Systematics and Diversity of Annelids

Systematics and Diversity of Annelids

Editors

Maria Capa
Pat Hutchings

MDPI • Basel • Beijing • Wuhan • Barcelona • Belgrade • Manchester • Tokyo • Cluj • Tianjin

Editors

Maria Capa
Biology Department
University of the Balearic Islands
Palma
Spain

Pat Hutchings
Australian Museum Research Institute,
Marine Invertebrates Collection
Australian Museum
Sydney
Australia

Editorial Office

MDPI
St. Alban-Anlage 66
4052 Basel, Switzerland

This is a reprint of articles from the Special Issue published online in the open access journal *Diversity* (ISSN 1424-2818) (available at: www.mdpi.com/journal/diversity/special_issues/systematics_annelids).

For citation purposes, cite each article independently as indicated on the article page online and as indicated below:

LastName, A.A.; LastName, B.B.; LastName, C.C. Article Title. *Journal Name* **Year**, *Volume Number*, Page Range.

ISBN 978-3-0365-1390-4 (Hbk)
ISBN 978-3-0365-1389-8 (PDF)

Contents

About the Editors

Maria Capa

Dr. Maria Capa's main line of research is the documentation of biodiversity in the oceans from an evolutionary perspective. She is interested in discovering forms and ways of life of current marine invertebrates (with a special predilection for annelids) and understanding the origins and processes responsible for shaping current diversity and distributions. Her line of research focuses on phylogeography and the delimitation of species limits in order to reveal possible cryptic species. Dr. Capa's interest in conservation has prompted her participation in projects related to invasive species and connectivity in anthropogenic environments, as well as the testing of new analytical tools for environmental monitoring.

Pat Hutchings

Dr. Pat Hutchings' research focuses on the systematics, phylogeny, and ecology of polychaetes and the role they play in benthic ecosystems. Dr. Hutchings has published extensively on Australian polychaete fauna, using both morphology and molecular techniques. This has included studying marine invasive species. Pat has also undertaken experimental studies on the bioerosion of coral substrates and how rates are changing with climate change. Pat is committed to mentoring and training the next generation of polychaete researchers; she has supervised PhD and postdoctoral students from Australia and from overseas.

Preface to "Systematics and Diversity of Annelids"

Annelida is a diverse and abundant group of invertebrates that populates all habitats on Earth, from the highest mountains to the abyssal depths. The origin of all extant annelids remains unsettled, but annelid-like fossils have been dated back to the Cambrian period, and genomic and transcriptomic studies have been key in placing them within Spiralia and relating them to molluscs, brachiopods, nemerteans, and phoronids. The enormous disparity in forms, lifestyles, and adaptions to different environments, most of which have occurred in a short period of time, have hindered relationships within Annelida. However, in the last two decades, the monophyly of most families has been assessed and, in some cases, the internal classifications have been reviewed. Approximately 20,000 species have been described to date; however, this number is increasing rapidly with the exploration of unsurveyed biogeographical regions, habitats, and depths, and also with new techniques in sampling, identifying, and analysing biodiversity. Special consideration needs to be given to the use of molecular data that allow for distinction between similarly looking, or identical, entities.

In this Special Issue, we address the state of the art of the systematics of the main annelid groups and the improvements in the diversity they hold, with special emphasis on the latest discoveries after the changes in fauna of well-studied areas, expeditions to unsurveyed areas or environments, or the use of novel techniques that allow for the improvement of biodiversity knowledge. We are hoping that this Special Issue will provide a platform facilitating a review of current knowledge on the subject, identifying the current research problems, as well as indicating directions and research trends for the future.

For this Special Issue, we have gathered 46 colleagues from 16 countries and 37 institutions at different stages in their careers, stressing the importance of collaboration (every chapter is multiauthored and multi-national); this is an amalgamation of different perspectives and sources of data, aiming at mentoring the next generation of annelid workers and highlighting the international and collaborative annelid community. Unfortunately, and due to various reasons, this Special Issue is not complete, and some relevant annelid groups and environments have not been included.

Maria Capa, Pat Hutchings

Editors

diversity

Review

Annelid Diversity: Historical Overview and Future Perspectives

María Capa [1,*] and Pat Hutchings [2,3]

[1] Departament de Biologia, Universitat de les Illes Balears, 07122 Palma, Spain
[2] Australian Museum Research Institute, Australian Museum, 1 William Street, Sydney 2010, Australia; pat.hutchings@Australian.Museum
[3] Biological Sciences, Macquarie University, North Ryde 2109, Australia
* Correspondence: maria.capa@uib.es

Abstract: Annelida is a ubiquitous, common and diverse group of organisms, found in terrestrial, fresh waters and marine environments. Despite the large efforts put into resolving the evolutionary relationships of these and other Lophotrochozoa, and the delineation of the basal nodes within the group, these are still unanswered. Annelida holds an enormous diversity of forms and biological strategies alongside a large number of species, following Arthropoda, Mollusca, Vertebrata and perhaps Platyhelminthes, among the species most rich in phyla within Metazoa. The number of currently accepted annelid species changes rapidly when taxonomic groups are revised due to synonymies and descriptions of a new species. The group is also experiencing a recent increase in species numbers as a consequence of the use of molecular taxonomy methods, which allows the delineation of the entities within species complexes. This review aims at succinctly reviewing the state-of-the-art of annelid diversity and summarizing the main systematic revisions carried out in the group. Moreover, it should be considered as the introduction to the papers that form this Special Issue on Systematics and Biodiversity of Annelids.

Keywords: Annelida; diversity; systematics; species; new developments; special issue

Citation: Capa, M.; Hutchings, P. Annelid Diversity: Historical Overview and Future Perspectives. *Diversity* **2021**, *13*, 129. https://doi.org/10.3390/d13030129

Academic Editor: Luc Legal

Received: 5 March 2021
Accepted: 14 March 2021
Published: 17 March 2021

1. Introduction

We entered the 21st century with a view about Annelida very different to what we have today, only twenty years later. In those days, we thought that the classification of the group was suffering a revolution because early molecular analyses placed clitellates within polychaetes [1–3]. Although there were also some indications that several of the then considered phyla, such as Sipuncula, Myzostomida, Vestimentifera, Pogonophora and Echiura, had annelid affinities (e.g., [4–11]), it was not until the advent of the phylogenomic methods that we were provided with strong enough evidence to consider these taxa within Annelida (e.g., [12–15]). This expansion of the Annelida concept greatly increased the diversity within the group, including aspects such as body plan, anatomy, reproductive biology, life history, feeding strategies, and species richness.

The fossil record evidenced that early annelids, provided with head appendages, birramous parapodia and simple chaetae were already present in the Early Cambrian [16] (although molecular clocks date the origin of annelids even earlier, e.g., [17]). These taxa do seem closely related to the extant annelid groups, which in most cases diversified rapidly during the Late Cambrian—Ordovician [16]. The deep relationships in the annelid radiation remain poorly resolved, in part due to the short basal branches as a consequence of this rapid diversification, but also due to the analysis of artifacts such as the long branch attraction of some groups [12,15,18–21]. Sister group relationships of Annelida are still being debated, but its placement within Lophotrochozoa and monophyly are now widely accepted (e.g., [12–15,22,23].

There have been several revisions of Annelida in the last 20 years. The volume *Polychaetes & Allies The Southern Synthesis* [24] (no longer in print but pdfs are available on the ABRS website, http://www.environment.gov.au/science/abrs/publications/fauna-of-australia/fauna-4a (accessed on 17 March 2021)), known as the *red book*, included an historical overview of not just polychaetes, but also of sipunculans, echiuroids, myzostomes and pogonophorans and, while oligochaetes were also originally to be included, this unfortunately did not happen. Its chapters include information on the biology and ecology of annelids, the higher classification based on the most recently available cladistics analysis by Rouse and Fauchald [4] and a chapter on each family then recognized with standardized content on morphology, diversity, physiology, reproduction, distribution, including how many species were present in Australian waters. While the title suggests it is Australia focused, the book is relevant to all parts of the world, as almost all polychaete families occur worldwide.

Shortly after the publication of *Polychaetes & Allies*, Rouse and Pleijel [8] published *Polychaetes*, also known as the *black book*. This book, whilst dealing with the anatomy, biology and ecology of polychaetes, it also focused on the phylogenetic relationships of the different clades and taxa. After the publication of these two books thorough reviews, some dramatic changes in the understanding of the systematics and classification of the annelids have taken place (e.g., [12–15]. Since 2014, an ambitious project aiming at updating the prestigious *Handbook of Zoology* [25–28] has gathered the interest of many Annelida experts with the aim of producing a comprehensive overview on different annelid groups, including updated information regarding the systematics, morphology, physiology, behavior, ecology and applied zoological research. The *Handbook of Zoology, Annelida* [29–31] appeared first online and later in book form, and it will eventually cover all clades and taxonomic annelid groups. It represents the third 21st century "must have" book series for annelid workers. In these volumes, it is highlighted the enormous efforts that have been put into resolving the phylogenetic relationships and the description of the diversity of forms and biological strategies exhibited within the group.

However, besides these comprehensive reviews, it is clear that further work is needed in order to pursue a better understanding of the diversification patterns in Annelida, to evaluate the current awareness on the species' richness, its distribution patterns and highlight where the major gaps of knowledge are in the different taxa.

Annelids are critically important in most marine ecosystems because of their diversity and abundance, especially in soft sediments from the intertidal to the deep-sea, as well as encrusting or attached to hard substrates. They exhibit a variety of feeding strategies ranging from deposit feeders, filter feeders, carnivores, herbivores and parasites, thus occupying all levels within the food chain. Some groups of polychaetes, fundamentally earthworms, are important bioturbators, turning over the sediment as well as breaking down organic matter. These mud swallowing feeders may also accumulate heavy metals and other contaminants in their body, and they are able to transfer these to other members of the trophic webs. Annelids are a major component of the marine benthos and terrestrial realm, and they comprise species with different tolerances to stress. Consequently, they have been considered as good bioindicators in environmental monitoring (e.g., [32–34]) and surrogates for marine biodiversity [35,36] biomarkers (e.g., [37–40]).

Annelids also exhibit a tremendous range of reproductive strategies ranging from mass spawning, brooding, laying of egg capsules as well as asexual reproduction [41]. Life spans range from a few weeks to many years, with some species spawning annually, whereas others only breed once and then die. Thus, while the biomass of polychaetes may not be high, in benthic communities they typically have a high productivity which is readily available to a wide range of organisms. Some species of annelids are widely collected as bait for recreational fishing (e.g., [42]), and some species are used as food for aquaculture practices, or are collected during the annual mass spawning associated with phases of the moon for human consumption. Annelid bioactive compounds are being used after showing properties compatible with anesthetics, fluorescent probes and even antibiotics and pesticides [43,44]. Negative ecological and economic impacts have also been reported because of annelids. For instance, an invasive species can cause problems such as blocking intake pipes to power stations, impact commercial value of mollusks by boring into shells, or modifying benthic habitats [45–47].

Until late in the last century, several polychaete workers assumed many polychaetes were widely distributed (e.g., [48–52]). Consequently, taxonomists and ecologists from around the world identified polychaetes using comprehensive and well-illustrated monographs such as those used by Fauvel [53,54] and Day [51,52], even though these were focused on the polychaete fauna of France and South Africa, respectively. The concept of "cosmopolitan" species was widely accepted with well-known examples such as *Capitella capitata* (Fabricius 1780), *Marphysa sanguinea* (Montagu, 1813), *Terebellides stroemii* Sars, 1835, *Owenia fusiformis* Delle Chiaje, 1844 or *Chaetozone setosa* Malmgren, 1867. However, the validity of this concept was questioned around the 1980s as detailed morphological studies, as well as molecular analyses, proved that "cosmopolitan species" were in fact siblings or cryptic species (e.g., [55–63]). As recently synthesized by several authors [64–66], most species natural distributions is discrete (although a few species from shallow waters and more from pelagic deep sea environments have been shown to exhibit wide distribution patterns), and these can be delimited with appropriate microscopic imagining techniques or molecular analyses. However, the number of translocated species (intentionally or unintentionally transported by anthropogenic means) outside of their natural range expands as we increase our surveys. Anthropogenic environments are more susceptible for the establishment of non-indigenous species, both in terrestrial [67] and in coastal areas, especially in ports and estuaries. Most recorded non-indigenous or invasive polychaete species are within the Serpulidae, Sabellidae and Spionidae [68].

2. Results and Discussion

2.1. Systematics

Since 1866 annelids (although not including clitellates) were divided in two groups: Annelidae erraticae and Annelidae sedentariae [69]. The terms Errantia and Sedentaria were widely used afterwards to refer to the more mobile or vagile forms and the tubiculous or sedentary forms, respectively. This division was originally made based on the type of segmentation: polychaetes with homonymous segmentation (with all segments similar) were placed in Errantia, and those with heteronomous segmentation (segments are grouped in morphology distinct series each with a different function) in Sedentaria. This was not strictly followed later on, and some annelid families were moved around between these two groups (e.g., [51–54,70–73]). Since the 1960s, a different classification system, where families were gathered in orders, started to be widely used (e.g., [74–78]). Some of these "orders" were shown to be natural clusters when cladistic analyses started to be implemented on annelids [79]. However, some of the early phylogenetic hypotheses of Annelida were conflicting, especially at interpreting the basal relationships. While some of the classifications, based on analyses of morphological data, divided Annelida into Clitellata and Polychaeta, and the later further split into Scolecida (parapodia with similar rami and the possession of two or more pairs of pygidial cirri) and Palpata (with palps and peristomium limited to the lips) [8,41,79,80], other studies contradicted this view based on methodological discrepancies (e.g., [81,82]). Later molecular phylogenies corroborated that clitellates, echiurids, myzostomids and sipunculids, were within the polychaetes and recovered some of the earlier considered taxa within Palpata (that is Canalipalpata) closely related to Scolecida, returning to the earlier concept of Errantia (see Figure 1) and Sedentaria (including Clitellata) (see Figure 2) (e.g., [12–15,83–85]). There is, however, a series of heterogeneous and basally branching annelids previously considered among the sedentarians or errants. These are the Palaeoannelida, Chaetopteridae, Amphinomida and Sipuncula (e.g., [13–15], Figure 1).

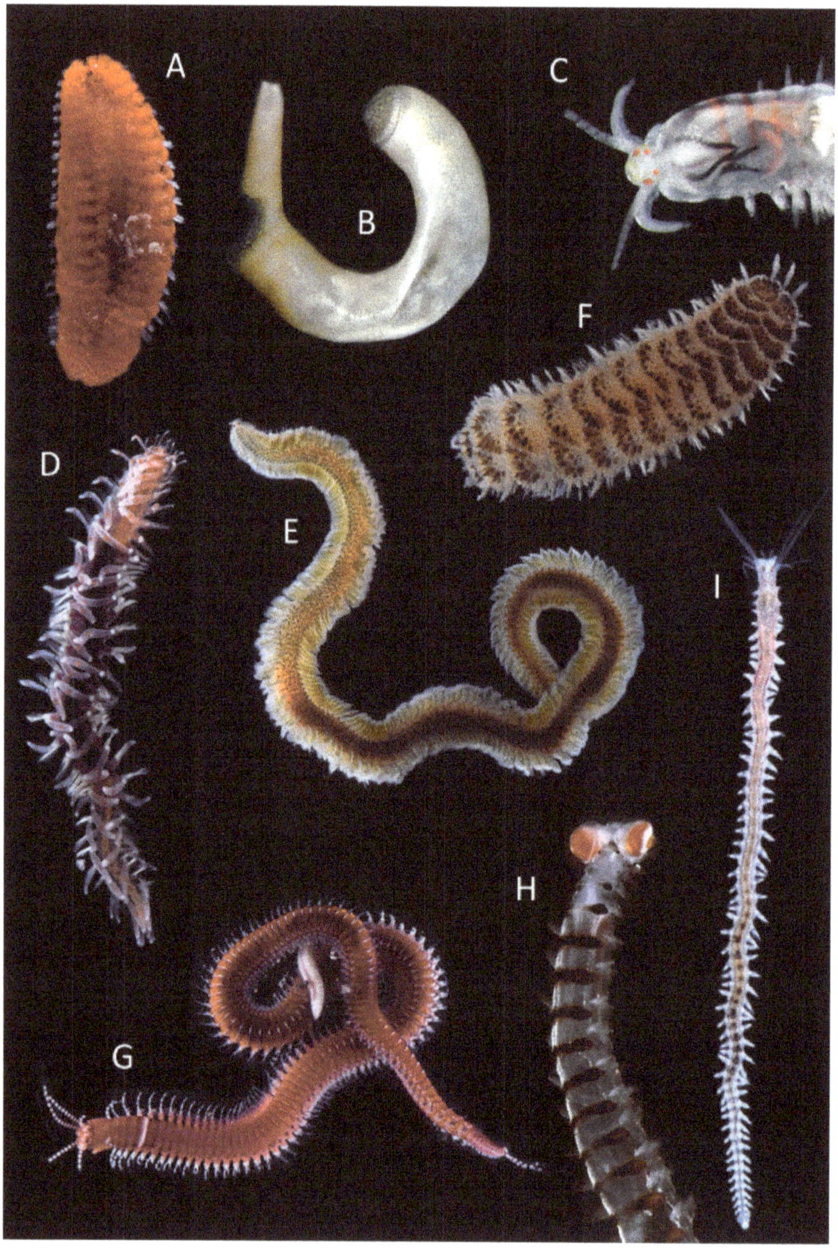

Figure 1. Small selection of some basal annelids (**A**,**B**) and Errantia (**C**–**I**). (**A**). *Euphrosine foliosa* Audouin & H Milne Edwards, 1833 (Amphinomida); (**B**). *Aspidosiphon muelleri* Diesing, 1851 (Sipuncula); **C**. *Dorvillea similis* (Crossland, 1924) (Eunicida); (**D**). *Amphiduros fuscescens* (Marenzeller, 1875) (Phyllodocida); (**E**). *Phyllodoce* sp. (Phyllodocida); (**F**). *Harmothoe areolata* (Grube, 1860) (Phyllodocida); (**G**). *Eunice* cf. *dubitata* Fauchald, 1974 (Eunicida); (**H**). *Vanadis Formosa* Claparède, 1870 (Phyllodocida); *Nereis* sp. (Phyllodocida). Photos: (**A**,**D**–**I**) by Xavier Salvador Costa; (**B**) by Daniel Martin; (**C**) by Alexander Semenov.

Figure 2. Small selection of the diversity encountered in Sedentaria. (**A**). *Amphictene auricoma* (O.F. Müller, 1776) (Terebelliformia); (**B**). *Maxmuelleria gigas* (M. Müller, 1852) (Echiura); (**C**). Maldanidae (Capitellida); (**D**). *Escalibregma* sp. (Scalibregmatidae); (**E**). *Armandia polyophthalma* Kükenthal, 1887 (Opheliida), (**F**). *Dasybranchus gajolae* Eisig, 1887 (Capitellida); (**G**). *Acromegalomma* sp. (Sabellida); (**H**). *Loimia tuberculata* Nogueira, Hutchings & Carrerette, 2015 (Terebelliformia). Photos: (**A**,**C**,**E**,**F**) by Xavier Salvador Costa; (**D**) by Dani Martin, (**G**,**H**) by Alexander Semenov.

2.2. Annelid Diversity

The current ranges of valid nominal species in the literature go from 14,000 to 20,000 [7,86–88], and databases such as WoRMS currently considers 23,774 accepted species of extant annelids [89]. Recounting the number of species after the latest revisions, such as the *Handbook of Zoology* chapters [29–31] and the present special issue ([90–101]), there seems to be around 20,000 currently accepted nominal species (Figure 3). There is a continuous documentation of new species and diversity patterns as new taxonomic surveys are carried out in poorly explored geographic areas and localities, in new environments, such as the deep-sea and, surprisingly, also in apparently well-known zones when using different collecting gear, sorting methods or identification techniques, such as SEM and molecular taxonomy.

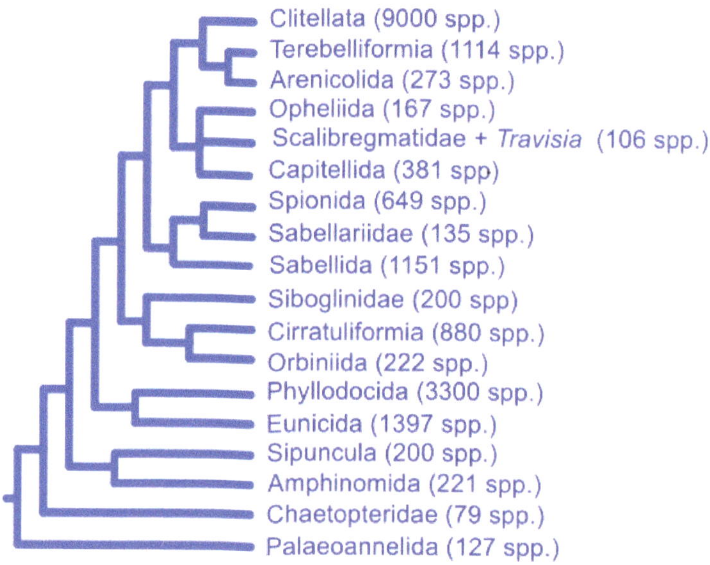

Figure 3. Metatree (based in [88,102]). The estimates for currently valid species were obtained from following studies or experts, although some minor groups are missing in the tree.; Clitellata, Erséus and Martinsson, pers. com. [94]; Terebelliformia [93]; Arenicolidae [89]; Opheliidae [90], Salibregmatidae and *Travisia* [89,90]; Capitellida [89]; Spionida [103–106]; Sabellariidae [89,107]; Sabellida [99]; Siboglinidae [108]; Cirratuliformia [89,100]; Orbiniida [95]; Phyllodocida [101]; Eunicida [91]; Sipuncula [97]; Amphinomida [89].

There are several factors influencing the discrepancy in the total numbers of annelid species considered as currently valid:

- Literature thorough scrutiny, including old and obscure publications. It is not uncommon to see how some taxonomic and nomenclatural mistakes have been passed on to most recent publications. Many annelid genera and families require a thorough follow up of synonyms (species that may have been synonymized with others or moved to a different genus or taxonomic group), and also way to rescue some old names that have been lost in most recent taxonomic lists. Of course, databases such as WoRMS are of invaluable help, but as much as they try to keep updated with taxonomic progress there is often a lag [109].
- Some groups require an exhaustive taxonomic revision that includes the examination of type material. This has become more difficult lately as shipping of preserved specimens has become an extremely regulated process, and unaffordable (for economic or time-related reasons) for some museums. Travelling to museums where the type

specimens of a genus are housed is also challenging as these may be spread around the globe.

- Taxonomists too often have a partial understanding of their group because we tend to specialize in particular families, genera, environments or geographic areas. This may lead to a non-confident or tentative perception of the overall diversity of larger groups.
- When referring to the total number of a species we generally mean that the currently accepted binomial names are based on morphologically recognizable entities. Taxonomists often hold knowledge of a wider number of morphospecies that are new but not formerly described. In fact, the average time it takes from the first collection of a specimen of a new species to its formal naming and description in a scientific paper is 21 years [110]. The reason for this may be the need for more specimens to account for intraspecific variability or to be preserved in a specific manner, revisions/examination of similar or related species are needed, the group requires a revision, a lack of funding or the need for collaboration with other (often lacking) experts in the group, e.g., taxonomic impediment, etc. [110,111]. Moreover, when the lines of evidence for delineating species are molecular (e.g., DNA sequences), these newly recognized lineages are not often accompanied by formal descriptions [112]. In these situations (no formal binomials), species will be missing from species lists in most cases.
- Cryptic species. The number of complexes of annelid species accounted for in recent years has vastly increased due to the use of molecular data. This has an impact in the real overall diversity present in the group. Although this is not necessary reflected in the total number of species accounted for, and this is because species delineated by molecular mean are often not accompanied by a formal species description (as discussed by Goldstein et al. [112] and exemplified by [61,113–116]).

According to some predictions, there are potentially still 13,000 to 24,000 of annelid species awaiting to be discovered and described [86].

2.3. *Gaps of Knowledge and Future Perspectives*

In this special issue, a considerable number of researchers have participated towards providing a summary of the current knowledge about the biology, systematics and diversity in a broad range of annelid taxonomic groups. Most of them acknowledge the improvement in the assessment of internal relationships within the groups after molecular phylogenetic analyses have been performed. However, they also indicate that these are still not fully settled in most cases, and the analyses require further data, analytical considerations or a combination of sources of information. Robust phylogenies with comprehensive taxon coverage are crucial for stablishing the backbone of classification and to trace the evolution. The review papers in this special issue point to some of the main gaps in the knowledge on each of the taxonomic groups they deal with, but there are large similarities between all of them. It seems clear that there are some geographic areas that have been scarcely studied. In the marine realm, these include the North Indo-Pacific, South America, polar waters, but mainly the African coastline. In the terrestrial and limnic realms, South America, Africa and Asia are in need of further taxonomic surveys (Figure 4, but see papers in this Special Issue for more information). Our knowledge about habitats such as extreme environments [92] and the deep sea is also very limited (e.g., [117]).

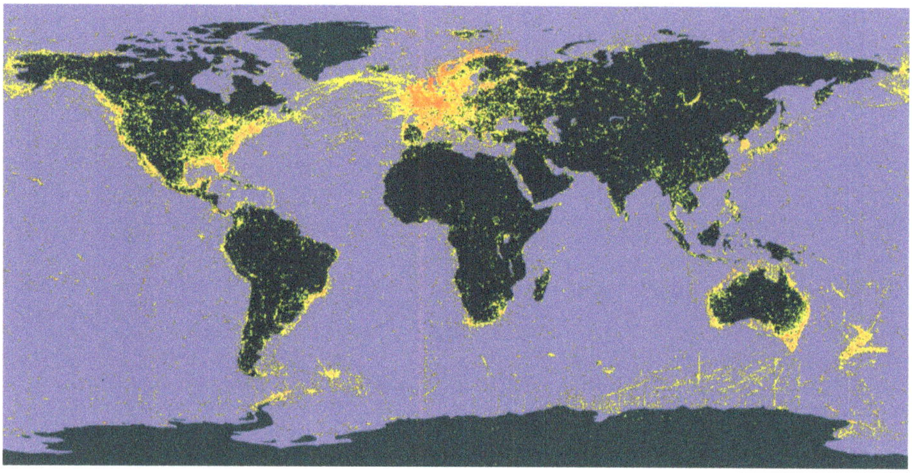

Figure 4. Occurrence data from Global Biodiversity Information Facility (GBIF) with 813,947 georeferenced records for Annelida in the marine and terrestrial environments. In red and orange, hotspots of occurrences, in yellow sites with one occurrence. This map does not show the real knowledge about annelid records, but those uploaded in this platform. However, it can be used as a proxy of the current number of undertaken surveys.

In the papers within this Special Issue, there are suggestions as to how to proceed in the future if we want to progress at a steady and efficient manner towards increasing our knowledge about Annelida and in discovering the real diversity held in this phylum. These can be summarized as follows:

- An increase in field work in the areas that have been poorly surveyed. There is also a large amount of material, specially coming from deep sea expeditions, that is awaiting to be studied in natural history museums. Therefore, only diving in poorly studied geographic areas, in extreme environments and in museum collections will provide us with a good understanding about the diversity of annelids.
- Most of the knowledge about the biology, anatomy and behavior of annelid groups is based on a limited number of species for each family. This is more obvious in modern biology that mainly focuses on model organisms, especially in the fields of comparative physiology and morphology. Studies based on a larger number of taxa and including some of the microscopic modern technics (phase contrast, scanning electron microscopy, transmission electron microscopy, confocal laser scanning microscopy, tomography and 3D reconstructions) are needed.
- It is recommended to undertake an integrative approach for delimiting and documenting species. This includes the combination of morphological, molecular and biological data. For several groups of annelids, understanding the reproductive and developmental features has also shown to be helpful for species delimitation.
- Species descriptions or re-description often requires re-examination of type specimens for species comparisons. Ideally, a thorough examination of specimens from different geographical areas and ecological features is needed in order to establish species boundaries and to assess intraspecific variability.
- Obtaining genetic information for at least the type species (preferably from type locality) of each genus is needed, as this would allow generic revisions. Specimens for genetic analysis should be collected from the type localities and vouchers deposited.
- For species with wide geographical or bathymetric distribution, population genetic studies are necessary to reveal potential cryptic species. In this line, the advent of high-throughput sequencing methods has a lot to offer for the generation of species delimitation datasets.

- Molecular taxonomy is revealing hidden diversity at a high speed, but formal taxonomic descriptions are lagging behind the molecular work. Therefore, we should increase the efforts made in describing the species encountered after molecular methods. Only trained taxonomist can undertake the development of a proper account and documentation of these lineages, as these activities often require revision of genera, revision of old literature, synonymies and varying terminology, re-examination of museum types, etc.
- Policy should take care of biodiversity, and governments should invest in systems and in training the next generation of taxonomists. The taxonomic work has been neglected for decades and we are suffering a loss of taxonomic knowledge. This has a direct impact on the speed with which species are described, but also on the quality of the biodiversity assessments and the studies based on those. We need to train and sustain more systematists able to discover, describe, identify and classify species.
- We need to promote, care and engage with museum collections and public databases. It is imperative that the type material (holotype, type series, additional specimens showing intraspecific variability, and DNA extractions) is always deposited in properly curated permanent museum collection(s) where it is maintained in optimal conditions. Museums and researchers need to commit to open-access databases such as World Register of Marine Species—WoRMS, Ocean Biodiversity Information System—OBIS, National Center for Biotechnology Information—Genbank, or the Global Biodiversity Information Facility—GBIF, that offer invaluable information for research projects.

2.4. This Special Issue

For the present special issue, we aimed at gathering updated information and discuss the recent advances in different diversity aspects and in recent systematic revisions of some of the major Annelida clades, including Palaeoannelida, Sipuncula, Phyllodocida, Eunicida, Orbiniida, Cirratuliformia, Sabellida, Opheliida and Scalibregmatidae, Terebelliformia, and Clitellata [90,91,93–97,99–101]. Other chapters take a more ecological approach, for example the papers on extreme or interstitial annelids [92,98]. By gathering this information, we aim to highlight the importance of annelids in biodiversity assessments and ecosystem functioning, to recapitulate differing diversity aspects about selected groups of annelids, and highlight the bridge between the written literature and the public databases and platforms regarding taxonomy, occurrence data and DNA sequences (e.g., WoRMS, GBIF, GenBank or DriloBASE Taxo). We also aim at revealing where the gaps of knowledge are and where the efforts should be concentrated if we want to progress towards a deeper understanding of the annelid diversity inhabiting our planet. We need to increase efforts in exploring understudied areas and in revisiting museum collections, in reviewing some neglected taxonomic groups, training the next generations of taxonomists and systematists, uncover hidden diversity, embrace methods for speeding up diversity assessments and taxonomic surveys, connect the updated taxonomic results with the more applied approach of ecology and pay special attention to the large number of species that have been translocated (e.g., [66]).

We have gathered 46 colleagues from 16 countries and 37 institutions at different stages in their careers, stressing the importance of collaboration (every chapter is multiauthored and multi-international), an amalgamation of different perspectives and sources of data, aiming at mentoring the next generation of annelid workers and highlighting the international and collaborative annelid community. Unfortunately, and due to different reasons, this special issue is not complete, and some relevant annelid groups and environments have not been included.

Author Contributions: Conceptualization, M.C. and P.H.; writing—original draft preparation, review and editing, M.C. and P.H.; visualization, M.C. Both authors have read and agreed to the published version of the manuscript.

Funding: M.C. was funded by the Ramón y Cajal program (RYC-2016-20799) funded by Spanish MINECO, Agencia Estatal de Investigación, Comunidad Autónoma de las Islas Baleares and the European Social Fund.

Acknowledgments: We would like to thank Dani Martin, Xavier Salvador and Alexander Semenov for allowing us to use the photographs to show the diversity of worms among annelids. It has been a privilege to be able to bring together this group of colleagues from all over the world gather some basic and updated knowledge about the diversity of Annelida. We are grateful to four anonymous reviewers for providing useful comments for the improvement of the manuscript.

Conflicts of Interest: The authors declare no conflict of interest.

References

1. Kojima, S. Paraphyletic Status of Polychaeta Suggested by Phylogenetic Analysis Based on the Amino Acid Sequences of Elongation Factor-1 Alpha. *Mol. Phylogenetics Evol.* **1998**, *9*, 255–261. [CrossRef] [PubMed]
2. Martin, P. On the Origin of the Hirudinea and the Demise of the Oligochaeta. *Proc. R. Soc. Lond. B* **2001**, *268*, 1089–1098. [CrossRef] [PubMed]
3. Brown, S.; Rouse, G.; Hutchings, P.; Colgan, D. Assessing the Usefulness of Histone H3, U2 SnRNA and 28S RDNA in Analyses of Polychaete Relationships. *Aust. J. Zool.* **1999**, *47*, 499. [CrossRef]
4. Rouse, G.W.; Fauchald, K. Cladistics and Polychaetes. *Zool. Scr.* **1997**, *26*, 139–204. [CrossRef]
5. McHugh, D. Molecular Evidence That Echiurans and Pogonophorans Are Derived Annelids. *Proc. Nat. Acad. Sci. USA* **1997**, *94*, 8006–8009. [CrossRef] [PubMed]
6. McHugh, D. Molecular Phylogeny of the Annelida. *Can. J. Zool.* **2000**, *78*, 1873–1884. [CrossRef]
7. Purschke, G.; Hessling, R.; Westheide, W. The Phylogenetic Position of the Clitellata and the Echiura-on the Problematic Assessment of Absent Characters. *J. Zool. Syst. Evol. Res.* **2000**, *38*, 165–173. [CrossRef]
8. Rouse, G.W.; Pleijel, F. *Polychaetes*; Oxford University Press: London, UK, 2001; pp. 1–354.
9. Bleidorn, C.; Vogt, L.; Bartolomaeus, T. New Insights into Polychaete Phylogeny (Annelida) Inferred from 18S RDNA Sequences. *Mol. Phylogenetics Evol.* **2003**, *29*, 279–288. [CrossRef]
10. Bleidorn, C.; Vogt, L.; Bartolomaeus, T. A Contribution to Sedentary Polychaete Phylogeny Using 18S RRNA Sequence Data. *J. Zool. Syst. Evol. Res.* **2003**, *41*, 186–195. [CrossRef]
11. Rousset, V.; Rouse, G.W.; Siddall, M.E.; Tillier, A.; Pleijel, F. The Phylogenetic Position of Siboglinidae (Annelida) Inferred from 18S RRNA, 28S RRNA and Morphological Data. *Cladistics* **2004**, *20*, 518–533. [CrossRef]
12. Struck, T.H.; Paul, C.; Hill, N.; Hartmann, S.; Hösel, C.; Kube, M.; Lieb, B.; Meyer, A.; Tiedemann, R.; Purschke, G.; et al. Phylogenomic Analyses Unravel Annelid Evolution. *Nature* **2011**, *471*, 95–98. [CrossRef] [PubMed]
13. Struck, T.H.; Golombek, A.; Weigert, A.; Franke, F.A.; Westheide, W.; Purschke, G.; Bleidorn, C.; Halanych, K.M. The Evolution of Annelids Reveals Two Adaptive Routes to the Interstitial Realm. *Curr. Biol.* **2015**, *25*, 1993–1999. [CrossRef] [PubMed]
14. Weigert, A.; Helm, C.; Meyer, M.; Nickel, B.; Arendt, D.; Hausdorf, B.; Santos, S.R.; Halanych, K.M.; Purschke, G.; Bleidorn, C.; et al. Illuminating the Base of the Annelid Tree Using Transcriptomics. *Mol. Biol. Evol.* **2014**, *31*, 1391–1401. [CrossRef] [PubMed]
15. Andrade, S.C.S.; Novo, M.; Kawauchi, G.Y.; Worsaae, K.; Pleijel, F.; Giribet, G.; Rouse, G.W. Articulating "Archiannelids": Phylogenomics and Annelid Relationships, with Emphasis on Meiofaunal Taxa. *Mol. Biol. Evol.* **2015**, *32*, 2860–2875. [CrossRef]
16. Parry, L.; Tanner, A.; Vinther, J. The Origin of Annelids. *Palaeontology* **2014**, *57*, 1091–1103. [CrossRef]
17. Peterson, K.J.; Cotton, J.A.; Gehling, J.G.; Pisani, D. The Ediacaran Emergence of Bilaterians: Congruence between the Genetic and the Geological Fossil Records. *Philos. Trans. R. Soc. B* **2008**, *363*, 1435–1443. [CrossRef]
18. Martin, P.; Kaygorodova, I.; Sherbakov, D.Y.; Verheyen, E. Rapidly Evolving Lineages Impede the Resolution of Phylogenetic Relationships among Clitellata (Annelida). *Mol. Phylogenetics Evol.* **2000**, *15*, 355–368. [CrossRef] [PubMed]
19. Bleidorn, C.; Eeckhaut, I.; Podsiadlowski, L.; Schult, N.; McHugh, D.; Halanych, K.M.; Milinkovitch, M.C.; Tiedemann, R. Mitochondrial Genome and Nuclear Sequence Data Support Myzostomida as Part of the Annelid Radiation. *Mol. Phylogenetics Evol.* **2007**, *24*, 1690–1701. [CrossRef]
20. Bleidorn, C.; Podsiadlowski, L.; Zhong, M.; Eeckhaut, I.; Hartmann, S.; Halanych, K.M.; Tiedemann, R. On the Phylogenetic Position of Myzostomida: Can 77 Genes Get It Wrong? *BMC Evol. Biol.* **2009**, *9*, 150. [CrossRef]
21. Struck, T.H. The Impact of Paralogy on Phylogenomic Studies—A Case Study on Annelid Relationships. *PLoS ONE* **2013**, *8*, e62892. [CrossRef]
22. Weigert, A.; Golombek, A.; Gerth, M.; Schwarz, F.; Struck, T.H.; Bleidorn, C. Evolution of Mitochondrial Gene Order in Annelida. *Mol. Phylogenetics Evol.* **2016**, *94*, 196–206. [CrossRef] [PubMed]
23. Laumer, C.E.; Bekkouche, N.; Kerbl, A.; Goetz, F.; Neves, R.C.; Sørensen, M.V.; Kristensen, R.M.; Hejnol, A.; Dunn, C.W.; Giribet, G.; et al. Spiralian Phylogeny Informs the Evolution of Microscopic Lineages. *Curr. Biol.* **2015**, *25*, 2000–2006. [CrossRef]
24. Beesley, P.L.; Ross, G.J.B.; Glasby, C.J. *Polychaetes & Allies: The Southern Synthesis. Fauna of Australia. Polychaeta, Myzostomida, Pogonophora, Echiura, Sipuncula*; CSIRO Publishing: Melbourne, Australia, 2000; Volume 44, pp. 1–465.
25. Reisinger, E. Zweiter Unterstamm der Vermes Polymera. Allgemeine Einleitung zur Naturgeschichte der Vermes Polymera. In *Handbuch der Zoologie*; Part 6, Vermes Polymera; DeGruyter & Co.: Leipzig/Berlin, Germany, 1928; Volume 2.2, pp. 1–32.

26. Hempelmann, F. Erste und Zweite Klasse der Vermes Polymera (Annelida). Archiannelida = UrRingelwürmer und Polychaeta = Borstenwürmer. In *Handbuch der Zoologie*; Part 7, Vermes Polymera; Kükenthal, W., Krumbach, T., Eds.; DeGruyter & Co.: Leipzig/Berlin, Germany, 1928; Volume 2.2, pp. 1–212.
27. Michaelsen, W. Dritte Klasse der Vermes Polymera (Annelida). Clitellata = Gürtelwürmer. 1. Ordnung der Clitellata: Oligochaeta = Regenwürmer und Verwandte. In *Handbuch der Zoologie*; Part 8, Vermes Polymera; Kükenthal, W., Krumbach, T., Eds.; DeGruyter & Co.: Leipzig/Berlin, Germany, 1928; Volume 2.2, pp. 1–118.
28. Scriban, I.A.; Autrum, H. 2. Ordnung Der Clitellata: Hirudinea = Egel. In *Handbuch der Zoologie*; Part 8, Vermes Polymera; Kükenthal, W., Krumbach, T., Eds.; DeGruyter & Co.: Leipzig/Berlin, Germany, 1928; Volume 2.2, pp. 119–352.
29. Purschke, G.; Böggemann, M.; Westheide, W. Basal Groups and Pleistoannelica, Sedentaria. In *Handbook of Zoology*; Annelida; De Gruyter: Berlin, Germany, 2019; Volume 1, pp. 1–472.
30. Purschke, G.; Böggemann, B.; Westheide, W. Pleistoannelida, Sedentaria II. In *Handbook of Zoology*; Annelida; De Gruyter: Berlin, Germany, 2019; Volume 2, pp. 1–465.
31. Purschke, G.; Böggemann, B.; Westheide, W. Pleistoannelida, Sedentaria III and Errantia I. In *Handbook of Zoology*; Annelida; De Gruyter: Berlin, Germany, 2021; Volume 3, pp. 1–480.
32. Belan, T.A. Marine Environmental Quality Assessment Using Polychaete Taxocene Characteristics in Vancouver Harbour. *Mar. Environ. Res.* **2004**, *57*, 89–101. [CrossRef]
33. Surugiu, V. The Use of Polychaetes as Indicators of Eutrophication and Organic Enrichment of Coastal Waters: A Study Case–Romanian Black Sea Coast. *An. Ştiinţifice ale Univ. "Al.I. Cuza" Iaşi s. Biol. Anim.* **2005**, *51*, 55–62.
34. Fründ, H.-C.; Graefe, U.; Tischer, S. Earthworms as Bioindicators of Soil Quality. In *Biology of Earthworms*; Springer: Berlin/Heidelberg, Germany, 2011; pp. 261–278.
35. Olsgard, F.; Brattegard, T.; Holthe, T. Polychaetes as Surrogates for Marine Biodiversity: Lower Taxonomic Resolution and Indicator Groups. *Biodiv. Conserv.* **2003**, *12*, 1033–1049. [CrossRef]
36. Gladstone, W.; Murray, B.R.; Hutchings, P. Promising yet Variable Performance of Cross-Taxon Biodiversity Surrogates: A Test in Two Marine Habitats at Multiple Times. *Biodiv. Conserv.* **2020**, *29*, 3067–3089. [CrossRef]
37. Bocchetti, R.; Fattorini, D.; Gambi, M.; Regoli, F. Trace Metal Concentrations and Susceptibility to Oxidative Stress in the Polychaete *Sabella spallanzanii* (Gmelin)(Sabellidae): Potential Role of Antioxidants in Revealing Stressful Environmental Conditions in the Mediterranean. *Arch. Environ. Contam. Toxicol.* **2004**, *46*, 353–361. [CrossRef] [PubMed]
38. Poirier, L.; Berthet, B.; Amiard, J.-C.; Jeantet, A.-Y.; Amiard-Triquet, C. A Suitable Model for the Biomonitoring of Trace Metal Bioavailabilities in Estuarine Sediments: The Annelid Polychaete *Nereis diversicolor*. *J. Mar. Biol. Assoc. UK* **2006**, *86*, 71–82. [CrossRef]
39. Dean, H.K. The Use of Polychaetes (Annelida) as Indicator Species of Marine Pollution: A Review. *Rev. Biol. Trop.* **2008**, *56*, 11–38.
40. Unrine, J.M.; Hunyadi, S.E.; Tsyusko, O.V.; Rao, W.; Shoults-Wilson, W.A.; Bertsch, P.M. Evidence for Bioavailability of Au Nanoparticles from Soil and Biodistribution within Earthworms (*Eisenia fetida*). *Environ. Sci. Technol.* **2010**, *44*, 8308–8313. [CrossRef]
41. Jamieson, B.G.M.; Rouse, G.W.; Pleijel, F. *Reproductive Biology and Phylogeny of Annelida*; Science Publishers: Enfield, UK, 2006; Volume 4, pp. 1–698.
42. Cole, V.J.; Chick, R.C.; Hutchings, P.A. A Review of Global Fisheries for Polychaete Worms as a Resource for Recreational Fishers: Diversity, Sustainability and Research Needs. *Rev. Fish Biol. Fish.* **2018**, *28*, 543–565. [CrossRef]
43. Datta, D.; Talapatra, S.N.; Swarnakar, S. Bioactive Compounds from Marine Invertebrates for Potential Medicines-an Overview. *Int. Let. Nat. Sci.* **2015**, *7*, 42–61. [CrossRef]
44. Rodrigo, A.P.; Costa, P.M. The Hidden Biotechnological Potential of Marine Invertebrates: The Polychaeta Case Study. *Environ. Res.* **2019**, *173*, 270–280. [CrossRef] [PubMed]
45. Bax, N.; Williamson, A.; Aguero, M.; Gonzalez, E.; Geeves, W. Marine Invasive Alien Species: A Threat to Global Biodiversity. *Mar. Pol.* **2003**, *27*, 313–323. [CrossRef]
46. Zenetos, A.; Çinar, M.; Pancucci-Papadopoulou, M.; Harmelin, J.; Furnari, G.; Andaloro, F.; Bellou, N.; Streftaris, N.; Zibrowius, H. Annotated List of Marine Alien Species in the Mediterranean with Records of the Worst Invasive Species. *Mediterr. Mar. Sci.* **2005**, *6*, 63–118. [CrossRef]
47. Molnar, J.L.; Gamboa, R.L.; Revenga, C.; Spalding, M.D. Assessing the Global Threat of Invasive Species to Marine Biodiversity. *Front. Ecol. Environ.* **2008**, *6*, 485–492. [CrossRef]
48. Ekman, S. *Zoogeography of the Sea*; Sidgwick & Jackson, Ltd.: London, UK, 1953; Volume 6, pp. 1–240. [CrossRef]
49. Knox, G. The Distribution of Polychaetes within the Indo-Pacific. In Proceedings of the 8th Pacific Science Congress, Quezon City, Philippines, 16–28 November 1957; National Research Council of the Phiuppines: Quezon City, Philippines, 1957; Volume 3, pp. 403–411.
50. Hartman, O. Catalogue of the Polychaeta Annelids of the World. *Allan Hancock Found. Publ.* **1959**, *23*, 1–628.
51. Day, J.H. *A Monograph on the Polychaeta of Southern Africa. Part 1. Errantia*; British Museum (Natural History): London, UK, 1967; pp. 1–458.
52. Day, J.H. *A Monograph on the Polychaeta of Southern Africa. Part II. Sedentaria*; British Museum (Natural History): London, UK, 1967; pp. 459–878.
53. Fauvel, P. Polychète Errantes. *Faune Fr.* **1923**, *5*, 1–488.

54. Fauvel, P. Polychètes Sédentaires. Addenda Aux Errantes, Archiannélides, Myzostomaires. *Faune Fr.* **1927**, *16*, 1–494.
55. Williams, S.J. The Status of *Terebellides stroemi* (Polychaeta; Trichobranchidae) as a Cosmopolitan Species, Based on a Worldwide Morphological Survey, Including Description of New Species. In Proceedings of the First International Polychaete Conference, Sydney, Australia, 4–9 July 1983; pp. 118–142.
56. Eckelbarger, K.; Grassle, J. Spermatogenesis, Sperm Storage and Comparative Sperm Morphology in Nine Species of *Capitella, Capitomastus* and *Capitellides* (Polychaeta: Capitellidae). *Mar. Biol.* **1987**, *95*, 415–429. [CrossRef]
57. Chambers, S.J.; Woodham, A. A New Species of *Chaetozone* (Polychaeta: Cirratulidae) from Deep Water in the Northeast Atlantic, with Comments on the Diversity of the Genus in Cold Northern Waters. *Hydrobiologia* **2003**, *496*, 41–48. [CrossRef]
58. Hutchings, P.; Karageorgopoulos, P. Designation of a Neotype of *Marphysa sanguinea* (Montagu, 1813) and a Description of a New Species of *Marphysa* from Eastern Australia. *Hydrobiologia* **2003**, *496*, 87–94. [CrossRef]
59. Koh, B.S.; Bhaud, M.R.; Jirkov, I.A. Two New Species of *Owenia* (Annelida: Polychaeta) in the Northern Part of the North Atlantic Ocean and Remarks on Previously Erected Species from the Same Area. *Sarsia* **2003**, *88*, 175–188. [CrossRef]
60. Ford, E.; Hutchings, P. An Analysis of Morphological Characters of *Owenia* Useful to Distinguish Species: Description of Three New Species of *Owenia* (Oweniidae: Polychaeta) from Australian Waters. *Mar. Ecol.* **2005**, *26*, 181–196. [CrossRef]
61. Nygren, A.; Parapar, J.; Pons, J.; Meißner, K.; Bakken, T.; Kongsrud, J.A.; Oug, E.; Gaeva, D.; Sikorski, A.; Johansen, R.A.; et al. A Mega-Cryptic Species Complex Hidden among One of the Most Common Annelids in the North East Atlantic. *PLoS ONE* **2018**, *13*, e0198356. [CrossRef] [PubMed]
62. Grosse, M.; Bakken, T.; Nygren, A.; Kongsrud, J.A.; Capa, M. Species Delimitation Analyses of NE Atlantic *Chaetozone* (Annelida, Cirratulidae) Reveals Hidden Diversity among a Common and Abundant Marine Annelid. *Mol. Phylogenetics Evol.* **2020**, *149*, 106852. [CrossRef]
63. Parapar, J.; Capa, M.; Nygren, A.; Moreira, J. To Name but a Few: Descriptions of Five New Species of *Terebellides* (Annelida, Trichobranchidae) from the North East Atlantic. *ZooKeys* **2020**, *992*, 1–58. [CrossRef]
64. Nygren, A. Cryptic Polychaete Diversity: A Review. *Zool. Scr.* **2014**, *43*, 172–183. [CrossRef]
65. Hutchings, P.; Kupriyanova, E. Cosmopolitan Polychaetes–Fact or Fiction? Personal and Historical Perspectives. *Invertebr. Syst.* **2018**, *32*, 1–9. [CrossRef]
66. Hutchings, P.; Lavesque, N. I Know Who You Are, but Do Others Know? Why Correct Scientific Names Are so Important for the Biological Sciences. *Zoosymposia* **2020**, *19*, 151–163. [CrossRef]
67. Hendrix, P.F.; Callaham Jr, M.A.; Drake, J.M.; Huang, C.-Y.; James, S.W.; Snyder, B.A.; Zhang, W. Pandora's Box Contained Bait: The Global Problem of Introduced Earthworms. *Ann. Rev. Ecol. Evol. Syst.* **2008**, *39*, 593–613. [CrossRef]
68. Çinar, M.E. Alien Polychaete Species Worldwide: Current Status and Their Impacts. *J. Mar. Biol. Ass.* **2013**, *93*, 1257–1278. [CrossRef]
69. De Quatrefages, A. *Histoire Naturelle Des Annéles Marine et d'eau Douce. Annélides et Géphyriens*; Librarie Encyclopédique de Roret: Paris, France, 1866; Volume 2, pp. 1–588.
70. Uschakov, P.V. *Polychaeta of the Far Eastern Seas of the U.S.S.R. Keys to the Fauna of the USSR. No. 56*; Zoological Institute of the Academy of Sciences of the USSR: Moscow, Russia, 1955; pp. 1–433.
71. Hartman, O. *Atlas of the Errantiate Polychaetous Annelids from California*; Allan Hancock Foundation, University of Southern California: Los Angeles, LA, USA, 1968; pp. 1–828.
72. Hartman, O. *Atlas of the Sedentariate Polychaetous Annelids from California*; Allan Hancock Foundation, University of Southern California: Los Angeles LA, USA, 1969; pp. 1–812.
73. Hartmann-Schröder, G. Annelida, Borstenwürmer, Polychaeta. *Tierwelt Dtschl.* **1971**, *58*, 1–594.
74. Dales, R.P. *Annelids*; Hutchinson University Library: London, UK, 1963; pp. 1–200.
75. Dales, R.P. The polychaete stomodeum and phylogeny. In *Essays on Polychaetous Annelids in Memory of Dr. Olga Hartman*; Reish, D.J., Fauchald, K., Eds.; Allan Hancock Foundation, University of Southern California: Los Angeles, LA, USA, 1977; pp. 525–546.
76. Fauchald, K. The Polychaete Worms. Definitions and Keys to the Orders, Families and Genera. *Nat. Hist. Mus. Los Angeles Cty. Sci. Ser.* **1977**, *28*, 1–188.
77. Pettibone, M.H. Annelida. In *Synopsis and Classification of living Organisms*; Parker, S.P., Ed.; McGraw-Hill Book Co: New York, NY, USA, 1982; Volume 2, pp. 1–43.
78. Hartmann-Schröder, G. Annelida, Borstenwürmer, Polychaeta. In *Die Tierwelt Deutschlands und der Angrenzenden Meeresteile*; 2nd revised ed.; Gustav Fischer: Jena, Germany, 1996; pp. 1–648.
79. Fauchald, K.; Rouse, G. Polychaete Systematics: Past and Present. *Zool. Scr.* **1997**, *26*, 71–138. [CrossRef]
80. Rouse, G.W.; Pleijel, F. Current Problems in Polychaete Systematics. *Hydrobiologia* **2003**, *496*, 175–189. [CrossRef]
81. Eibye-Jacobsen, D.; Nielsen, C. The Rearticulation of Annelids. *Zool. Scr.* **1996**, *25*, 275–282. [CrossRef]
82. Bartolomaeus, T.; Purschke, G.; Hausen, H. Polychaete Phylogeny Based on Morphological Data–a Comparison of Current Attempts. *Hydrobiologia* **2005**, *535*, 341–356.
83. Struck, T.H.; Schult, N.; Kusen, T.; Hickman, E.; Bleidorn, C.; McHugh, D.; Halanych, K.M. Annelida phylogeny and the status of Sipuncula and Echiura. *BMC Evol. Biol.* **2007**, *7*, 57. [CrossRef]
84. Struck, T.H.; Nesnidal, M.P.; Purschke, G.; Halanych, K.M. Detecting Possibly Saturated Positions in 18S and 28S Sequences and Their Influence on Phylogenetic Reconstruction of Annelida (Lophotrochozoa). *Mol. Phylogenetics Evol.* **2008**, *48*, 628–645. [CrossRef] [PubMed]

85. Zrzavý, J.; Říha, P.; Piálek, L.; Janouškovec, J. Phylogeny of Annelida (Lophotrochozoa): Total-Evidence Analysis of Morphology and Six Genes. *BMC Evol. Biol.* **2009**, *9*, 189. [CrossRef]
86. Appeltans, W.; Ahyong, S.T.; Anderson, G.; Angel, M.V.; Artois, T.; Bailly, N.; Bamber, R.; Barber, A.; Bartsch, I.; Berta, A. The Magnitude of Global Marine Species Diversity. *Curr. Biol.* **2012**, *22*, 2189–2202. [CrossRef] [PubMed]
87. Aguado, M.; Capa, M.; Oceguera-Figueroa, A.; Rouse, G. Annelida. In *The Tree of Life: Evolution and Classification of Living Organisms*; Vargas, P., Zardoya, R., Eds.; Sinauer Associates: Sunderland, MA, USA, 2014; pp. 254–269.
88. Weigert, A.; Bleidorn, C. Current Status of Annelid Phylogeny. *Org. Divers. Evol.* **2016**, *16*, 345–362. [CrossRef]
89. WoRMS Annelida. 2021. Available online: http://marinespecies.org/aphia.php?p=taxdetails&id=882 (accessed on 13 March 2021).
90. Parapar, J.; Martínez, A.; Moreira, J. On the Systematics and Biodiversity of the Opheliidae and Scalibregmatidae. *Diversity* **2021**, *13*, 87. [CrossRef]
91. Zanol, J.; Carrera-Parra, L.F.; Steiner, T.M.; Amaral, A.C.Z.; Wiklund, H.; Ravara, A.; Budaeva, N. The Current State of Eunicida (Annelida) Systematics and Biodiversity. *Diversity* **2021**, *13*, 74. [CrossRef]
92. Glasby, C.J.; Erséus, C.; Martin, P. Annelids in Extreme Aquatic Environments: Diversity, Adaptations and Evolution. *Diversity* **2021**, *13*, 98. [CrossRef]
93. Hutchings, P.; Carrerette, O.; Nogueira, J.M.; Hourdez, S.; Lavesque, N. The Terebelliformia-Recent Developments and Future Directions. *Diversity* **2021**, *13*, 60. [CrossRef]
94. Martinsson, S.; Erséus, C. Cryptic Clitellata: Molecular Species Delimitation of Clitellate Worms (Annelida): An Overview. *Diversity* **2021**, *13*, 36. [CrossRef]
95. Meca, M.A.; Zhadan, A.; Struck, T.H. The Early Branching Group of Orbiniida Sensu Struck et al., 2015: Parergodrilidae and Orbiniidae. *Diversity* **2021**, *13*, 29. [CrossRef]
96. Parapar, J.; Mortimer, K.; Capa, M.; Moreira, J. On the Systematics and Biodiversity of the Palaeoannelida. *Diversity* **2021**, *13*, 41. [CrossRef]
97. Schulze, A.; Kawauchi, G.Y. How Many Sipunculan Species Are Hiding in Our Oceans? *Diversity* **2021**, *13*, 43. [CrossRef]
98. Worsaae, K.; Kerbl, A.; Domenico, M.D.; Gonzalez, B.C.; Bekkouche, N.; Martínez, A. Interstitial Annelida. *Diversity* **2021**, *13*, 77. [CrossRef]
99. Capa, M.; Kupriyanova, E.; Nogueira, J.M.M.; Bick, A.; Tovar-Hernández, M.A. Fanworms: Yesterday, Today and Tomorrow. *Diversity* **2021**, in press.
100. Grosse, M.; Zhadan, A.; Langeneck, J.; Fiege, D.; Martínez, A. Still Digging: Advances and Perspectives in the Study of the Diversity of Several Sedentarian Annelid Families. *Diversity* **2021**, in press.
101. Martin, D.; Aguado, M.T.; Fernández Alamo, M.A.; Britayev, T.; Böggemann, M.; Capa, M.; Faulwetter, S.; Fukuda, M.V.; Helm, C.; Petti, M.A.V.; et al. An Approach to the Diversity of Phyllodocida (Annelida: Errantia), with a Focus on Glyceridae, Goniadidae, Nephtyidae, Polynoidae, Sphaerodoridae, Syllidae and the Holoplanktonic Families. *Diversity* **2021**, in press.
102. Struck, T. Phylogeny. Annelida, Basal Groups: Pleistoannelida, Sedentaria I. In *Handbook of Zoology*; De Gruyter: Berlin, Germany, 2019; Volume 1, pp. 37–68.
103. Blake, J.A.; Maciolek, N.J.; Meissner, K. Spionidae Grube, 1850. In *Handbook of Zoology*; Annelida; Volume 2: Pleistoannelida, Sedentaria II; De Gruyter: Berlin, Germany, 2019; pp. 1–102.
104. Blake, J.A.; Maciolek, N.J. Poecilochaetidae Hannerz, 1956. In *Handbook of Zoology*; Annelida; Volume 2: Pleistoannelida, Sedentaria II; De Gruyter: Berlin, Germany, 2019; pp. 103–119.
105. Blake, J.A.; Maciolek, N.J. Uncispionidae Green, 1982. In *Handbook of Zoology*; Annelida; Volume 2: Pleistoannelida, Sedentaria II; De Gruyter: Berlin, Germany, 2019; pp. 136–144.
106. Blake, J.A.; Maciolek, N.J. Trochochaetidae Pettibone,1963. In *Handbook of Zoology*; Annelida; Volume 2: Pleistoannelida, Sedentaria II.; De Gruyter: Berlin, Germany, 2019; pp. 120–135.
107. Capa, M.; Hutchings, P. Sabellariidae. In *Hanbook of Zoology*; Volume 2: Pleistoannelida, Sedentaria II; De Gruyter: Berlin, Germany, 2019; pp. 144–163.
108. Li, Y.; Kocot, K.M.; Whelan, N.V.; Santos, S.R.; Waits, D.S.; Thornhill, D.J.; Halanych, K.M. Phylogenomics of Tubeworms (Siboglinidae, Annelida) and Comparative Performance of Different Reconstruction Methods. *Zool. Scr.* **2017**, *46*, 200–213. [CrossRef]
109. Costello, M.J.; Wilson, S.; Houlding, B. Predicting Total Global Species Richness Using Rates of Species Description and Estimates of Taxonomic Effort. *Syst. Biol.* **2012**, *61*, 871. [CrossRef] [PubMed]
110. Fontaine, B.; Perrard, A.; Bouchet, P. 21 Years of Shelf Life between Discovery and Description of New Species. *Curr. Biol.* **2012**, *22*, R943–R944. [CrossRef]
111. Hutchings, P.A. Major Issues Facing Taxonomy—A Personal Perspective. *Megataxa* **2020**, *1*, 46–48. [CrossRef]
112. Goldstein, P.Z.; DeSalle, R. Integrating DNA Barcode Data and Taxonomic Practice: Determination, Discovery, and Description. *Bioessays* **2011**, *33*, 135–147. [CrossRef]
113. Capa, M.; Pons, J.; Hutchings, P. Cryptic Diversity, Intraspecific Phenetic Plasticity and Recent Geographical Translocations in *Branchiomma* (Sabellidae, Annelida). *Zol. Scr.* **2013**, *42*, 637–655. [CrossRef]
114. Seixas, V.C.; Zanol, J.; Magalhães, W.F.; Paiva, P.C. Genetic Diversity of *Timarete punctata* (Annelida: Cirratulidae): Detection of Pseudo-Cryptic Species and a Potential Biological Invader. *Estuar. Coast. Shelf. Sci.* **2017**, *197*, 214–220. [CrossRef]

115. Martinsson, S.; Erséus, C. Cryptic Diversity in Supposedly Species-Poor Genera of Enchytraeidae (Annelida: Clitellata). *Zool. J. Linn. Soc.* **2018**, *183*, 749–762. [CrossRef]
116. Marchán, D.F.; Fernández, R.; Domínguez, J.; Díaz Cosín, D.J.; Novo, M. Genome-Informed Integrative Taxonomic Description of Three Cryptic Species in the Earthworm Genus *Carpetania* (Oligochaeta, Hormogastridae). *Syst. Biodivers.* **2020**, *18*, 203–215. [CrossRef]
117. Gunton, L.M.; Kupriyanova, E.K.; Alvestad, T.; Avery, L.; Blake, J.A.; Biriukova, O.; Böggemann, M.; Borisova, P.; Budaeva, N.; Burghardt, I.; et al. Annelids of the Eastern Australian Abyss Collected by the 2017 RV 'Investigator' Voyage. *Zookeys* **2021**, *1020*, 1–198. [CrossRef] [PubMed]

diversity

MDPI

Review

On the Systematics and Biodiversity of the Palaeoannelida

Julio Parapar [1,*], Kate Mortimer [2], María Capa [3] and Juan Moreira [4]

1 Departamento de Bioloxía, Universidade da Coruña, 15001 A Coruña, Spain
2 Department of Natural Sciences, Amgueddfa Cymru—National Museum Wales, Cathays Park, Cardiff CF10 3NP, Wales, UK; Katie.Mortimer@museumwales.ac.uk
3 Departament de Biologia, Universitat de les Illes Balears, 07122 Palma, Spain; maria.capa@uib.es
4 Departamento de Biología (Zoología) and Centro de Investigación en Biodiversidad y Cambio Global (CIBC-UAM), Facultad de Ciencias, Universidad Autónoma de Madrid, 28049 Madrid, Spain; juan.moreira@uam.es
* Correspondence: julio.parapar@udc.es

Abstract: Palaeoannelida Weigert and Bleidorn, 2016 is an old clade branching off at the base of the Annelida radiation. It includes two morphologically and ecological divergent groups of sedentary burrowers and tube-dwellers: Magelonidae Cunningham and Ramage, 1888, and Oweniidae Rioja, 1917. Magelonids are characterised by a flattened, shovel-shaped prostomium and a pair of ventral papillated palps. Oweniids have simplified bodies lacking parapodia or appendages and are easily distinguished by the presence of oval patches of packed uncini, each with two distal curved teeth. The present review aims to summarise available information about the diversity of forms and life strategies displayed in the group, providing some guidelines for species identification and the techniques commonly used for their study. In addition, the assumed geographic distributions of some taxa are critically discussed. A brief introduction about the evolutionary relationships, systematics, and taxonomic history is given for both Magelonidae and Oweniidae. The motivation of this review is to highlight the main knowledge gaps from a taxonomic, methodological, and geographic perspective, aiming at stimulating further research into members of this clade.

Keywords: Oweniidae; Magelonidae; diversity; taxonomy; anatomy; biology

Citation: Parapar, J.; Mortimer, K.; Capa, M.; Moreira, J. On the Systematics and Biodiversity of the Palaeoannelida. *Diversity* **2021**, *13*, 41. https://doi.org/10.3390/d13020041

Received: 28 December 2020
Accepted: 15 January 2021
Published: 21 January 2021

1. Introduction

The term Palaeoannelida was proposed by Weigert and Bleidorn [1] for the clade formed by Oweniidae Rioja, 1917 and Magelonidae Cunningham and Ramage, 1888, branching of at the base of the annelid tree and sister to all the rest of annelids (see also [2,3]). The basal position of these two groups was assessed after analyses of phylogenomic data [4–6]. Previous analyses of just a few molecular markers had unveiled Oweniidae diverging early in the annelid radiation but only weakly supported (e.g., [7–15]). The position of these sedentary organisms at the base of the annelid tree generated some controversy with the fossil record, as most Cambrian annelid fossils showed morphologies corresponding to an epibenthic lifestyle. However, a recent finding of a tubicolous annelid dated in the early Cambrian [16], which was proposed within the Magelonidae, has suggested that a diversity of life modes, including sedentary and errant forms, may have inhabited the oceans at that time. The relationship between Magelonidae and Oweniidae, whilst supported by some morphological features such as the presence of a monociliated epidermis and lack of nuchal organs [17], still needs further work.

Magelonidae are known colloquially as the 'shovel head worms' due to their uniquely flattened and spade shaped prostomia (Figure 1) utilised in burrowing. The family is relatively small, containing 72 species worldwide [18]. Based on the diversity of species reported from relatively small geographic areas, the number of species is likely to be drastically underestimated [19]. The first species to be described was *Magelona papillicornis*

F. Müller, 1858 from Brazil, with the group being raised to the rank of family by Cunningham and Ramage [20]. Their unusual morphology has often led to difficulties in relating them to other annelid groups. However, in spite of this they are easily recognised by their characteristic prostomia, and two uniquely papillated and ventrally inserted palps (Figures 1A and 2A,B), characters which support the monophyly of the group [21] (and Mortimer et al., in preparation A). Magelonid species are relatively uniform in appearance and this has posed issues with species identification and the understanding of generic delineations within the family.

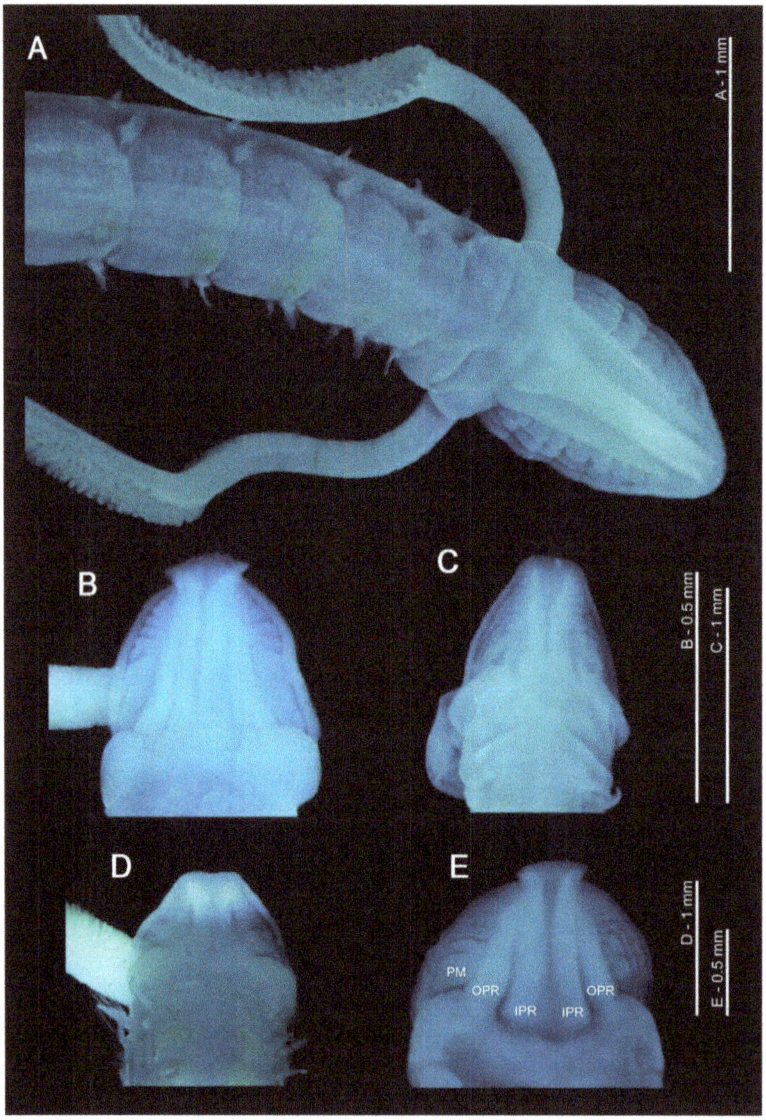

Figure 1. (**A**) Anterior end of *Magelona mirabilis*, showing prostomium, palps, achaetous first segment and first five chaetigers (dorsal view); (**B–E**) magelonid prostomia (dorsal views) of *M. crenulifrons*, *M. mahensis*, *M. symmetrica* and *M. wilsoni* respectively. All stained with methyl green. Abbreviations: IPR—inner prostomial ridge, OPR—outer prostomial ridge, PM—prostomial markings.

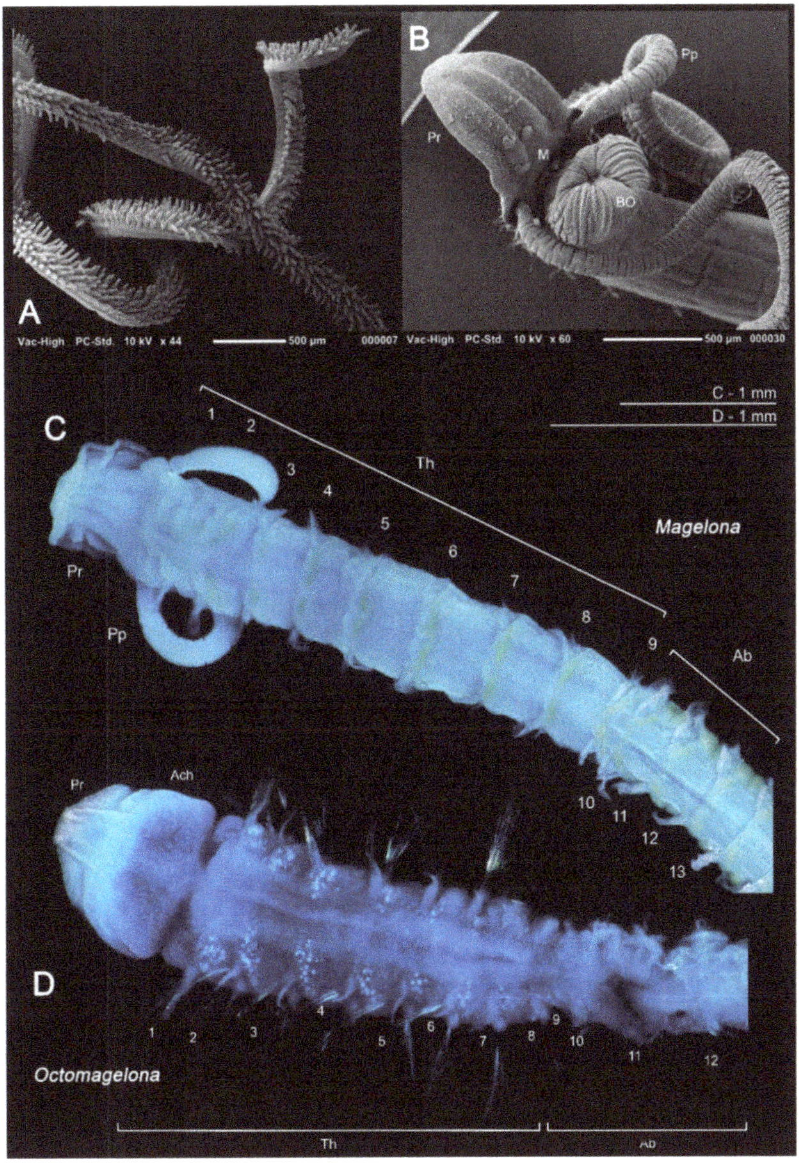

Figure 2. (**A**) Papillated palps of *Magelona johnstoni*; (**B**) anterior thorax of *M. johnstoni* showing partially everted burrowing organ, mouth at prostomium base, and basal portion of the palps (ventral view); (**C**) *M. gemmata* (dorsal view); (**D**) undescribed species of *Octomagelona* from Western Africa (dorsal view). Abbreviations: Ab—abdomen, Ach—achaetous first segment, BO—burrowing organ, M—mouth, Pp—palp, Pr—prostomium, Th—thorax. Numbers indicate chaetiger. (**C**,**D**) stained with methyl green.

At present, the family contains two genera: the type genus *Magelona* F. Müller, 1858 for species possessing thoracic regions of nine chaetigers (Figure 2C), and the monotypic *Octomagelona* Aguirrezabalaga, Ceberio and Fiege, 2001 for those possessing a thoracic region of only eight (Figure 2D). Whilst the latter genus contains only one described species, *Octomagelona bizkaiensis* Aguirrezabalaga, Ceberio and Fiege, 2001 from the Bay of Biscay

(North East Atlantic), several other undescribed species (from Mexico, Australia and West Africa) are known but have not yet been formally described [22]. All previously introduced generic names (*Maea* Johnston, 1865; *Rhynophylla* Carrington, 1865 and *Meredithia* Hernández-Alcántara and Solís-Weiss, 2000) have been synonymised with *Magelona*.

Several key researchers have worked on the family: McIntosh [23–28] provided much early knowledge on the morphology and anatomy of the group, and later Jones [29–33] published a series of taxonomic papers and a paper on the morphology, feeding and behaviour of an undescribed species from Woods Hole, Massachusetts. Jones [29] introduced unique terminology for the group, particularly relating to parapodial structures (e.g., lateral lamellae, ventral neuropodial lobes, dorsal medial and ventral medial lobes), which he subsequently modified [31,33]. The terms have been inconsistently applied [22] and their continued use may obscure potential homologies with conditions seen in other polychaete groups [19]. During a phylogenetic analysis of the family [21] standardisation of the terminology was attempted, with parapodial structures described in terms of shape, size and position. More recent papers include those from the Indian Ocean [34–39], Chinese waters [40,41], Gulf of Mexico [42–44]; California [45]; Brazil [46,47] and European waters [18,48–50]. Several observational papers looking at the behaviour of British species have also been published [51–53].

In general, magelonids are less than 1 mm wide (Figure 3A,B), but can reach over 100–150 mm in length [19,48]. Due to their fragility, average lengths of most species are unknown and thus total length is not generally a helpful characteristic for separation of species. However, body proportions are of use for identification. Whilst some species such as *Magelona minuta* Eliason, 1962, rarely attain widths greater than 0.5 mm, being slender animals (Figure 3A), others such as *Magelona alleni* Wilson, 1958 (Figure 3C), are more robust, attaining widths up to 1.5 mm (Mortimer et al., in preparation B). The terms slender, moderate and stout are terms often used in descriptions to describe overall body proportions and this is certainly something which warrants defining (Figure 3A–C).

Life span is unknown for most species. Although *Magelona sacculata* Hartman, 1961 is considered to be an annual species [54], animals of other species have been kept alive in tank environments for over two years (personal observations, KMJ). Magelonids are generally considered surface deposit-feeders [52], found in soft sediments, at a depth of less than 100 m [22]. However, several magelonid species have been recorded from deeper waters: 1000 to over 4000 m [55,56].

Regarding Oweniidae, there are around 60 species worldwide [57,58] included in four genera, namely *Owenia* Delle Chiaje, 1844, *Myriochele* Malmgren, 1867, *Galathowenia* Kirkegaard, 1959 and *Myriowenia* Hartman, 1960 [17].

In the last 50 years, there have been a number of studies on oweniid taxonomy in several regions across the world, namely Antarctica [59–63], Arctic and North Atlantic waters [64–68], Western Mediterranean [69], California [70,71], Gulf of Mexico and Caribbean Sea [72–74], Yellow Sea [75], Japan [76] and Australia [17,77,78]. However, updated information including revisions of old records are still lacking in many areas, such as eastern African coasts, across the Indian Ocean and most of the Pacific. Furthermore, some species, such as *Owenia fusiformis* Delle Chiaje, 1844 and *Myriochele heeri* Malmgren, 1867, have been for many years considered as having a large, cosmopolitan distribution (e.g., [67,71,79]). However, further morphological studies, focusing mainly on chaetal and crown tentacle morphology for *Owenia* species have revealed these are complexes of morphologically homogenous species [66,75,77,78,80,81].

The body length of oweniids usually ranges from 20 to 30 mm (e.g., *Myriochele*, *Galathowenia*) but species of *Owenia* may reach up to 10 cm [57]. Life span may be up to about 2–4 years [82]; development includes the planktonic larva mitraria that is unusual in having monociliated bands in comparison to the typical annelid trochophore [83]. Oweniids are deposit-feeders, tube dwellers usually found in sandy sediments, from shallow coastal habitats to the deep sea (e.g., [47,84]).

In this contribution we aim to provide an updated revision on current biodiversity knowledge in Palaeoannelida, focusing on species richness and distribution in different world sea regions, updates in taxonomy, classification, and systematics after recent advances in the knowledge of their anatomy, and evolutionary relationships and consequent classifications. We would like to emphasise major gaps in knowledge and where efforts should be made in terms of biodiversity surveys, analytical efforts, and the strengthening of taxonomic skills in order to increase our knowledge about the species inhabiting the planet.

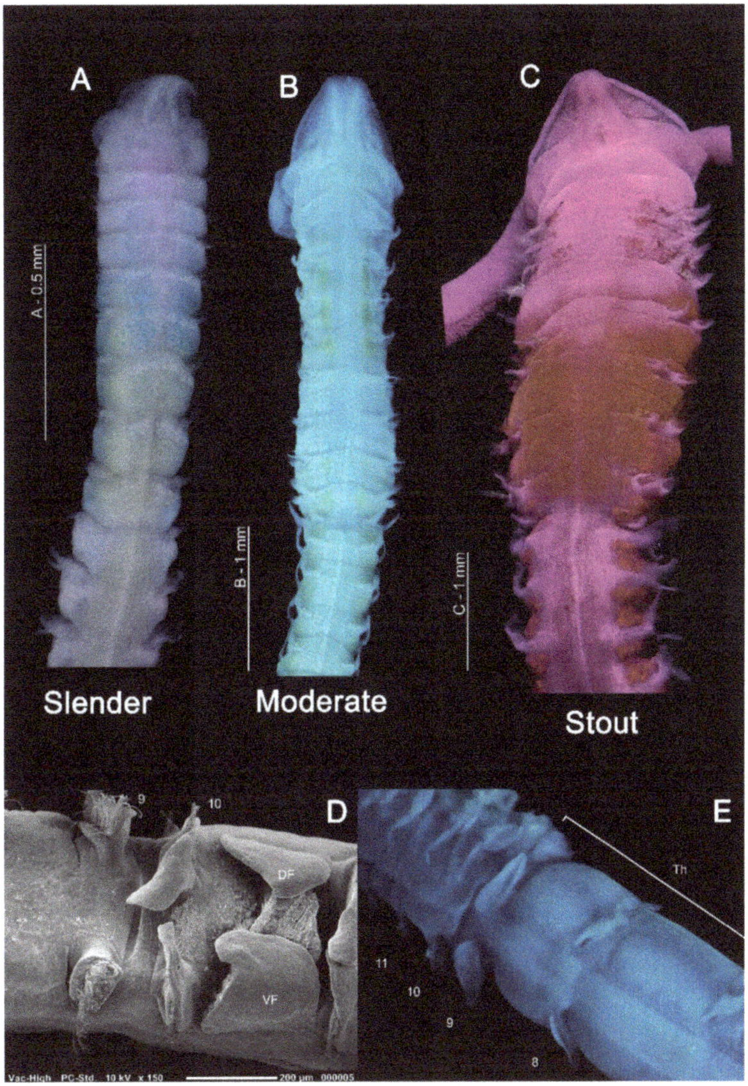

Figure 3. (**A**) Anterior of *Magelona minuta* (dorsal view, specimen stained with rose bengal and methyl green); (**B**) anterior of *M. mahensis* (dorsal view, stained with methyl green); (**C**) anterior of *M. alleni*, showing distinct pigment band between chaetigers 5–8 (dorsal view, stained with rose bengal); (**D**) thoracic/abdominal junction of *M. johnstoni* (lateral view), showing an anteriorly open pouch between chaetigers 10 and 11; (**E**) thoracic/abdominal junction of *M. johnstoni* (dorsal view). Abbreviations: DF—dorsal flap, Th—thorax, VF—ventral flap. Numbers indicate chaetiger.

2. Methods

A thorough literature review was performed in order to acquire information about species diversity, type locality, and geographic, bathymetric and ecological constrains of members of Palaeoannelida. Moreover, some recent advances in their anatomical study after Capa et al. [57] were incorporated due to their systematic significance. To accomplish this aim, the World register of Marine Species [58,85] database has been of great use, and some amendments herein made, (e.g., accepted synonymies according to latest references) have also been coordinated with WoRMS for its update. Tables with all considered currently valid nominal species alongside their type localities, depth (from original description) and marine realms (sensu Spalding et al. [86]) are provided (Tables A1 and A2). Additionally, a revision about methodologies and techniques used for species identification and characterisation is made, with the latest trends.

3. Results

3.1. Magelonidae Cunningham and Ramage, 1888

3.1.1. Systematics

There have been many difficulties relating magelonids to other annelid groups; a problem mirrored in the Oweniidae. For many years, magelonids have been considered among the spioniform polychaetes [19] and after phylogentic analyses of morphological data were placed in the Spionida (together with Apistobranchidae, Chaetopteridae, Longosomatidae, Poecilochaetidae, Spionidae, Trochochaetidae, Uncispionidae clades [87]), principally based on the presence of a pair of grooved palps and spiomorph parapodia [87], or related to Oweniidae [17]. Various other placements have since been suggested, for example, related to *Polydora* Bosc, 1802 (Spionidae), *Cirriformia* Hartman, 1936 and *Dodecaceria* Örsted, 1843 (Cirratulidae), or *Fauveliopsis* McIntosh, 1922 (Fauveliopsidae) [10] utilising combined morphological and DNA sequences of nuclear markers. Further, phylogenomic analyses placed Magelonidae as sister to Oweniidae branching off at the base of the annelid tree [4,6,88].

Although a sister-group relationship of Magelonidae and Oweniidae has been outlined, this and their position within Annelida are still the subject of intense debate [57]. Clarification is needed in regard to the extent to which the prostomium is fused to the peristomium in magelonids (see below) and also in the nature of the 'buccal organs' in the two families (see below in relation to the burrowing organ). It has been suggested that the head structures in the Palaeoannelida clade are heterogeneous, with *Magelona* possessing papillated peristomial palps whilst grooved prostomial palps are present in *Owenia* [89]. Further work is clearly needed, but perhaps further studies assessing similarities and differences between the two families may shed light on the subject.

There is only one cladistic analysis on the Magelonidae to date [21], which confirmed the monophyly of the group but from which no further proposals were made. A forth coming account of inferences of phylogenetic hypotheses within the Magelonidae (Mortimer et al., in preparation A) may add clarity. A fossil polychaete from the early Cambrian (*Dannychaeta tucolus* Chen, Parry, Vinther, Zhai, Hou and Ma, 2020) has recently been described within the Magelonidae based on phylogenetic analyses [16].

3.1.2. Taxonomic History

Taxonomically, Magelonidae received little attention prior to the 1930s (Figure 4), however, the number of taxonomic papers increased in the 1940s and 1950s with works by prolific polychaete researchers such as Hartman [90,91]. With the 1960s, the group saw a rapid expansion of taxonomic work, not only by notable magelonid workers such as Jones [29] but works by Gallardo [92], Eliason [93], Hartman [94,95], Glémarec [96], Kitamori [97], Reish [98], Hartmann-Schröder [99], Day [100] and Harmelin [101] (Table A1). This increase is likely to have been influenced by the publication of Jones, which set out a standard to which magelonid works should attain. The 1970s largely saw works by Jones, who by that time had been established as the magelonid expert [31–33]. His research on magelonids continued until 1979 where he switched his attention to the Vestimentifera (now Siboglinidae) [48]. The major taxonomic works of the 1980s and 1990s were those concentrating on Southeastern Brazil [46] and Thailand [34]. Since 2000, the number of researchers producing taxonomic works on the group, and the number of species being described has once again increased. Workers such as Mortimer and Mackie [35,36], Fiege et al. [48], Hernández-Alcántara and Solís-Weiss [43], Mortimer et al. [38,49] and Magalhães et al. [102] (see Table A1 for all references) contributing further to the taxonomic knowledge of the group. Certainly, with renewed interest in the group, it is likely that this trend in increasing publications will continue. However, outside of the key works by Jones there have been very few major revisions of the family [21,22,48]. A forthcoming account of inferences of phylogenetic hypotheses within Magelonidae (Mortimer et al., in preparation A) will add further information on the key morphological characters.

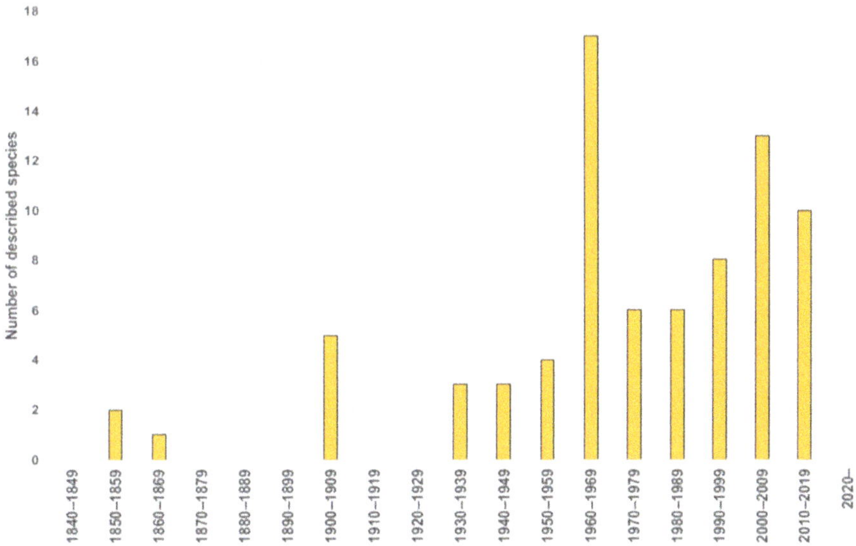

Figure 4. Number of magelonid species described per decade.

The only worldwide key to species was produced in 1963 [29]. However, over 50 species have been described since that time (WoRMS, Table A1). Whilst keys to local regions can be found in several papers: California [45], Eastern USA [95], Gulf of Mexico [42–44], Brazil [46], Europe [18,48,49], Japan [97,103] Viet Nam [92], Thailand [34], Western Indian Ocean [37,38], Seychelles [36], South Africa [104], updated versions are warranted in most regions.

3.1.3. Taxonomic Characters and External Morphology

Magelonids are generally fragile, being long and slender and for this reason they can be extremely difficult to collect whole. Over 60% of species have been described from posteriorly incomplete specimens. They are often pale in colour (white to cream), although the gut, which frequently has a darkish green hue, can often be seen through the body wall. Several species (*M. alleni*, *Magelona cincta* Ehlers, 1908, *Magelona equilamellae* Harmelin, 1964, *Magelona japonica* Okuda, 1937 and *Magelona variolamellata* Bolívar and Lana, 1986) are known to carry dark pigmentation occurring as a distinct band between chaetigers 5–8 [52,53] (Figure 3C), and pigmentation in other parts of the body such as the palps has additionally been reported, e.g., *Magelona mirabilis* (Johnston, 1865) [48,51]. In addition, specimens belonging to two undescribed species recently found off West Africa carry dark pigmentation of the parapodia, and distinct stripy pigmentation from head to toe, respectively (Mortimer et al., in preparation B).

The magelonid body is divided into distinct regions: the head (comprising the prostomium and the peristomium), the thorax (an achaetous first segment, and either eight or nine chaetigers), the abdomen comprising many chaetigers (approximately 50–160 chaetigers, see [22], although for most species this is unknown), and the pygidium (Figure 2C,D). The precise location and organisation of the peristomium has been debated [10,19]. Early works suggested the prostomium "merges posteriorly with the achaetous peristomium" [30], suggesting that the peristomium is represented by the achaetous first segment. However, magelonid larvae have been shown to possess two anterior segments bearing chaetae which subsequently become achaetous [105]. Subsequent workers regarded the peristomium to be limited to the buccal region [87]. Personal observations (KMJ) suggest that the peristomium is visible on the ventrum (see [22]: figure 4.2.5, posterior to buccal region) and this and the achaetous 'first segment' (Figure 1) warrant further investigation. Magelonids have long been regarded to possess an eversible ventral proboscis (e.g., [30,87]), however this heart shaped sac is not connected to the buccal region, plays no part in feeding and is involved only in burrowing (Figure 2B). The term burrowing organ has now been applied to this structure to avoid future confusions [18,22]. Between the thorax and abdomen, there is an often-marked constriction in the body (Figure 3D,E), and the two flanking regions are additionally discernible by a change in chaetal type (Figures 2C,D and 3) from capillary chaetae (Figure 5) to hooded hooks (Figure 6). Magelonids possess biramous parapodia, often carrying foliaceous flattened structures, for which the term '*lateral lamellae*' was coined [29]. However, these structures may be filiform to foliaceous and may be prechaetal, postchaetal, subchaetal or lateral in position. It is the identification of these features, particularly in the thoracic region which is a major diagnostic feature in separating species, and it is for this reason that all thoracic parapodia must be fully described and illustrated [21,31].

Several 'crucial morphological' characters for separation of species have been suggested [29,45]: (1) dentition of abdominal hooded hooks; (2) presence or absence of prostomial horns (distal projections of the anterior margin); (3) presence or absence of medial lamellae (DML, VML) in the posterior region (cirriform lobes adjacent to hooded hooks see Figures 6B and 7B,C); (4) presence or absence of specialised chaetae of the ninth chaetiger (mucronate, pennoned); (5) morphology of thoracic lamellae; (6) the relative dimensions of the prostomium (L:W ratio); (7) presence or absence of superior dorsal lobes on thoracic chaetigers; (8) presence and distribution of lateral abdominal pouches and (9) the presence or absence of postchaetal expansions behind chaetal rows in the abdomen. With the description of further taxa, several other characters are additionally noted: (10) pigmentation patterns and (11) overall body size and proportions (Figure 3).

Magelonid prostomia can be broadly divided into three categories: length approximately equal to width (Figure 1B,C), length greater than width (Figure 1A), or width greater than length (Figure 1E). The overall shape can aid identification, but care must be taken as the appearance can be modified if lateral or anterior margins become compressed. Prostomial horns may be present or absent. When present, they may be distinct (Figure 1B,E) such as in *Magelona montera* Mortimer, Cassà, Martin and Gil, 2012, or 'rudimentary' (Figures 2D and 3C) in which the anterior margin is straight and square, such as in *M. alleni*. The prostomial anterior margin may be smooth (e.g., *M. mirabilis* (Figure 1A), medially indented (e.g., *Magelona symmetrica* Mortimer and Mackie, 2006, Figure 1D) or crenulated (e.g., *Magelona crenulifrons* Gallardo, 1968) (Figure 1B,E). The prostomium may carry one (Figure 2D) or two pairs of dorsal muscular ridges (Figure 1A,B), and distinct patterned areas either side of the ridges (often as raised oblong or arched lines) may be present or absent (Figure 1).

The biramous parapodia of the thoracic region carry lamellae which may be equal in terms of size and shape in both rami, or noticeably larger in the notopodia. In some species the neuropodial lamellae may be marginally larger, however, this situation is generally infrequent. The lamellae may be filiform (e.g., *Magelona filiformis* Wilson, 1959) (Figure 5A) to foliaceous (e.g., notopodia of *Magelona sinbadi* Mortimer, Cassà, Martin and Gil, 2012) in shape (Figure 5B,D). The notopodial lamellae are generally postchaetal to subchaetal and encircle the chaetal bundle, confluent with a much lower prechaetal lamellae, almost cuff-like (Figure 5C). The upper edges of the lamellae may be smooth (e.g., *M. equilamellae*), crenulate (e.g., *M. johnstoni*) or bi-lobed (e.g., *Magelona obockensis* Gravier, 1905). At the top of the notopodia and in a slightly prechaetal position superior dorsal lobes (SDL) may be present (Figure 5B–D). These are smaller than the lamellae, and often digitiform in shape, although in some taxa they may be foliaceous. When present, they may occur on all thoracic chaetigers, from chaetigers 1–8 or only present in the posterior thorax (e.g., chaetigers 4–8). The neuropodial lamellae of the thoracic region may be prechaetal, subchaetal or postchaetal, the edges of which are generally smooth. Many species have filiform subchaetal lamellae attached to low pre- and post-chaetal ridges (Figure 5A,E). In these species the lamellae may occur in the same position along the thorax, or vary, often starting in a slightly prechaetal position, becoming ventral in the mid thorax and becoming postchaetal by the posterior thorax. The postchaetal portion may be more expanded in some species, particularly on the eighth and ninth chaetigers, where it is often triangular and of a similar length to the subchaetal lamellae (Figure 5D). The parapodia of the eighth and ninth chaetigers are particularly important to observe. In some species, the thoracic lamellae are similar on all thoracic chaetigers (e.g., *M. symmetrica*), whilst in others the first seven are similar with the lamellae of chaetigers eight and nine varying (e.g., *M. montera*), and in others it is the lamellae of chaetiger nine which varies in comparison to the remaining thoracic lamellae (e.g., *M. mirabilis*).

The thorax has unilimbate (e.g., *M. minuta*) or bilimbate capillary chaetae (e.g., *M. equilamellae*), the latter of which may have irregular blades [21,45,46]. More variation within thoracic chaetae may exist [18,50]. In most species, the ninth chaetiger has gently tapering chaetae like those on preceding chaetigers, however, several species have specialised chaetae on the 9th chaetiger. These may be mucronate (Figures 3D,E, 5F and 7B) or pennoned, in which the limbations broaden distally, culminating in an acute tip (see [33], p. 340).

The abdomen which starts at the 9th chaetiger for *Octomagelona* (Figure 2D) and the 10th chaetiger for *Magelona* (Figure 2C) is comprised of numerous chaetigers (~50–160 chaetigers). Parapodia are biramous, and as in the thorax, carry lateral lamellae (e.g., LL in Figure 7C) which are generally symmetrical in terms of size and shape between the two rami. However, in a few species such as *M. alleni* and *Magelona korena* Okuda, 1937 the notopodial lamellae are somewhat larger than those of the neuropodia. Lamellar shape varies but can be generally separated into those which are rounded (Figure 6B), carrying a basal constriction and those which are slender triangular, with no basal constriction (Figures 3D, 5A and 7C). Above the lamellae in the notopodia (Figure 6A), and below the lamellae in the neuropodia (Figure 7B), hooded hooks occur in a single row (N.B. the hooks of *M. equilamellae* were noted to occur in two rows towards the middle of the ramus [18]). In each ramus, hooks may be orientated in one direction, laterally towards the lamella (Figures 3D and 6A), or in two groups vis-à-vis (Figure 6B). Behind the chaetal row, a postchaetal expansion of the lamella may be present (Figure 6B). At the inner margins of the chaetal rows, small triangular to digitiform processes (dorsal (DML) and ventral medial lobes (VML) of Jones) may be present (Figures 6B and 7B,C).

Abdominal hooded hooks may be bidentate (with one secondary tooth above the main fang) (Figure 6C), tridentate (with two secondary teeth) (Figure 6D) or polydentate (Figure 6E,F). Quadridentate, pentadentate and hexodont hooks have all been reported to occur within the family. Enlarged hooks may be present, the form of which varies from spines, re-curved hooks or enlarged 'ordinary' hooks (e.g., *Magelona falcifera* Mortimer and Mackie, 2003 or *Magelona spinifera* (Hernández-Alcántara and Solís-Weiss, 2000)). Some species have a small hook emerging at the base of the lateral lamellae (e.g., *M. filiformis* see [18]: 104), whilst several curved support chaetae (aciculae) may also be present (e.g., *Magelona conversa* Mortimer and Mackie, 2003).

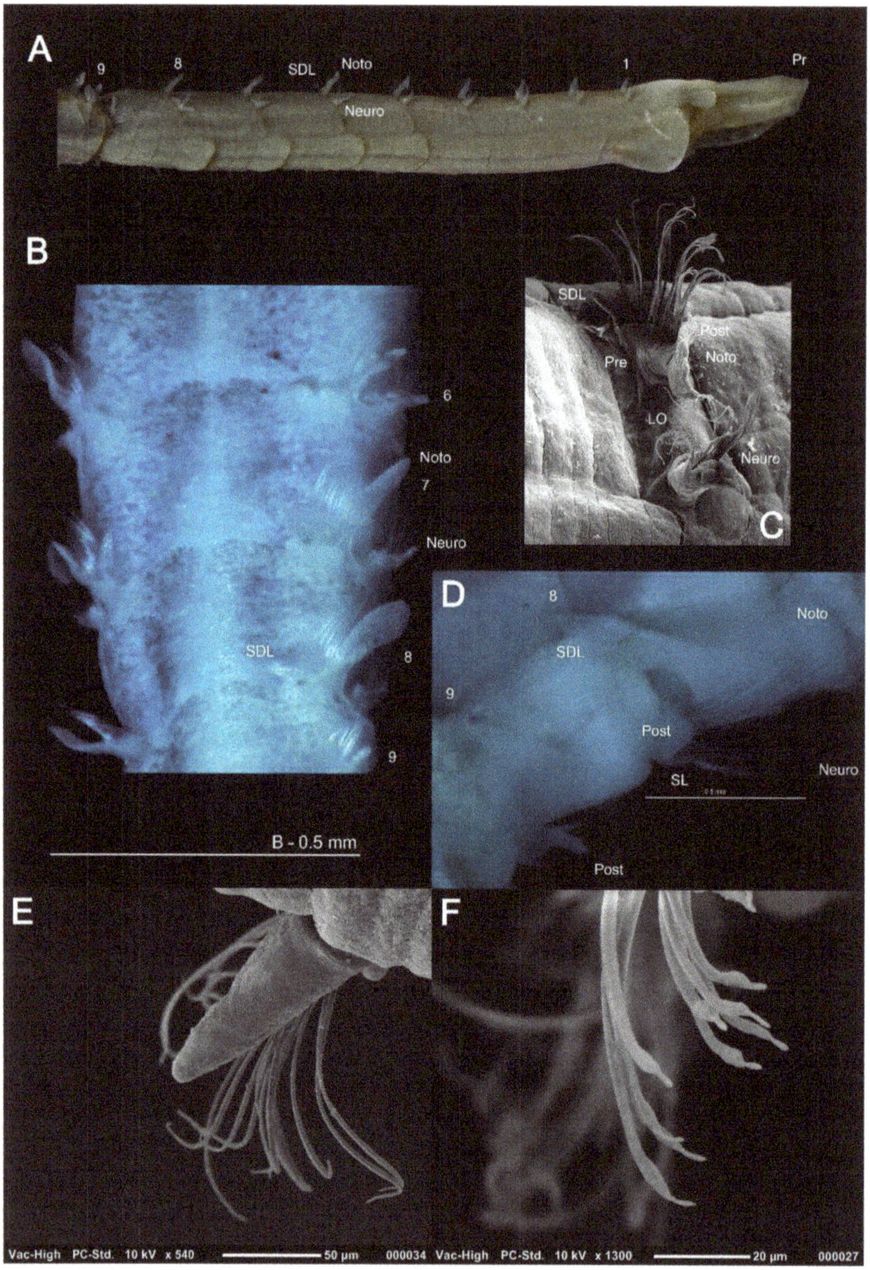

Figure 5. Thoracic parapodia and chaetae: (**A**) anterior of *Magelona filiformis* (lateral view) showing filiform lamellae; (**B**) *M. crenulifrons* from chaetiger 6–9 (dorsal view, stained with methyl green) showing foliaceous notopodial lamellae and superior dorsal lobes; (**C**) left-hand parapodia of chaetiger 8 of *M. johnstoni* (lateral view); (**D**) right-hand parapodia of chaetigers 8 and 9 of *M. wilsoni* (**E**) thoracic subchaetal neuropodial lamellae of *M. johnstoni* showing capillary chaetae; (**F**) mucronate capillary chaetae of chaetiger 9 of *M. johnstoni*. Abbreviations: LO—lateral organ, Noto—notopodia, Neuro—neuropodia, Post—postchaetal, Pr—prostomium, Pre—prechaetal, SDL—superior dorsal lobe, SL - subchaetal lamellae. Numbers indicate chaetiger.

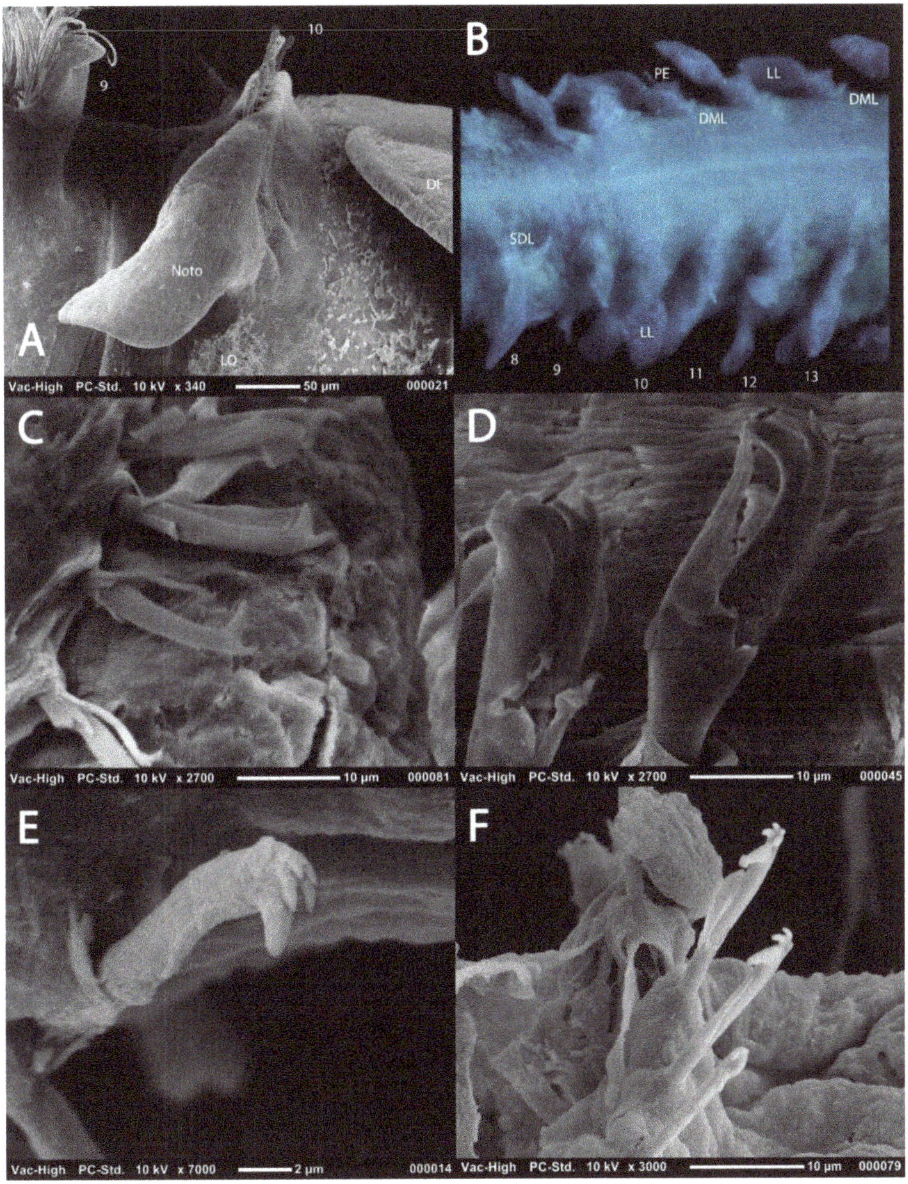

Figure 6. Abdominal parapodia and chaetae: (**A**) notopodia of chaetiger 10 of *Magelona johnstoni* showing abdominal hooded hoods in a vis-à-vis orientation (lateral view); (**B**), chaetigers 8–13 of *M. crenulifrons* (dorsal view, stained with methyl green); (**C**) bidentate abdominal hooded hook of *M. minuta* (lateral view); (**D**) tridentate abdominal hooded hook of *M. alleni* (lateral view); (**E**) pentadentate abdominal hooded hook of *M. fauchaldi* (oblique lateral view); (**F**) quadridentate and pentadentate abdominal hooded hooks of *M. fauchaldi* (oblique lateral view). (**C–F**), hoods broken via sonication prior to SEM. Abbreviations: DF—dorsal flap of the lateral pouch, DML—dorsal medial lobe, LL—lateral lamella, Noto—notopodia, PE—postchaetal expansion, SDL—superior dorsal lobe, VML—ventral medial lobe. Numbers indicate chaetiger. Figure F sourced from [39].

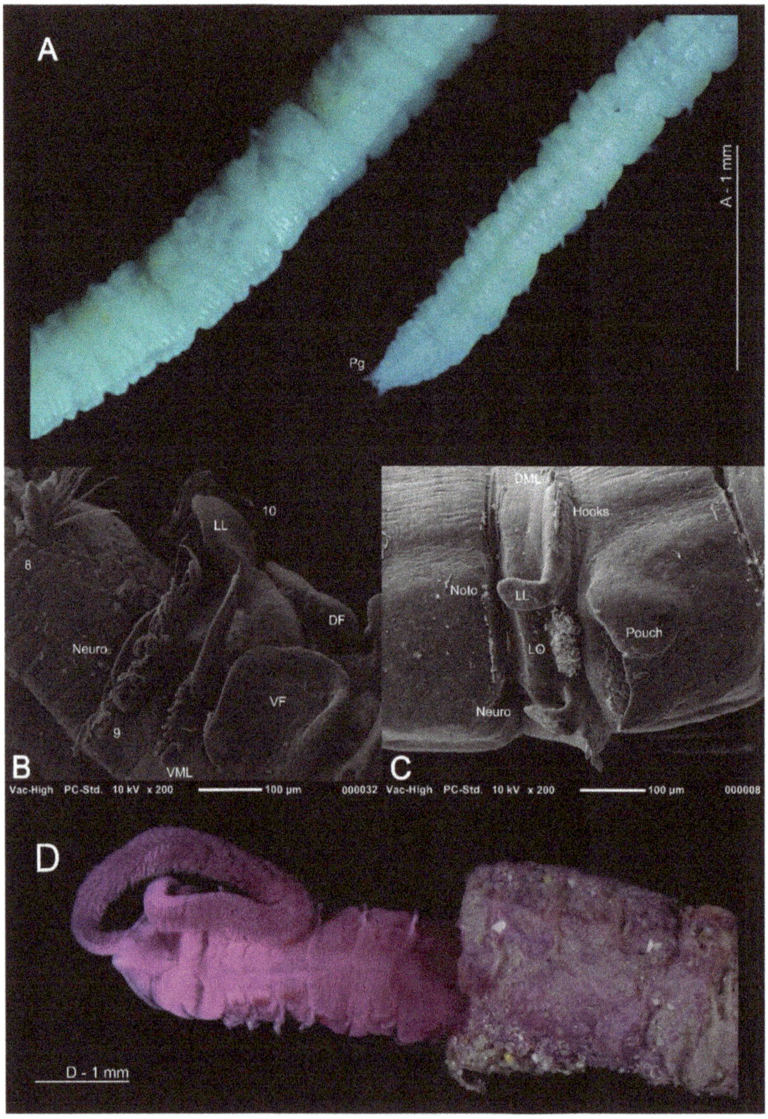

Figure 7. (**A**) Posterior abdomen and pygidium of *Magelona mahensis* (lateral and ventral views respectively), specimen ovigerous; (**B**) anteriorly open lateral pouch between chaetigers 10–11 of *M. johnstoni* (ventral view); (**C**) posteriorly open lateral pouch from posterior abdomen of *M. johnstoni* (lateral view); (**D**) anterior of *M. alleni* protruding from multi-layered sediment covered tube (dorsal view). (**A**) is stained with methyl green, (**D**) is stained with rose bengal. Abbreviations: DF—dorsal flap of the lateral pouch, DML—dorsal medial lobe, LL—lateral lamellae, LO—lateral organ, Neuro—neuropodia, Noto—notopodia, Pg—pygidium, VF—ventral flap of the lateral pouch, VML—ventral medial lobe. Numbers indicate chaetiger.

Lateral abdominal pouches are recorded in approximately half of all known species, although their function is currently unknown [51]. Whilst pouches may be divided roughly into two types [48], greater variation in morphology has been noted [22,37,51]. Pouches may be anteriorly (comprising of a dorsal and ventral flap, with a convoluted membrane in between) (Figures 3D and 7B) or posteriorly open (simple and pocket-like flaps) (Figure 7C). The former type generally occurs in pairs on either side of segments at the start of the abdomen, whilst the latter can be paired or unpaired, and may occur from median abdominal chaetigers, or only in the extreme posterior of the animal [51]. Posteriorly open pouches may occur on consecutive segments or alternating from segment to segment and from one side of the body to the other. They may be smooth or medially split.

For the 40% of species in which the pygidium is described, morphology is relatively uniform, possessing two slender cirri, placed laterally, with the anus being ventral (Figure 7A). However, variation has recently been highlighted [18,52] (Mortimer et al., in preparation B) with *M. alleni* and *M. equilamellae* possessing more robust projections either side of a terminal anus.

Several species have shown species-specific staining patterns using methyl green or methyl blue [34,39–41,50] and these methods can be extremely useful in identifying specimens in bulk samples.

3.1.4. Internal Morphology

In comparison to observations on the external morphology of magelonids the internal morphology has received less attention. Much of the early work on the anatomy of magelonids came with the publications of McIntosh [23–28] and Jones [30]. The latter author providing details of early studies, including the circulatory system [106], lateral organs [29], musculature [107,108], nervous system [109] and including his own observations on the nervous system, muscular system, septa, and circulatory system. Monociliated epidermal cells in the larvae of *M. mirabilis* were reported [110], although figure 1A of that publication is unlikely to be *M. mirabilis*.

The closed circulatory system consists of an anterior dorsal vessel, divided into a series of chambers set apart by valves. A heavily muscularised portion of the dorsal vessel is largely responsible for the movement of blood around the animal [30]. The circulatory system is vital in the eversion of the burrowing organ ([30], although note comment above about terminology) which can be clearly observed in live animals ([22]: figure 4.2.4). Notes on the unusual pale pink blood which contains numerous corpuscles have been made [111], the blood pigment is hemerythrin [112]. Palps contain a single blood vessel, clearly observed in live material ([51]: figure 1F) and it has been suggested that the palps have a secondary respiratory function [26,30].

Various authors have reviewed the structure and evolution of the nervous system [2,113,114]. The central nervous system of the Magelonidae has been considered to be simple in comparison to those of errant polychaetes [114]. The brain is composed of an anterior compact neuropil and posteriorly encircles the prostomial coelomic cavities. Thereafter, two lateral medullary cords branch off and fuse caudally. The ventral nerve cord comprises of two parallel cords of neurite bundles and paired neuropils which fuse between the 9th and 10th chaetigers [2]. Nuchal organs, ganglia and mushroom bodies are all considered to be absent [114], although the former are considered to occur in larvae [19]. Although eyes have been recorded as absent in adult magelonids, a species recently described from Hawaii, *Magelona cinthyae* Magalhães, Bailey-Brock and Watling, 2018 has distinct eyespots on the posterior prostomium.

The musculature of a species approaching *M. mirabilis* was investigated by phalloidin labelling and confocal laser scanning microscopy [115]. The prostomial muscles are anterior extensions of the ventral longitudinal muscles of the body and a complex of circumbuccal muscles are described ventral to the mouth opening. However, it is perhaps worth noting that this region is at the base of the burrowing organ rather than connected to the buccal region itself and the terminology should be reviewed. The palps possess muscles along their entire lengths, comprised of longitudinal fibres, which the authors [115] suggested would restrict movement to contraction and slight coiling. Whilst some authors have noted this to be the case for some species, including *M. mirabilis* [51,116] motile palps have been reported for other species [51,52] and will be discussed below. Each palp is moved by two palp retractor muscles. Dorsally and ventrally muscle strands run longitudinally, however, there is a distinct change in musculature between the thoracic and abdominal regions. Oblique and cross-striated muscles in *M. papillicornis* (likely to be *M. mirabilis* or *M. johnstoni*, see [48]) were investigated by Wissocq [117].

3.1.5. Species Diversity and Distribution

There are 72 magelonid species currently considered to be valid ([58], Table A1). Species have mainly been described from temperate and tropical environments, and the number described from the Temperate Northern Pacific, Temperate Northern Atlantic and Western Indo-Pacific is comparably larger than any other marine realm (Table A1, Figures 8 and 9A). There are no species described from the Arctic or the Southern Ocean (Table A1, Figures 8 and 9) although records from the Global Biodiversity Information Facility (GBIF [118] suggest magelonids have at least been found in these two regions but remain unverified at this time. These type locality occurrences perhaps reflect the influence of workers such as Jones [29], Hartman [95] and Wilson [119,120] in the Temperate North Atlantic, Hartman [90,94], and Jones [31,33] in the Temperate North Pacific, and Nateewathana and Hylleberg [34], Mortimer and Mackie [35,36] and Mortimer et al. [38] in the Western Indo-Pacific (Table A1), rather than differences in actual species numbers occurring in each region. However, further work is needed to corroborate this hypothesis.

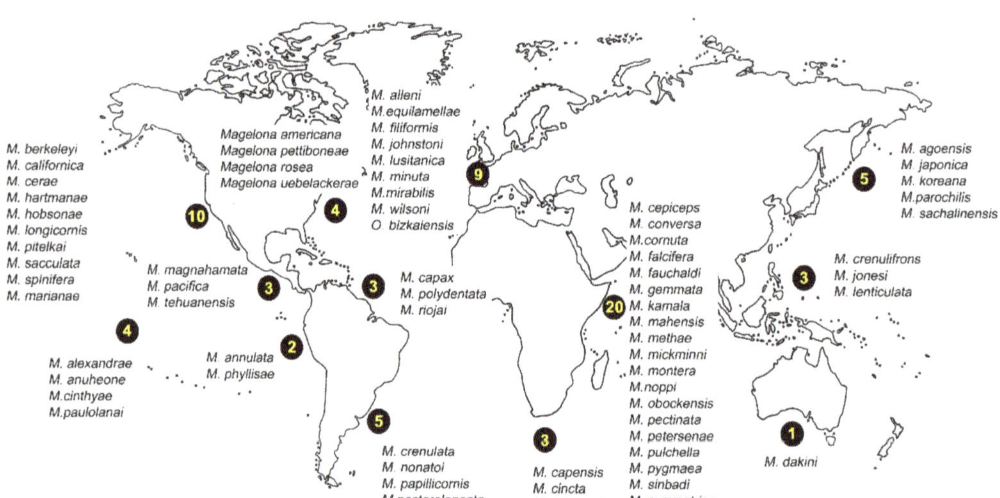

Figure 8. World map showing number of magelonid species described per realms (sensu [86]).

The diversity of magelonid species is generally high in relatively small geographical areas, e.g., 13 species in the Gulf of Mexico [42], nine species in European waters [18], 11 in the Arabian Gulf [38] and over 20 off west African waters (Mortimer et al., in preparation B). It was suggested that the high diversity of magelonid species observed off Phuket Island, Andaman Sea may be a direct result of sediment disturbance associated with monsoons [121], the authors hypothesising that catastrophic events affect interspecific competition by removing dominant species. However, areas with similar species diversities can be seen in non-monsoon affected areas. Abundance data on populations of *Magelona* in Monterey Bay indicate that population densities can be extremely variable form year to year [122].

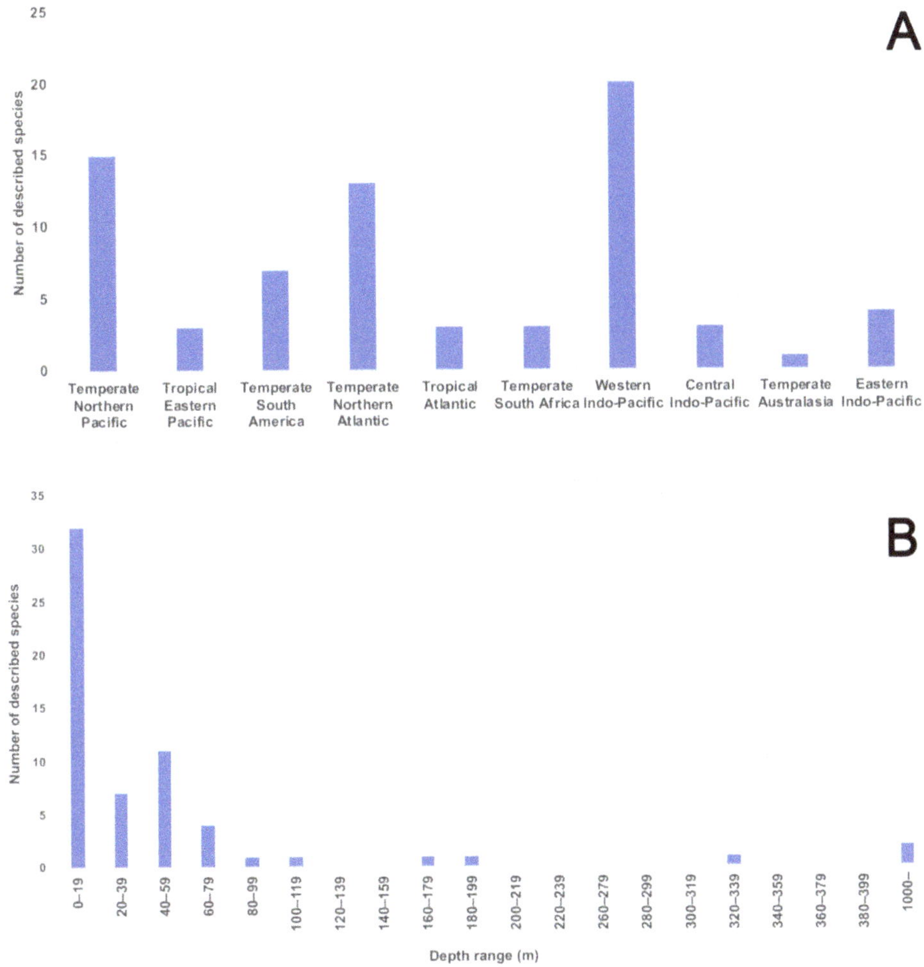

Figure 9. (**A**) All 72 magelonid species currently consider valid [58] listed under the marine realms (sensu [86]) from which they were described. Numbers indicate the number of species described from each bioregion; (**B**) number of species described from each depth range (depth in meters, data from the original descriptions of 61 species for which the data was provided).

Magelonids show a preference for shallow waters and data taken from the original descriptions of 61 magelonid species (for which data was provided) shows the number of species at depths of less than 20 m to be much greater than at any other depth range, with the number of species dropping off substantially after 60 m (Figure 9B). Although this information only includes data from the original descriptions, it does indicate the group's preference for shallower waters. Whilst the data may be biased by the ease of sampling onshore or in shallower waters it is likely that these depth patterns contrast with other annelid groups. However, further work is certainly needed to clarify depth ranges of individual species. Whilst some magelonid species are found both intertidally and subtidally, some are known to be distinct offshore species (e.g., *M. minuta*; [22]). Far fewer deep-water species have been recorded: *O. bizkaiensis* (1000–1040 m; [56]), *M. minuta* (1000 m; [48]), *M. capax* Hartman, 1965 (4769 m) and *Magelona* spp. (3753–5000 m) [55].

Magelona phyllisae Jones, 1963 (note incorrect spelling *M. physilia*) has been noted to occur in low oxygen habitats [123] in northern Chile. Sensitivity to hypoxic conditions was also noted in the "dead zone" of the Gulf of Mexico for a species approaching *M. phyllisae* [124].

Very little information about the distribution ranges of magelonid species exists, and those already documented need validation. *Magelona papillicornis* was believed to be a widespread species, occurring from regions such as America, Africa, Europe, New Zealand, and India [32]. It is now recognised that many of these records are erroneous and have since been referred to several other species such as *Magelona debeerei* Clarke, Paterson, Florence and Gibbons, 2010, *M. mirabilis* and *M. johnstoni* [32,48]. Current maps from GBIF [118] highlight many of these erroneous records still and should be verified to prevent further confusion. Recent work off the western coast of Africa (Mortimer et al., in preparation B) involving morphological and molecular work suggests that whilst some species have quite restricted distributions others may be far greater. Certainly, further work is needed but this relies heavily on increased taxonomic work to resolve the identity of species in many areas. Undoubtedly, species recorded large distances from the type locality should be treated with caution at this time, e.g., American and Brazilian species recorded off India [125].

3.1.6. Biology and Ecology

Magelonids are common in sands and muds, both intertidally and subtidally; most species occurring in waters less than 100 m deep. The group's preference for fine sediments has been linked to avoidance of sharp fragments that might damage the burrowing organ so vital for moving within sediments [27]. Whilst the broad sediment preferences are recognised for the family, very little is known about specific species habitats. Investigations into the distribution of four species (*M. alleni, M. filiformis, M. mirabilis* and *M. johnstoni*) in the German Bight showed sediment (median grain size diameter or mud content) to be the most important predictor in habitat suitability, but with salinity and water depth also of importance [126]. However, variations in habitat preference were observed between species (e.g., *M. filiformis* in sediments with less than 10% mud, whilst *M. alleni* occurred in sediments with elevated mud contents), something mirrored in observations in the UK, with *M. minuta* and *M. alleni* (personal observations, KMJ). Within the waters off the Arabian Peninsula individual species were noted to occur in sediments with distinct granulometric characteristics [38].

Whilst most magelonid species are believed to burrow more or less continuously through sediments [51,127], at least 10 species (e.g., *M. alleni, M. cincta, Magelona polydentata* Jones, 1963, *M. variolamellata*) are known to build distinct tubes (Figure 7D), which are often multi-layered paper-like tubes covered in sediment [22,52,53]. Individuals of *M. alleni*, have been observed to live for over a year in tube-lined burrows under tank conditions, being highly sedentary in comparison to previously observed species (Personal observations, KMJ).

Magelonids have been primarily described as surface deposit feeders [30,42,51,52], although suspension feeding [19,51,128,129], subsurface feeding and carnivory [116] have additionally been suggested. Varying feeding modes have been observed to predominate in different species observed simultaneously in the same laboratory setting [52]. Fauchald and Jumars [127] considered feeding to be selective, with selectivity increasing in nutrient-poor conditions. However, in contrast *M. alleni* was reported to be predominately non-selective [52] and magelonids feeding upon large volumes of sediment have been reported [116]. The feeding process has been described in detail for several species [30,51,52,130] and involves palps above the sediment surface, collecting particles (Figure 10A,B), which are then transferred along the palp to the mouth by cooperative movements of the papillae. A mucus thread has been postulated to aid transport of particles along the palps [30,130] and additionally reported to be involved in excretion in the tubicolous species *M. alleni* [52]. Given papillae involvement in particle transfer, differences in palp morphology between species, e.g., those species which possess slender palps with few papillae, in comparison to those that are broad carrying numerous rows of long papillae may indicate differences in feeding strategies and diet.

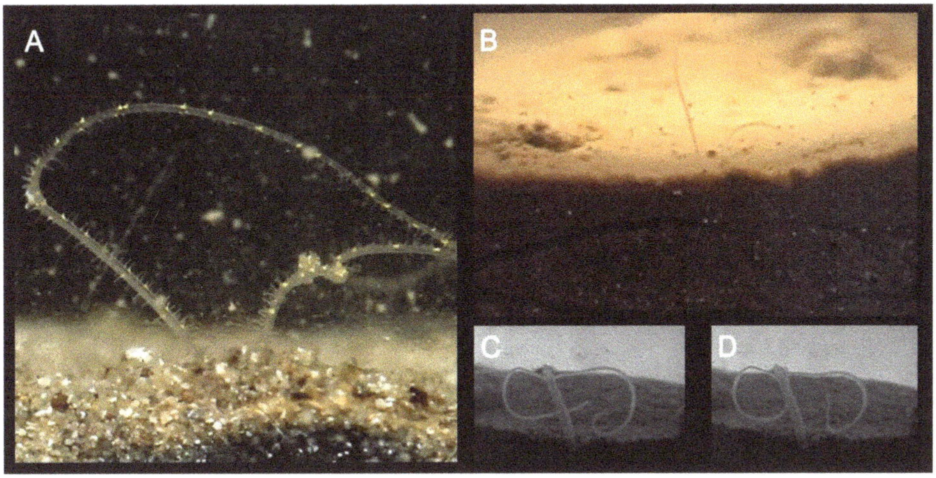

Figure 10. (**A**) Distal portion of the palps of *Magelona alleni* within the water column collecting food particles. Particles are transferred to the mouth via cooperative movements of the palp papillae (from [52]); (**B**) distal portion of the palps of *M. johnstoni* above the sediment surface. Animal can be seen in the burrow below, as well other burrowing tracks made by another individual; (**C**,**D**) anterior of an individual *M. johnstoni* protruding into the water column prior to initiation of lateral sinuous movements thought to be linked to reproduction.

Magelonids have been reported to consume crustaceans, crustacean larvae, sediment, detritus, diatoms, algal cysts, spores, tintinnids, molluscs, worms and other small animals [26,30,51,52,128,129,131–133]. Despite the number of papers recording diet, knowledge of species-specific diets is generally lacking, although interspecific variation is likely [51,52].

Very little is known about the reproductive biology of the group [19,22], although sexes are separate. Eggs are easily observed in the abdomen of live animals and range from 50 to 150 μm in size [22]. *Magelona sacculata* was recorded to possess small eggs, with a high fecundity, varying from 8 to 12,000 eggs [122]. Eggs first occurred generally between chaetigers 10 and 20, and in the most fecund individuals extended to the last chaetiger. The family are believed to have ect-aquasperm and fertilisation is thought to be external [22,134,135]. Studies on the life history and population dynamics of *M. sacculata* suggest it reaches maturity in approximately two months, spawning in its first year with death occurring at or shortly after [122]. Lateral sinuous movements of the thorax within the water column are believed to be associated with reproduction (Figure 10C,D), although further evidence is needed. Both planktotrophic and lecithotrophic larvae have been recorded to occur within the group [105,135,136], and it has been suggested that the planktotrophic larvae of *M. sacculata* are long-lived, surviving for at least two months [122]. Although little is known about larval development, several species have been described and illustrated [22,105,137]. The dispersal potential of four species of *Magelona* in the Southern California Bight was estimated to be high based on known and inferred life-history information [138]. Magelonid larvae (*Magelona* sp.) have been observed to perform reverse diel vertical migrations off Central Peru [139].

Magelonids as potential indicator species of marine pollution, have been suggested [140], based primarily on the observation that non-polluted mangrove areas in Brazil were characterised by the magelonid *M. papillicornis* [141].

3.2. Family Oweniidae Rioja, 1917

3.2.1. Systematics

The monophyly of Oweniidae is supported by embryos which do not go through the typical spiralian development, by having a unique mitraria larvae, a pair of protonephridia resembling those of deuterostomes, lacking the typical external cuticle of most annelids, possessing a monociliated epidermis and a nerve cord that is intraepidermal in some species [17,57,65,142–148]. Although robust and unambiguous characters, some of the aforementioned attributes are not easily observed. As a practical aid, members of this family are easily recognised by their elongate and cylindrical bodies, without protruding parapodia or appendages (except for those of the head, present in some species), and by bearing oval patches of transverse irregular rows of packed uncini provided with two distal teeth.

Due to the morphological and ontogenic peculiarities of members of this family, the relationships between oweniids and other annelids have been object of debate for decades (see [57] for a summary about the systematic history of the family).

Oweniidae currently contains four genera: *Owenia*, *Myriochele*, *Galathowenia*, and *Myriowenia*; a fifth genus, *Myrioglobula* Hartman, 1967 was considered as valid until its recent synonymisation with *Myriochele* [17]. Other genera such as *Ammochares* Grube, 1844, *Mitraria* Müller, 1851 and *Ops* Carrington, 1865 are now considered subjective synonyms of *Owenia*, and *Psammocollus* Grube, 1866 of *Myriochele* [85]. *Clymenia* Örsted, 1844 is still considered as a *nomen dubium* with uncertain homonymy with any of the aforementioned oweniid genera [85].

Monophyly of Oweniidae and the four currently considered valid genera has been assessed based on morphological information [17] but a phylogeny including molecular data has never been published for members of the family. The monophyly of *Owenia* is supported by the presence of their characteristic ramified tentacular crown on the anterior end and neuropodial uncini with teeth arranged side by side [17]. The monophyly of *Myriowenia* is also based on the presence of head appendages, but in this case, corresponding to a pair of grooved palps, as well as the arrangement of uncinal teeth, one on top of the other [17]. Members of *Galathowenia* are characterised by an anteriorly truncated head and the presence of a ventral cleft (e.g., [17,59,62,68,80,149,150]). The monophyly of *Myriochele*, as currently circumscribed, is supported by the presence of acicular chaetae with an elongated tapering distal end and a smooth surface and by the shape of the head, which is similar in width to the segments with a rounded anterior margin [17].

According to the latest phylogenetic hypothesis, oweniids possessing palps (*Myriowenia* and *Owenia*) branch off at the base of the family tree, with *Galathowenia* and *Myriochele* as sister groups [17,57] (Figure 11). If this hypothesis is corroborated, it would mean that the presence of palps is the plesiomorphic condition and have subsequently been lost in *Galathowenia* and *Myriochele*.

3.2.2. Taxonomic History

The first described oweniid was *Owenia fusiformis* Delle Chiaje, 1844. The genera *Owenia* and *Myriochele* are the most speciose containing 19 and 20 species respectively [85]. *Galathowenia* comprises 12 species with only one species described in the 19th century (*Galathowenia australis* (Grube, 1866), as *Psammocollus*). Finally, *Myriowenia* comprises only four species and was erected well into the 20th century.

Taxonomic studies on the family are closely related to significant historical events (Figures 12A and 13). For instance, a number of species were described from the mid-19th century to early-20th century (i.e., three *Owenia*, five *Myriochele*, one *Galathowenia*), followed by almost non-existent progress extending to the end of WWII (in which only *Myriochele picta* Southern, 1921 and *Galathowenia oculata* (Zachs, 1933) were described). From 1945, studies increased substantially; *Myriowenia* was erected by Hartman in 1960 and further species were described belonging to *Myriochele* and *Galathowenia*. However, the number of *Owenia* species did not increase in the same way and this is perhaps due to the belief that *O. fusiformis* was cosmopolitan. Finally, the 21st century is witnessing a substantial increase in number of new oweniids described, influenced heavily by the reconsideration of the status and past records of both *O. fusiformis* and *M. heeri*.

The taxonomic bibliography of the group includes several regional studies that include information about genera/species, keys and/or comparative tables of distinguishing characters. Among the most relevant are Fauvel [151] on Atlantic and Mediterranean French coasts, Ushakov [152] in the USRR, Hartman [153] in Antarctica, Day [154] in South Africa, Milligan [72] in the Gulf of Mexico, Nilsen and Holthe [65] in Scandinavia, Imajima and Morita [76] in Japan, Hartmann-Schröder [155] in Germany, and more recently Blake [71] in California, Cantone and Di Pietro [61] and Parapar [62,63] in Antarctica, Parapar [67,68] in Iceland, Gil [156] from European coasts and Capa et al. [17] in Australia.

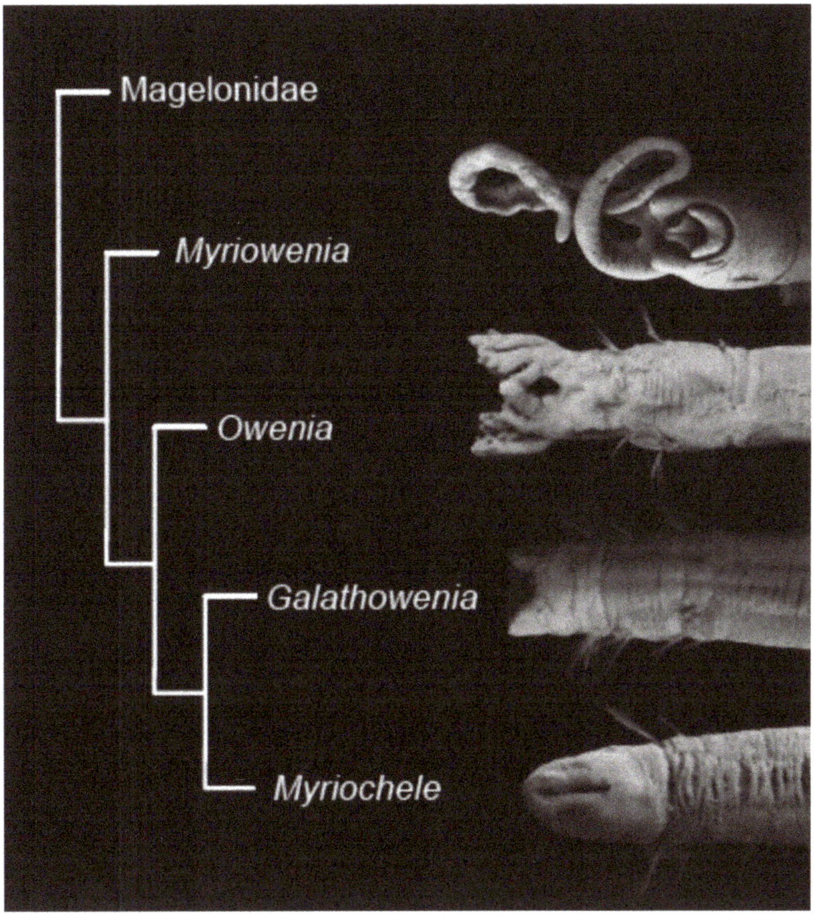

Figure 11. Phylogenetic hypothesis, based on morphological data, of relationships between oweniid genera. Modified after Capa et al. [17].

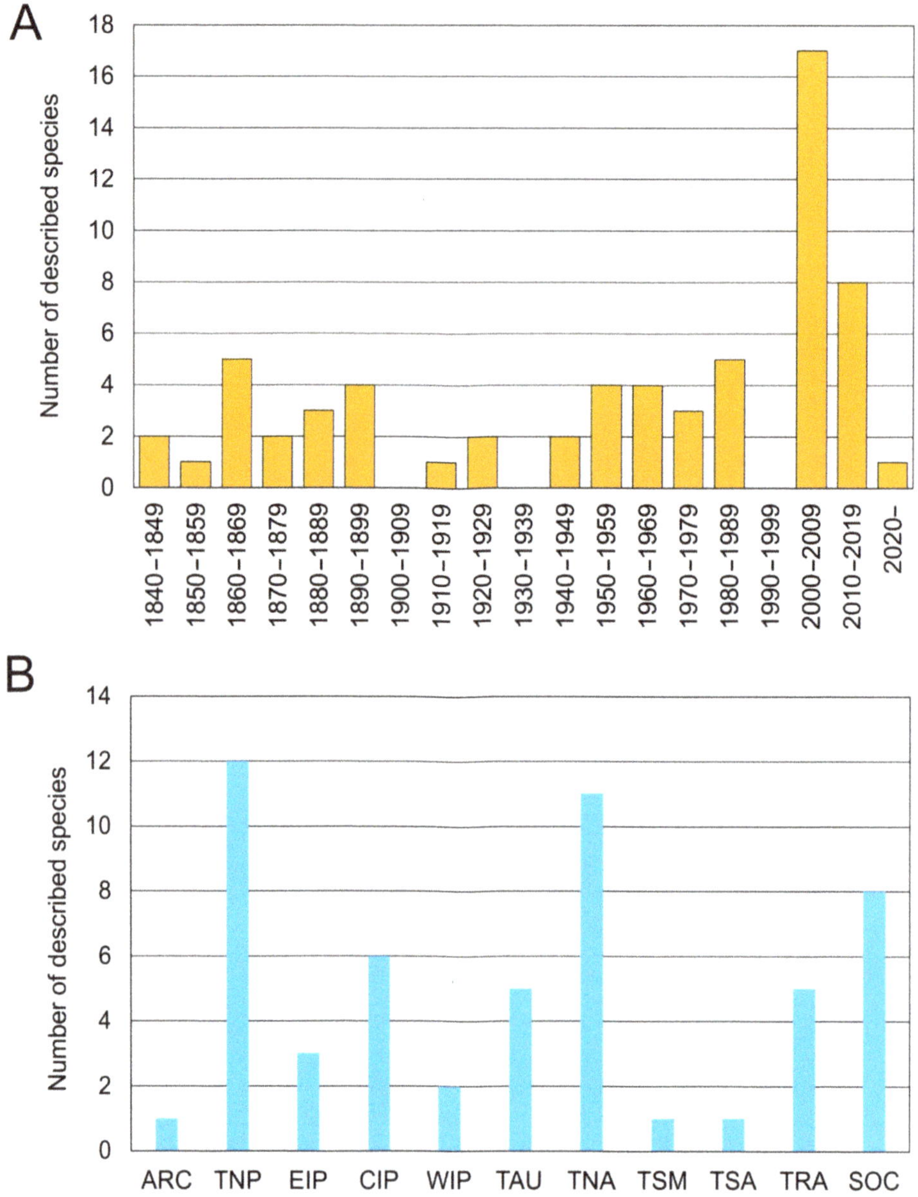

Figure 12. (**A**) Number of oweniid species described per decade; (**B**) all oweniid species currently consider valid [85] listed under the bioregion (sensu [86]) of their type locality. Abbreviations: ARC—Arctic; TNP—temperate North Pacific; EIP—Eastern Indo-Pacific; CIP—Central Indo-Pacific; WIP—Western Indo-Pacific; TAU—temperate Australasia; TNA—temperate Northern Atlantic; TSM—temperate South America; TSA—temperate South Africa; TRA—tropical Atlantic; SOC—Southern Ocean.

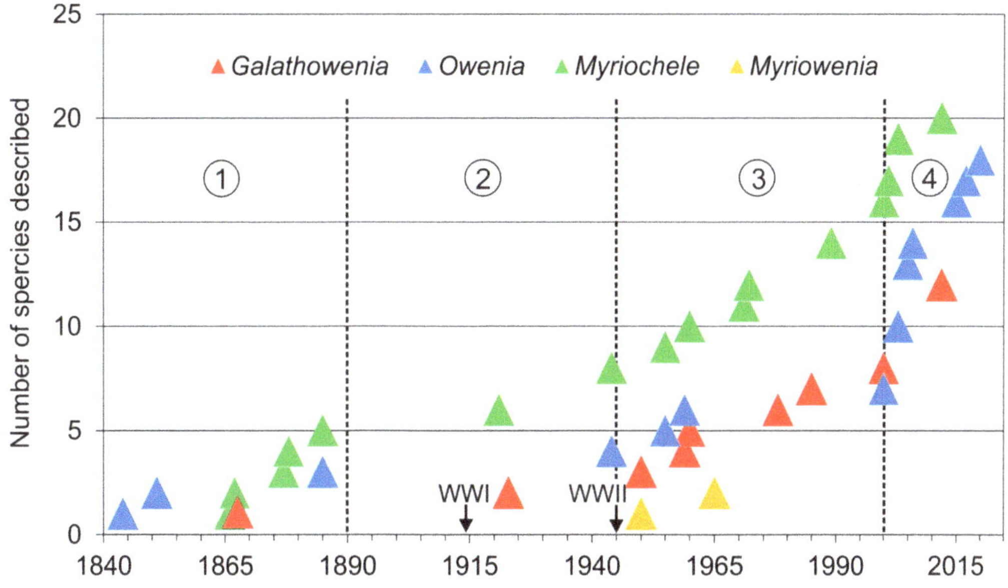

Figure 13. Number of described species (accumulated) of each oweniid genus. Encircled numbers indicate relevant historic moments (separated by discontinuous lines): (**1**) from first described species to 1890; (**2**) 1890 to the end of WWII in 1945; (**3**) 1945 to 2000; (**4**) 2000 to the present.

3.2.3. Taxonomic Characters

The shape of the body is consistent across members of the family, i.e., elongated, cylindrical and fairly rigid [71,157]. The prostomium and peristomium are fused forming the head [65,83] and its shape differs across the genera: distally truncated without palps (*Galathowenia*), distally rounded without palps (*Myriochele*), rounded and provided with a pair of grooved tentacles (*Myriowenia*) and with multi-lobed tentacles, in the shape of a tentacular crown (*Owenia*) (Figures 14 and 15). These head structures are generally used for particle selection and tube building (e.g., [158–161]), and are homologous between them and seemingly also to the palps of magelonids [161]. However, we consider further work is needed to assess the homology of the head appendages of members of these two groups.

The mouth and lips are differently developed among members of Oweniidae; *Myriochele*, *Myriowenia* and *Owenia* are provided with a button-hole mouth with shallow lips (Figure 14B–D); while *Galathowenia* bears a cleft directed backwards from the mouth, forming two large lateral lips (Figures 14A and 15A) (see [17,56]). An epithelial fold between the head and the first segment, often referred to as a collar is common in some oweniid species [60,68,71,76,79] (Figures 14E and 15D).

Posterior to the head, the trunk, consists of several chaetigers (up to 30 in species of *Galathowenia* and *Myriochele*, and up to 60 in species of *Owenia* and *Myriowenia*; [17]). The anterior three chaetigers are provided with uniramous parapodia whilst the following chaetigers are biramous (Figures 14 and 15). These regions have been previously referred to as thorax and abdomen, respectively (e.g., [65,71]), names that have recently been discouraged (e.g., [17,67,80]). The number and relative lengths of uniramous chaetigers is species specific and is one of the main taxonomic features used for species discrimination, as was shown by Parapar et al. [68] in the review of *Myriochele* (and the then accepted *Myrioglobula*) in Icelandic waters. It is common to find species descriptions including a formula indicating the relative length of uniramous (or else three anterior) segments, for example, for *Galathowenia annae* Capa, Parapar and Hutchings, 2012 the following formula is given RLUS = 1:2:1, where the second chaetiger is twice the length of the first and third [17]. Genera specific patterns can also be observed, especially in members of *Myriowenia* and *Owenia*, which always possess three uniramous chaetigers, while *Galathowenia* have two or, more commonly, three uniramous segments. Species of *Myriochele* show a wide range of variability for this feature from species with a single uniramous chaetiger (e.g., *Myriochele islandica* (Parapar, 2003)), others with two (e.g., *Myriochele olgae* Blake, 2000) and others with even three uniramous chaetigers (e.g., *M. heeri*). In some oweniids additional segments are added before the pygidium, thus chaetiger number is highly dependent on age. Biramous chaetigers are unequal in length, the anteriormost being longer and usually those towards the pygidium becoming progressively shorter (e.g., [17,67,80]).

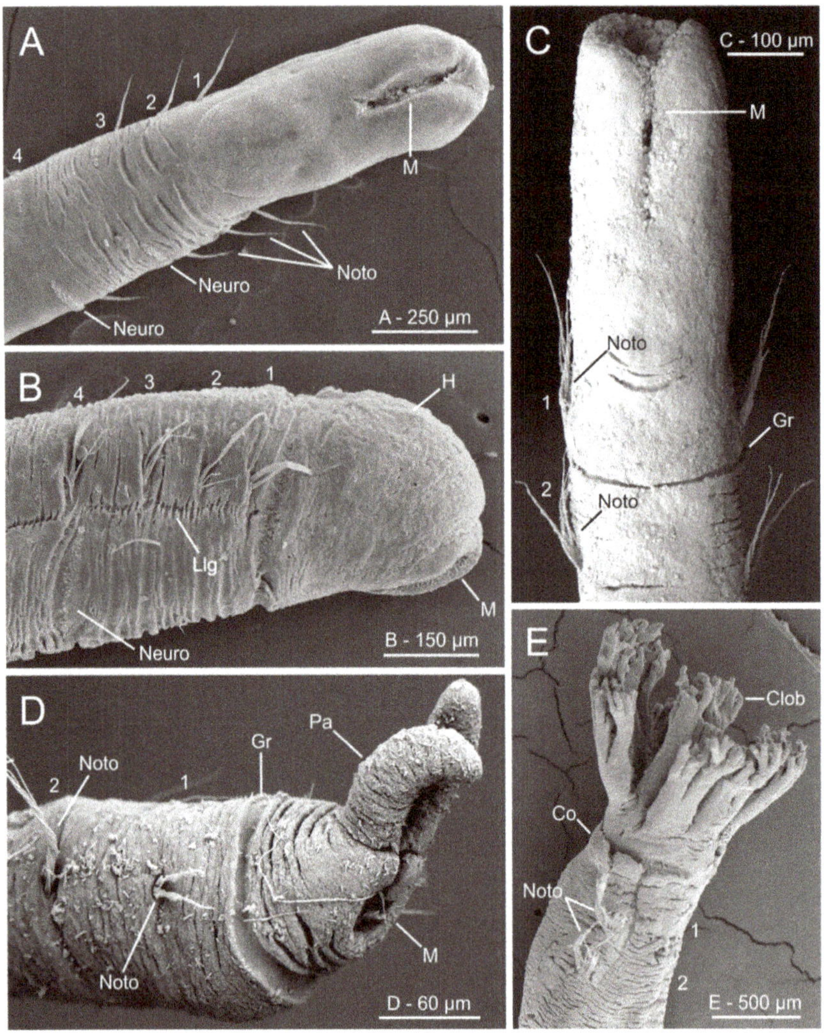

Figure 14. SEM micrographs of anterior end of representative species of all Oweniidae genera showing main diagnostic characters. (**A**) *Myriochele olgae*, ventral view; (**B**) *M. heeri*, lateral view; (**C**) *Galathowenia quelis*, ventral view; (**D**) *Myriowenia* sp., lateral view; (**E**) *Owenia fusiformis*, lateral view. (**A**,**B**) Modified after Parapar [68], (**C**,**D**) after Capa et al. [17]. Chaetigers number provided. Abbreviations: Clob—crown lobe, Co—collar, Gr—groove, H—head, Llg—longitudinal lateral groove, M—mouth, Neuro—neuropodium, Noto—notopodium, Pa—palps.

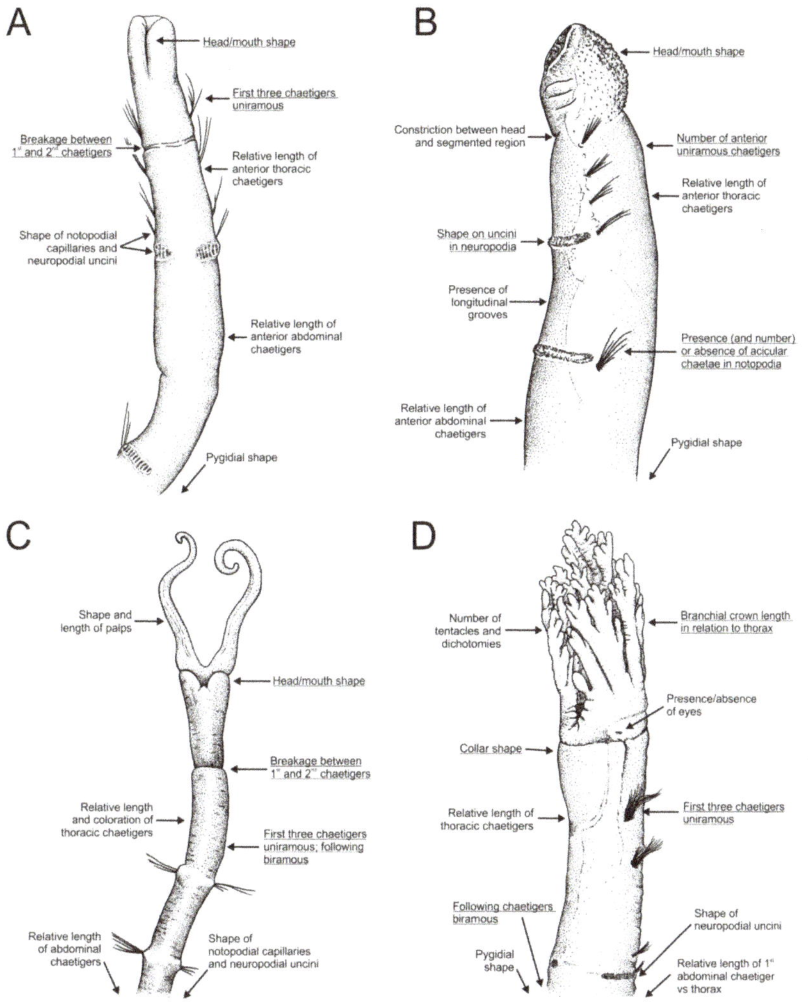

Figure 15. Stylised drawings of anterior region of representative species of all Oweniidae genera showing main diagnostic characters both of genus (underlined) and species. (**A**) *Galathowenia quelis* Capa, Parapar and Hutchings, 2012 in ventral view; (**B**) *Myriochele australiensis* Hartman, 1955 in lateral view; (**C**) *Myriowenia* sp. in ventral view; (**D**) *Owenia australis* Ford and Hutchings, 2005 in ventro-lateral view. (A) Redrawn after Parapar and Moreira ([78]; (**B**) after Blake [71]; (**C**) after Hartman [95]; (**D**) after Ford and Hutchings [77].

Oweniids possess ciliated, longitudinal lateral grooves, running between the noto- and neuropodia, homologous to the lateral organs of other annelids (Figure 14B, [17,161]). In addition, glandular longitudinal grooves (lacking cilia) are present on ventral body surfaces, while there is a non-glandular and non-ciliated groove running along the mid-dorsum [17,62], however, none of these characters are used in taxonomic descriptions.

Parapodia are conspicuously reduced in oweniids compared to other annelids, including the sister group Magelonidae. They lack structures such as cirri, branchiae or epidermal papillae (Figure 16A). The notochaetae emerge directly from the body wall and neurochaetae from low glandular transverse tori [162]. Notochaetae include fine distally tapering capillaries, known as "capillary chaetae" in *Galathowenia*, *Myriowenia* and *Owenia* (Figure 16A,B) [17,77], that coexist with the shorter and comparatively more robust chaetae, with a narrow distal tip, known as "acicular chaetae ", in *Myriochele* [17,62,65,67,68,76] (Figure 16A). Capillary chaetae show external structures under SEM like densely packed scales. Morphometric analyses of these scales have shown value for identification of species of *Owenia* (e.g., [163,164]), although some authors found the intra- and interspecific variation to overlap [77]. Neurochaetae come in the shape of dentate hooks with long shafts (i.e., uncini) provided with two distal teeth, homologous to the *capitium* of other annelids but lacking instead a true *rostrum* (Figure 14A,B, Figure 15A–C and Figure 16C,D) [57]. Uncini are arranged in multiple transverse rows per segments [146] forming the uncinal patches (Figure 16A). Uncinal morphology has been long used for generic assignment. Species of *Galathowenia* possess teeth arranged in an oblique row (with one tooth higher than the other). Teeth are generally arranged vertically (one on top of the other) in members of *Myriochele* and *Myriowenia,* and horizontally (teeth occurring side by side) in *Owenia* species [57]. Morphometric analyses with the aid of SEM have also been performed within members of *Owenia* in order to find distinct traits to separate species among this morphologically homogenous group [77,163,164].

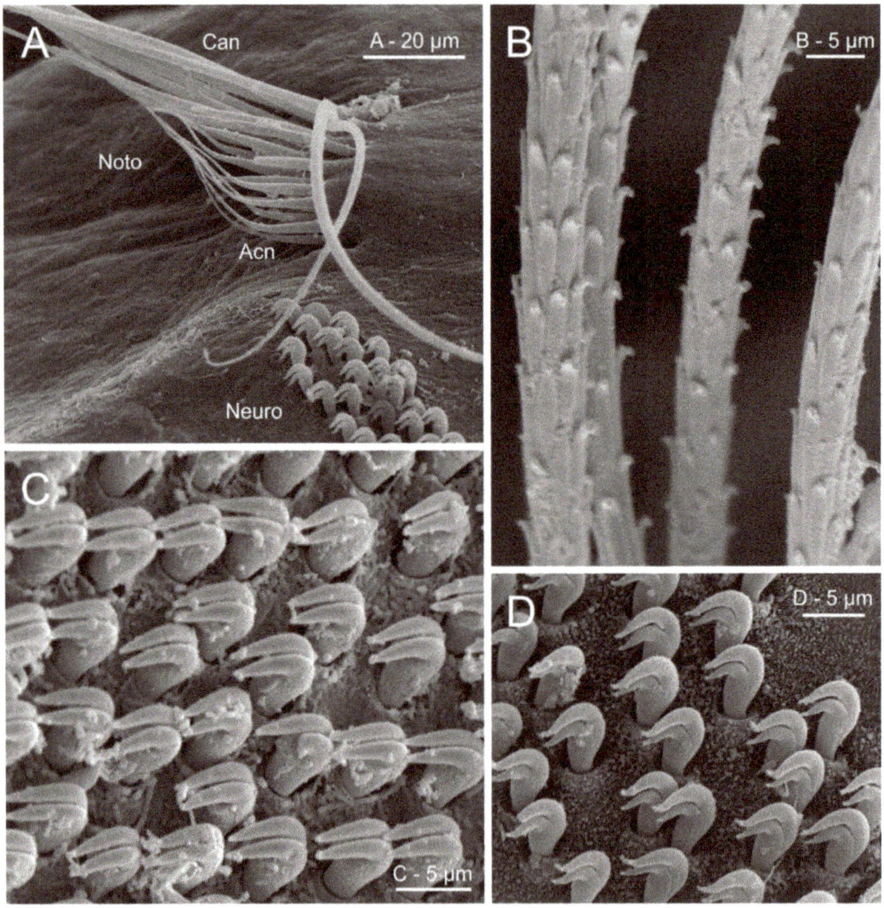

Figure 16. SEM micrographs illustrating several chaetal characters in Oweniidae. (**A**) Anterior abdominal chaetiger in *Myriochele olgae*; (**B**) detail of thoracic notochaetae scale covering of *Owenia fusiformis*; (**C**) neuropodial uncini of *O. fusiformis*; (**D**) neuropodial uncini of *M. olgae*. Abbreviations: Acn—acicular notochaetae, Can—capillary notochaetae, Neuro—neuropodium, Noto—notopodium.

Oweniids display a variability of pygidial shapes. They may be simple, bilobed, multilobed or pointed (Figure 17) [57] showing intrageneric variability but being species specific [62]. Even if reported, the presence of true pygidial cirri has not been confirmed yet and the structures reported for *Myriowenia gosnoldi* Hartman, 1965 may be related to body regeneration after breakage [17].

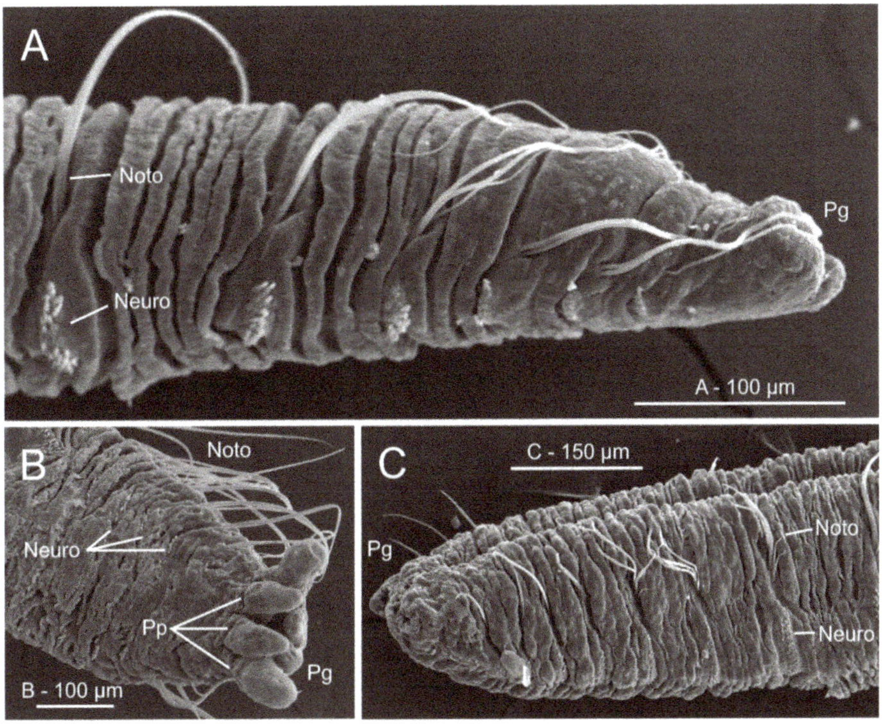

Figure 17. SEM micrographs of posterior end of (**A**) *Galathowenia oculata*; (**B**) *G. fragilis*; (**C**) *Myriochele heeri*. (**B**,**C**) modified from [68]. Abbreviations: Neuro—neuropodium, Noto—notopodium, Pg—pygidium, Pp—pygidial papillae.

With genera being fairly homogenous morphologically there has been a reconsideration of the number and identity of morphological characters used in oweniid taxonomy and characterisation of species compared with early studies (Figure 18). For instance, the unique combination of approximately eight characters were needed to identify species prior to 1970, which were visible under a stereo or optic microscope (but see [165]). In contrast, there are roughly twice as many characters needed to correctly identify oweniids to species currently (e.g., [65]). While the same characters are considered across the family due to the relative homogeneity in the oweniid body shape, the anterior end shows more variation among genera and is therefore more useful for species characterisation. However, for all members of Oweniidae, the use of a SEM is essential, and as mentioned before, very often morphometric analyses are needed for species discrimination [77,81,163,164]. Branching patterns of the head tentacles have also been shown to bear taxonomic information for some *Owenia* species [78].

The use of methyl blue stain and the different colouration body patterns in *Owenia* is being increasingly used (e.g., [71,164]) although its use as a taxonomic character in the characterisation of the species is not generally used in recent works (e.g., [74,78]).

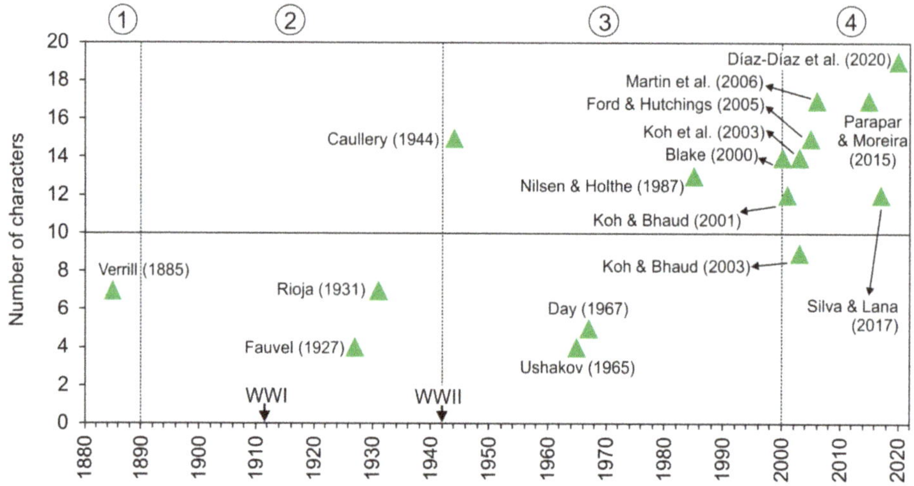

Figure 18. Number of characters (accumulated) used in the description of *Owenia* species. Encircled numbers denote relevant historic moments (separated by discontinuous lines): (**1**) from the first described species to 1890; (**2**) 1890 to the end of WWII in 1945; (**3**) 1945 to 2000; (**4**) 2000 to present. Species described in each paper corresponds to *O. fusiformis* or to the original description of the taxon. Verrill [166] (*O. artifex*); Fauvel [151] (*O. fusiformis*); Rioja [167] (*O. fusiformis*); Caullery [165] (*O. assimilis*); Uschakov [152] (*O. fusiformis*); Day [154] (*O. fusiformis*); Nilsen and Holthe [65] (*O. fusiformis*); Hartmann-Schröder [155] (*O. fusiformis*); Blake [71] (*O. johnstoni*); Koh and Bhaud [75] (*O. gomsoni*); Koh and Bhaud [66] (*O. petersenae*); Ford and Hutchings [77] (*O. australis*); Martin et al. [81] (*O. persica*); Parapar and Moreira [78] (*O. picta*); Silva and Lana [164] (*O. caissara*); Díaz-Díaz et al. [168] (*O. vieitezi*).

Iconography has also shown much variation across the literature through time (Figure 19), even showing evident differences in artistic quality. For instance, drawings and illustrations by Rioja [167] and Day [154] are generally superior to those found in some of the later works. In the case of *M. heeri*, accurate and artistic illustrations can, however, be found in early work (e.g., [169]).

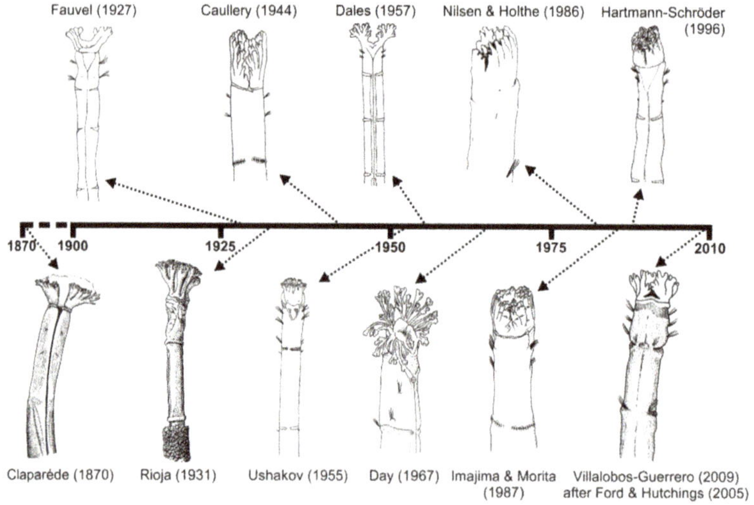

Figure 19. Selection of illustrations of anterior body end of *Owenia fusiformis* through time. All redrawn from original.

3.2.4. Internal Morphology

The first studies on internal anatomy date back to the beginning of the 20th century [158] but are still mostly limited to a handful of species such as those referred as *O. fusiformis* and *M. heeri* [158,170–172], the microanatomy of the central nervous system has received more attention due to its relevance in understanding the phylogeny of this group, and of annelids in general (see below).

The body wall in Oweniidae is composed of longitudinal muscles only, which are developed and grouped into bundles that gradually decrease in thickness [173,174]. The coelomic lining of the body wall is composed of a layer of monociliated peritoneal cells that cover the muscle cells except near the ventral and dorsal mesenteries, where longitudinal muscles form the lining [173,175].

The use of histological sectioning (HIS) (Figure 20A,B) and micro-computed X-ray tomography (micro-CT, or μCT) (Figures 20C,E and 21) has allowed for a re-examination and better assessment of the annelid internal anatomy [176]. Internal coelomic space is occupied mostly by the gut (Figures 20A,B and 21B–D), which is composed of three regions: foregut (esophagus and stomach), midgut (intestine) and hindgut (rectum), which can be easily distinguished histologically [177]. The anteriormost region of the gut of *Owenia*, *Myriochele* and *Myriowenia* has been described or illustrated as having a ventral pharyngeal organ [59,65,68,155,158–161,178]; this organ has often been reported incorrectly in previous work as a proboscis (e.g., [71]). Correspondingly, protractible ciliary folds on the dorsolateral walls of the foregut were observed in *Galathowenia*, *Myriochele* and *Myriowenia* [17,68,178]. The buccal region and the esophagus are lined by ciliated epithelial cells interspersed with glandular cells that secrete digestive enzymes; stomach walls are thick and involuted (Figure 20B) containing two types of numerous glandular epithelial cells [160]. The intestine lacks glandular cells and serves purely for absorption. The rectum is thin-walled and less convoluted; faecal pellets are mostly composed of sand grains and other material that are released when the animal reverses into its tube [160]. The gut musculature is responsible for the movement of material through the gut and is helped by contraction of the body wall muscles [57].

The nervous system shows some unusual features when compared to other annelids and is currently considered in phylogenetic studies [2]. It is largely intraepidermal with a well-developed nerve plexus, including many peripheral nerves but lacks segmental ganglia [179,180]. The absence of nuchal organs [172] a condition shared with the Magelonidae, could be regarded as plesiomorphic among the annelids and supporting their basal position at the base of the annelid tree [4]. However, further investigations are needed to truly verify their absence [57,181]. Eyespots are usually present in the head of adults and are closely associated to the brain [182] but it is still unclear whether they develop from the prostomium or the peristomium [17]. Helm et al. [2,183] compared the development of oweniid neuroarchitecture with that of other annelids based on histology, SEM and immunohistochemistry and found that development and metamorphosis of the mitraria larva is mostly similar to that of other annelids irrespective of the drastic changes in body shape during metamorphosis. The central nervous system in Oweniidae is medullary, with a ring-shaped and basiepidermal brain lacking higher brain centers, ganglia associated to ventral nerve cord or complex sensory organs (Figure 21A–C) [147,148,161]. In *Owenia borealis* Koh, Bhaud and Jirkov, 2003, the glandular structures located in the first two segments are parapodial organs (Figures 20B,C and 21D) which contain secretory cells producing the tube materials. Their relationship with the nephridial sacs located in first chaetigers of the abdominal region remains obscure and warrants further investigation [148].

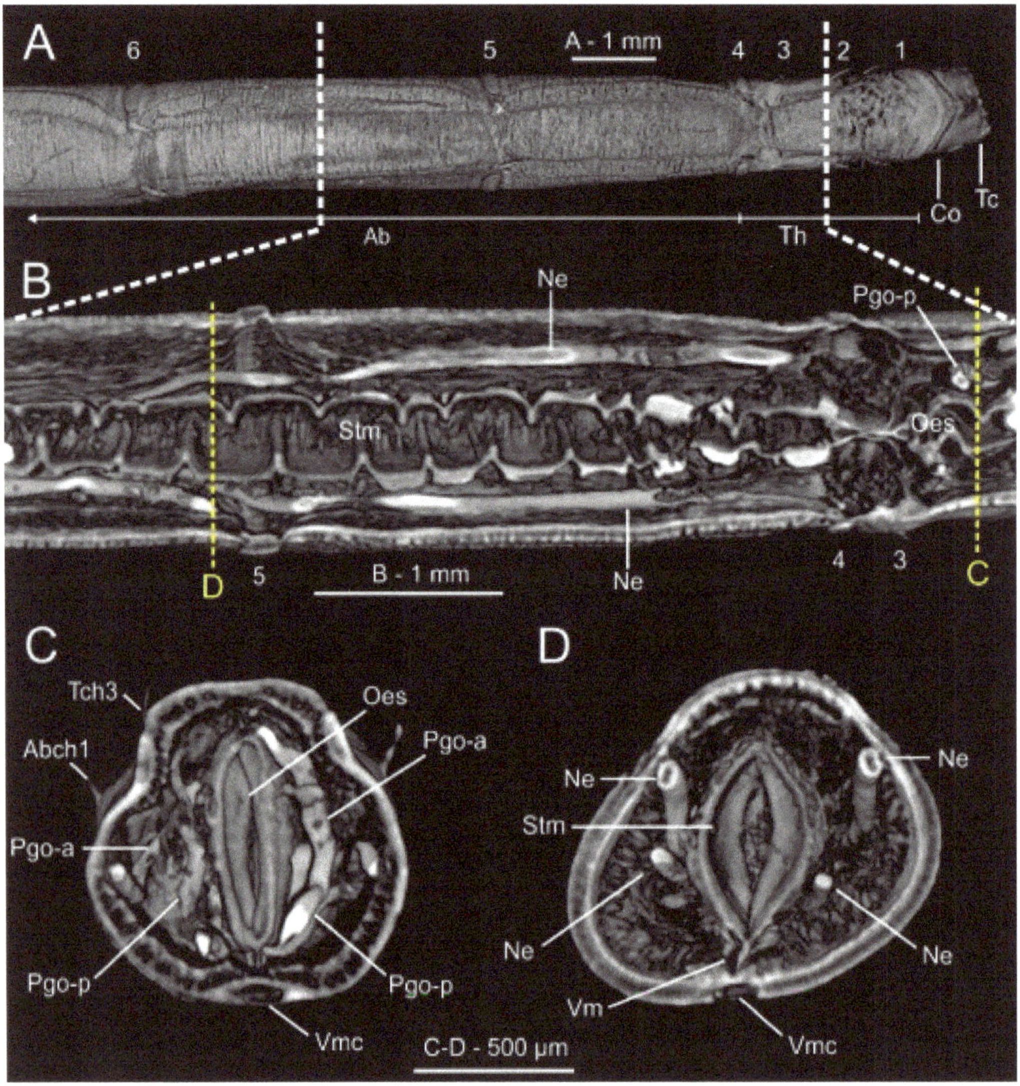

Figure 20. Micro-computed tomography (μCT) sections of *Owenia fusiformis* (MNCN 2866). (**A**) Dorsal view of body surface, (**B**) frontal and (**C**–**D**) transversal body sections showing internal anatomy. White discontinuous lines in (**A**) marking body region showed in (**B**). Yellow discontinuous lines in (**B**) marking position of transversal sections in (**C**) and (**D**) respectively. Chaetigers number provided. Abbreviations: Ab—abdomen, Abch1—abdominal chaetiger 1, Co—collar, Lml—longitudinal muscle layer, Ne—nephridial sac, Oes—oesophagus, Pgo-a/p—parapodial glandular organ—anterior/posterior, Stm—stomach, Tc—tentacular crown, Tch3—thoracic chaetiger 3, Th—thorax, Vm—ventral mesentery, Vmc—ventral medullary cord. Numbers indicate chaetigers.

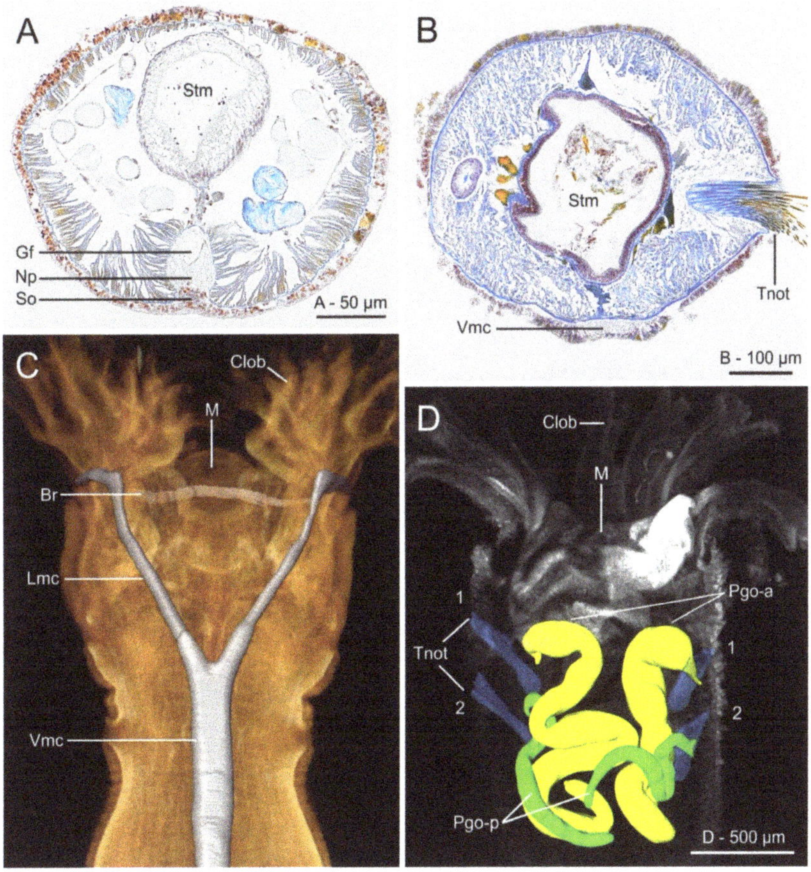

Figure 21. Histological cross sections (azan staining) at gut level of (**A**) *Myriowenia* sp. and (**B**) *Owenia fusiformis*; (**C**) micro-computed tomography (µCT) volume rendering and 3D-reconstruction of the nervous system of *O. fusiformis*; (**D**) 3D reconstruction of parapodial glandular organs (Pgo); anterior pair are in yellow, posterior ones are in green; notochaetae in blue. (**A–C**) From Beckers et al. [161]; (**D**) from Rimskaya-Korsakova et al. [148]. All modified from the original. Abbreviations: Br—brain ring, Clob—crown lobes, Gf—giant fibre, Lmc—lateral medullary cord, M—mouth, Np—neuropil, Pgo-a/p—parapodial glandular organ—anterior/posterior, So—somata, Stm—stomach, Tnot—thoracic notochaetae, Vmc—ventral medullary cord. In (**D**) first two thoracic chaetigers are indicated.

3.2.5. Species Diversity and Distribution

The family Oweniidae is currently composed of 55 valid species in four genera, 16 species in genus *Galathowenia*, 18 in *Myriochele*, 19 in *Owenia* (plus three *taxon inquirendum*) and two in *Myriowenia* [57,85]. The first species to be described was *Owenia fusiformis* Delle Chiaje, 1844 from Sicily and the group was raised to the rank of family by Rioja [184].

Oweniids have colonised a wide range of marine habitats across the world, from the intertidal fringe to the deep sea [185–187] (Figure 12B), occurring in all marine ecoregions (*sensu* [86]). Literature reviews show *Galathowenia*, *Myriochele* and *Owenia* to have broad distributions; *Myriowenia* appears restricted to the Atlantic and Pacific coasts of North and South America (Figure 22). There are, however, apparent gaps in the knowledge of diversity and distribution for all genera. According to the type localities of all described species, the temperate East Northern Atlantic (with nine species described thus far) appears to be as well-known as temperate Australasia (eastern coast of Australia); however, in contrast, data are still lacking for Eastern Indo-Pacific, Temperate South America and South Africa, and Western Indo-Pacific (Figure 22).

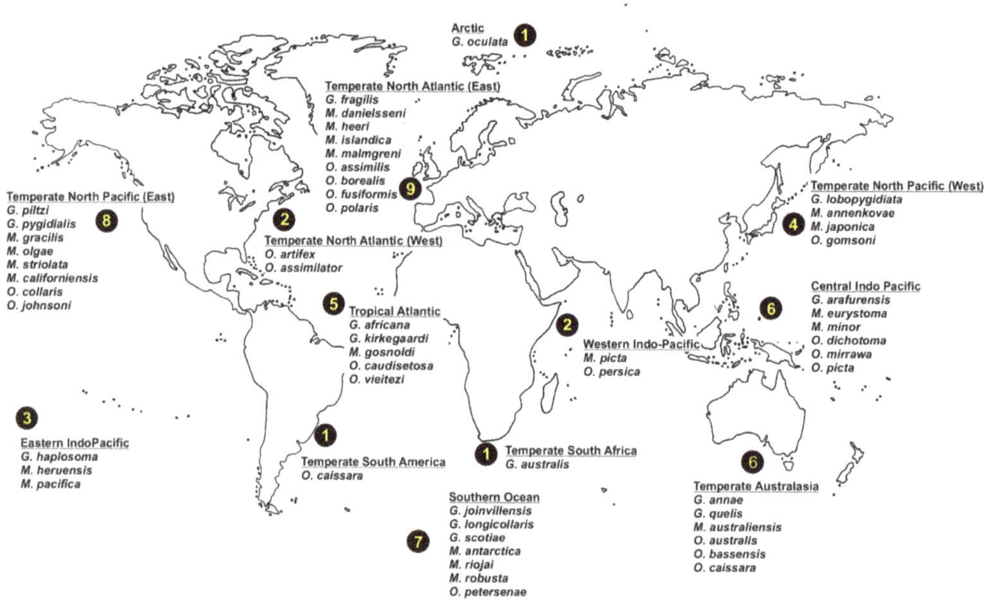

Figure 22. World map showing number of valid oweniid species (sensu [85]) described per bioregion (sensu [86]).

According to the international network and data infrastructure GBIF [118] there are up to 112,695 records for this family. Records are heterogeneously distributed, with many reports concentrated in Atlantic and Pacific North America, the Arctic and European Atlantic, the Australian Indo-Pacific and New Zealand; reports are fewer in comparison for the Subantarctic Scotia Arc, the Mediterranean Sea and some European and North American areas. These records correspond mostly to the genera *Galathowenia* (47%), *Myriochele* (17%) and *Owenia* (34%) (Figure 23A–D), and particularly to a few numbers of species considered previously as cosmopolitan (see below). *Myriowenia* records account for just 0.1% of all references and those are restricted to North American Atlantic and Australian Pacific (Figure 23C). Regarding *Galathowenia*, the most reported species are *Galathowenia fragilis* (Nilsen and Holthe, 1985) and *G. oculata*, particularly from North America and Europe (Figure 23A); a similar distribution pattern is found for *Myriochele*, mainly represented (83% of records) by *M. heeri*, *Myriochele danielsseni* Hansen, 1879 (both mostly in Europe), and, to a lesser extent, by *M. olgae*. Furthermore, *Galathowenia* and *Myriochele* are the only genera reported so far from subantarctic latitudes. *Owenia* is also often reported from the northern Hemisphere but also shows many records from Australia and New Zealand (Figure 21D); *O. fusiformis* accounts for up to 80% of the reports while *Owenia collaris* Hartman, 1955 is the most reported taxon from the North America Pacific. However, it is likely that further sampling and examination of oweniid collections from across Australasia and Pacific regions would reveal new undescribed taxa.

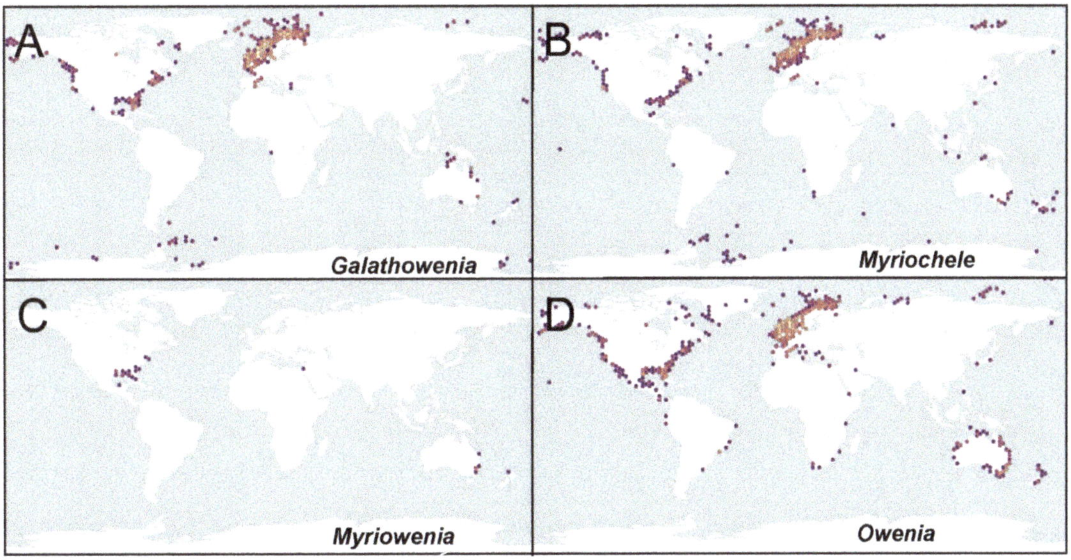

Figure 23. World distribution of all oweniid genera. (**A**) *Galathowenia*; (**B**) *Myriochele*; (**C**) *Myriowenia*; (**D**) *Owenia*. Source: GBIF [118] (https://www.gbif.org).

Several species such as *O. fusiformis*, *G. oculata* and *M. heeri*, have been traditionally reported as cosmopolitan or broadly distributed; recent available data suggests that most of these species fit better within the concept of complexes of morphologically similar species showing narrow ranges of distribution in regard to environmental features (e.g., [63,66,67,75,77,79,80,188–193]). However, the only study that has included molecular information to assess this hypothesis was that of Jolly et al. [194] who analysed several populations of *Owenia* in the northeast Atlantic, considering COI sequences and concluding that several species exist there that are, in turn, distinct from those occurring in the Mediterranean Sea.

3.2.6. Biology and Ecology

Knowledge about reproductive and developmental strategies is restricted to species of *Owenia* (summarised in [57]). Mature animals are dioecious [83]. It has been indicated that reproduction occurs once or several times a year depending on the geographical location and environmental conditions. For instance, some populations of *O. fusiformis* in NE Atlantic were found showing discrete seasonal spawning (e.g., [159,195]) while others breed throughout the year [196]. However, this disparity of reproductive modes should be considered with caution as some of the studied animals could have been misidentified as *O. fusiformis* and, in fact, belong to several different *Owenia* species inhabiting these waters (e.g., [66,194]). Maturity was reported after the first year of benthic life [195] and fecundity seems correlated with the size of females, but number of spawned eggs can reach up to 85,000 [195,197].

Owenia collaris and *O. fusiformis* embryos undergo holoblastic spiral cleavage to form a typical coeloblastula [83,198], where the anus is developed from the blastopore and the mouth is later formed from a secondary opening [83,159]. This development pattern shares more similarities to that seen in deuterostomes [83]. In addition, the larval cilial bands are formed later in the ontogenetic stages compared to other annelids, these are also short and simple and are differently arranged to the typical trochophore [83]. This mitraria larva develops completely 24 h after fertilisation in *Owenia*, having already a defined mouth and anus and the ability to swim. Several morphological changes take place in the larvae for about four weeks, and then metamorphosis starts with the protrusion of the trunk segments from the larval hyposphere, the reduction of the episphere as part of the head, and the disintegration of the ciliary bands [83]. Early juveniles have a head (fused prostomium and peristomium), a uniramous segment with two sets of chaetae, six or seven biramous segments each with a single pair of chaetae and one pair of uncinal tori, and a pygidium. Soon after metamorphosis the juvenile readily builds the tube by gathering materials with the anterior end; after two weeks, pairs of buds start to form the prostomial tentacles typical of adult *Owenia*. As noted above, Helm et al. [183] show that development and metamorphosis of the mitraria has many parallels to other annelids irrespective of the drastic changes in body shape during metamorphosis.

Asexual reproduction has been suggested in several species of *Galathowenia*, *Myriochele* and *Owenia* due to the presence of breakage grooves in anterior segments [17,57,62,67,199,200], bodies already broken at this level [62,63] or anterior/posterior ends showing different degrees of regeneration [57].

All oweniids are tube dwellers; the tube is flexible and made after mucous secretions binding foreign particles (i.e., sediment grains, foraminiferan thecae, shell fragments) [160,201,202]. Branchial tentacles and buccal lips are involved in tube building (e.g., [158–160]). It has been argued as to whether the materials utilised for the tube are pre-selected, and whether there is any preference for type and size of particles used [64,66,164] or whether the animal simply relies on any available material [67,68]. It appears that tube-building patterns depend on the species under consideration, development and the availability of materials. Experimental work with *Owenia caissara* Silva and Lana, 2017 has demonstrated that the animal is unable to rebuild the tube if completely released from it [164]; in turn, those partly removed were able to rebuild it using a wide range of sediment particles including filamentous debris, at least in laboratory conditions.

Oweniids are suspension or surface deposit feeders that collect particles at the sediment-water interface [71,160]. Species of *Owenia*, with a tentacular crown provided with cilia, create water currents that direct suspended particles into the crown and the mouth; *Owenia* and other oweniids bend over the sediment and collect sand and detritus directly from the bottom surface. Particles are sorted by the highly muscular lips helped by mucus [203]; the shape of the upper lip can be modified in order to facilitate sorting of materials, similarly to that observed for terebellids [204]. Feeding behaviour in *Myriowenia* has not been studied, though they may gather particles from sediment surface using their long palps [71].

Oweniids are quite common in subtidal soft seabeds, occurring across a wide range of depth, but very little is known about specific species habitats. Data taken from original descriptions of all oweniid species (i.e., Table A2) shows that: (1) only two species have been reported from the intertidal and the number of species at depths of less than 200 m is about 50% of the total; (2) about 20% of species show a wide depth range, from shallow waters to the continental slope realm, reaching 2000 m deep or more; (3) few species (e.g., *Myriochele malmgreni*) seem to be typical of deep waters. The second group includes the widely distributed *G. fragilis*, *G. oculata*, *Galathowenia scotiae* (Hartman, 1978) and *M. heeri*, but also recently described species such as *O. borealis* (Table A2); this suggests again the existence of species complexes or rather frequent misidentifications attributed to those species. In contrast to oweniids (see above), magelonids show a preference for shallow waters, with only handful of species that occur from 400–1000 m. Further work is certainly warranted to clarify depth ranges for most species.

Most ecological knowledge comes from studies done at European latitudes and about *O. fusiformis*, a species that has been attributed in the past to a cosmopolitan distribution but actually represents a complex of sibling species (e.g., [82,205–207]). *Owenia fusiformis* may reach high densities in sandy sediments, of up to 1000 indiv./m^2 [82] while those of the small-sized *G. oculata* range up to 500–700 indiv./m^2 [205,208]; an increase in the abundance of the latter has also been reported during early stages of eutrophication related to mariculture [209]. In the Northeast Atlantic, *O. fusiformis* may also be the dominant species in terms of biomass in shallow sandy-muddy sediments [210]. In general, the very presence of dense aggregations of oweniid tubes may serve for sediment stabilisation [211–213], and allows for an increase in local benthic diversity by favouring larval settlement and providing shelter for other invertebrates as other tube-building polychaetes do [213]. The ability of *O. fusiformis* to switch between deposit- and filter-feeding has been suggested as being a key factor for successful colonisation of habitats subjected to variations in hydrodynamism [206,214]. Finally, oweniids may serve as a relevant source of food for demersal fishes and flatfishes [155]; for instance, *Owenia* is among usual prey items for opportunistic preying fishes depending on the season [215].

4. Conclusions and Future Perspectives

Despite some efforts to increase the known biodiversity information of Magelonidae and Oweniidae (especially since the 60s in the former group and since 2000 in the latter), there are still large gaps in our taxonomic knowledge and the number of species world-wide, is most likely under reported, and in need of reappraisal. Additionally, the evolutionary relationships and subsequent classification for members of these two groups need to be assessed combining different sources of data.

Several marine realms and world seas should be further explored and sampled. Thus, Australia, South America, the Western Pacific and Africa at the very least for Magelonidae, while in East Atlantic, North West and West Indo-Pacific and also South America for Oweniidae. Whilst the forthcoming review of magelonids of Western Africa (Mortimer et al., in preparation B) should help in part to resolve the situation in that region, more work is needed, such as the description of the aforementioned undescribed *Octomagelona* species.

Most of our understanding of the biology, reproductive biology, anatomy and behaviour of both families have come from a relatively small number of studies on one or two species. Further work based on traditional and modern microscopy imaging technique (e.g., micro-CT, SEM) in combination with molecular methods (e.g., molecular phylogenies and species delimitation analyses) are needed to look at both assessing current species diversity and intraspecific variability. This is especially useful in morphologically homogenous groups, such as magelonids and oweniids, and crucial for a precise and effective assessment of their diversity, their distribution patterns and specific ecological preferences, and relevant for assessing putative cosmopolitan or broadly distributed species.

Although the latest molecular analyses suggest both families are part of the basal radiation of the Annelida (Palaeoannelida), morphological characters do not support a close relationship. Further investigation of potential homologies, understanding the morphological diversity found in each group. Assessment of the relationships within both Oweniidae and Magelonidae are also warranted, ideally integrating different sources of data.

Author Contributions: Conceptualisation and supervision, J.P., K.M., M.C. and J.M.; methodology, investigation, writing—original draft preparation, review and editing, illustrations, J.P., K.M., M.C. and J.M; funding acquisition, J.P. and M.C. All authors have read and agreed to the published version of the manuscript.

Funding: This research was partially funded by FAUNA IBÉRICA research project Polychaeta VII, Palpata, Canalipalpata II (PGC2018–095851–B–C64) funded by the Agencia Estatal de Investigación and coordinated by JP. MC was funded by the Ramón y Cajal program (RYC-2016-20799) funded by Spanish MINECO, Agencia Estatal de Investigación, Comunidad Autónoma de las Islas Baleares and the European Social Fund. Pat Hutchings (Australian Museum) also provided some funds to cover the publication costs.

Acknowledgments: Authors would thank Patrick Beckers (Institute of Evolutionary Biology, University of Bonn, Germany) and Elena Temereva (Department of Invertebrate Zoology, Moscow State University, Russia) for Figure 21A–D respectively. Also to Ada Castro and Catalina Sueiro (Servizos de Apoio á Investigación, Universidade da Coruña) and Sue Lidnsay (Australian Museum, Sydney) for SEM assistance, and to María Candás (Estación de Bioloxía Mariña da Graña–Ferrol, Universidade de Santiago de Compostela, Spain) for assistance with the micro-CT. Antón Taboada is thanked for help with line drawings. We would additionally like to thank Kimberley Mills (Amgueddfa Cymru and Cardiff University) for providing the image of magelonid palps feeding utilised in this paper. Authors deeply thank reviewers' comments on the manuscript.

Conflicts of Interest: The authors declare no conflict of interest. The funders had no role in the design of the study; in the collection, analyses, or interpretation of data; in the writing of the manuscript, or in the decision to publish the results.

Appendix A

Table A1. Magelonid valid species (according to WoRMS), type locality, assigned marine realms (sensu [86]), and depth of type locality. N.d. = no data.

SPECIES	TYPE LOCALITY	REALM	DEPTH (m)
Magelona agoensis Kitamori, 1967	Ago Bay, Japan	Temperate Northern Pacific	5
Magelona alexandrae Magalhães, Bailey-Brock and Watling, 2018	Hawaii, USA	Eastern Indo-Pacific	39.6
Magelona alleni Wilson, 1958	Plymouth, England	Temperate Northern Atlantic	Intertidal
Magelona americana Hartman, 1965	off New England, USA	Temperate Northern Atlantic	N.d.
Magelona annulata Hartmann-Schröder, 1962	Isla Blanca, Chimbote Bay, Peru	Temperate South America	8
Magelona anuheone Magalhães, Bailey-Brock and Watling, 2018	Easter Island	Eastern Indo-Pacific	Shallow
Magelona berkeleyi Jones, 1971	Puget Sound, Washington, USA	Temperate Northern Pacific	32–40
Magelona californica Hartman, 1944	Mission Bay, California, USA	Temperate Northern Pacific	?
Magelona capax Hartman, 1965	off the mouth of the Amazon River, Brazil	Tropical Atlantic	1500
Magelona capensis Day, 1961	Agulhas Bank, South Africa	Temperate South Africa	86
Magelona cepiceps Mortimer and Mackie, 2006	Mahé, Seychelles	Western Indo-Pacific	33–45
Magelona cerae Hartman and Reish, 1950	Coos Bay, Oregon, USA	Temperate Northern Pacific	37–73
Magelona cincta Ehlers, 1908	Algoa Bay, South Africa	Temperate South Africa	40
Magelona cinthyae Magalhães, Bailey-Brock and Watling, 2018	Hawaii, USA	Eastern Indo-Pacific	31–39
Magelona conversa Mortimer and Mackie, 2003	Mahé, Seychelles	Western Indo-Pacific	26–42
Magelona cornuta Wesenberg-Lund, 1949	E by N of Ras Jagin, Gulf of Oman	Western Indo-Pacific	12
Magelona crenulata Bolívar and Lana, 1986	Paranaguá Bay, Brazil	Temperate South America	10
Magelona crenulifrons Gallardo, 1968	Nha Trang Bay, Viet Nam	Central Indo-Pacific	6–48
Magelona dakini Jones, 1978	Careel Bay, New South Wales, Australia	Temperate Australasia	Shallow
Magelona debeerei Clarke, Paterson, Florence and Gibbons, 2010	Beverly Hill, Namibia	Temperate South Africa	37
Magelona equilamellae Harmelin, 1964	Rade de Villefranche and Golfe de Marseille, France	Temperate Northern Atlantic	13–18
Magelona falcifera Mortimer and Mackie, 2003	Mahé, Seychelles	Western Indo-Pacific	10–56
Magelona fauchaldi Shakouri, Mortimer and Dehani, 2017	Chabahar Bay, Iran, Gulf of Oman	Western Indo-Pacific	1.5–20
Magelona filiformis Wilson, 1959	Mill Bay, Salcombe, England	Temperate Northern Atlantic	Intertidal
Magelona gemmata Mortimer and Mackie, 2003	Mahé, Seychelles	Western Indo-Pacific	48–62
Magelona hartmanae Jones, 1978	near Port Hueneme and Oxnard, California, USA	Temperate Northern Pacific	12.7
Magelona hobsonae Jones, 1978	Eagle Cove, Puget Sound, Washington, USA	Temperate Northern Pacific	Intertidal
Magelona japonica Okuda, 1937	Jinsen, Korean Archipelago	Temperate Northern Pacific	N.d.
Magelona johnstoni Fiege, Licher and Mackie, 2000	St Andrews, Scotland	Temperate Northern Atlantic	N.d.
Magelona jonesi Hartmann-Schröder, 1980	Western Australia	Central Indo-Pacific	N.d.
Magelona kamala Nateewathana and Hylleberg, 1991	Kamala Bay, Phuket Island, Thailand	Western Indo-Pacific	10–20

Table A1. *Cont.*

SPECIES	TYPE LOCALITY	REALM	DEPTH (m)
Magelona koreana Okuda, 1937	Jinsen, Korean Archipelago	Temperate Northern Pacific	N.d.
Magelona lenticulata Gallardo, 1968	Nha Trang Bay, Viet Nam	Central Indo-Pacific	N.d.
Magelona longicornis Johnson, 1901	West Seattle, Puget Sound, USA	Temperate Northern Pacific	N.d.
Magelona lusitanica Mortimer, Gil and Fiege, 2011	SW Portuguese continental shelf	Temperate Northern Atlantic	105–327
Magelona magnahamata Aguado and San Martín, 2004	Granito de Oro, Panamá	Tropical Eastern Pacific	N.d.
Magelona mahensis Mortimer and Mackie, 2006	Mahé, Seychelles	Western Indo-Pacific	11–48
Magelona marianae Hernández-Alcántara and Solís-Weiss, 2000	Punta Mita, Mexico	Temperate Northern Pacific	23.5–46.4
Magelona methae Nateewathana and Hylleberg, 1991	Kamala Bay, Phuket Island, Thailand	Western Indo-Pacific	8.4–10
Magelona mickminni Nateewathana and Hylleberg, 1991	Patong Bay, Phuket Island, Thailand	Western Indo-Pacific	10
Magelona minuta Eliason, 1962	Öresund, Sweden	Temperate Northern Atlantic	15–18
Magelona mirabilis (Johnston, 1865)	St Andrews, Scotland, UK.	Temperate Northern Atlantic	N.d.
Magelona montera Mortimer, Cassà, Martin and Gil, 2012	Eilat, Gulf of Aqaba, Red sea	Western Indo-Pacific	Intertidal
Magelona nonatoi Bolívar and Lana, 1986	Rio de Janeiro, Brazil	Temperate South America	N.d.
Magelona noppi Nateewathana and Hylleberg, 1991	Nopparat-Tara Beach, Krabi, Thailand	Western Indo-Pacific	Intertidal
Magelona obockensis Gravier, 1905	Obock, Gulf of Aden, Red Sea	Western Indo-Pacific	Intertidal
Magelona pacifica Monro, 1933	Gorgona Island, Panamá	Tropical Eastern Pacific	Intertidal
Magelona papillicornis F. Müller, 1858	Santa Catarina Island, Brazil	Temperate South America	0–0.8
Magelona parochilis Zhou and Mortimer, 2013	Yellow Sea China	Temperate Northern Pacific	Intertidal
Magelona paulolanai Magalhães, Bailey-Brock and Watling, 2018	Hawaii, USA	Eastern Indo-Pacific	30–40
Magelona pectinata Nateewathana and Hylleberg, 1991	Kamala Bay, Phuket Island, Thailand	Western Indo-Pacific	10
Magelona petersenae Nateewathana and Hylleberg, 1991	Bang Tao Bay, Phuket Island, Thailand	Western Indo-Pacific	10
Magelona pettiboneae Jones, 1963	St Andrew Bay, Bay County, Florida, USA	Temperate Northern Atlantic	30
Magelona phyllisae Jones, 1963	Lamont Geological Observatory Station V-15-68, Santa, Peru	Temperate South America	181
Magelona pitelkai Hartman, 1944	Tomales Bay, California	Temperate Northern Pacific	Shallow
Magelona polydentata Jones, 1963	Bimini Lagoon, Bahamas	Tropical Atlantic	Intertidal
Magelona posterelongata Bolívar and Lana, 1986	Paraná Bay, Brazil	Temperate South America	Shallow
Magelona pulchella Mohammad, 1970	Kuwait, Arabian Gulf	Western Indo-Pacific	Intertidal
Magelona pygmaea Nateewathana and Hylleberg, 1991	Kamala Bay, Phuket Island, Thailand	Western Indo-Pacific	10
Magelona riojai Jones, 1963	Antón Lizardo, Veracruz, Mexico	Tropical Atlantic	Intertidal
Magelona rosea Moore, 1907	Buzzard's Bay, Wood's Hole, Massachusetts	Temperate Northern Atlantic	Intertidal
Magelona sacculata Hartman, 1961	San Pedro shelf area, California, USA	Temperate Northern Pacific	20–40
Magelona sachalinensis Buzhinskaja, 1985	Sakhalin Island, Sea of Japan	Temperate Northern Pacific	30–175
Magelona sinbadi Mortimer, Cassà, Martin and Gil, 2012	Iran, Persian Gulf	Western Indo-Pacific	20
Magelona spinifera (Hernández-Alcántara and Solís-Weiss, 2000)	Santa María Bay, Mexico	Temperate Northern Pacific	75
Magelona symmetrica Mortimer and Mackie, 2006	Mahé, Seychelles	Western Indo-Pacific	20

Table A1. *Cont.*

SPECIES	TYPE LOCALITY	REALM	DEPTH (m)
Magelona tehuanensis Hernández-Alcántara and Solís-Weiss, 2000	Western Salina Cruz, Mexico	Tropical Eastern Pacific	70
Magelona tinae Nateewathana and Hylleberg, 1991	Bang Tao Bay, Phuket Island, Thailand	Western Indo-Pacific	10
Magelona uebelackerae (Hernández-Alcántara and Solís-Weiss, 2000)	Texas, Gulf of Mexico	Temperate Northern Atlantic	3
Magelona variolamellata Bolívar and Lana, 1986	Paraná, São Paulo and Rio de Janeiro, Brazil	Temperate South America	Shallow
Magelona wilsoni Glémarec, 1966	South of Brittany, 'Grande Vasière', France	Temperate Northern Atlantic	60–110
Octomagelona bizkaiensis Aguirrezabalaga, Ceberio and Fiege, 2001	Capbreton Canyon, Bay of Biscay	Temperate Northern Atlantic	984–1040

Table A2. Oweniid valid species (partially according to WoRMS), type locality, assigned marine realms (sensu [86]), and depth of type locality. N.d.= No data.

SPECIES	TYPE LOCALITY	REALMS	DEPTH (m)
Galathowenia africana Kirkegaard, 1959	West Africa	Tropical Atlantic	30–42
Galathowenia annae Capa, Parapar and Hutchings, 2012	Botany Bay, Sydney, Australia	Temperate Australasia	16
Galathowenia arafurensis Capa, Parapar and Hutchings, 2012	Arafura Sea, Australia	Central Indo-Pacific	88
Galathowenia australis (Grube, 1866)	Saint Paul Island, Indian Ocean	Temperate Southern Africa	N.d.
Galathowenia eurystoma (Caullery, 1944)	Indonesia	Central Indo-Pacific	31–1570
Galathowenia fragilis (Nilsen and Holthe, 1985)	Norwegian Sea	Temperate Northern Atlantic	800–2600
Galathowenia haplosoma (Gibbs, 1972)	Cook Islands	Eastern Indo-Pacific	1–5
Galathowenia joinvillensis Hartmann-Schröder and Rosenfeldt, 1989	Antarctic Ocean	Southern Ocean	68–458
Galathowenia kirkegaardi De León-González and Sanchez-Hernández, 2012	Gulf of Mexico	Tropical Atlantic	1–4.1
Galathowenia lobopygidiata (Uschakov, 1950)	Okhotsk Sea	Temperate Northern Pacific	110–1366
Galathowenia longicollaris Hartmann-Schröder and Rosenfeldt, 1989	Antarctic Ocean	Southern Ocean	68
Galathowenia oculata (Zachs, 1923)	White Sea	Arctic	12–2500
Galathowenia piltzi Blake, 2000	Santa Maria Basin, California	Temperate Northern Pacific	92
Galathowenia pygidialis (Harman, 1960)	California and Baja California	Temperate Northern Pacific	>2000
Galathowenia quelis Capa, Parapar and Hutchings, 2012	North of Sydney, Australia	Temperate Australasia	1–60
Galathowenia scotiae (Hartman, 1978)	Weddell Sea	Southern Ocean	42–1592
Myriochele annenkovae Hartman, 1960 (1)	Northern Sea of Japan	Temperate Northern Pacific	45
Myriochele antarctica (Hartman, 1967)	South Orkney Islands	Southern Ocean	604–3816
Myriochele antarctica Cantone and Di Pietro, 2001	Ross Sea, Terra Nova Bay	Southern Ocean	126–136
Myriochele australiensis Capa, Parapar and Hutchings, 2012	East of Long Reef, Sydney, Australia	Temperate Australasia	60
Myriochele danielsseni Hansen, 1878	Norwegian Sea	Temperate Northern Atlantic	1187
Myriochele gracilis Hartman, 1955	Pacific Ocean, Southern California	Temperate Northern Pacific	115
Myriochele heeri Malmgren, 1867	Spitsbergen, Norway	Temperate Northern Atlantic	120–2600
Myriochele heruensis Gibbs, 1971	Solomon Islands	Eastern Indo-Pacific	16
Myriochele islandica (Parapar, 2003)	West Iceland	Temperate Northern Atlantic	1187–1407
Myriochele japonica (Imajima and Morita, 1987)	Japan	Temperate Northern Pacific	62–125
Myriochele malmgreni (Parapar, 2006)	Faroe Passage, Iceland	Temperate Northern Atlantic	1187–1407
Myriochele minor Caullery, 1944	Indonesia	Central Indo-Pacific	462–959

Table A2. *Cont.*

SPECIES	TYPE LOCALITY	REALMS	DEPTH (m)
Myriochele olgae Blake, 2000	off Half Moon Bay, California	Temperate Northern Pacific	100–200
Myriochele pacifica McIntosh, 1885	Central Pacific Ocean	Eastern Indo-Pacific	4754
Myriochele picta Southern, 1921	Chilka Lake, India	Western Indo-Pacific	1–3
Myriochele riojai Parapar, 2003	Bransfield Strait, Antarctic Ocean	Southern Ocean	647–1416
Myriochele robusta Parapar, 2003	Bransfield Strait, Antarctic Ocean	Southern Ocean	640–1500
Myriochele striolata Blake, 2000	Santa Maria Basin, California	Temperate Northern Pacific	91–400
Myriowenia californiensis Hartman, 1960	Southern California	Temperate Northern Pacific	106
Myriowenia gosnoldi Hartman, 1965	off Suriname	Tropical Atlantic	520–550
Owenia artifex (Verrill, 1885)	Massachusetts	Temperate Northern Atlantic	122
Owenia assimilator Caullery, 1944 (2)	n.d.	n.d.	N.d.
Owenia assimilis (Sars, 1851)	Norwegian Sea	Temperate Northern Atlantic	2–460
Owenia australis Ford and Hutchings, 2005	New South Wales, Australia	Temperate Australasia	<100
Owenia bassensis Ford and Hutchings, 2005	Victoria, East Bass Strait, Australia	Temperate Australasia	<100
Owenia borealis Koh, Bhaud and Jirkov, 2003	North East Atlantic	Temperate Northern Atlantic	41–1350
Owenia caissara Silva and Lana, 2017	Santa Catarina State, Brazil	Temperate South America	0–5
Owenia caudisetosa Hartmann-Schröder, 1959	El Salvador	Tropical Atlantic	N.d.
Owenia collaris Hartman, 1955	Santa Catalina Island	Temperate Northern Pacific	0–150
Owenia dichotoma Parapar and Moreira, 2015	Lizard Island, Australia	Central Indo-Pacific	12
Owenia fusiformis Delle Chiaje, 1844	Sicily, Mediterranean	Temperate Northern Atlantic	N.d.
Owenia gomsoni Koh and Bhaud, 2001	South Korea, Yellow Sea	Temperate Northern Pacific	0
Owenia johnsoni Blake, 2000	Tomales Bay, California	Temperate Northern Pacific	3–20
Owenia mirrawa Ford and Hutchings, 2005	Northern Territory, Australia	Central Indo-Pacific	<100
Owenia persica Martin, Koh, Bhaud, Dutrieux and Gil, 2006	Persian Gulf	Western Indo-Pacific	16
Owenia petersenae Koh and Bhaud, 2003	Wellington, New Zealand	Southern Ocean	20–30
Owenia picta Parapar and Moreira, 2015	Lizard Island, Australia	Central Indo-Pacific	2–12
Owenia polaris Koh, Bhaud and Jirkov, 2003	East Iceland	Temperate Northern Atlantic	12–930
Owenia vieitezi Díaz-Díaz, Parapar and Moreira, 2020	Gulf of Venezuela, Caribbean Sea	Tropical Atlantic	6–18

(1) Hartman [216] (p. 150) propose to rename *Myriochele pacifica* Annenkova [217] (p. 183) as *Myriochele annenkovae* for distinguishing it from *M. pacifica* McIntosh [218] (p. 413).

(2) Read and Fauchald [85] name it as *Owenia assimilator* Caullery, 1944 but Caullery [165] reports it as *O. assimilator* Andrews. Gil [156] affirms that Caullery was mistaken, because he was referring to *Ammochares aedificator* Andrews, 1891, a species that Read and Fauchald [85] report as synonym of *Owenia fusiformis*.

References

1. Weigert, A.; Bleidorn, C. Current Status of Annelid Phylogeny. *Org. Divers. Evol.* **2016**, *16*, 345–362. [CrossRef]
2. Helm, C.; Beckers, P.; Bartolomaeus, T.; Drukewitz, S.H.; Kourtesis, I.; Weigert, A.; Purschke, G.; Worsaae, K.; Struck, T.H.; Bleidorn, C. Convergent Evolution of the Ladder-like Ventral Nerve Cord in Annelida. *Front. Zool.* **2018**, *15*, 36. [CrossRef] [PubMed]
3. Struck, T. Phylogeny. In *Handbook of Zoology. Annelida. Basal Groups and Pleistoannelida, Sedentaria I*; Purschke, G., Böggemann, M., Westheide, W., Eds.; DeGruyter: Berlin, Germany; Boston, MA, USA, 2019; pp. 37–68.
4. Weigert, A.; Helm, C.; Meyer, M.; Nickel, B.; Arendt, D.; Hausdorf, B.; Santos, S.R.; Halanych, K.M.; Purschke, G.; Bleidorn, C.; et al. Illuminating the Base of the Annelid Tree Using Transcriptomics. *Mol. Biol. Evol.* **2014**, *31*, 1391–1401. [CrossRef] [PubMed]
5. Andrade, S.C.S.; Novo, M.; Kawauchi, G.Y.; Worsaae, K.; Pleijel, F.; Giribet, G.; Rouse, G.W. Articulating "Archiannelids": Phylogenomics and Annelid Relationships, with Emphasis on Meiofaunal Taxa. *Mol. Biol. Evol.* **2015**, *32*, 2860–2875. [CrossRef]
6. Struck, T.H.; Golombek, A.; Weigert, A.; Franke, F.A.; Westheide, W.; Purschke, G.; Bleidorn, C.; Halanych, K.M. The Evolution of Annelids Reveals Two Adaptive Routes to the Interstitial Realm. *Curr. Biol.* **2015**, *25*, 1993–1999. [CrossRef]
7. Kojima, S. Paraphyletic Status of Polychaeta Suggested by Phylogenetic Analysis Based on the Amino Acid Sequences of Elongation Factor 1-Alpha. *Mol. Phylogenet. Evol.* **1998**, *9*, 255–261. [CrossRef]

8. Eeckhaut, I.; McHugh, D.; Mardulyn, P.; Tiedemann, R.; Monteyne, D.; Jangoux, M.; Milinkovitch, C. Myzostomida: A Link between Trochozoans and Flatworms? *P. Roy. Soc. B Biol. Sci.* **2000**, *267*, 1383–1392. [CrossRef]
9. Bleidorn, C.; Vogt, L.; Bartolomaeus, T. New Insights into Polychaete Phylogeny (Annelida) Inferred from 18S RDNA Sequences. *Mol. Phylogenet. Evol.* **2003**, *29*, 279–288. [CrossRef]
10. Rousset, V.; Rouse, G.W.; Siddall, M.E.; Tillier, A.; Pleijel, F. The Phylogenetic Position of Siboglinidae (Annelida) Inferred from 18S RRNA, 28S RRNA and Morphological Data. *Cladistics* **2004**, *20*, 518–533. [CrossRef]
11. Rousset, V.; Pleijel, F.; Rouse, G.W.; Erséus, C.; Siddall, M.E. A Molecular Phylogeny of Annelids. *Cladistics* **2007**, *23*, 41–63. [CrossRef]
12. Struck, T.H.; Schult, N.; Kusen, T.; Hickman, E.; Bleidorn, C.; McHugh, D.; Halanych, K.M. Annelid Phylogeny and the Status of Sipuncula and Echiura. *BMC Evol. Biol.* **2007**, *7*, 57. [CrossRef] [PubMed]
13. Struck, T.H.; Nesnidal, M.P.; Purschke, G.; Halanych, K.M. Detecting Possibly Saturated Positions in 18S and 28S Sequences and Their Influence on Phylogenetic Reconstruction of Annelida (Lophotrochozoa). *Mol. Phylogenet. Evol.* **2008**, *48*, 628–645. [CrossRef] [PubMed]
14. Struck, T.H. The Impact of Paralogy on Phylogenomic Studies—A Case Study on Annelid Relationships. *PLoS ONE* **2013**, *8*, e62892. [CrossRef] [PubMed]
15. Zrzavý, J.; Říha, P.; Piálek, L.; Janouškovec, J. Phylogeny of Annelida (Lophotrochozoa): Total-Evidence Analysis of Morphology and Six Genes. *BMC Evol. Biol.* **2009**, *9*, 189. [CrossRef]
16. Chen, H.; Parry, L.A.; Vinther, J.; Zhai, D.; Hou, X.; Ma, X. A Cambrian Crown Annelid Reconciles Phylogenomics and the Fossil Record. *Nature* **2020**, *583*, 249–252. [CrossRef]
17. Capa, M.; Parapar, J.; Hutchings, P. Phylogeny of Oweniidae (Polychaeta) Based on Morphological Data and Taxonomic Revision of Australian Fauna. *Zool. J. Linn. Soc.* **2012**, *166*, 236–278. [CrossRef]
18. Mortimer, K.; Mills, K.; Jordana, E.; Pinedo, S.; Gil, J. A Further Review of European Magelonidae (Annelida), Including Redescriptions of *Magelona equilamellae* and *Magelona filiformis*. *Zootaxa* **2020**, *4767*, 89–114. [CrossRef]
19. Rouse, G.W. Magelona Müller, 1858. In *Polychaetes*; Rouse, G.W., Pleijel, F., Eds.; Oxford University Press: Oxford, UK, 2001; pp. 261–263. [CrossRef]
20. Cunningham, J.T.; Ramage, G.A. The Polychaeta Sedentaria of the Firth of Forth. *Trans. Roy. Soc. Edinburg* **1888**, *33*, 635–684. [CrossRef]
21. Dos Santos Brasil, A.C. Filogenia de Magelonidae Cunningham & Ramage, 1888 (Annelida–Polychaeta) com Base na Morfologia Externa. PhD Thesis, Universidade Federal do Paraná, Curitiva, Brazil, 2003.
22. Mortimer, K. 4.2 Magelonidae. In *Handbook of Zoology. Annelida. Basal Groups and Pleistoannelida, Sedentaria I*; Purschke, G., Böggemann, M., Westheide, W., Eds.; DeGruyter: Berlin, Germany; Boston, MA, USA, 2019; pp. 112–131.
23. McIntosh, W.C. On the Structure of *Magelona. Proc. R. Soc. Lond.* **1877**, *25*, 559–564.
24. McIntosh, W.C. Beiträge zur Anatomie von *Magelona. Z. Wiss. Zool.* **1878**, *31*, 401–472.
25. McIntosh, W.C. On the Circulatory System of *Magelona. J. Anat. Physiol., London* **1879**, *13*, 331–345.
26. McIntosh, W.C. On the Structure of *Magelona. Ann. Mag. Nat. Hist. (Ser. 8)* **1911**, *7*, 417–457. [CrossRef]
27. McIntosh, W.C. A monograph of the British marine annelids. In *Polychaeta. Opheliidae to Ammocharidae*; The Ray Society: London, UK, 1915; Volume 3, Part I. Text; pp. 1–368.
28. McIntosh, W.C. A monograph of the British marine annelids. In *Polychaeta. Opheliidae to Ammocharidae*; The Ray Society: London, UK, 1916; Volume 3, Part II. Plates.
29. Jones, M.L. Four New Species of *Magelona* (Annelida, Polychaeta) and a Redescription of *Magelona longicornis* Johnson. *Am. Mus. Novit.* **1963**, *2164*, 1–29.
30. Jones, M.L. On the Morphology, Feeding, and Behavior of *Magelona* sp. *Biol. Bull.* **1968**, *134*, 272–297. [CrossRef]
31. Jones, M.L. *Magelona berkeleyi* n. sp. from Puget Sound (Annelida: Polychaeta), with a Further Redescription of *Magelona longicornis* Johnson and a Consideration of Recently Described Species of *Magelona. J. Fish. Res. Bd. Can.* **1971**, *28*, 1445–1454. [CrossRef]
32. Jones, M.L. Redescription of *M. papillicornis* F. Müller. In *Essays on Polychaetous Annelids in Memory of Dr. Olga Hartman*; Reish, D.J., Fauchald, K., Eds.; Allan Hancock Foundation: University of Southern California: Los Angeles, CA, USA, 1977; pp. 247–266.
33. Jones, M.L. Three New Species of *Magelona* (Annelida, Polychaeta) and a Redescription of *Magelona pitelkai* Hartman. *Proc. Biol. Soc. Wash.* **1978**, *91*, 336–363.
34. Nateewathana, A.; Hylleberg, J. Magelonid Polychaetes from Thailand, the Andaman Sea, with Descriptions of Eight New Species. *Ophelia* **1991**, *5*, 169–184.
35. Mortimer, K.; Mackie, A.S.Y. The Magelonidae (Annelida: Polychaeta) from the Seychelles, with the Description of Three New Species. *Hydrobiologia* **2003**, *496*, 163–173. [CrossRef]
36. Mortimer, K.; Mackie, A.S.Y. The *Magelonidae* (Annelida: Polychaeta) from the Seychelles. 2. Description of Four Additional Species, Three New to Science. *Sci. Mar.* **2006**, *70*, 125–137. [CrossRef]
37. Mortimer, K. Magelonidae (Polychaeta) from the Arabian Peninsula: A Review of Known Species, with Notes on *Magelona tinae* from Thailand. *Zootaxa* **2010**, *2628*, 1–26. [CrossRef]
38. Mortimer, K.; Cassà, S.; Martin, D.; Gil, J. New Records and New Species of Magelonidae (*Polychaeta*) from the Arabian Peninsula, with a Re-Description of *Magelona pacifica* and a Discussion on the Magelonid Buccal Region. *Zootaxa* **2012**, *3331*, 1–43. [CrossRef]

39. Shakouri, A.; Mortimer, K.; Dehani, E. A New Species and New Records of *Magelona* (Annelida: Magelonidae) from Chabahar Bay, Gulf of Oman, South-Eastern Iran. *J. Mar. Biol. Ass. UK* **2017**, *97*, 1537–1552. [CrossRef]
40. Mortimer, K.; Mackie, A.S.Y. Magelonidae (Polychaeta) from Hong Kong, China; with Discussions on Related Species and Redescriptions of Three Species. *Zoosymposia* **2009**, *2*, 179–199. [CrossRef]
41. Zhou, J.; Mortimer, K. A New Species of *Magelona* (Polychaeta: Magelonidae) from Chinese Coastal Waters. *J. Mar. Biol. Ass. UK* **2013**, *93*, 1503–1510. [CrossRef]
42. Uebelacker, J.M.; Jones, M.L. Family Magelonidae. In *Taxonomic Guide to the Polychaetes of the Northern Gulf of Mexico*; Uebelacker, J.M., Johnson, P.G., Eds.; Barry A. Vittor & Associates: Mobile, AL, USA, 1984; Volume 2, pp. 7.1–7.29.
43. Hernández-Alcántara, P.; Solís-Weiss, V. Magelonidae from the Mexican Pacific and Northern Gulf of Mexico, with the Description of a New Genus (*Meridithia*) and Four New Species. *Bull. Mar. Sci.* **2000**, *67*, 625–644.
44. Hernández-Alcántara, P.; Solís-Weiss, V. Magelonidae Cunningham & Ramage, 1888. In *Poliquetos (Annelida: Poychaeta) de México y América Tropical*; León-González, J.A., Bastida-Zavala, J.R., Carrera Parra, L.F., García-Garza, M.E., Peña-Rivera, A., Salazar-Vallejo, S.I., Solís-Weiss, V., Eds.; Universidad Autónoma de Nuevo León: Nuevo León, Mexico, 2009; Volume 2, pp. 277–289.
45. Blake, J.A. Family Magelonidae Cunnigham & Ramage, 1888. In *Taxonomic Atlas of the Benthic Fauna of the Santa Maria Basin and Western Santa Barbara Channel*; Blake, J.A., Hilbig, B., Scott, P.H., Eds.; Santa Barbara Museum of Natural History: Santa Barbara, CA, USA, 1996; Volume 6, pp. 253–261.
46. Bolívar, G.A.; Lana, P.C. Magelonidae (Annelida: Polychaeta) do Litoral Sudeste do Brasil. *Neritica* **1986**, *1*, 131–147. [CrossRef]
47. Almeida, T.C.M.; Vivan, J.M.; Pesserl, B.H.; Lana, P. Polychaetes of the North-Central Santa Catarina State, Brazil. *Check List* **2012**, *8*, 204–206. [CrossRef]
48. Fiege, D.; Licher, F.; Mackie, A.S.Y. A Partial Review of the European Magelonidae (Annelida: Polychaeta): *Magelona mirabilis* Redefined and *M. johnstoni* sp. nov. Distinguished. *J. Mar. Biol. Assoc. UK* **2000**, *80*, 215–234. [CrossRef]
49. Mortimer, K.; Gil, J.; Fiege, D. Portuguese *Magelona* (Annelida: Magelonidae) with a Description of a New Species, a Re-Description of *Magelona wilsoni* Glémarec, and a Key to Adult Magelonidae from European Waters. *Ital. J. Zool.* **2011**, *78*, 124–139. [CrossRef]
50. Mills, K.; Mortimer, K. Redescription of *Magelona minuta* Eliason, 1962 (Annelida), with discussions on the validity of *Magelona filiformis minuta*. *Zootaxa* **2018**, *4527*, 541–559. [CrossRef]
51. Mortimer, K.; Mackie, A.S.Y. Morphology, Feeding and Behaviour of British *Magelona* (Annelida: Magelonidae), with Discussions on the Form and Function of Abdominal Lateral Pouches. *Mem. Mus. Vic.* **2014**, *71*, 177–201. [CrossRef]
52. Mills, K.; Mortimer, K. Observations on the Tubicolous Annelid *Magelona alleni* (Magelonidae), with Discussions on the Relationship between Morphology and Behaviour of European Magelonids. *J. Mar. Biol. Ass. UK* **2019**, *99*, 715–727. [CrossRef]
53. Mills, K.; Mortimer, K. Habitat and Distribution of the Shovelhead Worm *Magelona equilamellae* Harmelin, 1964 with Notes on the Morphologically Similar *Magelona alleni* Wilson, 1958. *Bull. Porcupine Mar. Nat. Hist. Soc.* **2019**, *12*, 63–67.
54. Fauchald, K. Life Diagram Patterns in Benthic Polychaetes. *Proc. Biol. Soc. Wash.* **1983**, *96*, 160–177.
55. Hartman, O. Abyssal Polychaetous Annelids from the Mozambique Basin off Southeast Africa, with a Compendium of Abyssal Polychaetous Annelids from World-Wide Areas. *J. Fish. Res. Board Can.* **1971**, *28*, 1407–1428. [CrossRef]
56. Aguirrezabalaga, F.; Ceberio, A.; Fiege, D. *Octomagelona bizkaiensis* (Polychaeta: Magelonidae) a New Genus and Species from the Capbreton Canyon (Bay of Biscay, North-East Atlantic). *J. Mar. Biol. Ass. UK* **2001**, *81*, 221–224. [CrossRef]
57. Capa, M.; Parapar, J.; Hutchings, P. 4.1. Oweniidae. In *Handbook of Zoology. Annelida. Basal Groups and Pleistoannelida, Sedentaria I*; Purschke, G., Böggemann, M., Westheide, W., Eds.; DeGruyter: Berlin, Germany; Boston, MA, USA, 2019; pp. 91–112. [CrossRef]
58. Read, G.; Fauchald, K. (Eds.) World Polychaeta Database. Magelonidae Cunningham & Ramage, 1888. 2021. Available online: http://marinespecies.org/aphia.php?p=taxdetails&id=975 (accessed on 12 January 2021).
59. Blake, J.A. Polychaeta Oweniidae from Antarctic Seas Collected by the United States Antarctic Research Program. In *Proceedings of the First International Polychaete Conference*; Hutchings, P.A., Ed.; Linnean Society of New South Wales: Sydney, Australia, 1984; pp. 112–117.
60. Hartmann-Schröder, G.; Rosenfeldt, P. Die Polychaten Der "Polarstern"- Reise ANT III/2 in Die Antarktis) 1984. Teil 2: Cirratulidae Dis Serpulidae. *Mitt. Hamb. Zool. Mus. Inst.* **1989**, *86*, 65–106.
61. Cantone, G.; Di Pietro, N. A New Species of *Myriochele* (Polychaeta, Oweniidae) from Antarctica, with Considerations on Antarctic Oweniids. *Polar Biol.* **1998**, *19*, 421–423. [CrossRef]
62. Parapar, J. Revision of Five Species Referred to *Myriochele* and *Galathowenia* (Polychaeta: Oweniidae) from the Antarctic Seas Based upon Type Material. *Proc. Biol. Soc. Wash.* **2001**, *114*, 403–413.
63. Parapar, J. Two New Species of *Myriochele* (Polychaeta: Oweniidae) from the Bransfield Strait (Antarctica). *Antarct. Sci.* **2003**, *15*, 219–226. [CrossRef]
64. Blake, J.A.; Dean, D. Polychaetous Annelids Collected by the R/V Hero from Baffin Island, Davis Strait, and West Greenland in 1968. *Bull. S. Calif. Acad. Sci.* **1973**, *72*, 31–39.
65. Nilsen, R.; Holthe, T. Arctic and Scandinavian Oweniidae (Polychaeta) with a Description of *Myriochele fragilis* sp.n., and Comments on the Phylogeny of the Family. *Sarsia* **1985**, *70*, 17–32. [CrossRef]
66. Koh, B.S.; Bhaud, M.R.; Jirkov, I.A. Two New Species of *Owenia* (Annelida: Polychaeta) in the Northern Part of the North Atlantic Ocean and Remarks on Previously Erected Species from the Same Area. *Sarsia* **2003**, *88*, 175–188. [CrossRef]

67. Parapar, J.O. Oweniidae (Annelida, Polychaeta) from Icelandic Waters, Collected by the BIOICE Project, with the Description of *Myrioglobula islandica* n. sp. *Sarsia* **2003**, *88*, 274–290. [CrossRef]
68. Parapar, J. The Genera *Myriochele* and *Myrioglobula* (Polychaeta, Oweniidae) in Icelandic Waters with the Revision of Type Material of *Myriochele heeri* Malmgren, 1867, and the Description of a New Species. *J. Nat. Hist.* **2006**, *40*, 523–547. [CrossRef]
69. Martin, D. Revisión de las especies de Owenidae (Annelida, Polychaeta) de la Peninsula Ibérica. *Sci. Mar.* **1989**, *53*, 47–52.
70. Hartman, O. *Atlas of the Sedentariate Polychaetous Annelids from California*; Allan Hancock Foundation, University of Southern California: Los Angeles, CA, USA, 1969.
71. Blake, J.A. Chapter 5. Family Oweniidae Rioja, 1917. In *Taxonomic Atlas of the Benthic Fauna of the Santa Maria Basin and the Western Santa Barbara Channel*; Blake, J.A., Hilbig, B., Scott, P.H., Eds.; Santa Barbara Museum of Natural History: Santa Barbara, CA, USA, 2000; Volume 7, pp. 97–127.
72. Milligan, M.R. Family Oweniidae Rioja, 1917. In *Taxonomic Guide to the Polychaetes of the Northern Gulf of Mexico*; Uebelacker, J.M., Johnson, P.G., Eds.; Barry A. Vittor & Associates: Mobile, AL, USA, 1984; Volume 6, pp. 46.1–46.12.
73. Villalobos-Guerrero, T.F. Oweniidae, Rioja 1917. In *Poliquetos (Annelida: Poychaeta) de México y América Tropical*; León-González, J.A., Bastida-Zavala, J.R., Carrera Parra, L.F., García-Garza, M.E., Peña-Rivera, A., Salazar-Vallejo, S.I., Solís-Weiss, V., Eds.; Universidad Autónoma de Nuevo León: Nuevo León, Mexico, 2009; Volume 2, pp. 391–402.
74. Díaz-Díaz, O.; Parapar, J.; Moreira, J. A New Species of Genus *Owenia* Delle-Chiaje, 1844 (Annelida; Oweniidae) from the Coast of Venezuela. *Cah. Biol. Mar.* **2018**, *59*, 589–597.
75. Koh, B.; Bhaud, M. Description of *Owenia gomsoni* n. sp. (Oweniidae, Annelida Polychaeta) from the Yellow Sea and Evidence That *Owenia fusiformis* Is Not a Cosmopolitan Species. *Vie Milieu* **2001**, *51*, 77–87.
76. Imajima, M.; Morita, Y. Oweniidae (Annelida, Polychaeta) from Japan. *Bull. Natl. Sci. Mus.* **1987**, *13*, 85–102.
77. Ford, E.; Hutchings, P. An Analysis of Morphological Characters of *Owenia* Useful to Distinguish Species: Description of Three New Species of *Owenia* (Oweniidae: Polychaeta) from Australian Waters. *Mar. Ecol.* **2005**, *26*, 181–196. [CrossRef]
78. Parapar, J.; Moreira, J. The Oweniidae (Annelida; Polychaeta) from Lizard Island (Great Barrier Reef, Australia) with the Description of Two New Species of *Owenia* Delle Chiaje, 1844. *Zootaxa* **2015**, *4019*, 604–620. [CrossRef] [PubMed]
79. Dauvin, J.-C.; Thiébaut, E. Is *Owenia fusiformis* Delle Chiaje a Cosmopolitan Species? *Mém. Mus. Nat. Paris* **1994**, *162*, 383–404.
80. Parapar, J. Resurrection of *Galathowenia australis* (Polychaeta, Oweniidae) Based upon Type Material. *Cah. Biol. Mar.* **2003**, *44*, 249–255.
81. Martin, D.; Koh, B.-S.; Bhaud, M.; Gil, J.; Dutrieux, E. The Genus *Owenia* (Annelida, Polychaeta) in the Persian Gulf, with Description of a New Species, *Owenia persica*. *Org. Divers. Evol.* **2006**, *6*, 325–326. [CrossRef]
82. Ménard, F.; Gentil, F.; Dauvin, J.-C. Population Dynamics and Secondary Production of *Owenia fusiformis* Delle Chiaje. Polychaeta from the Bay of Seine Eastern English Channel France. *J. Exp. Mar. Biol. Ecol.* **1989**, *133*, 151–168. [CrossRef]
83. Smart, T.I.; Von Dassow, G. Unusual Development of the Mitraria Larva in the Polychaete *Owenia collaris*. *Biol. Bull.* **2009**, *217*, 253–268. [CrossRef] [PubMed]
84. Kirkergaard, J.B. Benthic Polychaeta from Depths Exceeding 6000 m. *Galathea Rep.* **1956**, *2*, 63–78.
85. Read, G.; Fauchald, K. (Eds.) World Polychaeta Database. Oweniidae Rioja, 1917. 2021. Available online: http://marinespecies.org/aphia.php?p=taxdetails&id=975 (accessed on 12 January 2021).
86. Spalding, M.D.; Fox, H.E.; Allen, G.R.; Davidson, N.; Ferdaña, Z.A.; Finlayson, M.; Halpern, B.S.; Jorge, M.A.; Lombana, A.; Lourie, S.A.; et al. Marine Ecoregions of the World: A Bioregionalization of Coastal and Shelf Areas. *BioScience* **2007**, *57*, 573–583. [CrossRef]
87. Rouse, G.W.; Fauchald, K. Cladistics and Polychaetes. *Zool. Scr.* **1997**, *26*, 139–204. [CrossRef]
88. Weigert, A.; Golombek, A.; Gerth, M.; Schwarz, F.; Struck, T.H.; Bleidorn, C. Evolution of Mitochondrial Gene Order in Annelida. *Mol. Phylogenet. Evol.* **2016**, *94*, 196–206. [CrossRef]
89. Parry, L.; Vinther, J.; Edgecombe, G.D. Cambrian stem-group annelids and a metameric origin of the annelid head. *Biol. Lett.* **2015**, *11*, 20150763. [CrossRef] [PubMed]
90. Hartman, O. Polychaetous Annelids from California, Including the Descriptions of Two New Genera and Nine New Species. *Allan Hancock Pac. Exp.* **1944**, *10*, 239–307.
91. Hartman, O.; Reish, D.J. The Marine Annelids of Oregon. *Or. State Monogr. Stud. Zool.* **1950**, *6*, 1–64.
92. Gallardo, V.A. Polychaeta from the Bay of Nha Trang, South Viet Nam. *Naga Report* **1968**, *4*, 35–279.
93. Eliason, A. Undersökningar Över Öresund. XXXXI. Weitere Untersuchungen Über Die Polychaetenfauna Des Öresunds. *Acta Univ. Lund. Avd. 2* **1962**, *58*, 1–98.
94. Hartman, O. Polychaetous Annelids from California. *Allan Hancock Pac. Exp.* **1961**, *25*, 1–226.
95. Hartman, O. Catalogue of the Polychaetous Annelids of the World. Supplement 1960-1965 and Index. *Allan Hancock Found. Publ. Occas. Pap.* **1965**, *23*, 1–197.
96. Glémarec, M. Les Magelonidae des Coates de Bretagne. Description de *Magelona wilsoni* n. sp. *Vie Milieu* **1966**, *17*, 1077–1085.
97. Kitamori, R. Magelonidae (Polychaetous Annelids) from Japan, Including the Description of a New Species. *Bull. Tokai Reg. Fish. Res. Lab.* **1967**, *50*, 49–54.
98. Reish, D.J. Benthic Polychaetous Annelids from Bering, Chukchi, and Beaufort Seas. *Proc. US Nat. Mus.* **1965**, *117*, 131–157. [CrossRef]
99. Hartmann-Schröder, G. Zweiter Beitrag Zur Polychaetenfauna von Peru. *Kiel. Meeresforsch.* **1962**, *18*, 109–147.

100. Day, J.H. The Polychaet [Sic] Fauna of South Africa. Part 6. Sedentary Species Dredged off Cape Coasts with a Few New Records from the Shore. *J. Linn. Soc. London* **1961**, *44*, 463–560. [CrossRef]
101. Harmelin, J.G. Étude de l'endofaune des "mattes" d'herbiers de *Posidonia oceanica* Delile. *Recl. Trav. Stn. Mar. Endoume* **1964**, *35*, 43–105.
102. Magalhães, W.F.; Bailey-Brock, J.; Watling, L. Four New Species of *Magelona* (Annelida: Magelonidae) from Easter Island, Guam and Hawaii. *Zootaxa* **2018**, *4457*, 379–396. [CrossRef] [PubMed]
103. Okuda, S. Spioniform Polychaetes from Japan. *J. Fac. Sci. Hokkaido Univ. Ser.* **1937**, *6*, 217–254.
104. Clarke, D.T.; Paterson, G.L.J.; Florence, W.K.; Gibbons, M.J. A New Species of *Magelona* (Polychaeta: Magelonidae) from Southern Namibia. *Afr. Nat. Hist.* **2010**, *6*, 77–82.
105. Wilson, D.P. The Larval Development of Three Species of *Magelona* (Polychaeta) from Localities near Plymouth. *J. Mar. Biol. Assoc. UK* **1982**, *62*, 385–401. [CrossRef]
106. Mésnil, F. Études de morphologie externe chez les annélides. I. Les spionidens des côtes de la Manche. *Bull. Sci. Fr. Belg.* **1896**, *29*, 110–287. [CrossRef]
107. Romieu, M. Contribution à l'histologie comparée du muscle strié. *C. R. Acad. Sci. Paris* **1923**, *176*, 864–867.
108. Orrhage, L. Über das Vorkommen von Muskelzellen vom "Nematoden-Typus" bei Polychaeten als phylogenetisch-systematisches Merkmal. *Zool. Bidr. Upps.* **1962**, *35*, 321–327.
109. Orrhage, L. Über Die Anatomie Des Zentralen Nervensystemes Der Sedentären Polychaeten. *Ark. Zool.* **1966**, *19*, 99–133.
110. Bartolomaeus, T. Secondary Monociliarity in the Annelida: Monociliated Epidermal Cells in Larvae of *Magelona mirabilis* (Magelonidae). *Microfauna Mar.* **1995**, *10*, 327–332.
111. Benham, W.B. The Blood of *Magelona*. *Q. J. Microsc. Sci.* **1896**, *39*, 1–17.
112. Wells, R.M.; Dales, R.P. Oxygenational Properties of Haemerythrin in the Blood of *Magelona papillicornis* Muller (Polychaeta: Magelonidae). *Comp. Biochem. Physiol.* **1974**, *49A*, 57–64. [CrossRef]
113. Purschke, G. Annelida: Basal groups and Pleistoannelida. In *Structure and Evolution of Invertebrate Nervous Systems*; Schmidt-Rhaesa, A., Harzsch, S., Purschke, G., Eds.; Oxford University Press: Oxford, UK, 2016; pp. 254–312.
114. Beckers, P.; Helm, C.; Bartolomaeus, T. The Anatomy and Development of the Nervous System in Magelonidae (Annelida)—Insights into the Evolution of the Annelid Brain. *BMC Evol. Biol.* **2019**, *19*, 173. [CrossRef] [PubMed]
115. Filippova, A.; Purschke, G.; Tzetlin, A.B.; Müller, M.C.M. Reconstruction of the Musculature of *Magelona* cf. mirabilis (Magelonidae) and Prionospio cirrifera (Spionidae) (Polychaeta, Annelida) by Phalloidin Labeling and CLSM. *Zoomorphology* **2005**, *124*, 1–8. [CrossRef]
116. Jumars, P.A.; Dorgan, K.M.; Lindsay, S.M. Diet of Worms Emended: An Update of Polychaete Feeding Guilds. *Annu. Rev. Mar. Sci.* **2015**, *7*, 497–520. [CrossRef]
117. Wissocq, J.-C. Muscles à striation oblique et à striation transversale chez les Annélides Polychètes. *Ital. J. Zool.* **1981**, *48*, 57–74. [CrossRef]
118. GBIF.org. GBIF Home Page. 2021. Available online: https://www.gbif.org (accessed on 13 January 2021).
119. Wilson, D.P. The Polychaete *Magelona alleni* n. sp. and a Re- Assessment of *Magelona cincta* Ehlers. *J. Mar. Biol. Assoc. UK* **1958**, *37*, 617–626. [CrossRef]
120. Wilson, D.P. The Polychaete *Magelona filiformis* n. sp. and Notes on Other Species of *Magelona*. *J. Mar. Biol. Assoc. UK* **1959**, *38*, 547–556. [CrossRef]
121. Hylleberg, J.; Nateewathana, A. Temporal and Spatial Distribution of Subtidal Magelonid Polychaetes at Phuket Island, Thailand, Andaman Sea. *Ophelia* **1991**, *5*, 573–578.
122. Hannan, C.A. The Life Histories and Population Dynamics of the Polychaetes, *Nothria elegans* (Johnson) (Onuphidae) and *Magelona sacculata* Hartman (Magelonidae) in Monterey Bay, California. Master's Thesis, San Jose State University, San Jose, CA, USA, 1980.
123. Fajardo, M.; Andrade, D.; Bonicelli, J.; Bon, M.; Gómez, G.; Riascos, J.M.; Pacheco, A.S. Macrobenthic Communities in a Shallow Normoxia to Hypoxia Gradient in the Humboldt Upwelling Ecosystem. *PLoS ONE* **2018**, *13*, e0200349. [CrossRef] [PubMed]
124. Gaston, G. Effects of Hypoxia on Macrobenthos of the Inner Shelf off Cameron, Louisiana. *Estuar. Coast. Shelf Sci.* **1985**, *20*, 603–613. [CrossRef]
125. Sivadas, S.K.; Carvalho, R. Marine Annelida of India: Taxonomy and Status Evaluation and an Updated Checklist. *J. Threat. Taxa* **2020**, *12*, 16647–16714. [CrossRef]
126. Meißner, K.; Darr, A. Distribution of *Magelona* Species (Polychaeta: Magelonidae) in the German Bight (North Sea): A Modelling Approach. *Zoosymposia* **2009**, *2*, 567–586. [CrossRef]
127. Fauchald, K.; Jumars, P. The Diet of Worms: A Study of Polychaete Feedings Guilds. *Oceanogr. Mar. Biol. Ann. Rev.* **1979**, *17*, 193–284.
128. Wolff, W.J. The Estuary as a Habitat: An Analysis of Data on the Soft-Bottom Macrofauna of the Estuarine Area of the Rivers Rhine, Meuse, and Scheldt. *Zool. Verh.* **1973**, *126*, 1–242.
129. Hartmann-Schroder, G. Annelida, Borstenwurmer, Polychaeta. *Tierwelt Dtschl.* **1971**, *58*, 1–594.
130. McMahon, R.F.; Jones, M.L. Observations on Feeding in *Magelona* sp. *Biol. Bull.* **1967**, *133*, 476.
131. Hunt, O.D. The Food of the Bottom Fauna of the Plymouth Fishing Grounds. *J. Mar. Biol. Ass. UK* **1925**, *13*, 560–599. [CrossRef]

132. Mare, M.F. A Study of a Marine Benthic Community with Special Reference to the Micro-Organisms. *J. Mar. Biol. Ass. UK* **1942**, *25*, 517–554. [CrossRef]
133. Kühl, H. Über Vorkommen Und Nahrung der Larven von *Magelona papillicornis* O.F. Müller (Polychaeta Sedentaria) Im Mündungsgebiet von Elbe, Weser Und Ems. *Ber. Dtsch. Wiss. Komm. Meeresfirsch.* **1974**, *23*, 296–301.
134. Rouse, G.W. Polychaete Sperm: Phylogenetic and Functional Considerations. *Hydrobiologia* **1999**, *402*, 215–224. [CrossRef]
135. Rouse, G.W. Chapter 3. In *Annelid Sperm and Spermiogenesis*; Rouse, G.W., Pleijel, F., Eds.; Science Publishers Inc: Enfield, NH, USA, 2006; pp. 45–76.
136. Johnson, K.B.; Brink, L.A. Predation on Bivalve Veligers by Polychaete Larvae. *Biol. Bull.* **1998**, *194*, 297–303. [CrossRef]
137. Plate, S.; Husemann, E. Identification Guide to the Planktonic Larvae around the Island of Helgoland (German Bight). *Helgol. Meeresunters.* **1994**, *48*, 1–58. [CrossRef]
138. Carson, H.S.; Hentschel, B.T. Estimating the Dispersal Potential of Polychaete Species in the Southern California Bight: Implications for Designing Marine Reserves. *Mar. Ecol. Prog. Ser.* **2006**, *316*, 105–113. [CrossRef]
139. Criales-Hernández, M.I.; Schwamborn, R.; Graco, M.; Ayón, P.; Hirche, H.-J.; Wolff, M. Zooplankton Vertical Distribution and Migration off Central Peru in Relation to the Oxygen Minimum Layer. *Helgol. Mar. Res.* **2008**, *62*, 85–100. [CrossRef]
140. Dean, H.K. The Use of Polychaetes (Annelida) as Indicator Species of Marine Pollution: A Review. *Rev. Biol. Trop.* **2008**, *56*, 29.
141. Pagliosa, P.R. Another Diet of Worms: The Applicability of Polychaete Feeding Guilds as a Useful Conceptual Framework and Biological Variable. *Mar. Ecol.* **2005**, *26*, 246–254. [CrossRef]
142. Liwanow, N.A.; Porfirjewa, N.A. Die Organisation Der Pogonophoren Und Deren Beziehungen Zu Den Polychäten. *Biol. Zbl.* **1967**, *86*, 177–204.
143. Gardiner, S.L. Fine Structure of the Ciliated Epidermis on the Tentacles of *Owenia fusiformis* (Polychaeta, Oweniidae). *Zoomorphologie* **1978**, *91*, 37–48. [CrossRef]
144. Smith, P.R.; Ruppert, E.E.; Gardiner, S.L. A Deuterostome-like Nephridium in the Mitraria Larva of *Owenia fusiformis* (Polychaeta, Annelida). *Biol. Bull.* **1987**, *172*, 315–323. [CrossRef]
145. Westheide, W. The Direction of Evolution within the Polychaeta. *J. Nat. Hist.* **1997**, *31*, 1–15. [CrossRef]
146. Hausen, H. Chaetae and chaetogenesis in polychaetes (Annelida). *Hydrobiologia* **2005**, *535*, 37–52. [CrossRef]
147. Rimskaya-Korsakova, N.N.; Kristof, A.; Malakhov, V.V.; Wanninger, A. Neural Architecture of *Galathowenia oculata* Zach, 1923 (Oweniidae, Annelida). *Front. Zool.* **2016**, *13*, 1–19. [CrossRef]
148. Rimskaya-Korsakova, N.; Dyachuk, V.; Temereva, E. Parapodial Glandular Organs in *Owenia borealis* (Annelida: Oweniidae) and Their Possible Relationship with Nephridia. *J. Exp. Zool. Part B* **2020**, *334*, 88–99. [CrossRef]
149. Kirkegaard, J.B. The Polychaeta of West Africa Part I. Sedentary Species. *Atlantide Rep.* **1959**, *5*, 7–117.
150. Kirkegaard, J.B. Bathyal Benthic Polychaetes from the N.E. Atlantic Ocean, S.W. of the British Isles. *J. Mar. Biol. Ass. UK* **1983**, *63*, 593–608. [CrossRef]
151. Fauvel, P. Polychètes Sédentaires. Addenda aux Errantes, Archiannélides, Myzostomaires. *Faune France* **1927**, *16*, 1–494.
152. Uschakov, P. *Polychaeta of the Far Eastern Seas of the U.S.S.R.*; Israel Program for Scientific Translations: Jerusalem, Israel, 1965.
153. Hartman, O. Polychaeta Myzostomidae and Sedentaria of Antarctica. *Antarctic Res. Ser.* **1966**, *7*, 1–158.
154. Day, J.H. A Monograph on the Polychaeta of Southern Africa 2. Sedentaria. *Brit. Mus. (Nat. Hist.) Lond.* **1967**, 459–878.
155. Hartmann-Schröder, G. Annelida, Borstenwürmer, Polychaeta. *Tierwelt Dtschl.* **1996**, *58*, 1–648.
156. Gil, J. The European Fauna of Annelida Polychaeta. Ph.D. Thesis, University of Lisboa, Lisbon, Portugal, 2011.
157. Rouse, G.W. Oweniidae. In *Polychaetes*; Rouse, G.W., Pleijel, F., Eds.; Oxford University Press: Oxford, UK, 2001; pp. 185–188. [CrossRef]
158. Watson, A.T. On the structure and habits of the Polychaeta of the family Ammocharidae. *J. Linn. Soc.* **1901**, *28*, 230–260. [CrossRef]
159. Wilson, D.P. On the Mitraria Larva of *Owenia fusiformis* Delle Chiaje. *Philos. T. R. Soc. Lon. B.* **1932**, *22*, 231–334.
160. Dales, R.P. The Feeding Mechanism and Structure of the Gut of *Owenia fusiformis* Delle Chiaje. *J. Mar. Biol. Ass. UK* **1957**, *36*, 81–89. [CrossRef]
161. Beckers, P.; Helm, C.; Purschke, G.; Worsaae, K.; Hutchings, P.; Bartolomaeus, T. The central nervous system of Oweniidae (Annelida) and its implications for the structure of the ancestral annelid brain. *Front. Zool.* **2019**, *16*, 6. [CrossRef] [PubMed]
162. Hutchings, P.A. Family Oweniidae. In *Polychaetes & Allies: The Southern Synthesis*; Fauna of Australia. Vol. 4A Polychaeta, Myzostomida, Pogonophora, Echiura, Sipuncula; Beesley, P.L., Ross, G.J.B., Glasby, C.J., Eds.; CSIRO Publishing: Melbourne, Australia, 2000; pp. 173–176.
163. Koh, B.-S.; Bhaud, M. Identification of New Criteria for Differentiating between Populations of *Owenia fusiformis* (Annelida Polychaeta) from Different Origins: Rehabilitation of Old Species and Erection of New Species. *Vie Milieu* **2003**, *53*, 65–95.
164. Silva, L.; Lana, P. Strategies for tube construction in *Owenia caissara* (Oweniidae, Annelida) from southern Brazil. *Zoology* **2018**, *129*, 9–16. [CrossRef]
165. Caullery, M. Polychètes Sédentaire de l'Expédition Du Siboga: Ariciidae, Spionidae, Chaetopteridae, Chlorhaemidae, Opheliidae, Oweniidae, Sabellariidae, Sternaspidae, Amphictenidae, Ampharetidae, Terebellidae. In *Siboga-Expeditie Uitkomsten op Zoologisch, Bonatisch, Oceanographisch en Geologisch gebied Verzameld in Nederlandsch Oost-Indië 1899-1900*; World Register of Marine Species: Ostend, Belgium, 1944; Volume XXIV 2 bis, pp. 1–204.

166. Verrill, A.E. Notice of Recent Additions to the Marine Invertebrata of the Northeastern Coast of America, with Descriptions of New Genera and Species and Critical Remarks on Others. Part V. Annelida, Echinodermata, Hydroida, Tunicata. *Proc. US Nat. Mus.* **1885**, *8*, 424–448. [CrossRef]
167. Rioja, E. Estudio de Los Poliquetos de La Península Ibérica. *Mem. Real Acad. Ci. Exact. Madrid Ser. Cienc. Nat.* **1931**, *2*, 1–471.
168. Díaz-Díaz, O.F.; Parapar, J.; Moreira, J. Validation of *Owenia vieitezi* Díaz-Díaz, Parapar & Moreira, a polychaete from the coast of Venezuela (Annelida: Oweniidae). *Zootaxa* **2020**, *4729*, 145–146.
169. Malmgren, A.J. Annulata Polychaeta Spetsbergiae, Grönlandiae et Scandinaviae Hactenus Cognita. *Öfvers. K. Vet. Akad. Förh.* **1867**, *24*, 127–235.
170. Delle Chiaje, S. *Descrizione e Notomia degli Animali Invertebrati della Sicilia Citeriore osservati vivi negli anni 1822–1830. Tomo Ottavo. Appendice, Osservazioni Critiche, Indice Generale*; Stabilimento Tipografico di C. Batelli e Comp.: Naples, Italy, 1844.
171. Berkeley, E. Morphological characters of *Myriochele heeri* Malmgren. *Nature* **1949**, *164*, 239. [CrossRef]
172. Rullier, F. Étude Morphologique, Histologique et Physiologique de l'organe Nucal Chez Les Annélides Polychètes Sédentaires. *Ann. Inst. Ocean.* **1951**, *25*, 207–341.
173. Gardiner, S.L.; Rieger, R.M. Rudimentary Cilia in Muscle Cells of Annelids and Echinoderms. *Cell Tissue Res.* **1980**, *213*, 247–252. [CrossRef]
174. Tzetlin, A.B.; Filippova, A.V. Muscular system in polychaetes (Annelida). *Hydrobiologia* **2005**, *535*, 113–126. [CrossRef]
175. Fransen, M.E. Coelomic and Vascular Systems. Chapter XII. In *The Ultrastructure of Polychaeta. Microfauna Marina*; Westheide, W., Hermans, C.O., Eds.; Fischer Gustav Verlag GmbH & Co. KG.: Sttutgart, Germany, 1988; Volume 4, pp. 199–213.
176. Parapar, J.; Candás, M.; Cunha-Veira, X.; Moreira, J. Exploring annelid anatomy using micro-computed tomography: A taxonomic approach. *Zool. Anz.* **2017**, *270*, 19–42. [CrossRef]
177. Tzetlin, A.B.; Purschke, G. Pharynx and intestine. *Hydrobiologia* **2005**, *535*, 199–205. [CrossRef]
178. Purschke, G.; Tzetlin, A.B. Dorsolateral Ciliary Folds in the Polychaete Foregut: Structure, Prevalence and Phylogenetic Significance. *Acta Zool.* **1996**, *77*, 33–49. [CrossRef]
179. Bubko, O.V.; Minichev, Y.S. The Nervous System of Oweniidae (Polychaeta). *Zool. Zhurnal* **1972**, *51*, 1288–1299.
180. Coulon, J.; Bessone, R. Autoradiographic Detection of Indolamine and Catecholamine Neurons in the Nervous System of *Owenia fusiformis* (Polychaeta, Annelida). *Cell Tissue Res.* **1979**, *198*, 95–104. [CrossRef] [PubMed]
181. Purschke, G. Sense Organs in Polychaetes (Annelida). *Hydrobiologia* **2005**, *535*, 53–78. [CrossRef]
182. Purschke, G.; Arendt, D.; Hausen, H.; Müller, M.C.M. Photoreceptor Cells and Eyes in Annelida. *Arthropod Struct. Dev.* **2006**, *35*, 211–230. [CrossRef] [PubMed]
183. Helm, C.; Vöcking, O.; Kourtesis, I.; Hausen, H. *Owenia fusiformis*—A Basally Branching Annelid Suitable for Studying Ancestral Features of Annelid Neural Development. *BMC Evol. Biol.* **2016**, *16*. [CrossRef] [PubMed]
184. Rioja, E. Datos para el conocimiento de la fauna de anélidos poliquetos del Cantábrico. *Trab. Mus. Nac. Cienc. Nat. Ser. Zool.* **1917**, *29*, 1–111.
185. Day, J.H. *New Polychaeta from Beaufort, with a Key to All Species Recorded from North Carolina*; NOAA tech. Rep. NMFS, circ. # 375; U.S. Dept. of Commerce, NOAA, Marine Fisheries Service: Seattle, WA, USA, 1973; pp. 1–140.
186. Fauchald, K. The Polychaete Worms. Definitions and Keys to the Orders, Families and Genera. *Nat. Hist. Mus. Los Angel. Cty. Sci. Ser.* **1977**, *28*, 1–188.
187. Pettibone, M.H. Annelida. In *Synopsis and Classification of Living Organisms*; Parker, S.P., Ed.; McGraw-Hill Book Co: New York, NY, USA, 1982; Volume 2, pp. 1–43.
188. Hartman, O. Catalogue of the polychaetous annelids of the world. *Allan Hancock Found. Publ. Occas. Pap.* **1959**, *23*, 1–628.
189. Hutchings, P.A.; Rainer, S. The Polychaete Fauna of Careel Bay, Pittwater, New South Wales, Australia. *J. Nat. Hist.* **1979**, *13*, 745–796. [CrossRef]
190. Day, J.H.; Hutchings, P.A. Descriptive Notes on the Fauna and Flora of Merimbula, Pambula and Black Lakes, New South Wales. *Aust. Zool.* **1984**, *21*, 269–289.
191. Hutchings, P.; Murray, A. Taxonomy of Polychaetes from the Hawkesbury River and the Southern Estuaries of New South Wales, Australia. *Rec. Aust. Mus. Suppl.* **1984**, *3*, 1–118. [CrossRef]
192. Nygren, A. Cryptic Polychaete Diversity: A Review. *Zool. Scr.* **2014**, *43*, 172–183. [CrossRef]
193. Hutchings, P.; Kupriyanova, E. Cosmopolitan Polychaetes—Fact or Fiction? Personal and Historical Perspectives. *Invert. Syst.* **2018**, *32*, 1. [CrossRef]
194. Jolly, M.T.; Viard, F.; Gentil, F.; Thiébaut, E.; Jollivet, D. Comparative Phylogeography of Two Coastal Polychaete Tubeworms in the Northeast Atlantic Supports Shared History and Vicariant Events: Comparative Phylogeography Of Coastal Tubeworms. *Mol. Ecol.* **2006**, *15*, 1841–1855. [CrossRef] [PubMed]
195. Gentil, F.; Dauvin, J.-C.; Ménard, F. Reproductive Biology of the Polychaete *Owenia fusiformis* Delle Chiaje in the Bay of Seine (Eastern English Channel). *J. Exp. Mar Biol. Ecol.* **1990**, *142*, 13–23. [CrossRef]
196. McNulty, J.K.; López, N.N. Year-Round Production of Ripe Gametes by Benthic Polychaetes in Biscayne Bay, Florida. *Bull. Mar. Sci.* **1969**, *19*, 945–954.
197. Thiébaut, E.; Dauvin, J.C. Modelling of population dynamics of *Owenia fusiformis* (Annelida, Polychaeta) in eastern Bay of Seine. In *Report of the Workshop Modelling the Benthos Yerseke, The Netherlands*; Herman, P.M.J., Ed.; Delta Institute Communication, Delta Institute for Hydrobiological Research: Yerseke, The Netherlands, 1991; Volume 538, pp. 189–190.

198. Okada, K.A. Ch. 7. Annelida. In *Invertebrate Embryology*; Kume, M., Dan, K., Eds.; Bai Fu Kan Press: Tokyo, Japan, 1970; pp. 192–241.
199. Oliver, J.S. Selection for Asexual Reproduction in an Antarctic Polychaete Worm. *Mar. Ecol. Prog. Ser.* **1984**, *19*, 3–38. [CrossRef]
200. Aguirrezabalaga, F.; Gil, J.; Viéitez, J.M. Presence of *Myriochele danielsseni* Hansen, 1879 (Polychaeta, Oweniidae) en las costas de la Península Ibérica. *Bol. R. Soc. Esp. Hist. Nat.* **2000**, *96*, 57–68.
201. Signor, P.W.; McMenamin, M.A.S. The early Cambrian worm tube *Onuphionella* from California and Nevada. *J. Paleontol.* **1988**, *62*, 233–240. [CrossRef]
202. Gambi, M.C. Osservazioni su morfologia funzionale e comportamento trofico di *Owenia fusiformis* delle Chiaje (Polychaeta, Oweniidae) in rapporto ai fattori ambientali. *Oebalia* **1989**, *15*, 145–155.
203. Dales, R.P. The Nature of the Pigments in the Crown of Sabellid and Serpulid Polychaetes. *J. Mar. Biol. Assoc. UK* **1962**, *42*, 259–274. [CrossRef]
204. Ziegelmeier, E. Neue Untersuchungen über die Wohnröhren-Bauweise von *Lanice conchilega* (Polychaeta, Sedentaria). *Helgol. Wiss. Meeresunters.* **1969**, *19*, 216–229. [CrossRef]
205. Künitzer, A. Zur Verbreitung und Populationsdynamik der Bodenfauna der Zentralen Nordsee. PhD. Thesis, Universität Bremen, Bremen, Germany, 1990.
206. Dauvin, J.-C.; Gillet, P. Spatio-Temporal Variability in Population Structure of *Owenia fusiformis* Delle Chiaje (Annelida: Polychaeta) from the Bay of Seine (Eastern English Channel). *J. Exp. Mar. Biol. Ecol.* **1991**, *152*, 105–122. [CrossRef]
207. Pinedo, S.; Sardá, R.; Rey, C.; Bhaud, M. Effect of Sediment Particle Size on Recruitment of *Owenia fusiformis* in the Bay of Blanes (NW Mediterranean Sea): An Experimental Approach to Explain Field Distribution. *Mar. Ecol. Prog. Ser.* **2000**, *203*, 205–213. [CrossRef]
208. Deart, Y.V.; Britayev, T.A. "New" Benthic Community Dominated by Oweniidae (Polychaeta, Oweniidae) at the Murman Coast: Structure and Causes of Appearance. *Dokl. Biol. Sci.* **2014**, *454*, 34–38. [CrossRef] [PubMed]
209. Lee, H.W.; Bailey-Brock, J.H.; McGurr, M.M. Temporal Changes in the Polychaete Infaunal Community Surrounding a Hawaiian Mariculture Operation. *Mar. Ecol. Prog. Ser.* **2006**, *307*, 175–185. [CrossRef]
210. Kopp, D.; Le Bris, H.; Grimaud, L.; Nérot, C.; Brind'Amour, A. Spatial Analysis of the Trophic Interactions between Two Juvenile Fish Species and Their Preys along a Coastal–Estuarine Gradient. *J. Sea Res.* **2013**, *81*, 40–48. [CrossRef]
211. Fager, E.W. Marine Sediments: Effects of Tube-Dwelling Polychaete. *Science* **1964**, *143*, 356–359. [CrossRef]
212. Eckman, J.E.; Nowell, A.R.M.; Jumars, P.A. Sediment destabilization by animal tubes. *J. Mar. Res.* **1981**, *39*, 361–374.
213. Somaschini, A. A Mediterranean Fine-Sand Polychaete Community and the Effect of the Tube-Dwelling *Owenia fusiformis* Delle Chiaje on Community Structure. *Int. Rev. Gesamten Hydrobiol.* **1993**, *78*, 219–233. [CrossRef]
214. Ambrogi, R.; Fontana, P.; Gambi, M.C. Population dynamics and estimate of secondary production of *Owenia fusiformis* Delle Chiaje (Polychaeta, Oweniidae) in the coastal area of the Po river Delta (Italy). In *Biology and Ecology of Shallow Coastal Waters*; Eleftheriou, A., Smith, C.J., Ansell, A.D., Eds.; Olsen & Olsen: Fredensborg, Denmark, 1995; pp. 207–214.
215. Oyarzún, C.; Fernández, C.; Landaeta, M.; Cortés, N. Ontogenetic and Temporal Fluctuations in Feeding Habits of the Chilean Croaker *Cilus gilberti* (Perciformes, Sciaenidae) in Southern Chile. *Cah. Biol. Mar.* **2001**, *42*, 295–302.
216. Hartman, O. Systematic account of some marine invertebrate animals from the deep basins of southern California. In *The Benthic Fauna of the Deep Basins off Southern California. Part II*; Hartman, O., Barnard, J.L., Eds.; Allan Hancock Pacific Expeditions, University of Southern California Press: Los Angeles, CA, USA, 1960; Volume 22, pp. 69–215.
217. Annenkova, N.P. The Polychaete Fauna of the Northern Part of the Japan Sea. *Issled. Rus. Mor.* **1937**, *23*, 139–216.
218. McIntosh, W.C. Report on the Annelida Polychaeta Collected by H.M.S. Challenger during the Years 1873-1876. *Chall. Rep. (Zool.)* **1885**, *12*, 1–554.

diversity MDPI

Review

How Many Sipunculan Species Are Hiding in Our Oceans?

Anja Schulze [1,*] and Gisele Y. Kawauchi [2]

[1] Department of Marine Biology, Texas A&M University at Galveston, Galveston, TX 77554, USA
[2] Zoology Department, Institute of Biological Sciences, Federal University of Minas Gerais, Pampulha, Belo Horizonte, Minas Gerais 31270-901, Brazil; gykawauchi@ufmg.br
* Correspondence: schulzea@tamug.edu

Abstract: Sipuncula, long considered a separate phylum, are now commonly included in the Annelida based on phylogenomic analyses. The sipunculan body consists of an unsegmented trunk and a retractable introvert, usually with a set of tentacles at its anterior end. Unlike other annelids, they have no chaetae, but the introvert is often adorned with proteinaceous hooks that can be important taxonomic characters. Other external taxonomic characters include the tentacles (number, shape and arrangement), body papillae and, in some cases, hardened shields, as well as length ratios. Many species require dissection for correct identification to reveal internal characteristics, such as introvert retractor muscles, nephridia and contractile vessels. Here we summarize the state of the current knowledge of species diversity in sipunculans. We emphasize molecular studies, conducted over the past two decades, that have revealed multiple complexes of cryptic or pseudocryptic species. It has become obvious that diversity is significantly higher than the current taxonomic scheme accounts for, but formal species descriptions are lagging behind. Although the major branches in the sipunculan phylogeny have become increasingly consolidated, the internal relationships within most branches are still in flux.

Keywords: annelida; invertebrates; peanut worms; cryptic species; pseudocryptic; species complex

Citation: Schulze, A.; Kawauchi, G.Y. How Many Sipunculan Species Are Hiding in Our Oceans? *Diversity* **2021**, *13*, 43. https://doi.org/10.3390/d13020043

Academic Editor: Luc Legal
Received: 30 December 2020
Accepted: 20 January 2021
Published: 24 January 2021

1. Introduction

With their distinctive morphology, sipunculans have long been regarded as a phylum of their own (phylum Sipuncula Rafinesque, 1814). Sipuncula are commonly known by their vernacular names "peanut worms" or "star worms". Lacking typical annelid segmentation, they are more easily confused with sea cucumbers or anemones (especially sand anemones of the genus *Edwardsia*) than with more closely related annelids.

The recent placement of sipunculans within the annelid radiation is based on phylogenetic and phylogenomic analyses conducted since the early 2000s [1–7], although this view is still not universally accepted [8]. Their fossil record is sparse, but shows that their morphology has changed little since the Cambrian [9]. As adults, sipunculans show no trace of external or internal segmentation, but there has been disagreement on whether the nervous and muscular systems of the larvae are segmented [8,10,11]. The most recent analysis based on confocal laser scanning microscopy and whole mount in situ hybridization on *Themiste lageniformis* concluded that their ventral nerve cord is an unsegmented medullary cord and therefore rejects the hypothesis of larval segmentation [8].

Saiz Salinas [12] summarized the history of sipunculan research since the 1500s. Sipunculans have been associated with various animal groupings in the ever-changing animal classification schemes over the centuries. One notion that survived well into the middle of the 20th century was the Gephyrea (Greek, meaning "bridge") concept, originally proposed by de Quatrefages [13]. Apart from sipunculans, Gephyrea included taxa as diverse as holothurians, echiurans and priapulids. What all these taxa have in common is an elongated, unsegmented body with some type of appendages for food collection at the anterior end. This morphology was regarded as an intermediate form between segmented worms and echinoderms (although we recognize today that holothurians *are* echinoderms). Hyman [14] finally officially rejected the Gephyrea concept and, to our knowledge, the term has not been used since then, except in historical accounts.

The inclusion of Sipuncula in the Annelida may not be so surprising considering that other taxa without obvious segmentation (e.g., Siboglinidae, previously known as Pogonophora; Echiura and Myzostomida) have similarly been absorbed into the phylum Annelida. What is most baffling, however, is their inferred sister group relationship with amphinomids [6,7], or fireworms, as the two taxa starkly differ morphologically. In contrast to sipunculans, amphinomids are distinctly segmented and carry conspicuous calcareous chaetae on each segment.

The classification within the Sipuncula has also seen significant revisions over the past ~140 years. Selenka et al. in 1883 [15] proposed 10 genera, Fisher in 1952 [16] recognized 13 and Stephen in 1965 [17] 16. The two most recent monographs, Stephen & Edmonds in 1972 [18] and Cutler in 1994 [19], both recognize 17 genera, although there are some discrepancies with regard to which genera are considered valid and their definitions. The two monographs starkly differ in the number of recognized species: While Stephen and Edmonds [18] listed 320 species, Cutler [19] reduced the number to ~150 by assembling long lists of synonyms for many species. This practice of "taxonomic lumping" was based on the observations that many species have long-lived planktotrophic larval stages [20], theoretically enabling dispersal across ocean basins, leading to large geographic distribution ranges.

Since 1994, 12 additional species and one new subspecies have been described, and one previously synonymized species has been reinstated [21]. Based on morphological characters, Kawauchi and Rice [22] described two new species of *Nephasoma* from the western Atlantic. Hylleberg described a total of nine new species from Thailand (in the genera *Sipunculus*, *Xenosiphon*, *Siphonosoma* and *Aspidosiphon*) and Saiz Salinas described a new subspecies of *Phascolion* from Indian waters [23]. Relying primarily on molecular data, Silva-Morales et al. [24] described a new species of *Antillesoma* from the eastern Pacific.

Molecular phylogenetics has reshaped ideas about sipunculan relationships [25–30]. Although these studies, based on different sets of genes, do not necessarily agree in every aspect of the phylogeny, all strongly support the monophyly of the Sipuncula. Based on a phylogenetic tree derived from the analysis of five gene regions, Kawauchi et al. [29] established a new classification of sipunculans that recognizes 16 genera organized into six families, but no higher taxonomic ranks. Kawauchi et al.'s [29] major groupings were also confirmed in a later phylogenomic analysis of RNA Sequence data [30] (Figure 1).

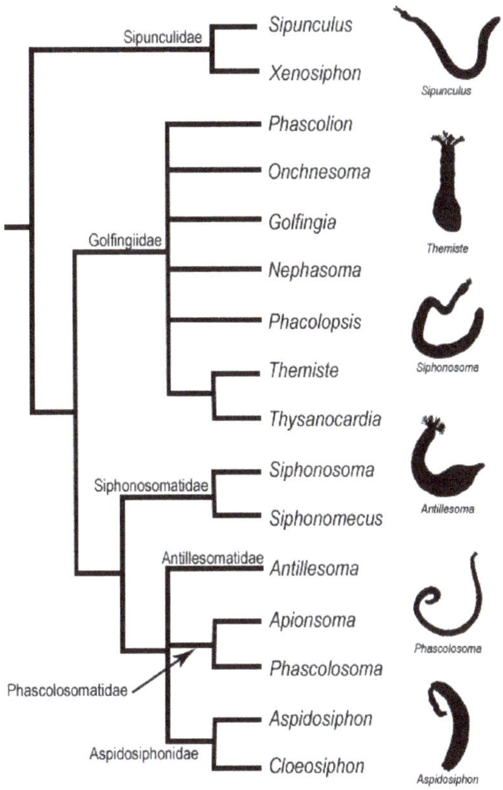

Figure 1. Phylogeny of the Sipuncula (modified from [29]). Right, exemplar shadows representing one genus from each family.

2. External Morphology

The simple body plan of sipunculans consists of a trunk region and a thinner introvert region that can be completely retracted into the trunk (Figures 2 and 3A). A number of tentacles are generally located at the anterior end of the introvert, either arranged in a circle around the mouth or in a horseshoe shape around the chemosensory nuchal organ on the dorsal margin of the oral disk. The introvert is often adorned with proteinaceous hooks. The remainder of the body wall may be smooth or covered with papillae of various shapes and sizes. Some species, particularly in the genera *Aspidosiphon* (Figure 2H) and *Cloeosiphon*, have hardened proteinaceous shields at the anterior and sometimes at the posterior end. There are no appendages, and the only more or less visible landmarks on the trunk are the dorsal anus near the anterior end and the (usually) paired openings of the nephridia on the ventral side, at roughly the same level of the anus. Many species are transparent or white without body pigmentation, while others are brown or gray in color, sometimes with black markings. A few species, especially in the genera *Antillesoma* (Figure 2G) and *Themiste* (Figure 2D), have some purple or greenish pigment on their tentacles or at the tentacle base. Sipunculans range in length from a few mm to about 30 cm.

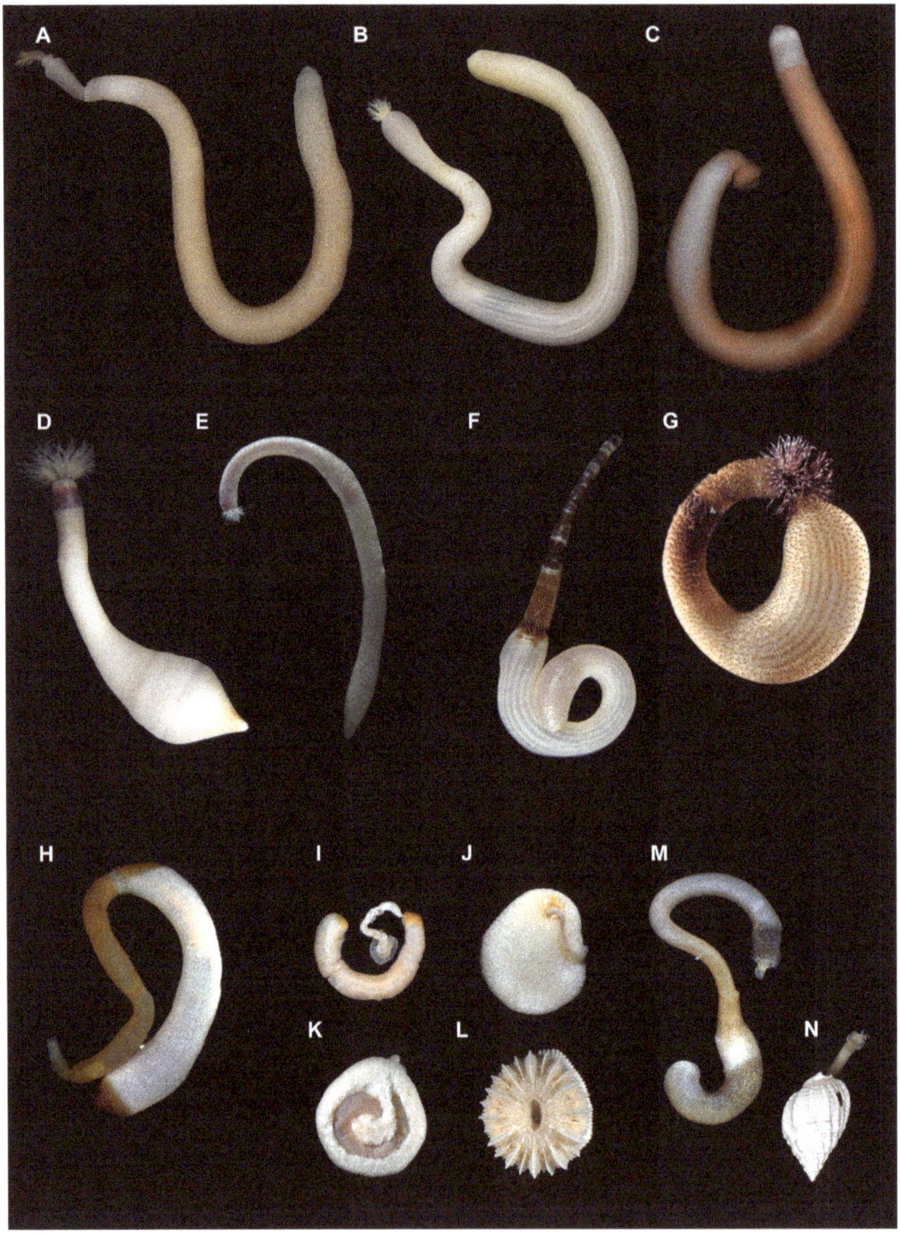

Figure 2. Diversity of sipunculan body plans (**A**) *Sipunculus phalloides*; (**B**) *Siphonosoma cumanense*; (**C**) *Xenosiphon branchiatus*; (**D**) *Themiste langeniformis*; (**E**) *Nephasoma pellucidum*; (**F**) *Phascolosoma perlucens*; (**G**) *Antillesoma antillarum*; (**H**) *Aspidosiphon fischeri*; (**I**) *Aspidosiphon muelleri*, a commensal sipunculan living in symbiosis with a solitary coral; (**J**) *A. mulleri* introvert coming out from a whole in the base of the coral made by the worm; (**K**) space inside a solitary coral inhabited by *A. mulleri*; (**L**) solitary coral from the genus *Heteropsammia*; (**M**) *Phascolion cryptum*; (**N**) *P. cryptum* inside a gastropod shell.

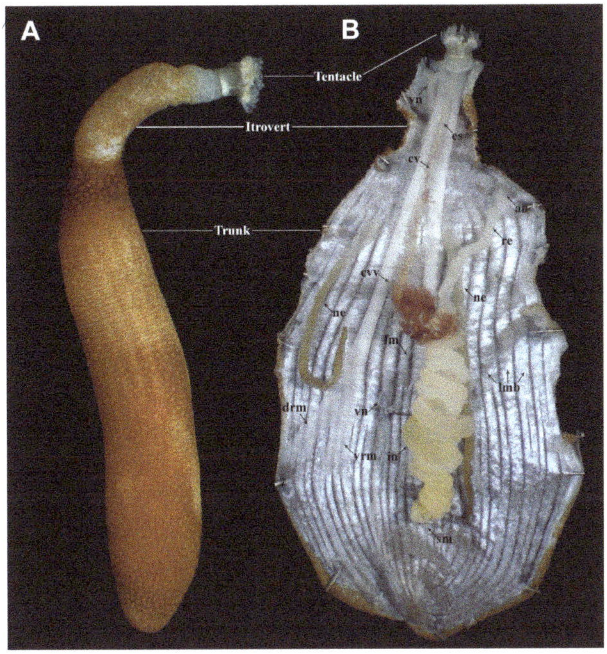

Figure 3. Sipuncula external and internal anatomy. (**A**) *Antillesoma antillarum*: external view. (**B**) *A. antillarum*: internal view. Abbreviations: an = anus, cv = contractile vessel, cvv = contractile vessel villi, drm = dorsal retractor muscle, es = esophagus, fm = fixing muscle, in = intestine, lmb = longitudinal muscle band, ne = nephridia, re = rectum, sm = spindle muscle, vn = ventral nerve cord, vrm = ventral retractor muscle.

3. Anatomy

Anatomically, one of the most distinctive features, present in most species, is an intestine that forms a double-helix in the trunk, consisting of a descending branch that stretches from the mouth at the tip of the introvert towards the posterior end of the trunk, and an ascending branch that runs back anteriorly towards the anus (Figure 3B). Also very distinctive are the well-developed introvert retractor muscles that connect the tip of the introvert to the body wall in the trunk. Four introvert retractors, divided into a dorsal and a ventral pair, appear to be the ancestral condition [31], but in many species they have been reduced to two or one, although separate insertion points in the body may remain visible. The central nervous system consists of a dorsal cerebral ganglion at the anterior end of the esophagus, a pair of circumesophageal connectives, and a single ventral nerve cord without obvious ganglia. A large, primary coelom constitutes the main body cavity. A secondary coelom extends into the tentacles and is connected to a pair of compensatory sacs or contractile vessels. The latter extend along the esophagus and sometimes carry hairlike villi, known as contractile vessel villi. The gonads are inconspicuous and located at the insertion points of the retractor muscles in the body wall. Oocytes or spermatocytes are released into the body coelom where they mature. Shortly before spawning, they are taken up by the nephridia and are then released through the nephridiopores. The nephridial system consists of one, or more often two, elongate sacs that extend into the body coelom along the ventral side.

4. Ecology

Sipunculans are exclusively marine and benthic as adults. They inhabit all oceans, ranging from polar waters to the tropics and from the intertidal zone to the abyss. Their presence is usually not obvious on superficial inspection of a benthic habitat, as they tend to be well hidden. While some sipunculan species inhabit sand or mud, others are associated with hard substrates, such as the underside of rocks, spaces between bivalve shells, the holdfasts of submerged vegetation or submerged wood. Although two species, *Phascolion psammophilum* Rice, 1993, and *Aspidosiphon exiguus* Edmonds, 1874, have been described as meiofaunal, adults of these species are larger than the generally recognized size thresholds for meiofauna of 500 µm or 1 mm. Most sipunculans found in meiofaunal samples actually represent postlarvae or juveniles of larger species [32]. Some species contribute to the erosion of hard substrate, such as coral rubble, mollusc shells, calcareous rock or sandstone (Figure 4). Other species, particularly in the genus *Phascolion,* inhabit abandoned mollusc shells (Figure 2M), polychaete tubes or foraminiferan tests. One species, *Phascolosoma turnerae* Rice, 1985, was originally described from submerged wood but has also been reported from methane and hydrocarbon seeps [33–35]. Another species in the same genus, *Phascolosoma saprophagicum* Gibbs, 1987, has been described from a whale carcass in abyssal depths. *Aspidosiphon* species can also inhabit empty molluscs shells, but one in particular has an interesting relationship with a solitary coral. As a juvenile *A. muelleri* looks for an empty mollusc shell as a shelter. Later, a planula larva from a solitary coral from the genera *Heteropsammia* and *Heterocyathus* can settle outside the occupied shell. The coral can overgrow and absorb the shell's material into its own tissue, leaving the sipunculan coiled inside the coral (Figure 2I–L). In this relationship the worm gains a portable shelter, and the coral benefits by being relocated if the local conditions deteriorate and by being kept upright [36]. Despite their cryptic nature, sipunculans can reach high densities. For example, Rice et al. [37] reported 8000 individuals/m^2 of *Phascolion cryptum* (Figure 2N) in the Indian River Lagoon in Florida.

Figure 4. Collecting Sipuncula. (**A**) Sand bar exposed by low tide, habitat for large sipunculans; (**A1**) Sipuncula holes; (**A2**) Digging deep into the sediment searching for sipunculans; (**B**) *Siphonosoma cumanenses* found in this habitat; (**C**) Beach with sabellariid mounds exposed by low tide; (**C1**) Pieces of polychaete colonies with sipunculans; (**C2**) breaking the pieces of polychaete colonies searching for sipunculans; (**D**) *Phascolosoma perlucens* found living among the polychate colony.

Sipunculan worms are either nonselective deposit feeders, surface grazers or suspension feeders [19,38–41]. As adults, they have a limited range of movement. While some species have been found burrowing up to about 1 m deep into sediment [42], not much is known about their horizontal movement, but it is most likely very limited. Many species are semi-sessile, although none physically attach to substrate.

5. Reproduction and Development

With a few exceptions [43–45], sipunculans are gonochoric. Reproduction is usually sexual via free-spawned gametes, although a few cases of parthenogenesis, budding and fission have been reported [45–47]. Embryonic and larval development has been exceptionally well-studied in some species, mostly through the work of Mary Rice on the northwest and southeast coasts of the U.S. and the Caribbean [48,49]. Sipunculan embryos undergo spiral cleavage [48–55]. Further development may include a trochophore larva and/or a unique larval form known as pelagosphera (Figure 5). While the trochophore larva is always lecithotrophic, the pelagosphera may be lecithotrophic or planktotrophic. Four developmental modes can be distinguished [56,57]: (I) direct development; (II) indirect development with a trochophore stage only; (III) indirect development with a trochophore and a lecithotrophic pelagosphera; and (IV) indirect development with a trochophore and a planktotrophic pelagospera. While there are still many sipunculan species for which development is unknown, it appears that the IV is the most common developmental mode.

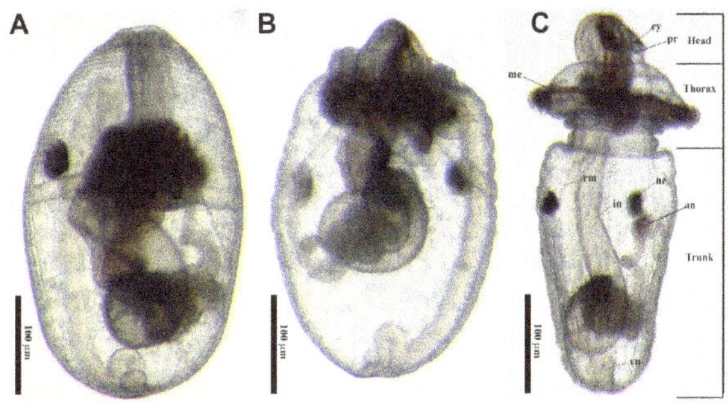

Figure 5. Pelagosphera larvae from *Sipunculus* sp. (**A**) Pelagosphera larva with the head and thorax retracted inside the trunk; (**B**) Pelagosphera lateral view of a larva in the process of extending the head from the trunk; (**C**) Dorsal view of the pelagosphera larva showing regions of the body and external and internal structures viewed through the translucent the body wall. Abbreviations: an = anus, ey = eye, in = intestine, me = metatroch, ne = nephridia, pr = prototroch, rm = retractor muscle, vn = ventral nerve cord.

Planktotrophic pelagosphera larvae can spend several months in the plankton [20] and are considered the primary dispersal mechanism for sipunculans. The body of the pelagosphera is spherical to elongate (Figure 5), with a distinctive constriction between the trunk and the thorax and a prominent metatrochal ciliary ring on the thorax used for swimming (Figure 5C). The head and thorax, with the metatroch, can be retracted into the trunk, temporarily rendering the larva incapable of swimming (Figure 5A). There are many morphologically distinct types of pelagosphera larvae, varying in size, color, body texture and head morphology [58]. Using DNA barcoding of adults and larvae, Schulze et al. [59] matched 11 of these morphotypes to a sipunculan species, and three additional ones to genus level. When discussing overall sipunculan diversity, the developmental mode and larval stages need to be taken into consideration as well.

6. Species Complexes

Compared to more "typical" annelids, sipunculans have few morphological characters that can be used for the diagnosis of species or higher-level taxa. While the relationships among the higher taxa are becoming increasingly consolidated, many species are ill-defined morphologically and molecularly. Since the 1990s, it has become easier and more affordable to generate molecular data that can be used in addition to morphological data to delineate species boundaries. This has helped identify several complexes of "cryptic" or "pseudocryptic" species within Sipuncula. Cryptic species are here defined as species that are morphologically indistinguishable but form evolutionary distinct units based on their genetic divergence [60]; pseudocryptic species that can be identified by morphological characters after a detailed comparative study of morphological and non-morphological features [61]. While two studies on sipunculan cryptic diversity relied on genetic fingerprinting techniques [62,63], most others used DNA sequencing of one or several mitochondrial or nuclear gene markers. The most commonly used marker has been the common "DNA barcoding" sequence [64], a ~650 bp sequence of the cytochrome c oxidase subunit I (COI) gene. Another mitochondrial marker is 16S ribosomal RNA (16S). Some studies have additionally used nuclear markers such as 28S ribosomal RNA (28S) or Histone H3 (H3). If genetic distances among clades are calculated, they are most commonly reported as average Kimura 2 Parameter (K2P) distances.

6.1. *Sipunculus nudus* Complex

Sipunculus nudus Linnaeus, 1766, is the most popular species among this group being used as bait for recreational fishing [65] and as delicatessen food in some Asian countries [66]. Specimens from this genus can reach 50 cm in trunk length, but *S. nudus* varies from 5 to 15 cm long. This species was described from "Europeans waters" and has been considered cosmopolitan in tropical, subtropical and temperate waters in all oceans [19]. However, a multilocus phylogenetic analysis using nuclear (28S and H3) and mitochondrial (COI and 16S) markers has shown a higher level of genetic variation among distant populations of putative species *S. nudus*, and morphological analyses corroborates phylogenetic discovery [67]. Average genetic distances among clades were not reported, but based on the phylogenetic trees, it is clear that *S. nudus* is a case of pseudocryptic species.

6.2. *Themiste pyroides/hexadactyla* Complex

Themiste pyroides (Chamberlin, 1920) is an intertidal species originally described from Laguna Beach, California, and has been widely reported from the northeast Pacific (Alaska through Baja California) [48] as well as the northwest Pacific [40,68]. A morphologically similar species, *Themiste hexadactyla*, was described from northern Japan [69] and reported from Hokkaido [70] as well as from California [70], but Cutler and Cutler [71] declared *T. hexadactyla* a junior synonym to *T. pyroides*. However, COI and 16S sequence data indicate significant genetic differences (11.5% average K2P distance for COI) between populations from British Columbia, Canada, and the Peter the Great Bay in the Sea of Japan [72]. Additionally, major differences in developmental mode and developmental timing have been reported between the eastern and western Pacific lineages, summarized in [72]. Whereas the lineage from British Columbia develops directly (developmental mode I), the lineage from the Sea of Japan includes lecithotrophic trochophore and pelagosphera stages (developmental mode III) [73]. Based on these differences, Maiorova and Adrianov [74] and Nishikawa [75] proposed the resurrection of *T. hexadactyla* in the western Pacific.

6.3. *Thysanocarida nigra/Golfingia pugettensis* Complex

Thysanocarida nigra (Ikeda, 1904) has a relative long (up to 10 cm), smooth trunk and an introvert up to twice as long. The type locality is listed as "Japan" [18]. The morphologically similar *Golfingia pugettensis* was described by Fisher [16] from Puget Sound, and has been the subject of investigations on reproduction, embryology and larval development [48,76]. The two species were synonymized by Gibbs et al. [77], under *Thysanocardia nigra* as the valid name. Genetic studies based on 16S sequences indicate that the eastern and western Pacific populations represent separate genetic lineages (29.3% K2P for 16S); moreover, there was indication that two clades exist in the eastern Pacific [72].

Based on [48,73], differences between the eastern and western Pacific lineages are also apparent in reproduction, development and larval behavior, summarized in [72].

6.4. *Phascolosoma agassizii* Complex

Phascolosoma agassizii agassizii Keferstein, 1866, is a relatively large and common shallow-water species that is abundant on the west coast of North America as well as the Northwest Pacific. There are also records from the Indian Ocean [78], tropical South Pacific [29], Atlantic [79] and Mediterranean Sea [80]. A second subspecies, *P. agassizii kurilense* (Satô, 1937) was described from the Kuril Islands. Schulze et al. [72] first reported a genetic break between the populations in the eastern and western north Pacific, based on COI and 16S sequence data. This was later confirmed with a larger sample size and broader geographic sampling range [81]. The average genetic distance between the eastern and western Pacific populations was over 32% for COI. Not only were the two taxa very genetically distinct, they were not even sister taxa that were separated by several other *Phascolosoma* species in the phylogenetic tree. Although there were some differences in the pigmentation patterns and hook morphology between the two clades, both fit the description of *P. agassizii agassizii* as given in Cutler's monograph [19]. The two specimens from the Kuril Islands included in the Johnson et al. study [81] showed no genetic divergence from other members of the western Pacific clade. Morphologically, *P. agassizii kurilense* differs from *P. agassizii agassizii* only by the presence of a small secondary lobe in the nephridia. The type locality for *P. agassizii agassizii* is Mendocino, California; it is therefore likely that the western Pacific represents a different, as yet undescribed species. DNA sequence data [81] and studies based on intersimple sequence repeat polymerase chain reaction (ISSR-PCR, a genetic fingerprinting technique) [63] indicated that there is slight genetic structure among the populations in the eastern Pacific and more substantial structure among the populations in the western Pacific, although there is no genetic break between the Kuril Island samples and those from the Sea of Japan. The larval development of both species has been studied extensively [48,73,82,83]. Both have planktotrophic pelagosphera larvae, which may possibly enable dispersal across the Pacific basin. Although some differences in developmental timing were observed between the larvae on both coasts [82], it is unclear whether these are examples of phenotypic plasticity or have a genetic basis.

6.5. *Phascolosoma nigrescens/varians* Complex

Keferstein [84] described *Phascolosoma nigrescens* from Fiji and *P. varians* from the St. Thomas in the Caribbean. The two species were synonymized with *P. nigrescens* as the senior synonym [19]. In addition to *P. varians*, Cutler [19] lists 12 other junior synonyms for *P. nigrescens*. However, Silva-Morales [21] recently provided morphological and molecular evidence to support the reinstatement of *P. varians*. The three Caribbean specimens included in their study diverged from the specimen from New Caledonia (the closest available sequence to Fiji, the type locality for *P. nigrescens*) by an average K2P distance of 24% for COI. They proposed that *P. nigrescens* (from the Indo-Pacific) and *P. varians* (from the Caribbean) can be distinguished by their hook morphology (rounded secondary tooth in *P. varians*; square in *P. nigrescens*), attachment of the nephridia to the body wall (partially in *P. varians*; fully in *P. nigrescens*) and the shape of the contractile vessel (simple in *P. varians*; with lateral sacs in *P. nigrescens*). Furthermore, Silva-Morales' molecular analyses suggest that *P. nigrescens* might consist of at least five additional species, from Spain, Australia, New Caledonia, Israel and South Africa, respectively [21].

6.6. *Phascolosoma perlucens* Complex

Phascolosoma perlucens Baird, 1868, is the third of the tree circumtropical and cosmopolitan species in the genus *Phascolosoma*. They are found in abundance in shallow waters inside coral rubble or in crevices of calcareous rocks [85] in the Caribbean, [86–88], western Pacific [15,89–91], eastern Pacific [16], several localities in Indian Ocean, [92–94], and few localities from eastern Atlantic, [95]. Described originally from Jamaica [87], the species is readily distinguished from other congeners by the presence of reddish, conical posteriorly directed, pre-anal papillae at the dorsal base of the introvert, and by the secondary round tooth at the concave side of the introvert hooks [96]. Cutler [19] reduced the confusion within *P. perlucens* but still lists 18 synonymies, resulting in putative *P. perlucens* species over-lumping. Following the tendency in testing the cosmopolitanism among invertebrates with few morphological characteristics, Kawauchi & Giribet [97] examined 56 putative individuals identified as *P. perlucens* from 13 localities throughout the tropics, using two mitochondrial genes (COI and 16S) in phylogenetic analyses. Although genetic distances among populations were not reported, the phylogenetic trees and haplotype networks show clear genetic separation of geographically distant populations of this species. More than that, the authors analyzed hook morphology, and identified at least four different lineages, corroborating part of the genetic analysis. However, to delimit species we need an effort in analyzing a larger sample of individuals, localities and more genes before a comprehensive taxonomic revision can be achieved.

6.7. *Apionsoma misakianum* Complex

Apionsoma (Apionsoma) misakianum (Ikeda, 1904) is a small species, with a trunk of 3-8 mm length and a thin introvert of up to 10 times the length of the trunk. The species has been widely reported in shallow water throughout the tropics and subtropics but is generally uncommon. The trunk morphology is relatively nondescript and the extreme length of the introvert makes it difficult to preserve specimens with their introverts fully extended. As a consequence, the distal rings of introvert hooks, which have distinctive characteristics, cannot easily be visualized. It is therefore possible that some records of this species are misidentifications. Using allozyme data, Staton and Rice [62] presented evidence of a genetic split between populations of this species between populations from the Florida Keys/Bahamas and a population from Sebastian Pinnacles, off the south-central Florida Atlantic coast. Moreover, although these populations could not be distinguished morphologically (in this case the introvert hooks were carefully examined), each group had distinctive pelagosphera larvae. Described in detail by Rice et al. [58,98], the larvae from the southern population have been dubbed "white blackhead" larvae, whereas those from the northern population are known as "spotted velvet" [58]. Not only can they be distinguished by their pigmentation patterns, but also in their slightly different developmental timing [56,58,98]. DNA sequencing of larvae and adults confirmed the original findings of genetic heterogeneity within *A. misakianum* [59] (average distance for all *A. misakianum* combined: 19% K2P). In addition to the population from Sebastian Pinnacles, the "spotted velvet" clade further included one adult specimen each from the Red Sea and New Caledonia, whereas the "white blackhead" clade included an additional specimen from Belize. White blackhead larvae have also been captured in surface waters of the central Atlantic (described as "type C" or *Baccaria oliva*) [99], indicating that they are long-lived and have high dispersal potential. The average within-clade genetic distance for COI was 16.2% for the "spotted velvet" clade, but only 0.7% for the "white blackhead" clade [59]. Based on these data and the wide geographic distribution of the "spotted velvet" clade, we suggest that this clade likely represents multiple species, although more extensive sampling would be necessary to resolve this clade. At the minimum, we can conclude that *Apionsoma misakianum* includes two species, but we cannot currently determine which, if either, is the nominal species. Cutler [19] synomized three species with *A. misakianum*: *Phascolosoma hespera* Chamberlain, 1920, *Golfingia (Phascolana) longirostris* Wesenberg-Lund, 1959, and *Golfingia (Phascolana) tenuissima* Wesenberg-Lund, 1959. Careful examination of type material and/or sequencing of topotypes would be necessary to evaluate whether any of these should be resurrected. The type locality for the species is in Japan [100]; it would therefore be of particular interest to study material from the western Pacific.

6.8. *Antillesoma antillarum* Complex

Antillesoma antillarum (Grube & Öersted, 1858) is another common and relatively large species reported from warm shallow-water habitats worldwide. Silva-Morales et al. [24] recently described a new species of *Antillesoma, A. mexicanum,* from the Mexican Pacific coast. This species differs morphologically from *A. antillarum* by having darker pigmentation, sparser distribution of body papillae and shorter trunk length. Moreover, the average genetic distance is 21% for COI between the Mexican Pacific and Caribbean-Florida clade. Silva-Morales et al. [24] also included a COI sequence from a specimen identified as *A. antillarum* from Thailand in their analysis that constituted its own distinct genetic lineage, but the specimen was not examined morphologically. Given the genetic divergence, we surmise that the Thailand sample constitutes a third species of *Antillesoma*. Addititional morphological and molecular studies covering a broader geographic scope might reveal that more species are present worldwide.

6.9. *Aspidosiphon* Species Complexes

No focused studies have yet been conducted on cryptic/pseudocryptic species complexes in the genus *Aspidosiphon*. However, based on phylogenetic analyses that included multiple representatives of some common species from different geographic locations [28,29], it is likely that several species complexes exist. For example, several specimens each of *Aspidosiphon (Paraspidosiphon) laevis* de Quatrefages, 1865, *Aspidosiphon (Paraspidosiphon) parvulus* Gerould, 1913, and *Aspidosiphon (Aspidosiphon) elegans* (Chamisso and Eysenhardt, 1821) from different locations were included in multigene phylogenetic reconstructions of the phylum and the resulting trees did not support the monophyly of these species.

7. Geographic Coverage

In general, sipunculan diversity data strongly reflect the amount of taxonomic effort for geographic areas. Due to easier accessibility, shallow-water species are better known than deep sea species. For example, of the ~70 lots of sipunculans collected from deep water along the eastern Australian margin by Gunton et al. [101], only 10 could reliably be identified to the species level and represented only two species. Murina [102] stressed the lack of knowledge of sipunculans all over the word, and pointed out that Indian, Pacific, Southwest Atlantic Oceans and China coast are the most unexplored regions. More than four decades later, only a few efforts in specific parts of the world have been made to fill this gap. The absence of specialists trained to identify Sipuncula contributes to this problem.

Sipunculans in Southwest Atlantic are still poorly known. In Brazil, the collection of this organism along the coast was mainly made by oceanographic expeditions at the continental shelf [103–106]. Besides that, the benthos from the intertidal zone and beyond the continental shelf of the Brazilian coast are still poorly or completely unexplored [107]. Lana, et al. [108] acknowledge that there is more information about the Brazilian southeastern and south marine fauna, and this is true for sipunculans also, because this situation is a consequence of the high level of the oceanographic expedition in this area and the consolidated researcher groups since the 1950s.

8. Conclusions

The global diversity of the Sipuncula is still poorly understood, but multiple studies using genetic techniques in combination with morphological and/or reproductive and developmental features have revealed that more diversity is present than the current taxonomic scheme [19] accounts for. Many of the lineages that were detected with molecular tools likely represent new species or species that were previously synonymized and should be reinstated. As in many other invertebrate taxa, formal taxonomic descriptions are lagging behind the molecular work. This gap is likely going to widen in the future due to the small number of sipunculan taxonomists worldwide, as well as lack of funding and recognition for taxonomic work. Although the major clades in the Sipuncula are well supported, the internal relationships within them need to be better resolved, especially in the Golfingiidae and the clade including Phascolosomatidae, Antillesomatidae, and Aspidosiphonidae.

Author Contributions: The authors contributed the following components to the manuscript: Conceptualization, A.S. and G.Y.K.; writing—original draft preparation, A.S.; writing—review and editing, A.S. and G.Y.K.; visualization, G.Y.K. Both authors have read and agreed to the published version of the manuscript.

Funding: This research received no external funding.

Acknowledgments: We thank the guest editors of this special edition, Maria Capa and Pat Hutchings, for inviting us to contribute this article.

Conflicts of Interest: The authors declare no conflict of interest.

References

1. Boore, J.L.; Staton, J.L. The mitochondrial genome of the sipunculid *Phascolopsis gouldii* supports its association with Annelida rather than Mollusca. *Mol. Biol. Evol.* **2002**, *19*, 127–137. [CrossRef]
2. Telford, M.J.; Wise, M.J.; Gowri-Shankar, V. Consideration of RNA secondary structure siginificantly improves likelihood-based estimates of phylogeny: Examples from the Bilateria. *Mol. Biol. Evol.* **2005**, *22*, 1129–1136. [CrossRef]
3. Dunn, C.W.; Hejnol, A.; Matus, D.Q.; Pang, K.; Browne, W.E.; Smith, S.A.; Seaver, E.; Rouse, G.W.; Obst, M.; Edgecombe, G.D.; et al. Broad phylogenomic sampling improves resolution of the animal tree of life. *Nature* **2008**, *452*. [CrossRef]
4. Mwinyi, A.; Meyer, A.; Bleidorn, C.; Lieb, B.; Bartolomaeus, T.; Podsiadlowski, L. Mitochondrial genome sequence and gene order of Sipunculus nudus give additional support for an inclusion of Sipuncula into Annelida. *BMC Genom.* **2009**, *10*, 27. [CrossRef]
5. Struck, T.H.; Paul, C.; Hill, N.; Hartmann, S.; Hosel, C.; Kube, M.; Lieb, B.; Meyer, A.; Tiedemann, R.; Purschke, G.; et al. Phylogenomic analyses unravel annelid evolution. *Nature* **2011**, *471*, 95–98. [CrossRef] [PubMed]
6. Weigert, A.; Golombek, A.; Gerth, M.; Schwarz, F.; Struck, T.H.; Bleidorn, C. Evolution of mitochondrial gene order in Annelida. *Mol. Phylogenet. Evol.* **2016**, *94*, 196–206. [CrossRef] [PubMed]
7. Weigert, A.; Helm, C.; Meyer, M.; Nickel, B.; Arendt, D.; Hausdorf, B.; Santos, S.R.; Halanych, K.M.; Purschke, G.; Bleidorn, C.; et al. Illuminating the base of the annelid tree using transcriptomics. *Mol. Biol. Evol.* **2014**, *31*, 1391–1401. [CrossRef] [PubMed]
8. Carrillo-Baltodano, A.M.; Boyle, M.J.; Rice, M.E.; Meyer, N.P. Developmental architecture of the nervous system in *Themiste lageniformis* (Sipuncula): New evidence from confocal laser scanning microscopy and gene expression. *J. Morphol.* **2019**, *280*, 1628–1650. [CrossRef] [PubMed]
9. Huang, D.; Chen, J.; Vannier, J.; Saiz Salinas, J.I. Early Cambrian sipunculan worms from southwest China. *Proc. R. Soc. Lond. Ser. B* **2004**, *271*, 1671–1676. [CrossRef]
10. Kristof, A.; Wollesen, T.; Wanninger, A. Segmental mode of neural patterning in Sipuncula. *Curr. Biol.* **2008**, *18*, 1129–1132. [CrossRef]
11. Kristof, A.; Wollesen, T.; Maiorova, A.S.; Wanninger, A. Cellular and muscular growth patterns during sipunculan development. *J. Exp. Zool. Part. B Mol. Dev. Evol.* **2011**, *316B*, 227–240. [CrossRef] [PubMed]
12. Saiz Salinas, J.I. Almost five centuries of systematic study of the enigmatic sipunculan worms. In *Proceedings of the Second International Symposium on the Biology of the Sipuncula*; Boyle, M.J., Kawauchi, G.Y., Eds.; Smithsonian Institution Scholarly Press: Washington, DC, USA, 2018; pp. 219–235.
13. De Quatrefages, A. Etudes sur les types inférieurs de l'embranchement des annelés. Mémoire sur l'echiure de Gaertner (Echiurus Gaertnerii NOB). *Ann. Sci. Nat. Zool. (Paris)* **1847**, *7*, 307–343.
14. Hyman, L.H. Phylum Sipunculida. In *Smaller Coelomate Groups*; McGraw-Hill: New York, NY, USA, 1959; Volume 5, pp. 610–690.
15. Selenka, E.; de Man, J.G.; Bülow, C. Die Sipunculiden, eine systematische Monographie. *Semper Reis. Archipel Phillippinen 2* **1883**, *4*, 1–131.
16. Fisher, W.K. The sipunculid worms of California and Baja California. *Proc. U. S. Natl. Mus.* **1952**, *102*, 371–450. [CrossRef]
17. Stephen, A.C. Echiura and Sipuncula from the Isreal South Red Sea Expedition. *Sea Fish. Res. Stn. Haifa Bull.* **1965**, *40*, 79–83.
18. Stephen, A.C.; Edmonds, S.J. *The Phyla Sipuncula and Echiura*; Trustees of the British Museum (Natural History): London, UK, 1972.
19. Cutler, E.B. *The Sipuncula. Their Systematics, Biology and Evolution*; Cornell University Press: Ithaca, NY, USA, 1994; p. 453.
20. Scheltema, R.S.; Hall, J.R. The dispersal of pelagosphera larvae by ocean currents and the geographical distribution of sipunculans. In *Proceedings of the International Symposium on the Biology of the Sipuncula and Echiura*; Rice, M.E., Todorovic, M., Eds.; Nauçno Delo: Belgrade, Serbia, 1975; pp. 103–116.
21. Silva-Morales, I. Reinstatement of Phascolosoma (Phascolosoma) varians Keferstein, 1865 (Sipuncula: Phascolosomatidae) based on morphological and molecular data. *PeerJ* **2020**, *8*, e10238. [CrossRef]
22. Kawauchi, G.Y.; Rice, M.E. Two New Species of Nephasoma (Sipuncula: Golfingiidae) from the Western Atlantic Ocean. *Proc. Biol. Soc. Wash.* **2009**, *122*, 1–13. [CrossRef]
23. Saiz Salinas, J.I.; Bustamante, M.; Tajadura, J.; Vijapure, T.; Sukumaran, S. A new subspecies of *Phascolion* Théel, 1875 (Sipuncula: Golfingiidae) from Indian waters. *Zootaxa* **2015**, *3931*, 433–437. [CrossRef]

24. Silva-Morales, I.; López-Aquino, M.J.; Islas-Villanueva, V.; Ruiz-Escobar, F.J.; Bastida-Zavala, R. Morphological and molecular differences between the Amphiamerican populations of *Antillesoma* (Sipuncula: Antillesomatidae), with the description of a new species. *Rev. Biol. Trop.* **2019**, *67*, S101–S109. [CrossRef]
25. Staton, J.L. Phylogenetic analysis of the mitochondrial cytochrome c oxidase subunit 1 gene from 13 sipunculan genera: Intra-and interphylum relationships. *Invertebr. Biol.* **2003**, *122*, 252–264. [CrossRef]
26. Maxmen, A.B.; King, B.F.; Cutler, E.B.; Giribet, G. Evolutionary relationships within the protostome phylum Sipuncula; a molecular analysis of ribosomal genes and histone H3 sequence data. *Mol. Phylogenet. Evol.* **2003**, *27*, 489–503. [CrossRef]
27. Schulze, A.; Cutler, E.B.; Giribet, G. Reconstructing the phylogeny of the Sipuncula. *Hydrobiologia* **2005**, *535/536*, 277–296. [CrossRef]
28. Schulze, A.; Cutler, E.B.; Giribet, G. Phylogeny of sipunculan worms: A combined analysis of four gene regions and morphology. *Mol. Phylogenet. Evol.* **2007**, *42*, 171–192. [CrossRef] [PubMed]
29. Kawauchi, G.Y.; Sharma, P.P.; Giribet, G. Sipunculan phylogeny based on six genes, with a new classification and the descriptions of two new families. *Zool. Scr.* **2012**, *41*, 186–210. [CrossRef]
30. Lemer, S.; Kawauchi, G.Y.; Andrade, S.C.S.; González, V.L.; Boyle, M.J.; Giribet, G. Re-evaluating the phylogeny of Sipuncula through transcriptomics. *Mol. Phylogenetics Evol.* **2015**, *83*, 174–183. [CrossRef]
31. Schulze, A.; Rice, M.E. Musculature in sipunculan worms: Ontogeny and ancestral states. *Evol. Dev.* **2009**, *11*, 97–108. [CrossRef]
32. Schulze, A.; Kawauchi, G.Y.; Migotto, A. 13 Sipuncula. In *Guide to the Identification of Marine Meiofauna*; Schmidt-Rhaesa, A., Ed.; Verlag Dr. Friedrich Pfeil: Munich, Germany, 2020; pp. 277–284.
33. Cordes, E.E.; Carney, S.L.; Hourdez, S.; Carney, R.S.; Brooks, J.M.; Fisher, C.R. Cold seeps of the deep Gulf of Mexico: Community structure and biogeographic comparisons to Atlantic equatorial belt seep communities. *Deep Sea Res. Part. I Oceanogr. Res. Pap.* **2007**, *54*, 637–653. [CrossRef]
34. MacAvoy, S.E.; Fisher, C.R.; Carney, R.S.; Macko, S.A. Nutritional associations among fauna at hydrocarbon seep communities in the Gulf of Mexico. *Mar. Ecol. Prog. Ser.* **2005**, *292*, 51–60. [CrossRef]
35. Cordes, E.E.; Hourdez, S.; Predmore, B.L.; Redding, M.L.; Fisher, C.R. Succession of hydrocarbon seep communities associated with the long-lived foundation species *Lamellibrachia luymesi*. *Mar. Ecol. Prog. Ser.* **2005**, *305*, 17–29. [CrossRef]
36. Yonge, C.M. A note on mutualism between sipunculans and scleractinian corals. In *Proceedings of the International Symposium on the Biology of Sipuncula and Echiura*; Rice, M.E., Todorovic, M., Eds.; Nauçno Delo: Belgrade, Serbia, 1975.
37. Rice, M.E.; Piraino, J.; Reichardt, H.F. Observations on the ecology and reproduction of the sipunculan *Phascolion cryptus* in the Indian River Lagoon. *Fla. Sci.* **1983**, *46*, 382–396.
38. Edmonds, S.J. Some notes on the abundance, environment, and nutrition of *Sipunculus nudus* L. (Sipunculoidea) at Morgat, Brittany. *Cah. Biol. Mar. III* **1962**, *3*, 183–190.
39. Pilger, J.F. Ultrastructure of the tentacles of *Themiste lageniformis* (Sipuncula). *Zoomorphology* **1982**, *100*, 143–156. [CrossRef]
40. Adrianov, A.V.; Malakhov, V.V.; Maiorova, A.S. Development of the tentacular apparatus in sipunculans (Sipuncula): I. *Thysanocardia nigra* (Ikeda, 1904) and *Themiste pyroides* (Chamberlin, 1920). *J. Morphol.* **2006**, *267*, 569–583. [CrossRef] [PubMed]
41. Jumars, P.A.; Dorgan, K.M.; Lindsay, S.M. Diet of worms emended: An update of polychaete feeding guilds. *Annu. Rev. Mar. Sci.* **2015**, *7*, 497–520. [CrossRef]
42. Maiorova, A.S.; Adrianov, A.V. Distribution of peanut worms (Sipuncula) in the West Pacific. In Proceedings of the China-Russia Bilateral Symposium on "Comparison on Marine Biodiversity in the Northwest Pacific Ocean", Qingdao, China, 10–11 October 2010.
43. Paul, G. Über *Petalostoma minutum* Keferstein und verwandte Arten. *Jahrb. Abtheilung Anat.* **1909**, *29*, 1–50.
44. Gibbs, P.E. Gametogenesis and spawning in a hermaphroditic population of *Golfingia minuta* (Sipuncula). *J. Mar. Biol. Assoc. U. K.* **1975**, *55*, 69–82. [CrossRef]
45. Pilger, J.F. Reproductive biology and development of *Themiste lageniformis*, a parthenogenic sipunculan. *Bull. Mar. Sci.* **1987**, *41*, 59–67.
46. Rice, M.E. Asexual reproduction in a sipunculan worm. *Science* **1970**, *167*, 1618–1620. [CrossRef]
47. Rajulu, G.S.; Krishnan, N. Occurrence of asexual reproduction by budding in Sipunculida. *Nature* **1969**, *223*, 186–207. [CrossRef]
48. Rice, M.E. A comparative study of the development of *Phascolosoma agassizii*, *Golfingia pugettensis*, and *Themiste pyroides* with a discussion of developmental patterns in the Sipuncula. *Ophelia* **1967**, *4*, 143–171. [CrossRef]
49. Rice, M.E. Observations on the development of six species of Caribbean Sipuncula with a review of development in the phylum. In *Proceedings of the International Symposium of Sipuncula and Echiura*; Rice, M.E., Todorovic, M., Eds.; Naucno Delo Press: Belgrade, Serbia, 1975; pp. 141–160.
50. Selenka, E. Eifurchung und Larvenbildung von *Phascolosoma elongatum*. *Z. Wiss. Zool.* **1875**, *25*, 442–450.
51. Gerould, J.H. The development of *Phascolosoma*. *Zool. Jahrb. Abtheilung Anat.* **1906**, *23*, 77–162.
52. Åkesson, B. A study of the nervous system of the sipunculoideae, with some remarks on the development of the two species *Phascolion strombi* Montagu and *Golfingia minuta* Keferstein. *Unders. Över Öresund* **1958**, *38*, 1–249.
53. Åkesson, B. The development of *Golfingia elongata* Keferstein (Sipunculidea) with some remarks on the development of neurosecretory cells in sipunculids. *Ark. Zool.* **1961**, *13*, 511–531.
54. Schulze, A.; Boyle, M.J.; Kawauchi, G.Y. Chapter 6.1: Sipuncula. In *Annelida*; Volume 1: Annelida Basal Groups and Pleistoannelida; Sedentaria, I., Purschke, G., Böggemann, M., Westheide, W., Eds.; De Gruyter: Berlin, Germany, 2019; Volume 1, pp. 177–201.

55. Boyle, M.J.; Seaver, E.C. Expression of FoxA and GATA transcription factors correlates with regionalized gut development in two lophotrochozoan marine worms: *Chaetopterus* (Annelida) and *Themiste lageniformis* (Sipuncula). *EvoDevo* **2010**, *1*, 2. [CrossRef] [PubMed]
56. Rice, M.E. Larvae adrift: Patterns and problems in life histories of sipunculans. *Am. Zool.* **1981**, *21*, 605–619. [CrossRef]
57. Boyle, M.J.; Rice, M.E. Comparative development and the evolution of life history diversity in Sipuncula. In *Proceedings of the Second International Symposium on the Biology of the Sipuncula*; Boyle, M.J., Kawauchi, G.Y., Eds.; Smithsonian Institution Scholarly Press: Washington, DC, USA, 2018.
58. Rice, M.E.; Piraino, J.; Reichardt, H.F.; Boyle, M.J. Observations on oceanic pelagosphera larvae (Sipuncula): Morphology, behavior, and metamorphosis. In *Proceedings of the Second International Symposium on the Biology of the Sipuncula*; Boyle, M.J., Kawauchi, G.Y., Eds.; Smithsonian Institution Scholarly Press: Washington, DC, USA, 2018; pp. 41–81.
59. Schulze, A.; Hipes, J.; Borda, E.; Rice, M.E. Who's who in the Sipuncula: Matching larvae and adults using DNA. In *Proceedings of the Second International Symposium on the Biology of the Sipuncula*; Boyle, M.J., Kawauchi, G.Y., Eds.; Smithsonian Institution Scholarly Press: Washington, DC, USA, 2018; pp. 83–93.
60. Knowlton, N. Sibling species in the sea. *Annu. Rev. Ecol. Syst.* **1993**, *24*, 189–216. [CrossRef]
61. Sáez, A.G.; Lozano, E. Body doubles. *Nature* **2005**, *433*, 111. [CrossRef]
62. Staton, J.; Rice, M.E. Genetic Differentiation despite teleplanic larval dispersal: Allozyme variation in sipunculans of the *Apionsoma misakianum* species complex. *Bull. Mar. Sci.* **1999**, *65*, 467–480.
63. Johnson, N.D.; Schulze, A. Genetic structure in two *Phascolosoma* species in the Pacific Ocean. *Mar. Biol. Res.* **2016**, *12*, 739–747. [CrossRef]
64. Hebert, P.D.N.; Ratnasingham, S.; DeWaard, J.R. Barcoding animal life: Cytochrome c oxidase subunit 1 divergences among closely related species. *Proc. R. Soc. B Biol. Sci.* **2003**, *270*, S96–S99. [CrossRef] [PubMed]
65. Costa, P.F.E.; Gil, J.; Passos, A.M.; Pereira, P.; Melo, P.; Batista, F.; Fonseca, L.C.D. The market features of imported nonindigenous polychaetes in Portugal and consequent ecological concerns. *Sci. Mar.* **2006**, *70*, 287–292. [CrossRef]
66. Kohn, A.J. Predation on Sipuncula. In *Proceedings of the International Symposium on the Biology of Sipuncula and Echiura*; Rice, M.E., Todorovic, M., Eds.; Nauçno Delo: Belgrade, Serbia, 1975; pp. 113–133.
67. Kawauchi, G.Y.; Giribet, G. *Sipunculus nudus* Linnaeus, 1766 (Sipuncula): Cosmopolitan or a group of pseudo-cryptic species? An integrated molecular and morphological approach. *Mar. Ecol.* **2014**, *35*, 478–491. [CrossRef]
68. Adrianov, A.V.; Maiorova, A.S.; Malakhov, V.V. Embryonic and larval development of the peanut worm *Themiste pyroides* (Sipuncula: Sipunculoidea) from the Sea of Japan. *Invertebr. Reprod. Dev.* **2008**, *52*, 143–151. [CrossRef]
69. Sato, H. Report on the biological survey of Mutsu Bay. 15. Sipunculoidea. *Sci. Rep. Ser. 4 Biol.* **1930**, *5*, 1–40.
70. Cutler, E.B.; Cutler, N.J. A reconsideration of Sipuncula named by I. Ikeda and H. Sato. *Publ. Seto Mar. Biol. Lab.* **1981**, *26*, 51–93.
71. Cutler, E.B.; Cutler, N.J. A revision of the genus *Themiste* (Sipuncula). *Proc. Biol. Soc. Wash.* **1988**, *101*, 741–766.
72. Schulze, A.; Maiorova, A.; Timm, L.E.; Rice, M.E. Sipunculan larvae and "cosmopolitan" species. *Integr. Comp. Biol.* **2012**, *52*, 497–510. [CrossRef]
73. Adrianov, A.; Maiorova, A. Reproduction and development of common species of peanut worms (Sipuncula) from the Sea of Japan. *Russ. J. Mar. Biol.* **2010**, *36*, 1–15. [CrossRef]
74. Maiorova, A.S.; Adrianov, A.V. Peanut worms of the phylum Sipuncula from the Sea of Japan with a key to species. *Deep Sea Res. Part. II Top. Stud. Oceanogr.* **2013**, *86–87*, 140–147. [CrossRef]
75. Nishikawa, T. Some comments on the taxonomy of the peanut worms (Annelida: Sipuncula) in Japanese waters toward a future revision. In *Species Diversity of Animals in Japan*; Motokawa, M., Kajihara, H., Eds.; Springer Japan: Tokyo, Japan, 2017; pp. 469–476. [CrossRef]
76. Rice, M.E. Gametogenesis in three species of Sipuncula: *Phascolosoma agassizii*, *Golfingia pugettensis*, and *Themiste pyroides*. *La Cell.* **1974**, *70*, 295–313.
77. Gibbs, P.E.; Cutler, E.B.; Cutler, N.J. A review of the genus *Thysanocardia* Fisher (Sipuncula). *Zool. Scr.* **1983**, *12*, 295–304.
78. Haldar, B.P. Sipunculans of the Indian coast. *Mem. Zool. Surv. India* **1991**, *17*, 1–169.
79. Murina, V.V. New records of Echiura and Sipuncula in the North Atlantic Ocean, with the description of a new species of Jacobia. *Mar. Biol. Res.* **2008**, *4*, 152–156. [CrossRef]
80. Açik, Ş. Checklist of Sipuncula from the coasts of Turkey. *Turk. J. Zool.* **2014**, *38*, 723–733.
81. Johnson, N.D.; Sanders, C.; Maiorova, A.; Schulze, A. Cryptic species in Pacific sipunculans (Sipuncula: Phascolosomatidae): East-west divergence between non-sister taxa. *Zool. Scr.* **2016**, *45*, 455–463. [CrossRef]
82. Adrianov, A.V.; Maiorova, A.S.; Malakhov, V.V. Embryonic and larval development of the peanut worm Phascolosoma agassizii (Keferstein 1867) from the Sea of Japan (Sipuncula: Phascolosomatidea). *Invertebr. Reprod. Dev.* **2011**, *55*, 22–29. [CrossRef]
83. Rice, M.E. Morphology, behavior and histogenesis of the pelagosphera larva of *Phascolosoma agassizii* (Sipuncula). *Smithson. Contrib. Zool.* **1973**, *132*, 1–51.
84. Keferstein, W. Beitrage zur anatomischen und systematischen Kenntniss der Sipunculiden. *Z. Fur Wiss. Zool.* **1865**, *15*, 404–445.
85. Rice, M.E. Survey of the Sipuncula of the coral and beach rock communities of the Caribbean Sea. In *Proceedings of the International Symposium of Sipuncula and Echiura*; Rice, M.E., Todorovic, M., Eds.; Naucno Delo Press: Belgrade, Serbia, 1975; pp. 35–49.
86. tenBroeke, A. Westindische Sipunculiden und Echuiriden. *Bijdr. Tot De Dierkd.* **1925**, *24*, 81–96.
87. Baird, W.B. Monograph on the species of worms belonging to the subclass Gephyreae. *Proc. Zool. Soc. Lond.* **1868**, *1868*, 77–114.

88. Murina, V.V. On the sipunculid fauna of the littoral of Cuba. *Zool. Zhurnal* **1967**, *46*, 35–47.
89. Shipley, A.E. Report on the Gephyrean worms collected by Mr. J. Stanley Gardiner at Rotuma and Funafuti. *J. Zool.* **1898**, *66*, 468–473. [CrossRef]
90. Edmonds, S.J. A revision of the systematics of Australian sipunculans (Sipuncula). *Rec. South. Aust. Mus.* **1980**, *18*, 1–74.
91. Cutler, E.B.; Cutler, N.J.; Nishikawa, T. The Sipuncula of Japan: Their systematics and distribution. *Publ. Seto Mar. Biol. Lab.* **1984**, *29*, 249–322. [CrossRef]
92. Fischer, W. Westindische Gephyreen. *Zool. Anz.* **1922**, *55*, 10–18.
93. Cutler, E.B.; Kirsteuer, E. Additional notes on some Sipuncula from Madagascar. Results of the Australian Indo-West Pacific Expedition, 1959–1960. Part 12. *Zool. Anz.* **1968**, *180*, 352–356.
94. Cutler, E.B.; Cutler, N.J. Madagascar and Indian Ocean Sipuncula. *Bull. Mus. Natl. D'hist. Nat. Paris* **1979**, *4*, 941–990.
95. Stephen, A.C. Echiuroidea and Sipunculoidea from Senegal, West Africa. *Bull. L'inst. Fr. D'afrique Noire* **1960**, *22A*, 512–520.
96. Cutler, N.J.; Cutler, E.B. A revision of the subgenus *Phascolosoma* (Sipuncula, Phascolosoma). *Proc. Biol. Soc. Wash.* **1990**, *3*, 691–730.
97. Kawauchi, G.; Giribet, G. Are there true cosmopolitan sipunculan worms? A genetic variation study within *Phascolosoma perlucens* (Sipuncula, Phascolosomatidae). *Mar. Biol.* **2010**, *157*, 1417–1431. [CrossRef]
98. Rice, M.E. Factors influencing larval metamorphosis in *Golfingia misakiana* (Sipuncula). *Bull. Mar. Sci.* **1986**, *39*, 362–375.
99. Hall, J.R.; Scheltema, R.S. Comparative morphology of open-ocean pelagosphera. In *Proceedings of the International Symposium on the Biology of the Sipuncula and Echiura*; Rice, M.E., Todorovic, M., Eds.; Nauçno Delo Press: Belgrade, Serbia, 1975; pp. 183–197.
100. Ikeda, I. The Gephyrea of Japan. *J. Coll. Sci. Imp. Univ. Tokyo* **1904**, *20*, 1–87.
101. Gunton, L.M.; Kupriyanova, E.; Alvestad, T.; Avery, L.; Blake, J.A.; Biriukova, O.; Böggemann, M.; Borisova, P.; Budaeva, N.; Burghardt, I.; et al. Annelids of the eastern Australian abyss collected by the 2017 RV Investigator voyage. *ZooKeys*, in press.
102. Murina, V.V. The geographic distribution of marine worms of the phylum Sipuncula of the world ocean. In *Proceedings of the International Symposium on the Biology of the Sipuncula and Echiura*; Rice, M.E., Todorovic, M., Eds.; Nauçno Delo: Bêlgrade, Serbia, 1975; Volume 1, pp. 3–18.
103. Cutler, E.B.; Cutler, N.J. Sipuncula from Southern Brazil. *Bol. Inst. Oceanogr. Sao Paulo* **1980**, *29*, 1–8. [CrossRef]
104. Migotto, A.E.; Ditadi, A.S.F. Aspidosiphonidae (Sipuncula) from the northern and northeastern coast of Brazil. *Rev. Bras. Biol.* **1988**, *48*, 245–263.
105. Kawauchi, G.Y.; Migotto, A. Filo Sipuncula. In *Biodiversidade Bentônica da Região Sudeste-Sul do Brasil*; Amaral, A.C.Z., Rossi-Wongtschowski, C.L.D.B., Eds.; USP Instituto Oceanográfico: São Paulo, Brasil, 2004; pp. 74–76.
106. Pires Vanin, A.M.S.; Corbisier, T.N.; Arasaki, E.; Möellmann, A.M. Composição e distribuição espaço-temporal da fauna bêntica no Canal de São Sebastião. *Realt. Técnico Inst. Oceanogr.* **1997**, *41*, 29–46.
107. Migotto, A.E.; Tiago, C.G. Síntese. In *Biodiversidade do Estado de São Paulo, Brasil: Síntese do Conhecimento ao Final do Século XX, Volume 3: Invertebrados Marinhos*; Migotto, A.E., Tiago, C.G., Eds.; Fundação de Amparo à Pesquisa do Estado de São Paulo, FAPESP: São Paulo, Brasil, 1999; pp. 301–310.
108. Lana, P.C.; Camargo, M.G.; Brogin, R.A.; Isaac, V.J. *O bentos da Costa Brasileira: Avaliação Crítica e Levantamento Bibliográfico (1858–1996)*; FEMAR: Rio de Janeiro, Brasil, 1996; p. 432.

diversity MDPI

Review

The Current State of Eunicida (Annelida) Systematics and Biodiversity

Joana Zanol [1], Luis F. Carrera-Parra [2], Tatiana Menchini Steiner [3], Antonia Cecilia Z. Amaral [3], Helena Wiklund [4], Ascensão Ravara [5] and Nataliya Budaeva [6,*]

1 Departamento de Invertebrados, Museu Nacional, Universidade Federal do Rio de Janeiro, Horto Botânico, Quinta da Boa Vista s/n, São Cristovão, Rio de Janeiro, RJ 20940-040, Brazil; joanazanol@mn.ufrj.br
2 Departamento de Sistemática y Ecología Acuática, El Colegio de la Frontera Sur, Chetumal, QR 77014, Mexico; lcarrera@ecosur.mx
3 Departamento de Biologia Animal, Instituto de Biologia, Universidade Estadual de Campinas, Campinas, SP 13083-862, Brazil; tatims@unicamp.br (T.M.S.); ceamaral@unicamp.br (A.C.Z.A.)
4 Department of Marine Sciences, University of Gothenburg, Carl Skottbergsgata 22B, 413 19 Gothenburg, Sweden; helena.wiklund@marine.gu.se
5 CESAM—Centre for Environmental and Marine Studies, Departamento de Biologia, Universidade de Aveiro, Campus de Santiago, 3810-193 Aveiro, Portugal; aravara@ua.pt
6 Department of Natural History, University Museum of Bergen, University of Bergen, Allégaten 41, 5007 Bergen, Norway
* Correspondence: Nataliya.Budaeva@uib.no

Abstract: In this study, we analyze the current state of knowledge on extant Eunicida systematics, morphology, feeding, life history, habitat, ecology, distribution patterns, local diversity and exploitation. Eunicida is an order of Errantia annelids characterized by the presence of ventral mandibles and dorsal maxillae in a ventral muscularized pharynx. The origin of Eunicida dates back to the late Cambrian, and the peaks of jaw morphology diversity and number of families are in the Ordovician. Species richness is heterogeneous among the seven recent families, with more than half of the valid species belonging to the Eunicidae + Onuphidae clade, one of the latest clades to diverge. Eunicidans inhabit soft and hard substrates from intertidal to deep waters in all oceans. The few freshwater species are restricted to Histriobdellidae, a family exclusively commensal/parasite of crustaceans. The reproductive biology, development and ecology of most families are poorly known and the information available suggests low dispersal ability. However, all families have records of widely distributed species. Scrutiny of these wide distributions has often revealed the presence of exotic species or more than one species. The exploration of the deep-sea and of new habitats has led to recent descriptions of new species. Furthermore, the revision of type specimens, the examination of new morphological features and the use of molecular data have revealed hidden biodiversity under unjustified synonyms, poor understanding of morphological features and incomplete descriptions. Molecular studies are still very few or nonexistent for the families Histriobdellidae, Hartmaniellidae, Lumbrineridae and Oenonidae. The integration of new methodologies for morphological and molecular study, along with information on biological and ecological traits appears to be the path to improve the knowledge on the diversity of Eunicida.

Keywords: marine invertebrate; diversity; distribution; jaw; maxillae; polychaete; Polychaeta; Dorvilleidae; Eunicidae; Hartmaniellidae; Histriobdellidae; Lumbrineridae; Oenonidae; Onuphidae

Citation: Zanol, J.; Carrera-Parra, L.F.; Steiner, T.M.; Amaral, A.C.Z.; Wiklund, H.; Ravara, A.; Budaeva, N. The Current State of Eunicida (Annelida) Systematics and Biodiversity. *Diversity* **2021**, *13*, 74. https://doi.org/10.3390/d13020074

Academic Editors: Bert W. Hoeksema and Maria Capa

Received: 30 December 2020
Accepted: 26 January 2021
Published: 9 February 2021

1. Introduction

Eunicida was raised to order level relatively recently (Uschacov [1] as Eunicimorpha, Dales [2] as Eunicida) despite being recognized as a taxon much earlier (superfamily Eunicea [3] or family Eunicidae [4]). The taxon identity is supported by the presence of eversible jaws, which comprise dorsal maxillae and ventral mandibles, in a ventral muscularized pharynx. The composition of Eunicida has been relatively stable including seven recent (Dorvilleidae Chamberlin, 1919; Eunicidae Berthold, 1827; Hartmaniellidae Imajima, 1977; Histriobdellidae Vaillant, 1890; Lumbrineridae Schmarda, 1861; Oenonidae Kinberg, 1865; and Onuphidae Kinberg, 1865) (Figure 1 and Figures 2–10) and 17 extinct families [5].

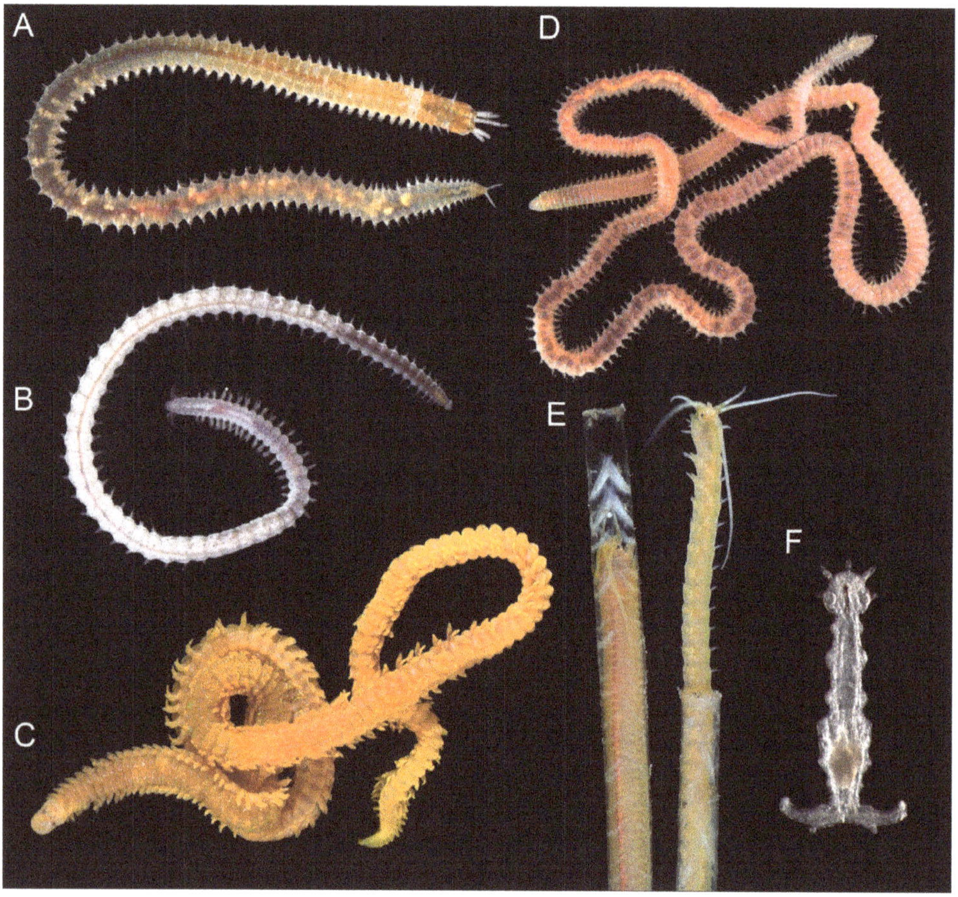

Figure 1. Live specimens representing eunicidan families: (**A**) *Nicidion* sp., Eunicidae, © A. Sememov; (**B**) Dorvilleidae gen. sp., © A. Sememov; (**C**) *Oenone fulgida* (Lamarck, 1818), Oenonidae, © A. Sememov; (**D**) Lumbrineridae gen. sp., © A. Sememov; (**E**) *Hyalinoecia tubicola* (O.F. Müller, 1776), Onuphidae, © N. Budaeva; (**F**) *Histriobdella homari* Van Beneden, 1858, Histriobdellidae, © C. Helm.

Eunicida is sister to Phyllodocida within the Errantia clade [6,7]. Phylogenetic analyses using multiple genes or genomic data recover monophyletic Eunicida and families, with exception of Dorvilleidae in analyses including the genus *Pettibonea* [6,8–12]. Eunicidae and Onuphidae are consistently recovered as sister groups in analyses with only recent taxa and molecular data but formed a polytomy with the extinct genus *Esconites* Thompson and Johnson, 1977 in analyses with extant and extinct taxa based on morphological data [9]. The relationship among the other families is inconsistent. Oenonidae, Dorvilleidae, Oenonidae + Dorvilleidae or Oenonidae+Lumbrineridae have resulted as sister groups of the Eunicidae + Onuphidae clade. Lumbrineridae and Dorvilleidae have also been recovered as a sister family to all other Eunicida [6,8,9]. The relationships recovered with phylogenomic analyses [6] agree, at least in part, with hypotheses drawn from Orensanz's [13] and Paxton´s [14] comprehension of the evolution of jaws. All of them recover the clade Eunicidae + Onuphidae and have Dorvilleidae as the sister group to all other Eunicida. These hypotheses contrast in the placement of Lumbrineridae as a sister group to Oenonidae [6,13] or in a clade with Hartmaniellidae, sister group to (Dorvilleidae (Oenonidae (Eunicidae, Onuphidae)) [14]. Orensanz [13] suggests the clade Lumbrineridae+Oenonidae is the sister to (Hartmaniellidae, (Eunicidae, Onuphidae)). Hartmaniellidae and Histriobdellidae have not been included in any formal phylogenetic analyses. Their monophyletic status and placement within Eunicida remain to be tested.

Species richness is heterogeneous among families. Eunicidae (12 genera, 453 spp.) bears the highest number of species followed by Onuphidae (22 genera, 340 spp.), Lumbrineridae (19 genera, 279 spp.), Dorvilleidae (32 genera, 200 spp.), Oenonidae (12 genera, 109 spp.), Histriobdellidae (3 genera, 13 spp.) and Hartmaniellidae (1 genus, 3 spp.) [15]. Despite being among one of the latest clades to diverge, the Eunicidae + Onuphidae clade comprises more than half of the recent species within Eunicida (1397 spp.). This could indicate a higher diversification rate, a lower extinction rate or just an underestimation of species diversity in other families, such as Dorvilleidae, for which 73% of the valid species were described in the last 50 years (Figure 2). In the families Eunicidae and Onuphidae, the species described in the last 50 years correspond, respectively, to 30% and 53% of the total of valid species. Many descriptions of new species in Eunicida are due to the discovery of new habitats and increase in exploration of the deep-sea (e.g., [16–19]) as well as the study of meiofaunal polychaetes (e.g., [20–23]). However, the revision of type specimens, the examination of new morphological features and the use of molecular data are revealing a hidden biodiversity under unjustified synonyms, poor understanding of morphological features and incomplete descriptions [24–26]. The lack of correspondence between the number of genera and the number of species per family (e.g., Eunicidae) illustrates more homogeneity in the external morphology of some families or just more conservative taxonomic decisions.

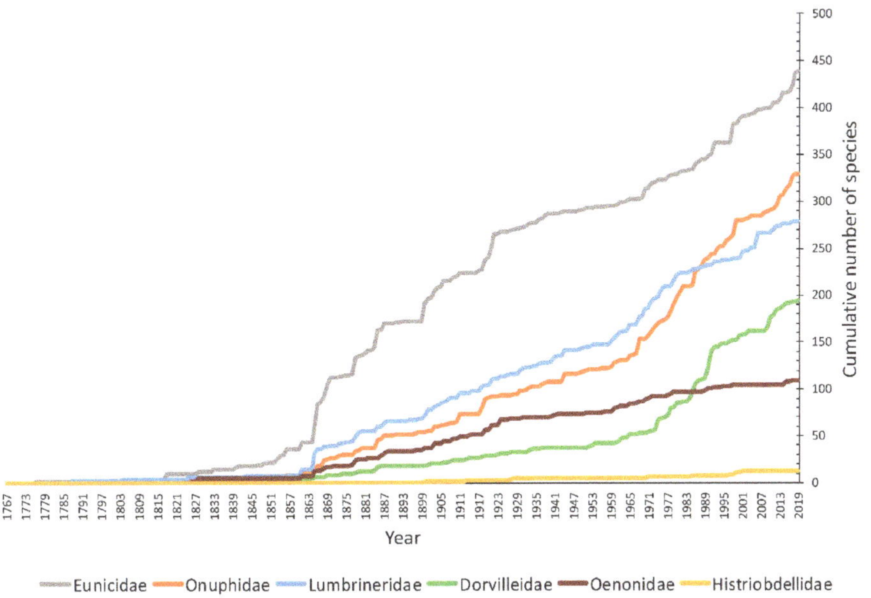

Figure 2. Cumulative number of species described in six eunicidan families, data on valid species only [15] (Hartmaniellidae not shown).

The origin and diversification of Eunicida date back to the late Cambrian, the age of the earliest fossil maxilla [27]. Maxillae compose the main fossil record of Eunicida and consist of a series of mostly paired sclerotized plates. Maxillae vary among families in relation to the number, morphology, symmetry and arrangement of the plates, and, when present, the shape of the carriers (Figure 3). The diversity of jaw morphology, which translates into the number of families, is more than twice as high in extinct than in extant taxa and is peaked in the Ordovician [27]. The general morphological patterns of maxillae are grouped in jaw-types fine-tuned along different studies considering both extinct and extant diversity [3,13,14,28,29]. Of the six jaw-types currently considered [14], two are exclusive of ten extinct families and all the other four jaw-types are shared among extinct and extant families [14]. Dorvilleidae is the recent family with the highest jaw diversity, motivating the proposition of its division into two different families (Dorvilleidae and Iphitimidae) [13,14,19] or into six informal groups [19], none supported by the only phylogenetic analyses including representatives of the dorvilleid jaw diversity [30].

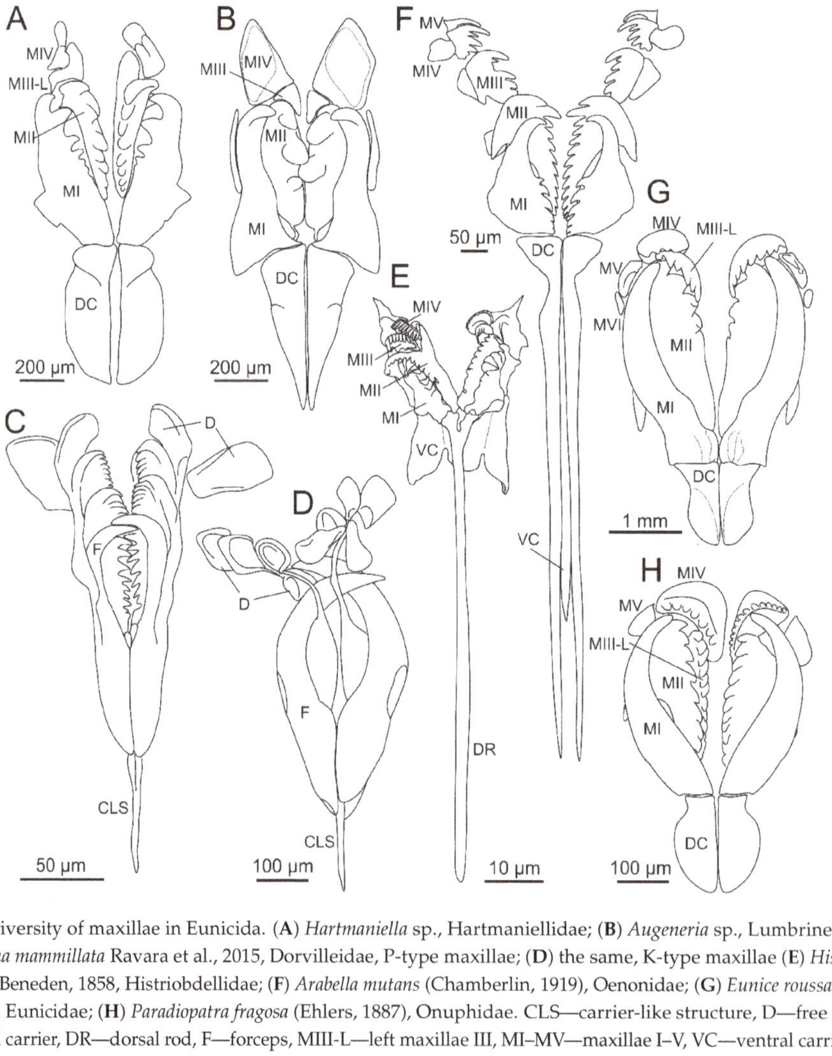

Figure 3. Diversity of maxillae in Eunicida. (**A**) *Hartmaniella* sp., Hartmaniellidae; (**B**) *Augeneria* sp., Lumbrineridae; (**C**) *Ophryotrocha mammillata* Ravara et al., 2015, Dorvilleidae, P-type maxillae; (**D**) the same, K-type maxillae (**E**) *Histriobdella homari* Van Beneden, 1858, Histriobdellidae; (**F**) *Arabella mutans* (Chamberlin, 1919), Oenonidae; (**G**) *Eunice roussaei* Quatrefages, 1866, Eunicidae; (**H**) *Paradiopatra fragosa* (Ehlers, 1887), Onuphidae. CLS—carrier-like structure, D—free denticles, DC—dorsal carrier, DR—dorsal rod, F—forceps, MIII-L—left maxillae III, MI–MV—maxillae I–V, VC—ventral carrier. (**C–D**) modified from Ravara et al. [31]; (**H**) modified from Budaeva and Fauchald [32].

The different jaw morphologies also have different biomechanics with implications for food capturing and shredding [33]. In maxillae bearing carriers of at least Lumbrineridae and Onuphidae (Figure 3B,H), those plates attached to the carriers are primary maxillae (maxillae I, II in both families and unpaired III in Onuphidae) responsible for catching and shredding (exclusive of serrated plates) food items. The unpaired serrated primary maxillae III in onuphid improves the breaking of food with different properties. The interlocking mechanism, present at the posterior end of maxillae I in both families, prevents slipping of the plates and it is an important mechanism in capturing moving food [33]. Feeding habits and food quality consumed are variable within recent and, probably, extinct Eunicida, including grazing, deposit-feeding, and predation, detritivory, herbivory, carnivory and omnivory [34]. Giant raptorial eunicidans were already present 400 million years ago [35].

Recent giant eunicidan species are among the longest annelids with up to 6 m in length (Eunicidae, *Eunice aphroditois* (Pallas, 1788) [36]). Length variation is great within Eunicida, since it also includes species among the shortest annelids with 255 μm (Dorvilleidae, *Neotenotrocha sterreri* Eibye-Jacobsen and Kristensen, 1994 [30]). The prostomium is well developed followed by a double (Dorvilleidae, Eunicidae, Hartmaniellidae, Lumbrineridae, and Oenonidae) or single (Onuphidae) ringed peristomium in all families, except for Histriobdellidae in which these structures are fused. Prostomial appendages, one to three antennae and a pair of palps, may be present on the prostomium of Dorvilleidae, Eunicidae, Histriobdellidae and Onuphidae (Figures 4, 5, 7 and 10). Lumbrineridae and Oenonidae generally have conical appendage-free prostomium with the exception of few genera bearing tiny dorsal appendages covered by the anterior fold of the peristomium or visible due to a peristomium dorsal incision (Figures 8 and 9). Buccal lips are conspicuous structures in the prostomium of Onuphidae (Figure 10E) and insconspicuous in Dorvilleidae, Eunicidae (Figure 5), Lumbrineridae, and Oenonidae [37]. The muscularized pharynx holding the maxillae contains at least homologous dorsolateral and pharyngeal folds in Dorvilleidae, Eunicidae, Lumbrineridae, Oenonidae and Onuphidae, and a dorsolateral fold anterior extension in the last four families [37,38]. Metamerism is conspicuous and the number of chaetigers is variable in all families but Histriobdellidae. Parapodia are reduced, uniramous or subbiramous with notopodia reduced to dorsal (notopodial) cirri. Capillary chaetae are present in all families, except Histriobdellidae, which lacks chaetae. Chaetae diversity increases along the evolution of Eunicida being highest in Onuphidae and Eunicidae [39]. The pygidium and two or four pygidial (anal) cirri, which may be absent, bear several nerve cell endings suggesting they have sensory function [40].

Eunicidans are mostly free-living and can be found in the majority of marine habitats in soft and hard substrates from the intertidal zone to the deep waters in all oceans. Hartmaniellidae is the only family that appears to be solely present in fine soft bottoms, sometimes with shell fragments. No other free-living family is exclusively present in one type of habitat or substrate. However, most members of some families, such as Lumbrineridae, Oenonidae and Onuphidae, prefer soft bottoms, while Eunicidae is most common in hard biogenic substrates [41,42]. Histriobdellidae is the only eunicidan family occurring in both marine and continental waters, with highest richness in the latter, and being exclusively commensals/parasites of crustaceans. Parasitic and symbiotic species are also present in the families Dorvilleidae, Eunicidae, and Oenonidae.

All families have records of widely distributed species, which deserve further investigation. Most widespread species studied in detail ended up as more than one species [43–45] or exotic species dispersed by human mediated transport [46,47]. However, some species appear to have naturally widespread distributions [48,49] despite the poor natural dispersal capability attributed to eunicidan species. Reproductive modes vary from viviparity, brooding of young in tubes or gelatinous egg masses to broadcast spawning. However, even in the latter the planktonic life span is expected to be relatively short, since all known eunicidan larvae are lecitotrophic [50]. Asexual reproduction happens at least by scissiparity (Oenonidae, [51]) and architomy (Dorvilleidae, [52]).

Eunicida includes species of economic and cultural importance. Eunicidae, Lumbrineridae, Oenonidae and Onuphidae comprise species used as baits for commercial and recreational fishing in diverse localities in Atlantic, Pacific, Adriatic, Mediterranean and Suez Canal coasts [53–58], and are also farmed to be used as bait and food supply in aquaculture (e.g., *Diopatra aciculata* Knox and Cameron, 1971, Onuphidae; [59]). Ripe swarming forms of *Palola* (Eunicidae) species are a traditional food and the source of festivals in South Pacific communities [49,60].

In this review, we analyze the current state of knowledge on extant Eunicida systematics, morphology, feeding, life history, habitat, ecology, distribution patterns, local diversity and exploitation, pointing out gaps of knowledge and future perspectives. The great diversity present in Eunicida prevents the presentation of a complete overview without considering each family separately. Thus, in the next sections, we present the current state of knowledge on the different aspects of the recent eunicidan families. Extinct families are not further considered because of the lack of information on most of the aspects considered here.

2. Dorvilleidae Chamberlin, 1919

2.1. Systematics and Phylogeny

The family Dorvilleidae comprises 32 genera and around 200 species. It includes the majority of the smaller forms of the eunicemorph polychaetes, with around 10% of the species described hitherto belonging to the interstitial fauna. Although the family name Dorvilleidae has been in common use since it was erected by Chamberlin [61], its validity remained questionable as its type genus name *Dorvillea* Parfitt, 1866 is a junior homonym of the mollusk genus name *Dorvillea* Leach in Gray, 1852. The alternative name Stauronereididae Verrill, 1900 was occasionally used (e.g., [62]), regardless the previous synonymy of its type genus *Stauronereis* Verrill, 1900 with the priority name *Dorvillea* [61]. Only recently, the genus name *Dorvillea* Parfitt, 1866 was validated as nomen protectum, based on its use in over 25 publications by more than 10 authors in the last 50 years (Article 23.9.2 of the ICZN code), while *Dorvillea* Leach, 1852 was deemed as nomen oblitum, since it has not been used as valid since 1899 [63]. This nomenclatural act also protects the family name Dorvilleidae Chamberlin, 1919.

The first dorvilleid species was described by Delle Chiaje [64] as *Nereis rudolphi* (posteriorly moved to the genus *Schistomeringos* Jumars, 1974). When Chamberlin erected the family, 21 of the currently valid dorvilleid species were already described. Two genera were included in the family: *Dorvillea* and *Ophryotrocha* Claparède and Mecznikow, 1869. Although already described, the genus *Iphitime* Marenzeller, 1902 was initially ascribed to a different family (first Lysaretidae Kinberg, 1865 and later Iphitimidae Fauchald, 1970, both currently unaccepted). Not until about four decades later were other genera described for the family. The remaining 29 genera were described following important revisions [13,19,21,30,65–68], among other works. An updated key and detailed diagnoses to all dorvilleid genera is available in Wiklund et al. [69]. Over two-thirds of the dorvilleid genera include only one or a couple of species, while the largest genera, *Ophryotrocha*, *Dorvillea* and *Schistomeringos*, encompass more than half of the existing species.

Most molecular studies within the family have dealt mainly with *Ophryotrocha* [70,71], often in combination with species descriptions (e.g., [31,72–74]). In several of these restricted molecular analyses, species from different dorvilleid genera fall within *Ophryotrocha*. This is the case for *Exallopus* Jumars, 1974, *Iphitime* and *Pseudophryotrocha* Hilbig and Blake, 1991 (e.g., [31,72–74]). However, since the respective type species for these genera have not been sequenced yet, the authors have chosen to await a formal family revision rather than renaming and moving only those species that have been used in those studies. The genus *Mammiphitime* Orensanz, 1990 might also belong to *Ophryotrocha* although no molecular data is available for it yet. The erection of *Mammiphitime* was justified by the large dorsal lobes in the median and posterior segments [13], but since then, another species with similar lobes has been described, *Ophryotrocha mammillata* Ravara, Marçal, Wiklund and

Hilário, 2015, and molecular data placed it within *Ophryotrocha* [31]. Due to these, and possibly other doubtful genera, the family Dorvilleidae is in need of a comprehensive revision using a combination of morphological characteristics and molecular data.

The mono- or paraphyletic condition of the family Dorvilleidae has been a matter of discussion depending on whether some taxa such as Dinophilidae Macalister, 1876, Diurodrilidae Kristensen and Niilonen, 1982 and *Pettiboneia* Orensanz, 1973 are included or not. In their phylogenetic study using morphological characteristics, Eibye-Jacobsen and Kristensen [30] refer to Iphitimidae and Dinophilidae as belonging to Dorvilleidae. However, subsequent studies using molecular methods or ultrastructure analyses have suggested that although Iphitimidae does indeed fall within Dorvilleidae, Dinophilidae and Diurodrilidae are not closely related to Dorvilleidae (e.g., [6,75–77]). The monophyly of Dorvilleidae is achieved when excluding the genus *Pettiboneia* [6,8,39]. In fact, Paxton [14] reported *Pettibonea* as lacking base dental plates and having two rows of free denticles, corroborating the unique morphology of the genus. However, in the molecular analysis performed by Struck et al. [8] the genus was represented only by two genetic markers (18S and 16S rDNA) of a single species, *Pettiboneia urcinensis* Campoy and San Martín, 1980, that was recovered as genetically closer to Lumbrineridae. Furthermore, there are very few sequences available for species in the other dorvilleid genera preventing a large-scale family analysis. Thus, phylogenetic analyses with better taxon coverage are required to test the monophyly of the family.

2.2. Morphology

Dorvilleids share with the other eunicidan families the outstanding jaw apparatus. It is mainly the jaw morphology that allows identifying the family and is also useful for delineation of genera and species. The jaws in Dorvilleidae consist of ctenognath maxillae (upper comb-like jaws), formed by two to four rows of symmetrical or subsymmetrical dentate elements and a posterior unpaired element (carrier-like structure) (Figure 3C,D), and mandibles (lower jaws). However, species in some genera lack jaws, e.g., *Ikosipodoides* Westheide, 2000 and *Parapodrilus* Westheide, 1965. Furthermore, the shape of the prostomium and its appendages as well as the shape of parapodia and chaetae is of importance to separate genera and species within the family (see [69]). Prostomial appendages, when present, are paired. Lateral antennae can be articulated or unarticulated while palps are simple or biarticulated (Figure 4A–D). Parapodia are missing, sub-biramous or uniramous, with or without dorsal and ventral cirri (Figure 4E). Some genera lack chaetae, but most species have simple supra-acicular and simple and/or compound subacicular chaetae (Figure 4F). A detailed description of the external and internal morphology of Dorvilleidae is available in Wiklund et al. [69]. Due to their small size and frequently minor morphological differences, dorvilleids can be difficult to identify correctly, and different species are often only detected following molecular analysis (e.g., [44]).

2.3. Feeding and Life History

The feeding habits vary within the family, with some species being reported as carnivores and others as omnivorous, grazing on algae or bacterial mats or living off detritus [34,78].

In most dorvilleid genera, the sexes are separated and the species lack sexual dimorphism. However, in the genus *Ophryotrocha*, some species exhibit sexual dimorphism and many species are consecutive or simultaneous hermaphrodites (e.g., [79,80]). Epitokal modifications may occur in some species, expressed by enlarged eyes and longer chaetae [81–83]. Some dorvilleid species are broadcast spawners with swarming phases [83], other species have pseudocopulation and possibly internal fertilization [84], and a few species have been reported to have asexual reproduction [52]. In *Ophryotrocha*, pseudocopulation and brooding of the offspring is common, and viviparity occurs in a few cases [79,84]. While some dorvilleids, e.g., *Schistomeringos rudolphi* (delle Chiaje, 1828), have planktonic larvae that can spend at least a week in the water column before settling [83], others brood their young, expressing direct development with the juveniles emerging from egg cases already resembling miniature adult worms [79].

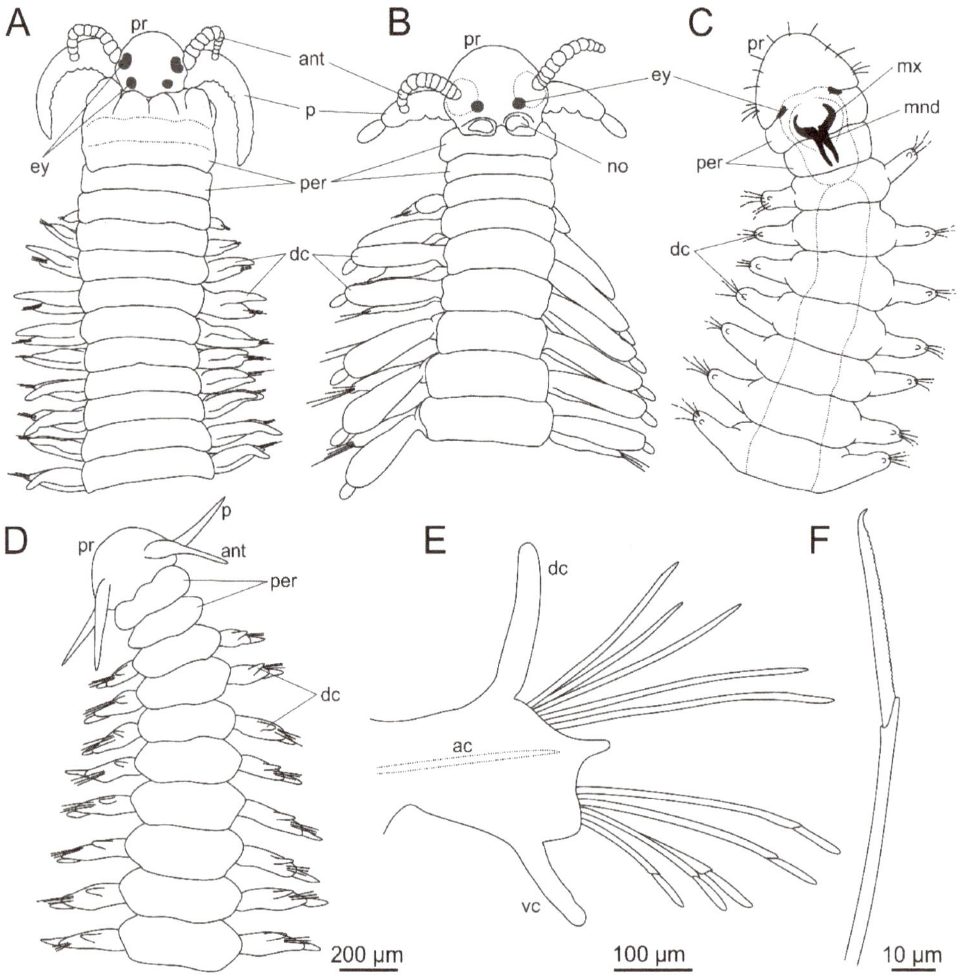

Figure 4. Morphology of Dorvilleidae. (**A**), (**B**) Dorvilleidade gen. sp. anterior end, dorsal view; (**C**) *Ophryotrocha sp.* Anterior end, dorsal view; (**D**) *Ophryotrocha langstrumpae* Wiklund, Altamira, Glover, Smith, Baco and Dahlgren, 2012, anterior end, dorsal view; (**E**) the same, midbody parapodium; (**F**) the same, compound chaeta. ac—acicula, ant—antenna, dc—dorsal cirrus, ey—eye, mnd—mandibles, mx—maxillae, no—nuchal organ, p—palp, per—peristomium, pr—prostomium, vc—ventral cirrus. (**D–F**) modified from Wiklund et al. [74].

2.4. Habitats and Ecology

Some of the smallest dorvilleid species are interstitial, while larger species are found on soft or hard substrates, and some are symbiotic/parasitic in e.g., crab gill chambers (species in *Iphitime* and two species in *Ophryotrocha*) or in the coelom of other polychaetes (*Veneriserva* Rossi, 1984). Free-living species can surround themselves with mucus, and some species construct tubes out of mucus [34].

Several studies indicate limited tolerance of some dorvilleid species to environmental factors such as temperature and salinity (e.g., [85,86]) although there are also records of euryhaline species [87,88]. Nevertheless, apart from a few records of estuarine occurrence, dorvilleids seem to be predominantly fully marine organisms. Species of the genera *Ophryotrocha* and *Parougia* Wolf, 1986 are particularly common and frequently highly abundant in organically enriched habitats, such as harbors, fish farms, sewage discharges and organic-falls (wood or the carcasses of large animals sunken to the seafloor) [16,31,44,72–74,89–92], being considered by some authors as nonspecialized opportunists (e.g., [48]). These two genera, along with *Exallopus*, include the majority of the species occurring in deeper waters [19,67,93–97], often in extreme habitats such as hydrothermal vents and cold seeps [18,97–102].

2.5. Distribution Patterns and Diversity

Dorvilleid worms are widely distributed and occur from the intertidal zone to the abyss. Significant family and genera revisions covering regional geographic areas include Fauchald [103], Orensanz [13,66], Oug [81], Wolf [68,104,105], Hilbig [95,106], Paiva and Nonato [107], Imajima [108], Núñez et al. [109] and Núñez [62] (as Stauronereididae). However, the majority of these works focus mainly on the Northern Hemisphere benthic fauna, while southern geographic areas such as the South Atlantic, Southeast Pacific and Indian Ocean are very poorly documented.

The exploration of organically enriched areas led to a great increase in the number of *Ophryotrocha* and *Parougia* species (e.g., [16,18,31,44,72–74,92,96–102,110]). The occurrence of several sympatric congeners is also frequently reported for both genera (e.g., [16,18,31,44,72–74,111–113]). Furthermore, with the upswing in deep-sea exploration, more species of these and other genera have been described also from areas that seemingly were not organically enriched (e.g., [19,93,94,114–116]). It is likely that there are many more species from these and other dorvilleid genera awaiting discovery, some still unknown in the ocean while others are already collected and stored, but not properly examined and described yet.

Despite the lack of planktonic larvae, some *Ophryotrocha* species have been shown to have quite a broad distribution range. For instance, *Ophryotrocha orensanzi* Taboada, Wiklund, Glover, Dahlgren, Cristobo and Ávila, 2013, *Ophryotrocha scutellus* Wiklund, Glover and Dahlgren, 2009 and *Ophryotrocha cyclops* Salvo, Wiklund, Dufour, Hamoutene, Pohle and Worsaae, 2014 have been reported from sites with hundreds to thousands kilometers between them, with species identities confirmed by molecular data [31,48,72,90]. Some species with wide distribution ranges are suspected to have been transferred with human activities such as in ballast waters (e.g., [78]). However, many dorvilleids are small and can be difficult to identify properly to species level, and thus the possibility of them having a continuous distribution between distant localities cannot be ruled out. More rigorous taxonomic investigations, preferably in combination with molecular information, are needed before we can make any assumptions about dorvilleid dispersal capacities and species real distribution.

2.6. Exploitation

Given their small size, Dorvilleids have never been an important commercial group. However, a recent study suggested the beneficial use of *Ophryotrocha* species in pisciculture by feeding on fish feces and in turn being fed to the fish [117]. In addition, because they have a short generation time and are easily kept in laboratory conditions, *Ophryotrocha* species may be suitable candidates as test model organisms for various kinds of studies [70,79,87,118].

3. Eunicidae Berthold, 1827

3.1. Systematics and Phylogeny

The family Eunicidae comprises 12 genera and approximately 453 species. It was first recognized as a taxon by Lamarck [119] under the name of Eunicea. However, until 1944 it was more inclusive and almost equivalent to the order Eunicida with variations depending on the author. The current definition of the family was established in Hartman [28] including four of the current valid genera, *Eunice* Cuvier, 1817 (242 species), *Lysidice* Lamarck, 1818 (30 species), *Marphysa* Quatrefages, 1865 (74 species), *Palola* Gray in Stair, 1847 (14 species). The composition of the family has been stable since then. However, currently it includes eight additional genera, one extinct, *Esconites* Thompson and Johnson, 1977 (one species), and seven extant, *Aciculomarphysa* Hartmann-Schröder, 1998 in Hartmann-Schröder and Zibrowius [120] (one species), *Euniphysa* Wesenberg-Lund, 1949 (11 species), *Fauchaldius* Carrera-Parra and Salazar-Vallejo, 1998 (two species), *Leodice* Lamarck, 1818 (33 species), *Nicidion* Kinberg, 1865 (14 species), *Paucibranchia* Molina-Acevedo, 2018 (19 species), *Treadwellphysa* Molina-Acevedo and Carrera-Parra, 2017 (eight species).

Phylogenetic analyses based solely on morphological data could not recover monophyletic Eunicidae and *Eunice* [121]. Combined molecular and morphological phylogenetic analyses recovered a monophyletic Eunicidae and nonmonophyletic *Eunice*, *Lysidice*, and *Marphysa* [10,11]. Based on these results, *Leodice* and *Nicidion* were resurrected to include species previously classified as *Eunice* or *Marphysa* that made both nonmonophyletic and *Nematonereis* Schmarda, 1861 was synonymized to *Lysidice*. It was still not possible to delimit a monophyletic *Eunice* because of the inconsistent placement of several of its species, including the type species *Eunice aphroditois*. *Eunice* species, which bear branchiae starting late, inflated base of the ventral cirri as thick ventral ridges, left MxIV with teeth restricted to the dorsal portion of the plate, and bidentate compound falcigerous chaetae with inconspicuous teeth in anteriormost chaetigers, consistently formed a clade which may represent another monophyletic genus. The further division of *Leodice* in two monophyletic taxa may also be justified in future analyses including more species. The genus has two main sister clades, one comprising species with dark subacicular hooks and the other species with light subacicular hooks. Thus, further analyses including more representatives of the diversity and other molecular markers are necessary.

A taxonomic key for the identification of genera valid at the time and considering the definitions based on eunicid current phylogenetic hypothesis is provided in Zanol et al. [11]. *Marphysa* is currently more restricted because some species have been transferred to *Paucibranchia* and *Treadwellphysa*. The key to genera in Molina-Acevedo and Carrera-Parra [122] includes *Treadwellphysa*. However, it considers *Nematonereis* valid and has more restricted morphological definitions for *Eunice*, *Leodice*, *Lysidice*, *Nicidion*, and *Marphysa*, not following genera definitions based on the phylogenetic hypothesis in Zanol et al. [11].

Taxonomic revisions for species are available for *Eunice* [36], *Euniphysa* [123], *Lysidice* from the Western Caribbean [124], *Marphysa* [25,122,125,126], *Palola* [127], *Paucibranchia* [128], and *Treadwellphysa* [129]. These revisions comprise more detailed descriptions, discussions on taxonomic acts and taxonomic keys, improving the knowledge of diagnostic morphological features, undoing unjustified synonymizations and facilitating identification. The revision of *Eunice* also includes species of *Leodice* and *Nicidion*, which were classified as *Eunice* at the time. Some of these have not had their genus officially changed yet. Thus, the number of *Leodice* and *Nicidion* species is probably much higher than the current figures and that of *Eunice* is much lower. *Palola* diversity is discussed in Schulze [49] and Schulze and Timm [116]. Despite the revisions, the identification of species of the genera *Eunice*, *Leodice*, *Lysidice*, *Nicidion* and *Palola* may still be confusing, thus, further revisions including more detailed descriptions with intraspecific variation would be of great value.

3.2. Morphology

Eunicidae is defined by two synapomorphies, dorsal buccal lip fused to the dorsal side of the prostomium and dorsolateral fold anterior extensions medially connected [11] (Figure 5A–E). Additional diagnostic features are present at prostomium, peristomium and jaws. These are ventral buccal lips with median transverse groove, one (single median antenna) to five prostomial appendages (a pair of palps, a pair of lateral antennae and a single median antenna), double ringed peristomium (Figures 1A and 5A–E), and asymmetric eulabidognath maxillae (Figure 3G). The family encompasses great length variation, bearing species from few millimeters to few meters long. All species have a variable number of chaetigers.

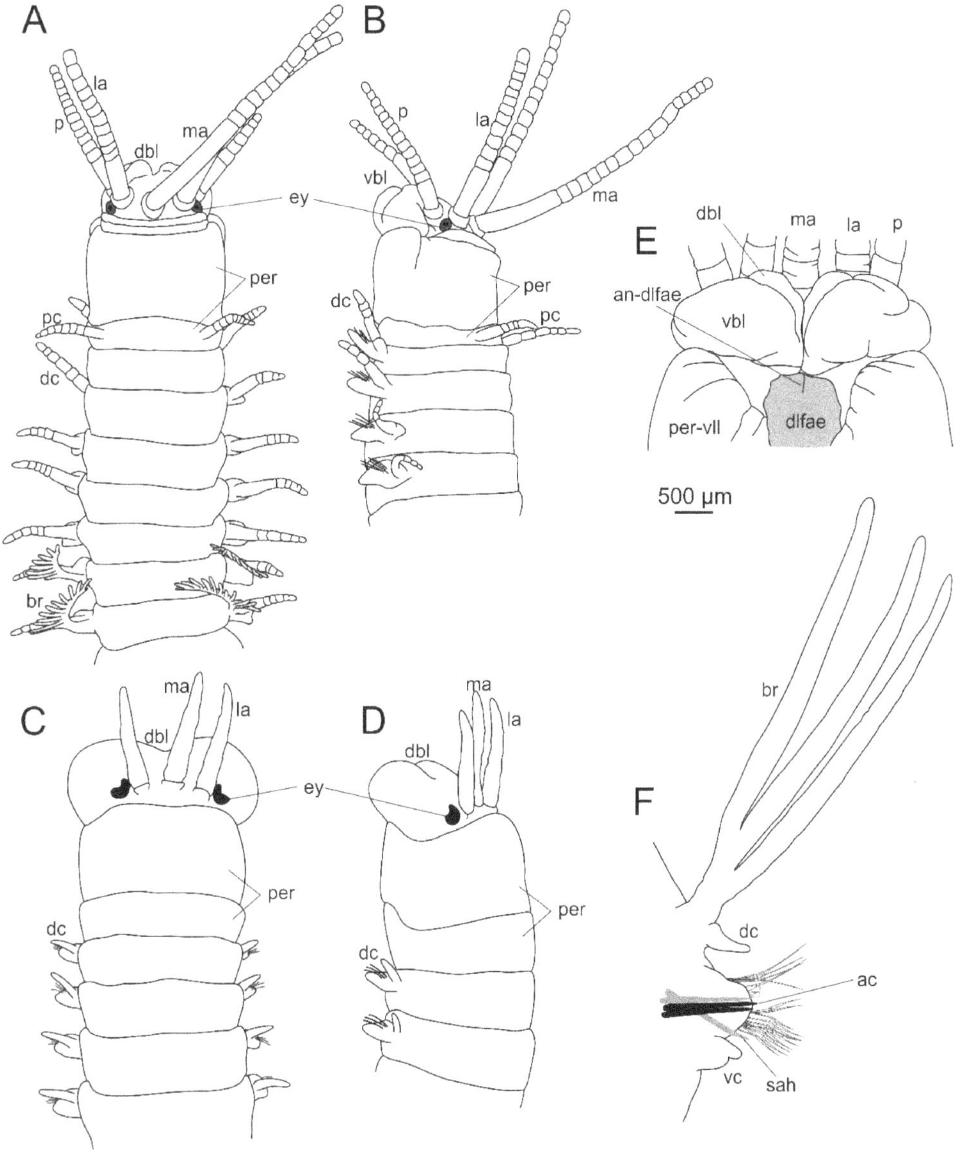

Figure 5. Morphology of Eunicidae. (**A**) *Leodice* sp., anterior end, dorsal view; (**B**) the same, lateral view; (**C**) *Lysidice* sp., anterior end, dorsal view; (**D**) the same, lateral view; (**E**) *Eunice violaceomaculata* Ehlers, 1887, prostomium, ventral view, dissected; (**F**) *Marphysa* sp., midbody parapodium. ac—acicula, an-dlfae—anterior notch of the dorsolateral fold anterior extension, br—branchia, dbl—dorsal buccal lip, dc—dorsal cirrus, dlfae—dorsolateral fold anterior extension, la—lateral antenna, ma—median antenna, p—palp, pc—peristomial cirrus, per—peristomium, per-vll—peristomial ventrolateral lips, sah—subacicular hook, vbl—ventral buccal lip, vc—ventral cirrus.

Diagnoses of extant genera take into account presence of lateral antennae and palps, articulation of prostomial appendages, presence of peristomial cirri, flat or curved mandibles, shape of maxillary plates, shape and distribution of branchiae, and presence and shape of pectinate chaetae and subacicular hooks (Figure 5F). Features present in chaetigers may vary in distribution and shape along the body. The complete set of chaetae in most genera includes limbate, pectinate, compound, subacicular hooks and aciculae. *Palola* is the only genus missing pectinate chaetae and subacicular hooks. In most genera, pectinate chaetae shafts are always thinner than limbate chaetae from the same parapodium (thin pectinate). However, in *Marphysa* and *Treadwellphysa*, they may have thick shafts (thicker than limbate chaetae in the same parapodium) and wide blades in mid and posterior chaetigers (thick and wide pectinate chaetae). The outer teeth of pectinate chaetae may be: both as long as inner teeth (anodont), just one longer (heterodont) or both longer than inner teeth (isodont *sensu* [130]).

In addition to these, features considered in the identification of species are the shape and size of prostomium, peristomium, prostomial appendages and peristomial cirri, dorsal and ventral cirri, prechaetal, chaetal, postchaetal lobes, pygidial cirri, the distribution and shape of branchiae, pectinate and compound chaetae, and the color, distribution, number and shape of aciculae and subacicular hooks. The detailed morphology of the chaetae varies along the body and is an important taxonomic feature, as in the case of pectinate chaetae in *Marphysa* species [131,132]. Compound chaetae are present in all species at least in part of the body with the exception of some species of *Marphysa*, which lack them completely. Only some species of *Euniphysa* bear pseudocompound chaetae in the anterior chaetigers.

Challenges in understanding the morphology of species are the difficulty in sampling complete specimens and the poor knowledge on intraspecific variation, known only for few species (e.g., *Leodice rubra* (Grube, 1856) in [133]; *Leodice vittata* (Chiaje, 1929) in [36,134]). Intraspecific variation, sometimes due to size variation, occurs at least in features such as the number of articles in the prostomial appendages, the presence/absence and distribution of subacicular hooks, the number of branchial filaments, the branchial distribution, and the end of inflated base of ventral cirri (e.g., [47,134–137]). In incomplete specimens, size is commonly estimated as length through chaetiger 10, and width at this chaetiger, however, at least in a few species the peristomium width is a better estimator [138]. An index of the product of the length through chaetiger 10 by the largest width at this region has also been used a size estimator [13]. Detailed descriptions of general morphological features are available in Ehlers [3], Treadwell [139], Fauchald [36], Zanol et al. ([11] in supplementary material, [121]), Carrera-Parra [140], Zanol and Budaeva [141], of branchial distribution patterns in Miura [135] and of maxillary features in Paxton [14], Molina-Acevedo and Carrera-Parra [122] and Molina-Acevedo [128].

3.3. Feeding and Life History

Eunicids have been reported as macrophagous or microphagous omnivorous, however, some species have more restricted diets, being herbivorous or carnivorous [34]. At least one *Eunice* species is an ambush predator. It captures the preys with the maxillae, which are kept open near the substrate while waiting for the prey [142].

All the species studied have separate sexes and sexual reproduction. Sexual dimorphism of unripe individuals, hermaphroditism and asexual reproduction have not been described for the family. The reproduction biology is known for less than 10% of the species [141]. Thus, it might be more variable than our current knowledge. Free spawning by benthic or epitokous forms followed by fertilization and larval development in the water column is the most common mode of reproduction. Some epitokous forms swarm. *Palola* swarmings in the South Pacific are the most well-known because they have cultural importance to local communities [60,143,144]. Exceptions to fertilizations in the water column happen in some *Leodice*, which fertilize eggs within their burrows, and estuarine *Marphysa* species, which produce jelly egg masses attached to the substrate [145–151].

All clearly described larvae are lecitotrophic and oocyte diameters also suggest lecitotrophic development [50]. Juvenile settlement may take 10–20 days from the beginning of the development [149–152], and eunicids are expected to have poor natural dispersal capabilities. However, they may disperse via rafting in drift plastic and wood (e.g., *Eunice* and *Marphysa;* [153,154]) and in association with oysters in human-mediated transport (e.g., *Marphysa victori* Lavesque, Daffe, Bonifácio and Hutchings, 2017; [47]). Commercial oysters can also host species of *Leodice*, *Lysidice*, *Nicidion* and *Marphysa* as epibionts [155].

3.4. Habitats and Ecology

Species of eunicids are present in soft and hard substrates in estuarine and marine habitats. They also occur associated with biogenic substrates, such as coralline algae, cnidarians, oyster shells, seagrass and sponges [139,155–158], and authigenic carbonate crusts formed in cold seeps [159]. Species live in burrows or tubes; some crawl outside, usually at night. Tube building appears to be restricted to some *Eunice* and *Leodice* species. They are most commonly composed of parchment-like material—soft or stiff. Some tube building species are symbionts with cnidarian species such as *Lophelia pertusa* (Linnaeus, 1758), *Madrepora oculata* Linnaeus, 1758 and *Epizoanthus* spp [160–162]. Species associated with corals act in their bioerosion, in assembling coral reefs and in determining the reef framework through tube building, which facilitates coral growth [162–164]. The mean density of eunicids in dead coral may be as high as 315 ind/m^2, while a single dominant species such as *Nicidion cariboea* (Grube, 1856) may reach a density of 1219 ind/m^2 [165]

3.5. Distribution Patterns and Diversity

Eunicids inhabit all oceans from the intertidal zone to around 3500 m in the deep-sea, being more common and diverse in shallow tropical and subtropical hard substrates. In the Antarctic and Artic regions, a maximum of four species has been recorded and only *Eunice* and *Leodice* are present [13,166,167]. Most genera are widespread; the exceptions are *Aciculomarphysa*, *Fauchaldius* and *Euniphysa*. *Aciculomarphysa* and *Fauchaldius* have not been recorded since their original description, thus their distribution is restricted to the depths they have been described from, respectively, 460–490 m and 150–250 m [120,130,168]. *Euniphysa* is mainly restricted to the shallow warm waters of the Northern Hemisphere, with most of the richness concentrated in the South China Sea and adjacencies [123]. The only known records for the South Atlantic Ocean are from the south and northeast of Brazil ([169], personal observation), which suggests that *Euniphysa* is widespread but not abundant.

Some local fauna studies focusing on eunicid taxa are available for the Antarctic and Subantarctic Seas [13], Argentina [170], Australia [24,131,171–175], Brazil [133,176–180], British Isles [181], the Canary Islands [182], the Caribbean Sea [25,42,122,124,130,136,139, 183–186], China [187,188], the eastern Pacific [28], the El Salvador Pacific Coast [189], Fiji and Samoa [190], the French Atlantic Coast [191], the Gulf of Mexico [192,193], Hong Kong [194], the Iberian Peninsula [132,195], India [196,197], Japan [135,198–202], Korea [203], Malaysia [204], the Mediterranean Sea [205–210], the Mexican Pacific Coast [103, 211,212], New Zealand [213], the Philippines [214,215], the Scandinavian and North Sea Coasts [167], South Africa [126,216,217], Taiwan [218,219], the USA's Atlantic Coast [148, 220,221], the USA's Pacific Coast [222], and Venezuela [223,224]. Carrera-Parra [140] includes a taxonomic key for species registered to the warm temperate Northeast Pacific, Tropical Eastern Pacific Ocean and Tropical Northwestern Atlantic Ocean. The eunicid fauna are most poorly known for the Southeast Atlantic, Southeast Pacific and Indian Ocean.

These studies on local fauna are initial guides to the eunicid diversity. However, some of the species included should be considered with care. Several species thought to be widespread at the time of publications are currently known [e.g., *Lysidice ninetta* Audouin and Milne Edwards, 1833a [124,225], *Marphysa sanguinea* (Montagu, 1813) [45], *Palola viridis* Gray in Stair, 1847 [116]] or suspected to have restricted distribution (e.g., *Eunice afra*

Peters, 1854, *Eunice aphroditois*, *Leodice antennata* Lamarck, 1818, *Leodice pennata* (Muller, 1776), *Leodice vittata*, *Lysidice collaris* Grube, 1870, *Lysidice hebes* (Verrill, 1900), *Lysidice unicornis* (Grube, 1870), *Nicidion cariboea*, *Palola siciliensis* (Grube, 1840), *Paucibranchia bellii* (Audouin and Milne Edwards, 1833)). *Eunice aphroditois* had several large (giant) species considered to be its junior synonyms, leading to the identification of most giant specimens as *E. aphroditois* [184]. The identity of most large specimens remains to be solved, as well as the diversity and evolution of giant species in the family [184,226].

The widespread status of most eunicid species is usually due to poor original descriptions, unjustified generalized synonymizations, and the use of local fauna monographs in the identification of the fauna of distinct regions as described for other polychaete groups [227]. *Palola* sp. (in [49,116,174]) and *Marphysa victori* are the only species with widespread distribution confirmed by both molecular and morphological data. *Palola* sp. is present in both Micronesia and the eastern Pacific and its means of dispersion are not fully understood [49,116,174]. Meanwhile, *M. victori* is an exotic species in France introduced in association with commercial oysters imported from Japan [47].

Taxonomic revisions of species based on type material (e.g., [36,123,127–129]) as well as designations of neotypes (e.g., *M. sanguinea* in [45]) provided the initial framework to improve the knowledge on the diversity (Figure 2) and distribution of eunicid species and to question records of species reported all over the world. Furthermore, more detailed examination of morphological features, such as jaws (e.g., [25,128]), chaetae variation along the body (e.g., [24,131]), and molecular data improved findings of misidentifications and morphologically similar species. These provide tools and encourage other authors to handle similar taxonomic issues generating a virtuous cycle as observed in *Marphysa*, eunicid genus with the highest number of species described since 2003 when the *M. sanguinea* neotype was designated in Hutchings and Karageorgopoulos [45].

Lewis and Karageorgopoulos [217] were the first to follow the lead to tackle the widespread issue in *Marphysa sanguinea* combining morphological and molecular data. They revealed high molecular diversity among species traditionally identified as *M. sanguinea*. The same pattern has also been found in *Palola* and *Lysidice*. *Palola* includes several deeply separated molecular lineages of mostly morphologically indistinct species and *Palola viridis* appears to have more restricted distribution than its records in the literature [49,116,174]. In *Lysidice*, molecular analyses [225,228] supported previously reported morphological differences between *Lysidice collaris* and *Lysidice ninetta* [229]. These are the only eunicid species subject to phylogeographic and population studies and depict different evolutionary histories. *L. collaris* has panmitic populations, while *L. ninetta* includes two genetic lineages corresponding to different color morphs separated at around 3.7–3.4 million years ago, which may be different species [225,228]. Molecular analyses using nucleotide sequences of cytochrome oxidase I (COI) and 16S rDNA have become common in papers on eunicid species [202,230]. These have aided in understanding intraspecific variation, such as the lack of a subacicular hook in large specimens of some *Marphysa* species [47], in improving species recognition and in revealing a hidden biodiversity even in well-studied places.

3.6. Exploitation

Species of *Eunice* and *Marphysa* are commonly used as bait worms. However, the species identity of many of these has just become clear in the last few years [24,132]. The clear understanding of this diversity is essential to avoid the overexploitation of species as well as the introduction of exotic species due to commercialization or the establishment of farming [24].

4. Hartmaniellidae Imajima 1977

4.1. Systematics and Phylogeny

Hartmaniellidae has only one genus and three species. The first species, *Hartmaniella erecta* Imajima, 1977, was described from Suruga Bay, Japan. A year later, Amoureux [168] described a new genus and species from Tuléar, Madagascar, *Pseudoninoe tulearensis*, which he placed within Lumbrineridae. However, as was suspected by Pettibone [231] and Orensanz [13], this species belongs to *Hartmaniella* [232]. The third species, *Hartmaniella fujianensis* He and Wu, 1986, was described from the Taiwan Strait. Three other potential new species were described but not formally named from the Gulf of Mexico [232,233], the Andaman Sea [234], and Natuna Island, Indonesia [235].

Hartmaniellids superficially resemble Orbiniidae by having parapodia displaced dorsally (Figure 6A,C). However, they have well-developed maxillae and mandibles (Figure 3A) which unambiguously place them within Eunicida. However, several authors have debated on the classification of hartmaniellid maxillae. They have diverging views on the maxillary apparatus symmetry, homology among maxillary plates, carrier connection to maxillae I, and, thus, on hartmaniellids' closest families. Imajima [236] mistakenly described a symmetric maxillary apparatus with four plates on both sides and illustrated the carriers as being poorly connected to maxillae I. Following Imajima's description, Orensanz [13] suggested Hartmaniellidae being closely related to the fossils *Synclinophora synclinalis* Eisenack, 1975 and the *Delosites*-like scolecodonts, and with extant Eunicidae and Onuphidae. Later, based on the poor connection between the carriers and maxillae I illustrated in Imajima's description, Fauchald and Rouse [237] considered the hartmaniellids maxillary apparatus more similar to the prionognath type and the family sister group of Oenonidae and Lumbrineridae. However, illustrations of *Hartmaniella erecta* by Szaniawski and Imajima [238] and the descriptions and illustrations of *Hartmaniella tuleransis* and *Hartmaniella* sp. by Carrera-Parra [232] show that the carriers are completely attached to the base of maxillae I. Szaniawski and Imajima [238] established a remarkable similarity of *Hartmaniella* jaws with fossil of *Palurites* Kozur, 1967 by the peculiar shape of maxillae I and II. They also proposed a close relationship between Hartmaniellidae extant Lumbrineridae and extinct Kielanoprionidae (asymmetric jaws), considering that the differences in structure (symmetric vs. asymmetric) of their maxillary apparatus are diagnostic at the family level. Eibye-Jacobsen and Oug [234] described an asymmetric maxillary apparatus in specimens from the Andaman Sea, and reported the right maxillae III and IV to be fused in contrast to the left separate plates. They hypothesized that asymmetry in maxillae could have evolved independently several times within Eunicida and supported the close relationships of hartmaniellids with Oenonidae or Lumbrineridae rather than with Eunicidae and Onuphidae. Carrera-Parra [232] stated that hartmaniellids have an asymmetrical maxillary apparatus with four maxillae on the left side and only three on the right side, reinforcing the earlier idea of close relationship with the asymmetrical fossils of Kielanoprionidae and with Eunicidae/Onuphidae. The latter was also supported by the presence of dorsal and ventral cirri in the three families. Lu [239], likewise based on the revision of Japanese specimens, described an asymmetric maxillary apparatus for *H. erecta*. Paxton [14] considered hartmaniellid jaws "subsymmetrical" and closest to the extant Lumbrineridae. For both families she proposed a new type of maxillae—symmetrognatha; however, the maxillary apparatus of hartmaniellids is asymmetric with four maxillae on the left side and only three on the right side [232,234,239] (Figure 3A).

A combined analysis of morphological characteristics and molecular data will help to clarify the placement of hartmaniellids within Eunicida, which, to date, has been hampered by the extremely scarce material and the lack of specimens suitable for molecular analyses.

No taxonomic revisions, key and phylogenetic analyses are available for the family.

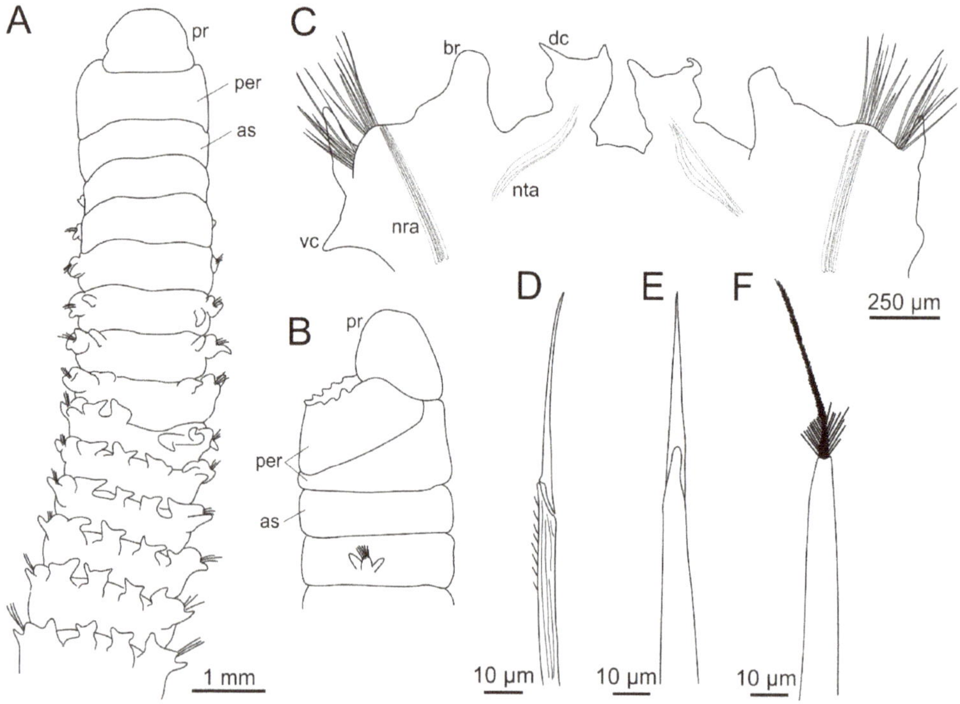

Figure 6. Morphology of Hartmaniellidae. (**A**) *Hartmaniella* sp., anterior end, dorsal view; (**B**) the same, anterior end, lateral view; (**C**) the same, parapodia from midbody region shifted dorsally; (**D**) *Hartmaniella tulearensis* (Amoureux, 1978), composite spiniger; (**E**) the same, acicular spine; (**F**) the same, acicula. as–achaetous segment, br–branchia, dc–dorsal cirrus, nra–neuropodal aciculae, nta–notopodial aciculae, per–peristomium, pr–prostomium, vc–ventral cirrus. (**B**), (**D**–**F**) modified from Carrera-Parra [232].

4.2. Morphology

Hartmaniellids are small worms up to 2.5 cm long; however, most of the few collected specimens are incomplete. The prostomium lacks eyes and appendages (Figure 6A). The peristomium is laterally projected over the prostomium, the first ring is dorsally reduced, and the second one is longer dorsally than ventrally (Figure 6B). The absence or presence of papillae in the peristomium may aid to species identification. The first segment is apodous. The body is divided into three regions by the morphology of parapodia. In the first region, the parapodia are lateral; the second region is characterized by having parapodia displaced over the dorsum with bifid dorsal cirri and foliose parapodial branchiae dorsal to prechaetal lobes; while in the third region, the dorsal cirri are digitiform, basally swollen and the branchiae are tapering. The relative length of the body regions might be of taxonomical value at species level; however, it should be taken with caution, as these characteristics may be size-dependent. From chaetiger 6–7, all segments are biannulated. The chaetae of hartmaniellids are limbate with a serrated margin, compound spinigers with distally spinulose shaft, internally with granulated base becoming canaliculated from the median to the distal end; acicular spines with a long mucro appear from about chaetigers 6 (Figure 6D,E). The aciculae are aristate, subdistally hirsute with a long mucro finely hairy (Figure 6F). The number of chaetae and aciculae per parapodium may be of diagnostic value at species level.

Hartmaniellids have an asymmetric maxillary apparatus with four maxillae on the left side and three on the right (Figure 3A). Maxillae I are mainly characterized by long internal teeth and a lateral spur located on the outer posterior margin, while maxillae II are characterized by being a narrow plate with a lateral projection on the anterior outer edge. The number of teeth in maxillae I and maxillae II are important to distinguish the species.

4.3. Feeding and Life History

Jumars et al. [34] suggested hartmaniellids could be predators of meiofauna based on their size and the presence of maxillary apparatus. The only information about reproduction in hartmaniellids is the presence of deep yellow oocytes with large amounts of yolk, 150–200 μm in diameter in median parapodia, suggesting the development via lecithotrophic larvae [234].

4.4. Habitat, Distribution and Diversity

Hartmaniellids have been found in clayey and sandy silt, soft mud, sandy mud, and sandy mud with shell fragments, from depths of 40 to 210 m in six regions of the Pacific Ocean (Japan, China, Indonesia, Thailand, Madagascar, and La Réunion), while in the Atlantic Ocean, they have only been reported in the Gulf of Mexico in clayey and sandy silt at depths of 175 to 189 m. They are considered as being "living fossils", abundant during the Mesozoic, but currently poorly represented [238].

There is no information about the ecology of the family.

5. Histriobdellidae Claus and Moquin-Tandon, 1884

5.1. Systematics and Phylogeny

The family Histriobdellidae is composed of highly specialized ectosymbionts known as "clown worms". The prefix "histrio" (from the Etruscan word "hister"), which means actor or dancer, alludes to the performative way in which these worms move, quite similar to the actor Charlie Chaplin, in his character "Carlitos". The species names *Stratiodrilus arreliai* Amaral and Morgado, 1997 and *Stratiodrilus circensis* Steiner and Amaral, 1999 also refer to the circus and its artists.

Histriobdellidae was first included in Hirudinea, which justifies the suffix "bdella" [240]. It has also been included among Rotifera and Archiannelida [241–244]. Mesnil and Caullery [245] have first suggested that maxillae of histriobdellids were homologous to other Eunicida, as a prionognath type. Tzetlin [246] considered the histriobdellid maxillae as of the ctenognath type. His views were recently corroborated by an extensive study of jaw morphology and ultrastructure in *Histriobdella homari* Van Beneden, 1858 showing the ultrastructure similarity of the maxillae with Dorvilleidae [247]. The inclusion of the family in Eunicida was followed by latter studies that considered histriobdellids as highly specialized commensals, lacking external resemblance to the traditional Errantia body plan [13,237,248–250].

This epizoic group is currently composed of three genera and 13 species that live within the branchial chambers (on branchiae, epipodites, or inner surfaces of carapace), in the egg masses, or, occasionally, on the general body surface (head, thorax, abdomen, or appendages) of marine and freshwater isopod and decapod crustaceans. *Histriobdella* Van Beneden, 1858 and *Steineridrilus* Zhang, 2014 are monospecific and exclusively associated with marine hosts, while *Stratiodrilus* Haswell, 1900 includes 11 species associated with freshwater hosts.

There is no phylogenetic hypothesis that deals with relationships between the genera and species within the family, only hypotheses regarding their marine or freshwater origin [251–255]. To date, most taxonomic and systematic studies on histriobdellids were based on external morphology with no molecular data available. A single study using SEM, TEM and confocal laser scanning microscopy described details of the jaw apparatus [247]. The only review was carried out by Steiner and Amaral [256] who proposed a nomenclature for some morphological structures, redescribed all species based on literature and proposed a new genus, *Dayus*, which, being preoccupied, was renamed for *Steineridrilus* [257]. Rosa et al. [255] updated the list of *Stratiodrilus* hosts based on published data and specialized databases discussing its distribution. An identification key for all described *Stratiodrilus* was published by Amato et al. [258].

5.2. Morphology

The worm-like translucent and delicate body is small, with the maximum length of adults varying between 0.5 to 1.5 mm, indistinctly and irregularly annulated, with constrictions more visible laterally, and divided into head, trunk and posterior region (Figures 1F and 7A,B,D,E). Histriobdellids have no true segments, parapodia, chaetae or aciculae, but the ventral dark sclerotized jaws can be easily seen by transparency (Figure 7D–G). Despite the absence of metamerism, internally histriobdellids have 9 to 11 ganglia (depending on the genus) in the ventral nerve cord [248,259].

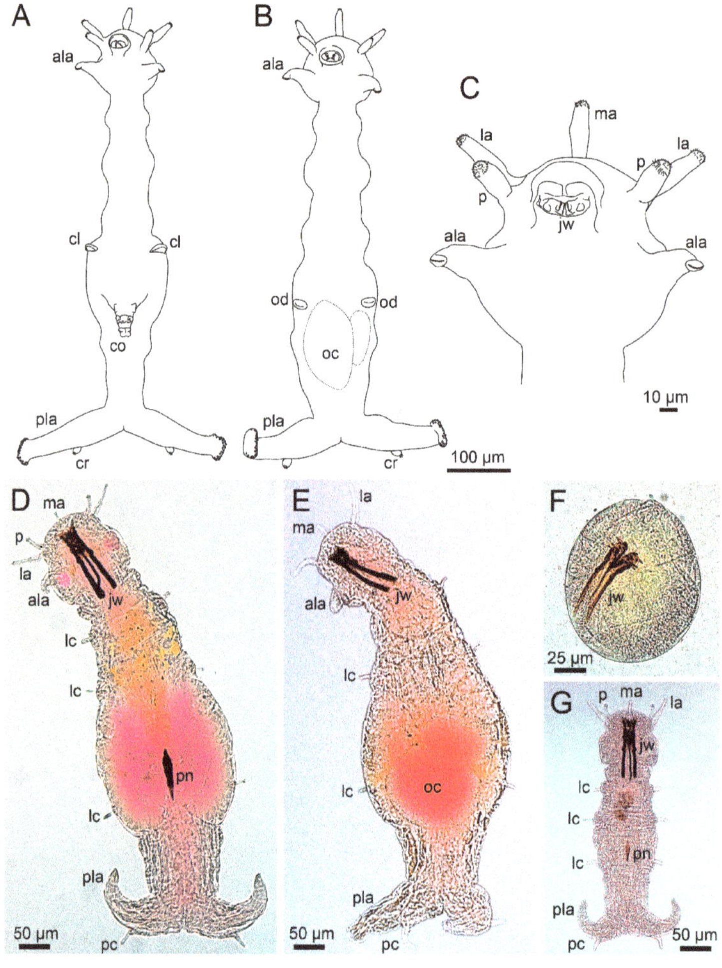

Figure 7. Morphology of Histriobdellidae. (**A**) *Histriobdella homari*, male, ventral view; (**B**) the same, female, ventral view; (**C**) the same, anterior end, ventral view; (**D**) *Stratiodrilus arreliai* Amaral and Morgado, 1997, male, ventral view; (**E**) the same, female, ventral view; (**F**) the same, early embryo; (**G**) the same, juvenile, ventral view. ala–anterior locomotory appendage, cl–clasper opening, co–copulatory organ, pc–posterior cirrus, jw–jaws, la–lateral antenna, lc–lateral cirrus, ma–median antenna, oc–oocyte, od–oviduct opening, p–palp, pla–posterior locomotory appendage, pn–penis. (**D–G**), Copyright © 2016, Editora Guanabara Koogan (Fransozo, A. & Negreiros-Fransozo, M.L., Zoologia dos Invertebrados, 1st edition, page 366), reproduced with permission of the publisher.

The head is composed of prostomium fused to peristomium, and the trunk has five segments, but only the second, third and fifth may have lateral cirri, which are probably homologous to dorsal and/or ventral cirri of other Eunicida [260] (Figure 7A–E). Nuchal organs were reported in some species [242,259]. The prostomium bears three antennae dorsally and two sensory palps frontally (Figure 7C,D). One pair of locomotory appendages with adhesive glands on the distal end is located on the lateral of the head, probably the peristomium (Figure 7A–E), retractable in most species. In the fourth trunk segment, males have one pair of lateral retractable claspers for holding females during copulation, and a penis on the ventral side of body (Figure 7A,D,G). The posterior region of the body, after the fifth segment, is composed of several fused segments without cirri or any kind of appendages and two lateral lobes, the posterior locomotory appendages, with the anus located between them (Figure 7A,B,D,E,G). These appendages have a duo-gland adhesive system [261] on their distal ends, in addition to cirri, lobes, and/or tubercles. The jaw apparatus is very complex, enclosed in the ventral pharyngeal organ. The paired mandibles are fused along the mandibular plates bearing about 10 denticles and attached to each other in the median part of the mandibular shafts, which are flattened and elongated. The maxillae consist of a single dorsal rod, a massive ventral carrier and four pairs of dentate plates–maxillae I–IV [247,249] (Figure 3E).

All known species are dioecious and sexually dimorphic when mature, one of the few examples in polychaetes. The chitinous black penis of *Stratiodrilus* is an eversible spine with a bevelled tip [260] like a hypodermic needle [262]. In *Histriobdella*, the penis is composed of two hard structures with a central duct [260], but not chinitized. Detailed descriptions of general morphological features are available in Shearer [242], Haswell [262,263], Roubaud [244], Jennings and Gelder [249], and Steiner and Amaral [256].

Histriobdella is characterized mainly by the presence of short and papillae-like antennae and palps and the absence of lateral cirri on trunk (Figure 7A–C). *Steineridrilus* and *Stratiodrilus* have lateral cirri and longer and narrower antennae and palps. In *Steineridrilus* the three antennae are biarticulated, while in *Stratiodrilus* only the lateral antennae bear two articles (Figure 7D,E). The shape of the lateral cirri and the presence or absence of cirri, lobes and tubercles in posterior locomotory appendages determine *Stratiodrilus* species, especially those from the Neotropical region. Studies that address intraspecific variations are only known for *Stratiodrilus circensis* [264] that has been widely recorded in the Neotropical region. These worms are relatively easy to collect, just by finding their hosts, which carry in their body, especially in the branchial chamber, a good amount of complete specimens.

The morphological adaptations to the commensal life habit are diverse, when compared to the Eunicida pattern. They range from the reduction in body size, absence of metamerism, parapodia, chaetae and aciculae, to the presence of a duo-gland adhesive system in the posterior region, which allows its attachment to different parts of the host's body. Sexual dimorphism and the penis also appear to be adaptations to the specific habitat of these worms.

5.3. Feeding and Life History

Histriobdellids find shelter in the host and feed on the microflora (bacteria, cyanophytic algae, diatoms, among others) that grows abundantly in the chamber and branchial filaments, bringing the benefit of always keeping these areas clean and free, so that gas exchange is carried out more efficiently. Jennings and Gelder [249] found that the relationship between *Histriobdella homari* and its crustacean host has a firm nutritional basis.

Regarding reproduction, the male hypodermically inserts its penis into the female and the sperm goes directly into the body cavity [265]. Whitish eggs in different stages of maturation are visible in the female's transparent body (Figure 7E). In all known species, there is a suppression of the larval stage, with egg-laying occurring in branchial chamber or in egg masses of the female host. The development, therefore, is direct, and immature miniatures of adult worms hatch from the eggs. Histriobdellid females attach their eggs to the host (branchial chamber, between eggs of host female, pleopods and uropods), which allows the young worms to continue their life cycle on the host. Shearer [242] and Simon [266] stated that *Histriobdella* can migrate from the branchial chamber to the egg mass of female host and, after attaching the eggs, return to chamber. *Histriobdella homari* also has the ability to migrate from one host to another [266].

5.4. Habitats and Ecology

Histriobdellids have been treated as parasites [267,268] or commensals [260,269]. According to Brattey and Campbell [270], in most infestations *Histriobdella homari* benefits the lobster. However, if the number of individuals is very high, the benefit of microfloral grazing may decrease, as worms can reduce the oxygen flow on the branchiae, also reducing success in hatching host's eggs. On the other hand, Lerch and Uglem [271] reported success in the hatching of female hosts' eggs, even with high numbers of *H. homari* in the branchial chamber. According to Martin and Britayev [269], a symbiotic species that completes the entire life cycle in a single host has a monoxenous pattern, and they consider that *H. homari* behaves more as a cleaning symbiont rather than as a parasite. Jennings and Gelder [249] classify the species as epizoic microphagous cleaning symbiont.

Studies on species biology are still few; the mechanisms of dispersion and selectivity in host choice are still unknown, as well as the real nature or extent of the association with their hosts.

5.5. Distribution Patterns and Diversity

The differentiation between the histriobdellid genera are also reflected in their geographical distribution. The first described genus, *Histriobdella*, occurs exclusively on lobsters (*Homarus*, *Nephrops*) in the Northern Hemisphere, on both sides of the Atlantic Ocean, in Europe (Norway, Ireland, England, the Netherlands, the Irish Sea, the Clyde Sea), Canada, and the United States (mid-Atlantic Bight). *Steineridrilus* is the marine group from the Southern Hemisphere, having an intertidal isopod (*Cirolana*) as its host. It is only known from South Africa, where it occurs from the Lagebaan Lagoon to East London [272]. *Stratiodrilus* is the most speciose genus and encompasses species living on freshwater decapods from the Southern Hemisphere. It has been identified on different hosts (*Astacopsis*, *Astacoides*, *Aegla*, *Cherax*, *Parastacus*, *Samastacus*, and *Trichodactylus*) that occur in diverse freshwater environments, such as streams, brooks, flowing rivers, lagoons and caves from Australia (including Tasmania), Madagascar, Chile, Argentina, Uruguay, and Brazil.

The distribution of histriobdellids reveals interesting information that contributes to the continental drift theory, with *Histriobdella* having a Laurasian distribution, while *Stratiodrilus* and *Steineridrilus* show a Gondwanan distribution, occupying the Neotropical, Afrotropical and Australasian zoogeographic regions [254,255]. Harrison [251] suggested that *Stratiodrilus* distribution demands former land connections between Australia, Madagascar and South America, and proposed a marine ancestor for the family. The co-occurrence of *Stratiodrilus* in parastacid crayfish suggests a single colonization in continental waters, prior to the breakup of Gondwana [254,260]. Thus, *Stratiodrilus* appears to have successfully invaded continental waters either in Palaeozoic or in Mesozoic [254]. Jennings and Gelder [249] suggested that the suppression of a free-swimming larval stage is an adaptation to the freshwater environment and that *Histriobdella* and *Steineridrilus* have become secondarily adapted to a marine habitat. On the other hand, one of the strategies of some commensal species is simplification of their life cycles, either reducing or eliminating the free-living larval stage, with the adult phase being responsible for dispersion [269]. Thus, the direct development could be an adaptation to the symbiotic lifestyle and not to the freshwater environment, which would support the marine origin theory for the family.

Up to now, 44 species of nine genera of four Decapoda families have been found to host *Histriobdella* and *Stratiodrilus* species. Only one species of Isopoda hosts *Steineridrilus*, which is known only from its original description [272], including some biological information, in addition to a photographic record made by Rouse [268], from an unknown locality. Some histriobdellids do not demonstrate specificity to a single host. *Histriobdella homari*, for example, is found in three different species of lobsters of the genera *Nephrops* and *Homarus*, on both sides of the Atlantic. *Stratiodrilus circensis* is also widely distributed, occurring in at least 10 species of Aeglidae and in some Parastacidae, in Chile, Argentina and Brazil [255]. Crabs from the genus *Aegla*, the hosts for several *Stratiodrilus* species, are widespread in the South America with 72 known species [273] of which only 20 have been studied for the histriobdellid symbionts [255]. Furthermore, other genera and species of decapods and isopods can potentially host histriobdellids. Thus, the diversity of Histriobdellidae could be highly underestimated due to the absence of systematic study covering the wide diversity of potential hosts.

It is not known how common cryptic speciation is within Histriobdellidae. Rosa et al. [255] proposed studies using molecular tools to clarify the taxonomic status of *S. circensis*, which may represent a complex of cryptic species, considering its wide distribution in four countries. Briggs et al. [274] also considered the possibility that *H. homari* from two localities of Europe are different species or subspecies. Considering that the species is reported from both sides of the Atlantic, it may represent a species complex.

Stratiodrilus gondwanian distribution suggests it may have speciated with host parastacid crayfish, which have similar distribution. However, despite early suggestions of this relationship [251,254,255] and of the connection between diversity of histriobdellids and their hosts, no study has tested these hypotheses yet. Biogeographic studies incorporating molecular analyses are necessary to study the coevolution of hosts and histriobdellid ectocommensals.

Histriobdellidae is a relatively poorly studied family with knowledge gaps in practically all areas, despite being relatively easy to collect and to observe. Until 1962, studies were limited to descriptions of new species based on external morphology [240,243,251,262, 263] and investigations of the internal anatomy [241,242,244,245,259,275]. More recently, the biology and behavior of several histriobdellid species [249,258,264–266,276,277], as well as relationships with their hosts and other ectosymbionts [252,271,278] received more attention. Few studies have been carried out on the ecology of the family [249,266,267,270, 271,278,279] as well as species distribution records [255,280,281]. Studies that focus on the life cycle, molecular techniques, ecology, and taxonomy will certainly help in reconstructing phylogenetic relationships among the species of the family, as well as their relationships with their hosts.

6. Lumbrineridae Schmarda, 1861

6.1. Systematics and Phylogeny

Lumbrineridae comprises 19 genera and about 279 species. The first species of Lumbrineridae was described by Müller in 1776 as *Lumbricus fragilis* (now *Scoletoma fragilis*). Previously, some authors considered the group as a subfamily of Eunicidae, but today it is regarded as an independent family [282].

Lumbrinerids have a very simple body shape, with a marked reduction of external morphological characteristics (Figure 1D). Thus, classifications based on external morphology only resulted in a simplified generic system that lumped all known species in only three or four genera [283]. Hartman [28] and Fauchald [103] treated genera *Augeneria* Monro, 1930, *Cenogenus* Chamberlin, 1919, *Eranno* Kinberg, 1865, and *Scoletoma* de Blainville, 1828 as synonyms of *Lumbrineris* de Blainville, 1828. The taxonomy of this family has changed after employing characteristics of the maxillary apparatus and mandibles, which allowed the reestablishment of all above-mentioned genera [13,283–285] and the description of 13 new genera [13,284–286]. A key to genera valid at the time is available in Carrera-Parra [286] and Carrera-Parra [282].

To date, there is only one phylogenetic study, which is based on morphological data, regarding the intergeneric relationships within Lumbrineridae. The family is divided into four main clades supported by the characteristics of the maxillary apparatus [286]. This study corroborated the inclusion of *Lysarete* Kinberg, 1865 into Lumbrineridae proposed by Orensanz [13], as well as the recognition of *Augeneria*, *Cenogenus*, *Eranno*, and

Scoletoma genera previously considered as junior synonyms of *Lumbrineris*. Furthermore, there is a phylogenetic study on interspecific relationships of *Lumbricalus* Frame, 1992 [287]. Only three genera of lumbrinerids underwent complete systematic revision: *Kuwaita* Mohammad, 1973 [288], *Lumbricalus* [289], and *Lumbrineris* [290]. The rest of the genera await revisions based on detailed study of the architecture of the maxillary apparatus and, possibly, molecular information.

6.2. Morphology

Lumbrinerids are worms of small to medium size, ranging from a few to over 60 centimeters [291,292]. Their prostomium can vary from round to conical; in most lumbrinerids, it lacks appendages (Figure 8A,B), except for *Augeneria*, *Lysarete*, and *Kuwaita*, which have three antennae (Figure 8C), as well as *Cenogenus* and *Sergioneris* Carrera-Parra, 2006, bearing a single antenna. Although *Augeneria* typically has three antennae, at least two species have multiple small antennae (*Augeneria polytentaculata* Imajima and Higuchi, 1975 and *Augeneria riojai* Aguirrezabalaga and Carrera-Parra 2007). Most lumbrinerids lack eyes, except *Lysarete*, which possess two pairs (Figure 8C). A pair of ventral pads of the dorsolateral folds anterior extension (*sensu* [37,38]; buccal lips *sensu* [284,293]), that may be well-developed, is present at the ventral posterior margin of the prostomium. The peristomium is composed of two rings and lacks appendages; dorsally, it is complete (Figure 8A,B) except in *Lysarete*, which have a dorsal incision reaching the first chaetiger (Figure 8C). Ventrally, the first peristomial ring is incomplete, and the second ring is projected forward. The parapodia are subbiramous with notopodium reduced to a small knob with notoaciculae in most lumbrinerids (Figure 8A), but well-developed dorsal cirri in *Kuwaita* and *Lysarete* (Figure 8C). Neuropodia are well developed in all lumbrinerids, constituted by prechaetal, chaetal, and postchaetal lobes, without ventral cirri (Figure 8D–F). Branchiae are absent in most genera; when present, they are associated with parapodial lobes of anterior chaetigers (Figure 8D); *Cenogenus* and *Ninoe* Kinberg, 1865) or over dorsum in posterior chaetigers (*Kuwaita*). All genera have limbate chaetae. The presence of other chaetae varies among genera: simple and compound multidentate hooded hooks (Figure 8G,H) and compound spinigers (*Lumbricalus*); simple and compound multidentate hooded hooks (*Augeneria*, *Gallardoneris* Carrera-Parra, 2006, *Hilbigneris* Carrera-Parra, 2006, *Loboneris* Carrera-Parra, 2006, *Lumbrineris*, and *Sergioneris* Carrera-Parra, 2006); simple multidentate hooded hooks and limbate simple multidentate hooded hooks (*Abyssoninoe* Orensanz, 1990); simple multidentate hooded hooks (*Cenogenus*, *Eranno*, *Gesaneris* Carrera-Parra, 2006, *Helmutneris* Carrera-Parra, 2006, *Kuwaita*, *Ninoe*, and *Scoletoma*); and simple bidentate hooded hooks (*Lumbrinerides* Orensanz, 1990 and *Lumbrineriopsis* Orensanz, 1990). *Arabellonereis* Hartmann-Schröder, 1979 and *Lysarete* lack hooks. The pygidium of lumbrinerids has two pairs of pygidial cirri, except in *Gallardoneris*, *Lumbrinerides*, and *Lumbrineriopsis*, which lack pygidial cirri. Until now, the pygidium shape of *Gesaneris* is unknown.

The study of lumbrinerids requires careful observation of the maxillary apparatus since the recognition of genera is based mainly on the differences in this complex structure. The maxillary apparatus is symmetrical of labidognath type (*sensu* [286]) or symmetrognath type (*sensu* [14]); the later was suggested to distinguish the lumbrinerid symmetrical labidognath maxillae from the asymmetrical labidognath maxillae of Eunicidae and Onuphidae (Figure 3). The number of maxillary plates varies among the genera as follows: six pairs in *Lysarete*; five pairs in *Arabelloneris*, *Eranno*, *Hilbigneris*, *Kuwaita*, *Ninoe*, *Lumbricalus*, *Lumbrineris*, and *Scoletoma*; and four pairs in *Abyssoninoe*, *Augeneria* (Figure 3B), *Cenogenus*, *Gallardoneris*, *Gesaneris*, *Helmutneris*, *Loboneris*, *Lumbrinerides*, *Lumbrineriopsis*, and *Sergioneris*. All maxillae can have either an accessory lamella, a connecting plate, or, in some cases, both structures. As a consequence, the maxillary apparatus of lumbrinerids has a great variety of shapes specific to each genus [286]. The mandibles are also an important characteristic for helping to recognize some genera such as *Gallardoneris* and *Lumbrinerides*, which have completely fused mandibles. In contrast, all other genera have mandibles fused up to 3/4 of their length [286].

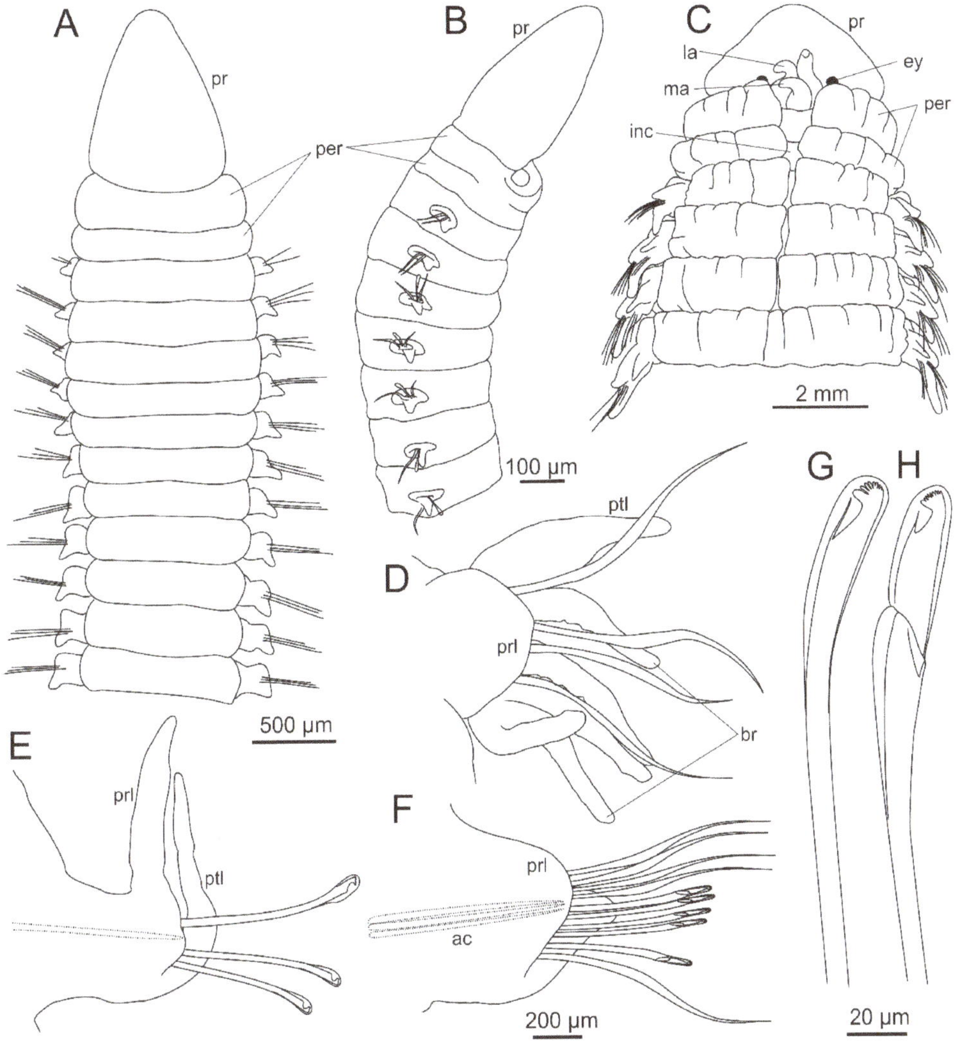

Figure 8. Morphology of Lumbrineridae. (**A**) *Lumbrineris mixochaeta* Oug, 1998, anterior end, dorsal view; (**B**) *Lumbrineriopsis* sp., anterior end, lateral view; (**C**) *Lysarete* sp., anterior end, dorsal view; (**D**) *Ninoe vargasi* Carrera-Parra, 2001, parapodium; (**E**) *Eranno lagunae* (Fauchald, 1970), parapodium; (**F**) *Lumbrineris* sp., parapodium; (**G**) *Scoletoma* sp., simple hook; (**H**) *Lumbrineris* sp., compound hook. ac–acicula, br–branchia, ey–eye, ins–dorsal incision, la–lateral antenna, ma–median antenna, per–peristomium, pr–prostomium, prl–prechaetal lobe, ptl–postchaetal lobe.

6.3. Feeding and Life History

In their feeding behavior, lumbrinerids were considered carnivores, feeding on various invertebrates, questionably herbivores [294–296] and deposit-feeders [297]. Based on the analysis of stable isotopes (δ^{15}N) in species of the genera *Hilbigneris*, *Lumbrineris*, *Lumbrinerides*, and *Scoletoma*, most lumbrinerids are carnivores with a few herbivorous [34].

There is little information on the reproductive aspects of this group. However, all species with known reproduction are gonochoric without sexual dimorphism. Mature specimens of *Lumbrineris latreilli* Audouin and Milne Edwards, 1834 have been reported swimming [298]. In the few existing studies [299–301], the morphology of the sperm is described as of the ect-aquasperm type (*sensu* [302]) suggesting broadcast spawning [84]. Osman et al. [57] described three maturity stages according to yolk deposition during the oogenesis of *Scoletoma funchalensis* (Kinberg, 1865), the first previtellogenic stage and the other two within the vitellogenic phase. Oocyte diameter is known from four species. It varies from around 100 to 300 μm in diameter, *S. funchalensis* (140 μm to 280 μm, [57]), *Gallardoneris iberica* Martins, Carrera-Parra, Quintino and Rodrigues, 2012 (129 μm to 188 μm, [303]), *Lumbrineris lusitanica* Martins, Carrera-Parra, Quintino and Rodrigues, 2012 (181 μm to 191 μm [303]) and *Lumbrineris pinaster* Martins, Carrera-Parra, Quintino and Rodrigues, 2012 (98 μm to 147 μm [303]).

Messina et al. [300] studied the first stages of the development of *Scoletoma impatiens* (Claparède, 1868). They observed the appearance of protrochophore larva in the 24 hours after fertilization. Three days later, it transforms into a metatrochophore with a duration of about seven days, in which time the first two segments are formed. By the eighth day, the postlarva benthic phase begins. Some species lack pelagic larvae; their gametes are deposited into the parental burrow or gelatinous masses attached to the substrate where the first development stages occur and benthic larvae emerge at different times, not allowing for a high dispersal capability [284,304–307].

6.4. Habitats and Ecology

Lumbrinerids are motile worms dwelling in soft sediment such as mud and sand; however, some species prefer hard substrates such as coral rocks. They can construct very fragile temporary tubes of sediment and mucus. Some species live in association with other invertebrates such as sponges, bryozoans, sea urchins [176,286,308,309]. The species *Helmutneris flabellicola* (Fage, 1936) and *Helmutneris corallicola* Carrera-Parra, 2006 are only found forming membranous transparent tubes in ahermatypic corals of the genera *Caryophyllia, Balanophyllia, Dendrophyllia,* and *Flabellum* [286,310]. Fossil corals with tube scars created by lumbrinerids were reported from the Miocene and Pliocene deposits in central Europe and in the Western Basin of the Mediterranean [310].

Very little is known about the ecology and population biology of the Lumbrineridae. It has been shown that some species of *Lumbrineris* and *Gallardoneris* prefer areas enriched with organic matter [303,311,312]. *Kuwaita hanneloreae* Arias and Carrera-Parra, 2014 was reported from a fine to medium sand with a mean organic matter content of 1.68%, reaching densities over 4–5 ind/m^2 and occurring together with other lumbrinerids, *Scoletoma impatiens* and *Lumbrineris latreilli* [313].

6.5. Distribution Patterns and Diversity

Lumbrinerids have a worldwide distribution, from intertidal to abyssal zones. They are more common in temperate and tropical regions, where all genera have been reported [282,286], while in the Antarctic and Artic regions only six genera have been found [13,290,308,314,315]. However, the taxonomic studies of lumbrinerids in the Antarctic and Arctic regions and the deep-sea habitats are scarce.

Some contributions to the regional faunas of lumbrinerid species are available for the Antarctic and Subantarctic seas [13], Argentina [284], Australia [316]; Brazil [176, 317,318], the British Isles [181]; the Canary Islands [319]; the Caribbean region [139,220, 308]; China [320–322]; Costa Rica [323]; the Eastern Pacific [28]; East Africa and the Red Sea [324]; El Salvador [325]; Fiji and Samoa [190]; the Gulf of Mexico [308,326–329]; the Iberian Peninsula [303]; India [330]; Japan [290,331–333]; the Mediterranean Sea [4,334–337]; New Zealand [338]; South California [291]; Thailand [286,339]; the USA East coast [285]; northern Europe [313,335,340,341]; western Mexico [94,103,290,342]; the Scandinavian and

North Sea coasts [315,343–346]; South Africa [216]; the southeastern Pacific [290,347–350]; Vietnam [286,351]; and West Africa [289,352–354].

Due to recent changes in the group's taxonomy, it is necessary to re-evaluate the lumbrinerid species worldwide. A number of new genera or species have been described, or other species have been re-established from their supposed synonyms even in well-studied areas [283,286,290,303,336,342]. Several lumbrinerid species have been regarded as having wide distribution (v. gr. *Scoletoma tetraura* (Schmarda, 1861), *Scoletoma impatiens*, *Lumbrineris latreilli*, *Lumbrineris inflata* Moore, 1911), most probably as a result of brief original descriptions and unjustified synonymies [283,290]. The revision of type materials with a detailed examination of the maxillary apparatus, chaetal morphology and shape of parapodia along the body is imperative to clarify these cases. Recently, some species have been re-established, and others were described as new species with restricted distributions [283,290,303,321,336].

The use of combined morphological and molecular data may help to improve the identification. Until now, only an Antarctic species identified as *Lumbrineris kerguelensis-cingulata* has been the focus of a molecular analysis [355]. The examined specimens exhibit a high range of genetic distance among 16S rDNA sequences (0–16%), suggesting the presence of more than one species. However, the poor morphological condition of the specimens and the lack of additional genetic markers in the analyses did not allow further conclusions and formal description of the obtained clades.

6.6. Exploitation

Lumbrinerids are used as bait for recreational or subsistence fishing. In the Suez Canal, *Scoletoma funchalensis* is under intensive harvesting due to its high price and local demand [57]. In the Gulf of Naples, *Scoletoma impatiens* is harvested from shallow sandy areas: this species is locally named "Esca rossa" or "Tremolina" [54]. In southwest Spain (Cádiz), sold specimens of *S. impatiens* are harvested in northwest Spain [356]; in the Canary Islands, *Lumbrineris latreilli* is a highly appreciated bait used by experienced fishers [357]. In northern Spain, *Kuwaita hanneloreae*, *S. impatiens*, and *L. latreilli* are highly appreciated as fishing bait and constitute an important natural resource from this area [313]; however, dermatitis problems have been documented when handling the latter species [358]. In Japan, *Kuwaita heteropoda* (Marenzeller, 1879) is commonly used as bait by Japanese fishers who observed that certain carnivorous insects die when they settle upon the worm. The insect death occurs by having contact with a neurotoxin called nereistoxin that is localized in the integument of the lumbrinerid [359].

7. Oenonidae Kinberg, 1865

7.1. Systematics and Phylogeny

Oenonidae comprises 12 genera and about 109 species. The first species of oenonids was described in 1804 by Montagu as *Nereis iricolor* (now *Arabella iricolor*) [360]. To date since the original family description, Oenonidae has undergone many changes in its composition, giving rise to a series of modifications and synonyms at the family level. Kinberg [361] proposed the family including four genera (*Aglaura* Savigny *in* Lamarck, 1818, *Andromache* Kinberg, 1865, *Danymene* Kinberg, 1865, and *Oenone* Savigny *in* Lamarck, 1818). Hartman [28] considered Oenonidae the synonym of Lysaretidae, another family proposed by Kinberg [361], which is currently included into Lumbrineridae. However, Colbath [362], based on the morphology and mineral composition of the maxillary apparatus, recognized Oenonidae as including the genera *Oenone*, *Halla* Costa, 1844, and *Tainoka* Knox and Green, 1972. One of the most notable changes in the delimitation of Oenonidae was the inclusion of the family Arabellidae Hartman, 1944 as its junior synonym, proposed by Orensanz [13]. Therefore, the family was expanded to 11 genera (*Arabella* Grube, 1850, *Biborin* Chamberlin, 1919, *Drilognathus* Day, 1960, *Drilonereis* Claparède, 1870, *Haematocleptes* Wirén, 1886, *Halla*, *Labrorostratus* de Saint-Joseph, 1888, *Notocirrus* Schmarda, 1861, *Oenone*, *Oligognathus* Spengel, 1882, and *Tainoka*). Two years later, Dean [363] proposed one more genus, *Pholadiphila*, bringing the total number of currently accepted genera to 12. A key to valid genera is available in Carrera-Parra [364].

Two genera, *Arabella* and *Drilonereis*, comprise 76% of the oenonid species; three genera are monospecific (*Biborin*, *Drilognathus*, and *Pholadiphila*), all parasites. Five genera have two species (*Haematocleptes*, *Halla*, *Oenone*, *Oligognathus*, and *Tainokia*), but *Oenone* includes another six species regarded as junior synonyms of the supposed cosmopolitan species *Oenone fulgida* Savigny in Lamarck, 1818 (Figure 1C). *Labrorostratus* and *Notocirrus* have six and nine species, respectively.

Phylogenetic studies based on molecular data supported the inclusion of Arabellidae into Oenonidae, finding a monophyletic group containing *Arabella*, *Drilonereis*, and *Oenone* [8]. No phylogenetic analyses focusing on the family have been carried yet.

7.2. Morphology

Oenonids are worms of variable length ranging from few centimeters to almost a meter—all of them are usually very thin. The prostomium usually lacks appendages (Figure 9A,B), except in *Tainokia*, which has one antenna, and *Halla* and *Oenone* with three antennae (Figure 9G). The peristomium is entire (Figure 9A,B,G) but can be dorsally incomplete with a median anterior incision poorly developed as in *Tainokia* or well-developed as in *Halla*. Usually, the parapodia have very small globular dorsal cirri (Figure 9C,D), but foliaceous in *Oenone* and *Halla* (Figure 9H). All oenonids lack ventral cirri. *Drilognathus* is the only genus lacking chaetae; in the other genera, all chaetae are simple, including smooth or denticulated limbate chaetae (Figure 9E,F). In some genera, there are acicular spines with or without hoods. Only *Oenone* species and one species of *Tainoika* (*Tainokia logachevae* Ravara and Cunha, 2018) have bidentate hooded hooks [364].

In addition to these important external characteristics, the study of this group of worms requires careful observation of the maxillary apparatus since its taxonomy is based mainly on the differences in this structure. The maxillary apparatus is of prionognath type, characterized by the presence of paired slender dorsal and unpaired ventral maxillary carriers, which are much longer than the maxillae themselves (Figure 3F). Due to the existence of free-living and parasitic species, the architecture of the maxillary apparatus in oenonids is highly variable. It is well developed in free-living species where several pairs of maxillae can be present, such as in *Oenone* or *Arabella*, or can be extremely reduced to only maxillary carriers and, sometimes, a pair of maxillae, as in parasitic *Drilognathus* and *Labrorostratus*.

The maxillary apparatus of oenonids is symmetrical, with the same number of maxillae on both left and right sides. In some taxa, there is a condition called asymmetric, which refers to the difference in the shape and size of maxillary elements, but not to the number of plates as in Eunicidae and Onuphidae. This type of asymmetry in the shape of the maxillary plates is usually observed in the first two maxillae. In some species, both maxillae I have similar size and falcate shape with a denticulated base, and they are similar in size to maxillae II (Figure 3F). In contrast, other species have the left maxilla I falcate with a denticulated base and the right maxilla I as a denticulated rectangular plate much shorter than maxillae II [14,364]. The mandibles of oenonids are not fused, and without calcified distal plates, lacking growth lines [13], except for *Oligognathus* (whose mandibles are fused in a single horseshoe-shaped piece), *Drilonereis* (where the mandibles may be absent in some species), or *Arabella aracaensis* Steiner and Amaral, 2009 (with growth lines on its mandibles).

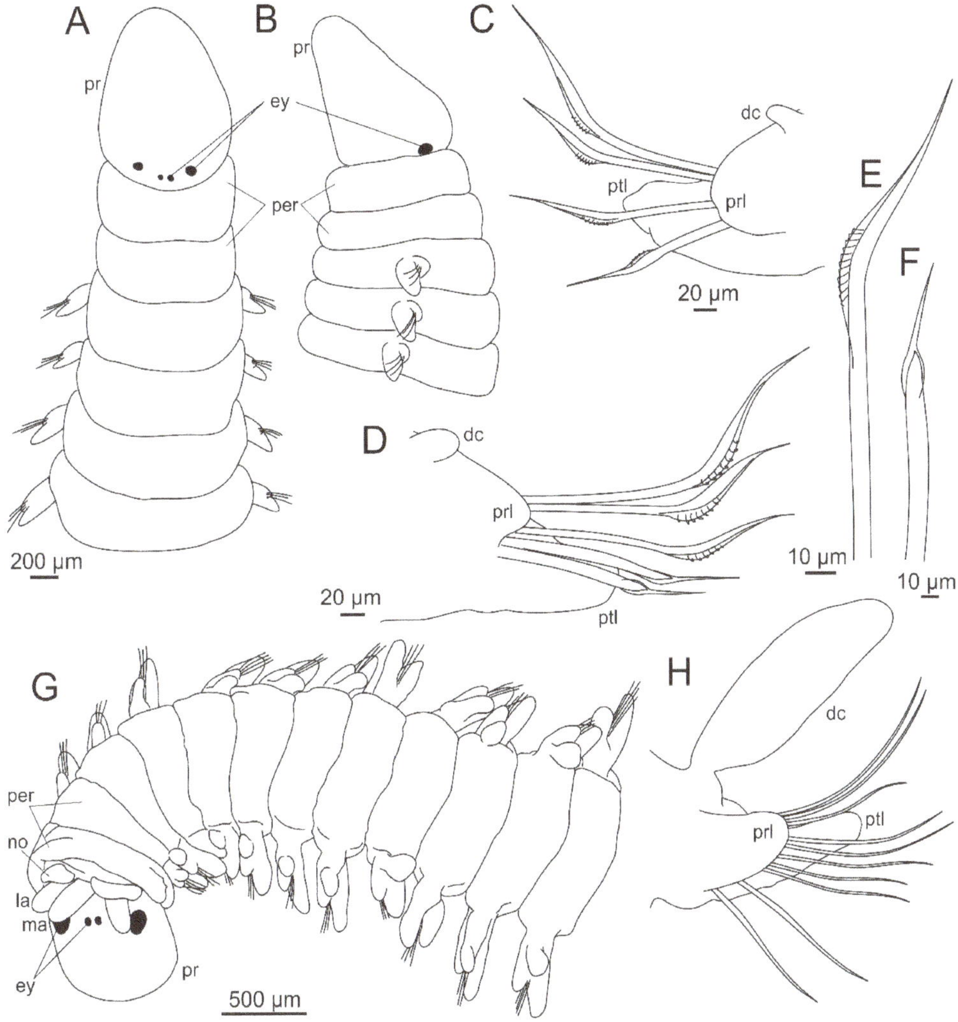

Figure 9. Morphology of Oenonidae. (**A**) *Arabella mutans* (Chamberlin, 1919), anterior end, dorsal view; (**B**) the same, lateral view; (**C**) the same, parapodium 5; (**D**) the same, parapodium 26; (**E**) the same, limbate chaeta; (**F**) the same, chaetae tapering abruptly to guards (**G**) *Oenone fulgida*, anterior end, dorsal view; (**H**) the same, parapodium from midbody region. dc– dorsal cirrus, ey–eye, la–lateral antenna, ma–median antenna, no–nuchal organ, per–peristomium, pr–prostomium, prl–prechaetal lobe, ptl–postchaetal lobe. (**B–F**) modified from Zanol and Ruta [365].

7.3. Feeding and Life History

Oenonids were regarded as selective carnivores or surface deposit-feeders [294]. Carnivorous activity in addition to the ability to feed on their hosts has been recently corroborated based on the results obtained with stable isotopes (δ^{15}N) [34]. Studies in species of the genus *Halla*, which have a marked preference for their prey, bivalves [58,366], indicate that these worms produce mucus that contains paralyzing and digestive substances that force the bivalve to open its valves, to later ingest the soft tissue [367]. Under stress conditions, *Halla* species secrete another type of purple mucus, which, when in contact with human skin, is difficult to eliminate [368]. This mucus is produced by glandular structures of the epidermis [369], and it is considered a chemical defense mechanism [368]. Purple mucus secretion also occurs in species of *Oenone*, when they are under handling stress [370].

Very little is known about reproductive biology in oenonids. Most of the studied species are gonochoric without sexual dimorphism. Amaral [371] and Steiner and Amaral [51] reported adult individuals of the parasitic species *Labrorotratus prolificus* Amaral, 1977 reproducing asexually by stolons. The study of the reproductive biology of *Halla parthenopeia* (delle Chiaje, 1828) showed that the appearance of mature sperm occurs in two periods, during late April and in late December; while spawning occurs in May and January [372]. The sperm of this species was described as of the ect-aquasperm type (*sensu* [302,372]). Osman et al. [58] described the oogenesis in this species with three maturity stages; the first one is a previtellogenic phase, with small racemelike oocytes, while the other two stages are vitellogenic phases, with oocytes free in the coelomic cavity.

7.4. Habitats and Ecology

Oenonidae includes free-living species, endoparasitic species, and species that, during their ontogenetic development, have an endoparasitic phase but are free-living as adults [51,371,373,374].

Free-living species are typically soft-sediment dwellers, and although they do not construct a tube, they secrete mucus to prevent their burrow from collapsing [375]. The parasitic forms mainly infest the polychaetes of different families, for example, Terebellidae, Nereididae, Onuphidae, Eunicidae, Arenicolidae, but they have also been found in spoon worms and bivalves [51,269,363–365,376,377]. Most of the parasitic species apparently have a high degree of specificity infecting a single host [269]. They can be found from one specimen to more than 50 infecting a single host, sometimes reaching larger sizes than their host [371,373,374,376]. Pettibone [373] suggested that infections occur during the early developmental stages of the parasite, but the mechanisms of infection and the stage at which it occurs are unknown.

Information about oenonid ecology is scarce. They are not frequently recorded, and usually occur in low abundance in mud, sand, coralline rock, among algae, or as parasites.

7.5. Distribution Patterns and Diversity

Oenonids have a worldwide distribution from the intertidal zone to the abyssal depths. They are more common in tropical and temperate regions. In the Arctic and Antarctic regions, no more than eight species of *Arabella*, *Driloneries*, and *Notocirrus* have been recorded [13,378]. Some contributions to regional faunas of oenonid species are available for the Antarctic and Subantarctic seas [13], Argentina [379], Australia [365,380]; Brazil [51,371,381], the British Isles [181]; the Caribbean region [139,220,377,382,383]; Chile [349,350]; Costa Rica, Pacific side [323,384]; the Eastern Pacific [28]; East Africa and the Red Sea [324]; Fiji and Samoa [190]; the Gulf of Cádiz [385]; the Gulf of Mexico [326,386–388]; India [330]; Japan [389–393]; the Mediterranean Sea [4,394–397]; New Zealand [398]; Norway [399]; South California [375]; Thailand [339]; western Mexico [103,376]; South Africa [216]; Vietnam [351]; and West Africa [353,354]. Some records of deep-sea oenonids in various areas of the world were reported by Chamberlin [61], Hartman [400], Hartman and Fauchald [401], and Kirkegaard [402].

Many regions of the world lack studies on oenonids, suggesting that oenonid diversity is still underestimated. Despite the low abundance of oenonids, new species have recently been found both in shallow-water and deep-sea habitats (*v gr.* [365,385]).

Some oenonid species have been regarded as widely distributed, but this distribution is usually the result of poor original description and unjustified synonymies. *Oenone fulgida*, originally described from the Red Sea, currently includes six species as junior synonyms: *Oenone diphyllidia* Schmarda, 1861 from Jamaica, *Oenone haswelli* Benham, 1915 from Australia, *Oenone pacifica* Fischli, 1900 from Ternate, *Oenone symmetrica* Fauvel, 1914 from the Gulf of Guinea, *Oenone telura* Chamberlin, 1919 from Marshall islands, and *Oenone vitiensis* Grube, 1871 from Fiji. *Arabella iricolor* (Montagu, 1804) originally described from

England has four junior synonyms: *Arabella lagunae* Chamberlin, 1919 from California, *Arabella maculosa* Verrill 1900 from the Bermudas, *Arabella multidentata* (Ehlers, 1887) from the Gulf of Mexico, and *Arabella setosa* Treadwell, 1921 from Tobago. Based on the revision of the type material, Perkins [326] found morphological differences to consider *A. multidentata* and *A. maculosa* as valid species, rejecting the synonymy with *A. iricolor*. The revision of type materials, including a detailed examination of maxillary apparatuses and chaetae, is critical to clarify the taxonomic status and distribution of oenonid species.

The use of molecular data may help to improve identification. However, until now, only seven species (*A. iricolor*, *Arabella semimaculata* (Moore, 1911), *Drilonereis longa* Webster, 1879, *Drilonereis filum* (Claparède, 1868), *Halla okudai* Imajima, 1967, *O. fulgida*, and *Tainokia logachevae*) have available molecular information in public databases. Other molecular information is available, but the specimens were identified just at family or genus level.

7.6. Exploitation

At least two species of the family are used as bait. *Halla parthenopeia* is commonly utilized for sport fishing in the Mediterranean Sea [58,372,403]. This species is locally named "cuc llobarrer" on the Catalan Maresme coast; fishers capture an average of 2.17 individuals per boat per day [403]. The second species is *Halla okudai*. It is locally named Tai-mushi in Japan, where considered an effective bait worm, reaching the highest prices among bait worms [404]. This species is also appreciated as bait in Malaysia; the specimens are harvested during low tide [405].

8. Onuphidae Kinberg, 1865

8.1. Systematics and Phylogeny

Onuphidae comprises 22 genera and approximately 340 species. Over their history, onuphids were recognized as a subfamily [4,216] or as a family [28,61] constituting 8–10 genera, with most of species belonging to *Onuphis* Audouin and Milne Edwards, 1833 and *Nothria* Malmgren, 1867. The revision of the family started from the work of Pettibone [406] who redefined *Nothria* and *Paradiopatra* Ehlers, 1887. Kucheruk [407] revised the generic diagnoses of 11 genera and described three new genera, one of which, *Aponuphis* Kucheruk, 1978, remains valid in the current system of the family. Fauchald [408] re-examined the type materials for all species of former *Onuphis*, *Nothria*, and *Paradiopatra* and described three new genera, of which *Mooreonuphis* Fauchald, 1982 and *Kinbergonuphis* Fauchald, 1982 are currently accepted. Paxton [409] provided the comprehensive revision of the whole family, redefining all accepted genera and describing five new genera. She subdivided onuphids into two monophyletic subfamilies: Onuphinae Kinberg, 1865 and Hyalinoeciinae Paxton, 1986 possessing morphological synapomorphies such as the position of subacicular hooks in the posterior parapodia and arrangement of the nurse cells in the oocytes (Figure 10D,H). Paxton's [409] system remained largely valid to date with few minor changes: *Neonuphis* Kucheruk, 1978 was synonymized with *Leptoecia* Chamberlin, 1919 [13,17]; monotypic *Fauchaldonuphis* Paxton, 2005 was described [410]; *Epidiopatra* Augener, 1918 and *Notonuphis* Kucheruk, 1978 were synonymized with *Diopatra* Audouin and Milne Edwards, 1833 and *Paradioptra*, respectively, and two more monotypic genera, *Paxtonia* Budaeva and Fauchald 2011 and *Protodiopatra* Budaeva and Fauchald 2011, were described [32]. A key to genera valid at the time is available in Paxton [409] and Carrera-Parra [411].

A number of onuphid genera have been revised and their taxonomy and identification are rather well resolved: *Australonuphis* Paxton, 1979 [412]; *Hirsutonuphis* Paxton, 1979 [413, 414]; *Brevibrachium* Paxton, 1986, *Longibrachium* Paxton, 1986, and *Rhamphobrachium* Ehlers, 1887 [415,416]; and *Paradiopatra* [32,417]. However, several species-rich genera, such as *Anchinothria* Paxton, 1986, *Diopatra*, *Hyalinoecia* Malmgren, 1867, *Kinbergonuphis*, *Nothria*, and *Onuphis* await detailed revision.

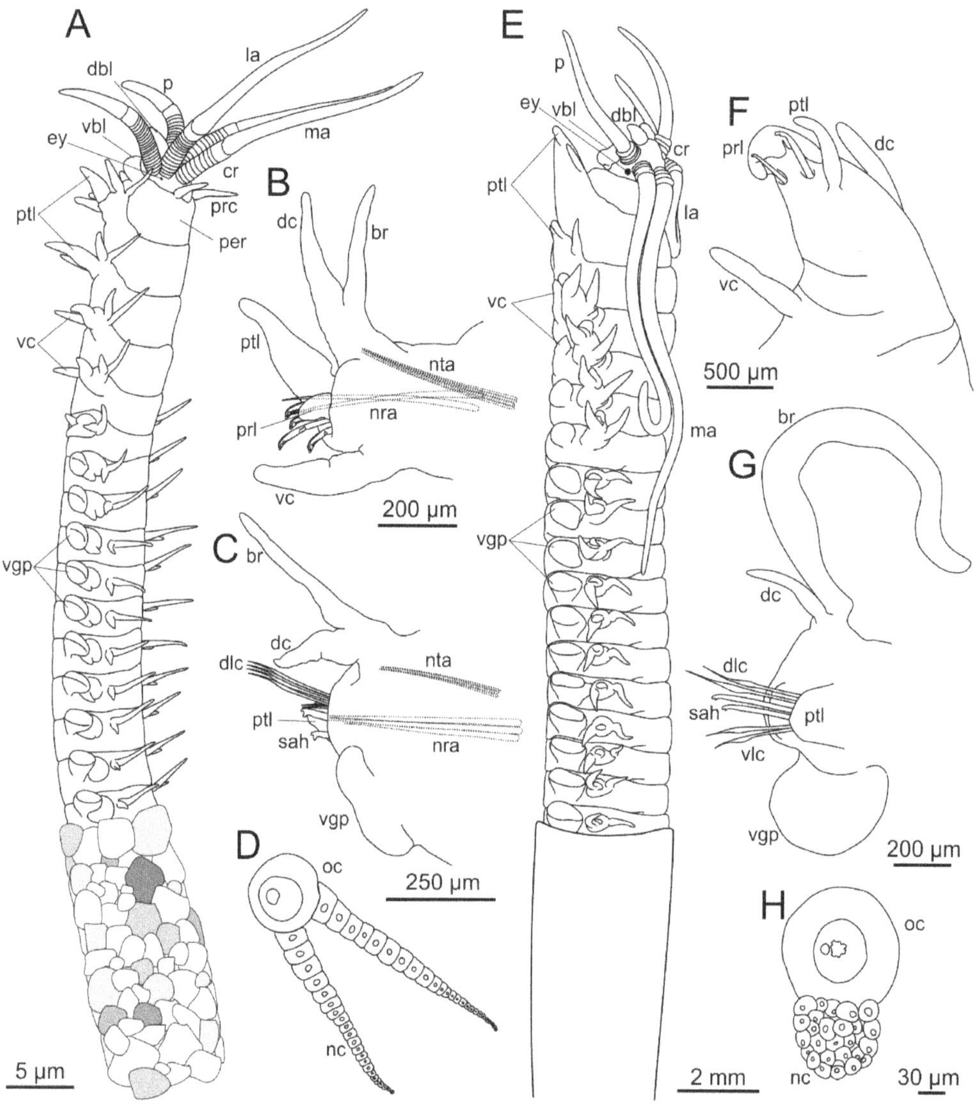

Figure 10. Morphology of Onuphidae. (**A**) *Onuphis* sp. (Onuphinae), anterior end, lateral view; (**B**) the same, anterior modified parapodium; (**C**) the same, posterior parapodium; (**D**) Onuphinae, oocyte; (**E**) *Hyalinoecia* sp. (Hyalinoeciinae), anterior end, lateral view; (**F**) the same, anterior modified parapodium, aciculae not shown; (**G**) the same, posterior parapodium, aciculae not shown; (**H**) *Leptoecia vivipara* Orensanz, 1990, oocyte. br–branchia, cr–ceratophore, dbl–dorsal buccal lip, dc–dorsal cirrus, dlc–dorsal limbate chaetae, ey–eye, la–lateral antenna, ma–median antenna, nc–nurse cells, nra–neuroaciculae, nta–notoaciculae, oc–oocyte, p, palp, per–peristomium, prc–peristomial cirrus, prl–prechaetal lobe, ptl–postchaetal lobe, sah–subacicular hook, vbl–ventral buccal lip, vc–ventral cirrus, vgp–ventral glandular pad, vlc–ventral limbate chaetae.

Phylogenetic hypotheses of Onuphidae were proposed based on morphological [409] and molecular [12] data. Both studies agreed in recognizing the two subfamilies; 10 of 22 genera described based on morphology were tested with molecular tools and were recovered as monophyletic, largely supporting the accepted system of the family provided by Paxton [409]. The hypotheses on the relationships between the genera and recognition of generic groups and complexes *sensu* Paxton [409] were not corroborated by the molecular data: nevertheless, taxon sampling was incomplete and only one nuclear (18S rDNA) and a fragment of a mitochondrial marker (16S rDNA) were used in the reconstruction [12]. Further detailed studies are required to reveal better resolution at the basal part of the onuphid tree and to assess interrelationships between the genera.

8.2. Morphology

Onuphids vary in size and number of chaetigers from 2 cm and about 100 chaetigers (e.g., *Leptoecia, Nothria,* and *Anchinothria*) to 70–300 cm and over 1000 chaetigers in beach worms *Australonuphis* and *Americonuphis*. The majority of species have about 200 chaetigers and are 10 cm in length [409].

Onuphids are defined by two main synapomorphies: the presence of well-delimited dorsal buccal lips (frontal lips) attached to the anterior margin of the prostomium and several pairs of modified anterior parapodia (Figure 10A,B,E,F). They have a rounded prostomium with dorsal and ventral buccal lips (frontal and upper lips), a pair of palps, a pair of lateral antennae and a single median antenna (Figure 10A,E). Both antennae and palps are placed on multiringed ceratophores consisting of 2–70 rings (Figure 10A,E). Many species possess eyes and dorsal nuchal organs in the shape of straight to semicircular ciliated grooves. The peristomium is a single ring, with (Figure 10A) or without (Figure 10E) peristomial cirri, ventrally forming a wide lower lip. The anterior 2–8 pairs of parapodia are modified, usually directed slightly anteriorly bearing special kind of chaetae: simple or pseudocompound falcigers (Figure 10B,F). Modified parapodia are used in locomotion and feeding and are highly adaptive to different lifestyles. In Hyalinoeciinae, the modified parapodia are enhanced and enlarged with auricular prechaetal lobes and short and stout falcigers (Figure 10E,F); they are used for crawling on the surface of the sea floor [418]. Species of *Rhamphobrachium*, *Longibrachium* and *Brevibrachium* have elongated modified parapodia with extremely long spiked and distally curved falcigers that can protrude anteriorly and are used in catching prey in an ambush-predator feeding strategy [419]. Australian beach worms (*Australonuphis*, *Hartmanonuphis* Paxton, 1986 and *Hirsutonuphis*) have massive anterior parapodia used for actively moving inside sandy sediments and making temporary borrows [53]. The modified parapodia may not be enlarged but always possess elongated ventral cirri (Figure 10B,F), while in following parapodia the ventral cirri are modified into glandular pads producing the organic secretion used in tube building (Figure 10C,G).

The majority of onuphid worms (except beach worms) are tubicolous dwellers and the shape and the composition of their tubes is quite diverse. Quill worms (*Hyalonoecia* and *Leptoecia*) build completely organic transparent light and firm tubes that they are able to carry along as epibenthic crawlers (Figures 1E and 10E; [418]). *Anchinothria* and *Nothria* have tubes made of a parchment-like inner layer covered by mud, sand of shell fragments; these tubes are dorsoventrally flattened, suggesting an epibenthic lifestyle for these two genera. Species of *Diopatra* construct vertical tubes with so-called "tube caps" protruding above the sediment surface and incrusted by fragments of shells, plant debris and other foreign objects. Many other species build cylindrical tubes covered by sand or mud particles, which usually are permanent and significantly exceed their inhabitants in length (Figure 10A).

The number and morphology (i.e., size and shape of various parapodia lobes) of the anterior modified parapodia, the arrangement of branchial filaments, number of rings on the ceratophores, and the presence or absence of specific chaetae are the main characteristics used in the identification of onuphid genera. The species level diagnostic characteristics are mainly the distribution of different types of chaetae and branchiae along the body, the number of denticles in the anterior falcigers, and the number of branchial filaments. Intraspecific variation in morphological characteristics is very common in onuphids, leading to difficulties in species identification and delimitation based on morphology. This is especially prominent in species-rich genera such as *Diopatra*, *Onuphis*, *Kinbergonuphis* and *Hyalinoecia* lacking complete revisions to date. It is probably impossible to revise these genera utilizing morphology only; a complete revision would require the inclusion of molecular data.

8.3. Feeding and Life History

Omnivorous or scavenging feeding modes have been documented in onuphids with some species feeding primarily on kelp [294]. Stable isotope analysis [418] and analysis of gut content [420] suggest that quill worms are secondary consumers. Being motile predators and/or scavengers, large-bodied quill worms are able to move around in search of sources of food on the sea floor and were many times recorded on underwater photographs and video attracted by fish bait or whale falls [421]. Large-bodied beach worms (*Australonuphis* spp.) were reported to aggregate in high numbers and feed on large carrion such as washed up whale or sheep's head [53]. Ambush predator behavior was described in detail for *Longibrachium arariensis* Nishi and Kato, 2009 utilizing its extremely modified, prolonged and retractable anterior parapodia and falcigers in catching small prey [419]. *Diopatra* species were shown to feed on foraminiferans [422] or minute invertebrates inhabiting their complex tubes protruding vertically from the sediment [423,424]. Little is known about the feeding habits of the small-bodied and largely sessile onuphids dwelling in permanent tubes inside the sediment.

Onuphids display a great variety of reproductive strategies. Several cases of protandric simultaneous [425,426] or sequential [427] hermaphroditism have been observed in onuphids as well as cases of sperm transfer via spermatophores [428] and sperm storage by females [428] or hermaphrodites [427]. All species investigated so far have lecitotrophic development but many provide parental care for their young by laying eggs inside their tubes or cocoons or egg masses attached to the tubes. The broadcast spawners are also known in the family; however, they are less common, possibly due to sampling bias. Few species were shown to have short-living pelagic stages [429,430] while the majority have direct development inside parental tubes or cocoons [430–435]. At least two cases of viviparity were observed in miniature deep-sea *Leptoecia* species, with juveniles developing inside the parental body cavity up to an advanced stage [13,436].

Large-bodied species (i.e., *Diopatra ornata* Moore, 1911) were shown to serve as hosts for oenonids, another eunicidan family of bristle worms [376]. Portuguese populations of *Diopatra neapolitana* Delle Chiaje, 1841 were reported to be infected by digeneid worms [437] and myxozoans [438]. Symbiosis between onuphids and ciliates attached to their parapodia was described by Arias et al. [439] and Arias and Paxton [440].

8.4. Habitats and Ecology

All known onuphid species are free-living. They are predominantly soft-bottom dwellers, either epibenthic or infaunal, commonly found in muddy or sandy sediments. Several species of *Diopatra* can form large aggregations and were described as ecosystem engineers—species physically creating or modifying the habitat [441]. Having vertical robust and complex tubes sticking above the sediment surface, the species act as sediment stabilizers and increase the abundance and richness of the local benthic fauna [442,443]. While *Diopatra cuprea* (Bosc, 1802) was shown to reach up to 250 ind/m^2 at the Atlantic coast of the US [443], the Hawaiian species, *Diopatra dexiognatha* Paxton and Bailey-Brock, 1986, is able to form intertidal mounds with extreme densities of up to 21800 ind/m^2. These mounds enhance organic matter accumulation and provide a complex tridimensional habitat for small invertebrate species [444]. No onuphids have been reported from hydrothermal vent fields; however, in cold seeps of the US Atlantic continental margin, *Hyalinoecia artifex* Verrill, 1880 occurs in high abundancies (up to 70 ind/m^2) [418].

8.5. Distribution Patterns and Diversity

Onuphids are widely distributed around the world, found in all oceans from the intertidal zone down to over 6000 m deep [445,446]. The diversity of Onuphidae in the Arctic is rather poor with only two species, *Nothria conchylega* (Sars, 1835) and *Paradiopatra pauli* (Annenkova, 1952), reported from the shelf areas [447], while the Antarctic onuphid fauna is diverse with many species found both in deep and shallow waters [13,166,448].

Regional onuphid faunas were described from Central America and adjacent waters of the East Pacific and West Atlantic [94,189,449–452], Antarctic and Subantarctic Seas [13,448]; Japan [453–456], Australia [53,413,416,417,430,457–459], South and West Africa [46,216,460–462]; South America [43,463,464] and the eastern Atlantic including Norway [343], southern Europe, the Mediterranean [440,465–468] and the Macaronesian region [469]. Large parts of the world's onuphid fauna remain poorly studied; for example, from the Indian Ocean and southeastern Asia as well as from most of the deep-sea habitats. Nonetheless, even relatively well-studied regions hold many undescribed species, which is demonstrated by the recent increase in the description of new species from Australia and the East Atlantic (Figure 2).

Deep-water onuphids were mostly described by McIntosh [445] after the HMS *Challenger* expedition, followed by the report from the USS *Albatross* Pacific expeditions by Chamberlin [61] and the USNS *Eltanin* Antarctic deep-sea expeditions by Hartman [448]. A series of works was published by Kucheruk [407,446,470,471] based on the material form the deep-sea Soviet expeditions. Deep-sea onuphids mostly belong to the genera *Anchinothria*, *Hyalinoecia*, *Leptoecia*, and *Paradiopatra*, and represent the fourth most diverse annelid family below 2000 m depth [472].

The most species-rich genera, *Onuphis*, *Nothria*, *Hyalinoecia*, and *Paradiopatra* display cosmopolitan distribution, while some others are restricted to more specific regions. Species of *Diopatra* are mostly found in intertidal and shelf tropical and subtropical areas with very few deep and temperate records. *Leptoecia* is almost exclusively a bathyal and abyssal genus distributed mainly in the Southern Hemisphere with few records in the North Atlantic and northeastern Pacific [13,17,446]. *Australonuphis*, a genus comprising seven species, is also known mostly form the Southern Hemisphere, being reported from Australia and South America and with one species known from western Mexico [412]. Until very recently, *Mooreonuphis* was restricted to the American waters; nonetheless, the detailed investigation of small-bodied onuphids from relatively well-known areas led to the description of two new species from the eastern Atlantic [461,465] and two new species from Australia [458]. Similarly, three new species of *Aponuphis*, the genus for long time known only from the North and East Atlantic, have been recently described from Australia [457].

Most onuphid species are known to have rather restricted geographical ranges; however, several examples of widely distributed species confirmed with genetic data were reported recently. *Diopatra neapolitana* originally described from the Mediterranean and widespread in Iberian Atlantic was subsequently reported from the Suez Canal, Egypt, the Brazilian Atlantic, and India [473,474]. Its sister species, *Diopatra aciculata* Knox and Cameron, 1971, common in Australian east coastal waters, has been recently reported from the South Africa estuaries and also from Egypt. Both species are broadcast spawners with larvae capable of spending some time in the water column. Thus, potential transport with ballast waters of via aquaculture could result in their recent spread due to human activities [46]. Another example of a species with a wide geographical range, but possibly not due to human mediation, is *Nothria conchylega*, a species commonly found in shelf and upper slope areas in the Arctic Ocean, the North Atlantic and the North Pacific [447]. Recent molecular investigations revealed very little genetic variation in mitochondrial markers in specimens collected across the whole range, indicating the presence of a large panmictic population in the Northern Hemisphere [475].

The vast majority of onuphid species are described based on morphological data and only a few recent studies have provided molecular support for species identification and description. This was almost exclusively done in *Diopatra* [43,476,477], the most specious genus of onuphids with high morphological variability in diagnostic characteristics. *Diopatra neapolitana* and *D. aciculata* are two very similar species, both genetically and morphologically, which display wide but also partly sympatric distribution. Analysis of 22 morphological characteristics and five genetic markers resulted in conclusion that these two species were in the grey speciation zone. Their validity was retained for the stability of the taxonomical system [474]. A similar case has been reported in newly described *Diopatra marinae* Steiner and Amaral, in Seixas et al., 2020 and *Diopatra victoriae* Steiner and Amaral, in Seixas et al., 2020 from Brazil [43] confirming previous observations in very rapid and recent speciation within *Diopatra* (Hektoen 2020 pers. comm.). *Diopatra cryptornata* Fauchald, Berke and Woodin, 2012 was synonymized with *D. neapolitana* based on molecular data. The species was originally described due to the presence of dorsal papillae, which were later shown to be sperm storage organs in simultaneous hermaphrodite individuals of *D. neapolitana* from the northern Spanish coast [427]. Similar papillae were later described in the Brazilian population of *D. neapolitana* [473].

Cryptic species diversity, although very common in some annelids [478], has not been described as a widespread phenomenon in Onuphidae, possibly due to their complex external morphology allowing distinguishing between closely related species. None of the studies provided clear evidence of cryptic species in onuphids with detailed morphological and genetic assessment of diversity; however, few cases of species complexes are known in *Aponuphis* [26], *Diopatra* [479], *Hyalinoecia* [480] and *Nothria* [481].

Population level studies are very scarce in onuphids. A study on the deep-sea *Hyalinoecia longibranchiata* McIntosh, 1885 based on two mitochondrial markers revealed significant population structure between the worms inhabiting several deep-sea sites west of New Zealand, which was hypothesized to be a result of complex current regimes that operate in the area as well as the potential limited ability for larval dispersal [482]. In contrast, the study of Single Nucleotide Polymorphism (SNP) in Australian beach worms, *Australonuphis teres,* showed no population structure along the eastern coast of Australia, possibly due natural disturbance events, changes in currents and temperature regimes, and/or anthropogenic disturbance [483].

8.6. Exploitation

Large-bodied onuphids are widely used as bait and thus harvested by professional and recreational fishermen from the intertidal zone habitats in Australia (*Australonuphis* spp. and *Hirsutonuphis* spp., [53]), South Africa (*D. aciculata*, [46]), the Mediterranean and Portuguese coasts (*D. neapolitana*, [54,56,484]). One species, *D. aciculata*, is commercially produced in aquaculture in New South Wales, Australia and sold as food in the conditioning of prawn broodstock and as dry bait for recreational fishing [59].

9. Future Perspectives

The phylogenetic relationships between eunicidan taxa and the evolution of jaw morphology are still open questions due to the varying placement of most families in different phylogenetic hypotheses, the absence of phylogenetic analyses including both extinct and extant taxa and the highly modified families Histriobdellidae and Hartmaniellidae, as well as lack of phylogenetic hypotheses for most families. More robust phylogenetic hypotheses for Eunicida could allow estimates of time of divergence, providing better understanding of eunicidan evolution as well as calibration points to time estimates within Annelida.

Homology between the different parts of the maxillae is one of the challenges to overcome in order to understand the evolution of jaw morphology and to allow the inclusion of the extinct taxa in the analyses. Paxton [14] undertook a very thorough analysis of morphology and proposed hypotheses of homology for carriers, maxillae I and basal plate, but homologies for other parts are still unclear and in some cases controversial (e.g., Hartmaniellidae). Information on the biomechanics of extant jaw types may also provide another source of information to understand homology among plates and the feeding biology of different species.

The reproductive biology, development and ecology of most families are poorly known despite being key information for understanding the dispersal and distribution of species. Generalizations are made based on knowledge built on few species, which sometimes makes explaining observed distribution patterns a challenge (e.g., [49]). Other factors that hinder the understanding of distribution patterns are the poor information available about the diversity of some regions, such as the South Atlantic, South Pacific and Indian Ocean, and the difficulty in species identification, which may hide continuous distributions or lead to flawed wide distributions. No information is available on the initial infection of the host by eunicidan parasitic and commensal species. Such information would also be of great importance in understanding the evolution of such life modes.

Despite the existence of poorly studied areas, well-studied regions also need to be better explored regarding synonymizations and widely distributed species. Reviews of type material have uncovered diversity hidden under unjustified synonymizations (e.g., [36,122]). Moreover, detailed analyses of morphology and molecular data, especially from type specimens or specimens collected from type localities, improve the understanding of species delimitation, correct use of names and thus of diversity. However, the most important factor of all in order to understand and to reveal the diversity and distribution patterns of taxa is the presence of specialists working on the groups, which generate large leaps in the number of known species (Figure 2). Therefore, the continuous formation of specialists coupled with new views of the morphology, knowledge on the biology and ecology and the use of molecular data appears to be the path to improve the knowledge on the diversity of Eunicida.

Author Contributions: Conceptualization, J.Z., N.B.; writing—sections Introduction, Eunicidae and future perspectives, J.Z.; Dorvilleidae, H.W., A.R.; Hartmaniellidae, Lumbrineridae and Oenonidae, L.F.C.-P.; Histriobdellidae, T.M.S., A.C.Z.A.; Onuphidae, N.B.; original draft preparation, review and editing, J.Z., N.B., L.F.C.-P., T.M.S., A.C.Z.A., H.W., A.R.; and illustrations, N.B. All authors have read and agreed to the published version of the manuscript.

Funding: J.Z. was funded by Fundação de Amparo à Pesquisa do Estado do Rio de Janeiro (FAPERJ Proc. E-26/200.071/2019). A.R. was supported by national funds (OE), through FCT (Fundação para a Ciência e a Tecnologia), in the scope of the framework contract foreseen in the Decree-Law 57/2016 (Nrs 4–6, art. 23), changed by Law 57/2017; and through CESAM by FCT/MCTES (UIDP/50017/2020+UIDB/50017/2020). NB was supported by a grant from the Russian Foundation for Basic Research (RFBR project number 19–04-00501).

Acknowledgments: We are thankful to María Capa and Pat Hutchings for organizing this issue and inviting us to contribute, to the three anonymous reviewers who helped to improve the manuscript, and to Alexander Semenov, Conrad Helm and Editora Guanabara Koogan for the permission to use their pictures.

Conflicts of Interest: The authors declare no conflict of interest.

References

1. Uschakov, P.V. Phylum Annelida, Class Polychaeta. In *Atlas of the Invertebrates of the Far Eastern Seas of the USSR*; Pavlovskii, E.N., Ed.; Stony Brook University: Stony Brook, NY, USA, 1955; pp. 109–128.
2. Dales, R.P. The polychaete stomodeum and the inter-relationships of the families of Polychaeta. *Proc. Zool. Soc. Lond.* **1962**, *139*, 389–428. [CrossRef]
3. Ehlers, E. *Die Borstenwürmer (Annelida Chaetopoda) Nach Systematischen und Anatomischen Untersuchungen Dargestellt. Volume II*; Wilhelm Engelmann: Leipzig, Germany, 1868; p. 748.
4. Fauvel, P. Polychètes Errantes. *Faune Fr.* **1923**, *5*, 1–488.
5. Budaeva, N.; Zanol, J. Eunicida Dales, 1962. In *Handbook of Zoology Annelida, Volume 3: Sedentaria III, Errantia I*; Purschke, G., Westheide, W., Böggemann, M., Eds.; De Gruyter: Berlin, Germany; Boston, MA, USA, 2021; pp. 353–360.
6. Struck, T.H.; Golombek, A.; Weigert, A.; Franke, F.A.; Westheide, W.; Purschke, G.; Bleidorn, C.; Halanych, K.M. The Evolution of Annelids Reveals Two Adaptive Routes to the Interstitial Realm. *Curr. Biol.* **2015**, *25*, 1993–1999. [CrossRef]
7. Weigert, A.; Helm, C.; Meyer, M.; Nickel, B.; Arendt, D.; Hausdorf, B.; Santos, S.R.; Halanych, K.M.; Purschke, G.; Bleidorn, C.; et al. Illuminating the base of the Annelid tree using transcriptomics. *Mol. Biol. Evol.* **2014**, *31*, 1391–1401. [CrossRef] [PubMed]

8. Struck, T.H.; Purschke, G.; Halanych, K.M. Phylogeny of Eunicida (Annelida) and exploring data congruence using a partition addition bootstrap alteration (PABA) approach. *Syst. Biol.* **2006**, *55*, 1–20. [CrossRef]
9. Parry, L.A.; Edgecombe, G.D.; Eibye-Jacobsen, D.; Vinther, J. The impact of fossil data on annelid phylogeny inferred from discrete morphological characters. *Proc. R. Soc. B Biol. Sci.* **2016**, *283*, 20161378. [CrossRef] [PubMed]
10. Zanol, J.; Halanych, K.M.; Struck, T.H.; Fauchald, K. Phylogeny of the bristle worm family Eunicidae (Eunicida, Annelida) and the phylogenetic utility of noncongruent 16S, COI and 18S in combined analyses. *Mol. Phylogenet. Evol.* **2010**, *55*, 660–676. [CrossRef]
11. Zanol, J.; Halanych, K.M.; Fauchald, K. Reconciling taxonomy and phylogeny in the bristleworm family Eunicidae (polychaete, Annelida). *Zool. Scr.* **2014**, *43*, 79–100. [CrossRef]
12. Budaeva, N.; Schepetov, D.; Zanol, J.; Neretina, T.; Willassen, E. When molecules support morphology: Phylogenetic reconstruction of the family Onuphidae (Eunicida, Annelida) based on 16S rDNA and 18S rDNA. *Mol. Phylogenet. Evol.* **2016**, *94*, 791–801. [CrossRef]
13. Orensanz, J.M. The Eunicemorph polychaete annelids from Antarctic and Subantarctic Seas. With addenda to the Eunicemorpha of Argentina, Chile, New Zealand, Australia, and the Southern Indian Ocean. *Antarct. Res. Ser.* **1990**, *52*, 1–183.
14. Paxton, H. Phylogeny of Eunicida (Annelida) based on morphology of jaws. *Zoosymposia* **2009**, *264*, 241–264. [CrossRef]
15. Read, G.; Fauchald, K. World Polychaeta Database. Eunicida. Available online: http://www.marinespecies.org/polychaeta/aphia.php?p=taxdetails&id=895 (accessed on 2 December 2020).
16. Wiklund, H.; Glover, A.G.; Dahlgren, T.G. Three new species of *Ophryotrocha* (Annelida: Dorvilleidae) from a whale-fall in the North-East Atlantic. *Zootaxa* **2009**, *2228*, 43–56. [CrossRef]
17. Budaeva, N. *Leptoecia midatlantica*, a new species of the deep-sea quill-worms (Polychaeta: Onuphidae: Hyalinoeciinae) from the Mid-Atlantic Ridge. *Zootaxa* **2012**, *3176*, 45–60.
18. Yen, N.K.; Rouse, G.W. Phylogeny, biogeography and systematics of Pacific vent, methane seep, and whale-fall *Parougia* (Dorvilleidae: Annelida), with eight new species. *Invertebr. Syst.* **2020**, *34*, 200–233.
19. Hilbig, B.; Blake, J.A. Dorvilleidae (Annelida: Polychaeta) from the U.S. Atlantic slope and rise. Description of two new genera and 14 new species, with generic revision of *Ophryotrocha*. *Zool. Scr.* **1991**, *20*, 147–183. [CrossRef]
20. Westheide, W.; Riser, N.W. Morphology and phylogenetic relationships of the neotenic interstitial polychaete *Apodotrocha progenerans* n.gen., n.sp. (Annelida). *Zoomorphology* **1983**, *103*, 67–87. [CrossRef]
21. Westheide, W.; von Nordheim, H. Interstitial Dorvilleidae (Annelida, Polychaeta) from Europe Australia and New Zealand. *Zool. Scr.* **1985**, *14*, 183–199. [CrossRef]
22. Westheide, W. *Ikosipodus carolensis* gen. et sp. n., an interstitial neotenic polychaete from North Carolina, U.S.A., and its phylogenetic relationships within Dorvilleidae. *Zool. Scr.* **1982**, *11*, 117–126. [CrossRef]
23. Núñez, J. *Marycarmenia lysandrae*, a new genus and interstitial species (Polychaeta: Dorvilleidae) from Madeira. *Bull. Mar. Sci.* **1998**, *62*, 115–119.
24. Zanol, J.; da Silva, T.D.S.C.; Hutchings, P. *Marphysa* (Eunicidae, polychaete, Annelida) species of the Sanguinea group from Australia, with comments on pseudo-cryptic species. *Invertebr. Biol.* **2016**, *135*, 328–344. [CrossRef]
25. Molina-Acevedo, I.C.; Carrera-Parra, L.F. Reinstatement of three species of the *Marphysa sanguinea* complex (Polychaeta: Eunicidae) from the Grand Caribbean Region. *Zootaxa* **2015**, *3925*, 37–55. [CrossRef] [PubMed]
26. Borisova, P.B.; Schepetov, D.M.; Budaeva, N.E. *Aponuphis* Kucheruk, 1978 (Annelida: Onuphidae) from western African waters. *Invertebr. Zool.* **2018**, *15*, 19–41. [CrossRef]
27. Hints, O.; Eriksson, M.E. Diversification and biogeography of scolecodont-bearing polychaetes in the Ordovician. *Palaeogeogr. Palaeoclim. Palaeoecol.* **2007**, *245*, 95–114. [CrossRef]
28. Hartman, O. Polychaetous Annelids. Part V. Eunicea. *Allan Hancock Pac. Exped.* **1944**, *10*, 1–237.
29. Kielan-Jaworowska, Z. Polychaete jaw apparatuses from the Ordovician and Silurian of Poland and a comparison with modern forms. *Palaeontol. Pol.* **1966**, *16*, 9–152.
30. Eibye-Jacobsen, D.; Kristensen, R.M. A new genus and species of Dorvilleidae (Annelida, Polychaeta) from Bermuda, with a phylogenetic analysis of Dorvilleidae, Iphitimidae and Dinophilidae. *Zool. Scr.* **1994**, *23*, 107–131. [CrossRef]
31. Ravara, A.; Marçal, A.R.; Wiklund, H.; Hilário, A. First account on the diversity of *Ophryotrocha* (Annelida, Dorvilleidae) from a mammal-fall in the deep-Atlantic Ocean with the description of three new species. *Syst. Biodivers.* **2015**, *13*, 875–890. [CrossRef]
32. Budaeva, N.; Fauchald, K. Phylogeny of the *Diopatra* generic complex with a revision of *Paradiopatra* Ehlres, 1887 (Polychaeta: Onuphidae). *Zool. J. Linn. Soc.* **2011**, *163*, 319–436. [CrossRef]
33. Clemo, W.C.; Dorgan, K.M. Functional morphology of eunicidan (Polychaeta) jaws. *Biol. Bull.* **2017**, *233*, 227–241. [CrossRef]
34. Jumars, P.A.; Dorgan, K.M.; Lindsay, S.M. Diet of worms emended: An update of polychaete feeding guilds. *Annu. Rev. Mar. Sci.* **2015**, *7*, 497–520. [CrossRef]
35. Eriksson, M.E.; Parry, L.A.; Rudkin, D.M. Earth's oldest 'Bobbit worm'—Gigantism in a Devonian eunicidan polychaete. *Sci. Rep.* **2017**, *7*, 43061. [CrossRef] [PubMed]
36. Fauchald, K. A review of the genus *Eunice* (Eunicidae: Polychaeta) based upon type material. *Smithson. Contrib. Zool.* **1992**, *523*, 1–422. [CrossRef]
37. Zanol, J. Homology of prostomial and pharyngeal structures in Eunicida (Annelida) based on innervation and morphological similarities. *J. Morphol.* **2010**, *271*, 1023–1043. [CrossRef] [PubMed]
38. Tzetlin, A.; Purschke, G. Pharynx and intestine. *Hydrobiologia* **2005**, *535*, 199–225.

39. Tilic, E.; Bartolomaeus, T.; Rouse, G.W. Chaetal type diversity increases during evolution of Eunicida (Annelida). *Org. Divers. Evol.* **2016**, *16*, 105–119. [CrossRef]
40. Starunov, V. Musculature and innervation of the pygidium in Eunicida (Annelida: Errantia). *Biol. Commun.* **2019**, *64*, 31–40. [CrossRef]
41. Fernández, P.V.; Bone, D.; Rodríguez, C. Cryptofaunal polychaetes from oceanic and continental reefs in Venezuela. *Bull. Mar. Sci.* **2012**, *88*, 339–352. [CrossRef]
42. Hernández-Alcántara, P.; Cruz-Pérez, I.N.; Solís-Weiss, V. Composition and diversity patterns of Eunicida and Amphinomida (Annelida) associated to dead coral in the Chinchorro Bank Biosphere Reserve, Caribbean Sea. *J. Mar. Biol. Assoc. UK* **2019**, *99*, 1547–1555. [CrossRef]
43. Seixas, V.C.; Steiner, T.M.; Solé-Cava, A.M.; Amaral, A.C.Z.; Paiva, P.C. Hidden diversity within the *Diopatra cuprea* complex (Annelida: Onuphidae): Morphological and genetics analyses reveal four new species in the south-west Atlantic. *Zool. J. Linn. Soc.* **2020**, *1840*, 1–35. [CrossRef]
44. Taboada, S.; Leiva, C.; Bas, M.; Schult, N.; McHugh, D. Cryptic species and colonization processes in *Ophryotrocha* (Annelida, Dorvilleidae) inhabiting vertebrate remains in the shallow-water Mediterranean. *Zool. Scr.* **2017**, *46*, 611–624. [CrossRef]
45. Hutchings, P.A.; Karageorgopoulos, P. Designation of a neotype of *Marphysa sanguinea* (Montagu, 1813) and a description of a new species of *Marphysa* from eastern Australia. *Hydrobiologia* **2003**, *496*, 87–94. [CrossRef]
46. van Rensburg, H.; Matthee, C.A.; Simon, C.A. Moonshine worms *Diopatra aciculata* (Onuphidae, Annelida) in the Knysna Estuary, South Africa; taxonomy and distribution. *J. Mar. Biol. Assoc. UK* **2020**, *100*, 897–907. [CrossRef]
47. Lavesque, N.; Hutchings, P.; Abe, H.; Daffe, G.; Gunton, L.M.; Glasby, C.J. Confirmation of the exotic status of *Marphysa victori* Lavesque, Daffe, Bonifácio & Hutchings, 2017 (Annelida) in French waters and synonymy of *Marphysa bulla* Liu, Hutchings & Kupriyanova, 2018. *Aquat. Invasions* **2020**, *15*, 355–366. [CrossRef]
48. Paxton, H.; Wiklund, H.; Alexander, F.; Taboada, S. Is the Antarctic *Ophryotrocha orensanzi* (Annelida: Dorvilleidae) a circumpolar non-specialized opportunist? *Syst. Biodivers.* **2017**, *15*, 105–114. [CrossRef]
49. Schulze, A. Phylogeny and genetic diversity of Palolo worms (*Palola*, Eunicidae) from the tropical North Pacific. *Biol. Bull.* **2006**, *210*, 25–37. [CrossRef] [PubMed]
50. Rouse, G.W. The epitome of hand waving? Larval feeding and hypotheses of metazoan phylogeny. *Evol. Dev.* **2000**, *2*, 222–233. [CrossRef]
51. Steiner, T.M.; Amaral, A.C.Z. *Arabella aracaensis*, a new species with growth rings on its mandibles, and some remarks on the endoparasitic *Labrorostratus prolificus* (Polychaeta: Oenonidae) from southeast Brazil. *J. Nat. Hist.* **2009**, *43*, 2537–2551. [CrossRef]
52. Åkesson, B.; Rice, S.A. Two new *Dorvillea* species (Polychaeta, Dorvilleidae) with obligate asexual reproduction. *Zool. Scr.* **1992**, *21*, 351–362. [CrossRef]
53. Paxton, H. Taxonomy and aspects of the life history of Australian beachworms (Polychaeta: Onuphidae). *Aust. J. Mar. Freshw. Res.* **1979**, *30*, 265–294. [CrossRef]
54. Gambi, M.C. Polychaetes of commercial and applied interest in Italy: An overview. *Mémoires Muséum Natl. D'histoire Nat.* **1994**, *162*, 593–601.
55. Bettoso, N.; Faverio, G.; Orel, G. Note sulla biologia e sulla pesca del verme de rimini (*Eunice aphroditois* Pallas) in alto Adriatico. *Hydrores* **1998**, *16*, 24–34.
56. Cunha, T.; Hall, A.; Queiroga, H. Estimation of the *Diopatra neapolitana* annual harvest resulting from digging activity in Canal de Mira, Ria de Aveiro. *Fish. Res.* **2005**, *76*, 56–66. [CrossRef]
57. Osman, I.; Gabr, H.; Saito, H.; El-Etreby, S. Reproductive biology of the highly commercial polychaetes in the Suez Canal. *J. Mar. Biol. Assoc. UK* **2010**, *90*, 281–290. [CrossRef]
58. Osman, I.H.; Gabr, H.R.; El-etreby, S.G. Rearing trials of *Halla parthenopeia* under laboratory conditions (Polychaeta: Oenonidae). *J. Exp. Mar. Biol. Ecol.* **2010**, *383*, 1–7. [CrossRef]
59. Safarik, M.; Redden, A.M.; Schreider, M.J. Density-dependent growth of the polychaete *Diopatra aciculata*. *Sci. Mar.* **2006**, *70*, 337–341.
60. Mondragón, C. Of Winds, Worms and Mana: The Traditional Calendar of the Torres Islands, Vanuatu. *Oceania* **2004**, *74*, 289–308. [CrossRef]
61. Chamberlin, R.V. The Annelida Polychaeta. *Mem. Mus. Comp. Zool. Harv. Coll.* **1919**, *48*, 1–514.
62. Núñez, J. Annelida Polychaeta: Staurocephalidae Verrill, 1900. In *Annelida Polychaeta V, Fauna Ibérica, Vol. 45*; Etal, P.J., Ed.; Museo Nacional de Ciencias Naturales. Consejo Superior de Investigaciones Cientificas: Madrid, Spain, 2018.
63. Salazar-Vallejo, S.; Carrera-Parra, L.F. Reversal of precedence for *Dorvillea* Parfitt, 1866 (Annelida, Errantia) over its senior homonym *Dorvillea* Leach, 1852 (Mollusca, Bivalvia): Maintenance of prevailing usage for *Dorvillea* Parfitt, 1866 and Dorvilleidae Chamberlin, 1919. *Bull. Zool. Nomencl.* **2020**, *77*, 105–111. [CrossRef]
64. delle Chiaje, S. *Memorie Sulla Storia e Notomia Degli Animali Senza Vertebre del Regno di Napoli. Volume 3*; Stamperia della Societá Tipografica: Naples, Italy, 1828; p. 232.
65. Pettibone, M.H. New species of polychaete worms from the Atlantic Ocean, with a revision of the Dorvilleidae. *Proc. Biol. Soc. Wash.* **1961**, *74*, 167–186.
66. Orensanz, J.M. Los anelidos poliquetos de la provincia biogeografica Argentina. III. Dorvilleidae. *Physis* **1973**, *32*, 325–342.

67. Jumars, P.A. A generic revision of the Dorvilleidae (Polychaeta), with six new species from the deep North Pacific. *Zool. J. Linn. Soc.* **1974**, *54*, 101–135. [CrossRef]
68. Wolf, P.S. Four new genera of Dorvilleidae (Annelida: Polychaeta) from the Gulf of Mexico. *Proc. Biol. Soc. Wash.* **1986**, *99*, 616–626.
69. Wiklund, H.; Purschke, G.; Ravara, A. Dorvilleidae Chamberlin, 1919. In *Handbook of Zoology Annelida, Volume 3: Sedentaria III, Errantia I*; Purschke, G., Böggemann, M., Westheide, W., Eds.; De Gruyter: Berlin, Germany; Boston, MA, USA, 2021; pp. 361–382. ISBN 978-3-11-029148-3.
70. Dahlgren, T.G.; Åkesson, B.; Schander, C.; Halanych, K.M.; Sundberg, P. Molecular Phylogeny of the Model Annelid *Ophryotrocha*. *Biol. Bull.* **2001**, *201*, 193–203. [CrossRef]
71. Heggoy, K.K.; Schander, C.; Åkesson, B. The phylogeny of the annelid genus *Ophryotrocha* (Dorvilleidae). *Mar. Biol. Res.* **2007**, *3*, 412–420. [CrossRef]
72. Ravara, A.; Wiklund, H.; Cunha, M.R. New species and records of Dorvilleidae (Annelida, Polychaeta) from deep-sea organic substrata, NE Atlantic. *Eur. J. Taxon.* **2020**, in press.
73. Taboada, S.; Wiklund, H.; Glover, A.G.; Dahlgren, T.G.; Cristobo, J.; Ávila, C. Two new Antarctic *Ophryotrocha* (Annelida: Dorvilleidae) described from shallow-water whale bones. *Polar Biol.* **2013**, *36*, 1031–1045. [CrossRef]
74. Wiklund, H.; Iris, V.; Glover, A.G.; Smith, C.R.; Baco, A.M.Y.R.; Dahlgren, T.G. Systematics and biodiversity of *Ophryotrocha* (Annelida, Dorvilleidae) with descriptions of six new species from deep-sea whale-fall and wood-fall habitats in the north-east Pacific. *Syst. Biodivers.* **2012**, *10*, 243–259. [CrossRef]
75. Struck, T.H.; Halanych, K.M.; Purschke, G. Dinophilidae (Annelida) is most likely not a progenetic Eunicida: Evidence from 18S and 28S rDNA. *Mol. Phylogenet. Evol.* **2005**, *37*, 619–623. [CrossRef] [PubMed]
76. Kristensen, R.M.; Eibye-Jacobsen, D. Ultrastructure of spermiogenesis and spermatozoa in *Diurodrilus subterraneus* (Polychaeta, Diurodrilidae). *Zoomorphology* **1995**, *115*, 117–132. [CrossRef]
77. Laumer, C.E.; Bekkouche, N.; Kerbl, A.; Goetz, F.; Neves, R.C.; Sørensen, M.V.; Kristensen, R.M.; Hejnol, A.; Dunn, C.W.; Giribet, G.; et al. Spiralian Phylogeny Informs the Evolution of Microscopic Lineages. *Curr. Biol.* **2015**, *25*, 2000–2006. [CrossRef]
78. Simonini, R.; Massamba-N'Siala, G.; Grandi, V.; Prevedelli, D. Distribution of the genus *Ophryotrocha* (Polychaeta) in Italy: New records and comments on the biogeography of Mediterranean species. *Vie Milieu-Life Environ.* **2009**, *59*, 79–88.
79. Åkesson, B. Reproduction and larval morphology of five *Ophryotrocha* species (Polychaeta, Dorvilleidae). *Zool. Scr.* **1973**, *2*, 145–155. [CrossRef]
80. Åkesson, B. Reproduction in the genus *Ophryotrocha* (Polychaeta, Dorvilleidae). *Pubbl. della Stn. Zool. di Napoli* **1975**, *39*, 377–398.
81. Oug, E. New and lesser known Dorvilleidae (Annelida, Polychaeta) from Scandinavian and northeast American waters. *Sarsia* **1978**, *63*, 285–303.
82. Purschke, G.; Nowak, K. Ultrastructure of pigmented eyes in Dorvilleidae (Annelida, Errantia, Eunicida) and their importance for understanding the evolution of eyes in polychaetes. *Acta Zool.* **2015**, *96*, 67–81. [CrossRef]
83. Richards, T.L. Reproduction and development of the polychaete *Stauronereis rudolphi*, including a summary of development in the superfamily Eunicea. *Mar. Biol.* **1967**, *1*, 124–133. [CrossRef]
84. Jamieson, B.G.M.; Rouse, G.W. The spermatozoa of the Polychaeta (Annelida): An ultrastructural review. *Biol. Rev.* **1989**, *64*, 93–157. [CrossRef]
85. Åkesson, B. A new *Ophryotrocha* species of the Labronica group (Polychaeta, Dorvilleidae) revealed in crossbreeding experiments. In Proceedings of the NATO Conference Series (Marine Science). 2 Marine Organisms, Venice, Italy, 24 March–4 April 1977; Battaglia, B., Beardmore, J., Eds.; Plenum Publishing: New York, NY, USA, 1978; pp. 573–590.
86. Prevedelli, D.; Simonini, R. Effect of temperature on demography of *Ophryotrocha labronica* (Polychaeta: Dorvilleidae). *Vie Milieu-Life Environ.* **2001**, *51*, 173–180.
87. Åkesson, B. *Ophryotrocha labronica* as test animal for the study of marine pollution. *Helgoländer Wiss. Meeresunters.* **1970**, *20*, 293–303. [CrossRef]
88. Simonini, R. Distribution and ecology of the genus *Ophryotrocha* (Polychaeta: Dorvilleidae) in Italian harbors and lagoons. *Vie Milieu-Life Environ.* **2002**, *2*, 59–65.
89. Paxton, H.; Davey, A. A new species of *Ophryotrocha* (Annelida: Dorvilleidae) associated with fish farming at Macquarie Harbour, Tasmania, Australia. *Zootaxa* **2010**, *2509*, 53–61. [CrossRef]
90. Salvo, F.; Wiklund, H.; Dufour, S.D.; Hamoutene, D.; Pohle, G.; Worsaae, K. A new annelid species from whalebones in Greenland and aquaculture sites in Newfoundland: *Ophryotrocha cyclops*, sp. nov. (Eunicida: Dorvilleidae). *Zootaxa* **2014**, *3887*, 555–568. [CrossRef] [PubMed]
91. Simonini, R.; Grandi, V.; Massamba-N'Siala, G.; Pia Martino, M.; Castelli, A.; Prevedelli, D. Diversity, habitat affinities and diet of *Ophryotrocha* species (Polychaeta, Dorvilleidae) living in Mediterranean harbour habitats. *Vie Milieu-Life Environ.* **2010**, *60*, 27–38.
92. Taboada, S.; Bas, M.; Ávila, C. A new *Parougia* species (Annelida, Dorvilleidae) associated with eutrophic marine habitats in Antarctica. *Polar Biol.* **2015**, *38*, 517–527. [CrossRef]
93. Detinova, N.N. Polychaetous worms from the Reykjanes Ridge (the North Atlantic) Bottom Fauna from Mid-Ocean Rises in the North Atlantic Donnaya fauna. *Otkryto-Okeanicheskikh Pod. Sev. Atl. Tr. Institua Okeanol.* **1985**, *120*, 96–136.
94. Fauchald, K. Benthic polychaetous annelids from deep water off western Mexico and adjacent areas in the Eastern Pacific Ocean. *Allan Hancock Monogr. Mar. Biol.* **1972**, *7*, 1–575.

95. Hilbig, B. Reconsideration of the genus *Exallopus* Jumars, 1974 (Polychaeta: Dorvilleidae) with descriptions of two new species from the western North Atlantic. *Ophelia* **1991**, *S5*, 101–110.
96. Thornhill, D.J.; Struck, T.H.; Ebbe, B.; Lee, R.W.; Mendoza, G.F.; Levin, L.A.; Halanych, K.M. Adaptive radiation in extremophilic Dorvilleidae (Annelida): Diversification of a single colonizer or multiple independent lineages? *Ecol. Evol.* **2012**, *2*, 1958–1970. [CrossRef]
97. Zhang, D.-S.; Zhou, Y.-D.; Wang, C.-S.; Rouse, G. A new species of *Ophryotrocha* (Annelida, Eunicida, Dorvilleidae) from hydrothermal vents on the Southwest Indian Ridge. *Zookeys* **2017**, *687*, 1–9.
98. Blake, J.A. Polychaeta from the vicinity of deep-sea geothermal vents in the eastern Pacific. I: Euphrosinidae, Phyllodocidae, Hesionidae, Nereididae, Glyceridae, Dorvilleidae, Orbiniidae and Maldanidae. *Bull. Biol. Soc. Wash.* **1985**, *6*, 67–101.
99. Blake, J.A.; Hilbig, B. Polychaeta from the vicinity of deep-sea hydrothermal vents in the Eastern Pacific Ocean. II. New species and records from the Juan de Fuca and Explorer Ridge systems. *Pac. Sci.* **1990**, *44*, 219–253.
100. Hilbig, B.; Fiege, D. A new species of Dorvilleidae (Annelida: Polychaeta) from a cold seep site in the northeast Pacific. *Proc. Biol. Soc. Wash.* **2001**, *114*, 396–402.
101. Paxton, H.; Morineaux, M. Three species of Dorvilleidae (Annelida: Polychaeta) associated with Atlantic deep-sea reducing habitats, with the description of *Ophryotrocha fabriae*, new species. *Proc. Biol. Soc. Wash.* **2009**, *122*, 14–25. [CrossRef]
102. Solís-Weiss, V.; Hilbig, B. Redescription of *Ophryotrocha platykephale* Blake (Polychaeta, Dorvilleidae) from the Guaymas Basin hydrothermal vents. *Bull. South. Calif. Acad. Sci.* **1992**, *91*, 92–96.
103. Fauchald, K. Polychaetous annelids of the families Eunicidae, Lumbrineridae, Iphitimidae, Arabellidae, Lysaretidae and Dorvilleidae from western Mexico. *Allan Hancock Monogr. Mar. Biol.* **1970**, *5*, 1–335.
104. Wolf, P.S. Family Dorvilleidae Chamberlin, 1919. In *Taxonomic Guide to the Polychaetes of the Northern Gulf of Mexico. Final Report to the Minerals Management Service, Contract 14-12-001-29091. Volume VI*; Uebelacker, J.M., Johnson, P.G., Eds.; Barry A. Vittor & Associates: Mobile, AL, USA, 1984; pp. 44.1–44.37.
105. Wolf, P.S. Three new species of Dorvilleidae (Annelida: Polychaeta) from Puerto Rico and Florida and a new genus for dorvillleids from Scandinavia and North America. *Proc. Biol. Soc. Wash.* **1986**, *99*, 627–638.
106. Hilbig, B. Family Dorvilleidae Chamberlin, 1919. In *Taxonomic Atlas of the Benthic Fauna of the Santa Maria Basin and the Western Santa Barbara Channel. Vol. 5. The Annelida Part 2, Polychaeta: Phyllodocida (Syllidae and Scale-Bearing Families), Amphinomida, and Eunicida*; Lake, J.A., Hilbig, B., Scott, P.H., Eds.; Santa Barbara Museum of Natural History Publications: Santa Barbara, CA, USA, 1995; pp. 341–364.
107. Paiva, P.C.; Nonato, E.F. On the genus *Iphitime* (Polychaeta: Iphitimidae) and description of *Iphitime sartorae* sp. nov. a commensal of brachyuran crabs. *Ophelia* **1991**, *34*, 209–215. [CrossRef]
108. Imajima, M. Dorvilleidae (Annelida, Polychaeta) from Japan. I. The genus *Dorvillea* (*Dorvillea*). *Bull. Natl. Sci. Mus. Tokyo* **1992**, *18*, 131–147.
109. Núñez, J.; Pascual, M.; Brito, M.C. Contribución al estudio de los Dorvilleidos (Annelida, Polychaeta) de Canarias. *Rev. Acad. Canar. Cienc.* **1996**, *8*, 139–151.
110. Hartmann-Schröder, G.; Rosenfeldt, P. Die Polychaeten der 'Polarstern'-Reise ANT V/1 in die Antarktis 1986. Teil 1: Euphrosinidae bis Iphitimidae. *Mitt. Hambg. Zool. Mus. Inst.* **1992**, *89*, 85–124.
111. Cunha, M.R.; Matos, F.L.; Génio, L.; Hilário, A.; Moura, C.J.; Ravara, A.; Rodrigues, C.F. Are organic falls bridging reduced environments in the deep sea?—Results from colonization experiments in the Gulf of Cádiz. *PLoS ONE* **2013**, *8*, e76688. [CrossRef] [PubMed]
112. Levin, L.A.; Ziebis, W.; Mendoza, G.F.; Growney, V.A.; Tryon, M.D.; Brown, K.M.; Mahn, C.; Gieskes, J.M.; Rathburn, A.E. Spatial heterogeneity of macrofauna at northern California methane seeps: Influence of sulfide concentration and fluid flow. *Mar. Ecol. Prog. Ser.* **2003**, *265*, 123–139. [CrossRef]
113. Prevedelli, D.; Massamba-N'Siala, G.; Simonini, R. The seasonal dynamics of six species of Dorvilleidae (Polychaeta) in the harbour of La Spezia (Italy). *Mar. Ecol. Evol. Perspect.* **2005**, *26*, 286–293. [CrossRef]
114. Fauchald, K.; Hancock, D.R. Deep-water polychaetes from a transect off central Oregon. *Allan Hancock Monogr. Mar. Biol.* **1981**, *11*, 1–73.
115. Hartmann-Schröder, G. Die polychaeten der Fahrten 11, 19, 21 und 26 (1967-1972) von F.S. "Meteor" in das Gebiet zwischen dem Golf von Biscaya und dem Auftriebsgebiet vor Westafrika. *Meteor Forsch.-Ergeb.* **1981**, *33*, 23–36.
116. Schulze, A.; Timm, L.E. Palolo and un: Distinct clades in the genus *Palola* (Eunicidae: Polychaeta). *Mar. Biodivers.* **2011**, *42*, 161–171. [CrossRef]
117. Nederlof, M.A.J.; Jansen, H.M.; Dahlgren, T.G.; Fang, J.; Meier, S.; Strand, Ø.; Sveier, H.; Verdegem, M.C.J.; Smaal, A.C. Application of polychaetes in (de)coupled integrated aquaculture: Production of a high-quality marine resource. *Aquac. Environ. Interact.* **2019**, *11*, 221–237. [CrossRef]
118. Åkesson, B. The use of certain polychaetes in bioassay studies. *Rapp. Proces Verbaux Reun.* **1980**, *179*, 315–321.
119. Lamarck, J.B. *Histoire Naturelle des Animaux Sans Vertèbres, Préséntant les Caractères Généraux et Particuliers de ces Animaux, Leur Distribution, Leurs Classes, Leurs Familles, Leurs Genres, et la Citation des Principales Espèces qui s'y Rapportent; Precedes d'une Introduction Offrant la Determination des Caracteres Essentiels de L'animal, sa Distinction du Vegetal et Desautres Corps Naturels, Enfin, l'Exposition des Principes Fondamentaux de la Zoologie. Vol. 5*; Deterville Libraire & Verdiere Libraire: Paris, France, 1818; p. 612.

120. Hartmann-Schröder, G.; Zibrowius, H. Polychaeta associated with Antipatharia (Cnidaria: Anthozoa): Description of Polynoidae and Eunicidae. *Mitt. Hambg. Zool. Mus. Inst.* **1998**, *95*, 29–44.
121. Zanol, J.; Fauchald, K.; Paiva, P.C. A phylogenetic analysis of the genus *Eunice* (Eunicidae, polychaete, Annelida). *Zool. J. Linn. Soc.* **2007**, *150*, 413–434. [CrossRef]
122. Molina-Acevedo, I.C.; Carrera-Parra, L.F. Revision of *Marphysa* de Quatrefages, 1865 and some species of *Nicidion* Kinberg, 1865 with the erection of a new genus (Polychaeta: Eunicidae) from the Grand Caribbean. *Zootaxa* **2017**, *4241*, 1–62. [CrossRef]
123. Lu, H.; Fauchald, K. A phylogenetic and biogeographic study of *Euniphysa* (Eunicidae, Polychaeta). *J. Nat. Hist.* **2000**, *34*, 997–1044. [CrossRef]
124. Carrera-Parra, L.F.; Fauchald, K.; Gambi, M.C. Revision of the taxonomic status of *Lysidice* (Polychaeta, Eunicidae) in the Western Caribbean Sea with observation on species reproductive features and habitat preference. *Ital. J. Zool.* **2011**, *78*, 27–40. [CrossRef]
125. Molina-Acevedo, I.C.; Idris, I. Reinstatement of species belonging *Marphysa sanguinea* complex (Annelida: Eunicidae) and description of new species from the mid-Pacific Ocean and the Adriatic Sea. *Zootaxa* **2020**, *4816*, 1–48. [CrossRef] [PubMed]
126. Kara, J.; Molina-Acevedo, I.C.; Zanol, J.; Simon, C.; Idris, I. Morphological and molecular systematic review of *Marphysa* Quatrefages, 1865 (Annelida: Eunicidae) species from South Africa. *PeerJ* **2020**, *8*, e10076. [CrossRef] [PubMed]
127. Fauchald, K. Review of the types of *Palola* (Eunicidae: Polychaeta). *J. Nat. Hist.* **1992**, *26*, 1177–1225. [CrossRef]
128. Molina-Acevedo, I.C. Morphological revision of the Subgroup 1 Fauchald, 1970 of *Marphysa* de Quatrefages, 1865 (Eunicidae: Polychaeta). *Zootaxa* **2018**, *4480*, 1–125. [CrossRef] [PubMed]
129. Molina-Acevedo, I.C. Redescription of two species previously regarded as *Marphysa* de Quatrefages, 1865, with the description of new species of *Treadwellphysa* Molina-Acevedo and Carrera-Parra, 2017. *J. Nat. Hist.* **2019**, *53*, 517–540. [CrossRef]
130. Carrera-Parra, L.F.; Salazar-Vallejo, S.I. A new genus and 12 new species of Eunicidae (Polychaeta) from the Caribbean Sea. *J. Mar. Biol. Assoc. UK* **1998**, *78*, 145–182. [CrossRef]
131. Glasby, C.J.; Hutchings, P.A. A new species of *Marphysa* Quatrefages, 1865 (Polychaeta: Eunicida: Eunicidae) including the genus *Nauphanta* Kinberg, 1865. *Zootaxa* **2010**, *45*, 29–45. [CrossRef]
132. Martin, D.; Gil, J.; Zanol, J.; Meca, M.A.; Portela, R.P. Digging the diversity of Iberian bait worms *Marphysa* (Annelida, Eunicidae). *PLoS ONE* **2020**, *15*, e0226749. [CrossRef] [PubMed]
133. Steiner, T.M.; Nogueira, J.M.D.M.; Amaral, A.C.Z. *Eunice rubra* Grube (Annelida, Polychaeta, Eunicidae), a redescription based on the holotype and Brazilian specimens. *Rev. Bras. Zool.* **2002**, *19*, 243–259. [CrossRef]
134. Parapar, J.; Freire, J.; Urgorri, V.; Besteiro, C. Morphological variability in *Eunice vittata* (Chiaje, 1828) (Polychaeta, Eunicidae) in the Ría de Ferrol (Galicia, NW Spain). *Ophelia* **1993**, *37*, 117–125. [CrossRef]
135. Miura, T. Japanese polychaetes of the genera *Eunice* and *Euniphysa*: Taxonomy and branchial distribution patterns. *Publ. Seto Mar. Biol. Lab.* **1986**, *31*, 269–325. [CrossRef]
136. Salazar-Vallejo, S.I.; Carrera-Parra, L.F. Eunícidos (Polychaeta) del caribe mexicano con claves para las especies del Gran Caribe: *Fauchaldius*, *Lysidice*, *Marphysa*, *Nematonereis* y *Palola*. *Rev. Biol. Trop.* **1998**, *45*, 1481–1498.
137. Parapar, J.; Harto, I. On the presence in the Iberian waters and morphological variability in *Eunice roussaei* Quatrefages, 1866 (Polychaeta, Eunicidae). *Bol. Real Soc. Esp. Hist. Nat. Secc. Biol.* **2001**, *96*, 165–174.
138. Costa-Paiva, E.M.; Paiva, P.C. A morphometric analysis of *Eunice* Cuvier (Annelida, Polychaeta) species. *Rev. Bras. Zool.* **2007**, *24*, 353–358. [CrossRef]
139. Treadwell, A.L. Leodicidae of the West Indian region. *Carnegie Inst. Wash. Publ.* **1921**, *293*, 1–131. [CrossRef]
140. Carrera-Parra, L.F. Eunicidae Berthold, 1827. In *Poliquetos (Annelida: Polychaeta) de México y América Tropical*; de León-González, J.Á., Bastida-Zavala, J.R., Parra, L.F.C., García-Garza, M.E., Peña-Rivera, A., Salazar-Vallejo, S.I., Solís-Weiss, V., Eds.; Universidad Autónoma de Nuevo León: Monterrey, Mexico, 2009; pp. 165–181.
141. Zanol, J.; Budaeva, N. Eunicidae. In *Handbook of Zoology Annelida, Volume 3: Sedentaria III, Errantia I*; Purschke, G., Westheide, W., Böggemann, M., Eds.; De Gruyter: Berlin, Germany; Boston, MA, USA, 2021; pp. 414–452.
142. Lachat, J.; Haag-Wackernagel, D. Novel mobbing strategies of a fish population against a sessile annelid predator. *Sci. Rep.* **2016**, *6*, 33187. [CrossRef]
143. Hofmann, D.K. Nachweise der Epitokie als Fortpflanzungsmodus des Polychaeten *Eunice siciliensis* (Polychaeta: Annelida). *Mar. Biol.* **1972**, *11*, 341–344. [CrossRef]
144. Pamungkas, J. Species richness and macronutrient content of wawo worms (Polychaeta, Annelida) from Ambonese waters, Maluku, Indonesia. *Biodiveristy Data J.* **2015**, *3*, e4251. [CrossRef] [PubMed]
145. Southern, R. Polychaeta of the Chilka Lake and also of fresh and brackish water in other parts of India. *Mem. Indian Mus.* **1921**, *4*, 563–659.
146. Aiyar, R.G. An account of the development and breeding habits of a brackish water Polychaete worm of the genus *Marphysa*. *J. Linn. Soc. Lond.* **1931**, *37*, 387–403. [CrossRef]
147. Pillai, T.G. Studies on a brackish-water polychaetous annelid, *Marphysa borradailei*, sp. n. from Ceylon. *Ceylon J. Sci. (Biol. Sci.)* **1958**, *1*, 94–106.
148. Pettibone, M.H. Marine polychate worms of the New England region Part 1. Aphroditidae through Trochochetidae. *United States Natl. Mus. Bull.* **1963**, *227*, 1–356.
149. Åkesson, B. The embryology of the Polychaete *Eunice kobiensis*. *Acta Zool.* **1967**, *48*, 141–192. [CrossRef]

150. Prevedelli, D.; N'siala, G.M.; Ansaloni, I.; Simonini, R. Life cycle of *Marphysa sanguinea* (Polychaeta: Eunicidae) in the Venice Lagoon (Italy). *Mar. Ecol.* **2007**, *28*, 384–393. [CrossRef]
151. Malathi, E.; Sunder Raj, S.K.; Mercy Bai, P. Larval development of *Marphysa gravelyi* (Polychaeta: Eunicidae) from Pulicat Lake, India. *Ital. J. Zool.* **2011**, *78*, 249–254. [CrossRef]
152. Herpin, R. Recherches biologiques sur la reproduction et le développement de quelques Annélides polychètes. *Bull. Société Sci. Nat. L'ouest Fr.* **1925**, *5*, 1–250.
153. Goldstein, M.C.; Carson, H.S.; Eriksen, M. Relationship of diversity and habitat area in North Pacific plastic-associated rafting communities. *Mar. Biol.* **2014**, *161*, 1441–1453. [CrossRef]
154. Jang, M.; Shim, W.J.; Han, G.M.; Song, Y.K.; Hong, S.H. Formation of microplastics by polychaetes (*Marphysa sanguinea*) inhabiting expanded polystyrene marine debris. *Mar. Pollut. Bull.* **2018**, *131*, 365–369. [CrossRef] [PubMed]
155. Díaz-Díaz, O.; Liñero-Arana, I. Epifaunal polychaetes on *Pinctada imbricata* Roding, 1798 (Bivalvia: Pteriidae) from the Gulf of Cariaco Venezuela. *Interciencia* **2003**, *28*, 298–301.
156. Hartman, O. Descriptions of new species and new generic records of polychaetous annelids from California of the families Glyceridae, Eunicidae, Stauronereidae, and Opheliidae. *Univ. Calif. Publ. Zool.* **1938**, *43*, 93–111.
157. Gambi, M.C. Spatio-temporal distribution and ecological role of polychaete borers of *Posidonia oceanica* (L.) Delile scales. *Bull. Mar. Sci.* **2002**, *71*, 1323–1331.
158. Molodtsova, T.; Budaeva, N. Modifications of corallum morphology in black corals as an effect of associated fauna. *Bull. Mar. Sci.* **2007**, *81*, 469–479.
159. Quiroga, E.; Levin, L.A. *Eunice pennata* (Polychaeta: Eunicidae) From Active and Passive Cold Seep Sites in Central and Southern Chile (36–46° S). *An. Inst. Patagon.* **2010**, *38*, 31–37. [CrossRef]
160. Mueller, C.E.; Lundälv, T.; Middelburg, J.J.; van Oevelen, D. The Symbiosis between *Lophelia pertusa* and *Eunice norvegica* stimulates coral calcification and worm assimilation. *PLoS ONE* **2013**, *8*, e58660. [CrossRef]
161. Kise, H.; Reimer, J.D. Unexpected diversity and a new species of *Epizoanthus* (Anthozoa, hexacorallia) attached to eunicid worm tubes from the Pacific Ocean. *Zookeys* **2016**, *2016*, 49–71. [CrossRef]
162. Roberts, J.M. Reef-aggregating behaviour by symbiotic eunicid polychaetes from cold-water corals: Do worms assemble reefs? *J. Mar. Biol. Assoc. UK* **2005**, *85*, 813–819. [CrossRef]
163. Hutchings, P.A. Biological destruction of coral reefs. *Coral Reefs* **1986**, *4*, 239–252. [CrossRef]
164. Chisholm, J.R.M.; Kelley, R. Worms start the reef-building process. *Nature* **2001**, *409*, 152. [CrossRef]
165. Ochoa-Rivera, V.; Granados-Barba, A.; Solís-Weiss, V. The polychaete cryptofauna from Cozumel Island, Mexican Caribbean. *Bull. Mar. Sci.* **2000**, *67*, 137–146.
166. Hartman, O. Polychaeta Errantia of Antarctica. *Antarct. Res. Ser.* **1964**, *3*, 1–131.
167. Winsnes, I.M. Eunicid polychaetes (Annelida) from Scandinavian and adjacent waters. Family Eunicidae. *Zool. Scr.* **1989**, *18*, 483–500. [CrossRef]
168. Amoureux, L. Annélides Polychètes profondes de Madagascar Description de deux nouvelles espèces (Collections Crosnier et Jouannie). *Bull. Muséum Natl. d'Histoire Nat.* **1978**, *344*, 1093–1108.
169. Nonato, E.F. Contribuição ao Conhecimento dos Anelídeos Poliquetas Bentônicos da Plataforma Continental Brasileira, Entre Cabo Frio e o Arroio Chuí. Livre Docência Thesis, Universidade de São Paulo, São Paulo, Brazil, 1981.
170. Orensanz, J.M. Los anelidos poliquetos de la provincia biogeografica Argentina. VII. Eunicidae y Lysaretidae. *Physis* **1975**, *34*, 85–111.
171. Fauchald, K. Review of the types and key to the species of *Eunice* (Eunicidae: Polychaeta) from the Australian region. *Rec. Aust. Mus.* **1986**, *38*, 241–262. [CrossRef]
172. Zanol, J.; da Silva, T.S.C.; Hutchings, P. One new species and two redescriptions of *Marphysa* (Eunicidae, Annelida) species of the Aenea-group from Australia. *Zootaxa* **2017**, *4268*, 411–426. [CrossRef] [PubMed]
173. Zanol, J.; Hutchings, P.A.; Fauchald, K. *Eunice sensu latu* (Annelida: Eunicidae) from Australia: Description of seven new species and comments on previously reported species of the genera *Eunice*, *Leodice* and *Nicidion*. *Zootaxa* **2020**, *4748*, 1–43. [CrossRef]
174. Schulze, A. Six genetically distinct clades of *Palola* (Eunicidae, Annelida) from Lizard Island, Great Barrier Reef, Australia. *Zootaxa* **2015**, *4019*, 695–706. [CrossRef] [PubMed]
175. Kott, P. Nereidae and Eunicidae of south Western Australia; also notes on the ecology of Western Australian limestone reefs. *J. R. Soc. West. Aust.* **1951**, *35*, 85–130.
176. Morgado, E.H.; Amaral, A.C.Z. Anelidos poliquetos associados a um briozoario I. Eunicidae, Lumbrineridae, Lysaretidae e Dorvilleidae. *Iheringia* **1981**, *60*, 33–54.
177. Steiner, T.; Amaral, A.C.Z. Two New Species of *Marphysa* Quatrefages, 1865 (Eunicidae, Polychaeta) From Intertidal Sandy Beaches of the São Sebastião Channel, State of São Paulo (Brazil). *Bull. Mar. Sci.* **2000**, *67*, 479–489.
178. Zanol, J.; Paiva, P.C.; Attolini, F.S. *Eunice* and *Palola* (Eunicidae: Polychaeta) from the eastern Brazilian Coast (13°00′–22°30′ S). *Bull. Mar. Sci.* **2000**, *67*, 449–463.
179. Nogueira, J.M.M.; Steiner, T.M.; Amaral, A.C.Z. Descriptions of two new species of *Eunice* Cuvier, 1817 (Polychaeta: Eunicidae) from coastal islands of the State of São Paulo, Brazil. *Sci. Mar.* **2001**, *65*, 47–57. [CrossRef]
180. Bergamo, G.; Carrerette, O.; Zanol, J.; Nogueira, J.M.M. Species of Eunicidae (Annelida, Errantia, Eunicida) from rhodolith beds off the states of Paraíba and Pernambuco, northeastern Brazilian coast. *Zootaxa* **2018**, *4521*, 376–390. [CrossRef] [PubMed]

181. George, J.D.; Hartmann-Schröder, G. Polychaetes: British Amphinomida, Spintherida and Eunicida. Keys and notes for the identification of the species. *Synop. Br. Fauna* **1985**, *32*, 1–221.
182. Núñez, J.; Pascual, M.; Delgado, J.D.; Brito, M.C. Anélidos poliquetos de Canarias: Familia Eunicidae. *Vieraea* **1997**, *26*, 47–75.
183. Carrera-Parra, L.F.; Salazar-Vallejo, S.I. Eunícidos (Polychaeta) del Caribe mexicano con claves para las especies del Gran Caribe: Eunice. *Rev. Biol. Trop.* **1998**, *45*, 1499–1521.
184. Salazar-Vallejo, S.I.; Carrera-Parra, L.F.; de León-González, J.A. Giant Eunicid Polychaetes (Annelida) in shallow tropical and temperate seas. *Rev. Biol. Trop.* **2011**, *59*, 1463–1474. [CrossRef]
185. Carrera-Parra, L.F.; Salazar-Vallejo, S.I. Redescriptions of *Eunice filamentosa* and *E. denticulata* and description of *E. tovarae* n. sp. (Polychaeta: Eunicidae), highlighted with morphological and molecular data. *Zootaxa* **2011**, *2880*, 51–64. [CrossRef]
186. de Léon-González, J.A. A new eunicid polychaete from the Caribbean Coast of Mexico. *Rev. Biol. Trop.* **1988**, *36*, 75–79.
187. Liu, Y.; Hutchings, P.; Sun, S. Three new species of *Marphysa* Quatrefages, 1865 (Polychaeta: Eunicida: Eunicidae) from the south coast of China and redescription of *Marphysa sinensis* Monro, 1934. *Zootaxa* **2017**, *4263*, 228–250. [PubMed]
188. Liu, Y.; Hutchings, P.; Kupriyanova, E. Two new species of *Marphysa* Quatrefages, 1865 (Polychaeta: Eunicida: Eunicidae) from northern coast of China and redescription for *Marphysa orientalis* Treadwell, 1936. *Zootaxa* **2018**, *4377*, 191. [CrossRef] [PubMed]
189. de León-González, J.A.; Rivera, C.G.; Romero, M.Y. Sublittoral Eunicidae and Onuphidae (Polychaeta) from softbottom of El Salvador, eastern Pacific. *J. Mar. Biol. Assoc. UK* **2004**, *84*, 93–101.
190. Treadwell, A.L. Leodicidae from Fiji and Samoa. *Carnegie Inst. Washingt. Publ.* **1922**, *312*, 127–170.
191. Lavesque, N.; Daffe, G.; Bonifácio, P.; Hutchings, P. A new species of the *Marphysa sanguinea* complex from French waters (Bay of Biscay, NE Atlantic) (Annelida, Eunicidae). *Zookeys* **2017**, *716*, 1–17. [CrossRef]
192. Gathof, J.M. Family Eunicidae Savigny, 1818. In *Taxonomic Guide to the Polychaetes of the Northern Gulf of Mexico. Final Report to the Minerals Management Service, Contract 14-12-001-29091. Volume VI*; Uebelacker, J.M., Johnson, P.G., Eds.; U.S. Department of the Interior Mineral Management Service Gulf of Mexico OCS.: Mobile, Italy, 1984; pp. 40.1–40.31.
193. de León-González, J.Á.; Díaz-Castañeda, V. Eunicidae (Annelida: Polychaeta) associated with *Phragmathopoma caudata* Morch, 1863 and some coral reefs from Veracruz, Gulf of Mexico. *Sci. Mar.* **2006**, *70*, 91–99. [CrossRef]
194. Wang, Z.; Zhang, Y.; Qiu, J. A New Species in the *Marphysa sanguinea* Complex (Annelida, Eunicidae) from Hong Kong. *Zool. Stud.* **2018**, *48*, 1–13. [CrossRef]
195. Arias, A.; Fernández-Álvarez, F.Á.; Martins, R.; Anadón, N. Rediscovery and redescription of *Leodice laurillardi* (Quatrefages, 1866) comb. nov. (Annelida: Eunicidae)—A rare European polychaete or just an overlooked species? *Zootaxa* **2015**, *3964*, 475–481. [CrossRef]
196. Hutchings, P.; Lavesque, N.; Priscilla, L.; Daffe, G.; Malathi, E.; Glasby, C.J. A new species of *Marphysa* (Annelida: Eunicida: Eunicidae) from India, with notes on previously described or reported species from the region. *Zootaxa* **2020**, *4852*, 285–308. [CrossRef] [PubMed]
197. Veeramuthu, S.; Ramadoss, R.; Subramaniyan, B.; Jeyaram, S.; Fernando, O.J. Abundance of the Boring Polychaetes of Eunicidae (Annelida) in Great Nicobar Islands. *Our Nat.* **2013**, *10*, 76–88. [CrossRef]
198. Miura, T. Eunicid polychaetous annelids from Japan—I. *La Mer (Bulletin la Société Fr. D'océanographie)* **1977**, *15*, 1–20.
199. Miura, T. Eunicid polychaetous annelids from Japan—II. *La Mer (Bulletin la Société Fr. D'océanographie)* **1977**, *15*, 61–81.
200. Miura, T. Eunicid polychaetous annelids from Japan—III. *La Mer (Bulletin la Société Fr. D'océanographie)* **1979**, *17*, 33–42.
201. Miura, T. New or little known species of the Family Eunicidae (Annelida, Polychaeta) from Japan. *Proc. Jpn. Soc. Syst. Zool.* **1987**, *36*, 1–9.
202. Abe, H.; Tanaka, M.; Taru, M.; Abe, S.; Nishigaki, A. Molecular evidence for the existence of five cryptic species within the Japanese species of *Marphysa* (Annelida: Eunicidae) known as "iwa-mushi". *Plankt. Benthos Res.* **2019**, *14*, 303–314. [CrossRef]
203. Choi, H.K.; Kim, J.G.; Kang, D.W.; Yoon, S.M. A new species of *Leodice* from Korean waters (Annelida, Polychaeta, Eunicidae). *Zookeys* **2017**, *2017*, 53–67. [CrossRef]
204. Idris, I.; Hutchings, P.; Arshad, A. Description of a new species of *Marphysa* Quatrefages, 1865 (Polychaeta: Eunicidae) from the west coast of Peninsular Malaysia and comparisons with species from *Marphysa* Group A from the Indo-West Pacific and Indian Ocean. *Mem. Mus. Vic.* **2014**, *71*, 109–121. [CrossRef]
205. Katsiaras, N.; Simboura, N.; Koutsoubas, D. The rare subgroup C1 of *Marphysa* (Polychaeta, Eunicidae): Re-description of species and first records in the Mediterranean Sea. *Zootaxa* **2014**, *3873*, 201–217. [CrossRef] [PubMed]
206. Kurt-Sahin, G.; Çinar, M.E. Eunicidae (Polychaeta) species in and around Iskenderun Bay (Levantine Sea, Eastern Mediterranean) with a new alien species for the Mediterranean Sea and a re-description of *Lysidice collaris*. *Turk. J. Zool.* **2009**, *33*, 331–347. [CrossRef]
207. Kurt-Sahin, G.; Çinar, M.E. Presence of *Marphysa disjuncta* (Polychaeta: Eunicidae) in the Mediterranean Sea. *Mediterr. Mar. Sci.* **2009**, *10*, 145–150. [CrossRef]
208. Kurt-Sahin, G. *Marphysa cinari*, a new species of Eunicidae (Polychaeta) from the coasts of Turkey (eastern Mediterranean) and re-descriptions of *Marphysa kinbergi* McIntosh, 1910 and *Marphysa disjuncta* Hartman, 1961. *J. Nat. Hist.* **2014**, *48*, 1989–2006. [CrossRef]
209. Kurt-Sahin, G.; Çinar, M.E. Distribution of Eunicidae (Annelida: Polychaeta) along the Levantine coast of Turkey, with special emphasis on alien species. *Mar. Biodivers.* **2017**, *47*, 421–431. [CrossRef]

210. Cantone, G. Censimento dei policheti dei mari italiani: Eunicidae Berthold, 1827. *Atti Soc. Toscana Sci. Nat. Resid. Pisa Mem. Ser. B* **1993**, *100*, 229–243.
211. de León-González, J.A. *Eunice orensanzi* n.sp. from the western coast of Baja California Sur, México and key to the Mexican *Eunice* (Polychaeta: Eunicidae). *Rev. Biol. Trop.* **1990**, *38*, 259–266.
212. Ruiz-Cancino, G.; Carrera-Parra, L.F.; Bastida-Zavala, R. Eunícidos (Polychaeta: Eunicidae) del Pacífico sur de México. *Ciencia y Mar* **2010**, *14*, 27–60.
213. Knox, G.A.; Green, K.M. The polychaetous annelids of New Zealand Part 4. Eunicidae. *J. R. Soc. N. Z.* **1972**, *2*, 459–470. [CrossRef]
214. Glasby, C.J.; Mandario, M.A.E.; Burghardt, I.; Kupriyanova, E.; Gunton, L.M.; Hutchings, P.A. A new species of the sanguinea-group Quatrefages, 1866 (Annelida: Eunicidae: *Marphysa*) from the Philippines. *Zootaxa* **2019**, *4674*, 264–282. [CrossRef]
215. Díaz, A.; López, E. Four New Records of Eunicidae (Annelida: Polychaeta) from the Western Philippines Islands. *Thalassas* **2020**, *36*, 321–332. [CrossRef]
216. Day, J.H. *A Monograph on the Polychaeta of Southern Africa. Part I. Errantia*; Trustees of the British Museum (Natural History): London, UK, 1967; p. 458.
217. Lewis, C.; Karageorgopoulos, P. A new species of *Marphysa* (Eunicidae) from the western Cape of South Africa. *J. Mar. Biol. Assoc. UK* **2008**, *88*, 277–287. [CrossRef]
218. Wu, Q.Q.; He, M.H. A new genus and new species of Eunicidae from Taiwan Strait. *Acta Zootaxon. Sin.* **1988**, *13*, 123–126.
219. Hsueh, P.-W.; Li, Y.-H. New species and new records of eunicids (Polychaeta, Eunicidae) from Taiwan. *Zootaxa* **2014**, *3802*, 151–172. [CrossRef] [PubMed]
220. Ehlers, E. Reports on the results of dredging, under the direction of L. F. Pourtalès, during the years 1868-1870, and of Alexander Agassiz, in the Gulf of Mexico (1877-78), and in the Caribbean Sea (1878-79), in the U.S. Coast Survey steamer "Blake", Lieut-Com. C. D. Sigsbee, U.S.N. and Commander J. R. Bartlett, U.S.N., commanding. XXXI. Report on the Annelids. *Mem. Mus. Comp. Zool. Harv. Coll.* **1887**, *15*, 1–335.
221. Ebbs, N.K. The coral-inhabiting polychaetes of the northern Florida reef tract. Part I. Aphroditidae, Polynoidae, Amphinomidae, Eunicidae and Lysaretidae. *Bull. Mar. Sci.* **1966**, *16*, 485–555.
222. Hartman, O. Polychaetous annelids from California, including the descriptions of two new genera and nine new species. *Allan Hancock Pac. Exped.* **1944**, *10*, 239–307.
223. Liñero-Arana, I. Poliquetos errantes bentónicos de la platforma continental nor-oriental de Venezuela. II. Eunicidae. *Bol. Inst. Oceanogr. Venez. Univ. Oriente* **1990**, *24*, 91–103.
224. Díaz-Díaz, O.F.; Bone, D.; Rodríguez, C.T.; López, A. Polychaetes associated to calcareous sediments, Venezuela: Eunicidae. *Panam. J. Aquat. Sci.* **2016**, *11*, 18–25.
225. Iannotta, M.A.; Gambi, M.C.; Patti, F.P. Molecular evidence of intraspecific variability in *Lysidice ninetta* (Polychaeta: Eunicidae) in the Mediterranean Sea. *Aquat. Biol.* **2009**, *6*, 121–132. [CrossRef]
226. Schulze, A. The Bobbit worm dilemma: A case for DNA. *Rev. Biol. Trop.* **2011**, *59*, 1475–1477.
227. Hutchings, P.; Kupriyanova, E. Cosmopolitan polychaetes—Fact or fiction? Personal and historical perspectives. *Invertebr. Syst.* **2017**, *32*, 1–9. [CrossRef]
228. Iannotta, M.A.; Patti, F.P.; Ambrosino, M.; Procaccini, G.; Gambi, M. Phylogeography of two species of *Lysidice* (Polychaeta, Eunicidae) associated to the seagrass *Posidonia oceanica* in the Mediterranean Sea. *Mar. Biol.* **2007**, *150*, 1115–1126. [CrossRef]
229. Martin, D. Anélidos poliquetos asociados a las concreciones de algas calcáreas del litoral Catalán. *Miscellània Zoològica Barc.* **1987**, *11*, 61–75.
230. Lavesque, N.; Daffe, G.; Grall, J.; Zanol, J.; Gouillieux, B.; Hutchings, P. Guess who? On the importance of using appropriate name: Case study of *Marphysa sanguinea* (Montagu, 1813). *Zookeys* **2019**, *859*, 1–15. [CrossRef] [PubMed]
231. Pettibone, M.H. Annelida. In *Synopsis and Classification of Living Organisms. 2*; Parker, S.P., Ed.; McGraw-Hill Book Co.: New York, NY, USA, 1982; pp. 1–43.
232. Carrera-Parra, L.F. Redescription of *Hartmaniella tulearensis* n. comb. (Amoureux, 1978) with comments on *Hartmaniella* sp and affinities of the family (Polychaeta: Hartmaniellidae). *J. Nat. Hist.* **2003**, *37*, 49–55. [CrossRef]
233. Gathof, J.M. Family Hartmaniellidae Imajima, 1977. In *Taxonomic Guide to the Polychaetes of the Northern Gulf of Mexico. Final Report to the Minerals Management Service, Contract 14-12-001-29091. Volume VII*; Uebelacker, J.M., Johnson, P.G., Eds.; Barry A. Vittor & Associates Inc.: Mobile, AL, USA, 1984; pp. 57.1–57.3.
234. Eibye-Jacobsen, D.; Oug, E. A report of Hartmaniellidae (Annelida: Polychaeta) from the Andaman Sea, Indian Ocean. *Phuket Mar. Biol. Cent. Spec. Publ.* **2002**, *24*, 101–114.
235. Al-Hakim, I.; Glasby, C.J. Polychaeta (Annelida) of the Natuna Islands, South China Sea. *Raffles Bull. Zool.* **2004**, *11*, 25–45.
236. Imajima, M. A new polychaete family, Hartmaniellidae, from Japan. In *Essays on Polychaetous Annelids in Memory of Dr. Olga Hartman*; Reish, D.J., Fauchald, K., Eds.; The Allan Hancock Foundation, University of Southern California: Los Angeles, CA, USA, 1977; pp. 211–216.
237. Fauchald, K.; Rouse, G.W. Polychaete systematics: Past and present. *Zool. Scr.* **1997**, *26*, 71–138. [CrossRef]
238. Szaniawski, H.; Imajima, M. Hartmaniellidae—Living fossils among polychaetes. *Acta Palaeontol. Pol.* **1996**, *41*, 111–125.
239. Lu, H. Systematics of Eunicida (Polychaeta: Annelida) Homology, Ontogeny, Morphometrics, Fossils and Phylogeny. Ph.D. Thesis, George Washington University, Washington, DC, USA, 2006.

240. Van Beneden, P.J. Histoire naturelle d'un animal nouveau designe sous le nom d'Histriobdella. *Bull. l'Académie R. Belg. Brux.* **1858**, *2*, 270–303.
241. Foettinger, A. Recherches sur l'organisation de *Histriobdella homari*, P.-J. Van Beneden rapportée aux Archiannélides. *Arch. Biol. (Liege)* **1884**, *5*, 435–516.
242. Shearer, C. On the anatomy of *Histriobdella homari*. *Q. J. Microsc. Sci.* **1910**, *55*, 17–20.
243. Cordero, E.H. Un nuevo Arquianélido, *Stratiodrilus platensis*, n. sp., que habita sobre *Aegla laevis* (Latr.). Nota preliminar. *Physis* **1927**, *8*, 574–578.
244. Roubaud, G. Recherches sur les *Stratiodrilus platensis* Cordero, Archiannelides parasites des *Aegla* des lacs de Patagonie. In *Biologie de l'Amerique Australe II. Études sur les Faune du sol*; Deboutteville, C.D., Rapport, E., Eds.; Éditions du Centre National de la Recherche Scientifique: Paris, France, 1962; pp. 31–54.
245. Mesnil, F.; Caullery, M. L'appareil maxillaire *d'Histriobdella homari*; affinités des Hitriobdellides avec les Euniciens. *Comptes Rendus l'Académie des Sci.* **1922**, *174*, 913–917.
246. Tzetlin, A.B. *Ophryotrocha schubravyi* sp.n. and the problem of evolution of the mouthparts in the Eunicemorpha (Polychaeta). *Zool. Zhurnal* **1980**, *59*, 666–676. (In Russian)
247. Tzetlin, A.; Budaeva, N.; Vortsepneva, E.; Helm, C. New insights into the morphology and evolution of the ventral pharynx and jaws in Histriobdellidae (Eunicida, Annelida). *Zool. Lett.* **2020**, *6*, 1–19. [CrossRef]
248. Gelder, S.R.; Jennings, J.B. The nervous system of the aberrant symbiotic polychaete *Histriobdella homari* and its implications for the taxonomic position of the Histriobdellidae. *Zool. Anz.* **1975**, *194*, 293–304.
249. Jennings, J.B.; Gelder, S.R. Observations on the Feeding Mechanism, Diet and Digestive Physiology of *Histriobdella homari* Van Beneden 1858: An Aberrant Polychaete Symbiotic with North American and European Lobsters. *Biol. Bull.* **1976**, *151*, 489–517. [CrossRef]
250. Fauchald, K. The polychaete worms. Definitions and keys to the orders, families and genera. *Nat. Hist. Mus. Los Angeles Cty. Sci. Ser.* **1977**, *28*, 1–188.
251. Harrison, L. On the genus *Stratiodrilus* (Archiannelida: Histriobdella) with a description of a new species from Madagascar. *Rec. Aust. Mus.* **1928**, *16*, 116–121. [CrossRef]
252. Dioni, W. *Didymorchis*, *Temnocephala* (Platyhelminthes) y *Stratiodrilus* (Annelida) vermes epizoicos sobre *Aegla* y *Parastacus* (Crustacea, Decapoda) de lagos Andino-Patagonicos: Notas taxonomicas y biogeograficas. *Acta Zool. Lilloana* **1972**, *29*, 167–179.
253. Rouse, G.W.; Pleijel, F. *Polychaetes*; Oxford University Press: New York, NY, USA, 2001; p. 354.
254. Glasby, C.J.; Timm, T. Global diversity of polychaetes (Polychaeta; Annelida) in freshwater. *Hydrobiologia* **2008**, *595*, 107–115. [CrossRef]
255. Rosa, J.J.D.S.; Marçal, I.C.; Teixeira, G.M.; Aguiar, A. Checklist of species of *Stratiodrilus* Haswell, 1900 (Annelida: Histriobdellidae), and new host records from Southern Brazil. *Zootaxa* **2018**, *4399*, 412–422. [CrossRef] [PubMed]
256. Steiner, T.M.; Amaral, A.C.Z. The family Histriobdellidae (Annelida, Polychaet) including descriptions of two new species from Brazil and a new genus. *Contrib. Zool.* **1999**, *68*, 1–17. [CrossRef]
257. Zhang, Y.T. New substitute name for the genus *Dayus* Steiner & Amaral, 1999 (Annelida: Polychaeta: Histriobdellidae). *Zootaxa* **2014**, *3802*, 400.
258. Amato, J.F.R.; Daudt, L.C.C.; Amato, S.B. New species of *Stratiodrilus* (Polychaeta, Histriobdellidae) from freshwater anomuran crustaceans of Southern Brazil. *Biociências* **2004**, *12*, 121–127.
259. Lang, K. A contribution to the morphology of *Stratiodrilus platensis* Cordero (Histriobdellidae). *Ark. Zool.* **1949**, *42*, 1–30.
260. Rouse, G.W. Histriobdellidae Vaillant, 1890. In *Polychaetes*; Rouse, G., Pleijel, F., Eds.; Oxford University Press: New York, NY, USA, 2001; pp. 160–163.
261. Gelder, S.R.; Tyler, S. Anatomical and cytochemical studies on the adhesive organs of the ectosymbiont *Histriobdella homari* (Annelida: Polychaeta). *Trans. Am. Microsc. Soc.* **1986**, *105*, 348–356. [CrossRef]
262. Haswell, W.A. On a new histriobdellid. *Q. J. Microsc. Sci.* **1900**, *43*, 299–335.
263. Haswell, W.A. Notes on the Histriobdellidae. *Q. J. Microsc. Sci.* **1913**, *59*, 197–226.
264. Daudt, L.C.C.; Amato, J.F.R. Morphological variation of *Stratiodrius circensis* (Polychaeta, Histriobdellidae) from a new host, *Aegla leptodactyla* (Crustacea, Anomura, Aeglidae) with identification of its type host species. *Zootaxa* **2007**, *1450*, 57–62. [CrossRef]
265. Jamieson, B.G.M.; Afzelius, B.A.; Franzén, Å. Ultrastructure of the acentriolar, aflagellate spermatozoa and the eggs of *Histriobdella homari* and *Stratiodrilus novaehollandiae* (Histriobdellidae, Polychaeta). *J. Submicrosc. Cytol.* **1985**, *17*, 363–372.
266. Simon, J.L. Incidence and behavior of *Histriobdella homari* (Annelida: Polychaeta) a commensal of the American lobster. *Bioscience* **1968**, *18*, 35–36.
267. Boghen, A.D. A parasitological survey of the American lobster *Homarus americanus* from the Northumberland Strait, southern Gulf of St. Lawrence. *Can. J. Zool.* **1978**, *56*, 2460–2462. [CrossRef]
268. Rouse, G.W. Polychaeta (Bristle Worms)—Chapter 5: Minor groups and fossils. In *Marine Parasitology*; Rohde, K., Ed.; CSIRO Publishing: Victoria, Australia, 2005; pp. 193–196.
269. Martin, D.; Britayev, T.A. Symbiotic polychaetes: Review of known species. *Oceanogr. Mar. Biol. Annu. Rev.* **1998**, *36*, 217–340.
270. Brattey, J.; Campbell, A. Occurrence of *Histriobdella homari* (Annelida: Polychaeta) on the American lobster in the Canadian Maritimes. *Can. J. Zool.* **1985**, *63*, 392–395. [CrossRef]

271. Lerch, F.; Uglem, I. High density of *Histriobdella homari* van Beneden, 1858 (Annelida, Polychaeta) on ovigerous female european lobsters (Decapoda, Nephropidae). *Crustaceana* **1996**, *69*, 916–921. [CrossRef]
272. Führ, I.M. A new histriobdellid on a marine isopod from South Africa. *S. Afr. J. Sci.* **1971**, *67*, 325–326.
273. Bond-Buckup, G.; Buckup, L. A Família Aeglidae (Crustacea, Decapoda, Anomura). *Arq. Zool.* **1994**, *32*, 159–346.
274. Briggs, R.P.; Atkinson, R.J.A.; McAliskey, M.; Rogerson, A. *Histriobdella homari* on *Nephrops norvegicus* from the Irish Sea and Clyde Sea area. *J. Mar. Biol. Assoc. UK* **1997**, *77*, 557–559. [CrossRef]
275. Haswell, W.A. On the embryology of *Stratiodrilus* (Histriobdellidae). *Q. J. Microsc. Sci.* **1916**, *61*, 301–311.
276. Amato, J.F.R.; Amato, S.B.; Daudt, L.C.C. The eggs of *Stratiodrilus* sp. (Polychaeta, Histriobdellidae). *Biociências* **2002**, *10*, 49–54.
277. Vila, I.; Véliz, D. Life cycle of a freshwater Gondwanic remnant polychaete *Stratiodrilus aeglaphilus* (Annelida: Eunicida: Histriobdellidae), commensal with *Aegla laevis* (Crustacea: Anomura). *Guyana* **2014**, *78*, 120–126. [CrossRef]
278. Cannon, L.R.G.; Jennings, J.B. Occurrence and nutritional relationships of four ectosymbiotes of the freshwater crayfishes *Cherax dispar* Riek and *Cherax punctatus* Clark (Crustacea: Decapoda) in Queensland. *Aust. J. Mar. Freshw. Res.* **1987**, *38*, 419–427. [CrossRef]
279. Pardo, R.; Vila, I.; Bustamante, R. Demographic response of *Stratiodrilus aeglaphilus* (Annelida, Histriobdellidae) to organic enrichment: Experimental assessement. *Hydrobiologia* **2008**, *598*, 11–16. [CrossRef]
280. Uzmann, J.R. *Histriobdella homari* (Annelida: Polychaeta) in the American lobster, *Homarus americanus*. *J. Parasitol.* **1967**, *53*, 210–211. [CrossRef]
281. Moyano, H.I.G.; Carrasco, F.D.; Gacitua, S. Sobre las especies chilenas de *Stratiodrilus* Haswell, 1900, (Polychaeta, Histriobdellidae). *Bol. Soc. Biol. Concepc.* **1993**, *64*, 147–157.
282. Carrera-Parra, L.F. Lumbrineridae Schmarda, 1861. In *Poliquetos (Annelida: Polychaeta) de México y América Tropical*; de León-González, J.Á., Bastida-Zavala, J.R., Carrera-Parra, L.F., García-Garza, M.E., Peña-Rivera, A., Salazar-Vallejo, S.I., Solís-Weiss, V., Eds.; Universidad Autónoma de Nuevo León: Monterrey, Mexico, 2009; pp. 263–275.
283. Carrera-Parra, L.F. Recognition of *Cenogenus* Chamberlin, 1919 (Polychaeta: Lumbrineridae) based on type material. *Proc. Biol. Soc. Wash.* **2001**, *114*, 720–724.
284. Orensanz, J.M. Los anelidos poliquetos de la provincia biogeografica Argentina. IV. Lumbrineridae. *Physis* **1973**, *32*, 343–393.
285. Frame, A.B. The lumbrinerids (Annelida: Polychaeta) collected in two northwestern Atlantic surveys with descriptions of a new genus and two new species. *Proc. Biol. Soc. Wash.* **1992**, *105*, 185–218.
286. Carrera-Parra, L.F. Phylogenetic analysis of Lumbrineridae Schmarda, 1861 (Annelida: Polychaeta). *Zootaxa* **2006**, *1332*, 1–36. [CrossRef]
287. Carrera-Parra, L.F. Phylogenetic relationship of *Lumbricalus* Frame, 1992 (Polychaeta: Lumbrineridae), with ecological and distribution data. *Mar. Ecol.* **2005**, *26*, 166–170. [CrossRef]
288. Carrera-Parra, L.F.; Orensanz, J.M. Revision of *Kuwaita* Mohammad, 1973 (Annelida, Polychaeta, Lumbrineridae). *Zoosystema* **2002**, *24*, 273–281.
289. Carrera-Parra, L.F. Revision of *Lumbricalus* (Polychaeta: Lumbrineridae). *J. Mar. Biol. Assoc. UK* **2004**, *84*, 81–91. [CrossRef]
290. Carrera-Parra, L.F. Revision of *Lumbrineris* de Blainville, 1828 (Polychaeta: Lumbrineridae). *Zootaxa* **2006**, *1336*, 1–64. [CrossRef]
291. Hilbig, B. Family Lumbrineridae Malmgren, 1867, emended Orensanz, 1990. In *Taxonomic Atlas of the Benthic Fauna of the Santa Maria Basin and Western Santa Barbara Channel. 5—The Annelida Part 2*; Blake, J.A., Hilbig, B., Scott, P.H., Eds.; Santa Barbara Museum of Natural History: Santa Barbara, CA, USA, 1995; pp. 279–313.
292. Pleijel, F. Lumbrineridae Schmarda, 1861. In *Polychaetes*; Rouse, G., Pleijel, F., Eds.; Oxford University Press: New York, NY, USA, 2001; pp. 164–165.
293. Pruvot, G. Recherches anatomiques et morphologiques sur le système nerveux des Annélides polychètes. *Arch. Zool. Expérimentale Générale* **1885**, *3*, 211–326.
294. Fauchald, K.; Jumars, P. The diet of worms: A study of polychaete feeding guilds. *Oceanogr. Mar. Biol. Annu. Rev.* **1979**, *17*, 193–284. [CrossRef]
295. Valderhaug, V.A. Population structure and production of *Lumbrineris fragilis* (Polychaeta, Lumbrineridae) in the Oslofjord (Norway) with a note on metal content of jaws. *Mar. Biol.* **1985**, *86*, 203–211. [CrossRef]
296. Carrasco, F.D.; Oyarzún, C. Diet of the polychaete *Lumbrineris tetraura* (Schmarda) (Lumbrineridae) in a polluted soft-bottom environmen. *Bull. Mar. Sci.* **1988**, *42*, 358–365.
297. Petch, D.A. Selective Deposit-Feeding by *Lumbrineris* cf. *latreilli (Polychaeta, Lumbrineridae), with a New Method for Assessing Selectivity by Deposit-Feeding Organisms. Mar. Biol.* **1986**, *93*, 443–448.
298. Gravier, C.; Dantan, J.L. Péchés nocturnes à la lumière dans la Baie d'Alger. Annélides polychètes. *Ann. l'Institut Océanographique* **1928**, *5*, 1–187.
299. Rouse, G.W. An ultrastructural study of the spermatozoa of *Eulalia* sp.(Phyllodocidae), *Lepidonotus* sp. (Polynoidae), *Lumbrineris* sp.(Lumbrineridae) and *Owenia fusiformis* (Oweniidae). *Helgoländer Wiss. Meeresunters.* **1988**, *42*, 67–78. [CrossRef]
300. Messina, P.; Di Filippo, M.; Gambi, M.C.; Zupo, V. In vitro fertilisation and larval development of a population of *Lumbrineris (Scoletoma) impatiens* (Claparede) (Polychaeta, Lumbrineridae) of the Gulf of Naples (Italy) in relation to aquaculture. *Invertebr. Reprod. Dev.* **2005**, *48*, 31–40. [CrossRef]
301. Conti, G.; Loy, F.; Diana, M.; Messina, P. The spermatozoon of *Lumbrineris* (*Scoletoma*) *impatiens* (Claparède, 1868) (Polychaeta, Lumbrineridae). *Ital. J. Zool.* **2011**, *78*, 215–218. [CrossRef]

302. Rouse, G.W.; Jamieson, B.G.M. An ultrastructural study of the spermatozoa of the polychaetes *Eurythoe complanata* (Amphinomidae), *Clymenella* sp. and *Micromaldane* sp. (Maldanidae), with definition of sperm types in relation to reproductive biology. *J. Submicrosc. Cytol.* **1987**, *19*, 573–584.
303. Martins, R.; Carrera-Parra, L.F.; Quintino, V.; Rodrigues, A.M. Lumbrineridae (Polychaeta) from the Portuguese continental shelf (NE Atlantic) with the description of four new species. *Zootaxa* **2012**, *3416*, 1–21. [CrossRef]
304. Hartman, O. New species of polychaetous annelids from southern California. With a discussion on the host of one of them. *Allan Hancock Pac. Exped.* **1939**, *7*, 157–172.
305. Nishihira, M.; Tsuchiya, M.; and Sato, M. Ecological aspect of the breeding of the polychaete, *Lumbrineris latreilli* (Audouin et Milne-Edwards) at Asamushi, northern Japan. *Bull. Mar. Biol. Stn. Asamushi* **1980**, *16*, 201–212.
306. Nishihira, M.; Tsuchiya, M.; Sato, M. Distribution and spawning site selection of the polychaete *Lumbrineris latreilli* (Audouin et Milne-Edwards) at Asamushi, northern Japan. *Bull. Mar. Biol. Stn. Asamushi* **1981**, *17*, 17–28.
307. Nishihira, M.; Tsuchiya, M.; Sato, M. Dispersal and recruitment of juveniles of the polychaete *Lumbrineris latreilli* (Audouin et Milne-Edwards). *Bull. Mar. Biol. Stn. Asamushi* **1984**, *17*, 191–203.
308. Carrera-Parra, L.F. Lumbrineridae (Annelida: Polychaeta) from the Grand Caribbean region with the description of six new species. *J. Mar. Biol. Assoc. UK* **2001**, *81*, 599–621. [CrossRef]
309. Ayyagari, A.; Kondamudi, R. Ecological significance of the association between *Stomopneustes variolaris* (Echinoidea) and *Lumbrineris latreilli* (Polychaeta) from Visakhapatnam Coast, India. *J. Mar. Biol.* **2014**, *2014*, 640784. [CrossRef]
310. Zibrowius, H.; Southward, E.C.; Day, J.H. New observations on a little known species of *Lumbrineris* (Polychaeta) living on various cnidarians, with notes on its recent and fossil scleractinian hosts. *J. Mar. Biol. Assoc. UK* **1975**, *55*, 83–108. [CrossRef]
311. García-Gómez, S.C.; Carrera-Parra, L.F.; Mas, F.A.; Freitas, R.; Martins, R. Novel insights on the diversity and ecology of the family Lumbrineridae (Polychaeta) along the Iberian Peninsula coasts. *J. Mar. Biol. Assoc. UK* **2016**, *96*, 1427–1435. [CrossRef]
312. Katsiaras, N.; Rousou, M.; Carrera-Parra, L.F.; García-Gómez, S.C.; Simboua, N.; Louizidou, P.; Chintiroglou, C.C.; Martins, R. Taxonomy, ecology and geographic distribution of *Gallardoneris iberica* (Polychaeta, Lumbrineridae) in southern Europe. *J. Mar. Biol. Assoc. UK* **2018**, *98*, 1609–1618. [CrossRef]
313. Arias, A.; Carrera-Parra, L.F. First record of the genus *Kuwaita* (Annelida: Lumbrineridae) in Europe with the description of a new species and new ultramorphological data for the genus. *Zootaxa* **2014**, *3887*, 68–78. [CrossRef] [PubMed]
314. Levenstein, R.Y. A new genus and species of Polychaeta (Family Lumbrineridae) from the deep-water trenches the the North Pacific. In *Essays on Polychaetous Annelids in Memory of Dr. Olga Hartman*; Reish, D.J., Fauchald, K., Eds.; The Allan Hancock Foundation, University of Southern California: Los Angeles, CA, USA, 1977; pp. 189–198.
315. Oug, E. A new small species of *Lumbrineris* from northern Norway and arctic waters, with comments on *L. minuta* (Theel, 1879) and *L. vanhoeffeni* (Michaelsen, 1898)—(Polychaeta: Lumbrineridae). *Ophelia* **1998**, *49*, 147–162. [CrossRef]
316. Hartmann-Schröder, G. Zur Kenntnis des Eulitorals der australischen Küsten unter besonderer Berücksichtigung der Polychaeten und Ostracoden, 13. Die Polychaeten der antiborealen Küste von Victoria (Australien) (Zwischen Warrnambool im Wasten und Port Welshpool im Osten). *Mitt. Hambg. Zool. Mus. Inst.* **1987**, *84*, 27–66.
317. Camargo, M.G.; da Lana, P.C. Lumbrineridae (Polychaeta: Eunicemorpha) da costa sul e sudeste do Brasil. I. *Iheringia Série Zool.* **1995**, *79*, 77–91.
318. Camargo, M.G.; Lana, P. da C. Lumbrineridae (Polychaeta: Eunicemorpha) da costa sul e sudeste do Brasil. II. *Lumbrineris. Iheringia* **1995**, *79*, 93–120.
319. Núñez, J.; Talavera, J.A.; Ocaña, O. Anélidos Poliquetos de Canarias: Familia Lumbrineridae. *Vieraea* **1991**, *20*, 89–100.
320. Paxton, H.; Chou, L.M. Polychaetous annelids from the South China Sea. *Raffles Bull. Zool.* **2000**, *48*, 209–232.
321. Cai, W.; Li, X. *Eranno carrerai* sp. nov. a new polychaete worm species (Polychaeta: Lumbrineridae) from coastal waters of China. *Chin. J. Oceanol. Limnol.* **2011**, *29*, 217–231. [CrossRef]
322. Cai, W.; Li, X. A new species and new recorded species of Lumbrineridae Schmarda, 1861 (Anellida: Polychaeta) from China. *Chin. J. Oceanol. Limnol.* **2011**, *29*, 356–365. [CrossRef]
323. Dean, K.D. Marine biodiversity of Costa Rica: Class Polychaeta (Annelida). *Rev. Biol. Trop.* **2004**, *52*, 131–181. [PubMed]
324. Crossland, C. Polychaeta of tropical East Africa, the Red Sea, and Cape Verde Islands collected by Cyril Crossland, and of the Maldive Archipelago collected by Professor Stanley Gardiner, M.A., F.R.S. The Lumbriconereidae and Staurocephalidae. *Proc. Zool. Soc. Lond.* **1924**, *1*, 1–106. [CrossRef]
325. Hartmann-Schröder, G. Zur Ökologie der Polychaeten des Mangrove-Estero-Gebietes von El Salvador. *Beiträge zur Neotropischen Fauna* **1959**, *1*, 69–183. [CrossRef]
326. Perkins, T.H. Lumbrineridae, Arabellidae, and Dorvilleidae (Polychaeta) from Florida, with descriptions of six new species. *Proc. Biol. Soc. Wash.* **1979**, *92*, 415–465.
327. Uebelacker, J.M. Family Lumbrineridae Malmgren, 1867. In *Taxonomic Guide to the Polychaetes of the Northern Gulf of Mexico. Final Report to the Minerals Management Service, Contract 14-12-001-29091. Volume VI*; Uebelacker, J.M., Johnson, P.G., Eds.; U.S. Department of the Interior Mineral Management Service Gulf of Mexico OCS.: Mobile, Italy, 1984; p. 41.
328. Granados-Barba, A.; Solís-Weiss, V. New records of polychaetous annelids (Order: Eunicida) from the southeastern Gulf of Mexico. *Bull. Mar. Sci.* **1994**, *52*, 420–427.

329. Solís-Weiss, V.; Granados-Barba, A.; Rodríguez-Villanueva, L.V.; Miranda-Vázquez, L.A.; Ochoa-Rivera, V.; Hernández-Alcántara, P. The Lumbrineridae of the continental shelf in mexican portion of the Gulf of Mexico. *Mitt. Hambg. Zool. Mus. Inst.* **1995**, *92*, 61–75.
330. Fauvel, P. Annelida Polychaeta of the Indian Museum, Calcutta. *Mem. Indian Mus.* **1932**, *12*, 1–262.
331. Imajima, M.; Higuchi, M. Lumbrineridae of polychaetous annelids from Japan, with descriptions of six new species. *Bull. Natl. Sci. Mus.* **1975**, *1*, 5–37.
332. Imajima, M. Six species of Lumbrinerides (Polychaeta, Lumbrineridae) from Japan. *Bull. Natl. Sci. Mus.* **1985**, *11*, 171–184.
333. Miura, T. Classification and morphological variations of the Japanese species of Lumbrinerides (Annelida: Lumbrineridae). *Species Divers.* **2017**, *22*, 7–27. [CrossRef]
334. Ramos, J.M. Lumbrineridae (Polychètes errantes) de Méditterranée. *Ann. l'Institut Océanograph.* **1976**, *52*, 103–137.
335. Miura, T. Lumbrineridae (Annélides Polychètes) abyssaux récoltés au cours de campagnes du Centre Océanologique de Bretagne dans l'Atlantique et la Mediterranée. *Bull. Muséum d'Histoire Nat.* **1980**, *2*, 1019–1057.
336. Carrera-Parra, L.F.; Çinar, M.E.; Dagli, E. Description of a new species of *Lumbrineris* (Polychaeta: Lumbrineridae) from the coasts of Turkey (eastern Mediterranean). *Mar. Biodivers.* **2011**, *41*, 343–347. [CrossRef]
337. D'Alessandro, M.; Cosentino, A.; Giacobbe, S.; Andaloro, F.; Romeo, T. Description of a new species of *Abyssoninoe* (Polychaeta: Lumbrineridae) from the NE Sicily (central Mediterranean Sea). *J. Mar. Biol. Assoc. UK* **2014**, *94*, 747–752. [CrossRef]
338. Knox, G.A.; Green, K.M. The Polychaetous Annelids of New Zealand Part 2: Lumbrineridae. *J. R. Soc. N. Z.* **1972**, *2*, 69–81.
339. Oug, E. Lumbrineridae from the Andaman Sea, Thailand, with notes on Oenonidae and Dorvilleidae (Annelida: Polychaeta). *Phuket Mar. Biol. Cent. Spec. Publ.* **2002**, *24*, 117–138.
340. Parapar, J.; Oconnor, B.; Besteiro, C.; Urgorri, V. *Abyssoninoe hibernica* (McIntosh) (Polychaeta: Lumbrineridae) a valid species from the Northeast Atlantic. *Sarsia* **1994**, *79*, 157–162. [CrossRef]
341. Aguirrezabalaga, F.; Carrera-Parra, L.F. Lumbrineridae (Polychaeta) from the Capbreton Canyon (Bay of Biscay, NE Atlantic) with the description of two new species. *Sci. Mar.* **2006**, *70*, 17–25. [CrossRef]
342. Hernández-Alcántara, P.; Pérez-Mendoza, A.Y.; Solís-Weiss, V. Description of three new species of *Ninoe* and *Cenogenus* (Polychaeta: Lumbrineridae) from the Mexican Pacific. *Sci. Mar.* **2006**, *70*, 81–90.
343. Fauchald, K. Deep-water errant polychaetes from Hardangerfjorden, western Norway. *Sarsia* **1974**, *57*, 1–32. [CrossRef]
344. Winsnes, I.M. A new species of *Lumbrineris* (Polychaeta) and a new subspecies of *Lumbrineris scopa* Fauchald from the coast of Norway. *Zool. Scr.* **1981**, *10*, 91–94. [CrossRef]
345. Winsnes, I.M. *Augeneria algida* (Wiren) comb. n., a deep-sea lumbrinerid from Spitzbergen with aberrant setae (Annelida, Polychaeta): Redescription of holotype. *Zool. Scr.* **1987**, *16*, 39–45. [CrossRef]
346. Oug, E. Guide to identification of Lumbrineridae (Polychaeta) in Norwegian and adjacent waters. *Nor. Polych. Forum Guid.* **2011**, 1–22.
347. Hartmann-Schröder, G. Zur polychaeten-fauna von Peru. *Beiträge zur Neotropischen Fauna* **1960**, *2*, 1–44. [CrossRef]
348. Hartmann-Schröder, G. Die Polychaeten des Eulitorals. Zur Kenntnis des Eulitorals der chilenischen Pazifikküste und der argentinischen Küste Südpatagoniens unter besonderer Berücksichtigung der Polychaeten und Ostracoden. *Mitt. Hambg. Zool. Mus. Inst.* **1962**, *60*, 57–167.
349. Rozbaczylo, N.; Vásquez-Yáñez, P.; Moreno, R.A.; Díaz-Díaz, O. Poliquetos bentónicos Amphinomida, Phyllodocida y Eunicida (Annelida: Polychaeta) de la región de fiordos y canales australes de Chile recolectados durante los cruceros CIMAR 13 al 20 fiordos. *An. Inst. Patagon.* **2018**, *45*, 51–91. [CrossRef]
350. Rozbaczylo, N.; Moreno, R.A.; Díaz-Díaz, O. Poliquetos bentónicos submareales de fondos blandos de la región de Aysén, Chile: Clados Amphinomida, Eunicida, Spionida, Sabellida y Scolecida (Annelida, Polychaeta). *Investig. Mar.* **2006**, *34*, 43–62. [CrossRef]
351. Gallardo, V.A. Polychaeta from the bay of Nha Trang, South Viet Nam. *NAGA Rep.* **1968**, *4*, 35–279.
352. Fauvel, P.; Rullier, F. Annélides Polychètes. Resultats Scientifiques des Campagnes de Calypso IV. *Ann. l'Institut Océanograph.* **1959**, *37*, 143–205.
353. Amoureux, L. Quelques annélides polychètes de l'Afrique occidentale et équatoriale. *Cah. ORSTOM sér Océnnogr* **1973**, *11*, 41–65.
354. Intes, A.; Le Loeuff, P. Les annélides polychètes de Côte d'Ivoire. 1. Polychètes Errantes—Compte rendu systématique. *Cah. ORSTOM sér Océnnogr* **1975**, *13*, 267–321.
355. Brasier, M.J.; Wiklund, H.; Neal, L.; Jeffreys, R.; Linse, K.; Ruhl, H.; Glover, A.G. DNA barcoding uncovers cryptic diversity in 50% of deep-sea antarctic polychaetes. *R. Soc. Open Sci.* **2016**, *3*, 160432. [CrossRef] [PubMed]
356. Font, T.; Gil, J.; Lloret, J. The commercialization and use of exotic baits in recreational fisheries in the north-western Mediterranean: Environmental and management implications. *Aquat. Conserv. Mar. Freshw. Ecosyst.* **2018**, *28*, 651–661. [CrossRef]
357. Núñez, J.; Núñez, L.; Maggio, Y. *Invertebrados que se Comercializan en Canarias Como cebo vivo para la Pesca Deportiva*; Gobierno de Canarias: La Laguna, Spain, 2011.
358. Usamentiaga, P.; Rodríguez, F.; Martin-Gil, D.; Saiz, N.; Morchon, E.; Alonso, S.; Jerez, J.; Bartolomé, B. Protein contact dermatitis by fishing bait (*Lumbrinereis latreilli*). *Contact Dematitis* **2005**, *53*, 236–237. [CrossRef]
359. Ruggeri, G.D. Drugs from the Sea. Marine organisms with novel chemical constituents are excellent sources of new drugs. *Science* **1976**, *194*, 491–497.
360. Montagu, G. Description of several marine animals found on the south coast of Devonshire. *Trans. Linn. Soc. Lond.* **1804**, *7*, 80–84. [CrossRef]

361. Kinberg, J.G.H. Annulata nova. Öfversigt af Königlich Vetensk. *Förhandlingar Stock* **1865**, *21*, 559–574.
362. Colbath, G.K. Revision of the family Lysaretidae, and recognition of the family Oenonidae Kinberg, 1865 (Eunicida: Polychaeta). *Proc. Biol. Soc. Wash.* **1989**, *102*, 116–123.
363. Dean, H.K. A new arabellid polychaete living in the mantle cavity of deep-sea wood boring bivalves (family Pholadidae). *Proc. Biol. Soc. Wash.* **1992**, *105*, 224–232.
364. Carrera-Parra, L.F. Oenonidae Kinberg, 1865. In *Poliquetos (Annelida: Polychaeta) de México y América Tropical*; de León-González, J.Á., Bastida-Zavala, J.R., Carrera-Parra, L.F., García-Garza, M.E., Peña-Rivera, A., Salazar-Vallejo, S.I., Solís-Weiss, V., Eds.; Universidad Autónoma de Nuevo León: Monterrey, Mexico, 2009; pp. 355–362.
365. Zanol, J.; Ruta, C. New and previously known species of Oenonidae (Polychaeta: Annelida) from Lizard Island, Great Barrier Reef, Australia. *Zootaxa* **2015**, *4019*, 745–772. [CrossRef]
366. Saito, H.; Imabayashi, H.; Kawai, K. Growth of the bivalve-feeder *Halla okudai* (Polychaete: Lysaretidae) under wild and rearing conditions, in relation to species and abundance of prey organisms. *Fish. Sci.* **1999**, *65*, 230–234. [CrossRef]
367. Kawai, K.; Kunitake, H.; Saito, H.; Imabayashi, H. Paralytic and digestive activities of jelly-like substances secreted by a lysaretid polychaete, *Halla okudai*. *Benthos Res.* **1999**, *54*, 1–7. [CrossRef]
368. Iori, D.; Forti, L.; Massamba-N'Siala, G.; Prevedelli, D.; Simonini, R. Toxicity of the purple mucus of the polychaete *Halla parthenopeia* (Oenonidae) revealed by a battery of ecotoxicological bioassays. *Sci. Mar.* **2014**, *78*, 589–595.
369. Bielig, H.J.; Möllinger, H. Über hallochrom, den epithelfarbstoff von *Halla parthenopeia* Costa. *Physiol. Chem.* **1960**, *321*, 276–289. [CrossRef]
370. Carrera-Parra, L.F. Oenonidae Kinberg, 1865. In *Poliquetos (Annelida: Polychaeta) de México y América Tropical*; de León-González, J.A., Bastida-Zavala, J.R., Carrera-Parra, L.F., García-Garza, M.E., Salazar-Vallejo, S.I., Solís-Weiss, V., Tovar-Hernández, M.A., Eds.; unpublished.
371. Amaral, A.C. Um poliqueto endoparasita, *Labrorostratus prolificus* sp. nov. em nereídeo. *Bol. Inst. Oceanogr.* **1977**, *26*, 285–292. [CrossRef]
372. Abd-Elnaby, F.A. Observations on Spermatogenesis in Polychaete. Species, *Halla parthenopeia* (Family: Oenonidae, Polychaeta). *World Appl. Sci. J.* **2009**, *6*, 144–153.
373. Pettibone, M.H. Endoparasitic polychaetous annelids of the family Arabellidae with descriptions of new species. *Biol. Bull. Biol. Lab.* **1957**, *113*, 170–187. [CrossRef]
374. Pleijel, F. Oenonidae Kinberg, 1865. In *Polychaetes*; Rouse, G., Pleijel, F., Eds.; Oxford University Press: New York, NY, USA, 2001; pp. 166–168.
375. Hilbig, B. Family Oenonidae Kinberg, 1865, emended Orensanz, 1990. In *Taxonomic Atlas of the Benthic Fauna of the Santa Maria Basin and Western Santa Barbara Channel. 5—The Annelida Part 2*; Blake, J.A., Hilbig, B., Scott, P.H., Eds.; Santa Barbara Museum of Natural History: Santa Barbara, CA, USA, 1995; pp. 315–339.
376. Hernández-Alcántara, P.; Sólis-Weiss, V. Parasitism among polychaetes: A rare case illustrated by a new species: *Labrorostratus zaragozensis*, n. sp. (Oenonidae) found in the Gulf of California, Mexico. *J. Parasitol.* **1998**, *84*, 978–982. [CrossRef] [PubMed]
377. Hernández-Alcántara, P.; Cruz-Pérez, I.N.; Sólis-Weiss, V. *Labrorostratus caribensis*, a new oenonid polychaete from the Grand Caribbean living in the body cavity of a nereidid, with emendation of the genus. *Zootaxa* **2015**, *4048*, 127–139. [CrossRef]
378. Carr, C.M. Polychaete diversity and distribution patterns in Canadian marine waters. *Mar. Biodivers.* **2012**, *42*, 93–107. [CrossRef]
379. Orensanz, J.M. Los anelidos poliquetos de la provincia biogeografica Argentina. VI. Arabellidae. *Physis* **1974**, *33*, 381–408.
380. Hartmann-Schröder, G. Die Polychaeten der tropischen Nordwestküste Australiens (zwischen Derby im Norden und Port Hedland im Süden). Teil 2. Zur Kenntnis des Eulittorals der australischen Küsten unter besonderer Berücksichtigung der Polychaeten und Ostracoden. *Mitteilungen aus dem Hamburgischen Zoologischen Museum und Institut* **1979**, *76*, 75–218.
381. Rizzo, A.E.; Almeida dos Santos, S.; da Conceição Guerreiro Couto, E. First report of *Labrorostratus caribensis* (Annelida, Oenonidae) as endoparasite of *Haplosyllis rosenalessoae* (Annelida, Syllidae) from Brazil. *Int. J. Parasitol. Parasites Wildl.* **2020**, *12*, 64–66. [CrossRef]
382. Ibarzabal, D.R. Poliquetos de Punta del Este, Isla de la Juventud, Cuba. IV Orden Eunicida. *Poeyana* **1989**, *384*, 1–28.
383. Díaz-Díaz, O.; Liñero-Arana, I. Poliquetos asociados a substratos artificiales sumergidos en la costa noroerental de Venezuela 3. Eunicida (Annelida: Polychaeta). *Bol. del Inst. Oceanogr. Venez.* **2002**, *41*, 3–14.
384. Dean, K.D.; Sibaja-Cordero, J.A.; Cortés, J. Polychaetes (Annelida: Polychaeta) of Cocos Island National Park, Pacific Costa Rica. *Pac. Sci.* **2012**, *66*, 347–386. [CrossRef]
385. Ravara, A.; Cunha, M.R. An unusual new species of Oenonidae (Polychaeta, Eunicida) from the North Atlantic Ocean. *Mar. Biodivers.* **2018**, *48*, 1037–1043. [CrossRef]
386. Uebelacker, J.M. A new parasitic polychaetous annelid (Arabellidae) from the Bahamas. *J. Parasitol.* **1978**, *64*, 151–154. [CrossRef]
387. Uebelacker, J.M. Family Arabellidae Hartman, 1944. In *Taxonomic Guide to the Polychaetes of the Northern Gulf of Mexico. Final Report to the Minerals Management Service, Contract 14-12-001-29091. Volume VI*; Uebelacker, J.M., Johnson, P.G., Eds.; U.S. Department of the Interior Mineral Management Service Gulf of Mexico OCS.: Mobile, Italy, 1984; pp. 42.1–42.29.
388. Gilbert, K. Family Lysaretidae Kinberg, 1865. In *Taxonomic Guide to the Polychaetes of the Northern Gulf of Mexico. Final Report to the Minerals Management Service, Contract 14-12-001-29091. Volume VI*; Uebelacker, J.M., Johnson, P.G., Eds.; U.S. Department of the Interior Mineral Management Service Gulf of Mexico OCS.: Mobile, Italy, 1984; pp. 43.1–43.7.

389. Imajima, M. Errant polychaetous annelids from Tsukumo Bay and vicinity of Noto Penisula, Japan. *Bull. Natl. Sci. Mus.* **1967**, *10*, 403–441.
390. Imajima, M. Polychaetous annelids of Suruga Bay, central Japan. *Natl. Sci. Mus. Monogr.* **1997**, *12*, 149–228.
391. Imajima, M. Deep-sea benthic polychaetous annelids of Tosa Bay, southwestern Japan. *Natl. Sci. Mus. Monogr.* **2001**, *20*, 31–100.
392. Imajima, M. Polychaetous annelids from Sagami Bay and the Sagami Sea, central Japan. *Mem. Natl. Sci. Mus.* **2006**, *40*, 317–408.
393. Kobayashi, G.; Mukai, R.; Itoh, H. New record of *Halla okudai* Imajima, 1967 (Annelida, Eunicida, Oenonidae) from Fukue Island in the Goto Islands, Japan. *Check List* **2020**, *16*, 1199–1203.
394. Ramos, J.M. Arabellidae (Polychètes Errantes) de Méditerranée. *Ann. l'Institut Océanograph.* **1976**, *52*, 247–257.
395. San Martín, G.; Sardá, R. Sobre la presencia de un Arabélido (Polychaeta: Arabellidae) parásito de Sílidos (Polychaeta: Syllidae), del género *Labrorostratus* en las costas espanolas. *Boletín la Real Soc. Española Hist. Nat.* **1986**, *82*, 141–146.
396. Cantone, G. Censimento de policheti de mari Italiani: Arabellidae Hartman, 1944, Oenonidae Kinberg, 1865. *Atti Soc. Toscana Sci. Nat. Mem. Ser. B* **1996**, *103*, 99–103.
397. Mikac, B. A sea of worms: Polychaete checklist of the Adriatic Sea. *Zootaxa* **2015**, *3943*, 1–172. [CrossRef]
398. Knox, G.A.; Green, K.M. The polychaetes of New Zealand. Part 3. Lysaretidae. *J. R. Soc. N. Z.* **1972**, *2*, 431–434. [CrossRef]
399. Fauchald, K. Some Polychaetous Annelids From the Deep Basins in Sognefjorden, Western Norway. *Sarsia* **1972**, *49*, 89–106. [CrossRef]
400. Hartman, O. Deep-water benthic polychaetous annelids off New England to Bermuda and other North Atlantic areas. *Occas. Pap. Allan Hancock Found.* **1965**, *28*, 1–378.
401. Hartman, O.; Fauchald, K. Deep-water benthic polychaetous annelids off New England to Bermuda and other North Atlantic Areas. Part II. *Allan Hancock Monogr. Mar. Biol.* **1971**, *6*, 1–327.
402. Kirkegaard, J.B. Bathyal and abyssal polychaetes (errant species). *Galathea Repor* **1995**, *17*, 7–56.
403. Baeta, M.; Navarrete, A.; Ballesteros, M. First description of a new worm bait fishery in the NW Mediterranean Sea. *Mediterr. Mar. Sci.* **2019**, *20*, 302–308.
404. Saito, H.; Kawai, K.; Umino, T.; Imabayashi, H. Fishing bait worm supplies in Japan in relation to their physiological traits. *Mem. Mus. Vic.* **2014**, *71*, 279–287. [CrossRef]
405. Idris, I.; Arshad, A. Checklist of polychaetous annelids in Malaysia with redescription of two commercially exploited species. *Asian J. Anim. Vet. Adv.* **2013**, *8*, 409–436. [CrossRef]
406. Pettibone, M.H. Polychaeta Errantia of the Siboga Expedition. Part IV. Some additional polychaetes of the Polynoidae, Hesionidae, Nereidae, Goniadidae, Eunicidae, and Onuphidae, selected as new species by the late Dr. Hermann Augener with remarks on other related species. *Siboga-Exped. Monogr.* **1970**, *24*, 199–270.
407. Kucheruk, N.V. Deep-water Onuphidae (Polychaeta) from the collections of 16th cruise of R/V Dmitry Mendeleev (to the generic classification of the family Onuphidae). *Tr. Institut. Okeanol. Akad. Nauk SSSR* **1978**, *113*, 88–106. (In Russian)
408. Fauchald, K. Revision of *Onuphis*, *Nothria*, and *Paradiopatra* (Polychaeta: Onuphidae) based upon type material. *Smithson. Contrib. Zool.* **1982**, *356*, 1–109. [CrossRef]
409. Paxton, H. Generic Revision and Relationships of the Family Onuphidae (Annelida: Polychaeta). *Rec. Aust. Mus.* **1986**, *38*, 1–74. [CrossRef]
410. Paxton, H. *Fauchaldonuphis*, a new genus for the enigmatic polychaete *Diopatra paradoxa* Quatrefages, 1866. *Mar. Ecol.* **2005**, *26*, 209–215. [CrossRef]
411. Carrera-Parra, L.F. Onuphidae Kinberg, 1865. In *Poliquetos (Annelida: Polychaeta) de México y América Tropical*; de León-González, J.Á., Bastida-Zavala, J.R., Carrera-Parra, L.F., García-Garza, M.E., Peña-Rivera, A., Salazar-Vallejo, S.I., Solís-Weiss, V., Eds.; Universidad Autónoma de Nuevo León: Monterrey, Mexico, 2009; pp. 363–378.
412. de León-González, J.A.; Cornejo-Rodríguez, M.H.; Degraer, S. A new species of *Australonuphis* (Polychaeta: Onuphidae) from the eastern Pacific. *J. Mar. Biol. Assoc. UK* **2008**, *88*, 739–742. [CrossRef]
413. Paxton, H. *Hirsutonuphis* (Polychaeta: Onuphidae) from Australia, with a discussion of setal progression in juveniles. *Invertebr. Taxon.* **1996**, *10*, 77–96. [CrossRef]
414. Estrella-Ruiz, A.; Hernández-Alcántara, P.; Solís-Weiss, V. Description of a new species of *Hirsutonuphis* (Annelida: Onuphidae) from the Mexican Pacific, with notes on the world distribution of the genus. *J. Mar. Biol. Assoc. UK* **2013**, *93*, 1777–1783. [CrossRef]
415. Paxton, H. Revision of the *Rhamphobrachium* complex (Polychaeta: Onuphidae). *Rec. Aust. Mus.* **1986**, *38*, 75–104. [CrossRef]
416. Paxton, H.; Budaeva, N. *Minibrachium*, a new subgenus of *Rhamphobrachium* (Annelida: Onuphidae) from Australia with the description of three new species. *Zootaxa* **2015**, *4019*, 621–634. [CrossRef]
417. Paxton, H.; Budaeva, N. *Paradiopatra* (Annelida: Onuphidae) from eastern Australian waters, with the description of six new species. *Zootaxa* **2013**, *3686*, 140–164.
418. Meyer, K.S.; Wagner, J.K.S.; Ball, B.; Turner, P.J.; Young, C.M.; Van Dover, C.L. *Hyalinoecia artifex*: Field notes on a charismatic and abundant epifaunal polychaete on the US Atlantic continental margin. *Invertebr. Biol.* **2016**, *135*, 211–224. [CrossRef]
419. Nishi, E.; Kato, T. *Longibrachium arariensis*, a new species of Onuphidae (Annelida: Polychaeta) from the shallow water of Izu Peninsula, central Japan, with notes on its feeding behavior. *Zootaxa* **2009**, *2081*, 46–56. [CrossRef]
420. Gaston, G.R.G. Benthic Polychaeta of the Middle Atlantic Bight: Feeding and distribution. *Mar. Ecol. Prog. Ser.* **1987**, *36*, 251–262. [CrossRef]

421. Dayton, P.K.; Hessler, R.R. Role of biological disturbance in maintaining diversity in the deep sea. *Deep Sea Res.* **1972**, *19*, 199–208. [CrossRef]
422. Lipps, J.H.; Ronan, T.E. Predation on Foraminifera by the polychete worm *Diopatra*. *J. Foraminifer. Res.* **1974**, *4*, 139–143. [CrossRef]
423. Mangum, C.P.; Santos, S.L.; Rhodes, W.R., Jr. Distribution and feeding in the onuphid polychaete, *Diopatra cuprea* (Bosc). *Mar. Biol.* **1968**, *2*, 33–40. [CrossRef]
424. Brenchley, G.A.; Tidball, J.G. Tube-cap orientations of *Diopatra cuprea* (Bosc) (Polychaeta): The compromise between physiology and foraging. *Mar. Behav. Physiol.* **1980**, *7*, 1–13. [CrossRef]
425. Arias, A.; Richter, A.; Anadón, N.; Paxton, H. Evidence of simultaneous hermaphroditism in the brooding *Diopatra marocensis* (Annelida: Onuphidae) from northern Spain. *J. Mar. Biol. Assoc. UK* **2013**, *93*, 1533–1542. [CrossRef]
426. Arias, A.; Paxton, H. The cryptogenic bait worm *Diopatra biscayensis* Fauchald et al., 2012 (Annelida: Onuphidae)—Revisiting its history, biology and ecology. *Estuar. Coast. Shelf Sci.* **2015**, *163*, 22–36. [CrossRef]
427. Arias, A.; Paxton, H.; Budaeva, N. Redescription and biology of *Diopatra neapolitana* (Annelida: Onuphidae), a protandric hermaphrodite with external spermaducal papillae. *Estuar. Coast. Shelf Sci.* **2016**, *174*, 1–17. [CrossRef]
428. Hsieh, H.L.; Simon, J.L. The sperm transfer system in *Kinbergonuphis simoni* (Polychaeta: Onuphidae). *Biol. Bull.* **1990**, *178*, 85–93. [CrossRef] [PubMed]
429. Blake, J.A. The larval development of Polychaeta from the nothern California coast. II. *Nothria elegans* (Family Onuphidae). *Ophelia* **1975**, *13*, 43–61. [CrossRef]
430. Paxton, H. *Diopatra* Audouin and Milne Edwards (Polychaeta, Onuphidae) from Australia, with a discussion of developmental patterns in the genus. *Beagle Rec. North Territ. Mus. Arts Sci.* **1993**, *10*, 115–154.
431. Carrasco, F.D. Description of adults and larvae of a new deep water species of *Hyalinoecia* (Polychaeta, Onuphidae) from the southeastern Pacific Ocean. *J. Nat. Hist.* **1983**, *17*, 87–93. [CrossRef]
432. Hsieh, H.L.; Simon, J.L. Larval development of *Kinbergonuphis simoni*, with a summary of development patterns in the family Onuphidae (Polychaeta). *Bull. Biol. Soc. Wash.* **1987**, *7*, 194–210.
433. Fadlaoui, S.A.; Lechapt, J.P.; Retière, C. Larval development of the onuphid *Diopatra marocensis* (Annelida: Polychaeta) from the Atlantic coast of Morocco. *J. Mar. Biol. Assoc. UK* **1995**, *75*, 957–966. [CrossRef]
434. Budaeva, N.; Fauchald, K. *Diopatra tuberculantennata*, a new species of Onuphidae (Polychaeta) from Belize with a key to onuphids from the Caribbean Sea. *Zootaxa* **2008**, *1795*, 29–45. [CrossRef]
435. Budaeva, N.; Fauchald, K. Larval development of *Mooreonuphis stigmatis* (Treadwell 1922) (Polychaeta: Onuphidae) from the north-east Pacific. *Mar. Biol. Res.* **2010**, *6*, 6–24. [CrossRef]
436. Budaeva, N.; Pyataeva, S.; Meißner, K. Development of the deep-sea viviparous quill worm *Leptoecia vivipara* (Hyalinoeciinae, Onuphidae, Annelida). *Invertebr. Biol.* **2014**, *133*, 242–260. [CrossRef]
437. Rangel, L.; Santos, M. *Diopatra neapolitana* (Polychaeta: Onuphidae) as a second intermediate host of *Gymnophallus choledochus* (Digenea: Gymnophallidae) in the Aveiro Estuary (Portugal): Distribution within the host and histopathology. *J. Parasitol.* **2009**, *95*, 1233–1236. [CrossRef]
438. Rangel, L.F.; Cech, G.; Székely, C.; Santos, M.J. A new actinospore type unicapsulactinomyxon (Myxozoa), infecting the marine polychaete, *Diopatra neapolitana* (Polychaeta: Onuphidae) in the Aveiro Estuary (Portugal). *Parasitology* **2011**, *138*, 698–712. [CrossRef] [PubMed]
439. Arias, A.; Anadón, N.; Paxton, H. New records of *Diopatra marocensis* (Annelida: Onuphidae) from northern Spain. *Zootaxa* **2010**, *2691*, 67–68. [CrossRef]
440. Arias, A.; Paxton, H. Hidden diversity within the polychaete *Onuphis eremita sensu lato* (Annelida: Onuphidae)—Redescription of *O. eremita* Audouin & Milne-Edwards, 1833 and reinstatement of *Onuphis pancerii* Claparède, 1868. *Zootaxa* **2014**, *3861*, 145–169. [PubMed]
441. Jones, C.; Lawton, J.; Shachak, M. Organisms as ecosystem engineers. *Oikos* **1994**, *69*, 373–386. [CrossRef]
442. Berke, S.K.; Mahon, A.R.; Lima, F.P.; Halanych, K.M.; Wethey, D.S.; Woodin, S.A. Range shifts and species diversity in marine ecosystem engineers: Patterns and predictions for European. *Glob. Ecol. Biogeogr.* **2010**, *19*, 223–232. [CrossRef]
443. Berke, S.K. Biogeographic variability in ecosystem engineering: Patterns in the abundance and behavior of the tube-building polychaete *Diopatra cuprea*. *Mar. Ecol. Prog. Ser.* **2012**, *447*, 1–13. [CrossRef]
444. Bailey-Brock, J. Ecology of the tube-building polychaete *Diopatra leuckarti* Kinberg, 1865 (Onuphidae) in Hawaii: Community structure, and sediment stabilizing properties. *Zool. J. Linn. Soc. Lond.* **1984**, *80*, 191–199. [CrossRef]
445. McIntosh, W.C. Report on the Annelida Polychaeta collected by H.M.S. Challenger during the years 1873–1876. *Rep. Sci. Results Voyag. H.M.S.Chall. Dur. Years 1872–76* **1885**, *12*, 1–554.
446. Kucheruk, N.V. On the specific composition and distribution of the deep-sea genus *Paraonuphis* (sic) (Polychaeta, Eunicidae). *Tr. Institut. Okeanol. Akad. Nauk SSSR* **1977**, *108*, 44–51. (In Russian)
447. Jirkov, I.A. *Polychaeta of the Arctic Ocean*; Yanus-K: Moscow, Russia, 2001.
448. Hartman, O. Polychaetous annelids collected by the USNS Eltanin and Staten Island cruises, chiefly from Antarctic Seas. *Allan Hancock Monogr. Mar. Biol.* **1967**, *2*, 1–387.
449. Fauchald, K. Onuphidae (Polychaeta) from Western Mexico. *Allan Hancock Monogr. Mar. Biol.* **1968**, *3*, 1–82.
450. Fauchald, K. Polychaetes from Central American sandy beaches. *Bull. South Calif. Acad. Sci.* **1973**, *72*, 19–31.
451. Fauchald, K. Onuphidae (Polychaeta) from Belize, Central America, with notes on related taxa. *Biol. Soc. Wash.* **1980**, *93*, 797–829.

452. de León-González, J.A. de Soft bottom polychaetes from the western coast of Baja California Sur, México. 4. Onuphidae. *Cah. Biol. Mar.* **1994**, *35*, 57–67.
453. Imajima, M. Two New Species of *Nothria* (Polychaeta, Onuphidae) from off Sanriku, Northeastern Honshu, Japan. *Bull. Natl. Sci. Mus.* **1989**, *15*, 1–10.
454. Imajima, M. Onuphidae (Annelida, Polychaeta) from Japan, excluding the genus *Onuphis*. *Natl. Sci. Mus. Monogr.* **1999**, *16*, 1–115.
455. Maekawa, N.; Hayashi, I. Onuphid polychaetes from Wakasa Bay, Sea of Japan. *Mem. Coll. Agric.* **1989**, *134*, 61–93.
456. Maekawa, N.; Hayashi, I. Taxonomic study on the genus *Onuphis* (Polychaeta, Onuphidae) from Japan and adjacent seas, with descriptions of six new species. *Bull. Natl. Sci. Mus.* **1999**, *25*, 163–214.
457. Paxton, H. Three new species of *Aponuphis* (Annelida: Onuphidae) from eastern Australia. *Zootaxa* **2017**, *4344*, 246–260. [CrossRef]
458. Paxton, H. Two new species of *Mooreonuphis* (Annelida: Onuphidae) from Australia. *Zootaxa* **2018**, *4486*, 589–600. [CrossRef]
459. Budaeva, N.; Paxton, H. *Nothria* and *Anchinothria* (Annelida: Onuphidae) from eastern Australian waters, with a discussion of ontogenetic variation of diagnostic characters. *J. Mar. Biol. Assoc. UK* **2013**, *93*, 1487–1502. [CrossRef]
460. Augener, H. Polychaeta. *Beiträge zur Kennt. der Meeresfauna Westafrikas* **1918**, *2*, 67–625.
461. Arias, A. *Onuphis* and *Mooreonuphis* (Annelida: Onuphidae) from West Africa with the description of three new species and the reinstatement of *O. landanaensis* Augener, 1918. *Zootaxa* **2016**, *4168*, 481–511. [CrossRef] [PubMed]
462. Paxton, H. A new species of *Diopatra* (Annelida: Onuphidae) from Namibia, southwestern Africa. *Mar. Biodivers.* **2016**, *46*, 889–895. [CrossRef]
463. Orensanz, J.M. Los anelidos poliquetos de la provincia biogeografica Argentina. V. Onuphidae. *Physis* **1974**, *33*, 75–122.
464. Paxton, H. The *Diopatra chiliensis* confusion—Redescription of *D. chiliensis* (Polychaeta, Onuphidae) and implicated species. *Zool. Scr.* **1998**, *27*, 31–48. [CrossRef]
465. Arias, A.; Paxton, H.; Anadón, N. *Mooreonuphis vespa*, a new brooding species of Onuphidae (Annelida) from northern Spain (Bay of Biscay). *Zootaxa* **2013**, *3741*, 583–592. [CrossRef]
466. Arias, A.; Núñez, J.; Paxton, H. Onuphid polychaetes associated with the *Cymodocea nodosa* meadows of La Gomera (Canary Islands, NW Africa)—New species and new records from the eastern North Atlantic. *J. Mar. Biol. Assoc. UK* **2017**, *97*, 857–869. [CrossRef]
467. Arias, A.; Paxton, H. *Onuphis* and *Aponuphis* (Annelida: Onuphidae) from southwestern Europe, with the description of a new species. *Zootaxa* **2015**, *3949*, 345–369. [CrossRef]
468. Arias, A.; Paxton, H. *Paradiopatra* Ehlers, 1887 (Annelida: Onuphidae) from southwestern Europe with the description of a new species and new ultramorphological data for the genus. *Zootaxa* **2015**, *4040*, 149–168. [CrossRef]
469. Paxton, H.; Arias, A. Unveiling a surprising diversity of the genus *Diopatra* Audouin & Milne Edwards, 1833 (Annelida: Onuphidae) in the Macaronesian region (eastern North Atlantic) with the description of four new species. *Zootaxa* **2017**, *4300*, 505–535.
470. Kucheruk, N.V. New species of the family Onuphidae (Polychaeta) from the Pacific. *Zool. Zhurnal* **1979**, *58*, 1227–1230. (In Russian)
471. Kucheruk, N.V. A new genus and species of polychaetes (Onuphidae). *Zool. Zhurnal* **1979**, *58*, 1585–1587. (In Russian)
472. Paterson, G.L.J.; Glover, A.G.; Frojan, C.B.; Whitaker, A.; Budaeva, N.; Chimonides, J.; Doner, S. A census of abyssal polychaetes. *Deep-Sea Res. II* **2009**, *56*, 1739–1746. [CrossRef]
473. Bergamo, G.; Carrerette, O.; Nogueira, J.M.M. Continuous and non-seasonal reproductive cycle of the alien species *Diopatra neapolitana* (Onuphidae, Annelida) in a tropical bay of SW Atlantic. *Estuar. Coast. Shelf Sci.* **2019**, *231*, 106479. [CrossRef]
474. Elgetany, A.H.; van Rensburg, H.; Hektoen, M.; Matthee, C.; Budaeva, N.; Simon, C.A.; Struck, T.H. Species delineation in the speciation grey zone—The case of *Diopatra* (Annelida, Onuphidae). *Zool. Scr.* **2020**, *49*, 516–534.
475. Budaeva, N.; Borisova, P.; Willassen, E. Nothria conchylega (Onuphidae, Annelida)—An example of a widespread species of polychaetes in the Northern Hemisphere. unpublished.
476. Rodrigues, A.M.; Pires, A.; Mendo, S.; Quintino, V. *Diopatra neapolitana* and *Diopatra marocensis* from the Portuguese coast: Morphological and genetic comparison. *Estuar. Coast. Shelf Sci.* **2009**, *85*, 609–617. [CrossRef]
477. Pires, A.; Paxton, H.; Quintino, V.; Rodrigues, A.M. *Diopatra* (Annelida: Onuphidae) diversity in European waters with the description of *Diopatra micrura*, new species. *Zootaxa* **2010**, *2395*, 17–33. [CrossRef]
478. Nygren, A. Cryptic polychaete diversity: A review. *Zool. Scr.* **2014**, *43*, 172–183.
479. Hektoen, M.M. (Norwegian University of Science and Technology, Trondheim, Norway). *Personal communication*. 2020.
480. Budaeva, N.; Paxton, H.; Ribeiro, P.; Sellanes, J.; Willassen, E. Systematics and diversity of quill worms (Hyalinoeciinae, Onuphidae, Annelida). unpublished.
481. Paxton, H.; Budaeva, N. Nothria (Onuphidae, Annelida) from deep Australian waters. unpublished.
482. Bors, E.K.; Rowden, A.A.; Maas, E.W.; Clark, M.R.; Shank, T.M. Patterns of Deep-Sea Genetic Connectivity in the New Zealand Region: Implications for Management of Benthic Ecosystems. *PLoS ONE* **2012**, *7*, e49474. [CrossRef] [PubMed]
483. Padovan, A.; Chick, R.C.; Cole, V.J.; Dutoit, L.; Hutchings, P.A.; Jack, C.; Fraser, C.I. Genomic analyses suggest strong population connectivity over large spatial scales of the commercially important baitworm, *Australonuphis teres* (Onuphidae). *Mar. Freshw. Res.* **2020**, *71*, 1549–1556. [CrossRef]
484. Dağli, E.; Ergen, Z.; Çinar, M.E. One-year observation on the population structure of *Diopatra neapolitana* Delle Chiaje (Polychaeta, Onuphidae) in Izmir Bay (Aegean Sea, eastern Mediterranean). *Mar. Ecol.* **2005**, *26*, 265–272. [CrossRef]

diversity

MDPI

Review

On the Diversity of Phyllodocida (Annelida: Errantia), with a Focus on Glyceridae, Goniadidae, Nephtyidae, Polynoidae, Sphaerodoridae, Syllidae, and the Holoplanktonic Families

Daniel Martin [1,*], Maria Teresa Aguado [2,*], María-Ana Fernández Álamo [3], Temir Alanovich Britayev [4], Markus Böggemann [5], María Capa [6], Sarah Faulwetter [7,8], Marcelo Veronesi Fukuda [9], Conrad Helm [2], Monica Angelica Varella Petti [10], Ascensão Ravara [11] and Marcos A. L. Teixeira [12,13]

1 Centre d'Estudis Avançats de Blanes (CEAB-CSIC), 17300 Blanes, Spain
2 Animal Evolution & Biodiversity, Georg-August-Universität, 37073 Göttingen, Germany; chelm@uni-goettingen.de
3 Laboratorio de Invertebrados, Facultad de Ciencias, Universidad Nacional Autónoma de México, Ciudad de México 04510, Mexico; mafa@ciencias.unam.mx
4 A. N. Severtzov Institute of Ecology and Evolution (RAS), 119071 Moscow, Russia; britayev@yandex.ru
5 Fakultät II-Natur- und Sozialwissenschaften Department, University of Vechta, Fach Biologie, Driverstraße 22, 49377 Vechta, Germany; markus.boeggemann@uni-vechta.de
6 Departament de Biologia, Universitat de les Illes Balears, 07122 Palma, Spain; maria.capa@uib.es
7 Department of Geology, University of Patras, 26504 Patras, Greece; sarifa@hcmr.gr
8 Hellenic Centre for Marine Research, Institute of Oceanography, 19013 Anavyssos, Greece
9 Museu de Zoologia, Universidade de São Paulo, São Paulo 04263-000, Brazil; mvfukuda@usp.br
10 Instituto Oceanográfico, Universidade de São Paulo, São Paulo 05508-120, Brazil; mapetti@usp.br
11 Centre for Environmental and Marine Studies (CESAM), Departamento de Biologia, Campus de Santiago, Universidade de Aveiro, 3810-193 Aveiro, Portugal; aravara@ua.pt
12 Centre of Molecular and Environmental Biology (CBMA), Departamento de Biologia, Universidade do Minho, Campus de Gualtar, 4710-057 Braga, Portugal; mark-us_teixeira@hotmail.com
13 Campus de Gualtar, Institute of Science and Innovation for Bio-Sustainability (IB-S), Universidade do Minho, 4710-057 Braga, Portugal
* Correspondence: dani@ceab.csic.es (D.M.); aguadomolina@uni-goettingen.de (M.T.A.); Tel.: +34-636-046-003 (D.M.); +49-(0)551-39-25536 (M.T.A.)

Citation: Martin, D.; Aguado, M.T.; Fernández Álamo, M.-A.; Britayev, T.A.; Böggemann, M.; Capa, M.; Faulwetter, S.; Fukuda, M.V.; Helm, C.; Petti, M.A.V.; et al. On the Diversity of Phyllodocida (Annelida: Errantia), with a Focus on Glyceridae, Goniadidae, Nephtyidae, Polynoidae, Sphaerodoridae, Syllidae, and the Holoplanktonic Families. *Diversity* **2021**, *13*, 131. https://doi.org/10.3390/d13030131

Academic Editor: Bert W. Hoeksema

Received: 5 February 2021
Accepted: 8 March 2021
Published: 17 March 2021

Abstract: Phyllodocida is a clade of errantiate annelids characterized by having ventral sensory palps, anterior enlarged cirri, axial muscular proboscis, compound chaetae (if present) with a single ligament, and of lacking dorsolateral folds. Members of most families date back to the Carboniferous, although the earliest fossil was dated from the Devonian. Phyllodocida holds 27 well-established and morphologically homogenous clades ranked as families, gathering more than 4600 currently accepted nominal species. Among them, Syllidae and Polynoidae are the most specious polychaete groups. Species of Phyllodocida are mainly found in the marine benthos, although a few inhabit freshwater, terrestrial and planktonic environments, and occur from intertidal to deep waters in all oceans. In this review, we (1) explore the current knowledge on species diversity trends (based on traditional species concept and molecular data), phylogeny, ecology, and geographic distribution for the whole group, (2) try to identify the main knowledge gaps, and (3) focus on selected families: Alciopidae, Goniadidae, Glyceridae, Iospilidae, Lopadorrhynchidae, Polynoidae, Pontodoridae, Nephtyidae, Sphaerodoridae, Syllidae, Tomopteridae, Typhloscolecidae, and Yndolaciidae. The highest species richness is concentrated in European, North American, and Australian continental shelves (reflecting a strong sampling bias). While most data come from shallow coastal and surface environments most world oceans are clearly under-studied. The overall trends indicate that new descriptions are constantly added through time and that less than 10% of the known species have molecular barcode information available.

Keywords: polychaetes; species diversity; molecular diversity; phylogeny; ecology; distribution

1. Introduction

The order Phyllodocida was first formulated as monophyletic by Dales [1], based on the muscular pharynx shared by the members of the clade. Later it was supported based on characters such as the ventral position of sensory palps, the presence of anterior enlarged cirri, the loss of dorsolateral folds (i.e., protrusible dorsolateral walls of the foregut), the presence of an axial muscular proboscis and the presence of compound chaetae with a single ligament [2]. It was further considered paraphyletic [3], but its monophyly has been recovered with strong support by modern molecular approaches [4,5]. Phyllodocida, like most soft-bodied polychaetes, is not well represented in the fossil record, but fossil specimens have been described from several families [3,6]. Most extant groups first appear in the Carboniferous [7], although *Arkonips topororum* Farrell & Briggs, 2007 [8], from the Devonian, seems to group within Phyllodocida, thus possibly constituting the earliest known member of the clade.

As currently delimited, Phyllodocida is one of the largest and most diverse polychaete clades. It includes Aphroditiformia, Glyceriformia, Nereidiformia, and Phyllodociformia, each with several generally well-defined clades interpreted with the taxonomic hierarchical level of family, but also several unplaced taxa (as Phyllodocida *incertae sedis*). In total, Phyllodocida holds more than 6600 species-level taxa, of which around 4627 are currently considered valid in the World Polychaeta Database (WPD) [6], where 28 families (excluding Pholoidae, a synonym of Sigalionidae [9]) and 566 valid genera are listed. However, higher taxa are constantly being revised as more insights from molecular methods and new morphological data (e.g., through microCT, confocal imaging, etc.) become available [10,11].

Phyllodocida is a ubiquitous group of annelids. Most members are typically marine benthic, but a small fraction also inhabit brackish waters, freshwater, and even terrestrial environments, and a few are holoplanktonic [12]. Benthic forms live as in- or epifauna in muddy and sandy bottoms, mixed sediments, under rocks, or hiding in crevices in hard surfaces, from shallow littoral to the deepest marine bottoms [3], including extreme environments such as hydrothermal vents [13–15]. Most species are free living (especially within Nereidiformia and Phyllodociformia), some burrowing in sediments (e.g., Glyceriformia), and some are tubicolous (e.g., some Aphroditiformia) [12]. Most species are 'active-searching' or 'sit-and-wait' predators, feeding on other invertebrates (e.g., among Nereidiformia, Phyllodociformia, Glyceriformia, or Aphroditiformia); some may be carrion-feeders and herbivorous, rarely alternate these trophic guilds with filter feeding (e.g., among Nereidiformia) [12]. Moreover, a large number of species (particularly within Aphroditiformia) live symbiotically with other benthic species (including echinoderms, cnidarians, decapods and, even, other polychaetes) [16,17].

Phyllodocida are among the most phylogenetically diverse groups of organisms [18,19], while the key roles they play in marine ecosystems lead them to be a demanding component for morphology-based biomonitoring [20]. Moreover, molecular tools are also being increasingly integrated in regular and large-scale biomonitoring initiatives thanks, for instance, to high-throughput sequencing technologies [21,22]. However, to achieve their full potential, the creation and constant improvement of DNA barcode libraries is an essential task to support species identification. Together with the emergence of DNA metabarcoding and eDNA-based approaches for ecological and biological research [23], the need to update molecular libraries becomes crucial [24] not only for already known species, but also for the remarkable hidden diversity that is being continuously revealed with the support of molecular data [25–27]. Taking this into account, the first part of this paper aims at analyzing all public Barcode Of Life Data System (BOLD) [28] data to assess the worldwide DNA barcode coverage for the species of Phyllodocida. This will allow us to evaluate taxonomic uncertainties, as well as to analyze species phylogenetic diversity, to improve DNA metabarcoding studies at the taxonomic assignment step [24] and to highlight the existing knowledge gaps and the main still-pending taxonomic revisions.

The literature on the species of the order is immense, as can be partly shown just by going through the list of authors describing new species compiled in the WPD [6]. In addition, there are also numerous papers on anatomy, biology, ecology, and other disciplines (some of them, but certainly not all, listed in the present paper). Such a vast amount of information makes reviewing diversity-related aspects in the group challenging. In this sense, we will also focus on providing wide-spectrum information on different topics dealing with a selected group of representative families (and here we use Linnean ranks for practical purposes) within Phyllodocida, mainly targeting diversity-related aspects. Relying on the different expertise of the involved authors, the groups included in this paper are the two families of Glyceriformia (i.e., Goniadidae and Glyceridae), the holoplanktonic Alciopini, Iospilidae, Lopadorrhynchidae, Pontodoridae, Tomopteridae, Typhloscolecidae and Yndolaciidae, and the families Nephtyidae, Polynoidae, Sphaerodoridae, and Syllidae.

2. Materials and Methods

2.1. Species List

Species and subspecies for Phyllodocida were downloaded from the WPD [6], a subset of the World Register of Marine Species (WoRMS) [29] on 2020-09-06, using the *Worrms* library [30] in R 3.6.1 (R Foundation, Vienna, Austria) [31]. Subsequently, taxa with an unclear taxonomic status (*nomen nudum*, *interim* unpublished, temporary name, uncertain, *taxon inquirendum*) were excluded. Alternative representations of names were treated as objective synonyms (all data and scripts available via figshare—DOI: https://doi.org/10.6084/m9.figshare.13678570, posted on 2 March 2020).

2.2. Occurrence Records

Distribution records were downloaded on 2020-04-21 from the Ocean Biodiversity Information System (OBIS) [32] using the library *Robis* [33] in R and additional records were downloaded from the Global Biodiversity Information Facility (GBIF) via the online search interface [34]. The two datasets were joined and duplicate records (same species sampled at the same location on the same date) were excluded. Only species-level records were considered, and they were matched against the WPD [6] to obtain the currently accepted names. Species with an unclear taxonomic status (i.e., *nomen nudum*, *interim* unpublished, temporary name, uncertain, *taxon inquirendum*) and fossil records were excluded. Coordinates were rounded to two decimal places to obtain the same precision for all coordinates. Points falling on land were removed with the Obistools package [35] in R (with a buffer of 5 km to allow for imprecision in the coastline and to include lagoons and swamps). For depth analyses, points deeper than 10,902 m were excluded. Depths in OBIS are given as positive numbers, any negative numbers were checked manually and converted to true depths. This resulted in a dataset of 538,599 records, covering all families of Phyllodocida, all geographic regions and latitudes and all depths.

2.3. Biogeographic Distribution

We used a system of biogeographic regions dividing the world's ocean into 32 realms [36]. As these realms do not constitute regions of similar sizes or depths, for a finer resolution, a hexagon grid with equally sized cells of ca. 69,967.85 km^2 surface was created using the *dggridR* package in R [37]. For analyses of latitudinal patterns, data were grouped into 5° latitudinal bands. Depth distributions were analyzed within depth zones from intertidal to hadal (intertidal and shallow subtidal: <10 m, high subtidal: 10–100 m, low subtidal: 100–200 m, continental slope: 200–1000 m, bathyal: 1000–4000 m, abyssal: 4000–6000 m, hadal: >6000 m). No standardization of the ocean surface or water column volume corresponding to these intervals was performed, as the scope of the analysis was merely descriptive.

As the data exhibit a strong sampling bias towards the coasts of Europe and North America and species richness is strongly correlated with the number of records, the expected number of species for a random sample of 50 occurrence records (ES50) [38] for each observational unit (hexagon cell/region/latitudinal band/depth interval) was calculated using the *calc* function by OBIS in R (https://github.com/iobis/ebsa/blob/master/lib.R, accessed on 20 April 2020) using frequencies of species per observational unit in place of abundances.

Uniqueness of species was calculated by determining which percentage of the total number of species in a biogeographic area were unique for that area (i.e., did not occur in any other biogeographic area).

2.4. Non-Indigenous Species

A list of non-indigenous species and their regions of introduction was compiled from the World Register of Introduced Marine Species (WriMS) [39] and additional literature sources [40–45] and plotted using shapefiles for regional seas [46]. Non-indigenous species are defined as "species introduced outside of their natural range (past or present) and outside of their natural dispersal potential" [47].

2.5. Data Mining and BOLD Dataset Creation

We uploaded a list of selected taxa of Phyllodocida to BOLD [28] (CL-MTVPP, DOI: https://doi.org/10.6084/m9.figshare.13678570, posted on 2 March 2020), comprising 27 families, 566 genera, 4680 species, and 161 subspecies. We also uploaded a list of species considered non-indigenous (CL-MTAPP, DOI: https://doi.org/10.6084/m9.figshare.13678570, posted on 2 March 2020), containing 13 families, 44 genera, 62 species, and one subspecies (Table 1). We used the BOLD platform to search for all the publicly available COI-5P sequences from Phyllodocida and GenBank to create the dataset DS-MTAPP (DOI: https://doi.org/10.5883/DS-MTAPP, posted on 2 March 2020) for the analysis. A species was considered successfully barcoded if at least one COI-5P sequence (>300 bp) was available. COI sequences without information on species name and with less than 300 base pairs, lacking BINs and flagged for contamination, stop codons or indels were subsequently removed. We started from a dataset containing 11,799 sequences corresponding to 1418 species. However, only 7831 barcodes (from 830 species) were publicly available. Using them we apply the methods described above to obtain a final dataset (also used for statistical analyses) that included 6361 DNA barcodes from 620 species (3509 exclusive to BOLD and 2852 mined from GenBank making). Since most GenBank records lack metadata (e.g., GPS coordinates, depth), we excluded GenBank-only records from the species list to generate a new dataset with 3509 records that was also uploaded to BOLD (DS-MTBPP, DOI: https://doi.org/10.5883/DS-MTBPP, posted on 2 March 2020) to analyze bathymetric patterns in barcode availability.

Table 1. Non-native species. PO: possible origin; AOI: area of introduction; BNN: number of sequences (seq) barcoded non-native areas; BOA: number of sequences (seq) barcoded in other areas.

Family	Species	TL	AOI	BNN	BOA	Source
Chrysopetalidae	*Bhawania goodei*	Bermuda	Mediterranean Sea		Florida Keys, USA (1 seq)	[39]
Glyceridae	*Glycera capitata*	Greenland	Black Sea		Artic Russia (9 seq)—Kandalaksha Bay, Velikaya Salma Strait; NAmerica (10 seq)—Hudson Bay, Canada; Saglek Fiord, Canada; Devon Island, Canada; Alaska, USA; Ratnagiri coast, India (2 seq)	[40]
Hesionidae	*Podarkeopsis capensis*	South Africa	E Mediterranean		N Spain (1 seq)	[39]
Nephtyidae	*Inermonephtys inermis* (1)	Florida Keys, USA	Red Sea		China (3 seq)—Laizhou Bay	[40]
Nereididae	*Alitta succinea*	Cuxhaven, Germany	Australian Exclusive Economic Zone; Argentina; Caribbean Sea; Hawaii; Japan; South Africa; USA Pacific		USA Atlantic (8 seq)	[39,40]
Nereididae	*Alitta virens* (2)	Manger (N Bergen, Norway)	Baltic Sea, North Sea		Kandalaksha Bay, Velikaya Salma Strait, Russia Arctic (3 seq) St. Andrews, Blockhouse, USA Atlantic (50 seq)	[40]
Nereididae	*Leonnates decipiens*	Gulf of Mannar, Indian Ocean	Mediterranean		India (1 seq)—Mumbai coast	[40]
Nereididae	*Namalycastis abiuma*	Santa Catarina Island, Brazil	Hawaii		China (2 seq)—Yuandang Lake in Xiamen; India (5 seq)—Kadinamkulam estuary	[40]
Nereididae	*Neanthes acuminata* (3)	Gulf of Naples, Italy	USA Pacific	S California, USA (54 seq)	Portugal (5 seq); Hawai (1 seq), USA Atlantic, Connecticut (5 seq); Baja California, Pacific Mexico (6 seq)	[40]
Nereididae	*Perinereis aibuhitensis*	Palau, Pacific Ocean	Japan, Portugal		No GPS data (7 seq); Zhoushan Zhujiajian, China (1 seq)	[40]
Nereididae	*Perinereis nuntia*	Gulf of Suez	Mediterranean Sea		Pari Island, Indonesia (1 seq)	[40]
Nereididae	*Pseudonereis anomala*	Gulf of Aden	Mediterranean Sea		Queensland, Australia (21 seq)	[39]
Paralacydoniidae	*Paralacydonia paradoxa* (4)	Monaco, Mediterranean Sea	Red Sea		Bohai Sea, China (18 seq)	[39,40]
Phyllodocidae	*Eumida sanguinea* (5)	Denmark (Kattegat)	Hawaii		NE Atlantic (29 seq)—Scilly islands, Great Britain; Finnmark, Norway; Bergen, Norway; Bohuslän, Sweden; Helsingör, Denmark	[40]

Table 1. *Cont.*

Family	Species	TL	AOI	BNN	BOA	Source
Pilargidae	*Sigambra parva*	S South Africa	E Mediterranean		Ratnagiri coast, India (2 seq)	[39]
Polynoidae	*Paralepidonotus ampulliferus*	Philippines	New Zealand		No GPS data (3 seq)	[40]
Polynoidae	*Subadyte pellucida* (6)	Croatia, Adriatic Sea	Red Sea		Cádiz—Spain (2 seq)	[40]
Sigalionidae	*Pisione guanche*	La Gomera, Spain	E Mediterranean		Lanzarote, Spain (1 seq)	[39]
Syllidae	*Amblyosyllis speciosa* (7)	Misaki, Japan	USA Pacific	Port of Los Angeles, San Pedro Marina (3 seq)		[40]
Syllidae	*Branchiosyllis exilis* (8)	Djibouti, Red Sea	Aegean Sea, USA Pacific		Shark Bay, Western Australia (1 seq)	
Syllidae	*Eusyllis kupfferi*	Madeira island, Portugal	E Mediterranean		Kalbarry, Western Australia (1 seq)	[39]
Syllidae	*Syllis bella* (9)	Laguna Beach, Pacific USA	E Mediterranean		Philipines (1 seq)	[39]
Syllidae	*Syllis gracilis* (10)	Gulf of Naples, Mediterranean Sea	Argentina		Peru (5 seq); Australia (2 seq); Los Angeles, Pacific USA (12); Phillipines (8 seq); Italy (2 seq); Galicia, Spain (4 seq)	[40]
Syllidae	*Megasyllis nipponica*	Sea of Japan	USA Pacific		Manazuru Peninsula, Pacific coast of Japan (1 seq)	[40]

(1) Red sea specimens differ slightly from Western Atlantic specimens and thus require confirmation [48]. (2) Natural distribution not clearly delimited, reported areas of introduction are near the type locality, so status in these areas unclear. In addition, North Atlantic and North Pacific specimens comprise at least two cryptic lineages [49]. (3) Different clades (possibly species) in different worldwide localities have been confirmed with molecular methods [50]. (4) Red Sea record in Fauvel (1933) [51] and should be reviewed. (5) Known species complex, comprised of at least 10 different cryptic species in Europe alone [25]. Thus worldwide locations likely host undescribed species, too. (6) Several specimens previously assigned to *S. pellucida* have been re-examined [52] and assigned to *Subadyte albanyensis* Hanley & Burke, 1990 [53]. Thus, the presence of *S. pellucida* in the Red Sea needs confirmation. (7) See details in Section 3.2.5 and in Aguado et al. [54]. (8) A species complex in need of revision. Mediterranean specimens could belong to the native *Branchiosyllis cirropunctata* (Michel, 1909) [55], which previously considered a synonym of *Branchiosyllis exilis* (Gravier, 1900) [56] by Faulwetter [41]. Thus, worldwide locations likely host undescribed species, too. (9) Minor morphological differences between specimens from California and Lebanon reported [57]. (10) Known species complex, comprised of at least three different cryptic species in the Mediterranean alone [58]. Thus, worldwide locations likely host undescribed species, too.

2.6. Data Processing and Analyses

We conducted a global gap-analysis by comparing the available barcoded species of Phyllodocida by 4 April 2020 and its congruence with the total number of valid species [24,59,60]. We compared the species list CL-MTVPP with all publicly available COI-5P sequence records using the BOLD checklist tool to obtain the percentage of barcoded species. We only included records identified at the species level and discarded those with tag codes added by BOLD users. Tag codes are often used either to distinguish lineages within cryptic complexes or between different populations in certain BOLD projects. As such, these records are considered as different species by the Checklist Progress Report (CPR) tool in BOLD. Thus, they will not match with the corresponding species found by the CPR tool (e.g., '*Nereis pelagica* CMC01' will be considered a different species from '*Nereis pelagica*').

All species in the dataset had a Barcode Index Number (BIN). We annotate them with one of four possible taxonomic congruency grades: Discordant (i.e., more than one nominal species assigned to the same BIN, which often include conflicts with sequences of species labelled with tag codes), complex (i.e., one nominal species assigned to more than one BIN), concordant (i.e., one species assigned to a single BIN), and singletons (nominal species with just one available sequence). We carefully inspected discordant BINs by checking their placement in neighboring-joining (NJ) phenograms, looking for valid species names, synonyms or contaminations, and by inspecting BINs' content on BOLD database. We considered BINs as "complex" when the same species had more than two sequences for at least two different BINs and were close to each other in the phylogenetic tree. Additionally, if the same species have two BINs with more than two sequences and a third BIN with one sequence, we would consider the third BIN as part of the complex as well, instead of a singleton. The BIN system clusters COI sequence data into molecular operational taxonomic units (MOTUs) independent of prior taxonomic assignment. As such, allows us confirming barcode sequence clusters vs. species designations concordance. We performed this validation by comparing the taxonomy on input records against all others in the same BINs, including those submitted and managed by other users [28].

We built the worldwide barcode map based on georeferenced data with the *dggridR* package in R (see Section 2.3 for details) and we used the BOLD Accumulation Curve tool to visualize the total number of sequences, species and BINs over time, for the whole order and for each family of Phyllodocida. We represented further data analyses by histogram and pie charts created with Microsoft Excel.

2.7. Analyses at the Family Level

The cumulative curve of accepted species along the temporal axis [61] for each selected taxon was calculated based on the year of description according to the WPD [6]. We have considered recent, currently accepted species names since the first species attributed to a given family. The information for each targeted family has been based on a selection (i.e., according to the authors criteria) of the respective most relevant references that are listed directly in the reference list.

3. Results and Discussion

3.1. Species Diversity Trends among Phyllodocida

Based on the data from the WPD (see Section 2.1. for details), the most species-rich family by far is Syllidae, with over 1100 currently accepted species, followed by Polynoidae, Nereididae, and Phyllodocidae (Figure 1a). In most families, about half of all described species names are currently considered accepted, though in some species the number of synonymized species names is considerably higher or lower (Figure 1b). The first species in what is today Phyllodocida was described in pre-Linnean times as *Aphrodita aculeata* Linnaeus, 1758 [62], previously known as *Eruca echinata* Barrelier, 1714 [63]. Since then, the number of described species has been growing at first slowly and more rapidly after the

1860s, when several major works were published [64–67]. To date, there is no indication that the number of newly described species is slowing (Figure 2a,b).

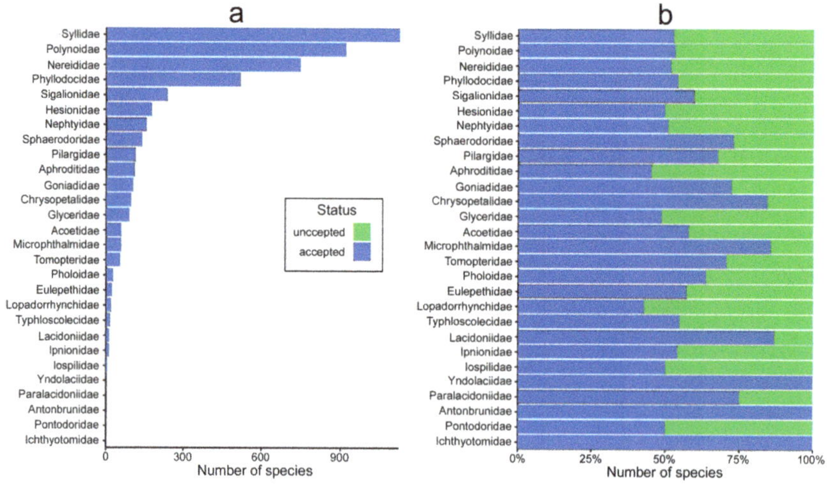

Figure 1. (**a**) The number of currently accepted species per family; (**b**) percent of accepted and unaccepted species names per family. Based on the WPD.

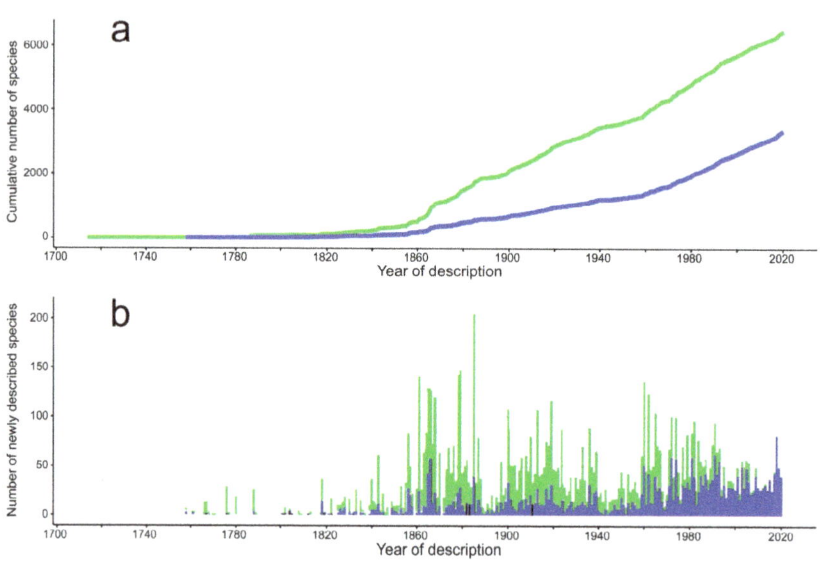

Figure 2. Description rates of species in Phyllodocida, based on the WPD. (**a**) Cumulative number of newly described species; (**b**) number of species described each year. Green: all species names including those currently unaccepted, blue: currently accepted species names.

3.1.1. Biogeographic Distribution Patterns

Phyllodocida are distributed globally, although the highest numbers of species have been reported from European, North American (especially the Gulf of Mexico), and Australian waters (Figure 3a). However, these numbers are biased by an increased sampling effort in these regions and do not reflect true species richness, as species richness was highly correlated with the number of records (Spearman's ρ = 0.737332, $p \leq 0.001$ for hexagons, ρ = 0.6774194, $p \leq 0.001$ for biogeographic regions). The five species with the most occurrence records were *Nephtys hombergii* (Savigny in Lamarck, 1818) [68], *Hediste diversicolor* (O.F. Müller, 1776) [69], *Nephtys cirrhosa* Ehlers, 1868 [65], *Eteone longa* (Fabricius, 1780) [70], and *Goniada maculata* Örsted, 1843 [71]. The five families with the most occurrence records were Nephtyidae, Phyllodocidae, Syllidae, Nereididae, and Polynoidae. The five species with the widest distribution (highest number of hexagon cells) were the holopelagic *Pelagobia longicirrata* Greeff, 1879 [72], *Tomopteris septentrionalis* Steenstrup, 1849 [73], and *Typhloscolex muelleri* Busch, 1851 [74], as well as the benthic *Glycera capitata* Örsted, 1843 [71] and *Harmothoe imbricata* (Linnaeus, 1767) [75]. The five taxa with the widest distribution were Phyllodocidae (including the holoplanktonic Alciopini), Polynoidae, Nereididae, Syllidae, and Lopadorrhynchidae.

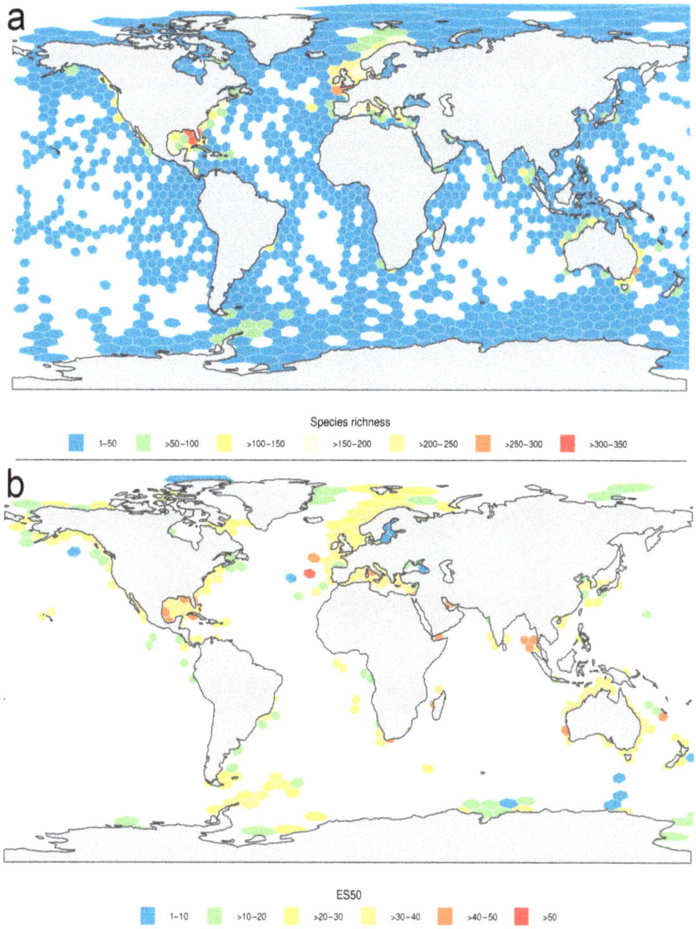

Figure 3. Global species richness of Phyllodocida. (**a**) The number of species; (**b**) expected number of species (ES50).

ES50 was slightly less biased than the number of species. While still showing the highest values in European, North American, and Australian coasts, those of South America and Antarctica also hosted a considerable diversity (Figure 3b). However, ES50 assumes sufficiently large sample sizes and the same sampling methods. In our analysis, many cells do not have a species number large enough for ES50 to be calculated, thus preventing acceptable estimates of the respective number of expected species. When analyzing larger biogeographic areas [36], the most species-rich were the Caribbean/Gulf of Mexico and the coasts of the Indian Ocean (Figure 4a). However, being large areas, sampling efforts were slightly more balanced and the ES50 gave what looked like a more accurate picture, with hotspots of species diversity in the Red Sea and the coastal Indian Ocean (Figure 4b).

Some biogeographic areas—notably Antarctica and the Pacific coasts of the America and Asia—had very unique species compositions, with 40–50% of endemism (i.e., species reported only from a given area). Similarly, circumtropical areas had a high number of unique species (with the exception of the offshore Indian and Pacific Ocean). Endemism was less frequent in the temperate areas of the Atlantic and the Arctic Oceans (Figure 4c). Latitudinal patterns of sampling effort appeared strongly biased towards the northern hemisphere (Figure 5a). However, raw species richness did not reflect this sampling bias, showing a secondary peak in the southern hemisphere (Figure 5b). ES50 appeared less biased, with lower numbers in temperate latitudes and several distinct peaks (Figure 5c). The number of occurrence records was correlated with the number of species (Spearman's $\rho = 0.7413102$, $p \leq 0.001$) but not with ES50 (Spearman's $\rho = 0.3011364$, $p \leq 0.08883$). The weak latitudinal gradient, with a rather uniformly high diversity across tropical and temperate latitudes and a drop of species richness only at extreme latitudes has been observed also in amphipods [76] as well as in a variety of other taxonomic groups [77,78].

The distribution patterns of Phyllodocida showed similarities to those found for polychaetes overall [79], which showed the highest species richness in the North Atlantic, Indonesia and Australia. These, plus the Mediterranean, Caribbean, and Red Sea and the Indian Ocean coasts, were also the regions where most species of Phyllodocida were expected (ES50). This may be in part due to correlation of the two datasets (as Phyllodocida form a large clade within Annelida). However, these trends have been identified by independent authors with a different methodology, thus supporting our results. Similarly, the bimodal latitudinal pattern we observed has been found not only for Annelida overall but seems to hold true for a variety of marine taxa [77,80] and may be related to sea temperature [79,80].

Regarding bathymetric patterns, the majority of the world's ocean are under-sampled, and the vast majority of information is available from the shallower coastal and surface areas [81], and Phyllodocida is no exception. Our results showed: (1) most records (i.e., 88.5%) coming from 10 to 100 m depth (Figure 6a); (2) a similar distribution (although more balanced) for species number, which was correlated with sampling effort (Spearman's $\rho = 0.964285$, $p \leq 0.002778$) (Figure 6b); and (3) ES50 not correlated with the number of records (Spearman's $\rho = 0.2857143$, $p \leq 0.556$), being similar at all depths, except for the most extreme (i.e., very shallow and very deep) (Figure 6c). Although depth zones are not equally sized sampling areas and, thus, our results have to be interpreted with care, we are nevertheless showing that members of Phyllodocida seemed well adapted to live in deeper waters. However, in most depth zones half of the occurrence records belonged to less than five families, particularly to Polynoidae, while Nereididae, Syllidae, and Nephtyidae dominated in shallow waters (Figure 6d).

Our analysis also revealed that only 74 over the total number of species of Phyllodocida known to date have been reported as occurring non-natively in certain parts of the world. Most of them (i.e., >30) occur in Mediterranean waters, while in other coasts the number of non-native species is lower than 10 (Figure 7a). Interestingly, over half of the non-native species belong to Nereididae and Syllidae (Figure 7b). Although this would be expected given these families are among the most species rich, it seems that Polynoidae did not follow the same pattern and this is certainly something that merits further studies. As for barcoding, some non-native species of Phyllodocida listed in CL-MTAPP have been upload to BOLD indicating that they are considered invasive in certain areas (Table 1). However, the total number of barcoded alien Phyllodocida is relatively low (24, ca. 40%) (Figure 7c, Table 1).

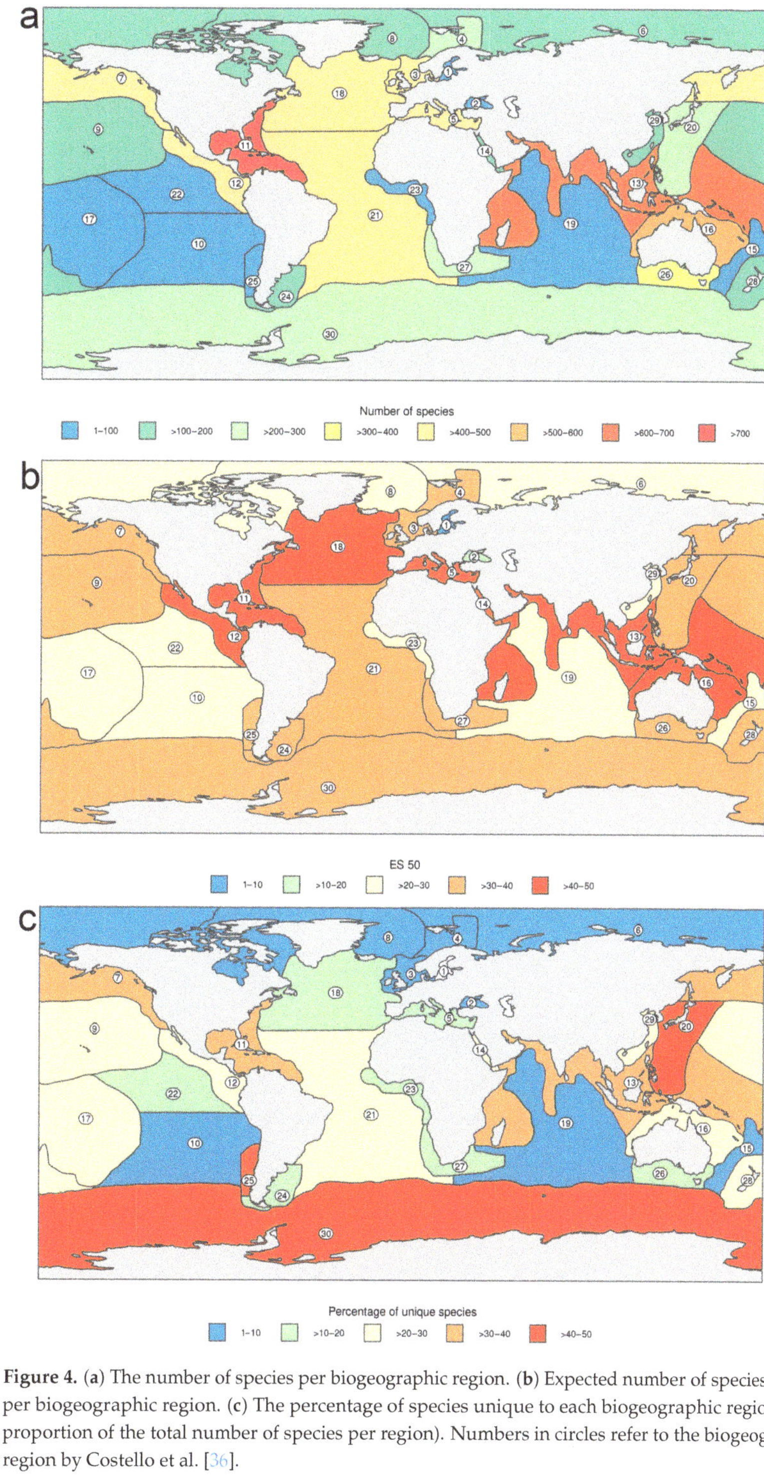

Figure 4. (**a**) The number of species per biogeographic region. (**b**) Expected number of species (ES50) per biogeographic region. (**c**) The percentage of species unique to each biogeographic region (as a proportion of the total number of species per region). Numbers in circles refer to the biogeographic region by Costello et al. [36].

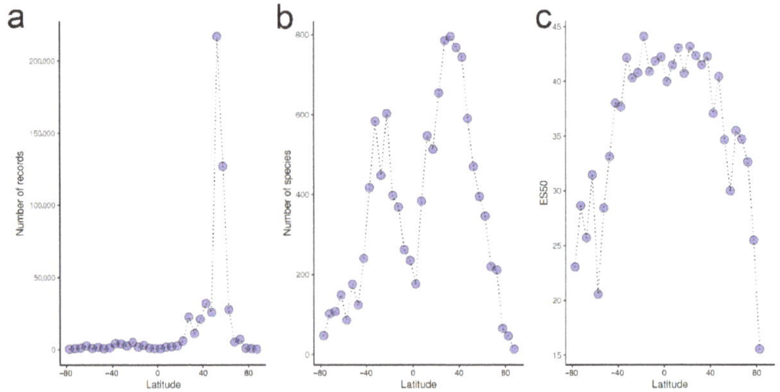

Figure 5. (**a**) The number of records, (**b**) number of species, and (**c**) expected number of species per 5° latitude bands.

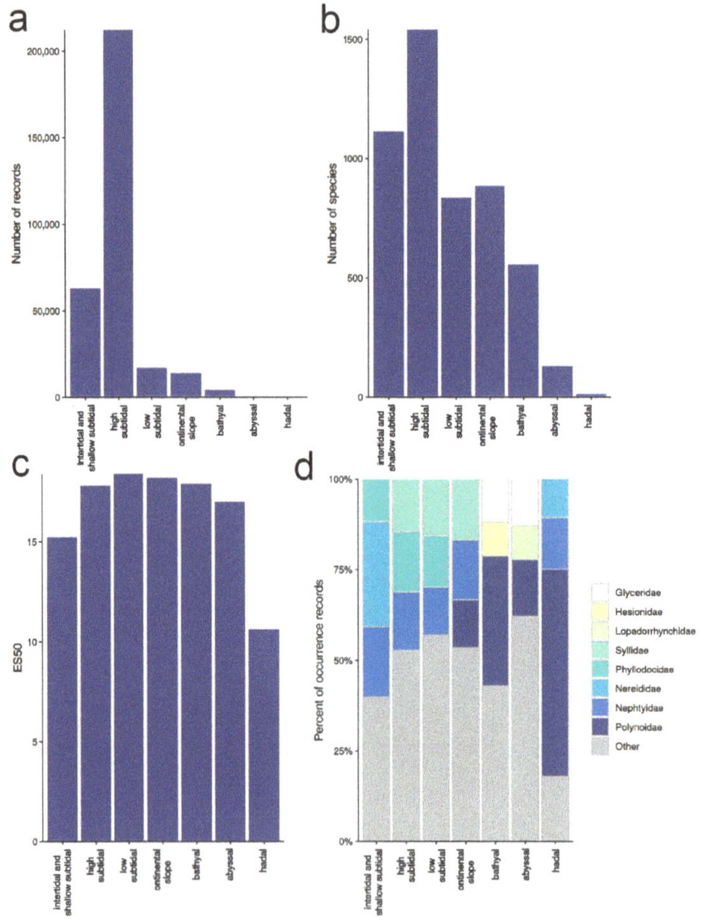

Figure 6. Depth distribution of Phyllodocida: (**a**) The number of records, (**b**) number of species, (**c**) ES50, and (**d**) the percentage of occurrence records per family and depth zone.

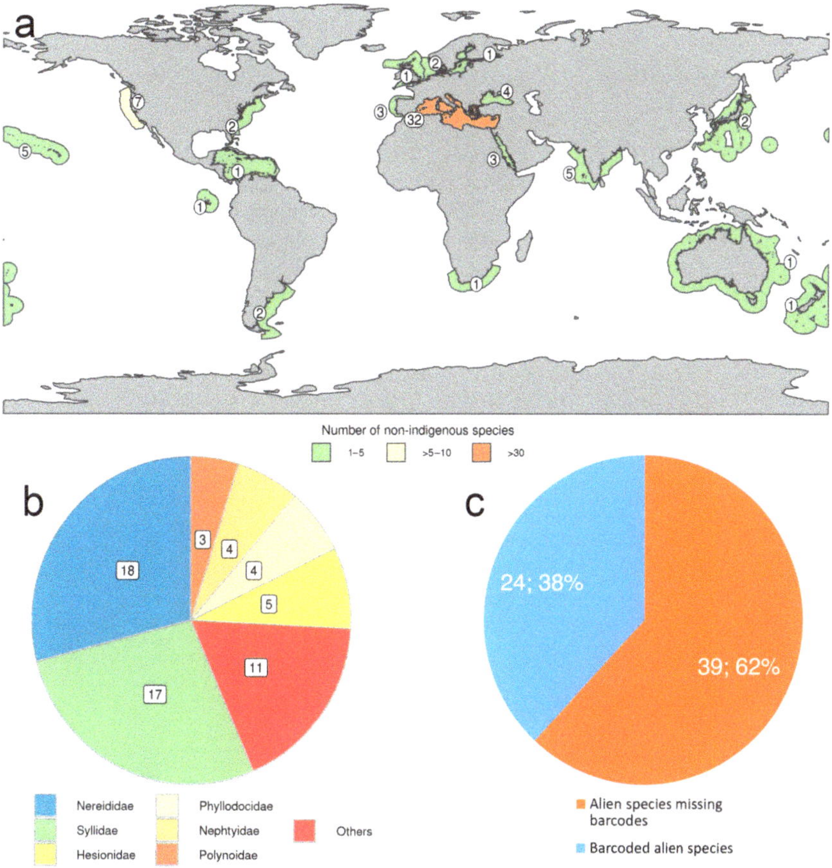

Figure 7. (**a**) The number of non-indigenous Phyllodocida species in various areas of the world. (**b**) Families with the most non-indigenous species, independent of region. (**c**) The number of barcoded species belonging to reported alien species found in literature.

As mentioned, much care must be paid to the information included in the databases, particularly concerning taxa identification. An example of how complex the situation can be, for instance, occurs with the report of *Amblyosyllis speciosa* Izuka, 1912 [82] from the USA (Table 1). The specimens of this species from Dorsey (California) were identified as part of *Amblyosyllis hectori* Aguado, Capa, Lago-Barcia et al., 2019 [54], while those from Washington in Pernet [83] belong to *Amblyosyllis anae* Aguado, Capa, Lago-Barcia et al., 2019 [54]. *Amblyosyllis nigrolineata* Okada, 1934 [84], which occurs in Japan, Australia, and California, was synonymized with *A. speciosa* [85] and later considered valid [54], while *Amblyosyllis finmarchica* (Malmgren, 1867) [86] occur in Norway, Russia, and Canada [54]. Therefore, the specimens reported from the USA could belong to one of four species, more likely *A. nigrolineata*, but certainly we may state that they do not belong to *A. speciosa*, as the species does not occur in the USA.

In addition, reports of alien species in the past were often not accompanied by detailed species descriptions—only molecular data—and many of these reports were propagated through publications and databases of non-indigenous species. While some of these species can be clearly identified as non-native or even invasive (even pests), many of them may in fact be undescribed species native to the area that simply constitute misidentifications. A good practice for publishing future reports of non-native species occurrences should, therefore, provide detailed taxonomic descriptions and barcoding data if possible. In addition, the specimen in question should be deposited in a reference collection for future consultation. Once barcodes of the species from the type locality or native area become available, these "aliens" can be assessed much more accurately.

3.1.2. Global Gap-Analysis

A total of 620 species of Phyllodocida have sequences published in BOLD, while the total number of BINS is 1215 (Figure 8a), with the most advanced libraries coming from North America. The relationship between sequences and BINs is caused by sequences being assigned to higher taxonomic ranks (genus or family), but also to wrong taxonomy assignment. In terms of number of sequences per family, Polynoidae took the largest share (24%), followed by Nereididae (20%), Phyllodocidae and Syllidae (11% each), Hesionidae (10%), Nephtyidae and Glyceridae (4% each), and only 2% are identified at order level only. All remaining families (except Nautiliniellidae—to date within Chrysopetalidae–and Pisionidae that are currently not accepted in WoRMS) [6] represent a 14% (Figure 8b). However, Syllidae held the highest number of sequenced species (Figure 8c) and Polynoidae, Nereididae, Phyllodocidae, and Syllidae also appeared as the most afflicted with multiple BINs.

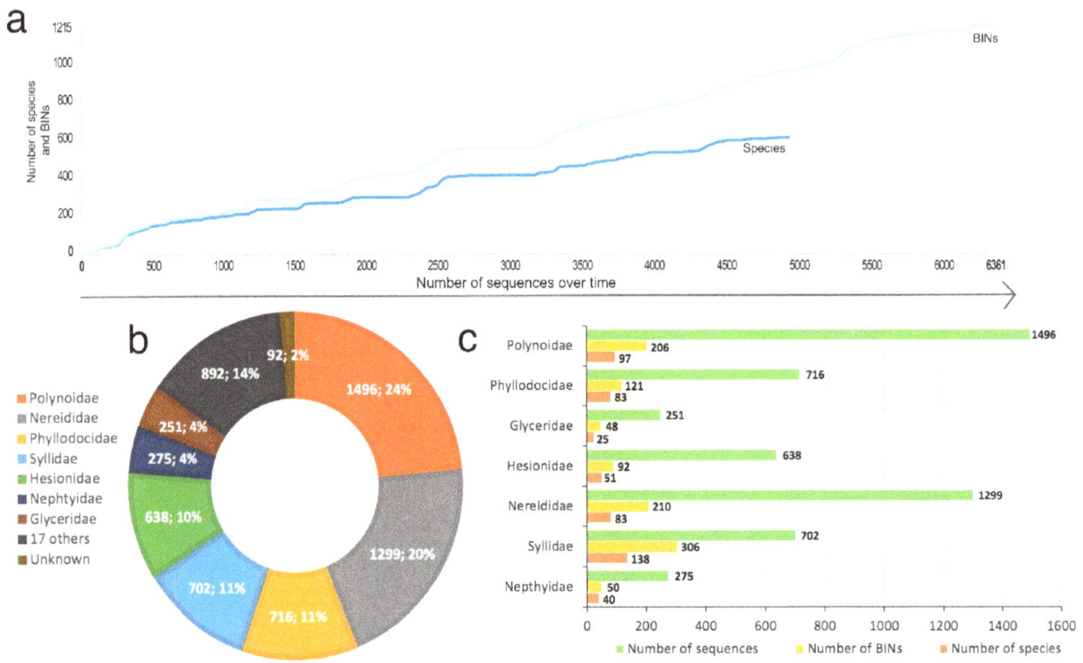

Figure 8. (**a**) Accumulation curve using all records from the dataset. The number of species and number of BINs by number of published/public sequences submitted to BOLD over time from 2008–2019. (**b**) The number of available sequences per family; records lacking family assignations (unknown) correspond to sequences only identified at the order level. (**c**) The number of species, BINs, and the total number of sequences for the most represented families.

The number of DNA barcodes assigned to different taxa levels among Phyllodocida was highly variable (Figure 9a), with 3787 (59.5%) having species names, 754 (11.9%) having only genus names, 559 (8.8%) having family or subfamily names, and 94 (1.4%) having just the order assigned. In turn, 1169 (18.4%) barcodes had tag codes added to the species name. However, only 59.5% over the total 6361 sequences found in BOLD had species names. Thus, only these records could be compared against the worldwide Phyllodocida species-level list (CL-MTAPP), which results in only 10.26% of the species (480) and 0.62% of the subspecies (1) from the species list having barcodes by April 4, 2020 (Figure 10a). Using the same approach, 32.63% (185) of the genera and 78.57% (22) of the families were represented with DNA barcodes (Figure 10b,c). Overall, from the 6361 sequences, it was only possible to analyze 4917 barcodes, which imply that there are at least 1400 sequences misidentified and/or with invalid, misspelled, or synonymized names.

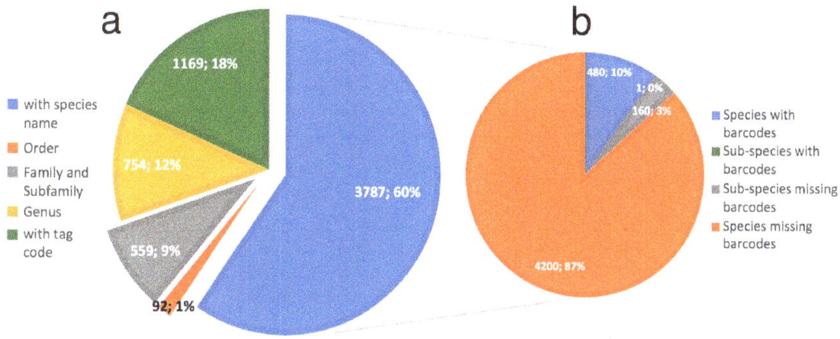

Figure 9. (**a**) The number of DNA barcodes with species names, barcodes identified only at the genus, family and subfamily, order and barcodes with tag codes added to the species name. (**b**) The number of barcoded records with species name present in the list of Phyllodocida (CL-MTAPP).

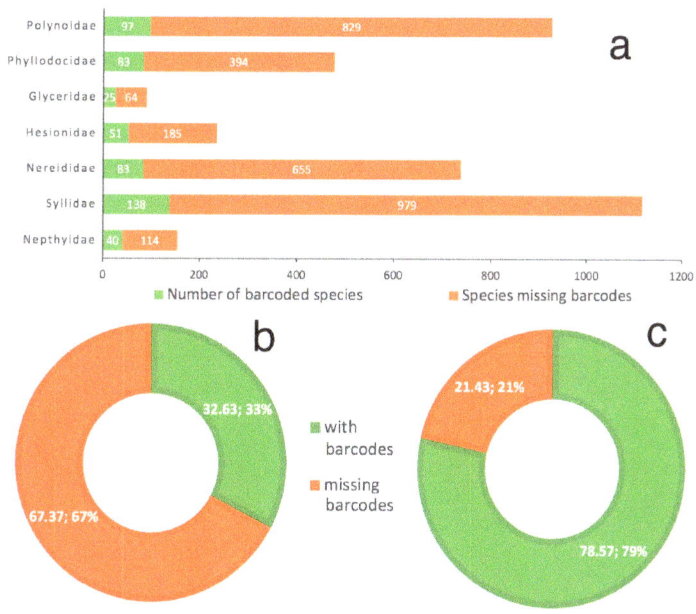

Figure 10. (**a**) The percentage of barcoded species and species still missing molecular data for the most represented families based on the list of Phyllodocida (CL-MTAPP). (**b**) The percentage of barcoded genera. (**c**) The percentage of families with DNA barcodes. Records identified only at the order were discarded.

As mentioned above, Polynoidae and Nereididae had the highest number of representative sequences. However, at the same time they are also by far the families showing the lowest level of completion (Figure 10a, 10.5% and 11.2%, respectively). Conversely, Glyceridae and Nephtyidae doubled these numbers (28% and 26%, respectively). When comparing the total number of barcoded species (527) among these families, Syllidae was the richest family, with 26.7% (138) of the sequenced species, while Glyceridae was the poorest (4.8%, 25 species). These data are still more informative and the lack of knowledge may be better assessed, if taking into account the extremely disparate number of valid taxa of these families: 1117 for Syllidae, 926 for Polynoidae, 736 for Nereididae, 89 for Glyceridae, and 154 for Nephtyidae [6].

As for the biogeographic distribution, although the total number of sequenced species in the DS-MTAPP dataset having georeferenced coordinates is certainly still very low (only 4145 records), barcoding in Phyllodocida showed similar biogeographic trends (Figure 11a) as those reported for the taxa and a similar bias. Most records came from North America (2382), followed by South East Asia (688) and Europe (484), there is also a considerable amount that have unspecific locations (358). As for the number of BINs (Figure 11b), from a total of 1215, 34% (corresponding to 220 species) showed no apparent taxonomic conflict (i.e., concordant), while 6.7% (81 species) showed taxonomic conflicts (i.e., discordant). Moreover, although 44.1% of the records (i.e., 500) were singletons (i.e., having just a single barcode), a significative number of them were identified only at the genus/family level or had tag codes. Thus, our analysis proved that there were only 257 species identified at the species level and having a single available sequence, while 35 "species" (15.2% of the BINs) were possible cryptic complexes.

From the DS-MTBPP dataset (3509 barcodes from 277 species), only 1666 sequences were identified at the species level (and had no tag codes) allowing to analyze the respective bathymetric trends. Accordingly, barcoding appeared to be mostly available for shallow areas (Figure 6e), while deep-sea species showed a significantly low number of sequences after 100 m depth. Not only it is more costly to sample in such locations, but also it is often exceptionally hard to identify deep-sea specimens because of tissue degradation due to the combined effect of different environmental pressures and sampling techniques [19]. Indeed, most deep sea records of sequenced Phyllodocida, correctly identified at the species level, came from a few papers, e.g., [19,49], which certainly indicates that further efforts must be addressed in barcoding deep-sea members of the group. In addition, from the few species having specimens collected from significant different depth levels (more than 100 m apart), three showed again possible evidence of cryptic complexes with lineages specific to each depth layer: *Phyllodoce madeirensis* Langerhans, 1880 [87] (BINs: BOLD:AAZ1549, BOLD:AAZ0051 and BOLD:AAZ0052 at 246, 392 and 660 m depth, respectively); *Glycera kerguelensis* McIntosh, 1885 [88] (BINs: BOLD:AAA8690 and BOLD:AAA8688 at 5000 and 2000 m depth, respectively) and *Eunereis longissima* (Johnston, 1840) [89] (BINs: BOLD:AAY3565 and BOLD:AAZ1159 at 300 and 700 m depth, respectively).

Indeed, there is a still unknown number of possible cryptic species complexes, which we have inferred, in part, from BINs and records having "tag codes" usually attributed by BOLD users to differentiate between cryptic lineages. For instance, "*Nereis pelagica* CMC01" and "*Nereis pelagica* CMC03", which display high COI intraspecific divergence appearing on different BINs. Over the last decade, cryptic species have been increasingly reported, thereby emerging as a substantial fraction of biodiversity and as a much more widespread and frequent phenomenon than previously thought, especially in marine invertebrates [18,90–93]. Dedicated studies about this topic can highly increase the representativeness of sequences belonging to these groups in genetic databases. Thirty-five species were considered possible cryptic species complexes, corresponding in total to 185 BINs. Some notorious examples are *Platynereis bicanaliculata* (Baird, 1863) [94] (six BINs), *Treptopale homalos* Watson, 2010 [95] (seven BINs) and *Pseudonereis anomala* Gravier, 1899 [96] (seven BINs). *Syllis gracilis* Grube, 1840 [97] (six BINs) was already a target study for cryptic diversity [98], with the authors refraining from naming the new species due to the existence of multiple lineages in the same type locality with no apparent morphological differences and inability to access the holotype for sequencing. An extreme case with a unique genetic fragmentation by presenting intraspecific divergence higher than usual compared to other annelids (>3%) but still not enough to be considered different species for most cases (<8%) [49,99], is that of *Hediste diversicolor* (O.F. Müller, 1776) [69]. It was already documented [100,101] and in our study, we have found 37 BINS in 140 sequences. *Hediste diversicolor*, together with *Hediste atoka* Sato & Nakashima, 2003 [102] (10 BINs in our analysis), seem to be outliers where the number of MOTUs clearly and far surpasses the number of possible species within the complex [103]. Overall, Polynoidae and Nereididae showed the highest number of representative sequences while having lower levels of completion (10.5% and 11.2%, respectively; Figure 10a), which might be underestimated due to possible hidden diversity. Integrative taxonomy is thus essential to solve this kind of situations and to allow naming the involved undescribed species. Otherwise, most molecular data providing enough support for species hypothesis [104] will continue to be unused, and large biodiversity sections would remain unnoticed [105].

The problem of cryptic species is, to some extent, intrinsically linked to the detection of exotic species. In some cases, supposedly non-indigenous or introduced species belonging to cryptic complexes. These complexes require detailed morphological studies, often combined with molecular data, to resolve the delimitation of the involved species, often leading to new species descriptions. Obviously, Phyllodocida is not an exception [54,58,90,106–108]. An obvious advantage of metabarcoding studies is the ability to easily detect invasive species in certain locations or even to report species in previously undocumented locations. However, a relatively low number (24, ca. 40%) of Phyllodocida have been uploaded to BOLD with indications that they are considered invasive in certain areas (Table 1), while only two (i.e., only one syllid, one nereidid) have been sequenced in the location reported as being "invaded" (Figure 7c, Table 1). In some cases, the populations from the invaded area or nearby have different sequences in each of these areas, which also differ from that in type locality. This certainly raises the question whether these species are actually non-native or just overlooked cryptic complexes, which certainly merits further analyses.

To assess this and other complex taxonomic and biogeographic problems, recent tools, like the R-based application Barcode, Audit and Grade System (BAGS), may potentially be a valuable addition to forthcoming DNA metabarcoding studies, as it may long-term contribute to globally improve the quality and reliability of the public reference libraries. BAGS can quickly screen reference libraries to gauge data congruence and to facilitate the triage of ambiguous records for posterior review, allowing researchers to obtain the most useful and reliable data by highlighting and segregating records according to their congruency) [109]. Our analyses show the key importance of keeping libraries adequately curated, together with the need of adding metadata (e.g., GPS coordinates, depth) to public databases. This is especially critical as the library we have analyzed still has considerable gaps, numerous poorly represented species, and potential misidentifications or other errors in barcode generation. Certainly, this opens the door to future works that will allow to obtain a more precise picture of the biodiversity within Phyllodocida and, by extension, through the whole tree of life.

Even though the number of sequences and barcoded species have grown almost exponentially since 2008, our results highlight the apparent difficulty of having molecular data with correct identifications among Phyllodocida, with less than 60% of the records being usable to species-level in statistical analysis. Additionally, less than 11% of the compiled worldwide Phyllodocida list had barcodes. This might be partly justified by other factors, such as possible contaminations, misidentifications, outdated taxonomic identifications and synonyms. For example, two families in the BOLD dataset are now invalid [6]: Nautiliniellidae and Pisionidae, with the accepted names being Calamyzinae Hartmann-Schröder, 1971 [110] (subfamily for Chrysopetalidae Ehlers, 1864 [64]) and Sigalionidae Kinberg, 1856 [111], respectively, or the species *Glycera tridactyla* Schmarda, 1861 [112], identified as "*Glycera convoluta*", a subjective synonym, or the species *Sphaerodoridium minutum* (Webster

& Benedict, 1887) [113], being also identified as "*Sphaerodoropsis minuta*", a superseded subsequent combination. Indeed, less than 80% of the species were found barcoded in the list (i.e., 481 of 620), while there where 81 discordant BINs and 535 singletons (Figure 11b). The latter are subject to high uncertainty and low confidence due to the lack of comparable sequences and sources from multiple studies. Even if all species from the analyzed dataset could be found in the list, it still is a far cry compared to the current 4627 valid species of Phyllodocida [6]. This could be due to the marine biodiversity assessment challenge caused by the large-scale geographical sampling effort required, which can affect community richness outcomes [114]. However, the number of studies dedicated to this annelid group and, consequently, that of the associated barcoding projects must also be taken into account [24]. For example, in the case of fishes, the amount of dedicated projects is significantly higher and, thus, the barcode library closer to completion [115–117], which is not the case for macroinvertebrate barcoding projects and the current state of its molecular libraries [60].

3.2. *Selected Taxa*

3.2.1. Glyceriformia

Glyceriformia Fauchald, 1977 [118] is a monophyletic clade within Phyllodocida that includes only Glyceridae Grube, 1850 [119] and Goniadidae Kinberg, 1865 [3,120]. All species are characteristically elongated, slender, cylindrical polychaetes that can reach considerable sizes up to 1 m long [121,122]. Their pointed, usually annulated prostomium with two pairs of terminal appendages, and their long, muscular, eversible axial proboscis, which is densely covered with papillae and provided with terminal jaws, are unique characters among Annelida [123].

Morphology based analyses confirmed the sister-group relationship of glycerids and goniadids and suggested a common ancestor with a few small and equal jaws and ciliated proboscidial papillae [121,123]. However, different molecular markers supported alternative hypotheses: sister-group relationship [124,125], monophyletic Glyceridae nested within a paraphyletic Goniadidae [124] or separated taxa [124,126]. The later indicates that the morphological similarities might be also the result of convergent evolution due to similar selection pressure, which is not very likely. However, both groups are distinguishable from each other, especially by the proboscidial armature (jaws and papillae) and the prostomial ciliation pattern [121–123,127,128].

Glyceriformia is one of the best investigated clades within Annelida as, in addition to two monographic reviews [121,122], there are many other dedicated papers dealing with special topics as well as with ultrastructural, phylogenetic and molecular aspects, which are summarized in the following paragraphs. Prior to the monographs, 165 glycerid and 111 goniadid species or subspecies (not counting fossil forms) and, afterwards, 12 additional species, had been described [124,129–136]. However, only 46 Glyceridae (40 *Glycera* Lamarck, 1818 [68], one *Glycerella* Arwidsson, 1899 [137], five *Hemipodia* Kinberg, 1865 [120]), and 64 Goniadidae taxa are currently accepted as valid (five *Bathyglycinde* Fauchald, 1972 [138], 12 *Glycinde* F. Müller, 1858 [139], 31 *Goniada* Audouin & Milne Edwards, 1833 [140], six *Goniadella* Hartman, 1950 [141], five *Goniadides* Hartmann-Schröder, 1960 [142], three *Goniadopsis* Fauvel, 1928 [143], one *Ophiogoniada* Böggemann, 2005 [122], 1 *Progoniada* Hartman, 1965 [144]) [128]. The current state of the taxonomic diversity of the two families is perfectly reflected by the large disparity between all described taxa and valid taxa (i.e., 286 vs. 110). Overall, the increase of valid species descriptions within Glyceriformia has been slow and more or less constant, with a highest rhythm during the last years of the 18th century, a small but significant peak during the early 20th century and a markedly sharp peak at the beginning of the 21th century, with a maximum of seven in 2001 (Figure 12a–d).

Today the main diagnostic characters allowing to identify the typical species of Glyceridae (Figures 13a,b and 14a) and Goniadiae (Figure 13c,d and Figure 14b,c) are the chaetae, the jaw apparatus, the parapodia, and the proboscidial papillae, whereas, e.g., annulation, color, presence or absence of eyes are of little taxonomic use [121,122]. Scanning electron microscope studies demonstrated that the proboscidial papillae provide one of the most reliable characters for identification [121,122,145–148]. Attempts to use the finer structures of glycerid jaws failed due to intra-specific variation [149]. However, this might be the result of a misidentification [121]. Therefore, further scanning electron microscope examinations are needed to clarify the diagnostic value of glycerid jaws. This is less important to separate recent species, but it might be useful to increase the value of scolecodonts (jaws of fossil annelids) as index fossils for geology [123].

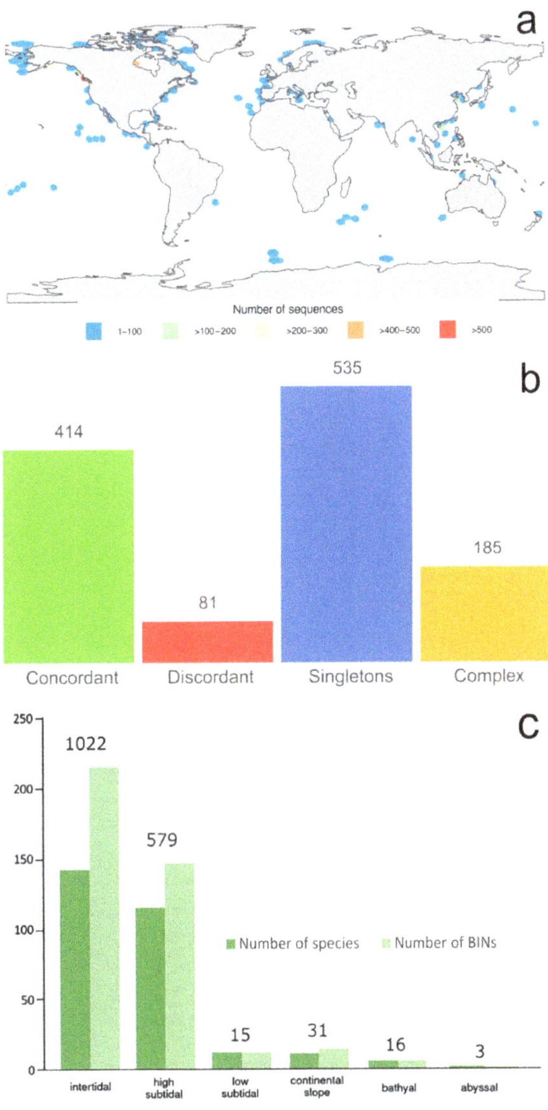

Figure 11. (**a**) Worldwide barcode distribution. (**b**) The number of barcode index numbers (BINs) according to congruency grades. Concordant: The number of BINs with no apparent taxonomic conflict; Discordant: taxonomic conflict within BINs; Singletons: BINs with just one single barcode record; Complex: one species assigned to more than one BIN. (**c**) The number of species with barcode and number of BINs—numbers on the top of each bar refer to the total number of sequences.

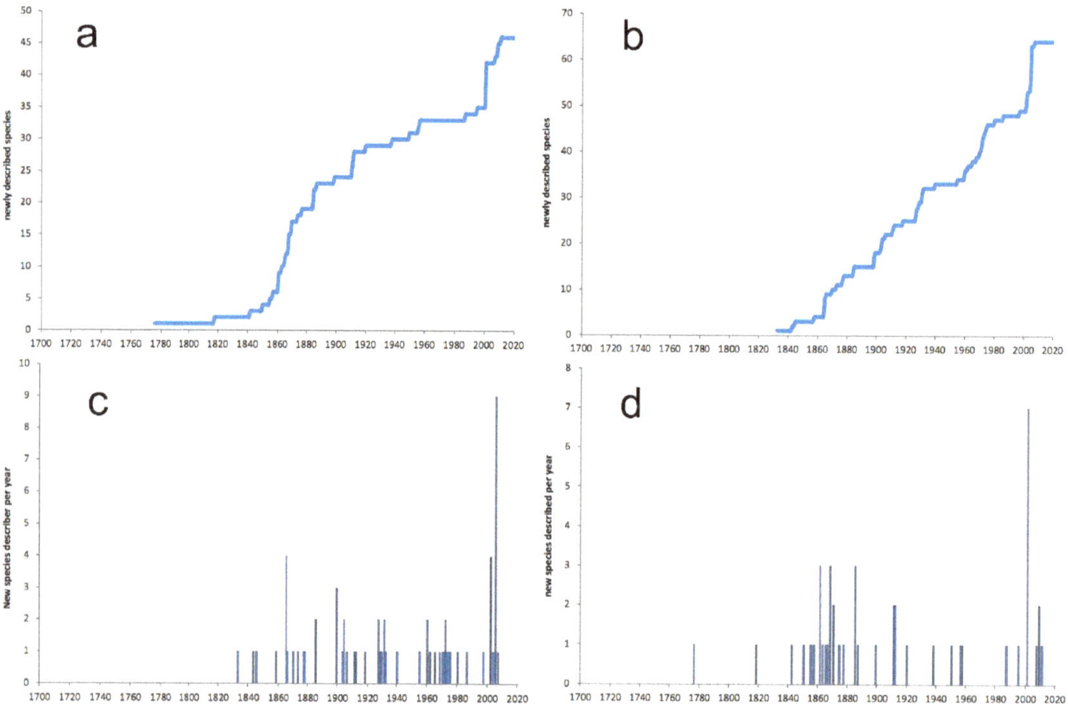

Figure 12. Cumulative curve of newly described species through time (**a**,**c**) and number of new species described per year (**b**,**d**) in Glyceriformia. (**a**,**b**) Glyceridae, and (**c**,**d**) Goniadidae.

Armed with a powerful, eversible proboscis, glycerids (Figure 13a,b and Figure 14a) and goniadids (Figure 13c,d and Figure 14b,c) are important but usually not very abundant benthic predators. They are common burrowers in soft and sandy bottom sediments, forming complex semi-permanent burrow systems with a series of openings [150] or being motile [134,151–153]. Glyceriformia themselves are predated by some birds [154], as well as benthic fishes and crustaceans [155–162]. Furthermore, larger species of glycerids are used as fish bait for sport and commercial fisheries [152,154,163–170].

Glyceriformia have a worldwide distribution from intertidal to abyssal depths [121,122,124]. Glycerids are even present in extreme environments like hydrothermal vents [171,172] and cold seeps [134]. Widespread taxa were very common among polychaetes before using molecular data in species delineation, and some Glyceriformia were even regarded as cosmopolitan [121,122]. Therefore, detecting a complex of cryptic species using two molecular markers was not a surprise [173]. These species supposedly being morphologically "identical" were only investigated via light microscopy and obviously belonged to different taxa [174]. Genetic cryptic species are of course also possible within Glyceriformia [92,124,175], while previous research also confirmed widespread distributions of some deep-sea taxa [174]. The distribution patterns of these species demonstrated that some taxa have a high dispersal capability and show an extended level of eurybathy, whereas other species are restricted to the deep sea. Nevertheless, the underlying processes to become cosmopolitan are unknown.

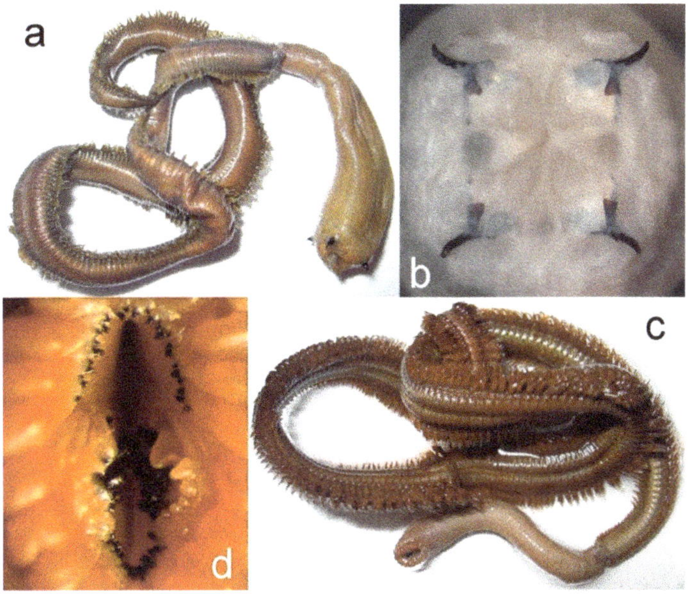

Figure 13. (**a**) An example of an entire preserved specimen of *Glycera* (Glyceridae), with fully everted pharynx; (**b**) detail of the buccal armature in an everted pharynx of *Glycera*; (**c**) an example of an entire preserved specimen of *Goniada* (Goniadidae), with fully everted pharynx; and (**d**) detail of the buccal armature in an everted pharynx of *Goniada*.

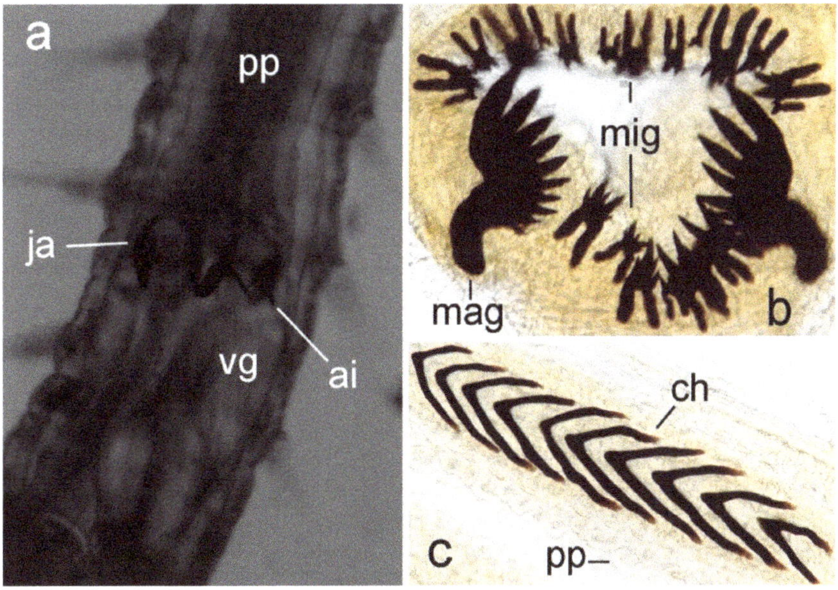

Figure 14. (**a**) *Glycera* sp. (Glyceridae), part of a retracted anterior gut; ai: aileron; ja: jaw; pp: proboscidial papillae; vg: venom gland; (**b**) *Goniada* sp. (Goniadidae), ring of macro- (mag) and micrognaths (mig). (**c**) Part of the proboscis with chevrons (ch) and proboscidial papillae (pp).

Concerning predominate acceptance, passive transport by bottom ocean currents is the main factor for the dispersal [176,177], whereas own studies suggested surface currents as the main driving factor [124,174,178]. There is evidence to support this alternative hypothesis. For example, the glycerid *Glycera capitata* has planktotrophic larvae [179] with a fairly long pelagic larval duration. Not surprisingly, it was present in all investigated basins of the South Atlantic and Southern Ocean, showing almost identical (99–100%) mitochondrial (COI, 16S) and nuclear DNA sequences (18S, ITS 1 + 2). By contrast, the goniadid *Progoniada regularis* Hartman, 1965 [144], has lecithotrophic planktonic larvae [179], which have yolk that must settle before depleting their food source. Species like *P. regularis* are also present in all investigated areas, but populations from different basins show a higher diversity in terms of gene markers (COI, ITS1, ITS2). Molecular analyses at population level revealed an interesting distribution pattern. Those from Northern Brazil and Southern Angola Basins were nearly identical and separated from those of Argentine, Southern Brazil, Guinea, and Cape Basin. This distribution cannot be explained by deep-water currents in this area [180]. Conversely, a presumed influence by surface or subsurface currents [181] makes much more sense. However, these studies were limited in scope as they were based on only a few specimens from each locality and the question still merits further efforts.

Large ocean basins are some of the least explored parts of the world and we know almost nothing about their inhabitants such as benthic animals [182]. For a long time, these areas were postulated to be only muddy deserts, but they revealed to be great reservoirs of biodiversity [183,184] with strong connections to the surface [185,186]. To assess their real biodiversity, extensive and well-preserved material for morphological and molecular studies is needed. The combination of classical taxonomic techniques with modern aspects of biodiversity research allows the analysis of factors influencing the distribution and migration of species as well as the investigation of the background of biogeographic zonation. Furthermore, the results will be useful to clarify the phylogenetic relationships within Glyceriformia and to determinate their position within Phyllodocida. However, the "well"-known shelf and coastal areas have to be newly investigated to identify for example alien species, species complexes, and cryptic or sibling species, and to gain knowledge about changes in biodiversity due to climate change and increasing human activity.

3.2.2. Holoplanktonic Taxa

Many holoplanktonic polychaetes (i.e., those completing the entire life cycle in the plankton) were considered as aberrant [2,187–189]. Their phylogenetic relationships have been largely discussed [188–193] and their taxonomy still remains complex and controversial. In fact, the holoplanktonic species have either been grouped as minor taxa within Phyllodocidae [3,189,192,194–196] or as independent families within Phyllodocida [64,119,188,191,197–201]. Even at the phylogenomics' age, the holoplanktonic polychaetes are still enigmatic, poorly known and with uncertain phylogenetic relationships, thus requiring further research to advance in their knowledge. Nevertheless, the currently accepted situation includes Poeobiidae (including *Poeobius* Heath, 1930 [202]) and Flotidae (including *Flota* Hartman, 1967 [203] but now synonymized with Flabelligeridae by Osborn [204]) being related to Terebellida and all other holoplanktonic families being related within Phyllodocida [205,206]. The latter, the subject of this paper, are Alciopini Ehlers, 1864 [64], Iospilidae Bergström, 1914 [190], Lopadorrhynchidae Claparède, 1868 [66], Pontodoridae Bergström, 1914 [190], Tomopteridae Grube, 1850 [119], Typhloscolecidae Uljanin, 1878 [207], and Yndolaciidae Støp-Bowitz,1987 [193] (Figure 15a–f).

Figure 15. Examples of holoplanktonic Phyllodocida. (**a**) Tomopteridae: Living *Tomopteris* sp.; (**b**,**c**) Alciopini: (**b**) preserved *Vanadis* sp.; (**c**) detail of the anterior end of a living *Vanadis* sp; (**d**) Iospilidae: preserved *Phalacrophorus* sp.; (**e**) Lopadorrhynchidae: living *Lopadorrhynchus* sp.; (**f**) Typhloscolecidae: *Typhloscolex* sp.; scale bars are 1 mm. Images from Gabriel Monteiro, provided by ColBIO-IOUSP (**b**,**d**,**f**) and Xavi Salvador Costa (**a**,**c**,**e**).

Holoplanktonic polychaetes are relatively less diverse than their benthic relatives, from which they can be clearly distinguished by specific adaptations and distinctive morpho-physiological traits. All of them are assumed to be derived from benthic ancestors, even though there is no real evidence supporting this hypothesis [2]. Their relative simplification more likely resulted from an adaptation to a pelagic mode of life, pointing on a basally-branching position. Alternatively, it was hypothesized that "pelagic groups could have given rise to other pelagic groups. And even if their sister taxa are benthic, they could have members which, secondarily, have returned to a benthic life" [194]. In this sense, it is important to consider that, although rare, there are fossil records of pelagic polychaetes assigned to Tomopteridae dating back to the Lower Carboniferous [208,209].

Most species were described in the last part of the 19th century and the first half of the twentieth century, with very few posterior new species, e.g., [193,210–212] (Figure 16a,b). All synonymies and the main systematic information are listed in Dales [213] and Dales and Peter [214]. Currently there are around 143 nominal species within Phyllodocida. Among them, 12 nominal genera and approximately 49 species, most of them belonging to *Rhynchonereella* Costa, 1864 [215] and *Vanadis* Claparède, 1870 [67], are included within Alciopini. Among Iospilidae, there are currently five nominal species included in three genera: *Iospilus* Viguier, 1886 [216], *Paraiospilus* Viguier, 1911 [217], and *Phalacrophorus* Greeff, 1879 [72,218,219]. The first described species of Lopadorrhynchidae was *Lopadorrhynchus brevis* Grube, 1855 [220], which was included within Phyllodocidae [119,220].

Since then, only six nominal genera have been described, one of them very recently, *Bathypelagobia* Kolbasova in [218]. They include 21 species, most of them within *Lopadorrhynchus* Grube, 1855 [220] and *Maupasia* Viguier, 1886 [216]. Pontodoridae, in turn, only includes *Pontodora* Greeff, 1879 [72], with *Pontodora pelagica* Greeff, 1879 [72]. *Epitoka pelagica* Treadwell, 1943 [221], initially placed within Syllidae, was considered a junior synonym of *P. pelagica* [189].

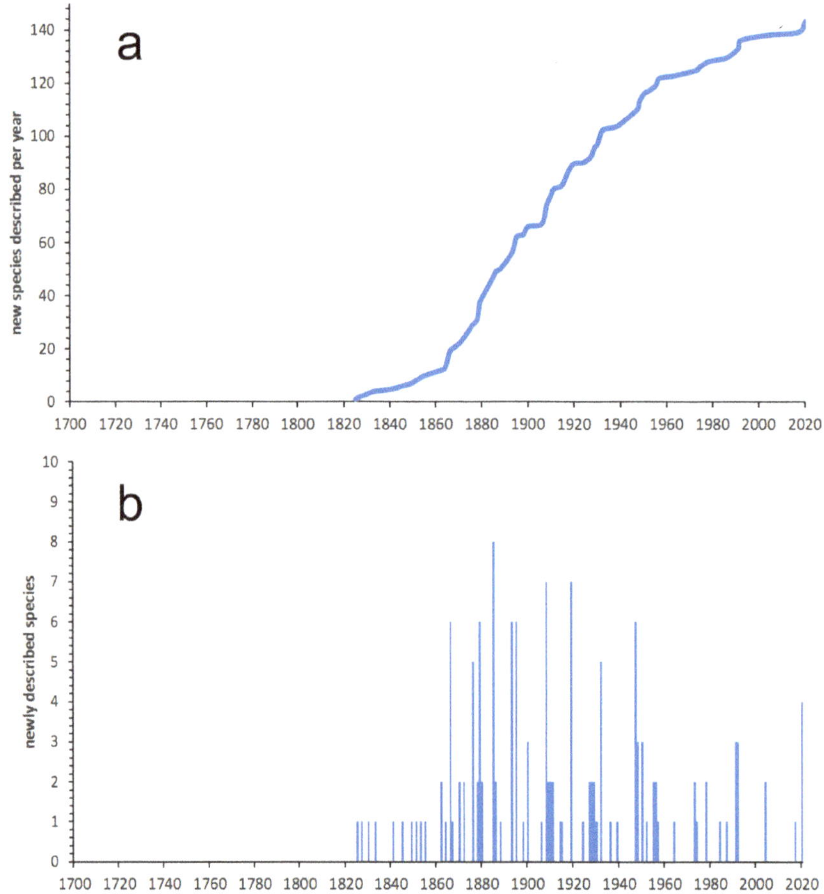

Figure 16. Holoplanktonic polychaetes. (**a**) Cumulative curve of newly described species through time. (**b**) The number of new species described per year.

The number of species of Tomopteridae is still uncertain, because many of them have not been recorded after their original descriptions, which in many cases were also incomplete. Therefore, it is difficult to know the real extent of the existing synonyms, which are not considered in the current databases on the family (Fernández-Álamo, in preparation). There are currently about 53 known species within three nominal genera, *Tomopteris* Eschscholtz, 1825 [222], *Briaraea* Quoy & Gaimard, 1827 [223], and *Enapteris* Rosa, 1908 [224]. Most of them belong to *Tomopteris*, as the other two genera are monospecific. In fact, some authors considered the existence of *Tomopteris* as the single genus of the family [225,226]. Typhloscolecidae also includes three genera, *Typhloscolex* Busch, 1851 [74],

Sagitella Wagner, 1872 [227], and *Travisiopsis* Levinsen, 1885 [228], but the number of species is significantly lower (i.e., about 16).

Alciopini

Members of this group were first considered as a subfamily of Phyllodocidae [189,196,229], but also as an independent family [2,64,210,230–232]. They were also considered as the tribe Alciopini [3,194,195], a criterion apparently confirmed by molecular analysis [205] and nowadays accepted as valid [6,233]. Alciopini have always been considered close to phyllodocids, either within Phyllodocoidea by Chamberlin [234], Phyllodocemorpha by Uschakov [235], Phyllodocida (without special category) by Dales [1], Phyllodocida: Phyllodociformia by Fauchald [118], Phyllodocida: Phyllodociformia: Phyllodocidacea by Pettibone [187] and in Phyllodocida: Phyllodocoidea by Pleijel and Dales [236]. Within the family, the genera have been divided in two morpho-groups [189,191,198,210,237] which, based on a cladistic analysis, derived into the subfamilies Alciopinae and Watelinae [238].

Alciopini (Figure 15b,c) are usually classified as macrophagous visual hunters that feed on copepods, euphausiids and zooplankton, although some small-sized members also feed on diatoms and other phytoplanktonic organisms [12,239]. The recently described genus *Ctenophoricola* San Martín, Álvarez-Campos et al. 2021 (including two species) shows a characteristic parasitic behavior, feeding and living on ctenophores [240]. Most species are epipelagic, with a limited range of vertical migration [239]. However, some occur below the photic zone, deeper than 500 m depth [241] and, infrequently, in neritic environments [239]. Many Alciopini stand out by their very wide distributions, in warm waters of tropical and temperate areas of the open ocean [189]. However, some species occur in cold waters, always in the Antarctic, while they have never been recorded in the Arctic [242].

Iospilidae

Iospilids (Figure 15d) were first included in Phyllodocidae [243,244], which was then divided into Phyllodocides (benthic) and Lopadorrhynchoides (pelagic) [66]. The later included the iospilids, which were defined as subfamily (Iospilinae) within Phyllodocidae [189,191,201,244–247], and as an independent family (Iospilidae) [2,118,190,214,225,231,232,248]. Their systematic position has also been controversial, being placed within Phyllodocoidea [234] and in Phyllodocida, within a "not recognized" suborder [118], within the superfamily Phyllodocidacea (aberrant) [187] and within the superfamily Iospiloidea [236]. More recently it has been suggested that only a rough placement of the iospilids (together with pontodorids and typhloscolecids) within Phyllodocida can be supported, while their actual phylogenetic relationships are still uncertain [249].

Members of this family seem to be motile herbivores feeding mainly on diatoms, as inferred based on body size and the lack of pharyngeal armature, while others apparently behave as ambush predators, as inferred from their appearance, which does not suggest a powerful swimming capacity [12]. In general, however, they are considered as relatively good swimmers compared to similar small-sized holopelagic polychaetes [205]. They are relatively common in open surface waters, but may occur down to 200 m depth and appear to be common in the Southern Ocean near the Antarctic Peninsula [205]. However, they have in fact a cosmopolitan distribution, with the scarce records in other ocean areas likely being caused by their small size and fragility leading the animals to be easily broken during towing, with their fragments being often collected but commonly overlooked in plankton samples [231,246,247,250,251].

Lopadorrhynchidae

As already mentioned, members of this group (Figure 15e) were considered a pelagic tribe within Phyllodocidae [66]. They were raised to subfamily [188–190,200,201,225,244–247,252–254], and to family [248], which is the currently accepted situation [6]. However, the family showed separate clades within Phyllodocidae, with *Pelagobia* Greeff, 1879 [72] being related to *Eteone* Savigny, 1822 [255], and *Maupasia* Viguier, 1886 [216], and *Pedi-*

nosoma Reibisch, 1895 [244] having unclear relationships [3,194], and its whole phylogenetic relationships are still uncertain [249]. Accordingly, lopadorrhynchids were considered within Phyllodocida: Phyllodociformia [118], within Phyllodocidaea [187], and within Phyllodocoidea [236].

Members of this family seem to be either motile herbivores, feeding on microalgae, or predators feeding on small protists and animals, grasping the preys with their enlarged, muscular anterior parapodia and gently curved spines [12,191]. Diurnal vertical migrations from surface to deeper waters during daylight have been reported for *L. brevis* [247]. Many members are considered cosmopolitan. For instance, *P. longicirrata*, one of the most abundant species that has been collected from surface to below 4000 m depth [256] and throughout the world [188,189,214,232,245–247,257,258], including Arctic [256] and Antarctic [245,259–261] waters. However, its cosmopolitan status has been recently questioned. Based on morphological and molecular data, it might represent a species complex [218]. Further comparative molecular studies on material from different areas of the world are required, including specimens such as those attributed to *P.* cf. *longicirrata* from mid Pacific waters.

Pontodoridae

The single known pontodorid, *P. pelagica*, was originally placed within Phyllodocidae [72], and then raised to subfamily (Pontodorinae) [189,201], and to family (within the Tribe Phyllodocidiformia) [2,6,118,144,190,230,231,248,254,262]. Due to the pharyngeal shape, the family was first considered to be related with Syllidae [225] and Nephtyidae [230]. However, the shape, size and arrangement of the pharyngeal papillae more closely resemble those of the phyllodocids. Pontodorids were placed in Phyllodocoidea [234] and in Phyllodocida, within Phyllodociformia [118], Phyllodocidacea [187], and Phyllodocoidea [236]. Nevertheless, their phylogenetic relationships are still unclear [194,249].

Pontodora pelagica seems to be a motile or discretely motile microphagous, but its real diet is still unknown [12]. The species is probably bioluminescent [189] and, although it may be relatively common in warm waters of the world oceans, it could have been overlooked in plankton samples due to its small size and transparent body [213,263]. The species was described from subtropical areas of the Atlantic Ocean (Canary Islands) and the Mediterranean [189] and later reported from the equatorial Atlantic, the Northwestern and Eastern Pacific (from Canada to Peru and Chile) [189,221,264,265], the Indian Ocean [266] and South Africa [225]. There are no records from Antarctic and Arctic waters.

Tomopteridae

Tomopterids are typically enigmatic holoplanktonic polychaetes [194], whose monophyly was the absence of chaetae except the acicular structures supporting the first two parapodia that are parts of the head [230]. Their unique morpho-physiological traits, derived from their adaptations to the holoplanktonic mode of cycle, difficult the establishment of their relationships with the other polychaetes, leading them to be considered as a clade with uncertain phylogenetic affinities [190,194,249]. Nevertheless, they are generally associated with Phyllodocida [3,188]. Their achaetous parapodia led to create the order Gymnocopa to include a single family, Tomopteridea [119]. Subsequently, the family was named as Tomopteridae [267,268], habitually within Phyllodocida, but sometimes as "unrecognized suborder" [118] or as a superfamily, either Phyllodocidacea (aberrant) [187], Tomopteroidea [236] or Tomopterimorpha [269]. The aberrant character of the family seems to be accepted by many authors. They were particularly considered a very old and highly specialized group placed in Phyllodocemorpha: Phyllodociformia, emphasizing its ancient character by pointing out that it was separated very early from all other polychaetes and followed a differential evolutionary path than, for instance, Alciopini [189,235]. Overall, the relationships of this family with Phyllodocida seem to be scarce and even more obscure than those of the other holoplanktonic families. Accordingly, possible advances in their phylogenetic knowledge could be acquired through molecular analyses, which could even result in leaving them out of Phyllodocida [194]. In fact, the family has been recently placed in close relationship with Glyceriformia [4].

Members of the family are often highly motile organisms, looking like delicate transparent feathers [225] (Figure 15a), relatively easy to be imaged in their natural environment with the modern tools used in oceanographic research [270]. They feed with a short, unarmed, eversible pharynx, being either primarily herbivorous feeding mainly on diatoms (small species), primarily carnivorous (large species) or mixed feeders feeding on diatoms, protozoans and other animals (intermediate-sized species) [12]. Some of them seem to be either specialized predators or parasites of chaetognaths, which show sucking marks due to the feeding activity of tomopterids [271]. They are frequent in plankton samples, mainly from surface waters down to a few hundred meters deep, although they have also been reported as deep as 3000 m [189,225,266,272,273]. In fact, they are used in zoogeographical studies since their distributions seemed to be related to particular water masses [199,232,245,246,274]. For instance, there has been a recent notable rise in the abundance of *Tomopteris (Johnstonella) pacifica* (Izuka, 1914) [275] and *Tomopteris planktonis* Apstein, 1900 [198] related to an upwelling in the thermal dome of Costa Rica [276] and of *T. (J.) pacifica* related to the entry of warmer and saltier waters in the Southern Adriatic Sea [277].

Tomopterids are distributed worldwide, living in oceanic and near-shore waters from polar to equatorial areas [245]. However, most species are known from only one area and only a few appear to be cosmopolitan (ex. *Tomopteris apsteini* Rosa, 1908 [224], *Tomopteris elegans* Chun, 1887 [278], *T. planktonis*) [279] or considered as cosmopolitan in cold waters (*T. septentrionalis*) [274].

Typhloscolecidae

Typhloscolecids (Figure 15f) have particularly complex phylogenetic relationships with the rest of polychaetes, as a result of combining the adaptations to pelagic and ectoparasitic modes of life. The name of the family was proposed by Uljanin [207], who relate it with *Phyllodoce* due to a supposed similarity of its "swimming paddles" with the foliose dorsal cirrus characteristic of this genus. Later, they were considered as possible neotenic forms of benthic organisms [191] or, more generally, as an independent family within Phyllodocida [1,280], either within an "*unrecognized suborder* [118], within Phyllodocidacea, as Typhloscolecidae (aberrant) [187] or within Phyllodocida: Typhloscolecoidea [236]. More recently, Typhloscolecidae were included within Phyllodocidae [281], although this was later considered unacceptable owing to analytical problems [282].

Members of the family are motile or discretely motile parasitoids on chaetognaths [12,283]. They may occur attached behind the head of their host chaetognaths [284], which may even lose their heads as a result of the activity of the parasites [283,285]. They may live from surface waters to abyssal depths [3,189,286]. In some cases (e.g., the species of *Travisiopsis*), the adults live in the deep sea, but the juveniles may sometimes be found near the surface [225].

Yndolaciidae

Yndolaciids form a small family whose phylogenetic relationships are still uncertain. They were proposed to be raised to order level [287], but Buzhinskaja [288] considered this as premature due to the poor existing knowledge, thus keeping them as a family within Phyllodocida. Although not proved, they have been considered as carnivores [12]. Moreover, they are only known from deep waters and just known from specific sites, such as the Gulf of Guinea, Southeastern Atlantic [193], the Cape Verde archipelago, Northeastern Atlantic [289], the Mid-Atlantic ridge [290], and the Arctic Ocean [288].

3.2.3. Nephtyidae

Nephtyidae Grube, 1850 [119] is a morphologically well-defined and a monophyletic group within Phyllodocida [2,4,291]. Its monophyly is based mainly on the presence of interramal branchiae attached ventrally to the notopodia (although absent in some taxa) and the single median pygidial cirrus, which are both unique features within the order [230,292]. The family was at first included within 'Phyllodociformia' along with Phyllodocidae and Alciopini [293]. Later, most errant families were regrouped within 'Nereimorpha' (excluding only Amphinomidae and 'Spiomorpha') [294]. 'Phyllodocemorpha' and 'Nereimorpha' were again distinguished [235]. After the erection of polychaete orders, Nephtyidae remained within Phyllodocida, although with an uncertain position [1]. Different analyses placed them close to Phyllodocidae, Paralacydoniidae, Glyceridae Goniadidae, and *Pisione* Grube, 1857 [2,295], or as sister group to Glyceridae and Goniadidae [292] or to Hesionidae, Pilargidae, and Nereididae [4,126,291].

The family includes at present four genera and almost 150 accepted species [6]. Most of them are within *Aglaophamus* Kinberg, 1866 [296] (Figure 17a,b) and *Nephtys* Cuvier, 1817 [297] (Figure 17c,d), while *Inermonephtys* Fauchald, 1968 [298] and *Micronephthys* Friedrich, 1939 [299] include only 10 and 15 species, respectively. A large number of these species (over 50%) were described between the mid-19th and 20th centuries (1865–1972), following important oceanographic expeditions, such as the SMS Gazelle (1874–1876), the dredging by the U.S. coast survey steamer "Blake", the Challenger (1872–1876), the Hirondelle and Princesse-Alice oceanographic campaigns (1885–1910), the F.I.S. "Endeavour" (1909–1914), the Allan Hancock Pacific Expedition, the Swedish Deep-sea Expedition (1947–1948), and the USNS Eltanin and Staten Island Cruises [88,138,203,300–305].

Many taxonomic revisions and major works covering regional geographic areas were subsequently published [48,141,144,201,225,306–341]. Overall, this gave rise to a regular trend of species description with six major peaks during the 20th century (Figure 18a,b), when a significant number of species were described mainly from the East Pacific [141,319], Australia [325,337,338], and tropical West Pacific areas [298,334].

Many taxonomic problems have already been solved, but doubts still remain concerning the boundaries among genera and the correct allocation of many species. Molecular phylogenetic analyses recently started to address these questions [342,343]. The two most speciose genera, *Nephtys* and *Aglaophamus*, appears to be clearly monophyletic, while poor representation of the two other genera did not allow a proper assessment [343]. In this analysis, only 18% of the total known species were included, which was not fully representative of the diversity of the family. For instance, (1) the species of *Aglaophamus* having lyriform chaetae were not included, which did not allow to check whether they would consistently group with the species lacking those chaetae; (2) *Nephtys* was subdivided in two morphologically distinct clades; (3) *Micronephthys* was only represented by two abranchiate and one branchiate species that group in different, quite distant clades, for which the erection of a new genus has been a matter of discussion [307,343]; and (4) *Inermonephtys* is represented by only one species falling outside the nephtyid clade, although with very low support. All these results denoted a clear lack of stability and the need for further reassessment of the phylogenetic relationships within the family at the genus level.

Members of the family are mainly predators, although they may exceptionally be deposit-feeders in highly dense populations [12,344,345]. Most often, they hunt within the sediment, capturing any other motile invertebrates (including smaller conspecifics), although they may also be highly selective [12,344–346]. They are common in most coastal and offshore environments around the world and at all depths, including the deep-sea [138,203,234,301,347,348] and extreme habitats, such as methane seeps, despite the apparently poor tolerance to sulphide [349–351]. However, comprehensive studies on deep-sea species are lacking for all the oceanic basins. Its abundance is highly variable, but may be so high that, especially in coastal environments, large-bodied species are important components of the trophic network and are often exploited as fish baits [167,345,352,353].

Figure 17. Nephtyidae. (**a**) *Aglaophamus trissophyllus* (Grube, 1877) [300]. (**b**) *Aglaophamus elamellatus* (Eliason, 1951) [348]. (**c**) *Nephtys hombergii* Savigny in [68]. (**d**) *Nephtys capensis* Day, 1953 [354].

Habitat preferences may vary considerably, mainly depending on water depth, salinity, grain size and mud content [355]. Different species may co-occur, sharing resources although slightly segregated by small differences in life-traits [355–358], such as the periodic reproductive failures allowing the sympatric distribution of *N. hombergii* and *Nephtys caeca* (Fabricius, 1780) [70] reported by Olive and co-authors [356,359]. Life-stages can be environmentally segregated, such as the adult *N. hombergii* living in intertidal mudflats and the juveniles inhabiting the sublittoral [360]. Shallow water estuarine species may tolerate a wide range of salinities and a few occur in nearly freshwater (e.g., *Micronephthys oligobranchia* (Southern, 1921) [361], *Nephtys fluviatilis* Monro, 1937 [362], and *Nephtys polybranchia* Southern, 1921 [361]) [298,332]. Most species prefer sandy bottoms with variable mud and clay contents, where they move freely to as deep as 20 cm within the sediment [363].

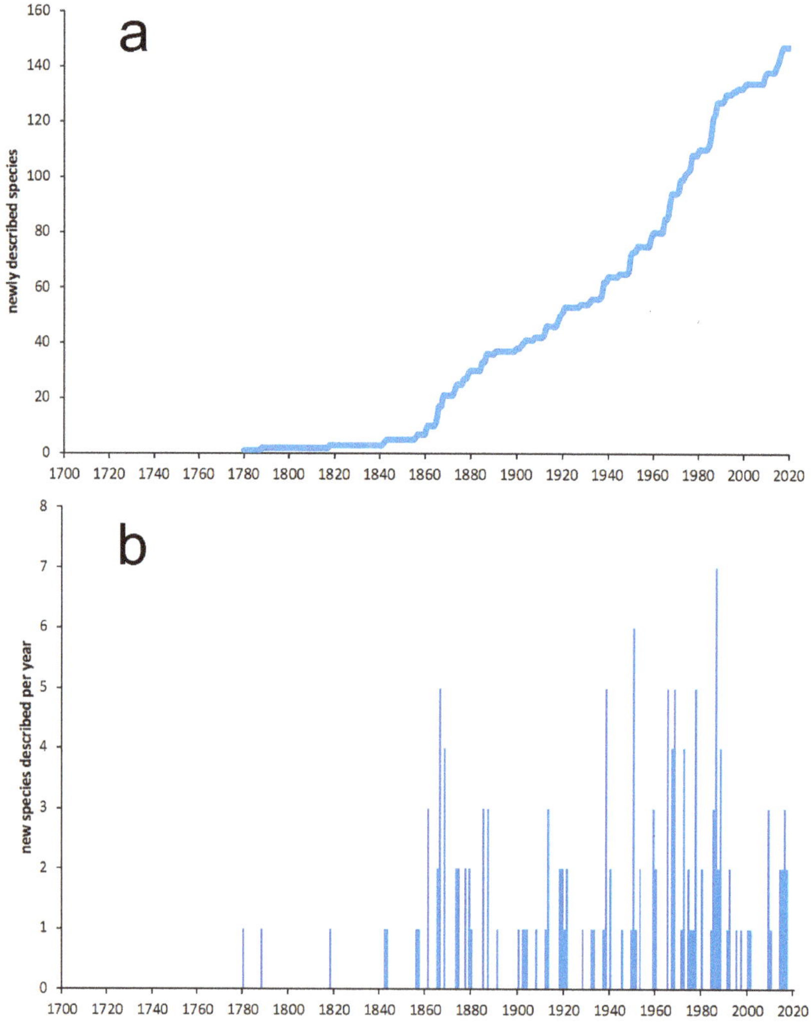

Figure 18. Nephtyidae. (**a**) Cumulative curve of newly described species through time. (**b**) The number of new species described per year.

Despite the multiple areas of the world (e.g., Arctic, Atlantic, Pacific, and Antarctic Oceans, Mediterranean, Black, and Red Seas) covered by the taxonomic revisions mentioned above, there are still geographic areas very poorly documented, such as the Indian Ocean and the African and South American coasts. Furthermore, comprehensive deep-sea studies are lacking for all the oceanic basins. Only 13 deep-sea species are known to date, from the E Pacific, N Atlantic, Antarctic and Tasman Sea, the majority of which described during the 20th century [138,203,234,301,347,348]. As a consequence, many currently accepted species are known only from their original descriptions, that are often very short, incomplete and may lack illustrations (or are of low quality). This casts doubts on their validity, particularly when trying to compare them with others described for the same location. To even complicate the situation, the existence of cryptic species has already been pointed out (e.g., *Micronephthys longicornis* (Perejaslavzeva, 1891) [364], *Micronephthys sphaerocirrata* (Wesenberg-Lund, 1949) [365], *Nephtys cirrosa* Ehlers, 1868 [65], and

Nephtys hombergii Savigny 1822 [366]) [318,333,367]. Therefore, some (if not many) widely widespread taxa may likely involve more than one species (e.g., *Aglaophamus verrilli* (McIntosh, 1885) [88], *Inermonephtys inermis* (Ehlers, 1887) [301], *Micronephthys sphaerocirrata*, *Nephtys caeca* (Fabricius, 1780) [70], and *Nephtys paradoxa* Malm, 1874 [368]) [318,322,333]. This, together with the fact that several species still need to be correctly allocated within the family, denote that further local works and comprehensive revisions, preferably including molecular tools, are still required.

3.2.4. Polynoidae

Polynoidae is included within the clade Aphroditiformia (Aphroditoidea) and is the largest polychaete family after Syllidae [6]. Members of the family typically have more or less similar segments all along the body and, together with all other members of the clade, are commonly known as scale-worms, due to the key trait of having the dorsum more or less covered by elytra or scales (Figures 19a–d and 20a,b).

Figure 19. Polynoidae. (**a**,**b**) Two different color morphs of *Harmothoe imbricata* Linnaeus, 1767) [75]. (**c**). *Lepidonotus tenuisetosus* (Gravier, 1902) [369]. (**d**). *Hololepidella* sp. from a sea urchin *Toxopneustes pileolus* (Lamarck, 1816) [370]. (3) *Pottsiscalisetosus praelongus* (Marenzeller, 1902) [371], symbiont of the starfish *Luidia maculata* Müller et Troschel, 1842 [372]. (**d**,**e**) Photos by J.V. Deart.

Despite its relevancy among polychaetes, the phylogenetic and taxonomic relationships within the family have been seldom approached. Early studies based on morphological characters [373,374] and general analyses dealing with higher taxa relationships [2,292] supported the monophyly of Aphroditiformia, but the position of Polynoidae remained unresolved. Conversely, Struck, et al. [375] placed Pisionidae within Aphroditiformia and, while refusing the monophyly of the suborder, strongly supported the placement of Polynoidae as a well-supported clade within them. Further combined molecular and morphological approaches placed the family as sister group to Acoetidae and Iphionidae [376,377]. More recently, the monophyly was recovered both for the whole Aphroditiformia, with a strong support and having the presence of elytra as apomorphic trait, and for Polynoidae, although some shallow-water (e.g., Lepidastheniinae) and deep-sea (e.g., Branchinotogluminae and Macellicephalinae) subfamilies revealed to be paraphyletic [377,378]. Particularly among deep-sea polynoids, there seems to be at least two contrasting interpretations, which include both a small number of species placed within a large number of subfamilies [377,378] and the synonymy of up to ten subfamilies within Macellicephalinae, which formed an homogeneous clade defined by the absence of lateral antennae [379]. Interestingly, the later agreed with an early postulate based on morphological characteristics, which included five deep-sea subfamilies established by Pettibone [380] (i.e., Bathyedithinae, Bathymacellinae, Macellicephaloidinae, Macelloidinae, and Polaruschakovinae) within the Macillicephalinae as one natural group [381].

Figure 20. Polynoidae. (**a**) *Malmgrenia mcintoshi* (Tebble & Chambers, 1982) [382] in its natural habitat. (**b**) *Asterophilia culcitae* Britayev & Fauchald, 2005 [383] (brown) and the shrimp *Zenopontonia soror* (Nobili, 1904) [384] (white) on the surface of a starfish of the genus *Nardoa* Gray, 1840 [385]. Photos by: (**a**) T.I. Antokhina, (**b**) J.V. Deart.

The first still valid species of the family, *Lepidonotus squamatus* (Linnaeus, 1758) [62], was described as *Aphrodita squamata* Linnaeus, 1758 [62]. Since then, 1486 species have been described, of which 871 are currently considered as valid and grouped into 154 genera within nine subfamilies, together with 13 genera that were not placed in any of these subfamilies [6]. Almost half of all known species (i.e., 433) belong to the Polynoinae, while the Admetellinae, Eulagescinae and Uncopolynoinae consist on one to eight species [6]. The three largest genera, *Harmothoe* Kinberg, 1856 [111], (150 species), *Lepidonotus* Leach, 1816 [386] (79) and *Eunoe* Malmgren, 1865 [387] (43) account for about 32% of all known species, while nearly half of all known genera (i.e., 48%) are monotypic [6].

The rhythm of new species descriptions (Figure 21a) mirrors that of the whole "Polychaeta" [61] and Phyllodocida (this paper). From 1758 to 1850 (with 26 species), a limited number of generalist taxonomists described 0.3 species per year; from 1850 to 1939, about 400 species were described. There was a ten-year parenthesis during the second world war (with eight species described, less than one per year) and then a highly productive reprise, which involves a high number of specialized taxonomists allowing to describe almost 400 more new species to date (i.e., 5.7 new species per year) (Figure 21b). Moreover, as the number of newly described species did not show traces of stabilization (Figure 21b), it is expected a great increase of the species diversity of the family.

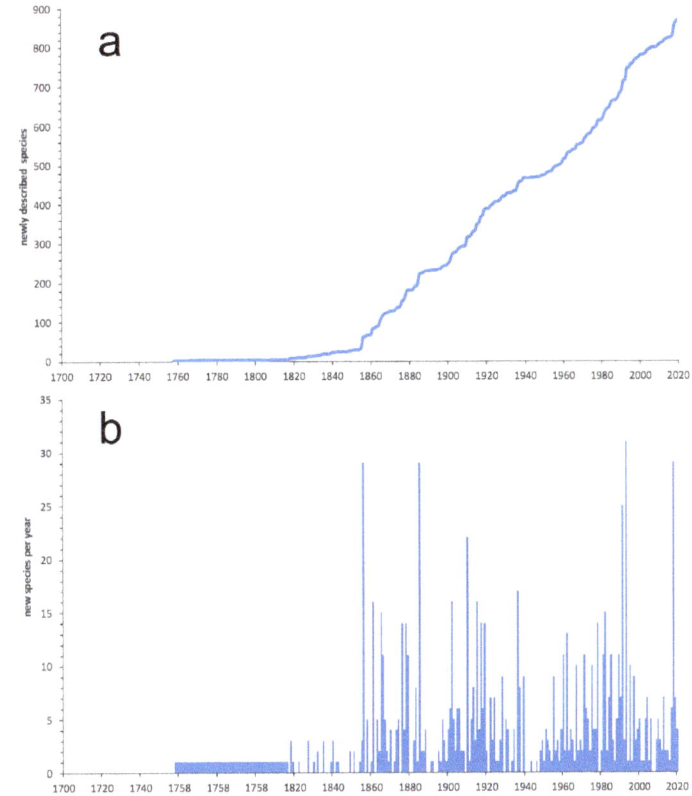

Figure 21. Polynoidae. (**a**) Cumulative curve of newly described species through time. (**b**) The number of new species described per year.

An additional fact contributes to increase the complexity of the taxonomy of Polynoidae, as one can also expect the discovery of new sibling or pseudo-sibling complexes. Although poorly documented to date, some cases have been reported for free-living polynoids with large distribution ranges, like *Harmothoe imbricata* (Linnaeus, 1767) [75]. The species is well-known to show numerous distinct color morphotypes (Figure 20a,b) and the 10 color morphs inhabiting Scandinavian and Svalbard coasts revealed to belong to a single polymorphic species [388]. However, an analysis of a wider geographic range revealed a species complex composed of at least four diverging lineages [107]. Similar situations occur within the symbiotic taxa, particularly with the polyxenous species (i.e., those showing a wide host-ranges). The number of species of *Branchipolynoe* Pettibone, 1984 [389], which was four (with wide distributions and associated to different deep sea mytilids) until 2019, was recently doubled after a study of Pacific methane seeps and hydrothermal vents [106]. However, still nowadays, the most spectacular case is that of *Harmothoe lunulata* (Delle Chiaje, 1841) [390]. The species used to be known as associated with numerous species of echinoderms, cnidarians, polychaetes, sipunculans and balanoglossids. However, it is currently accepted that these partnerships involve at least 15 different species from three different genera (i.e., *Malmgreniella* Hartman, 1967 [203], *Lepidonopsis* Pettibone, 1977 [391] and *Wilsoniella* Pettibone, 1993 [392], the latter currently accepted as *Pettibonesia* Nemésio, 2006 [393]), showing clear biogeographic segregation and, in many cases, specific adaptations, such as color mimicry [392].

To date, molecular data available at GenBank refer only to about 80 species of Polynoidae, which represents no more than 10% of the total number of known species, which certainly gives room to the discovery of new species complexes. Surprisingly, despite the number of known shallow-water species is higher than that of deep-sea relatives, about half of the molecular studies dealing with polynoids refer to deep-sea species. Therefore, to improve our understanding on the phylogeny and classification of Polynoidae, further efforts must be addressed to increase the number of molecular analyses on shallow-water species (especially Arctonoinae and Lepidasteniinae), while continuing with the studies on deep-sea species.

A particular aspect that has not been addressed for the whole family, e.g., [376,377] is the possible origin of the symbiotic mode of life typically occurring in many polynoids [17]. Conversely, based on a limited approach mainly including deep sea species, it has been suggested that: (1) most basal species of the subfamily Polynoinae are obligate symbionts showing specific morphological adaptations; (2) commensal and free-livings modes of life have evolved several times; and (3) the obligate coral commensals were monophyletic [394]. Based on a still more limited dataset, Shields et al. [395] just confirmed the commensal association of *Eunoe bathydomus* (Ditlevsen, 1917) [396] with holothurians, in addition to supporting the specific difference of two geographically distant (non-symbiotic) morphotypes of what could be *Harmothoe* cf. *glabra* (Malmgren, 1865) [387].

At present, nine subfamilies are recognized within Polynoidae [6], which partially coincide with the structure proposed by Bonifácio and Menot [379]. Overall, systematics of polynoids at subfamily and genera levels is among the worst developed within polychaetes, and there are no recent revisions on the family. The number of subfamilies varied through time from 9 to 21 [3,6,52] and the boundaries between most of them are not clear, while there are a number of genera whose assignment to a given subfamily is not possible. Some subfamilies (e.g., Uncopolynoinae, Branchipolynoinae, Polaruschakovinae) include one or few species, while among genera many are monotypic (e.g., *Capitulatinoe* Hanley & Burke, 1989 [397], *Gastrolepidia* Schmarda, 1861 [112], *Pottsiscalisetosus* Pettibone, 1969 [398], *Alentiana* Hartman, 1942 [399]), and some are giant (including 100 species or more) and paraphyletic with indistinct boundaries (e.g., *Harmothoe*, *Lepidonotus*).

The vast majority of polynoids are highly motile, typically benthic organisms that can be found in any oxic marine environment. However, there are about ten species (belonging to *Drieschia* Michaelsen, 1982 [400] and *Podarmus* Chamberlin, 2019 [234]) known to be holopelagic [12] and a few may inhabit brackish environments [401] or anchialine caves [402]. They usually occur in very low densities, but local episodes of high abundance have been reported, likely in connection with recruitment events [403] or introduced/invasive species (Marwa Chaibi, personal communication). Most polynoids are typical free-living organisms, being basically sit-and-wait predating or active hunting carnivores, although some may feed on algae or on vent and seep bacterial mats [12]. Moreover their diets may also include different organisms that could either be real targets or captured as bycatches of the real preys (e.g., diatom thecae, algal filaments, crustacean exoskeletons, hydroid thecae, bivalve shells, gastropod radulae, bryozoan skeletons, sponge spiculae, and ciliate loricae), and often include remnants of other polychaetes, such as nereidids or phyllodocids, but also other polynoids, even members of the own species [404]. Whether this may be a real cannibalistic behavior or the result of territorial aggressive interactions is still unknown. However, the fact that territoriality leads to strong and very complex aggressive intra-specific interactions is well known among Polynoidae. For instance, this behavior has been recently reported in free-living species of the deep sea genus *Peinaleopolynoe* Desbruyères & Laubier, 1988 [405], where the purpose was not described (but certainly looks like territoriality) [406] but mainly in symbiotic [16,17] species. Their species diversity highly depends on the studied region, varying from very high (e.g., 114 in the Caribbean and Gulf of Mexico) to very low (e.g., four in the Baltic Sea or seven in Chilean coasts). The most speciose regions are the North Pacific and Tropical Indo-Pacific, including the coastal

Indian Ocean (94 species and 42 genera each), the North Atlantic (79 species and 38 genera) and the Antarctic Ocean (71 species and 32 genera) (Supplementary Figure S1a).

Polynoids are highly diverse in almost all marine environments, from cold Arctic and Antarctic to tropical waters, and from intertidal to deep-sea habitats, although they are absent in fresh waters and their diversity also declines in brackish waters [381]. For instance, while 47 species have been recorded in the Mediterranean Sea [407] (salinity = 36–40‰), in the adjacent Black Sea (salinity = 18‰) there are only four known species [408]. Among the family, the most widely distributed species are the free-living *H. imbricata* and *L. squamatus*, known from coastal boreal and arctic Atlantic waters, as well as from the Pacific and Arctic Oceans (Supplementary Figure S2a,b). However, the former revealed to be a cryptic species-complex [107], while the status of the later has not yet been checked. Moreover, concerning *L. squamatus*, data from OBIS do not reflect its real distribution, as most Pacific records, derived from Uschakov [381], are no included. Symbiotic species also seem to have very wide distributions across tropical Indo-Pacific waters, such as *Hololepidella nigropunctata* (Horst, 1915) [409] or *Paradyte crinoidicola* (Potts, 1910) [410–412]. However, at least the first seems to be a species-complex (Britayev and Fiege, unpublished data). Overall, further efforts (including molecular analyses) are needed to clarify the status of most polynoids having such wide distributions.

The latitudinal diversity gradient traditionally assumed for most benthic macrofaunal organisms seems not to be evident in polynoids, which show a similar bimodal pattern to Phyllodocida overall (Figure 5b), with that in the northern one being significantly higher (Supplementary Figure S1b,c). There is a maximum of diversity in boreal waters that obeys to a much higher sampling effort, particularly in the high latitudes of the Northern Hemisphere, while a lower effort probably explains the low diversity of equatorial seas (Supplemental Figure S1a,c). In fact, recent surveys in the tropical West-Pacific have shown that the diversity of these tropical waters is comparable or even higher than in temperate latitudes [79,413,414]. Moreover, as much as 76 species have been recorded in the South China Sea [413,414], which is comparable with the entire North Atlantic (79 species). Up to 57 species have been reported along Indonesian coasts [79], which is close to or even higher than the diversity of the well-studied Mediterranean, North, or Norwegian Seas. Taking this into account, together with high habitat diversity of the tropical areas, we certainly expect a significant increase in the number of species of polynoid being reported from these regions.

Polynoids have revealed to be particularly conspicuous in the deep-sea. At least 140 species (i.e., ca. 16%) have been found living deeper than 1000 m, some of them reaching the 10,000 m depth at the Kurile-Kamchatka Trench [235]. They are frequent inhabitants of vent and seep sites, organic falls and other chemosynthesis-based habitats from the Atlantic and Pacific Oceans, where numerous new species are being described thanks to new observation and sampling technologies used in recent expeditions e.g., [13,14,379,406]. Other deep sea habitats are also being explored and proved to be inhabited by polynoids, such as sea-mounts in the Indian Ocean [394] or mud volcanoes in the Atlantic Ocean [415]. Therefore, deep-sea habitats seem certainly to be underestimated in terms of polynoid diversity. In line with this, it is not surprising that the number of new deep-sea polynoids described over the past 10 years almost doubled that from shallow waters (i.e., 37 vs. 18, respectively). Many more species remain to be described, particularly in the intertidal Antarctic and tropical waters, which are much poorly explored than the shallow subtidal Arctic Ocean and temperate European and North American waters, but certainly also in deep-sea environments.

Some deep sea polynoids observed thanks to the numerous images captured in recent expeditions are holoplanktonic species with specific adaptations to this particular environment (e.g., https://www.youtube.com/watch?v=yrlSmxG5yZY, accessed on 18 November 2020). However, like their shallow-water relatives, most of them are benthic, many likely correspond to undescribed species, and quite a lot seem to be symbionts. Among them, new discoveries include species associated with carnivorous sponges [416,417], black corals [418], cold-water corals [419,420], holothurians [395,421], and acorn worms (https://www.youtube.com/watch?v=yrlSmxG5yZY, accessed on on 18 November 2020).

Actually, polynoids are not only the family including the most symbiotic species within Phyllodocida (e.g., Figure 19d,e and Figure 20b), but also within the whole 'Polychaeta', with 220 (ca. 25%) species involved in about 600 (36%) relationships [16,17]. The vast majority are roughly considered as commensals, particularly in taxonomic papers, while basic biological, ecological and ethological knowledge is often lacking [16]. Indeed, some widely accepted "commensals" revealed to be mutualists when deeply studied, and only four (i.e., *Gastrolepidia clavigera* Schmarda, 1861 [112], *Branchipolynoe seepensis* Pettibone, 1986 [422], *Eunoe opalina* McIntosh, 1885 [88], and *Thormora johnstoni* (Kinberg, 1856) [111]) turned out to be parasites [16].

Still, nowadays, nothing clearly allows to explain the astonishing dominance of polynoids among symbiotic polychaetes. A possible reason is the use of free-living traits [12] as adaptations to a symbiotic life [16]. For instance, the pharyngeal biting during intraspecific fighting known in free-living species, such as the deep-sea, hydrothermal vent *Peinaleopolynoe orphanae* Hatch & Rouse, 2020 in [406]. This behavior acquires a dramatic intensity in the symbiotic *Ophthalmonoe pettiboneae* Petersen and Britayev, 1997 [423],which use violent attacks to discourage their conspecifics when trying to occupy their hosts [424] Britayev & Martin, unpublished results]. Similarly, there is a remarkably absence of key taxonomic differences between many symbionts and their free-living relatives, while "free-living genera" often include symbiotic species and vice-versa. For instance, 20 over the 150 known species of *Harmothoe* are symbionts, or six over 40 in *Eunoe*.

However, there are specific adaptations to a symbiotic mode of life. Some affect the color which may vary in symbionts with respect to their free-living relatives, even if they belong to the same species. In *Lepidonotus glaucus* (Peters, 1854) [425] the specimens living with eunicid hosts were almost black, while the free-living ones were much paler [53]. Coloring may also vary among symbiotic con-specifics living with different hosts, as in *Gastrolepidia clavigera* Schmarda, 1861 [112,426]. The possible origin of the color mimicry, as well as whether an individual may or not change its color depending on that of the host [17] has not yet been solved.

Mimicry may also be achieved by morphological adaptations (e.g., in antennae, palps, tentacular and dorsal cirri, and elytra) simulating parts of the host body. Additionally, some parts of the body tend to be simplified (e.g., small elytra lacking ornamentation, subbiramous parapodia with reduced notopodia and a few or no notochaetae), typically in bivalve endosymbionts and tube dwelling species [17], attachment structures may be developed (e.g., simplified, hooked chaetae or sucker-like parapodial lobes) or sensory organs may be modified (e.g., eye size and position) [16,17]. Specialized behavioral traits have been developed, such as host/symbiont co-ordination in movements and feeding [427] or distant host-recognition mechanisms, likely chemically mediated [17]. The study of the chemically mediated behavioral interactions between hosts and symbionts lead also to discover that some hosts may also recognize the symbionts [428]. This imply that hosts also gain some benefits (e.g., defense against predators or cleaning), which led to recent reconsideration of some commensal associations as mutualisms [16].

The problem of lacking information also makes it difficult to assess the real extent of the associations established by symbiotic polynoids. Many scale-worms are known to be associated with one (or a few, closely related) host species. For instance, *Medioantenna variopinta* Di Camillo, Martin and Britayev, 2011 [429] associated with the hydrozoan *Solanderia secunda* (Inaba, 1892) [430] or *Adyte hyalina* (G.O. Sars, 1873) [431] associated with two related species of echinoids (*Echinus esculentus* Linnaeus, 1758 [62] and *Gracilechinus acutus* (Lamarck, 1816) [370]) [17,429]. In the other extreme, a few scale-worms appear to be associated with many hosts that frequently belong to the same taxa (i.e., class, order or even family), such as *Paradyte crinoidicola* (Potts, 1910) [410] known to live on ca. 20 species of unstalked crinoids [432], or *G. clavigera* living with 13 species of tropical holothurians [426,433,434]. In addition, these widely polyxenous species often show specialized morphological adaptations, like the hooked ventral chaetae of *P. crinoidicola* [435] or the ventral sucker-like lobes of *G. clavigera* [426,436]. The degree of polyxeny of some species is still not well-known and for instance, *Asterophilia culcitae* Britayev & Fauchald, 2005 [383], was always reported as exclusive associate of asteroids until found on crinoids [412] and holothurians [437].

From the point of view of biodiversity, however, the most interesting species are those showing polyxenous associations involving hosts from different taxonomic groups, like some species of *Hololepidella* Willey, 1905 [438] and *Arctonoe*. *Hololepidella nigropunctata* (Horst, 1915) [409] is known to live with sixteen echinoderm species (mostly ophiuroids, but also asteroids and one echinoid), one sponge and one cnidarian. *Arctonoe pulchra* (Johnson, 1897) [439] with six echinoderms, two mollusks and one polychaete, and *Arctonoe vittata* (Grube, 1855) [220] with as many as 40 species including cnidarians, polychaetes, mollusks, crustaceans, and echinoderms [16,17]. This large number of host species may sometimes (but not always) be explained by combined effect of hosts' biogeographic distributions and the development of symbiont's preferences for a given host species, e.g., in the case of *A. vittata* [440–443], but the possibility of hiding cryptic species-complexes cannot be discarded.

On the other hand, it must be highlighted that, although less reported than the inverse situation, polynoids may act as hosts. For instance, they are known to host epizootic kamptozoan entoprocts, which occur between parapodia and below elytra in *Lepidonotus clava* (Montagu, 1808) [444] or to be parasitized by numerous species of copepods [52,445].

3.2.5. Sphaerodoridae

Recent molecular phylogenetic analyses assessed the monophyly of Sphaerodoridae [446], confirming a previous assumption based on the presence of conspicuous epithelial tubercles arranged in rows (e.g., [446–449]). As a result of these analyses, a new classification was proposed, gathering long and short bodied forms in two major clades, respectively [450].

The family currently includes 131 species classified in eight genera [6,446,447,450–453]. As currently delineated, the most speciose genus is *Sphaerephesia* Fauchald, 1972 [138] (36 species), followed by *Sphaerodorum* Örsted, 1843 [71] (24), *Sphaerodoridium* Lützen, 1961 [454] (24), and *Geminofilum* Capa et al. 2019 [453] (19). In turn, *Commensodorum* Fauchald, 1974 [455] is monospecific, with *Commensodorum commensalis* (Lützen, 1961) [454] having a particularly remarkable external morphology, with reduced epithelial tubercles [453].

In general, sphaerodorids are poorly studied, as evidenced by the fact that several species have been reported once or are only known from the holotype e.g., [327,455–459]. Likely, this was due to their small size and collecting difficulties [447]. In fact, they were overlooked until the 1970s, when a high number of species were discovered and described, a trend that continues today (Figure 22a,b). Indeed, 85% of the known species have been described in the last 50 years.

Long-bodied sphaerodorids are morphologically homogeneous, show obvious segmentation and macrotubercles with terminal papillae above parapodia forming two longitudinal rows, together with reduced dorsal cirri or microtubercles [446,450,455,460], and belong all them to *Sphaerodorum* [450] (Figure 23a). They were long time considered as belonging to three distinct genera: *Ephesiella* Chamberlin, 1919 [234], *Ephesiopsis* Hartman & Fauchald, 1971 [327], and *Sphaerodorum*, based on their chaetal morphology (only compound, both compound and simple, and only simple, respectively) (e.g., [327,450,455]). The presence of pseudocompound chaetae—apparently simple but with an oblique indentation—had been reported in some species of the three genera (e.g., [450,461–463]). Doubts on the usefulness of chaetal arrangement to distinguish between genera were early indicated [464]. Then, the results of the analyses of nuclear and mitochondrial markers confirmed the synonymy of the genera and proved chaetal morphology as being highly variable within the group, but valuable for species discrimination [450].

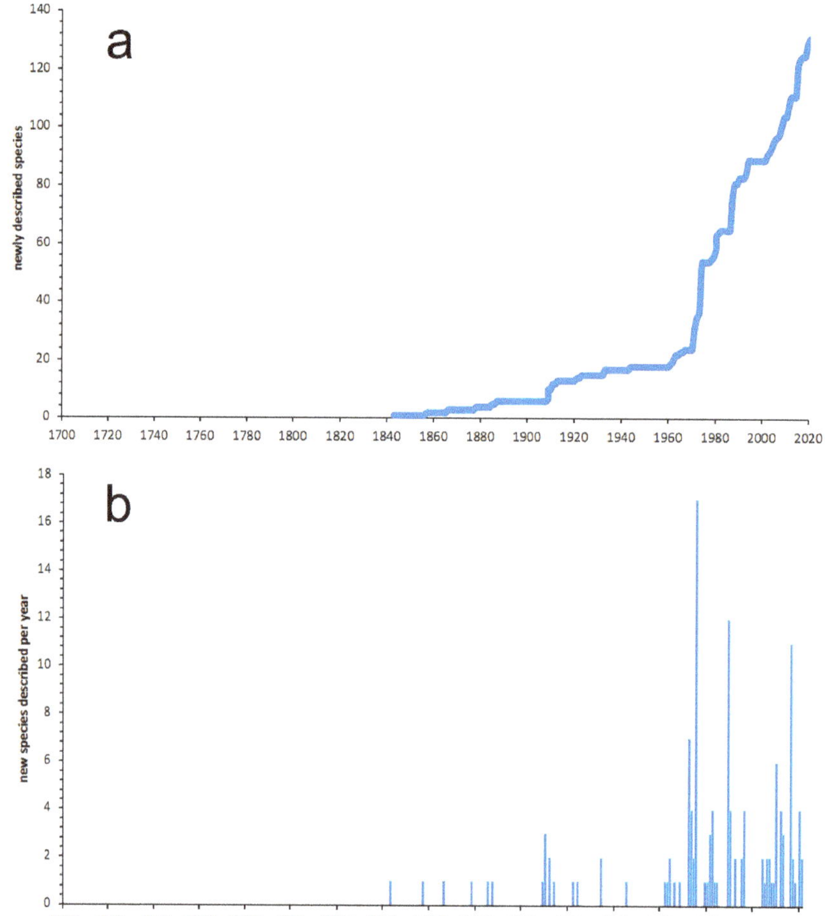

Figure 22. Sphaerodoridae. (**a**) Cumulative curve of newly described species through time. (**b**) The number of new species described per year.

Short bodied sphaerodorids show a broader morphological diversity, especially in number and arrangement of dorsal epithelial tubercles [446,447,455]. The classification and nomenclature have changed with time, with the most recent versions reflecting evolutionary relationships. Before this systematic revision, the morphology, number and arrangement of dorsal epithelial tubercles were considered among the most taxonomically informative features: (1) *Euritmia* Sardá-Borroy, 1987 [458] and *Amacrodorum* Kudenov, 1987 [465] lacked large tubercles (macrotubercles) but were completely covered by small papillae, *Commensodorum* also lacked large macrotubercles and had four dorsal longitudinal rows of small papillae; (2) *Sphaerephesia* was characterized by having four longitudinal rows of dorsal macrotubercles with terminal papillae; (3) *Sphaerodoropsis*, thought to be the most speciose, showed from four dorsal longitudinal rows of macrotubercles in a single, transverse, segmental row, to numerous dorsal longitudinal and transverse rows of tubercles (Figure 23b); and (4) *Clavodorum* and *Sphaerodoridium* showed six or more dorsal longitudinal rows of stalked macrotubercles [447].

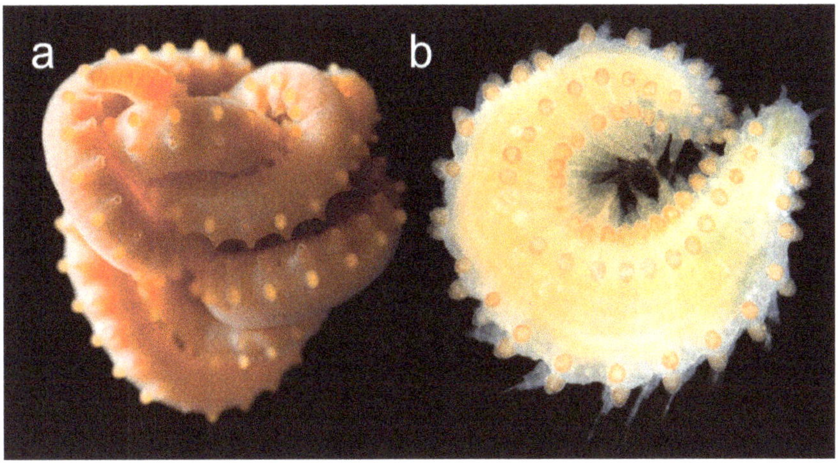

Figure 23. Sphaerodoridae. (**a**) *Spaeherodorum* sp. (**b**) *Sphaerodoropsis* cf. *philippi* (Fauvel, 1911) [464]. (Photos by A. Nygren).

Although still incomplete (i.e., representatives of some genera were missing), the new classification based on phylogenetic analyses of nuclear and mitochondrial DNA showed *Sphaerodoropsis* as being split in three major clades: (1) *Sphaerodoropsis* Group 1 *sensu* Borowski [466], should be synonymized with *Sphaerephesia* [453] to group all sphaerodorids with four longitudinal rows of sessile macrotubercles (regardless if they are smooth or with a terminal papilla, spherical or hemispherical) arranged in a single transverse row per segment; (2) *Sphaerodoropsis* Group 3 *sensu* Borowski [466] is characterized by having two transverse rows of sessile macrotubercles per segment, leading to the erection of *Geminofilum* [453]; and (3) *Sphaerodoropsis* Group 2 *sensu* Borowski [466] that has sessile macrotubercles arranged in more than six longitudinal rows (in a single transverse row per segment) was found nested with other sphaerodorids with stalked macrotubercles and a similar arrangement. Consequently, it was included in *Sphaerodoridium*. Finally, the species having exactly six rows of macrotubercles were recovered as a paraphyletic clade and left under *Clavodorum*.

Members of *Sphaerodoropsis* with small macrotubercles in three or more transverse rows per segment (Group 4 *sensu* Borowski [466]) were not included in the existing phylogenetic studies, neither were those sphaerodorids without large tubercles (i.e., *Amacrodorum* and *Euritmia*). However, since members of these two taxa shared the arrangement of epithelial papillae and the presence of only simple chaetae, *Amacrodorum* was synonymized under *Euritmia* [451].

Although the sphaerodorids have been reported as not common in benthic samples [467], some recent data also show their distribution is often patchy at shelf sediments and can actually be abundant [453,468]. In fact, they are exclusively benthic marine organisms, and are present in all oceans and at all depths. They are often reported from deep-sea sediments, but also occur in shallow waters, including hard substrates and algae (as epibionts) [447,452,453,455,458,468–470], where they are often considered as members of the meiofauna [471,472].

The number of species described at the different marine benthic zones has now been reviewed. The continental shelf (intertidal to ca. 200 m depth), with 61 species, is the richest benthic zone, although more than a half (i.e., 35) were described from shallow waters (<40 m depth). The bathyal zone (200–4000 m depth), with 48 species, and the abyssal (4000–6000 m depth) with 11, led a total of 59 deep sea species, almost as much as in shallower waters. Given that the continental shelf is better studied than the deep-sea, these numbers may confirm the deep-sea as more specious.

Most members of the family have been described from Atlantic waters, both Temperate Northern (31 species) and Tropical (18 species) realms *sensu* Spalding et al. [473], followed by the Southern Ocean (16 species), the Central Indo-Pacific (15 species) and the Arctic (13 species). The highest number of georeferenced occurrences [34] are also reported in the Northern Atlantic, the Southern Ocean, and the Arctic, which actually indicate that these areas are better studied than others worldwide. The Temperate Northern Pacific seems also to be well surveyed and show a high number of records [34], but does not hold a high number of species (one three were described in this area and a few others have been reported there). However, by considering georeferenced reports as a direct measure of how well surveyed an area is, the most surprising situation occurs in the Central Indo-Pacific, where several species have been recently described species, but the overall number of records is very low. Conversely, the Temperate South America, especially the north-western coast, the Temperate South Africa and the Indo-Pacific (Eastern and Western) appear to be understudied, with(seven or less species described from these realms and few occurrences reported [34]. Thus, they would need to be prioritized in future surveys.

3.2.6. Syllidae

Syllidae is the largest polychaete family, comprising ca. 1100 species distributed in 79 genera [6]. Its intricate taxonomic history is reflected in a maze of doubtful, invalid, and synonymized names. The family Amytidea Grube, 1850 [119] was first proposed to include several genera that were later invalidated and recognized as reproductive forms (epitokes), mainly based on the pivotal work describing the reproductive cycle of *Autolytus cornutus* Agassiz, 1862 [474], today included in *Proceraea* Ehlers, 1864 [64]. The tribes Syllideae Grube, 1850 [119], Exogoneae Langerhans, 1879 [475], and Autolyteae Langerhans, 1879 [475] were then proposed to better accommodate the syllids [475], which were later complemented by adding the Eusylleae Malaquin, 1893 [476]. In fact, this work was the first large monograph on the family, where the author described new species, proposed new classifications and addressed numerous biological aspects [476]. In the 20th century, the tribes were raised to the subfamily level [201,477], a recognition still followed today, as Eusyllinae Malaquin, 1893 [476], Syllinae Grube, 1850 [119], Exogoninae Langerhans, 1879 [475], and Autolytinae Langerhans, 1879 [475] (Figures 24a–d and 25a–d). It was not until the 21th century that the fifth subfamily, the Anoplosyllinae Aguado & San Martín, 2009 [478], was erected together with the publication of major works synthesizing the current knowledge on the family [479,480].

The first undoubtful description of a syllid, *Nereis armillaris* O.F. Müller, 1771 [481], dates back to the 18th century using the then newly proposed Linnean classification. Today the species is allocated in *Syllis* Savigny *in* Lamarck, 1818 [68]. Like in this case, a first phase of syllids history was characterized by numerous animals (now syllids) being described within different genera [482]. In fact, most species lacked a rigorous taxonomic position, often with the descriptions only highlighting their similarity with *Nereis* Linnaeus, 1758 [62]. It was only as a result of a major review of Annelida [119] that the family was described, implicitly recognizing the consistency of the group formed by animals that had begun to be described less than 100 years before. Since then, the number of species never stopped to increase and there is a continuous dripping of new descriptions that has no traces of deceleration still to date (Figure 26a,b).

During the last decades, several phylogenetic analyses have been performed on Syllidae, with the pioneer dealing with the morphology of *Typosyllis* Langerhans, 1879 [475,483], reproductive modes and subfamilies based on morphology [484] and Autolytinae based on molecular data [485]. The two latter studies used a limited number of terminal taxa, but their results were overall corroborated by analyzing a large number of taxa based on morphological and molecular data [486,487]. Other studies focused on Autolytinae [488] or on *Pionosyllis* Malmgren, 1867 [86,489], but also on the systematics of the whole family including a higher number of terminal taxa and based on combining three molecular markers (18S, 16S and COI) and/or morphology [478,486,487,490]. The overall congruent results of these studies revealed the monophyly of Anoplosyllinae, Eusyllinae, Autolytinae, Exogoninae and Syllinae, and reorganized the Eusyllinae, while *Anguillosyllis* Day, 1963 [491], *Amblyosyllis* Grube, 1857 [295], and *Perkinsyllis* San Martín, López & Aguado, 2009 [489] were considered independent genera. The evolution of certain morphological characters (e.g., the pharyngeal structures), as well as the reproductive modes, could be traced in the obtained phylogenies. Epigamy is the plesiomorphic condition, while schizogamy has appeared twice in Syllinae and in Autolytinae [484,486,487,492].

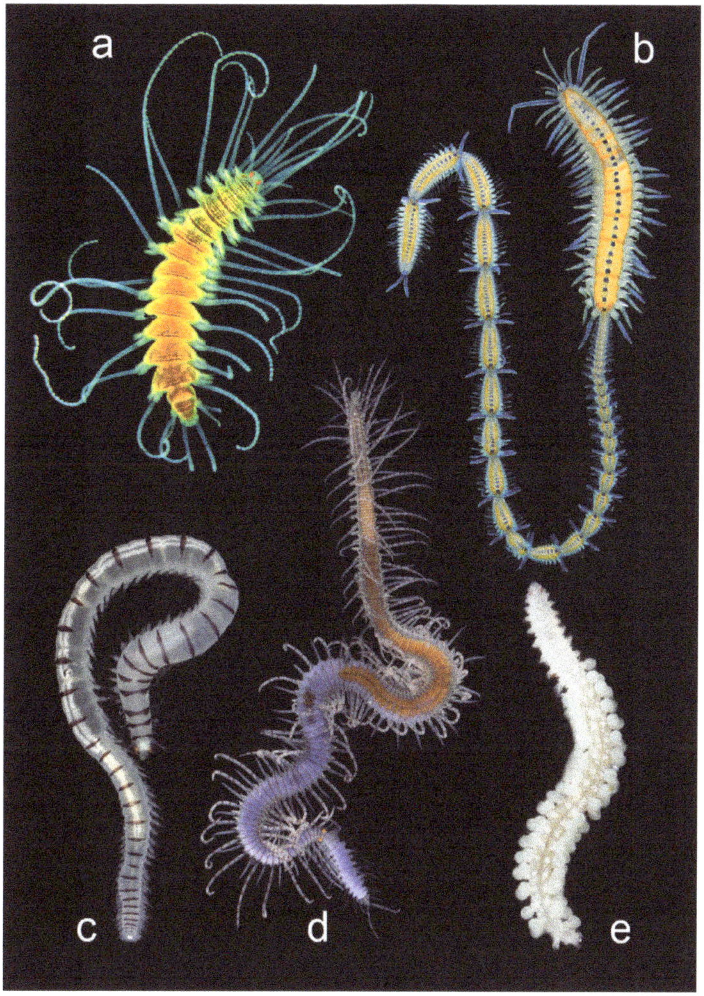

Figure 24. Syllidae. (**a**) Eusyllinae: *Amblyosyllis anae* Aguado, Capa, Lago-Barcia et al., 2019 [54]; photo by A. Nygren in [54]). (**b**) Autolytinae: *Myrianida pachycera* (Augener, 1913) [493]; photo by L. Harris. (**c**) Eusyllinae: *Odontosyllis marombibooral* San Martín & Hutchings, 2006 [494]; photo by A. Semenov in Aguado et al. [495]. (**d**) Syllinae: *Syllis maganda* Martínez & San Martín, 2020 [496]; photo by A. Semenov. (**e**) Exogoninae: *Prosphaerosyllis longipapillata* (Hartmann-Schröder, 1979) [497]; photo by A. Semenov in Aguado et al. [495].

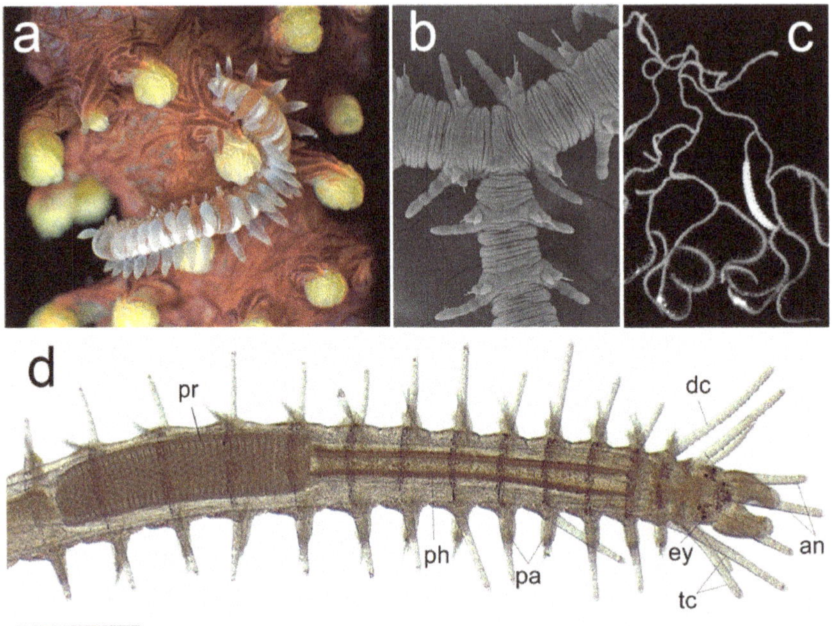

Figure 25. Syllidae. (**a**) *Alcyonosyllis phili* Glasby & Watson, 2001 [498] on its alcyonacean host; photo by C. Glasby in [498]. (**b**) SEM of a *Ramisyllis multicaudata* Glasby, Schroeder and Aguado, 2012 [499] branching point. (**c**) *Ramisyllis multicaudata*, mid-posterior body showing terminal branches and developing female stolon, after [499]. (**d**) The anterior end of *Syllis malaquini*, Ribeiro, Ponz-Segrelles, Helm, Egger & Aguado, 2020; photo by R. P Ribeiro in [500]; an: antenna, ey: eye, dc: dorsal cirrus, pa: parapodia, ph: pharynx, pl: palp, pr: proventricle, py: pygidium, tc: tentacular cirri.

The systematics of the family were corroborated by a phylogenetic analysis of complete mitochondrial genome, revealing also a high variability in the mitochondrial gene order, particularly in Syllinae [501]. Almost in parallel: (1) the existence of a ′ribbon-clade′ lineage, which includes *Ramisyllis* Glasby, Schroeder & Aguado, 2012 [499] and the flattened body syllids was proposed [492], and (2) *Typosyllis* and the previous division of the large genus *Syllis* in subgenera were proposed as non-valid [502]. Lastly, the monophyly of the family was supported based on all available sequences and markers, although subdivided in two different lineages: the ′ribbon-clade′ (clade A) and clade B (including *Syllis* as paraphyletic genus) [500]. Accordingly, all hypotheses on the monophyly of the main groups within Syllidae (i.e., the five subfamilies and the main lineages within them) are highly congruent. This stable phylogenetic scenario allowed to trace the evolution of biological processes such as the reproductive modes and a large variety of regenerative abilities [500,503–505].

Syllidae is a complex family including large genera, such as *Syllis* [502,506] that is currently considered paraphyletic [500], together with very small taxa, with only few known species, or even only one [507,508]. The family includes species with a large diversity of morphologies (Figures 24a–e and 25a–c), from meiofaunal organisms with few chaetigers and less than 1 mm like in *Neopetita* San Martín, 2003 [479] and *Erinaceusyllis* San Martín, 2003 [479], to relatively large animals like in *Trypanosyllis* Claparède, 1864 [509], which may reach impressive lengths of 15 cm and hundreds of segments, or the even longer *Syllis ramosa* McIntosh, 1879 [510] and *Ramisyllis multicaudata* Glasby, Schroeder and Aguado, 2012 [499], which also show an unusual branched body pattern (Figure 25b,c). To some extent, species level identifications tend to be difficult, but the members of the family are relatively easily recognizable by the presence of the proventricle, a muscular, barrel-shaped structure associated to the pharynx [511] (Figure 25d).

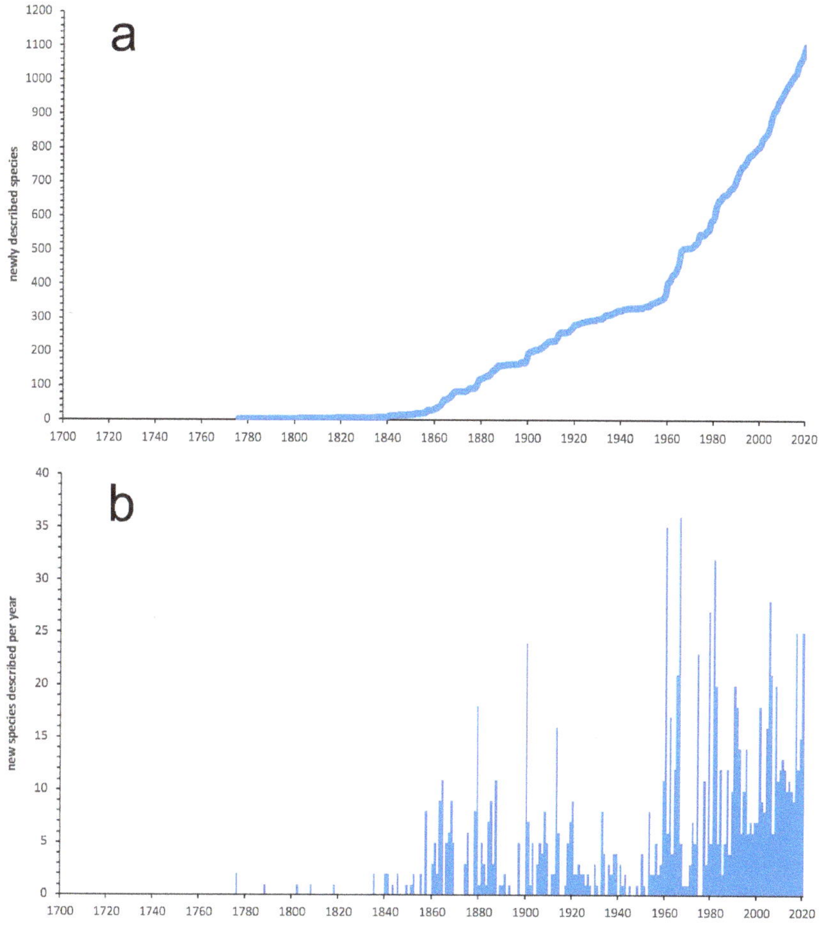

Figure 26. Syllidae. (**a**) Cumulative curve of newly described species through time. (**b**) The number of new species described per year.

Traditionally, species descriptions have been based on morphology, largely aided since the 1980s by detailed observations under scanning electron microscope [512]. Modern careful observations, complemented by molecular techniques, revealed the existence of numerous cryptic or pseudo-cryptic complexes *sensu* Nygren [18], hidden among the so-called "cosmopolitan" species [176], having very wide geographical distributions and bathymetric ranges [54,58,513].

Members of the family can be found in great abundance and diversity in almost any marine environment, although they are slightly rarer at greater depths [479]. Their planktonic epitoke and benthic atoke forms occur in a wide range of habitats, including small interstitial forms inhabiting different types of soft sediments [513–516] to moderately large cryptofaunal forms [517–519] living in among boulders, coral reefs, sponges, algae, and mollusk banks, among others, from the mesolittoral and shallow intertidal to the deep sea.

Syllids present a worldwide distribution, but non-surprisingly showing a skewed pattern towards showing more occurrences in better studied regions [520], such as the Mediterranean, North Atlantic European and American coasts, Gulf of Mexico, and Australia. Instead of reflecting a likely true scenario, this rather 'Wallacean shortfall' [521], in the case of Syllidae, has been mitigated in the aforementioned areas by numerous contributions spanning from the 19th century to recent years. Among these, it is worth mentioning some large taxonomic efforts either concentrated on, or having the family as one of their focuses, in European waters [119,475,476,479,522], North American Atlantic coast [523], the Gulf of Mexico [524–529], and Australia [469,495,530–534].

Despite their nearly ubiquity in marine environments, syllids have been the subject of few biogeographic studies to date, most of them concentrated in the Mediterranean. A thorough account of the syllids from the Balearic Islands (Spain), found most species as having disjunct or Atlantic-Mediterranean distributions [522], a trend also found for the whole Mediterranean syllids [535,536], which allowed corroborating previous propositions of dividing the Mediterranean in northern and southern basins [537]. In a narrower cut, the fauna of Rovinj (Croatia, Northern Adriatic Sea) revealed numerous new records while pointing to a possible "meridionalization" of the region, which might be explained by global warming effects [538].

It is worth mentioning that syllids can be found in nearly all marine environments, being particularly abundant in rocky shore intertidal communities. This leads to the species being frequently identified by non-specialists and, non-rarely, based on outdated European keys (cf. [539]). Thorough revisions (e.g., [54,58,540,541] have already demonstrated that, in many instances, careful examinations may reveal a hidden diversity larger than previously thought, not only in poorly studied locations but also among "well-known" species. As already pointed out [522,535,536,538], the lack of long-term taxonomical data in many regions, frequently with large knowledge voids, preclude more sound biogeographic conclusions, a fact that may help to explain the abovementioned lack of studies within the group.

Syllids present a wide range of feeding strategies, usually using a single tooth to graze, and/or a crown of denticles associated to the pharynx to punch on other animals. Food ingestion is assisted by a pumping action of the proventricle. Traditionally, Syllinae and Autolytinae are considered as carnivores, Eusyllinae omnivorous and Exogoninae selective deposit feeders and/or opportunistic carnivores [151,542,543]. However, this classification was a generalization based on a few cases, some of which were also based on assumptions by previous authors. A more mixed-up scenario proposed the species of Syllinae, Eusyllinae and Exogoninae to be herbivores, omnivores, and detritivores [544]. More recently, trophic guilds have been updated, but no clear subfamily-level trends have been proposed [12]. Instead, a rather omnivore habit has been postulated across the family, as indicated by the frequent findings of recognizable detritus (e.g., fragments of diatoms, forams, algae and copepods) in guts and fecal pellets of different species [12].

Members of the family are also often found living in symbiosis with many other organisms (Figure 24a–c), either as commensals of sponges, alcyonaceans, gorgonians, scleractinian corals, bryozoans, tunicates, asteroids, ophiuroids, and crustaceans, or (more rarely) as parasites of sponges, cnidarians, other polychaetes, crustaceans, and tunicates (see [16] and references therein) or as hosts, for instance of epibiotic ciliophoran protozoa [545] or an haplosporidian internal parasites [546]. In fact, they are the second most diverse polychaete family, after Polynoidae, in terms of symbiotic relationships [16]. Living animals, especially symbionts, frequently display camouflage colorings, which are usually easily lost in preserved material. The most widespread symbiotic association within the family probably occurs in *Haplosyllis* Langerhans, 1879 [475]. The so-called sponge-associated type species *Haplosyllis spongicola* Langerhans, 1879 [475] is in fact a specialized sponge predator whose distribution is restricted to European coasts, while all other reports of this species revealed to be numerous strictly sponge-symbiotic new species with locally-restricted distributions, e.g., [547,548]. Other interesting associations are those of *Alcyonosyllis*, which was named after its alcyonacean octocoral original hosts [498], although some species were later found living with gorgonians and hexacorals [549], and those of *S. ramosa* and *R. multicaudata*, whose complex branching bodies move inside the channels of their host sponges [492] (Figure 25a–c). Very likely, taking into account the great proportion of cryptic species within Syllidae, some of the symbionts currently accepted as being polyxenous could certainly be different species still requiring to be properly described.

4. Conclusions

1. The highest numbers of species of Phyllodocida have been reported from European, North American, and Australian waters, although these numbers are biased by an increased sampling effort in these regions and do not reflect true species richness. DNA barcode data show similar patterns, but also similar bias.
2. At the family level, the highest number of distribution records are for Nephtyidae, Phyllodocidae, Syllidae, Nereididae, and Polynoidae and widest distribution ranges were for Phyllodocidae, Polynoidae, Nereididae, Syllidae, and Lopadorrhynchidae.
3. Overall, there is a weak latitudinal gradient in species richness, with a rather uniformly high diversity across tropical and temperate latitudes and a drop only in extreme latitudes.
4. Antarctic and Pacific coasts of America and Asia, together with the circumtropical areas worldwide show the highest level of endemism, while the lowest numbers occur in temperate Atlantic areas and in the Arctic Ocean.
5. Most records of Phyllodocida and the highest species number and barcode data come from the high subtidal, where Nereididae, Syllidae, and Nephtyidae dominate. However, members of Phyllodocida seems to be well adapted to deep waters, particularly polynoids.
6. Less than 3% of the known species have been reported as occurring non-natively in certain parts of the world, most of them in Mediterranean waters, and more than half belong to Nereididae and Syllidae. However, many "non-native" or "introduced" species, particularly those belonging to critic species-complexes, turned to be native species with locally restricted populations when carefully examined. None of them has been considered as invasive or as pest to date.
7. There is a still unknown number of possible cryptic species complexes, this being a recurrent trend in most examined families.
8. Most examined families except to some extent Glyceridae and Goniadidae, show no traces of stabilization of the accumulative curve of species description, indicating that more new species are expected to be described in the coming years. Sources of new species diversity are mainly related with cryptic species complexes, but also with sampling in poorly explored regions and environments, with the deep-sea being particularly promising.
9. Only 620 species of Phyllodocida have sequences published in BOLD, for 1215 BINS as a consequence of having sequences (1) assigned to higher taxonomic ranks (genus or family), and (2) with wrong taxonomy assignments, the latter representing 22% and including sequences either misidentified and/or with invalid, misspelled, or synonymized names.
10. Our analyses show the key importance of keeping barcode libraries adequately curated, together with the need of adding metadata, while highlighting the apparent difficulty of having molecular data with correct identifications among Phyllodocida, with less than 60% of the records being usable at the species-level in statistical analysis.
11. Despite the amount of knowledge on the systematics of Phyllodocida, we would like to stress that there are still many open questions regarding the correct phylogenetic placement of most taxa (at different levels) so that further efforts must be dedicated to collecting new materials, allowing precise morphological descriptions in parallel with sequences.
12. We would like to highlight that there is a similar lack of knowledge with respect to the ecology of most species of Phyllodocida, as well as on their functional role in marine ecosystems all over the world oceans.

13. Taking into account that we are entering in the 2020s Oceans Decade, during which marine ecosystems have to be re-evaluated from many different points of view (from basic science to sustainable ecosystem services and derived benefits), having a real and accurate picture of the world oceans emerges as a strategic pillar, with the knowledge on the diversity they hold being keystone.

Supplementary Materials: The following are available online at https://www.mdpi.com/1424-2818/13/3/131/s1, Figure S1: Polynoidae. (a) Biogeographic diversity distribution showing the number of species/genera record. Polynoidae. (b) Number of records, (c) number of species and (d) expected number of species per 5° of latitude, Figure S2: Polynoidae. Geographical distribution of two the most wide spread poly-noids: (a) Harmothoe imbricata and (b) Lepidonotus squamatus. Blrue circles: Data from OBIS; Greesn area: data from [351].

Author Contributions: Conceptualization: D.M., M.T.A., M.C.; methodology: M.A.L.T., S.F., D.M.; writing—sections, Analyses of Phyllodocida: S.F., M.A.L.T.; Glyceriformia: M.B.; Holoplanktonic families: M.-A.F.Á., M.A.V.P., C.H.; Nephtyidae: A.R.; Polynoidae: D.M., T.A.B.; Sphaerodoridae: M.C.; Syllidae: M.T.A., M.V.F.; writing—original draft: D.M.; writing—review and editing: D.M., M.T.A., M.-A.F.Á., T.A.B., M.B., M.C., S.F., M.V.F., C.H., M.A.V.P., A.R., M.A.L.T.; illustrations: D.M., M.T.A., M.-A.F.Á., T.A.B., M.B., M.C., S.F., M.V.F., C.H., M.A.V.P., A.R., M.A.L.T.; tables: S.F. All authors have read and agreed to the published version of the manuscript.

Funding: We acknowledge support of the publication fees by the CSIC Open Access Publication Support Initiative through its Unit of Information Resources for Research (URICI) and the Open Access Publication Funds of the Georg-August-Universität Göttingen. This research was funded by the Spanish "Agencia Estatal de Investigación" (AEI) and the European Funds for Regional Development (FEDER), Research Project PopCOmics (CTM2017-88080) to DM; the Russian Scientific Foundation for Basic Research, grant no. RFBR 18-05-00459 to TAB; Fundação para a Ciência e a Tecnologia (FCT), contract foreseen in the Decree-Law 57/2016 (Nrs 4-6, art. 23), changed by Law 57/2017 to AR and FCT/MCTES to CESAM (UIDP/50017/2020+UIDB/50017/2020) through national funds; São Paulo Research Foundation (FAPESP), fellowship proc. 2007/53040-9 to MVF; Spanish MINECO, AEI, Comunidad Autónoma de las Islas Baleares, European Social Funds and Ramón y Cajal program, RYC-2016-20799 to MC; FCT and ESF (SFRH/BD/131527/2017) through a PhD grant to MALT.

Data Availability Statement: All data used in this study have been made available on-line, and the corresponding links are indicated in the text.

Acknowledgments: The images of preserved holoplanktonic Phyllodocida have been provided by the Biological Collection "Edmundo F. Nonato" (ColBIO)—Oceanographic Institute, University of São Paulo, SP, Brazil, while those form living specimens have been provided by Xavi Salvador Costa. Some images of the Syllidae were taken by L. Harris, A. Nygren, A. Semenov, R. P. Ribeiro and C. Glasby, while some of Sphaerodoridae were taken by A. Nygren. The authors would like to thank the anonymous reviewers, who greatly appreciate our work and whose comment substantially contributed to increase the quality and readability of our paper.

Conflicts of Interest: The authors declare no conflict of interest. The funders had no role in the design of the study; in the collection, analyses, or interpretation of data; in the writing of the manuscript, or in the decision to publish the results.

References

1. Dales, R.P. The polychaete stomodeum and the inter-relationships of the families of Polychaeta. *Proc. Zool. Soc. Lond.* **1962**, *139*, 389–428. [CrossRef]
2. Rouse, G.W.; Fauchald, K. Cladistics and polychaetes. *Zool. Scr.* **1997**, *26*, 139–204. [CrossRef]
3. Rouse, G.W.; Pleijel, F. *Polychaetes*; Oxford University Press Inc.: New York, NY, USA, 2001.
4. Weigert, A.; Bleidorn, C. Current status of annelid phylogeny. *Org. Divers. Evol.* **2016**, *16*, 345–362. [CrossRef]
5. Struck, T.H.; Paul, C.; Hill, N.; Hartmann, S.; Hosel, C.; Kube, M.; Lieb, B.; Meyer, A.; Tiedemann, R.; Purschke, G.; et al. Phylogenomic analyses unravel annelid evolution. *Nature* **2011**, *471*, 95–98. [CrossRef] [PubMed]
6. Read, G.; Fauchald, K. World Polychaeta Database. 2020. Available online: http://www.marinespecies.org/polychaeta (accessed on 6 September 2020).
7. Parry, L.; Tanner, A.; Vinther, J. The origin of annelids. *Palaeontology* **2014**, *57*, 1091–1103. [CrossRef]

8. Farrell, U.C.; Briggs, D.E. A pyritized polychaete from the Devonian of Ontario. *Proc. R. Soc. B Biol. Sci.* **2007**, *274*, 499–504. [CrossRef] [PubMed]
9. Gonzalez, B.C.; Martínez, A.; Borda, E.; Iliffe, T.M.; Eibye-Jacobsen, D.; Worsaae, K. Phylogeny and systematics of Aphroditiformia. *Cladistics* **2018**, *34*, 225–259. [CrossRef]
10. Parapar, J.; Caramelo, C.; Candás, M.; Cunha-Veira, X.; Moreira, J. An integrative approach to the anatomy of *Syllis gracilis* Grube, 1840 (Annelida) using micro-computed X-ray tomography. *PeerJ* **2019**, *7*, e7251. [CrossRef]
11. Watson, C.; Faulwetter, S. Stylet jaws of Chrysopetalidae (Annelida). *J. Nat. Hist.* **2017**, *51*, 2863–2924. [CrossRef]
12. Jumars, P.A.; Dorgan, K.M.; Lindsay, S.M. Diet of worms emended: An update of polychaete feeding guilds. *Annu. Rev. Mar. Sci.* **2015**, *7*, 497–520. [CrossRef] [PubMed]
13. Wu, X.; Zhan, Z.; Xu, K. Two new and two rarely known species of *Branchinotogluma* (Annelida: Polynoidae) from deep-sea hydrothermal vents of the Manus Back-Arc basin, with remarks on the diversity and biogeography of vent polynoids. *Deep Sea Res. Part I Oceanogr. Res. Pap.* **2019**, *149*, 103051. [CrossRef]
14. Zhang, Y.; Chen, C.; Qiu, J.-W. Sexually dimorphic scale worms (Annelida: Polynoidae) from hydrothermal vents in the Okinawa Trough: Two new species and two new sex morphs. *Front. Mar. Sci.* **2018**, *5*, 112. [CrossRef]
15. McCowin, M.F.; Rouse, G.W. Phylogeny of hydrothermal vent Iphionidae, with the description of a new species (Aphroditiformia, Annelida). *ZooKeys* **2018**, *779*, 89–107. [CrossRef] [PubMed]
16. Martin, D.; Britayev, T.A. Symbiotic polychaetes revisited: An update of the known species and relationships (1998–2017). *Oceanogr. Mar. Biol. Annu. Rev.* **2018**, *56*, 371–448. [CrossRef]
17. Martin, D.; Britayev, T.A. Symbiotic polychaetes: Review of known species. *Oceanogr. Mar. Biol. Annu. Rev.* **1998**, *36*, 217–340.
18. Nygren, A. Cryptic polychaete diversity: A review. *Zool. Scr.* **2014**, *43*, 172–183. [CrossRef]
19. Ravara, A.; Ramos, D.; Teixeira, M.A.L.; Costa, F.O.; Cunha, M.R. Taxonomy, distribution and ecology of the order Phyllodocida (Annelida, Polychaeta) in deep-sea habitats around the Iberian margin. *Deep Sea Res. Part II Top. Stud. Oceanogr.* **2017**, *137*, 207–231. [CrossRef]
20. Borja, A.; Franco, J.; Pérez, V. A marine biotic index to establish the ecological quality of soft-bottom benthos within European estuarine and coastal environments. *Mar. Pollut. Bull.* **2000**, *40*, 1100–1114. [CrossRef]
21. Pennisi, E. DNA barcodes jump-start search for new species. *Science* **2019**, *364*, 920–921. [CrossRef] [PubMed]
22. Leese, F.; Bouchez, A.; Abarenkov, K.; Altermatt, F.; Borja, Á.; Bruce, K.; Ekrem, T.; Čiampor, F.; Čiamporová-Zaťovičová, Z.; Costa, F.O.; et al. Chapter Two—Why We Need Sustainable Networks Bridging Countries, Disciplines, Cultures and Generations for Aquatic Biomonitoring 2.0: A Perspective Derived From the DNAqua-Net COST Action. *Adv. Ecol. Res.* **2018**, *58*, 63–99. [CrossRef]
23. Deiner, K.; Lopez, J.; Bourne, S.; Holman, L.; Seymour, M.; Grey, E.K.; Lacoursière, A.; Li, Y.; Renshaw, M.A.; Pfrender, M.E.; et al. Optimising the detection of marine taxonomic richness using environmental DNA metabarcoding: The effects of filter material, pore size and extraction method. *Metabarcoding Metagenom.* **2018**, *2*, e28963. [CrossRef]
24. Weigand, H.; Beermann, A.J.; Čiampor, F.; Costa, F.O.; Csabai, Z.; Duarte, S.; Geiger, M.F.; Grabowski, M.; Rimet, F.; Rulik, B.; et al. DNA barcode reference libraries for the monitoring of aquatic biota in Europe: Gap-analysis and recommendations for future work. *Sci. Total Environ.* **2019**, *678*, 499–524. [CrossRef] [PubMed]
25. Nygren, A.; Pleijel, F. From one to ten in a single stroke–resolving the European *Eumida sanguinea* (Phyllodocidae, Annelida) species complex. *Mol. Phylogenet. Evol.* **2011**, *58*, 132–141. [CrossRef]
26. Delić, T.; Trontelj, P.; Rendoš, M.; Fišer, C. The importance of naming cryptic species and the conservation of endemic subterranean amphipods. *Sci. Rep.* **2017**, *7*, 3391. [CrossRef] [PubMed]
27. Fišer, C.; Robinson, C.T.; Malard, F. Cryptic species as a window into the paradigm shift of the species concept. *Mol. Ecol.* **2018**, *27*, 613–635. [CrossRef]
28. Ratnasingham, S.; Hebert, P.D.N. Barcoding: Bold: The Barcode of Life Data System (http://www.barcodinglife.org). *Mol. Ecol. Notes* **2007**, *7*, 335–345. [CrossRef]
29. Horton, T.; Kroh, A.; Ahyong, S.; Bailly, N.; Boyko, C.B.; Brandão, S.N.; Gofas, S.; Hooper, J.N.A.; Hernandez, F.; Holovachov, O.; et al. World Register of Marine Species (WoRMS). 2021. Available online: http://www.marinespecies.org (accessed on 7 January 2021).
30. Chamberlain, S. Worrms: World Register of Marine Species (WoRMS) Client. R Package Version 0.4.0. 2019. Available online: https://CRAN.R-project.org/package=worrms (accessed on 20 April 2020).
31. R Core Team. *R: A Language and Environment for Statistical Computing*; R Core Team: Vienna, Austria, 2013. Available online: https://www.R-project.org/ (accessed on 20 April 2020).
32. OBIS. Ocean Biodiversity Information System. Intergovernmental Oceanographic Commission of UNESCO. 2020. Available online: http://www.iobis.org (accessed on 20 April 2020).
33. Provoost, P.; Bosch, S. "Robis: R Client to access data from the OBIS API." Ocean Biogeographic Information System. Intergovernmental Oceanographic Commission of UNESCO. R Package Version 2.1.8. 2020. Available online: https://cran.r-project.org/package=robis (accessed on 20 April 2020).
34. "GBIF.org". GBIF Occurrence Download. 2020. Available online: https://doi.org/10.15468/dl.yntzgf (accessed on 20 April 2020).

35. Provoost, P.; Bosch, S. "Obistools: Tools for Data Enhancement and Quality Control." Ocean Biogeographic Information System. Intergovernmental Oceanographic Commission of UNESCO. R Package Version 0.0.9. 2020. Available online: https://cran.r-project.org/package=obistools (accessed on 20 April 2020).
36. Costello, M.J.; Tsai, P.; Wong, P.S.; Cheung, A.K.L.; Basher, Z.; Chaudhary, C. Marine biogeographic realms and species endemicity. *Nat. Commun.* **2017**, *8*, 1057. [CrossRef]
37. Barnes, R. DggridR: Discrete Global Grids. R Package Version 2.0.4. 2020. Available online: https://github.com/r-barnes/dggridR/ (accessed on 20 April 2020).
38. Hurlbert, S.H. The nonconcept of species diversity: A critique and alternative parameters. *Ecology* **1971**, *52*, 577–586. [CrossRef] [PubMed]
39. Ahyong, S.; Costello, M.J.; Galil, B.S.; Gollasch, S.; Hutchings, P.A.; Katsanevakis, S.; Lejeusne, C.; Marchini, A.; Occhipinti, A.; Pagad, S.; et al. World Register of Introduced Marine Species (WRiMS). **2020**. Available online: http://www.marinespecies.org/introduced (accessed on 6 September 2020). [CrossRef]
40. Çinar, M.E. Alien polychaete species worldwide: Current status and their impacts. *J. Mar. Biol. Assoc. UK* **2013**, *93*, 1257–1278. [CrossRef]
41. Faulwetter, S.; Simboura, N.; Katsiaras, N.; Chatzigeorgiou, G.; Arvanitidis, C. Polychaetes of Greece: An updated and annotated checklist. *Biodivers. Data J.* **2017**, *5*, e20997. [CrossRef] [PubMed]
42. Keppel, E.; Keith, I.; Ruiz, G.M.; Carlton, J.T. New records of native and non-indigenous polychaetes (Annelida: Polychaeta) in the Galapagos Islands. *Aquat. Invasions* **2019**, *14*, 59–84. [CrossRef]
43. Keppel, E.; Tovar-Hernandez, A.M.; Ruiz, G. First record and establishment of *Branchiomma coheni* (Polychaeta: Sabellidae) in the Atlantic Ocean and review of non–indigenous species of the genus. *Zootaxa* **2015**, *4058*, 499–518. [CrossRef] [PubMed]
44. Langeneck, J.; Lezzi, M.; Del Pasqua, M.; Musco, L.; Gambi, M.C.; Castelli, A.; Giangrande, A. Non-indigenous polychaetes along the coasts of Italy: A critical review. *Mediterr. Mar. Sci.* **2020**, 238–275. [CrossRef]
45. López, E.; Richter, A. Non-indigenous species (NIS) of polychaetes (Annelida: Polychaeta) from the Atlantic and Mediterranean coasts of the Iberian Peninsula: An annotated checklist. *Helgol. Mar. Res.* **2017**, *71*, 19–36. [CrossRef]
46. Flanders Marine Institute. MarineRegions.org. 2020. Available online: www.marineregions.org (accessed on 1 November 2020).
47. Olenin, S.; Alemany, F.; Cardoso, A.C.; Gollasch, S.; Goulletquer, P.; Lehtiniemi, M.; McCollin, T.; Minchin, D.; Miossec, L.; Occhipinti-Ambrogi, A.O.H.; et al. *Marine Strategy Framework Directive–Task Group 2 Report. Non-Indigenous Species*; Office for Official Publications of the European Communities, EU/ICES: Luxembourg, 2010. [CrossRef]
48. Ravara, A.; Carvalho, S. Nephtyidae (Polychaeta, Phyllodocida) from the Red Sea, with record of a new species. *J. Mar. Biol. Assoc. UK* **2017**, *97*, 843–856. [CrossRef]
49. Carr, C.M.; Hardy, S.M.; Brown, T.M.; Macdonald, T.A.; Hebert, P.D.N. A tri-oceanic perspective: DNA barcoding reveals geographic structure and cryptic diversity in Canadian polychaetes. *PLoS ONE* **2011**, *6*, e22232. [CrossRef]
50. Reish, D.J.; Anderson, F.E.; Horn, K.M.; Hardege, J. Molecular phylogenetics of the Neanthes acuminata (Annelida: *Nereididae*) species complex. *Mem. Mus. Vic.* **2014**, *71*, 271–278. [CrossRef]
51. Fauvel, P. Mission Robert Ph. Dollfus en Egypte. Annélides Polychètes. *Mémoires Présentes A L'institut D'egypte Et Publes Sous Les Auspices De Sa Majesté Fouad Ierroi D'egypte* **1933**, *21*, 31–83.
52. Wehe, T. Revision of the scale worms (Polychaeta: Aphroditoidea) occurring in the seas surrounding the Arabian Peninsula. Part I: Polynoidae. *Fauna Arab.* **2006**, *22*, 23–197.
53. Hanley, J.R.; Burke, M. Scaleworms (Polychaeta: Polynoidae) of Albany, Western Australia. In *Proceedings of the Third International Marine Biological Workshop: The Marine Fauna of Albany, Western Australia*; Wells, F.E., Walker, D.I., Kirkman, H., Lethbridge, R., Eds.; Western Australian Museum: Perth, Australia, 1990; Volume I, pp. 203–236.
54. Aguado, M.T.; Capa, M.; Lago-Barcia, D.; Gil, J.; Pleijel, F.; Nygren, A. Species delimitation in *Amblyosyllis* (Annelida, Syllidae). *PLoS ONE* **2019**, *14*, e0214211. [CrossRef]
55. Michel, A. Sur les divers types de stolons chez les syllidiens, specialement sur une nouvelle espèce (*Syllis cirropunctata* n.sp.) à stolons acephales et sur la reobservation du stolon tetracère de *Syllis amica* Qfg. *C. R. L'acad. Sci. Paris* **1909**, *148*, 318–320.
56. Gravier, C. Contribution à l'étude des Annélides Polychètes de la Mer Rouge. Première partie. *Nouv. Arch. Mus. D'hist. Nat. Paris* **1900**, *2*, 137–282.
57. Aguado, M.T.; San Martín, G. Syllidae (Polychaeta) from Lebanon with two new reports for the Mediterranean Sea. *Cah. Biol. Mar.* **2007**, *48*, 207–224.
58. Álvarez-Campos, P.; Giribet, G.; Riesgo, A. The *Syllis gracilis* species complex: A molecular approach to a difficult taxonomic problem (Annelida, Syllidae). *Mol. Phylogenet. Evol.* **2017**, *109*, 138–150. [CrossRef] [PubMed]
59. Leite, B.; Vieira, P.; Teixeira, M.; Lobo-Arteaga, J.; Hollatz, C.; Borges, L.; Duarte, S.; Troncoso, J.; Costa, F. Gap-analysis and annotated reference library for supporting macroinvertebrate metabarcoding in Atlantic Iberia. *Reg. Stud. Mar. Sci.* **2020**, 101307. [CrossRef]
60. Duarte, S.; Vieira, P.E.; Costa, F.O. Assessment of species gaps in DNA barcode libraries of non-indigenous species (NIS) occurring in European coastal regions. *Metabarc. Metagen.* **2020**, *4*, e55162. [CrossRef]
61. Pamungkas, J.; Glasby, C.J.; Read, G.B.; Wilson, S.P.; Costello, M.J. Progress and perspectives in the discovery of polychaete worms (Annelida) of the world. *Helgol. Mar. Res.* **2019**, *73*, 4. [CrossRef]

62. Linnaeus, C. *Systema Naturae per Regna Tria Naturae, Secundum Classes, Ordines, Genera, Species, cum Characteribus, Differentiis, Synonymis, Locis. Tomus I. Editio Decima, Reformata;* Laurentii Salvii: Stockholm, Sweden, 1758; p. 824.
63. Barrelier, J. *Icones Plantarum per Galliam, Hispaniam et Italiam Observatur ad Vivum Exivitarum;* Opus Posthumum Editum Cura et Studio Antonii de Jussieu: Paris, France, 1714; p. 148.
64. Ehlers, E. *Die Borstenwürmer (Annelida Chaetopoda) nach Systematischen und Anatomischen Untersuchungen Dargestellt;* Wilhelm Engelmann: Leipzig, Germany, 1864; Volume I, pp. 1–268.
65. Ehlers, E. *Die Borstenwürmer (Annelida Chaetopoda) nach Systematischen und Anatomischen Untersuchungen Dargestellt;* Wilhelm Engelmann: Leipzig, Germany, 1868; Volume II, pp. 269–748. [CrossRef]
66. Claparède, É. Les Annélides Chétopodes du Golfe de Naples. *Mém. Soc. Phys. Hist. Nat. Genève* **1868**, *19*, 313–584.
67. Claparède, É. Les annélides chétopodes du Golfe de Naples. Supplément. *Mém. Soc. Phys. Hist. Nat. Genève* **1870**, *20*, 365–542.
68. Lamarck, J.B. *Histoire Naturelle des Animaux sans Vertèbres, Présentant les Caractères Généraux et Particuliers de ces Animaux, Leur Distribution, Leurs Classes, Leurs Familles, Leurs Genres, et la Citation des Principales Espèces qui s'y Rapportent; Précédée d'une Introduction Offrant la Détermination des Caractères Essentiels de l'Animal, sa Distinction du Végétal et des Autres Corps Naturels, Enfin, L'exposition des Principes Fondamentaux de la Zoologie. Vol. 5;* Déterville & Verdière: Paris, France, 1818; Volume 5, p. 612.
69. Müller, O.F. *Zoologicae Danicae Prodromus, seu Animalium Daniae et Norvegiae Indigenarum Characteres, Nomina et Synonyma Imprimis Popularium;* Hallageriis: Copenhagen, Denmark, 1776; p. 282. [CrossRef]
70. Fabricius, O. *Fauna Groenlandica, Systematice Sistens, Animalia Groenlandiae Occidentalis Hactenus Indagata, Quoad Nomen Specificum, Triviale, Vernaculumque Synonyma Auctorum Plurium, Descriptionem, Locum, Victum, Generationem, Mores, Usum, Capturamque Singuli Prout Detegendi Occasio Fuit, Maximaque Parte Secundum Proprias Observationes;* Impensis Ioannis Gottlob Rothe: Copenhagen, Denmark; Leipzig, Germany, 1780; p. 452.
71. Örsted, A.S. *Annulatorum Danicorum Conspectus. Fasc. 1 Maricolae;* Librariae Wahlianae: Hafniae, Denmark, 1843; p. 52.
72. Greeff, R. Über pelagische Anneliden von der Küste der canarischen Inseln. *Z. Wiss. Zool.* **1879**, *32*, 237–284.
73. Steenstrup, J.S. Af Slaegten *Tomopteris. Vidensk. Medd. Dan. Nat. Foren. I Köbenhavn* **1850**, *1850*, 75–77.
74. Busch, W. *Beobachtungen ueber Anatomie und Entwicklung Einiger Wirbellosen Seethiere;* Aug. Hirschwald: Berlin, Germany, 1851; p. 143.
75. Linnaeus, C. *Systema Naturae per Regna Tria Naturae, Editio Duodecima, Reformata, Tomus I, Pars II. Regnum Animale;* Laurentii Salvii: Stockholm, Sweden, 1767.
76. Arfianti, T.; Costello, M.J. Global biogeography of marine amphipod crustaceans: Latitude, regionalization, and beta diversity. *Mar. Ecol. Prog. Ser.* **2020**, *638*, 83–94. [CrossRef]
77. Chaudhary, C.; Saeedi, H.; Costello, M.J. Marine species richness is bimodal with latitude: A reply to Fernandez and Marques. *Trends Ecol. Evol.* **2017**, *32*, 234–237. [CrossRef] [PubMed]
78. Hillebrand, H. Strength, slope and variability of marine latitudinal gradients. *Mar. Ecol. Prog. Ser.* **2004**, *273*, 251–267. [CrossRef]
79. Pamungkas, J.; Glasby, C.J.; Costello, M.J. Biogeography of polychaete worms (Annelida) of the world. *Mar. Ecol. Prog. Ser.* **2021**, *657*, 147–159. [CrossRef]
80. Chaudhary, C.; Saeedi, H.; Costello, M.J. Bimodality of latitudinal gradients in marine species richness. *Trends Ecol. Evol.* **2016**, *31*, 670–676. [CrossRef]
81. Webb, T.J.; Berghe, E.V.; O'Dor, R. Biodiversity's big wet secret: The global distribution of marine biological records reveals chronic under-exploration of the deep pelagic ocean. *PLoS ONE* **2010**, *5*, e10223. [CrossRef] [PubMed]
82. Izuka, A. The errantiate Polychaeta of Japan. *J. Coll. Sci. Imp. Univ. Tokyo* **1912**, *30*, 1–262.
83. Pernet, B. Benthic egg masses and larval development of Amblyosyllis speciosa (Polychaeta: Syllidae). *J. Mar. Biol. Assoc. UK* **1998**, *78*, 1369–1372. [CrossRef]
84. Okada, Y.K. *Amblyosyllis nigrolineata*, une nouvelle variété de l'*A. speciosa* Izuka. *Annot. Zool. Jpn.* **1934**, *14*, 317–320.
85. Imajima, M.; Hartman, O. The Polychaetous Annelids from Japan. Pt. I. *Allan Hancock Found. Spec. Publ.* **1964**, *26*, 1–237.
86. Malmgren, A.J. *Annulata Polychaeta Spetsbergiae, Groelanlandiae, Islandiae et Scandinaviae Hactenus Cognita. Cum XIV. Tabulis;* Ex Officina Frenckelliana: Helsinki, Finland, 1867; p. 127.
87. Langerhans, P. Die Wurmfauna von Madeira. III. *Z. Wiss. Zool.* **1880**, *34*, 87–143.
88. McIntosh, W.C. Report on the Annelida Polychaeta collected by the H.M.S. Challenger during the years 1873–1876. *Rep. Sci. Res. Voy. H.M.S. Chall. 1872-76* **1885**, *12*, 1–554.
89. Johnston, G. Miscellanea Zoologica. [Continued from vol. iv. p. 375.] Contributions towards a history of Irish Annelids (1). *Ann. Mag. Nat. Hist.* **1840**, *5*, 168–179. [CrossRef]
90. Teixeira, M.A.L.; Vieira, P.E.; Pleijel, F.; Sampieri, B.R.; Ravara, A.; Costa, F.O.; Nygren, A. Molecular and morphometric analyses identify new lineages within a large *Eumida* (Annelida) species complex. *Zool. Scr.* **2020**, *49*, 222–235. [CrossRef]
91. Vieira, P.E.; Desiderato, A.; Holdich, D.M.; Soares, P.; Creer, S.; Carvalho, G.R.; Costa, F.O.; Queiroga, H. Deep segregation in the open ocean: Macaronesia as an evolutionary hotspot for low dispersal marine invertebrates. *Mol. Ecol.* **2019**, *28*, 1784–1800. [CrossRef]
92. Brasier, M.J.; Wiklund, H.; Neal, L.; Jeffreys, R.; Linse, K.; Ruhl, H.; Glover, A.G. DNA barcoding uncovers cryptic diversity in 50% of deep-sea Antarctic polychaetes. *R. Soc. Open Sci.* **2016**, *3*, 160432. [CrossRef]

93. Nygren, A.; Parapar, J.; Pons, J.n.; Meißner, K.; Bakken, T.; Kongsrud, J.A.; Oug, E.; Gaeva, D.; Sikorski, A.; Johansen, R.A.; et al. A mega-cryptic species complex hidden among one of the most common annelids in the North East Atlantic. *PLoS ONE* **2018**, *13*, e0198356. [CrossRef]
94. Baird, W. Descriptions of several new species of worms belonging to the Annelida Errantia and Sedentaria or Tubicola of Milne Edwards. *Proc. Zool. Soc. Lond.* **1863**, *1863*, 106–110.
95. Watson, C. Revision of the pantropical genus *Treptopale* (Annelida: Phyllodocida: Chrysopetalidae): Redescription of *Treptopale rudolphi* Perkins, 1985 and description of two new species including comparison of *Treptopale* populations in northern Australia. *Beagle Rec. Mus. Art Gall. North. Territ.* **2010**, *26*, 37–55.
96. Gravier, C. Contribution à l'étude des Annélides Polychètes de la Mer Rouge (Suite). *Bull. Mus. D'hist. Nat. Paris* **1899**, *6*, 288–298.
97. Grube, A.E. *Actinien, Echinodermen und Wurmen des Adriatischen und Mittelmeers nach Eigenen Sammlungen Beschrieben*; J. H. Bon: Königsberg, Russia, 1840; p. 92. [CrossRef]
98. Langeneck, J.; Scarpa, F.; Maltagliati, F.; Sanna, D.; Barbieri, M.; Cossu, P.; Mikac, B.; Curini Galletti, M.; Castelli, A.; Casu, M. A complex species complex: The controversial role of ecology and biogeography in the evolutionary history of *Syllis gracilis* Grube, 1840 (Annelida, Syllidae). *J. Zool. Syst. Evol. Res.* **2020**, *58*, 67–78. [CrossRef]
99. Lobo, J.; Teixeira, M.A.L.; Borges, L.M.S.; Ferreira, M.S.G.; Hollatz, C.; Gomes, P.T.; Sousa, R.; Ravara, A.; Costa, M.H.; Costa, F.O. Starting a DNA barcode reference library for shallow water polychaetes from the southern European Atlantic coast. *Mol. Ecol. Resour.* **2016**, *16*, 298–313. [CrossRef] [PubMed]
100. Audzijonyte, A.; Ovcarenko, I.; Bastrop, R.; Vainola, R. Two cryptic species of the *Hediste diversicolor* group (Polychaeta, Nereididae) in the Baltic Sea, with mitochondrial signatures of different population histories. *Mar. Biol.* **2008**, *155*, 599–612. [CrossRef]
101. Virgilio, M.; Fauvelot, C.; Costantini, F.; Abbiati, M.; Backeljau, T. Phylogeography of the common ragworm *Hediste diversicolor* (Polychaeta: Nereididae) reveals cryptic diversity and multiple colonization events across its distribution. *Mol. Ecol.* **2009**, *18*, 1980–1994. [CrossRef] [PubMed]
102. Sato, M.; Nakashima, A. A review of Asian *Hediste* species complex (Nereididae, Polychaeta) with descriptiuons of two new species and a redescription of *Hediste japonica* (Izuka,1908). *Zool. J. Linn. Soc. Lond.* **2003**, *137*, 203–445. [CrossRef]
103. Tosuji, H.; Bastrop, R.; Götting, M.; Park, T.; Hong, J.-S.; Sato, M. Worldwide molecular phylogeny of common estuarine polychaetes of the genus *Hediste* (Annelida: Nereididae), with special reference to interspecific common haplotypes found in southern Japan. *Mar. Biodivers.* **2019**, *49*, 1385–1402. [CrossRef]
104. Fujita, M.K.; Leaché, A.D.; Burbrink, F.T.; McGuire, J.A.; Moritz, C. Coalescent-based species delimitation in an integrative taxonomy. *Trends Ecol. Evol.* **2012**, *27*, 480–488. [CrossRef]
105. Fontaneto, D.; Flot, J.-F.; Tang, C.Q. Guidelines for DNA taxonomy, with a focus on the meiofauna. *Mar. Biodivers.* **2015**, *45*, 433–451. [CrossRef]
106. Lindgren, J.; Hatch, A.S.; Hourdez, S.; Seid, C.A.; Rouse, G.W. Phylogeny and biogeography of *Branchipolynoe* (Polynoidae, Phyllodocida, Aciculata, Annelida), with descriptions of five new species from methane seeps and hydrothermal vents. *Diversity* **2019**, *11*, 153. [CrossRef]
107. Gastaldi, A. *Harmothoe Imbricata: Species Complex or Complex Species?* University of Alaska Fairbanks: Fairbanks, AK, USA, 2019.
108. Álvarez-Campos, P.; Giribet, G.; San Martín, G.; Rouse, G.W.; Riesgo, A. Straightening the striped chaos: Systematics and evolution of *Trypanosyllis* and the case of its pseudocryptic type species *Trypanosyllis krohnii* (Annelida, Syllidae). *Zool. J. Linn. Soc.* **2017**, *179*, 492–540. [CrossRef]
109. Fontes, J.T.; Vieira, P.E.; Ekrem, T.; Soares, P.; Costa, F.O. BAGS: An automated Barcode, Audit & Grade System for DNA barcode reference libraries. *Mol. Ecol. Resour.* **2021**, *21*, 573–583. [CrossRef] [PubMed]
110. Hartmann-Schröder, G. Annelida, Borstenwürmer, Polychaeta. *Tierwelt Dtschl.* **1971**, *58*, 1–594.
111. Kinberg, J.G.H. Nya slägten och arter af Annelider. *Öfvers. Kongliga Vetensk. Akad. Förhandlingarstockholm* **1856**, *12*, 381–388.
112. Schmarda, L.K. *Neue wirbellose Thiere Beobachtet und Gesammelt auf Einer Reise un Die Erdr 1853 bis 1857. Erster Band (Zweite Halfte) Turbellarian, Rotatorien un Anneliden*; Wilhelm Engelmann: Leipzig, Germany, 1861; p. 164.
113. Webster, H.E.; Benedict, J.E. The Annelida Chaetopoda, from Eastport, Maine. *Annu. Rep. United States Comm. Fish Fish.* **1887**, *12*, 707–758.
114. Bergsten, J.; Bilton, D.T.; Fujisawa, T.; Elliott, M.; Monaghan, M.T.; Balke, M.; Hendrich, L.; Geijer, J.; Herrmann, J.; Foster, G.N.; et al. The effect of geographical scale of sampling on DNA barcoding. *Syst. Biol.* **2012**, *61*, 851–869. [CrossRef] [PubMed]
115. Cariani, A.; Messinetti, S.; Ferrari, A.; Arculeo, M.; Bonello, J.J.; Bonnici, L.; Cannas, R.; Carbonara, P.; Cau, A.; Charilaou, C.; et al. Improving the Conservation of Mediterranean Chondrichthyans: The ELASMOMED DNA Barcode Reference Library. *PLoS ONE* **2017**, *12*, e0170244. [CrossRef]
116. Oliveira, L.M.; Knebelsberger, T.; Landi, M.; Soares, P.; Raupach, M.J.; Costa, F.O. Assembling and auditing a comprehensive DNA barcode reference library for European marine fishes. *J. Fish Biol.* **2016**, *89*, 2741–2754. [CrossRef]
117. Costa, F.O.; Landi, M.; Martins, R.; Costa, M.H.; Costa, M.E.; Carneiro, M.; Alves, M.J.; Steinke, D.; Carvalho, G.R. A ranking system for reference libraries of DNA barcodes: Application to marine fish species from Portugal. *PLoS ONE* **2012**, *7*, e35858. [CrossRef]
118. Fauchald, K. The polychaete worms. Definitions and keys to the orders, families and genera. *Nat. Hist. Mus. Los Angel. County. Sci. Ser.* **1977**, *28*, 1–188.

119. Grube, A.E. Die Familien der Anneliden. *Arch. Nat. Berl.* **1850**, *16*, 249–364.
120. Kinberg, J.G.H. Annulata nova. *Öfvers. Kongliga Vetensk. Akad. Förhandlingarstockholm* **1865**, *21*, 559–574.
121. Böggemann, M. Revision of the Glyceridae Grübe 1850 (Annelida: Polychaeta). *Abh. Senckenbergischen Nat. Ges.* **2002**, *555*, 1–249.
122. Böggemann, M. Revision of the Goniadidae (Annelida, Polychaeta). *Abh. Nat. Ver. Hambg. (Neue Folgen)* **2005**, *39*, 1–354.
123. Böggemann, M. Worms that might be 300 million years old. *Mar. Biol. Res.* **2006**, *2*, 130–135. [CrossRef]
124. Böggemann, M. Polychaetes (Annelida) of the abyssal SE Atlantic. *Org. Divers. Evol.* **2009**, *9*, 251–428. [CrossRef]
125. Richter, S.; Schwarz, F.; Hering, L.; Böggemann, M.; Bleidorn, C. The utility of genome skimming for phylogenomic analyses as demonstrated for glycerid relationships (Annelida, Glyceridae). *Genome Biol. Evol.* **2015**, *7*, 3443–3462. [CrossRef]
126. Rousset, V.; Pleijel, F.; Rouse, G.W.; Erseus, C.; Siddall, M.E. A molecular phylogeny of annelids. *Cladistics* **2007**, *23*, 41–63. [CrossRef]
127. Böggemann, M. Chapter 7.11.8 Glyceridae Grube, 1850. In *Handbook of Zoology Online*; Westheide, W., Purschke, G., Böggemann, M., Eds.; De Gruyter: Berlin, Germany, 2014.
128. Böggemann, M. Chapter 7.11.9 Goniadidae Kinberg, 1865. In *Handbook of Zoology Online*; Westheide, W., Purschke, G., Böggemann, M., Eds.; De Gruyter: Berlin, Germany, 2014.
129. Imajima, M. Polychaetous Annelids from Sagami Bay and Sagami Sea Collected by the Emperor Showa of Japan and Deposited at the Showa Memorial Institute, National Science Museum, Tokyo (II): Orders included within the Phyllodocida, Amphinomida, Spintherida and Eunicida. *Natl. Sci. Mus. Monogr.* **2003**, *23*, 1–221.
130. Barrett, B.M.; Bailey-Brock, J.H. *Progoniada oahuensis*, a new species from Oahu, Hawaii (Annelida: Polychaeta: Goniadidae). *Proc. Biol. Soc. Wash.* **2005**, *118*, 251–258. [CrossRef]
131. Böggemann, M.; Orensanz, J.M. Glyceriformia Fauchald, 1977 (Annelida:"Polychaeta") from the SW Atlantic Shelf, between 30° and 45° S. *Mitt. Aus Dem Hambg. Zool. Mus. Und Inst.* **2007**, *104*, 11–59.
132. Rizzo, A.E.; Steiner, T.M.; Amaral, A.C.Z. Glyceridae Grube 1850 (Annelida: Polychaeta) from southern and southeastern brazil, including a new species of *Glycera*. *Biota Neotrop.* **2007**, *7*, 41–59. [CrossRef]
133. Imajima, M. Deep-sea benthic polychaetes off Pacific coast of the northern Honshu, Japan. *Natl. Mus. Nat. Sci. Monogr.* **2009**, *39*, 192.
134. Böggemann, M.; Bienhold, C.; Gaudron, S.M. A new species of Glyceridae (Annelida:"Polychaeta") recovered from organic substrate experiments at cold seeps in the eastern Mediterranean Sea. *Mar. Biodivers.* **2012**, *42*, 47–54. [CrossRef]
135. Magalhaes, W.F.; Rizzo, A.E. Glyceridae (Annelida: Polychaeta) from Guam, Mariana Islands with description of a new species of *Glycera* Savigny in Lamarck, 1818. *Zootaxa* **2012**, *3338*, 60–68. [CrossRef]
136. Hossain, M.B.; Hutchings, P.A. *Glycera sheikhmujibi* n. sp.(Annelida: Polychaeta: Glyceridae): A new species of Glyceridae from the saltmarsh of Bangladesh. *Diversity* **2020**, *12*, 213. [CrossRef]
137. Arwidsson, I. Studien über die Familien Glyceridae und Goniadidae. *Bergen. Mus. Årb.* **1899**, *2*, 1–69.
138. Fauchald, K. Benthic polychaetous annelids from deep water off western Mexico and adjacent areas in the Eastern Pacific Ocean. *Allan Hancock Monogr. Mar. Biol.* **1972**, *7*, 1–575.
139. Müller, F. Einiges über die Annelidenfauna der Insel Santa Catharina an der brasilianischen Küste. *Arch. Für Nat.* **1858**, *24*, 211–220.
140. Audouin, J.V.; Milne Edwards, H. Classification des Annélides, et description de celles qui habitent les côtes de la France. *Ann. Sci. Nat. Paris* **1833**, *30*, 411–425. [CrossRef]
141. Hartman, O. Goniadidae, Glyceridae and Nephtyidae. *Allan Hancock Pac. Exped.* **1950**, *15*, 1–181.
142. Hartmann-Schröder, G. Polychaeten aus dem Roten Meer. *Kiel. Meeresforsch.* **1960**, *16*, 69–125.
143. Fauvel, P. Annélides polychètes nouvelles de l'Inde. I. *Bull. Mus. D'hist. Nat. Paris* **1928**, *34*, 90–96.
144. Hartman, O. Deep-water benthic polychaetous annelids off New England to Bermuda and other North Atlantic areas. *Occas. Pap. Allan Hancock Found.* **1965**, *28*, 1–378.
145. Böggemann, M. Polychaeten der Deutschen Küstengewässer (Auswertung einer "Senckenberg"-Fahrt im Sommer 1990). Ph.D. Thesis, Westfälischen Wilhelms-Universität Münster, Münster, Germany, 1995; pp. 1–281.
146. Böggemann, M.; Fiege, D. Proboscidial papillae of the Glyceridae (Annelida, Polychaeta)—A useful taxonomic character. *Verh. Dtsch. Zool. Ges. Kurzpublikationen* **1996**, *89*, 4.
147. Smith, L.E.; Trabanino, S.; Baerwald, R.J. Scanning electron microscopical observations of the proboscideal papillae of *Glycinde armigera* (Annelida: Polychaeta). *Invertebr. Biol.* **1995**, *14*, 46–50. [CrossRef]
148. Fiege, D.; Böggemann, M. Scanning electron microscopy of the proboscidal papillae of some European Glyceridae. *Bull. Mar. Sci.* **1997**, *60*, 559–563.
149. Wolf, G. Kieferorgane von Glyceriden (Polychaeta) ihre Funktion und ihr taxonomischer Wert. *Senckenbergiana Marit.* **1977**, *9*, 261–283.
150. Ockelmann, K.W.; Vahl, O. On the biology of the polychaete *Glycera alba*, especially its burrowing and feeding. *Ophelia* **1970**, *8*, 275–294. [CrossRef]
151. Fauchald, K.; Jumars, P.A. The diet of worms: A study of polychaete feeding guilds. *Oceanogr. Mar. Biol. Annu. Rev.* **1979**, *17*, 193–284.
152. Klawe, W.L.; Dickie, L.M. Biology of the bloodworm, *Glycera dibranchiata* Ehlers, and its relation to the bloodworm of the Maritime Provinces. *Bull. Fish. Res. Board Can.* **1957**, *115*, 1–37.

153. Mattson, S. Burrowing and feeding of *Goniada maculata* Ørsted (Polychaeta). *Sarsia* **1981**, *66*, 49–51. [CrossRef]
154. Brown, B. Maine's baitworm fisheries: Resources at risk? *Am. Zool.* **1993**, *33*, 568–577. [CrossRef]
155. McIntosh, W.C. A monograph of the British annelids. Polychaeta. Syllidae to Ariciidae. *Ray Soc. Lond.* **1910**, *2*, 233–524. [CrossRef]
156. McIntosh, W.C. Notes from the Gatty Marine Laboratory, St. Andrews. No. 26. 1. On the Pacific, Atlantic and Japanese *Palolo*. 2. On the British Goniadidae and Ariciidae. 3. On the same groups found in the 'Porcupine' Expeditions of1869 and 1870. 4. On the same groups dredged by Dr. Whiteaves of Canadain 1872 and 1873. 5. On the same groups procured by Canon Norman inNorway and Finmark. 6. On some Japanese Glyceridae. 7. On the form described as *Hemipodus magellanicus* in the Challenger. *Ann. Mag. Nat. Hist.* **1905**, *715*, 33–57.
157. Ben-Eliahu, M.N.; Golani, D. Polychaetes (Annelida) in the gut contents of goatfishes (Muliidae), with new polychaete records for the Mediterranean coast of Israel and the Gulf of Elat (Red Sea). *Mar. Ecol.* **1990**, *11*, 193–205. [CrossRef]
158. Ben-Eliahu, M.N.; Golani, D.; Ben-Tuvia, A. On predation of polychaetes (Annelides) by the squirrel- fish *Adioryx ruber* (Holocentridae), with a new polychaete record for the Mediterranean coast of Israel. *Téthys* **1983**, *11*, 15–19.
159. Benham, W.B. Notes on Polychaeta: Two new species of the genus *Goniada*. *Bull. Soc. Linn. Normandie* **1932**, *9*, 553–566.
160. Creaser, E.P., Jr. Reproduction of the bloodworm *Glycera dibranchiata* in the Sheepscot Estuary, Maine. *J. Fish. Res. Board Can.* **1973**, *30*, 161–166. [CrossRef]
161. Pettibone, M.H. Marine polychaete worms of the New England region. Part 1. Families Aphroditidae through Trochochaetidae. *Bull. United States Natl. Mus.* **1963**, *227*, 1–356.
162. Treadwell, A.L. Polychaetous annelids collected by Captain Robert A. Bartlett in Greenland, Fox Basin, and Labrador. *J. Wash. Acad. Sci.* **1937**, *27*, 23–36.
163. Brown, B. The worms crawl in, the worms crawl out, what's the bait worm fishery all about. *Am. Zool.* **1992**, *32*, 174A.
164. Creaser, E.P.; Clifford, D.A.; Hogan, M.J.; Sampson, D.B. A commercial sampling program for sandworms, *Nereis virens* Sars, and bloodworms, *Glycera dibranchiata* Ehlers, harvested along the Maine coast. *Noaa Tech. Rep.* **1983**, *NMFS SSRF-767*, 1–56.
165. Dow, R.L. Chances in abundance of the marine worm, *Glycera dibranchiata*, associated with seawater temperature fluctuations. *Commer. Fish. Rev.* **1964**, 7–9.
166. Dow, R.L. Fluctuations in marine species abundance during climatic cycles. *Mar. Tech. Soc. J.* **1973**, *7*, 38–42.
167. Olive, P.J.W. Polychaeta as a world resource: A review of exploitation as a sea angling baits, and the potential for aquaculture based production. *Mém. Mus. Natl. D'hist. Nat. Paris* **1994**, *162*, 603–610.
168. Taylor, P.M.; Saloman, C.H. Some effects of dredging and coastal development in Boca Ciega Bay, Florida. *United States Fish Wild Life Serv. Fish. Bull.* **1968**, *67*, 213–241.
169. Watson, G.J.; Murray, J.M.; Schaefer, M.; Bonner, A. Bait worms: A valuable and important fishery with implications for fisheries and conservation management. *Fish Fish.* **2017**, *18*, 374–388. [CrossRef]
170. Wilson, W.H., Jr.; Ruff, R.E. Species profiles: Life histories and environmental requirements of coastal fishes and invertebrates (North Atlantic)—Sandworm and bloodworm. *United States Fish Wildl. Serv. Biol. Rep.* **1988**, *82*, 1–23.
171. Böggemann, M. Annelida, Polychaeta, Phyllodocida, Glyceridae, *Glycera branchiopoda* Moore, 1911. In *Handbook of Deep-Sea Hydrothermal Vent Fauna. Denisia 18*; Desbruyères, D., Segonzac, M., Bright, M., Eds.; IFREMER: Brest, France, 2006; p. 195.
172. Desbruyères, D. Annelida, Polychaeta, Phyllodocida, Glyceridae, Glycera tesselata Grube, 1863. In *Handbook of Deep-Sea Hydrothermal Vent Fauna. Denisia 18*; Desbruyères, D., Segonzac, M., Bright, M., Eds.; IFREMER: Brest, France, 2006; pp. 196–197.
173. Schüller, M. Evidence for a role of bathymetry and emergence in speciation in the genus *Glycera* (Glyceridae, Polychaeta) from the deep Eastern Weddell Sea. *Polar Biol.* **2011**, *34*, 549–564. [CrossRef]
174. Böggemann, M.; Dietz, A. Glyceriformia (Annelida) from the deep sea of the Atlantic sector of the Southern Ocean. *Polar Biol.* **2016**, *39*, 1505–1510. [CrossRef]
175. Brasier, M.J.; Harle, J.; Wiklund, H.; Jeffreys, R.M.; Linse, K.; Ruhl, H.A.; Glover, A.G. Distributional patterns of polychaetes across the West Antarctic based on DNA barcoding and particle tracking analyses. *Front. Mar. Sci.* **2017**, *4*, 356. [CrossRef]
176. Hutchings, P.A.; Kupriyanova, E. Cosmopolitan polychaetes—Fact or fiction? Personal and historical perspectives. *Invertebr. Syst.* **2018**, *32*, 1–9. [CrossRef]
177. Menzel, L.; George, K.H.; Arbizu, P.M. Submarine ridges do not prevent large-scale dispersal of abyssal fauna: A case study of Mesocletodes (Crustacea, Copepoda, Harpacticoida). *Deep Sea Res. Part I Oceanogr. Res. Pap.* **2011**, *58*, 839–864. [CrossRef]
178. Böggemann, M. Glyceriformia (Annelida) of the abyssal SW Atlantic and additional material from the SE Atlantic. *Mar. Biodivers.* **2016**, *46*, 227–241. [CrossRef]
179. Rouse, G.W. Bias? What bias? The evolution of downstream larval-feeding in animals. *Zool. Scr.* **2000**, *29*, 213–236. [CrossRef]
180. Stramma, L.; England, M. On the water masses and mean circulation of the South Atlantic Ocean. *J. Geophys. Res. Ocean.* **1999**, *104*, 20863–20883. [CrossRef]
181. Boeckel, B.; Baumann, K.-H.; Henrich, R.; Kinkel, H. Coccolith distribution patterns in South Atlantic and Southern Ocean surface sediments in relation to environmental gradients. *Deep Sea Res. Part I Oceanogr. Res. Pap.* **2006**, *53*, 1073–1099. [CrossRef]
182. Clarke, A. Chapter 8. The Polar Deep Sea. In *Ecosystems of the Deep Oceans. Ecosystems of the World 28*; Tyler, P.A., Ed.; Elsevier: Amsterdam, The Netherlands, 2003; pp. 239–260.
183. Martinez Arbizu, P.; Brix, S. Bringing light into deep-sea biodiversity. *Zootaxa* **1866**, *5*, 2008.
184. Martínez Arbizu, P.; Schminke, H.K. DIVA-1 expedition to the deep sea of the Angola Basin in 2000 and DIVA-1 workshop in 2003. *Org. Divers. Evol.* **2005**, *5*, 11–12. [CrossRef]

185. Ebbesmeyer, C.C.; Ingraham, W.J., Jr. Pacific toy spill fuels ocean current pathways research. *Eostrans. Am. Geophys. Union* **1994**, *75*, 425–430. [CrossRef]
186. Hohn, D. *Moby-Duck. The True Story of 28,800 Bath Toys Lost at Sea and of the Beachcombers, Oceanographers, Environmentalists, and Fools, Including the Author, Who Went in Search of Them*; Viking: New York, NY, USA, 2011.
187. Pettibone, M.H. Annelida. Polychaeta. In *Synopsis and Classifiction of Living Organisms*; Parker, S.P., Ed.; McGraw-Hill Book Company: New York, NY, USA, 1982; Volume 2, pp. 1–43.
188. Støp-Bowitz, C. Polychaeta from the Michael Sars North Atlantic deep-sea expedition 1910. *Rep. Sci. Results Michael Sars North Atl. Deep-Sea Exped.* **1948**, *5*, 1–91.
189. Uschakov, P.V. Fauna of the USSR. Polychaetes. Vol. I. Polychaetes of the suborder Phyllodociformia of the Polar Basin and the North-Western part of the Pacific. (Family Phyllodocidae, Alciopidae, Tomopteridae, Typhloscolecidae and Lacydoniidae). *Akad. Nauk Sssr* **1972**, *102*, 1–272. (In Russian)
190. Bergström, E. Zur Systematik der Polychætenfamilie der Phyllodociden. *Zool. Bidr. Upps.* **1914**, *3*, 37–224.
191. Dales, R.P. The evolution of the pelagic alciopid and phyllodocid polychaetes. *Proc. Zool. Soc. Lond.* **1955**, *125*, 411–420. [CrossRef]
192. Southern, R. Polychaeta of the coasts of Ireland. Pelagic Phyllodocidae. *Fish. Irel. Sci. Investig.* **1909**, *3*, 1–11.
193. Støp-Bowitz, C. A new genus and species (*Yndolacia lopadorrhynchoides*) of pelagic polychaetes representative of a new family, Yndolaciidae. *Bull. Biol. Soc. Wash.* **1987**, *7*, 128–130.
194. Rouse, G.W.; Pleijel, F. Problems in polychaete systematics. *Hydrobiologia* **2003**, *496*, 175–189. [CrossRef]
195. Rouse, G.W.; Pleijel, F. *Reproductive Biology and Phylogeny of Annelida*; Science Publishers Inc.: Enfield, NH, USA, 2006.
196. Southern, R. Polychaeta of the coasts of Ireland. 3. The Alciopidae, Tomopteridae and Typhloscolecidae. *Fish. Irel. Sci. Investig.* **1911**, *3*, 1–37.
197. Apstein, C. *Vorbericht ueber Die Alciopiden und Tomopteriden der Plankton-Expedition 1*; Lipsius & Tischer: Kiel, Germany, 1899; pp. 135–138.
198. Apstein, C. *Die Alciopiden und Tomopteriden der Plankton-Expedition 2*; Lipsius: Kiel, Germany; Tischer: Leipzig, Germany, 1900; p. 62.
199. Dales, R.P. The pelagic Polychaetes of Monterey Bay, California. *Ann. Mag. Nat. Hist.* **1955**, *12*, 434–444. [CrossRef]
200. Dales, R.P. Pelagic polychaetes of the Pacific Ocean. *Bull. Scripss Inst. Oceanogr. Univ. Calif.* **1957**, *7*, 99–167.
201. Fauvel, P. Faune de France. Polychètes Errantes. *Faune Fr.* **1923**, *5*, 1–488.
202. Heath, H. A connecting link between the Annelida and the Echiuroidea *Gephyrea armata*. *J. Morphol.* **1930**, *49*, 223–249. [CrossRef]
203. Hartman, O. Polychaetous annelids collected by the USNS Eltanin and Staten Island cruises, chiefly from Antarctic Seas. *Allan Hancock Monogr. Mar. Biol.* **1967**, *2*, 1–387.
204. Osborn, K.J.; Rouse, G.W. Multiple origins of pelagicism within Flabelligeridae (Annelida). *Mol. Phylogenet. Evol.* **2008**, *49*, 386–392. [CrossRef] [PubMed]
205. Halanych, K.M.; Cox, L.N.; Struck, T.H. A brief review of holopelagic annelids. *Integr. Comp. Biol.* **2007**, *47*, 872–879. [CrossRef]
206. Struck, T.H.; Schult, N.; Kusen, T.; Hickman, E.; Bleidorn, C.; McHugh, D.; Halanych, K.M. Annelid phylogeny and the status of Sipuncula and Echiura. *BMC Evol. Biol.* **2007**, *7*, 57. [CrossRef]
207. Uljanin, B. Sur le Genre *Sagitella* N. Wagner. *Arch. Zool. Exp. Gén.* **1878**, *7*, 1–32.
208. Brigss, D.E.G.; Clarkson, E.N.K. The first tomopterid, a polychaete from the Carboniferous of Scotland. *Lethaia* **1987**, *20*, 257–262. [CrossRef]
209. Clark, N. Shrimping at Granton-the Muirhouse'shrimp-bed'revisited. *Edinb. Geol.* **2014**, *55*, 14–19.
210. Rice, S.A. Reproductive biology, systematics, and evolution in the polychaete family Alciopidae. *Bull. Biol. Soc. Wash.* **1987**, *7*, 114–127.
211. Støp-Bowitz, C. Some new or rare species of pelagic polychaetes from the Gulf of Guinea. *Ophelia* **1991**, (Suppl. 5), 261–270.
212. Støp-Bowitz, C. Polychètes pélagiques des campagnes de l'Ombango dans les eaux équatoriales et tropicales Ouest-Africanes. *Ed. L'orstrom Collect. Études Thèses* **1992**, *1992*, 1–115.
213. Dales, R.P. An annotated list of the pelagic Polychaeta. *Ann. Mag. Nat. Hist. Ser. 12* **1956**, *9*, 289–304. [CrossRef]
214. Dales, R.P.; Peter, G. A synopsis of the Pelagic Polychaeta. *J. Nat. Hist.* **1972**, *6*, 55–92. [CrossRef]
215. Costa, A. Illustrazione Iconografica degli Anellidi rari o poco conosciuti del Golfo di Napoli. *Annu. Dell'instituto Mus. Zool. Dell'univ. Napoli* **1864**, *2*, 159–168.
216. Viguier, C. Etudes sur les Animaux inferieurs de la Baie d'Alger. Recherches sur les Annelides pelagiques. *Arch. Zool. Exp. Gén.* **1886**, *4*, 347–442.
217. Viguier, C. Nouvelles etudes sur le plankton de la baie d'Alger. *Ann. Sci. Nat. Paris* **1911**, *13*, 187–267.
218. Kolbasova, G.; Kosobokova, K.; Neretina, T. Bathy-and mesopelagic annelida from the Arctic Ocean: Description of new, redescription of known and notes on some "cosmopolitan" species. *Deep Sea Res. Part I Oceanogr. Res. Pap.* **2020**, *165*, 103327. [CrossRef]
219. Núñez, J.; Hernández, F.; Ocaña, O.; Jiménez, S. Poliquetos pélagicos de Canarias: Familias Iospilidae y Lopadorrhynchidae. *Vieraea* **1992**, *21*, 101–108.
220. Grube, A.E. Beschreibungen neuer oder wenig bekannter Anneliden. *Arch. Nat. Berl.* **1855**, *21*, 81–136.
221. Treadwell, A.L. Scientific Results of Cruise VII of the Carnegie during 1928–1929 under Command of Captain J.P. Ault. Biology—IV. Biological results of the last cruise of the Carnegie. III. Polychaetous annelids. *Carnegie Inst. Wash. Publ.* **1943**, *55*, 30–59.

222. Eschscholtz, F. Berich ueber zoologische Ausbeute waehrend der Reise von Kronstadt bis St. Peter und Paul. *Isis Von Oken* **1825**, *16*, 733–747.
223. Quoy, J.R.; Gaimard, P. Observations zoologiques faites a bord de l'Astrolabe en mai 1826, dans le detroit de Gibraltar. *Ann. Sci. Nat. Paris* **1827**, *10*, 5–239.
224. Rosa, D. Nouve specie di Tomopteridi. Diagnosi preliminari. *Boll. Mus. Zool. Anat. Comp. Torino* **1908**, *23*, 1.
225. Day, J.H. A monograph on the Polychaetes of Southern Africa. Part 1. Errantia. *Trustees Br. Mus. (Nat. Hist.)* **1967**, *656*, 1–656. [CrossRef]
226. Fernández-Alamo, M.A. Tomopterids (Annelida: Polychaeta) from the Eastern Tropical Pacific Ocean. In: Proceedings of the VI International Polychaete Conference. *Bull. Mar. Sci.* **2000**, *67*, 45–53.
227. Wagner, N. Nouveau groupe d'Annelides. *Tr. Obsch. Estest St. Petersburg* **1872**, *3*, 344–347.
228. Levinsen, G.M.R. Spolia atlantica. Om nogle pelagiske Annulata. *Skr. Fra Vidensk. I Kjoebenhavnmath.-Nat. Rekke* **1885**, *3*, 321–344.
229. Krohn, A. Zoologische und anatomische Bemerkungen über die Alciopen. *Arch. Nat. Berl.* **1845**, *11*, 171–184.
230. Fauchald, K.; Rouse, G.W. Polychaete systematics: Past and present. *Zool. Scr.* **1997**, *26*, 71–138. [CrossRef]
231. Fernández-Álamo, M.A. Los poliquetos pelágicos (Annelida-Polychaeta) del Pacífico Tropical Oriental: Sistemática y Zoogeografía. Ph.D. Thesis, Universidad Nacional Autónoma de Mexico, Ciudad de México, Mexico, 1983.
232. Orensanz, J.M.; Ramirez, F.C. Taxonomía y distribución de los poliquetos pelágicos del Atlántico Sudoccidental. *Bol. Inst. Biol. Mar. Mar Plata* **1973**, *21*, 1–122.
233. Costello, M.J.; Bouchet, P.; Boxshall, G.; Arvanitidis, C.; Appeltans, W. European Register of Marine Species. Alciopini Ehlers. 1864. Available online: http://www.marbef.org/data/aphia.php?p=taxdetails&id=932 (accessed on 8 December 2020).
234. Chamberlin, R.V. The annelida polychaeta (In: Reps. Sci. Res. Expeds. U. S. Fish. Steamer "Albatross", 1891, 1899–1900 and 1904–1905). *Mem. Mus. Comp. Zool. Harv. Coll.* **1919**, *48*, 1–514.
235. Uschakov, P.V. Polychaeta of the Far Eastern seas of the Soviet Union. *Akad. Nauk Sssr* **1955**, *56*, 1–445. (In Russian)
236. Pleijel, F.; Dales, R.P. *Polychaetes: British Phyllodocoideans, Typhloscolecoideans and Tomopteroideans*; Universal Book Services/Dr. W. Backhuys: Avon, OH, USA, 1991; p. 202.
237. Hering, E. Zur Kenntnis der Alciopiden von Messina. *Sber. Akad. Wiss. Wien* **1892**, *101*, 713–768.
238. Wu, B.L.; Hua, L. Phylogeny of Alciopidae (pelagic polychaetes): A cladistic analysis. *Mém. Mus. Natl. D'hist. Nat. Paris* **1994**, *162*, 317–322.
239. Jiménez-Cueto, S.; Suárez-Morales, E. An account of *Alciopina*, *Torrea*, and *Rhynconereella* (Polychaeta: Alciopidae) of the western Caribbean Sea. *Belg. J. Zool.* **2008**, *138*, 70.
240. San Martín, G.; Álvarez-Campos, P.; Kondo, Y.; Núñez, J.; Fernández-Álamo, M.A.; Pleijel, F.; Goetz, F.E.; Nygren, A.; Osborn, K.J. New symbiotic association in marine annelids: Ectoparasites of comb jellies. *Zool. J. Linn. Soc.* **2020**, *191*, 672–694. [CrossRef]
241. Støp-Bowitz, C. Polychaeta. In *Atlas del zooplancton del Atlántico Sudoccidental y Métodos de Trabajo con el Zooplancton Marino*; Boltovskoy, D., Ed.; Instituto Nacional de Investigacion y Desarrollo Pesquero: Mar del Plato, Argentina, 1981; pp. 471–492.
242. Maurer, D.; Reish, D.J. Pelagic polychaetes from ice stations (Arlis I and II) in the Arctic Basin. *J. Nat. Hist.* **1984**, *18*, 381–387. [CrossRef]
243. Granata, L. Annelidi, Alciopidi e Fillidocidi. Raccolte planctoniche fatte dalla R. Nave Liguria nel viaggio di circonnavigazione del 1903-05 sotto il commando di S.A.R. Luigi di Savioa, Duca degli Abruzzi. *Pub. R. Inst. Stud. Sup. Prat. Perfez. Firenze Sez. Sci. Fis. Nat.* **1911**, *2*, 245–327.
244. Reibisch, J.G.F. *Die Pelagischen Phyllodociden und Typhloscoleciden der Plankton-Expedition*; Lipsius & Tischer: Kiel, Germany, 1895.
245. Tebble, N. The distribution of pelagic polychaetes in the South Atlantic Ocean. *Dis. Rep.* **1960**, *30*, 161–299. [CrossRef]
246. Tebble, N. The distribution of pelagic polychaetes across the North Pacific Ocean. *Bull. Br. Mus. (Nat. Hist.)* **1962**, *7*, 371–492.
247. Dales, R.P. Pelagic polychaetes from the Bay of Biscay. *Ann. Mag. Nat. Hist.* **1957**, *9*, 661–664. [CrossRef]
248. Hartman, O. Catalogue of the polychaetous Annelids of the World, parts 1-2. *Allan Hancock Found. Publ. Occas. Pap.* **1959**, *23*, 1–628.
249. Eklöf, J.; Pleijel, F.; Sundberg, P. Phylogeny of benthic Phyllodocidae (Polychaeta) based on morphological and molecular data. *Mol. Phylogenet. Evol.* **2007**, *45*, 261–271. [CrossRef]
250. Fernández-Álamo, M.A. Iospilidae Bergström, 1914. In *Poliquetos (Annelida: Polychaeta) de México y América Tropical*; León-González, J.A.B.-Z., Bastida Zavala, J.R., Carrera-Parra, L.F., García-Garza, M.E., Peña-Rivera, A., Salazar-Vallejo, S.I., Solís-Weiss, V., Eds.; Universidad Autónoma de Nuevo León: Monterrey, Mexico, 2009; pp. 245–250.
251. Tovar-Faro, B.; Leocádio, M.; de Paiva, P.C. Distribution of Iospilidae (Annelida) along the eastern Brazilian coast (from Bahia to Rio de Janeiro). *Lat. Am. J. Aquat. Res.* **2013**, *41*, 323–334.
252. Benham, W.B. The pelagic Polychaeta. *Br. Antarct. Terra Nova Exped. Nat. Hist. Rep. Zool.* **1929**, *7*, 183–201.
253. Berkeley, C. A checklist of Polychaeta recorded from British Columbia since 1923, with references to name changes, descriptions, and synonymies. I. Errantia. *Can. J. Zool.* **1967**, *45*, 1049–1059. [CrossRef] [PubMed]
254. Berkeley, E.; Berkeley, C. Some further records of pelagic Polychaeta from the northeast Pacific north of latitude 40° N and east of longitude 175° W, together with records of Siphonophora, Mollusca and Tunicata from the same region. *Can. J. Zool.* **1960**, *38*, 787–799. [CrossRef]
255. Savigny, J.-C. Système des annelides, principalement de celles des côtes de l'Égypte et de la Syrie, offrant les caractères tant distinctifs que naturels des Ordres, Familles et Genres, avec la description des espèces. *Des* **1822**, *1*, 1–128. [CrossRef]

256. Yingst, D.R. The vertical distribution and reproductive biology of *Pelagobia longicirrata* (Annelida) in the central Arctic Ocean. *Biol. Bull.* **1974**, *147*, 457–465. [CrossRef] [PubMed]
257. Fernández-Álamo, M.A. Lopadorrhynchidae Claparède, 1868. In *Poliquetos (Annelida: Polychaeta) de México y América Tropical*; León-González, J.A.B.-Z., Bastida Zavala, J.R., Carrera-Parra, L.F., García-Garza, M.E., Peña-Rivera, A., Salazar-Vallejo, S.I., Solís-Weiss, V., Eds.; Universidad Autónoma de Nuevo León: Monterrey, Mexico, 2009; pp. 255–261.
258. Guglielmo, R.; Gambi, M.C.; Granata, A.; Guglielmo, L.; Minutoli, R. Composition, abundance and distribution of holoplanktonic polychaetes within the Strait of Magellan (southern America) in austral summer. *Polar Biol.* **2014**, *37*, 999–1015. [CrossRef]
259. Lana, P.C.; Blankensteyn, A. Poliquetas pelágicos coletados pelo N. Ap. Oc. "Barão de Teffé", durante a I Expedição Antártica Brasileira. *Ann. Acad. Ciênc.* **1986**, *58*, 13–22.
260. Lana, P.C.; Blankensteyn, A. Distribution patterns of pelagic polychaetes in the Southern Drake Passage and Bransfield Strait (January-February 1984). *Nerítica* **1987**, *2*, 37–64.
261. Sicinski, J. Pelagic Polychaeta in the Scotia Front west of Elephant Island (BIOMASS III, October-November 1986). *Pol. Polar Res.* **1988**, *9*, 277–282.
262. Bellan, G. Polychaeta. In *European Register of Marine Species: A Check-List of the Marine Species in Europe and a Bibliography of Guides to Their Identification*; Costello, M.J., Emblow, C., White, R.J., Eds.; Muséum National d'Histoire Naturelle: Paris, France, 2001; pp. 214–231.
263. Wilson, R.S. Family Pontodoridae. In *Polychaetes & Allies: The Southern Synthesis. Fauna of Australia. Vol. 4A. Polychaeta, Myzostomida, Pogonophora, Echiura, Sipuncula*; Beesley, P.L.G., Ross, J.B., Glasby, C.J., Eds.; CSIRO Publishing: Melbourne, Australia, 2000; pp. 156–157.
264. Berkeley, E.; Berkeley, C. Notes on some pelagic and some swarming polychaeta taken off the coast of Perú. *Can. J. Zool.* **1964**, *42*, 121–133. [CrossRef]
265. Fernández-Álamo, M.A. Pontodoridae Bergström, 1914. In *Poliquetos (Annelida: Polychaeta) de México y América Tropical*; León-González, J.A.B.-Z., Bastida Zavala, J.R., Carrera-Parra, L.F., García-Garza, M.E., Peña-Rivera, A., Salazar-Vallejo, S.I., Solís-Weiss, V., Eds.; Universidad Autónoma de Nuevo León: Monterrey, Mexico, 2009; pp. 473–476.
266. Day, J.H. Zooplancton de la région de Nosy-Bé. X. The Biology of planktonic Polychaeta near Nosy-Bé, Madagascar. *Cah. Orstom (Off. Rech. Sci. Tech. Outre-Mer) Ser. Océanogr.* **1975**, *13*, 197–216.
267. Malaquin, A.; Carin, F. Note preliminaire sur les annelides pelagiques provenant des Campganes de l'Hirondelle et de la Princesse-Alice. *Bull. L'inst. Océanogr.* **1911**, *205*, 1–16.
268. Malaquin, A.; Carin, F. Tomopterides provenant des campagnes de l'Hirondelle et de la Princesse-Alice (1888–1910). *Result. Camp. Sci. Monaco* **1922**, *61*, 1–52.
269. Mileikovsky, S.A. On the systematic interrelationships within the Polychaeta and Annelida. An attempt to create an integrated system based on their larval morphology. In *Essays on Polychaetous Annelids in Memory of Dr. Olga Hartman*; Reish, D.J., Fauchald, K., Eds.; The Allan Hancock Foundation, University of Southern California: Los Angeles, CA, USA, 1977; pp. 503–524.
270. Fernández-Álamo, M.A. Tomopteridae Grube, 1850. In *Poliquetos (Annelida: Polychaeta) de México y América Tropical*; León-González, J.A.B.-Z., Bastida Zavala, J.R., Carrera-Parra, L.F., García-Garza, M.E., Peña-Rivera, A., Salazar-Vallejo, S.I., Solís-Weiss, V., Eds.; Universidad Autónoma de Nuevo León: Monterrey, Mexico, 2009; pp. 657–663.
271. Rakusa-Suszczewski, S. Predation of chaetognatha by *Tomopteris helgolandica* Greff. *Ices J. Mar. Sci.* **1968**, *32*, 226–231. [CrossRef]
272. Fernández-Álamo, M.A.; Thuese, A.B. Polychaeta. In *South Atlantic Zooplankton*; Boltovskoy, D., Ed.; Backhuys Publishers: Leiden, The Netherlands, 1999.
273. Fernández-Álamo, M.A. Composition, abundance and distribution of holoplanktonic polychaetes from the expedition "El Golfo 6311-12" of Scripps Institution of Oceanography. *Sci. Mar.* **2006**, *70*, 209–215. [CrossRef]
274. Ekman, S. *Zoogeography of the Sea*; Sidgwick & Jackson, Ltd.: London, UK, 1953; p. 417.
275. Izuka, A. On the pelagic annelids of Japan. *J. Coll. Sci. Imp. Univ. Tokyo* **1914**, *36*, 1–14.
276. Fernández-Álamo, M.A. Composición, densidad y distribución de los poliquetos planctónicos (Phyllodocida: Tomopteridae) en el Domo Térmico de Costa Rica, Pacífico Tropical Oriental. *Rev. Biol. Trop.* **2020**, *68*, S238–S247. [CrossRef]
277. Guglielmo, R.; Bergamasco, A.; Minutoli, R.; Patti, F.P.; Belmonte, G.; Spanò, N.; Zagami, G.; Bonanzinga, V.; Guglielmo, L.; Granata, A. The Otranto Channel (South Adriatic Sea), a hot-spot area of plankton biodiversity: Pelagic polychaetes. *Sci. Rep.* **2019**, *9*, 19490. [CrossRef] [PubMed]
278. Chun, C. Die pelagische Thierwelt in groesseren Meerestiefen. *Bibl. Zool. Cassel* **1887**, *1*, 1–66.
279. Wilson, R.S. Family Tomopteridae. In *Polychaetes & Allies: The Southern Synthesis. Fauna of Australia. Vol. 4A. Polychaeta, Myzostomida, Pogonophora, Echiura, Sipuncula*; Beesley, P.L.G., Ross, J.B., Glasby, C.J., Eds.; CSIRO Publishing: Melbourne, Australia, 2000; pp. 167–168.
280. Glasby, C.J.; Hutchings, P.A.; Fauchald, K.; Paxton, H.; Rouse, G.W.; Watson Russell, C.; Wilson, R.S. Class Polychaeta. In *Polychaetes & Allies: The Southern Synthesis. Fauna of Australia. Vol. 4A. Polychaeta, Myzostomida, Pogonophora, Echiura, Sipuncula*; Beesley, P.L.G., Ross, J.B., Glasby, C.J., Eds.; CSIRO Publishing: Melbourne, Australia, 2000; pp. 1–296.
281. Struck, T.H.; Halanych, K.M. Origins of holopelagic Typhloscolecidae and Lopadorhynchidae within Phyllodocidae (Phyllodocida, Annelida). *Zool. Scr.* **2010**, *39*, 269–275. [CrossRef]
282. Nygren, A.; Pleijel, F. Chimaeras and the origins of the holopelagic annelids Typhloscolecidae and Lopadorhynchidae: A reply to Struck & Halanych (2010). *Zool. Scr.* **2011**, *40*, 112–114. [CrossRef]

283. Feigenbaum, D. Predation on chaetoganths by typhloscolecid polychaetes, one explanation for headless specimens. *J. Mar. Biol. Assoc. United Kingd.* **1979**, *59*, 631–633. [CrossRef]
284. Øresland, V.; Pleijel, F. An ectoparasitic typhloscolecid polychaete on the chaetognath *Eukronia hamata* from the Antarctic Peninsula. *Mar. Biol.* **1991**, *108*, 429–432. [CrossRef]
285. Øresland, V.; Bray, R.A. Parasites and headless chaetognaths in the Indian Ocean. *Mar. Biol.* **2005**, *147*, 725–734. [CrossRef]
286. Fernández-Álamo, M.A. Typhloscolecidae Uljanin, 1878. In *Poliquetos (Annelida: Polychaeta) de México y América Tropical*; León-González, J.A.B.-Z., Bastida Zavala, J.R., Carrera-Parra, L.F., García-Garza, M.E., Peña-Rivera, A., Salazar-Vallejo, S.I., Solís-Weiss, V., Eds.; Universidad Autónoma de Nuevo León: Monterrey, Mexico, 2009; pp. 671–675.
287. Chambers, S.J. Polychaetes: British Chrysopetaloidea, Pisionoidea, and Aphroditoidea. *Synop. Br. Fauna* **1997**, *54*, 1–202.
288. Buzhinskaja, G. Two new genera of the pelagic family Yndolaciidae (Polychaeta) from the Arctic Ocean with an addition to the description of *Yndolacia lopadorrhynchoides* Stöp-Bowitz. *Sarsia* **2004**, *89*, 338–345. [CrossRef]
289. Fernández-Álamo, M.A.; Tejera, E.; León, M.E. Poliquetos pelágicos de las Islas de Cabo Verde: Resultados de la campaña TFMCBM/98, Proyecto Macaronesia 2000. *Rev. Acad. Canar. Cienc. Folia Canar. Acad. Sci.* **2003**, *15*, 87–98.
290. Zhadan, A.E.; Tzetlin, A.B. Polychaetes from deep pelagic zone of the Mid-Atlantic Ridge. *Invertebr. Zool.* **2008**, *5*, 97–109. [CrossRef]
291. Zrzavý, J.; Ríha, P.; Piálek, L.; Janouskovec, J. Phylogeny of Annelida (Lophotrochozoa): Total-evidence analysis of morphology and six genes. *BMC Evol. Biol.* **2009**, *9*, 189. [CrossRef] [PubMed]
292. Pleijel, F.; Dahlgren, T.G. Position and delineation of Chrysopetalidae and Hesionidae (Annelida, Polychaeta, Phyllodocida). *Cladistics* **1998**, *14*, 129–150. [CrossRef]
293. Levinsen, G.M.R. Systematisk geografisk Oversigt over do Nordiske Annulata, Gephyrea, Choetognathi, og Balanoglossi; II. *Vidensk. Medd. Fra Dan. Nat. Foren. I Kjφbenhavn* **1883**, *4*, 92–350.
294. Hatschek, B. System der Anneliden, ein vorläufiger Bericht. *Lotos* **1893**, *13*, 123–126.
295. Grube, A.E. Annulata Örstediana. Enumeratio Annulatorum, quae in itinere per Indiam occidentalem et Americam centralem annis 1845–1848 suscepto legit cl. A. S. Örsted, adjectis speciebus nonnullis a cl. H. Kröyero in itinere ad Americam meridionalem collectis. (Fortsaettelse). [Part 2]. *Vidensk. Medd. Fra Dan. Nat. Foren. I Köbenhavn* **1857**, *1857*, 158–186.
296. Kinberg, J.G.H. Annulata nova. Continuatio. [various errantia & sedentaria]. *Ofvers. Afk. Vetensk. Forh. Stockh.* **1867**, *22*, 239–258.
297. Cuvier, G.L. *Le Règne Animal Distribué D'après Son Organisation, pour Servir de Base à L'histoire Naturelle des Animaux et D'introduction à L'anatomie Comparée. Vol. 4 Les Zoophytes, les Tables, et les Planches*; Deterville: Paris, France, 1817.
298. Fauchald, K. Nephtyidae (Polychaeta) from the Bay of Nha Trang, South Vietnam. *Sci. Res. Mar. Invert. S. China Sea G. Thailand, 1959–61. NAGA Rep.* **1968**, *4*, 5–33.
299. Friedrich, H. Polychaetenstudien. IV- Zur polychaetenfauna der Barents-See. *Kiel. Meeresforsch.* **1939**, *3*, 122–132.
300. Grube, A.E. Anneliden-Ausbeute S.M.S. Gazelle. *Mon. K. Preuss. Akad. Wiss. Zu Berl.* **1877**, 509–554.
301. Ehlers, E. Report on the annelids of the dredging expedition of the U.S. coast survey steamer Blake. *Mem. Mus. Comp. Zool. Harv. Coll.* **1887**, *15*, 1–335.
302. Fauvel, P. Annélides polychètes non-pélagiques provenant des campagnes de l'Hirondelle et de la Princesse-Alice (1885–1910). *Résultats Des. Camp. Sci. Accompl. Sur Son Yatch Par Albert Ier Prince Souver. Monaco* **1914**, *46*, 1–432.
303. Benham, W.B. Report on the Polychaeta obtained by the F.I.S. 'Endeavour' on the coasts of New South Wales, Victoria, Tasmania and South Australia. Part 2. In *Fisheries. Zoological (and Biological) results of the Fishing Experiments carried out by F.I.S. "Endeavour", 1909–1914*; H.C. Dannevig: Sydney, Australia, 1916; Volume 4, pp. 127–169.
304. Hartman, O. Polychaetous annelids. Part II. Chrysopetalidae to Goniadidae. *Allan Hancock Pac. Exped.* **1940**, *7*, 173–287.
305. Petterson, H.; Jerlov, N.G.; Kullenberg, B. *Reports of the Swedish Deep-Sea Expedition 1947–1948*; Swedish National Scientific Research Council: Stockholm, Sweden, 1951.
306. Dnestrovskaya, N.Y.; Jirkov, I.A. Nephtyidae Grube, 1850. In *Polikhety Severnogo Ledovitogo Okeana [Polychaeta of the Arctic Ocean]*; Jirkov, I.A., Ed.; Yanus-K Press: Moscow, Russia, 2001; pp. 181–212.
307. Dnestrovskaya, N.Y.; Jirkov, I.A. *Micronephthys* (Polychaeta: Nephtyidae) of Northern Europe and Arctic. *Invertebr. Zool.* **2010**, *7*, 107–121. [CrossRef]
308. Dnestrovskaya, N.Y.; Jirkov, I.A. Identification key for Nephtyidae (Polychaeta) of the eastern Atlantic and the North Polar Basin. *Invertebr. Zool.* **2012**, *9*, 143–150. [CrossRef]
309. Fauchald, K. Nephtyidae (Polychaeta) from Norwegian waters. *Sarsia* **1963**, *13*, 1–32. [CrossRef]
310. Foret-Montardo, P. Étude systématique et écologique des Nephtyidae (Polychètes Errantes) des parages de Marseille. *Téthys* **1969**, *1*, 807–832.
311. Jirkov, I.A.; Paraketsova, N. Review of the species of the genus Micronephthys (Polychaeta: Nephthyidae) from the white sea. *Zool. Zhurnal* **1996**, *75*, 831–840. (In Russian)
312. Laborda, A.J. Annelida, Polychaeta I. In *Fauna Iberica*, 25th ed.; Viéitez, J.M., Alós, C., Parapar, J., Besteiro, C., Moreira, J., Nuñez, J., Laborda, A.J., San Martin, G., Eds.; Museo Nacional de Ciencias Naturales (MNCN–CSIC): Madrid, Spain, 2004; pp. 390–419.
313. Rainer, S.F. *Nephtys pente* sp. nov. (Polychaeta: Nephtyidae) and a Key to *Nephtys* from Northern Europe. *J. Mar. Biol. Assoc. UK* **1984**, *64*, 899–907. [CrossRef]
314. Rainer, S.F. Redescription of *Nephtys assimilis* and *N. kersivalensis* (Polychaeta: Phyllodocida) and a key to Nephtys from Northern Europe. *J. Mar. Biol. Assoc. UK* **1989**, *69*, 875–889. [CrossRef]

315. Rainer, S.F. The genus *Nephtys* (Polychaeta: Phyllodocida) in Northern Europe: Redescription of *N. hystricis* and *N. incisa*. *J. Nat. Hist.* **1990**, *24*, 361–372. [CrossRef]
316. Rainer, S.F. The genus *Nephtys* (Polychaeta: Phyllodocida) of northern Europe: A review of species, including the description of *N. pulchra* sp. n. and a key to the Nephtydae. *Helgoländer Meeresunters.* **1991**, *45*, 65–96. [CrossRef]
317. Rainer, S.F. Nephtyid Polychaetes from the Faroe Islands. *Northurlandahsth Arsrit* **1992**, *1991–1992*, 80–81.
318. Ravara, A.; Cunha, M.R.; Pleijel, F. Nephtyidae (Annelida, Polychaeta) from southern Europe. *Zootaxa* **2010**, *2682*, 1–68. [CrossRef]
319. Hartman, O. Review of the annelid worms of the family Nephtyidae from the Northeast Pacific, with descriptions of five new species. *Proc. United States Natl. Mus.* **1938**, *85*, 143–158. [CrossRef]
320. Hartman, O. Polychaeta Errantia of the Antarctica. *Antarct. Res. Ser.* **1964**, *3*, 1–131.
321. Hartman, O. Polychaeta from the Weddel Sea quadrant, Antarctica. *Antarct. Res. Ser.* **1978**, *26*, 125–223.
322. Alalykina, I.L.; Dnestrovskaya, N.Y.; Jirkov, I.A. Identification key to Nephtyidae (Annelida) of the Sea of Okhotsk. *ZooKeys* **2017**, *684*, 1–18. [CrossRef]
323. Dixon-Bridges, K.; Gladstone, W.; Hutchings, P. One new species of *Micronephthys* Fredrich, 1939 and one new species of *Nephtys* Cuvier, 1817 (Polychaeta: Phylllodocidae: Nephtyidae) from eastern Australia with notes on *Aglaophamus australiensis* (Fauchald, 1965) and a key to all Australian species. *Zootaxa* **2014**, *3872*, 513–540. [CrossRef]
324. Dnestrovskaya, N.Y. Identification key to Nephtyidae (Annelida) of the Black Sea. *ZooKeys* **2020**, *908*, 1–17. [CrossRef]
325. Fauchald, K. Some Nephtyidae (Polychaeta) from Australian waters. *Rec. Aust. Mus.* **1965**, *26*, 333–340. [CrossRef]
326. Franco, N.B.; Rizzo, A.E. Nephtyidae (Annelida: Polychaeta) from the Campos Basin, including two new species and a new record. *Zootaxa* **2016**, *4114*, 291–308. [CrossRef] [PubMed]
327. Hartman, O.; Fauchald, K. Deep-water benthic polychaetous annelids off New England to Bermuda and other North Atlantic Areas. Part II. *Allan Hancock Monogr. Mar. Biol.* **1971**, *6*, 1–327.
328. Hilbig, B. The annelida. Part 1. Oligochaeta: Phyllodocida (Phyllodocidae to Paralacydoniidae). In *Taxonomic Atlas of the Benthic Fauna of the Santa Maria Basin and the Western Santa Barbara Channel*; Blake, J.A., Hilbig, B., Scott, P.H., Eds.; US Department of the Interior, Minerals Management Service, Pacific OCS Region: Santa Barbara, CA, USA, 1997; Volume 42, pp. 317–349.
329. Imajima, M.; Takeda, Y. Nephtyidae (Polychaeta) from Japan. I. The genera *Inermonephtys*, *Micronephtys* and *Aglaophamus*. *Bull. Natl. Sci. Mus. Tokyo* **1985**, *11*, 57–90.
330. Imajima, M.; Takeda, Y. Nephtyidae (Polychaeta) from Japan. II. The genera *Dentinephtys* and *Nephtys*. *Bull. Natl. Sci. Mus. Tokyo* **1987**, *13*, 41–77.
331. Jung, R.H.; Hong, J.S. Nephtyidae (Annelida: Polychaeta) from the Yellow Sea. *Bull. Mar. Sci.* **1997**, *60*, 371–384.
332. Lana, P.C. Nephtyidae (Annelida: Polychaeta) do litoral do estado do Paraná (Brasil). *Nerítica* **1986**, *1*, 135–154. [CrossRef]
333. Murray, A.; Wong, E.; Hutchings, P.A. Nephtyidae (Annelida: Phyllodocida) of Lizard Island, Great Barrier Reef, Australia. *Zootaxa* **2015**, *4019*, 414–436. [CrossRef]
334. Nateewathana, A.; Hylleberg, J. Nephtyid polychaetes from the west coast of Phuket Island, Andaman Sea, Thailand with description of five new species. *Proc. Linn. Soc. New South. Wales* **1986**, *108*, 195–215.
335. Paxton, H. Contribution to the study of Australian Nephtyidae (Polychaeta). *Rec. Aust. Mus.* **1974**, *29*, 197–208. [CrossRef]
336. Pérez-Torrijos, J.; Hernández-Alcántara, P.; Solís-Weiss, V. Nephtyidae (Polychaeta) from the Gulf of California (Mexican Pacific) with the description of two new species of *Aglaophamus*. *J. Mar. Biol. Assoc. UK* **2009**, *89*, 697–710. [CrossRef]
337. Rainer, S.F.; Hutchings, P. Nephtyidae (Polychaeta: Errantia) from Australia. *Rec. Aust. Mus.* **1977**, *31*, 307–347. [CrossRef]
338. Rainer, S.F.; Kaly, U.L. Nephtyidae (Polychaeta: Phyllodocida) of Australia: New species from the North West Shelf, and a key to Australian species. *J. Nat. Hist.* **1988**, *22*, 685–703. [CrossRef]
339. Rizzo, A.E.; Amaral, A.C.Z. Nephtyidae (Annelida: Polychaeta) from São Paulo state, Brazil, including a new record for the Brazilian coast. *Biota Neotrop.* **2007**, *7*, 253–263. [CrossRef]
340. Rozbaczylo, N.; Castilla, J.C. La familia Nephtyidae en Chile (Annelida, polychaeta). *Stud. Neotrop. Fauna* **1974**, *9*, 179–206. [CrossRef]
341. Taylor, J.L. Chapter 35. Family Nephtyidae Grube, 1850. In *Taxonomic Guide to the Polychaetes of the Northern of Gulf of Mexico*; Uebelacker, J.M., Johnson, P.G., Eds.; Barry A. Vittor & Associates, Inc.: Mobile, AL, USA, 1984; pp. 1–20.
342. Pattaratumrong, M.S.; Wongkhamhaeng, K. The molecular Identification of *Nephtys species* (Polychaeta: Phyllodocida) from Songkhla Lake, Southern Thailand. *Naresuan Univ. J. Sci. Technol.* **2019**, *27*, 1–9. [CrossRef]
343. Ravara, A.; Wiklund, H.; Cunha, M.R.; Pleijel, F. Phylogenetic relationships within Nephtyidae (Polychaeta, Annelida). *Zool. Scr.* **2010**, *39*, 394–405. [CrossRef]
344. Clark, R.B. Observations on the food of *Nephtys*. *Limnol. Oceanogr.* **1962**, *7*, 380–385. [CrossRef]
345. Schubert, A.; Reise, K. Predatory effects of *Nephtys hombergii* on other polychaetes in tidal flat sediments. *Mar. Ecol. Prog. Ser.* **1986**, *34*, 117–124. [CrossRef]
346. Würzberg, L.; Peters, J.; Schüller, M.; Brandt, A. Diet insights of deep-sea polychaetes derived from fatty acid analyses. *Deep Sea Res. Part. II Top. Stud. Oceanogr.* **2011**, *58*, 153–162. [CrossRef]
347. Benham, W.B. *Report on the Polychaeta obtained by the F.I.S. 'Endeavour' on the coasts of New South. Wales, Victoria, Tasmania and South. Australia, Part. 1*; H.C. Dannevig: Sydney, Australia, 1915; pp. 171–237.
348. Eliason, A. Polychaeta. In *Reports of the Swedish Deep-Sea Expedition 1947–1948*; Petterson, H., Jerlov, N.G., Kullenberg, B., Eds.; Swedish National Scientific Research Council: Stockholm, Sweden, 1951; Volume 2.

349. Demopoulos, A.W.J.; Bourque, J.R.; Durkin, A.; Cordes, E.E. The influence of seep habitats on sediment macrofaunal biodiversity and functional traits. *Deep Sea Res. Part. I Oceanogr. Res. Pap.* **2018**, *142*, 77–93. [CrossRef]
350. Grupe, B.M.; Krach, M.L.; Pasulka, A.L.; Maloney, J.M.; Levin, L.A.; Frieder, C.A. Methane seep ecosystem functions and services from a recently discovered southern California seep. *Mar. Ecol.* **2015**, *36*, 91–108. [CrossRef]
351. Levin, L.A.; Ziebis, W.; Mendoza, G.F.; Growney, V.A.; Tryon, M.D.; Brown, K.M.; Mahn, C.; Gieskes, J.M.; Rathaburn, A.E. Spatial heterogeneity of macrofauna at northern California methane seeps: Influence of sulfide concentration and fluid flow. *Mar. Ecol. Prog. Ser.* **2003**, *265*, 123–139. [CrossRef]
352. Serrano, A.; Velasco, F.; Olaso, I. Polychaete annelids in the diet of demersal fish from the southern shelf of the Bay of Biscay. *J. Mar. Biol. Assoc. United Kingd.* **2003**, *83*, 619–623. [CrossRef]
353. Varadharajan, D.; Soundarapandian, P. Contribution of polychaetes in feeding capability of commercially important crabs, South East coast of India. *Mar. Sci. Res. Dev.* **2013**, *3*, 2–6. [CrossRef]
354. Day, J.H. The Polychaete fauna of South Africa. Part 2. Errant species from Cape shores and estuaries. *Ann. Natal Mus.* **1953**, *12*, 397–441.
355. Meißner, K.; Darr, A.; Rachor, E. Development of habitat models for *Nephtys* species (Polychaeta: Nephtyidae) in the German Bight (North Sea). *J. Sea Res.* **2008**, *60*, 276–291. [CrossRef]
356. Olive, P.J.W.; Garwood, P.R.; Bentley, M.G. Reproductive failure and oosorption in Polychaeta in relation to their reproductive strategies. *Bull. Soc. Zool. Fr.* **1981**, *106*, 263–268.
357. Olive, P.J.W.; Morgan, P.J. The reproductive cycles of four British intertidal *Nephtys* species in relation to their geographical distribution (Polychaeta: Nephtyidae). In: Systematics, biology and morphology of world polychaeta. Proceedings of the Second International Polychaeta Conference. *Ophelia* **1991**, *5*, 351–361.
358. Rainer, S.F. Distribution, growth and production of *Nephtys hombergii* and *N. assimilis* (Polychaeta: Nephtyidae) in benthic communities of the North Sea. *Bull. Mar. Sci.* **1991**, *48*, 330–345.
359. Olive, P.J.W.; Garwood, P.R.; Bentley, M.G.; Wright, N.H. Reproductive success, relative abundance and population structure of two species of Nephtys in an estuarine beach. *Mar. Biol.* **1981**, *63*, 189–196. [CrossRef]
360. Desroy, N.; Retière, C. The influence of structure and dynamics of infaunal predator populations on predatory activity: The example of *Nephtys hombergii. Estuar. Coast. Shelf Sci.* **2003**, *58*, 445–453. [CrossRef]
361. Southern, R. Fauna of the Chilka Lake and also of fresh and brackish waters in other parts of India. *Mem. Indian Mus. Calcutta* **1921**, *5*, 563–659.
362. Monro, C.A.A. On some freshwater polychaetes from Uruguay. *Ann. Mag. Nat. Hist. Mus. Lond.* **1937**, *10*, 241–250. [CrossRef]
363. Wilson, R.S. Polychaetes & Allies: The southern synthesis. Fauna of Australia. Vol. 4a Polychaeta, Myzostomida, Pogonophora, Echiura, Sipuncula. In *Family Nephtyidae*; Beesley, P.L., Ross, G.J.B., Glasby, C.J., Eds.; CSIRO Publishing: Melbourne, Australia, 2000; pp. 136–137.
364. Perejaslavzeva, S.M. Supplementations to the fauna of the Black Sea. *Proc. Soc. Nat. Kharkov Univ.* **1891**, *25*, 235–274.
365. Wesenberg-Lund, E. Polychaetes of the Iranian Gulf. *Dan. Sci. Investig. Iran.* **1949**, *4*, 247–400.
366. Savigny, J.C. Troisième partie. Systèmes de diverses classes d'animaux sans vertèbres. Système des annélides, principalement de celles des côtes de l'Égypte et de la Syrie, offrant les charactères tant distinctifs que naturels des Ordres, Familles et Genres, avec la description des espèces. *Descr. L'egyptehist. Nat.* **1822**, *3*, 1–128.
367. Böggemann, M. Polychaeten aus der Deutschen Bucht. Taxonomische Beabeitung und Dokumentation der vom Forschungsinstitut Senckenberg hauptsachlich in der Deutschen Bucht gesammelten Polychaeten. *C. Forschung. Senckenb.* **1997**, *202*, 91–96.
368. Malm, A.W. Annulata i hafvet utmed Sveriges westkust och omkring Göteborg. *Göteborgs K. Vetensk. Och Vitterh. Handl.* **1874**, *14*, 67–105.
369. Gravier, C. Contribution à l'étude des Annélides Polychètes de la Mer Rouge. Familles des Nereidiens, Aphroditiens, tribus des Polynoina, Sigalionina, familles des Amphinomiens, Palmyriens. *Nouv. Arch. Mus. Paris Sér. 4* **1902**, *3*, 147–268.
370. Lamarck, J.B. Ordre Second. Radiaires Échinodermes. *Hist. Nat. Des. Animaux Sans Vertèbres* **1816**, *2*, 522–568.
371. Marenzeller, E. Südjapanische Anneliden.III. *Denkschr. Akad. Wiss. Wien.* **1902**, *72*, 563–582.
372. Müller, J.X.; Troschel, F.H. *System der Asteriden*; F. Papier, Druck und Verlag von Friedrich Vieweg und Son: Brünswick, Germany, 1842.
373. Seidler, H.J. Beitrage zur kenntnis der Polynoiden, I. *Arch. Nat. Berl.* **1923**, *89*, 1–217.
374. Uschakov, P.V.; Reish, D.J.; Fauchald, K. Phylogenetic relationships in the family Polynoidae (Polychaeta). In *Essays on Polychaetous Annelids in Memory of Dr. Olga Hartman*; University of Southern California, Allan Hancock Press: Los Angeles, CA, USA, 1977; pp. 29–38.
375. Struck, T.H.; Purschke, G.; Halanych, K.M. A scaleless scale worm: Molecular evidence for the phylogenetic placement of *Pisione remota* (Pisionidae, Annelida). *Mar. Biol. Res.* **2005**, *1*, 243–253. [CrossRef]
376. Wiklund, H.; Nygren, A.; Pleijel, F.; Sundberg, P. Phylogeny of Aphroditiformia (Polychaeta) based on molecular and morphological data. *Mol. Phylogenet. Evol.* **2005**, *37*, 494–502. [CrossRef]
377. Norlinder, E.; Nygren, A.; Wiklund, H.; Pleijel, F. Phylogeny of scale-worms (Aphroditiformia, Annelida), assessed from 18SrRNA, 28SrRNA, 16SrRNA, mitochondrial cytochrome c oxidase subunit I (COI), and morphology. *Mol. Phylogenet. Evol.* **2012**, *65*, 490–500. [CrossRef] [PubMed]

378. Zhang, Y.; Sun, J.; Rouse, G.W.; Wiklund, H.; Pleijel, F.; Watanabe, H.K.; Chen, C.; Qian, P.-Y.; Qiu, J.-W. Phylogeny, evolution and mitochondrial gene order rearrangement in scale worms (Aphroditiformia, Annelida). *Mol. Phylogenet. Evol.* **2018**, *125*, 200–231. [CrossRef] [PubMed]
379. Bonifácio, P.; Menot, L. New genera and species from the Equatorial Pacific provide phylogenetic insights into deep-sea Polynoidae (Annelida). *Zool. J. Linn. Soc.* **2018**, *185*, 555–635. [CrossRef]
380. Pettibone, M.H. Revision of the genus *Macellicephala* McIntosh and the subfamily Macellicephalinae Hartmann-Schröder (Polychaeta: Polynoidae). *Smithson. Contrib. Zool.* **1976**, *229*, 1–71.
381. Uschakov, P.V. Polychaetes of the Suborder Aphroditiformia of the Arctic Ocean and the Northwestern Part of the Pacific, Families Aphroditidae and Polynoidae. *Fauna Sssrmnogoshchetinkovyye Chervi* **1982**, *2*, 1–272. (In Russian)
382. Tebble, N.; Chambers, S. *Polychaetes from Scottish Waters, Part. 1. Family Polynoidae*; Royal Scottish Museum: Edinburgh, UK, 1982; p. 73.
383. Britayev, T.A.; Fauchald, K. New species of symbiotic scaleworms *Asterophilia* (Polychaeta, Polynoidae) from Vietnam. *Invertebr. Zool.* **2005**, *2*, 15–22. [CrossRef]
384. Nobili, G. Diagnoses préliminaires de vingt-huit espèces nouvelles de stomatopodes et décapodes macroures de la Mer Rouge. *Bull. Mus. D'hist. Nat. Paris* **1904**, *10*, 228–238.
385. Gray, J.E. XXXII. A synopsis of the genera and species of the class Hypostoma (*Asterias*, Linnaeus). *Ann. Mag. Nat. Hist.* **1840**, *6*, 275–290. [CrossRef]
386. Leach, W.E. Annulosa. In *Encyclopedia Britannica, Supplement to the Fourth, Fifth and Sixth Editions*; Encyclopedia Britannica, Inc.: London, UK, 1816; Volume 1, pp. 401–453.
387. Malmgren, A.J. Nordiska Hafs-Annulater. [part one of three]. *Öfvers. Königlich Vetensk. Förhandlingarstockholm* **1865**, *22*, 51–110.
388. Nygren, A.; Norlinder, E.; Panova, M.; Pleijel, F. Colour polymorphism in the polychaete *Harmothoe imbricata* (Linnaeus, 1767). *Mar. Biol. Res.* **2011**, *7*, 54–62. [CrossRef]
389. Pettibone, M.H. A new scale-worm commensal with deep-sea mussels on the Galapagos hydrotermal vent (Polychaeta: Polynoidae). *Proc. Biol. Soc. Wash.* **1984**, *97*, 226–239.
390. Delle Chiaje, S. *Descrizione et Notomia Degli Animali Invertebrati de la Sicilia Cateriore Oservati Vivi Negli Anni 1822–1830*; Batelli & Co.: Naples, Italy, 1841; Volume 5, pp. 1–137.
391. Pettibone, M.H.; Reish, D.J.; Fauchald, K. Review of *Halosydnopsis* and related genera (Polychaeta: Polynoidae: Lepidonotinae). In *Essays on Polychaetous Annelids in Memory of Dr. Olga Hartman*; The Allan Hancock Foundation, University of Southern California: Los Angeles, CA, USA, 1977; pp. 39–62.
392. Pettibone, M.H. Scaled polychaetes (Polynoidae) associated with ophiuroids and other invertebrates and review of species referred to *Malmgrenia* McIntosh and replaced by *Malmgeniella* Hartman, with descriptions of new taxa. *Smithson. Contrib. Zool.* **1993**, *538*, 1–92. [CrossRef]
393. Nemésio, A. A replacement name for *Wilsoniella* Pettibone, 1993 (Annelida: Polynoidae), junior homonym of *Wilsoniella* Khalfin, 1939 (Brachiopoda: Rhynchonellida), and revalidation of *Pessoaiella* Guimarães, 1940 over *Wilsoniella* Eichler, 1940 (Phthiraptera: Ischnocera: Philopteridae) also a junior homonym of *Wilsoniella* Khalfin, 1939. *Zootaxa* **2006**, *1260*, 67–68.
394. Serpetti, N.; Taylor, M.L.; Brennan, D.; Green, D.H.; Rogers, A.D.; Paterson, G.L.J.; Narayanaswamy, B.E. Ecological adaptations and commensal evolution of the Polynoidae (Polychaeta) in the Southwest Indian Ocean Ridge: A phylogenetic approach. *Deep Sea Res. Part. II Top. Stud. Oceanogr.* **2017**, *137*, 273–281. [CrossRef]
395. Shields, M.A.; Glover, A.G.; Wiklund, H. Polynoid polychaetes of the Mid-Atlantic Ridge and a new holothurian association. *Mar. Biol. Res.* **2013**, *9*, 547–553. [CrossRef]
396. Ditlevsen, H. *Annelids. I*; H. Hagerup: Copenhagen, Denmark, 1917; Volume VI, p. 71.
397. Hanley, J.R.; Burke, M. A new genus and species of commensal scaleworm (Polychaeta, Polynoidae) from Broome, Western Australia. *Beagle Rec. Mus. Art Gall. North. Territ.* **1989**, *6*, 97–103.
398. Pettibone, M.H. Review of some species referred to *Scalisetosus* McIntosh (Polychaeta, Polynoidae). *Proc. Biol. Soc. Wash.* **1969**, *82*, 1–30.
399. Hartman, O. A review of the types of polychaetous annelids at the Peabody Museum of Natural History, Yale University. *Bull. Bingham Oceanogr. Coll.* **1942**, *8*, 1–98.
400. Michaelsen, W. Polychaeten von Ceylon. *Jb. Hamb. Wiss. Anst.* **1892**, *9*, 1–23.
401. Hutchings, P.A.; Murray, A. Taxonomy of Polychaetes from the Hawkesbury River and the Southern Estuaries of New South Wales, Australia. *Rec. Aust. Mus.* **1984**, *36*, 1–119. [CrossRef]
402. Gonzalez, B.C.; Martínez, A.; Borda, E.; Iliffe, T.M.; Fontaneto, D.; Worsaae, K. Genetic spatial structure of an anchialine cave annelid indicates connectivity within—But not between—Islands of the Great Bahama Bank. *Mol. Phylogenet. Evol.* **2017**, *109*, 259–270. [CrossRef] [PubMed]
403. Sarvala, J. Ecology of *Harmothoe sarsi* (Malmgren) (Polychaeta, Polynoidae) in the northern Baltic area. *Ann. Zool. Fenn.* **1971**, *8*, 231–309.
404. Plyuscheva, M.; Martin, D.; Britayev, T.A. Diet analyses of the scale-worms *Lepidonotus squamatus* and *Harmothoe imbricata* (Polychaeta, Polynoidae) in the White Sea. *Mar. Biol. Res.* **2010**, *6*, 271–281. [CrossRef]

405. Desbruyères, D.; Laubier, L. Exploitation d'une source de matière organique concentrée dans l'océan profond: Intervention d'une annélide polychète nouvelle [Exploitation of a concentrated organic matter source in the deep sea: Role of a new polychaetous annelid]. *C. R. Hebdomaidaire Des. Séances L'acad. Des. Sci.* **1988**, *307*, 329–336.
406. Hatch, A.S.; Liew, H.; Hourdez, S.; Rouse, G.W. Hungry scale worms: Phylogenetics of *Peinaleopolynoe* (Polynoidae, Annelida), with four new species. *ZooKeys* **2020**, *932*, 27–74. [CrossRef]
407. Coll, M.; Piroddi, C.; Steenbeek, J.; Kaschner, K.; Ben Rais Lasram, F.; Aguzzi, J.; Ballesteros, E.; Bianchi, C.N.; Corbera, J.; Dailianis, T.; et al. The biodiversity of the Mediterranean Sea: Estimates, patterns and treats. *PLoS ONE* **2010**, *5*, e11842. [CrossRef] [PubMed]
408. Kiseleva, M.I. *Mnogoshchetinkovye Chervi (Polychaeta) Chernogo i Azovskogo Morei [Polychaeta from the Black and Azov Seas]*; Russian Academy of Sciences, Kola Scientific Center, Murmansk Marine Biological Institute: Murmansk, Russian; National Academy of Sciences of Ukraine, A.O. Kovalevsky Institute of Biology of the South Seas, Apatity, KSCRAN Publications: Kiev, Ukraine, 2004.
409. Horst, R. On new and little-known species of Polynoinae from the Netherland's East Indies. *Zool. Meded. Leiden* **1915**, *1*, 2–20.
410. Potts, F.A. Polychaeta of the Indian ocean. Pt. 2. The Palmyridae, Aphroditidae, Polynoidae, Acoetidae and Sigalionidae. *Trans. Linn. Soc. Zool.* **1910**, *16*, 325–353. [CrossRef]
411. Barnich, R.; Fiege, D.; Sun, R. Polychaeta (Annelida) of Hainan Island, South China Sea Part III. Aphroditoidea. *Species Divers.* **2004**, *9*, 285–329. [CrossRef]
412. Britayev, T.A.; Antokhina, T.I. Chapter 1. Symbiotic polychaetes of from the Bay of Nhatrang Bay, Vietnam. In *Benthic fauna of the Bay of Nhatrang, Southern Vietnam. Volume 2*; Britayev, T.A., Pavlov, D.S., Eds.; KMK Scientific Press Ltd.: Moscow, Russia, 2012; Volume 2, pp. 11–54.
413. Salazar-Vallejo, S.I.; Carrera-Parra, L.F.; Muir, A.I.; León-González, J.A.d.; Piotrowsky, C.; Sato, M. Polychaete species (Annelida) described from the Philippine and China Seas. *Zootaxa* **2014**, *3842*, 1–68. [CrossRef]
414. Glasby, C.J.; Lee, Y.-L.; Hsueh, P.-W. Marine Annelida (excluding clitellates and siboglinids) from the South China Sea. *Raffles Bull. Zool.* **2016**, (Suppl. 34), 178–234. Available online: https://www.researchgate.net/profile/Pan-Wen-Hsueh-2/publication/306216590_Marine_Annelida_excluding_clitellates_and_siboglinids_from_the_South_China_Sea/links/57f3505908ae280dd0b56b0c/Marine-Annelida-excluding-clitellates-and-siboglinids-from-the-South-China-Sea.pdf (accessed on 4 February 2021).
415. Ravara, A.; Cunha, M.R. Two new species of scale worms (Polychaeta: Aphroditiformia) from deep-sea habitats in the Gulf of Cadiz (NE Atlantic). *Zootaxa* **2016**, *4097*, 422–450. [CrossRef] [PubMed]
416. Taboada, S.; Silva, A.S.; Neal, L.; Cristobo, J.; Ríos, P.; Álvarez-Campos, P.; Hestetun, J.T.; Koutsouveli, V.i.; Sherlock, E.; Riesgo, A. Insights into the symbiotic relationship between scale worms and carnivorous sponges (Cladorhizidae, Chondrocladia). *Deep Sea Res. Part. I Oceanogr. Res. Pap.* **2020**, *156*, 103191. [CrossRef]
417. Taboada, S.; Serra Silva, A.; Díez-Vives, C.; Neal, L.; Cristobo, J.; Ríos, P.; Hestetun, J.T.; Clark, B.; Rossi, M.E.; Junoy, J.; et al. Sleeping with the enemy: Unravelling the symbiotic relationships between the scale worm *Neopolynoe chondrocladiae* (Annelida: Polynoidae) and its carnivorous sponge hosts. *Zool. J. Linn. Soc.* **2020**, zlaa146. [CrossRef]
418. De Assis, J.E.; de Souza, J.R.B.; de Lima, M.M.; de Lima, G.V.; Cordeiro, R.T.S.; Pérez, C.D. Association between deep-water scale-worms (Annelida: Polynoidae) and black corals (Cnidaria: Antipatharia) in the Southwestern Atlantic. *Zoologia* **2019**, *36*, 1–13. [CrossRef]
419. Barnich, R.; Beuck, L.; Freiwald, A. Scale worms (Polychaeta: Aphroditiformia) associated with cold-water corals in the eastern Gulf of Mexico. *J. Mar. Biol. Assoc. UK* **2013**, *93*, 2129–2143. [CrossRef]
420. Miranda, V.d.R.; Brasil, A.C.S. Two new species and a new record of Scale-worms (Polychaeta) from Southwest Atlantic deep-sea coral mounds. *Zootaxa* **2014**, *3856*, 211–226. [CrossRef]
421. Schiaparelli, S.; Alvaro, M.C.; Bohn, J.; Albertelli, G. 'Hitchhiker' polynoid polychaetes in cold deep waters and their potential influence on benthic soft bottom food webs. *Antarct. Sci.* **2010**, *22*, 399–407. [CrossRef]
422. Pettibone, M.H. A new scale worm commensal with deep-sea mussels in the seep-sites at the Florida Escarpment in the Eastern Gulf of Mexico (Polychaeta: Polynoidae: Branchipolynoidae). *Proc. Biol. Soc. Wash.* **1986**, *99*, 444–451.
423. Petersen, M.E.; Britayev, T.A. A new genus and species of polynoid scaleworm commensal with *Chaetopterus appendiculatus* Grube from the Banda Sea (Annelida: Polychaeta), with a review of commensals of Chaetopteridae. *Bull. Mar. Sci.* **1997**, *60*, 261–276.
424. Britayev, T.A.; Mekhova, E.; Deart, Y.; Martin, D. Do syntopic host species harbour similar symbiotic communities? The case of *Chaetopterus* spp. (Annelida: Chaetopteridae). *PeerJ* **2017**, *5*, e2930. [CrossRef]
425. Peters, W.C.H. Über die Gattung *Bdella*, Savigny, (Limnatis, Moquin-Tandon) und die in Mossambique beobachteten Anneliden, wovon hier eine Mittheilung folgt [informal title in meeting report]. *Ber. Über Die Zur Bekanntm. Geeigneten Verh. Königlichen Preuss. Akad. Wiss. Zu Berl.* **1854**, *21*, 607–614.
426. Britayev, T.A.; Zamyshliak, E.A. Association of the commensal scaleworm *Gastrolepidia clavigera* (Polychaeta: Polynoidae) with holothurians near the coast of South Vietnam. *Ophelia* **1996**, *45*, 175–190. [CrossRef]
427. Dales, R.P. Interrelations of organisms. A. Commensalism. In Treatise on marine ecology and palaeecology (Ed. J.W. Hedgpeth). *Mem. Geol. Soc. Am.* **1957**, *67*, 391–412.
428. Wagner, R.H.; Phillips, D.W.; Standing, J.D.; Hand, C. Commensalism or mutualism: Atraction of a sea star towards a symbiotic polychaete. *J. Exp. Mar. Biol. Ecol.* **1979**, *39*, 205–210. [CrossRef]

429. Di Camillo, C.G.; Martin, D.; Britayev, T.A. Symbiotic association between *Solanderia secunda* (Cnidaria, Hydrozoa, Solanderiidae) and *Medioantenna variopinta* sp. nov. (Annelida, Polychaeta, Polynoidae) from North Sulawesi (Indonesia). *Helgol. Mar. Res.* **2011**, *65*, 495–511. [CrossRef]
430. Inaba, M. Soshu, Miura, Misaki ni oide edaru Hydroidea. [The hydroids collected at Miura and Misaki in Soshu.]. *Zool. Mag.* **1892**, *4*, 124–131.
431. Sars, G.O. Bidrag til Kundskaben om Dyrelivet paa vore Havbanker. *Forh. Vidensk. Selsk. I Christiania* **1873**, *1872*, 73–119.
432. Britayev, T.A.; Mekhova, E.S. Assessment of hidden diversity of crinoids and their symbionts in the Bay of Nhatrang, Vietnam. *Org. Divers. Evol.* **2011**, *11*, 275–286. [CrossRef]
433. Marudhupandi, T.; Kumar, T.T.A.; Prakash, S.; Gopi, M.; Balasubramanian, T. A first report of symbiotic polychaete scale worm *Gastrolepidia clavigera* Schmarda, 1861 (Phyllodocida: Polynoidae) from Lakshadweep Archipelago, India. *J. Treatened Taxa* **2014**, *6*, 3685–6388. [CrossRef]
434. Britayev, T.A.; Lyskin, S.A. Feeding of the symbiotic polychaete *Gastrolepidia clavigera* (Polynoidae) and its interactions with its hosts. *Dokl. Biol. Sci.* **2002**, *385*, 352–356. [CrossRef]
435. Hanley, J.R. A new host and locality records of the commensal *Adyte crinoidicola* (Polychaeta, Polynoidae). *Beagle Rec. Mus. Art Gall. North. Territ.* **1984**, *1*, 87–92.
436. Hanley, J.R. Revision of the scaleworm genera *Arctonoe* Chamberlin and *Gastrolepidia* Schmarda (Polychaeta, Polynoidae) with the erection of a new subfamily, Arctonoinae. *Beagle Rec. Mus. Art Gall. North. Territ.* **1989**, *6*, 1–34.
437. Sugiyama, T.; Jimi, N.; Goto, R. Widening the host range of the ectosymbiotic scale-worm *Asterophilia culcitae* (Annelida: Polynoidae) to three echinoderm classes, with data on its body color variation. *Plankton Benthos Res.* **2020**, *15*, 289–295. [CrossRef]
438. Willey, A. Report on the Polychaeta collected by Professor Herdman, at Ceylon, in 1902. *Rep. Gov. Ceylon Pearl Oys* **1905**, *30*, 212–324.
439. Johnson, H.P. A preliminary account of the marine annelids of the Pacific coast, with descriptions of new species. *Proc. Calif. Acad. Sci.* **1897**, *1*, 153–199.
440. Britayev, T.A.; Ivashchenko, N.I.; Litvinov, E.G. Peculiarities of the commensal complex in the polychaete *Arctonoe vittata*. *Biol. Morya* **1977**, *1978*, 76–78. (In Russian)
441. Park, T.; Lee, S.; Kim, W. New record of commensal scale worms, *Arctonoe vittata* (Grube, 1855) and *Hyperhalosydna striata* (Kinberg, 1856)(Polychaeta: Polynoidae) from Korean waters. *J. Species Res.* **2016**, *5*, 517–529. [CrossRef]
442. Tokaji, H.i.; Nakahara, K.; Goshima, S. Host switching improves survival rate of the symbiotic polychaete *Arctonoe vittata. Plankton Benthos Res.* **2014**, *9*, 189–196. [CrossRef]
443. Davenport, D. Studies in the physiology of commensalism. I. The polynoid genus *Arctonoë*. *Biol. Bull.* **1950**, *98*, 81–93. [CrossRef] [PubMed]
444. Montagu, G., II. Description of several Marine Animals found on the South Coast of Devonshire. *Trans. Linn. Soc. Lond.* **1808**, *9*, 81–114. [CrossRef]
445. Conradi, M.; Bandera, M.E.; Marin, I.; Martin, D. Polychaete-parasitizing copepods from the deep-sea Kuril-Kamchatka Trench (Pacific Ocean), with the description of a new species of *Ophelicola*. *Deep Sea Res. Part II Top. Stud. Oceanogr.* **2015**, *111*, 147–165. [CrossRef]
446. Capa, M.; Aguado, M.T.; Bakken, T. Phylogenetic hypothesis of Sphaerodoridae Malmgren, 1867 (Annelida) and its position within Phyllodocida. *Cladistics* **2016**, *32*, 335–350. [CrossRef]
447. Capa, M.; Bakken, T. Revision of the Australian Sphaerodoridae (Annelida) including the description of four new species. *Zootaxa* **2015**, *4000*, 227–267. [CrossRef] [PubMed]
448. Ruderman, L. Recherches sur *Ephesia gracilis* Rathke, Annélide polychète de la famille des sphaerodorides; morphologie, anatomie, histologie. *Mém. Soc. Zool. Fr.* **1911**, *24*, 1–96.
449. Reimers, H. Morphologie der Polychaetengattung *Sphaerodorum*. Monographie. *Zool. Jahrbücherabteilung Syst. Ökologie Und Geogr. Tiere* **1933**, *64*, 41–110.
450. Capa, M.; Bakken, T.; Meißner, K.; Nygren, A. Three, two, one! Revision of the long-bodied sphaerodorids (Sphaerodoridae, Annelida) and synonymization of *Ephesiella*, *Ephesiopsis* and *Sphaerodorum*. *PeerJ* **2018**, *6*, e5783. [CrossRef]
451. Capa, M.; Osborn, K.J.; Bakken, T. Sphaerodoridae (Annelida) of the deep Northwestern Atlantic, including remarkable new species of *Euritmia* and *Sphaerephesia*. *ZooKeys* **2016**, *615*, 1–32. [CrossRef]
452. Capa, M.; Rouse, G.W. Sphaerodoridae (Annelida) from Lizard Island, Great Barrier Reef, Australia, including the description of two new species and reproductive notes. *Zootaxa* **2015**, *4019*, 168–183. [CrossRef] [PubMed]
453. Capa, M.; Nygren, A.; Parapar, J.; Bakken, T.; Meißner, K.; Moreira, J. Systematic re-structure and new species of Sphaerodoridae (Annelida) after morphological revision and molecular phylogenetic analyses of the North East Atlantic fauna. *ZooKeys* **2019**, *845*, 1–97. [CrossRef]
454. Lützen, J. Sur une nouvelle espèce de Polychète, *Sphaerodoridium commensalis* n. gen., n. spec. (Polychaeta Errantia, famille des Sphaerodoridae) vivant en commensal de *Terebellides stroemi* Sars. *Cah. Biol. Mar.* **1961**, *2*, 409–416.
455. Fauchald, K. Sphaerodoridae (Polychaeta: Errantia) from world-wide areas. *J. Nat. Hist.* **1974**, *8*, 257–289. [CrossRef]
456. Desbruyères, D. Sphaerodoridae (Annélides Polychètes) profonds du Nord-Est Atlantique. *Bull. Mus. D'hist. Nat. Paris* **1980**, *4*, 109–128.

457. Kudenov, J.D. Five new species of Sphaeroridae (Annelida: Polychaeta) from the Gulf of Mexico. *Proc. Biol. Soc. Wash.* **1987**, *100*, 927–935.
458. Sardá-Borroy, R. Sphaerodoridae (Annelida, Polychaeta) from the region of the Gibraltar strait with description of *Euritmia hamulisetosa* gen. et sp. n. *Zool. Scr.* **1987**, *16*, 47–50. [CrossRef]
459. Rizzo, A.E. A new species of *Ephesiopsis* (Polychaeta: Sphaerodoridae) from off southeastern Brazil. *Zootaxa* **2009**, *2307*, 61–67. [CrossRef]
460. Helm, C.; Capa, M. Comparative analyses of morphological characters in Sphaerodoridae and allies (Annelida) revealed by an integrative microscopical approach. *Front. Mar. Sci.* **2015**, *1*, 1–15. [CrossRef]
461. Martin, D.; Alvà, V. *Sphaerodorum ophiurophoretos* n. sp., une nouvelle éspece de Sphaerodoridae (Annelida, Polychaeta) commensal sur *Amphipholis squamata* (Echinodermata, Ophiuridae). *Bull. L'inst. R. Des. Sci. Nat. Belg. Biol.* **1988**, *58*, 45–49.
462. Moore, J.P. Polychaetous annelids from Monterey Bay and San Diego, California. *Proc. Acad. Nat. Sci. Phila.* **1909**, *61*, 235–295.
463. Moreira, J.; Parapar, J. Sphaerodoridae (Annelida: Polychaeta) from the Bellingshausen Sea (Antarctica) with the description of two new species. *Polar Biol.* **2011**, *34*, 193–204. [CrossRef]
464. Fauvel, P. *Annélides Polychètes. Duc d'Orleans' Campagne Arctique de 1907*; Charles Bulens: Brussels, Belgium, 1911; p. 45.
465. Kudenov, J.D. Four species of Sphaerodoridae (Anellida: Polychaeta) including one new genus and three new species from Alaska. *Proc. Biol. Soc. Wash.* **1987**, *100*, 917–926.
466. Borowski, C. Three new deep-sea species of Sphaerodoridae (Annelida, Polychaeta) from the eastern tropical South Pacific. *Zool. Scr.* **1994**, *23*, 193–203. [CrossRef]
467. Pleijel, F. Sphaerodoridae Malmgren, 1867. In *Polychaetes*; Rouse, G.W., Pleijel, F., Eds.; Oxford University Press: Oxford, UK, 2001; pp. 136–138.
468. Capa, M.; Bakken, T.; Purschke, G. *Sphaerodoridae Malmgren, 1867*; De Gruyter: Ösnabruck, Germany, 2014.
469. Hartmann-Schröder, G. Zur kenntnis des eulitorals der Australischen küsten unter besonderer berücksichigung de Polychaeten und Ostracoden. Teil 6. Die Polychaeten der tropisch-subtropischen westküste Australiens (zwischen Exmouth im Norden und Cervantes im Süden). *Mitt. Hambg. Zool. Mus. Inst.* **1981**, *78*, 19–96.
470. Moreira, J.; Parapar, J. Two new species of *Sphaerodoropsis* Hartman & Fauchald, 1971 (Polychaeta: Sphaerodoridae) from Iceland (BIOICE programme). *Mar. Biol. Res.* **2012**, *8*, 584–593. [CrossRef]
471. Villora-Moreno, S. Environmental heterogeneity and the biodiversity of interstitial Polychaeta. *Bull. Mar. Sci.* **1997**, *60*, 494–501.
472. Worsaae, K.; Kristensen, R.M. Evolution of interstitial Polychaeta (Annelida). In *Morphology, Molecules, Evolution and Phylogeny in Polychaeta and Related Taxa*; Springer: Berlin/Heidelberg, Germany, 2005; pp. 319–340.
473. Spalding, M.D.; Fox, H.E.; Allen, G.R.; Davidson, N.; Ferdaña, Z.A.; Max, F.; Halpern, B.S.; Jorge, M.A.; Lombana, A.; Lourie, S.A.; et al. Marine ecoregions of the world: A bioregionalization of coastal and shelf areas. *Bioscience* **2007**, *57*, 573–583. [CrossRef]
474. Agassiz, A. On alternate generation of annelids and the embryology of *Autolytus cornutus*. *J. Boston Soc. Nat. Hist.* **1862**, *7*, 384–409.
475. Langerhans, P. Die Wurmfauna von Madeira [part I]. *Z. Wiss. Zool.* **1879**, *32*, 513–592.
476. Malaquin, A. Recherches sur les syllidiens: Morphologie, anatomie, reproduction, développement. *Mém. Soc. Des. Sci. L'agric. Des. Arts Lille4e Sér.* **1893**, *18*, 1–477.
477. Rioja, E. Anélidos poliquetos de San Vicente de la Barquera (Cantábrico). *Trab. Mus. Nac. Cienc. Nat. Ser. Zool.* **1925**, *53*, 1–62.
478. Aguado, M.T.; San Martín, G. Phylogeny of Syllidae (Polychaeta) based on morphological data. *Zool. Scr.* **2009**, *38*, 379–402. [CrossRef]
479. San Martín, G. Annelida, Polychaeta II: Syllidade. In *Fauna Iberica., vol. 21*; Ramos, M.A., Alba, J., Bellés, X., Gosálbez, J., Guerra, A., Macpherson, E., Serrano, J., Templado, J., Eds.; CSIC: Madrid, Spain, 2003; Volume 21, p. 554.
480. San Martín, G.; Aguado, M.T. Family Syllidae. In *Handbook of Zoology Online, Annelida*; De Gruyter: Berlin, Germany, 2014; pp. 1–52.
481. Müller, O.F. *Von Würmern des Süssen und Salzigen Wassers*; H. Mumme and Faber: Copenhagen, Denmark, 1771; p. 200.
482. Audouin, J.V.; Milne-Edwards, H. *Recherches pour Servir a L'histoire Naturelle du Littoral de la France, ou, Recueil de Mémoires sur L'anatomie, la Physiologie, la Classification et les Moeurs des Animaux des nos Côtes: Ouvrage Accompagné de Planches Faites D'après Nature*; Crochard: Paris, France, 1834.
483. Licher, F. Revision der Gattung *Typosyllis* Langerhans, 1879 (Polychaeta: Syllidae). Morphologie, Taxonomie und Phylogenie. *Abh. Senckenbergischen Nat. Ges.* **1999**, *551*, 1–336.
484. Nygren, A. Phylogeny and reproduction in Syllidae (Polychaeta). *Zool. J. Linn. Soc. Lond.* **1999**, *126*, 365–386. [CrossRef]
485. Nygren, A.; Sundberg, P. Phylogeny and evolution of reproductive modes in Autolytinae (Syllidae, Annelida). *Mol. Phylogenet. Evol.* **2003**, *29*, 235–249. [CrossRef]
486. Aguado, M.T.; Martin, S.; Siddall, G. Systematics and evolution of syllids. *Annelidasyllidae. Cladistics* **2012**, *28*, 234–250. [CrossRef]
487. Aguado, M.T.; Nygren, A.; Siddall, M.E. Phylogeny of Syllidae (Polychaeta) based on combined molecular analysis of nuclear and mitochondrial genes. *Cladistics* **2007**, *23*, 552–564. [CrossRef]
488. Nygren, A. Revision of Autolytinae (Syllidae: Polychaeta). *Zootaxa* **2004**, *680*, 1–314. [CrossRef]
489. San Martín, G.; López, E.; Aguado, M.T. Revision of the genus *Pionosyllis* (Polychaeta: Syllidae: Eusyllinae), with a cladistic analysis, and the description of five new genera and two new species. *J. Mar. Biol. Assoc. UK* **2009**, 1–44. [CrossRef]

490. Aguado, M.T.; Bleidorn, C. Conflicting signal within a single gene confounds syllid phylogeny (Syllidae, Annelida). *Mol. Phylogenet. Evol.* **2010**, *55*, 1128–1138. [CrossRef] [PubMed]
491. Day, J.H. The polychaete fauna of South Africa Part 8: New species and records from grab samples and dredgings. *Bull. Br. Mus. Nat. Hist.* **1963**, *10*, 381–445. [CrossRef]
492. Aguado, M.T.; Glasby, C.J.; Schroeder, P.l.C.; Weigert, A.; Bleidorn, C. The making of a branching annelid: An analysis of complete mitochondrial genome and ribosomal data of *Ramisyllis multicaudata*. *Sci. Rep.* **2015**, *5*, 12072. [CrossRef]
493. Augener, H. Polychaeta I. Errantia. In *Die Fauna Südwest-Australiens. Ergebnisse der Hamburger Südwest-Australischen Forschungsreise 1905*; Michaelsen, W., Hartmeyer, R., Eds.; Gustav Fischer: Jena, Germany, 1913; Volume IV, pp. 65–304.
494. San Martín, G.; Hutchings, P. Eusyllinae (Polychaeta: Syllidae) from Australia with the description of a new genus and fifteen new species. *Rec. Aust. Mus.* **2006**, *58*, 257–370. [CrossRef]
495. Aguado, M.T.; Murray, A.; Hutchings, P.A. Syllidae (Annelida: Phyllodocida) from Lizard Island, Great Barrier Reef, Australia. *Zootaxa* **2015**, *4019*, 35–60. [CrossRef]
496. Martínez, M.J.; San Martin, G. Syllidae (Annelida) from East Timor and the Philippines (Pacific Ocean), with the description of three new species of *Syllis* Savigny in Lamarck, 1818. *Zootaxa* **2020**, *4834*, 231–263. [CrossRef]
497. Hartmann-Schröder, G. Die Polychaeten der tropischen Nordwestküste Australiens (Zwischen Port Samson in Norden und Port Hedland in Süden). Teil 2. In: Hartmann-Schröder, G. and Hartmann, G. (eds.), Zur Kenntnis des Eulitorals der australischen Küsten unter besonder Berücksichtigung der Polychaeten und Ostracoden. *Mitt. Hambg. Zool. Mus. Inst.* **1979**, *76*, 75–218.
498. Glasby, C.J.; Watson, C. A new genus and species of Syllidae (Annelida: Polychaeta) commensal with octocorals. *Beagle Rec. Mus. Art Gall. North. Territ.* **2001**, *17*, 43–51.
499. Glasby, C.J.; Schroeder, P.C.; Aguado, M.T. Branching out: A remarkable new branching syllid (Annelida) living in a *Petrosia* sponge (Porifera: Demospongiae). *Zool. J. Linn. Soc.* **2012**, *164*, 481–497. [CrossRef]
500. Ribeiro, R.P.; Ponz-Segrelles, G.; Helm, C.; Egger, B.; Aguado, M.T. A new species of *Syllis* Grube, 1850 including transcriptomic data and an updated phylogeny of Syllinae (Annelida: Syllidae). *Mar. Biodivers.* **2020**, *50*, 31. [CrossRef]
501. Aguado, M.T.; Richter, S.; Sontowski, R.; Golombek, A.; Struck, T.H.; Bleidorn, C. Syllidae mitochondrial gene order is unusually variable for Annelida. *Gene* **2016**, *594*, 89–96. [CrossRef] [PubMed]
502. Álvarez-Campos, P.; Riesgo, A.; Hutchings, P.; San Martin, G. The genus *Syllis* Savigny in Lamarck, 1818 (Annelida, Syllidae) from Australia. Molecular analysis and re-description of some poorly-known species. *Zootaxa* **2015**, *4052*, 297–331. [CrossRef]
503. Ponz-Segrelles, G.; Ribeiro, R.P.; Bleidorn, C.; Aguado, M.T. Sex-specific gene expression differences in reproducing *Syllis prolifera* and *Nudisyllis pulligera* (Annelida, Syllidae). *Mar. Genom.* **2020**, 100772. [CrossRef] [PubMed]
504. Ponz-Segrelles, G.; Bleidorn, C.; Aguado, M.T. Expression of vasa, piwi, and nanos during gametogenesis in Typosyllis antoni (Annelida, Syllidae). *Evol. Dev.* **2018**, *20*, 132–145. [CrossRef]
505. Ribeiro, R.P.; Ponz-Segrelles, G.; Bleidorn, C.; Aguado, M.T. Comparative transcriptomics in Syllidae (Annelida) indicates that posterior regeneration and regular growth are comparable, while anterior regeneration is a distinct process. *Bmc Genom.* **2019**, *20*, 855. [CrossRef]
506. San Martín, G.; Álvarez-Campos, P.; Hutchings, P. The genus *Syllis* Savigny in Lamarck, 1818 (Annelida: Syllidae: Syllinae) from Australia (second part): Four new species and re-description of twelve previously described species. *Zootaxa* **2017**, *4237*, 201–243. [CrossRef] [PubMed]
507. Aguado, M.T.; San Martín, G. Re-description of some enigmatic genera of Syllidae (Phyllodocida: Polychaeta). *J. Mar. Biol. Assoc. United Kingd.* **2008**, *88*, 35–56. [CrossRef]
508. Fukuda, M.V.; San Martin, G.; Carrerette, O.; Paresque, K. On a new species of the rare syllid genus *Exogonoides* (Annelida, Phyllodocida, Syllidae). *Zootaxa* **2016**, *4144*, 291–295. [CrossRef]
509. Claparède, É. Glanures zootomiques parmi les annélides de Port-Vendres (Pyrénées Orientales). *Mém. Soc. Phys. D'hist. Nat. Gen.* **1864**, *17*, 463–600.
510. McIntosh, W.C. On a remarkably branched *Syllis* dredged by H.M.S. Challenger. *J. Linn. Soc.* **1879**, *14*, 720–724. [CrossRef]
511. Pleijel, F. Syllidae Grube, 1850. In *Polychaetes*; Rouse, G.W., Pleijel, F., Eds.; Oxford University Press: Oxford, UK, 2001; pp. 102–105.
512. San Martín, G.; Aguado, M.T. Contribution of Scanning Electron Microscope to the study of morphology, biology, reproduction, and phylogeny of the family Syllidae (Polychaeta). In *Scanning Electron Microscopy*; Kazmiruk, V., Ed.; IntechOpen: Rijeka, Croatia, 2012; Volume 8, pp. 129–145.
513. Westheide, W.; Hass-Cordes, E. Molecular taxonomy: Description of a cryptic *Petitia* species (Polychaeta: Syllidae) from the island of Mahé (Seychelles, Indian Ocean) using RAPD markers and ITS2 sequences. *J. Zool. Syst. Evol. Res.* **2001**, *39*, 103–111. [CrossRef]
514. Di Domenico, M.; da Cunha Lana, P.; Garraffoni, A.R.S. Distribution patterns of interstitial polychaetes in sandy beaches of southern Brazil. *Mar. Ecol.* **2009**, *30*, 47–62. [CrossRef]
515. Nogueira, J.M.M.; San Martín, G.; Fukuda, M.V. On some exogonines (Polychaeta, Syllidae, Exogoninae) from the northern coast of the State of São Paulo, southeastern Brazil. Results of BIOTA/FAPESP/Bentos Marinho Project. *Meiofauna Mar.* **2004**, *13*, 45–77.
516. Riera, R.; Núñez, J.; Brito, M.C. A new species of the interstitial genus *Neopetitia* (Polychaeta, Syllidae, Eusyllinae) from Tenerife, with modified acicular chaetae in males. *Helgol. Mar. Res.* **2007**, *61*, 221–223. [CrossRef]
517. Barroso, R.; Paiva, P.C.D.; Nogueira, J.M.D.M.; Fukuda, M.V. Deep sea Syllidae (Annelida, Phyllodocida) from Southwestern Atlantic. *Zootaxa* **2017**, *4221*, 401–430. [CrossRef] [PubMed]

518. Böggemann, M.; Purschke, G. Abyssal benthic Syllidae (Annelida: Polychaeta) from the Angola Basin. *Org. Divers. Evol.* **2005**, *5*, 221–226. [CrossRef]
519. San Martín, G. Deep sea Syllidae from the Pacific Ocean collected during cruises with the RV Sonne (Annelida, Polychaeta, Syllidae). *Senckenbergiana Biol.* **2004**, *84*, 13–25.
520. Hortal, J.; Jiménez-Valverde, A.; Gómez, J.F.; Lobo, J.M.; Baselga, A. Historical bias in biodiversity inventories affects the observed environmental niche of the species. *Oikos* **2008**, *117*, 847–858. [CrossRef]
521. Lomolino, M.V. *Conservation Biogeography. Frontiers of Biogeography: New Directions in the Geography of Nature*; Lomolino, M.V., Heaney, L.R., Eds.; Sinauer Associates: Sunderland, MA, USA, 2004; pp. 293–296.
522. San Martín, G. Estudio Biogeográfico, Faunístico y Sistemático de los Poliquetos de la Familia Sílidos (Syllidae, Polychaeta) en Baleares. Ph.D. Thesis, Universidad Complutense, Madrid, Spain, 1984.
523. Perkins, T.H. Syllidae (Polychaeta), principlally from Florida, with descriptions of a new genus and twenty-one new species. *Proc. Biol. Soc. Wash.* **1980**, *93*, 1080–1172.
524. Uebelacker, J.M. Chapter 30. Family Syllidae. In *Taxonomic Guide to the Polychaetes of the Northern Gulf of Mexico*; Uebelacker, J.M., Johnson, P.G., Eds.; Barry A. Vitor & Associated: Mobile, AL, USA, 1984; pp. 1–151.
525. San Martín, G. Eusyllinae (Syllidae, polychaeta) from Cuba and Gulf of Mexico. *Bull. Mar. Sci.* **1990**, *46*, 590–619.
526. San Martín, G. Grubeosyllis and Exogone (Exogoninae, Syllidae, Polychaeta) from Cuba, the Gulf of Mexico, Florida and Puerto Rico, with a revision of Exogone. *Bull. Mar. Sci.* **1991**, *49*, 715–740.
527. San Martín, G. Syllis Savigny in Lamarck, 1818 (Polychaeta: Syllidae: Syllinae) from Cuba, The Gulf of Mexico, Florida and North Carolina, with a revision of several species described by Verrill. *Bull. Mar. Sci.* **1992**, *51*, 167–196.
528. Capa, M.; San Martín, G.; López, E. Syllinae (Syllidae: Polychaeta) del Parque Nacional de Coiba, Panamá. *Rev. Biol. Trop.* **2001**, *49*, 103–115.
529. Capa, M.; San Martín, G.; López. E. Autolytinae, Eusyllinae y Exogoninae (Syllidae: Polychaeta) del Parque Nacional de Coiba, Panamá. *Rev. Biol. Trop.* **2001**, *49*, 621–628.
530. Hartmann-Schröder, G. Zur kenntnis des eulitorals des Australischen küsten unter besonderer berücksichtigung der Polycheten und Ostracoden. Teil 2. Die Polychaeten der tropischen Nord westküste Australiens (zwischen Derby im Nordem und Port Hedland im Südem). *Mitt. Hambg. Zool. Mus. Inst.* **1979**, *76*, 75–218.
531. Hartmann-Schröder, G. Die Polychaeten der subtropisch-tropischen bis tropischen Ostkueste Australiens zwischen Maclean (New South Wales) und Gladstone (Queensland) sowie von Heron Island (Grosses Barrier Riff) IN: Hartmann-Schröder, Gesa and Gerd Hartmann. Zur Kenntnis des eulitorals der australischen Kuesten unter besounderer Beruecksichtigung des Polychaeten und Ostracoden. Teil 16. *Mitt. Hambg. Zool. Mus. Inst.* **1991**, *88*, 17–71.
532. San Martin, G. Exogoninae (Polychaeta:Syllidae) from Australia with the description of a new genus and twenty-two new species. *Rec. Aust. Mus.* **2005**, *57*, 39–152. [CrossRef]
533. San Martín, G.; Hutchings, J.A.; Aguado, M.T. Syllinae (Polychaeta: Syllidae) from Australia. Part 1. Genera *Branchiosyllis*, *Eurysyllis*, *Karroonsyllis*, *Parasphaerosyllis*, *Plakosyllis*, *Rhopalosyllis*, *Tetrapalpia* n.gen., and *Xenosyllis*. *Rec. Aust. Mus.* **2008**, *60*, 119–160. [CrossRef]
534. San Martín, G.; Hutchings, P.; Aguado, M.T. Syllinae (Polychaeta: Syllidae) from Australia. Part 3. Genera *Alcyonosyllis*, Genus A, *Parahaplosyllis*, and *Trypanosyllis* (*Trypanobia*). *Zootaxa* **2010**, *2493*, 35–48. [CrossRef]
535. Musco, L.; Giangrande, A. A new sponge-associated species, *Syllis mayeri* n. sp.(Polychaeta: Syllidae), with a discussion on the status of *S. armillaris* (Müller, 1776). *Sci. Mar.* **2005**, *69*, 467–474. [CrossRef]
536. Musco, L.; Giangrande, A. Mediterranean Syllidae (Annelida: Polychaeta) revisited: Biogeography, diversity and species fidelity to environmental features. *Mar. Ecol. Prog. Ser.* **2005**, *304*, 143–153. [CrossRef]
537. Bianchi, C.N. Biodiversity issues for the forthcoming tropical Mediterranean Sea. *Hydrobiologia* **2007**, *580*, 7–21. [CrossRef]
538. Mikac, B.; Musco, L. Faunal and biogeographic analysis of Syllidae (Polychaeta) from Rovinj (Croatia, northern Adriatic Sea). *Sci. Mar.* **2010**, *74*, 353–370. [CrossRef]
539. Hutchings, P.A.; Lavesque, N. I know who you are, but do others know? Why correct scientific names are so important for the biological sciences. *Zoosymposia* **2020**, *19*, 151–163. [CrossRef]
540. Lattig, P.; San Martín, G.; Martin, D. Taxonomic and morphometric analyses of the *Haplosyllis spongicola* complex (Polychaeta: Syllidae: Syllinae) from Spanish seas, with re-description of the type species and descriptions of two new species. *Sci. Mar.* **2007**, *71*, 551–570. [CrossRef]
541. Lattig, P.; Martin, D. A taxonomic revision of the genus *Haplosyllis* Langerhans, 1887 (Polychaeta: Syllidae: Syllinae). *Zootaxa* **2009**, *2220*, 1–40. [CrossRef]
542. Muniz, P.; Pires, A.M.S. Trophic structure of polychaetes in the São Sebastião Channel (southeastern Brazil). *Mar. Biol.* **1999**, *134*, 517–528. [CrossRef]
543. Pagliosa, P.R. Another diet of worms: The applicability of polychaete feeding guilds as a useful conceptual framework and biological variable. *Mar. Ecol.* **2005**, *26*, 246–254. [CrossRef]
544. Giangrande, A.; Licciano, M.; Pagliara, P. The diversity of diets in Syllidae (Annelida: Polychaeta). *Cah. Biol. Mar.* **2000**, *41*, 55–65.
545. Álvarez-Campos, P.; Fernández-Leborans, G.; Verdes, A.; San Martín, G.; Martin, D.; Riesgo, R. The tag-along friendship: Epibiotic protozoans and syllid polychaetes. Implications for the taxonomy of Syllidae (Annelida), and description of three new species of *Rhabdostyla* and *Cothurnia* (Ciliophora, Peritrichia). *Zool. J. Linn. Soc.* **2014**, *17*, 265–281. [CrossRef]

546. Siddall, M.E.; Aguado, M.T. Molecular phylogenetic evidence of a haplosporidian parasite infecting the polychaete *Syllis nipponica* (Imajima, 1966). *Parasitol. Res.* **2006**, *99*, 309–312. [CrossRef]
547. Lattig, P.; Martin, D. Sponge-associated *Haplosyllis* (Polychaeta: Syllidae: Syllinae) from the Caribbean Sea, with the description of four new species. *Sci. Mar.* **2011**, *75*, 733–758. [CrossRef]
548. Paresque, K.; Nogueira, J.M.M. The genus *Haplosyllis* Langerhans, 1879 (Polychaeta: Syllidae) from northeastern Brazil, with descriptions of two new species. *Mar. Biol. Res.* **2014**, *10*, 554–576. [CrossRef]
549. Glasby, C.J.; Aguado, M.T. A new species and new records of the anthozoan commensal genus *Alcyonosyllis* (Polychaeta: Syllidae). *Beagle Rec. Mus. Art Gall. North. Territ.* **2009**, *25*, 55–63.

diversity

MDPI

Review

The Early Branching Group of Orbiniida Sensu Struck et al., 2015: Parergodrilidae and Orbiniidae

Miguel A. Meca [1,*], Anna Zhadan [2] and Torsten H. Struck [3]

1 Department of Natural History, University Museum of Bergen, University of Bergen, Postboks 7800, 5020 Bergen, Norway
2 N.A. Pertsov White Sea Biological Station, Biological Faculty, M.V. Lomonosov Moscow State University, Leninskie Gory 1–12, 119234 Moscow, Russia; azhadan@wsbs-msu.ru
3 Frontiers of Evolutionary Zoology Research Group, Natural History Museum, University of Oslo, 0562 Oslo, Norway; t.h.struck@nhm.uio.no
* Correspondence: Miguel.Meca@uib.no

Abstract: This review addresses the state of the art of the systematics and the improvements in the biology, ecology and species diversity of the two annelid taxa Parergodrilidae and Orbiniidae, the early branching group of Orbiniida sensu Struck et al., 2015 according to molecular studies. An effort to identify gaps of knowledge is given to understand the distribution, dispersal and the diversity Parergodrilidae and Orbiniidae hold, as well as to give several directions for future research. Parergodrilidae is a taxon of interstitial annelids constituted by the terrestrial *Parergodrilus heideri* (monotypic genus up to date), reported throughout Europe but also in Korea and North America, and the genus *Stygocapitella*, which includes eleven species from the upper shore of sandy beaches distributed along Europe and other regions of the world. Orbiniidae contains more than 200 described species spread over 20 valid genera, varying in size from a few millimeters up to 30 cm, distributed globally and living in a wide variety of soft bottoms. Improving the knowledge on these two sister-taxa is crucial for the understanding of the evolution to interstitial forms by progenesis in Annelida.

Keywords: biodiversity; biology; distribution; diversity; ecology; phylogeny; taxonomy

Citation: Meca, M.A.; Zhadan, A.; Struck, T.H. The Early Branching Group of Orbiniida Sensu Struck et al., 2015: Parergodrilidae and Orbiniidae. *Diversity* **2021**, *13*, 29. https://doi.org/10.3390/d13010029

Received: 17 November 2020
Accepted: 21 December 2020
Published: 13 January 2021

1. Introduction

While abounding the concepts of Sedentaria and Errantia, Fauchald [1] erected Orbiniida comprising Orbiniidae, Paraonidae and Questidae due to the lack of antennae and palps and possession of an eversible pharynx and biramous parapodia with simple chaetae. All three taxa had before been regarded as part of Sedentaria, e.g., [2–4]. Based on their morphological-cladistic analyses, Rouse and Fauchald [5] regarded all three taxa also as a monophyletic group within Scolecida. Molecular data placed Questidae within Orbiniidae but Paraonidae as probably more closely related to Cirratulida see [6,7]. Struck et al. [8] re-erected the groups Sedentaria and Errantia but with changes in their taxon composition; Orbiniidae was in this study part of Errantia. However, a following study showed that this was caused by a single erroneously assigned paralogous gene and instead Orbiniidae had to be placed in Sedentaria [9]. Subsequent phylogenomic studies generally found Orbiniidae as sister to or being part of the sistergroup to the remaining Sedentaria [7,10,11].

As part of this placement within Sedentaria, Orbiniida has been recognized again, but with a different taxon composition [12]. Besides Orbiniidae (including Questidae), it now comprises several interstitial annelid groups, each of which were at one point assigned to the polyphyletic Archiannelida (see [7,13] and the article on interstitial annelids in this issue for a more detailed discussion). Namely, these taxa belonging to Orbiniida are Parergodrilidae, Dinophilidae, *Diurodrilus*, *Apharyngtus* and Nerillidae [12]. According to the authors, the evolution of these groups is best explained by recurrent independent events of progenesis, which most likely happened in an orbiniid-like ancestor. However, depending on the analytical strategy applied some recent analyses showed that Dinophilidae are placed outside Orbiniida and together with *Lobatocerebrum* as sister to Pleistoannelida, while independent of the strategy Parergodrilidae, *Apharyngtus* and *Diurodrilus* remained closely related to Orbiniidae [11]. Nerillidae were not included in the analyses and, hence, no data have been provided yet excluding them from Orbiniida. Of all interstitial taxa possibly placed within Orbiniida, the close relationship of Parergodrilidae and Orbiniidae is the best supported by molecular data [12,14–21].

In conclusion, Orbiniida as of today consists of Orbiniidae, Parergodrilidae, Nerillidae, *Apharyngtus* and *Diurodrilus* as well as possibly Dinophilidae [7,10–12]. However, for reasons of consistency all interstitial taxa except Parergodrilidae are treated in the article on interstitial annelids in this issue. Here, we will review the recent advances of our knowledge about the biology, ecology, taxonomy and species diversity as well as identify relevant gaps of knowledge to understand the species diversity, distribution and dispersal of Orbiniidae and Parergodrilidae.

2. Parergodrilidae Reisinger, 1925

Parergodrilidae is a taxon of small-sized, stout annelids of 0.8 to 2.8 mm body length to 0.1 to 0.25 mm body width, which until very recently consisted of only two species in two genera, namely *Stygocapitella subterranea* Knöllner, 1934 and *Parergodrilus heideri* Reisinger, 1925. The latter is a terrestrial polychaete, while the former lives in the transition zone from the marine environment to the terrestrial habitat. Parergodrilidae was in its first description placed within Archiannelida [22], while others regarded them as clitellates [23]. Moreover, before Karling [24] it was not recognized that *S. subterranea* and *P. heideri* constitute one taxon and *S. subterranea* was regarded as closely related to Capitellidae [25]. After Karling [24] Parergodrilidae was often considered as closely related to Ctenodrilidae but without the indication of morphological autapomorphies, e.g., [1,26]. Alternatively, a closer relationship to *Hrabeiella periglandulata*, another terrestrial polychaete species, has been suggested [27]. However, none of the relationships was strongly supported by morphological data, e.g., [28–30]. Accordingly, the morphological-cladistic analyses by Rouse and Fauchald [5] found Parergodrilidae as *in certae sedis* within polychaetes. Nowadays, however, as mentioned above molecular data unequivocally support a sistergroup relationship to Orbiniidae.

2.1. *Parergodrilus* Reisinger, 1925

Parergodrilus heideri was first described from leaf litter samples in Austrian beech forests by Reisinger [22] (Figure 1D). It had been assumed for some time that the species is restricted to this kind of habitat, specifically the leaf-litter, organic soil layer of montane beech forests, which is humid, but not water-logged, and slightly alkaline to moderately acid, e.g., [30–35]. However, the species has now been recorded from different habitats including different forest types like spruce and maple forests, low-land, wetter and more acidic ones, e.g., [31,36–40]. Hence, the original restricted distribution could reflect more biases in sampling than restriction of the species. Nonetheless, the species seems to occur only in the humus-rich upper layer of soil, often within the first three to six centimeters.

Interestingly, there is a strong difference in the sex ratio favoring females with ratios of 1:8 to 1:170 [28–30,33,41]. Accordingly, in the beginning only female specimens were found and considered to be hermaphrodites [22]. However, later these were recognized as females and *P. heideri* is generally regarded as being gonochoristic, while some doubts remain about the latter conclusion [24,33,42–45]. Different reasons for this pronounced sex ratio imbalance have been suggested, which include differences in behavior, longevity and size [33]. In Austria, males have only been found during the summer months, while in a German beech forest, mature females with vitellogenic oocytes and spermatozoa in the receptacula seminis have been found throughout the year [33,45]. Hence, it is uncertain if reproduction occurs throughout the year, only during the summer period or only insemination occurs during the summer months with spermatozoa stored in the receptacula seminis throughout the year. Eggs are deposited in cocoons, which are attached to the substrate. The eight to ten day-long development is direct and four-chaetiger stages of 60–80 µm body length hatch from the cocoon. Early cleavage resembles an unequal spirale cleavage [33].

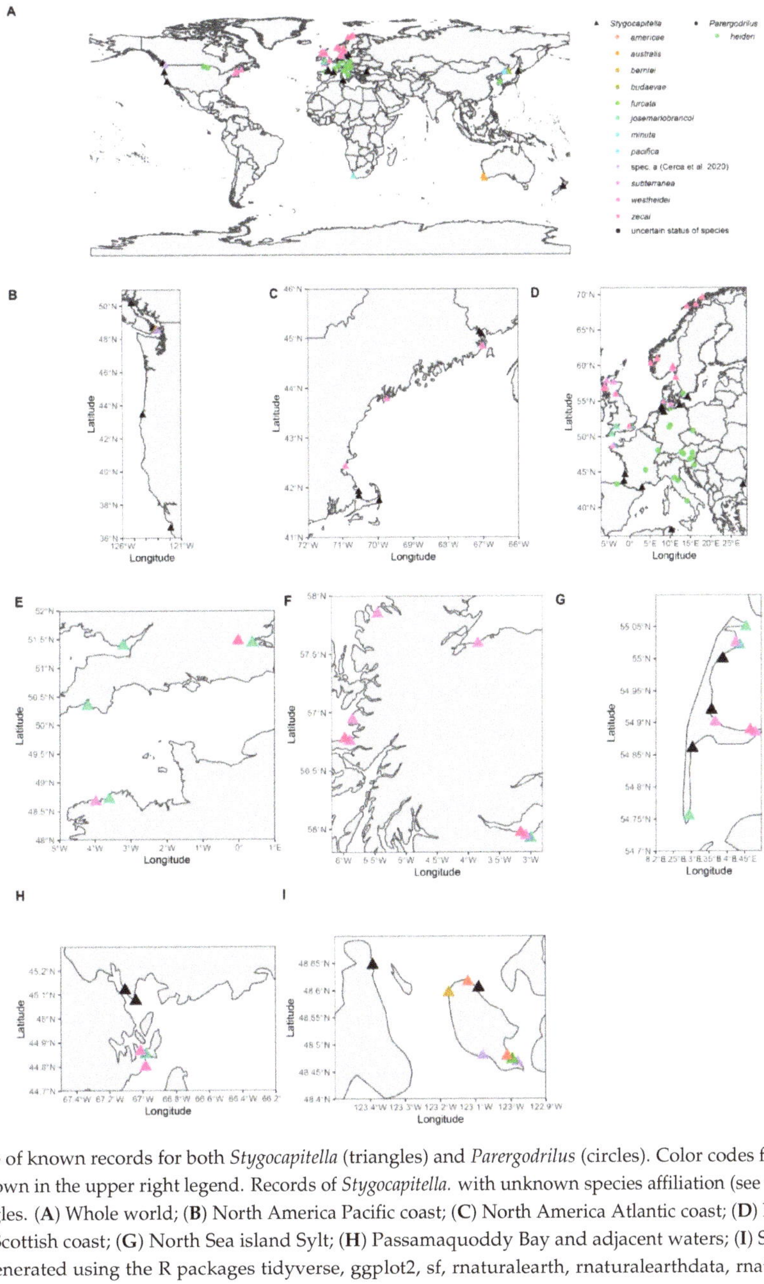

Figure 1. Map of known records for both *Stygocapitella* (triangles) and *Parergodrilus* (circles). Color codes for the different species are shown in the upper right legend. Records of *Stygocapitella.* with unknown species affiliation (see text) are shown as black triangles. (**A**) Whole world; (**B**) North America Pacific coast; (**C**) North America Atlantic coast; (**D**) Europe; (**E**) The Channel; (**F**) Scottish coast; (**G**) North Sea island Sylt; (**H**) Passamaquoddy Bay and adjacent waters; (**I**) San Juan Island. Maps were generated using the R packages tidyverse, ggplot2, sf, rnaturalearth, rnaturalearthdata, rnaturalearthhires, ggmap and ggpubr.

In *P. heideri*, males possess ten chaetigers (Figure 2), while females have only eight. Both also have an additional achaetigerous segment after the peristomium [33]. Each chaetigerous segment has a pair of bundles, which contain two very simple, oligochaete-like chaetae (Figure 2). For a more detailed description of the morphology and internal anatomy please see Purschke [45]. As there is at present only a single species recognized in the genus *Parergodrilus*, no morphological diagnostic features for species identification are known. However, the possible presence of cryptic species has been suggested. *Parergodrilus* can be differentiated from *Stygocapitella*, based on several characters, which among others include the presence of simple chaetae, the lack of bilimbate and whipped chaetae as well as of two praepygidial achaetigerous segments (Figure 2).

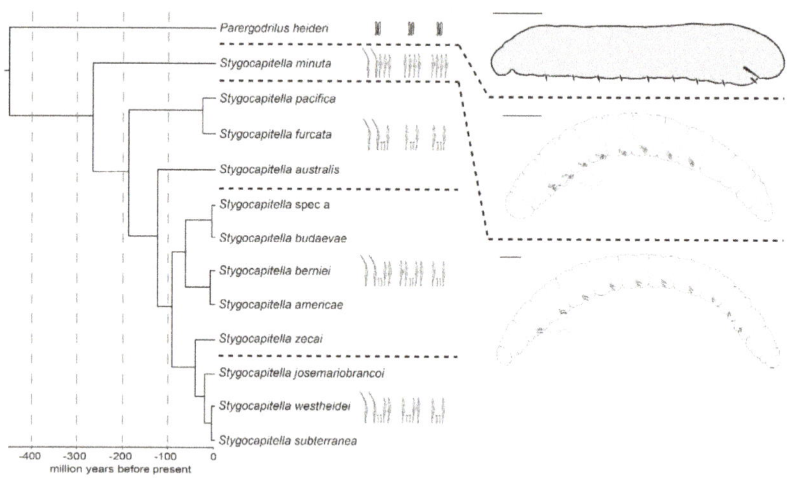

Figure 2. Parergodrilidae based on different sources (Cerca et al. [46]; Cerca et al. [47]; Struck et al. [48]). Besides the species names the different chaetal patterns are shown for the 1st, 2nd and 3rd chaetiger. All following chaetigers have the same pattern as shown for the 3rd. Schematic drawings of the general morphology of *Parergodrilus* and the two different ones for *Stygocapitella* are based on Reisinger [33] and Struck et al. [48]. The scale bars equal 125 μm.

Distribution

Besides the first records from Austria, *P. heideri* has been found throughout Europe including records from Spain, Italy and Croatia in the South to Sweden in the North (Figure 1D), e.g., [28,30–32,34,36,38–41,49,50]. Recently, also first findings outside Europe have been reported, specifically in Korea and North America (Figure 1A) [37,51]. Hence, the known distribution of *P. heideri* has been strongly expanded in recent years. Even though *P. heideri* is very well studied with respect to morphology, anatomy, and early development by comparison to other meiofaunal species for more details see [45] and is relatively well documented from Europe, many questions concerning the genus' species composition and diversity still remain uncertain or unanswered. The increasing records from atypical habitats indicate that the habitat restrictions are less rigorous than originally considered. Hence, what are the habitat requirements for this species? How sensitive is it truly to desiccation and flooding, different forest types and soil conditions? The center of distribution thus far seems to be Europe and the distribution is seemingly restricted to the Northern hemisphere. Is the lack of *Parergodrilus* species from the Southern hemisphere and the sparse distribution outside Europe true or does it just reflect a sampling bias with stronger efforts having taken place in Europe? Finally, a thorough taxonomic revision of the species including also molecular data from all records around the world and from different habitats has not been conducted yet. Hence, it is uncertain if cryptic species are present within this species and if such cryptic species could be associated with different habitat preferences and/or different regions. For example, do the records outside Europe truly represent the same species? In the same vein, the molecular data will also allow one to assess if the records outside Europe could indicate recent invasions, for example, alongside earthworm introductions or if the species are native species in these areas. Schlaghamerský and Frelich [37] regard the species as native as they regard transport of such small annelids as part of commercial fish bait of anglers (i.e., earthworms) as not very likely. However, they also state, "an initial introduction to North America with soil from Europe cannot be fully excluded".

2.2. *Stygocapitella* Knöllner, 1934

Stygocapitella was first described by Knöllner [25] from a wide sandy beach with medium coarse sediments at the German Baltic coast (Figure 1D). Following records also found *Stygocapitella* at similar sandy beaches with medium coarse sediments (Figure 3A–C), e.g., [24,52–56]. However, more extensive sampling also at beaches, which are not wide with only medium coarse sediments, revealed that *Stygocapitella* also occurred at them [30,43,52,57], sometimes in high abundance (personal observations). These beaches could be wide or narrow, with or without stones and pebbles on top of or intermixed with medium-coarse sediment (Figure 3D–G). Usually the specimens can be found around or up to 18 m above high tide level, but at some beaches of the Thames (i.e., Cutty Sark, UK) and the Nærøyfjord (Bakka, Norway) they are found substantially below high tide level or even at low tide level (personal observations). In the beaches, they occur usually in the first 20 cm of depth but can also be found as deep as 1 m [30,58,59]. Throughout the year the abundance is around 10–20 specimens/100 cm^3 in most beaches investigated in more detail thus far, but it can occasionally be three to four times higher [30,58,59]. During winter months, the specimens migrate to deeper layers to avoid freezing and the same migration pattern can be observed to avoid desiccation in the uppermost layers [30,58,59]. Generally, *Stygocapitella* seems to prefer the zone of damp sand in the beach with 3% water content in relation to the dry weight of the sediment, but they seem to sustain also water content up to 21.5% [30]. The latter point is further supported by records closer to low tide level, which means that for at least a few hours each day these animals experience complete water saturation. In summary, *Stygocapitella* is usually found in medium coarse sediments in the damp zone of the upper shore of sandy beaches but occasionally also in sediments with higher water content. These beaches can be differently exposed to tides, small or wide, with or without stones and pebbles.

In contrast to *P. heideri*, the sex ratio in *Stygocapitella* is 1:1 and no dimorphism is visible. Reproduction seems to take place throughout the year as all juvenile stages can be observed in all months. The direct development resembles the one of *P. heideri* with eggs being deposited in cocoons and hatching at the four-chaetiger stage. However, the hatchlings are larger with a body length of about 0.75 mm and do not possess the full complement of chaetae in the first two segments, which develop in the later juvenile stages. Cleavage pattern and duration of development are unknown. In general, 70–90% of the populations are immature adults [45].

The first described species of *Stygocapitella* was *S. subterranea* by Knöllner [25]. *Stygocapitella subterranea* possesses ten chaetigers plus one achaetigerous segment after the peristomium and two achaetigerious before the pygidium. Three different types of chaetae are present: whipped-like chaetae, bilimbate chaetae and forked chaetae (Figure 2). The chaetal composition at the chaetigers shows a specific pattern in each pair of bundles. At the first chaetiger, each bundle possesses two whipped-like, two forked and two bilimbate chaetae. The second one has one bilimbate, two forked and two bilimbate chaetae. Finally, the third and all following ones have one bilimbate, two forked and one bilimbate chaetae (Figure 2). For a more detailed description of the morphology and internal anatomy please see Purschke [45].

Figure 3. Pictures of the different beaches with known records of *Stygocapitella*. (**A**) Medium coarse sediment from Hausstrand, North Sea island Sylt; (**B**) Ellenbogen, North Sea island Sylt; (**C**) Sarge Bay, Australia; (**D**) Bakka, Norway; (**E**) Reid State Park, San Juan Island; (**F**) 4th of July beach, San Juan Island; (**G**) Roche Harbor, San Juan Island.

2.2.1. Distribution

After the first description of *S. subterranea* from the German Baltic Sea [25], this species has also been found at numerous additional beaches in Europe including the North, Baltic, Mediterranean, and Black Sea and the European Atlantic coast (Figure 1D) [24,30,44,50,53,55,58–67]. It has also been found at several beaches outside Europe, in specific North America, Australia, New Zealand and Japan (Figure 1A–C) [30,52,54,56,57,68]. Interestingly, all records in both hemispheres are so far restricted to the boreal and temperate zones barely reaching into the subtropical zone (Figure 1A). As there have been several attempts focusing on this species, the distribution gap in the tropics and subtropics seem not to be a sampling effect [45]. All of these records were assigned to *S. subterranea*. Therefore, *Stygocapitella* was often considered a typical example of the Meiofauna paradox as it had a cosmopolitan distribution but seemingly no active or passive long-distance dispersal stage [57,60,69,70]. However, a first molecular study with one population each from the US Pacific, US Atlantic and European Atlantic coast indicated that no gene flow between these three populations occurs [57], but no taxonomic action was taken. Additionally, analyses of sperm ultrastructure revealed no differences between the same three populations (Purschke personal communication).

2.2.2. Taxonomy

The first record of *Stygocapitella*, which was not assigned to *S. subterranea*, was *S. minuta* Struck et al., 2017, which was found in South Africa (Figure 1A) [48]. The specimens are clearly different from *S. subterranea* as *S. minuta* has only eight chaetigers but four achaetigerous segments in front of the pygdium (Figure 2). Moreover, mature adults are only about 1.5 mm long and hence much smaller than *S. subterranea*. Forked chaetae are lacking altogether and two additional bilimbate chaetae are present instead. As part of this study specimens from the Australian localities were investigated anew and it could be shown that even though they were overall similar to specimens from the type locality they differed in the chaetal composition. They have one bilimbate chaeta less in the first and second chaetiger (Figure 2). Therefore, these specimens were described as *S. australis* Struck et al., 2017 (Figures 1A and 2). The morphological results were supported by molecular data showing that there were deep divergences between the three species dating back about 260 and 80 million years, respectively.

This study was followed by a more thorough revision of the genus *Stygocapitella* by Cerca et al. [46] including several populations from both North American coastlines, Europe and a new record from the Russian Pacific coast. This study found that besides the *S. minuta*-, *S. australis*- and *S. subterranea*-morphotypes a fourth morphotype can be observed, which is very similar to the *S. subterranea*-morphotype but differs in the chaetal composition of the second chaetiger by possessing one more bilimbate chaeta (Figure 2). In addition, a few species are smaller than others. However, species delimitation in this study had to mostly rely on molecular data using two mitochondrial and two nuclear markers. In total, twelve different species could be differentiated of which eight were described anew. One species could not be described as no holotype could be assigned. Two of these eight species, *S. pacifica* and *S. budaeva*, were completely new to science as they were collected from the Russian Pacific coast, which has not been sampled previously (Figure 1A). Along the North American Pacific coast, three species, *S. furcata*, *S. berniei* and *S. americae*, were described (Figure 1A,B). Hence, including the undescribed species four species previously assigned to *S. subterranea* occur on San Juan island alone (Figure 1I). At the Northern European coastlines, two additional species, *S. josemariobrancoi* and *S. zecai*, are now recognized besides *S. subterranea* (Figure 1A,D,E–G). The European species *S. josemariobrancoi* also has a record from the North American Atlantic coast (Figure 1H), where *S. westheidei* has also been found (Figure 1H). Additional support for these delimitations is gained from the fact that species occur in sympatry at each of these coastlines, which in this case means that they occur within the same beach, often within the same 50 cm^3 of sediment (Figure 1A,F–H). Hence, the number of formally described species has increased from one to eleven in recent days and only three of them are due to new records, while the remaining ones have been recorded as *S. subterranea* before. All species are cryptic species except for *S. minuta*, as it is the only one, which can unaminously be distinguished from all other species based on its morphology [71]. Accordingly, the previously assumed cosmopolitan distribution of *S. subterranea* is no longer given. While several species still show a relatively broad distribution along European, American or Australian coastlines, other species (i.e., *S. pacifica*, *S. budaeva*, *S. furcata*, *S. berniei*, *S. americae* and *S. minuta*) seem to have a very restricted distribution. However, in these latter cases only one population (*S. pacifica*, *S. budaeva*, and *S. minuta*) or very few populations in close vicinity to each other (*S. furcata*, *S. berniei*, and *S. americae*) have been studied. Finally, all previous records, which have been assigned to *S. subterranea* (black triangles in Figure 1), need to be investigated again to assign them to one of these species or even identify a new one. As the morphological characters to differentiate the species from each other are very limited and except for *S. minuta* restricted to the chaetal pattern at the first two chaetigers and to some degree body size, these new analyses need to include molecular data to be more reliable. As long as this has not been done these records should be treated as *Stygocapitella* sp.

Besides the geographic distribution, the temporal distribution has also been studied in this complex of cryptic species [47]. It was confirmed that the radiation of *Stygocapitella* took place about 260 million years ago (Figure 2). More interestingly, the three morphotypes each represented by several species have not changed for at least 18 million (*S. subterranea*-morphotype) to 140 million years (*S. australis*-morphotype). [47]. This can also be shown when the morphological disparity is compared to genetic distance. The morphological disparity in this group is about five to eight times lower than, for example, in its sistergroup Orbiniidae or another interstitial group, Nerillidae (Figure 4) [47]. Even at very high genetic distance, it is only half the disparity of these two groups. Hence, these species are clearly affected by morphological stasis and include species with the longest known period of stasis namely *S. pacifica*, *S. furcata* and *S. australis* [46,47].

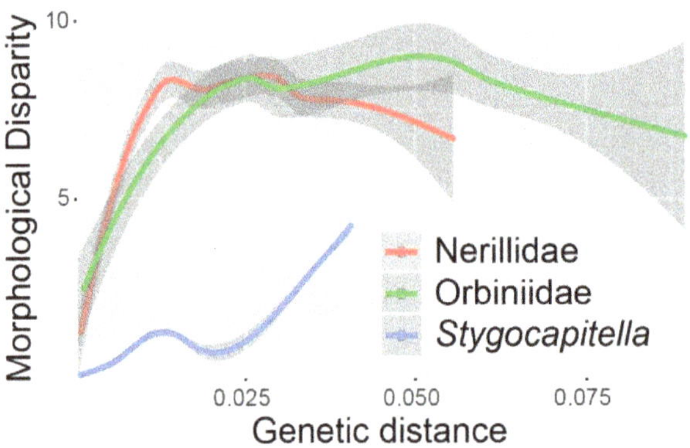

Figure 4. A simplified re-drawing of Figure 5C of Cerca et al. [47]. The plot shows the relationship of morphological disparity measured by pairwise MMD indices in relation to genetic distance measured by pairwise genetic distances of the nuclear 18S rRNA marker. The values for *Stygocapitella* are compared to the sistergroup of Parergodrilidae, Orbiniidae, and another interstitial annelid group, Nerillidae. The grey area indicates the 95% confidence interval of the fit (line) to the data points (not shown).

2.2.3. Conclusions

Besides large-scale differences in the distribution along different oceanic coastlines, no other differences, for example in macroecological factors like annual average temperature, could be detected so far between the species [46,47]. Hence, it is not certain yet, which factors drive distribution, speciation and stasis in these species. However, detailed studies of their different interstitial environments are lacking. The first studies in Europe in this respect [30,58,59,61,62] can only be starting points to investigate these relationships and dependencies and in the light of the new taxonomy the sampling is not sufficient, because what was thought before to be the result of one species is now representing populations from three different species and also includes some sympatric populations. For example, the beaches on the North Sea Island sampled in these studies contain only very rarely *S. subterranea*, but *S. josemariobrancoi* or species of uncertain status (Figure 1G) [30,58,59].

The dates listed above for divergence of morphotypes or species are too young to be able to explain the recent day distribution of *Stygocapitella* species by vicariance [46,48]. Hence, this distribution must have been established by occasional long-distance dispersal events across oceans including at least two times dispersal across the equator leading to the present-day distribution gap [46]. However, how this was accomplished is still uncertain. Moreover, the colonization of *S. josemariobrancoi* of the North American Atlantic coast must have happened very recently as these specimens share identical haplotypes with European specimens across all molecular markers [46]. This could have possibly been human-mediated distribution, maybe by ballast sand. However, this scenario seems not to occur otherwise in *Stygocapitella* [72].

In summary, this whole group is very well suited to study the factors causing morphological stasis as well as the drivers of speciation in cryptic species as it allows studying change and stasis at different time scales using morphological, ecological, biogeographic, reproductive biological, developmental, physiological, population dynamic and genomic data. These kinds of data will also help to understand if the different species occupy different microniches allowing to survive in sympatry. However, these kinds of data are also lacking to a very large degree and research in this direction is only just beginning. Moreover, given the recent results there is a high probability that the number of species in this genus will still substantially increase. Finally, due to strong sampling efforts many records are known from Europe and North America, but this is not the case on the other continents. Given the recent findings in South Africa and Russia it is very likely that they also occur in the temperate zones of the Southern hemisphere and along the Western Pacific coastlines is East Asia.

3. Orbiniidae Hartman, 1942

Orbiniidae are sedentary annelids which can be distinguished generally from the rest of Annelida by a rounded or pointed prostomium without appendages, one or two achaetous rings of the peristomium (Figure 5A), body separated into thorax and abdomen bearing biramous parapodia (Figure 5B) and branchia disposed dorsally in the abdomen (Figure 5C). This general scheme fits with the medium- to large-sized genera like *Scoloplos, Leitoscoloplos, Leodamas, Naineris, Orbinia* and *Phylo* (Figure 5D–F), being at the same time the most frequently reported genera and also the most species-rich genera. The adults of most of the species of these genera measure between 3.5 and 20 cm in length, though others can be up to 30 cm long (e.g., some *Orbinia*). On the other hand, in the small-sized taxa (with adults measuring few mm) like in *Orbiniella* or *Pettibonella* (Figure 5G), the distinction of the body regions is weak or lacking and some species do not possess branchia (Figure 5H). Moreover, *Orbiniella branchiata* Hartman, 1967 [73] bears three achaetous rings in the peristomium. The presence of crenulations in the capillary chaetae, the autapomorphy of the family, is characteristic to all genera independent of their size and defines the attribution of a species to Orbiniidae [6,74] (see the part on chaetae typology in the discussion on taxonomical characters from this subchapter for details).

The most relevant regional identification keys for orbiniids are those of Day [3] for South Africa, Day [75] for North Carolina (USA), Day [76] and Zhadan [77] for Australia and New Zealand, Kirkegaard [78] for the North Sea, Blake [79] for Southern California (USA), Lopez [80] for the Western Mediterranean and Diaz-Diaz et al. [81] for the Caribbean Sea. A number of keys for the different orbiniid groups are available, such as Gillet [82] and Parapar et al. [83] for *Orbiniella*, Solis-Weiss and Fauchald [84] for Protoaricinae, Sun and Li [85] for 22 species of *Orbinia*, Sun et al. [86] for 30 species of *Leodamas* and Blake [87] for the seven deep-water species of *Leitoscoloplos* from the Eastern Pacific.

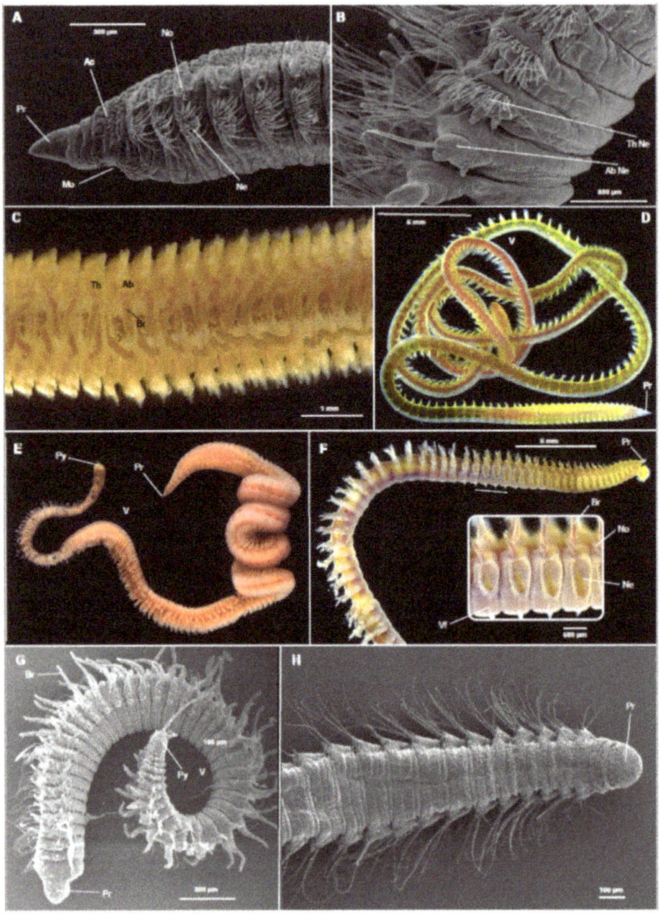

Figure 5. Scanning electron microscopy (SEM) and light microscopy images of different orbiniids. (**A**). *Scoloplos armiger*, anterior end. (**B**). *Orbinia* cf. *armandi*, transition thorax-abdomen. (**C**). *Naineris laevigata*, detail of the dorsum with branchiae. (**D**). *Leodamas chevalieri*, general view of a living specimen. (**E**). *Scoloplos armiger*, general view of a living specimen. (**F**). *Phylo kuwaitica*, anterior-mid body view of a living specimen and detail of posterior thoracic chaetigers. (**G**). *Pettibonella multiuncinata*, general view. (**H**). *Orbiniella* sp., dorsum of the anterior end. Ab. Abdomen; Ab Ne. Abdominal Neuropodium; Ac. Achaetous ring; Br. Branchiae; Mo. Mouth; Ne. Neuropodium; No. Notopodium; Pr. Prostomium; Py. Pygidium; V. 1478 Ventrum; Vf. Ventral fringe. Photo E is not scaled.

3.1. Systematics

3.1.1. Morphological Era

Originally orbiniids carried the name Ariciidae Malmgren, 1867 [88] which has been subsequently changed by Hartman [89] to Orbiniidae, as the type genus of the family, *Aricia* Savigny 1822 [90], was preoccupied in Lepidoptera. *Orbinia* Quatrefages 1866 [91] had been designated as the type genus few years earlier [92]. The first important taxonomic work on Orbiniidae systematics belongs to Eisig [93], who reviewed previous works and suggested most of the currently accepted terminology. Later, Hartman [94] performed the revision of orbiniid systematics, in which she redefined all genera, reviewed most of the common species and divided the family into two subfamilies based on the number of the peristomial rings: Orbiniinae (one peristomial ring) and Protoariciinae (two peristomial rings). In the same year, Pettibone [95] reviewed several genera and species based on the material from the east coast of North America, describing three new species. The system of Orbiniidae suggested by Hartman [94] was largely accepted until the end of the 20th century and it is still the most comprehensive and detailed work on orbiniids. Nonetheless, there were important contributions to the orbiniid systematics. Day [75] reviewed the generic system for the subfamily Orbiniinae and Solis-Weiss and Fauchald [84] did the same for the subfamily Protoariciinae. Day [76] erected the genus *Leitoscoloplos* and Mackie [96] revised the genus, transferring some species to the genus *Scoloplos*. Gillet [82], Parapar et al. [83], and Blake [87] reviewed *Orbiniella* and Badalamenti and Castelli [97] *Schroederella*. Blake [98] described *Methanoaricia dendrobranchiata*, a new species and genus (monotypic up to date) reported from the seeps of the Gulf of Mexico. The discovery of this genus led to the new revision of the whole family. This resulted in the division of Orbiniidae into three subfamilies: Orbiniinae, Microrbiniinae and Methanoariciinae, established for the unusual *M. dendrobranchiata*. Blake [98] demonstrated that the number of rings in peristomium can vary during the ontogeny in several genera and excluded it from orbiniid taxonomy. Following Blake [79,98], many protoariciins represent the juveniles of other orbiniid genera suggesting synonymization of *Pararicia* with *Protoariciella* or *Naineris* in Blake [79].

3.1.2. Genetics Era

The first molecular phylogenetic study of Orbiniidae was conducted by Bleidorn et al. [99]. It included eight orbiniid species among other annelid genera and received strong support for a close relationship between orbiniids and Questidae Hartman, 1966 [100]. Currently, questids are considered part of Orbiniidae [6,101], constituting the genus *Questa*. Three molecular phylogenetic reconstructions focusing on relationships between orbiniid genera reported several genera as paraphyletic [19,101,102] without support for Hartman's [94] or Blake's [98] classifications. Bleidorn [19] confirmed the placement of *Methanoaricia dendrobranchiata* within Orbiniidae based on two genes (16S and 18S). Twenty species belonging to 11 different genera were included in the analyses, in which *Scoloplos, Leitoscoloplos, Orbinia* and *Phylo* were recovered as paraphyletic. An extension of this phylogenetic reconstruction was made by Bleidorn et al. [101] (Figure 6A) adding to the analyses four more genes (cox1, cox3, nad1 and nad4), six more species and a morphological character matrix. Similar results (i.e., paraphyly of the former genera except for *Naineris*, which was monophyletic) were obtained even with the expanded dataset, and most of the traditional characters revealed to be useless for phylogeny. The authors suggested that repeated loss of characters and progenesis were the main processes in the evolution of orbiniids. In addition, some of the small-sized species with two peristomial rings (i.e., belonging to Protoariciinae sensu Hartman, 1957) included in the analyses showed significant genetic divergence and, thus, good evidence of being valid species. However, one specimen identified as *Protoariciella uncinata* Hartmann-Schröder, 1962 [103] was found to be genetically identical with the individuals of *Leodamas tribulosus* (Ehlers, 1897) [104]. This has partly corroborated the hypothesis by Blake [79,98] that species/genera of Protoariciinae might be misidentified juvenile stages of Orbiniidae. The latest phylogeny focused on Orbiniidae by Zhadan et al. [102] included 53 specimens belonging to 33 species from 12 genera. Three NJ analyses performed independently for COI, 16S and 18S (Figure 6B–D) recovered all genera tested by Bleidorn et al. [101] as paraphyletic; however, the tree topologies were different in the two studies, with *Naineris* being paraphyletic in Zhadan et al. [102].

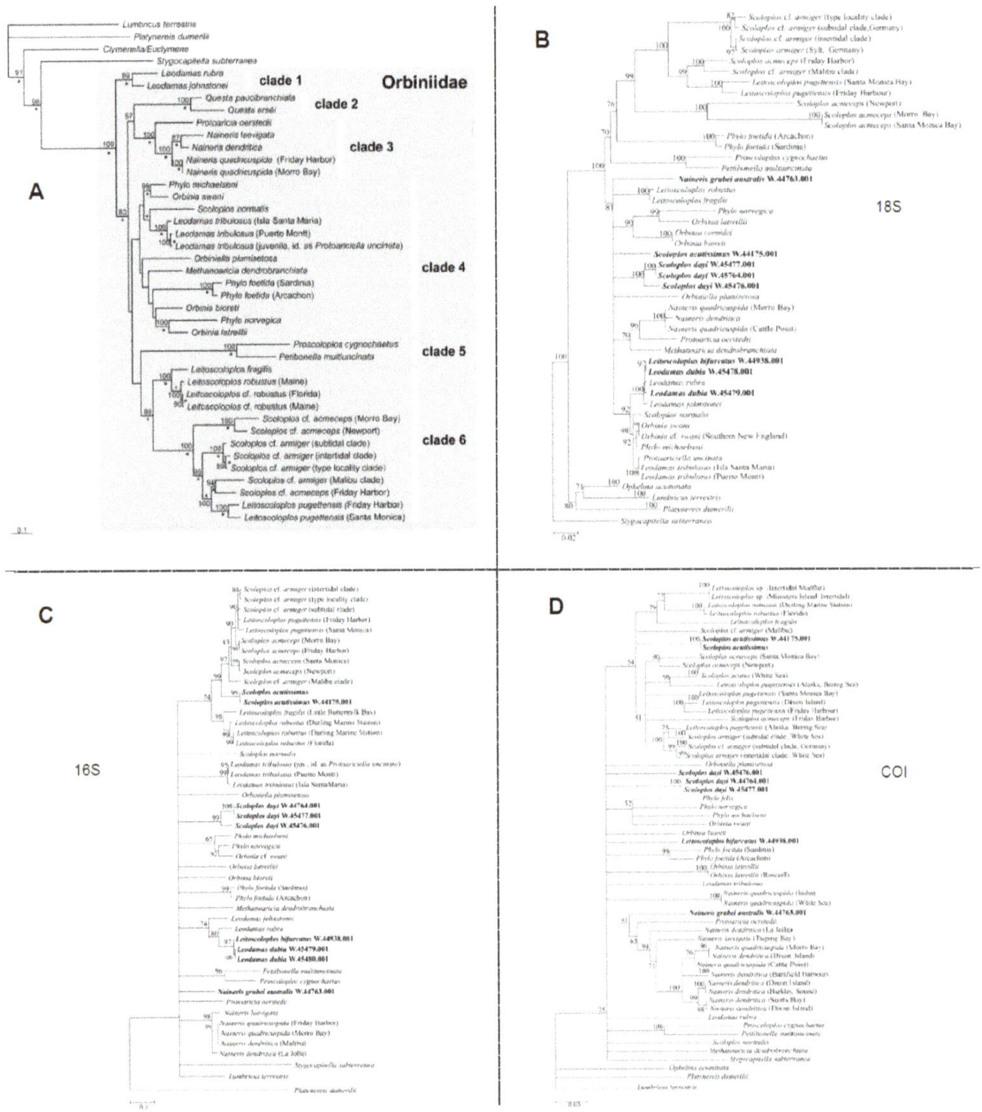

Figure 6. The four phylogenetic inferences performed for Orbiniidae in Bleidorn et al. [101] (**A**) and Zhadan et al. [102] (**B–D**). **A**. Maximum likelihood analyses based on six concatenated gene fragments (18S, 16S, cox1, cox3, nad1 and nad4), resulting in six orbiniid clades. Numbers at the nodes show bootstrap values from 1000 replicates and asterisks (*) indicate nodes with a posterior probability of >0.95 in the Bayesian analysis. (**B–D**). Three Neighbor Joining analyses based on individual genes (18S, 16S and COI) marked in each tree. The bootstrap values were obtained from 1000 replicates. Most of the clades from **A** changed in adding species in (**B–D**).

3.1.3. Current State: Traditional Taxonomy vs. Genetics

Currently, Orbiniidae comprise more than 200 accepted species within 20 valid genera (see the Tables S1 and S2 in Supplementary Materials). Following Bleidorn and Helm [6] these numbers should be taken "with a grain of salt" given the high number of upcoming papers describing new orbiniid fauna [87] and the paraphyly of most of the genera as shown by the molecular analyses. We consider the monotypic genus *Scoloplella* as nonvalid following Bleidorn and Helm [6], as it is undistinguishable from the juveniles of *Leitoscoloplos* and consequently synonymized. With his monographs of 2017 [74] and 2020 [87], Blake has updated the previously accepted system for Orbiniidae [98], emending the two subfamilies Orbiniinae (*Berkeleyia, Califia, Leitoscoloplos, Leodamas, Naineris, Orbinia, Phylo, Protoaricia, Schroederella, Scoloplella, Scoloplos* and *Uncorbinia*) and Microrbiniinae (*Microrbinia, Orbiniella, Proscoloplos* and *Pettibonella*), and reviewing the definition of ten genera. *Protoariciella* was included in Orbiniinae in Blake [74] as problematic and not considered in Blake [87] since only the type species *Protoariciella uncinata* bore thickened notopodial chaetae in the posteriormost segments, the diagnostic character for the genus. Furthermore, one individual identified as *P. uncinata* was shown to be genetically identical to *Leodamas tribulosus* in Bleidorn et al. [101]. The genera *Pararicia, Paraorbiniella* and *Questa* were not considered by Blake [74,87] in his redefinition of the orbiniid sufamilies, whilst *Scoloplella* was included into Orbiniinae.

The morphological characters used in the Blake's generic system generally were the same that Hartman used in her revision from 1957, namely: the shape of the prostomium, the number of thoracic segments, the position of the first pair of branchiae, the shape of the parapodia, the presence/absence of subpodial papillae and their number, and the chaetal characteristics. The characters show high degree of variation among orbiniid species and genera with some species having ambiguous generic placement. This makes the orbiniid system unresolved which has been suggested by Mackie [96] in his revision of *Leitoscoloplos* even before the molecular analyses emerged. *Leitoscoloplos multipapillatus* Alcántara and Solís-Weiss, 2014 [105] is an example of a species that combines diagnostic characters of two genera. The species does not have hooks in the thoracic neuropodia characteristic to *Leitoscoloplos* but bears subpodial and stomach papillae, which, together with the absence of the spear-like spines, suggests its placing within *Orbinia*. Another example of ambiguous attribution of a species to a genus is *Naineris setosa* (Verrill, 1900) [106], a large orbiniid with a *Naineris*-like rounded prostomium bearing only camerated capillaries in the thoracic neuropodia, the diagnostic character of *Leitoscoloplos*. Furthermore, *Orbinia sagitta* Leão and Santos, 2016 [107] was erroneously placed into *Orbinia* lacking subpodal lobes and having *Leodamas*-like uncini, being similar to *Leodamas sinensis* Sun, Sui and Li, 2018 [86] (authors' observations).

A phylogeny with sufficient taxon coverage and amount of genetic data is required as a first step for the revision of the orbiniid system. Once the robust phylogeny of the family is constructed, the morphological revision of the well-supported monophyletic clades, in search of synapomorphies, is necessary in order to provide supported system of orbiniids. It is worth saying that none of the known orbiniid species was originally described, both traditionally and currently, incorporating molecular analyses and the only available genetic data in public databases (as GenBank and BOLD) comes from the commented phylogeny papers of Bleidorn, Zhadan and collaborators [19,101,102], other few works as Kruse et al. [108], Meyer et al. [109] or Carr et al. [110] or from unpublished data.

3.2. Discussion on Taxonomical Characters

Naineris and *Protoaricia* are sister groups forming well supported clade on most phylogenetic trees [101,102]. This is the rare case in orbiniid phylogeny when morphological characters are congruent with molecular data. Besides the presence of statocysts mentioned in Bleidorn et al. [101], these two genera also share the dorsal ciliated ridges between branchial bases in abdominal segments and special chaetae in thoracic—subuluncini and cauduncini [93] (p. 216, Figure XIV); [94] (p. 219–221). Progenetic evolution is the best explanation for the similarities between *Protoaricia* and *Naineris* [19,101]. Phylogenetic studies with bigger taxon coverage of both genera are needed to confirm this statement. In the following, we discuss some taxonomical characters which are often overlooked or not described well but can be useful for phylogenetic analysis of Orbiniidae.

Dorsal organs are paired ciliated patches usually with a nonciliated mound between them, situated on all abdominal and sometimes on posterior thoracic segments. They are serially homologous to nuchal organs [6,94]. In living and sometimes in preserved worms, dorsal organs are marked with pigmentation; however, often the shape of the pigmented spots differs from the shape of ciliated dorsal organs seen with SEM. The dorsal organs are rarely illustrated and described but potentially can serve as genus- or species-specific character. For example, *Scoloplos* and *Leitoscoloplos* species have dorsal organs as strait or curved ciliated stripes, whereas *Naineris* bear five pairs of round ciliated spots [102].

The most promising source of characters useful for phylogenetic analysis of Orbiniidae are shape, number and distribution of chaetae, especially in thoracic neuropodia. Thoracic neurochaetae are organized in several transverse rows in most orbiniids. Hoffmann and Hausen [111] investigated the chaetal arrangement of three species of Orbiniidae belonging to different clades both in morphology-based and sequence-based phylogenies: *Scoloplos armiger*, *Orbinia latreillii* and *Pettibonella multiuncinata*. SEM and 3D reconstructions based on series of histological sections were used to reveal the position of different chaetae, their organizations in rows, the depth of their bases inside the parapodial tissue and the locations of formative sites of the chaetal rows. All three investigated species showed very similar patterns in the thoracic neuropodia: they had anterior main and posterior secondary transverse rows; main rows had a dorsally located formative site, which indicates homology with the transverse rows of other sedentary polychaete taxa. The number of chaetae and the depth of the chaetal bases decreased from frontal to caudal rows. Two additional secondary rows had their own formative site in the dorsoventral middle of the chaetal patch, and they were located along the caudoventral edge of the posteriormost main row and bent frontally. Interestingly, only capillary chaetae were present in the posterior main row and the posterior secondary row in all three investigated species whereas other rows could contain uncini as well as capillaries. The author supposed the described design of the thoracic neuropodia with main and secondary transverse rows represents the ancestral pattern for Orbiniidae [111]. This parapodial pattern is illustrated here on example of juvenile *Scoloplos armiger* (Figure 7A).

This approach suggested for the study and description of chaetal arrangement seems to be very promising. There were no such detailed investigations of other orbiniid genera, so it is hard to confirm or refuse that the general scheme works for all Orbiniidae; but good quality SEM photos sometimes are enough to reveal the chaetal position in other orbiniid taxa. The species of the genus *Leodamas* are the most convenient for SEM investigations of the thoracic neuropodia, as their chaetae usually are short and oriented more or less perpendicularly to a body wall. Unfortunately, in most orbiniids SEM images do not allow recording the chaetal arrangement in details, when anterior chaetae are long, oriented in caudal direction and cover the bases of posterior chaetae; also, chaetal rows can be not obvious, like in many *Scoloplos* and *Leitoscoloplos* species. In this situation special efforts are necessary to illustrate the chaetal position in neuropodia; it can be parapodia, mounted with different angles, parapodia with broken chaetae to show only their bases, serial histological sections with 3D reconstructions or confocal laser scanning microscopy (cLSM), which allows one to see chaetae themselves by autofluorescence as well as formative and degenerative zones with specific markers.

Here we present the attempt to analyze chaetal arrangement in thoracic neuropodia of different orbiniid taxa and compare them with Hoffmann and Hausen's [111] scheme using published SEM photos. *Leodamas acutissimus* (Hartmann-Schröder, 1991) [112] has thoracic neuropodia which are in good agreement with the orbiniid general pattern, including the presence of capillaries only in posterior main row and posterior secondary row. It has three of four main rows consisting of uncini, then a short main row consisting of capillaries which is located only in upper half of the neuropodia; and two secondary rows, anterior consisting of uncini and posterior row of capillaries, both located only in lower part of the neuropodia. The length of uncini increases from anterior to posterior rows. Unlike other orbiniids discussed here the last row is situated behind the papilla of postchaetal lamellae (Figure 7B). A similar pattern is found in thoracic neuropodia of *Orbinia camposiensis* Leão and Santos, 2016 [107]. It differs by a longer first secondary row which is bent frontally under the main rows and a reduced posterior secondary row with two capillaries only (Figure 7C). *Leodamas verax* Kinberg, 1866 [113] (the type species of the genus *Leodamas*) was redescribed with designation of the holotype by Blake [74]. Its thoracic neuropodia bear three main rows of uncini and one short secondary row, consisting of longer uncini; it is curved down frontally under the main rows; capillary chaetae are totally absent. Unlike general pattern of orbiniids sensu Hoffmann and Hausen [111], the length of the chaetae in the main rows decrease from anterior to posterior (Figure 7D). The same neuropodial arrangement was found in *Leodamas cirratus* (Ehlers, 1897) [104] (Figure 7E). Unlike *L. acutissimus* and *L. verax*, *Leodamas dubia* (Tebble, 1955) [114] from Australia had four main rows containing uncini and a tuft of two capillaries in the uppermost position of the third row. Uncini of the posterior row were the longest, which correspond with Hoffmann and Hausen's [111] description but no sign of secondary rows was seen; also, capillaries were located in the third, and not in the posterior row (Figure 7F). *Califia bilamellata* Blake, 2017 [74] has three main rows of uncini and no secondary rows (Figure 3G). The given examples show that the scheme suggested by Hoffmann and Hausen [111] has many modifications among Orbiniidae. When the phylogenetic tree of Orbiniidae is obtained, tracing of various states of this character will help one to understand the evolution of chaetal arrangement in this group. Especially interesting to investigate are the spear-shape chaetae in the genus *Phylo*. According to Hartman [94] (p. 222), the oldest chaetae in the row is the uppermost one. It means the formative site should be ventral, so these specialized chaetae supposedly grow in the opposite direction to uncini of other studied genera.

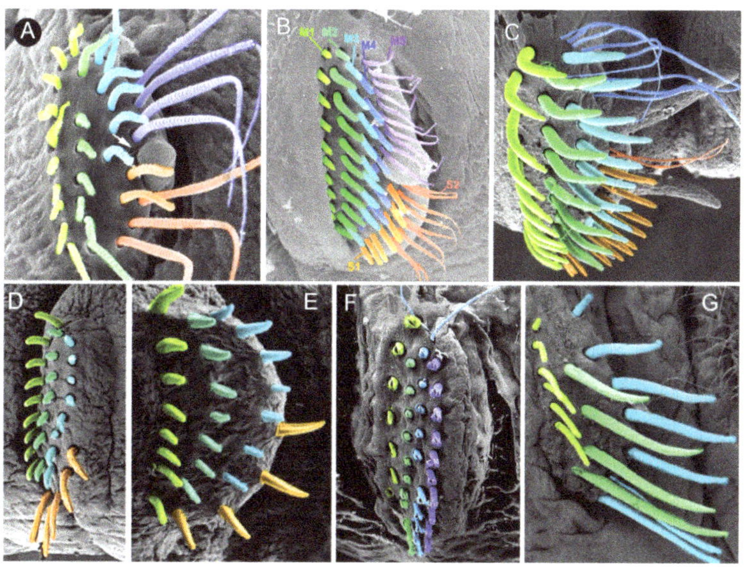

Figure 7. Thoracic neuropodia of Orbiniidae with colored rows of chaetae, SEM. (**A**). *Scoloplos armiger*, juvenile, Ch 1; (**B**). *Scoloplos acutissimus*, Ch 11; (**C**). *Orbinia camposiensis*, Ch 18; (**D**). *Leodamas verax*; (**E**). *Leodamas cirratus*; (**F**). *Leodamas dubia*, Ch 15; (**G**). *Califia bilamellata*, Ch 3. Rows designations are given according to Hoffmann and Hausen [111]: green, blue and purple colors indicate main rows (M1–M5), orange and red—secondary rows (S1, S2; white arrow points formative site). The anterior end is on the left in all photos. **A**: after Hoffmann and Hausen [111]; (**B**,**F**): after Zhadan et al. [102]; (**C**): after Leão and Santos [107]; (**D**–**G**): after Blake [74]. Photos are not in scale.

In addition to the arrangement of chaetae in parapodia, the ratio of different types of chaetae is important for orbiniid taxonomy. For example, *Leitoscoloplos* bear capillaries only, different *Scoloplos* species—both uncini and capillaries in different proportions, and some *Leodamas* species—uncini only. Cryptic species of *Scoloplos armiger* complex can be distinguished by the number of uncini in thoracic neuropodia (unpublished data).

The most common chaetal type for Orbiniidae is crenulated capillaries. They can be present in notopodia and neuropodia and in thorax and abdomen. The crenulations are formed by transverse rows of barbs (Figure 8A,B,I). On capillaries, these structures are also called camerations, which may be single, double or in a form of interlocking transverse ribs [74]. Similar crenulations are also present in other types of chaetae (Figure 8D,G–I,L,N–Q,S–V). Some more types of specialized chaetae are known in orbiniids. Acicular spines are present in neuropodia and sometimes in notopodia in *Orbiniella* (Figure 8C). Forked chaetae occur in abdominal notopodia in many orbiniid species [74] and they have two unequal tynes with blunt or tapered tips. Those with a blunt tip have a distinct hole in the tip. Flattened filaments arise from inner part of the tynes (Figure 8D,E). Swan-shape chaetae are present in both noto-and neuropodia in posterior body of *Proscoloplos* and *Pettibonella* (Figure 8F) and most probably evolved convergently to uncini [111]. Flail chaetae are similar with capillaries but have abruptly tapering tips (Figure 8G); they occur in abdominal neuropodia (sometimes in notopodia) in many orbiniid genera. Subuluncini are intermediate between uncini and capillaries having thick bases and pointed tips (Figure 8H). Abdominal parapodia bear thick protruding aciculae that can be smooth or serrated, straight or strongly hooked (Figure 8I–K). Other types of chaetae not illustrated here are brush-tipped uncini of *Califia*, bifid and trifid crochets of *Questa*, spear-shape chaetae of *Phylo*, and hirsute spines of *Orbiniella spinosa* Blake, 2017 [74] (pp. 112–113, Figure 54F,G).

Uncini (also called hooks, crochets, spines, blunt-tipped chaetae) differ by curvature, serration, comparative thickness and length, development of a hood, shape of tips and presence of grooves. They vary from notably (Figure 8L–N) to slightly bent (Figure 8O–R) or are straight (Figure 8S–V). They can be coarsely (Figure 8L), moderately (Figure 8O–Q,U) or slightly (Figure 8N,S,T) serrated or smooth (Figure 8R). The number of denticles varies from three (Figure 8L) to 10–15 (Figure 8O,P). Uncini can be long and have the same thickness as capillaries (Figure 8O,P) or be short and much thicker (Figure 8V). They can have a clearly visible hood (Figure 8N,Q,T), a thin, poorly visible hood (Figure 8L,O,S) or lack a hood (Figure 8R). Sometimes the hood is as thick as the chaeta itself, so the impression of bidentate tips arises (Figure 8P,V). The tips of uncini are always more or less rounded, but the exact shape varies from conical to obtuse (Figure 8O,Q,R–U). In some species, uncini bear a longitudinal groove on the distal end, which can be short and shallow (Figure 8S,Q) or long and deep, so the tip of the chaetae is deeply notched (Figure 8N). Often uncini have different shape and size in anterior and posterior thoracic segments (Figure 8L,M) or even in one parapodia in different rows (Figure 8V).

The functional role of different types of chaetae in Orbiniidae is not clear. Generally, capillaries are believed to be important in locomotion, stabilization during peristalsis and sensing the environment; they also help to irrigate borrows and tubes [116]. Most likely, capillaries in Orbiniidae perform all these functions. For large orbiniids with dorsally elevated abdominal parapodia, it can be that notopodia with notochaetae form a channel covering branchiae, which produce a water current inside it with cilia. Hooks (including spines and uncini) play a role in anchoring polychaetes [116]. As orbiniids are not tube dwellers, they possibly use uncini for preventing backward slipping during burrowing in sediment, but this statement needs to be proven in experiments and live observations. The function of forked, flail and other types of orbiniid chaetae is unknown.

A promising approach which can be used for species delimitation when the morphological differences between lineages are subtle is morphometrics. The characters more suitable for morphometry in orbiniids are from the cephalic regions (i.e., prostomium, achateous segments), parapodia (i.e., neuropodial and notopodial lobes, both thoracic and abdominal, and subpodal papillae), branchiae and the ventral side. Morphometrics has proven useful in annelid groups such as Hesionidae [117,118], Phyllodocidae [119], Syllidae [120] or Nereididae [121]. The only morphometric study performed in Orbiniidae was implemented by Hernandez-Alcantara and Solis-Weiss [105] to distinguish *Leitoscoloplos multipapillatus* among its coexistent *Leitoscoloplos panamensis* (Monro, 1933) [122] and a third taxon, *Leitoscoloplos* sp., in the Gulf of California.

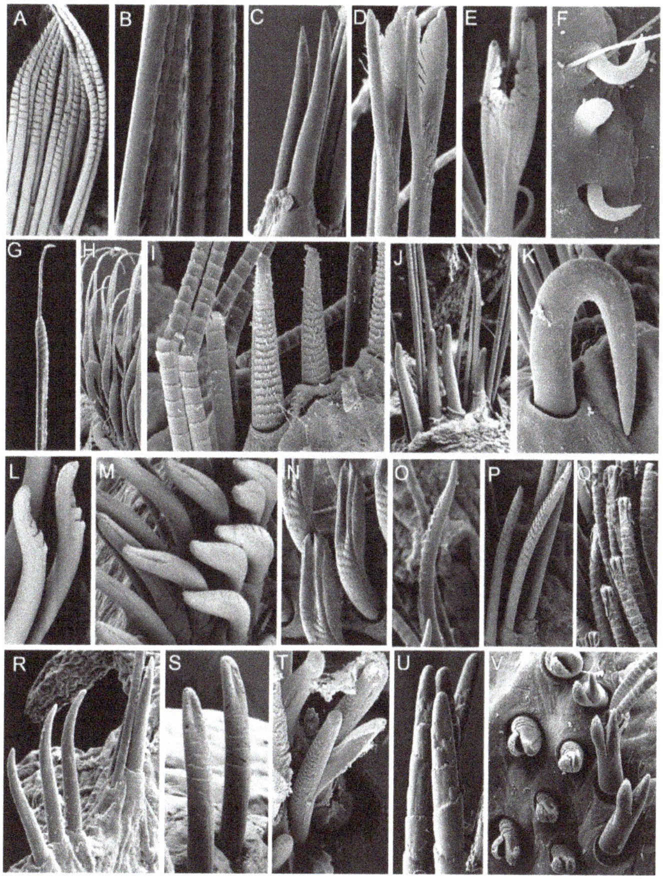

Figure 8. Chaetae of Orbiniidae. (**A**). *Orbinia orensanzi*, thoracic notopodial capillaries; (**B,C**). *Orbiniella andeepia*, anterior capillary notochaetae and notopodial spines respectively; (**D**). *Leodamas dubia*, forked chaetae of abdominal notopodia; (**E**). *Scoloplos bathytatus*, forked chaeta of abdominal notopodia; (**F**). *Proscoloplos cygnochaetus*, posterior parapodium; (**G**). *Orbinia camposiensis*, flail chaetae from abdominal neuropodium; (**H–J**). *Naineris grubei australis*, subuluncini of thoracic neuropodia and abdominal neuropodial capillaries and spines, respectively; (**K**). *Leodamas dubia*, abdominal neuropodial acicula, protruding and strongly hooked; (**L–V**): thoracic neuropodial uncini. (**L,M**). *Orbinia orensanzi*, from middle and posterior neuropodia respectively; (**N**). *Leodamas tribulosus*; (**O**). *Scoloplos bathytatus*; (**P**). *Scoloplos suroestense*; (**Q**). *Naineris grubei*; (**R**). *Scoloplos maranhensis*; (**S**). *Leodamas verax*; (**T**). *Scoloplos dayi*; (**U**). *Leodamas hyphalos*; (**V**). *Leodamas dubia*. (**A–C,E,J,L–Q,S,U**): after Blake [74]; (**D,H,I,K,T,V**): after Zhadan et al. [102], (**F**): after Meyer et al. [109]; (**G**): after Leão and Santos [107]; (**R**): after Oliveira et al. [115]. Photos are not in scale.

3.3. *Diversity*

3.3.1. Species Numbers

Since the first described orbiniid, *Scoloplos armiger* (Müller 1776) [123], the number of reported species was increasing slowly until Hartman's revision in 1957 [94] accepting 74 valid species (Figure 9). Several new orbiniids were described in different studies following Hartman's system [73,124–127] and, as a result, the species number reached around

120 [1]. During the following years, several local faunistic studies of Orbiniidae were performed [76,82,84,96,97,112,128–142]. Rouse [143] listed around 150 species of Orbiniidae; twice as much as Hartman [94] defined in her monograph less than 50 years earlier. The 21st century brought several important contributions to the knowledge about orbiniid diversity [83,85,86,98,107,144–151]. The two recent works by Blake [74,87] described 37 new orbiniids from which 23 were from the deep-sea (deeper than 500 m), and among them, 18 were from more than 1500 m depth. These studies aid a better understanding of the deep-sea orbiniid diversity, increasing the number of reported deep-sea species from 28 to 51. Bleidorn and Helm [6] listed 204 valid species belonging to 20 genera and Blake [87] reported 240 species of orbiniids; however, here we report 222 species including all described subspecies (commented below).

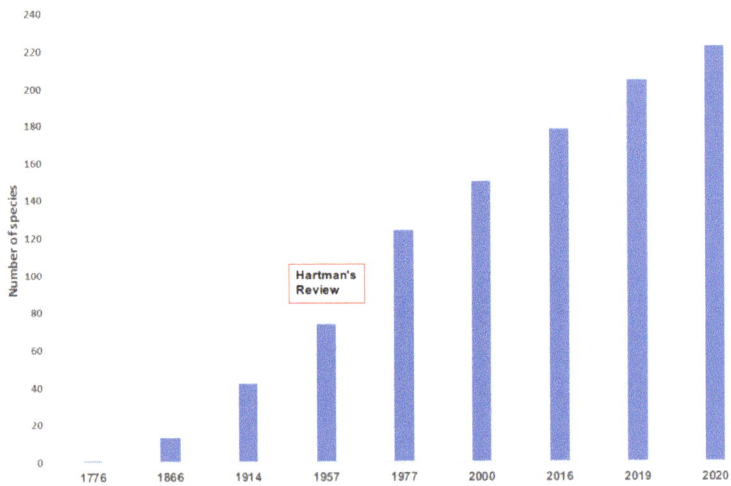

Figure 9. Bar diagram showing the evolution of the number of known orbiniid species from the first described species, *Scoloplos armiger*, in Müller [123] to this study. Note the exponential growth in species discoveries after Eisig [93] and, specially, during the subsequent decades to the Hartman's review [94].

3.3.2. Species Distribution

Orbiniidae occur world-wide in all marine depths, being most common, sometimes forming dense aggregations, in intertidal and shallow waters down to 500 m. Their diversity, depending on the region, can vary from having a great number of species (e.g., 14 in Southern California) to few (e.g., three in the White Sea). The most diverse areas in terms of number of reported species are Asia (with 45 species and 10 genera), the Pacific North America (with 42 species and 10 genera), the Atlantic North America (with 33 species and 16 genera) and the Atlantic South America (with 33 species and 9 genera) (Figure 10). The variation on the orbiniid diversity in certain areas possibly reflects the systematic efforts done by different researchers (e.g., Hartman, Fauchald or Solis-Weiss in the North East Pacific, the Gulf of Mexico and the Caribbean Sea; Lana, Almeida or Pagliosa in the South West Atlantic; Day in South Africa; Hartmann-Schröder or Hutchings in Australia). The poorly studied areas (e.g., East Coast of USA and Canada, South Eastern Pacific, Western Africa or Indo-West Pacific) may have a great number of still undiscovered species.

The deep-sea orbiniids (below 500 m) are rarely encountered and poorly known [87]. The 51 species of Orbiniidae from the deep-sea habitats are from *Leitoscoloplos* (12), *Orbiniella* (11), *Leodamas* (6), *Scoloplos* (6), *Berkeleyia* (4), *Califia* (4), *Phylo* (4), *Naineris* (3) and *Microrbinia* (1). The genera *Orbiniella, Berkeleyia, Califia* and *Microrbinia* are composed mainly by deep-water species. From all these species, only *Microrbinia linea* Hartman, 1965 was reported as dominant at 2000 m depth off North Carolina (Western North Atlantic) during a monitoring program conducted from 1983 to 1987 [152]. Relatively dense populations of two other species were reported by Blake [87]: *Leitoscoloplos gordaensis* Blake, 2020 (with 31 specimens in a single HOV Alvin core from the Gorda Ridge, Escanaba Trough, 3271m) and *Naineris uncinata* Hartman, 1957 (with 30 specimens from a multicore deployment at a gas hydrate site on the Cascadia Subduction zone off Oregon, 786 m). The deepest recorded orbiniid is *Berkeleyia hadala* Blake, 2017, which was found in 6143 m in the abyssal plain from the Peru-Chile Trench.

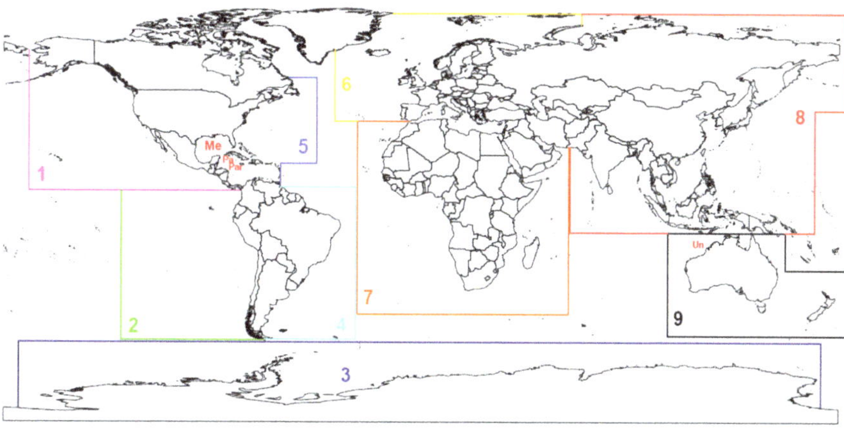

Figure 10. The nine biogeographic areas used for the review of the Orbiniidae diversity distribution with, first, the number of species records and, second, the genera reported for each area. (**1**). Pacific North America (42, 10); (**2**). Pacific South America (23, 10); (**3**). Antarctica (18, 7); (**4**). Atlantic South America (33, 9); (**5**). Atlantic North America (33, 16); (**6**). Europe (31, 11); (**7**). Africa (32, 13); (**8**). Asia (45, 10); (**9**). Australia and New Zealand (29, 7). The monotypic genera based on the original report are represented: Pa. *Paraorbiniella paucibranchiata*; Par. *Pararicia belizensis*; Un. *Uncorbinia brevibranchiata*. The restricted distribution of the monotypic genus *Methanoaricia* is also marked: Me. *M. dendrobranchiata*. Map was generated using the R packages ggplot2 and sf.

The most widely distributed species are *Scoloplos armiger* and *Naineris laevigata*. The first revealed to be a complex of cryptic species (see details below), while the status of the second has not been confirmed yet with molecular data. *Naineris laevigata* was first described from Nice (France) and later reported from different localities around the Mediterranean Sea as well as in other distant locations in North and South America, South Africa and Asia. Other widely distributed orbiniids are *Naineris quadricuspida* (Fabricius, 1780) [153], *Phylo felix* Kinberg, 1866 [113], *Phylo norvegicus* (M. Sars in G. O. Sars, 1872) [154], and *Protoaricia oerstedii* (Claparède, 1864) [155], the status of the last three has not been tested yet. For the first, a significant genetic differentiation was reported in Zhadan et al. [102] between three genetically identical specimens from three locations (i.e., Friday Harbor, Morro Bay and Cattle Point) in the North East Pacific and two genetically identical specimens from two distant places (i.e., India and White Sea) (Figure 6D), but more data is required for confirming a nonworldwide distribution pattern. An interesting case of a confirmed nearly cosmopolitan distribution was reported in *Proscoloplos cygnochaetus* Day, 1954 [156]. Analysis of morphology and genetics of three different species of *Proscoloplos* from Australia, South Africa and France revealed a single species with a wide geographical range [109]. The authors suggested a possible human transport by vessels and, thus, an invasive nature of the species. A similar case is that of *Naineris setosa*, an alien orbiniid in the Mediterranean Sea. The species showed an American subtropical-tropical distribution (see Table S1 in Supplementary Materials for details) and was reported, first, from an aquiculture facility in the Adriatic Sea (Italy) by Blake and Giangrande [157] and, later, in Tunisia by

Khedhri et al. [158]. However, the status of the "alien" populations vs. the "native" ones in *N. setosa* has yet to be confirmed. *Scoloplos capensis* (Day, 1961) has been recorded as an alien species in the Bay of Bengal (India) with a South-African origin [159,160]. *Naineris quadraticeps* Day, 1965 [161] was described from the Red Sea and afterwards noticed in the Aegean Sea [162] based on two incomplete specimens in poor conditions. Harmelin [162] considered the specimens as juveniles of *N. quadraticeps*, with some minor differences attributed to their juvenile condition. As the record is based on two juveniles, it is considered doubtful [160]. *Leitoscoloplos kerguelensis* (McIntosh, 1885) [163] has been recorded globally but considered as an Antarctic and sub-Antarctic exclusive species by Blake [74] and as an alien species in the Thyrrenian Sea [164] based on an incomplete specimen. Mackie [96] compared this specimen with the type material and concluded that they have some differences, considering the record in the Mediterranean as doubtful.

3.3.3. Cryptic Diversity and Subspecies

Cryptic species have been reported for *Scoloplos armiger*, *Phylo foetida* (Claparède, 1868) [165], *Scoloplos acmeceps* Chamberlin, 1919 [166], *Leitoscoloplos pugettensis* (Pettibone, 1957) [95] and *Naineris dendritica* (Kinberg, 1866) [101,110,167,168]. Another potential sibling species issue is known for *Leitoscoloplos acutus* (Verrill, 1873) [169] and *Leitoscoloplos mammosus* Mackie, 1987, two similar species with overlapping distributions (see below for details).

Scoloplos armiger was originally described from Kristiansand (Norway) and later reported in several ecological studies [79,170–175] being a dominant macrofaunal species and showing a cosmopolitan distribution with records at the Pacific North American coast, in Europe, including the Arctic and the White Seas, and Japan. It has been reported to be a complex of cryptic species based on, first, RAPD DNA markers [108] and, later, mitochondrial markers [168,174], with at least two clades in the Pacific region (designated as "Malibu clade" and "San Diego Clade") and four clades in the North Sea (named "Type locality clade", "Subtidal clade", "Intertidal clade" and "Intertidal clade 2"). Recent morphological studies performed on the *Scoloplos* populations from Southern California revealed three morphotypes of *S. armiger*-like species and two of *S. acmeceps*-like species (Haggin, pers. comm.). The study of the genetic clustering among these morphotypes and their relationships with the reported clades by Bleidorn et al. [168] is in process with the aim to describe the species recovered in phylogenetic analyses. Similarly, the species descriptions referred to the North Sea's clades is in preparation [6]. In addition, the populations of *S. armiger* from the White Sea have been shown to constitute two clades (Zhadan, unpubl.). The eight documented clades of *S. armiger* are accompanied by subtle diagnostic morphological characters and, in the case of the European populations, along with separation by depth. In the North Sea, two clades are known from the intertidal and the other two from the subtidal and, in the White Sea, one is from the intertidal and the other from the subtidal and close to the "Type locality clade". Moreover, two distinct reproductive modes occur in the North Sea's populations [176]. Intertidal females produce egg cocoons, but no pelagic larvae, while subtidal females produce pelagic larvae, but no egg cocoons. Furthermore, the intertidal males have spermatozoa with heads twice as long as those in subtidal males and also a significantly shorter flagellum. The subtidal clades showed an additional autumn spawning [177]. The authors suggested that the difference in the sperm morphology could cause the reproductive isolation at the fertilization stage.

Leitoscoloplos pugettensis is a common macrofaunal component in the West Coast of North America, from Alaska to Costa Rica, being the most abundant orbiniid in intertidal and subtidal bottoms in California [79]. In the phylogenetic works by Bleidorn and collaborators [101,168] two distinct clades of *L. pugettensis* were reported, one from Friday Harbor and another from Santa Monica, Northern and Southern California, respectively. Furthermore, Carr et al. [110] found five MOTUs among the Canadian populations of *L. pugettensis* using the DNA barcoding approach. More recently, Haggin (pers. comm.) detected at least six morphotypes showing a different pattern of depth distribution. Another widespread species from Western North America (i.e., *Scoloplos acmeceps*) showed to have at least three clades (Morro Bay and Friday Harbor from Northern California, and Newport from Rhode Island) by means of the molecular analyses of Bleidorn et al. [101]. Similarly to *L. pugettensis*, Haggin (pers. comm.) detected at least two different morphotypes of *S. acmeceps* coexisting in the intertidal and the shelf (<200 m) areas in Southern California.

Naineris dendritica is frequently sampled in the intertidal sands and muds along the Californian coast [79,178] and is also recorded in the Pacific North America, the Arctic Canada and the Gulf of Mexico (see Table S1 in Supplementary Materials for details). Carr et al. [110] reported four MOTUs among the British Columbia and the Arctic Canada, with a high divergent clustering in the British Columbia. Additionally, Zhadan et al. [102], using some of the published sequences from Carr et al. [110] and previous works of Bleidorn and collaborators [19,101], found significant genetic dissimilarity between two Southern Californian specimens according to the 16S NJ tree (Figure 6C) and a high genetic distance between one Southern Californian specimen and the Canadian populations according to the COI NJ Tree (Figure 6D).

Leitoscoloplos mammosus, described from Loch Creran (Scotland) by Mackie [96], is morphologically very similar to *Leitoscoloplos acutus*, described from Massachusetts, USA, and was later recorded from North Carolina to Canadian Arctic and also in the White Sea, Russia [179,180]. In recent years, several records of both *L. mammosus*, from Iceland, The Channel and the North and the Norwegian Seas, and *L. acutus*, from the Barents and the Kara Seas, have been reported (pers. obs. by the authors; Andy Mackie, pers. comm.; OBIS, https://obis.org/taxon/607421 and https://obis.org/taxon/130514; GBIF, https://www.gbif.org/species/4289003 and https://www.gbif.org/species/2319969). Morphological comparison of materials from the type locality together with molecular studies should help to resolve the relationships and validity of the two species.

Several subspecies were recorded for Orbiniidae, but none of them has been verified by genetic analyses. *Phylo foetida* is the orbiniid with the most documented subspecies (seven, see Table S1 in Supplementary Materials). The species was initially described from the Gulf of Naples (Italy) and later reported, together with the distinct subspecies, from several localities around the Mediterranean Sea, Gulf of Biscay, English Channel as well as from Mozambique, Madagascar and Libia in Africa [181]. Bleidorn et al. [101] found significant genetic dissimilarity between two distinct specimens of *P. foetida* from Sardinia (Mediterranean) and Arcachon (Gulf of Biscay). The remaining reported subspecies are *Scoloplos acmeceps profundus* Hartman, 1960 [182], *Phylo felix asiaticus* Wu, 1962 [124] and *Leodamas chevalieri candensis* Harmelin, 1969 [162].

3.4. Ecology

3.4.1. Habitat

Orbiniids are burrowing deposit-feeding annelids which live in soft bottoms rich in organic matter. They are strictly known from marine waters though some taxa, such as *Naineris laevigata* (Grube, 1855) [183], *Scoloplos capensis* (Day, 1961) [184] or *Scoloplos marsupialis* (Southern, 1921) [185], are quite tolerant to low salinity [159,186,187]. Orbiniidae can be found in a wide variety of sediments, as in mud (Figure 11A,B), sand (Figure 11C,D), mud or sand with shell fragments, mixed bottoms with stones or annelid reefs (Figure 11E), bivalve beds or in algal meadows (Figure 11F,G). The large-sized genera (as *Scoloplos* and *Leitoscoloplos*) are usually recorded from intertidal and shallow muddy bottoms, except for *Orbinia*, which prefers sandy sediments, and *Califia*, mostly reported from deep waters. The small-sized genera (as *Protoariciella* and *Protoaricia*), together with *Naineris*, are often encountered among algal holdfasts, with the exception of *Questa* and the deep-sea genera *Berkeleyia* and *Microrbinia*, which have never been reported in algal congregations. Although *Orbiniella* is mainly deep sea, some shallow species inhabits algal groupings, as *O. spinosa* or *O. plumisetosa* Buzhinskaya, 1993 [139]. The interstitial *Questa* live in coarse, fine and coralline sands from shallow waters. The monotypic genera *Paraorbiniella*, *Pararicia* and *Uncorbinia* were reported from littoral zones and have not been documented again since their original description (see Table S1 in Supplementary Materials for details). The species inhabiting the intertidal are tolerant to hypoxic conditions through several adaptations. However, *S.* cf. *armiger* showed only moderate resistance to anoxia in the Wadden Sea, and individuals actively ascends to oxidative layers during low tides [188]. In contrast, the coexistent *Arenicola marina* (Linnaeus, 1758) [189] can inhabit deeper sediment layers by means of a more efficient anaerobic metabolism. *Leitoscoloplos mammosus* is a dominant macrofauna species in the anoxic mud of the Upper Basin of Loch Creran (Scotland), an area with strong accumulations of detritus (Andy Mackie, pers. comm.). The most hypoxic habitats inhabited by Orbiniidae are abyssal plains (with 16 recorded species), hydrothermal vents (i.e., *Leitoscoloplos sahlingi* Blake, 2020, *Orbiniella hobsonae* Blake and Hilbig, 1990 [138] and *Orbiniella grasslei* Blake, 2020) and hydrocarbon seeps (i.e., *Methanoaricia dendrobranchiata*). This last species lives in association with the bivalve *Gigantidas childressi* (Gustafson, R. D. Turner, Lutz and Vrijenhoek, 1998) [190] as an adaptation to the hydrocarbon enriched sediments on the Louisiana continental slope (Gulf of Mexico). Other adaptations of this orbiniid to this extreme habitat are described and compiled elsewhere [6,191–193].

Figure 11. Photographs of some common and special habitats in Orbiniidae. (**A**). Muddy intertidal. (**B**). Muddy subtidal. (**C**). Sandy intertidal. (**D**). Sandy subtidal. (**E**) Mixed muddy/rocky intertidal. (**F**). Intertidal algal meadow, with detail on the algal species *Ulva* sp. (**G**). Subtidal algal meadow (different filamentous algae). All photos were taken from the White Sea, Russia, except **C** which was taken from Oxwich Bay, Wales.

Traditionally, the description of a new orbiniid species is not accompanied by the description of its habitat, which are relevant data for species delimitation according to the Integrative Taxomony concept [118,194–196]. Furthermore, it is crucial to resample certain areas, especially type localities. We encourage researchers to include data on the habitat of the species in ongoing work.

3.4.2. Relation to Pollution

Although orbiniids have not been traditionally used as bioindicators, some species (e.g., *Scoloplos* cf. *armiger* and *Scoloplos madagascarensis* Fauvel, 1919 [197] in Reunion Islands, Eastern Madagascar) increase in density under the presence of pollutants, showing the potential of their use as indicators of disturbance or nonpollution. These species were referred to ecological group III "Species tolerant to excess of organic matter" according to the AMBI biotic index [198], as their populations were stimulated by discharges of industrial wastes in their natural habitat [199]. Similarly, *Leitoscoloplos pugettensis* in Southern California marine bays was referred to group III by Teixeira et al. [200]. An interesting case is that of *Leitoscoloplos fragilis* (Verrill, 1873) [169], which showed tissue accumulation of the hydrocarbon Benzo(a)pyrene under high exposures without effects on mortality [201], making it tolerant to and indicator of Polycyclic Aromatic Hydrocarbons contamination.

Other orbiniids are sensitive to high concentrations of pollutants, decreasing their density or disappearing completely from the environment. For instance, *Leitoscoloplos foliosus* (Hartman, 1951) [202] revealed to be sensitive and an indicator of low metals and low/moderate organic pollution in Northern Gulf of Mexico estuaries [203]. *Scoloplos* cf. *armiger* and *Scoloplos typicus* (Eisig, 1914) [93] are referred to ecological group I "Species very sensitive to organic enrichment and present under unpolluted conditions" in the Cantabrian Sea (Northern Spain) [198]. In a domestic sewage impact study in the macrofaunal composition of several intertidal mussel beds of Mar de la Plata (Argentina) [204], *Protoariciella uncinata* was subdominant in a control unpolluted station and disappeared in the most impacted station.

3.5. Conclusions

Orbiniidae are common and often abundant in bottom communities. In spite of a long history of investigations, many aspects of their morphology, biology, systematics and phylogeny remain poorly known. We can suggest several directions of topical future research. The main approach should be integrative taxonomy combining morphological, molecular and biological data. For many species, re-examination of type specimens or designating of neotypes is required as well as thorough investigations of specimens from different geographical areas. Preferably these studies should include scanning electron microscopy for revealing fine details of chaetal and ciliation patterns. Other techniques, such as histological sections or confocal microscopy can be useful for examination of chaetal arrangement. The phylogenetic system of Orbiniidae requires advanced genetic studies, including new technologies such as next generation sequencing, in combination with morphological examinations of the well-supported clades in search of possible new informative characters. Taxon coverage of molecular based studies should be expanded considerably. Obtaining genetic information for type species of each genus is crucial; specimens for genetic analysis should be collected from the type localities. For species with wide geographical distribution, population genetic studies are necessary to reveal cryptic species. Studies of postlarval development will help researchers to understand the transformations of morphological structures during ontogenesis and the progenetic evolution of orbiniids. Morphofunctional observations and experimental research of burrowing will possibly shed light on the function of different types of chaetae and other parapodial structures.

Supplementary Materials: The following are available online at https://www.mdpi.com/1424-2818/13/1/29/s1, Table S1: Species table, Table S2: Summary table.

Author Contributions: The three authors contributed to the conceptualization, investigation and writing (both for the original draft and review and editing) in the following order: M.A.M., T.H.S. and A.Z. All authors have read and agreed to the published version of the manuscript.

Funding: A.Z. was funded by grants from the Russian Foundation of Basic Research no. 18-05-60158, 19-04-00501.

Data Availability Statement: No new data were created or analyzed in this study. Data sharing is not applicable to this article.

Acknowledgments: We greatly acknowledge the text revision and comments made on previous versions of this Manuscript by Nataliya Budaeva. Christoph Bleidorn made inputs on orbiniid knowledge and provided Figure 5G. We also would like to thank the original photographs provided by Katrine Kongshavn (Figure 5H), Kate Mortimer and Andrew S. Y. Mackie (Figure 11C), the bibliography contributed by João Gil and Verônica Oliveira, and the shared knowledge on the Californian orbiniid morphotypes and on the orbiniid species distribution from NE Pacific by Brent Haggin and Leslie Harris.

Conflicts of Interest: The authors declare no conflict of interest.

References

1. Fauchald, K. The polychaete worms. Definitions and keys to the orders, families and genera. *Nat. Hist. Mus. Los Angeles Cty.* **1977**, *28*, 188.
2. Fauvel, P. Polychètes Sédentaires. Addenda aux Errantes, Archiannélides, Myzostomaires. *Faune Fr.* **1927**, *16*, 1–494.
3. Day, J.H. *A Monograph on the Polychaeta of South Africa. Part 2. Sedentaria.*; British Museum (Natural History): London, UK, 1967.
4. Hartmann-Schröder, G. *Teil 58. Annelida, Borstenwürmer, Polychaeta*; VEB Gustav Fischer Verlag: Jena, Germany, 1971.
5. Rouse, G.W.; Fauchald, K. Cladistics and polychaetes. *Zoöl. Scr.* **1997**, *26*, 139–204. [CrossRef]
6. Bleidorn, C.; Helm, C. 7.1.4 Orbiniidae Hartman, 1942. In *Annelida Basal Groups and Pleistoannelida*; Sedentaria, I., Purschke, G., Böggemann, M., Westheide, W., Eds.; De Gruyter: Berlin/Boston, Germany, 2019; pp. 237–250. [CrossRef]
7. Struck, T.H. 7.2 Phylogeny. In *Annelida Basal Groups and Pleistoannelida*; Sedentaria, I., Purschke, G., Böggemann, M., Westheide, W., Eds.; De Gruyter: Berlin/Boston, Germany, 2019; pp. 37–68. [CrossRef]
8. Struck, T.H.; Paul, C.; Hill, N.; Hartmann, S.; Hösel, C.; Kube, M.; Lieb, B.; Meyer, A.; Tiedemann, R.; Purschke, G.; et al. Phylogenomic analyses unravel annelid evolution. *Nat. Cell Biol.* **2011**, *471*, 95–98. [CrossRef]
9. Struck, T.H. The Impact of Paralogy on Phylogenomic Studies—A Case Study on Annelid Relationships. *PLoS ONE* **2013**, *8*, e62892. [CrossRef]
10. Weigert, A.; Bleidorn, C. Current status of annelid phylogeny. *Org. Divers. Evol.* **2016**, *16*, 345–362. [CrossRef]
11. Martín-Durán, J.M.; Vellutini, B.C.; Marlétaz, F.; Cetrangolo, V.; Cvetesic, N.; Thiel, D.; Henriet, S.; Grau-Bové, X.; Carrillo-Baltodano, A.M.; Gu, W.; et al. Conservative route to genome compaction in a miniature annelid. *Nat. Ecol. Evol.* **2020**, 1–12. [CrossRef]
12. Struck, T.H.; Golombek, A.; Weigert, A.; Franke, F.A.; Westheide, W.; Purschke, G.; Bleidorn, C.; Halanych, K.M. The Evolution of Annelids Reveals Two Adaptive Routes to the Interstitial Realm. *Curr. Biol.* **2015**, *25*, 1993–1999. [CrossRef]
13. Worsaae, K.; Kristensen, R.M. Evolution of interstitial Polychaeta (Annelida). *Hydrobiologia* **2005**, *535*, 319–340. [CrossRef]
14. Struck, T.H.; Purschke, G.; Dordel, J.; Hösel, C.; Nesnidal, M.P.; Diersing, F.; Bleidorn, C.; Paul, C.; Hill, N.; Tiedemann, R.; et al. Phylogeny and evolution of Annelida based on molecular data. In *Deep Metazoan Phylogeny: The Backbone of the Tree of Life—New Insights from Analyses of Molecules, Morphology, and Theory of Data Analysis*; Wägele, J.W., Bartolomaeus, T., Eds.; De Gruyter: Berlin, Germany, 2014; pp. 143–160.
15. Struck, T.H.; Nesnidal, M.; Purschke, G.; Halanych, K.M. Detecting possibly saturated positions in 18S and 28S sequences and their influence on phylogenetic reconstruction of Annelida (Lophotrochozoa). *Mol. Phylogenet. Evol.* **2008**, *48*, 628–645. [CrossRef]
16. Jordens, J.; Struck, T.; Purschke, G. Phylogenetic inference regarding Parergodrilidae and *Hrabeiella periglandulata* ('Polychaeta', Annelida) based on 18S rDNA, 28S rDNA and COI sequences. *J. Zoöl. Syst. Evol. Res.* **2004**, *42*, 270–280. [CrossRef]
17. Struck, T.H.; Hessling, R.; Purschke, G. The phylogenetic position of the Aeolosomatidae and Parergodrilidae, two enigmatic oligochaete-like taxa of the 'Polychaeta', based on molecular data from 18S rDNA sequences. *J. Zoöl. Syst. Evol. Res.* **2002**, *40*, 155–163. [CrossRef]
18. Bleidorn, C. The role of character loss in phylogenetic reconstruction as exemplified for the Annelida. *J. Zoöl. Syst. Evol. Res.* **2007**, *45*, 299–307. [CrossRef]
19. Bleidorn, C. Phylogenetic relationships and evolution of Orbiniidae (Annelida, Polychaeta) based on molecular data. *Zoöl. J. Linn. Soc.* **2005**, *144*, 59–73. [CrossRef]
20. Bleidorn, C.; Vogt, L.; Bartolomaeus, T. New insights into polychaete phylogeny (Annelida) inferred from 18S rDNA sequences. *Mol. Phylogenet. Evol.* **2003**, *29*, 279–288. [CrossRef]
21. Zrzavý, J.; Riha, P.; Piálek, L.; Janouškovec, J. Phylogeny of Annelida (Lophotrochozoa): Total-evidence analysis of morphology and six genes. *BMC Evol. Biol.* **2009**, *9*, 189. [CrossRef]

22. Reisinger, E. Ein landbewohnender archiannelide. *Zoomorphology* **1925**, *3*, 197–254. [CrossRef]
23. Meyer, A. Ist Parergodrilus heideri (Reisinger) ein Archiannelide? *Zool. Anz.* **1927**, *72*, 19–35.
24. Karling, T.G. Zur Kenntnis von Stygocapitella subterranea Knöllner und Parergodrilus heideri Reisinger. *Ark. Zool.* **1958**, *1*, 307–324.
25. Knöllner, F. Stygocapitella subterranea nov.gen., nov.spec. *Schriften Naturwissenschaftlichen Ver. Schlesw. Holst.* **1934**, *20*, 468–472.
26. Hartmann-Schröder, G. *Teil 58. Annelida, Borstenwürmer, Polychaeta*, 2nd ed.; Gustav Fischer: Stuttgart, Germany, 1996.
27. Pizl, V.; Chalupsky, J. Hrabeiella periglandulata gen. et sp. n.(Annelida): A curious worm from Czechoslovakia. *Veštn. Českoslov. Spol. Zool.* **1984**, *48*, 291–295.
28. Rota, E. Morphology and adaptations of Parergodrilus Reisinger and Hrabeiella Pizl & Chalupsky, two enigmatic soil-dwelling annelids. *Ital. J. Zoöl.* **1998**, *65*, 75–84. [CrossRef]
29. Purschke, G. Male genital organs, spermatogenesis and spermatozoa in the enigmatic terrestrial polychaete Parergodrilus heideri (Annelida, Parergodrilidae). *Zoomorphology* **2002**, *121*, 125–138. [CrossRef]
30. Purschke, G. Terrestrial polychaetes—Models for the evolution of the Clitellata (Annelida)? *Hydrobiologia* **1999**, *406*, 87–99. [CrossRef]
31. Rota, E.; De Nicola, F.; Bargagli, R. Parergodrilus heideri Reisinger, 1925 (Annelida: Polychaeta) from a holm oak wood in an extinct volcano of southern Italy. *Zootaxa* **2010**, *2687*, 65–68. [CrossRef]
32. Rota, E. First Italian record of the terrestrial polychaeteparergodrilus heiderireisinger, with anatomical and ecological notes. *Ital. J. Zoöl.* **1997**, *64*, 91–96. [CrossRef]
33. Reisinger, E. Die Lösung des Parergodrilus-Problems. *Z. Morphol. Ökol. Tiere* **1960**, *48*, 517–544. [CrossRef]
34. Graefe, U. Die Gliederung der Zersetzergesellschaften für die standortsökologische Ansprache. *Mitteilungen Dtsch. Bodenkd. Ges.* **1993**, *69*, 95–98.
35. Graefe, U.; Schmelz, R.M. Indicator values, strategy types and life forms of terrestrial Enchytraeidae and other microannelids. In *Newsletter on Enchytraeidae 6, Proceedings of the 3rd International Symposium on Enchytraeidae, Osnabrück, Germany, 3–4 July 1998*; Schmelz, R.M., Sühlo, K., Eds.; Universitätsverlag Rasch: Osnabrück, Germany, 1999; pp. 59–67.
36. Martinez-Ansemil, E.; Parapar, J. Primera cita en la península Ibérica del poliqueto terrestre Parergodrilus heideri Reisinger, 1925 (Polychaeta, Parergodrilidae). *Graellsia* **2009**, *65*, 235–240. [CrossRef]
37. Schlaghamerský, J.; Frelich, L.E. First records of Parergodrilus heideri (Annelida: "Polychaeta") from North America. *Zootaxa* **2012**, *3498*, 81–86. [CrossRef]
38. Rota, E.; De Jong, Y. Fauna Europaea: Annelida—Terrestrial Oligochaeta (Enchytraeidae and Megadrili), Aphanoneura and Polychaeta. *Biodivers. Data J.* **2015**, *3*, e5737. [CrossRef] [PubMed]
39. Römbke, J.; Jans, W. On terrestrial Polychaeta (Annelida) from forest soils of southern Germany. *Verhandlungen Dtsch. Zool. Ges.* **1991**, *84*, 447–448.
40. Beylich, A.; Graefe, U. Investigations on the enchytraeid fauna in floodplain soils of the Lower Middle Elbe. *Folia Fac. Sci. Nat. Univ. Masaryk. Brun. Brno Biol.* **2007**, *110*, 107–122.
41. Chalupský, J. Terrestrial Enchytraeidae (Oligochaeta) and Parergodrilidae (Polychaeta) from Sweden, with description of a new enchytraeid species. *Zool. Scr.* **1992**, *21*, 133–150. [CrossRef]
42. Rouse, G.W.; Pleijel, F. *Polychaetes*; Oxford University Press: New York, NY, USA, 2001; p. 354.
43. Purschke, G. Problematic Annelid Groups. In *Reproductive Biology and Phylogeny of Annelida*; Rouse, G.W., Pleijel, F., Eds.; Science Publisher: Plymouth, UK, 2006; Volume 4, pp. 639–667.
44. Westheide, W. *Polychaetes: Interstitial Families*, 2nd ed.; Field Studies Council: Shrewsbury, UK, 2008.
45. Purschke, G. 7.1.3 Parergoldrilidae Reisinger, 1925. In *Annelida Basal Groups and Pleistoannelida*; Sedentaria, I., Purschke, G., Böggemann, M., Westheide, W., Eds.; De Gruyter: Berlin/Boston, Germany, 2019; pp. 237–250. [CrossRef]
46. Cerca, J.; Meyer, C.; Purschke, G.; Struck, T.H. Delimitation of cryptic species drastically reduces the geographical ranges of marine interstitial ghost-worms (Stygocapitella; Annelida, Sedentaria). *Mol. Phylogenet. Evol.* **2020**, *143*, 106663. [CrossRef] [PubMed]
47. Cerca, J.; Meyer, C.; Stateczny, D.; Siemon, D.; Wegbrod, J.; Purschke, G.; Dimitrov, D.; Struck, T.H. Deceleration of morphological evolution in a cryptic species complex and its link to paleontological stasis. *Evolution* **2019**, *74*, 116–131. [CrossRef]
48. Struck, T.H.; Koczula, J.; Stateczny, D.; Meyer, C.; Purschke, G. Two new species in the annelid genus Stygocapitella (Orbiniida, Parergodrilidae) with comments on their biogeography. *Zootaxa* **2017**, *4286*, 301. [CrossRef]
49. Jans, W.; Römbke, J. Funde eines terrestrischen Polychaeten (Annelida) in Wäldern Baden-Württembergs. *Carolinea* **1989**, *47*, 158–162.
50. Rota, E.; Martin, P.; Erséus, C. Soil-dwelling polychaetes: Enigmatic as ever? Some hints on their phylogenetic relationships as suggested by a maximum parsimony analysis of 18S rRNA gene sequences. *Contrib. Zool.* **2001**, *70*, 127–138. [CrossRef]
51. Dózsa-Farkas, K.; Hong, Y. Three new Hemienchytraeus species (Enchytraeidae, Oligochaeta, Annelida) from Korea, with first records of other enchytraeids and terrestrial polychaetes (Annelida). *Zootaxa* **2010**, *2406*, 29–56. [CrossRef]
52. Riser, N.W. The aberrant polychaete Stygocapitella from some American beaches. *Wasmann J. Biol.* **1980**, *38*, 10–17.
53. Westheide, W. Zur Polychaetenfauna des Eulitorals der Nordseeinsel Sylt. *Helgol. Mar. Res.* **1966**, *13*, 203–209. [CrossRef]
54. Riser, N.W. General observations on the intertidal interstitial fauna of New Zealand. *Tane* **1984**, *30*, 239–250.

55. Worsfold, T.M.; Dyer, M.F. *UK Marine SACs Project. Monitoring Methods Workshop at Plymouth (April 1997) and Millport (May 1997). (Contractor: Unicomarine, Letchworth)*; Joint Nature Conservation Committee: Peterborough, UK, 1997.
56. Hartmann-Schröder, G. Teil 9—Die Polychaeten der antiborealen Südwestküste Australiens (zwischen Dunsborough im Norden und Denmark im Süden). *Mitteilungen Hambg. Zool. Mus. Inst.* **1983**, *80*, 123–167.
57. Schmidt, H.; Westheide, W. Are the meiofaunal polychaetes Hesionides arenaria and Stygocapitella subterranea true cosmopolitan species?—Results of RAPD-PCR investigations. *Zool. Scr.* **2000**, *29*, 17–27. [CrossRef]
58. Schmidt, P. Zonation of the interstitial polychaeteStygocapitella subterranea (Stygocapitellidae) in European sandy beaches. *Mar. Biol.* **1970**, *7*, 319–323. [CrossRef]
59. Schmidt, P. Die quantitative Verteilung und Populationsdynamik des Mesopsammons am Gezeiten-Sandstrand der Nordsee-Insel Sylt1). II. Quantitative Verteilung und Populationsdynamik einzelner Arten. *Int. Rev. Hydrobiol.* **1969**, *54*, 95–174. [CrossRef]
60. Westheide, W. The geographical distribution of interstitial polychaetes. *Mikrofauna Meeresbod.* **1977**, *61*, 287–302.
61. Schmidt, P. Zonierung und jahreszeitliche Fluktuationen des Mesopsammons im Sandstrand von Schilksee (Kieler Bucht). *Mikrofauna Meeresbod.* **1972**, *10*, 1–60.
62. Schmidt, P. Zonierung und jahreszeitliche Fluktuationen der interstitiellen Fauna in Sandstränden des Gebiets von Tromsø (Norwegen). *Mikrofauna Meeresbod.* **1972**, *10*, 81–164.
63. Renaud-Debysier, J. Recherches écologiques sur la faune interstitielle des sables. Bassin d'Arcachon, Île de Bimini, Bahamas. *Vie Milieu Suppl.* **1963**, *15*, 1–157.
64. Valkanov, A. Zwei neue Polychaeten für das Schwarze Meer [Bulgar]. *Trud. Chernomorsk. Biol. Sta. Varna* **1954**, *18*, 55–58.
65. Westheide, W. La faune des Polychètes et des Archiannélides dans les plages à Ressac de la côte méditerranéenne de la Tunisie. *Bull. Insti. Natl. Sci. Technol. Mer* **1972**, *2*, 449–468.
66. Worsfold, T.M. British records of the interstitial polychaete Stygocapitella subterranea (Annelida: Parergodrilidae). *Mar. Biodivers. Rec.* **2008**, *1*, 40. [CrossRef]
67. Brinck, P.; Dahl, E.; Wieser, W. On the littoral subsoil fauna of the Simrishamn Beach in Eastern Scania. *K. Fysiogr. Sällsk. Lund Förh.* **1955**, *25*, 109–129.
68. Itô, T. Studies on the interstitial animals in the Ishikari beach, Hokkaido, northern Japan-A preliminary report. *BENTHOS Res.* **1984**, 1–14. [CrossRef]
69. Westheide, W. Meiofauna geographic distribution: Vicariance and dispersal. *Meiofauna Mar.* **2005**, *14*, 201–207.
70. Cerca, J.; Purschke, G.; Struck, T.H. Marine connectivity dynamics: Clarifying cosmopolitan distributions of marine interstitial invertebrates and the meiofauna paradox. *Mar. Biol.* **2018**, *165*, 123. [CrossRef]
71. Struck, T.H.; Feder, J.L.; Bendiksby, M.; Birkeland, S.; Cerca, J.; Gusarov, V.I.; Kistenich, S.; Larsson, K.-H.; Liow, L.H.; Nowak, M.D.; et al. Finding Evolutionary Processes Hidden in Cryptic Species. *Trends Ecol. Evol.* **2018**, *33*, 153–163. [CrossRef]
72. Gerlach, S.A. Means of meiofauna dispersal. *Mikrofauna Meeresbod.* **1977**, *61*, 89–103.
73. Hedgpeth, J.W. Polychaeta Errantia of Antarctica. Volume 3, Antarctic Research Series.Olga HartmanPolychaeta Myzostomidae and Sedentaria of the Antarctica. Volume 7, Antarctic Research Series.Olga HartmanPolychaetous Annelids Collected by the USNS Eltanin and Staten Island Cruises Chiefly from Antarctic Seas. Allan Hancock Monographs in Biology, No. 2.Olga Hartman. *Q. Rev. Biol.* **1968**, *43*, 206. [CrossRef]
74. Blake, J.A. Polychaeta Orbiniidae from Antarctica, the Southern Ocean, the Abyssal Pacific Ocean, and off South America. *Zootaxa* **2017**, *4218*, 1–145. [CrossRef] [PubMed]
75. Day, J.H. New polychaeta from Beaufort: With a key to all species recorded from North Carolina. *Biodivers. Herit. Libr.* **1973**, *375*, 1–140. [CrossRef]
76. Day, J. A review of the Australian and New Zealand Orbiniidae (Annelida: Polychaeta). In *Essays on Polychaetous Annelids in Memory of Dr. Olga Hartman*; Reish, D.J., Fauchald, K., Eds.; Allan Hancock Foundation: Los Angeles, CA, USA, 1977; pp. 217–243.
77. Zhadan, A. Review of Orbiniidae (Annelida, Sedentaria) from Australia. *Zootaxa* **2020**, *4860*, 451–502. [CrossRef] [PubMed]
78. Kirkegaard, J. *Havborsteorme II. Sedentaria*; Dansk Naturhistorisk Forening: Copenhagen, Denmark, 1996; Volume 86, p. 451.
79. Blake, J.A. Family Orbiniidae Hartman, 1942. In *Taxonomic Atlas of the Benthic Fauna of Santa Maria Basin and Western Santa Barbara Channel. The Annelida Part 3—Polychaeta: Orbiniidae to Cossuridae*; Blake, J.A., Hilbig, B., Scott, P.H., Eds.; Santa Barbara Museum of Natural History: Santa Barbara, CA, USA, 1996; Volume 6, pp. 1–26.
80. López, E. Familia Orbiniidae Hartman, 1942. In *Fauna Iberica. Annelida Polychaeta III*; Parapar, J., Alós, C., Núñez, J., Moreira, J., López, E., Aguirrezabalaga, F., Besteiro, C., Martinez, A., Eds.; Museo Nacional de Ciencias Naturales. CSIC: Madrid, Spain, 2012; Volume 36, pp. 96–160.
81. Díaz-Díaz, O.; Vanegas-Espinosa, V.; Cárdenas-Oliva, A.; Liñero-Arana, I. Orbiniidae Hartman, 1942 (Annelida: Polychaeta) de las costas de Venezuela. *Biota Colomb.* **2012**, *13*, 18–34.
82. Gillet, P. A new species of Orbiniella (Orbiniidae: Polychaeta) from Marion Island, Indian Ocean. *Proc. Biol. Soc. Wash.* **1999**, *112*, 592–597.
83. Parapar, J.; Moreira, J.; Helgason, G.V. First record of genus Orbiniella Day, 1954 (Polychaeta: Orbiniidae) in North Atlantic Ocean with the description of a new species. *Zootaxa* **2015**, *4006*, 330–346. [CrossRef]
84. Solis-Weiss, V.; Fauchald, K. Orbiniidae (Annelida: Polychaeta) from mangrove root-mats in Belize, with a revision of protoariciin genera. *Proc. Biol. Soc. Wash.* **1989**, *102*, 772–792.

85. Sun, Y.; Li, X. Orbinia wui, a new species from China, with redescription of O. dicrochaeta Wu, 1962 (Annelida, Orbiniidae). *Zootaxa* **2018**, *4403*, 351–364. [CrossRef]
86. Sun, Y.; Sui, J.; Li, X. A new species of Leodamas Kinberg, 1866 (Polychaeta: Orbiniidae) from the Yellow Sea and the East China Sea. *Acta Oceanol. Sin.* **2018**, *37*, 130–135. [CrossRef]
87. Blake, J.A. New species and records of deep-water Orbiniidae (Annelida, Polychaeta) from the Eastern Pacific continental slope, abyssal Pacific Ocean, and the South China Sea. *Zootaxa* **2020**, *4730*, 1–61. [CrossRef] [PubMed]
88. Malmgren, A.J. Annulata Polychaeta Spetsbergiae, Groenlandiae, Islandiae et Scandinaviae hactenus cognita. *Öfvers. K. Vetenskapsakademiens Forh.* **1867**, *24*, 127–235.
89. Hartman, O. A review of the types of polychaetous annelids at the Peabody Museum of Natural History, Yale University. *Bull. Bingham Oceanogr. Collect. Yale Univ.* **1942**, *8*, 1–98.
90. Savigny, J.-C.; Benedict, J.E. Systèmes de Diverses Clases D'animaux Sans Vertèbres. Principalement de Celles des Côtes de l'Égypte et de la Syrie, Offrant les Caractères Tant Distinctifs que Naturels des Ordres, Familles et Genres, avec la Description des espèces. Available online: https://www.biodiversitylibrary.org/bibliography/66284#/summary (accessed on 18 December 2020).
91. Quatrefages, A. Note sur la Classification des Annelides. *Ann. Sci. Nat. Paris* **1865**, *5*, 253–296.
92. Hartman, O. Nomenclatural changes involving California polychaete worms. *J. Wash. Acad. Sci.* **1936**, *26*, 31–32.
93. Eisig, H. Zur Systematik, Anatomie und Morphologie der ariciiden nebst beiträgen zur generellen Systematik. *Mitteilungen Zool. Station Neapel* **1914**, *21*, 153–500.
94. Hartman, O. Orbiniidae, Apistobranchidae, Paraonidae and Longosomidae. *Allan Hancock Pac. Exped.* **1957**, *15*, 211–393.
95. Pettibone, M.H. North American genera of the family Orbiniidae (Annelida: Polychaeta), with descriptions of new species. *J. Wash. Acad. Sci.* **1957**, *47*, 159–167.
96. Mackie, A.S. A review of species currently assigned to the genus Leitoscoloplos Day, 1977 (Polychaeta: Orbiniidae), with descriptions of species newly referred to Scoloplos Blainville, 1828. *Sarsia* **1987**, *72*, 1–28. [CrossRef]
97. Badalamenti, F.; Castelli, A. Schroederella laubieri, a new species of the subfamily protoariicinae (polychaeta, orbiniidae), with some notes on the genusschroederellalaubier, 1962. *Boll. Zool.* **1991**, *58*, 95–98. [CrossRef]
98. Blake, J.A. A new genus of polychaete worm (Family Orbiniidae) from methane seeps in the Gulf of Mexico, with a review of the systematics and phylogenetic interrelationships of the genera of Orbiniidae. *Cah. Biol. Mar.* **2000**, *4*, 435–449.
99. Bleidorn, C.; Vogt, L.; Bartolomaeus, T. A contribution to sedentary polychaete phylogeny using 18S rRNA sequence data. *J. Zool. Syst. Evol. Res.* **2003**, *41*, 186–195. [CrossRef]
100. Hartman, O. Quantitative survey of the benthos of SanPedro Basin, Southern California. Part 11. Final results and conclusions. *Allan Hancock Pac. Exped.* **1966**, *19*, 187–456.
101. Bleidorn, C.; Hill, N.; Erséus, C.; Tiedemann, R. On the role of character loss in orbiniid phylogeny (Annelida): Molecules vs. morphology. *Mol. Phylogenet. Evol.* **2009**, *52*, 57–69. [CrossRef] [PubMed]
102. Zhadan, A.; Stupnikova, A.N.; Neretina, T. Orbiniidae (Annelida: Errantia) from Lizard Island, Great Barrier Reef, Australia with notes on orbiniid phylogeny. *Zootaxa* **2015**, *4019*, 773–801. [CrossRef] [PubMed]
103. Hartmann-Schroeder, G. Zweiter Beitrag zur Polychaetenfauna von Peru. *Kiel. Meeresforsch.* **1962**, *18*, 109–147.
104. Ehlers, E. *Ergebnisse der Hamburger Magalhaensischen Sammelreise 1892/93*; L. Friederichsen & Co.: Hamburg, Germany, 1897; p. 148.
105. Hernández-Alcántara, P.; Solís-Weiss, V. Anatomical and morphometric analysis of a new species of Leitoscoloplos (Annelida: Orbiniidae) with numerous stomach papillae, from the Gulf of California, Eastern Pacific. *Contrib. Zool.* **2014**, *83*, 133–150. [CrossRef]
106. Verrill, A.E. Additions to the Turbellaria, Nemertina, and Annelida of the Bermudas, with revisions of some New England genera and species. *Trans. Connect. Acad. Arts Sci.* **1900**, *10*, 595–671. [CrossRef]
107. Leão, L.S.D.; Santos, C.S.G. Orbinia (Polychaeta: Orbiniidae) from the Brazilian coast: Two new species and two new records. *Zootaxa* **2016**, *4105*, 145–158. [CrossRef]
108. Kruse, I.; Reusch, T.B.; Schneider, M.V. Sibling species or poecilogony in the polychaete Scoloplos armiger? *Mar. Biol.* **2003**, *142*, 937–947. [CrossRef]
109. Meyer, A.; Bleidorn, C.; Rouse, G.W.; Hausen, H. Morphological and molecular data suggest a cosmopolitan distribution of the polychaete Proscoloplos cygnochaetus Day, 1954 (Annelida, Orbiniidae). *Mar. Biol.* **2007**, *153*, 879–889. [CrossRef]
110. Carr, C.M.; Hardy, S.M.; Brown, T.M.; Macdonald, T.A.; Hebert, P.D.N. A Tri-Oceanic Perspective: DNA Barcoding Reveals Geographic Structure and Cryptic Diversity in Canadian Polychaetes. *PLoS ONE* **2011**, *6*, e22232. [CrossRef] [PubMed]
111. Hoffmann, S.; Hausen, H. Chaetal arrangement in Orbiniidae (Annelida, Polychaeta) and its significance for systematics. *Zoomorphology* **2007**, *126*, 215–227. [CrossRef]
112. Hartmann-Schröder, G. Zur Kenntnis des eulitorals der australischen Kuesten unter besounderer Beruecksichtigung des Polychaeten und Ostracoden. Teil 16. Die Polychaeten der subtropisch-tropischen bis tropischen Ostkueste Australiens zwischen Maclean (New South Wales) und Gladstone (Queensland) sowie von Heron Island (Grosses Barrier Riff). *Mitteilungen Hamburgischen Zool. Mus. Inst.* **1991**, *88*, 17–71.
113. Kinberg, J. Annulata nova. *Öfversight K. Vetensk.-Adakemiens Forh. Stockh.* **1866**, *22*, 239–258.
114. Tebble, N. The polychaete fauna of the Gold Coast [=Ghana]. Bulletin of the British Museum (Natural History), Ser. *Zoology* **1955**, *3*, 59–148.

115. De Oliveira, V.M.; Cutrim, A.S.T.; Vieira, A.K.M.; Ferreira, C.N.; Almeida, Z.S.; Júnior, M.N. Description of Scoloplos maranhensis sp. nov. (Orbiniidae, Annelida) from tropical Brazilian mangrove. *Iheringia. Série Zool.* **2019**, *109*. [CrossRef]
116. Merz, R.A. Polychaete chaetae: Function, fossils, and phylogeny. *Integr. Comp. Biol.* **2006**, *46*, 481–496. [CrossRef]
117. Martin, D.; Meca, M.A.; Gil, J.; Drake, P.; Nygren, A. Another brick in the wall: Population dynamics of a symbiotic species of Oxydromus (Annelida, Hesionidae), described as new based on morphometry. *Contrib. Zool.* **2017**, *86*, 181–211. [CrossRef]
118. Meca, M.A.; Drake, P.; Martin, D. Does polyxenous symbiosis promote sympatric divergence? A morphometric and phylogeographic approach based on Oxydromus okupa (Annelida, Polychaeta, Hesionidae). *Contrib. Zool.* **2019**, *88*, 1–28. [CrossRef]
119. Teixeira, M.A.L.; Vieira, P.E.; Pleijel, F.; Sampieri, B.R.; Ravara, A.; Costa, F.O.; Nygren, A. Molecular and morphometric analyses identify new lineages within a large Eumida (Annelida) species complex. *Zool. Scr.* **2019**, *49*, 222–235. [CrossRef]
120. Lattig, P.; Martín, G.S.; Martin, D. Análisis toxinómico y morfométrico del completo Haplosyllis spongicola (Polychaeta: Syllidae: Syllinae) de los mares de España, con la re-descripción de la especie tipo y la descripción de dos nuevas especies. *Sci. Mar.* **2007**, *71*, 551–570. [CrossRef]
121. Coutinho, M.C.L.; Paiva, P.C.; Santos, C.S.G. Morphometric analysis of two sympatric species ofPerinereis (Annelida: Nereididae) from the Brazilian coast. *J. Mar. Biol. Assoc. U. K.* **2015**, *95*, 953–959. [CrossRef]
122. Monro, C.C.A. The Polychaeta Sedentaria collected by Dr. C. Crossland at Colón, in the Panama Region, and the Galapagos Islands during the Expedition of the S.Y. 'St. George'. *Proc. Zool. Soc. Lond.* **2009**, *103*, 1039–1092. [CrossRef]
123. Müller, O.F.; Dall, W.H. Zoologiae Danicae Prodromus, seu, Animalium Daniae et Norvegiae Indigenarum Characteres, Nomina, et Synonyma Imprimis Popularium. Available online: https://www.biodiversitylibrary.org/bibliography/13268#/summary (accessed on 18 December 2020).
124. Wu, B. New species of polychaete worms of the family Orbiniidae and Paraonidae from the Yellow Sea. *Acta Zool. Sin.* **1962**, *14*, 421–426.
125. Hartmann-Schröder, G. Die Polychaeten des Sublitorals. zur Kentnnis des Sublitorals der chilenischen Küste unter besonderer Berücksichtigung der Polychaeten und Ostracoden (Mit Bemerkungen über den Einfluss sauerstoffarmer strömungen auf die Besiedlung von marinen Sedimenten). *Mitteilungen Hamburg. Zool. Mus. Inst.* **1965**, *62*, 59–305.
126. Gallardo, V.A. Polychaeta from the bay of Nha Trang, South Viet Nam. *NAGA Rep.* **1968**, *4*, 91–96.
127. Fauchald, K. Benthic polychaetous annelids from deep water off western Mexico and adjacent areas in the eastern Pacific Ocean. *Allan Hancock Monogr. Mar. Biol.* **1972**, *7*, 162–171.
128. Maciolek, N.J.; Holland, J. Scoloplos texana: A new orbiniid polychaete from south Texas, with notes on the related species Scoloplos treadwelli Eisig. *Contrib. Mar. Sci.* **1978**, *21*, 163–169.
129. Hartmann-Schröder, G. Die Polychaeten der tropischen Nordwestküste Australiens (zwischen Derby im Norden und Port Hedland im Süden). *Mitteilungen Hamburgisch. Zool. Mus. Inst.* **1979**, *76*, 77–218.
130. Hartmann-Schröder, G. des Eulitorals der australischen Küsten unter besonder Berücksichtigung der Polychaeten und Ostracoden. *Mitteilungen Hamburgisch. Zool. Mus. Inst.* **1980**, *77*, 67–68.
131. Hartmann-Schröder, G. Zur Kenntnis des Eulitorals der australischen Kusten unter besonderer Berucksichtigung der Polychaeten und Ostracoden. Teil 6. Die Polychaeten der tropisch-subtropischen Westkuste Australiens (zwischen Exmouth im Norden und Cervantes im Suden). *Mitteilungen Hamburgisch. Zool. Mus. Inst.* **1981**, *78*, 48–49.
132. Hartmann-Schröder, G. Zur Polychaetenfauna der Polynesischen Inseln Huahiné (Gesellschafts inseln) und Rangiroa (Tuamotu-Inseln). *Mitteilungen Hamburgisch. Zool. Mus. Inst.* **1992**, *89*, 68.
133. Mohammad, M. Polychaete annelids from Kuwaitian islands, Arabian Gulf, with descriptions of four new species. *Zool. J. Linn. Soc.* **1980**, *69*, 31–42. [CrossRef]
134. Westheide, W. Interstitielle Fauna von Galapagos. XXVI. Questidae, Cirratulidae, Acrocirridae, Ctenodrilidae (Polychaeta). *Mikrofauna Meeresbod.* **1981**, *82*, 6–8.
135. Jamieson, B.; Webb, R. The morphology, spermatozoal ultrastructure and phylogenetic affinities of a new species of questid (Polychaeta, Annelida). In Proceedings of the 1st International Polychaete Conference, Sydney, Australia, 1–5 August 2016; Linnean Society of New South Wales: Sydney, Australia, 2016; pp. 21–34.
136. Blake, J.A. Polychaeta from the vicinity of deep-sea geothermal vents in the eastern Pacific. I. Euphrosinidae, Phyllodocidae, Hesionidae, Nereididae, Glyceridae, Dorvilleidae, Orbiniidae, and Maldanidae. *Bull. Biol. Soc. Wash.* **1985**, *6*, 67–101.
137. Mackie, A.S. A new species of Scoloplos (Polychaeta: Orbiniidae) from Hong Kong and a comparison with closely related Scoloplos marsupialis Southern, 1921 from India. *Asian Mar. Biol.* **1991**, *8*, 35–44.
138. Blake, J.A.; Hilbig, B. Polychaeta from the vicinity of deep-sea hydrothermal vents in the Eastern Pacific Ocean. II. New species and records from the Juan de Fuca and Explorer Ridge systems. *Pac. Sci.* **1990**, *44*, 219–253.
139. Buzhinskaya, G.N. Occurrence of Orbiniella plumisetosa sp. n. (Polychaeta: Orbiniidae) in the north-western Pacific Ocean and a key to species of the genus Orbiniella. *Explor. Fauna Seas* **1993**, *43*, 76–81.
140. Branch, M.L. Four new species of Polychaeta from subantarctic Marion Island. *Ann. S. Afr. Mus.* **1998**, *105*, 249–265.
141. Giere, O.; Erséus, C. A systematic account of the Questidae (Annelida, Polychaeta), with description of new taxa. *Zool. Scr.* **1998**, *27*, 345–360. [CrossRef]
142. Mirsa, A. Polychaete [Fauna of West Bengal]. In *Zoological Survey of India, Calcutta, Series 3*; State Fauna Series, Fauna of West Bengal, Part 10, Invertebrates; Gosh, A.K., Ed.; Zoological survey of India: Calcuttam, India, 1999; pp. 125–225.
143. Rouse, G.W. Orbiniidae Hartman, 1942. In *Polychaetes*; Rouse, G., Pleijel, F., Eds.; University Press: Oxford, UK, 2001; pp. 57–60.

144. Eibye-Jacobsen, D. The Orbiniidae (Annelida: Polychaeta) of the BIOSHELF Project, Andaman Sea, Thailand. *Phuket Mar. Biol. Cent. Spec. Publ.* **2002**, *24*, 77–99.
145. López, E.; Cladera, P.; Martín, G.S.; Lpez, E.; Martn, G.S. Two new species of the genus Leodamas (Orbiniidae: Scolecida: Polychaeta) from the Pacific coast of Panama. *J. Mar. Biol. Assoc. U. K.* **2003**, *83*, 367–374. [CrossRef]
146. López, E.; Cladera, P.; San Martín, G. Orbiniidae polychaetes (Polychaeta: Scolecida) from Coiba Island, eastern Pacific of Panama, with description of a new species. *Rev. Biol. Trop.* **2006**, *54*, 1307–1318. [CrossRef] [PubMed]
147. Narayanaswamy, B.E.; Blake, J.A. A new species of Orbiniella (Polychaeta: Orbiniidae) from deep basins of Antarctica. *J. Mar. Biol. Assoc. U. K.* **2005**, *85*, 843–846. [CrossRef]
148. Giere, O.; Ebbe, B.; Erseus, C. Questa (Annelida, Polychaeta, Orbiniidae) from Pacific regions—New species and reassessment of the genus Periquesta. *Org. Divers. Evol.* **2008**, *7*, 304–319. [CrossRef]
149. Imajima, M. Deep-sea benthic polychaetes off Pacific coast of the northern Honshu, Japan. In *Deep-Sea Fauna and Pollutants off Pacific Coast of Northern Japan*; Fujita, T., Ed.; National Museum of Nature and Science Monographs: Tokyo, Japan, 2009; Volume 39, pp. 39–192.
150. Gopal, A.; Useph, A.J.K.; Varghese, S.A.; Narayana, S.V. A new species of polychaete, Pettibonella shompenssp. nov. (Orbiniidae), from the Nicobar Islands, North Indian Ocean. *Mar. Biol. Res.* **2014**, *10*, 1033–1037. [CrossRef]
151. Dean, H.K.; Blake, J.A. The Orbiniidae (Annelida: Polychaeta) of Pacific Costa Rica. *Zootaxa* **2015**, *3956*, 183–198. [CrossRef]
152. Blake, J.A. Life history analysis of five dominant infaunal polychaete species from the continental slope off North Carolina. *J. Mar. Biol. Assoc. U. K.* **1993**, *73*, 123–141. [CrossRef]
153. Fabricius, O. Fauna Groenlandica, systematice sistens, Animalia Groenlandiae occidentalis hactenus indagata, quoad nomen specificum, triviale, vernaculumque; synonyma auctorum plurium, descriptionem, locum, victum, generationem, mores, usum, capturamque singuli; prout detegendi occasion fuit, maximaque parti secundum proprias observationes. *Hafniae Lipsiae* **1780**, *i-xvi*, 1–452.
154. Sars, G.O. Diagnoser af nye Annelider fra Christianiaforden, efter Professor M. Sar's efterladte Manuskripter. *Forh. Vidensk. Christiania* **1872**, *1871*, 406–417.
155. Claparède, É. Glanures zootomiques parmi les annélides de Port-Vendres (Pyrénées Orientales). *Mém. Société Phys. d'Histoire Nat. Genève* **1864**, *17*, 463–600.
156. Day, J.H. The Polychaeta of Tristan da Cunha. *Results Nor. Sci. Exped. Tristan Cunha* **1954**, *29*, 1–35.
157. Blake, J.A.; Giangrande, A. Naineris setosa (Verrill) (Polychaeta, Orbiniidae), an American subtropical–tropical polychaete collected from an aquaculture facility in Brindisi (Adriatic Sea, Italy): A possible alien species. *Ital. J. Zool.* **2011**, *78*, 20–26. [CrossRef]
158. Khedhri, I. First record of Naineris setosa (Verrill, 1900) (Annelida: Polychaeta: Orbiniidae) in the Western Mediterranean Sea. *BioInvasions Rec.* **2014**, *3*, 83–88. [CrossRef]
159. Raman, A.; Ganapati, P. Occurrence of Scolaricia capensis Day, 1961 (Polychaeta: Orbiniidae) off the east coast of India, Bay of Bengal. *Indian J. Mar. Sci.* **1977**, *6*, 185–186.
160. Zenetos, A.; Gofas, S.; Verlaque, M.; Cinar, M.; Raso, J.G.; Bianchi, C.; Morri, C.; Azzurro, E.; Bilecenoglu, M.; Froglia, C.; et al. Alien species in the Mediterranean Sea by 2010. A contribution to the application of European Union's Marine Strategy Framework Directive (MSFD). Part I. Spatial distribution. *Mediterr. Mar. Sci.* **2010**, *11*, 381. [CrossRef]
161. Day, J.H. Israel South Red Sea Expedition, 1962, Reports no. 7. Some Polychaeta from the Israel South Red Sea Expedition, 1962. *Bull. Sea Fish. Res. Stn. Haifa* **1965**, *38*, 15–27.
162. Harmelin, J.G. Contribution à l'étude de l'endofaune des prairies d'Halophila stipulacea de Méditerranée Orientale. *Recl. Trav. Stn. Mar. d'Endoume* **1969**, *45*, 305–316.
163. McIntosh, W.C. Report on the Annelida Polychaeta collected by HMS "Challenger" during the years 1873–1876. Reports on the Scientific Results of the Voyage of HMS "Challenger". *Zoology* **1885**, *12*, 1–554.
164. Ramos, J.M. Haploscoloplos kerguelensis McIntosh, 1885. Nouvel Orbiniidae en Méditerranée Occidentale. *Vie Milieu* **1976**, *26*, 1–9.
165. Claparède, R.-É. Royal College of Physicians of Edinburgh. In *Les Annelides Chétopodes du Golfe de Naples*; Smithsonian Institution: Washington, DC, USA, 1870; Volume 19, pp. 313–584.
166. Chamberlin, R.V. New polychaetous annelids from Laguna Beach, California. *Pomona Coll. J. Entomol. Zool.* **1919**, *11*, 1–23.
167. Kinberg, J. Annulata Nova. *Öfversigt Königlich Vetensk. Förhandlingar* **1866**, *22*, 337.
168. Bleidorn, C.; Kruse, I.; Albrecht, S.; Bartolomaeus, T. Mitochondrial sequence data expose the putative cosmopolitan polychaete Scoloplos armiger (Annelida, Orbiniidae) as a species complex. *BMC Evol. Biol.* **2006**, *6*, 47. [CrossRef] [PubMed]
169. Verrill, A.E. XVIII. Report upon the Invertebrate Animals of Vineyard Sound and the Adjacent Waters, with an Account of the Physical Characters of the Region. Available online: https://www.biodiversitylibrary.org/bibliography/11688#/summary (accessed on 18 December 2020).
170. Holte, B.; Gulliksen, B. Common macrofaunal dominant species in the sediments of some north Norwegian and Svalbard glacial fjords. *Polar Biol.* **1998**, *19*, 375–382. [CrossRef]
171. Carvalho, S.; Ravara, A.; Quintino, V.; Rodrigues, A. Macrobenthic community characterisation of an estuary from the western coast of Portugal (Sado estuary) prior to dredging operations. *Boletín Inst. Español Oceanogr.* **2001**, *17*, 179–190.
172. Belan, T.A. Marine environmental quality assessment using polychaete taxocene characteristics in Vancouver Harbour. *Mar. Environ. Res.* **2004**, *57*, 89–101. [CrossRef]

173. Kraan, C.; Dekinga, A.; Piersma, T. Now an empty mudflat: Past and present benthic abundances in the western Dutch Wadden Sea. *Helgol. Mar. Res.* **2010**, *65*, 51–58. [CrossRef]
174. Luttikhuizen, P.C.; Bol, A.; Cardoso, J.F.; Dekker, R. Overlapping distributions of cryptic Scoloplos cf. armiger species in the western Wadden Sea. *J. Sea Res.* **2011**, *66*, 231–237. [CrossRef]
175. Faulwetter, S.; Simboura, N.; Katsiaras, N.; Chatzigeorgiou, G.; Arvanitidis, C. Polychaetes of Greece: An updated and annotated checklist. *Biodivers. Data J.* **2017**, *5*, e20997. [CrossRef]
176. Kruse, I.; Reise, K. Reproductive isolation between intertidal and subtidal Scoloplos armiger (Polychaeta, Orbiniidae) indicates sibling species in the North Sea. *Mar. Biol.* **2003**, *143*, 511–517. [CrossRef]
177. Kruse, I.; Strasser, M.; Thiermann, F. The role of ecological divergence in speciation between intertidal and subtidal Scoloplos armiger (Polychaeta, Orbiniidae). *J. Sea Res.* **2004**, *51*, 53–62. [CrossRef]
178. Francoeur, A.A.; Dorgan, K.M. Burrowing Behavior in Mud and Sand of Morphologically Divergent Polychaete Species (Annelida: Orbiniidae). *Biol. Bull.* **2014**, *226*, 131–145. [CrossRef] [PubMed]
179. Blake, J.A.; Hilbig, B. Dense infaunal assemblages on the continental slope off Cape Hatteras, North Carolina. *Deep. Sea Res. Part II Top. Stud. Oceanogr.* **1994**, *41*, 875–899. [CrossRef]
180. Zhadan, A. Taxonomy of the genus Scoloplos (orbiniidae) in the white, barents and north seas. *ooo a* **1998**, *77*, 177–190.
181. Fauvel, P. Sur l'Aricia foetida, Claparède et ses variétés. *Bull. Société Zool. Fr.* **1924**, *49*, 518–528.
182. Hartman, O. Systematic account of some marine invertebrate animals from the deep basins off southern California. *Allan Hancock Pac. Exped.* **1960**, *22*, 69–216.
183. Grube, A.E. Beschreibungen neuer oder wenig bekannter Anneliden. *Archiv für Naturgeschichte* **1855**, *21*, 81–136. [CrossRef]
184. Day, F.J.H. The polychaet fauna of south africa. part 6. sedentary species dredged off cape coasts with a few new records from the shore. *J. Linn. Soc. Lond. Zool.* **1961**, *44*, 463–560. [CrossRef]
185. Southern, R. Polychaeta of the Chilka Lake and also of fresh and brackish waters in other parts of India. *Mem. Indian Mus.* **1921**, *5*, 563–659.
186. Giangrande, A.; Fraschetti, S. A population study of Naineris laevigata (Polychaeta, Orbiniidae) in a fluctuating environment (Mediterranean Sea). *Sci. Mar.* **1995**, *59*, 39–48.
187. Khan, S.A.; Murugesan, P. Polychaete diversity in Indian estuaries. *Indian J. Mar. Sci.* **2005**, *34*, 114–119.
188. Grieshaber, M. Adaptation of the polychaete worm Scoloplos armiger to hypoxic conditions. *Mar. Biol.* **1988**, *99*, 215–222. [CrossRef]
189. Linnaeus, C. Systema Naturae per regna tria naturae secundum classes, ordines, genera, species cum characteribus, differentiis, synonymis, locis. Editio decima, reformata. *Laurentius Salvius Holmiae* **1758**, *2*, 824.
190. Gustafson, R.G.; Turner, R.D.; Lutz, R.; Vrijenhoek, R.C. A new genus and five new species of mussels (Bivalvia, Mytilidae) from deep-sea sulfide/hydrocarbon seeps in the Gulf of Mexico. *Malacologia* **1998**, *40*, 63–112.
191. Hourdez, S.; Schernecke, A.; Fisher, C.R.; Frederick, L.-A. Functional respiratory anatomy of a deep-sea orbiniid polychaete from the Brine Pool NR-1 in the Gulf of Mexico. *Invertebr. Biol.* **2005**, *120*, 29–40. [CrossRef]
192. Hourdez, S.; Weber, R.E.; Green, B.N.; Kenney, J.M.; Fisher, C.R. Respiratory adaptations in a deep-sea orbiniid polychaete from Gulf of Mexico brine pool NR-1: Metabolic rates and hemoglobin structure/function relationships. *J. Exp. Biol.* **2002**, *205*, 1669–1681.
193. Van Gaest, A.L.; Young, C.M.; Young, J.J.; Helms, A.R.; Arellano, S.M. Physiological and behavioral responses of Bathynerita naticoidea (Gastropoda: Neritidae) and Methanoaricia dendrobranchiata (Polychaeta: Orbiniidae) to hypersaline conditions at a brine pool cold seep. *Mar. Ecol.* **2007**, *28*, 199–207. [CrossRef]
194. Will, K.W.; Mishler, B.D.; Wheeler, Q.D. The Perils of DNA Barcoding and the Need for Integrative Taxonomy. *Syst. Biol.* **2005**, *54*, 844–851. [CrossRef]
195. Padial, J.M.; Miralles, A.; De La Riva, I.; Vences, M. The integrative future of taxonomy. *Front. Zool.* **2010**, *7*, 16. [CrossRef]
196. Giribet, G. Morphology should not be forgotten in the era of genomics—A phylogenetic perspective. *Zool. Anz.-A J. Comp. Zool.* **2015**, *256*, 96–103. [CrossRef]
197. Fauvel, P. Annélides Polychètes nouvelles de l'Afrique orientale: 2ième note. *Bull. Mus. D'histoire Nat.* **1919**, *1*, 33–39.
198. Borja, A.; Franco, J.; Pérez, V. A Marine Biotic Index to Establish the Ecological Quality of Soft-Bottom Benthos Within European Estuarine and Coastal Environments. *Mar. Pollut. Bull.* **2000**, *40*, 1100–1114. [CrossRef]
199. Bigot, L.; Grémare, A.; Amouroux, J.-M.; Frouin, P.; Maire, O.; Gaertner, J.C. Assessment of the ecological quality status of soft-bottoms in Reunion Island (tropical Southwest Indian Ocean) using AZTI marine biotic indices. *Mar. Pollut. Bull.* **2008**, *56*, 704–722. [CrossRef] [PubMed]
200. Teixeira, H.; Weisberg, S.B.; Borja, A.; Ranasinghe, J.A.; Cadien, D.B.; Velarde, R.G.; Lovell, L.L.; Pasko, D.; Phillips, C.A.; Montagne, D.E.; et al. Calibration and validation of the AZTI's Marine Biotic Index (AMBI) for Southern California marine bays. *Ecol. Indic.* **2012**, *12*, 84–95. [CrossRef]
201. Driscoll, S.K.; McElroy, A.E. Bioaccumulation and metabolism of benzo[A]pyrene in three species of polychaete worms. *Environ. Toxicol. Chem.* **1996**, *15*, 1401–1410. [CrossRef]
202. Hartman, O. The littoral marine annelids of the Gulf of Mexico. *Publ. Inst. Mar. Sci. Port Aransas Texas* **1951**, *2*, 7–124.
203. Rakocinski, C.F.; Brown, S.S.; Gaston, G.R.; Heard, R.W.; Walker, W.W.; Summers, J.K. Macrobenthic Responses to Natural and Contaminant-Related Gradients in Northern Gulf of Mexico Estuaries. *Ecol. Appl.* **1997**, *7*, 1278. [CrossRef]

204. Elías, R.; Rivero, M.S.; Vallarino, E.A. Sewage impact on the composition and distribution of Polychaeta associated to intertidal mussel beds of the Mar del Plata rocky shore, Argentina. *Iheringia. Série Zool.* **2003**, *93*, 309–318. [CrossRef]

diversity

MDPI

Review

Still Digging: Advances and Perspectives in the Study of the Diversity of Several Sedentarian Annelid Families

Maël Grosse [1,*], Anna Zhadan [2], Joachim Langeneck [3], Dieter Fiege [4] and Alejandro Martínez [5]

1 Tjärnö Marine Laboratory, Department of Marine Sciences, University of Gothenburg, SE 452 96 Strömstad, Sweden
2 N.A. Pertsov White Sea Biological Station, Biological Faculty, M.V. Lomonosov Moscow State University, 119234 Moscow, Russia; azhadan@wsbs-msu.ru
3 Department of Biology, University of Pisa, 56126 Pisa, Italy; jlangeneck@biologia.unipi.it
4 Sektion Marine Evertebraten II, Forschungsinstitut und Naturmuseum Senckenberg, 60325 Frankfurt, Germany; dieter.fiege@senckenberg.de
5 Molecular Ecology Group (MEG), Water Research Institute (IRSA), National Research Council of Italy (CNR), 28922 Pallanza, Italy; alejandro.martinezgarcia@cnr.it
* Correspondence: maelgrosse@gmail.com

Abstract: Sedentarian annelids are a diverse and heterogeneous group of marine worms representing more than 8600 species gathered in ca. 43 families. The attention brought to these organisms is unevenly distributed among these families, and the knowledge about them sometimes scarce. We review here the current knowledge about the families Acrocirridae, Cirratulidae (including Ctenodrilidae), Cossuridae, Longosomatidae, Paraonidae, and Sternaspidae in terms of biodiversity as well as the evolution of the taxonomy and systematics of each group. We present the challenges faced when studying these organisms and compare methodologies across groups and perspectives in future research.

Keywords: Annelida; systematics; biodiversity; taxonomy; Acrocirridae; Cirratulidae; Sternaspidae; Longosomatidae; Paraonidae; Cossuridae

Citation: Grosse, M.; Zhadan, A.; Langeneck, J.; Fiege, D.; Martínez, A. Still Digging: Advances and Perspectives in the Study of the Diversity of Several Sedentarian Annelid Families. *Diversity* **2021**, *13*, 132. https://doi.org/10.3390/d13030132

Academic Editors: Michael Wink, Pat Hutchings and Maria Capa

Received: 14 January 2021
Accepted: 22 February 2021
Published: 18 March 2021

1. Introduction

Sedentarian annelids represent a morphologically heterogeneous clade of worms comprising ca. 43 families and more than 8600 species [1]. Although the relationships among annelids are far from being understood, different independent sources of evidence suggest that the clades Cirratulida, Sabellida, and Spionida might represent the sister clade of the remaining Sedentaria, which at least include the orders Terebellida, Arenicolida, Opheliida, Capitellida, Echiura, and Clitellata [2–4]. However, relationships among these groups remain to be investiged using extensive datasets with both molecular and morphological characters and a larger taxon sampling. Within this incomplete picture, there are several lineages whose relationships with the remaining Sedentaria remain particularly obscure, including Cossuridae Day, 1963 [5], Paraonidae Cerruti, 1909 [6], Siboglinidae Caullery, 1914 [7], *Hrabeiella* Pizl, 1984 [8], *Aeolosoma* Ehrenberg, 1828 [9], *Potamodrilus* Lastochkin, 1935 [10], Scalibregmatidae Malmgren, 1867 [11], and Travisiidae Hartmann-Schröder, 1971 [12], among many others. Many recent studies have reviewed the current knowledge of several of these sedentarian lineages [13–16], including several articles within this issue [17–21], while others have received less attention (See [4]). In this article we focus on the cirratuliform families Acrocirridae Banse, 1969 [22], Cirratulidae Ryckholt, 1851 [23], and Sternaspidae Carus, 1863 [24], as well as the families Longosomatidae Hartman, 1944 [25], Paraonidae, and Cossuridae, with the goal of providing an updated overview of our current knowledge on their diversity and systematics.

Acrocirridae, Cirratulidae, and Sternaspidae have been placed within the Cirratuliformia along with Flabelligeridae Saint-Joseph, 1984 [26] and Fauveliopsidae Hartman,

1971 [27]. Most families within this clade consist of worms with few external morphological characters, typically thin and elongated, with little body regionalization and reduced parapodia [28,29]. They often exhibit some kind of more or less filamentous branchiae and/or tentacles. Those are sometimes represented by a few, segmentally arranged long filamentous branchiae, as in Acrocirridae, and other times consisting of numerous tentacles, as in Cirratulidae. However, there are exceptions, such as Sternaspidae and Fauveliopsidae, which are represented by species with thick, short bodies lacking prostomial appendages [30,31]. The clade mostly consists of benthic organisms, present mostly in sediments from the intertidal area to the abyssal plains. However, some Cirratulidae are hard-bottom dwellers and a few species of acrocirrids and flabelligerids have conquered the pelagic realm, exhibiting staggering divergent morphologies [32,33]. Among the cirratuliform families described in detail within this article, the most species-rich is Cirratulidae, with approximately 355 described species and 16 genera; followed by Acrocirridae, with 45 species and 10 genera; and Sternaspidae, with four genera and 43 species [1].

The Cossuridae and the Paraonidae are small, threadlike animals [29,34]. The Cossuridae bear a characteristic unpaired branchial filament attached to the dorsal side of one of the anterior segments. Paraonids are also simple-looking, but exhibit a certain heteronomy in the body, with the anterior part wider and slightly flattened, bearing a variable number of segmentally arranged branchiae. Species in both families are benthic and nearly exclusively found in sediments of various granulometries all over the world. In terms of number of species, both families are considered relatively poor. More than 150 species among eight genera of Paraonidae are currently known [35], while only 29 species have been described for the monotypic Cossuridae [36]. These numbers are likely affected, though, by the relatively simple, conservative morphology shown in both groups that makes the identification at the species level challenging. Therefore, they are likely to increase as integrative studies, including detailed morphological analyses and inclusion of molecular data, are performed within both families. Paraonidae and Cossuridae remain unplaced in the annelid tree as analyses of different sources provide mutually exclusive hypothesis, and so far, no comprehensive analyses have been performed. Despite that, some authors have suggested an affinity of these two families with the Cirratuliformia [4]. Initially, *Cossura* was placed within the family Cirratulidae, until Day established the Cossuridae [5]. In a phylogenetic hypothesis based on morphological characteristics, Cossuridae was placed in the clade Scolecida as a basally branching lineage [37]. In contrast, molecular data have grouped Cossuridae with Fauveliopsidae and Paraonidae, in a clade sister to other Cirratulida [38]. A close relationship between Paraonidae and Spionidae or Orbiniidae was suggested during a series of comprehensive cladistic studies [37,39], although it seems more likely that these results were misled by the presence of shared plesiomorphic morphological characters in some of these families [40]. Analyses of DNA sequences have recovered Paraonidae nested within Cirratuliformia, somehow related to Cirratulidae and Cossuridae [41], or as the sister group of Sternaspidae [42].

Finally, Longosomatidae is a family of small, cylindrical, elongated worms, characterized by their distinct body regionalization with an enlarged anterior and posterior end, and thin, very long middle segments [43]. Longosomatidae is a small family, with eight species, all belonging to *Heterospio* Ehlers, 1874 [44], although many species remain undescribed. The phylogenetic affinities of this family remain controversial. Based on morphology, Longosomatidae have long been thought related to spioniform polychaetes, but they also share many morphological similarities with Cirratulidae, and COI analyses recover Longosomatidae somehow related to Cirratulidae, yet acknowledging the limitation that this marker presents for deep phylogenetic reconstruction [45].

This review follows a family-by-family approach. For the families Acrocirridae, Cirratulidae (including Ctenodrilidae), Cossuridae, Longosomatidae, Paraonidae, and Sternaspidae, we summarize our current knowledge on species diversity, systematics, morphological diversity, distribution patterns and ecological preferences. These topics have also been discussed for Fauveliopsidae in a recent review by Salazar-Vallejo et al. [46].

The main gaps of knowledge relative to each of these aspects and the technical and practical aspects of systematic study of these polychaetes in general are discussed along with perspectives for future research.

2. Acrocirridae Banse, 1969

Annelids belonging to the family Acrocirridae Banse, 1969 [22] are cryptic, yet locally abundant worms, characterized by cylindrical bodies with conspicuous falcigerous compound chaetae, and up to four pairs of segmentally arranged, filiform branchiae on the anteriormost body segments [28] (Figure 1). There are benthic and pelagic forms, inhabiting a variety of environments. Whereas this family has historically received little attention, it currently benefits from a regain of interest among marine biologists and even the general public, due to the recent discovery of several deep-sea pelagic species presenting quite peculiar morphologies [32].

2.1. Morphology

Acrocirridae is a morphologically rather heterogenous family, exhibiting quite a large variation in morphology and habitat across species within different clades. Important characters for species identification include the position, size, and shape of the prostomium, shape of the nuchal organs, presence of eyes, presence of head branchiae, number and shape of achaetous first segments, number of branchiae, arrangement and shape of interramal papillae, and the shape and arrangements of the chaetae.

The prostomium is usually dorsal to the peristomium and varies in size and shape between genera and species. The prostomium is rounded and well developed in species of *Acrocirrus* Grube, 1873 [47], *Macrochaeta* Grube, 1850 [48], *Flabelligena* Gillet, 2001 [49], *Actaedrilus* Jimi, Fujimoto and Inamura, 2020 [50] and *Flabelligella* Hartman, 1965 [51]; triangular and smaller in *Chauvinelia* Laubier, 1974 [52] and *Helmetophorus* Hartman, 1978 [51]; and reduced to small regions encircling the nuchal organs in *Swima* Osborn Haddock, Pleijel, Madin, and Rouse, 2009 [53] and *Teuthidodrilus* Osborn, Madin, and Rouse, 2011 [32]. When present, eyes can be one to three pairs in *Acrocirrus* and *Macrochaeta*. Head branchiae are present in *Chauviniela*, *Heteromorphus*, and *Swima*, and their number and shape (elongated or short and stout) are important in recognizing some species [52,54]. Nuchal organs exhibit great variation between genera, varying from simple straight ridges in *Acrocirrus* and *Macrochaeta*, to complex looped rings with branching appendages in *Teuthidodrilus*. Acrocirridae have a variable number of anterior achaetous segments, which can also vary greatly between species, in terms of number, length, distinctiveness, or modification in a "cephalic hood" [28]. The first of these is considered homologous to the peristomium. Segmentally arranged branchiae, also referred to as main branchiae, can be found as one to four pairs on the first anteriormost segments, e.g., ref. [55]. However, they can be easily lost upon collection, and their notable absence in *Flabelligella* may be an artefact. Main branchiae are elongated cirriform appendages in most species but take the shape of small rounded bioluminescent bombs in the species of *Swima* [53].

Parapodia are typically protruding lobes, except in *Macrochaeta*, where they are very reduced [56]. Chaetae may be compound, pseudocompound chaetae, hooks, paddle-like chaetae, and spinous chaetae. The nature and arrangements of chaetae are of systematic importance [57].

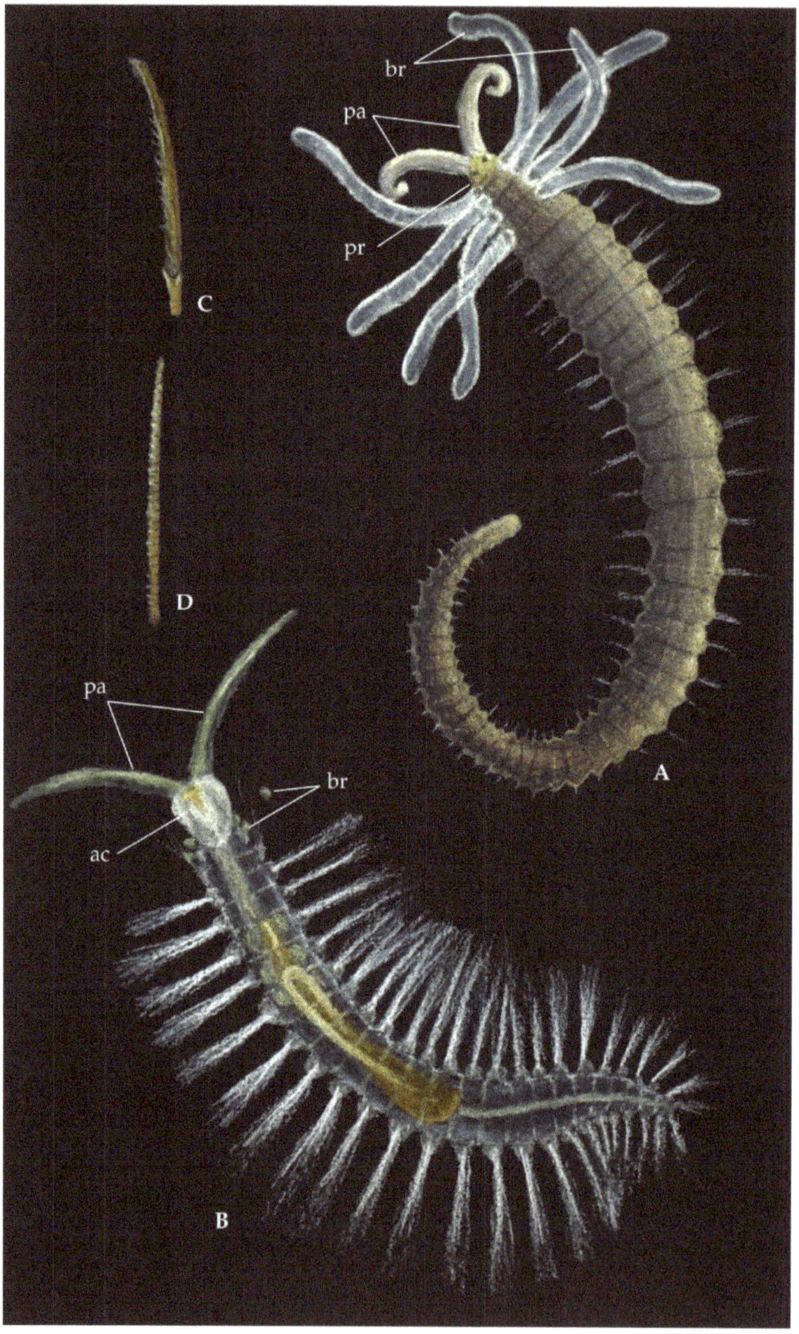

Figure 1. Family Acrocirridae, artistic representation. (**A**) *Acrocirrus validus*, approx. 5 cm. (**B**) *Swima bombiviridis*, approx. 3.5 cm. (**C**) Example of compound chaetae, approx. 100 μm. (**D**) Example of simple chaetae, approx. 100 μm. **ac**, anterior cephalic region; **br**, branchiae; **pa**, palps; **pr**, prostomium. Pastels and colour pencils from Maël Grosse ©. (**B**) after Osborn et al. [53] and (**C**,**D**) after Hartman [58].

2.2. Diversity and Phylogeny

The family Acrocirridae has historically received little attention, with most species described around 1970 (Figure 2A). However, while this has not yet led to a significant rise in species descriptions, renewed interest in deep-sea pelagic habitats and cave habitats opens new perspectives for increased species descriptions in the future, e.g., refs. [53,59]. There are currently 45 described species of Acrocirridae in ten genera [32,50] and at least two additionally new, undescribed species have been reported [59–61]. The majority of these species are described from the Pacific and Atlantic Oceans (Figure 3A) and belong to the benthic genera *Acrocirrus* (13) and *Macrochaeta* (12).

Phylogenetic analyses have inferred three clades within the family, congruent with previous systematic arrangements of the group [50,57], with a clade including *Macrochaeta, Actaedrilus* and *Acrocirrus*, splitting next to a monophylum, which includes two sister clades: *Flabelligella-Flabelliseta-Flabelligena* and the pelagic forms *Swima-Teuthidodrilus*. Whereas each of these clades is well supported by morphological characters, the genera within are sometimes poorly defined and described based on a combination of, often plesiomorphic, characters, emphasizing the need for a systematic review of the family [22,49,62]. Furthermore, several species, potentially important to understand character evolution within the group, remain to be formally described [53,59–61].

2.3. Biogeography

Despite the increasing attention that the group has received lately, knowledge of the global distribution of Acrocirridae remains fragmentary, and it is difficult to infer generalized distribution patterns for most species. The better understood are, by far, members of *Acrocirrus* and *Macrochaeta*, which have been reported mostly in shallow waters around the world, e.g., refs. [55,63,64]. Members of *Macrochaeta* are often recorded from shallow sandy bottoms in the North Atlantic Ocean, especially in the North Sea and Scandinavia [12,65], but also in the Mediterranean and the Canary Islands [28,63]. In contrast, species of *Acrocirrus* are more common in the Pacific Ocean, with 11 species found around Japan, Australia, New Zealand, Hawaii, and the east coast of North America, typically intertidally under rocks [55,64]. There are, of course, exceptions such as *A. frontifilis*, which is fairly common in the Mediterranean and nearby areas of the Atlantic [66]; or *M. pege* and *M. multipapillata* from Washington State and the Galapagos Islands, respectively [22,67]. Members of *Flabelligena* and *Flabelligella* are often found deeper and prefer mud, with several records both in the Atlantic and the Pacific Ocean [49,58,62], as well as the morphologically bizarre *M. polyonyx*, whose generic status demands closer examination [68]. Finally, the actual distribution of other acrocirrids can barely be inferred, since they have been mostly known from anecdotical evidence until recently, when technological advances have started to shed light on their actual habitats in the deep sea.

2.4. Biology and Ecology

Acrocirridae is an exception among other families within Cirratuliformia as it contains both benthic and pelagic species, a feature only shared by Flabelligeridae. Benthic species are usually found in sandy or muddy sediments and sometimes on hard bottoms on or within which they crawl [28]. Pelagic species have for now only been recorded from deep sea, and ecological preferences are difficult to ascertain at this stage. Three undescribed species are notable for being exclusively found in anchialine caves [59,69]. These species are, however, benthic and morphologically similar to their marine relatives, in contrast to other annelids colonizing cave habitats that exhibit secondarily adaptations to swim in the water column [15,70–72].

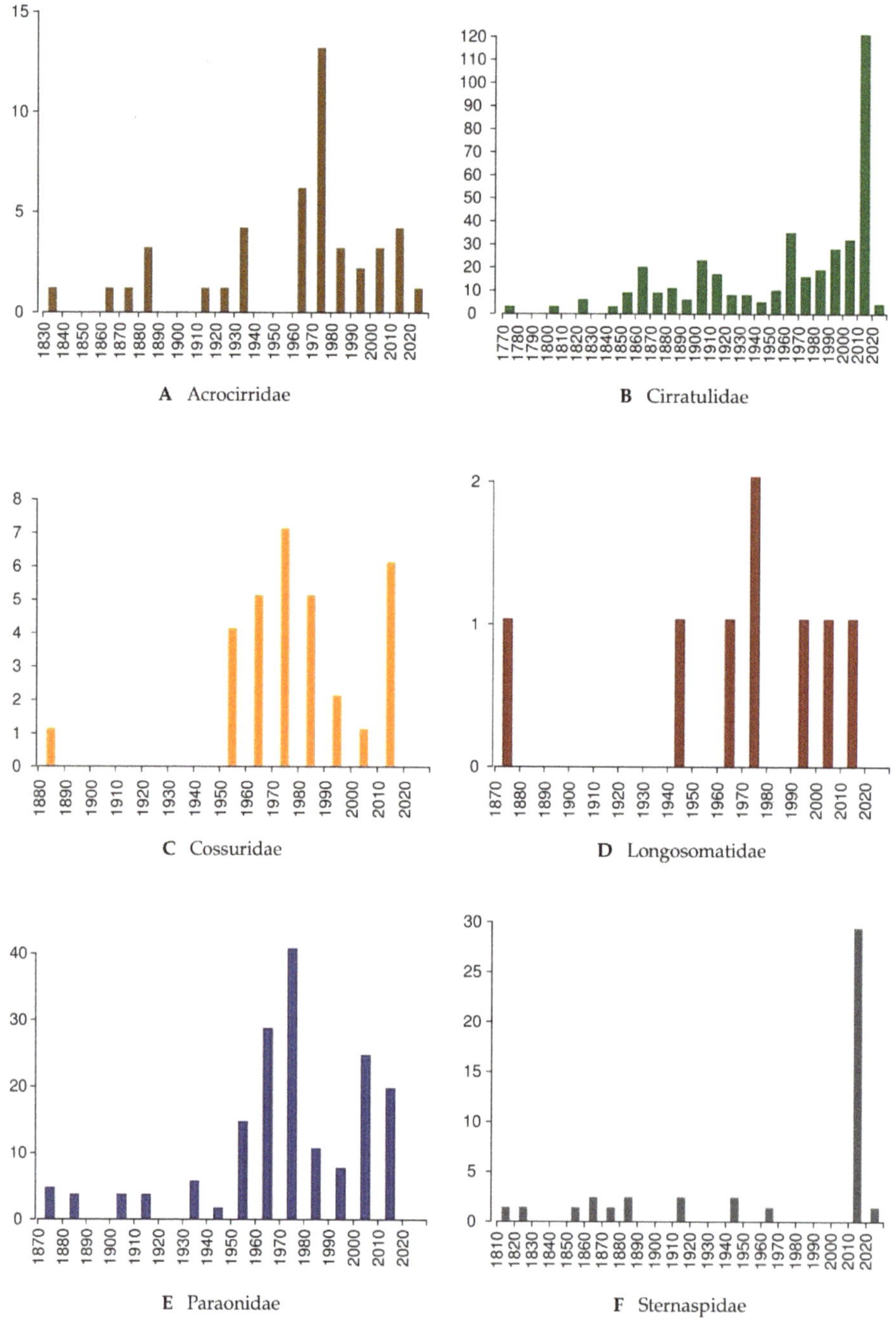

Figure 2. Number of species described per decade per family: (**A**) Acrocirridae; (**B**) Cirratulidae; (**C**) Cossuridae; (**D**) Longosomatidae; (**E**) Paraonidae; (**F**) Sternaspidae.

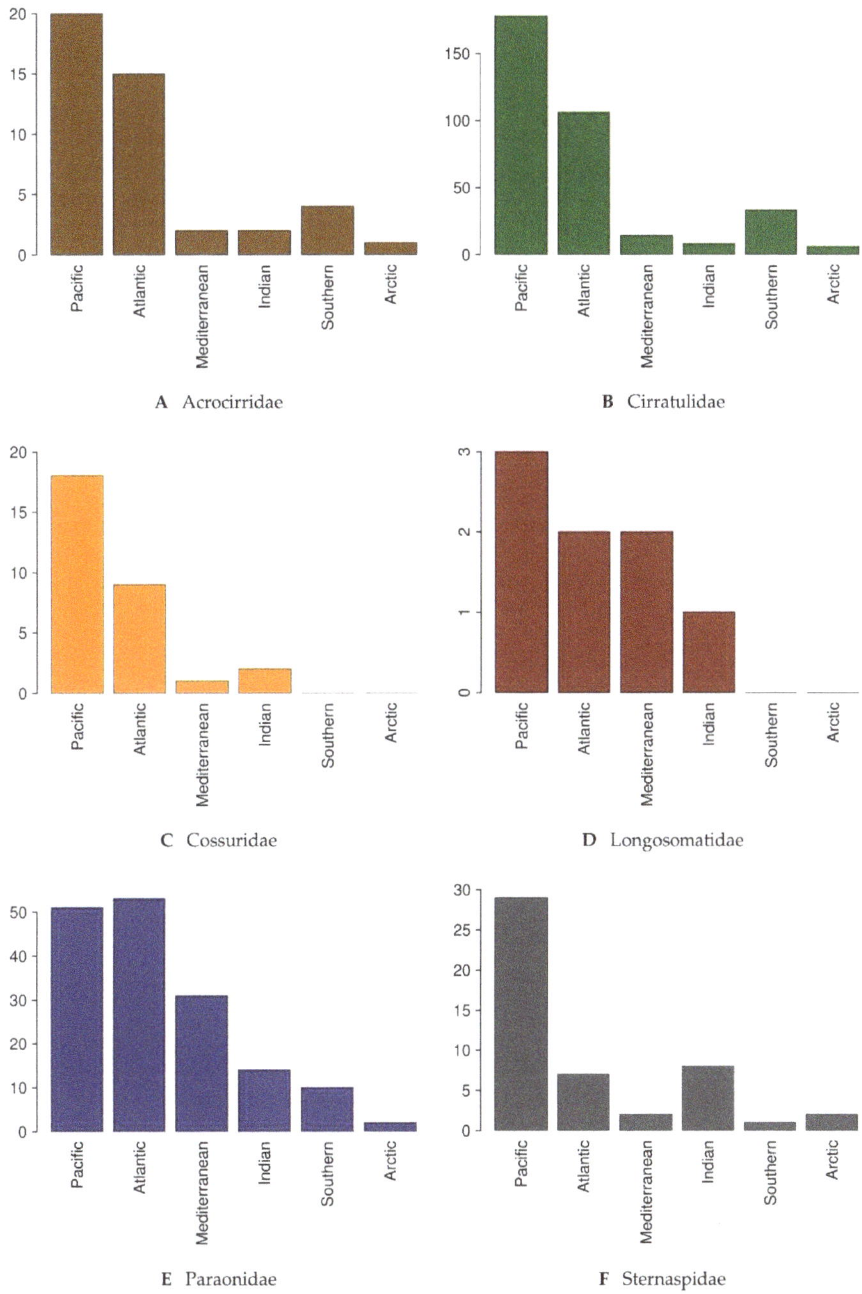

Figure 3. Number of species described (types, localities) per ocean/sea per family: (**A**) Acrocirridae; (**B**) Cirratulidae; (**C**) Cossuridae; (**D**) Longosomatidae; (**E**) Paraonidae; (**F**) Sternaspidae.

Most benthic acrocirrids are classified as selective deposit feeders [73], while pelagic species are probably suspension feeders [32].

Little is known about the reproduction and development of acrocirrids, but the few species studied all reproduce sexually (e.g., *Macrochaeta clavicornis* (Sars, 1835) [74]). Details on sperm transfer, fertilization, and larval development are still unknown in Acrocirridae

2.5. Conclusions

While we expect an increased interest in Acrocirridae in the coming years with the study of new habitats, little is still known about this relatively small, but diverse family of annelids. The wide morphological variation within the family, presumably correlated with the secondary colonization of pelagic and deep-sea habitats, together with the existence of fossils, highlights the potential of Acrocirridae for future ecological and evolutionary studies.

3. Cirratulidae (Ryckholt, 1851) Including Ctenodrilidae (Kennel, 1882)

Annelids belonging to the Cirratulidae Ryckholt, 1851 [23] form a rather morphologically homogenous group. They are elongated, cylindrical worms, with numerous segments and parapodia reduced or absent [29]. The anterior thoracic region is sometimes expanded, with narrower segments, as are sometimes the last few segments. Their grooved tentacles and long filamentous branchiae, which most species bear on many segments, are characteristic features (Figure 4). There are four main groups within Cirratulidae, although they do not necessarily reflect monophyletic groups: (1) multitentaculate cirratulids (*Cirratulus* Lamarck, 1818 [75], *Cirriformia* Hartman, 1936 [76], *Timarete* Kinberg 1866 [77], *Fauvelicirratulus* Çinar and Petersen, 2011 [78], and *Protocirrineris* Czerniavsky, 1881 [79]) are characterized by numerous filamentous dorsal tentacles; (2) bitentaculate cirratulids (*Aphelochaeta* Blake, 1991 [80], *Caulleriella* Chamberlin, 1919 [81], *Chaetocirratulus* Blake, 2018 [82], *Chaetozone* Malmgren, 1867 [11] (Figure 5A), *Kirkegaardia* Blake, 2016 [83], and *Tharyx* Webster and Benedict, 1887 [84]) are characterized by a single pair of thick dorsal tentacles; (3) *Dodecaceria* Ørsted, 1843 [85] is characterized by a single pair of thick lateral tentacles and a comparatively smaller number of branchiae; (4) ctenodrilids (*Ctenodrilus* Claparède, 1863 [86], *Aphropharynx* Wilfert, 1974 [87], *Raphidrilus* Monticelli, 1910 [88], and *Raricirrus* Hartman 1961 [89] (Figure 5B)) are characterized by their lack of tentacles. The status of Ctenodrilidae Kennel 1882 [90] as a distinct family has been controversial [91], but several morphological characters as well as recent molecular phylogenetic analyses support its inclusion within Cirratulidae (e.g., [92,93]. All cirratulids are benthic, and most live in sediments, except for *Dodecaceria*, which is a strictly hard-bottom tubicolous genus, as well as many multitentaculate species.

3.1. Morphology

The identification of cirratulids can be challenging, since most species are small (ranging between a few millimetres and a couple of centimetres) and present characters that can be difficult to observe and interpret. Important species characters include the shape of the prostomium and peristomium, presence of eyes, pattern of tentacles (paired or multiple) and branchiae, presence of a first achaetous segment, shapes and distribution of chaetae, shapes of potentially modified segments, presence of body regionalization, presence of dorsal/ventral grooves/ridges, shape of the pygidium, and methyl green/blue staining patterns [29].

The prostomium varies from narrow and conical (most bitentaculate species) to broad and wedge-shaped (most multitentaculate species) but exhibits little variation in shape and length. A pair of eyespots or two rows of multiple eyespots (the number of which can vary through ontogeny) may be present in some species. The length of the peristomium, number of annulations, and their shape and degree of completeness are of systematic importance [29]. A problematic character for bitentaculate cirratulids is the possible presence of one or two achaetous segments between the peristomium and the first chaetiger, since interpreting an achaetigerous ring as a segment or as a peristomial annulation remains ambiguous without histological studies, and therefore has been interpreted differently among systematists, affecting how the placement of the first branchiae and tentacles are described [94]. Tentacles in cirratulids take two shapes, a single pair of thick appendages in bitentaculate cirratulids and *Dodecaceria* (thicker than the branchiae and easily distinguished from them) or as a variable number of thin filaments in multitentaculate cirratulids (similar to branchiae in shape and length). The position of tentacles (either paired tentacles or tentacular filaments) is important in recognizing species. Among multitentaculate genera, the position of tentacles is used in recognizing different genera. Tentacular filaments arise from a single anterior segment, accompanying the first pair of branchiae in *Cirratulus;* from one or two anterior segments, or accompanying the first pair of branchiae or posterior to it in *Cirriformia;* from two or more anterior segments, posteriorly to the first pair branchiae in *Timarete;* from two or more anterior segments, usually accompanying the first pair branchiae and in small numbers in *Protocirrineris;* and from the first segment, accompanying the first pair of branchiae in *Fauvelicirratulus* [78]. Branchiae are also useful for recognizing both genera and species. Branchiae arise in pairs (two branchiae per segment) typically just above the notopodia in most species, but shift to mid-dorsum posteriorly in *Timarete* and are more numerous in *Fauvelicirratulus*. Like the position of the tentacles, the position of the first pair of branchiae is of systematic importance.

Chaetae include different types of capillaries and acicular spines. Compound chaetae are absent. Shape, number, arrangement, and distribution are important in determining species and genera. Smooth capillaries are present in all genera except *Raphidrilus* and are the only type of chaetae found in *Aphelochaeta* and *Protocirrineris* [95,96]. Serrated capillaries are characteristic of *Kirkegaardia* and *Raphidrilus* [95,96]. Knobby-tipped spines are present in small numbers in *Tharyx* [80]. Bidentate hooks in well-separated neuro- and notopodia are characteristics of *Caulleriella* and can sometimes be found in *Chaetozone* [82]. Unidentate spines are characteristics of *Chaetozone* when arranged in spread fascicles on elevated parapodial membranes creating disctinctive cinctures (Figure 5C), and characteristic of *Chaetocirratulus* when few and not arranged in cinctures [82]. Unidentate spines are also found in *Timarete, Fauvelcirratulus, Cirratulus,* and *Cirriformia*. Short serrated spines are found in *Raricirrus, Ctenodrilus,* and *Aphropharynx* (Figure 5D). Stout spoon- or chisel-shaped chaetae are characteristic of *Dodecaceria*. However, the use of chaetae presents several practical complications. First, complete specimens are needed to be certain of the presence or absence of certain types of chaetae, in particular the spines of bitentaculate genera. Second, the nature and distribution of chaetae can vary with ontogeny, making these characters difficult to assess in immature specimens [97]. Finally, there is no unified terminology for the different types of chaetae found in cirratulids, e.g., ref. [98].

The pygidium is usually simple and rounded, sometimes pointed (e.g., *Chaetozone elakata* Blake and Lavesque 2017 [99]), and can present terminal cirri in a few species (e.g., *Caulleriella venefica* Doner and Blake 2006 [100]). However, cirratulids often break upon collection, so the pygidium can rarely be examined.

Methyl green or methylene blue staining can reveal distinct patterns for some species, mostly of bitentaculate genera, e.g., refs. [99,100]. However, not all species exhibit this pattern, and staining success can vary with the specimens' condition and preservation method.

3.2. *Diversity and Phylogeny*

There is a total of 355 species of Cirratulidae, belonging to 16 genera [1,29,101]. The bitentaculate Cirratulidae comprise approximately 227 species. Nearly 142 new species of bitentaculate Cirratulidae were described in the past 20 years, compared to 15 multitentaculate species, four species of *Dodecaceria*, and four Ctenodrilidae (Figure 2B). Most species were described from the Pacific and Atlantic Oceans, with 168 and 106 type localities, respectively, while the Mediterranean Sea, Indian Ocean, and Arctic Ocean are considerably less studied, with less than 10 species described from the Indian Ocean (Figure 3B). More recently, more than 40 species have been described from the Clarion-Clipperton Fracture Zone in the North-East Pacific, the Southern Ocean, Antarctica, and South-East America [82,94,102], which are otherwise poorly known areas. Other areas where cirratulids have recently been studied and new species described include the Caribbean Sea, Hawaii, Korea, North-East Atlantic, and South Africa [103–106]. In particular, the recent description of five new species of *Protocirrineris* from South Africa and South America is remarkable as it represents important progress in the study of a little-known genus in a poorly studied geographical area [97,107,108].

Although several studies have inferred phylogenetic hypothesis of part or all of Cirratulidae, the monophyly of each genus has not yet been assessed, and the relationships between members of the family are not fully resolved. Several species of *Cirratulus* have been recovered as a monophyletic group based on COI, 16S, and 28S data [91,93]. *Cirriformia* and *Timarete* have not been recovered monophyletic based on COI and 16S data [91,93]. *Raricirrus* is recovered monophyletic using COI and 16S data with two species sequenced [91]. The genera *Chaetozone* and *Aphelochaeta* were also recovered monophyletic using COI and 28S data, although with few reliably identified species of *Aphelochaeta* included [93]. The monophyly of multitentaculate cirratulids as a whole is supported by COI, 28S, and 16S data [91,93]. However, the monophyly of bitentaculate cirratulids as a whole is not supported, and relationships between larger groups (multitentaculate, bitentaculate, *Dodecaceria*, and ctenodrilids) remain unresolved [93].

3.3. Biogeography

Bitentaculate cirratulids show the broadest distribution range, occurring in all oceans and from the intertidal zone to the abyss. Species of Ctenodrilidae show a similar distribution pattern, although based on fewer records, probably reflecting their smaller size and difficulties in identification. *Ctenodrilus serratus* (Schmidt, 1857) [109] is well known for being often found in aquaria, and *Aphropharynx heterochaeta* Wilfert, 1974 [87] was described from an aquarium in Germany. The multitentaculate cirratulids and *Dodecaceria* species mostly occur in coastal shallow water around the globe. As for many other groups, the North American and European coasts have the most records for the family, reflecting sampling efforts in these areas.

As for many other species, early described cirratulids have been recorded worldwide. Species such as *Chaetozone setosa*, *Cirratulus cirratus*, or *Timarete punctata* (Grube, 1858) [110] have long been thought as cosmopolitan. Recent examination of these records and of specimens collected worldwide revealed species complexes in each case [98,111,112]. *Ctenodrilus serratus* was one of the early annelids investigated as potential cryptic species [113]. However, cosmopolitan species also exist. *Raricirrus jennae* Magalhães, Linse, and Wiklund, 2017 [91] was recorded from sunken wood in the northeast Pacific and hydrothermal vents in the Atlantic sector of the Southern Ocean. One species within the *Timarete punctata* complex has been recorded from the South Atlantic Ocean off Brazil, the Caribbean Sea, and Hawaii [112]. *Chaetozone corona* Berkeley and Berkeley, 1941 [114] has been described from California but subsequently reported from Brazil, Turkey, Greece, and France, e.g., refs. [115–117].

3.4. Biology and Ecology

Bitentaculate cirratulids mostly live in sediments, burrowing just under the surface, from intertidal areas (e.g., *Chaetozone christiei* Chambers, 2000 [118]) to abyssal plains (e.g., *Aphelochaeta abyssalis* Blake, 2019 [102]), although some can be found on hard substrates or coralline algae (e.g., *Caulleriella viridis* Langerhans, 1881 [119]). Multitentaculate cirratulids live in more varied habitats, including soft bottoms (e.g., *Cirriformia tentaculata* Montagu, 1808 [120]), rocks and crevices (e.g., *Protocirrineris socialis* Blake, 1996 [95]), or coralline algae (e.g., *Protocirrineris strandloperarum* Elías, Simon, and Sarracho-Bottero, 2019 [107]). *Dodecaceria* species burrow into mollusc shells and calcareous rocks (e.g., *Dodecaceria concharum*), or build their own tubes, creating rocklike colonies (e.g., *Dodecaceria fewkesi* Berkeley and Berkeley, 1954 [121]). Ctenodrilids, like bitentaculate cirratulids, occur from the intertidal to the deep ocean (e.g., *Raricirrus variabilis* Dean, 1995 [122]).

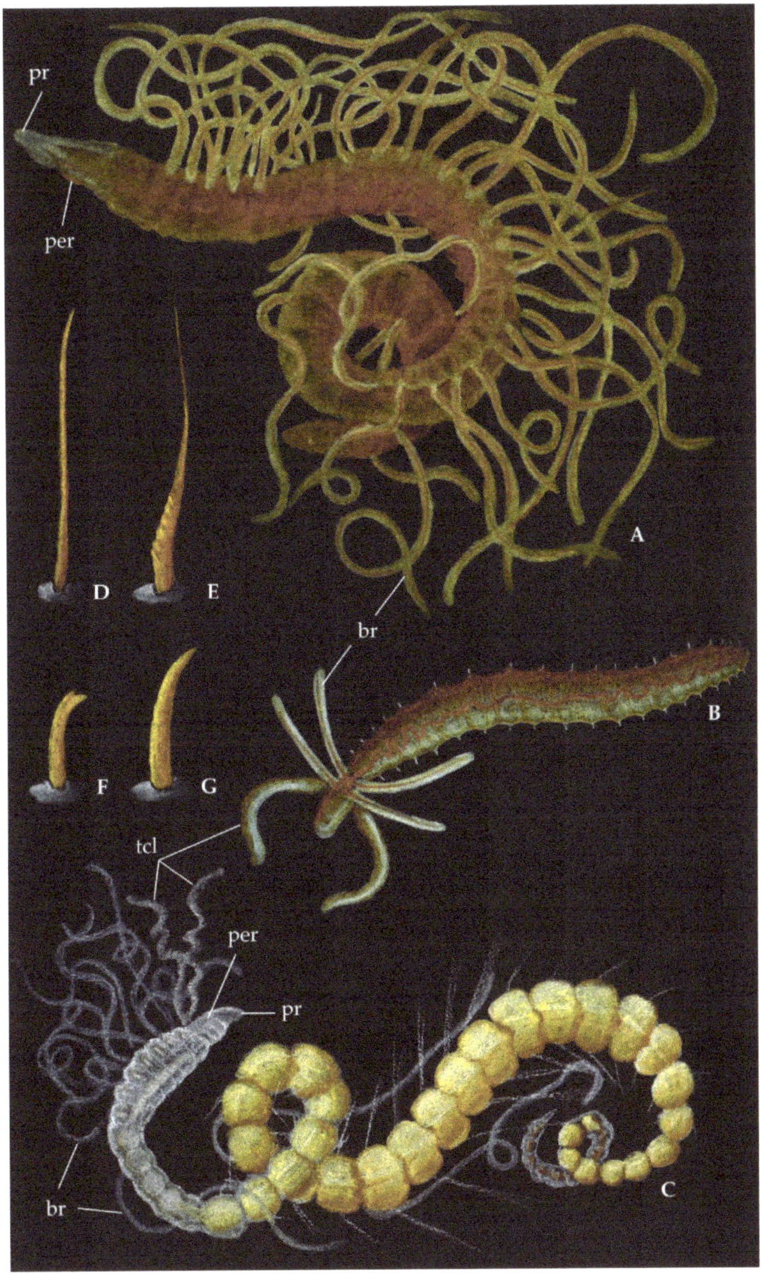

Figure 4. Family Cirratulidae, artistic representation. (**A**) *Cirratulus* sp., approx. 2 cm; (**B**) *Dodecaceria concharum*, approx. 8 mm; (**C**) *Aphelochaeta* sp., approx. 3 mm; (**D**) example of smooth capillary chaeta, approx. 70 μm; (**E**) example of serrated capillary chaeta, approx. 70 μm; (**F**) example of bidentate hook, approx. 50 μm; (**G**) example of unidentate spine, approx. 100 μm. **br**, branchiae; **per**, peristomium; **pr**, prostomium; **tcl**, tentacles. Pastels and colour pencils from Maël Grosse ©.

Cirratulids can occur in high numbers in sediments (e.g., up to 10,000 inviduals per square meter [123]). Several studies also highlight the preference of some species for organically enriched sediments [122,124–127], making them potential bioindicators. Cirratulids and ctenodrilids also exhibit a certain tolerance to some heavy metals, such as arsenic, copper, chromium, cadmium, and zinc [128–131].

Cirratulids are surface or subsurface deposit feeders, using their tentacles to collect particles [73].

Reproductive strategies can vary greatly between (and sometimes within) species of cirratulids and ctenodrilids, and that of Cirratulidae have last been reviewed by Petersen [132]. Species may reproduce exclusively sexually (e.g., *Cirratulus cirratus*), or both sexually and asexually (e.g., *Dodecaceria concharum*). Most species are gonochoristic, although sexual dimorphism has rarely been observed, and a few are known to be simultaneous hermaphrodites (e.g., *Chaetozone vivipara* (Christie, 1984) [133]). Larval development includes direct lecithotrophic, and occasionally viviparous development.

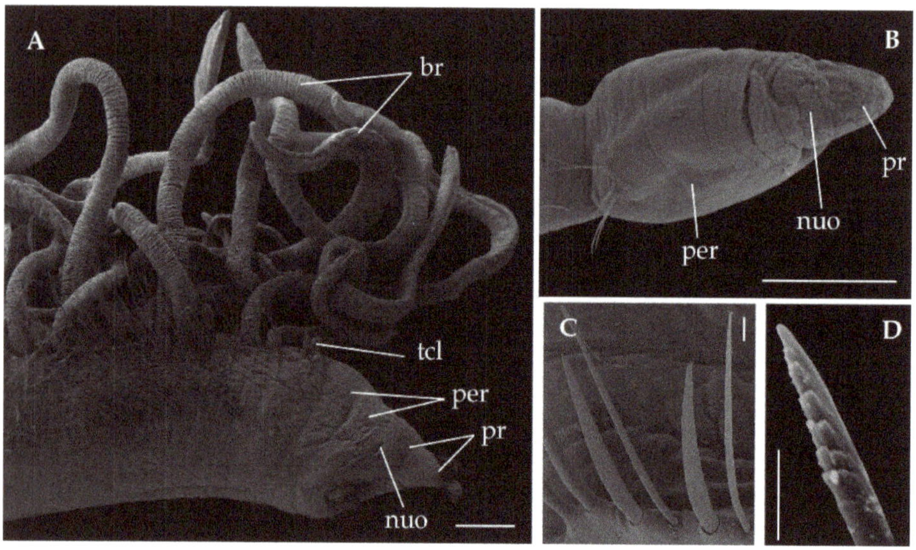

Figure 5. Scanning electron micrographs of Cirratulidae and Ctenodrilidae. (**A**) *Chaetozone* sp., anterior in lateral view; (**B**) *Raricirrus beryli* Petersen and George, 1991 [124], anterior in lateral view; (**C**) *Chaetozone* sp., neuropodial chaetae; (**D**) *Raricirrus beryli*, neuropodial chaeta. **br**, branchiae; **nuo**, nuchal organ; **per**, peristomium; **pr**, prostomium; **tcl**, tentacle base. Scale bars: (**A**,**B**) 300 μm; (**C**) 20 μm; (**D**) 10 μm.

3.5. Conclusions

Considering that we still find new species even in well-studied areas, the presence of cryptic diversity, and the number of areas still unexplored, we can say that even if Cirratulidae is already a very diverse group, there are still a great many species left to be discovered and described. In such a diverse group like, characterized by morphological characters that are difficult to interpret, molecular data can be a great tool for species discovery and description in support of morphological work. A better understanding of species boundaries will also help better understand their distribution and all the other life history traits of the species.

4. Cossuridae Day, 1963

Annelids belonging to the family Cossuridae Day, 1963 [5] are small, threadlike worms, easily recognizable owing to a single unpaired branchial filament attached to the dorsal side of one of the anterior segments (Figure 6). Their body is cylindrical, and the anterior part is more muscular and frequently slightly flattened. Anterior (thoracic) and posterior-most segments are shorter than they are wide, and middle segments are longer and often beaded [34]. Cossurids lack head appendages; their parapodia are reduced, and chaetae emerge directly from the body wall. The pygidium typically bears three cirri.

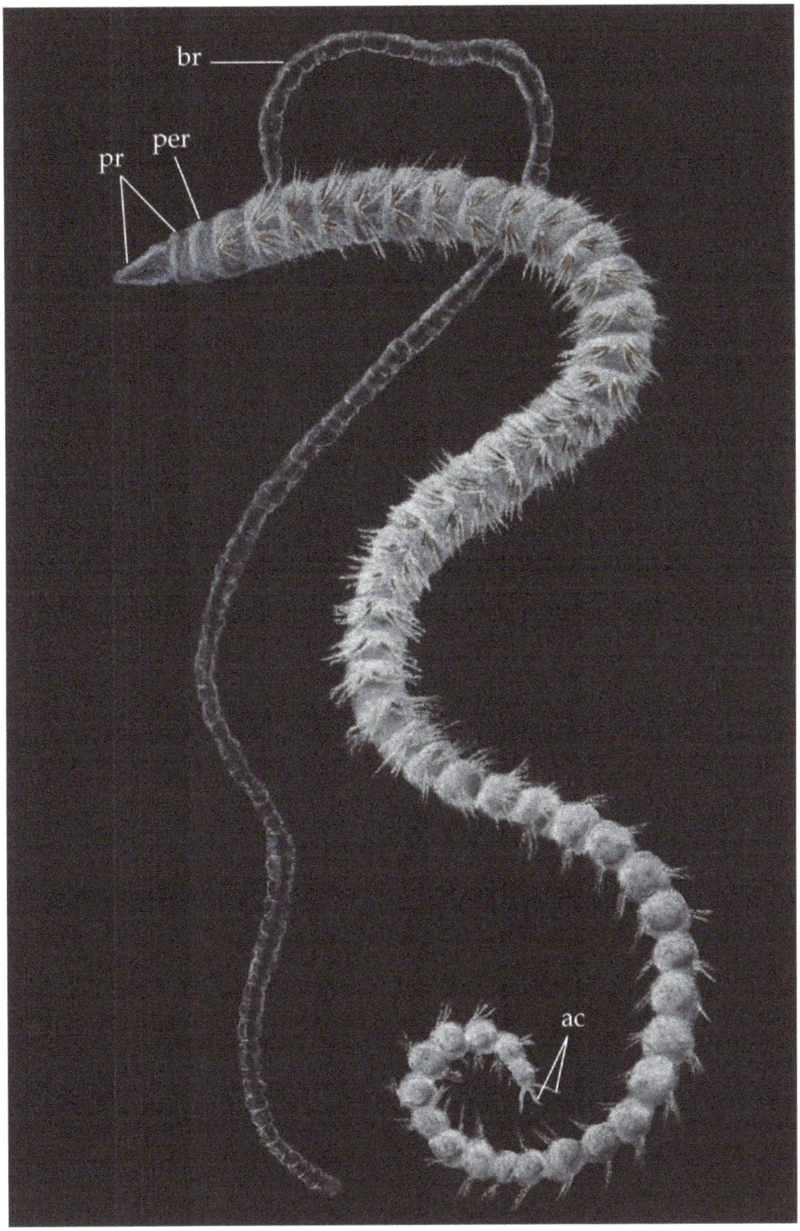

Figure 6. Family Cossuridae, artistic representation. *Cossura soyeri*, 4 mm. **ac**, anal cirri; **br**, branchial filament; **per**, peristomium; **pr**, prostomium. Pastels from Maël Grosse ©.

4.1. Morphology

Cossuridae specimens are difficult to distinguish morphologically and do not have many taxonomically informative characteristics. Their appearance is simple and quite uniform: low body regionalization, simple head without appendices, soft unarmed proboscis, absence of parapodial lobes, and simple uniform chaetae (Figure 7). The most important characteristics used for cossurid identification are the shape of the prostomium, position of the branchial filament, number of thoracic chaetigers, chaetal shapes, and shape of the pygidium. Methyl green staining patterns are also useful for identifying Cossuridae species [134]. Characteristics such as the number of anterior achaetous segments and the number of uniramous parapodia are not useful for cossurid systematics.

The shape of the prostomium is typically conical or triangular from a dorsal view (which is the most common shape in Cossuridae) to round, trapezium-shaped, and almost quadrangular; *Cossura ginesi* Liñero-Arana and Díaz-Díaz, 2010 [135] shows an unusually shaped prostomium with anterior horns. The branchial filament typically arises from chaetiger 2 or 3 or the border between them (Figure 7A,B). Exceptionally, it might arise from chaetiger 4 or 5. The exact position of the branchial filament is typically difficult to establish and requires strictly lateral positioning of the worm during investigation or detailed scanning electron microscopy (SEM) images. Furthermore, the previous segment may have dorsal extensions, from which the branchial filament arises [136,137]. This difficulty produced different interpretations regarding the position of the branchial filament in the same species. For example, *C. consimilis* Read, 2000 [138] was described with the filament arising from the anterior border of chaetiger 3; however, in a redescription, the filament was found to be attached to the segmental border between chaetigers 2 and 3 [136,138].

The number of anterior achaetous rings is not systematically informative, as the first of them has been shown to be of prostomial origin and not seen in relaxed specimens [139]. The number of thoracic segments is often difficult to count. Generally, thoracic segments are short and dorsoventrally flattened, and chaetae emerge from their anterior borders; abdominal segments are longer, frequently appearing beadlike, and chaetae emerge from their central areas (Figure 7C,D). In some species previously assigned to *Cossurella* Hartman, 1976 [62], the transition from thorax to abdomen is very distinct because the abdominal segments bear spinelike or acicular-like chaetae, whereas in other species, variations in the shape of the segments and positions of chaetae are gradual, the border indistinct, and accurate counting difficult [137], which has questioned the validity of species separation based on these characteristics [138]. The position of chaetal bundles has been proposed as a main indicator of transition [136], the first abdominal segment being the first segment which bears chaetae situated in its centre. The number of thoracic segments increases with the worm size and varies considerably between individuals of the same species. Thus, for accurate identification, it is necessary to examine several adult specimens.

Chaetal types and arrangement are very similar across cossurids, yet important for species identification. The first chaetiger is uniramous, and all others are biramous. In early descriptions, the number of segments with uniramous parapodia was erroneously used as a characteristic for distinguishing species, e.g., ref. [140]. Thoracic chaetae are arranged in two vertical rows on both rami (Figure 7B). All thoracic chaetae are capillaries of different lengths and thicknesses, with a cylindrical shaft and more or less flattened hirsute blades. Typically, chaetae of the anterior row are shorter and thicker than those of the posterior row, and neurochaetae are frequently thicker than notochaetae. In some species, coarse thickened chaetae tend to be short, curved, and abruptly tapered. Differences in size and shape of chaetae gradually diminish towards the abdomen. SEM is the most reliable method for examining these chaetae.

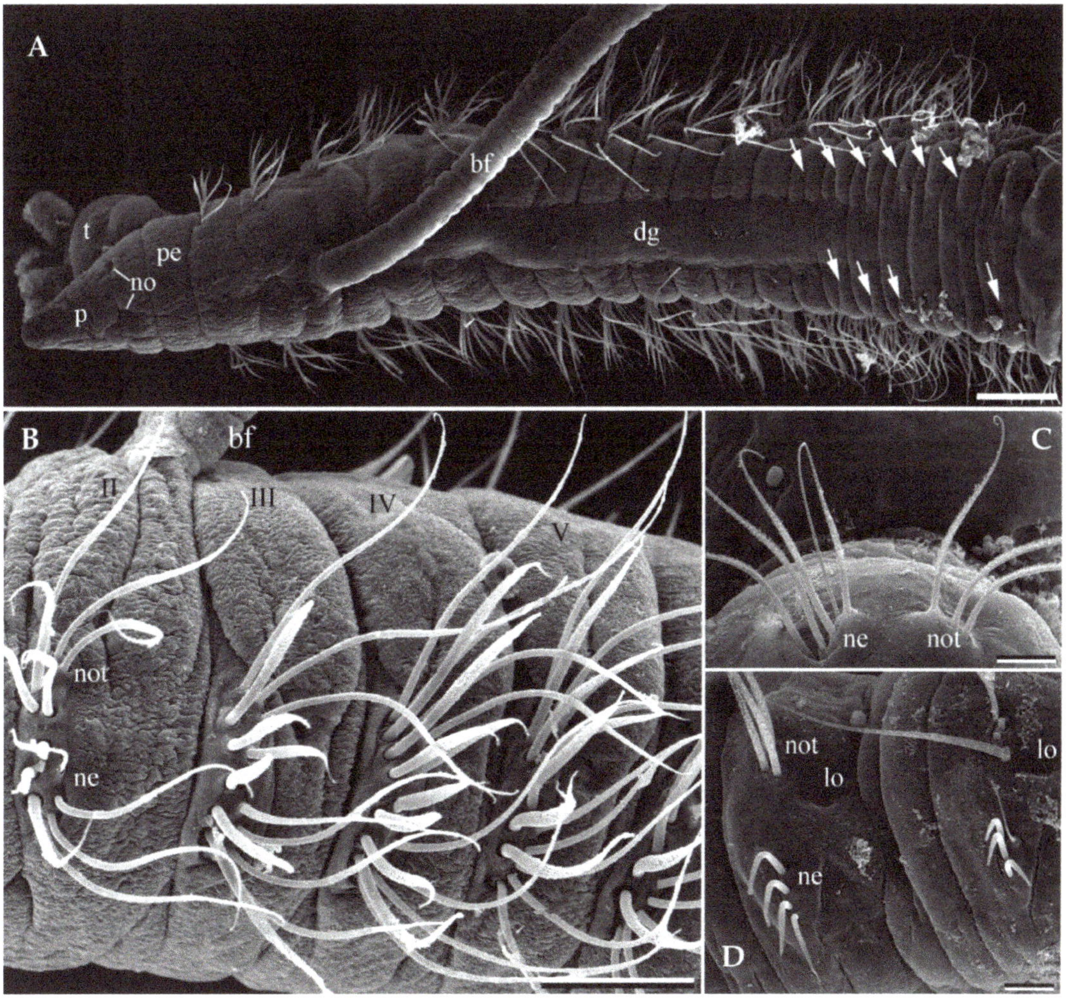

Figure 7. *Cossura pygodactylata*. (**A**) Thorax and beginning of the abdomen, dorsal view; arrows indicate additional transversal furrows dividing segments; (**B**) anterior thorax, lateral view; (**C**) abdominal chaetiger, anterior view; (**D**) two abdominal chaetigers, lateral view. **bf**, branchial filament; **dg**, dorsal groove; **lo**, lateral organ; **ne**, neuropodia; **no**, nuchal organs; **not**, notopodia; **p**, prostomium; **pe**, peristomium; **t**, buccal tentacles. Scale bars: (**A**) 100 µm; (**B**) 60 µm; (**C**,**D**) 20 µm. All reproduced from Zhadan et al. [139] with permission.

Pygidial appendages are important for species identification. Most cossurid species have three long anal cirri, one of which is ventral and two are dorsal. In addition to anal cirri, *Cossura pygodactylata* has 12–20 fingerlike intercirral processes, whereas juveniles of this species do not possess cirri or appendages [139,141]. In *C. pseudakaina* Ewing, 1987 [142], the pygidial rim is scalloped and bears only one short ventral cirrus; in *C. coasta* Kitamori, 1960 [143], three pygidial cirri are branched at their ends. Unfortunately, cossurids are very fragile, and usually only anterior ends are found in collections. In a size assessment study using standard procedures for benthic samples, only 0.3% of Cossuridae were represented by complete specimens [144] In many species, the pygidium remains unknown [138].

4.2. Diversity and Phylogeny

The Cossuridae family comprises only one genus with 29 species. The taxonomic history was described by Ebbe and Blake [34]. During the past 20 years, eight new species have been described. These descriptions were produced by revisions of museum collections [136,138] or by exploring new areas such as northeastern Venezuela [135], the Congo deep-sea fan [145], and northern and northeastern Brazilian tropical estuaries [36]. The number of species described per decade is shown in Figure 2C. The largest number of cossurid species was described in the Pacific Ocean (*n* = 18), and more than one third of this number was from the Pacific coast of North America (Table 1). There are 10 species described in the Atlantic Ocean; most of these also have type locality in the North and South American coasts, and only one is from the Mediterranean Sea (Table 1). Only two species have type locality in the Indian Ocean, and none in the Arctic and in the Southern Ocean (Figure 3C). This can be explained, in part, by different levels of scrutiny of studies on marine benthos in different regions; however, this does not apply generally, since European waters were investigated in detail and for longer periods of time than any other.

Table 1. Species of Cossuridae described from different regions.

Geographic Distribution	No. of Species	Species
Pacific Ocean	**18**	
NE Pacific, North American coast	7	*C. bansei, C. brunnea, C. candida, C. modica, C. pygodactylata, C. rostrata, C. sima*
SE Pacific, South American coast	4	*C. abyssalis, C. alba, C. chilensis, C. laeviseta*
NW Pacific, Japan, China	3	*C. coasta, C. duplex, C. aciculata*
SW Pacific, Australia, New Zeeland, Fiji	4	*C. consimilis, C. hutchingsae, C. keablei, C. queenslandensis*
Atlantic Ocean	**10**	
NW Atlantic, North American coast	3	*C. delta, C. longocirrata, C. pseudokaina*
NW Atlantic, Caribbean Sea, South American coast	2	*C. pettibonea, C. ginesi*
NE Atlantic, Mediterranean Sea	1	*C. soyeri*
SW Atlantic, South American coast	2	*C. heterochaeta, C. yacy*
SE Atlantic, Congo deep-sea fan	2	*C. flabelligera, C. platypus*
Indian Ocean	**2**	*C. dayi, C. dimorpha*

Currently, there is no study on phylogenetic relationships within Cossuridae. One of the reasons for this may be the deficiency of information on species morphology and incompleteness of descriptions. Many species are known only from single or a few incomplete specimens; descriptions are frequently insufficient and sometimes use invalid characteristics (such as the number of achaetous segments or segments with uniramous parapodia), whereas the most important characteristics of those species remain unknown (see [36,138] for review). The second reason is the lack of molecular genetic data for cossurids. GenBank contains only a few sequences of two *Cossura* species. It is unclear why molecular genetics of these annelids have not been studied so far.

Therefore, the family Cossuridae requires revision. Re-investigation of type material and a large number of specimens per species are needed to reveal intraspecific variability and to obtain comprehensive morphological character data on species. Sequence data of mitochondrial and nuclear genetic markers are required to produce molecular phylogenetic trees of this family. Numerous questions need to be answered—e.g., What is the plesiomorphic condition of different characteristics? Which are the most basally splitting cossurids? Do morphological characteristics and molecular genetic data give congruent phylogenetic information? Do former *Cossurella* species form a monophyletic group? Are there clear patterns of character evolution within the family, suggesting the delineation of other genera?

4.3. Biogeography

Examining distribution patterns of cossurids is frequently associated with difficulties regarding species identification. Most cossurid species are known only from type localities or the same biogeographic region, whereas other species were reported far from type localities, such as *Cossura ginesi* or *C. soyeri* Laubier, 1963 [135,136,146]. *Cossura longocirrata* is the first described species of the family, and for many years, all cossurid findings were assigned to this species. Re-examination of material from different localities has revealed much confusion with *Cossura pygodactylata*, and *C. soyeri*, *C. ginesi*, *C. pygodactylata*, and *C. coasta* are other putative examples of cossurid species complexes that may need re-examination. The reported wide distibutions of several cossurid species can be probably ascribed to both cryptic species and misidentification, and comparative analysis of morphology and genetic investigations are needed to confirm these records.

A few notes on cossurids as introduced species have been made. Callier et al. [147] suggested that *Cossura pygodactylata* was introduced at the Southsea Marina (south coast of England) by boats, subsequently colonizing the area and probably extending its distribution range from within the marina to its environs. *Cossura coasta* was suggested as a Lessepsian migrant, which entered the Mediterranean Sea through the Suez Canal [148]. This species is included in lists of Mediterranean alien species as a cryptogenic and questionable species [149,150] and also listed as invasive in India [151,152].

4.4. Biology and Ecology

Members of Cossuridae inhabit sediment, such as silt, clay, and sometimes fine sand, usually occurring within the upper 10 cm layer. They can be found from the intertidal to the abyss (e.g., *Cossura flabelligera* Zhadan, 2017 [145], at a maximum depth of 4719 m). Cossuridae species differ based on their preferred layer of sediment. For example, *C. chilensis* was found mainly at 5–10 cm depth [153], and *Cossura* sp. were found near the surface (0–5 cm) and mostly at 0–2 cm sediment depth [154]. Cossurids do not build permanent tubes; however, worms typically inhabit temporary tubes consisting of mucus with adhered sediment particles. In *C. flabelligera*, this mucous sheath is quite strong, persists during washing, and resembles the tunic of flabelligerids [145].

Ecological preferences also vary among Cossuridae species. Cossurids seemed rare within benthic communities before the second half of the 20th century, probably reflecting the methods used for washing sediments and the use of mesh size of 1.5 mm, too large to retain most cossurid species. The current use of mesh sizes of 500 µm (or even finer) have revealed that cossurids are actually abundant and often the dominant species in many areas such as the Atlantic coast of France and Spain [141,155]. All cossurids are marine, but some species inhabit estuaries and tolerate reduced salinity as well as substantial changes in salinity, e.g., *Cossura coasta* [156,157], *C. delta* [140,158–160], *C. yaci* [36], and *C. pygodactylata* [161]. Ecological differences can indicate presence of cryptic species. Some species exhibit a broad depth range, from shallow subtidal to lower slope, and it has been suggested that separate populations of these species occupy narrower depth zones [34]. Cossurid populations are resilient and may even respond positively to bioturbation [162,163]. Some cossurids are also indicators of different types of pollutants, such as heavy metals [147,154,160,164–169] or organic pollutions [156,170–172], and are potentially oportunistic. Some species, such as *C. coasta*, have been considered to belong to groups III and IV of the AZTI's Marine Biotic Index (AMBI) [173–177], which is commonly used to evaluate pollution based on ecologic groups of different sensitivity to organic matter. In contrast, *C. candida* in Los Angeles Harbor was found to be a characteristic species of healthy environments [178,179]. Similarly, in New Zealand, *C. consimilis* was predominant under pristine conditions and was thus categorized in group I: species which are very sensitive to organic enrichment and occur in unpolluted environments [180].

Cossurids are motile subsurface deposit feeders, collecting food particles with their short ciliary tentacles from the walls of their burrows [73].

The reproduction biology of Cossuridae is poorly known. Annual recruitment with spring peaks has been shown in several species [139,181]. Planktonic larvae are most probably absent in Cossuridae, as they have never been observed. However, there have been a few sightings of juvenile cossurids with natatory chaetae in the water column [138,182], suggesting a semi-planktonic dispersal phase in some species, which must be taken into account during morphological studies.

4.5. Conclusions

Cossuridae—despite the small number of described species—remains poorly known. The most pressing problems are insufficient species descriptions, gaps in the knowledge of reproduction and development, questionable species distributions, numerous misidentifications, possible cryptic species, lack of genetic and morphological data, and no phylogenetic hypotheses.

5. Longosomatidae (Hartman, 1944)

Annelids belonging to the family Longosomatidae Hartman, 1944 [25] are small, cylindrical, elongated worms, with few segments. They are characterized by their distinct body regionalization, with enlarged anterior and posterior ends, and thin, very long middle segments (Figure 8).

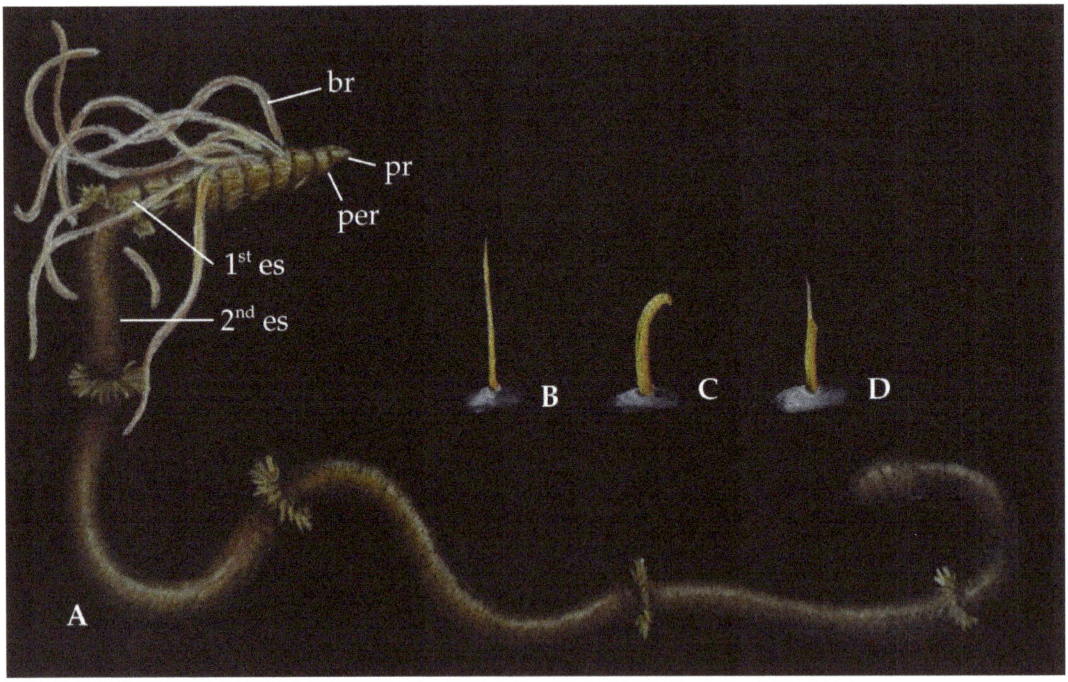

Figure 8. Family Longosomatidae, artistic representation. (**A**) *Heterospio* sp., 4 cm (imaginary colours—there is no record of observation of any live specimen—and shown without palps as most species are presented); (**B**) capillary chaetae, 100 µm; (**C**) hook, 80 µm; (**D**) aristate spine, 80 µm. **1st es**, first elongated segment; **2nd es**, second elongated segment; **br**, branchiae; **per**, peristomium; **pr**, prostomium. Pastels and colour pencils from Maël Grosse ©. (**B–D**) after Hartman [183,184].

5.1. Morphology

The prostomium is usually conical and rounded, with a pair of nuchal organs. The peristomium is short and the presence of palps is confirmed for several species [185]. The thorax is usually six to eight chaetigers with a variable number of branchiae. The number of thoracic and abdominal segments and pairs of branchiae are of systematic relevance. The abdominal segments are narrow and greatly elongated, with many chaetae typically arranged in a cincture anteriorly. The length of the first elongated segment is an important character to differentiate species. The posterior end, which is usually lost during the collecting process, is typically composed of inflated segments.

Chaetae include capillaries of various lengths and acicular spines, whose shape and arrangement are taxonomically relevant, although the lack of a unified terminology for chaetal morphology complicates the assessment of this character [185].

The pygidium is a simple ring.

5.2. Taxonomy and Phylogeny

Longosomatidae is a small family, with eight species in the monotypic genus *Heterospio* Ehlers, 1874 [44] (Figure 2D), although many species probably remain undescribed. Longosomatids are known from the Atlantic, Indian, and Pacific Oceans, as well as from the Mediterranean Sea, with respectively two, one, three, and two species described from each area (Figure 3D).

5.3. Biogeography

Longosomatids have been found from intertidal sediments to abyssal plains. *Heterospio longissima* Ehlers, 1874 [44] was originally described from the Atlantic Ocean and has been subsequently recorded from the Indian and Pacific Oceans [45]. However, as for many early described species, some of these records might concern other species and the distribution range of *H. longissima* might be more restricted. In particular, records from the Indian Ocean have been shown to belong to another species, *H. indica*, which is so far known to be only from this part of the world [45].

5.4. Biology and Ecology

Little is known of the biology and ecology of this family, as most of the species have rarely been encountered and no individuals described alive. They inhabit soft sediments and are probably subsurface deposit feeders [73].

5.5. Conclusions

Longosomatidae is a small family, and its members are rarely encountered. Therefore, it is probably the least studied member of Cirratuliformia, and very little is known about its biology. It is certain that more species are still to be described.

6. Paraonidae Cerruti, 1909

Annelids belonging to the family Paraonidae Cerruti, 1909 [6] are usually small to medium-sized, and characterized by a relatively simple, conservative morphology, which makes identification at the species level challenging (Figure 9). The body shows a certain heteronomy, with an anterior part wider and often slightly flattened, bearing a variable number of branchiae; and a posterior part, cylindrical, with reduced parapodia and often possessing modified chaetae in either the notopodium or the neuropodium. The pygidium is usually semicircular, subtriangular, or bilobed, and bears two or three anal cirri.

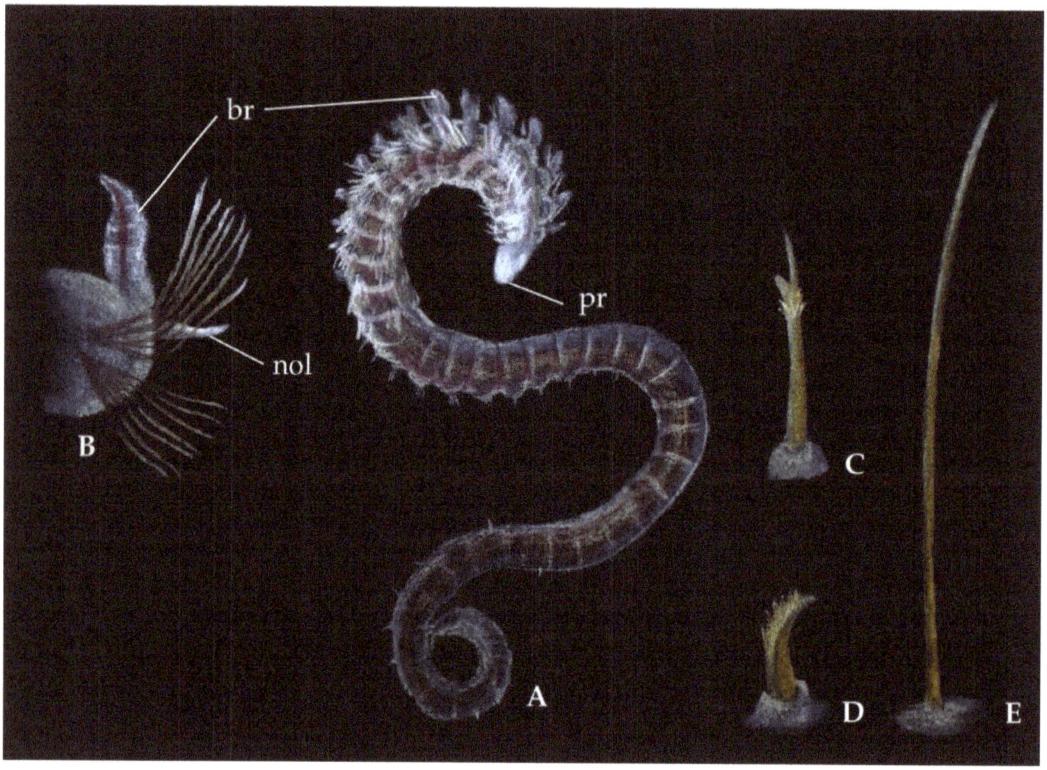

Figure 9. Family Paraonidae, artistic representation. (**A**) *Aricidea assimilis*, 5 mm; (**B**) transversal view of a parapodium, 400 μm; (**C**,**D**) neurochaetae, 30 μm; (**E**) capillary chaetae, 40 μm. **br**, branchiae; **pr**, prostomium; **nol**, notopodial lamellae. Pastels and colour pencils from Maël Grosse ©. (**B–E**) after Blake [35,186].

6.1. Morphology

Paraonidae are characterized by a higher degree of regionalization in comparison with other Cirratuliformia, showing a distinction between anterior (thoracic) and posterior (abdominal) regions. Nonetheless, the transition between thoracic and abdominal regions is gradual, with intermediate segments showing transitionnal features. Diagnostic morphological characters include shape of the prostomium, presence or absence of an antenna, number of pre-branchial and branchial chaetigers, pattern of notopodial and neuropodial post-chaetal lobes, and presence of modified chaetae. Despite these characters, the identification of paraonid species is often hindered by their fragility, making it impossible to appreciate several morphological features, as well as by the scarce information on the ontogenetic variation in key morphological characters [37,187].

The prostomium is usually triangular, sub-trapezoidal, or rounded, but in some species of the genera *Aricidea* (e.g., *Aricidea trilobata* Imajima, 1973 [188]) and *Sabidius* (e.g., *Sabidius cornatus* (Hartman, 1965) [58]), it is more or less strongly trilobed. *Sabidius* species are peculiar in having a thickened cuticle on the anterior margin of the prostomium [189]. The presence or absence of a single antenna has been originally used to separate different genera [35], but it has been shown that all monophyletic lineages within Paraonidae include both species with and without antenna [14]. The usefulness of the presence and shape of antennae in species descriptions has been instead confirmed by several systematics works [190–192], although the occurrence of a branched antenna, originally considered diagnostic [193,194], is most likely the result of issues in the regeneration or pathological conditions [14,195]. Eyes are usually simple eyespots located in the ventral part of the prostomium, and their presence is not always easy to identify in preserved material, as these structures can fade after preservation in ethanol. Examination of live material showed, however, that in conspecific individuals at different growth stages, eyes may be present or absent, and that juvenile individuals usually show eyespots, while adults lack them. The peristomium is largely fused with the prostomium, and difficult to sort out; the alleged presence of a distinct, achaetous peristomial ring in the original descriptions of some species [84,196] has later been proved to be a preservation artefact [39,194]. Prostomial and peristomial sensory structures often include ciliary bands, a terminal retractile organ and paired nuchal organs[39]. These features might represent useful taxonomic characters, but their informativeness remains to be determined. The proboscis is soft and unarmed.

The first chaetigers are devoid of branchiae and, usually, of modified chaetae. The majority of species has three pre-branchial chaetigers, but four pre-branchial chaetigers have been observed in members of the genera *Aricidea*, *Cirrophorus*, and *Paradoneis*, and adult individuals of *Aricidea simonae* Laubier and Ramos, 1973 [197] have branchiae starting at chaetiger 3. The genus *Levinsenia* has typically five to eight pre-branchial chaetigers [39]. Although the first descriptions suggested a certain degree of intraspecific variation in the number of pre-branchial chaetigers [39,198,199], the variability of this character has been re-evaluated [194,200], and the number of pre-branchial chaetigers seems to be informative even at the supraspecific level [14]. Ontogenetic change in the number of pre-branchial chaetigers observed in *A. simonae* [187] is in fact an exception to a general trend of intraspecific stability of this character. Branchiae are paired and are usually flattened and ciliated, showing the presence of two blood vessels, and limited to a certain number of anterior segments. The number of branchiae in the vast majority of species increases with the body size [39,201], and the branchial region does not usually exceed the first third of the body length. An exception is represented by *Cirrophorus nikebianchii* Langeneck, Barbieri, Maltagliati, and Castelli, 2017 [201], showing a very high number of branchial pairs, which might represent an adaptation to hypoxic environments [192]. Abranchiate species are known in almost all genera, usually from abyssal environments [39].

Parapodia in Paraonidae are reduced, biramous, and the majority of chaetae are thick, curved capillaries, becoming thinner and straighter posteriorly. In addition, several modified chaetae, usually restricted either to the neuropodia or to the notopodia, occur in both parapodial rami in *Paradoneis eliasoni* Mackie, 1991 [202], *Paradoneis strelzovi* de León-González and Diaz-Castañeda, 2011 [203], *Paradoneis andreae* López and Sikorski, 2017 [204], and *C. nikebianchii*. Modified notochaetae occur in the genera *Cirrophorus* and *Paradoneis* and are usually lyrate or derived from typical lyrate chaetae through thickening of one of the rami; *Paradoneis spinifera* (Hobson, 1972) [205] and *Paradoneis drachi* Laubier and Ramos, 1974 [197] are characterized by unbranched notopodial spines. Modified notochaetae usually start either at the last pre-branchial chaetiger or in the branchial region, and in species with thickened, acicular chaetae, they are usually lyrate in the branchial region, becoming thicker afterwards. Modified neurochaetae exhibit a wide variation in shape ranging from thickened capillaries to strong hooks or spines, sometimes with additional hairs or a sub-terminal hyaline sheath. The shapes of modified neuro chaetae were employed by Strelzov [39] to diagnose infrageneric groups, although without real phylogenetic meaning [187,201]. However, the shape and distribution of modified noto- and neurochaetae remain as useful diagnostic features. Chaetigers usually carry notopodial post-chaetal lobes, as cirriform, tubercular, or digitiformprotrusions of the body wall, whose shape, variation, and distribution along the body are often useful to discern species, particularly within *Cirrophorus* and *Paradoneis* [39,206]. Neuropodial post-chaetal lobes are limited to pre-branchial and branchial regions and usually inconspicuous, except for a few *Aricidea* species. Additional papillary structures have been reported in a few species and are usually considered as systematically informative [187,194,207].

The pygidium is usually semicircular or bilobed and bears three elongated anal cirri, except for species of *Levinsenia,* characterized by triangular pygidium with two cirri. Additional pairs of cirri have been reported in a few species, mostly interpreted as paedomorphic traits [208,209]; based on personal observations, additional pairs of pygidial cirri actually represent the notopodial post-chaetal lobes of differentiating chaetigers in the proliferation zone between metastomium and pygidium.

6.2. Diversity and Phylogeny

Despite the high diversity and abundance displayed in seas all over the world, Paraonidae have been the subject of relatively few comprehensive studies, and its evolution, species diversity, and ecology remain largely unknown [35,39]. The wrong assignment of early described species to Orbiniidae [210], Spionidae [211–214], and Cirratulidae [215], poor early descriptions [84,216], and nomenclatural debates (See for example ICZN opinion 1139) have been sources of much systematic confusion.

For a long time, the family was considered species-poor and consisting of relatively few species with broad distributions. Currently, more than 150 Paraonidae species have been described, belonging to eight genera: *Aparaonis* Hartman, 1965 [58], *Aricidea* Webster, 1879 [210], *Cirrophorus* Ehlers, 1908 [215], *Levinsenia* Mesnil, 1897 [217], *Paradoneis* Hartman, 1965 [58], *Paraonella* Strelzov, 1973 [39], *Paraonis* Grube, 1872 [211], and *Sabidius* Strelzov, 1973 [35,39]. Descriptions of new species peaked in the 1960s and 1970s [39,58,188,197,199,207], and increased again in the 2000s and 2010s (Figure 2E). Nonetheless, the number of undescribed species might well be comparable with that of currently described ones [35], which most likely comprise large complexes of cryptic and pseudocryptic species [14,218]. Approximately half of the described Paraonidae species is associated with deep-sea environments (up to 6000 m depth), which are expected to host a high number of undescribed species [219].

Recent phylogenetic analyses suggest that the systematics of Paraonidae requires a thorough revision. The phylogenetic inference based on morphological data by Reuscher [220] confirmed Strelzov's synonymy between *Cirrophorus* and *Paradoneis* [39], while a molecular phylogeny [14] showed several inconsistencies between accepted genera and monophyletic clades retrieved. Species of the genus *Paraonis* were recovered nested within *Aricidea*, while *Cirrophorus* and *Paradoneis* species were intermixed in two different clades. A part of *Aricidea* represented a separate, morphologically homogeneous clade, sister to *Levinsenia*; for this clade, Langeneck et al. [14] described *Blakeia*, which is a junior homonym of *Blakeia* Kiriakoff, 1967 (Lepidoptera) and thus unavailable. A synthesis between contrasting morphological and molecular data and a systematic revision based on total evidence are still lacking for this family, while molecular sequences are lacking for the majority of the species, and several genera (such as *Paraonella*, *Sabidius*, and the subgenus *Aedicira*).

6.3. Biogeography

Paraonidae have been reported worldwide, from the intertidal zone [221] to bathyal and abyssal environments [39]. A similar number of Paraonidae species have been described from the Atlantic and Pacific Oceans (53 vs. 51); the Mediterranean accounts for a relatively high number of taxa (31), while only a few species have the type localities in the Indian Ocean, the Southern Ocean, and the Arctic Ocean (Figure 3E). The scarce information on Paraonidae diversity and distribution in some geographical areas, such as the tropical Eastern Atlantic Ocean and Indian Ocean, most likely reflects limited sampling efforts [39].

Some species have an alleged cosmopolitan distribution [39,186,222], whereas according to Strelzov [39] other species show a bipolar distribution, which could reflect the scarce knowledge on the diversity of paraonids in tropical and equatorial areas. Individuals of *Levinsenia gracilis* sampled in the Pacific Ocean were morphologically identical to Eastern Atlantic individuals [223], but another allegedly widespread species, *Aricidea catherinae* Laubier, 1967 [207], turned out to include several pseudocryptic lineages, with the taxon *A. catherinae s s.* likely restricted to the Mediterranean Sea [14]. On the other hand, several deep-sea species described for the Mediterranean Sea, and until recently considered endemic [197,199,224], have been recently reported for the North-Eastern Atlantic Ocean [194,204], suggesting that the restricted distribution of several deep-sea species might be an artefact due to limited sampling efforts. Many deep-sea species are in fact known only from type material, and other taxa, especially recently described ones, are under-reported due to the rather complex systematics of the group and the use of outdated identification keys.

Only two species of Paraonidae have been suggested to represent non-indigenous species. *Aricidea fragilis* Webster, 1879 [210], with type locality in the western Atlantic Ocean, has been proposed as potentially invasive in the Mediterranean [195,225]. *Aricidea bulbosa* Hartley, 1984 [226] was reported from the Sea of Marmara as a possible Lessepsian immigrant [227]. Reports of species far from their type locality are relatively frequent, but they are usually just names in species inventories and general ecological works [66,228], without descriptions of specimens. Given the high number of pseudocryptic lineages found in this family [14], it is far more likely that these records are misidentifications of native species [150] or refer to undescribed taxa [14].

6.4. Biology and Ecology

Paraonidae live on the surfaces of soft sediments, or in their uppermost layer, where they can reach high densities [186,229]. Many species build temporary mucous tubes, covered by sediment particles, which are difficult to observe due to their brittleness. Living animals reside in the sediment in a characteristic "corkscrew" position with the middle and posterior parts spiralled beyond the axis delineated by the anterior, which remains straight and bears the branchiae. A few species have been sporadically reported from hard-bottom assemblages [230], and some species show a strict association with seagrass and *Caulerpa* beds [231–233]. Sediment grain size and origin seem to affect distributions of congeneric species. In shallow environments, individuals occur in gravel, coarse sand, fine sand, and silt; infralittoral environments typically show a low Paraonidae diversity, with few co-occurring species, one of which is often dominant [196]. Deeper environments are characterized by the prevalence silt and clay and often show a higher species diversity, although in lower abundances [234]. Several species have been reported from a wide bathymetric range; for instance, *Cirrophorus branchiatus* Ehlers, 1908 [215] has been reported from 8 to 2700 m. Other species are more strictly related to bathyal and abyssal depths [194]. Extremely wide bathymetric ranges are not uncommon among polychaetes, but this feature could actually be an artefact due to incorrect identification, or absence of reliable morphological features. The majority of Paraonidae species is typically marine and stenohaline, and only few species have been reported from brackish estuarine environments and coastal lagoons [190,235,236]. The occurrence of the same species in very different habitats could suggest the existence of cryptic species, rather than an extreme adaptability of a single species [14]. The same hypothesis can account also for the high number of species assumed to be cosmopolitan, or with very wide geographical distribution.

Paraonidae occurring in brackish environments usually exhibit a low tolerance to wide variations of salinity, and usually occur in enclosed environments with limited salinity variation, as well as in marine environments characterized by organic enrichment [201,237]. For this reason, Paraonidae occurring in brackish-water environments can be considered marine species, but with high tolerance towards eutrophic conditions, rather than brackish-water species.

Paraonidae were traditionally considered as subsurface deposit feeders [73]. However, a few detailed studies have demonstrated that *Paraonis* spp. feed on microbenthos and microbial films growing on the surface layer of the sediment without leaving their tubes. An active selection of food items has also been observed [238,239].

Reproduction remains largely unknown. Paraonids are usually gonochoristic, with extremely sporadic reports of hermaphrodite individuals [240]. In most species, mature individuals are found all year round [240]. Larvae are unknown, but brooding of juveniles is known [241]. This, together with the occurrence of epitokes [242] and the large size of eggs, suggests that this family might have lecithotrophic larvae or even direct development [243].

6.5. Conclusions

Although Paraonidae have been the subject of several detailed taxonomic studies, and are partially investigated with molecular tools, there are still several gaps in the knowledge of this family. The available molecular data support the widespread belief that a large number of species (perhaps the majority) is still undescribed [35,39]. Furthermore, some genera are still unknown from the molecular point of view, and the current systematic view is largely inconsistent with the phylogenetic reconstruction based on molecular data. The uncertain identity of some early-described species has consequences for the systematics of the group and represents one of the first issues that should be tackled. Lastly, the ecology and the biology have been studied only in a few species, and in particular, the reproduction of Paraonidae is still largely unknown.

7. Sternaspidae (Carus, 1863)

Sternaspidae Carus, 1863 [24] is a taxon easily recognized by the presence of a bilaterally symmetrical ventro-caudal shield (Figures 10 and 11). The masklike shape and prominent colouring of this shield resembles the eye region of owls and earned Sternaspidae the common name "mud owls" [244]. Present knowledge on the morphology, biology, and systematics of the group has been summarized by Sendall and Salazar-Vallejo [244] and (including anatomy) by Fiege [31,245].

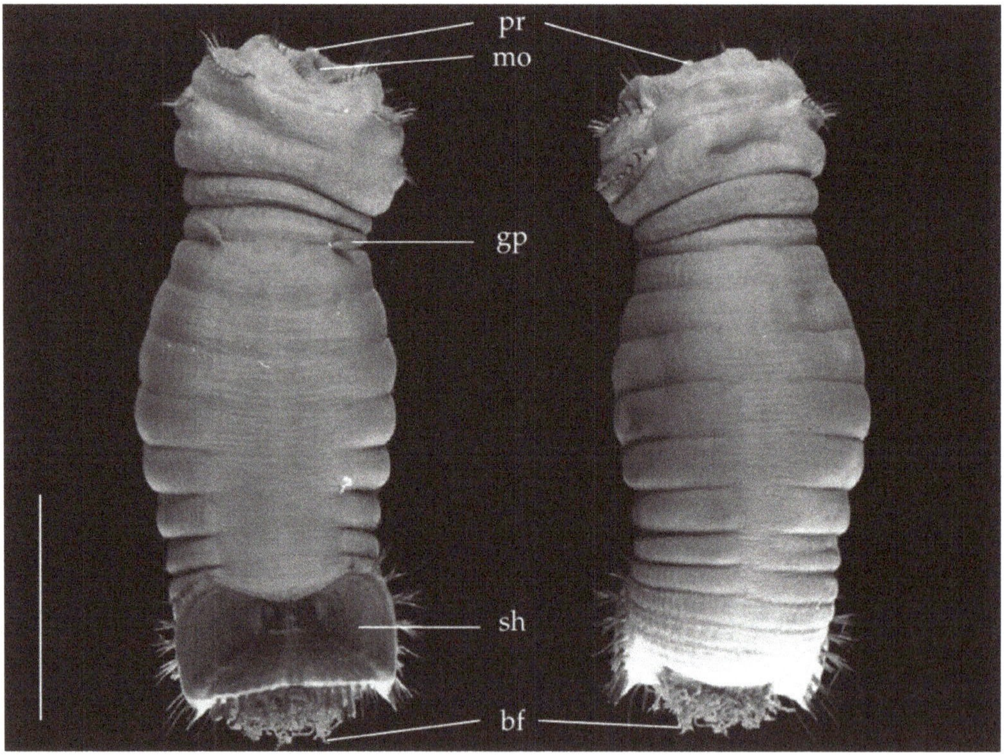

Figure 10. *Sternaspis* sp. Ventral (**left**) and dorsal (**right**) view. **bf**, branchial filament; **gp**, genital papillae; **mo**, mouth; **pr**, prostomium; **sh**, shield. Scale bar: 1 cm.

7.1. Morphology

Sternaspidae are relatively short-bodied worms, appearing grub- or peanut-shaped. The body colour is whitish to yellowish with the exception of a body maculated with numerous black spots in *Sternaspis lindae* Salazar-Vallejo, 2017 [246]. An anterior retractable region, including rows of strong hooks on segments 1–3, can be completely withdrawn into the abdomen, resulting in an inflated ovoid body shape. In extended individuals, the following body regions can be distinguished: (1) an anterior introvert comprising prostomium, peristomium, and segments 1–5/6; (2) a constricted region comprising segments 4–6, and (3) an abdominal region.

Figure 11. *Mauretanaspis longichaeta*. Holotype, MNHN-IA-TYPE 2005, ventral view; **pch**, posterior shield chaetae; * body damaged with oesophagus protruding; arrowheads point at gonopodial lobes. Scale bar: 1 mm. Modified from Fiege and Barnich [247], licensed under CC BY 4.0.

The anterior end lacks appendages. The prostomium is a small, smooth hemispherical cupule, and the peristomium is reduced to a mound surrounding the mouth and is often at least partly covered by papillae. Eyes are absent, except perhaps in a juvenile specimen of *S. fossor* Stimpson, 1853 [248] from California [249]. Nuchal organs have not yet been identified with certainty.

The constricted region exhibits a pair of genital papillae protruding ventrally, usually from the intersegmental groove between segments 5 and 6 (7 and 8 if prostomium and peristomium are errouneously included in the counting). They often stay at least partly visible when the introvert is retracted.

The abdominal region consists of the segmented pre-shield region comprising 7 segments in *Sternaspis* Otto, 1821 [250] and *Caulleryaspis* Sendall and Salazar-Vallejo, 2013 [244] and 8 segments in *Petersenaspis* Sendall and Salazar-Vallejo, 2013 [244] and *Mauretanaspis* Fiege and Barnich, 2020 [247], as well as the posterior region bearing the ventral shield, and the dorsal branchial region with threadlike branchial filaments and filiform papillae originating from branchial cushions next to the anus. Segmentation is distinct along the body, though sometimes best visible laterally. Parapodia are reduced, i.e., parapodial lobes, appendages, and aciculae are absent. Morphological characters of the shield are of primary importance for species distinction. These characters include the shield's general shape, colour, surface sculpture (ribs, concentric rings), subdivisions and relative proportions, shape of the median suture, and presence/absence and firmness of attachment of sediment cover. Shields are stiff with loose sediment easily brushed off in all species of *Sternaspis* and *Petersenaspis*, but soft in *Caulleryaspis* with sediment firmly adherent. In the latter case, the sediment cover obscures structures of the shield surface from observation. The stiffness of the shields can be tested by bending its margin [251], but as free margins are sometimes absent and rather merge into the epidermis, this character is not always easy to evaluate and therefore considered of limited value for generic distinction. The adherence of sediment to the ventrocaudal shield, i.e., "firmly adherent" (not removable) versus "loosely attached" (easily brushed off), appears more useful in diagnosing sternaspid genera in combination with other characters [247,252].

Total length and abdomen length have been examined as diagnostic characters, but retraction can prevent these measurements. Shield measurements have been performed on several species, in an attempt to find a relation with body length, with varying results showing no clear correlation [253,254].

Segments 1–3 bear a single row of stout spinelike chaetae, increasing dorsally in length and thickness. Their shape (tapering or subdistally expanded), colour, number, relative size, and possibly the relative length of rows, and degree of spacing among spines in extended specimens are of systematic importance. However, they cannot be observed in retracted specimens. All remaining chaetae are simple capillaries, mainly situated along the margins of the shield in adult animals, with numerous lateral and a lower number of caudal fascicles with shorter capillaries. Chaetae in lateral and posterior fascicles might be covered by a feltlike sheath in some species [31,245,252,255,256]. Chaetae of lateral fascicles increase in length caudally, reaching about four times the body length in *Mauretanaspis longichaeta* Fiege and Barnich, 2020 [247]. Peg chaetae—a dense group of numerous short chaetae embedded in a fibrous matrix and covered by a common sheath [256]—are situated at the posterolateral corners of the shield in species of *Sternaspis* and some species of *Caulleryaspis*. They are absent in *Petersenaspis* and *Mauretanaspis*. If present, peg chaetae are often accompanied on the inner side towards the body midline by a few very thin capillary chaetae. However, they can be missing or present on only one side and are therefore of limited systematic value. For juveniles, chaetae have been reported in the anterior abdomen (pre-shield) segments, they are usually missing in adults probably due to abrasion. The chaetae surrounding the shield might be a useful character with regard to number of chaetal fascicles and respective number of individual chaetae. The lengths of posterior chaetae relative to the size of the shield as well as the arrangement of chaetae in lateral fascicles (e.g., *Sternaspis affinis* Stimpson, 1864 [257], see [252]) may be of systematic value and should be further explored.

The shape and distribution of different types of cuticular papillae that can be found in varying densities in different body regions, including the ventrocaudal shield [244], might also be informative for species distinction. They can best be found along intersegmental folds or at the anterior end. Last but not least, the shape of the branchial cushions might also be useful for species distinction [31,244,245,249]. However, sediment cover as well as abrasion can hamper the study of these characters.

A pygidium is not clearly defined, and pygidial appendages are absent except for an eversible anal peduncle that has been described for *Sternaspis costata* von Marenzeller, 1879 [258] and *S. papillosa* Zhadan, Tzetlin, and Salazar-Vallejo, 2017 [256].

Sendall and Salazar-Vallejo [244] established morphological characters for species distinction, which led to a rapid increase in the number of species described. However, discrepancies between morphology and molecular results raised questions about the diagnostic power of morphological shield characters for sternaspid identification and points to a much more complex taxonomy comprising cryptic species and possibly the presence of widely distributed deep-water species. Considering the shield characters Kobayashi et al. [259] and Drennan et al. [252] suggested revising the systematic position of *Caulleryaspis nuda* Salazar-Vallejo and Buzhinskaja, 2013 [260] and possibly the validity of the genus per se because the original character combination, "soft shield with firmly adhered layer of sediment", for *Caulleryaspis* had later been emended by Salazar-Vallejo and Buzhinskaja [260] to incorporate species with "soft shields with loosely adhered layer of sediment", such as *C. nuda*, as well as species with "soft shields without sediment attached" such as *C. villamari* Salazar-Vallejo and Buzhinskaja, 2013 [260] and *C. nana* Zhadan, Tzetlin, and Salazar-Vallejo, 2017 [256].

Juvenile specimens of Sternaspidae can apparently not per se be distinguished by small body size because at least in some species individuals of only 2–8 mm length have been found with oocytes [253,256,261]. Following the assumption that gonopodial lobes are instrumental in releasing sexual products, their absence as observed in some small specimens of *Mauretanaspis longichaeta* might indicate that these specimens are indeed not yet sexually mature [247].

Typically, juvenile characters for Sternaspidae appear to be a more distinctive papillation of the body, including the surface of the ventrocaudal shield. Moreover, in small (juvenile) specimens, the shield has been described as generally thinner and less sclerotinized, shield margins and surface structures not well defined, but anterior keels more pronounced. With increasing body size, shield plates become less rounded and ribs more defined. Shield colour appears to become darker (*Sternaspis williamsae* [260]; *S. sendalli* [262]; *S. chinensis*, *S. liui* [263]; *Caulleryaspis villamari* [246]; *S. radiata*, *S. spinosa* [264]).

7.2. Taxonomy and Phylogeny

To date, 43 valid species of Sternaspidae are known to belong to four genera: *Sternaspis* (32 species), *Caulleryaspis* (5), *Petersenaspis* (5), and *Mauretanaspis* (1) [31,245,247]. Including the revision of Sendall and Salazar-Vallejo [244], 31 species were described during the past decade, more than doubling the number of known species [246,247,256,260,262–266] (Table 2; Figure 2F). A key to species of *Sternaspis*, *Caulleryaspis* and *Petersenaspis* is provided by Salazar-Vallejo [246] and a table of characters for distinction of genera by Fiege and Barnich [247].

Table 2. List of Sternaspidae species currently accepted.

Species	Distribution	Depth (m)	Type Locality
Caulleryaspis fauchaldi Salazar-Vallejo and Buzhinskaja, 2013	NE Pacific, Oregon to Southern California	1387–2800	Oregon, Cascadia Abyssal Plain, W of Yaquina Bay, 44°30.0′ N, 125°34.0′ W to 44°39.0′ N, 125°33.2′ W, 2800 m
Caulleryaspis gudmundssoni Sendall and Salazar-Vallejo, 2013	N Atlantic, Iceland	452–1207	Iceland, southeast of Vestmannaey jar, 63°02.30′ N, 21°50.80′ W, 1072 m
Caulleryaspis laevis (Caullery, 1944)	Andaman Sea to South China Sea, Southeastern Australia	39–147	Sumbawa Island, Bay of Bima, 08°27.5′ S, 118°43.5′ E, 55 m

Table 2. *Cont.*

Species	Distribution	Depth (m)	Type Locality
Caulleryaspis nuda Salazar-Vallejo and Buzhinskaja, 2013	NE Pacific, off Oregon NW Pacific, Sea of Okhotsk, off Kuril Islands as *C. nuda*	2519	Off Oregon, 48°9.0′ N, 127°4.2′ W, 2519 m
Caulleryaspis villamari Salazar-Vallejo, 2017	E Pacific, off northern Peru	160	Peru, off Isla Lobos de Tierra, 06°27′ S, 80°56′ W to 06°23′ S, 80°55′ W, 160 m
Mauretanaspis longichaeta Fiege and Barnich, 2020	CE and SE Atlantic, off Mauritania and Angola	2700–4400	Off Mauretania; 18°21.27′ N, 17°57.44′ W; 2712 m
Petersenaspis capillata (Nonato, 1966)	SW Atlantic, Brazil	(intertidal) 52–138	Brazil, Vitoria Island, 23°45′18″ S, 44°00′54″ W, 52 m
Petersenaspis deani Salazar-Vallejo, 2017	E Pacific, Costa Rica	22	Costa Rica, Golfo de Nicoya, 09°44′00″ N, 84°59′25″ W, 22 m
Petersenaspis harrisae Salazar-Vallejo, 2017	SW Indian Ocean, between South Africa and Madagascar	15–128	South Africa, off Durban, 29°11′ S, 31°37′ E, 15 m
Petersenaspis palpallatoci Sendall and Salazar-Vallejo, 2013	Philippine Islands to Malaysia	5.5–93	Philippine Islands, Sibuyan Sea, E off Kalibo, 11°42.6′ N, 122°31.5′ E, 93 m
Petersenaspis salazari Wu and Xu, 2017	Northern South China Sea	174–182	South China Sea, 19°30′ N, 112°30′ E, 174 m
Sternaspis affinis Stimpson, 1864	NE Pacific Ocean, Alaska to Gulf of California	95–154	Canada, British Columbia, Strait of Georgia
Sternaspis Africana Augener, 1918	E Atlantic, W African coast from Ghana to Angola	20–70	Angola, St. Paul Loanda
Sternaspis andamanensis Sendall and Salazar-Vallejo, 2013	Indian Ocean, Andaman Sea and Malaysia, South China Sea	5–45	Andaman Sea, Thailand, 45 m
Sternaspis annenkovae Salazar-Vallejo and Buzhinskaja, 2013	NW Pacific, east of northern Kuril I lands SW Pacific, Australia, off E Gippsland	3980–4070	East of northern Kuril Island, 49°46′01″ N, 157°48′06″ E, 3980–4070 m
Sternaspis britayevi Zhadan, Tzetlin and Salazar-Vallejo, 2017	South China Sea, Gulf of Tonkin	65	South China Sea, Gulf of Tonkin, 19°16′ N, 107°19′ E, 65 m
Sternaspis buzhinskajae Salazar-Vallejo, 2014	Arctic Ocean, Okhotsk, Chukchi, Bering and Beaufort Seas	3.5–90	Alaska, Beaufort Sea, off Point Barrow, 3.5–5.0 m
Sternaspis chilensis Díaz-Díaz and Rozbaczylo, 2017	SE Pacific, Austral Chilean channel and fjord region	50–515	Chile, Magellan Province, Moraleda Channel, 45°06′ S, 73°38′ W, 250 m
Sternaspis chinensis Wu, Salazar-Vallejo and Xu, 2015	Yellow and East China Sea, Bohai Sea	8–63.7; 500	Bohai Sea, 38°38′ N, 120°23′ E, 39 m
Sternaspis costata von Marenzeller, 1879	NW Pacific, Southern Sakhalin Island to Philippines, South China Sea	15–70	Japan, Honshu Island, Chiba, Boso Peninsula
Sternaspis fossor Stimpson, 1853	NW Atlantic, Canada to northeastern United States	20–153	Canada, Nova Scotia, 88 km E Cape Sable, 153 m
Sternaspis islandica Malmgren, 1867	NE Atlantic, Norwegian Sea, Iceland and Faroe Islands	7–226	Iceland, Berufjord, 64°48′ N, 14°30′ W, 27–55 m
Sternaspis lindae Salazar-Vallejo, 2017	E Pacific, Gulf of Panama to NW Colombia	119–275	Panama, 08°39.3′ N, 79°31.7′ W, 22–27 m
Sternaspis liui Wu, Salazar-Vallejo and Xu, 2015	Yellow Sea	21–43.3	Yellow Sea, 35°59′ N, 122°00′ E; 43.3 m

Table 2. *Cont.*

Species	Distribution	Depth (m)	Type Locality
Sternaspis londognoi Salazar-Vallejo, 2017	Gulf of Mexico and Caribbean Sea	intertidal-344	SW Caribbean, Panama, Caledonia Bay, intertidal
Sternaspis maior Chamberlin, 1919	E Pacific, Central Gulf of California	180–1158	Gulf of California, Sinaloa, Off Isla Altamura, 24°52′ N, 108°58′ W, 830 m
Sternaspis maureri Salazar-Vallejo and Buzhinskaja, 2013	E Pacific and SW Pacific	795–6489	Peru, W of Trujillo, 08°16′ S, 81°05′ W, 6156–6489 m
Sternaspis nana Zhadan, Tzetlin and Salazar-Vallejo, 2017	Vietnam, Nha Trang Bay, South China Sea	15–40	South China Sea, Nha Trang Bay, 12°11′50″ N, 109°15′06.6″ E, 23 m
Sternaspis papillosa Zhadan, Tzetlin and Salazar-Vallejo, 2017	Vietnam, Nha Trang Bay, South China Sea	11–40	South China Sea, Nha Trang Bay, 12°10′24″ N, 109°15′42″ E, 21.1 m
Sternaspis piotrowskiae Salazar-Vallejo, 2014	Philippine Islands	272–636	Philippines, Luzon Island, Balayan Bay, 13.82°N, 120.75°E, 272–291 m
Sternaspis princeps Selenka, 1885	SW Pacific, New Zealand, off North Island	1274	New Zealand, North Island, NE off Gisborne, 37°34′ S, 179°22′ E, 1274 m
Sternaspis radiate Wu and Xu, 2017	Northern South China Sea	16–105.5	South China Sea 22°15′ N, 115°00′ E, 41 m
Sternaspis rietschi Caullery, 1944	Arafura Sea, Indonesia, off Wokam Island	1788	Indonesia, W of Wokam Island, 5°46′ S, 134°00′ E, 1788 m
Sternaspis scutata (Renier in Ranzani, 1817)	Mediterranean Sea	9–36	Aegean Sea, Izmir Bay, 38°30′00″ N, 26°50′00″E, 33 m
Sternaspis sendalli Salazar-Vallejo, 2014 incl. *S. monroi* Salazar-Vallejo, 2014	Southern Ocean, South Georgia to Antarctic Peninsula and Weddell Sea	160–1500	South Orkneys, Scotia Sea
Sternaspis sherlockae Salazar-Vallejo, 2017	Red Sea	Presumed shallow water	Red Sea
Sternaspis spinosa Sluiter, 1882	W Pacific from South China Sea to Andaman Sea and Coral Sea off Australia	7–110	Indonesia, Java, Bay of Batavia, "Batavia Roads", outside Jakarta, 30 m
Sternaspis sunae Wu and Xu, 2017	South China Sea	224	South China Sea, Nansha Islands, 224 m
Sternaspis thalassemoides Otto, 1821	NE Atlantic and Mediterranean Sea	n.d.	Tyrrhenian Sea, Naples
Sternaspis thorsoni Sendall and Salazar-Vallejo, 2013	Indian Ocean, Arabian/Iranian Gulf to Andaman Sea	3–110	Gulf of Oman, off Jask, 25°45′ N, 57°12′ E, 110 m
Sternaspis uschakovi Salazar-Vallejo and Buzhinskaja, 2013	NW and E Pacific, Okhotsk Sea, California, Gulf of California, W Mexico	592–2548	Okhotsk Sea, 55°13′ N, 146°52′ E, 592 m
Sternaspis williamsae Salazar-Vallejo and Buzhinskaja, 2013	NE Pacific, Oregon to California NW Pacific, off Kuril Islands as S. cf. williamsae	1000–2800	Oregon, W of Yaquina Bay, 44°30.0′ N, 125°34.0′ W to 44°39.0′ N, 125°33.2′ W, 2800 m
Sternaspis wui Wu and Xu, 2017	Northern South China Sea	173–230	South China Sea, 17°45′ N, 110°30′ E, 230 m

For better resolution of sternaspid systematics, more comprehensive molecular data for more species would be desirable to complement known characters and suitably treated material needs to be collected. As a first example using next generation sequencing (NGS) techniques, the complete mitochondrial genome of specimens identified as *Sternaspis scutata* Renier in Ranzani, 1817 [267] collected from Jiazhou Bay, near Qingdao (Yellow Sea), was sequenced [268].

7.3. Biogeography

Most sternaspid species have been recorded from low intertidal to shelf depths (200 m), but 14 have been found between 200 m and 6500 m deep (Table 3). The majority of species occurs in the Pacific (29 species), nine are recorded for the Atlantic, eight for the wider Indian Ocean, and only two and one for the Arctic and the Antarctic Ocean, respectively (Table 4, Figure 3F). The South China Sea has been found to be an area of high sternaspid diversity with 11 species reported, followed by the central eastern Pacific with nine species recorded. For the northwestern Pacific, with four species currently recorded, five distinct but yet undescribed species might have to be added due to recent molecular studies [252, 259]. Among the species occurring in the Atlantic and its neighbouring seas, there are two species recorded in the Mediterranean Sea and two or possibly three species in the North-East Atlantic. Six species are widely recorded from the Indian Ocean and one species each in the Arabian/Persian Gulf and the Red Sea. Maps showing the type localities of known species were provided by Fiege [31,245], Díaz-Díaz and Rozbaczylo [265], and Drennan et al. [252].

Table 3. Deep-water Sternaspidae (predominantly recorded deeper than 200 m; *Sternaspis chinensis* Wu, Salazar-Vallejo and Xu, 2015 [263] is not included as only a single record from 500 m exists [252]).

Species	Distribution	Depth (m)
Caulleryaspis fauchaldi Salazar-Vallejo and Buzhinskaja, 2013	NE Pacific, Oregon to Southern California	1387–2800
Caulleryaspis gudmundssoni Sendall and Salazar-Vallejo, 2013	North Atlantic, Iceland	452–1207
Caulleryaspis nuda Salazar-Vallejo and Buzhinskaja, 2013	NE Pacific, off Oregon	2519
Mauretanaspis longichaeta Fiege and Barnich, 2020	CE and SE Atlantic, off Mauritania and Angola	2700–4400
Sternaspis annenkovae Salazar-Vallejo and Buzhinskaja, 2013	NW Pacific, east of northern Kurile Island	3980–4070
Sternaspis chilensis Díaz-Díaz and Rozbaczylo, 2017	SE Pacific, Austral Chilean channel and fjord region	50–515
Sternaspis maior Chamberlin, 1919	E Pacific, Central Gulf of California	180–1158
Sternaspis maureri Salazar-Vallejo and Buzhinskaja, 2013	E and SW Pacific	795–6489
Sternaspis piotrowskiae Salazar-Vallejo, 2014	Philippine Islands	272–636
Sternaspis princeps Selenka, 1885	SW Pacific, New Zealand	1274
Sternaspis rietschi Caullery, 1944	Arafura Sea, off Wokam Island	1788
Sternaspis sendalli Salazar-Vallejo, 2014	Southern Ocean and Antarctic Peninsula	160–1500
Sternaspis uschakovi Salazar-Vallejo and Buzhinskaja, 2013	N Pacific, Okhotsk Sea, California, Gulf of California	592–2548
Sternaspis williamsae Salazar-Vallejo and Buzhinskaja, 2013	NE Pacific, Oregon to California	1000–2800

The distributional range varies greatly among species, and if genetic data are considered besides morphological distinctness, the picture becomes even more complex. It appears that narrow distribution ranges are found in shallow-water species, while some deep-water species inhabit much larger areas [252], e.g., nine species are known only from their respective type locality (Table 2). Seven species (*Petersenaspis salazari* Wu and Xu, 2017 [264], *Sternaspis britayevi* Zhadan, Tzetlin, and Salazar-Vallejo, 2017 [256], *S. nana*, *S. papillosa*, *S. radiata* Wu and Xu, 2017 [264], *S. sunae* Wu and Xu, 2017 [264], and *S. wui* Wu and Xu, 2017 [264]) described from the South China Sea are known only from this area. Some species overlap in distribution, e.g., adults of *S. nana* and juveniles of *S. papillosa* were found close together in Nha Trang Bay, South China Sea [238]. *Sternaspis monroi* Salazar-Vallejo, 2014 [262] and *S. sendalli* have both been described from close localities off the South Orkney Islands in the Scotia Sea [262] and were recently found synonymous due to morphological and genetical studies [252]. *Caulleryaspis fauchaldi* Salazar-Vallejo and Buzhinskaja, 2013 [260] and *S. williamsae* Salazar-Vallejo and Buzhinskaja, 2013 [260] have been described from the same sample, off Oregon (NE Pacific) [260].

Table 4. Geographic distribution of sternaspid species including some putative records of *Sternaspis scutata* (bold: species known only from type locality; underlined: species known predominantly from deep water; depth >> 200 m; possible but unconfirmed records of *S. scutata* are preceded by a question mark).

Geographic Distribution	No. of Species	Species
Arctic Ocean	**2**	*S. bushinskajae, S. fossor*
Pacific	**29**	
NE Pacific	5	*C. fauchaldi, C. nuda, S. affinis, S. uschakovi, S. williamsae*
NW Pacific	4 (+5)	*S. annenkovae, S. bushinskajae, S. costata, S. uschakovi,* (+ *C.* cf. *nuda, S.* cf. *williamsae,* and 3 unidentified species of *Sternaspis*; see [259] and [252]
CE Pacific	9	*C. fauchaldi,* ***C. villamari, P. deani,*** *S. affinis, S. lindae, S. maior, S. maureri, S. uschakovi, S. williamsae*
CW Pacific	3	*P. palpallatoci, S. costata, S. piotrowskiae*
W Pacific		
E China Sea, Yellow Sea	2	*S. chinensis, S. liui,* (+? *S. scutata* in Jiaozhou Bay, see Wang et al. [268])
South China Sea	11	*C. laevis,* ***P. salazari,*** *S. andamanensis,* ***S. britayevi,*** *S. costata, S. nana, S. papillosa, S. radiata, S. spinosa,* ***S. sunae, S. wui,***
SW Pacific	5	*C. laevis, S. annenkovae, S. maureri,* ***S. princeps,*** *S. spinosa*
SE Pacific	1	*S. chilensis*
Atlantic	**9**	
NE Atlantic	2 (+1)	***C. gudmundssoni,*** *S. islandica,* (+ *S.* cf. *scutata* in English Channel; see [252])
Mediterranean Sea	2	*S. scutata, S. thalassemoides*
NW Atlantic	1	*S. fossor*
CE Atlantic	2	*M. longichaeta, S. africana*
CW Atlantic		
Gulf of Mexico	1	*S. londognoi*
SW Atlantic	1	*P. capillata*
SE Atlantic	2	*M. longichaeta, S. africana*
Indian Ocean	**8**	
Indian Ocean	6	*C. laevis, P. palpallatoci, S. andamanensis,* ***S. rietschi,*** *S. spinosa, S. thorsoni,* (+? *S. scutata* in Sundarban Mangroves, Gulf of Bengal, see [269], Cochin Estuary, SW India, see [270]; West coast of India, see [271])
NW Indian Ocean		
Arabian/Persian Gulf; Gulf of Oman	1	*S. thorsoni*
Red Sea	1	*S. sherlockae,* (+? *S. scutata* Gulf of Suez; see [272])
SW Indian Ocean	1	*P. harrisae*
Antarctic Ocean	**1**	*S. sendalli*

On the other hand, phylogeographic studies using mitochondrial and nuclear genes revealed examples of low genetic differences over large distances, e.g., 1500 km among populations from lower bathyal to abyssal depths in the NW Pacific for *Sternaspis* sp. cf.

S. williamsae [259]. Low genetic variation was also found among populations of *Sternaspis affinis* over a geographic distance of 1800 km along the west coast of North America (Washington to California), raising the question of only this single species being present in this area. Genetic data also show close relationships among specimens of *Sternaspis* cf. sp. *S. annenkovae* from off SE Australia with a clade of *Sternaspis* sp. cf. *S. williamsae* and *Sternaspis* sp. cf. *S. nuda* in the NW Pacific covering a distance of about 9600 km across the Pacific Ocean [252]. *Sternaspis scutata* is apparently the most widely reported name in the past but was recorded also more recently from the Cochin Estuary (SW India) [270], the English Channel [273], where it was considered an invasive species, the west coast of India (Arabian Sea) [271], the Sundarban Mangroves (Bengal Sea) [269], Jiaouzhou Bay (Yellow Sea) [268,274], and the Gulf of Suez (Red Sea) [272]. The latter record could be the result of Anti-Lessepsian migration, i.e., Mediterranean species entering the Red Sea via the Suez Canal [275]. Due to this wide geographic range of records far away from the type locality, it seems likely that cryptic or undescribed species are present, which need to be distinguished [259]. Molecular data support morphological observations regarding the genetic distinctness of *Sternaspis* sp. cf. *S. scutata* reported from the English Channel, from *S. scutata* present in the Mediterranean and a molecular record from the Bay of Biscay (North East Atlantic), which was found to be the first record of *S. scutata* outside the Mediterranean. Other records of sternaspids identified as *S. scutata* in GenBank from the Bering Sea, the Bay of Bengal, and Southern Chile were shown to be different from each other and from Mediterranean *S. scutata*, lending further support against assumptions of a wide distribution of *S. scutata* outside the Mediterranean Sea [252].

Lecithotrophic larvae have been observed for some sternaspid species, although with a rather short planktonic period. Cold temperatures in deep water might inhibit development of larvae and in combination with favourable currents it might result in a wide distribution in the deep sea. This hypothesis, however, requires confirmation by studies on reproduction. So far planktonic phases for sternaspid larvae have only been found to last for less than two days in shallow water among *S. scutata* [276].

7.4. Biology and Ecology

The biology and ecology of Sternaspidae is only poorly known. They live burrowing in the sediment, mostly in muddy bottoms, but have also been recorded from a variety of sediment types [244]. Individuals are reported to bury in the sediment head first, mostly with their dorsal side up, leaving the gills exposed at the sediment surface while the anterior part of the body is retracted [276,277]. Sternaspids are able to move quickly in the sediment by extending and retracting their introvert [276,278,279], but they have also been found to stay in the same position for prolonged periods of time [73]. Sternaspids are usually not represented in samples in high numbers, but they can be locally abundant and dominant in benthic communities with up to 1335 spms/m^2 [271]. They have been classified as microphagous subsurface deposit feeders [73] preferring particles of small size [280].

Sternaspids are gonochoristic and show no signs of sexual dimorphism. Year-round recruitment with a peak in summer and lecitotrophic larvae with a short planktonic phase (less than two days) have been observed [276,281].

8. Discussion: Methods and Perspectives

Rigorous implementation of diverse methods of morphological observation and their integration with molecular methods has played a major part in systematic understanding and biodiversity discovery in the families Acrocirridae, Cirratulidae, Cossuridae, Longosomatidae, Paraonidae, and Sternaspidae, e.g., refs. [32,93]. Continued rigour in morphological studies, cooperation, and further integration of molecular data will be key to resolving the current issues in diversity and systematics in these groups, such as the resolution of their classification and a unified terminology for morphological characters.

8.1. Washing Samples

To accurately assess small benthic polychaete abundance and avoid the excessive fragmentation and loss of important diagnostic features, it is necessary to gently wash samples with sieves with fine mesh sizes not exceeding 500 µm; 150–300 µm screens produce more accurate results (Figure 12); 500 µm sieves are better suited for shelf samples, while in deep environments the use of 250–300 µm sieves is advisable. Sternaspid species have a stouter shape and are more robust polychaetes. Finer sieves also allow for retaining early developmental stages [83]. Ohwada [282] recommended preserving samples before washing to retain the maximum number of specimens. However, this can be problematic when processing large samples in ethanol for future DNA extraction, as care must be ensured to maintain a sufficient concentration of ethanol and a homogenization of the sample.

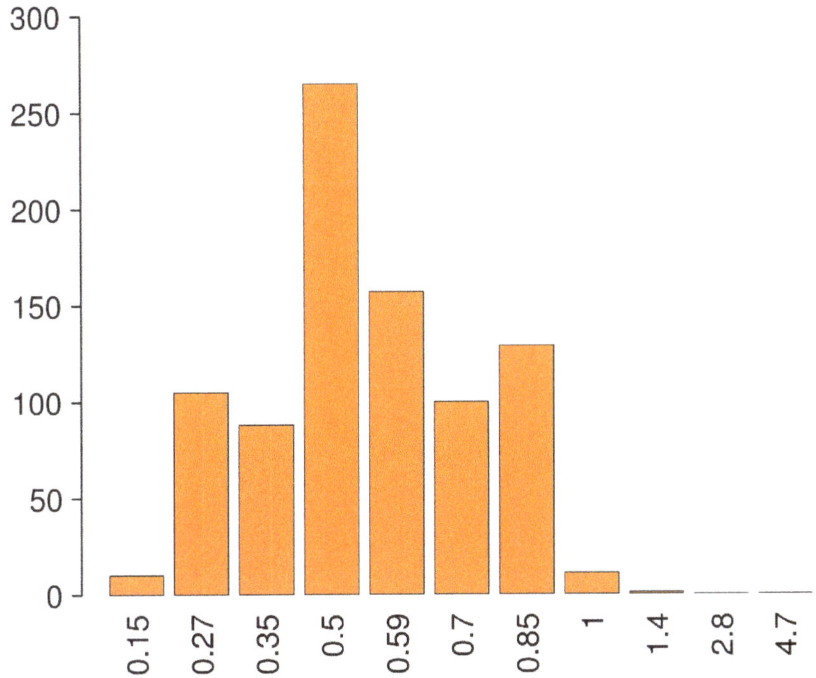

Figure 12. Diagram showing number of specimens of *Cossura candida* on sieves with different mesh sizes in mm (data obtained from [282] Table 1).

8.2. Fixation

Individuals used for DNA investigation should be ideally sorted alive and fixed in 96% ethanol and preserved at −20 °C until DNA extraction. Bulk-fixed samples give a variable extraction and amplification success, based on the concentration of the ethanol and time elapsed between fixation and sorting. Samples fixed with 96% ethanol and stored in a cold room allow for obtaining useful genetic material up to at least 10 years after sampling, even if in old samples the amplification of nuclear markers might be difficult. Samples fixed with 70% ethanol are useful for genetics for a limited time and should be sorted and transferred to clean ethanol within a few months after sampling. Sample storage at cold, stable temperature allows for obtaining useful DNA extracts for a longer time. Samples fixed with denatured ethanol or stained with Rose Bengal to ease the sorting are unsuitable for genetics.

8.3. Species Description

Ideally, species descriptions should include light microscopy and scanning electron microscopy, and molecular genetic investigation. Light microscopy should be used for dorsal and lateral view to display the shape of the prostomium and the thorax–abdomen transition (e.g., in Cossuridae), to display the location of the branchial filament(s), eventual palps/tentacles, and chaetal characteristics. Different kinds of staining are highly recommended for improving the visibility of morphological characteristics [283]. Molecular data are also very welcome, as integrative systematics is necessary to disentangle the diversity of most families, especially in the case of pseudocryptic species complexes, which are likely to be common [93]. Ecological preferences, such as depth, granulometry, and habitat type, are not strictly necessary for the description, but might be useful for parataxonomists in the identification of taxa, and might suggest the occurrence of cryptic species in the case of discrepancies. Biological features such as feeding, reproductive mode(s), and egg or larvae characteristics are in most cases difficult to obtain, but can also prove important in discriminating species and inferring phylogenetic hypotheses.

8.3.1. Staining

Methyl green and methylene blue are commonly used for polychaete systematics, and both produce similar results. These stains can be used for two purposes. First, for contrasting external structures, such as chaetae, parapodial lobes, papillae, head appendages, cirri, branchiae, and scars of these structures when they were lost. This method requires transferring worms to water before staining and examination in water. The concentration of water-soluble pigment is low for this method. Examples of methylene blue staining have been published previously [136,145]. Second, these stains can be used to stain glandular structures. For this, a saturated ethanol solution of dye is used, and worms are stained for some time, after which they are examined in clean ethanol after partial de-staining and differentiation. This method is especially useful for cossurids, as their staining patterns are species-specific, which accelerates mass identification, even for short anterior fragments [134]. This method is also often used for Cirratulidae species, but not so much in Longosomatidae, Acrocirridae or Paraonidae systematics as the majority of these taxa do not present any specific pattern. The exception to this is represented by members of *Levinsenia*, in Paraonidae, where the anterior part of the body shows a distinctive and often diagnostic staining pattern [223]. It has not yet been tested for members of Sternaspidae.

Shirlastain A is also a commonly used dye in polychaete systematics, e.g., ref. [284]. Its water solution makes external structures more distinguishable, similar to methylene blue. Shirlastain A can be used to highlight external structures, such as branchiae, prostomial antenna, and parapodial lobes, and to identify broken structures through their scars, making the examination and identification of individuals easier. It is also routinely used for members of Cirratulidae [83,283].

8.3.2. Scanning Electron Microscopy

Scanning electron microscopy (SEM) illustrations are highly recommended for species descriptions to document the exact position of the branchial filament of cossurids or tentacles of cirratulids, and especially for illustrating chaetae, including their exact number, position, shape, length, and thickness [57,98]. They are also helpful in revealing broken branchiae by making the scars clearly visible. The preparation of the sample for SEM imaging, however, might lead to artefacts in the shape of the soft structures, and also the shape of chaetae should be carefully evaluated, as in recent descriptions, broken chaetae have been treated as entire and considered as diagnostic characters [221]. Dehydrating the specimens using a series of baths from 100%EtOH to 100%HMDS before air drying has been proposed as a gentler alternative to the most fragile specimens compared to critical point drying [285,286].

8.4. Molecular Genetic Investigations

This approach requires ethanol-preserved specimens and is extremely important for resolving systematic and phylogenetic problems. It is recommended to target several genetic markers, both mitochondrial and nuclear [4,287]. For exact species identification, reference sequences from the type locality of each species are desired, and at higher taxonomic levels, sequences from type localities of at least the type species are wanted.

DNA barcoding of Cirratulidae and Ctenodrilidae has proven relatively easy, and the use of molecular data to study this group is increasing. The markers COI, 16S rRNA (16S), and 28S rRNA (28S) have in particular been used for taxonomic and phylogenetic studies. Several new species of *Cirriformia* and *Timarete* were described, including data on COI and 16S [104,105]. The *Timarete punctata* species complex was also revealed thanks to COI and 16S data. Regarding these genera, it is important to note that the complete mitochondrial genomes of *Timarete posteria* and *Cirriformia* cf. *tentaculata* were made available [288,289]. The diversity of the bitentaculate Cirratulidae in North-East Atlantic was recently reviewed using the markers COI and 28S with species delimitation analyses, revealing both cryptic diversity, and new species that presented distinct morphologies [93].

Currently, the majority of sequence data available for Paraonidae consists of partial sequences referred to two mitochondrial markers, COI and 16S, and one nuclear marker, 18S rRNA (18S) [14]. Both mitochondrial markers turned out to be useful for species delimitation, but they are scarcely suitable for the resolution of deeper branches; 18S, instead, allows to disentangle phylogenetic relationships between different paraonid groups, but it is less useful to identify species, especially in the genus *Aricidea*, which seems to have radiated rather recently [14]. A fragment of the internal transcribed spacer 2 (ITS-2) represents a promising candidate as a nuclear gene that can be employed for species delimitation [290], but this fragment is highly variable, might be present in different copies, and sequences from somewhat distant species can be difficult to align, complicating subsequent analyses (see [291]). While universal primers are usually suitable for 16S, 28S, and 18S amplification, both COI and ITS-2 show a variable amplification success across different genera and species and might need the design of customized primer pairs.

The few sequences of Cossuridae and Acrocirridae available for phylogenetic analyses include Histone H3, 28S, 16S, and COI, e.g., ref. [50]. Even fewer sequences are available for Longosomatidae, of COI and 18S, but the amplification of all these markers has proven relatively easy to amplify for both of these families.

More sequences are available for Sternaspidae, mainly of the mitochondrial markers COI and 16S, as well as a few sequences of the nuclear markers 28S, 18S, 5.8S, and Histone H3 [252].

8.5. Internal Morphology

Histological studies of cirratulids have been scarce and rarely used for systematic purposes, e.g., refs. [292,293]. While it might not be relevant for every single description, histology can provide useful information on the nature of some characters and how to describe some features such as the presence of an achaetous first segment in some bitentaculate cirratulids [94]. Histology also reveals details of reproductive biology, e.g., refs. [292,294], which can be another tool in species delimitation and important to make phylogenetic hypotheses. While confocal microscopy has been employed to study the internal morphology of acrocirrids and cossurids, it has yet to find an application in systematics [28,295]. Internal morphology of sternaspids has been studied using SEM, and TEM has been used to study the ultrastructure of sensory organs [31]. Confocal microscopical study of innervation of the branchial filament in *Cossura pygodactylata* suggested that the filament is homologous to branchiae of other polychaetes. Ultrastructural and confocal studies of the same cossurid species, including juvenile stages, revealed the nature of the anterior achaetous ring as a part of the prostomium, showed the origin of buccal tentacles from the soft axial pharynx, and illustrated some unique features such as the muscular system in the prostomium and an unusual heart body [296]. Micro-CT investigation of

C. pygodactylata together with histological studies revealed details of internal anatomy, which can be useful for phylogenetic studies [295]. Micro-CT has not yet been used for other cirratuliform polychaetes.

9. Conclusions

The Acrocirridae, Cirratulidae, Cossuridae, Longosomatidae Paraonidae, and Sternaspidae are all examples of seemingly understudied groups of polychaetes, due to their rarity, difficulty to work with, or even a past lack of interest. However, more than 200 species have been described worldwide in the past 20 years, and significant progress has been made in understanding phylogenetic relationships to each other and to other annelid families with the advent of molecular phylogenetics and phylogenomics. Many species are still undescribed [297], and the distribution of many others, especially so-called cosmopolitan species, is still unknown. A continued interest in these groups and further integration of molecular data with more and more detailed morphological studies will be key to achieving these goals.

Author Contributions: Conceptualization and supervision, M.G., A.Z., J.L., D.F. and A.M.; methodology, writing—original draft preparation, review and editing and visualization M.G., A.Z., J.L., D.F. and A.M. All authors have read and agreed to the published version of the manuscript.

Funding: M.G. benefited from support by the Norwegian Taxonomy Initiative (http://www.biodiversity.no/Pages/135523 (accessed on 5 March 2021) (Cryptic polychaetes in Norwegian waters, knr 49-13, pnr 70184228). A.Z. was supported by grants from the Russian Foundation of Basic Research (no. 18-05-60158 and 19-04-00501).

Institutional Review Board Statement: Not applicable.

Informed Consent Statement: Not applicable.

Data Availability Statement: Not applicable.

Acknowledgments: The authors would like to thank James Blake for discussion about Cirratuliade and collection methods, Julio Parapar for helpful discussions about Longosomatidae, and Leslie Harris for helpful discussion and advice on working with Cossuridae and staining methods. Thanks are also due to three anonymous reviewers who greatly improved the quality of this article. The first author is particularly grateful to Arne Nygren and Maria Capa for their support and the opportunity to write this article.

Conflicts of Interest: The authors declare no conflict of interest.

References

1. WoRMS—World Register of Marine Species. Available online: http://www.marinespecies.org/index.php (accessed on 10 March 2021).
2. Struck, T.H.; Golombek, A.; Weigert, A.; Franke, F.A.; Westheide, W.; Purschke, G.; Bleidorn, C.; Halanych, K.M. The Evolution of Annelids Reveals Two Adaptive Routes to the Interstitial Realm. *Curr. Biol.* **2015**, *25*, 1993–1999. [CrossRef] [PubMed]
3. Weigert, A.; Helm, C.; Meyer, M.; Nickel, B.; Arendt, D.; Hausdorf, B.; Santos, S.R.; Halanych, K.M.; Purschke, G.; Bleidorn, C.; et al. Illuminating the Base of the Annelid Tree Using Transcriptomics. *Mol. Biol. Evol.* **2014**, *31*, 1391–1401. [CrossRef]
4. Struck, T.H. Phylogeny. In *Hanbook of Zoology, Annelida*; Purschke, G., Böggemann, M., Westheide, W., Eds.; De Gruyter: Berlin, Germany; Boston, MA, USA, 2019; Volume 1: Annelida Basal Groups and Pleistoannelida, Sedentaria I; pp. 37–68.
5. Day, J.H. The polychaete fauna of South Africa. Part 8: New species and records from grab samples and dredgings. *Bull. Br. Mus. (Nat. Hist.) Ser. Zool.* **1963**, *10*, 381–445. [CrossRef]
6. Cerruti, A. Contributo all'anatomia, biologia e sistematica delle Paraonidae (Levinsenidae) con particolare riguardo alle specie del golfo di Napoli. *Mittheilungen aus der Zoologischen Station zu Neapel* **1909**, *19*, 459–512.
7. Caullery, M. Sur les Siboglinidae, type nouveau d'invertébrés receuillis par l'expédition du Siboga. *Comptes Rendus Hebdomadaires des Séances de l'Académie des Sciences* **1914**, *158*, 2014–2017.
8. Pizl, V.; Chalupský, J. *Hrabeiella periglandulata* gen. et sp. n. (Annelida) a curious worm from Czechoslowakia. *Vestn. Cesk. Spol. Zool.* **1984**, *48*, 291–295.
9. Ehrenberg, C.G. Animalia evertebrata exclusis Insectis. Series prima. In *Symbolae Physicae, Seu Icones et Descriptiones Mammalium, Avium, Insectorum et Animalia Evertebra, Quae ex Itinere per Africam Borealem et Asiam Occidentalem Studio nova aut Illustrata Redierunt*; Hemprich, F.G., Ehrenberg, C.G., Eds.; Officina Academica, ed.: Berlin, Germany, 1828; p. 126.

10. Lastochkin, D.A. New name [*Potamodrilus*] for the genus *Stephensoniella*. *Ann. Mag. Nat. Hist.* **1935**, *16*, 488. [CrossRef]
11. Malmgren, A.J. Annulata Polychaeta Spetsbergiæ, Grœnlandiæ, Islandiæ et Scandinaviæ. *Öfversigt af Kongliga Vetenskaps-Akademien Förhandlingar* **1867**, *24*, 127–235.
12. Hartmann-Schröder, G. Annelida, Borstenwürmer, Polychaeta. In *Die Tierwelt Deutschlands 58*; Dahl, M., Peus, F., Eds.; Gustav Fischer: Jena, Germany, 1971; p. 594.
13. Struck, T.H.; Hessling, R.; Purschke, G. The phylogenetic position of the Aelosomatidae and Parergodrilidae, two enigmatic oligochaete-like taxa of the 'Polychaeta', based on molecular data from 18S rDNA sequences. *J. Zool. Syst. Evol. Res.* **2002**, *40*, 155–163. [CrossRef]
14. Langeneck, J.; Barbieri, M.; Maltagliati, F.; Castelli, A. Molecular phylogeny of Paraonidae (Annelida). *Mol. Phylogenet. Evol.* **2019**, *136*, 1–13. [CrossRef] [PubMed]
15. Martínez, A.; Di Domenico, M.; Worsaae, K. Evolution of cave *Axiokebuita* and *Speleobregma* (Scalibregma, Annelida). *Zool. Scr.* **2013**, *42*, 623–636.
16. Li, Y.; Kocot, K.M.; Whelan, N.W.; Santos, S.R.; Waits, D.S.; Thornhill, D.J.; Halanych, K.M. Phylogenomics of tube worms (Siboglinidae, Annelida) and comparative performance of different reconstruction methods. *Zool. Scr.* **2017**, *46*, 200–213. [CrossRef]
17. Parapar, J.; Martínez, A.; Moreira, J. On the systematics and biodivesity of the Opheliidae and Scalibregmatidae. *Diversity* **2021**, *13*, 87. [CrossRef]
18. Hutchings, P.; Carrerette, O.; Nogueira, J.M.M.; Hourdez, S.; Lavesque, N. The Terebelliformia-Recent Developments and Future Directions. *Diversity* **2021**, *13*, 60. [CrossRef]
19. Worsaae, K.; Kerbl, A.; Di Domenico, M.; Gonzalez, B.C.; Bekkouche, N.; Martínez, A. Interstitial Annelida. *Diversity* **2021**, *13*, 77. [CrossRef]
20. Martinsson, S.; Erséus, C. Cryptic Clitellata: Molecular Species Delimitation of Clitellate Worms (Annelida): An Overview. *Diversity* **2021**, *13*, 36. [CrossRef]
21. Meca, M.A.; Zhadan, A.E.; Struck, T.H. The Early Branching Group of Orbiniida Sensu Struck et al., 2015: Parergodrilidae and Orbiniidae. *Diversity* **2021**, *13*, 29. [CrossRef]
22. Banse, K. Acrocirridae n.fam. (Polychaeta Sedentaria). *J. Fish. Res. Board Can.* **1969**, *26*, 2595–2620. [CrossRef]
23. Ryckholt [Baron de], P. Mélanges paléontologiques, première partie. *Mémoires Couronnés des Savants Étrangers de l'Académie Royale des Sciences, des Lettres et des Beaux Arts de Belgique* **1851**, *24*, 1–176.
24. Carus, J.V. Vermes. In *Handbuch der Zoologie*; Kükenthal, W., Ed.; Wilhelm Engelmann: Leipzig, Germany, 1863; Volume 2, pp. 422–454.
25. Hartman, O. Polychaetous annelids. Part 6. Paraonidae, Magelonidae, Longosomidae, Ctenodrilidae, and Sabellariidae. *Allan Hancock Pac. Exped.* **1944**, *10*, 311–389.
26. Saint Joseph [Baron de], A.A. Les annélides polychètes des côtes de Dinard. Troisième partie. *Ann. des Sci. Nat. Zool. et Paléontol.* **1894**, *17*, 1–395.
27. Hartman, O. Abyssal polychaetous annelids from the Mozambique Basin off southeast Africa, with a compendium of abyssal polychaetous annelids from world-wide areas. *J. Fish. Res. Board Can.* **1971**, *28*, 1407–1428. [CrossRef]
28. Martínez, A.; Worsaae, K.; Núñez, J. Acocirridae Banse, 1968. In *Handbook of Zoology. Annelida*; Westheide, W., Böggemann, M., Purschke, G., Eds.; De Gruyter: Berlin, Germany; Boston, MA, USA, 2019; Volume 1: Annelida Basal Groups and Pleistoannelida, Sedentaria I; pp. 422–439.
29. Blake, J.A.; Magalhães, W. Cirratulidae Ryckholt, 1851. In *Handbook of Zoology. Annelida*; Purschke, G., Westheide, W., Böggemann, M., Eds.; De Gruyter: Berlin, Germany; Boston, MA, USA, 2019; Volume 1: Annelida Basal Groups and Pleistoannelida, Sedentaria I; pp. 339–397.
30. Zhadan, A.E.; Salazar-Vallejo, S. Fauveliopsidae Hartman, 1971. In *Hanbook of Zoology. Annelida*; Purschke, G., Böggemann, M., Westheide, W., Eds.; De Gruyter: Berlin, Germany; Boston, MA, USA, 2019; Volume 1: Annelida Basal Groups and Pleistoannelida, Sedentaria I; pp. 317–328.
31. Fiege, D. Sternaspidae Carus, 1863. In *Handbook of Zoology. Annelida*; Purschke, G., Böggemann, M., Westheide, W., Eds.; De Gruyter: Berlin, Germany; Boston, MA, USA, 2019; Volume 1: Annelida Basal Groups and Pleistoannelida, Sedentaria I; pp. 440–457.
32. Osborn, K.J.; Madin, L.P.; Rouse, G.W. The remarkable squidworm is an example of discoveries that await in deep-pelagic habitats. *Biol. Lett.* **2011**, *7*, 449–453. [CrossRef] [PubMed]
33. Salazar-Vallejo, S.I.; Carrera-Parra, L.F.; Fauchald, K. Phylogenetic affinities of the Flabelligeridae (Annelida, Polychaeta). *J. Zool. Syst. Evol. Res.* **2008**, *46*, 203–215. [CrossRef]
34. Ebbe, B.; Blake, J.A. Cossuridae Day, 1963. In *Handbook of Zoology. Annelida*; Purschke, G., Böggemann, M., Westheide, W., Eds.; De Gruyter: Berlin, Germany; Boston, MA, USA, 2019; Volume 1: Annelida Basal Groups and Pleistoannelida, Sedentaria I; pp. 309–317.
35. Blake, J.A. Paraonidae Cerruti 1909. In *Handbook of Zoology. Annelida*; Purschke, G., Böggemann, M., Westheide, W., Eds.; De Gruyter: Berlin, Germany; Boston, MA, USA, 2019; Volume 1: Annelida Basal Groups and Pleistoannelida, Sedentaria I; pp. 281–308.

36. Sousa, L.K.S.; Júnior, M.N.; Cutrim, M.V.J.; De Oliveira, V.M. *Cossura yacy* sp. Nov. (cossuridae, annelida) from a tropical brazilian estuary. *Iheringia Ser. Zool.* **2019**, *109*. [CrossRef]
37. Rouse, G.W.; Fauchald, K. Cladistics and polychaetes. *Zool. Scr.* **1997**, *26*, 139–204. [CrossRef]
38. Zrzavý, J.; Říha, P.; Piálek, L.; Janouškovec, J. Phylogeny of Annelida (Lophotrochozoa): Total-evidence analysis of morphology and six genes. *BMC Evol. Biol.* **2009**, *9*, 189. [CrossRef]
39. Strelzov, V.E. *Polychaete Worms of the Family Paraonidae Cerruti, 1909 (Polychaeta, Sedentaria)*; Akademia Nauk SSSR: Leningrad, Russia, 1973; pp. 1–169.
40. Bleidorn, C. The role of character loss in phylogenetic reconstruction as exemplified for the Annelida. *J. Zool. Syst. Evol. Res.* **2007**, *45*, 299–307. [CrossRef]
41. Rousset, V.; Pleijel, F.; Rouse, G.W.; Erséus, C.; Siddall, M.E. A molecular phylogeny of annelids. *Cladistics* **2007**, *23*, 41–63. [CrossRef]
42. Bleidorn, C. Phylogenetic relationships and evolution of Orbiniidae (Annelida, Polychaeta) based on molecular data. *Zool. J. Linn. Soc.* **2005**, *144*, 59–73. [CrossRef]
43. Blake, J.; Maciolek, N.J. Longosomatidae Hartman, 1944. In *Handbook of Zoology. Annelida*; Purschke, G., Böggemann, M., Westheide, W., Eds.; De Gruyter: Berlin, Germany; Boston, MA, USA, 2019; Volume 1: Annelida Basal Groups and Pleistoannelida, Sedentaria I; pp. 457–466.
44. Ehlers, E. Annulata nova vel minus cognita in Expeditione 'Porcupine' capta. *Ann. Mag. Nat. Hist.* **1874**, *13*, 292–298. [CrossRef]
45. Vijapure, T.; Sukumaran, S.; Manohar, C.S. Molecular characterization and phylogenetics of Indian polychaete fauna: Scope for implementation in ecological monitoring. *Aquat. Ecol.* **2019**, *53*, 665–677. [CrossRef]
46. Salazar-Vallejo, S.I.; Zhadan, A.E.; Rizzo, A.E. Revision of Fauveliopsidae Hartman, 1971 (Annelida, Sedentaria). *Zootaxa* **2019**, *4637*, 1–67. [CrossRef]
47. Grube, A.E. Die Familie der Cirratuliden. *Jahres Bericht der Schlesischen Gesellschaft für Vaterländische Cultur Breslau* **1873**, *50*, 59–66.
48. Grube, A.E. Die Familien der Anneliden. *Arch. für Nat. Berl.* **1850**, *16*, 249–364.
49. Gillet, P. *Flabelligena amoureuxi* new genus, new species (Polychaeta: Acrocirridae) from Crozet Islands (Indian Ocean). *Bull. Mar. Sci.* **2001**, *68*, 125–131.
50. Jimi, N.; Fujimoto, S.; Imura, S. A new interstitial genus and species of Acrocirridae from Okinawa-jima Island, Japan. *Zoosymposia* **2020**, *19*, 164–172. [CrossRef]
51. Hartman, O. Polychaeta from the Weddell Sea quadrant, Antarctica. *Antarct. Res. Ser.* **1978**, *26*, 125–223.
52. Laubier, L. *Chauvinelia biscayensis* gen. sp. nov., un Flabelligeridae (annélide polychète sédentaire) aberrant de l'étage abyssal du Golfe de Gascogne. *Bull. Soc. Zool. Fr.* **1974**, *99*, 391–399.
53. Osborn, K.J.; Haddock, S.H.D.; Pleijel, F.; Madin, L.P.; Rouse, G.W. Deep-Sea, Swimming Worms with Luminescent "Bombs". *Science* **2009**, *325*, 964. [CrossRef] [PubMed]
54. Salazar-Vallejo, S.; Gillet, P.; Carrera-Parra, L.F. Revision of *Chauviniella*, redescriptions of *Flabelliseta incrusta*, and *Helmetophorus rankini*, and their recognition as acrocirrids (Polychaeta: Acrocirridae). *J. Mar. Biol. Assoc. U. K.* **2007**, *87*, 465–477. [CrossRef]
55. Kudenov, J.D. Polychaeta from southeaster Australia 1. Acrocirridae Banse, 1969, from Victoria and New South Wales. *Rec. Aust. Mus.* **1976**, *30*, 137–151. [CrossRef]
56. Martínez, A.; Di Domenico, M.; Leasi, F.; Curini-Galletti, M.; Todaro, M.A.; Zotto, M.D.; Gobert, S.; Artois, T.; Norenburg, J.; Jörger, K.M.; et al. Patterns of diversity and endemism of soft-bodied meiofauna in an oceanic island, Lanzarote, Canary Islands. *Mar. Biodivers.* **2019**, *49*, 2033–2055. [CrossRef]
57. Osborn, K.J.; Rouse, G.W. Phylogenetics of Acrocirridae and Flabelligeridae (Cirratuliformia, Annelida). *Zool. Scr.* **2011**, *40*, 204–219. [CrossRef]
58. Hartman, O. Deep-Water Benthic Polychaetous Annelids off New England to Bermuda and Other North Atlantic Areas. *Occas. Pap. Allan Hancock Found.* **1965**, *28*, 1–378.
59. Gonzalez, B.C.; Borda, E.; Carbalho, R.; Schulze, A. Polychaetes from the Mayan underworld: Phylogeny, evolution, and cryptic diversity. *Nat. Croat. Period. Musei Hist. Nat. Croat.* **2012**, *21*, 51–53.
60. Nunez, J.; Ocana, O.; Brito, M.C. Two New Species (Polychaeta: Fauveliopsidae and Nerillidae) and other Polychaetes from the Marine Lagoon Cave of Jameos Del Agua, Lanzarote (Canary Islands). *Bull. Mar. Sci.* **1997**, *60*, 252–260.
61. Riera, R.; Monterroso, Ó.; Núñez, J.; Martínez, A. Distribution of meiofaunal abundances in a marine cave complex with secondary openings and freshwater filtrations. *Mar. Biodivers.* **2018**, *48*, 203–215. [CrossRef]
62. Hartman, O. Polychaetous annelids of the Indian Ocean including an account of species collected by members of the International Indian Ocean Expeditions, 1963-1964 and a catalogue and bibliography of the species from India. *J. Mar. Biol. Assoc. India* **1976**, *16*, 191–252.
63. Santos, P.J.P.; Silva, V.M.A.P. *Macrochaeta westheidi* n. sp.: First record of Acrocirridae for the Brazilian coast (annelida Polychaeta). *Nerítica Curitiba* **1993**, *7*, 7–12.
64. Magalhães, W.; Bailey-Brock, J.H. A new species of *Acrocirrus* (Polychaeta: Acrocirridae) from Coconut Island, Oahu, Hawaii. *J. Mar. Biol. Assoc. U. K.* **2012**, *92*, 1019–1102. [CrossRef]
65. Friedrich, H. Polychaetenstudien, I-III. *Kiel. Meeresforsch.* **1937**, *1*, 343–351.
66. Fauvel, P. *Polychètes Sédentaires. Addenda aux Errantes, Archiannélides, Myzostomaires*; Paul Chevalier: Paris, France, 1927; Volume 16, 494p.

67. Westheide, W. Interstitielle Fauna von Galapagos. XXVI. Questidae, Cirratulidae, Acrocirridae, Ctenodrillidae (Polychaeta). *Mikrofauna des Meeresbod.* **1981**, *82*, 1–23.
68. Parapar, J.; Moreira, J. Polychaeta of the 'DIVA-Artabria I' project (cruise 2002) in the continental shelf and upper slope off Galicia (NW Spain). *Cah. de Biol. Mar.* **2009**, *50*, 57–78.
69. Martínez, A.; Gonzalez, B.C. Volcaline anchialine habitats of Lanzarote. In *Cave Ecology*; Moldovan, O.T., Kováč, L., Halse, S., Eds.; Ecological Studies; Springer: Cham, Switzerland, 2018; Volume 235, pp. 399–414.
70. Martínez, A.; Kvindebjerg, K.; Lliffe, T.; Worsaae, K. Evolution of cave suspension feeding in Protodrilidae (Annelida). *Zool. Scr.* **2017**, *46*, 214–226. [CrossRef]
71. Worsaae, K.; Gonzalez, B.C.; Kerbl, A.; Holdfold Nielsen, S.; Terp Jørgensen, J.; Armenteros, M.; Iliffe, T.M.; Martínez, A. Diversity and evolution of the stygobitic *Speleonerilla* nom. nov. (Nerillidae, Annelida) with description of three new species from anchialine caves in the Caribbean and Lanzarote. *Mar. Biodivers.* **2019**, *49*, 2167–2192. [CrossRef]
72. Gonzalez, B.C.; Worsaae, K.; Fontaneto, D.; Martínez, A. Anophtalmia and elongation of body appendages in cave scale worms (Annelida: Aphroditiformia). *Zool. Scr.* **2018**, *47*, 106–121. [CrossRef]
73. Jumars, P.A.; Dorgan, K.M.; Lindsay, S.M. Diet of Worms Emended: An Update of Polychaete Feeding Guilds. *Annu. Rev. Mar. Sci.* **2015**, *7*, 497–520. [CrossRef]
74. Sars, M. *Beskrivelser og Iagttagelser over nogle Moerkelige eller nye i Havet ved den Bergenske Kyst Levende Dyr af Polypernes, Acalephernes, Radiaternes, Annelidernes og Molluskernes Classer, med en kort Oversigt over de Hidtil af Forfatteren Sammesteds Fundne Arter og deres Forekommen*; Thorstein Hallagers: Bergen, Norway, 1835; 81p.
75. Lamarck, J.B. *Histoire Naturelle des Animaux sans Vertèbres, Préséntant les Caractères Généraux et Particuliers de ces Animaux, leur Distribution, leurs Classes, leurs Familles, leurs Genres, et la Citation des Principales Espèces qui s'y Rapportent; Précedés d'une Introduction Offrant la Determination des Caracteres Essentiels de l'Animal, sa Distinction du Vegetal et des Autres Corps Naturels, Enfin, l'Exposition des Principes Fondamentaux de la Zoologie*; Deterville: Paris, France, 1818.
76. Hartman, O. Nomenclatural changes involving California polychaete worms. *J. Wash. Acad. Sci.* **1936**, *26*, 31–32.
77. Kinberg, J.G.H. Annulata Nova. *Öfversigt af Königlich Vetenskapsakademiens Förhandlingar Stockholm* **1866**, *22*, 239–258.
78. Çinar, M.E.; Petersen, M.E. Re-description of *Cirratulus dollfusi* (Polychaeta: Cirratulidae), and *Fauvelicirratulus* as a new genus. *J. Mar. Biol. Assoc. U. K.* **2011**, *91*, 415–418. [CrossRef]
79. Czerniavski, V. Materialia ad zoographiam Ponticam comparatam. *Bull. de la Société Impériale des Naturalistes de Moscou* **1881**, *56*, 338–420.
80. Blake, J. Revision of some genera and species of Cirratulidae from the Western North Atlantic. *Ophelia Suppl.* **1991**, *5*, 17–30.
81. Chamberlin, R.V. The Annelida Polychaeta [Albatross Expeditions]. *Mem. Mus. Comp. Zool. Harv. Coll.* **1919**, *48*, 1–514.
82. Blake, J.A. Bitentaculate Cirratulidae (Annelida, Polychaeta) collected chiefly during cruises of the R/V Anton Bruun, USNS Eltanin, USCG Glacier, R/V Hero, RVIB Nathaniel B. Palmer, and R/V Polarstern from the Southern Ocean, Antarctica, and off Western South America. *Zootaxa* **2018**, *4537*, 1–130. [CrossRef]
83. Blake, J.A. *Kirkegaardia* (Polychaeta, Cirratulidae), new name for *Monticellina* Laubier, preoccupied in the Rhabdocoela, together with new records and descriptions of eight previously known and sixteen new species from the Atlantic, Pacific, and Southern Oceans. *Zootaxa* **2016**, *4166*, 1–93. [CrossRef] [PubMed]
84. Webster, H.E.; Benedict, J.E. *The Annelida Chaetopoda from Eastport, Maine*; Report of the United States Commissioner of Fisheries; U.S. Commission of Fish & Fisheries: Washington, DC, USA, 1887; pp. 707–758.
85. Ørsted, A.S. *Annulatorum Danicorum Conspectus. Auctore A.S. Örsted. Fasc. I. Maricolæ. (Quæstio ab Universitate Hafniensi ad Solvendum Proposita et Proemio Ornata)*; Sumtibus Librariæ Wahlianæ: Copenhagen, Denmark, 1843.
86. Claparède, A.R.E. *Beobachtungen über Anatomie und Entwicklungsgeschichte Wirbelloser Thiere: An der Küste von Normandie Angestellt*; W. Engelmann: Leipzig, Germany, 1863; p. 120, 118 plates.
87. Wilfert, M. *Aphropharynx heterochaeta* nov. gen. nov. spec., ein neuer polychaet aus der familie Ctenodrilidae Kennel 1882. *Cah. de Biol. Mar.* **1974**, *15*, 495–504.
88. Monticelli, F.S. Di un nuovo Ctenodrilide del Golfo di Napoli. Nota preliminare riassuntiva del socio ordinario. *Rend. dell'Accad. delle Sci. Fische e Mat.* **1910**, *16*, 61–64.
89. Hartman, O. Polychaetous annelids from California. *Allan Hancock Pac. Exped.* **1961**, *25*, 1–226.
90. Kennel, J. Über *Ctenodrilus pardalis* Clpd. Ein Beitrag zur Kenntniss der Anatomie und Knospung der Anneliden. *Arb. aus dem Zool. Zootomischen Inst. Würzbg.* **1882**, *5*, 373–429.
91. Magalhães, W.F.; Linse, K.; Wiklund, H. A new species of *Raricirrus* (Annelida: Cirratuliformia) from deep-water sunken wood off California. *Zootaxa* **2017**, *4353*, 51–68. [CrossRef] [PubMed]
92. Weidhase, M.; Bleidorn, C.; Simon, C.A. On the taxonomy and phylogeny of *Ctenodrilus* (Annelida: Cirratulidae) with a first report from South Africa. *Mar. Biodivers.* **2016**, *46*, 243–252. [CrossRef]
93. Grosse, M.; Bakken, T.; Nygren, A.; Kongsrud, J.A.; Capa, M. Species delimitation analyses of NE Atlantic *Chaetozone* (Annelida, Cirratulidae) reveals hidden diversity among a common and abundant marine annelid. *Mol. Phylogenet. Evol.* **2020**, *149*, 106852. [CrossRef]
94. Elias, R.; Rivero, M.S.; Orensanz, J.M.L. New species of *Monticellina* and *Chaetozone* (Polychaeta: Cirratulidae) in the SW Atlantic, and a review of *Monticellina* species. *J. Mar. Biol. Assoc. U. K.* **2017**, *97*, 1553–1563. [CrossRef]

95. Blake, J. Family Cirratulidae Ryckholdt, 1851. Including a revision of the Genera and Species from the Eastern North Pacific. In *Taxonomic Atlas of the Benthic Fauna of the Santa Maria Bassin and Western Santa Barbara Channel*; Blake, J., Hilbig, B., Scott, P.H., Eds.; Santa Barbara Museum of Natural History: Santa Barbara, CA, USA, 1996; Volume 6—The Annelida Part 3. Polychaeta: Orbiniidae to Cossuridae; pp. 27–70.
96. Magalhães, W.F.; Bailey-Brock, J.H.; Davenport, J.F. On the genus *Raphidrilus* Monticelli, 1910 (Polychaeta: Ctenodrilidae) with description of two new species. *Zootaxa* **2011**, *2804*, 1–14. [CrossRef]
97. Elías, R.; Rivero, M.S. Two new species of Cirratulidae (Annelida: Polychaeta) from Mar del Plata, Argentina (SW Atlantic). *Zoosymposia* **2009**, *2*, 139–148. [CrossRef]
98. Dean, H.K.; Blake, J. *Monticellina* (Polychaeta: Cirratulidae) from the Pacific coast of Costa ica with descriptions of six new species. *Zoosymposia* **2009**, *2*, 105–126. [CrossRef]
99. Blake, J.A.; Lavesque, N. A new species of *Chaetozone* (Polychaeta, Cirratulidae) from the Bay of Biscay offshore France, together with a review of *Chaetozone* species from the North Atlantic and adjacent waters. *Zootaxa* **2017**, *4312*, 565. [CrossRef]
100. Doner, S.A.; Blake, J.A. New species of Cirratulidae (Polychaeta) from the northeastern United States. *Sci. Mar.* **2006**, *70*, 65–73. [CrossRef]
101. Dean, H.K.; Blake, J.A. Ctenodrilidae Kennel, 1882. In *Handbook of Zoology. Annelida*; Westheide, W., Purschke, G., Böggemann, M., Eds.; De Gruyter: Berlin, Germany; Boston, MA, USA, 2019; Volume 1: Annelida Basal Groups and Pleistoannelida, Sedentaria I; pp. 328–338.
102. Blake, J.A. New species of Cirratulidae (Annelida, Polychaeta) from abyssal depths of the Clarion-Clipperton Fracture Zone, North Equatorial Pacific Ocean. *Zootaxa* **2019**, *4629*, 151–187. [CrossRef]
103. Magalhães, W.F.; Bailey-Brock, J.H. Bitentaculate Cirratulidae (Annelida: Polychaeta) from the northwestern Pacific Islands with description of nine new species. *Zootaxa* **2013**, *3630*, 80–116. [CrossRef]
104. Magalhães, W.F.; Seixas, V.C.; Paiva, P.C.; Elias, R. The Multitentaculate Cirratulidae of the Genera *Cirriformia* and *Timarete* (Annelida: Polychaeta) from Shallow Waters of Brazil. *PLoS ONE* **2014**, *9*, e112727. [CrossRef]
105. Choi, H.K.; Kim, H.; Yoon, S.M. *Timarete posteria*, a new cirratulid species from Korea (Annelida, Polychaeta, Cirratulidae). *ZooKeys* **2018**, *806*, 1–15. [CrossRef] [PubMed]
106. Blake, J.A.; Dean, H.K. New Species of Cirratulidae (Annelida, Polychaeta) from the Caribbean Sea. *Zootaxa* **2019**, *4671*, 301–338. [CrossRef]
107. Elías, R.; Saracho-Bottero, M.A.; Simon, C.A. *Protocirrineris* (Polychaeta:Cirratulidae) in South Africa and description of two new species. *Rev. de Biol. Trop.* **2019**, *67*, S70–S80. [CrossRef]
108. Elías, R.; Saracho-Bottero, M.A.; Magalhães, W.F. Two new species of *Protocirrineris* (Polychaeta: Cirratulidae) from Brazil. *Rev. de Biol. Trop.* **2019**, *67*, S81–S91. [CrossRef]
109. Schmidt, O. Zur Kenntnis der Turbellaria, Rhabdocoela und einiger anderer Wuermer des Mittelmeeres. *Sitzungsberichte der Kaiserliche Akademie der Wissenschaften Wien Mathematisch Naturwissenschaftliche Klasse* **1857**, *23*, 347–366.
110. Grube, A.E. Annulata Örstediana. Enumeratio Annulatorum, quae in itinere Indiam Occidentalem et Americam Centralem annis 1845-1848 suscepto legit cl. A.S. Oersted, adjectis speciebus nonnullis a cl. H. Kroyero in itinere ad Americam meridionalem collectis. *Videnskabelige Meddelelser fra Dansk naturhistorisk Forening i Köbenhavn* **1858**, *3*, 105–120.
111. Blake, J.A. New species of *Chaetozone* and *Tharyx* (Polychaeta: Cirratulidae) from the Alaskan and Canadian Arctic and the Northeastern Pacific, including a description of the lectotype of *Chaetozone setosa* Malmgren from Spitsbergen in the Norwegian Arctic. *Zootaxa* **2015**, *3919*, 501. [CrossRef]
112. Seixas, V.C.; Zanol, J.; Magalhães, W.F.; Paiva, P.C. Genetic diversity of *Timarete punctata* (Annelida: Cirratulidae): Detection of pseudo-cryptic species and a potential biological invader. *Estuar. Coast. Shelf Sci.* **2017**, *197*, 214–220. [CrossRef]
113. Westheide, W.; Haß-Cordes, E.; Krabusch, M. *Ctenodrilus serratus* (Polychaeta: Ctenodrilidae) is a truly amphi-Atlantic meiofauna species—Evidence from molecular data. *Mar. Biol.* **2003**, *142*, 637–642. [CrossRef]
114. Berkeley, E.; Berkeley, C. On a collection of Polychaeta from Southern California. *Bull. South. Calif. Acad. Sci.* **1941**, *40*, 16–60.
115. Omena, E.; Creed, J.C. Polychaete fauna of seagrass beds (*Halodule wrightii* Ascherson) along the coast of Rio de Janeiro (southeast Brazil). *P. S. Z. N. Marine Ecology* **2004**, *25*, 273–288. [CrossRef]
116. Çinar, M.E.; Ergen, Z. The presence of *Chaetozone corona* (Polychaeta: Cirratulidae) in the Mediterranean Sea: An alien or a native species? *Cah. de Biol. Mar.* **2007**, *48*, 339–346.
117. Le Garrec, V.; Grall, J.; Chevalier, C.; Guyonnet, B.; Jourde, J.; Lavesque, N.; Bonifácio, P.; Blake, J.A. *Chaetozone corona* (Polychaeta, Cirratulidae) in the Bay of Biscay: A new alien species for the North-east Atlantic waters? *J. Mar. Biol. Assoc. U. K.* **2017**, *97*, 433–445. [CrossRef]
118. Chambers, S.J. A redescription of *Chaetozone setosa* Malmgren, 1867 including a definition of the genus, and a description of a new species of *Chaetozone* (Polychaeta: Cirratulidae) from the Northeast Atlantic. *Bull. Mar. Sci.* **2000**, *67*, 10.
119. Langerhans, P. Die Wurmfauna von Madeira. III. *Z. für Wiss. Zool.* **1880**, *34*, 87–143.
120. Montagu, G. Description of several marine animals found on the south coast of Devonshire. *Trans. Linn. Soc. Lond.* **1808**, *9*, 81–114. [CrossRef]
121. Berkeley, E.; Berkeley, C. Notes on the life-history of the polychaete *Dodecaceria fewkesii* (nom.n.). *J. Fish. Res. Board Can.* **1954**, *11*, 326–334. [CrossRef]

122. Dean, H.K. A new species of *Raricirrus* (Polychaeta: Ctenodrilidae) from wood collected in the Tongue of the Ocean, Virgin Islands. *Proc. Biol. Soc. Wash.* **1995**, *108*, 169–179.
123. Hily, C. Spatio-temporal variability of *Chaetozone setosa* (Malmgren) populations on an organic gradient in the Bay of Brest, France. *J. Exp. Marine Biol. Ecol.* **1987**, *112*, 201–216. [CrossRef]
124. Petersen, M.E.; George, J.D. A new species of *Raricirrus* from Northern Europe, with notes on its biology and a discussion of the affinities of the genus (polychaeta: Ctenodrilidae). *Ophelia Suppl.* **1991**, *5*, 185–208.
125. Pearson, T.H.; Rosenberg, R. Macrobenthic succession in relation to organic enrichment and pollution of the marine environment. *Oceanogr. Mar. Biol. Annu. Rev.* **1978**, *16*, 229–311.
126. Elías, R.; Rivero, M.S.; Palacios, J.R.; Vallarino, E.A. Sewage-induced disturbance on polychaetes inhabiting intertidal mussel beds of *Brachidontes rodriguezii* off Mar del Plata (SW Atlantic, Argentina). *Sci. Mar.* **2006**, *70*, 187–196. [CrossRef]
127. Elías, R.; Rivero, M.S.; Vallarino, E.A. Sewage impact on the composition and distribution of Polychaeta associated to intertidal mussel beds of the Mar del Plata rocky shore, Argentina. *Iheringia Ser. Zool.* **2003**, *93*, 309–318. [CrossRef]
128. Gibbs, P.; Langston, W.; Burt, G.; Pascoe, P. *Tharyx marioni* (Polychaeta): A remarkable accumulator of arsenic. *J. Mar. Biol. Assoc. U. K.* **1983**, *63*, 313–325. [CrossRef]
129. Milanovich, F.P.; Spies, R.; Guram, M.S.; Sykes, E.E. Uptake of copper by the polychaete *Cirriformia spirabrancha* in the presence of dissolved yellow organic matter of natural origin. *Estuar. Coast. Mar. Sci.* **1976**, *4*, 585–588. [CrossRef]
130. Moore, D.C. *Raricirrus beryli* Petersen and George (Ctenodrilidae): A new polychaete indicator species for hydrocarbon-polluted sediments. *Ophelia Suppl.* **1991**, *5*, 477–486.
131. Reish, D.J.; Gerlinger, T.V. A Review of the Toxicological Studies with Polychaetous Annelids. *Bull. Mar. Sci.* **1997**, *60*, 584–607.
132. Petersen, M.E. Reproduction and development in Cirratulidae (Annelida: Polychaeta). In *Reproductive Strategies and Developmental Patterns in Annelids*; Dorresteijn, A.W.C., Westheide, W., Eds.; Springer: Dordrecht, The Netherlands, 1999; pp. 107–128.
133. Christie, G. A new species of *Tharyx* (Polychaeta: Cirratulidae) from the estuaries in north-east England. *Sarsia* **1984**, *69*, 69–73.
134. Hilbig, B. Family Cossuridae Day, 1963. In *Taxonomic Atlas of the Benthic Fauna of the Santa Maria Basin and Western Santa Barbara Channel*; Blake, J.A., Hilbig, B., Scott, P.H., Eds.; Santa Barbara Museum of Natural History: Santa Barbara, CA, USA, 1996; Volume 6—The Annelida Part 3. Polychaeta: Orbiniidae to Cossuridae; pp. 385–402.
135. Liñero-Arana, I.; Díaz-Díaz, Ó. A new species of Cossuridae (Annelida: Polychaeta) from Venezuela. *Interciencia* **2010**, *35*, 789–792.
136. Zhadan, A. Cossuridae (Annelida: Polychaeta: Sedentaria) from australian and adjacent waters: The first faunistic survey. *Rec. Aust. Mus.* **2015**, *67*, 1–24. [CrossRef]
137. Fournier, J.A.; Petersen, M.E. *Cossura longocirrata*: Redescription and distribution, with notes on reproductive biology and a comparison of described species of *Cossura* (Polychaeta: Cossuridae). *Ophelia Suppl.* **1991**, *5*, 63–80.
138. Read, G.B. Taxonomy and distribution of a new *Cossura* species (Annelida: Polychaeta: Cossuridae) from New Zealand. *Proc. Biol. Soc. Wash.* **2000**, *113*, 1096–1110.
139. Zhadan, A.E.; Vortsepneva, E.V.; Tsetlin, A.B. Redescription and biology of *Cossura pygodactylata* Jones, 1956 (Polychaeta: Cossuridae) in the White Sea. *Invertebr. Zool.* **2012**, *9*, 115–125. [CrossRef]
140. Reish, D.J. Description of a new species of *Cossura* (Annelida: Polychaeta) from the Mississippi Delta. *J. Wash. Acad. Sci.* **1958**, *48*, 53–55.
141. Bachelet, G.; Laubier, L. Morphology, ecology and juvenile development of *Cossura pygodactylata* Jones (Polychaeta, Cossuridae) in Arcachon Bay, SW France, with a reassessment of the geographical distribution of *C. pygodactylata* and *C. soyeri* Laubier. *Mem. Mus. Natl. D'histoire Nat.* **1994**, *162*, 355–369.
142. Ewing, R.M. Review of the genus *Cossurella* (Polychaeta: Cossuridae) including descriptions of two new species and a key to the species of the world. *Bull. Biol. Soc. Wash.* **1987**, *7*, 3–10.
143. Kitamori, R. Two new species of cirratulid and Nephthydidae (Annelida: Polychaeta. *Bull. Jpn. Soc. Sci. Fish.* **1960**, *26*, 1082–1085. [CrossRef]
144. Górska, B.; Gromisz, S.; Włodarska-Kowalczuk, M. Size assessment in polychaete worms—application of morphometric correlations for common North Atlantic taxa. *Limnol. Oceanogr. Methods* **2019**, *17*, 254–265. [CrossRef]
145. Zhadan, A. Two new species of *Cossura* (Cossuridae, Annelida) from the terminal lobes of the Congo River deep-sea fan. *J. Mar. Biol. Assoc. U. K.* **2017**, *97*, 881–888. [CrossRef]
146. Laubier, L. Découverte du genre *Cossura* (polychète, Cossuridae) en Méditerranée: *Cossura soyeri* sp. n. *Vie et Milieu* **1963**, *14*, 833–842.
147. Callier, M.D.; Fletcher, R.L.; Thorp, C.H.; Fichet, D. Macrofaunal community responses to marina-related pollution on the south coast of England and west coast of France. *J. Mar. Biol. Assoc. U. K.* **2009**, *89*, 19–29. [CrossRef]
148. Bogdanos, C.; Fredj, G. Sur la présence de *Cossura coasta* Kitamori, 1960 (Polychaeta, Cossuridae) dans les eaux cotieres Grecques (Mer Égée, Mer Ionienne). *Thalassographica* **1983**, *6*, 5–15.
149. Zenetos, A.; Gofas, S.; Verlaque, M.; Cinar, M.E.; Garcia Raso, J.E.; Bianchi, C.N.; Morri, C.; Azzurro, E.; Bilecenoglu, M.; Froglia, C.; et al. Alien species in the Mediterranean Sea by 2010. A contribution to the application of European Union's Marine Strategy Framework Directive (MSFD). Part I. Spatial distribution. *Mediterr. Mar. Sci.* **2010**, *11*, 381. [CrossRef]
150. Faulwetter, S.; Simboura, N.; Katsiaras, N.; Chatzigeorgiou, G.; Arvanitidis, C. Polychaetes of Greece: An updated and annotated checklist. *Biodivers. Data J.* **2017**, *22*, e20997. [CrossRef]
151. Raghunathan, C.; Gopi, K.C.; Chandra, K. Invasive alien faunal species in India. *J. Gov.* **2019**, *18*, 124–140.

152. Gaonkar, C.A.; Sawant, S.S.; Anil, A.C.; Venkat, K.; Harkantra, S.N. Mumbai harbour, India: Gateway for introduction of marine organisms. *Environ. Monit. Assess.* **2010**, *163*, 583–589. [CrossRef] [PubMed]
153. Gutiérrez, D.; Gallardo, V.A.; Mayor, S.; Neira, C.; Vásquez, C.; Sellanes, J.; Rivas, M.; Soto, A.; Carrasco, F.; Baltazar, M. Effects of dissolved oxygen and fresh organic matter on the bioturbation potential of macrofauna in sublittoral sediments off Central Chile during the 1997/1998 El Niño. *Mar. Ecol. Prog. Ser.* **2000**, *202*, 81–99. [CrossRef]
154. Ryu, J.; Khim, J.S.; Kang, S.-G.; Kang, D.; Lee, C.-h.; Koh, C.-h. The impact of heavy metal pollution gradients in sediments on benthic macrofauna at population and community levels. *Environ. Pollut.* **2011**, *159*, 2622–2629. [CrossRef] [PubMed]
155. Moreira, J.; Lourido, A.; Troncoso, J.S. Temporal dynamics of the benthic assemblage in the muddy sediments of the harbour of Baiona (Galicia, NW Iberian Peninsula). *Thalassas* **2010**, *26*, 9–22.
156. Balachandar, K.; Sundaramanickam, A.; Kumaresan, S. Spatial and seasonal variation of macrobenthos from Puducherry Coast, Southeast Coast of India. *Int. J. Curr. Microbiol. Appl. Sci.* **2016**, *5*, 33–49. [CrossRef]
157. Chainho, P.; Chaves, M.L.; Costa, J.L.; Costa, M.J.; Dauer, D.M. Use of multimetric indices to classify estuaries with different hydromorphological characteristics and different levels of human pressure. *Mar. Pollut. Bull.* **2008**, *56*, 1128–1137. [CrossRef]
158. Khan, S.A.; Murugesan, P. Polychaete diversity in Indian estuaries. *Indian J. Geo-Mar. Sci.* **2005**, *34*, 114–119. [CrossRef]
159. Murugesan, P.; Ajmal Khan, S.; Ajithkumar, T.T. Temporal changes in the benthic community structure of the marine zone of Vellar estuary southeast coast of India. *J. Mar. Biol. Assoc. India* **2007**, *49*, 154–158.
160. Rakocinski, C.F.; Brown, S.S.; Gaston, G.R.; Heard, R.W.; Walker, W.W.; Summers, J.K. Macrobenthic responses to natural and contaminant-related gradients in northern Gulf of Mexico estuaries. *Ecol. Appl.* **1997**, *7*, 1278–1298. [CrossRef]
161. Méndez, N. Annelid assemblages in soft bottoms subjected to human impact in the Urías estuary (Sinaloa, Mexico). *Oceanol. Acta* **2002**, *25*, 139–147. [CrossRef]
162. Lohrer, A.M.; Chiaroni, L.D.; Hewitt, J.E.; Thrush, S.F. Biogenic Disturbance Determines Invasion Success In A Subtidal Soft-Sediment System. *Ecology* **2008**, *89*, 1299–1307. [CrossRef]
163. Widdicombe, S.; Austen, M.C. Mesocosm investigation into the effects of bioturbation on the diversity and structure of a subtidal macrobenthic community. *Mar. Ecol. Prog. Ser.* **1999**, *189*, 181–193. [CrossRef]
164. Rygg, B. Effect of sediment copper on benthic fauna. *Mar. Ecol. Prog. Ser.* **1985**, *25*, 83–89. [CrossRef]
165. Olsgard, F.; Hasle, J.R. Impact of waste from titanium mining on benthic fauna. *J. Exp. Mar. Biol. Ecol.* **1993**, *172*, 185–213. [CrossRef]
166. Burd, B.J. Evaluation of mine tailings effects on a benthic marine infaunal community over 29 years. *Mar. Environ. Res.* **2002**, *53*, 481–519. [CrossRef]
167. Ellis, D.V. The concept of "sustainable ecological succession"; and its value in assessing the recovery of sediment seabed biodiversity from environmental impact. *Mar. Pollut. Bull.* **2003**, *46*, 39–41. [CrossRef]
168. Ellis, D.V. Biodiversity reduction on contaminated sediments in the sea: Contamination or sediment mobility? *Mar. Pollut. Bull.* **2005**, *50*, 244–246. [CrossRef]
169. Lancellotti, D.A.; Stotz, W.B. Effects of shoreline discharge of iron mine tailings on a marine soft-bottom community in northern Chile. *Mar. Pollut. Bull.* **2004**, *48*, 303–312. [CrossRef]
170. Raman, A.V.; Ganapati, P.N. Pollution effects on ecobiology of benthic polychaetes in Visakhapatnam Harbour (Bay of Bengal). *Mar. Pollut. Bull.* **1983**, *14*, 46–52. [CrossRef]
171. Varadharajan, D.; Soundarapandian, P.; Gunalan, B.; Babu, R. Seasonal abundance of macro benthic composition and diversity along the south east coast of India. *Eur. J. Appl. Sci.* **2010**, *2*, 1–5.
172. Natesan, U.; Kalaivani, S.; Kalpana, G. Pollution assessment of Ennore (India) creek using macrobenthos. *J. Environ. Geol.* **2017**, *1*, 9–16. [CrossRef]
173. Shokat, P.; Nabavi, S.M.B.; Savari, A.; Kochanian, P. Ecological quality of Bahrekan coast, by using biotic indices and benthic communities. *Transit. Waters Bull.* **2010**, *4*, 25–34.
174. Grall, J.; Glémarec, M. Using biotic indices to estimate macrobenthic community perturbations in the Bay of Brest. *Estuar. Coast. Shelf Sci.* **1997**, *44*, 43–53. [CrossRef]
175. Borja, A.; Franco, J.; Pérez, V. A Marine Biotic Index to Establish the Ecological Quality of Soft-Bottom Benthos Within European Estuarine and Coastal Environments. *Mar. Pollut. Bull.* **2000**, *40*, 1100–1114. [CrossRef]
176. Teixeira, H.; Weisberg, S.B.; Borja, A.; Ranasinghe, J.A.; Cadien, D.B.; Velarde, R.G.; Lovell, L.L.; Pasko, D.; Phillips, C.A.; Montagne, D.E.; et al. Calibration and validation of the AZTI's Marine Biotic Index (AMBI) for Southern California marine bays. *Ecol. Indic.* **2012**, *12*, 84–95. [CrossRef]
177. Bigot, L.; Grémare, A.; Amoureux, J.-M.; Frouin, P.; Maire, O.; Gaertner, J.C. Assessment of the ecological quality status of soft-bottoms in Reunion Island (tropical Southwest Indian Ocean) using AZTI marine biotic indices. *Mar. Pollut. Bull.* **2008**, *56*, 704–722. [CrossRef] [PubMed]
178. Reish, D. Benthic Invertebrates as Indicators of Marine Pollution: 35 Years of Study. In Proceedings of the OCEANS'86, Washington, DC, USA, 23–25 September 1986; pp. 885–888. [CrossRef]
179. Reish, D.J. The relation of polychaetous Annelids to harbor pollution. *Public Health Rep.* **1955**, *70*, 1168–1174. [CrossRef]
180. Keeley, N.B.; Macleod, C.K.; Forrest, B.M. Combining best professional judgement and quantile regression splines to improve characterisation of macrofaunal responses to enrichment. *Ecol. Indic.* **2012**, *12*, 154–166. [CrossRef]

181. Blake, J.A. Life history analysis of five dominant infaunal polychaete species from the continental slope off North Carolina. *J. Mar. Biol. Assoc. U. K.* **1993**, *73*, 123–141. [CrossRef]
182. Berkeley, E.; Berkeley, C. Some Further Records of Pelagic Polychaeta from the Northeast Pacific North Of Latitude 40° N. and East of Longitude 175° W., Together with Records of Siphonophora, Mollusca, and Tunicata from the Same Region. *Can. J. Zool.* **1960**, *38*, 787–799. [CrossRef]
183. Hartman, O. Endemism in the North Pacific Ocean, with emphasis on the distribution of marine annelids, and descriptions of new or little known species. In *Essays in the Natural Sciences in Honor of Captain Allan Hancock on the Occasion of His Birthday July 26*; Allan Hancock Foundation, Ed.; University of Southern California Press: Los Angeles, CA, USA, 1955; pp. 39–60.
184. Hartman, O. Polychaetous annelids collected by the USNS Eltanin and Staten Island cruises, chiefly from Antarctic Seas. *Allan Hancock Monogr. Mar. Biol.* **1967**, *2*, 1–387.
185. Parapar, J.; Vijapure, T.; Moreira, J.; Sukumaran, S. A new species of *Heterospio* (Annelida, Longosomatidae) from the Indian Ocean. *Eur. J. Taxon.* **2016**. [CrossRef]
186. Blake, J.A. Family Paraonidae Cerruti 1909. In *Taxonomic Atlas of the Benthic Fauna of the Santa Maria Basin and Western Santa Barbara Channel*; 6—The Annelida Part 3. Polychaeta: Orbiniidae to Cossuridae; Blake, J.A., Hilbig, B., Scott, P.H., Eds.; Santa Barbara Museum of Natural History: Santa Barbara, CA, USA, 1996; pp. 27–70.
187. Hartley, J.P. The family Paraonidae (Polychaeta) in british waters: A new species and new records with a key to species. *J. Mar. Biol. Assoc. U. K.* **1981**, *61*, 133–149. [CrossRef]
188. Inajima, M. Paraonidae (Polychaeta) from Japan. *Bull. Natl. Sci. Mus. Tokyo* **1973**, *16*, 253–292.
189. Ranauro, N.; Barroso, R.; Paiva, P.C.; Nogueira, J.M.d.M. On the genus *Sabidius* Strelzov, 1973 (Annelida: Paraonidae), with a redescription of the type species and the description of a new species. *PLoS ONE* **2020**, *15*, e0229717. [CrossRef]
190. Arriaga-Hernández, S.; Hernández-Alcántara, P.; Solís-Weiss, V. Description and distribution of two new species of Paraonidae (Annelida: Polychaeta) from a lagoon-estuarine ecosystem in the Southern Gulf of Mexico. *Zootaxa* **2013**, *3686*, 51–64. [CrossRef] [PubMed]
191. Zhou, L.; Reuscher, M.G. Description of a new species of *Aricidea* (Polychaeta: Paraonidae) from Chinese Seas. *Proc. Biol. Soc. Wash.* **2013**, *126*, 276–283. [CrossRef]
192. Erdoğan-Dereli, D.; Çinar, M.E.; Dağli, E. *Cirrophorus* species (Annelida: Polychaeta: Paraonidae) in the Sea of Marmara, with description of a new species. *Zootaxa* **2017**, *4347*, 137–150. [CrossRef] [PubMed]
193. Lovell, L. Paraonidae (Annelida: Polychaeta) Of The Andaman Sea, Thailand. *Phuket Mar. Biol. Cent. Spec. Publ.* **2002**, *24*, 33–56.
194. Aguirrezabalaga, F.; Gil, J. Paraonidae (Polychaeta) from the Capbreton Canyon (Bay of Biscay, NE Atlantic) with the description of eight new species. *Sci. Mar.* **2009**, *73*, 631–666. [CrossRef]
195. Langeneck, J.; Mazziotti, C.; Mikac, B.; Scirocco, T.; Castelli, A. *Aricidea fragilis* (Annelida: Paraonidae) in the Mediterranean Sea: Overlooked native or alien species? *Eur. Zool. J.* **2018**, *85*, 267–272. [CrossRef]
196. Castelli, A. Paraonidae (Annelida, Polychaeta) des fonds meubles infralittoraux des cotes toscanes. *Cah. de Biol. Mar.* **1985**, *26*, 267–279.
197. Laubier, L.; Ramos, J. Paraonidae (Polychètes Sédentaires) de Méditerranée. *Bull. du Mus. Natl. d'Histoire Nat. Ser. 3 Zool.* **1974**, *168*, 1097–1148.
198. Hartman, O. Orbiniidae, Apistobranchidae, Paraonidae and Longosomidae. *Allan Hancock Pac. Exped.* **1957**, *15*, 211–393.
199. Katzmann, W.; Laubier, L. Paraonidae (Polychètes Sédentaires) de l'Adriatique. *Ann. der Nat. Mus. Wien* **1975**, *79*, 567–588.
200. Çinar, M.E.; Dagli, E.; Açik, S. Annelids (Polychaeta and Oligochaeta) from the Sea of Marmara, with descriptions of five new species. *J. Nat. Hist.* **2011**, *45*, 2105–2143. [CrossRef]
201. Langeneck, J.; Barbieri, M.; Maltagliati, F.; Castelli, A. A new species of *Cirrophorus* (Annelida: Paraonidae) from Mediterranean organically enriched coastal environments, with taxonomic notes on the family. *J. Mar. Biol. Assoc. U. K.* **2017**, *97*, 871–880. [CrossRef]
202. Mackie, A.S.Y. *Paradoneis eliasoni* sp. nov. (Polychaeta: Paraonidae) from northern European waters, with a redescription of *Paradoneis lyra* (Southern, 1914). *Ophelia Suppl.* **1991**, *5*, 147–155.
203. de Leon-Gonzáles, J.A.; Díaz-Castañeda, V. A new species of Paraonidae (Polychaeta: Paraonidae) from the western coast of Baja California, Mexico. *Proc. Biol. Soc. Wash.* **2011**, *124*, 40–44. [CrossRef]
204. López, E.; Sikorski, A. The Paraonidae (Annelida: Sedentaria) from Norway and adjacent seas, with two new species, four new records, and a redescription of *Paraonides nordica* Strelzov, 1968 based on type material. *Zootaxa* **2017**, *4320*, 41–67. [CrossRef]
205. Hobson, K.D. Two new species and two new records of the Family Paraonidae (Annelida, Polychaeta) from the northeastern Pacific Ocean. *Proc. Biol. Soc. Wash.* **1972**, *85*, 549–556.
206. Quintanar-Retama, O.; Hernández-Alcántara, P.; Solís-Weiss, V. Distribution patterns of three new species of *Paradoneis* (Annelida: Paraonidae) from the southern Gulf of Mexico, with a dichotomous key for the Grand Caribbean species. *Mar. Biodivers.* **2019**, *49*, 2851–2870. [CrossRef]
207. Laubier, L. Sur quelques *Aricidea* (Polychètes, Paraonidae) de Banyuls-sur-Mer. *Vie et Milieu Ser. A* **1967**, *18*, 99–132.
208. Jones, M.L. *Paraonis pygoenigmatica* new species, a new annelid from Massachusetts (Polychaeta: Paraonidae). *Proc. Biol. Soc. Wash.* **1968**, *81*, 323–334.
209. Mclelland, J.; Gaston, G. 2 New Species Of *Cirrophorus* (Polychaeta, Paraonidae) From The Northern Gulf of-Mexico. *Proc. Biol. Soc. Wash.* **1994**, *107*, 524–531.

210. Webster, H.E. The Annelida Chaetopoda of the Virginian coast. *Trans. Albany Inst.* **1879**, *9*, 202–269.
211. Grube, A.E. Über ein paar neue Anneliden aus der Familie der Spiodeen. *Jahres Bericht der Schlesischen Gesellschaft für Vaterländische Cultur* **1872**, *50*, 57–58.
212. McIntosh, W.C. On the Annelida obtained during the Cruise of H.M.S. 'Valorous' to Davis Strait in 1875. *Trans. Linn. Soc. Lond.* **1878**, *1*, 499–511. [CrossRef]
213. Tauber, P. *Annulata Danica. En Kritisk Revision af de in Danmark Fundne Annulata Chaetognatha, Gephyrea, Balanoglossi, Discophoreae, Oligochaeta, Gymnocopa og Polychaeta*; Reitzer: Kobenhavn, Denmark, 1879; p. 143.
214. Levinsen, G.M.R. Systematisk-geografisk Oversigt over de nordiske Annulata, Gephyrea, Chaetognathi og Balanoglossi. *Vidensk-abelige Meddelelser fra den Naturhistoriske Forening i Kjöbenhavn* **1884**, *45*, 92–350.
215. Ehlers, E. Die bodensässigen Anneliden aus den Sammlungen der deutschen Tiefsee-Expedition. In *Wissenschaftlichte Ergebnisse der Deutschen Tiefsee-Expedition auf dem Dampfer 'Valdivia' 1898–1899*; Chun, C., Ed.; Gustav Fischer: Jena, Germany, 1908; Volume 16, pp. 1–168.
216. Annenkova, N.P. Paraoniden der Meeren des Fernen Osten der USSR. *Doklady Akademii Nauk SSSR* **1934**, *3*, 656–658.
217. Mesnil, F. Études de morphologie externe chez les Annélides. II. Remarques complémentaires sur les Spionidiens. La famille nouvelle des Disomidiens. La place des Aonides (sensu Tauber, Levinsen). *Bull. Sci. Fr. Belg.* **1897**, *30*, 83–100.
218. Brasier, M.J.; Wiklund, H.; Neal, L.; Jeffreys, R.; Linse, K.; Ruhl, H.; Glover, A.G. DNA barcoding uncovers cryptic diversity in 50% of deep-sea Antarctic polychaetes. *R. Soc. Open Sci.* **2016**, *3*, 160432. [CrossRef] [PubMed]
219. Hilbig, B.; Blake, J.A. Deep-sea polychaete communities in the northeast Pacific Ocean off the Gulf of the Farallones, California. *Bull. Mar. Sci.* **2006**, *78*, 243–269.
220. Reuscher, M.G. Biodiversity, Ecology, and Natural History of Polychaetous Annelids from the Gulf of Mexico. Ph.D. Thesis, Corpus Christi University, Corpus Christi, TX, USA, 2013.
221. Ribeiro, R.P.; Alves, P.R.; de Almeida, Z.d.S.; Ruta, C. A new species of *Paraonis* and an annotated checklist of polychaetes from mangroves of the Brazilian Amazon Coast (Annelida, Paraonidae). *ZooKeys* **2018**, 1–34. [CrossRef]
222. Zhou, L.; Li, X.-Z. A report of the family Paraonidae (Annelida, Polychaeta) from China seas. *Acta Zootaxon. Sin.* **2007**, *32*, 275–282.
223. Lovell, L.L.; Fitzhugh, K. Taking a closer look: An SEM review of *Levinsenia* species (Polychaeta: Paraonidae) reported from California. *Zootaxa* **2020**, *4751*, 256–275. [CrossRef] [PubMed]
224. Langeneck, J.; Busoni, G.; Aliani, S.; Castelli, A. Deep-sea polychaetes (Annelida) from the Malta Escarpment (western Ionian Sea). *Eur. Zool. J.* **2017**, *84*, 142–152. [CrossRef]
225. Langeneck, J.; Lezzi, M.; Pasqua, M.D.; Musco, L.; Gambi, M.C.; Castelli, A.; Giangrande, A. Non-indigenous polychaetes along the coasts of Italy: A critical review. *Mediterr. Mar. Sci.* **2020**, *21*, 238–275.
226. Hartley, J.P. Cosmopolitan polychaete species: The status of Aricidea belgicae (Fauvel, 1936) and notes on the identity of A. suecica Eliason, 1920 (Polychaeta; Paraonidae). In *Proceedings of the First International Polychaete Conference, Sydney, Australia, 4–9 July 1983*; The Linnean Society of New South Wales: Sydney, Australia, 1984; pp. 7–20.
227. Erdoğan-Dereli, D.; Çinar, M.E. The diversity of the genus *Aricidea* (Polychaeta: Paraonidae) from the Sea of Marmara, with descriptions of two new species and two new records for the Mediterranean fauna. *Zootaxa* **2020**, *4844*, 1–73. [CrossRef]
228. Bakalem, A.; Gillet, P.; Pezy, J.-P.; Dauvin, J.-C. Inventory and the biogeographical affinities of Annelida Polychaeta in theAlgerian coastline (Western Mediterranean). *Mediterr. Mar. Sci.* **2020**, *21*, 157–182. [CrossRef]
229. Gibbs, P.E. Recent additions to the marine fauna of Whitstable, with a description of *Aricidea minuta* southward (Annelida: Polychaeta). *Ann. Mag. Nat. Hist.* **1965**, *8*, 33–36. [CrossRef]
230. Martín, D. La comunidad de Anélidos Poliquetos de las concreciones de algas calcáreas del litoral catalán. Caracterización de las especies. *Publicaciones del Departamiento de Zoologia de Barcelona* **1987**, *13*, 45–54.
231. Box, A.; Martin, D.; Deudero, S. Changes in seagrass polychaete assemblages after invasion by *Caulerpa racemosa* var. *cylindracea (Chlorophyta: Caulerpales): Community structure, trophic guilds and taxonomic distinctness. Sci. Mar.* **2010**, *74*, 317–329. [CrossRef]
232. Brito, M.C.; Martin, D.; Núñez, J. Polychaetes associated to a *Cymodocea nodosa* meadow in the Canary Islands: Assemblage structure, temporal variability and vertical distribution compared to other Mediterranean seagrass meadows. *Mar. Biol.* **2005**, *146*, 467–481. [CrossRef]
233. Çinar, M.E.; Dagli, E. Polychaetes (Annelida: Polychaeta) from the Aegean and Levantine coasts of Turkey, with descriptions of two new species. *J. Nat. Hist.* **2013**, *47*, 911–947. [CrossRef]
234. Langeneck, J.; Busoni, G.; Aliani, S.; Lardicci, C.; Castelli, A. Distribution and diversity of polychaetes along a bathyal escarpment in the western Mediterranean Sea. *Deep Sea Res. Part I Oceanogr. Res. Pap.* **2019**, *144*, 85–94. [CrossRef]
235. Lardicci, C.; Como, S.; Corti, S.; Rossi, F. Changes and recovery of macrozoobenthic communities after restoration measures of the Orbetello Lagoon (Tyrrhenian coast, Italy). *Aquat. Conserv. Mar. Freshw. Ecosyst.* **2001**, *11*, 281–287. [CrossRef]
236. Morri, C.; Castelli, A.; Diviacco, G.; Mori, M.; Bianchi, C.N. Zonazione di comunità bentiche lungo l'estuario della Magra (Mar Ligure). *Atti Soc. Toscana Sci. Nat. Ser. B* **1990**, *97*, 311–327.
237. Magalhães, W.F.; Bailey-Brock, J.H.; Barroso, R. A new species of *Paradoneis* Hartman, 1965 (Annelida: Paraonidae) from Hawaii with notes on its reproductive strategy. *Mar. Biodivers.* **2019**, *49*, 1617–1625. [CrossRef]
238. Gaston, G.R.; McLelland, J.A.; Heard, R.W. Feeding Biology, Distribution, and Ecology of Two Species of Benthic Polychaetes: *Paraonis fulgens* and *Paraonis pygoenigmatica* (Polychaeta: Paraonidae). *Gulf Res. Rep.* **1992**, *8*, 395–399. [CrossRef]

239. Röder, H. Gangsysteme von *Paraonis fulgens* Levinsen 1883 (Polychaeta) in ökologischer, ethologischer und aktuopaläontologischer Sicht. *Senckenbergiana Marit.* **1971**, *3*, 3–51.
240. Blake, J.A.; Baptiste, E.M. Life history studies on dominant polychaete species from Georges Bank. In *Georges Bank Benthic Infauna Monitoring Program, Final Report for Third Year of Sampling, U.S.*; Maciolek-Blake, N., Grassle, J.F., Neff, J.M., Eds.; Department of the Interior, Minerals Management Service: Herndon, VA, USA, 1985; Chapter 6; pp. 140–178.
241. Levin, L.A.; Plaia, G.R.; Huggett, C.L. The influence of natural organic enhancement on life histories and community structure of bathyal polychaetes. In *Reproduction, Larval biology and Recruitment of the Deep-Esa Benthos*; Young, C.M., Eckelbarger, K.J., Eds.; Columbia University Press: New York, NY, USA, 1994; pp. 261–283.
242. López-Jamar, E.; O'Connor, B.; González, G. Demography and gametogenic cycle of *Paradoneis armata* Glémarec (Polychaeta, Paraonidae) in Ría de la Coruña, North-West Spain. *Ophelia* **1987**, *27*, 127–136. [CrossRef]
243. Giangrande, A. Polychaete reproductive patterns, life cycles and life histories: An overview. *Oceanogr. Mar. Biol. Annu. Rev.* **1997**, *35*, 323–386.
244. Sendall, K.; Salazar-Vallejo, S. Revision of *Sternaspis* Otto, 1821 (Polychaeta, Sternaspidae). *ZooKeys* **2013**, *286*, 1–74. [CrossRef]
245. Fiege, D. Sternaspidae Carus, 1863. In *Handbook of Zoology Online. Annelida: Polychaetes*; Westheide, W., Purschke, G., Eds.; De Gruyter: Berlin, Germany; Boston, MA, USA, 2016; Available online: https://db.degruyter.com/view/Zoology/bp_029146-9_24 (accessed on 17 March 2021).
246. Salazar-Vallejo, S.I. Six New Tropical Sternaspid Species (Annelida, Sternaspidae) with Keys to Identify Genera and Species. *Zool. Stud.* **2017**, *56*. [CrossRef]
247. Fiege, D.; Barnich, R. A new genus and species of Sternaspidae (Annelida: Polychaeta) from the deep eastern Atlantic. *Eur. J. Taxon.* **2020**, *699*, 1–13. [CrossRef]
248. Stimpson, W. Synopsis of the marine Invertebrata of Grand Manan: Or the region about the mouth of the Bay of Fundy, New Brunswick. *Smithson. Contrib. Knowl.* **1853**, *6*, 28–37.
249. Petersen, M.E. Family Sternaspidae Carus, 1863. In *Taxonomic Atlas of the Benthic Fauna of the Santa Maria Basin and Western Santa Barbara Channel*; Volume 7–The Annelida Part 4. Polychaeta: Flabelligeridae to Sternaspidae; Blake, J.A., Hilbig, B., Scott, P.V., Eds.; Santa Barbara Museum of Natural History: Santa Barbara, CA, USA, 2000; pp. 331–336.
250. Otto, A.G. Animalium maritimorum, nondum editorum, genera duo. *I. Sternaspis thalassemoides. Verhandlungen der Kaiserlichen Leopoldinisch Carolinischen Academie der Naturforscher* **1821**, *2*, 619–627.
251. Sendall, K. Review and Revision of the Genus *Sternaspis* (Polychaeta: Sternaspidae) Using Cladistics on Morphological Characters. Master's Thesis, University of Victoria, Greater Victoria, BC, Canada, 2006.
252. Drennan, R.; Wiklund, H.; Rouse, G.W.; Georgieva, M.N.; Wu, X.; Kobayashi, G.; Yoshino, K.; Glover, A.G. Taxonomy and phylogeny of mud owls (Annelida: Sternaspidae), including a new synonymy and new records from the Southern Ocean, North East Atlantic Ocean and Pacific Ocean: Challenges in morphological delimitation. *Mar. Biodivers.* **2019**, *49*, 2659–2697. [CrossRef]
253. Méndez, N.; Yáñez-Rivera, B. Distribution and morphometry of the deep-sea sternaspids, *Sternaspis maior, Sternaspis uschakovi,* and *Caulleryaspis fauchaldi* (Polychaeta), in Mexican Pacific waters. *Bull. Mar. Sci.* **2015**, *91*, 457–467. [CrossRef]
254. Yoshino, K.; Nagayoshi, M.; Sato, M.; Katano, T.; Ito, Y.; Fuji, N.; Hamada, T.; Hayami, Y. Life history of *Sternaspis costata* (Sternaspidae: Polychaeta) in Ariake Bay, Japan. *J. Mar. Biol. Assoc. U. K.* **2016**, *96*, 647–655. [CrossRef]
255. Fiege, D.; Bütfering, B. SEM-Investigations on the morphology and anatomy of *Sternaspis scutata* (Ranzani, 1817) (Annelida: Polychaeta). In Proceedings of the 6th International Polychete Conference, Held at Curitiba, Paraná, Brazil, 2–7 August 1998. *Bull. Mar. Sci.* **2000**, *67*, 662–663.
256. Zhadan, A.E.; Tzetlin, A.B.; Salazar-Vallejo, S.I. Sternaspidae (Annelida, Sedentaria) from Vietnam with description of three new species and clarification of some morphological features. *Zootaxa* **2017**, *4226*, 75–92. [CrossRef] [PubMed]
257. Stimpson, W. Descriptions of new species of marine Invertebrata from Puget Sound, collected by the Naturalists of the North-west Boundary Commission, A.H. Campbell, Esq., Commissioner. *Proc. Acad. Nat. Sci. Phila.* **1864**, *16*, 153–161.
258. von Marenzeller, E. Südjapanische Anneliden I. *Denkschriften der Kaiserlichen Akademie der Wissenschaften. Mathematisch Naturwissenschaftliche Classe* **1879**, *41*, 109–154.
259. Kobayashi, G.; Mukai, R.; Alalykina, I.; Miura, T.; Kojima, S. Phylogeography of benthic invertebrates in deep waters: A case study of *Sternaspis* cf. *williamsae (Annelida: Sternaspidae) from the northwestern Pacific Ocean. Deep Sea Res. Part II Top. Stud. Oceanogr.* **2018**, *154*, 159–166. [CrossRef]
260. Salazar-Vallejo, S.; Buzhinskaja, G. Six new deep-water sternaspid species (Annelida, Sternaspidae) from the Pacific Ocean. *ZooKeys* **2013**, *348*, 1–27. [CrossRef] [PubMed]
261. Southern, R. Fauna of the Chilka Lake. Polychaeta of the Chilka Lake and also of fresh and brackish waters in other parts of India. *Mem. Indian Mus. Calcutta* **1921**, *5*, 563–693.
262. Salazar-Vallejo, S.I. Three new polar species of *Sternaspis* Otto, 1821 (Polychaeta: Sternaspidae). *Zootaxa* **2014**, *3861*, 333. [CrossRef]
263. Wu, X.; Salazar-Vallejo, S.; Xu, K. Two new species of *Sternaspis* Otto, 1821 (Polychaeta: Sternaspidae) from China seas. *Zootaxa* **2015**, *4052*, 373. [CrossRef]
264. Wu, X.; Xu, K. Diversity of Sternaspidae (Annelida: Terebellida) in the South China Sea, with descriptions of four new species. *Zootaxa* **2017**, *4244*, 403–415. [CrossRef]
265. Díaz-Díaz, O.; Rozbaczylo, N. Article *Sternaspis chilensis* n. sp., a new species from austral Chilean channels and fjords (Annelida, Sternaspidae). *Zootaxa* **2017**, *4254*, 269–276. [CrossRef]

266. Salazar-Vallejo, S.I. Sternaspis piotrowskiae sp. nov. (Polychaeta: Sternaspidae) from the Philippine Islands. In *The Coral Triangle: The 2011 Hearst Philippine Biodiversity Expedition*; Williams, G.C., Terrence, M., Eds.; California Academy of Sciences: San Francisco, CA, USA, 2014; pp. 165–169.
267. Ranzani, C. Descrizione di una nova specie del genere *Thalassema*. *Opusculi Sci.* **1817**, *1*, 112–116.
268. Wang, Z.; Liu, C.; Liu, H.; Xu, Z.; Wang, B.; Zheng, F. Complete mitochondrial genome of the *Sternaspis scutata* (Sternaspidae, Annelida). *Mitochondrial DNA Part B* **2019**, *4*, 1700–1701. [CrossRef]
269. Preetha Mini Jose, H.M.; Muthuvelu, S.; Sivaraj, S.; Murugesan, P. A note on the occurrence of *Sternaspis scutata* (Renier 1807)—A sternaspid polychaete from Sundarban Mangroves, West Bengal, India. *J. Bombay Nat. Hist. Soc.* **2014**, *111*, 62–64. [CrossRef]
270. Pillai, N.G.K. On some benthic polychaetes from Cochin estuary. *J. Mar. Biol. Assoc. of India* **2001**, *43*, 120–135.
271. Joydas, T.V.; Damodaran, R. Infaunal macrobenthos along the shelf waters of the west coast of India, Arabian Sea. *Indian J. Mar. Sci.* **2009**, *38*, 14.
272. Abdelnaby, F.A. First record of *Sternaspis scutata* (Polychaeta, Sternaspidae) in Egyptian waters. *Egypt. J. Aquat. Biol. Fish.* **2020**, *24*, 317–326. [CrossRef]
273. Townsend, M.; Worsfold, T.M.; Smith, P.R.J.; Martina, L.J.; McNeill, C.L.; Kendall, M.A. Occurrence of *Sternaspis scutata* (Polychaeta: Sternaspidae) in the English Channel. *Cah. Biol. Mar.* **2006**, *47*, 281–285.
274. Wang, J.-B.; Li, X.-Z.; Wang, H.-F. Ecological characteristics of dominant polychaete species from the Jiaozhou Bay. *Acta Zool. Sin.* **2006**, *52*, 63–69.
275. Por, F.D. *Lessepsian Migration. The influx of Red Sea Biota into the Mediterranean by Way of the Suez Canal*; Springer: Berlin, Germany, 1978; p. 228.
276. Rietsch, M. Étude sur le Sternaspis scutata. *Ann. des Sci. Nat. Sixieme Sér. Zool.* **1882**, *13*, 1–84.
277. Fauchald, K.; Jumars, P.A. The Diet of Worms: A Study of Polychaete Feeding Guilds. *Oceanogr. Mar. Biol. Annu. Rev.* **1979**, *17*, 193–284.
278. Dales, R.P. The polychaete stomodeum and the inter-relationships of the families of Polychaeta. *Proc. Zool. Soc.* **1962**, *39*, 389–428. [CrossRef]
279. Dorgan, K.M.; D'Amelio, C.; Lindsay, S.M. Strategies of burrowing in soft muddy sediments by diverse polychaetes. *Invertebr. Biol.* **2016**, *135*, 287–301. [CrossRef]
280. Self, R.F.L.; Jumars, P.A. Cross-phyletic patterns of particle selection by deposit feeders. *J. Mar. Res.* **1988**, *46*, 119–143. [CrossRef]
281. Lim, H.S.; Hong, J.S. Distribution and growth pattern of *Sternaspis scutata* (Polychaeta: Sternaspidae) in Chinhae Bay, Korea. *J. Korean Fish. Soc.* **1996**, *29*, 537–545.
282. Ohwada, T. Mesh size for settlement and recruitment studies of polychaetes with reference to the effect of fixation on washing bottom samples. *Asian Mar. Biol.* **1988**, *5*, 35–40.
283. Blake, J.A.; Göransson, P. Redescription of *Tharyx killariensis* (Southern) from Ireland and description of two new species of *Tharyx* from the Kattegat, Sweden (Polychaeta, Cirratulidae). *Zootaxa* **2015**, *4039*, 501. [CrossRef] [PubMed]
284. Jägersten, G. Studies on the morpgology, larval development and biology of *Protodrilus*. *Zool. Bidrag från Uppsala* **1952**, *29*, 426–512.
285. Helm, C.; Capa, M. Comparative analyses of morphological characters in Sphaerodoridae and allies (Annelida) revealed by an integrative microscopical approach. *Front. Mar. Sci.* **2015**, *1*, 82. [CrossRef]
286. Surugiu, V.; Capa, M. The Occurrence of *Amphiglena mediterranea* (Leydig, 1851) (Annelida: Sabellidae) at the Romanian Coast of the Black Sea: A Case on an Unsuccessful Invasion? *Russ. J. Biol. Invasions* **2020**, *11*, 293–299. [CrossRef]
287. Nygren, A. Cryptic polychaete diversity: A review. *Zool. Scr.* **2014**, *43*, 172–183. [CrossRef]
288. Choi, H.K.; Kim, H.J.; An, Y.R.; Kim, H. The complete mitochondrial genome of *Cirriformia* cf. *tentaculata (Polychaeta, Cirratulidae) from Dok-Do, Korea. Mitochondrial DNA B* **2017**, *2*, 847–848.
289. Kim, H.; Min, G.-S.; Choi, H.K. The complete mitochondrial genome of Korean endemic species, *Timarete posteria* (Polychaeta, Cirratulidae). *Mitochondrial DNA Part B* **2019**, *4*, 501–502. [CrossRef]
290. Di Giuseppe, G.; Barbieri, M.; Vallesi, A.; Luporini, P.; Dini, F. Phylogeographical pattern of *Euplotes nobilii*, a protist ciliate with a bipolar biogeographical distribution. *Mol. Ecol.* **2013**, *22*, 4029–4037. [CrossRef] [PubMed]
291. Flot, J.-F.; Couloux, A.; Tillier, S. Haplowebs as a graphical tool for delimiting species: A revival of Doyle's "field for recombination" approach and its application to the coral genus *Pocillopora* in Clipperton. *BMC Evol. Biol.* **2010**, *10*, 372. [CrossRef] [PubMed]
292. Christie, G. A Comparative Study of the Reproductive Cycles of Three Northumberland Populations of *Chaetozone Setosa* (Polychaeta:Cirratulidae). *J. Mar. Biol. Assoc. U. K.* **1985**, *65*, 239–254. [CrossRef]
293. Day, J.H. *A Monograph on the Polychaeta of Southern Africa. Part 2. Sedentaria*; Trustees of the British Museum (Natural History): London, UK, 1967; p. 419.
294. Olive, P.J.W. Ovary structure and oogenesis in *Cirratulus cirratus* (Polychaeta: Cirratulidae). *Mar. Biol.* **1971**, *8*, 243–259. [CrossRef]
295. Parapar, J.; Zhadan, A.; Tzetlin, A.B.; Vortsepneva, E.V.; Moreira, J. Exploring the anatomy of *Cossura pygodactylata* Jones, 1956 (Annelida, Cossuridae) using micro-computed tomography, with special emphasis on gut architecture. *Mar. Biodivers.* **2018**, *48*, 751–761. [CrossRef]

296. Zhadan, A.; Vortsepneva, E.; Tzetlin, A. Three-dimensional reconstruction of the musculature of *Cossura pygodactylata* Jones, 1956 (Annelida: Cossuridae). *Zool. Anz.* **2014**, *253*, 181–191. [CrossRef]
297. Pamungkas, J.; Glasby, C.J.; Read, G.B.; Wilson, S.P.; Costello, M.J. Progress and perspectives in the discovery of polychaete worms (Annelida) of the world. *Helgol. Mar. Res.* **2019**, *73*, 10. [CrossRef]

diversity

MDPI

Review

Fanworms: Yesterday, Today and Tomorrow

María Capa [1,*], Elena Kupriyanova [2], João Miguel de Matos Nogueira [3], Andreas Bick [4] and María Ana Tovar-Hernández [5]

1 Departament de Biologia, Universitat de les Illes Balears, 07122 Palma, Spain
2 Australian Museum Research Institute, Australian Museum, Sydney, NSW 2010, Australia; elena.kupriyanova@australian.museum
3 Departamento de Zoologia, Instituto de Biociências, Universidade de São Paulo, São Paulo 05508-090, Brazil; nogueira@ib.usp.br
4 Institut für Biowissenschaften, Allgemeine und Spezielle Zoologie, Universität Rostock, Universitätsplatz 2, D-18055 Rostock, Germany; andreas.bick@uni-rostock.de
5 Facultad de Ciencias Biológicas, Universidad Autónoma de Nuevo León, Nuevo León 66455, Mexico; maria_ana_tovar@yahoo.com
* Correspondence: maria.capa@uib.es

Abstract: Sabellida Levinsen, 1883 is a large morphologically uniform group of sedentary annelids commonly known as fanworms. These annelids live in tubes made either of calcareous carbonate or mucus with agglutinated sediment. They share the presence of an anterior crown consisting of radioles and the division of the body into thorax and abdomen marked by a chaetal and fecal groove inversion. This study synthesises the current state of knowledge about the diversity of fanworms in the broad sense (morphological, ecological, species richness), the species occurrences in the different biogeographic regions, highlights latest surveys, provides guidelines for identification of members of each group, and describe novel methodologies for species delimitation. As some members of this group are well-known introduced pests, we address information about these species and their current invasive status. In addition, an overview of the current evolutionary hypothesis and history of the classification of members of Sabellida is presented. The main aim of this review is to highlight the knowledge gaps to stimulate research in those directions.

Keywords: Sabellida; Sabellidae; Serpulidae; Fabriciidae; Annelida; polychaetes; biodiversity assessment; systematics; methods; gaps of knowledge

Citation: Capa, M.; Kupriyanova, E.; Nogueira, J.M.d.M.; Bick, A.; Tovar-Hernández, M.A. Fanworms: Yesterday, Today and Tomorrow. *Diversity* **2021**, *13*, 130. https://doi.org/10.3390/d13030130

Academic Editor: Luc Legal

Received: 30 December 2020
Accepted: 10 March 2021
Published: 17 March 2021

Publisher's Note: MDPI stays neutral with regard to jurisdictional claims in published maps and institutional affiliations.

1. Introduction

Sabellida Levinsen, 1883 is a morphologically uniform clade of sedentary annelids historically given a rank of Order. Sabellida currently includes members of Fabriciidae Rioja, 1923, Sabellidae Latreille, 1825, and Serpulidae Rafinesque, 1815 [1–3]. They are commonly known as fanworms, feather-duster worms, or flowers of the sea, because their radioles are arranged in a crown, protruding from the tube made of calcium carbonate or mucus with agglutinated sediment (Figure 1). In addition to the presence of protective tube and the prostomial crown made of radioles with secondary ramifications (generally referred as pinnules, but see [4] for Fabriciidae), which are mainly used for feeding and respiration, all members of Sabellida share the presence of chaetal inversion [thoracic chaetigers with simple chaetae on notopodia and uncini (hooks) on neuropodia, and abdominal chaetigers with opposite arrangement]. Sabellida includes about 1200 species distributed world-wide, from tropical to polar waters and found in all habitats, from freshwater to fully marine conditions, and intertidal to deepest ocean trenches.

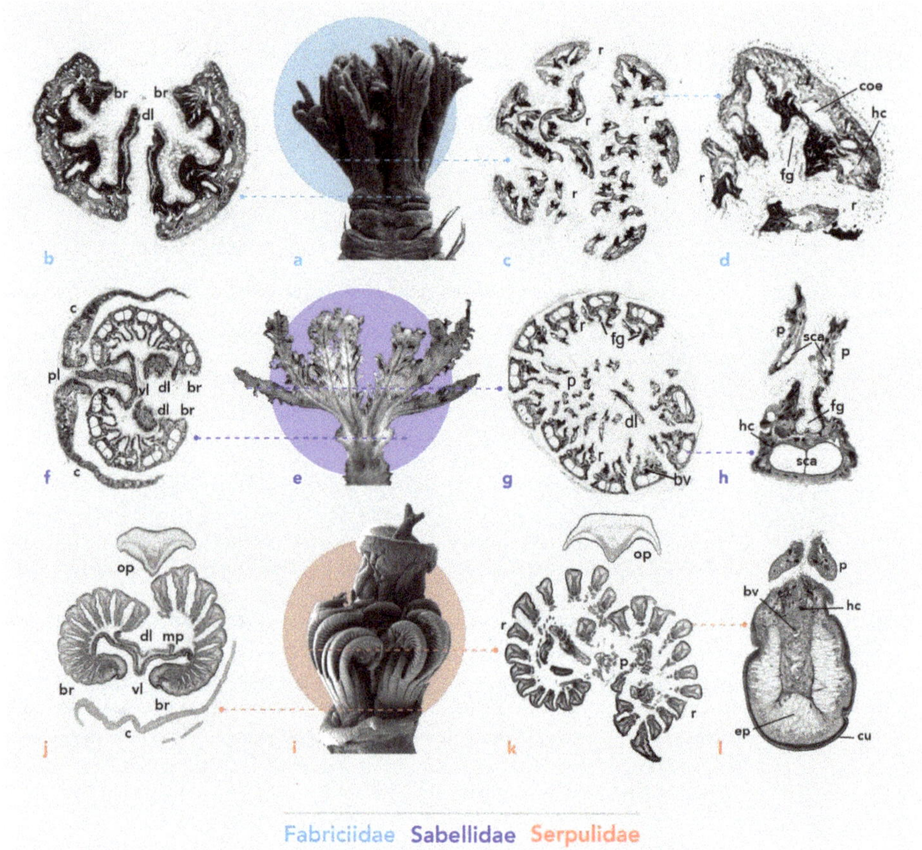

Figure 1. Comparison of the radiolar crown structure among the three families of Sabellida. (**a**) Anterior end of *Fabricia stellaris*, dorsal view; (**b**) Histological section of crown of *F. stellaris* at base; (**c**) Histological section of crown of *F. stellaris* at mid-length; (**d**) Histological section of one radiole and presumed pinnules of *F. stellaris*; (**e**) Anterior end of *Laonome xeprovala*, dorsal view; (**f**) Histological section of crown of *Laonome xeprovala* at base; (**g**) Histological section of crown of *L. xeprovala* at mid-length; (**h**) Histological section of one radiole and pinnules of *L. xeprovala*; (**i**) Anterior end of *Spirobranchus lamarki*, ventral view; (**j**) Histological section of crown of *S. lamarcki* at base; (**k**) Histological section of crown of *S. lamarcki* at mid-length; (**l**) Histological section of one radiole and pinnules of *S. lamarcki*. Abbreviations: br: base of radioles; bv: blood vessel; c: collar; coe: coelom; cu: cuticle; dl: dorsal lips; ep: epithelium; fg: faecal groove; hc: hyaline cartilage; mp: mouth palp; op: opercular peduncle; p: pinnules; pl: parallel lamellae; r: radioles; sca: supporting cellular axis; vl: ventral lips.

The Sabellida concept and even the group name has changed over time. Since their erection in the early 19th century, sabellids (including fabriciids), building soft sediment tubes, and serpulids, building calcareous tubes, have always been considered related, based on their general morphology and grouped into the section Amphitrites sabelliennes [5], the family Serpulacei [6], family Serpulacea [7–9], Serpulidae [10], and finally Sabellida [11,12].

With the advent of cladistic analyses, close relationships of sabellids and serpulids with Sabellariidae Johnston, 1865, Siboglinidae Caullery, 1914 and Oweniidae Rioja, 1917 have been suggested, and consequently the composition of Sabellida expanded to incorporate these three taxa (e.g., [13]) (Figure 2). However, subsequent molecular analyses using increasing number of taxa and DNA markers revealed that the three late incorporated taxa were neither closely related to fanworms, nor to each other [14–16].

Morphological lines of evidence related to the ontogeny, internal anatomy, position of the ciliated groove, as well as chaetal morphology and arrangement also supported this lack of close relationship [17–26]. Consequently, Sabellida now again includes only fanworms (Figure 2), but the former sabellid subfamily Fabriciinae has been elevated to Fabriciidae based on DNA evidence [1–3,24,27]. The sister group of Sabellida, according to the latest phylogenomic studies, is a clade including Spionidae Grube, 1850 and Sabellariidae [14,28,29].

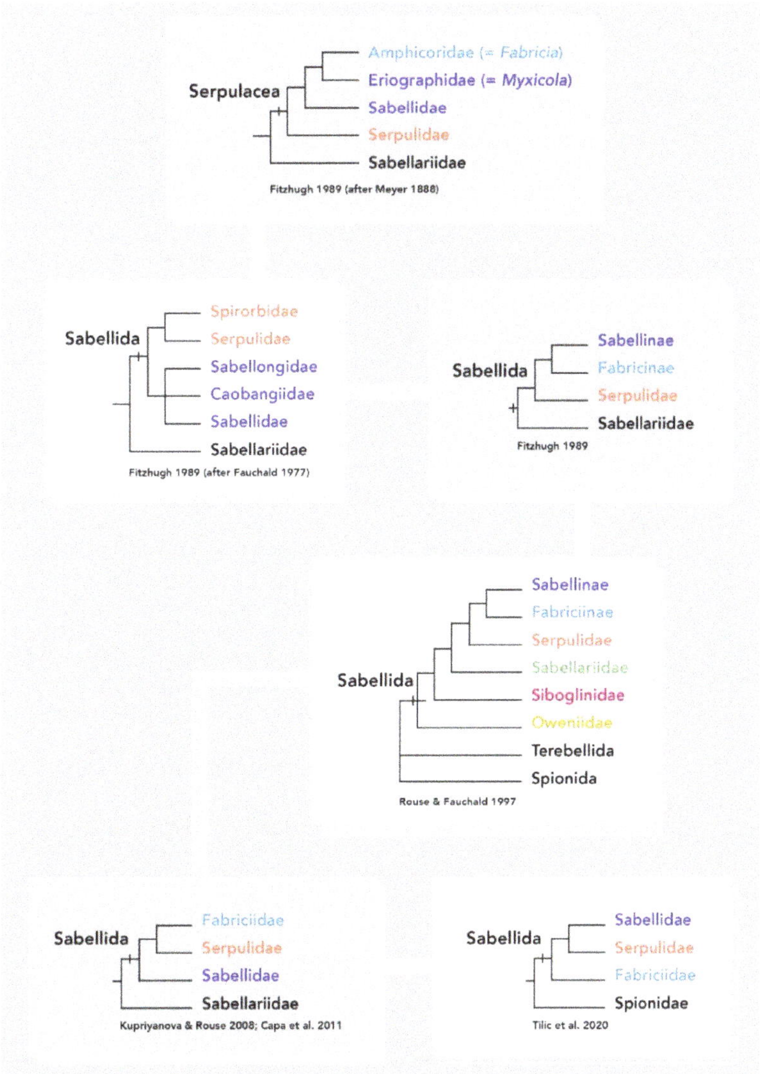

Figure 2. History of the evolutionary hypotheses within Sabellida. The coloured taxa indicate the groups that had been considered as members of Sabellida (or Serpulacea) according to the represented phylogenetic hypotheses. In light blue are members of the currently accepted Fabriciidae, in dark blue are Sabellidae and Serpulidae are in orange. In green, pink and yellow are other taxa previously considered as Sabellida.

Most of the taxonomic work in the Sabellida has aimed to document the regional species diversity, rather than to provide comprehensive world-wide generic revisions. These regional studies have been geographically unevenly distributed because of their strong association to political, economic, historical and traditional context. Moreover, available georeferenced databases (i.e., in Global Biodiversity Information Facility, GBIF, https://www.gbif.org (accessed on 3 March 2021)) are also biased, as not all biodiversity information from museum collections and research institutions is shared with this service. The information available in GBIF database not necessarily adequately reflects true species richness and abundance in natural habitats, as more conspicuous, well-known or easier to identify species are predominant in such datasets.

The main aim of this study is to synthesise in a single, easily accessible publication the current knowledge of evolutionary relationships, classification, species diversity and distribution of Sabellida, paying special attention to the latest surveys and novel methodologies used for species delimitation. As some members of Sabellida are easily translocated by anthropogenic means and establish outside of their native ranges, updated information about their invasive status is provided. Another important goal was to highlight the knowledge gaps in order to stimulate research in those directions.

2. Materials and Methods

The present study is a literature review of the information on the Sabellida species diversity, distribution and ecology, highlighting the geographic areas that need further scientific effort and the taxonomic groups that need revisions. The characters and methods used for species delineation have been revised. We summarise the phylogenetic position of Sabellida and the relationships within the group from a historic perspective.

The World Register of Marine Species (WoRMS) [30] database has been key for accounting the number of current valid taxa and analyses of species richness. However, this list has been further revised, including synonymies, new combinations, corrections of names for gender agreement, years of publication, specification of habitats, type localities, and assigning statuses such as *inquirenda* or *indeterminable* to taxa (Appendix A).

Scientific names for all taxa are followed by the authority the first time a taxon is mentioned in the text and in the supplementary material tables (Tables S1–S3). However, due to the high number of taxa dealt with in this review, the citations of authorships have not been included in the reference list. Since many species are still reported as having suspiciously wide distributions, generalised type localities (not details of precise collection locations) are included in Tables S1–S3). Original descriptions, details of type localities and synonymies can be found in WoRMS [30].

Biodiversity information (occurrence data) is referred to the geographic regions (realms) proposed by Spalding et al. [31] for the marine and Udvardy [32] for limnic environments with terminological changes by Olson et al. [33]. Type localities instead of currently reported distributions were used to assign each species to a biogeographic realm. Available biodiversity records have been downloaded as global maps with georeferenced occurrences for Fabriciidae, Sabellidae and Serpulidae separately from the Global Biodiversity Information Facility (GBIF platform, [34]). Distribution of the most common taxa at global scale and species richness in each of the marine realms have been analysed and discussed. In order to delimit these realms in the GBIF maps, polygons following the boundaries of these realms have been drawn with the tools given at the GBIF platform.

DNA sequences available at the National Center for Biotechnology Information (NCBI, [35]) and Barcode of Life Data System (BOLDSystems [36]) have been used to assess the state of the genetic information available for each of the three groups of fanworms. Moreover, currents trends, including phylogenomic data are discussed.

3. Results

3.1. Systematics

The history of the Sabellida as Fabriciidae + Sabellidae + Serpulidae) has been convoluted. Rafinesque [37] grouped the worms with calcareous tubes as Serpularia, now Serpulidae. The subfamily Spirorbinae was established for small-bodied serpulids with spirally coiled tubes [38] and the subfamily Filograninae for the taxa with pinnulated operculum-bearing radioles or lacking opercula [39]. Thus, Serpulidae was subdivided into Filograninae, Serpulinae, and Spirorbinae until Pillai [40] elevated the Spirorbinae to the family status. However, even first morphology-based phylogenetic analyses suggested that Spirorbinae are more closely related to Serpulinae than to Filograninae [19,41,42] and that Filograninae is paraphyletic [41,43,44]. Moreover, further analyses integrating molecular data [45–49] unequivocally found both traditional Serpulinae and Filograninae paraphyletic, and Spirorbinae nested within Serpulidae. Thus, consensus that recognition of Spirorbidae would make Serpulidae a paraphyletic group has prevailed and the rank of the spirorbids was lowered back to Spirorbinae [50].

Initially, Sabellidae included the large-bodied species of fanworms. Rioja, in 1923 [39], divided Sabellidae in three subfamilies: Fabriciinae Rioja (1923), gathering species with acicular uncini; Myxicolinae (only *Myxicola* Koch *in* Renier, 1847), having radioles joined by a membrane for most of their length, abdominal uncinal tori forming almost complete cinctures and tubes made of thick gelatinous mucus; and Sabellinae, characterised by avicular thoracic uncini, often with companion chaetae, distinct faecal groove and ventral shields. Johansson [51] included *Myxicola* in Fabriciinae. Fauchald [12] followed Rioja's arrangement and recognized Sabellongidae Harman, 1969 (with *Sabellonga* Hartman, 1969) and Caobangiidae Chamberlin, 1919 (with *Caobangia* Giard, 1893) as valid taxa.

As a result of the first morphology-based phylogenetic analysis of Sabellidae (as perceived at the time), using Serpulidae as the outgroup [42], only subfamilies Fabriciinae and Sabellinae were recognised and their composition changed dramatically (Figure 2). *Caobangia*, 1893 was included in Fabriciinae, while *Myxicola*, *Sabellonga* and some genera previously considered as fabriciins (*Chone* Krøyer, 1856, *Desdemona* Banse, 1957, *Euchone* Malmgren, 1866, *Fabrisabella* Hartman, 1969 and *Jasmineira* Langerhans, 1880) were recovered in Sabellinae. The Sabellinae was defined by the presence of the "radiolar skeleton" composed of at least two rows of vacuolated cells and dorsally fused radiolar lobes [42]. The subfamily Fabriciinae was characterised by the presence of rasp-shaped (having several rows of teeth) abdominal uncini, absence of ventral lips, separated radiolar lobes (except in *Caobangia*) and the absence of "radiolar skeleton" (except in *Caobangia*, with one longitudinal row of vacuolated cells). Later *Caobangia* was transferred to Sabellinae, simplifying the definition for Fabriciinae [52].

Analyses of molecular data have changed the understanding of the relationships within Sabellida. Molecular data provided evidence of Serpulidae being sister to Fabriciinae, and Sabellinae the sister group of this clade [1]. Consequently, Fabriciinae was raised to the family rank (Fabriciidae, Figure 2). Further studies supported the validity of the three families, although relationships among them were not consistently supported [2,24,27]. The latest study using transcriptomes for a broad range of sabellids recovered Fabriciidae as sister to a clade of Sabellidae and Serpulidae [3] (Figure 2). Now it is generally accepted that Sabellidae does not include Fabriciidae [1,2,24,27] and is more closely related to Serpulidae than to Fabriciidae [3].

3.1.1. Fabriciidae

Monophyly of Fabriciidae is supported by the branching patterns of the radiolar crown, the absence of ventral lips, abdominal uncini with elongate handle referred to as manubrium, and presence of radiolar hearts [4] (Table 1, Figure 3), as well as by reproductive characters [27]. Relationships among fabriciids were first explored using morphological data [42,53–60]. Although the phylogenies were not fully resolved, these studies recovered two main groups: one paraphyletic, branching off at the base of the tree and including

Fabriciola Friedrich, 1939, *Manayunkia* Leidy, 1859, *Monroika* Hartman, 1951, and a clade with the remaining nine genera considered valid at the time. The position of *Pseudofabriciola* Fitzhugh, 1990b varied with the different analyses [57].

Table 1. Morphological diagnostic features of the three taxa of Sabellida.

Feature	Fabriciidae	Sabellidae	Serpulidae
Tube material	mucus and sediment/none	mucus and sediment *	calcium carbonate
Radiolar lobes	separated	fused	separated
Vacuolated cells supporting radioles	absent	present	absent
Operculum	absent	absent	absent or present
Thoracic membrane	absent	absent	present
Thoracic uncini	acicular	avicular **	avicular
Number of abdominal chaetigers	usually three (exceptionally two or four)	more than three	more than three
Abdominal uncini	with elongate and wide handle (manubrium)	with short handle or lacking handles	lacking handles
Branchial hearts	present	absent	absent

* *Glomerula* is an exception with a calcareous tube; ** *Terebrasabella* is an exception with three types of thoracic uncini: acicular, avicular and palmate.

A recent comprehensive study assessing phylogenetic relationships within Fabriciidae incorporated DNA sequence and reproductive data into a morphological dataset [27]. The results corroborated the apomorphies proposed earlier: the absence of ventral lips, modified abdominal uncini with elongate manubrium and presence of radiolar hearts, together with six apomorphic reproductive traits: (1) spermatogenesis occurring only in the thorax, (2) spermatids developed in large clusters with a central cytophore, (3) presence of a single dorsal sperm duct, (4) presence of a sperm nuclear projection, (5) sperm nuclear membrane thickening, and (6) sperm extra-axonemal sheath. The study recovered the two main groups already revealed by Fitzhugh [42,56,57,61,62]. One clade contained *Manayunkia* and *Echinofabricia* Huang, Fitzhugh and Rouse, 2011 (Genus A in [57]), and branched off basally, sister to all the other Fabriciidae. The second larger clade showed *Rubifabriciola* Huang, Fitzhugh and Rouse, 2011 (the red-eyed '*Fabriciola*'), branching off at the base, sister to six other genera. Some taxa need further study to assess their monophyly since *Novafabricia labrus* Fitzhugh, 1998 was not found nested within other species in the genus (as in [57]), *Monroika africana* (Monro, 1939) formed a polytomy with the *Manayunkia* species, and there were not enough characters to support *Augeneriella* Banse, 1957 and *Pseudoaugeneriella* Fitzhugh, 1988 as distinct genera [27].

With one exception, genera of Fabriciidae have been revised and most of the revisions were accompanied by phylogenies (Table 2). *Brandtika* Jones, 1974 was not included in phylogenetic analyses [27,59] due to poorly preserved types and incomplete descriptions [63].

Figure 3. Some fabriciid representatives showing a range of the diversity of forms found in the group. (**a**) cf. *Fabriciola* sp., deep-sea basin of the southwest Atlantic; (**b**,**c**,**h**) *Fabricia stellaris*, Baltic Sea; (**d**,**f**) *Manayunkia athalassia*, South Australia; (**e**) *Fabriciola* sp., Brazil (**g**) *Monroika* sp., Argentina ((**c**) by A. Dietrich, (**d**) by G. Rouse, (**g**) by L. Armendariz).

Table 2. Fabriciid currently accepted genera that have undergone major or partial revisions and phylogenetic analyses (based on morphological features, molecular data or both).

Genera	Number of Species *	Taxonomic Revision	Phylogenetic Studies
Augeneriella Banse, 1957	5	Banse 1957, Fitzhugh 1983, 1990a, 1993	Fitzhugh 1991a-b, 1992, 1993, 2010
Bansella Fitzhugh, 2010	1	Fitzhugh 2010	Fitzhugh 2010
Brandtika Jones, 1974	1	Fitzhugh 2001	NO
Brifacia Fitzhugh, 1998	2	Fitzhugh 1998; Giangrande et al. 2014	Fitzhugh 2010
Echinofabricia Huang, Fitzhugh and Rouse, 2011	4	Huang et al. 2011	Fitzhugh 1991a, 1992, 1993, 1998, 2010 (as genus A)
Fabricia de Blainville, 1828	1	Fitzhugh 1991b, 2010	Fitzhugh 1991a, 1992, 1993, 1998, 2010
Fabricinuda Fitzhugh, 1990b	7	Fitzhugh 1983, 1990b, 2002a; López and Rodríguez 2008	Fitzhugh 1991a, 1992, 1993, 1998, 2002a, 2010; López and Rodríguez 2008
Fabriciola Friedrich, 1939	6	Fitzhugh 1991b, 1992, 1999; Bick 2005	Fitzhugh 1991a, 1992, 1993, 1998, 1999, 2010
Manayunkia Leidy, 1859	10	Sitnikova et al. 2014, Atkinson et al. 2020	Fitzhugh 1991a, 1992, 1993, 1998, 2010; Sitnikova et al. 2014; Pudovkina et al. 2016
Monroika Hartman, 1951	1	Fitzhugh 1992	Fitzhugh 1992, 1998, 2010
Novafabricia Fitzhugh, 1990c	11	Fitzhugh 1983, 1990c, 1998; Bick 2005	Fitzhugh 1991a, 1992, 1993, 1998, 2010
Parafabricia Fitzhugh, 1992	2	Fitzhugh 1992	Fitzhugh 1992, 1993, 1998, 2010
Pseudoaugeneriella Fitzhugh, 1998	5	Fitzhugh 1998	Fitzhugh 1998, 2010
Pseudofabricia Cantone, 1972	1	Fitzhugh 1995	Fitzhugh 1992, 1998, 2010
Pseudofabriciola Fitzhugh, 1990b	13	Fitzhugh 1990c, 1991b, 1993, 1996, 2002a; Fitzhugh et al. 1994, Fitzhugh and Simboura 1995	Fitzhugh 1991a-b, 1993, 1994, 1996, 1998, 2002a, 2010; Fitzhugh et al. 1994, Fitzhugh and Simboura 1995
Raficiba Fitzhugh, 2001	1	Fitzhugh 2001	Fitzhugh 2010
Rubifabriciola Huang, Fitzhugh and Rouse, 2011	10	Huang et al. 2011	Huang et al. 2011

* Excluding subspecies.

3.1.2. Sabellidae

Monophyly of Sabellidae is supported by the presence of dorsal and ventral lips, the presence of vacuolated cells supporting radioles and pinnules, and the dorsal fusion of the radiolar lobes (Table 1, Figure 4) [2,3,24,42,52]. Currently, monophyletic clades Sabellinae and Myxicolinae are recognized within Sabellidae, the latter is composed of Amphiglenini and Myxicolini [3]. The current composition of the Sabellinae (now Sabellidae) has not changed significantly since the early cladistic analyses [42,52], although nine genera have been erected since (Table 3), and *Megalomma* Johansson, 1925 was replaced by *Acromegalomma* Gil and Nishi, 2017 because the name was preoccupied by a group of carabid beetles. Fitzhugh [42] provided diagnoses of all genera accepted at the time, with their potential apomorphies. The most recent review of morphology and diagnostic features of genera and species identification is found in [24].

Figure 4. Some sabellid representatives showing a range of the diversity of forms found in the group. (**a**) *Stylomma palmatum*, Lizard Island, Australia; (**b**) *Acromegalomma* spp., Lizard Island, Australia; (**c**) *Paradialychone ambigua*, Lizard Island, Australia; (**d**) *Pseudobranchiomma paraemersoni*, São Paulo, Brazil; (**e**) *Laonome xeprovala*, Sea of Azov; (F) *Branchiomma* sp., Mexico; (**g**) *Bispira brunnea*, Caribbean; (**h**) *Notaulax* sp., Lizard Island, Australia; (**i**) *Anamobaea orstedii*., Mexico; (**j**) *Sabellastarte magnifica*, Mexico. ((**a**,**b**,**g**), by M. Bok; (**e**) by V. Syomin, (**i**,**j**) by H. Bahena).

Table 3. Sabellid currently accepted genera that have gone through major or partial revisions and phylogenetic analyses (either considering morphological features, molecular data or both).

Genera	Number of Species *	Revision	Phylogenetic Analyses
Acromegalomma Gil and Nishi, 2017	38	Perkins 1984, Tovar-Hernández and Salazar-Vallejo 2008; Capa and Murray 2009; Tovar-Hernández and Carrera-Parra 2011, Gil and Nishi 2017	Capa and Murray 2009, Tovar-Hernández and Carrera-Parra 2011
Amphicorina Claparède, 1864	46	Rouse 1990 (as *Oriopsis*), Cochrane 2003	Cochrane 2003
Amphiglena Claparède, 1864	14	Capa and Rouse 2007; Tilic et al. 2019	Capa and Rouse 2007, Tilic et al. 2019
Anamobaea Krøyer, 1856	2	Tovar-Hernández et al. 2020	NO
Aracia Nogueira, Fitzhugh and Rossi, 2004	3	Nogueira et al. 2004; Tovar-Hernández 2014	Nogueira et al. 2010
Bispira Krøyer, 1856	24	Knight-Jones and Perkins 1998; Capa 2008	Capa 2008
Branchiomma Kölliker, 1859	30	Tovar-Hernández and Knight-Jones 2006	Capa et al. 2013; del Pasqua et al. 2018
Caobangia Giard, 1893	7	Jones 1974	NO
Chone Krøyer, 1856	20	Cochrane 2003, Tovar-Hernández 2005, 2006, 2007a, b, c, 2008	Tovar-Hernández 2008
Claviramus Fitzhugh, 2002	5	Fitzhugh 2002; Nishi et al. 2019	NO
Dialychone Claparède, 1869	19	Tovar-Hernández 2008	Tovar-Hernández 2008
Euchone Malmgren, 1866	35	Cochrane 2003, Giangrande and Licciano 2006, Giangrande et al. 2017	Cochrane 2003
Euchoneira Licciano, Giangrande and Gambi, 2009	1	Licciano et al. 2009	Licciano et al. 2009
Eudistylia Bush, 1905	5	Hartman 1938, Banse 1979	NO
Hypsicomus Grube, 1870	1	Perkins 1984	NO
Jasmineira Langerhans, 1880	19	Cochrane 2003, Capa and Murray 2015	Cochrane 2003
Laonome Malmgren, 1866	10	Fitzhugh 2002, Capa 2007, Bick et al. 2018	Capa 2007
Notaulax Tauber, 1879	26	Perkins 1984	NO
Paradialychone Tovar-Hernández, 2008	16	Tovar-Hernández 2008	Tovar-Hernández 2008
Parasabella Bush, 1905	29	Perkins 1984, Giangrande 1994, Tovar-Hernández and Harris 2010, Capa and Murray 2015b, Keppel et al. 2020	Capa and Murray 2015b
Perkinsiana Knight-Jones, 1983	16	Knight-Jones 1983, Giangrande and Gambi 1997, Tovar-Hernández et al. 2012	NO
Potamethus Chamberlin, 1919	11	Knight-Jones 1983	NO
Potamilla Malmgren, 1866	10	Knight-Jones 1983	NO
Potaspina Hartman, 1969	2	Capa 2007	Capa 2007
Pseudobranchiomma Jones, 1962	19	Knight-Jones 1994, Knight-Jones and Giangrande 2003	Capa and Murray 2016
Pseudopotamilla Bush, 1904	23	Knight-Jones et al. 2017	Capa 2007
Sabella Linnaeus, 1767	39	Knight-Jones and Perkins 1998	NO
Sabellastarte Krøyer, 1856	8	Knight-Jones and Mackie 2003	Capa et al. 2010
Sabellomma Nogueira, Fitzhugh and Rossi, 2010	4	Nogueira et al. 2010, Capa and Murray 2015	Nogueira et al. 2010
Stylomma Knight-Jones, 1997	2	Knight-Jones and Perkins 1998, Capa 2007	Capa 2007
Terebrasabella Fitzhugh and Rouse, 1999	3	Murray and Rouse 2007	Murray and Rouse 2007

* Excluding subspecies.

Phylogeny of Sabellidae has been largely assessed [2,64–70]. Monophyly of *Acromegalomma, Amphiglena* Claparède, 1864, *Branchiomma* Kölliker, 1859, *Chone, Dialychone* Claparède, 1869, *Paradialychone* Tovar-Hernández, 2008 and *Pseudobranchiomma* Jones, 1962) have been confirmed [68,71–74]. Members of the genera *Chone, Dialychone* and *Paradialychone* are still problematic due to their small size and because genera and species are delineated based on combination of features, such as details of uncini dentition (anterior and posterior abdominal chaetigers) and radiolar crown structures (lips and pinnular appendages) that are often difficult to interpret. As a result, the position of some species within either *Dialychone, Paradialychone* or *Chone* based on morphology is uncertain (e.g., *P. ambigua* Capa and Murray, 2015) and a molecular approach to this group is needed. The genera *Bispira* Krøyer, 1856 [2,69,74], *Euchone* [67,75,76] and *Perkinsiana* Knight-Jones, 1983 [68] appear to be paraphyletic, whereas monophyly of *Laonome* Malmgren, 1866, *Parasabella* Bush, 1905 and *Sabellastarte* Krøyer, 1856 should be assessed.

In the last two decades a number of sabellid genera have undergone major or partial taxonomic revisions (Table 3) that included morphological comparisons of congeners, as well as examination and re-description of types to detect potential synonyms or undescribed species. In a few of them, monophyly has been assessed through phylogenetic analyses of mainly morphological data.

3.1.3. Serpulidae

Monophyly of Serpulidae is supported by the presence of calcareous tubes with complex ultrastructures, distinct from the simple structure found in calcareous tubes of the unique sabellid *Glomerula piloseta* (Perkins, 1991). The serpulid thorax is surrounded by the thoracic membranes, which are absent in sabellids and fabriciids. Most serpulids have an operculum (or several), a modification of the distal part of a radiole, acting as a plug when animals hide in their tubes (Table 1, Figure 5).

The first formal phylogenetic analysis based on morphological data [43] recovered monophyletic Spirorbinae (as sister group to Serpulinae, including *Chitinopoma* Levinsen, 1884, *Crucigera* Benedict, 1887, *Serpula* Linnaeus, 1758, *Hydroides* Gunnerus, 1768, *Ficopomatus* Southern, 1921, *Galeolaria* Lamarck, 1818, *Spirobranchus* Blainville, 1818) and paraphyletic Filograninae (*Filograna* Berkeley, 1835, *Microprotula* Uchida, 1978, *Protula* Risso, 1826). Phylogenetic studies using DNA data inferred two major clades within Serpulidae, e.g., [45–49]. The clade A comprised two clades: the *Serpula-Crucigera-Hydroides* (Clade AI '*Serpula*-group') and the *Spirobranchus-Ficopomatus-Ditrupa* (Clade AII '*Spirobranchus*-group'). The Clade B included a monophyletic Spirorbinae as sister group to the *Protis-Protula-Vermiliopsis-Chitinopoma* (clade BI '*Protula*-group'). Position of serpulin genera, such as *Vermiliopsis* and *Chitinopoma* within clade BI along with typical filogranins, made both traditional Filograninae and Serpulidae paraphyletic.

Within Clade A, further assessment of AI '*Serpula*-group' (*Serpula, Crucigera, Hydroides*) [77] supported monophyly of *Hydroides*, but *Serpula* was recovered as paraphyletic basal grade and *Crucigera* was polyphyletic. Later studies assessed relationships within the largest serpulid genus *Hydroides* [78–80]. Within AII '*Spirobranchus*-group' several studies examined relationships within the genus *Spirobranchus* [81–84] and demonstrated sister group relationship between brackish-water genus *Ficopomatus* and freshwater monotypic *Marifugia* Absolon and Hrabě, 1930 [46].

Figure 5. Diversity within Serpulidae. (**a**) *Vermiliopsis glandigera/pygidialis*-complex sp., Lizard Island, Australia; (**b**) *Serpula* sp., Lizard Island, Australia; (**c**) *Hydroides lirs*, Lizard Island, Australia (**d**) *Spirobranchus corniculatus,* Lizard Island, Australia; (**e**) *Pomatostegus actinoceras,* Lizard Island, Australia.; (**f**) *Protula* sp., Lizard Island, Australia. ((**a–e**) by A. Semenov, (**f**) by G. Rouse).

Within clade B, studies focused on Spirorbinae, classification of which is based [85] on the six distinct types of brooding, two opercular (Pileolariinis and Januini) and four tubular (Romanchellini, Paralaeospirini, Circeini, Spirorbini). It has been repeatedly argued that tube incubation precedes opercular brooding [85–88], but Thorp and Segrove [89] advocated for an ancestral opercular incubation. Results of the first morphology-based phylogenetic analysis of spirorbins [44] confirmed the ancestry of tube brooding, but suggested that the opercular brooding arose once and the brooding cup of Januini is a simplification of the brooding structure of Pileolariini. Another analysis of morphological data [90] confirmed that opercular brooding is derived, but suggested that the two types arose independently. No molecular spirorbin phylogeny is available to test this arrangement. As neither traditional Serpulinae, nor Filograninae are monophyletic, re-classification based on a comprehensive integrative analysis and re-formulation of the subfamily diagnoses are needed. Meanwhile Spirorbinae is accepted as nested within Serpulidae, but other serpulid genera are not assigned into subfamilies. Most serpulid genera have not been revised (Table 4).

Table 4. Serpulid currently accepted genera that have undergone major or partial revisions and phylogenetic analyses (using morphological features, molecular data or both).

Genera	Number of Species *	Revision	Phylogenetic Analyses
Bathyditrupa Kupriyanova, 1993	1	Kupriyanova and Ippolitov 2015	NO
Bathyvermilia Zibrowius, 1973	7	Zibrowius 1973	NO
Crucigera Benedict, 1887	5	ten Hove and Jansen-Jacobs 1984	Kupriyanova et al. 2008
Ditrupa Berkeley, 1835		ten Hove and Smith 1990	NO
Ficopomatus Southern, 1921	6	ten Hove and Weerdenburg 1978	Kupriyanova et al. 2009; Styan et al. 2017
Galeolaria Lamarck, 1818	3	NO	Halt et al. 2009; Smith et al. 2012
Hydroides Gunnerus, 1768	99	Bastida-Zavala and ten Hove 2002, 2003; Sun et al. 2015; Sun et al. 2018	Sun et al. 2018
Laminatubus ten Hove and Zibrowius, 1986	3	Rouse and Kupriyanova 2021	Rouse and Kupriyanova 2021
Marifugia Absolon and Hrabe, 1930	1	Kupriyanova et al. 2009	Kupriyanova et al. 2009
Metavermilia Bush, 1905	15	Zibrowius 1971; Nishi et al. 2007	NO
Pseudochitinopoma Zibrowius, 1969	5	Kupriyanova et al. 2012	NO
Pyrgopolon de Montfort, 1808	3	ten Hove 1973	NO
Serpula Linnaeus, 1758	30	NO	Kupriyanova et al. 2008
Spirobranchus de Blainville, 1818	36	ten Hove 1970	Willette et al. 2015; Perry et al. 2019; Pazoki et al. 2020
Spiraserpula Regenhardt, 1961	18	Pillai and ten Hove 1994	NO
Spirodiscus Fauvel, 1909	2	Kupriyanova and Nishi 2011, Kupriyanova and Ippolitov 2015	NO

* Excluding subspecies.

3.2. *Diversity and Species Discovery*

3.2.1. Number of Genera and Species

Within Fabriciidae, 17 genera, 82 species and four subspecies are currently considered valid (Table S1). This revised dataset differs from that of Pamungkas et al. [91], who listed 21 genera and 91 species, but erroneously counted the genera *Eriographis* Grube, 1850, *Leiobranchus* Quatrefages, 1850, *Leptochone* Claparède, 1870 and *Tuba* Renier, 1804, all of them already synonymised with the sabellid genus *Myxicola* (Fitzhugh 1989). Discrepancy in the number of species is due to synonymisations and new combinations, as recently updated in WoRMS.

The first fabriciid genus, *Fabricia* Blainville, 1828, was established to accommodate *Tubularia stellaria* Müller, 1774, a species with only 12 chaetigers and smaller than the sabellids described at that time [90]. In mid-19th century the genus *Manayunkia* was erected and *Fabriciola* was established only in the 20th century. In the second half of the 20th century, 11 genera were established and four genera were erected at the beginning of the 21st century (Figure 6). It is remarkable that among the 17 currently valid Fabriciidae genera, eight were established by Fitzhugh alone (*Bansella* Fitzhugh, 2010, *Brifacia* Fitzhugh, 1998, *Fabricinuda* Fitzhugh, 1990a, *Novafabricia* Fitzhugh, 1990b, *Parafabricia* Fitzhugh, 1992, *Pseudoaugeneriella*, *Pseudofabriciola* and *Raficiba* Fitzhugh, 2001) or with collaborators (*Echinofabricia* and *Rubifabriciola*) [27].

According to Pamungkas et al. [91], whose data were collected in 2016, Sabellidae comprises 42 genera and 493 valid species. Since 2016, 20 new species have been described [74,92–103] and a new monotypic genus *Euchonoides* Magalhães, Bailey-Brock and Tovar-Hernández, 2020 was established. After the WoRMS database has been updated (see Materials and Methods section), the current count is 42 genera and 512 valid species in Sabellidae (Table S2). The first described genus was *Sabella* Linnaeus, 1767. From 1801 to 1850 only the genus *Myxicola* was established (Koch *in* [104]). The second half of the 19th century was a productive period, with 15 genera established. In the 20th century, six genera were erected in the first half, and nine in the second. Finally, from 2001, eight genera have been erected (Figure 6).

According to Pamungkas et al. [91], Serpulidae comprises 576 species in 77 genera, as they mistakenly included several fossil taxon names in the count of extant species. Serpulidae now comprises 562 species in 69 genera, which includes 48 genera with 374 extant species of Serpulinae *sensu lato* and 23 genera with 188 extant species of Spirorbinae (Table S3). These numbers, however, do not include those of fossil taxa (180 species, in 53 genera), not considered here. Out of 69 species of the genus *Spirorbis* Daudin, 1800, 46 were described before Bailey [85] re-classified spirorbins according to the incubation methods and never revised, so some of them upon a revision can be re-assigned to other genera.

The first serpulids described within the newly erected genus *Serpula* by Linnaeus [105] were *Spirobranchus triqueter* (non Linnaeus, *sensu* Fabricius, 1780), as *S. triquetra*, *Spirorbis spirorbis* (Linnaeus, 1758), as *S. spirorbis*, and *Circeis spirillum* (Linnaeus, 1758), as *S. spirillum*. *Serpula vermicularis* Linnaeus, 1767, the type species of the type genus, was described only nine years later. The monotypic genus *Hydroides*, with *H. norvegica* Gunnerus, 1768, was the second serpulid genus to be described in the 18th century. In the 19th century, 28 genera, eight spirorbins and 20 serpulins, were described. In 1900–1959, 22 genera (including three spirorbin) were described, while 48 genera were erected in the second half of the 20th century (1960–2000). Finally, from 2001, four genera, including three serpulin and one spirorbin, have been erected [49,106,107] (Figure 6).

3.2.2. Taxonomists and Species Discovery

Kirk Fitzhugh is the most productive author in terms of the number of discovered fabriciid species, with 27 species described alone and four in collaborations. Other authorities of fabriciid species include Karl Banse, Gesa Hartmann-Schröder and Greg Rouse, with six species described by each, all as single authors (Table S1).

The most prolific author in terms of sabellid species discovery is Adolph Eduard Grube, who described 40 species before 1881. Other productive taxonomists are María Ana Tovar-Hernández (42 species: 36 as first or single author, 6 as co-author); María Capa (29 species, all as first author or alone), Adriana Giangrande (29 species: 14 as first author or alone, and 15 as co-author), Olga Hartman (18 species) and Gesa Hartmann-Schröder (16 species) (Table S2).

The most productive serpulid taxonomist is Harry ten Hove who so far described 49 species, including seven alone. Gottfried Pillai described 46 species, including 31 species described alone and 15 in collaboration with ten Hove. Phyllis and Wynn Knight-Jones described 41 species, mostly spirorbins. Katherine Bush described 33 species, including 25 alone. Helmut Zibrowius described 31 species, including 29 alone and Elena Kupriyanova authored 29 species, including eight alone. Other productive serpulid taxonomists (over 15 species described) are Minoru Imajima (21 species) and Alexander Rzhavsky (17 species, mostly spirorbins) (Table S3).

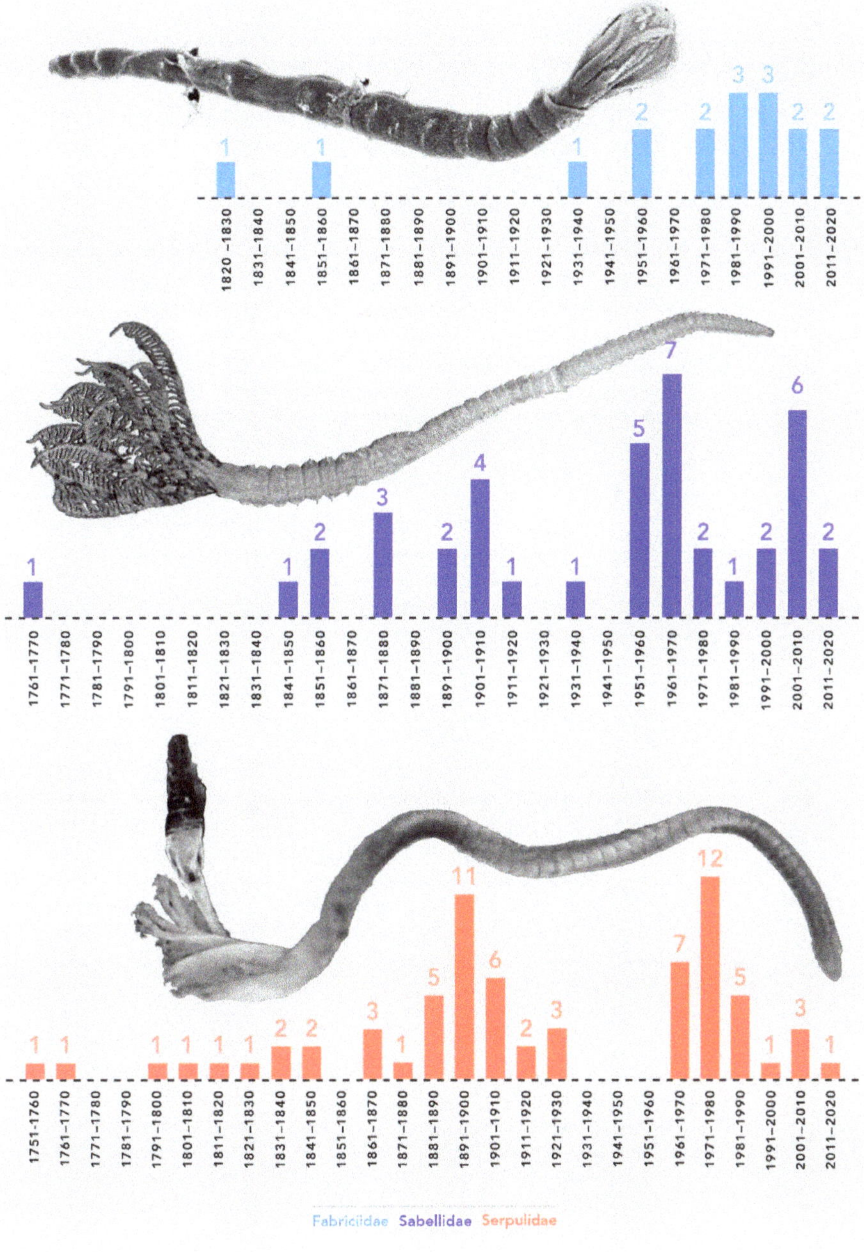

Figure 6. Number of genera erected each decade in Fabriciidae, Sabellidae, and Serpulidae.

3.2.3. Identification Keys, Diversity Assessments, and Recent Regional Taxonomic Studies

The key to all polychaete genera by Fauchald [12] includes all sabellids and serpulids considered valid at the time, but it is outdated and not recommended for taxonomic work anymore. A key to Fabriciidae genera recognized until 1998 was provided by Fitzhugh [57]. In that study, the currently accepted *Echinofabricia* was named Genus A. The most recently updated key to all fabriciid genera is that by Tovar-Hernández and Fitzhugh, in press. The keys to Sabellidae genera were provided by Fitzhugh [42], Tovar-Hernández [108] and most recently Tovar-Hernández and Fitzhugh [109]. The review by ten Hove and Kupriyanova [110] includes diagnoses and a key to all serpulid genera (excluding spirorbins) valid at the time.

Revision of literature reveals that intensive fieldwork and continuous taxonomic studies by a single scientist in a specific area have had a great impact in biodiversity knowledge of a region. However, large geographic regions have been scarcely studied, not only in difficult to access deep-sea environments, but even in the intertidal and subtidal zone either because not enough work has been put into taxonomic surveys, or members of Sabellida were not among targeted groups. It is expected that our understanding of species diversity will improve after efforts (financial, logistical, technological and taxonomic expertise) are devoted to fill those gaps. Herein, information about the most recent regional surveys and taxonomic revisions is provided, and the number of species described in these areas given as a rough approximation of their biodiversity knowledge.

The regions with the overall highest number of type localities of described species are the coastal areas of Europe, both coasts of North America, and Central and Western Indo-Pacific, while the areas with lower number of original descriptions are the majority of Africa, South America, as well as Tropical Eastern Pacific and Eastern Indo-Pacific (Figure 7).

The current state of biodiversity knowledge and a summary of the most recent (defined here as last 20 years) comprehensive checklists, faunistic and taxonomic regional studies (excluding single species descriptions) are organised below by marine realms. References to main comprehensive taxonomic studies are provided as recommendations for getting started with faunas of each realm.

Arctic

This realm covers the Arctic Ocean down to Newfoundland in the western Atlantic, including the northern half of Iceland, northern Russia, from the White Sea to the Bering strait, and all northern Alaska and Canada. One fabriciid, 16 sabellid species, and 16 serpulids have been described from the Arctic, most of them from the Western sector of the Arctic Ocean (Figure 7, Tables S1–S3). Knight-Jones et al. [111] reviewed species of Pseudopotamilla from Iceland, Greenland and the Canadian Arctic. Jirkov's book [112] on Arctic polychaetes, that includes diagnoses, illustrations, and keys to sabellids (including fabriciids) and serpulids (including spirorbins), as well as the recent comprehensive illustrated revisions with taxonomic keys to all Arctic Serpulidae (including Spirorbinae) by Rzhavsky et al. [113,114] are recommended for studies in this region.

Temperate Northern Atlantic

This realm is delimited in the north with the Arctic realm, and in the south reaches the coasts of Florida, including the northern half of the Gulf of Mexico, and is delimited in the east by the Cape Verde archipelago and the coasts of Mauritania. It also includes the Mediterranean and the Black Sea. Sixteen fabriciids, 124 sabellids, and 108 serpulids have been described from this realm (Figure 7). Of these, one fabriciid, 18 sabellid, and 12 serpulid species were from the Western coasts of the Atlantic Ocean and three sabellids were from northwest Africa. All the rest were described from European waters and the Mediterranean (Tables S1–S3).

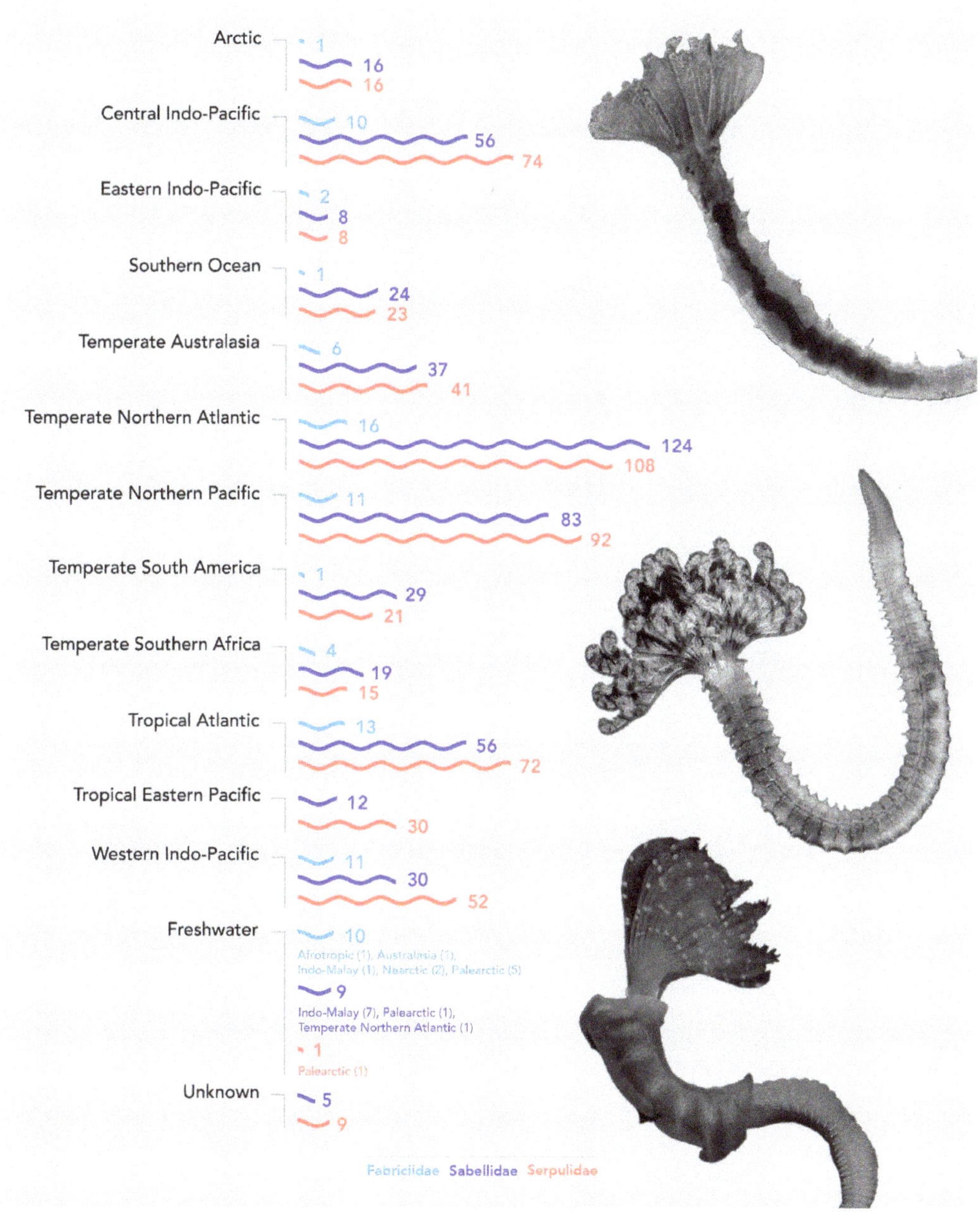

Figure 7. Number of species described for each marine realm (in alphabetical order, according to Spalding et al. [31]) and Udvardy 1975 for limnic realms.

The book by Fauvel on sedentary polychaetes of France [115] provide descriptions and illustrations of the common species of the north eastern Atlantic and western Mediterranean and is still been widely used despite being outdated. Relatively more updated sources of serpulid diversity data in the Mediterranean are books by Zibrowius [116] and Bianchi [117] that include keys, descriptions and illustrations. The illustrated key by Knight-Jones [118] to the British Isles and North-West Europe is recommended as an initial source of data on Sabellidae, Fabriciidae and Serpulidae. The incoming book on Fauna Ibérica includes chapters on Fabriciidae and Sabellidae and will be a new standard reference for the region [119,120].

Despite the overall high number of species described from this region, only a few taxonomic and faunistic studies have been carried out on Sabellida in the last 20 years in the Atlantic provinces of this realm. In particular, fabriciids were studied by Bick [121], species of *Euchone* by Bick and Randel [122], *Chone* by Tovar-Hernández et al. [123,124] and *Pseudopotamilla* by Knight-Jones et al. [111]. However, the notable exception has been the Mediterranean, where much taxonomic activity, with particular emphasis on introduced species, has taken place recently. Some representative contributions dealing with Mediterranean fanworms include those by Çinar (on serpulids [125], on non-indigenous species [126]), the general annelid checklist of polychaetes from Turkey [127], Selim et al. (on *Dialychone* and *Paradialychone* [128]), Giangrande et al. (sabellids of the Ionian Sea [129], on *Acromegalomma* [130]), the general papers on annelid diversity of the Adriatic Sea [131,132] and the fabriciids and sabellids of the Adriatic [133], the checklist of Iberian species [134] and the most recent one by Tilic et al. [103] dealing with *Amphiglena*. A checklist of the polychaetes from the Black Sea includes three fabriciids, six sabellids, and 11 serpulids [135].

Temperate Northern Pacific

The western side of this realm is demarcated by the Bering Strait in the north, and an imaginary line from Taiwan to the south of Baja California Peninsula, in the south. The number of fabriciids described in this realm is 11, eight of them from the Eastern side (Table S1). The number of described sabellids is 83, 37 of which from the Western side, and 46 from the Eastern side (Table S2). Out of 92 serpulid species described from this region, 54 are from the western side (Table S3).

The catalogue of sedentary polychaetes from California [136], still widely used for the region, is outdated and is not recommended. The monograph of polychaetes of the Russian Far-East [136] and its English translation [137] that remains the main source of keys and information on polychaetes of the region, including Sabellida, is also outdated and thus should be used with caution.The revision of spirorbins from the east Pacific coast [137] is still the most recent source of information on this group.

The most recent literature-based annotated checklist of polychaetes from Pacific coasts of Russia lists 37 sabellids (including fabriciids) and 40 serpulids [138]. Sabellids from Japan were recently reported by Nishi et al. [95,139] and Yoshihara et al. [140]. Taxonomic studies on serpulids of Japan are summarized in the book by Imajima [141] that provides an illustrated key to 55 species of Serpulidae, but does not include sabellids. The most recent account of Chinese polychaetes [142] provides diagnoses and keys to 64 species of sabellids and 98 serpulid species, and is recommended as a source of biodiversity data and taxonomic keys for China. The complementary revision [143] includes a checklist of most annelid groups from the South China Sea, and lists three fabriciid, 33 sabellid and 72 serpulid species. Recent revisionary studies, including taxonomic keys, from the Pacific coast of North America have reported seven species of sabellids and more than 40 serpulids in the following contributions dealing with species of *Hydroides* from Northern Mexico [144], serpulids from the Eastern Pacific [145,146], sabellids and serpulids from northern Mexico [147,148], and sabellid *Acromegalomma* [149] and *Notaulax* species also from Northern Mexico [101].

Tropical Atlantic

This realm is delimited in the north by the Temperate Northern Atlantic realm and in the south by an imaginary line from Rio de Janeiro in the west to the southern border of Angola in the east. It also includes the southern half of the Gulf of Mexico and the Caribbean. Thirteen fabriciid species have been described from western side of the realm, specifically from the Caribbean, and none from the eastern Atlantic (Table S1). Out of 56 sabellids described in the region, only one came from the African coasts (Table S2) and of the 72 serpulids, only 10 were described from Africa (Table S3).

Zibrowius [150] made the first study on Brazilian serpulids. Other recommended taxonomic studies of serpulids (other than *Hydroides*) of Caribbean are those by ten Hove [151–154].

Recent studies of the Caribbean fanworms included those describing fabriciids [27,155]; revisions of species in *Chone* [156] and *Branchiomma* [157]; records and new species of sabellids [155,158]; and selected serpulids, such as *Hydroides* [159], *Serpula* and *Spiraserpula* [160]. The tropical coasts of South America have been scarcely studied. The checklist of polychaetes of Brazilian Tropical Atlantic region reports 11 sabellid and 24 serpulid species for the area [161–164], with no fabriciids registered so far. However, several of those are records of species described from Europe, North America and South Africa, demanding further study. Additionally, Amaral et al. [163] checklist treated many already synonymised species as valid.

The illustrated key of Sabellidae and Fabriciidae by Tovar-Hernández and Fitzhugh [105] includes all species currently known for the Caribbean, whereas Caribbean Serpulidae are available in [165].

Western Indo-Pacific

This large realm covers most of East coast of Africa, Madagascar, Arabian (Persian) Gulf, the Red Sea, shelf of Bay of Bengal and Andaman Sea. The number of species described in this realm is 93, including 11 fabriciids, 30 sabellids, and 52 serpulids (Figure 7, Tables S1–S3). Out of 52 serpulids, only eight were described from African coasts.

An influential book on Indian polychaetes [166], unfortunately, lists European species (and even illustrates specimens collected in France) and therefore, is not recommended as an identification tool beyond the generic level. The most recent checklist by Sivadas and Carvalho [167] includes two fabriciids, 11 sabellids and 34 serpulids from India and critically evaluated annelid species richness in the region, stressing that native species diversity of India is severely underestimated. The relatively large number of serpulids described from Sri Lanka is due to the intense work of Pillai [40,168,169]. These publications still remain as the only source of faunal information for that region.

An annotated literature-based checklist by Wehe and Fiege [170] is the best compilation of annelid diversity in the area surrounding the Arabian Peninsula. The most recent checklist of intertidal polychaetes of Kuwait by Al-Kadari et al. [171], based on newly collected material, reported seven species of Sabellidae and 12 Serpulidae. A monograph on Serpulidae from the Suez Canal by Ben-Eliahu and ten Hove [172] included 16 species. In the Red Sea, Perry at al. [81] reviewed of serpulids of the genus *Spirobranchus* and sabellids have not been studied since Knight-Jones [173].

Central Indo-Pacific

This realm comprising the largest number of ecoregions (40) includes part of the coast of South-East Asia, from Taiwan to Malaysia, down to Tropical Australia from Coral Bay, in the West, to Brisbane, in the East. The eastern boundary of this region is delimited by an imaginary line from Fiji up to the south of Japan. The realm includes the Coral Triangle recognized as the global centre of marine biodiversity [174], and fanworms are also diverse in this region. Ten fabriciids, 56 sabellids, and 74 serpulids have been described from this realm (Figure 7), mainly from the Philippines and tropical Australia (Tables S1–S3). Other than newly described species, 78 species of Sabellidae have been reported only from the Gulf of Thailand, Indonesian Archipelago and the Philippine Seas [98,169], and at least 11 taxa are awaiting formal description [175,176]. Serpulidae from Hong Kong were most recently revised by Sun et al. [177], who provided illustrations, diagnoses and taxonomic keys. Tropical Australian sabellids belonging to 12 genera have been documented in a series of recent studies [68,69,72–74,178]. Serpulids from Kimberley (Western Australia) were revised by Pillai [106] and those from Lizard Island (Queensland) by Kupriyanova et al. [179], whereas the revision of the genus *Hydroides* in Australia [180] includes both tropical and temperate species. The most comprehensive treatment of Australian sabellids and serpulids is still the interactive key by Wilson et al. [181], but it is outdated in the light of the recent studies. The digital guide [182,183] allows distinguishing 38 native and non-indigenous species of Serpulidae and 14 Sabellidae from Australia, and includes a glossary with main diagnostic features for members of both groups.

Eastern Indo-Pacific

This small in terms of the coastline length Pacific realm includes Hawaii, Marshall, Gilbert and Ellis Islands, Central and Southeast Polynesia, Marquesas, Eastern Island, and the shelf around them. It hosts type localities of 18 fanworms, including two fabriciids, eight sabellids, and eight serpulids (Figure 7, Tables S1–S3).

Fauna of this realm is poorly known beyond that of Hawaii. Out of eight serpulids described from this realm, five were described from Hawaii. The latest studies include those on serpulids of Hawaii (*Hydroides* [140], a key and records excluding spirorbins [145], records of 16 species with a key [142], and serpulids from Cross Seamounts in the Hawaiian chain [184]. Out of eight sabellids, five were described from Hawaii, and two species of *Branchiomma* were reported most recently [185]. A recent study of serpulids from atolls of Marshall Islands [186] reported 29 serpulids (including spirorbins). Small number of publications dealing with members of Sabellida highlights the need for taxonomic work in the area.

Tropical Eastern Pacific

This realm is delimited in the North by the Cape San Lucas, Baja California, by the northern border of Peru in the South and includes the Galapagos Islands in the West. No fabriciids have been described in this region (Table S1) and the number of described sabellids is 12, five of them originally reported from coast of Panama and the rest from further north (Figure 7, Table S2). The number of described serpulids is 30, most of them from Galapagos, Panama and Mexico (Figure 7, Table S3).

Recent studies of Sabellida from the region include those dedicated to Panamanian sabellids [187], some *Acromegalomma* [149] and *Chone* [124], and those focused on sabellids and serpulids from Mexico [145,148,188]. Three sabellids and two serpulid species introduced in the Galapagos Islands were reported by Keppel et al. [189].

Temperate South America

The realm covers both Pacific and Atlantic coasts of South America, from Peru to Rio de Janeiro, respectively. Only one fabriciid was described from this region (Table S1). Twenty-nine sabellids have been described (Figure 7), mainly from Chile and Argentina (Table S2), and 21 serpulids were described at a variety of localities from Brazil to Peru (Figure 7, Table S3).

The publication by Zibrowius [150] remains the most comprehensive study on Brazilian serpulids in both Tropical Atlantic and Temperate South America realms. More recent publications reported three species of fabriciids, 29 of sabellids and 22 species of serpulids for the Brazilian part of the Temperate South America realm [163] and 27 species of serpulids and 17 sabellids for Argentina [190]. Tovar-Hernández et al. [99] studied sabellids mainly from Argentina and Chile. A key to Sabellidae and Serpulidae from continental Chile is available in [191].

Temperate South Africa

This realm includes the coastline of Namibia and South Africa as well as Amsterdam and St. Paul Islands. It hosts type localities of 38 species of Sabellida, including four fabriciids, 19 sabellids and 15 serpulids (Figure 7, Tables S1–S3). Of these, 30 have been described in the littoral zone in South Africa, indicating that less taxonomic effort has been devoted to other areas within this region. The recent papers re-described two serpulid species based on type material (e.g., [192,193].

Although South African polychaetes, including fanworms, were summarized in the influential book by Day [194], most of Sabellida included in the monograph are 'cosmopolitan species' with European type localities (e.g., of 27 serpulids only six have type localities in South Africa), so native Sabellida species diversity is severely underestimated. Spirorbins from South Africa have not been reviewed since studies of Knight-Jones [195] and Knight-Jones and Knight-Jones [196].

Temperate Australasia

The realm includes coasts of Southern Australia and New Zealand hosting type localities of 86 species of Sabellida. This number includes six fabriciids from Australia (Table S1), 37 sabellids (28 described from the Australia and nine from New Zealand (Table S2) and 41 serpulids (28 from Australia and 15 from New Zealand, Table S3) (Figure 7).

Spirorbins from southern Australia have not been studied since they were reviewed by Knight-Jones et al. [197]. In New Zealand, spirorbins were studied by Vine [198], who reported 24 species, nine of them new to science, while a list of sabellids and serpulids was provided by Glasby and Read [199].

Sabellids from Australian temperate waters have been well documented in a series of recent papers [68,69,72–74,200–202] along with records of temperate species. Most recent papers on temperate Serpulidae are Sun et al. [203] and Styan et al. [204], whereas the Australian *Hydroides* revision [180] also includes temperate species.

Southern Ocean

This large realm covers coasts of Antarctica and sub-Antarctic Islands. It hosts type localities of a single fabriciid, 24 sabellid and 23 serpulid species (six serpulins and 17 spirorbins) (Figure 7, Tables S1–S3). Many of these species were discovered and described as a result of Antarctic expeditions of the 19th and early 20th century (e.g., [205–210]).

The most recent contributions from the region are descriptions of two spirorbin species from Kerguelen and Bouvet Islands [211]; species of *Perkinsiana* [212] and reports of 19 still undescribed sabellids from Falkland Islands [213],demonstrating the underestimated diversity. There is no contribution summarizing Sabellida species diversity of this region.

3.3. Diagnostic Characters and Techniques Used for Species Discrimination

Most species within Sabellida are characterised by a unique combination of morphological features. The most useful morphological characters used for Sabellidae species identification are summarised in Capa et al. [24]; for Fabriciidae, see Bick [4]; for Serpulidae, see ten Hove and Kupriyanova [110] and Kupriyanova et al. [214]. Since detailed information is provided in these thorough revisions, only succinct identification guidelines are given below for each family.

3.3.1. Shortcuts to Identification of Fabriciidae

The small body size (most species are less than 5 mm long) and the absence of significant diagnostic characters make the identification of Fabriciidae difficult. All fabriciid species possess a radiolar crown with three pairs of radioles (Figure 8a–c). The branches are formed by successive longitudinal splitting of the radioles. The symmetrical branching of the radioles leads to bi-pectinated radioles, as found in most genera (Figure 8c), whereas pectinated radioles result in asymmetrical branching (Figure 8b), as in *Manayunkia* and *Monroika* only [4]. Ventral filamentous appendages, present in some genera, are associated with the radiolar crown (Figure 8b). These appendages have been described as non-vascularized (e.g., in all species of *Fabriciola*, *Pseudofabricia* Cantone, 1972 and *Rubifabriciola*) or vascularized (e.g., in all species of *Augeneriella*, *Echinofabricia*, *Manayunkia*, *Monroika* and *Pseudoaugeneriella*, and also some species of *Fabricinuda* and *Pseudofabriciola*) [57]. These appendages are branched only among species of *Augeneriella*.

Figure 8. Main diagnostic characters for members of Fabriciidae. (**a**) *Fabricia stellaris*, dorsal view; (**b**) Pectinated radiolar branching pattern, *Manayunkia athalassia*; (**c**) Bi-pectinated radiolar branching pattern, *F. stellaris*; (**d**) Thoracic uncini of *Manayunkia zenkewitschii*; (**e**) Gamete bearing region in thorax; (**f**) Branchial heart (bh), spermathecal (s) and peristomial eye (pe), in this order; (**g**) Thoracic chaetae, *M. athalassia*; (**h**) Abdominal uncini, *Pseudoaugeneriella nigra*; (**i**) Pygidial eye (py); (**j**) Abdominal chaetae, *M. athalassia*; (**k**) Thoracic transitional chaetae (below), *Manayunkia godlewskii*; (**l**) Chaetal inversion, *M. athalassia*; (**m**) Abdominal pin-ead chaeta, *Rubifabriciola tonerella*. ((**a**,**c**) by A. Dietrich; (**b**) by G. Christie).

Peristomial eyes (Figure 8f) are developed among most species of Fabriciidae, black in most fabriciids, or red, as in *Echinofabricia* and *Rubifabriciola* [27].

The presence of thoracic pseudospatulate and transitional (=pilose, after Jones 1974) chaetae is of taxonomic significance [4] (Figure 8g,k). However, the distribution of pseudospatulate chaetae is not consistent within the genera. These chaetae occur on chaetigers 2–5 (some *Manayunkia* species), 2–8 (*Raficiba barryi* Fitzhugh, 2001), 3–5 (*Monroika africana* and most *Novafabricia* species), 3–6 (*Pseudoaugeneriella*, some species of *Augeneriella* and *Novafabricia*), 3–7 (*Brifacia metastellaris* Fitzhugh, 1998, *Fabricia stellaris*, *Parafabricia ventricingulata* Fitzhugh, 1992, and some species of *Augeneriella* and *Fabricinuda*), or 3–8 (most *Fabricinuda* species) [4]. Pseudospatulate chaetae are absent in *Bansella, Echinofabricia, Fabriciola, Pseudofabriciola* and *Rubifabriciola* [4]. Species of *Rubifabriciola* have pin-head chaetae on the abdominal neuropodia [27] (Figure 8m). These chaetae have a blunt tip and a number of small teeth apically. Transitional chaetae (Figure 8k) replacing thoracic uncini occur on the last thoracic chaetigers (chaetigers 6–8) of *Brandtika* spp., *Manayunkia godlewskii* (Nusbaum, 1901), females of *M. occidentalis* Atkinson, Bartholomew and Rouse, 2020 and *M. zenkewitschii* Sitnikova, Shcherbakov and Kharchenko, 1997 [4,215–217].

The thoracic uncini are characterized by a long manubrium (homologous to handle in sabellids) and a main fang surmounted by a series of smaller teeth (Figure 8d). A slightly offset medium-sized tooth occurs between the large main fang and the smaller apical teeth in *Augeneriella, Fabricia, Fabricinuda, Monroika, Novafabricia, Parafabricia, Pseudofabricia, Pseudoaugeneriella* and some species of *Pseudofabriciola* [42]. The apical teeth can also be approximately of the same size in *Echinofabricia*, or may gradually decrease in size away from the main fang as in *Fabriciola, Manayunkia* and some species of *Pseudofabriciola* [57]. The abdominal uncini usually exhibit multiple rows of equal-sized teeth (Figure 8h). Only members of *Novafabricia chilensis* (Hartmann-Schröder, 1962) and *N. gerdi* (Hartmann-Schröder, 1974) have uncini with a single row of teeth.

Fabriciids usually have three abdominal chaetigers (Figure 8j). However, *Brandtika* spp., *Fabriciola minuta* Rouse, 1996, and *Monroika africana* have two abdominal chaetigers, while *Echinofabricia* spp. has four [215,218,219].

The pygidium is triangular or bluntly rounded in most species, but has a ventral depression in *Pseudofabriciola analis* Fitzhugh, Giangrande and Simboura, 1994. A pair of black or dark brown pygidial eyes is present in most species of Fabriciidae (Figure 8j). They are red in *Echinofabricia* (disappearing after fixation) and *Rubifabriciola* (persisting after fixation), but are always absent in all species of *Manayunkia* and *Monroika*, *Fabriciola parvus* Rouse, 1993 and two undescribed deep-sea species [220]. Unique among members of Sabellida, emergent spicules are present in the epithelium of *Echinofabricia* species [27].

3.3.2. Shortcuts to Identification of Sabellidae

Sabellids are relatively easily to visually identify to the generic level because genera are provided with unique and conspicuous diagnostic features (Figure 9). The diversity of radiolar eyes within members of Sabellidae is remarkable (e.g., Figure 9a), and the eye number, type and arrangement offer a very powerful taxonomic aid for genera and species identification [24,69,221]. The large compound eyes located in the tips of dorsal radioles are unequivocally characteristic of members of *Acromegalomma*, whereas the large single and bulging compound eyes arranged, in a longitudinal row on the outer margin of the radioles, are typical of members of *Pseudopotamilla* Bush, 1904 (Figure 9a). *Anamobaea* Krøyer, 1856 and *Notaulax* Tauber, 1879 are easily recognized due to the presence of long radiolar lobes with dorsal and ventral flanges. Other generic synapomorphies related to the radiolar morphology are the dichotomously branching radioles, found only in *Schizobranchia* Bush, 1905 and *Eudistylia* Bush, 1905 (most likely due to a regeneration processes [111,222]), and the external paired radiolar appendages, called stylodes, in members of *Branchiomma* (Figure 9e). *Euchone* is recognisable by the presence of a typical pre-pygidial depression with lateral wings, but this character is only visible in adults. Species of *Claviramus* Fitzhugh, 2002 have radiolar tips with expanded flanges, rolled inwards or bilobed, this feature is easily seen if radioles are complete. *Potamethus* Chamberlin, 1919 is recognizable due the very long collar (2–4 times the length of next thoracic segment). Other genera are recognised by unique traits, which are not evident to a naked eye and require optic aids. These are the typical companion chaetae of members of *Parasabella*, the absence of posterior peristomial ring collar in members of *Amphiglena*, or the presence of a broad, oblique glandular (clitellum-like) ring on third abdominal segment in *Euchonoides*. Internal structures, such as the rows of vacuolated cells supporting radioles, dorsal lips and radiolar appendages, are of taxonomic significance and often used for species discrimination [64,102,122,223–226].

3.3.3. Serpulidae Diagnostic Characters

Within Serpulidae, body symmetry separates serpulins from spirorbins, as serpulins are bilaterally symmetrical, while spirorbins are curved in the direction of the tube coil. Specific identification has been based on a combination of characters such as morphology of the operculum and opercular peduncle, degree of development of the collar and thoracic membranes, and chaetal structures (Figure 10). Tube morphology and ultrastructure are important for identification of both extant and fossil taxa [227] (Figure 10a–c). Serpulid genera have been described on the basis of unique characters or on unique combinations of characters (or absence of characters), rather than on presence of shared derived characters. Mentioned here morphological characters used for serpulid identification have been described in details and illustrated in ten Hove and Kupriyanova [110], Wong et al. [228], Kupriyanova et al. [214].

The operculum that is present in most serpulins and in all spirorbins is one of the most important diagnostic characters (Figure 10d–h). The shape of the operculum varies significantly, ranging from soft transparent vesicles to complex structures reinforced with chitinous or calcareous endplates and spines. The distinct funnel-shaped opercula of *Crucigera* and *Serpula* are composed of numerous radii (Figure 10d), while in *Hydroides* the funnel is topped with a verticil of chitinous spines (Figure 10e). The operculum-bearing radiole can be identical to others (e.g., *Filograna*, *Apomatus* Philippi, 1844), but usually is modified into a smooth peduncle (Figure 10j). In cross-section, the peduncle is commonly cylindrical, but it is flat ribbon-like in members of the genus *Metavermilia* Bush, 1905. Below the operculum, the peduncle may bear diagnostic distal wings (e.g., *Spirobranchus*) (Figure 10h).

The collar segment bears only notopodial (collar) chaetae that may be absent (e.g., *Ditrupa* Berkeley, 1835, *Marifugia*, *Placostegus* Philippi, 1844). The collar chaetae may bear four types of diagnostic modified chaetae: bayonet-type (e.g., *Serpula*, *Hydroides*, Figure 10i), fin-and-blade (e.g., *Chitinopoma*, *Protis* Ehlers, 1887), *Spirobranchus*-type (e.g., *Spirobranchus*, *Laminatubus* ten Hove and Zibrowius, 1986) and *Ficopomatus*-type (see [214]). Tonguelets, located between collar lobes, are diagnostic of the genera *Spirobranchus* and *Pyrgopolon* de Montfort, 1808. The thoracic membranes (Figure 10l) may be ending at the first (*Ditrupa*) or the second thoracic chaetiger (*Chitinopoma*), or may continue to the mid-thorax (e.g., *Pomatostegus*, *Vermiliopsis* Saint-Joseph, 1894), to the last thoracic chaetiger (some *Spiraserpula* Regenhardt, 1961 and *Metavermilia* spp.), or past the end of the thorax forming the ventral apron (e.g., *Ficopomatus*, *Serpula*, *Hydroides*, *Protula*, *Spirobranchus*) (Figure 10m). In most genera the thorax consists of seven chaetigerous segments (first with collar chaetae only and six with both notopodia and neuropodia). The number of thoracic segments varies from five (*Tanturia* Ben-Eliahu, 1976 and *Bathyditrupa* Kupriyanova, 1993) or six (*Laminatubus*, *Hyalopomatus* Marenzeller, 1878, *Spirodiscus* Fauvel, 1909) to 10 (*Kimberleya* Pillai, 2009), while spirorbins have three to five thoracic segments.

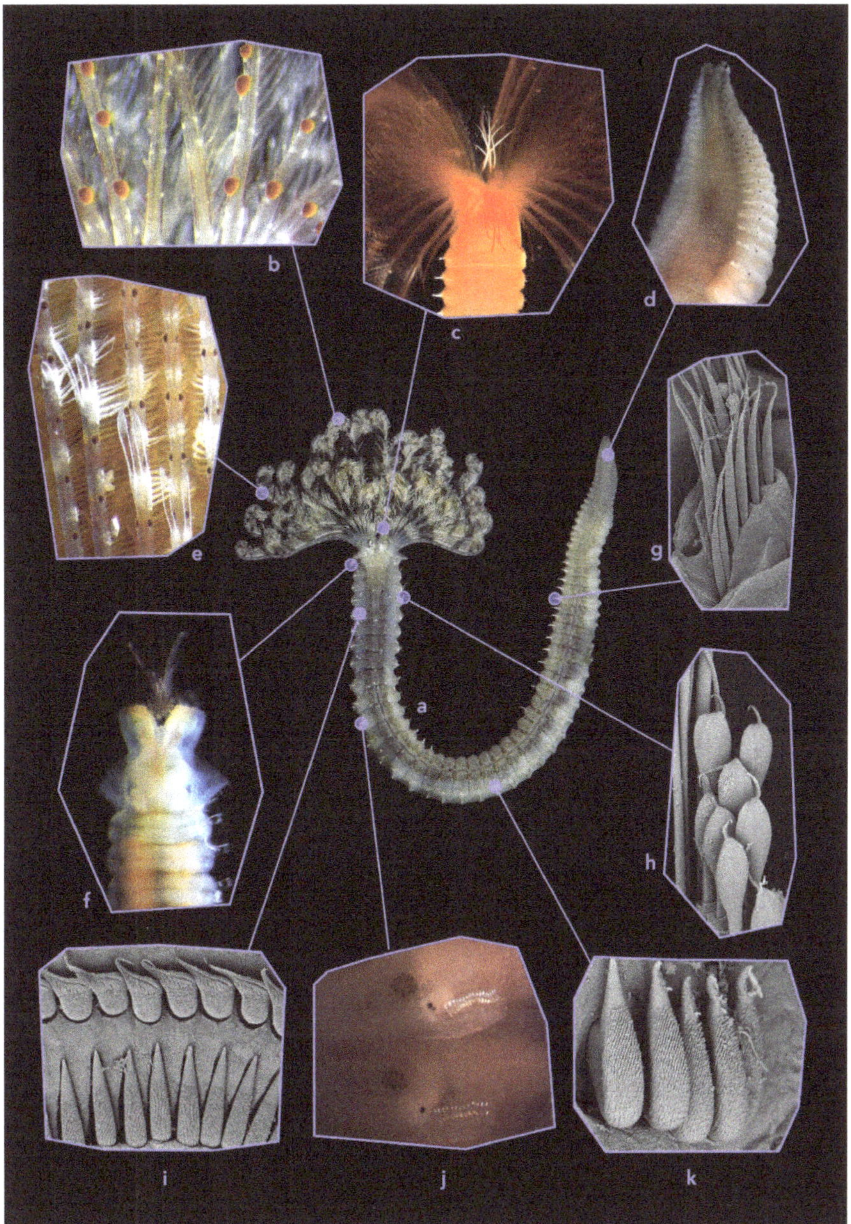

Figure 9. Main diagnostic characters for members of Sabellidae. (**a**) *Parasabella microphtalma*, ventral view; (**b**) Compound eyes on dorsal radioles, *Pseudopotamilla* sp.; (**c**) Radiolar internal structures, *Chone infundibiliformis*; (**d**) Pygidial morphology, *Bispira* sp.; (**e**) Radiolar stylodes, *Branchiomma* sp.; (**f**) Collar, radiolar lobes and glandular girdle in second chaetiger, *Jasmineira* sp.; (**g**) Abdominal chaetae, *Parasabella* sp.; (**h**) Thoracic chaetae, *Parasabella* sp.; (**i**) Thoracic uncini and companion chaetae, *Notaulax* sp.; (**j**) Parapodia morphology and arrangement of chaetae, *Bispira* sp.; (**k**) Abdominal uncini, *Notaulax* sp.

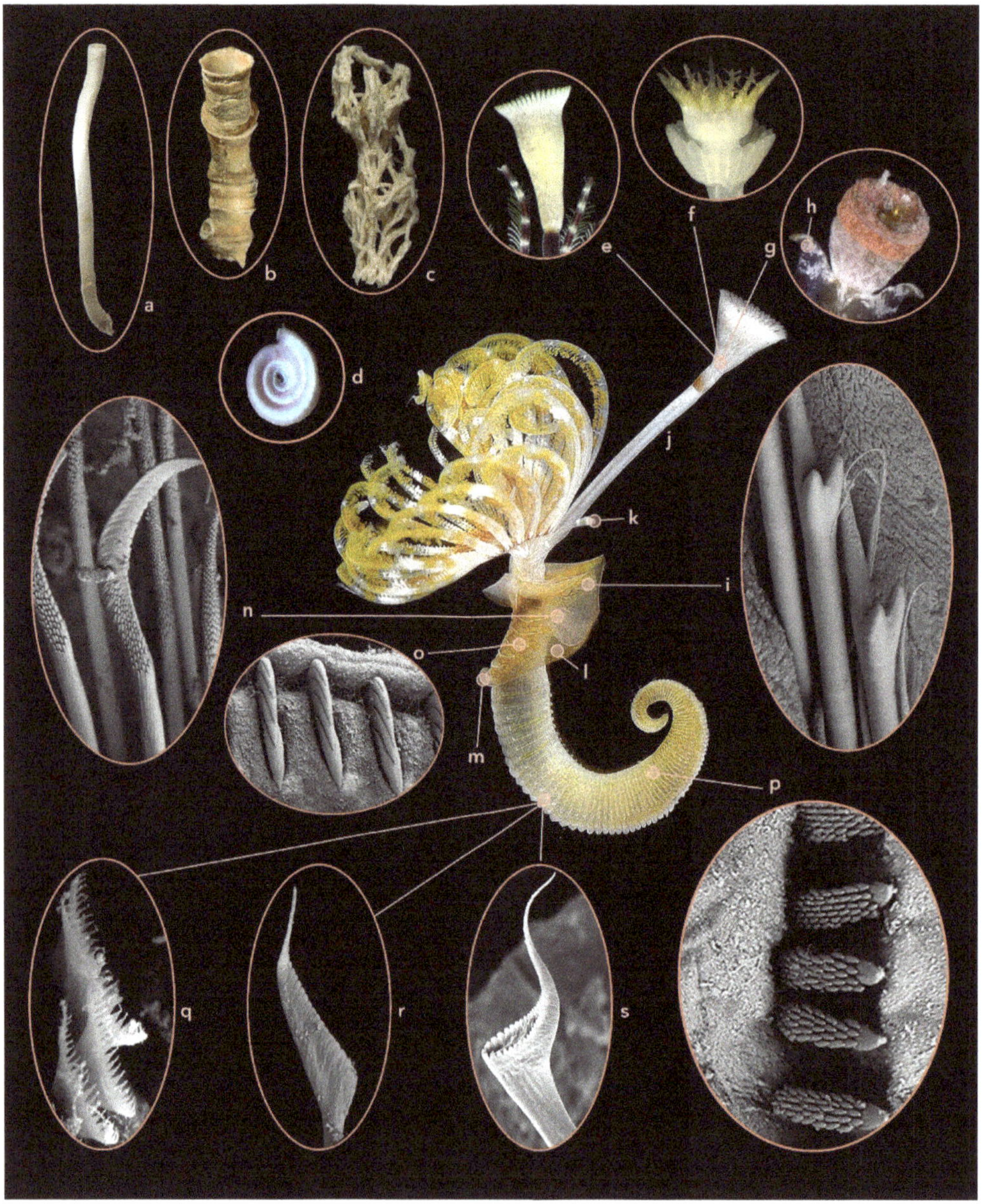

Figure 10. Main diagnostic characters for members of Serpulidae. Centre: lateral view of a *Serpula* specimen removed from the tube; (**a**–**d**) Tubes: *Hyalopomatus biformis* (**a**), *Serpula vermicularis* (**b**), *Filograna implexa* (**c**), *Circeis armoricana* (**d**); (**e**–**h**) Opercula: *Serpula* (**e**), *Hydroides* (**f**), *Spirobranchus* (**g**) showing distal wings (**h**); (**i**) Special (bayonet) collar chaetae of *Hydroides*; (**j**) Peduncle; (**k**) Pseudoperculum; (**l**) Thoracic membranes; (**m**) Apron; (**n**) Thoracic *Apomatus* (sickle-shaped) chaetae of *Vermiliopsis*; (**o**) Thoracic saw-shaped uncini of *Hydroides*; (**p**) Posterior abdominal rasp-shaped uncini of *Hydroides*; (**q**–**s**) Abdominal chaetae: flat trumpet-shaped of *Serpula;* (**q**), flat geniculate of *Vermiliopsis;* (**r**) true trumpet-shaped of *Spirobranchus* (**s**). (**a**–**c**,**f**) by E. Wong, (**d**) by A. Rzhavsky, (**e**) by G. Rouse, (**g**,**h**) by A. Semenov, (**j**–**m**) (central photo of *Serpula*) by F. Verbiest, (**n**–**s**) (SEM images) by S. Lindsay).

In the posterior thoracic segments chaetae are supplemented by diagnostically important *Apomatus* or sickle-shaped chaetae (Figure 10n). The number of vertical teeth rows in the thoracic and anterior abdominal uncini (saw-shaped, with one row of teeth, e.g., *Hydroides*, *Serpula* (Figure 10o); rasp-shaped, with several rows of teeth, e.g., *Hyalopomatus*, *Placostegus*, *Marifugia*; or saw-to rasp-shaped, from with one tooth distally to a row of up to five teeth near the peg, e.g., *Filogranula* Langerhans, 1884) is diagnostic. Posterior abdominal uncini are always rasp-shaped (Figure 10p). Even more important is the shape of the anterior tooth of uncini. The anterior teeth are either pointed fangs (e.g., *Filograna*, *Hydroides*, and *Serpula* (Figure 10o,p), and pileolariin spirorbins), or a wide variety of blunt 'wedge' shaped pegs (e.g., *Pseudovermilia* Bush, 1907, *Spirobranchus*, *Galeolaria*, *Ficopomatus*, *Hyalopomatus*, *Chitinopoma*, *Pyrgolopon*, *Vermiliopsis*, *Protula*).

The shape of abdominal chaetae is very important for generic diagnostics (Figure 10q–s). The simplest forms are capillary (*Bathyditrupa*) and acicular (*Paumotella* Chamberlin, 1919). The flat trumpet-shaped chaetae with a single row of teeth are found in *Crucigera*, *Hydroides*, *Serpula* (Figure 10q). Abdominal chaetae previously referred to as 'geniculate' are two distinct types of chaetae, true trumpet-shaped, typical for, e.g., *Ficopomatus*, *Galeolaria*, *Placostegus*, *Spirobranchus* (Figure 10s) that lack thoracic '*Apomatus*' chaetae, and flat geniculate are found in taxa with *Apomatus* chaetae, e.g., spirorbins, *Apomatus*, *Chitinopoma*, *Vermiliopsis* (Figure 10r).

3.3.4. Data and Techniques Used for Species Identification and Systematics

Initial Collection, Observation and Fixation in the Field

To identify individuals to the species level, specimens are first examined under a stereomicroscope. If conditions in the field permit, they should be examined and photographed alive to document colouration. Removing individuals from their tubes, specially serpulids, without any tube or specimen damage is rarely possible unless 0.05% phenol-seawater solution is used for several hours [229–231]. This method does not work in spirorbins (Bick pers. obs.) and it is unknown whether DNA is affected by phenol. It is important to examine and photograph intact tubes if they are broken to extract animals.

Fixation and preservation of specimens vary depending on further purposes of samples. Specimens aimed for a morphological study only are commonly fixed in a 4% solution of formaldehyde in sea water for 24–48 h, if possible after relaxation of individuals (in magnesium chloride). Animals are then rinsed in distilled water and preserved in 70–80% ethanol. For scanning electron microscopy (SEM) osmium tetraoxide is preferable as a fixative.

Specimens aimed for genetic sequencing should avoid all contact with formaldehyde as it degrades DNA, impeding amplification of the usual size fragments to be sequenced. Best procedures for DNA sequencing include fixation of fresh specimens in high concentration ethanol (the higher the better) and storing samples at 4–6 °C, or at least not in direct sunlight at room temperature. The ethanol should be changed at least once, preferably more often. RNA sequencing may need other protocols such as fixing with RNAlater. For integrative (morphological and molecular) studies, a tissue sample taken from a specimen should be fixed in ethanol and stored in a fridge or a freezer, while the rest of the specimen should be fixed in formalin and preserved in ethanol. Further reading on fixation and preservation of samples is found at Rouse and Pleijel [232].

Morphological Studies of Preserved Specimens

When examining freshly preserved or museum material lacking natural pigmentation, staining with methyl (or methylene) green (blue) helps to increase contrast and to reveal glandular patterns, including thoracic ventral shields and glandular girdles [24]. Examination of chaetae requires higher magnification (>100x), therefore, chaetae and noto- and neuropodia are dissected from the specimen, placed on a slide in a drop of ethanol, glycerin or permanent media, and covered with a cover glass. Applying gentle pressure on the cover glass ensures that uncini and chaetae lay in a lateral position. Using SEM is essential to reveal details of external features, such as ciliation, chaetal morphology and body wall microstructure, as well as anatomical features not easily distinguished in small specimens (appendages of the radiolar crown, for example). SEM is also an indispensable tool to examine tube ultrastructure, important for taxonomy of serpulids (reviewed by Ippolitov et al. [227]).

Structures of the radiolar crown, e.g., the rows of vacuolated cells supporting the radioles, dorsal and ventral lips, dorsal radiolar and pinnular as well as ventral radiolar appendages, and a parallel lamella in Sabellidae, dorsal lips and ventral filamentous appendages in Fabriciidae, are examined after fine sections with a sharp blade or, better, histological semi-thin sections are made and mounted on temporary or permanent slides, and stained with solutions, such as Mallory or Cason (e.g., [226]).

Morphometric characters such as counts (e.g., numbers of radioles or pinnules), measurements and proportions of soft body parts (e.g., thorax to abdomen ratio, length of dorsal lips) have traditionally been considered diagnostic for some taxa, but individuals may show sexual dimorphism [216], size and age-related variability ([70,233] or their size may be affected by anaesthetization and fixation techniques [234], which needs to be considered when comparing material.

For drawings to be made to scale, a camera Lucida attached to both the stereo- and the compound scopes is used. Traditional ink drawing (pencil sketching followed by India ink tracing) is currently supplemented or replaced by digital tracing of scanned pencil sketches, using a drawing pad (e.g., [235]). Advances in digital photography and universal availability of microscope-mounted digital cameras and Z-stacking software also resulted in photographs of live or preserved specimens, rather than line drawings being used in species descriptions. Use of SEM micrographs helps to illustrate both chaetal and soft body diagnostic features characters with precision and objectivity. Micro-computed tomography techniques have been proved useful in studies of internal anatomy in sabellids and serpulids ([49,236,237], and may offer taxonomically useful information.

Genetic Data

Genetic methods have been used in studies of Sabellida for nearly 20 years. The earliest publication by Patty et al. [238] used the C1 regions of 28S (123 bp) of 16 species to assess evolutionary relationships among Sabellidae. Kupriyanova et al. [45] published the first phylogeny of Serpulidae based on analyses of 18S rDNA, 28S rDNA, and morphological characters of 29 taxa. Combination of molecular (18S rDNA, the D1 region of 28S rDNA, and histone H3) and morphological datasets have been used to assess for the first time the relationships within Fabriciidae [27]. Other studies of Sabellida have used a limited number of molecular markers commonly used in polychaete systematics [239]. Currently, the number of sequences in GenBank is 246 for Fabriciidae, 814 for Sabellidae, and 2880 for Serpulidae, figures that indicate the relative larger effort put into molecular studies in serpulids compared to the other two families (Figure 11). In addition, "BOLD Systems [36] includes 443 barcodes (fragments of COI gene) of Serpulidae, 349 of these records are mined from GenBank and 19 diferent BOLD records are also shared with GenBank. Similarly, BOLD holds 692 COI sabellid barcodes, 399 of which are mined from GenBank and 148 BOLD records that were also uploaded in GenBank. Among the unidentified 105 sabellids there is an unknown number of sequences that belong to fabriciids as this database still follows the old clasitication. Sumarising, there are 75 serpulid and 145 sabellid/fabriciid COI sequences available in BOLD, in addition to those found in GenBank.

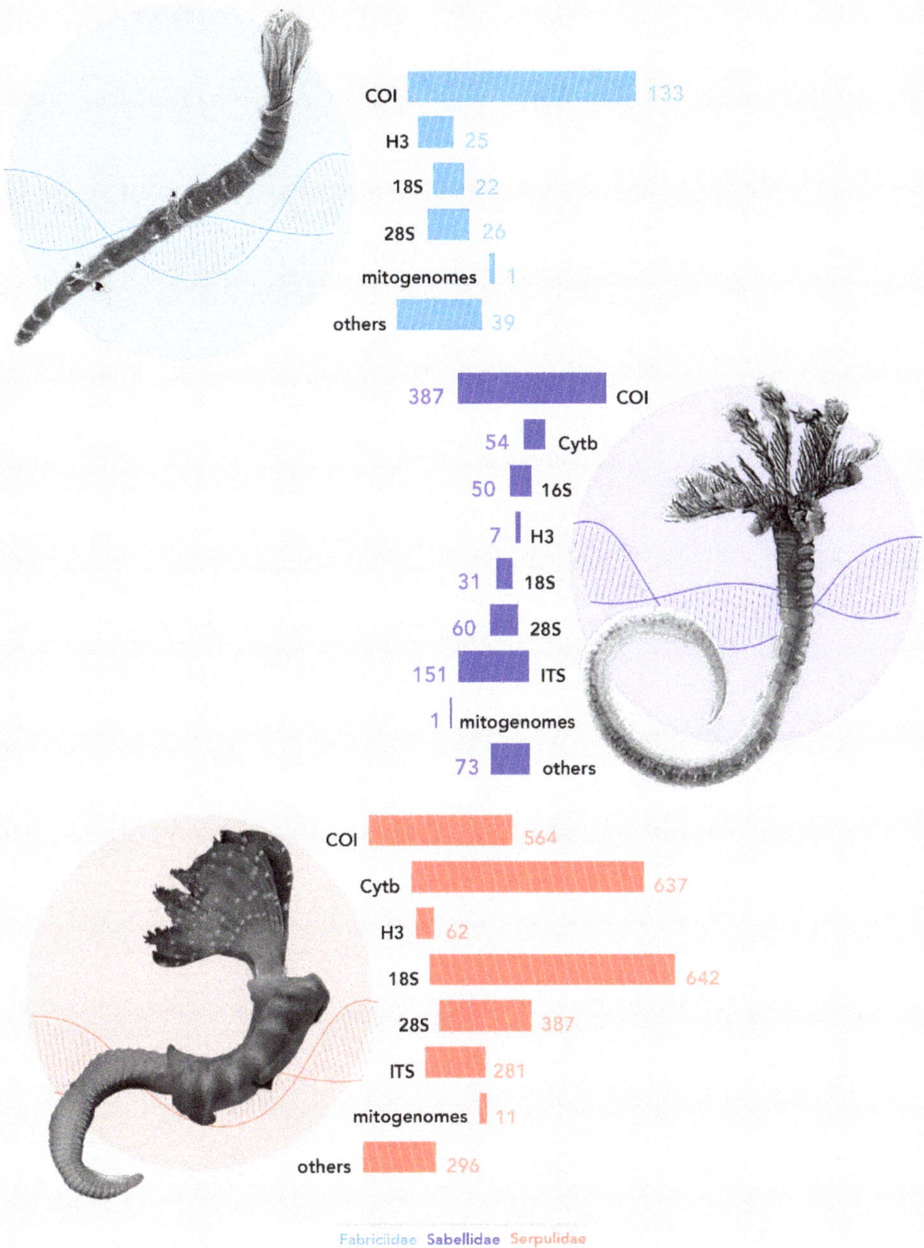

Figure 11. Number of sequences available in Genbank for Fabriciidae, Sabellidae; and Serpulidae.

The universal DNA barcoding fragment of COI gene is by far the most popular marker for the Fabriciidae, accounting for 54% (133 sequences) of the sequences available for this group (Figure 11), and Sabellidae, 38% (387 sequences) (Figure 11). For Serpulidae, despite all efforts (reviewed in Sun et al. [78]), no COI sequences had been available until

Carr et al. [240] reported six. However, Sun et al. [76], who developed genus-specific primers to generate COI sequences for 11 species of *Hydroides*, showed that "serpulid" sequences in Carr et al. [240] are likely from bacteria. Progress in COI barcoding in serpulids is mainly a result of new primer development [77]. Currently, the number of COI sequences for members of Serpulidae in Genbank is 564 (not including those mentioned problematic sequences [240], Figure 11).

Other markers widely used in systematics studies of Sabellida are nuclear 18S (642 sequences for Serpulidae, 31 for Sabellidae, 22 for Fabriciidae), 28S RNA (387 for Serpulidae, 60 for Sabellidae, 26 for Fabriciidae), mitochondrial cytochrome b (cytb) (637 for Serpulidae and 54 for Sabellidae), nuclear internal transcribed spacer (ITS2), ATP synthase, and Histone H3 (Figure 11). Mitochondrial 16S, widely used in sabellids (50 sequences in Genbank), has not been successfully amplified for serpulids.

Manayunkia is the fabriciid genus with most sequences in GenBank (179). The Sabellidae genera with most available sequences are *Branchiomma* (259), followed by *Sabella* (86) and *Sabellastarte* (61). Among the serpulids, *Hydroides* is the genus with the highest number of sequences in GenBank (1366), followed by *Spirobranchus* (601); all the rest have at least one order of magnitude less sequences available (Figure 12).

Although Sabellida are still behind other annelids in terms of genomic approach, several studies recently reported mitochondrial genomes and used transcriptomes for resolving systematics and evolutionary questions within this group. The first Sabellida mitochondrial genome was published for the serpulid *Spirobranchus giganteus* (Pallas, 1776) [241]. Mitochondrial genome sequences of ten *Hydroides* species have been reported in Sun et al. [242]. The mitochondrial genomes of *Sabella spallanzanii* (Gmelin, 1791) and freshwater fabriciid *Manayunkia occidentalis* were recently published [243,244]. The phylogeny by Tilic et al. [3] includes transcriptome sequences of 20 species of Sabellida (three fabriciids, 15 sabellids, and two serpulids), containing up to 3015 orthologous genes. Several other studies have also dealt with expressed sequence tag (EST) libraries and transcriptomes to address molecular mechanisms of larval settlement, or gene order and loss [245] in serpulids, adding up to 4205 sequences of mRNAs of larval cDNA library. With current fast adoption of the genomic approach, the number of sequences is expected to raise dramatically in the near future.

Species Delimitation and Identification

Application of molecular methods in combination with traditional morphological techniques or alone have expanded in the last two decades with regard to species delimitation in annelids in general, and in Sabellida, in particular. These methods are sustained by the definition of species as independently evolving entities (metapopulations), that are genetically (and often phenotypically) distinct [246,247]. Thus, species are expected to be reciprocally monophyletic clusters, morphologically distinct and/or genetically divergent, as a result of evolutionary forces applied to closely related lineages. Molecular-based approaches have not only improved species delimitation by providing additional evidence to morphological taxonomy, but also helped to reveal cryptic (only genetically distinct) species [248].

In Fabriciidae, even though species appear morphologically similar due to the animal's small size and the diagnostic features being difficult to recognize by non-specialists, molecular approach to species identification and delimitation is still uncommon. Only one species, *Manayunkia occidentalis*, has been described based mainly on genetic data [216].

In Sabellidae, boundaries between species within the genera *Amphiglena*, *Branchiomma*, *Parasabella*, *Pseudobranchiomma* Jones, 1962, *Sabellastarte* and *Sabellomma* Nogueira, Fitzhugh and Rossi, 2010 were assessed with molecular and morphological data [70,72–74]. Results revealed cryptic diversity hidden in species complexes and helped to assess the diagnostic features traditionally used for morphological species identification [2,27,72–74,103,140].

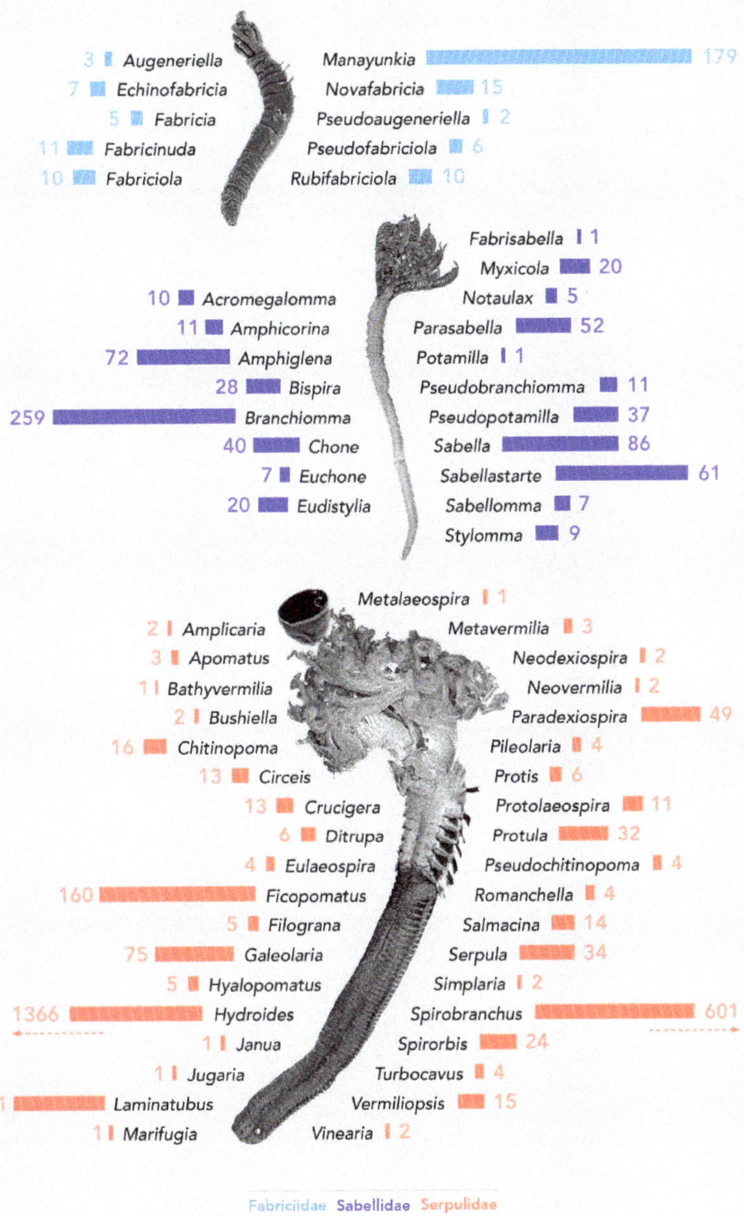

Figure 12. Number of DNA sequences available in GenBank for genera in Fabriciidae, Sabellidae, and Serpulidae.

In Serpulidae, the study of Halt et al. [249] was the first to name a new species in Sabellida without morphological indicators, after analyses of DNA revealed two cryptic species with non-overlapping distributions within *Galeolaria caespitosa* Lamarck, 1818. Another study revealed three genetic species with overlapping distributions within *Ficopomatus enigmaticus* (Fauvel, 1923), two cryptic and one morphologically distinct [204]. A combination of molecular and morphological data helped to partially resolve species complexes within the genera *Hydroides* [203,250,251] and *Spirobranchus* [84,193,252].

The idea of 'DNA barcoding' is that a species can be uniquely characterised by a short DNA fragment and then identified by comparing such a fragment from an unknown specimen to a reference DNA sequence [253]. Initially, a 650 base pair fragment of the mitochondrial COI was proposed as a standard barcoding gene for animals [253] because of its variability among closely related taxa and supposed ease of amplification. Later, however, a number of other mitochondrial markers (cytb, e.g., [81–84,193,204,249,252] and nuclear (ITS, [82,249]) have been used, especially in serpulids where amplification of COI proved to be challenging. The first attempts to use DNA data alone while ignoring any morphological and biogeographic evidence to identify potentially invasive serpulids [254], were rather a failure. The authors mistakenly claimed discovery of an Australian species *Spirobranchus taeniatus* (Lamarck, 1818) (mostly likely *Spirobranchus triqueter*) and North American *Serpula columbiana* Johnson, 1901 (almost certainly *Serpula vermicularis*) attached to drifting marine litter in the Mediterranean, after comparing (minimum 97% nucleotide identity was accepted) partial sequences of the conservative (thus unsuitable for species-level barcoding) 18S gene with the limited set of sequences available in GenBank. Similarly, Langeneck et al. [255] criticised another paper by the same authors (Rech et al. [256]) who identified specimens associated with floating debris in the Lagoon of Venice as *Hydroides sanctaecrucis*, suggesting its presence in the Mediterranean might have been overlooked due to misidentification as common *H. dianthus*. Rech et al. [256] again did not examine the morphology of the specimens and used 18S sequences for identification, accepting an identity ≥97% with sequences of *H. sanctaecrucis* in GenBank. However, 18S rDNA sequences are ill-suited for molecular identification because they have identity close to 100% in closely related species. The low sequence identity shows that the specimens in Rech et al. [256] study did not belong to *H. sanctaecrucis*, and the species name should be removed from checklists of species non-indigenous for the Mediterranean [255].

3.4. Ecology, Distribution and Biogeography

3.4.1. Ecology

Fabriciidae

A review of the ecology and biology of Fabriciidae was published recently [4]. Most fabriciids occur in intertidal and subtidal zones, mainly in sheltered areas on sandy, muddy or rocky sediments, in mangroves, on red and green algal mats, and in seagrass beds, with low benthic species richness.

Fabriciids are mainly distributed in marine and brackish ecosystems worldwide, but species of the genus *Manayunkia* are also common in freshwater, and even hypersaline lakes, where they survive salinities of 82 psu for several months [257]. The abundance of some species tends to be very high in habitats with low biodiversity. The highest abundances of *Fabricia stellaris* and *Manayunkia aestuarina* (over 10^6 ind. m^{-2}) have been reported in physiologically stressful conditions, such as sediments with a high organic matter content and waters of highly variable salinities [258–261]. A reduction in organic matter content from 1.8% to 1.0% in the Baltic Sea was followed by a reduction in fabriciid abundance from 16 000 to 6000 ind. m^{-2} [262]. Giangrande et al. [263] found five fabriciid species in a coastal Mediterranean system naturally acidified by carbon dioxide vent emissions. Among these, *Parafabricia mazzellae* Giangrande, Gambi, Micheli and Kroeker, 2014 and *Brifacia aragonensis* Giangrande, Gambi, Micheli and Kroeker, 2014 were most abundant even in the extremely low pH zone (pH 6.6–7.2).

Fabriciidae species produce flexible tubes consisting of the finest sediment particles stabilized by mucus. Detritus might also be deposited on the outside of the tubes. Fabriciids are not obligatory tube dwellers and they can voluntarily leave their tubes and build new ones. When outside, they crawl with the posterior end in front, while the radiolar crown is folded up and dragged behind [264,265].

Fabriciids are suspension-feeders like other Sabellida, but *Manayunkia* spp. are deposit feeders. In addition to detritus, they ingest bacteria, heterotrophic protozoa, cyanophyceans and diatoms. The sizes of the ingested particles range from 1–2 μm to 2–7 μm and occasionally reach up to 20 μm [265].

Some fabriciids are commensals of molluscs, e.g., the freshwater *Brandtika asiatica* Jones, 1974 and *Monroika africana* [215,266], or the marine *Rubifabriciola tonerella* (Banse, 1959) and *Novafabricia infratorquata* (Fitzhugh, 1983) [231], but these species have been also found in other substrates. Another example of commensalism is the occurrence of peritrichous ciliates on anterior chaetigers in *Manayunkia aestuarina* [4]. *Manayunkia speciosa* Leidy, 1859 is an obligate invertebrate host of the myxozoan parasites *Ceratonova shasta* (Noble, 1950) and *Parvicapsula minibicornis* Kent, Whitaker and Dawe, 1997, which cause ceratomyxosis in salmon and trout in North America [267,268].

Sabellidae

A review of ecology and biology of sabellids was recently published [24]. Sabellids are able to inhabit either hard or soft sediments. Species of *Amphiglena, Bispira, Perkinsiana, Pseudobranchiomma, Sabellomma, Sabellastarte* and *Stylomma* Knight-Jones, 1997 mainly inhabit littoral hard substrates, as epibionts of algae, or associated with biogenic structures, including live coral or rubble [2,68,73,74,269,270]. Some species of *Perkinsiana* and *Pseudopotamilla* associated with dead coral and limestone sediments are capable of actively boring into the calcium carbonate [271,272].

Several species of *Acromegalomma, Amphiglena, Branchiomma, Eudistylia, Notaulax, Parasabella, Pseudobranchiomma* and *Sabella* are abundant in biofouling communities [72, 74,101,147,273–276]. Some of the soft bottom species need large and stable enough surface (shell, rock, holdfast or root) to attach to build their tubes. This is characteristic of species of *Branchiomma, Parasabella, Bispira manicata* (Grube, 1878), *Acromegalomma*, or the Mediterranean *Sabella spallanzanii*, a species associated, in natural conditions, to *Posidonia* K. Koenig seagrass roots [73,129,133,277,278]. However, other soft-bottom species can build tubes within the sediment grains, like *Euchonoides meone* Magalhães, Bailey-Brock and Tovar-Hernández, 2020. This species is found near a sewage outfall in Hawaii, reaching 141,046 ind/m^2, the highest densities ever reported for Sabellidae [279].

Sabellids have been only recently found in chemosynthesis-based environments, such as hydrothermal vents, methane seeps and organic falls, and their diversity in such habitats is poorly understood. *Bispira wireni* (Johansson, 1922) was reported from a hydrothermal vent from Okinawa, Japan [280]. *Jasmineira* sp. and an undetermined sabellid colonized bone and sunken wood in the southwestern Indian Ridge [281]. Unidentified sabellids have been reported from methane seeps in the Gulf of Mexico [282] and Chile [283]. Recently, an undescribed species of *Bispira* was found at a deep-sea cold seep off the Pacific coast of Costa Rica [284].

Although the group is typically marine, a few exceptional species have adapted to brackish and even fresh water environments. The most remarkable example is the exclusively freshwater genus *Caobangia*. Euryhaline sabellids are, for example, the Australian *Desdemona aniara* Hutchings and Murray, 1984 and *Laonome triangularis* Hutchings and Murray, 1984, Indian *Potamilla leptochaeta* Southern, 1921, American *Aracia sinaloae* Tovar-Hernández, 2014, and the cryptogenic *Desdemona ornata* Banse, 1957, *Euchone limnicola* Reish, 1959, and *Laonome xeprovala* Bick and Bastrop, 2018 [68,102,285–290]. Some species typically found in fully marine conditions are tolerant to brackish water conditions, e.g., members of *Euchone, Branchiomma* and *Parasabella* [147,291], while *Laonome calida* Capa, 2007 and *L. albicingillum* Hsieh, 1995 have been reported in environments ranging from fully marine to freshwater [68,292–294].

With the exception of *Glomerula piloseta*, that inhabits calcareous tubes, all sabellids build tubes by secreting the mucous base and enforcing it with different size particles they attach, including mud, sand, feces or biogenic fragments [24]. Smaller species are more liable to leave the tubes if disturbed and can build new ones [295], but larger species, even if capable to build new tubes, tend to inhabit the same one for longer periods or their whole lives [270,296,297].

Until very recently, sabellids have been found mostly in areas of high productivity and assumed to be obligatory suspension feeders [298]. However, a sabellid-bacterial symbiosis, fueled by methane, between a still undescribed species of *Bispira* and methane-oxidizing Methylococcales bacteria, has recently been reported from a methane seep [284]. This makes *Bispira* a new addition to the list of annelids (including Siboglinidae and two new serpulids of the genus *Laminatubus*, see below) relaying on chemosynthetic symbionts for nutrition.

Associations of sabellids with other organisms relate to their ability to bore into calcium carbonate. The seven species of *Caobangia* are commensals or parasites of freshwater gastropods and bivalves in rivers of southeastern Asia [289,299]. *Terebrasabella heterouncinata* Fitzhugh and Rouse, 1999 bores into the shells of marine gastropods, including abalones and limpets, in South Africa and California [66,300,301]. *Notaulax montiporicola* Tovar-Hernández and ten Hove, 2020 associated with the living coral *Montipora nodosa* (Dana, 1846) does not bore into coral, but uses crevices to settle and allows coral tissue to grow around its tube [101].

Serpulidae

A review of ecology and biology of serpulids was recently published [214]. Serpulids are typical on hard substrates in all marine environments. Inhabitants of areas with predominantly soft-sediments always attach to rocky outcrops, stones and shells, and can deal with high sedimentation rate by building their tubes upwards to avoid being buried in the sediment [302]. Many serpulids are notorious opportunistic foulers, capable of colonising any available hard substrates. The ability to settle and build large aggregations on human-made surfaces makes serpulids important and troublesome members of fouling communities. However, some show high habitat selectivity, resulting from non-random larval settlement and juvenile survival (reviewed by Kupriyanova et al. [303]). A few unusual serpulids are pre-adapted to living unattached on soft substrates in subtidal-shelf (*Ditrupa*) [110] or bathyal-abyssal (*Bathyditrupa* and *Spirodiscus*) environments [107]. However, larvae of free-living *Ditrupa* need to attach initially to small particles during settlement and metamorphosis to start building the tube [304].

Serpulids are some of recognizable animals to inhabit the periphery of seeps and hydrothermal vents. *Laminatubus alvini* ten Hove and Zibrowius, 1986 and *Protis hydrothermica* ten Hove and Zibrowius, 1986 were the first serpulids to be formally described from vent communities of East Pacific Rise and *Laminatubus* is known from other seeps (e.g., Pescadero Transform Fault, Gulf of California) and vents, e.g., Alarcon Rise, Gulf of California [305]. *Hyalopomatus mironovi* Kupriyanova, 1993 and *Protis* sp. were reported from hydrothermal vents of North Fiji [306]. Less is known about seep-associated serpulids, as more species have been reported from fossil than modern hydrocarbon seeps [307]. Serpulids (tentatively identified as members of *Neovermilia* Day, 1961) have been reported from cold seep communities in Nankai Trough [308], the Peruvian active margin [309], the Terevaka ridge [310], the Peru Trench, Middle American Trench off Mexico [311], and the Barbados prism [312].

Although serpulids are predominately marine, some species of *Hydroides* tolerate mixohaline conditions (e.g., [313], for *H. elegans* (Haswell, 1883)), while representatives of *Ficopomatus* can cope with a wide range of salinities and are common in brackish-water environments world-wide [110]. *Marifugia cavatica* Absolon and Hrabě, 1930, closely related to *Ficopomatus*, is the only known truly fresh-water serpulid, inhabitant of subterranean caves of the Dinaric Alps [46].

Serpulids produce their calcareous tubes using a pair of calcium carbonate secreting glands located on the collar. As obligate tube dwellers, they never leave their tubes and cannot build new ones if removed. Adult serpulids lie in the tube with their dorsum facing the substrate and locomotion is limited to partial emergence from and withdrawal into the tube [314]. Hiding behaviour is a common antipredator tactic, and animals may adjust the durations of such behaviour to current benefits and costs [315].

As suspension-feeders, serpulids show varying abilities in particle sorting and clearance rates. High planktonic biomass removal (>50% of initial standing stock) and significant differences in clearance for different components of the community by reef-building *Ficopomatus enigmaticus* indicate that the serpulid can regulate planktonic biomass and promote changes in plankton community structure [316]. *Ditrupa arietina* (Müller, 1776) lives unattached in soft sediments and ingests diatoms, haptophytes, bacteria and cyanobacteria ranging from 1 to 50 μm in size, and the origin of the food is both planktonic and benthic [317]. Recently metanotrophy (similar to that found for members of *Bispira* sp.) has been reported, as a result of symbiosis between two species of the genus *Laminatubus* (*L. joicebrooskae* Rouse and Kupriyanova, 2021 and *L. paulbrooksi* Rouse and Kupriyanova, 2021) and methane-oxidizing Methylococcales bacteria, from a seep off Costa Rica [284].

While a large number of serpulids is found in coral reefs on coral rubble (e.g., [318]: Fiji, [319]: Okinawa, Japan, [179]: Queensland, Australia), only some taxa, such as *Floriprotis* Uchida, 1978, *Pseudovermilia*, *Spirobranchus* and *Vermiliopsis* spp., are found in association with live corals. Some *Spirobranchus* species are reported as obligate associates of corals to the extent that their successful settlement occurs only on live corals (e.g., [320–322]), although recent observations indicate that while *Spirobranchus* larvae have a preference for live corals, they will survive on other substrates [83]. Many serpulids form epizootic associations with other invertebrates, mostly molluscs, crustaceans, bryozoans, and sponges. For example, *Hydroides spongicola* Benedict, 1887 occurs symbiotically in the chemically aggressive do-not-touch-me sponge *Neofibularia nolitangere* (Duchassin and Michelotti, 1864), while *Circeis paguri* Knight-Jones and Knight-Jones, 1977 is associated with hermit crabs (reviewed in [303]). Spirorbins are commonly found in specific epiphytic associations with macrophytes and their settlement can be stimulated by algal extracts [323,324].

3.4.2. Biogeography, Distribution and Bathymetry

Members of the Sabellida are found world-wide and, like most polychaetes, for much of the 20th century were assumed to have naturally wide, even cosmopolitan, distributions (e.g., [166,194,325]). Darling and Carlton [326] use the term eucosmopolitan to refer to the species with naturally broad distribution (found in two or more oceans). However, recent studies overwhelmingly show that 'cosmopolitan' taxa represent complexes of either morphospecies or cryptic species (reviewed in [327]). All evidence to date suggests that polychaetes have restricted natural geographic and bathymetric distributions, thus taxa reported with wide ranges should be treated as potential species complexes. Two general exceptions to the rule of restricted distributions are deep-sea and invasive species. Ranges of deep-sea polychaetes are expected to be wider than those found in shallow seas as a result of stable environmental conditions over wide distances, a traditional view (e.g., [328]) also supported by recent studies (e.g., [329,330]). Annelids that are easily translocated by anthropogenic means can establish and become invasive in remote localities, consequently expanding their ranges [327]).

Biodiversity patterns may be influenced not only by intrinsic ecological and historical factors, but also by 'extrinsic factors' *sensu* Giangrande and Licciano [331]. When a group is studied by a few specialists working in a particular area, species distribution may correlate with that of the specialists ('author effect'). The concentration of taxonomic expertise in some regions may increase the number of species in those areas compared to less studied areas ('concentration effect').

Fabriciidae

Fabriciids have been described from all marine realms except for the Tropical Eastern Pacific (Table S1). Temperate Northern Atlantic is the province with the highest number of fabriciids (16 species), followed by Tropical Atlantic (13 species), Temperate Northern Pacific and Western Indo-Pacific (11 species each), Central Indo-Pacific (10 species) and Temperate Australasia (six species), Temperate Southern Africa (four species), Eastern Indo-Pacific (two species), and Actic, Southern Ocean, and Temperate Southern America, each with a single species (Table S1). Most of the 4739 georeferenced occurrences in GBIF are also from the Temperate North Atlantic Realm and the Tropical Atlantic (Figure 13) and refer to members of *Fabricia*, *Fabricinuda* and *Manayunkia* [34]. There are records from neither Tropical Eastern Pacific nor from Western Indo-Pacific, the west coast of South America, and the west coast of Africa. The historical records until 1999, compiled by Giangrande and Licciano [331], showed 55.8% of the total fabriciid species are found in the tropics. Interestingly, the currently available information, based on species type localities show sthat most species were described from the Atlantic Ocean, at all latitudes. Explanation to latest results could be 'concentration effect', with taxonomic expertise accumulated at both sides of the Atlantic, and poor state of knowledge elsewhere.

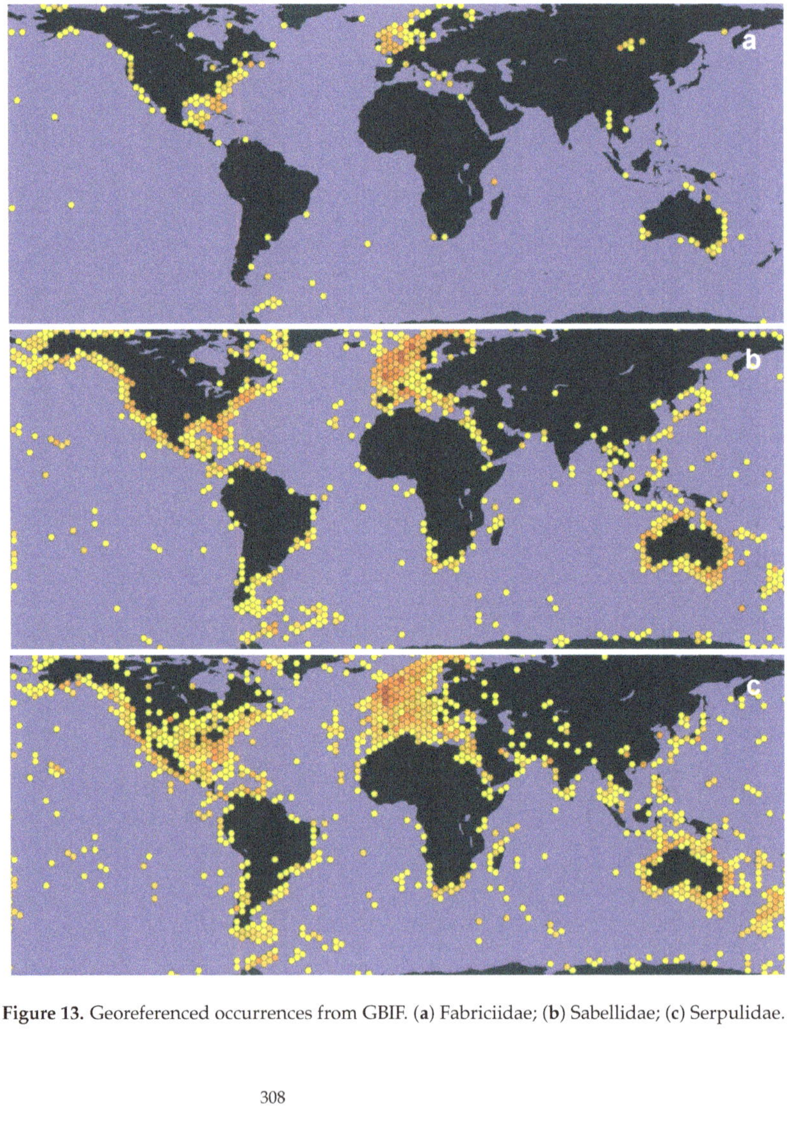

Figure 13. Georeferenced occurrences from GBIF. (**a**) Fabriciidae; (**b**) Sabellidae; (**c**) Serpulidae.

The species of most genera are distributed almost worldwide. The adaptability of Fabriciidae to different environments and the wide distribution of some taxa can be shown by the example of the genus *Manayunkia*. The ten extant *Manayunkia* species occur worldwide in marine, brackish and freshwater habitats, as well as hypersaline lakes. Their common ancestor was most likely already present in marine habitats [4]. There are ten species adapted to freshwater conditions: *M. speciosa* and *M. occidentalis* (Nearctic); *M. zenkewitschii, M. baicalensis* (Nusbaum, 1901) and *M. godlewskii* (Palearctic) [216,217]. One species, *M. athalassia* Hutchings, Dekker and Geddes, 1981, was found in hypersaline lakes in Australia [257], another species, *M. mizu* Rouse, 1996, in marine habitats [219,332] and three species, *M. aestuarina*, *M. caspica* Annenkova, 1928), *M. brasiliensis* Banse, 1956, in brackish environments [333,334] (Nogueira pers. obs.).

Most Fabriciidae occur in intertidal and subtidal waters. Only certain *Pseudofabriciola* species (e.g., *P. californica* Fitzhugh, 1991, *P. filamentosa* (Day, 1963), *P. filaris* Fitzhugh, 2002, and *P. longipyga* Fitzhugh, Giangrande and Simboura, 1994), *Fabricinuda longilabrum* Fitzhugh, 2002 and *Raficiba barryi* occur between depths of 50 m and 335 m [53,58,63,335]. The record of *Fabricia sabella* (Ehrenberg, 1937) reported by Hartman (1965) from 1000 m off New England needs a revision because the nominal *F. stellaris* is a brackish-water species from the Baltic Sea. In the southwest Atlantic, exceptionally, two yet undescribed species provisionally assigned to the genera *Fabriciola* and *Novafabricia* have been found at 4500 m [220] and one other of *Fabriciola* from the Okhotsk Sea below 2000 m (Table 5).

Table 5. Deepest records for members of Sabellida below 1000 m.

Taxon	Depth (m)	References
FABRICIIDAE		
Fabriciola sp. (Okhotsk Sea)	>2000	Alalykina 2020
Fabriciola sp.	4600	Baumhaker 2012
Novafabricia sp.	4600	Baumhaker 2012
SABELLIDAE		
Jasmineira bermudensis Hartman, 1965	1000	Original description
Potaspina australiensis Capa, 2007	1000	Original description
Euchone magna (Fauchald, 1972)	1071	Original description
Perkinsiana assimilis (McIntosh, 1885)	1100	Original description
Chone gracilis Moore, 1906	1244	Méndez 2006
Bispira wireni (Johansson, 1922)	1335	Capa et al. 2013
Potamethus filiformis Hartmann-Schröder, 1977	1430	Original description
Pseudopotamilla intermedia Moore, 1905	1682	Original description
Bispira sp. (Costa Rica)	1887	Goffredi et al. 2020
Chone sp. (Okhotsk Sea and N. Pacific abyss)	>2000	Alalykina 2020
Euchone sp. (Okhotsk Sea and N. Pacific abyss)	>2000	Alalykina 2020
Jasmineira sp. (Okhotsk Sea and N. Pacific abyss)	>2000	Alalykina 2020
Potamethus sp. 1 (Okhotsk Sea and N. Pacific abyss)	>2000	Alalykina 2020
Potamethus sp. 2 (Okhotsk Sea)	>2000	Alalykina 2020
Potamethus singularis Hartman, 1965	2000	Original description
Potamilla neglecta (Sars, 1851)	2030	Hansen 1882
Potamethus malmgreni (Hansen, 1878)	2222	Original description
Euchone cf. *incolor* Hartman, 1965	2500	Alalykina 2020
Fabrisabella similis Fauchald, 1972	2520	Original description
Potamethus scotiae (Pixell, 1913)	2578	Original description
Euchone papillosa (Sars, 1851)	2900	Uschakov 1955; Levenstein 1969
Jasmineria pacifica Annenkova, 1937	2900	Original description
Jasmineira schaudinni Augener, 1912	3500	Augener 1912 (abyssal, no depth given); Jirkov 1982, 2001
Chone infundibuliformis Krøyer, 1856	3521	Wesenberg-Lund 1950
cf Sabellidae species 1 (Clarion-Clipperton Zone)	4029	Amon et al. 2017
Potamethus mucronatus (Moore, 1923)	4131	Original description, as *Notaulax*

Table 5. *Cont.*

Taxon	Depth (m)	References
Potamilla abyssicola Uschakov, 1952	4200	Original description; Levenstein 1961 1969; Alalykina 2020
Potamethus spathiferus (Ehlers, 1887)	4360	Fauvel 1914
Euchone incolor Hartman, 1965	4862	Original description; Hartman 1971
?*Potamethus* sp. Mozambique Basin	5068	Hartman 1971
Potamehus dubius (Eliason, 1951)	5860	Original description
Jasmineira sp. Japan	6207	Levenstein 1961b
Potamethus singularis Hartman, 1965	6023	Original description; Hartman 1971
Sabellidae sp. Pacific Ocean	8042	Levenstein 1969; Lemche et al. 1976
Potamilla sp. Kurile-Kamchatka Trench	8100	Uschakov 1952; Belyaev 1989
Jasmineira sp. Kermadec Trench Trench	8300	Kirkegaard 1956; Hartman and Fauchald 1971; Belayev 1972
Potamethus sp. Izu-Bonin Trench	8735	Belayev 1989
Jasmineira filitovae Levenstein, 1961 (as *Potamethus*)	9735	Levenstein 1969, 1973; Belyaev 1989
SERPULIDAE		
Laminatubus joicebrooksae Rouse and Kupriyanova 2021	1011	Original description
Hyalopomatus madreporae Sanfilippo, 2009	1146	Original description
Neovermilia falcigera (Roule, 1898)	1580	Zibrowius and ten Hove 1987
Metavermilia ogasawaraensis Nishi, Kupriyanova and Tachikawa, 2007	1603	Original description
Zibrovermilia zibrowii Kupriyanova and Ippolitov, 2015	1710	Original description
Hyalopomatus dieteri Kupriyanova and Ippolitov, 2015	1980	Original description
Hyalopomatus biformis (Hartman, 1960)	1982	Kupriyanova and Nishi 2010
Metavermilia zibrowii Bailey-Brock and Magalhães, 2012	2013	Original description
Vermiliopsis notialis Monro, 1930	2016	Averintsev 1974
Bushiella (Jugaria) atlantica (Knight-Jones, 1978)	2100	Original description
Bathyvermilia islandica Sanfilippo, 2001	2399	Original description
Filogranula stellata (Southward, 1963)	2464	Ben-Eliahu and Fiege 1996
Hyalopomatus variorugosus Ben-Eliahu and Fiege, 1996	2474	Original description
Laminatubus paulbrooksi Rouse and Kupriyanova, 2021	2478	Original description
Protis browni (Pixell, 1913)	2585	Original description
Protis hydrothermica ten Hove and Zibrowius 1986	2620	ten Hove and Zibrowius 1986
Hyalopomatus marenzelleri Langerhans, 1884	2800	Zibrowius 1968, 1969, 1977
Laminatubus alvini ten Hove and Zibrowius, 1986	2842	Original description
Neovermilia cf. *sphaeropomata* (Benham, 1927)	3261	Rouse and Kupriyanova 2021
Apomatus globifer Théel, 1878	3384	Uschakov 1957
Hyalopomatus claparedii Marenzeller, 1878	3550	Kupriyanova and Jirkov 1997
Hyalopomatus jirkovi Kupriyanova, 1993	3949	Kupriyanova et al. 2011
Spirodiscus grimaldii Fauvel, 1909	4124	ten Hove and Kupriyanova 2009
Bathyvermilia kupriyanovae Bastida-Zavala, 2008	4190	Original description
Apomatus similis Marion and Bobretzky, 1875	4400	Fauvel (1909), Fauvel (1914)
Spirodiscus groenlandicus (McIntosh, 1877)	4440	Kupriyanova and Ippolitov 2015
Bathyvermilia zibrowiusi Kupriyanova, 1993	4550	Kupriyanova et al. 2011
Hyalopomatus sikorskii Kupriyanova, 1993	4550	Kupriyanova et al. 2011
Protis simplex Ehlers, 1887	4810	Knight-Jones et al. (1997)
Protis polyoperculata Kupriyanova, 1993	5110	Original description
Hyalopomatus mironovi Kupriyanova, 1993	5216	Rouse and Kupriyanova 2021
Protis arctica (Hansen, 1879)	5300	Zibrowius 1969
Bathyvermilia challengeri Zibrowius, 1973	5719	Original description
Bathyvermilia langerhansi Fauvel, 1909	5987	Eliason 1951
Bathyvermilia gregrousei Kupriyanova and Ippolitov, 2015	6050	Original description
Nidificaria levensteinae (Bailey-Brock and Knight-Jones, 1977)	6096	Original description
Bathyditrupa hovei Kupriyanova, 1993	6330	Kupriyanova et al. 2011
Protis sp. 2	8345	Kupriyanova et al. 2014
Protis sp. 1	9735	Kupriyanova et al. 2014

Sabellidae

Sabellids have been described from all marine realms and all seven members of *Caobangia* are known so far are from the Indo-Malay limnic realm. In the current analysis, the Temperate Northern Atlantic is the realm with the highest number of sabellids described (125 species), which represents 24% of the sabellids described worldwide, followed by the Temperate Northern Pacific (79 species, 15%), Central Indo-Pacific (58 species, 11%), Tropical Atlantic (56 species, 11%), Temperate Australasia (37), Western Indo-Pacific (20 species), Temperate South America (28 species), Southern Ocean (24 species), Temperate Southern Africa (18), Arctic (16 species), Tropical Eastern Pacific (15 species), Eastern Indo-Pacific (8), 23 species non-marine and eight species with unknown type locality (Table S2). Most of the 117 073 georeferenced occurrences in GBIF are from the Temperate Northern Atlantic and the Tropical Atlantic realms, and records belong to members of genera *Euchone*, *Jasmineira*, *Chone* and *Sabella* [34] (Figure 13). Other realms with large representation of sabellids records are Tropical Atlantic, the Arctic and Temperate Australasia. The number of described species and number of records is higher in the Atlantic than in any other ocean, the result that contradicts the patterns showing an increase in sabellid species richness towards the tropics, and mainly in the Indo-Pacific [331].

Some genera with few species, such as *Anamobaea* or *Stylomma*, and the freshwater *Caobangia*, are exclusive of tropical environments [69,158,229], and others, such as *Branchiomma*, *Bispira*, *Acromegalomma*, *Notaulax*, *Sabellastarte*, and *Sabellonga*, have either tropical or temperate distribution [149,157,336–338]. A few genera show a significant preference for colder waters and are either well represented at greater depths or in higher latitudes (e.g., members of *Chone*, *Euchone*, and *Jasmineira*). The Antarctic region is richer in number of genera and species than the Arctic [24] (Figure 7), and genera, such as *Perkinsiana*, are mainly distributed in the Southern Ocean [212].

Sabellids found below 6000 m have all been reported from the Western Pacific Ocean (Table 5) and include species of the genera *Jasmineira*, *Potamethus* and *Potamilla* Malmgren, 1866 [339–344]. Species reported in the abyssal zone (2000–6000 m deep) include members of the genera *Chone*, *Euchone*, *Fabrisabella*, *Jasmineira*, *Potamilla*, and *Potamethus* (Table 5). The genus *Potamethus* is the most speciose deep-sea taxon (Table 5). *Jasmineira filitovae* Levenstein, 1961 is the deepest record (9735 m). Sabellids reported from between 1000 and 2000 m depths include species of *Bispira*, *Chone*, *Euchone*, *Jasmineira*, *Perkinsiana*, *Potamethus*, *Potaspina* and *Pseudopotamilla* [280,284,345–348]. Among all sabellids recorded below 1000 m, 13 have been identified to the genus level only, most probably constituting new species.

Serpulidae

Serpulids have been described from all marine realms. The realm with highest number of species is the Temperate Northern Atlantic (108 species), followed by the Temperate Northern Pacific (92), Central Indo-Pacific (74), Tropical Atlantic (71), Western Indo-Pacific (52), Temperate Australasia (41), Tropical Eastern Pacific (28), Southern Ocean (23), Temperate South America (21), Arctic (16), Temperate Southern Africa (15), Eastern Indo-Pacific (8), and nine species with unknown type locality (Table S3). Of the 107 859 georeferenced records in GBIF (2020), more than half are within the Temperate Northern Atlantic (mainly identified as *Spirobranchus*, *Hydroides*, *Ditrupa* and *Spirorbis*) and following realms with highest occurrences are the Temperate Australasia (mainly *Spirobranchus*, *Hydroides*, *Galeolaria* and *Serpula*) and the Tropical Atlantic (*Hydroides*, *Spirobranchus*, *Vermiliopsis* and *Pseudovermilia*) (Figure 13). There were no previous analyses of global serpulid distribution patterns to compare with these data.

Serpulids are common inhabitants of intertidal, subtidal and shelf locations, but they can occur at all latitudes from intertidal to hadal depths. Spirorbin bathymetric distribution ranges from littoral to abyssal depths, but they are most commonly found in the sublittoral zone. The best known representatives of the genera *Ficopomatus*, *Galeolaria*, *Hydroides*, *Salmacina* Claparède, 1870, *Serpula*, *Spirobranchus* and *Vermiliopsis* are inhabitants of shallower waters (below 500 m), and so are representatives of less known and more cryptic genera, such as, for example, *Chitinopoma, Floriprotis, Josephella* Caullery and Mesnil, 1896, *Metavermilia, Pomatostegus, Pseudochitinopoma* Zibrowius, 1969, *Pseudovermilia, Rhodopsis* Bush, 1905, *Semivermilia* ten Hove, 1975 and *Spiraserpula.* Some genera, e.g., *Apomatus, Filogranula, Neovermilia* and *Protula* may include both subtidal and bathyal species.

Serpulids found below 2000 m were reviewed by Zibrowius [349], who corrected Hartman's [339] compendium of abyssal polychaetes by removing taxa typical of subtidal and shelf depths (*Hydroides, Ditrupa, Placostegus, Serpula, Spirobranchus*). As a result, he listed 25 species, including one unidentifiable specimen from Kermadec Trench (6620–6730 m, [350]). Belyaev [344] added two unidentified hadal specimens from 6410–6757 m (Aleutian Trench) to 9715–9735 m (Izu-Bonin Trench), the latter being the deepest record for a serpulid. In their review, Paterson et al. [351] list only 26 serpulids from over 2000 m world-wide, including five species from depths beyond 3500 m, all described by Kupriyanova [352,353] from Kuril-Kamchatka Trench alone. Sanfilippo [354], Kupriyanova et al. [355,356] Kupriyanova and Nishi [357], Bailey-Brock and Magalhães [184], Kupriyanova and Ippolitov [107], and Rouse and Kupriyanova [358] recently provided additional records and descriptions of new deep-sea serpulid taxa. Kupriyanova et al. [356] reviewed and revised hadal records below 5000 m, demonstrating that the deepest serpulid records (8345 and 9735 m) reported in Belyaev's book [344]) belong to the genus *Protis.*

Currently, 36 named species have published records below 1000 m, 28 of them are reported from below 2000 m (Table 5), although Kupriyanova et al. [356] lists eight records in open nomenclature (*Bathyvermilia* sp., *Hyalopomatus* sp., *Protis* sp., Serpulidae gen. sp. A and B). In summary, serpulids from bathyal and abyssal depths (>1000 m) are found in the genera *Apomatus, Bathyditrupa, Bathyvermilia* Zibrowius, 1973, *Bushiella (Jugaria)* Knight-Jones, 1978, *Filogranula, Hyalopomatus, Laminatubus, Metavermilia, Neovermilia, Nidificaria* Knight-Jones, 1984, *Spirodiscus, Protis,* and *Zibrovermilia* Kupriyanova and Ippolitov, 2015, but only species of *Bathyditrupa, Bathyvermilia, Hyalopomatus,* and *Protis* are found in the abyss, also penetrating into the hadal zone. Non-operculate *Protula* and operculate *Apomatus* are often confused with non-operculate and operculate *Protis* sp., so that abyssal records of supposed *Protula* and *Apomatus* might belong to *Protis* [356].

3.5. Non-Indigenous and Invasive Species

Shallow-water Sabellida, due to their sedentary tubicolous lifestyle, are common members of biofouling communities and are easily translocated by anthropogenic means, i. e., on ship hulls and floating marine debris [147]. Distribution as larvae in ballast water has been suggested [359–362] and is the most plausible hypothesis for *Sabella spallanzanii* [363,364], but larvae of sabellids and serpulids have never been reported from ballast water samples. If become established in remote localities, such translocated taxa significantly expand their ranges. However, some reported broad distributions are a reflection of uncertain taxonomic status. Such species remain as widely distributed or even 'cosmopolitan' only as long taxonomic uncertainty persists, and a molecular investigation usually split hem into a number of geographically restricted species.

According to the literature survey of polychaetes reported outside their natural ranges, Serpulidae and Sabellidae account for 22% of the total number of non-native polychaetes world-wide [365]. However, a critical assessment of non-indigenous species records is needed. The number of such species is a function of the research effort put into distinguishing non-native and native taxa, which in turn depends on the knowledge of native diversity and the state of taxonomy of a group. Integrative taxonomic revisions of species with reportedly global distributions are important because such taxa may include cryptic invaders that are particularly difficult to track because they are often assumed to be native species or wrongly assigned to other invasive species [366].

Several criteria for distinguishing non-indigenous from native species have been proposed as lines of indirect evidence, such as a new record for an area, a new localised occurrence showing a population explosion, species with disjoint distribution, with insufficient natural dispersal capabilities to account for observed distribution range, or species associated with means of human-made transport, e.g., found on ship hulls (e.g., [172,367,368]). In the last two decades, molecular genetics tools have proved indispensable direct lines of evidence for assessing the status of a reportedly invasive taxon. The population genetics studies allow testing if distant populations belong to same species, and also if genetic variability of suspected new arrivals is lower than that observed in the native range, which indicates a recent translocation (e.g., [72,74,204,294,368–370]). Molecular data can also help to determine the origin of non-native populations. For example, haplotype variability analysis suggests that although serpulid *Hydroides dianthus* (Verrill, 1873) was originally described from New England, its native range may be the Mediterranean [250].

A good example of a sabellid with an uncertain invasive status is *Laonome calida* described from the Calliope River, Australia, and later reported as introduced in the Netherlands [294], Odra River and the Sea of Azov [371,372]. Simultaneously, the morphologically similar *Laonome xeprovala* was described from the Baltic Sea, and DNA sequences from specimens from the Netherlands and Sea of Azov showed that they belong to the same species [102]. Neither *L. calida* nor *L. xeprovala* had been reported from Europe before 2014, and increased occurrences suggest a recent invasion [373]. Molecular analyses, however, are needed to determine whether the European populations belong to the same species found in Australia (implying that *L. xeprovala* could be a junior synonym of *L. calida*) or a distinct non-indigenous species of unknown origin is found in Europe (in which case *L. xeprovala* would be valid). Similarly, *Branchiomma* species are easily translocated outside of their native ranges [72,129,147,374–379], but members of this genus are so morphologically homogenous that species identification using morphological characters only is problematic. Therefore, assessing the identity of *Branchiomma* spp. with invasive potential requires a comprehensive generic revision, including DNA-based species delimitation analyses [72,379]. Other records still to be confirmed by molecular studies are the sabellids *Euchone variabilis* Hutchings and Murray, 1984, *Laonome triangularis, Desdemona aniara* and the three species of *Pseudobranchiomma* reported from Australia [74,126,380,381].

Table 6. List of translocated species, with presumned origin indicated by provinces or ecoregions (*sensu* Spalding et al. [31]). Those species that are included as unresolved, require confirmation and two are not cosmopolitan (*Branchiomma curtum* and *Parasabella pallida*). NA: not applicable.

Species	Type Locality	Presumed Origin	Category	New Occurrences (Outside of Expected Natural Range)	Molecular Markers	Comments
Acromegalomma claparedei (Gravier, 1906)	Djibouti, Gulf of Aden	Red Sea	? Non-indigenous	Western Mediterranean	none	Listed in the Mediterranean [382] but needs confirmation [130].
Amphicorina pectinata (Banse, 1957)	Masked Island, New Zealand	Southern Island New Zealand	? Non-indigenous	Western and Central Mediterranean	none	Redescribed from Italy and Spain without examination of types from New Zealand [383]. Included in the Mediterranean checklists [365,381,384] but requires confirmation.
Bispira polyomma Giangrande and Faasse in Faasse and Giangrande, 2012	Yerseke Marina, the Netherlands	Unknown	? Non-indigenous	NA	none	Found among cultured oysters, presumably as an introduction [385].
Branchiomma bairdi (McIntosh, 1885)	Bermuda	Tropical Northwestern Atlantic	Non-indigenous	Gulf of California, Australia, Mediterranean, Suez Canal, northeastern Atlantic Ocean, California, Hawaii, Galapagos, Madeira	COI, Cytb, 16S, 28S, ITS	Widest distribution reported for a sabellid [72,126,129,147,185,189, 376,377,379,386–388]. Some records need confirmation [185].
Branchiomma boholense (Grube, 1878)	Bohol, Philippines	Western Coral Triangle	Non-indigenous	Mediterranean	COI	Reported from the Mediterranean [375,389,390]. Çinar [126] corrected his records as *B. bairdi*. Many records of *B. bairdi* in the Mediterranean are *B. boholense* [379].

Table 6. *Cont.*

Species	Type Locality	Presumed Origin	Category	New Occurrences (Outside of Expected Natural Range)	Molecular Markers	Comments
Branchiomma coheni Tovar-Hernández and Knight-Jones, 2006	Panama, Pacific	Tropical Eastern Pacific	Non-indigenous	Florida, Gulf of California	none	Records from Tampa Bay and Gulf of California [391] first since description. In Florida it was an early detected non-indigenous species [391].
Branchiomma conspersum (Ehlers, 1887)	Key West, Florida, USA	Floridian	Non-indigenous	Hawaii	ITS	Reported from Hawaii [185]
Branchiomma curtum (Ehlers, 1901)	Juan Fernández Island, Chile	Juan Fernández	Not translocated	? Mexican Caribbean	none	Caribbean records may be erroneous [391]. Both syntypes of *B. curtum* and Caribbean specimens were juveniles produced by fission [392]. See comments in [391].
Branchiomma luctuosum (Grube, 1870)	Red Sea	Red Sea	Non-indigenous	Mediterranean, Brazil	none	Common in the Mediterranean as Lessepsian migrant [129,389,390,393–401], and Brazil [234,402]
Desdemona ornata Banse, 1957	South Africa	Agulhas	? Non-indigenous	Iberian Peninsula, UK, Marmara Sea, Portugal	none	Reported as introduced in Australia [403], Spain [404], UK [405], Marmara Sea [406], Portugal [407] and the Netherlands [408], but types of *D. ornata* were not examined.

Table 6. *Cont.*

Species	Type Locality	Presumed Origin	Category	New Occurrences (Outside of Expected Natural Range)	Molecular Markers	Comments
Euchone limnicola Reish, 1959	Los Angeles, California, USA	Warm Temperate Northeast Pacific	? Non-indigenous	Dunkerke, Australia	none	Reported from Australia by McArthur [409] and included in Hewitt et al. [380] based on McArthur dissertation. Reported by Guyonnet and Borg [410] from the French coast of North Sea.
Laonome calida Capa, 2007	Queensland, Australia	Unknown	? Non-indigenous	Australia (Queensland, Northern Territory, Western Australia), ? Europe (the Netherlands, Sea of Azov, Baltic Sea, Mosel River)	none	Reported in Australian in fully marine but also estuarine conditions, in both pristine and port environments [68]. Bick et al. [102] suggest that European records belong to *Laonome xeprovala*, not *L. calida*.
Laonome elegans Gravier, 1906	Red Sea	Red Sea	? Non-indigenous	East Mediterranean	none	Listed in Zenetos et al. [383]. Presence in the Mediterranean area as Lessepsian migrant needs to be confirmed.
Laonome triangularis Hutchings and Murray, 1984	New South Wales, Australia	East Central Australian Shelf	? Non-indigenous	Turkey	none	Reported as introduced in Turkey [126].
Laonome xeprovala Bick and Bastrop in Bick et al. 2018	Estonia, Baltic Sea	Unknown	Non-indigenous	the Netherlands, Sea of Azov, Baltic Sea, Mosel River	COI, 16S, 18S	Specimens from the Baltic Sea, the Netherlands and the Sea of Azov possessed identical genotypes, but unknown origin [102]. Confirmation that it is not the same as *L. calida* is needed.

Table 6. *Cont.*

Species	Type Locality	Presumed Origin	Category	New Occurrences (Outside of Expected Natural Range)	Molecular Markers	Comments
Myxicola infundibulum (Renier *in* Meneghini, 1847)	Devon, UK	Northern European Seas	Non-indigenous	Australia, ?North America	COI, 16S	Analysed sequences of specimens from European and Australian waters belong to same species; sequences from Maine showed some differences [411].
Parasabella fullo Grube, 1878)	Northern Japan	Temperate Northwest Pacific	Non-indigenous	Santa Barbara and San Diego, California, USA	none	Collected on ship hulls in California, and a resident population appears to exist in the region [412].
Parasabella pallida Moore, 1923	California, USA	Warm Temperate Northeast Pacific	Not translocated	NA	none	Included in the list of translocated species by mistake [294], as it was described from California [143] not the Caribbean, as later fixed in [73].
Parasabella rugosa (Moore, 1904)	San Diego, California, USA	Warm Temperate Northeast Pacific	Non-indigenous	Australia	none	It was reported in Australia near an international port as sp. cf. *P. rugosa* Capa and Murray [73].
Pseudobranchiomma emersoni Jones, 1962	Port Jackson, Jamaica	Tropical Northwestern Atlantic	Non-indigenous	Australia	none	As cf. *P. emersoni* in Capa and Murray [74].
Peudobranchiomma orientalis (McIntosh, 1885)	Hong Kong	Unknown	Non-indigenous	Australia	none	As cf. *P. orientalis* in Capa and Murray [74]. Ethanol fixed specimens are need for molecular analysis.

Table 6. *Cont.*

Species	Type Locality	Presumed Origin	Category	New Occurrences (Outside of Expected Natural Range)	Molecular Markers	Comments
Pseudobranchiomma schizogenica Tovar-Hernández and Dean, 2014	La Paz, Mexico	Tropical Northwestern Atlantic	Non-indigenous	Australia, Galapagos	ITS, Cytb	As cf. *P. schizogenica* in Capa and Murray [74] from Australia, also reported in Galapagos [183].
Sabella spallanzanii (Gmelin, 1791)	Malta	Mediterranean	Non-indigenous	Australia, New Zealand	COI, H3, 18S, 28S, 16S, ITS	Considered an invasive pest in Australia and New Zealand [202,364,370,413].
Sabellastarte spectabilis (Grube, 1878)	Bohol, Masalac, Philippines and Singapore	Western Coral Triangle	? Non-indigenous	Sri Lanka, Solomon Islands, Mauritius, Japan, Taiwan, Hawaii, Malaysia, Saipan, Pakistan	COI, 16S	Reports of accidental introductions to Hawaii [414–416] rely on invalid morphological features [70]. Evidence for wide distribution exists [70].
Terebrasabella heterouncinata Fitzhugh and Rouse, 1999	South Africa	Unknown	Non-indigenous	California, Chile	none	Parasite of red abalone. Reported from abalone farms from California and Chile [66,300–419].
Ficopomatus enigmaticus (Fauvel, 1923)	France, native range is unknown, likely southerm Australia	Unknown	Non-indigenous	Europe, New Zealand, Japan, USA (both coasts), Argentina, Tunisia, Egypt, Côte d'Ivoire, South Africa	Cytb, COI	Reviewed by Dittmann [420]. Styan et al. [204] revealed three species (not formally described) with overlapping ranges in Australia, one of which is morphologically distinct from the other two. Grosse et al. [421] found two other species within the complex Spain.

Table 6. *Cont.*

Species	Type Locality	Presumed Origin	Category	New Occurrences (Outside of Expected Natural Range)	Molecular Markers	Comments
Ficopomatus uschakovi (Pillai, 1960)	Sri Lanka	Unknown	? Non-indigenous	Indo-Pacific, Western Africa (Nigeria, Ivory Coast), Brazil, Venezuela, Colombian Caribbean and southern Mexican Pacific ?Australia	none	Several records given outside the native range [422–425]. Likely a species complex [204].
Hydroides brachyacantha Rioja, 1941	Pacific coast of Mexico	Warm Temperate Northeast Pacific	Native	Likely restricted to Pacific coast of Mexico, no evidence of translocations	18S, cytb, ITS	Belongs to a complex of species, records from warm-temperate and tropical localities world-wide likely belong to other species of *H. brachacantha* complex [203].
Hydroides dianthus (Verrill, 1873)	New England, USA	Either East Coast of the USA or the Mediterranean	Non-indigenous	Brazil, China, Japan, West Africa, the Mediterranean (or US East Coast), the Black Sea, Texas	COI	Observed higher haplotypes diversity in the Mediterranean contradicts the accepted native range of *H. dianthus* in the USA. The cryptic lineage found in Texas was evidently introduced to the Black Sea recently [250].
Hydroides dirampha Mörch, 1863	St. Thomas Island, US Virgin Islands	Tropical Northwestern Atlantic	Non-indigenous	Australia, Brazil, Japan, Hawaii, New Zealand, Panama, the Pacific from Mexico and California, and the Mediterranean	18S, cytb, COI, 28S, ITS	Several records given outside the native range [426–429]. Molecular data from Australia, Brazil, Panama [78,79].

Table 6. *Cont.*

Species	Type Locality	Presumed Origin	Category	New Occurrences (Outside of Expected Natural Range)	Molecular Markers	Comments
Hydroides elegans (Haswell, 1883)	Sydney, Australia	Unknown	Non-indigenous	Sub-tropical world-wide: Mediterranean-Atlantic, Indo-West Pacific, tropical Pacific America, West Atlantic, East Atlantic, South Africa	Microsatellites, 18S, cytb, COI, 28S	See [106]. Biofouling has been shown as a major mode of dispersal for this species [430]. Molecular data from Panama, California, Australia, Brazil, Italy and Spain [79,421].
Hydroides ezonesis Okuda, 1934	Northern Japan	Warm Temperate Northwest Pacific	Non-indigenous	France, UK, Australia	18S, 28S	Imported from Japan with oysters to the Atlantic coast of France [431,432] then the UK and Australia [359,432,433]. Molecular data from China, Japan [79].
Hydroides operculata (Treadwell, 1929)	Gulf of Aden	Western Indo-Pacific	Native, some species in the complex maybe non-indigenous	A complex of species, no evidence of translocations for any species of the complex	18S, cytb, COI, 28S, ITS	Reports from South and East Africa, India, Pakistan, Sri Lanka, Hong Kong, tropical Australia, and the eastern Mediterranean likely belong to other species of the *H. operculata* complex [251].
Hydroides sanctaecrusis Krøyer [in] Mörch, 1863	Saint Croix, Virgin Islands	Eastern Caribbean	Non-indigenous	Singapore, tropical Australia, Hong Kong, Taiwan, Florida, India. Not found in the Mediterranean [240]	18S, COI, 28S, ITS	Records outside of the native range include [78,180,362,434]. Molecular data from Florida [77], Panama, Australia, India [79], Pacific Mexico [80].

Table 6. *Cont.*

Species	Type Locality	Presumed Origin	Category	New Occurrences (Outside of Expected Natural Range)	Molecular Markers	Comments
Spirobranchus kraussii (Baird, 1865)	Cape of Good Hope, South Africa	Agulhas	Native, some species in the complex non-indigenous, but not *S. kraussi*	Restricted to South Africa, no evidence of translocations	18S, cytb	Reports from warm-temperate and tropical localities in the Indo-Pacific and Mediterranean Sea belong to other species of *S. kraussii* complex [84,193].
Spirobranchus tetraceros (Schmarda, 1861)	NSW, Australia	East Central Australian Shelf	Native, some species in the complex likely non-indigenous, but not *S. tetraceros*	Likely restricted to south-eastern Australia, no evidence of translocations.	cytb	Reports from warm-temperate and tropical localities world-wide belong to other species of the *S. tetraceros* complex. At least one of these species was introduced and established in the Mediterranean [252].

One of the best examples of a sabellid with the invasive status confirmed through a combination of morphological and genetic data is *Sabella spallanzanii*. This large conspicuous Mediterranean native was introduced to Australia in 1965 and to New Zealand in 2008 [364,435,436]. Analyses of COI sequences from the native and non-indigenous populations proved that the New Zealand incursion originated from Australia rather than from the Mediterranean [370]. Other examples of sabellids with invasive status (but ofunknown origin) confirmed with DNA analyses are *Parasabella crassichaeta* Capa and Murray, 2015 and *Pseudobranchiomma* cf. *schizogenica* Tovar-Hernández and Dean, 2014, reported from both Hawaii and Australia [73,74]. Out of seven nominal species of the serpulid genus *Hydroides* reported as translocated outside of their natural range, five are confirmed invaders (Table 6). One of them, *H. elegans*, is the best-known cryptogenic polychaete, reported from most sub-tropical locations world-wide [110], and biofouling as the major mode of its dispersal was supported by DNA data [430]. *Ficopomatus enigmaticus* is another cryptogenic serpulid, because, although it was described from France, its native range is enigmatic (hence the name), likely to be southern Australia [420]. This typical species has invaded warm-temperate estuaries world-wide, as confirmed by DNA studies [204,437,438].

Ficopomatus uschakovi (Pillai, 1960), described from Sri Lanka, a tropical species with supposedly wide distribution in Indo-Pacific, was recently reported as introduced to South America [422–425]. The invasive status of the species has not been examined with DNA, but preliminary molecular data (Kupriyanova unpubl.) suggest that this taxon is a complex of species. Two nominal *Hydroides* species, *H. brachyacantha* Rioja, 1941 and *H. operculata* (Treadwell, 1929), are examples of complexes of morphologically similar species [180,251]. Similarly, the invasive status attributed to serpulids *Spirobranchus kraussii* (Baird, 1865) and *S. tetraceros* (Schmarda, 1861) [365,439] is unjustified, as both are members of species complexes [84,193,252].

The Mediterranean Sea leads the rank when it comes to reported introductions, with 13 serpulid and 10 sabellid non-indigenous species reported, mainly as a result of Lessepsian migration from the Red Sea [125,365,440,441]. In this region, 11 species of sabellids and serpulids, mainly of the genera *Branchiomma, Ficopomatus,* and *Hydroides*, have been listed among the top 100 worst invasive species, based on their economic and ecological impacts [442]. However, taxonomic and invasive status of many of these taxa needs to be revised.

3.6. Fanworms Are Important: Some Applications

3.6.1. Nuisance Fouling Species

Several serpulid species, predominantly of the genera *Hydroides, Ficopomatus,* and *Spirobranchus*, are capable of colonizing a wide range of natural and artificial substrates and settling gregariously, which makes them economically and ecologically important fouling nuisance species.

Serpulid foulers constitute a significant financial burden to due to costs associated with the removal of tubes from artificial structures. Millions of dollars are spent annually to prevent the fouling of marine organisms, especially of *Hydroides*, on human-made structures [443]. Dense tube aggregates attach to underwater seawater intake pipes of power plants reducing water flow and causing blockages. Fouled docks require cleaning maintenance in harbours around the world. Fouling interferes with navigation and shipping industries by decreasing ship speed, while increasing the weight and drag of buoys [444,445].

In marine aquaculture the key impact is the direct fouling of stock causing physical damage, biological competition and environmental modification, while infrastructure, such as aquaculture nets and cages, is also damaged. The conservative estimates of economic loss to the aquaculture industry are 5–10% of production costs attributed to biofouling [446].

Manayunkia speciosa and/or *M. occidentalis* (see [216]) are obligate hosts of the myxozoan parasites *Ceratonova shasta* and *Parvicapsula minibicornis*, which cause ceratomyxosis in salmon and trout in North America [267,268,447]. Management actions, such as flow manipulations to increase the mortality of *M. speciosa* and disturbance of its habitat, have been implemented [268].

3.6.2. Non-Indigenous and Invasive Species

Non-native to a region species translocated to another region can expand and have significant impact on human health, economic interests or environmental values. Such translocations of fouling species of Sabellida are well documented (e.g., [448–452]). Countries around the world have established biosecurity systems, aimed to prevent the introduction and/or spread of non-indigenous organisms. Some Sabellida have been listed in individual countries' Laws and Regulations, indicating its status as unwanted non-indigenous species (invasive, pests, parasites, pathogens). For example, in New Zealand, *Sabella spallanzanii* has been registered as a notifiable organism, subject to targeted surveillance work, including study of population dynamics and reproduction, under the New Zealand Biosecurity Act 1993 [364]. In Australia, non-indigenous marine species already found and those not yet found but have demonstrated significant impacts elsewhere are ranked according to their invasive impact and potential. For example, *S. spallanzanii* is regarded as a high impact, notifiable invasive species, while *Hydroides dirampha* Mörch, 1863, *H. dianthus*, *H. sanctaecrucis* Krøyer in Mörch, 1863, *H. ezoensis* Okuda, 1934 are listed as medium or low priority species [433]. In Mexico, sabellids *Branchiomma bairdi* (McIntosh, 1885) and *Terebrasabella heterouncinata* and six serpulids (*Ficopomatus enigmaticus*, *F. miamiensis* (Treadwell, 1934), *F. uschakovi*, *Hydroides elegans*, *H. bispinosa* Bush, 1910 and *H. dirampha*) are regulated under the Diario Oficial de la Federación [453]. *Ficopomatus enigmaticus* is the only annelid registered in the Spanish Catalogue of Exotic Invasive Species [421]. In Brazil, the only species of Sabellida reported as invasive is *Branchiomma luctuosum* (Grube, 1870) [401].

3.6.3. Indicators of Pollution

Manayunkia speciosa is an indicator of moderate organic pollution, but is intolerant of severe pollution or anoxic sediments [447,454,455]. Decrease of the organic content of the sediment from 1.8% to 1.0% leads to reduction in abundance of its congener, *M. aestuarina*, from 16 000 to 6000 ind.m^2 in the Baltic Sea [262]. *Euchonoides moeone* was proposed as an indicator of sediment organic enrichment in a sewage outfall in Hawaii [279].

Sabella spallanzanii can trap anthropogenic micro-particles and glue these to their tubes, and it has been proposed as an indicator of microlitter pollution in sheltered and polluted environments such as ports [456]. Larvae of *Hydroides elegans* have been used as indicators for biomonitoring and ecotoxicology tests (e.g., [457–459]).

Some sabellids and serpulids have been suggested as bioindicators of heavy metal pollution. For example, the tube of *Sabella spallanzanii* is an important compartment in metal retention and suitable for evaluation of the pollution by traced elements [460], while *Branchiomma bairdi* and *B. luctuosum*, invasive sabellids in the Mediterranean, can accumulate high concentrations of arsenic (As), cadmium (Cd), chromium (Cr) and lead (Pb), considered to be priority toxic or ubiquitous persistent, bioaccumulative and toxic (PBT) substances under the EU Water Framework Directive [461]. Some studies have focused on the effects of heavy metals on larval development and metamorphosis using serpulid larvae (*Hydroides elegans*: [462,463]); *Galeolaria caespitosa:* [464]).

3.6.4. Bioremeditators

As suspension feeders, *Sabella spallanzanii*, *Branchiomma luctuosum* and *B. bairdi* have been tested as bioremediators for aquaculture waste-water treatment in polluted coastal areas [273,274,465–470]. However, these three taxa are invasive in some areas, and may pose a threat to native ecosystems. Nevertheless, the use of non-indigenous species as bioremediator may allow to transform a potential risk into a benefit, with high potential commercial gain and economic feasibility [470]. Due to their important role in organic sediment bioremediation, the Food and Agriculture Organization of the United Nations

(FAO, Roma, Italy) recommended *Sabella* as one of organisms with most potential for the development of integrated multi-trophic aquaculture systems [471].

Ficopomatus enigmaticus is a dominant species in estuaries and lagoons, where it can affect the community structure and contribute to the invertebrate biomass [472,473]. Due to its ability to build extensive reefs, *F. enigmaticus* is considered an ecosystem engineer that can modify estuarine ecosystem, changing water flow, sedimentation rates, and creating a structured hard substrate habitat in a soft-sediment environment. Large aggregations of *F. enigmaticus* remove suspended particulate matter, reduce excess nutrient loads and improve oxygen levels in enclosed waters, thereby improving the water quality and environmental conditions for other benthic species (reviewed in [420]). Davies et al. [474] stressed that because of the fundamental role *F. enigmaticus* played in the maintenance of water quality of an enclosed system near Cape Town, South Africa, eradication of this non-indigenous species should not be a management option.

3.6.5. Models Organisms in Research

Sabellids are used as models in regeneration biology, most notably in studies examining the developmental basis and functional ecology of regeneration [222,475–477]. Members of the genus *Myxicola* are known for the giant axon [478] that directly innervates the worm's muscles, presumably aiding in super-fast retraction into the tubes [479]. The outsized nerves make this species a model organism for studies of neuroanatomy, neuroactivity and electrophysiology [480–482]. *Myxicola*'s giant axons were also used for testing the effects of the anticonvulsant Carbamazepine on the ionic conductance [483]. Moreover, the mucus of *Myxicola infundibulum* (Renier *in* Meneghini, 1847), with natural antibacterial and antioxidant compounds, showed potential for drug prospection [470].

Hydroides elegans is easily adapted for laboratory research because of its rapid generation time (three weeks) and ease of propagation. The adults spawn and eggs easily fertilise, their larvae become metamorphically competent in several days and readily settle in the laboratory. Thus, *H. elegans* has been declared an important model organism [484] and has been used routinely during last two decades in hundreds of experimental embryological, larval ecology and biofouling studies, including tests of newly formulated marine coatings (e.g., [485–490]). Other *Hydroides* species, such as *H. ezoensis* and *H. dianthus*, have also acted as model organisms for larval ecology research (e.g., studies of mechanisms of gregarious settlement by [491–493]. *Spirobranchus lamarcki* (Quatrefages, 1866) has provided an important model system for molecular and embryological work, including studies on the organization and expression of its developmental genes (e.g., [494–497]). Recently *H. elegans* and *S. triqueter* have served as models in ocean acidification and biomineralization research (e.g., [498–500]).

3.6.6. Objects of Ornamental Trade

Sabellida includes some of the most beautiful marine invertebrates due to their colourful radiolar crowns. They are listed among the ten most imported ornamental invertebrates [501]) and are amongst the most photographed polychaetes found in marine guides and featured on postcards, stamps, calendars, T-shirts and even tattoos. Largest sabellids (*Anamobaea, Bispira, Notaulax, Sabella* and *Sabellastarte*) and serpulids such as Christmas tree worms (*Spirobranchus*) and coco worms (*Protula*) are popular in the aquarium trade. The vast majority of ornamental sabellids and serpulids are tropical species, although a market for cold-water species has been growing [502,503]. Efforts to culture sabellids (e.g., *Sabellastarte spectabilis* (Grube, 1878) [504–506], *Sabella pavonina* Savigny, 1822 [270,507] and *Bispira brunnea* (Treadwell, 1917) [508] are well under way. Aquaculture can provide environmental benefits by reducing collecting pressure on highly traded species.

3.7. Future Perspectives in Fanworm Research

As it is clear from this review, knowledge of Sabellida biodiversity is incomplete and the reported species numbers appear to be an underestimation of the true diversity.

This review highlights that some of the lesser known coastal and continental shelf areas including Hudson complex, the Atlantic coast of South America (especially the tropical Atlantic region, excluding the Caribbean Sea), the coastlines along the Arabic Sea and Gulf of Bengal, and the Far East of Russia. However, Africa, with the exception of South Africa and Morocco, is by far the most neglected continent when it comes to taxonomic studies.

More surveys into deep-sea (abyssal and hadal), chemosynthesis-based (hydrothermal vents, methane seeps and organic falls) and freshwater habitats are needed for a better understanding of the Sabellida diversity and adaptations to these habitats. The fact that undescribed species have been collected in recent deep-sea cruises along several worldwide regions (e.g., [284,509,510]) provides evidences for deep sea fanworms still awaiting to be discovered. Studies of symbiotic/commensal relationships with other organisms, e.g., molluscs, corals, sponges, or examinations of bacterial microbiomes may reveal not only new taxa, but also new ecological relationships and trophic networks (e.g., [284,490]).

Importantly, the diversity of some remote areas, including deep sea environments, is poorly known not only because of the obvious logistical difficulties with collecting, but also due the insufficient number of experts and their unbalanced distribution across the globe known as 'taxonomic impediment' (e.g., [94,511]). We need to train and sustain more systematists able to discover, describe, identify and classify species and also to increase efforts directed to manage and curate existing research collections [355].

As many more species are yet to be discovered, either in the field or in museum collections, particular attention should be paid to setting a high standard for the new species descriptions, which would include use of modern microscopic techniques (e.g., SEM, Phase Contrast), assessment of intra- and interspecific variability, and preparation of quality informative illustrations (digital drawings and high-quality digital photographs of stained fixed or live specimens, when possible). Exploration of both new characters, e.g., ultrastructure of calcareous serpulid tubes that proved useful for species delimitation [107,227,512], and new techniques to examine existing morphological characters, e.g., tomography and 3D reconstructions (e.g., [237]) should significantly improve species descriptions in the future and aid species delimitations.

Although multivariate morphometrics have been used to analyse differences among annelid species and populations (e.g., [513,514]), this technique is not very common for species delimitation because body shapes of these soft-bodied organisms vary depending on the fixation or anaesthetization methods [234]. We suggest that application of morphometrics to chaetal or opercular traits should be explored. However, regardless of availability DNA data, morphological studies must include statistical assessment of intraspecific variability and its sources (such as size-dependent, ontogenetic, environmental) (e.g., [72,278,515]).

Understanding of true species diversity of Sabellida requires world-wide revisionary studies of existing species and their distribution ranges. As early species descriptions are often very short and sometimes poorly illustrated, further re-descriptions of older species (especially described before mid-20th century) are needed, ideally based on topotypical material (as, e.g., done for serpulid *Spirobranchus kraussii* by Simon et al. [193]). For the species with lost or lacking types, neotypes should designated (e.g., as for *Pseudopotamilla reniformis* (Bruguière, 1789) by Knight-Jones et al. [111]), preferably accompanied by DNA sequence data (e.g., as done for *Hydroides brachyacantha* by Sun et al. [197]. It is imperative that the type material (holotype, type series, additional specimens showing intraspecific variability, and DNA extractions) is always deposited in properly curated permanent museum collection(s) where it is maintained in optimal conditions [516].

Contrary to previous conceptions that the ocean has no boundaries and that polychaetes more often than other organisms have cosmopolitan distributions [327], it now became clear that the genetic and species diversity of marine invertebrates is highly structured geographically and significant species diversity is hidden in former 'cosmopolitan polychaete species'. Therefore, resolution of 'cosmopolitan species' should be one of the main goals of revisionary studies of Sabellida. This goal is only achieved with application of fast-developing molecular tools such as DNA sequencing and genomics/transcriptomics. Molecular tools and analytical methods are indispensable to further improve our understanding of the species diversity, but also to trace the pathways and origins of invasive species, to determine biogeographic boundaries between species, and to provide reliable phylogenic hypotheses. Robust well-resolved phylogenies with significant taxon coverage using transcriptome and mitochondrial genome data are important to address important character evolution questions (e.g., photoreceptor evolution and evolution of the reproductive and larval strategies in Sabellidae and Serpulidae). Finally, in the future molecular identification of species by non-specialists might replace morphology-based identifications only if reliable databases of reference sequences supported by voucher specimen depositories are built.

Supplementary Materials: The following are available online at https://www.mdpi.com/1424-2818/13/3/130/s1, Table S1: Fabriciidae, species and type localities; Table S2: Sabellidae, species and type localities; Table S3: Serpulidae, species and type localities.

Author Contributions: Conceptualization, M.C. and M.A.T.-H.; validation, M.C., E.K., A.B., J.M.d.M.N. and M.A.T.-H.; writing—original draft preparation M.C., E.K., A.B., J.M.d.M.N. and M.A.T.-H.; writing—review and editing, M.C., E.K., A.B., J.M.d.M.N. and M.A.T.-H.; visualization, M.C., E.K., A.B., J.M.d.M.N. and M.A.T.-H.; coordination, M.C., E.K., A.B., J.M.d.M.N. and M.A.T.-H.; funding acquisition, M.C. and E.K. All authors have read and agreed to the published version of the manuscript.

Funding: M.C. was funded by the Ramón y Cajal program (RYC-2016-20799) funded by Spanish MINECO, Agencia Estatal de Investigación, Comunidad Autónoma de las Islas Baleares and the European Social Fund. EK was funded by the Australian Biological Research Study (ABRS) grant RG18-21. J.M.M.N. was funded by a productivity grant from "CNPq—Conselho Nacional de Desenvolvimento Científico e Tecnológico, Brazil, level 2".

Institutional Review Board Statement: Not applicable.

Informed Consent Statement: Not applicable.

Data Availability Statement: Not applicable.

Acknowledgments: We would like to thank Mike Bok (Lund University, Sweden), Alexander Semenov (Moscow University, Russia), Greg Rouse (Scripps Institution of Oceanography, UCSD, USA), Graeme Christie (Australia), Anna Dietrich (Institute for Applied Ecosystem Research, Germany), Vitaly Syomin (P.P. Shirshov Institute of Oceanology), Humberto Bahena (Ecosur, Mexico), Eunice Wong (University of Texas at Austin, USA), Alexander Rzhavsky (Russian Academy of Sciences, Russia), Orlemir Carrete (Universidade de São Paolo, Brazil)and Floor Verbiest (Rijksmuseum van Natuurlijke Historie, Leiden) for letting us use some their fantastic pictures of live animals. Laura C. Armendáriz (Instituto de Limnología "Dr Raúl A. Ringuelet", CONICET UNLP, Argentina) and Sue Lindsay (Macquarie University) provided with SEM micrographs. Alisson Ricardo da Silva (Museu da Casa Brasileira, Brazil), helped with the elaboration of the figures. We appreciate Geoff Read (NIWA New Zealand) for updating WoRMS and also Vollrath Wiese (Haus der Natur, Cismar) for helping us to clarify identity of *Sabella* species described by Gmelin. Further thanks are due to the publisher De Gruyter for permission to use the following images from chapter 7.4.8 Fabriciidae Rioja, 1823, from Handbook for Zoology, vol. 3 (Eds. G. Purschke, M. Böggemann and W. Westheide), De Gruyter GmbH, Berlin/Boston, 2021: 7.4.8.1A; 7.4.8.2 A, C; 7.4.8.6 D. We appreciate helpful comments from three anonymous reviewers.

Conflicts of Interest: The authors declare no conflict of interest.

Appendix A

Fabriciidae and Sabellidae species with doubtful identity requiring further investigation (*inquirenda*), indeterminable or incorrect assignation.

Fabriciidae

1. *Manayunkia siaukhu* Annenkova, 1938 *inquirenda*. Based on the description, *M. siaukhu* has pygidial eyes [517] and thus does not fulfill diagnostic features for the genus.

Sabellidae

2. *Clymeneis* Rathke, 1843 *inquirenda*

Clymeneis and its type species *Clymeneis stigmosa* Rathke, 1843 are of doubtful identity requiring further investigation. Description was based in specimens inquirenda (?) apparently without crown and types have not been found. It has not been reported over more than a century, but recently mentioned in the paper about original specimens and type localities of early described polychaete species from Norway [518].

3. *Sabella aculeata* Gmelin in Linnaeus, 1788. Insecta: Trichoptera.
4. *Sabella ammonita* Gmelin in Linnaeus, 1788. Insecta: Trichoptera.
5. *Sabella arenaria* Montagu, 1803 *indeterminable*

Described based on the tube only, the worm is unknown [42] (pp. 552).

6. *Sabella arundinacea* Gmelin in Linnaeus, 1788. Insecta: Trichoptera.
7. *Sabella clavata* Gmelin in Linnaeus, 1788. Insecta: Trichoptera.
8. *Sabella compressa* Montagu, 1803 *indeterminable*

Original description was based only in the tube. Hartman [325] (pp.559) suggested that the tube is perhaps from a pectinariid.

9. *Sabella conica* Gmelin in Linnaeus, 1788. Insecta: Trichoptera.
10. *Sabella corticalis* Gmelin in Linnaeus, 1788. Insecta: Trichoptera.
11. *Sabella dimidiata* Gmelin in Linnaeus, 1788. Insecta: Trichoptera.
12. *Sabella flabellata* Savigny, 1820 *inquirenda*

Declared as *inquirenda* by Knight-Jones and Perkins [337] (pp. 398).

13. *Sabella fixa* Gmelin in Linnaeus, 1788. Insecta: Trichoptera.
14. *Sabella grossa* Baird, 1865 *inquirenda*

Declared as *inquirenda* by Knight-Jones and Mackie [338] (pp. 2296).

15. *Sabella helicina* Gmelin in Linnaeus, 1788. Insecta: Trichoptera.
16. *Sabella nigra* Gmelin in Linnaeus, 1788. Insecta: Trichoptera.
17. *Sabella sabulosa* Gmelin in Linnaeus, 1788. Insecta: Trichoptera.
18. *Sabella setiformis* Montagu, 1803 *indeterminable*

The tube was the only structure described, the worm is unknown [519] (pp.553).

19. *Sabella stagnalis* Gmelin in Linnaeus, 1788. Insecta: Trichoptera.
20. *Sabella subcylindrica* Montagu, 1803 *indeterminable*

Only the tube was described, animal unknown [519] (pp. 552-553).

21. *Sabella teredula* Chiereghini *in* Siebold, 1850 *indeterminable*

Only the tube was described [520] (pp. 369).

22. *Sabella trigona* Chiereghini *in* Siebold, 1850 *indeterminable*

Only the tube was described [520] (pp. 369).

23. *Sabella uncinata* Gmelin in Linnaeus, 1788. Insecta: Trichoptera.
24. *Sabella vegetabilis* Gmelin in Linnaeus, 1888. Insecta: Trichoptera.
25. *Sabella zonalis* Stimpson, 1854 *inquirenda*

Declared *inquirenda* by Knight-Jones and Perkins [337] (pp. 405).

References

1. Kupriyanova, E.K.; Rouse, G.W. Yet another example of paraphyly in Annelida: Molecular evidence that Sabellidae contains Serpulidae. *Mol. Phylogenet. Evol.* **2008**, *46*, 1174–1181. [CrossRef]
2. Capa, M.; Hutchings, P.; Aguado, M.T.; Bott, N.J. Phylogeny of Sabellidae (Annelida) and relationships with other taxa inferred from morphology and multiple genes. *Cladistics* **2011**, *27*, 449–469. [CrossRef]

3. Tilic, E.; Sayyari, E.; Stiller, J.; Mirarab, S.; Rouse, G.W. More is needed—thousands of loci are required to elucidate the relationships of the 'Flowers of the Sea' (Sabellida, Annelida). *Mol. Phylogenet. Evol.* **2020**, *151*, 106892. [CrossRef]
4. Bick, A. Fabriciidae Rioja, 1923. In *Handbook of Zoology Annelida, Volume 3: Pleistoannelida, Sedentaria III and Errantia I*; Purschke, G., Westheide, W., Böggemann, M., Eds.; De Gruyter: Berlin, Germany; Boston, MA, USA, 2021; pp. 1–33.
5. Savigny, J.-C. *Systèmes des Annélides, Principalement de Celles des Côtes de l'Égypte et de la Syrie, Offrant les Caractères tant Distinctifs que Naturels des Ordres, Familles et Genres, Avec la Description des Espèces Système des Annelides (I)*; Commission des sciences et arts d'Egypte: Paris, France, 1822; Volume 1, pp. 1–128.
6. Burmeister, H. *Handbuch Der Naturgeschichte. Zum Gebrauch bei Voresungen*; Zweite Abt. Zoologie.: Berlin, Germany, 1837; pp. 1–858.
7. Grube, A.E. Die Familien der Anneliden. *Archiv Für Naturgeschichte* **1850**, *16*, 249–364.
8. Grube, A.E. Beschreibungen Einiger von Georg Ritter von Frauenfeld Gesammelter Anneliden und Gephyreen des Rothen Meeres. *Verhandlungen der Kaiserlich Königlichen Zoologisch-Botanischen Gesellschaft in Wien* **1868**, *18*, 629–650.
9. Malmgren, A.J. Nordiska Hafs-Annulater. *Öfversigt af Königlich Vetenskapsakademiens Förhandlingar* **1866**, *22*, 355–410.
10. Hatschek, B. System Der Anneliden, ein vorläufiger Bericht. Sitzungsberichte des Deutschen Naturwissenschaftlich-Medicinischen Vereines für Böhmen"Lotos" in Prag. *Neue Folge* **1893**, *13*, 123–126.
11. Dales, R.P. The nature of the pigments in the crown of sabellid and serpulid polychaetes. *J. Mar. Biol. Assoc. UK* **1962**, *42*, 259–274. [CrossRef]
12. Fauchald, K. The polychaete worms. Definitions and keys to the orders, families and genera. *Nat. Hist. Mus. LA County Sci. Ser.* **1977**, *28*, 1–188.
13. Fauchald, K.; Rouse, G.W. Polychaete systematics: Past and present. *Zool. Scr.* **1997**, *26*, 71–138. [CrossRef]
14. Andrade, S.C.S.; Novo, M.; Kawauchi, G.Y.; Worsaae, K.; Pleijel, F.; Giribet, G.; Rouse, G.W. Articulating "Archiannelids": Phylogenomics and annelid relationships, with emphasis on meiofaunal taxa. *Mol. Biol. Evol.* **2015**, *32*, 2860–2875. [CrossRef] [PubMed]
15. Weigert, A.; Bleidorn, C. Current status of annelid phylogeny. *Org. Divers. Evol.* **2016**, *16*, 345–362. [CrossRef]
16. Helm, C.; Bok, M.J.; Hutchings, P.; Kupriyanova, E.; Capa, M. Developmental studies provide new insights into the evolution of sense organs in Sabellariidae (Annelida). *BMC Evol. Biol.* **2018**, *18*, 149–162. [CrossRef] [PubMed]
17. Wilson, D.P. The development of the sabellid *Branchiomma vesiculosum*. *J. Cell Sci.* **1936**, *78*, 534–603.
18. Berrill, N. Functional morphology and development of segmental inversion in sabellid polychaetes. *Biol. Bull.* **1977**, *153*, 453–467. [CrossRef]
19. Smith, R.S. Relationships within the Order Sabellida (Polychaeta). *Ophelia Suppl.* **1991**, *5*, 249–260.
20. Bartolomaeus, T. Structure and formation of thoracic and abdominal uncini in *Fabricia stellaris* (Müller, 1774)–implication for the evolution of Sabellida (Annelida). *Zool. Anz. J. Comp. Zool.* **2002**, *241*, 1–17. [CrossRef]
21. Kieselbach, D.; Hausen, H. Chaetal arrangement provides no support for a close relationship of Sabellidae and Sabellariidae (Annelida). *J. Morphol.* **2008**, *269*, 104–117. [CrossRef] [PubMed]
22. Capa, M.; Hutchings, P.; Peart, R. Systematic revision of Sabellariidae (Polychaeta) and their relationships with other polychaetes using morphological and DNA sequence data: Systematics of Sabellariidae. *Zool. J. Linn. Soc.* **2012**, *164*, 245–284. [CrossRef]
23. Capa, M.; Parapar, J.; Hutchings, P. Phylogeny of Oweniidae (Polychaeta) based on morphological data and taxonomic revision of Australian fauna: Oweniid phylogeny and Australian fauna. *Zool. J. Linn. Soc.* **2012**, *166*, 236–278. [CrossRef]
24. Capa, M.; Giangrande, A.; Nogueira, J.M.M.; Tovar-Hernández, M.A. Sabellidae Latreille, 1825. In *Handbook of Zoology, Annelida, Volume 2: Pleistoannelida, Sedentaria II*; Purschke, G., Böggemann, M., Westheide, W., Eds.; De Gruyter: Berlin, Germany, 2019; pp. 164–212.
25. Capa, M.; Hutchings, P. Sabellariidae Johnston, 1865. In *Handbook of Zoology, Annelida, Volume 2: Pleistoannelida, Sedentaria II*; Purschke, G., Böggemann, M., Westheide, W., Eds.; De Gruyter: Berlin, Germany, 2019; pp. 144–163.
26. Capa, M.; Parapar, J.; Hutchings, P.; Mortimer, K. *4. Palaeoannelida. Annelida*; Purschke, G., Böggemann, M., Westheide, W., Eds.; De Gruyter: Berlin, Germany, 2019; pp. 91–132. ISBN 978-3-11-029158-2.
27. Huang, D.; Fitzhugh, K.; Rouse, G.W. Inference of phylogenetic relationships within Fabriciidae (Sabellida, Annelida) using molecular and morphological data. *Cladistics* **2011**, *27*, 356–379. [CrossRef]
28. Weigert, A.; Helm, C.; Meyer, M.; Nickel, B.; Arendt, D.; Hausdorf, B.; Santos, S.R.; Halanych, K.M.; Purschke, G.; Bleidorn, C.; et al. Illuminating the base of the annelid tree using transcriptomics. *Mol. Biol. Evol.* **2014**, *31*, 1391–1401. [CrossRef]
29. Struck, T.H.; Golombek, A.; Weigert, A.; Franke, F.A.; Westheide, W.; Purschke, G.; Bleidorn, C.; Halanych, K.M. The evolution of annelids reveals two adaptive routes to the interstitial realm. *Curr. Biol.* **2015**, *25*, 1993–1999. [CrossRef]
30. WoRMS 2020, Sabellida. Available online: http://www.marinespecies.org/aphia.php?p=taxdetails&id=882 (accessed on 2 March 2021).
31. Spalding, M.D.; Fox, H.E.; Allen, G.R.; Davidson, N.; Ferdaña, Z.A.; Finlayson, M.; Halpern, B.S.; Jorge, M.A.; Lombana, A.; Lourie, S.A.; et al. Marine ecoregions of the World: A bioregionalization of coastal and shelf areas. *BioScience* **2007**, *57*, 573–583. [CrossRef]
32. Udvardy, M.D. *A Classification of the Biogeographical Provinces of the World*; IUCN Occasional Paper 18: Morges, Switzerland, 1975; pp. 1–49.

33. Olson, D.M.; Dinerstein, E.; Wikramanayake, E.D.; Burgess, N.D.; Powell, G.V.; Underwood, E.C.; D'amico, J.A.; Itoua, I.; Strand, H.E.; Morrison, J.C. Terrestrial ecoregions of the world: A new map of life on Earth a new global map of terrestrial ecoregions provides an innovative tool for conserving biodiversity. *BioScience* **2001**, *51*, 933–938. [CrossRef]
34. GBIF. 2020. Available online: https://www.gbif.org/es/ (accessed on 3 March 2021).
35. NCBI. 2020. Available online: https://www.ncbi.nlm.nih.gov (accessed on 3 March 2021).
36. Ratnasingham, S.; Hebert, P.D.N. BOLD: The Barcode of Life Data System (http://www.barcodinglife.org). *Mol. Ecol. Notes* **2007**, *7*, 355–364. [CrossRef]
37. Rafinesque, C.S. *Analyse de la Nature ou Tableau de l'Universe et Des Corps Organize*; De l'imprimerie de Jean Barravecchia: Palermo, Italy, 1815.
38. Chamberlin, R.V. The Annelida Polychaeta. *Mem. Mus. Comp. Zool.* **1919**, *48*, 1–518.
39. Rioja, E. Estudio sistemático de las especies Ibéricas del Suborden Sabelliformia. *Trab. Mus. Nac. Cien. Nat. (Ser. Zool.)* **1923**, *48*, 1–144.
40. Pillai, G. Studies on a collection of spirorbids from Ceylon, together with a critical review and revision of spirorbid systematics and an account of their phylogeny and zoogeography. *Ceylon J. Sci. (Biol. Sci.)* **1970**, *8*, 100–172.
41. Ten Hove, H.A. Towards a phylogeny in serpulids (Annelida; Polychaeta). In *Proceedings of the First International Polychaete Conference*; Hutchings, P.A., Ed.; Linnean Society of New South West: Sydney, Australia, 1984; pp. 181–196.
42. Fitzhugh, K. A systematic revision of the Sabellidae-Caobangiidae-Sabellongidae complex (Annelida: Polychaeta). *Bull. Am. Mus. Nat. Hist.* **1989**, *192*, 1–104.
43. Kupriyanova, E.K. Life history evolution in serpulimorph polychaetes: A phylogenetic analysis. *Hydrobiologia* **2003**, *496*, 105–114. [CrossRef]
44. Macdonald, T.A. Phylogenetic relations among spirorbid subgenera and the evolution of opercular brooding. *Hydrobiologia* **2003**, *496*, 125–143. [CrossRef]
45. Kupriyanova, E.K.; Macdonald, T.A.; Rouse, G.W. Phylogenetic relationships within Serpulidae (Sabellida, Annelida) inferred from molecular and morphological data. *Zool. Scr.* **2006**, *35*, 421–439. [CrossRef]
46. Kupriyanova, E.K.; ten Hove, H.A.; Sket, B.; Zakšek, V.; Trontelj, P.; Rouse, G.W. Evolution of the unique freshwater cave-dwelling tube worm *Marifugia cavatica* (Annelida: Serpulidae). *Syst. Biodivers* **2009**, *7*, 389–401. [CrossRef]
47. Lehrke, J.; Harry, A.; Macdonald, T.A.; Bartolomaeus, T.; Bleidorn, C. Phylogenetic relationships of Serpulidae (Annelida: Polychaeta) based on 18S rDNA sequence data, and implications for opercular evolution. *Org. Divers Evol.* **2007**, *7*, 195–206. [CrossRef]
48. Kupriyanova, E.K.; Nishi, E. Serpulidae (Annelida, Polychaeta) from Patton-Murray Seamounts, Gulf of Alaska, North Pacific Ocean. *Zootaxa* **2010**, *2665*, 51–68. [CrossRef]
49. Prentiss, N.K.; Vasileiadou, K.; Faulwetter, S.; Arvanitidis, C.; ten Hove, H.A. A new genus and species of Serpulidae (Annelida, Polychaeta, Sabellida) from the Caribbean Sea. *Zootaxa* **2014**, *3900*, 204–222. [CrossRef]
50. Rzhavsky, A.V.; Kupriyanova, E.K.; Sikorski, A.V. Two new species of serpulid polychaetes (Annelida) from the Barents Sea. *Fauna Norv.* **2013**, *32*, 27–38. [CrossRef]
51. Johansson, K.E. Beiträge Zur Kenntnis der Polychaeten-Familien Hermellidae Sabellidae und Serpulidae. *Zool. Bidr. Uppsala* **1927**, *11*, 1–184.
52. Fitzhugh, K. Further revisions of the Sabellidae subfamilies and cladistic relationships among the Fabriciinae (Annelida: Polychaeta). *Zool. J. Linn. Soc.* **1991**, *102*, 305–332. [CrossRef]
53. Fitzhugh, K. Systematics of several fabriciin fan worms (Polychaeta: Sabellidae: Fabriciinae) previously referred to *Fabricia* or *Fabriciola*. *J. Nat. Hist.* **1991**, *25*, 1101–1120. [CrossRef]
54. Fitzhugh, K. On the systematic position of *Monroika africana* (Monro) (Polychaeta: Sabellidae: Fabriciinae) and a description of a new fabriciin genus and species from Australia. *Proc. Biol. Soc. Wash.* **1992**, *105*, 116–131.
55. Fitzhugh, K. Additions to the description of the fanworm genus *Pseudofabricia* Cantone, 1972 (Polychaeta: Sabellidae: Fabriciinae). *Contrib. Sci.* **1995**, *456*, 1–6.
56. Fitzhugh, K. New fanworm species (Polychaeta: Sabellidae: Fabriciinae) in the genus *Pseudofabriciola* Fitzhugh. *J. Nat. Hist.* **1996**, *30*, 1267–1286. [CrossRef]
57. Fitzhugh, K. New fan worm genera and species (Polychaeta, Sabellidae, Fabriciinae) from the Western Pacific, and cladistic relationships among genera. *Zool. Scr.* **1998**, *27*, 209–245. [CrossRef]
58. Fitzhugh, K. New species of *Fabricinuda* Fitzhugh and *Pseudofabriciola* Fitzhugh (Polychaeta: Sabellidae: Fabriciinae), with an emendation of *Pseudofabriciola australiensis* (Hartmann-Schröder). *J. Nat. Hist.* **2002**, *36*, 893–925. [CrossRef]
59. Nogueira, J.M.D.M.; Fitzhugh, K.; Rossi, M.C.S. A new genus and new species of fan worms (Polychaeta: Sabellidae) from Atlantic and Pacific Oceans—the formal treatment of taxon names as explanatory hypotheses. *Zootaxa* **2010**, *2603*, 1–52. [CrossRef]
60. Fitzhugh, K.; Giangrande, A.; Simboura, N. New species of *Pseudofabriciola* Fitzhugh, 1990 (Polychaeta: Sabellidae: Fabriciinae), from the Mediterranean Sea. *Zool. J. Linn. Soc.* **1994**, *110*, 219–241. [CrossRef]
61. Fitzhugh, K. *Fabricinuda*, a new genus of Fabriciinae (Polychaeta: Sabellidae). *Proc. Biol. Soc. Wash.* **1990**, *103*, 161–178.
62. Fitzhugh, K. Two new genera of the Subfamily Fabriciinae (Polychaeta: Sabellidae). *Am. Mus. Novit.* **1990**, *2967*, 1–19.
63. Fitzhugh, K. A new deep-water genus and species of Fabriciinae fanworm (Polychaeta: Sabellidae) from Antarctica. *Contrib. Sci.* **2001**, *491*, 1–8.

64. Fitzhugh, K. A new species of *Megalomma* Johansson, 1927 (Polychaeta: Sabellidae: Sabellinae) from Taiwan, with comments on sabellid dorsal lip classification. *Zool. Stud.* **2003**, *42*, 106–134.
65. Rouse, G.; Fitzhugh, K. Broadcasting fables: Is external fertilization really primitive? sex, size, and larvae in sabellid polychaetes. *Zool Scr.* **1994**, *23*, 271–312. [CrossRef]
66. Fitzhugh, K.; Rouse, G.W. A remarkable new genus and species of fan worm (Polychaeta: Sabellidae: Sabellinae) associated with marine gastropods. *Invertebr. Biol.* **1999**, *118*, 357–390. [CrossRef]
67. Cochrane, S. Snowflakes and feather-dusters–some challenges for soft-bottom fanworm systematics. *Hydrobiologia* **2003**, *496*, 49–62. [CrossRef]
68. Capa, M. Taxonomic revision and phylogenetic relationships of apomorphic sabellids (Polychaeta) from Australia. *Invertebr. Syst.* **2007**, *21*, 537. [CrossRef]
69. Capa, M. The genera *Bispira* Krøyer, 1856 and *Stylomma* Knight-Jones, 1997 (Polychaeta, Sabellidae): Systematic revision, relationships with close related taxa and new species from Australia. *Hydrobiologia* **2008**, *596*, 301–327. [CrossRef]
70. Capa, M.; Bybee, D.R.; Bybee, S.M. Establishing species and species boundaries in *Sabellastarte* Krøyer, 1856 (Annelida: Sabellidae): An integrative approach. *Org. Divers. Evol.* **2010**, *10*, 351–371. [CrossRef]
71. Tovar-Hernández, M.A. Phylogeny of *Chone* Krøyer, 1856 (Polychaeta: Sabellidae) and related genera. *J. Nat. Hist.* **2008**, *42*, 2193–2226. [CrossRef]
72. Capa, M.; Pons, J.; Hutchings, P. Cryptic diversity, intraspecific phenetic plasticity and recent geographical translocations in *Branchiomma* (Sabellidae, Annelida). *Zool. Scr.* **2013**, *42*, 637–655. [CrossRef]
73. Capa, M.; Murray, A. Integrative taxonomy of *Parasabella* and *Sabellomma* (Sabellidae: Annelida) from Australia: Description of new species, indication of cryptic diversity, and translocation of some species out of their natural distribution range: Australian *Parasabella* and *Sabellomma*. *Zool. J. Linn. Soc.* **2015**, *175*, 764–811. [CrossRef]
74. Capa, M.; Murray, A. Combined morphological and molecular data unveils relationships of *Pseudobranchiomma* (Sabellidae, Annelida) and reveals higher diversity of this intriguing group of fan worms in Australia, including potentially introduced species. *ZooKeys* **2016**, *622*, 1–36. [CrossRef]
75. Giangrande, A.; Licciano, M. The genus *Euchone* (Polychaeta, Sabellidae) in the Mediterranean Sea, addition of two new species and discussion on some closely related taxa. *J. Nat. Hist.* **2006**, *40*, 1301–1330. [CrossRef]
76. Licciano, M.; Giangrande, A.; Gambi, M.C. A new genus of Sabellidae (Annelida, Polychaeta) from Antarctica, with discussion of relationships among plesiomorphic genera within Sabellinae. *Zootaxa* **2009**, *2226*, 28–42. [CrossRef]
77. Kupriyanova, E.K.; Bastida-Zavala, R.; Halt, M.N.; Lee, M.S.Y.; Rouse, G.W. Phylogeny of the *Serpula-Crucigera-Hydroides* clade (Serpulidae: Annelida) using molecular and morphological data: Implications for operculum evolution. *Invertebr. Syst.* **2008**, *22*, 425–437. [CrossRef]
78. Sun, Y.; Kupriyanova, E.K.; Qiu, J.W. COI Barcoding of *Hydroides*: A road from impossible to difficult. *Invertebr. Syst.* **2012**, *26*, 539–547. [CrossRef]
79. Sun, Y.; Wong, E.; Ahyong, S.T.; Williamson, J.E.; Hutchings, P.A.; Kupriyanova, E.K. Barcoding and multi-locus phylogeography of the globally distributed calcareous tubeworm genus *Hydroides* Gunnerus, 1768 (Annelida, Polychaeta, Serpulidae). *Mol. Phylogenet. Evol* **2018**, *127*, 732–745. [CrossRef] [PubMed]
80. Tovar-Hernández, M.A.; Villalobos-Guerrero, T.F.; Kupriyanova, E.K.; Sun, Y. A new fouling *Hydroides* (Annelida, Sabellida, Serpulidae) from Southern Gulf of California. *J. Mar. Biol. Assoc. UK* **2016**, *96*, 693. [CrossRef]
81. Smith, A.M.; Henderson, Z.E.; Kennedy, M.; King, T.M.; Spencer, H.G. Reef formation versus solitariness in two New Zealand serpulids does not involve cryptic species. *Aquat. Biol* **2012**, *16*, 97–103. [CrossRef]
82. Willette, D.A.; Iniguez, A.R.; Kupriyanova, E.K.; Starger, C.J.; Varman, T.; Toha, A.H.; Maralit, B.A.; Barber, P.H. Christmas tree worms of Indo-Pacific coral reefs: Untangling the *Spirobranchus corniculatus* (Grube, 1862) complex. *Coral Reefs* **2015**, *34*, 899–904. [CrossRef]
83. Perry, O.; Bronstein, O.; Simon-Blecher, N.; Atkins, A.; Kupriyanova, E.; ten Hove, H.; Levy, O.; Fine, M. On the genus *Spirobranchus* (Annelida, Serpulidae) from the Northern Red Sea, and a description of a new species. *Inverterb. Syst.* **2018**, *32*, 605. [CrossRef]
84. Pazoki, S.; Rahimian, H.; Struck, T.H.; Katouzian, A.R.; Kupriyanova, E.K. A new species of the *Spirobranchus kraussii*-complex (Annelida, Serpulidae) from the Persian Gulf and Gulf of Oman. *Zootaxa* **2020**, *4748*, 401–430. [CrossRef] [PubMed]
85. Bailey, J.H. Methods of brood protection as a basis for reclassification of the Spirorbinae (Serpulidae). *Zool. J. Linn. Soc.* **1969**, *48*, 387–407. [CrossRef]
86. Thorp, C.H. The structure of the operculum in *Pileolaria granulata* (L.) (Polychaeta: Serpulidae) and related species. *J. Exp. Mar. Biol. Ecol.* **1975**, *20*, 215–235. [CrossRef]
87. Knight-Jones, P.; Thorp, C.H. The opercular brood chambers of Spirorbidae. *J. Linn. Soc. (Zool.)* **1984**, *80*, 121–133. [CrossRef]
88. Nishi, E. On the origin of brooding characteristics in spirorbids with the phylogeny of sabellids and serpulids (Annelida, Polychaeta, Sedentaria). *Proc. Jap. Soc. Syst. Zool.* **1993**, *49*, 6–12.
89. Thorp, C.; Segrove, F. The opercular moult in *Spirorbis spirorbis* (L.) and *S. pusilloides* Bush (Polychaeta: Serpulidae). *J. Exp. Mar. Biol. Ecol.* **1975**, *19*, 117–143. [CrossRef]
90. Rzhavsky, A.V.; Kupriyanova, E.K. Evolution of spirorbin brooding: A phylogenetic analysis and a test of an oxygen limitation hypothesis. *Invertebr. Zool.* **2019**, *16*, 409–430. [CrossRef]

91. Pamungkas, J.; Glasby, C.J.; Read, G.B.; Wilson, S.P.; Costello, M.J. Progress and perspectives in the discovery of polychaete worms (Annelida) of the world. *Helgol. Mar. Res.* **2019**, *73*, 4. [CrossRef]
92. Fitzhugh, K. A Revision of the genus *Fabricia* Blainville, 1828 (Polychaeta: Sabellidae: Fabriciinae). *Sarsia* **1990**, *75*, 1–16. [CrossRef]
93. Cepeda, D.; Lattig, P. New reports and description of a new species of Sabellidae (Annelida) for the Iberian Peninsula and Balearic Archipelago. *Mar. Biol. Res.* **2017**, *13*, 832–853. [CrossRef]
94. Giangrande, A.; Wasson, B.; Lezzi, M.; Licciano, M. Description of *Euchone anceps* sp. nov. (Annelida: Sabellidae) from the Mediterranean Sea and Northeast Atlantic, with remarks on the difficulty of generic definition. *Eur. Zool. J.* **2017**, *84*, 193–207. [CrossRef]
95. Çinar, M.E.; Giangrande, A. A new species of *Pseudobranchiomma* (Sabellidae, Polychaeta) from the Sea of Marmara (Turkey). *Mar. Biodivers* **2018**, *48*, 1563–1569. [CrossRef]
96. López, E.; Olivier, F.; Grant, C.; Archambault, P.A. new species and four new records of sedentary polychaetes from the Canadian High Arctic. *J. Mar. Biol. Assoc. UK* **2017**, *97*, 1685–1694. [CrossRef]
97. Nishi, E.; Gil, J.; Tanaka, K.; Kupriyanova, E.K. *Notaulax yamasui* sp. n. (Annelida, Sabellidae) from Okinawa and Ogasawara, Japan, with notes on its ecology. *ZooKeys* **2017**, *660*, 1–16. [CrossRef]
98. Nishi, E.; Tanaka, K.; Tovar-Hernández, M.A. A new species of *Claviramus* (Annelida, Sabellida, Sabellidae) from the Ariake Inland Sea, Kyushu, Japan. *ZooKeys* **2019**, *880*, 25–32. [CrossRef] [PubMed]
99. Tovar-Hernández, M.A.; de León-González, J.A.; Bybee, D.R. Sabellid worms from the Patagonian Shelf and Humboldt Current System (Annelida, Sabellidae): Phyllis Knight-Jones' and José María Orensanz's collections. *Zootaxa* **2017**, *4283*, 1–64. [CrossRef]
100. Tovar-Hernández, M.A.; Harry, A.; Vinn, O.; Zatoń, M.; de León-González, J.A.; García-Garza, M.E. Fan worms (Annelida: Sabellidae) from Indonesia collected by the Snellius II Expedition (1984) with descriptions of three new species and tube microstructure. *PeerJ* **2020**, *8*, e9692. [CrossRef]
101. Tovar-Hernández, M.A.; García-Garza, M.E.; de León-González, J.A. Sclerozoan and fouling sabellid worms (Annelida: Sabellidae) from Mexico with the establishment of two new species. *Biodivers. Data J.* **2020**, *8*, e57471. [CrossRef]
102. Bick, A.; Bastrop, R.; Kotta, J.; Meißner, K.; Meyer, M.; Syomin, V. Description of a new species of Sabellidae (Polychaeta, Annelida) from fresh and brackish waters in Europe, with some remarks on the branchial crown of *Laonome*. *Zootaxa* **2018**, *4483*, 349–364. [CrossRef]
103. Tilic, E.; Feerst, K.G.; Rouse, G.W. Two new species of *Amphiglena* (Sabellidae, Annelida), with an assessment of hidden diversity in the Mediterranean. *Zootaxa* **2019**, *4648*, 337–353. [CrossRef]
104. Renier, S.A. Osservazioni postume di Zoologia Adriatica, del Professore Stefano Andrea Renier, Membro Effettivo dell'Instituto Italiano, Pubblicate per Cura dell'IR Istituto Veneto Di Scienze, Lettere er Arti a Studio Del Membro Effettivo Prof. G. Meneghini. Giovanni Cecchini: Venice, Italia, 1847; pp. 1–122.
105. Linnaeus, C. *Systema Naturae*, 10th ed.; L. Salvius: Holmiae, Sweden, 1758.
106. Pillai, T.G. A revision of the genera *Galeolaria* and *Pyrgopolon* (Polychaeta: Serpulidae), with discussions on opercular insertion as a character in their taxonomy and relationships, and their zoogeography. *Zootaxa* **2009**, *2060*, 47–58. [CrossRef]
107. Kupriyanova, E.K.; Ippolitov, A.P. Deep-sea serpulids (Annelida: Polychaeta) in tetragonal tubes: On a tube convergence path from the Mesozoic to Recent. *Zootaxa* **2015**, *4044*, 151–200. [CrossRef]
108. Tovar-Hernández, M.A. Taxonomic update of the sabellids (Polychaeta: Sabellidae) from Chile and taxa established by prof. Ernst Ehlers, with a key to genera of Sabellinae. *An. Inst. Patagon.* **2010**, *38*, 7–29. [CrossRef]
109. Tovar-Hernández, M.A.; Fitzhugh, K. Sabellidae Latreille, 1825. In *Poliquetos (Annelida: Polychaeta) de México y América Tropical*, 2nd ed.; De León-González, J.A., Bastida-Zavala, J.R., Carrera-Parra, L.F., García-Garza, M.E., Salazar-Vallejo, S.I., Solís-Weiss, V., Tovar-Hernández, M.A., Eds.; Universidad Autónoma de Nuevo León: Monterrey, Mexico, 2021; Volume 3.
110. Ten Hove, H.A.; Kupriyanova, E.K. Taxonomy of Serpulidae (Annelida, Polychaeta): The state of affairs. *Zootaxa* **2009**, *2036*, 1–126. [CrossRef]
111. Knight-Jones, P.; Darbyshire, T.; Petersen, M.E.; Tovar-Hernández, M. What is *Pseudopotamilla reniformis* (Sabellidae)? Comparisons of populations from Britain, Iceland and Canada with comments on *Eudistylia* and *Schizobranchia*. *Zootaxa* **2017**, *4254*, 201–220. [CrossRef] [PubMed]
112. Jirkov, I. *Polikhety Severnogo Ledovitogo Okeana [Polychaeta of the Arctic Ocean.]*; Janus-K Press: Moscow, Russia, 2001; pp. 1–632. (In Russian)
113. Rzhavsky, A.V.; Kupriyanova, E.K.; Sikorski, A.V.; Dahle, S. *Calcareous Tubeworms (Polychaeta, Serpulidae) of the Arctic Ocean*; KMK Scientific Press: Moscow, Russia, 2014; pp. 1–187.
114. Rzhavsky, A.; Kupriyanova, E.; Sikorski, A. *Field guide to Calcareous Tubeworms (Polychaeta, Serpulidae) of the Arctic Ocean*; KMK Scientific Press: Moscow, Russia, 2018; pp. 1–184.
115. Fauvel, P. Polychètes sédentaires. Addenda aux errantes, Arachiannélides, Myzostomaires. In *Faune de France*; Paul Lechevalier: Paris, France, 1927; Volume 16, pp. 1–494.
116. Zibrowius, H. Étude morphologique, systématique et écologique des Serpulidae (Annelida Polychaeta) de La région de Marseille. *Recl. Trav. Stn. Mar. Endoume Bull.* **1968**, *43*, 81–252.
117. Bianchi, C.N. Les espèces de Serpuloidea (Annélides, Polychaètes) des lagunes côtières Italiennes. *Rapp. Commis. Int. Mer. Médit. Monaco* **1981**, *27*, 195–196.

118. Knight-Jones, P. Sabellidae and Serpulidae. In *The Marine Fauna of the British Isles and North West Europe*; Hayward, P.J., Ryland, J.S., Eds.; Clarendon Press: Oxford, UK, 1990; Volume 1, pp. 271–287.
119. López, E.; Cepeda, D. Familia Fabriciidae Rioja, 1923. In *Fauna Ibérica*; Museo Nacional de Ciencias Naturales, Consejo Superior de Investigaciones Científicas: Madrid, Spain, 2020.
120. Cepeda, D.; López, E. Familia Sabellidae Latreille, 1825. In *Fauna Ibérica*; Museo Nacional de Ciencias Naturales, Consejo Superior de Investigaciones Científicas: Madrid, Spain, 2020.
121. Bick, A. Redescription of *Fabriciola tonerella* Banse, 1959, and a new record of *Novafabricia infratorquata* (Fitzhugh, 1983) from the Mediterranean Sea, with a key for the Fabriciinae (Annelida: Polychaeta) of the Mediterranean Sea and the North-East Atlantic. *Zool. Anz. J. Comp. Zool.* **2005**, *244*, 137–152. [CrossRef]
122. Bick, A.; Randel, N. Ontogenetic variations in characters of *Euchone analis* (Krøyer, 1856) (Polychaeta, Sabellidae, Sabellinae) from Spitsbergen, and new assignments of *Oriopsis ingelorae* Plate, 1995 and *O. liefdefjordensis* Plate, 1995. *Acta Zool.* **2005**, *86*, 145–157. [CrossRef]
123. Tovar-Hernández, M.A.; Licciano, M.; Giangrande, A. Revision of *Chone* Krøyer, 1856 (Polychaeta: Sabellidae) from the Eastern Central Atlantic and Mediterranean Sea with descriptions of two new species. *Sci. Mar.* **2007**, *71*, 315–338. [CrossRef]
124. Tovar-Hernández, M.A. Revision of *Chone* Krøyer, 1856 (Polychaeta: Sabellidae) from North America and descriptions of four new species. *J. Nat. Hist.* **2007**, *41*, 511–566. [CrossRef]
125. Çinar, M. Serpulid species (Polychaeta: Serpulidae) from the Levantine coast of Turkey (Eastern Mediterranean), with special emphasis on alien species. *Aquat. Invasions* **2006**, *1*, 223–240. [CrossRef]
126. Çinar, M.E. Alien polychaete species (Annelida: Polychaeta) on the Southern coast of Turkey (Levantine Sea, Eastern Mediterranean), with 13 new records for the Mediterranean Sea. *J. Nat. Hist.* **2009**, *43*, 2283–2328. [CrossRef]
127. Çinar, M.E.; Dağli, E.; Şahin, G.K. Checklist of Annelida from the coasts of Turkey. *Turk. Zool. Derg.* **2014**, *38*, 734–764. [CrossRef]
128. Selim, S.A.; Rzhavsky, A.V.; Britayev, T.A. *Dialychone* and *Paradialychone* (Polychaeta: Sabellidae) from the Mediterranean Coast of Egypt with description of *Dialychone egyptica* sp. n. *Invertebr. Zool.* **2012**, *9*, 105–114. [CrossRef]
129. Giangrande, A.; Cosentino, A.; Presti, C.L.; Licciano, M. Sabellidae (Annelida) from the Faro coastal lake (Messina, Ionian Sea), with the first record of the invasive species *Branchiomma bairdi* along the Italian Coast. *Mediterr. Mar. Sci.* **2012**, *13*, 283–293. [CrossRef]
130. Giangrande, A.; Caruso, L.; Mikac, B.; Licciano, M. The genus *Megalomma* (Annelida: Sabellidae) in the Mediterranean Sea, with description of two new species from Italian and Croatian Coasts. *Ital. J. Zool.* **2015**, *82*, 521–534. [CrossRef]
131. Mikac, B.A. Sea of worms: Polychaete checklist of the Adriatic Sea. *Zootaxa* **2015**, *3943*, 1–172. [CrossRef]
132. Mikac, B.; Licciano, M.; Jaklin, A.; Iveša, L.; Giangrande, A.; Musco, L. Diversity and distribution patterns of hard bottom polychaete assemblages in the North Adriatic Sea (Mediterranean). *Diversity* **2020**, *12*, 408. [CrossRef]
133. Mikac, B.; Giangrande, A.; Licciano, M. Sabellidae and Fabriciidae (Polychaeta) of the Adriatic Sea with particular retrospect to the Northern Adriatic and the description of two new species. *J. Mar. Biol. Assoc. UK* **2013**, *93*, 1511–1524. [CrossRef]
134. López, E.; Cepeda, D. Catálogo de las especies de Sabellidae Latreille, 1825 y Fabriciidae Rioja, 1923 (Annelida; Sabellida) citadas en la Península Ibérica e Islas Baleares y Chafarinas. *Graellsia* **2017**, *73*, e064. [CrossRef]
135. Kurt Şahin, G.; Çınar, M.E. A check-list of polychaete species (Annelida: Polychaeta) from the Black Sea. *J. Black Sea/Medit. Environ.* **2012**, *18*, 10–48.
136. Hartman, O. *Atlas of the Sedentariate Polychaetous Annelids from California*; Allan Hancock Foundation, University of Southern California: Los Angeles, CA, USA, 1969; pp. 1–812.
137. Knight-Jones, P.; Knight-Jones, E.W.; Dales, R.P. Spirorbidae (Polychaeta Sedentaria) from Alaska to Panama. *J. Zool.* **1979**, *189*, 419–458. [CrossRef]
138. Buzhinskaja, G. *Polychaetes of the Far East. Seas of Russia and Adjacent Waters of the Pacific Ocean.: Annotated Checklist and Bibliography*; KMK Scientific Press: Moscow, Russia, 2013; pp. 1–131.
139. Nishi, E.; Tanaka, K.; Tovar-Hernández, M.A.; Giangrande, A. *Dialychone*, *Jasmineira* and *Paradialychone* (Annelida: Polychaeta: Sabellidae) from Japan and adjacent waters, including four new species descriptions. *Zootaxa* **2009**, *2167*, 1–24. [CrossRef]
140. Yoshihara, T.; Hiruta, S.; Katoh, T.; Kajihara, H. Three species of *Amphicorina* (Annelida, Sabellida, Sabellidae) from Japan, with descriptions of two new species. *Zookeys* **2012**, *187*, 45–62. [CrossRef]
141. Imajima, M. *Kankeidoubutsu Tamourui (Annelida Polychaeta)*; Biological Research Co.: Tokyo, Japan, 1996; pp. 1–530. (In Japanese)
142. Sun, R.; Yang, D. Polychaeta III. Sabellida. In *Fauna Sinica: Invertebrata*; Science Press: Beijing, China, 2014.
143. Glasby, C.J.; Lee, Y.-L.; Hsueh, P.-W. Marine Annelida (excluding clitellates and siboglinids) from the South China Sea. *Raffles Bull. Zool.* **2016**, *34*, 178–234.
144. Bastida-Zavala, J.R.; ten Hove, H.A. Revision of *Hydroides* Gunnerus, 1768 (Polychaeta: Serpulidae) from the Eastern Pacific Region and Hawaii. *Beaufortia* **2003**, *53*, 67–110.
145. Bastida-Zavala, J.R. Serpulids (Annelida: Polychaeta) from the Eastern Pacific, including a brief mention of Hawaiian serpulids. *Zootaxa* **2008**, *1722*, 1–61.
146. Bastida-Zavala, J.R.; McCann, L.D.; Keppel, E.; Ruiz, G.M. The fouling serpulids (Polychaeta: Serpulidae) from United States coastal waters: An overview. *Eur. J. Taxon.* **2017**, *344*, 1–76. [CrossRef]
147. Tovar-Hernández, M.A.; Méndez, N.; Villalobos-Guerrero, T.F. Fouling polychaete worms from the Southern Gulf of California: Sabellidae and Serpulidae. *Syst. Biodivers* **2009**, *7*, 319–336. [CrossRef]

148. Bastida-Zavala, J.R.; Buelna, A.S.R.; De León-González, J.A.; Camacho-Cruz, K.A.; Carmona, I. New records of sabellids and serpulids (Polychaeta: Sabellidae, Serpulidae) from the Tropical Eastern Pacific. *Zootaxa* **2016**, *4184*, 401–457. [CrossRef]
149. Tovar-Hernández, M.A.; Carrera-Parra, L.F. *Megalomma* Johansson, 1925 (Polychaeta: Sabellidae) from America and other world-wide localities, and phylogenetic relationships within the genus. *Zootaxa* **2011**, *2861*, 1–71. [CrossRef]
150. Zibrowius, H. Contribution a l'etude des Serpulidae (Polychaeta Sedentaria) du Bresil. *Bol. Inst. Oceanogr.* **1970**, *19*, 1–32. [CrossRef]
151. Ten Hove, H.A. Serpulinae (Polychaeta) from the Caribbean: I-the genus *Spirobranchus*. *Stud. Fauna Curaçao* **1970**, *32*, 1–57.
152. Ten Hove, H.A. Serpulinae (Polychaeta) from the Caribbean: II-the genus *Sclerostyla*. *Stud. Fauna Curaçao* **1973**, *43*, 1–21.
153. ten Hove, H.A. Serpulinae (Polychaeta) from the Caribbean: III-the genus *Pseudovermilia*. *Stud. Fauna Curaçao* **1975**, *47*, 46–101.
154. Ten Hove, H.A. Serpulinae (Polychaeta) from the Caribbean: IV *Pseudovermilia madracicola* sp. n., a symbiont of corals. In *Natuurwetenschappelijke Studiekring voor Suriname en de Nederlandse Antillen*; Van Der Steen, L.J., Ed.; 1989; Volume 123, pp. 135–144.
155. Giangrande, A.; Licciano, M.; Gambi, M.C. A collection of Sabellidae (Polychaeta) from Carrie Bow Cay (Belize, Western Caribbean Sea) with the description of two new species. *Zootaxa* **2007**, *1650*, 41–53. [CrossRef]
156. Tovar-Hernández, M.A. Redescription of *Chone americana* Day, 1973 (Polychaeta: Sabellidae) and description of five new species from the Grand Caribbean Region. *Zootaxa* **2005**, *1070*, 1–30. [CrossRef]
157. Tovar-Hernández, M.A.; Knight-Jones, P. Species of *Branchiomma* (Polychaeta: Sabellidae) from the Caribbean Sea and Pacific Coast of Panama. *Zootaxa* **2006**, *1189*, 1–37. [CrossRef]
158. Tovar-Hernández, M.A.; Salazar-Vallejo, S.I. Sabellids (Polychaeta: Sabellidae) from the Grand Caribbean. *Zool. Stud.* **2006**, *45*, 24–66.
159. Bastida-Zavala, J.R.; ten Hove, H.A. Revision of *Hydroides* Gunnerus, 1768 (Polychaeta: Serpulidae) from the Western Atlantic Region. *Beaufortia* **2002**, *52*, 103–173.
160. Bastida-Zavala, R. *Serpula* and *Spiraserpula* (Polychaeta, Serpulidae) from the Tropical Western Atlantic and Gulf of Guinea. *ZooKeys* **2012**, *198*, 1–23. [CrossRef] [PubMed]
161. Nogueira, J.; Knight-Jones, P.A. New species of *Pseudobranchiomma* Jones (Polychaeta: Sabellidae) found amongst Brazilian coral, with a redescription of *P. punctata* (Treadwell, 1906) from Hawaii. *J. Nat. Hist.* **2002**, *36*, 1661–1670. [CrossRef]
162. Nogueira, J.M.D.M.; Abbud, A. Three new serpulids (Polychaeta: Serpulidae) from the Brazilian Exclusive Economic Zone. *Zoosymposia* **2009**, *2*, 201–227. [CrossRef]
163. Amaral, A.C.Z.; Nallin, S.A.H.; Steiner, T.M.; Forroni, T.D.O.; Gomes-Filho, D. *Catálogo das Espécies de Annelida Polychaeta do Brasil*; Unicamp: Campinas, Brazil, 2013; pp. 1–159.
164. Costa, D. de A.; Fernandes, H.F.; da Silva, F. de A.; Christoffersen, M.L. Checklist de espécies de Polychaeta (Annelida) da Praia do Seixas, João Pessoa, Estado Da Paraíba, Nordeste do Brasil. *Rev. Bras. Gestão Amb. Sust.* **2017**, *4*, 313–320. [CrossRef]
165. Bastida-Zavala, J.R.; Sánchez Ovando, J.P. Serpulidae Rafinesque 1815. In *Poliquetos (Annelida: Polychaeta) de México y América Tropical*, 2nd ed.; De León-González, J.A., Bastida-Zavala, J.R., Carrera-Parra, L.F., García-Garza, M.E., Salazar-Vallejo, S.I., Solís-Weiss, V., Tovar-Hernández, M.A., Eds.; Universidad Autónoma de Nuevo León: Monterrey, Mexico, 2021; Volume 3.
166. Fauvel, P. Annelida Polychaeta. In *Fauna of India, Pakistan, Ceylon, Burma and Malaya*; The Indian Press Ltd: Allahabad, India, 1953; pp. 1–507.
167. Sivadas, S.K.; Carvalho, R. Marine Annelida of India: Taxonomy and status evaluation and an updated checklist. *J. Threat. Taxa* **2020**, *12*, 16647–16714.
168. Pillai, G. Some marine and brackish-water serpulid polychaetes from Ceylon, including new genera and species. *Ceylon J. Sci. (Biol. Sci.)* **1960**, *3*, 1–40.
169. Pillai, T.G. Annelida Polychaeta of Tambalagam Lake, Ceylon. *Ceylon J. Sci. (Biol. Sci.)* **1961**, *4*, 1–40.
170. Wehe, T.; Fiege, D. Annotated checklist of the polychaete species of the seas surrounding the Arabian Peninsula: Red Sea, Gulf of Aden, Arabian Sea, Gulf of Oman, Arabian Gulf. *Fauna Arabia* **2002**, *19*, 7–238.
171. Al-Kandari, M.; Sattari, Z.; Hussain, S.; Radashevsky, V.I.; Zhadan, A. Checklist of intertidal Polychaetes (Annelida) of Kuwait, Northern Part of the Arabian Gulf. *Reg. Stud. Mar. Sci.* **2019**, *32*, 100872. [CrossRef]
172. Ben-Eliahu, M.N.; ten Hove, H.A. Serpulidae (Annelida: Polychaeta) from the Suez Canal—from a Lessepsian migration perspective (a monograph). *Zootaxa* **2011**, *2848*, 1–147. [CrossRef]
173. Knight-Jones, E.W.; Knight-Jones, P.; Oliver, P.G.; Mackie, A.S.Y. A.S.Y. A new species of *Hyalopomatus* (Serpulidae: Polychaeta) which lacks an operculum: Is this an adaptation to low oxygen. In *Interactions and Adaptation Strategies of Marine Organisms*; Naumov, A.D., Hummel, H., Sukhotin, A.A., Ryland, J.S., Eds.; Springer: Dordrecht, The Netherlands, 1997; pp. 145–151.
174. Veron, J.E.; Devantier, L.M.; Turak, E.; Green, A.L.; Kininmonth, S.; Stafford-Smith, M.; Peterson, N. Delineating the Coral Triangle. *Galaxea* **2009**, *11*, 91–100. [CrossRef]
175. Fitzhugh, K. Fan worm polychaetes (Sabellidae: Sabellinae) collected during the Thai-Danish BIOSHELF Project. *Phuket Mar. Biol. Center Spec. Publ.* **2002**, *24*, 353–424.
176. Al-Hakim, I.; Glasby, C.J. Polychaeta (Annelida) of the Natuna Islands, South China Sea. *Raffles Bull. Zool.* **2004**, *21*, 25–45.
177. Sun, Y.; ten Hove, H.A.; Qiu, J.W. Serpulidae (Annelida: Polychaeta) from Hong Kong. *Zootaxa* **2012**, *3424*, 1–42. [CrossRef]
178. Murray, A.; Rouse, G.W. Two new species of *Terebrasabella* (Annelida: Sabellidae: Sabellinae) from Australia. *Zootaxa* **2007**, *1434*, 51–68. [CrossRef]

179. Kupriyanova, E.K.; Sun, Y.; ten Hove, H.A.; Wong, E.; Rouse, G.W. Serpulidae (Annelida) of Lizard Island, Great Barrier Reef, Australia. *Zootaxa* **2015**, *4019*, 275–353. [CrossRef]
180. Sun, Y.; Wong, E.; ten Hove, H.A.; Hutchings, P.A.; Williamson, J.E.; Kupriyanova, E.K. Revision of the genus *Hydroides* (Annelida: Serpulidae) from Australia. *Zootaxa* **2015**, *4009*, 1–99. [CrossRef] [PubMed]
181. Wilson, R.S.; Hutchings, P.A.; Glasby, C.J. *Polychaetes: An. Interactive Identification Guide*; CSIRO Publishing: Collingwood, VIC, Australia, 2003; pp. 1–24.
182. Kupriyanova, E.K.; Wong, E.; Hutchings, P.A. Invasive Polychaete Identifier–an Australian Perspective. Version 1.1 2013. Available online: http://polychaetes.australianmuseum.net.au/ (accessed on 3 March 2021).
183. Kupriyanova, E.; Hutchings, P.; Wong, E. A fully illustrated web-based guide to distinguish native and introduced polychaetes of Australia. *Manag. Biol. Invasions* **2016**, *7*, 305–312. [CrossRef]
184. Bailey-Brock, J.H.; Magalhães, W.F. A new species and record of Serpulidae (Annelida: Polychaeta) from Cross Seamount in the Hawaiian Chain. *Zootaxa* **2012**, *3192*, 49–58. [CrossRef]
185. Keppel, E.; Tovar-Hernández, M.A.; Ruiz, G. New records of the non-indigenous species *Branchiomma bairdi* and *B. conspersum* (Polychaeta: Sabellidae) on the Pacific Coast of North America. *Bioinvasions Rec.* **2018**, *7*, 229–236. [CrossRef]
186. Bailey-Brock, J.H.; Magalhães, W.F.; Brock, R.E. Coral reef inhabiting tubeworms (Polychaeta: Serpulidae) from Enewetak, Kwajalein, Rongelap and Utirik Atolls, Marshall Islands. *J. Mar. Biol. Assoc. UK* **2012**, *92*, 967–988. [CrossRef]
187. Capa, M.; López, E. Sabellidae (Annelida: Polychaeta) living in blocks of dead coral in the Coiba National Park, Panama. *J. Mar. Biol. Assoc. UK* **2004**, *84*, 63–72. [CrossRef]
188. Yáñez-Rivera, B.; Tovar-Hernández, M.A.; Galván-Villa, C.M.; Ríos-Jara, E. Tubicolous polychaete worms (Annelida) from Bahía de Chamela Islands Sanctuary, Mexico, with the description of a new bamboo worm. *Biodivers. Data J.* **2020**, *8*, e57572. [CrossRef]
189. Keppel, E.; Keith, I.; Ruiz, G.M.; Carlton, J.T. New records of native and non-indigenous Polychaetes (Annelida: Polychaeta) in the Galapagos Islands. *Aquat. Invasions* **2019**, *14*, 59–84. [CrossRef]
190. Elías, R.; Jaubet, M.L.; Ferrando, A.; Saracho, M. Historia y perspectivas de los estudios sobre poliquetos en Argentina. In *Poliquetos de Sudamérica*; Díaz-Díaz, O., Bone, D., Rodríguez, C.T., Delgado-Blas, V.H., Eds.; Poliquetos de Sudamérica. Volumen especial del Boletín del Instituto Oceanográfico de Venezuela: Cumaná, Venezuela, 2017; pp. 3–23.
191. Rozbaczylo, O.; Moreno, R.A.; Díaz-Díaz, Ó. Poliquetos Bentónicos en Chile. In *Poliquetos de Sudamérica*; Díaz-Díaz, Ó., Bone, D., Rodríguez, C.T., Delgado-Blas, V.H., Eds.; Volumen especial del Boletín del Instituto Oceanográfico de Venezuela: Cumaná, Venezuela, 2017; pp. 51–70.
192. Kupriyanova, E.K.; ten Hove, H.A.; Nishi, E.A. Taxonomic revision of *Pseudochitinopoma* Zibrowius, 1969 (Annelida, Serpulidae) with description of two new species. *Zootaxa* **2012**, *3507*, 57–78. [CrossRef]
193. Simon, C.A.; van Niekerk, H.H.; Burghardt, I.; ten Hove, H.A.; Kupriyanova, E.K. Not out of Africa: *Spirobranchus kraussii* (Baird, 1865) is not a global fouling and invasive serpulid of Indo-Pacific Origin. *Aquat. Invasions* **2019**, *14*, 221–249. [CrossRef]
194. Day, J.H. A Monograph on the Polychaeta of Southern Africa. Part II. *Brit. Mus. Nat. Hist. Publ.* **1967**, *656*, 459–878.
195. Knight-Jones, P. New species and a new subgenus of Spirorbinae (Serpulidae: Polychaeta) from Kenya. *J. Zool.* **1972**, *166*, 1–18. [CrossRef]
196. Knight-Jones, P.; Knight-Jones, E.W. Spirorbinae (Polychaeta: Serpulidae) from South Africa, including three new species. *Mar. Biol.* **1974**, *25*, 253–261. [CrossRef]
197. Knight-Jones, E.W.; Knight-Jones, P.; Llewellyn, L.C. Spirorbinae (Polychaeta: Serpulidae) from Southeastern Australia. Notes on their taxonomy, ecology and distribution. *Rec. Aust. Mus.* **1974**, *29*, 107–151. [CrossRef]
198. Vine, P.J. The marine fauna of New Zealand. Spirorbinae (Polychaeta: Serpulidae). *NZOI Mem.* **1977**, *68*, 1–66.
199. Glasby, C.J.; Read, G.B. A chronological review of polychaete taxonomy in New Zealand. *J. R. Soc. N. Z.* **1998**, *28*, 347–374. [CrossRef]
200. Capa, M.; Rouse, G.W. Phylogenetic relationships within *Amphiglena* Claparède, 1864 (Polychaeta: Sabellidae), description of five new species from Australia, a new species from Japan, and comments on previously described species. *J. Nat. Hist.* **2007**, *41*, 327–356. [CrossRef]
201. Capa, M.; Murray, A. Review of the genus *Megalomma* (Polychaeta: Sabellidae) in Australia with description of three new species, new records and notes on certain features with phylogenetic implications. *Rec. Aust. Mus.* **2009**, *61*, 201–224. [CrossRef]
202. Murray, A.; Keable, S.J. First report of *Sabella spallanzanii* (Gmelin, 1791) (Annelida: Polychaeta) from Botany Bay, New South Wales, a northern range extension for the invasive species within Australia. *Zootaxa* **2013**, *3670*, 391–395. [CrossRef]
203. Sun, Y.; Wong, E.; Tovar-Hernández, M.A.; Williamson, J.E.; Kupriyanova, E.K. Is *Hydroides brachyacantha* (Serpulidae: Annelida) a widespread species? *Invertebr. Syst.* **2016**, *30*, 41–59. [CrossRef]
204. Styan, C.; McCluskey, C.; Sun, Y.; Kupriyanova, E. Cryptic sympatric species across the Australian range of the global estuarine invader *Ficopomatus enigmaticus* (Fauvel, 1923) (Serpulidae, Annelida). *Aquat. Invasions* **2017**, *12*, 53–65. [CrossRef]
205. Baird, W. Description of a new species of annelids belonging to the Family Amphinomidae. *Trans. Linn. Soc. London* **1864**, *24*, 449–450. [CrossRef]
206. Baird, W. Description of several new species and varieties of tubicolous annelids (= Tribe Limivora of Grube), in the collection of the British Museum. *J. Linn. Soc. Zool.* **1865**, *8*, 10–22. [CrossRef]
207. Ehlers, E. Polychaeten der Hamburger Magalhaenischen Sammelreise. *Ergeb. Hambg. Magalhaenischen Sammelreise Liefersung* **1897**, *2*, 1–147.

208. Gravier, C. Sur les Annélides Polychètes Recueillies par l'Expédition Antarctique Française (Terebelliens, Serpuliens). *Bull. Mus. d'Hist. Nat. Paris* **1907**, *13*, 46–52.
209. Pixell, H.L.M. Polychaeta of the Families Serpulidae and Sabellidae, collected by the Scottish National Antarctic Expedition. *Trans. R. Soc. Edinburgh* **1913**, *49*, 347–358. [CrossRef]
210. Benham, W.B. Polychaeta. British Antarctic "Terra Nova" Expedition, 1910. *Nat. Hist. Rep. Zool.* **1927**, *7*, 47–182.
211. Rzhavsky, A.V. Two new species of *Pileolaria* (Polychaeta: Spirorbidae) from the Southern hemisphere with a brief review of related species. *Inverterb. Zool.* **2010**, *7*, 81–91. [CrossRef]
212. Tovar-Hernández, M.A.; Yáñez-Rivera, B.; Giangrande, A.; Gambi, M.C. Notes on the species of *Perkinsiana* (Polychaeta: Sabellidae) from Antarctica with the description of *P. brigittae* sp. nov. *Zootaxa* **2012**, *3485*, 56–68. [CrossRef]
213. Neal, L.; Paterson, G.L.; Blockley, D.; Scott, B.; Sherlock, E.; Huque, C.; Glover, A.G. Biodiversity data and new species descriptions of polychaetes from offshore waters of the Falkland Islands, an area undergoing hydrocarbon exploration. *ZooKeys* **2020**, *938*, 1–86. [CrossRef]
214. Kupriyanova, E.K.; Rzhavsky, A.; ten Hove, H.A. Serpulidae Rafinesque, 1815. In *Handbook of Zoology, Annelida, Volume 2: Pleistoannelida, Sedentaria II*; Purschke, G., Böggemann, M., Westheide, W., Eds.; De Gruyter: Berlin, Germany, 2019; pp. 213–275.
215. Jones, M.L. *Brandtika asiatica* new genus, new species, from Southeastern Asia and a redescription of *Monroika africana* (Monro) (Polychaeta: Sabellidae). *Proc. Biol. Soc. Wash.* **1974**, *87*, 217–230.
216. Atkinson, S.D.; Bartholomew, J.L.; Rouse, G.W. The invertebrate host of salmonid fish parasites *Ceratonova shasta* and *Parvicapsula minibicornis* (Cnidaria: Myxozoa), is a novel fabriciid annelid, *Manayunkia occidentalis* sp. nov. (Sabellida: Fabriciidae). *Zootaxa* **2020**, *4751*, 310–320. [CrossRef]
217. Sitnikova, T.Y.; Shcherbakov, D.Y.; Kharchenko, V.V. On taxonomic status of polychaetes of the genus *Manayunkia* (Sabellidae, Fabriciinae) from the open Lake Baikal. *Zool. Zhurnal* **1997**, *76*, 16–27.
218. Monro, C.C.A. On a collection of Polychaeta from near the Mouth of the River Congo. *Rev. Zool. Bot. Afr.* **1939**, *32*, 213–225.
219. Rouse, G. New *Fabriciola* and *Manayunkia* species (Fabriciinae: Sabellidae: Polychaeta) from Papua New Guinea. *J. Nat. Hist.* **1996**, *30*, 1761–1778. [CrossRef]
220. Baumhaker, H. Taxonomische Untersuchungen an Sabellidae und Fabriciidae (Annelida: Polychaeta) der Tiefsee des Südwest-Atlantiks. Master's Thesis, University of Rostock, Rostock, Germany, 2012.
221. Bok, M.J.; Capa, M.; Nilsson, D.-E. Here, there and everywhere: The radiolar eyes of fan worms (Annelida, Sabellidae). *Integr. Comp. Biol.* **2016**, *56*, 784–795. [CrossRef]
222. Brown, S.D.; Emlet, R.B. Natural radiole damage and regeneration in the feather duster worm *Schizobranchia insignis*. *Invertebr. Biol.* **2020**, e12307. [CrossRef]
223. Orrhage, L. On the structure and homologues of the anterior end of the polychaete families Sabellidae and Serpulidae. *Zoomorphology* **1980**, *96*, 113–167. [CrossRef]
224. Giangrande, A. The genus *Chone* (Polychaeta, Sabellidae) in the Mediterranean Sea with description of *C. longiseta* n. sp. *Boll. Zool.* **1992**, *59*, 517–529. [CrossRef]
225. Tovar-Hernández, M.A.; Sosa-Rodríguez, T. Redescription of *Chone infundibuliformis* Krøyer, 1856 (Polychaeta: Sabellidae) and histology of the branchial crown appendages, collar and glandular ridge. *Zootaxa* **2006**, *1115*, 31–59. [CrossRef]
226. Capa, M.; Nogueira, J.M.D.M.; Rossi, M.C.S. Comparative internal structure of dorsal lips and radiolar appendages in Sabellidae (Polychaeta) and phylogenetic implications. *J. Morphol.* **2011**, *272*, 302–319. [CrossRef] [PubMed]
227. Ippolitov, A.P.; Vinn, O.; Kupriyanova, E.K.; Jäger, M. Written in stone: History of serpulid polychaetes through time. *Mem. Mus. Vic.* **2014**, *71*, 123–160. [CrossRef]
228. Wong, E.; Kupriyanova, E.K.; Hutchings, P.; Capa, M.; Radashevsky, V.I.; ten Hove, H.A. A graphically illustrated glossary of polychaete terminology: Invasive species of Sabellidae, Serpulidae and Spionidae. *Mem. Mus. Vic.* **2014**, *71*, 327–342. [CrossRef]
229. Handley, S.J. Spionid polychaetes in Pacific oysters, *Crassostrea gigas* (Thunberg) from Admiralty Bay, Marlborough Sounds, New Zealand. *N. Z. J. Mar. Freshw. Res.* **1995**, *29*, 305–309. [CrossRef]
230. Bick, A. Morphology and ecology of *Dipolydora armata* (Polychaeta, Spionidae) from the Western Mediterranean Sea. *Acta Zool.* **2001**, *11*, 177–187. [CrossRef]
231. Bick, A. Polychaete communities associated with gastropod shells inhabited by the hermit crabs *Clibanarius erythropus* and *Calcinus tubularis* from Ibiza, Mediterranean Sea. *J. Mar. Biol. Assoc. UK* **2006**, *86*, 83–92. [CrossRef]
232. Rouse, G.W.; Pleijel, F. *Polychaetes*; Oxford University Press: London, UK, 2001; pp. 1–354.
233. Bick, A. *Bedeutung Postlarvaler Ontogenetischer Merkmalsvariationen für die Taxonomie der Spionidae und Sabellidae*; Habilitationsschrift an der Mathematisch-Naturwissenschaftliche Fakultät Fakultät der Universität Rostock: Rostock, Germany, 2007.
234. Costa-Paiva, E.M.; Paiva, P.C. Anaesthetization and fixation effects on the morphology of sabellid polychaetes (Annelida: Polychaeta: Sabellidae). *J. Mar. Biol. Assoc. UK* **2007**, *87*, 1127–1132. [CrossRef]
235. Coleman, C.O. "Digital inking": How to make perfect line drawings on computers. *Org. Div. Evol. Electr. Suppl.* **2003**, *14*, 1–14. [CrossRef]
236. Paterson, G.L.; Sykes, D.; Faulwetter, S.; Merk, R.; Ahmed, F.; Hawkins, L.E.; Dinley, J.; Ball, A.D.; Arvanitidis, C. The pros and cons of using micro-computed tomography in gross and micro-anatomical assessments of polychaetous annelids. *Mem. Mus. Vic.* **2014**, *71*, 237–246. [CrossRef]

237. Parapar, J.; Candás, M.; Cunha-Veira, X.; Moreira, J. Exploring annelid anatomy using micro-computed tomography: A taxonomic approach. *Zool. Anz.* **2017**, *270*, 19–42. [CrossRef]
238. Patti, F.P.; Gambi, M.C.; Giangrande, A. Preliminary study on the systematic relationships of Sabellinae (Polychaeta, Sabellidae), based on the C1 domain of the 28S rDNA, with discussion of reproductive features. *Ital. J. Zool.* **2003**, *70*, 269–278. [CrossRef]
239. Halanych, K.M. A review of molecular markers used for annelid phylogenetics. *Integr. Comp. Biol.* **2006**, *46*, 533–543. [CrossRef]
240. Carr, C.M.; Hardy, S.M.; Brown, T.M.; Macdonald, T.A.; Hebert, P.D.N. A Tri-Oceanic perspective: DNA barcoding reveals geographic structure and cryptic diversity in Canadian polychaetes. *PLoS ONE* **2011**, *6*, e22232. [CrossRef] [PubMed]
241. Seixas, V.C.; de Moraes Russo, C.A.; Paiva, P.C. Mitochondrial genome of the Christmas Tree Worm *Spirobranchus giganteus* (Annelida: Serpulidae) reveals a high substitution rate among annelids. *Gene* **2017**, *605*, 43–53. [CrossRef]
242. Sun, Y.; Daffe, G.; Zhang, Y.; Qiu, J.-W.; Kupriyanova, E.K. Another blow to the conserved gene order in Annelida: Mitogenomics reveals extensive gene rearrangement in the tubeworm genus *Hydroides* (Serpulidae). *Mol. Phylogenet. Evol.* **2021**, 107124. [CrossRef] [PubMed]
243. Tilic, E.; Atkinson, S.D.; Rouse, G.W. Mitochondrial genome of the freshwater annelid *Manayunkia occidentalis* (Sabellida: Fabriciidae). *Mitochondrial DNA B* **2020**, *5*, 3313–3315. [CrossRef] [PubMed]
244. Daffe, G.; Sun, Y.; Ahyong, S.T.; Kupriyanova, E.K. Mitochondrial genome of *Sabella spallanzanii* (Sabellida: Sabellidae). *Mitochondrial DNA B* **2020**, *6*, 499–501. [CrossRef] [PubMed]
245. Takahashi, T.; McDougall, C.; Troscianko, J.; Chen, W.-C.; Jayaraman-Nagarajan, A.; Shimeld, S.M.; Ferrier, D.E. An EST screen from the annelid *Pomatoceros lamarckii* reveals patterns of gene loss and gain in animals. *BMC Evol. Biol.* **2009**, *9*, 1–17. [CrossRef]
246. De Queiroz, K. Species concepts and species delimitation. *Syst. Biol.* **2007**, *56*, 879–886. [CrossRef]
247. Barraclough, T.G. Evolving entities: Towards a unified framework for understanding diversity at the species and higher levels. *Phil. Trans. R. Soc. B Biol. Sci.* **2010**, *365*, 1801–1813. [CrossRef] [PubMed]
248. Nygren, A. Cryptic polychaete diversity: A review. *Zool. Scr.* **2014**, *43*, 172–183. [CrossRef]
249. Halt, M.N.; Kupriyanova, E.K.; Cooper, S.J.B.; Rouse, G.W. Naming species with no morphological indicators: Species status of *Galeolaria caespitosa* (Annelida: Serpulidae) inferred from nuclear and mitochondrial gene sequences and morphology. *Invertebr. Syst.* **2009**, *23*, 205–222. [CrossRef]
250. Sun, Y.; Wong, E.; Keppel, E.; Williamson, J.E.; Kupriyanova, E.K. A global invader or a complex of regionally distributed species? clarifying the status of an invasive calcareous tubeworm *Hydroides dianthus* (Verrill, 1873) (Polychaeta: Serpulidae) Using DNA Barcoding. *Mar. Biol.* **2017**, *164*, 28-1–28-12. [CrossRef]
251. Sun, Y.; Al-Kandari, M.; Kubal, P.; Walmiki, N.; Kupriyanova, E.K. Cutting a Gordian knot of tubeworms with DNA data: The story of the *Hydroides operculata*-complex (Annelida, Serpulidae). *Zootaxa* **2017**, *4323*, 39–48. [CrossRef]
252. Palero, F.; Torrado, H.; Perry, O.; Kupriyanova, E.K.; Ulman, A.; Genis-Armero, R.; ten Hove, H.A.; Capaccioni-Azzati, R. Following the Phoenician example: Western Mediterranean colonization by *Spirobranchus* cf. *tetraceros* (Annelida: Serpulidae). *Sci. Mar.* **2020**, *84*, 83–92. [CrossRef]
253. Hebert, P.D.; Ratnasingham, S.; De Waard, J.R. Barcoding animal life: Cytochrome c oxidase subunit 1 divergences among closely related species. *Proc. R. Soc. London Biol. Sci.* **2003**, *270*, S96–S99. [CrossRef]
254. Rech, S.; Borrell Pichs, Y.J.; García-Vazquez, E. Anthropogenic marine litter composition in coastal areas may be a predictor of potentially invasive rafting fauna. *PLoS ONE* **2018**, *13*, e0191859. [CrossRef] [PubMed]
255. Langeneck, J.; Lezzi, M.; Del Pasqua, M.; Musco, L.; Gambi, M.C.; Castelli, A.; Giangrande, A. Non-indigenous polychaetes along the coasts of Italy: A critical review. *Mediterr. Mar. Sci.* **2020**, *21*, 238–275. [CrossRef]
256. Rech, S.; Salmina, S.; Pichs, Y.J.B.; García-Vazquez, E. Dispersal of alien invasive species on anthropogenic litter from European mariculture areas. *Mar. Poll. Bull.* **2018**, *131*, 10–16. [CrossRef]
257. Hutchings, P.; De Decker, P.; Geddes, M.C. A new species of *Manayunkia* (Polychaeta) from ephemeral lakes near the Coorong, South Australia. *Trans. R. Soc. S. Aust.* **1981**, *105*, 25–28.
258. Schütz, L. Über Verbreitung, Ökologie und Biologie des Brackwasserpolychaeten *Manayunkia aestuarina* (Bourne) Insbesondere an den Küsten Schleswig-Holsteins. *Faun. Mitt. Norddeut.* **1965**, *9*, 226–234.
259. Lewis, D.B. Some aspects of the ecology of *Fabricia sabella* (Ehr.) (Annelida, Polychaeta). *J. Linn. Soc. (Zool.)* **1968**, *47*, 515–526. [CrossRef]
260. Light, W.J. Extension of range for *Manayunkia aestuarina* (Polychaeta: Sabellidae) to British Columbia. *J. Fish. Board Canada* **1969**, *26*, 3088–3091. [CrossRef]
261. Bagheri, E.; McLusky, D. Population dynamics of oligochaetes and small polychaetes in the polluted forth estuary ecosystem. *Neth. J. Sea Res.* **1982**, *16*, 55–66. [CrossRef]
262. Bick, A. Reproduction and larval development of *Manayunkia aestuarina* (Bourne, 1883) (Polychaeta, Sabellidae) in a coastal region of the Southern Baltic. *Helgoländer Meeresunters* **1996**, *50*, 287–298. [CrossRef]
263. Giangrande, A.; Gambi, M.; Micheli, F.; Kroeker, K. Fabriciidae (Annelida, Sabellida) from a naturally acidified coastal system (Italy) with description of two new species. *J. Mar. Biol. Assoc. UK* **2014**, *94*, 1417. [CrossRef]
264. Muus, B.J. The Fauna of Danish estuaries and lagoons: Distribution and ecology of dominating species in the shallow reaches of the mesohaline zone. *Medd. Danmarks Fisk. Havunders (n. ser.)* **1967**, *5*, 1–316.
265. Lewis, D.B. Feeding and tube-building in the Fabriciinae (Annelida, Polychaeta). *Proc. Linn. Soc. Lond.* **1968**, *179*, 37–49. [CrossRef]

266. Hartman, O. Fabricinae (Featherduster Polychaetous Annelids) in the Pacific. *Pac. Sci.* **1951**, *5*, 379–391.
267. Bartholomew, J.L.; Atkinson, S.D.; Hallett, S.L. Involvement of *Manayunkia speciosa* (Annelida: Polychaeta: Sabellidae) in the life cycle of *Parvicapsula minibicornis*, a myxozoan parasite of Pacific salmon. *J. Parasitol.* **2006**, *92*, 742–748. [CrossRef]
268. Alexander, J.D.; Hallett, S.L.; Stocking, R.W.; Xue, L.; Bartholomew, J.L. Host and parasite populations after a ten-year flood: *Manayunkia speciosa* and *Ceratonova* (Syn *Ceratomyxa*) *shasta* in the Klamath River. *Northwest. Sci.* **2014**, *88*, 219–233. [CrossRef]
269. Giangrande, A.; Gambi, M.C. The genus *Perkinsiana* (Polychaeta, Sabellidae) from Antarctica, with descriptions of the new species *P. milae* and *P. borsibrunoi*. *Zool. Scr.* **1997**, *26*, 267–278. [CrossRef]
270. Murray, J.; Watson, G.; Giangrande, A.; Licciano, M.; Bentley, M. Regeneration as a novel method to culture marine ornamental sabellids. *Aquaculture* **2013**, *410*, 129–137. [CrossRef]
271. Chughtai, I.; Knight-Jones, W.E. Burrowing into limestone by sabellid polychaetes. *Zool. Scr.* **1988**, *17*, 231–238. [CrossRef]
272. Fonseca, E.A.; Dean, H.K.; Cortés, J. Non-colonial coral macro-borers as indicators of coral reef status in the South Pacific of Costa Rica. *Rev. Biol. Trop.* **2006**, *54*, 101–115. [CrossRef] [PubMed]
273. Licciano, M.; Stabili, L.; Giangrande, A. Clearance rates of *Sabella spallanzanii* and *Branchiomma luctuosum* (Annelida: Polychaeta) on a pure culture of *Vibrio alginolyticus*. *Water Res.* **2005**, *39*, 4375–4384. [CrossRef]
274. Licciano, M.; Terlizzi, A.; Giangrande, A.; Cavallo, R.A.; Stabili, L. Filter-feeder macroinvertebrates as key players in culturable bacteria biodiversity control: A case of study with *Sabella spallanzanii* (Polychaeta: Sabellidae). *Mar. Environ. Res.* **2007**, *64*, 504–513. [CrossRef]
275. Stabili, L.; Schirosi, R.; Licciano, M.; Giangrande, A. The mucus of *Sabella spallanzanii* (Annelida, Polychaeta): Its involvement in chemical defense and fertilization success. *J. Exp. Mar. Biol. Ecol.* **2009**, *374*, 144–149. [CrossRef]
276. Giangrande, A.; Pierri, C.; Arduini, D.; Borghese, J.; Licciano, M.; Trani, R.; Corriero, G.; Basile, G.; Cecere, E.; Petrocelli, A. An innovative IMTA System: Polychaetes, sponges and macroalgae co-cultured in a southern Italian in-shore mariculture plant (Ionian Sea). *J. Mar. Sci. Eng.* **2020**, *8*, 733. [CrossRef]
277. Selim, S.A. New records of sabellid species (Polychaeta: Sabellinae) from the coastal Egyptian waters. *Egypt J. Aquat. Res.* **2008**, *34*, 108–128.
278. Giangrande, A.; Caruso, L.; Musco, L.; Licciano, M. Variability among Mediterranean populations of *Sabella pavonina* (Annelida: Sabellidae). *Ital. J. Zool.* **2014**, *81*, 100–111. [CrossRef]
279. Magalhães, W.F.; Bailey-Brock, J.H.; Tovar-Hernández, M.A. An abundant new genus and species of fan worms (Polychaeta: Sabellidae) from Hawaii. *Zootaxa* **2020**, *4763*, 85–98. [CrossRef]
280. Capa, M.; Nishi, E.; Tanaka, K.; Fujikura, K. First record of a *Bispira* species (Sabellidae: Polychaeta) from a hydrothermal vent. *Mar. Biodivers. Rec.* **2013**, *6*, e68. [CrossRef]
281. Amon, D.J.; Glover, A.G.; Wiklund, H.; Marsh, L.; Linse, K.; Rogers, A.D.; Copley, J.T. The discovery of a natural whale fall in the Antarctic deep sea. *Deep Sea Res. Part. II Top. Stud. Ocean.* **2013**, *92*, 87–96. [CrossRef]
282. Cordes, E.E.; Carney, S.L.; Hourdez, S.; Carney, R.S.; Brooks, J.M.; Fisher, C.R. Cold seeps of the deep Gulf of Mexico: Community structure and biogeographic comparisons to Atlantic Equatorial belt seep communities. *Deep Sea Res. Part. I Ocean. Res. Pap.* **2007**, *54*, 637–653. [CrossRef]
283. Sellanes, J.; Quiroga, E.; Neira, C. Megafauna community structure and trophic relationships at the recently discovered Concepción methane seep area, Chile,~ 36° S. ICES. *Mar. Sci.* **2008**, *65*, 1102–1111. [CrossRef]
284. Goffredi, S.K.; Tilic, E.; Mullin, S.W.; Dawson, K.S.; Keller, A.; Lee, R.W.; Wu, F.; Levin, L.A.; Rouse, G.W.; Cordes, E.E. Methanotrophic bacterial symbionts fuel dense populations of deep-sea feather duster worms (Sabellida, Annelida) and extend the spatial influence of methane seepage. *Sci. Adv.* **2020**, *6*, eaay8562. [CrossRef]
285. Hutchings, P.; Murray, A. Taxonomy of the polychaetes from the Hawkesbury River and the Southern estuaries of New South Wales, Australia. *Rec. Aust. Mus.* **1984**, *36*, 1–119. [CrossRef]
286. Lardicci, C.; Castelli, A. *Desdemona ornata* Banse, 1957 (Polychaeta, Sabellidae, Fabricinae) New record in the Mediterranean Sea. *Oebalia* **1986**, *13*, 195–201.
287. Castelli, A.; Galletti, C.; Lardicci, C. Structure of benthic communities of brackish-water microhabitats: Spatial and temporal variations. *MAP Tech. Rep. Ser. (UNEP)* **1988**, *22*, 19–46.
288. Panagopoulos, D.; Nicolaidou, A. A population of *Desdemona ornata* Banse, 1957 (Polychaeta, Sabellidae) settled in a fully marine habitat of the Mediterranean. *Oebalia* **1989**, *16*, 35–39.
289. Glasby, C.J.; Timm, T.; Muir, A.I.; Gil, J. Catalogue of non-marine Polychaeta (Annelida) of the World. *Zootaxa* **2009**, *2070*, 1–52. [CrossRef]
290. Hossain, M.B.; Marshall, D.J. Benthic infaunal community structuring in an acidified tropical estuarine system. *Aquat. Biosyst.* **2014**, *10*, 1–12. [CrossRef] [PubMed]
291. Hernández-Alcántara, P.; Cortés-Solano, J.D.; Medina-Cantú, N.M.; Avilés-Díaz, A.L.; Solís-Weiss, V. Polychaete diversity in the estuarine habitats of Términos lagoon, Southern Gulf of Mexico. *Mem. Mus. Vic.* **2014**, *71*, 97–107. [CrossRef]
292. Hsieh, H.L. *Laonome albicingillum*, a new fan worm species (Polychaeta, Sabellidae, Sabellinae) from Taiwan. *Proc. Biol. Soc. Wash.* **1995**, *108*, 130–135.
293. Hsieh, H.L. Self-fertilization-a potential fertilization mode in an estuarine sabellid polychaete. *Mar. Ecol. Prog. Ser.* **1997**, *147*, 143–148. [CrossRef]

294. Capa, M.; van Moorsel, G.; Tempelman, D. The Australian feather-duster worm *Laonome calida* Capa, 2007 (Annelida: Sabellidae) introduced into European inland waters? *Bioinvasions Rec.* **2014**, *3*, 1–11. [CrossRef]
295. Bonar, D.B. Feeding and tube construction in *Chone mollis* Bush (Polychaeta, Sabellidae). *J. Exp. Mar. Biol. Ecol.* **1972**, *9*, 1–18. [CrossRef]
296. Fitzsimons, G. Feeding and tube building in *Sabellastarte magnifica* (Shaw) (Sabellidae: Polychaeta). *Bull. Mar. Sci.* **1965**, *15*, 642–671.
297. Licciano, M.; Murray, J.M.; Watson, G.J.; Giangrande, A. Morphological comparison of the regeneration process in *Sabella spallanzanii* and *Branchiomma luctuosum* (Annelida, Sabellida). *Invertebr. Biol.* **2012**, *131*, 40–51. [CrossRef]
298. Jumars, P.A.; Dorgan, K.M.; Lindsay, S.M. Diet of worms emended: An update of polychaete feeding guilds. *Ann. Rev. Mar. Sci* **2015**, *7*, 497–520. [CrossRef]
299. Jones, M.L. On the Caobangiidae, a new family of the Polychaeta, with a redescription of *Caobangia billeti* Giard. *Smithson. Contr. Zool.* **1974**, 1–55. [CrossRef]
300. Kuris, A.M.; Culver, C.S. An introduced sabellid polychaete pest infesting cultured abalones and its potential spread to other California gastropods. *Invertebr. Biol.* **1999**, *118*, 391–403. [CrossRef]
301. Ruck, K.R.; Cook, P.A. Sabellid infestations in the shells of South African molluscs: Implications for abalone mariculture. *J. Shellfish Res.* **1998**, *17*, 693–699.
302. Kupriyanova, E.K.; Badyaev, A.V. Ecological correlates of Arctic Serpulidae (Annelida, Polychaeta) distributions. *Ophelia* **1998**, *49*, 181–193. [CrossRef]
303. Kupriyanova, E.K.; Nishi, E.; ten Hove, H.A.; Rzhavsky, A.V. A review of life history patterns in serpulimorph polychaetes: Ecological and evolutionary perspectives. *Ocean. Mar. Biol. Ann. Rev.* **2001**, *39*, 1–101.
304. Charles, F.; Jordana, E.; Amouroux, J.-M.; Gremare, A.; Desmalades, M.; Zudaire, L. Reproduction, recruitment and larval metamorphosis in the serpulid polychaete *Ditrupa arietina* (O.F. Muller). *Estuar. Coast. Shelf Sci.* **2003**, *57*, 435–443. [CrossRef]
305. Goffredi, S.K.; Johnson, S.; Tunnicliffe, V.; Caress, D.; Clague, D.; Escobar, E.; Paduan, J.B.; Rouse, G.; Salcedo, D.L.; Soto, L.A.; et al. Hydrothermal vent fields discovered in the Southern Gulf of California clarify role of habitat in augmenting regional diversity. *Proc. R. Soc. B Biol. Sci.* **2017**, *284*, 20170817. [CrossRef] [PubMed]
306. Kupriyanova, E.K.; Nishi, E.; Kawato, M.; Fujiwara, Y. New records of Serpulidae (Annelida, Polychaeta) from hydrothermal vents of North Fiji, Pacific Ocean. *Zootaxa* **2010**, *2389*, 57–68. [CrossRef]
307. Vinn, O.; Kupriyanova, E.K.; Kiel, S. Systematics of serpulid tubeworms (Annelida, Polychaeta) from Cretaceous and Cenozoic hydrocarbon-seep deposits in North America and Europe. *Neues Jahrb. Geol. Paläontol.* **2012**, *265*, 315–325. [CrossRef]
308. Juniper, S.K.; Sibuet, M. Cold seep benthic communities in Japan subduction zones: Spatial organisation, trophic strategies and evidence for temporal evolution. *Mar. Ecol. Prog. Ser.* **1987**, *40*, 115–126. [CrossRef]
309. Olu, K.; Duperret, A.; Sibuet, M.; Foucher, J.-P.; Fiala-Médioni, A. Structure and distribution of cold seep communities along the Peruvian Active Margin: Relationship to geological and fluid patterns. *Mar. Ecol. Progr. Ser.* **1996**, *132*, 109–125. [CrossRef]
310. Segonzac, M.; Hekinian, R.; Auzende, J.; Francheteau, J. Recently discovered animal communities on the South East Pacific Rise (17-19 S and the Eastern Microplaque Region). *Cah. Biol. Mar.* **1997**, *38*, 140–141.
311. López-González, P.; Rodríguez, E.; Gili, J.-M.; Segonzac, M. New records on sea anemones (Anthozoa: Actiniaria) from hydrothermal vents and cold seeps. *Zool. Verh.* **2003**, *345*, 215–243.
312. Sibuet, M.; Olu, K. Biogeography, biodiversity and fluid dependence of deep-sea cold-seep communities at active and passive margins. *Deep Sea Res. Part. II Top. Stud. Ocean.* **1998**, *45*, 517–567. [CrossRef]
313. Mak, P.; Huang, Z. The salinity tolerance of the serpulid polychaete, *Hydroides elegans* (Haswell, 1883), and its possible applications in bio-antifouling. The marine flora and fauna of Hong Kong and Southern China. *Ecol. Morph. Behav. Phys.* **1982**, *1*, 817–823.
314. Knight-Jones, P. Behaviour, setal inversion and phylogeny of Sabellida (Polychaeta). *Zool. Scr.* **1981**, *10*, 183–202. [CrossRef]
315. Dill, L.M.; Fraser, A.H.G. The worm re-turns: Hiding behavior of a tube-dwelling marine polychaete, *Serpula vermicularis*. *Behav. Ecol.* **1997**, *8*, 186–193. [CrossRef]
316. Pan, J.; Marcoval, M.A. Top-down effects of an exotic serpulid polychaete on natural plankton assemblage of estuarine and brackish systems in the SW Atlantic. *J. Coast. Res.* **2014**, *30*, 1226–1235. [CrossRef]
317. Jordana, E.; Charles, F.; Gremare, A.; Amouroux, J.-M.; Chretiennot-Dinet, M.-J. Food sources, ingestion and absorption in the suspension-feeding polychaete, *Ditrupa arietina* (O. F. Muller). *J. Exp. Mar. Biol. Ecol.* **2001**, *266*, 219–236. [CrossRef]
318. Bailey-Brock, J.H. Polychaetes from Fijian coral reefs. *Pac. Sci.* **1985**, *39*, 195–220.
319. Nishi, E. Serpulid polychaetes associated with living and dead corals at Okinawa Island, Southwest Japan. *Publ. Seto Mar. Biol. Lab.* **1996**, *36*, 305–318. [CrossRef]
320. Marsden, J.R. Coral preference behaviour by planktotrophic larvae of *Spirobranchus giganteus corniculatus* (Serpulidae: Polychaeta). *Coral Reefs* **1987**, *6*, 71–74. [CrossRef]
321. Hunte, W.; Conlin, B.E.; Marsden, J.R. Habitat selection in the tropical polychaete *Spirobranchus giganteus*. I. Distribution on corals. *Mar. Biol.* **1990**, *104*, 87–92. [CrossRef]
322. Marsden, J.R. Light responses of the planktotrophic larva of the serpulid polychaete *Spirobranchus polycerus*. *Mar. Ecol. Prog. Ser.* **1990**, *58*, 225–233.
323. Williams, G.B. The effect of extracts of *Fucus serratus* in promoting the settlement of larvae of *Spirorbis borealis* (Polychaeta). *J. Mar. Biol. Assoc. UK* **1964**, *44*, 397–414. [CrossRef]

324. Gee, J.M. Chemical stimulation of settlement in larvae of *Spirorbis rupestris* (Serpulidae). *Anim. Behav.* **1965**, *13*, 181–186. [CrossRef]
325. Hartman, O. Catalogue of the polychaetous annelids of the world. Part II. *Occ. Pap. Allan Hancock Found.* **1959**, *23*, 355–628.
326. Darling, J.A.; Carlton, J.T. A framework for understanding marine cosmopolitanism in the Anthropocene. *Front. Mar. Sci.* **2018**, *5*, 1–293. [CrossRef]
327. Hutchings, P.; Kupriyanova, E. Cosmopolitan polychaetes–fact or fiction? Personal and historical perspectives. *Invertebr. Syst.* **2018**, *32*, 1–9. [CrossRef]
328. Briggs, J.C. *Marine Zoogeography*; McGraw-Hill: New York, NY, USA, 1974; pp. 1–475.
329. McClain, C.R.; Schlacher, T.A. On some hypotheses of diversity of animal life at great depths on the sea floor. *Mar. Ecol.* **2015**, *36*, 849–872. [CrossRef]
330. Higgs, N.D.; Attrill, M. Biases in biodiversity: Wide-ranging species are discovered first in the deep sea. *Front. Mar. Sci.* **2015**, *2*, 61. [CrossRef]
331. Giangrande, A.; Licciano, M. Factors influencing latitudinal pattern of biodiversity: An example using Sabellidae (Annelida, Polychaeta). *Biodiv. Conserv.* **2004**, *13*, 1633–1646. [CrossRef]
332. Banse, K. Beiträge zur kenntnis der gattungen *Fabricia, Manayunkia* und *Fabriciola* (Sabellidae, Polychaeta). *Zool. Jahrb. (Abt. Syst. Okol. Geogr. Tiere)* **1956**, *84*, 415–438.
333. Bourne, A.G. On *Haplobranchus*, a new genus of Capitobranchiate annelids. *J. Cell Sci.* **1883**, *23*, 168–176.
334. Annenkova, N.P. Über Die Pontokaspischen Polychaeten. II. Die Gattungen *Hypaniola, Parhypania, Fabricia* und *Manajunkia*. *Ann. Mussée Zool. L'Academie des Sci. L'URSS, Leningrad* **1928**, *30*, 13–20.
335. Day, J.H. The polychaete fauna of South Africa Part 8: New species and records from grab samplings and dreddings. *Bull. Brit. Mus. (Nat. Hist.) Zool.* **1963**, *10*, 381–445. [CrossRef]
336. Perkins, T.H. Revision of *Demonax* Kinberg, *Hypsicomus* Grube, and *Notaulax* Tauber, with a review of *Megalomma* Johansson from Florida (Polychaeta: Sabellidae). *Proc. Biol. Soc. Wash.* **1984**, *97*, 285–368.
337. Knight-Jones, P.; Perkins, T.H. A revision of *Sabella* and *Bispira* and *Stylomma*. (Sabellidae: Polychaeta). *Zool. J. Linn. Soc.* **1998**, *123*, 385–467. [CrossRef]
338. Knight-Jones, P.; Mackie, A.S. A revision of *Sabellastarte* (Polychaeta: Sabellidae). *J. Nat. Hist.* **2003**, *37*, 2269–2301. [CrossRef]
339. Hartman, O. Abyssal polychaetous annelids from the Mozambique Basin off Southeast Africa, with a compendium of abyssal polychaetous annelids from world-wide Areas. *J. Fish. Res. Board Can.* **1971**, *28*, 1407–1428. [CrossRef]
340. Uschakov, P.V. Bathypelagic and deepwater forms of marine polychaetes from Kamchatka Area of the Pacific. *Issled. Fauny Morej USSR* **1952**, *3*, 103–112.
341. Levenstein, R.Y. Polychaeta from deep water in the Bering Sea. *Proc. Inst. Oceanol. USSR Acad. Sci* **1961**, *46*, 147–178.
342. Levenstein, R.Y. Polychaeta. In *Pacific Ocean. Biology. Deep-Sea Fauna*; Zenkevitch, L.A., Ed.; Nauka: Moscow, Russia, 1969; pp. 52–62. (In Russian)
343. Belyaev, G.M. *Hadal Bottom Fauna of the World Ocean*; Institute of Oceanology, USSR Academy of Sciences: Moscow, Russia, 1972; pp. 1–199.
344. Belyaev, G.M. *Glubokovodnye Okeanicheskie Zheloba i Ikh Fauna [Deep-Sea Oceanic Trenches and Their Fauna]*; Nauka: Moscow, Russia, 1989; pp. 1–256. (In Russian)
345. Moore, J.P. Additional new species of Polychaeta from the North Pacific. *Proc. Acad. Nat. Sci. Phil.* **1906**, *58*, 217–260.
346. Moore, J.P. The Polychætous annelids Dredged by the U. S. S. "Albatross" off the Coast of Southern California in 1904. IV. Spionidae to Sabellariidae. *Proc. Acad. Nat. Sci. Phil.* **1923**, *7*, 179–259.
347. Hartman, O. Deep-water benthic polychaetous annelids off New England to Bermuda and other North Atlantic Areas. *Occ. Pap. Allan Hancock Found.* **1965**, *28*, 1–378.
348. Méndez, N. Trophic categories of soft-bottom epibenthic deep-sea polychaetes from the Southeastern Gulf of California (Mexico) in relation with environmental variables. *Pan Am. J. Aquat. Sci.* **2013**, *8*, 299–311.
349. Zibrowius, H. Review of Serpulidae (Polychaeta) from depths exceeding 2000 meters. In *Essays on Polychaetous Annelids in Memory of Dr. Olga Hartman*; Reish, D.J., Fauchald, K., Eds.; Allan Hancock Press: Los Angeles, CA, USA, 1977; pp. 289–306.
350. Kirkergaard, J.B. Benthic Polychaeta from depths exceeding 6000 m. *Galathea Rep.* **1956**, *2*, 63–78.
351. Paterson, G.L.; Glover, A.G.; Froján, C.R.B.; Whitaker, A.; Budaeva, N.; Chimonides, J.; Doner, S. A Census of abyssal polychaetes. *Deep Sea Res. Part. II Top. Stud. Ocean.* **2009**, *56*, 1739–1746. [CrossRef]
352. Kupriyanova, E.K. Deep-water Serpulidae (Annelida, Polychaeta) from the Kurile-Kamchatka Trench. 2. Genera *Bathyditrupa, Bathyvermilia* and *Protis*. *Zool. Zhurnal* **1993**, *72*, 21–28.
353. Kupriyanova, E.K. Deep-water Serpulidae (Annelida, Polychaeta) from the Kurile-Kamchatka Trench. 1. Genus *Hyalopomatus*. *Zool. Zhurnal* **1993**, *72*, 145–152.
354. Sanfilippo, R. New species of *Hyalopomatus* Marenzeller, 1878 (Annelida, Polychaeta, Serpulidae) from Recent Mediterranean deep-water coral mounds and comments on some congeners. *Zoosystema* **2009**, *31*, 147–161. [CrossRef]
355. Kupriyanova, E.K.; Bailey-Brock, J.; Nishi, E. New records of Serpulidae (Annelida, Polychaeta) collected by R/V "Vityaz" from bathyal and abyssal depths of the Pacific Ocean. *Zootaxa* **2011**, *2871*, 43–60. [CrossRef]
356. Kupriyanova, E.K.; Vinn, O.; Taylor, P.D.; Schopf, J.W.; Kudryavtsev, A.B.; Bailey-Brock, J. Serpulids living deep: Calcareous tubeworms beyond the abyss. *Deep Sea Res. Part. I Ocean. Res.* **2014**, *90*, 91–104. [CrossRef]

357. Kupriyanova, E.K.; Nishi, E. New records of the deep-sea *Nogrobs grimaldii* (Serpulidae: Annelida). *Mar. Biodivers. Rec.* **2011**, *4*, e74. [CrossRef]
358. Rouse, G.W.; Kupriyanova, E.K. *Laminatubus* (Serpulidae, Annelida) from Eastern Pacific hydrothermal vents and methane seeps, with description of two new species. *Zootaxa* **2021**, *4915*, 1–27. [CrossRef]
359. Thorp, C.H.; Pyne, S.; West, S.A. *Hydroides ezoensis* Okuda, a fouling serpulid new to British coastal waters. *J. Nat. Hist.* **1987**, *21*, 863–877. [CrossRef]
360. Zibrowius, H. A propos des pretendud "Recifs de Serpulidae" de l'Ile Rousse, Corse (Mediterranee Nord-Occidentale). *Mesogee* **1991**, *50*, 71–74.
361. Zibrowius, H. Assessing scale and impact of ship-transported alien fauna in the Mediterranean. In *Alien Marine Organisms Introduced by Ships in the Mediterranean and Black Seas*; Briand, F., Ed.; CIESM Workshop Monographs: Monaco, France, 2002; Volume 20, pp. 62–68.
362. Lewis, J.A.; Watson, C.; Harry, A. Establishment of the Caribbean serpulid tubeworm *Hydroides sanctaecrucis* Krøyer [in] Mörch, 1863, in Northern Australia. *Biol. Invasions* **2006**, *8*, 665–671. [CrossRef]
363. Giangrande, A.; Licciano, M.; Pagliara, P.; Gambi, M.C. Gametogenesis and larval development in *Sabella spallanzanii* (Polychaeta: Sabellidae) from the Mediterranean Sea. *Mar. Biol.* **2000**, *136*, 847–861. [CrossRef]
364. Read, G.; Inglis, G.; Stratford, P.; Ahyong, S. Arrival of the alien fanworm *Sabella spallanzanii* (Gmelin, 1791) (Polychaeta: Sabellidae) in two New Zealand harbours. *Aquat. Invasions* **2011**, *6*, 273–279. [CrossRef]
365. Çinar, M.E. Alien polychaete species worldwide: Current status and their impacts. *J. Mar. Biol. Assoc. UK* **2013**, *93*, 1257–1278. [CrossRef]
366. Morais, P.; Reichard, M. Cryptic invasions: A review. *Sci. Total Environ.* **2018**, *613*, 1438–1448. [CrossRef]
367. Chapman, J.W.; Carlton, J.T. A test of criteria for introduced species: The global invasion by the isopod *Synidotea laevidorsalis* (Miers, 1881). *J. Crust. Biol.* **1991**, *11*, 386–400. [CrossRef]
368. Wolff, W.J. Non-Indigenous marine and estuarine species in the Netherlands. *Zool. Med. Leiden* **2005**, *79*, 1–116.
369. Patti, F.; Gambi, M. Phylogeography of the invasive polychaete *Sabella spallanzanii* (Sabellidae) based on the nucleotide sequence of internal transcribed spacer 2 (ITS2) of nuclear rDNA. *Mar. Ecol. Prog. Ser.* **2001**, *215*, 169–177. [CrossRef]
370. Ahyong, S.T.; Kupriyanova, E.; Burghardt, I.; Sun, Y.; Hutchings, P.A.; Capa, M.; Cox, S.L. Phylogeography of the invasive Mediterranean fan worm, *Sabella spallanzanii* (Gmelin, 1791), in Australia and New Zealand. *J. Mar. Biol. Assoc. UK* **2017**, *97*, 985–991. [CrossRef]
371. Pabis, K.; Krodkiewska, M.; Cebulska, K. Alien freshwater polychaetes *Hypania invalida* (Grube 1860) and *Laonome calida* Capa, 2007 in the Upper Odra River (Baltic Sea catchment area). *Knowl. Manag. Aquat. Ecosyst.* **2017**, *418*, 46. [CrossRef]
372. Boltachova, N.; Lisitskaya, E.; Frolenko, L.; Kovalev, E.; Barabashin, T. The finding of polychaete *Laonome calida* Capa, 2007 (Annelida: Sabellidae) in the Southeast Sea of Azov. *Russ. J. Biol. Invasions* **2017**, *8*, 303–306. [CrossRef]
373. Tamulyonis, A.Y.; Gagaev, S.Y.; Stratanenko, E.A.; Zuyev, Y.A.; Potin, V.V. Invasion of the Polychaeta *Laonome xeprovala* Bick & Bastrop, 2018 (Sabellidae, Polychaeta) into the estuary of the Luga and Khabolovka Rivers (Luga Bay, Gulf of Finland). *Russ. J. Biol. Invasions* **2020**, *11*, 148–154. [CrossRef]
374. Tovar-Hernández, M.A.; Yáñez-Rivera, B.; Bortolini-Rosales, J.L. Reproduction of the invasive fan worm *Branchiomma bairdi* (Polychaeta: Sabellidae). *Mar. Biol. Res.* **2011**, *7*, 710–718. [CrossRef]
375. Roman, S.; Perez-Ruzafa, A.; Lopez, E. First record in the Western Mediterranean Sea of *Branchiomma boholense* (Grube, 1878) (Polychaeta: Sabellidae), an alien species of Indo-Pacific origin. *Cah. Biol. Mar.* **2009**, *50*, 241–250.
376. Arias, A.; Giangrande, A.; Gambi, M.C.; Anadón, N. Biology and new records of the invasive species *Branchiomma bairdi* (Annelida: Sabellidae) in the Mediterranean Sea. *Mediterr. Mar. Sci.* **2013**, *10*, 162–171. [CrossRef]
377. Ramalhosa, P.; Camacho-Cruz, K.; Bastida-Zavala, R.; Canning-Clode, J. First record of *Branchiomma bairdi* McIntosh, 1885 (Annelida: Sabellidae) from Madeira Island, Portugal (Northeastern Atlantic Ocean). *Bioinvasions Rec.* **2014**, *3*, 235–239. [CrossRef]
378. Cepeda, D.; Rodríguez-Flores, P. First record of the invasive worm *Branchiomma bairdi* (Annelida: Sabellidae) in the Balearic Sea (Western Mediterranean). *J. Mar. Biol. Assoc. UK* **2018**, *98*, 1955–1963. [CrossRef]
379. Del Pasqua, M.; Schulze, A.; Tovar-Hernández, M.A.; Keppel, E.; Lezzi, M.; Gambi, M.C.; Giangrande, A. Clarifying the taxonomic status of the alien species *Branchiomma bairdi* and *Branchiomma boholense* (Annelida: Sabellidae) using molecular and morphological evidence. *PLoS ONE* **2018**, *13*, e0197104. [CrossRef]
380. Hewitt, C.L.; Campbell, M.L.; Thresher, R.E.; Martin, R.B.; Boyd, S.; Cohen, B.F.; Currie, D.R.; Gomon, M.F.; Keough, M.J.; Lewis, J.A.; et al. Introduced and cryptogenic species in Port Phillip Bay, Victoria, Australia. *Mar. Biol.* **2004**, *144*, 183–202. [CrossRef]
381. Zenetos, A.; Çinar, M.E.; Pancucci-Papadopoulou, M.A.; Harmelin, J.G.; Furnari, G.; Andaloro, F.; Bellou, N.; Streftaris, N.; Zibrowius, H. Annotated list of marine alien species in the Mediterranean with records of the worst invasive species. *Mediterr. Mar. Sci.* **2005**, *6*, 63–118. [CrossRef]
382. Gravili, C.; Belmonte, G.; Cecere, E.; Denitto, F.; Giangrande, A.; Guidetti, P.; Longo, C.; Mastrototaro, F.; Moscatello, S.; Petrocelli, A. Nonindigenous species along the Apulian Coast, Italy. *Chem. Ecol.* **2010**, *26*, 121–142. [CrossRef]
383. Giangrande, A.; Montanaro, P.; Castelli, A. On some *Amphicorina* (Polychaeta, Sabellidae) species from the Mediterranean coast, with the description of *A. grahamensis*. *Ital. J. Zool.* **1999**, *66*, 195–203. [CrossRef]

384. Zenetos, A.; Gofas, S.; Verlaque, M.; Çinar, M.E.; Garcia Raso, J.E.; Bianchi, C.N.; Morri, C.; Azzurro, E.; Bilecenoglu, M.; Froglia, C.; et al. Alien species in the Mediterranean Sea by 2010. A contribution to the application of European Union's Marine Strategy Framework Directive (MSFD). Part I. Spatial distribution. *Mediter. Mar. Sci.* **2010**, *11*, 381. [CrossRef]
385. Faasse, M.A.; Giangrande, A. Description of *Bispira polyomma* n. sp. (Annelida: Sabellidae): A probable introduction to the Netherlands. *Aquat. Invasions* **2012**, *7*, 591–598. [CrossRef]
386. Capa, M.; Murray, A. A taxonomic guide to the fanworms (Sabellidae, Annelida) of Lizard Island, Great Barrier Reef, Australia, including new species and new records. *Zootaxa* **2015**, *4019*, 98–167. [CrossRef] [PubMed]
387. Khedhri, I.; Tovar-Hernández, M.A.; Bonifácio, P.; Ahmed, A.; Aleya, L. First report of the invasive species *Branchiomma bairdi* McIntosh, 1885 (Annelida: Sabellidae) along the Tunisian Coast (Mediterranean Sea). *Bioinvasions Rec.* **2017**, *6*, 139–145. [CrossRef]
388. Abdelnaby, F. Alien polychaete species and the first record of *Branchiomma bairdi* (McIntosh, 1885) from the Suez Canal and the Mediterranean coast of Egypt. *Egypt. J. Aquat. Biol. Fish.* **2020**, *24*, 13–32. [CrossRef]
389. Knight-Jones, P.; Knight-Jones, W.; Ergen, Z. Sabelliform Polychaetes, mostly from Turkey's Aegean coast. *J. Nat. Hist.* **1991**, *25*, 837–858.
390. Çinar, M.E. Polychaetes from the coast of Northern Cyprus (Eastern Mediterranean Sea), with Two new records for the Mediterranean Sea. *Cah. Biol. Mar.* **2005**, *46*, 143–160.
391. Keppel, E.; Tovar Hernández, M.A.; Ruiz, G.M. First record and establishment of *Branchiomma coheni* (Polychaeta: Sabellidae) in the Atlantic Ocean and review of non indigenous species of the genus. *Zootaxa* **2015**, *4058*, 499–518. [CrossRef]
392. Tovar-Hernández, M.A.; Dean, H.A. New gregarious sabellid worm from the Gulf of California reproducing by dpontaneous fission (Polychaeta, Sabellidae). *J. M. Biol. Assoc. UK* **2014**, *94*, 935–946. [CrossRef]
393. Bianchi, C. Serpuloidea (Annelida, Polychaeta) Delle Lagune Costiere Laziali e Campane. *Annali del Museo Civico di Storia Naturale di Genova* **1983**, *84*, 231–243.
394. Arvanitides, C. Polychaete fauna of the Aegean Sea: Inventory and new information. *Bull. Mar. Sci.* **2000**, *66*, 73–96.
395. Simboura, N.; Nicolaidou, A. *The Polychaetes (Annelida, Polychaeta) of Greece: Checklist, distribution and ecological characteristics. Monographs on Marine Sciences, Ser. 4*; National Center for Marine Research: Athens, Greece, 2001; pp. 1–115.
396. Licciano, M.; Giangrande, A.; Gambi, M.C. Reproduction and simultaneous hermaphroditism in *Branchiomma luctuosum* (Polychaeta, Sabellidae) from the Mediterranean Sea. *Invertebr. Biol.* **2002**, *121*, 55–65. [CrossRef]
397. Mastrototaro, F.; Petrocelli, A.; Cecere, E.; Matarrese, A. Non indigenous species settle down in the Taranto Seas. *Biogeographia* **2004**, *25*, 47–54. [CrossRef]
398. Çinar, M.E.; Bilecenoglu, M.; Öztürk, B.; Can, A. New records of alien species on the Levantine coast of Turkey. *Aquat. Invasions* **2006**, *1*, 84–90. [CrossRef]
399. El Haddad, M.; Capaccioni Azzati, R.; García-Carrascosa, A.M. *Branchiomma luctuosum* (Polychaeta: Sabellidae): A non-indigenous species at Valencia Port (Western Mediterranean Sea, Spain). *Mar. Biodivers. Rec.* **2008**, *1*, e61. [CrossRef]
400. Giangrande, A.; Licciano, M. Revision of the species of *Megalomma* (Polychaeta, Sabellidae) from the Mediterranean Sea, with the description of *M. messapicum* n. sp. *Ital. J. Zool.* **2008**, *75*, 207–217. [CrossRef]
401. Tanduo, V.; Golemaj, A.; Crocetta, F. Citizen-science detects the arrival and establishment of *Branchiomma luctuosum* (Grube, 1870) (Annelida: Polychaeta: Sabellidae) in Albania. *Biodivers. Data J.* **2020**, *8*, e54790. [CrossRef]
402. De Matos Nogueira, J.M.; Rossi, M.C.S.; López, E. Intertidal species of *Branchiomma* Kolliker and *Pseudobranchiomma* Jones (Polychaeta: Sabellidae: Sabellinae) occurring on rocky shores along the state of Sao Paulo, Southeastern Brazil. *Zool. Stud.* **2006**, *45*, 586–610.
403. Hartmann-Schröder, G. Teil 8. Die Polychaeten der Subtropisch-Antiborealen Westküste Australiens (Zwischen Cervantes im Norden und Cape Naturliste im Suden). *Mitt. Hamb. Zool. Mus. Inst.* **1982**, *79*, 51–118.
404. Ceberio, A.; Martínez, J.; Aguirrezabalaga, F. Presencia de *Desdemona ornata* Banse, 1957 (Polychaeta, Sabellidae) en las costas de La Península Ibérica, Golfo de Vizcaya. *Munibe* **1998**, *50*, 37–41.
405. Smith, P.; Perrett, J.; Garwood, P.; Moore, G. Two additions to the UK marine fauna: *Desdemona ornata* Banse, 1957 (Polychaeta, Sabellidae) and *Grandidierella japonica* Stephensen, 1938 (Amphipoda, Gammaridea). *Newsl. Porcupine Mar. Nat. Hist. Soc.* **1999**, *2*, 8–11.
406. Çinar, M.E.; Dagli, E.; Açik, S. Annelids (Polychaeta and Oligochaeta) from the Sea of Marmara, with descriptions of five new species. *J. Nat. Hist.* **2011**, *45*, 2105–2143. [CrossRef]
407. Carvalho, S.; Pereira, P.; Pereira, F.; de Pablo, H.; Vale, C.; Gaspar, M.B. Factors structuring temporal and spatial dynamics of macrobenthic communities in a eutrophic coastal lagoon (Óbidos Lagoon, Portugal). *Mar. Environ. Res.* **2011**, *71*, 97–110. [CrossRef]
408. Faasse, M. Dispersal of the invasive tubeworms *Desdemona ornata* and *Pseudopolydora paucibranchiata* to the Netherlands (Polychaeta: Sedentaria). *Nederl. Faun. Med.* **2016**, *46*, 49–56.
409. McArthur, M. Systematics and Introduced Status of *Euchone* (Polychaeta: Sabellidae) in Port Phillip Bay, and the Feeding Biology of *Euchone limnicola*. Bachelor's Thesis, Victoria University of Technology, Melbourne, Australia, 1997.
410. Guyonnet, B.; Borg, D. Premier signalement de l'espèce introduite *Euchone limnicola* Reish, 1959 (Polychaeta: Sabellidae) sur ses Côtes Françaises de la Mer du Nord (Grand Port Maritime de Dunkerque). *Cah. Nat. Observat. Mar.* **2015**, *4*, 15–23.
411. 411. Dane, L Morphological and Genetic Variation in the Cryptic Species Complex *Myxicola infundibulum* (Polychaeta, Sabellidae), and its Introduction to Australia. Master's Thesis, University of Melbourne, Melbourne, Australia, 2008.

412. Keppel, E.; Ruiz, G.; Tovar–Hernández, M. Re-description of *Parasabella fullo* (Grube, 1878) (Polychaeta: Sabellidae) and diagnostic characteristics for detection in California. *Eur. Zool. J.* **2020**, *87*, 105–115. [CrossRef]
413. Currie, D.; McArthur, M.; Cohen, B. *Exotic Marine Pests in the Port. of Geelong, Victoria. Report 8*; Marine and Freshwater Resources Institute: Queenscliff, Australia, 1998; pp. 1–57.
414. Coles, S.L. Historical and recent introductions of non-indigenous marine species into Pearl Harbor, Oahu, Hawaiian Islands. *Mar. Biol.* **1999**, *135*, 147–158. [CrossRef]
415. Eldredge, L.G.; Smith, C.A. *Guidebook of Introduced Marine Species in Hawaii*; Bishop Museum Technical Report 21; Bishop Museum: Honolulu, Hawaii, USA, 2001.
416. Coles, S.; Eldredge, L.G. Nonindigenous species introductions on coral reefs: A need for information. *Pac. Sci.* **2002**, *56*, 191–209. [CrossRef]
417. Culver, C.S.; Kuris, A.M.; Beede, B. *Identification and Management of the Exotic Sabellid Pest in California Cultured Abalone*; Californian Sea Grant College System: La Jolla, CA, USA, 1997.
418. Moore, J.D.; Juhasz, C.I.; Robbins, T.T.; Grosholz, E.D. The introduced sabellid polychaete *Terebrasabella heterouncinata* in California: Transmission, methods of control and survey for presence in native gastropod populations. *J. Shellfish Res.* **2007**, *26*, 869–876. [CrossRef]
419. Moreno, R.A.; Neill, P.E.; Rozbaczylo, N. Native and non-indigenous boring polychaetes in Chile: A threat to native and commercial mollusc species. *Rev. Chil. Hist. Nat.* **2006**, *79*, 263–278. [CrossRef]
420. Dittmann, S.; Rolston, A.; Benger, S.N.; Kupriyanova, E.K. *Habitat Requirements, Distribution and Colonisation of the Tubeworm Ficopomatus Enigmaticus in the Lower Lakes and Coorong*; Report for the South Australian Murray-Darling Basin Natural Resources Management Board; Department of Environment and water: Adelaide, Australia, 2009; pp. 1–99.
421. Grosse, M.; Pérez, R.; Juan-Amengual, M.; Pons, J.; Capa, M. The elephant in the room-first record of invasive gregarious species of serpulids (Calcareous Tube Annelids) in Majorca (Western Mediterranean). *Sci. Mar.* **2021**, in press.
422. Assis, J.D.; Alonso, C.; Christoffersen, M.L. First record of *Ficopomatus uschakovi* (Pillai, 1960) Serpulidae (Polychaeta: Annelida) for the Western Atlantic. *Rev. Nord. Biol.* **2008**, *19*, 51–58.
423. Bastida-Zavala, R.; García-Madrigal, S. First record in the Tropical Eastern Pacific of the exotic species *Ficopomatus uschakovi* (Polychaeta, Serpulidae). *ZooKeys* **2012**, *238*, 45–55. [CrossRef]
424. Liñero-Arana, I.; Díaz-Díaz, Ó. Presencia del poliqueto exótico *Ficopomatus uschakovi* (Polychaeta: Serpulidae) en Venezuela: Descripción y comentarios sobre su distribución. *Interciencia* **2012**, *37*, 234–237.
425. Arteaga-Flórez, C.; Fernández-Rodríguez, V.; Londoño-Mesa, M.H. First record of the polychaete *Ficopomatus uschakovi* (Pillai, 1960) (Annelida, Serpulidae) in the Colombian Caribbean, South America. *Zookeys* **2014**, *371*, 1–11. [CrossRef] [PubMed]
426. Zibrowius, H. Les espèces Méditerrannéennes du genre *Hydroides* (Polychaeta Serpulidae). Remarques sur le prétendu polymorphisme de *Hydroides uncinata*. *Téthys* **1971**, *2*, 691–746.
427. Zibrowius, H. Introduced invertebrates: Examples of success and nuisance in the European Atlantic and in the Mediterranean. Introduced species in European coastal waters. *Ecosyst. Res. Rep.* **1994**, *8*, 44–49.
428. Cohen, A.N.; Harris, L.H.; Bingham, B.L.; Carlton, J.T.; Chapman, J.W.; Lambert, C.C.; Lambert, G.; Ljubenkov, J.C.; Murray, S.N.; Rao, L.C. *Project Report for the Southern California Exotics Expedition*; A Rapid Assessment Survey of Exotic Species in Sheltered Coastal Waters California Department of Fish and Game, Sacramento CA State Water Resources Control Board, Sacramento CA National Fish and Wildlife Foundation: San Francisco, CA, USA, 2002; pp. 1–37.
429. Carlton, J.T.; Eldredge, L.G. *Marine bioinvasions of Hawai'i*; Bernice, P., Ed.; Bishop Museum Press: Honolulu, HI, USA, 2009; pp. 1–202.
430. Pettengill, J.; Wendt, D.; Schug, M.D.; Hadfield, M. Biofouling likely serves as a major mode of dispersal for the polychaete tubeworm *Hydroides elegans* as inferred from microsatellite loci. *Biofouling* **2007**, *23*, 161–169. [CrossRef]
431. Gruet, Y.; Hèral, M.; Robert, J.M. Premières observations sur l'introduction de la faune associée au naissain d'huîtres Japonaises *Crassostrea gigas* (Thunberg), importé sur la Côte Atlantique Française. *Cah. Biol. Mar.* **1976**, *17*, 173–184.
432. Zibrowius, H. Introduction du polychete Serpulidae Japonais *Hydroides ezoensis* sur la Cote, Atlantique Frnacoise et remarques sur la propagation d'autre espeies de Serpulidae. *Tethys* **1978**, *8*, 141–150.
433. Hayes, K.; Sliwa, C.; Migus, S.; McEnnulty, F.; Dunstan, P. National Priority Pests: Part II Ranking of Australian Marine Pests. In *An independent report undertaken for the Department of Environment and Heritage by CSIRO Marine Research*; Australian Government Department of the Environment and Heritage: Canberra, Australia, 2005; pp. 1–106.
434. Ferrario, J.; Minchin, D. Spread of the non-indigenous serpulid *Hydroides sanctaecrucis* Krøyer in Mörch, 1863 in the Pacific Ocean: A new record from Taiwan. *Bioinvasions Rec.* **2017**, *6*, 33–38. [CrossRef]
435. Lee, A.L.; Dafforn, K.A.; Hutchings, P.A.; Johnston, E.L. Reproductive strategy and gamete development of an invasive fanworm, *Sabella spallanzanii* (Polychaeta: Sabellidae), a field study in Gulf St Vincent, South Australia. *PLoS ONE* **2018**, *13*, e0200027.
436. Clapin, G.; Evans, D.R. The status of the introduced marine fanworm *Sabella spallanzanii* in Western Australia: A preliminary investigation. *Cent. Res. Introd. Mar. Pests Tech. Rep.* **1995**, *2*, 1–34.
437. Yee, A.; Mackie, J.; Pernet, B. The distribution and unexpected genetic diversity of the non-indigenous annelid *Ficopomatus enigmaticus* in California. *Aquat. Invasions* **2019**, *14*, 250–266. [CrossRef]

438. Oliva, M.; De Marchi, L.; Sanches, M.V.; Pires, A.; Cuccaro, A.; Baratti, M.; Chiellini, F.; Morelli, A.; Freitas, R.; Pretti, C. Atlantic and Mediterranean populations of the widespread serpulid *Ficopomatus enigmaticus*: Developmental responses to carbon nanotubes. *Mar. Pollut. Bull.* **2020**, *156*, 111265. [CrossRef] [PubMed]
439. Zibrowius, H. Serpulidae (Annelida Polychaeta) de l'Ocean Indien arrives sur des Coques de Bateaux a Toulon (France, Mediterranee). *Rapports et Proces Verbaux des Reunions Commission Internationale pour l'Exploration Scientifique de la Mer Mediterranee Monaco* **1979**, *25–26*, 133–134.
440. Ben-Eliahu, M.N. Red Sea serpulids (Polychaeta) in the Eastern Mediterranean. *Ophelia Suppl.* **1991**, *5*, 515–528.
441. Zenetos, A.; Çinar, M.E.; Crocetta, F.; Golani, D.; Rosso, A.; Servello, G.; Shenkar, N.; Turon, X.; Verlaque, M. Uncertainties and validation of alien species catalogues: The Mediterranean as an example. *Estuar. Coast. Shelf Sci.* **2017**, *191*, 171–187. [CrossRef]
442. Streftaris, N.; Zenetos, A. Alien marine species in the Mediterranean-the 100 'worst invasives' and their impact. *Mediterr. Mar. Sci.* **2006**, *7*, 87–118. [CrossRef]
443. Dürr, S.; Watson, D.I. Biofouling and antifouling in aquaculture. *Biofouling* **2010**, *12*, 267–287.
444. Wang, J.; Huang, Z. Fouling polychaetes of Hong Kong and adjacent saters. *Asian Mar. Biol.* **1993**, *10*, 1–12.
445. Mohan, P.C.; Aruna, C. The biology of serpulid worms in relation to biofouling. In *Recent Developments in Biofouling Control*; Thompson, M.F., Nagabhushanam, R., Sarojini, R., Fingerman, M., Eds.; Oxford & IBH Publishing Company: New Delhi, India, 1994; pp. 59–64.
446. Fitridge, I.; Dempster, T.; Guenther, J.; De Nys, R. The impact and control of biofouling in marine aquaculture: A review. *Biofouling* **2012**, *28*, 649–669. [CrossRef] [PubMed]
447. Schloesser, D.W.; Malakauskas, D.M.; Malakauskas, S.J. Freshwater polychaetes (*Manayunkia speciosa*) near the Detroit River, Western Lake Erie: Abundance and life-history characteristics. *J. Great Lakes Res.* **2016**, *42*, 1070–1083. [CrossRef]
448. Read, G.B.; Gordon, D.P. Adventive occurrence of the fouling serpulid *Ficopomatus enigmaticus* (Polychaeta) in New Zealand. *N. Z. J. Mar. Freshw. Res.* **1991**, *25*, 269–273. [CrossRef]
449. Probert, P.K. First record of the introduced fouling tubeworm *Ficopomatus enigmaticus* (Polychaeta: Serpulidae) in Hawke Bay, New Zealand. *N. Z. J. Zool.* **1993**, *20*, 35–36. [CrossRef]
450. Pernet, B.; Barton, M.; Fitzhugh, K.; Harris, L.; Lizárraga, D.; Ohl, R.; Whitcraft, C. Establishment of the reef-forming tubeworm *Ficopomatus enigmaticus* (Fauvel, 1923) (Annelida: Serpulidae) in Southern California. *Bioinvasions Rec.* **2016**, *5*, 13–19. [CrossRef]
451. Heiman, K.W.; Micheli, F. Non-native ecosystem engineer alters Estuarine communities. *Integr. Comp. Biol.* **2010**, *50*, 226–236. [CrossRef] [PubMed]
452. Ulman, A.; Ferrario, J.; Forcada, A.; Seebens, H.; Arvanitidis, C.; Occhipinti-Ambrogi, A.; Marchini, A. Alien species spreading via biofouling on recreational vessels in the Mediterranean Sea. *J. Appl. Ecol.* **2019**, *56*, 2620–2629. [CrossRef]
453. Secretaría de Medio Ambiente y Recursos Naturales, Estados Unidos Mexicanos. *Acuerdo por el que se Determina la Lista de las Especies Exóticas Invasoras Para México*; Secretaría de Medio Ambiente y Recursos Naturales: Ciudad de México, Mexico, 2016.
454. Mackie, G.L.; Qadri, S.U. A polychaete, *Manayunkia speciosa*, from the Ottawa River, and Its North American distribution. *Canad. J. Zool.* **1971**, *49*, 780–782. [CrossRef]
455. Stocking, R.W.; Bartholomew, J.L. Distribution and habitat characteristics of *Manayunkia speciosa* and infection prevalence with the parasite *Ceratomyxa shasta* in the Klamath River, Oregon–California. *J. Parasitol.* **2007**, *93*, 78–88. [CrossRef] [PubMed]
456. Piazzolla, D.; Cafaro, V.; Mancini, E.; Scanu, S.; Bonamano, S.; Marcelli, M. Preliminary investigation of microlitter pollution in low-energy hydrodynamic basins using *Sabella spallanzanii* (Polychaeta: Sabellidae) tubes. *Bull. Environ. Contam. Toxicol.* **2020**, *104*, 345–350. [CrossRef]
457. Gopalakrishnan, S.; Thilagam, H.; Raja, P.V. Comparison of heavy metal toxicity in life stages (spermiotoxicity, egg toxicity, embryotoxicity and larval toxicity) of *Hydroides elegans*. *Chemosphere* **2008**, *71*, 515–528. [CrossRef] [PubMed]
458. Thilagam, H.; Gopalakrishnan, S.; Vijayavel, K.; Raja, P.V. Effluent toxicity test using developmental stages of the marine polychaete *Hydroides elegans*. *Arch. Environ. Contam. Toxicol.* **2008**, *54*, 674–683. [CrossRef]
459. Vijayaragavan, S.; Raja, P.V. The toxic effects of phorate on early embryonic stages of sedentary polychaete *Hydroides elegans* (Haswell, 1883). *J. Basic Appl. Zool.* **2018**, *79*, 45. [CrossRef]
460. Bellante, A.; Piazzese, D.; Cataldo, S.; Parisi, M.G.; Cammarata, M. Evaluation and comparison of trace metal accumulation in different tissues of potential bioindicator organisms: Macrobenthic filter feeders *Styela plicata*, *Sabella spallanzanii*, and *Mytilus galloprovincialis*. *Environ. Toxicol. Chem.* **2016**, *35*, 3062–3070. [CrossRef]
461. Giangrande, A.; Licciano, M.; Del Pasqua, M.; Fanizzi, F.P.; Migoni, D.; Stabili, L. Heavy metals in five Sabellidae species (Annelida, Polychaeta): Ecological implications. *Environ. Sci. Pollut. Res.* **2017**, *24*, 3759–3768. [CrossRef]
462. Xie, Z.-C.; Wong, N.C.; Qian, P.-Y.; Qiu, J.-W. Responses of polychaete *Hydroides elegans* life stages to copper stress. *Mar. Ecol. Prog. Ser.* **2005**, *285*, 89–96. [CrossRef]
463. Gopalakrishnan, S.; Thilagam, H.; Raja, P. Toxicity of heavy metals on embryogenesis and larvae of the marine sedentary polychaete *Hydroides elegans*. *Arch. Environ. Contamin. Toxicol.* **2007**, *52*, 171–178. [CrossRef] [PubMed]
464. Ross, K.E.; Bidwell, J.R. A 48-h larval development toxicity test using the marine polychaete *Galeolaria caespitosa* Lamarck (Fam. Serpulidae). *Arch. Environ. Contamin. Toxicol.* **2001**, *40*, 489–496.
465. Giangrande, A.; Cavallo, A.; Licciano, M.; Mola, E.; Pierri, C.; Trianni, L. Utilization of the filter feeder polychaete *Sabella*. *Aquac. Int.* **2005**, *13*, 129–136. [CrossRef]

466. Licciano, M.; Stabili, L.; Giangrande, A.; Cavallo, R.A. Bacterial accumulation by *Branchiomma luctuosum* (Annelida: Polychaeta): A tool for biomonitoring marine systems and restoring polluted waters. *Mar. Environ. Res.* **2007**, *63*, 291–302. [CrossRef]
467. Pierri, C.; Fanelli, G.; Giangrande, A. Experimental co-culture of low food-chain organisms, *Sabella spallanzanii* (Polychaeta, Sabellidae) and *Cladophora prolifera* (Chlorophyta, Cladophorales), in Porto Cesareo area (Mediterranean Sea). *Aquac. Res.* **2006**, *37*, 966–974. [CrossRef]
468. Stabili, L.; Licciano, M.; Giangrande, A.; Longo, C.; Mercurio, M.; Marzano, C.N.; Corriero, G. Filtering Activity of *Spongia officinalis* var. *adriatica* (Schmidt) (Porifera, Demospongiae) on bacterioplankton: Implications for bioremediation of polluted seawater. *Water Res.* **2006**, *40*, 3083–3090. [CrossRef]
469. Stabili, L.; Licciano, M.; Lezzi, M.; Giangrande, A. Microbiological accumulation by the Mediterranean invasive alien species *Branchiomma bairdi* (Annelida, Sabellidae): Potential tool for bioremediation. *Mar. Pollut. Bull.* **2014**, *86*, 325–331. [CrossRef]
470. Stabili, L.; Licciano, M.; Giangrande, A.; Gerardi, C.; De Pascali, S.A.; Fanizzi, F.P. First insight on the mucus of the annelid *Myxicola infundibulum* (Polychaeta, Sabellidae) as a potential prospect for drug discovery. *Mar. Drugs* **2019**, *17*, 396. [CrossRef]
471. Soto, D. *Integrated Mariculture: A Global Review.*; Food and Agriculture Organization of the United Nations (FAO): Rome, Italy, 2009.
472. Stewart, B.; Davies, B. Effects of macrophyte harvesting on invertebrates associated with *Potamogeton pectinatus* L. in the Marina Da Gama, Zandvlei, Western Cape. *Trans. R. Soc. S. Africa* **1986**, *46*, 35–49. [CrossRef]
473. Joyce, C.B.; Vina-Herbon, C.; Metcalfe, D.J. Biotic variation in coastal water bodies in Sussex, England: Implications for saline lagoons. *Estuar. Coast. Shelf Sci.* **2005**, *65*, 633–644. [CrossRef]
474. Davies, B.; Stuart, V.; De Villiers, M. The filtration activity of a serpulid polychaete population (*Ficopomatus enigmaticus* (Fauvel) and its effects on water quality in a coastal marina. *Estuar. Coast. Shelf Sci.* **1989**, *29*, 613–620. [CrossRef]
475. Licciano, M.; Watson, G.J.; Murray, J.M.; Giangrande, A. Evidence of regenerative ability in *Myxicola infundibulum* (Annelida, Sabellida): Evolutionary and systematic implications. *Invertebr. Biol.* **2015**, *134*, 48–60. [CrossRef]
476. Bely, A.E. Distribution of segment regeneration ability in the Annelida. *Integr. Comp. Biol.* **2006**, *46*, 508–518. [CrossRef] [PubMed]
477. Prentiss, N.K.; Tyler, M.S.; Dean, D.A. Morphological and histological investigation of the regeneration in *Myxicola infundibulum* (Montagu, 1808) (Sabellida, Annelida). *Mar. Biol. Assoc. UK* **2017**, *97*, 1155–1165. [CrossRef]
478. Nicol, J. The giant nerve-fibres in the central nervous system of *Myxicola* (Polychaeta, Sabellidae). *Q. J. Micros. Sci.* **1948**, *3*, 1–45.
479. Roberts, M.B.V. The Rapid response of *Myxicola infundibulum* (Grube). *J. Mar. Biol. Assoc. U.K.* **1962**, *42*, 527–539. [CrossRef]
480. Gilbert, D. Axoplasm architecture and physical properties as seen in the *Myxicola* giant axon. *J. Physiol.* **1975**, *253*, 257–301. [CrossRef]
481. Barry, J.F.; Turner, M.J.; Schloss, J.M.; Glenn, D.R.; Song, Y.; Lukin, M.D.; Park, H.; Walsworth, R.L. Optical magnetic detection of single-neuron action potentials using quantum defects in diamond. *Proc. Natl. Acad. Sci. USA* **2016**, *113*, 14133–14138. [CrossRef]
482. Purschke, G. Annelida: Basal groups and Pleistoannelida. In *Structure and Evolution of Invertebrate Nervous Systems*; Schmidt-Rhaesa, A., Harzsch, S., Purschke, G., Eds.; Oxford University Press: Oxford, UK, 2016; pp. 254–312.
483. Schauf, C.; Davis, F.A.; Marder, J. Effects of carbamazepine on the ionic conductances of *Myxicola* giant axons. *J. Pharmacol. Exp. Ther.* **1974**, *189*, 538–543.
484. Nedved, B.T.; Hadfield, M.G. Hydroides elegans (Annelida: Polychaeta): A model for biofouling research. In *Marine and Industrial Biofouling*; Flemming, H.-S., Sriyutha Murthy, P., Venkatesan, R., Cooksey, K., Eds.; Springer Series on Biofilms: Los Angeles, CA, USA, 2008; Volume 4, pp. 203–218.
485. Hadfield, M.G. The D.P. Wilson Lecture. Research on settlement and metamorphosis of marine invertebrate larvae: Past, present and future. *Biofouling* **1998**, *12*, 9–29. [CrossRef]
486. Hadfield, M.G. Biofilms and marine invertebrate larvae: What bacteria produce that larvae use to choose settlement sites. *Ann. Rev. Mar. Sci* **2011**, *3*, 453–470. [CrossRef] [PubMed]
487. Qian, P.-Y. Larval settlement of polychaetes. *Hydrobiologia* **1999**, *402*, 239–253. [CrossRef]
488. Hadfield, M.G.; Nedved, B.T.; Wilbur, S.; Koehl, M. Biofilm cue for larval settlement in *Hydroides elegans* (Polychaeta): Is contact necessary? *Mar. Biol.* **2014**, *161*, 2577–2587. [CrossRef]
489. Shikuma, N.J.; Antoshechkin, I.; Medeiros, J.M.; Pilhofer, M.; Newman, D.K. Stepwise metamorphosis of the tubeworm *Hydroides elegans* is mediated by a bacterial inducer and MAPK Signaling. *Proc. Natl. Acad. Sci. USA* **2016**, *113*, 10097–10102. [CrossRef] [PubMed]
490. Vijayan, N.; Lema, K.A.; Nedved, B.T.; Hadfield, M.G. Microbiomes of the polychaete *Hydroides elegans* (Polychaeta: Serpulidae) across its life-history stages. *Mar. Biol.* **2019**, *166*, 19. [CrossRef]
491. Toonen, R.J.; Pawlik, J.R. Foundations of gregariousness. *Nature* **1994**, *370*, 511–512. [CrossRef]
492. Toonen, R.J.; Pawlik, J.R. Settlement of the gregarious tube worm *Hydroides dianthus* (Polychaeta: Serpulidae): Cues for gregarious settlement. *Mar. Biol.* **1996**, *126*, 725–733. [CrossRef]
493. Toonen, R.; Pawlik, J. Settlement of the gregarious tube worm *Hydroides dianthus* (Polychaeta: Serpulidae). I. gregarious and nongregarious settlement. *Mar. Ecol. Prog. Ser.* **2001**, *224*, 103–114. [CrossRef]
494. Kenny, N.J.; Shimeld, S.M. Additive multiple k-mer transcriptome of the keelworm *Pomatoceros lamarckii* (Annelida; Serpulidae) reveals annelid trochophore transcription factor cassette. *Dev. Genes Evol.* **2012**, *222*, 325–339. [CrossRef]
495. Barton-Owen, T.B.; Szabó, R.; Somorjai, I.M.; Ferrier, D.E. A revised spiralian homeobox gene classification incorporating new polychaete transcriptomes reveals a diverse TALE class and a divergent hox gene. *Genome Biol. Evol.* **2018**, *10*, 2151–2167. [CrossRef] [PubMed]

496. Szabó, R.; Ferrier, D.E.K. Cell proliferation dynamics in regeneration of the operculum head appendage in the annelid *Pomatoceros lamarckii*. *J. Exp. Zool. Part. B Mol. Dev. Evol.* **2014**, *322*, 257–268. [CrossRef] [PubMed]
497. Szabó, R.; Ferrier, D.E. Another biomineralising protostome with an *msp130* gene and conservation of *msp130* gene structure across Bilateria. *Evol. Dev.* **2015**, *17*, 195–197. [CrossRef]
498. San Chan, V.B.; Li, C.; Lane, A.C.; Wang, Y.; Lu, X.; Shih, K.; Zhang, T.; Thiyagarajan, V. CO_2-driven ocean acidification alters and weakens integrity of the calcareous tubes produced by the serpulid tubeworm, *Hydroides elegans*. *PLoS ONE* **2012**, *7*, e42718. [CrossRef]
499. Lane, A.C.; Mukherjee, J.; Chan, V.B.; Thiyagarajan, V. Decreased pH does not alter metamorphosis but compromises juvenile calcification of the tube worm *Hydroides elegans*. *Mar. Biol.* **2013**, *160*, 1983–1993. [CrossRef]
500. Díaz-Castañeda, V.; Cox, T.E.; Gazeau, F.; Fitzer, S.; Delille, J.; Alliouane, S.; Gattuso, J.-P. Ocean acidification affects calcareous tube growth in adults and reared offspring of serpulid polychaetes. *J. Exp. Biol.* **2019**, *222*, jeb196543. [CrossRef]
501. Wabnitz, C. *From Ocean to Aquarium: The Global Trade in Marine Ornamental Species*; UNEP/Earthprint: Cambridge, UK, 2003.
502. Murray, J.M.; Watson, G.J.; Giangrande, A.; Licciano, M.; Bentley, M.G. Managing the marine aquarium trade: Revealing the data gaps using ornamental polychaetes. *PLoS ONE* **2012**, *7*, e29543. [CrossRef] [PubMed]
503. Bybee, D.R.; Murray, J.M. Polychaetes. *Mar. Ornam. Spec. Aquacult.* **2017**, *1*, 565–579.
504. Bybee, D.R.; Bailey-Brock, J.H.; Tamaru, C.S. Larval development of *Sabellastarte spectabilis* (Grube, 1878) (Polychaeta: Sabellidae) in Hawaiian waters. *Sci. Mar.* **2006**, *70*, 279–286. [CrossRef]
505. Bybee, D.R.; Bailey-Brock, J.H.; Tamaru, C.S. Evidence for sequential hermaphroditism in *Sabellastarte spectabilis* (Polychaeta: Sabellidae) in Hawai'i. *Pac. Sci.* **2006**, *60*, 541–547. [CrossRef]
506. Bybee, D.R.; Bailey-Brock, J.H.; Tamaru, C.S. Gametogenesis and spawning periodicity in the fan worm *Sabellastarte spectabilis* (Polychaeta: Sabellidae). *Mar. Biol.* **2007**, *151*, 639–648. [CrossRef]
507. Murray, J.M. Regeneration and Reproduction in *Sabella pavonina* (Savigny): Developing a Novel Method to Culture Marine Ornamental Sabellids. Ph.D. Thesis, University of Portsmouth, Portsmouth, UK, 2010.
508. Dávila-Jiménez, Y.; Tovar-Hernández, M.A.; Simões, N. The social feather duster worm *Bispira brunnea* (Polychaeta: Sabellidae): Aggregations, morphology and reproduction. *Mar. Biol. Res.* **2017**, *13*, 782–796. [CrossRef]
509. Alalykina IL, Polychaeta. A review of the deep-sea benthic polychaetes along the NW Pacific. In *Biogeographical Atlas of the Deep NW Pacific Fauna*; Pensoft Publishers: Sofia, Bulgaria, 2020; pp. 175–212.
510. Gunton, L.M.; Kupriyanova, E.K.; Alvestad, T.; Avery, L.; Blake, J.A.; Biriukova, O.; Böggemann, M.; Borisova, P.; Budaeva, N.; Burghardt, I.; et al. Annelids of the eastern Australian abyss collected by the 2017 RV 'Investigator'voyage. *ZooKeys* **2021**, *1020*, 1–198. [CrossRef]
511. Simon, C. The wonderful world of worms. *Quest* **2014**, *10*, 32–34.
512. Ippolitov, A.; Rzhavsky, A. Tube morphology, ultrastructures and mineralogy in recent Spirorbinae (Annelida; Polychaeta; Serpulidae). I. General introduction. Tribe Paralaeospirini. *Invertebr. Zool.* **2014**, *11*, 293–314. [CrossRef]
513. Sigvaldadóttir, E.; Mackie, A.S. *Prionospio steenstrupi*, *P. fallax* and *P. dubia* (Polychaeta, Spionidae): Re-evaluation of identity and status. *Sarsia* **1993**, *78*, 203–219. [CrossRef]
514. Martin, D.; Britayev, T.A.; San Martín, G.; Gil, J. Inter-population variability and character description in the sponge-associated *Haplosyllis spongicola* complex (Polychaeta: Syllidae). In *Advances in Polychaete Research*; Springer: Dordrecht, The Netherlands, 2003; pp. 145–162.
515. Ten Hove, H.A.; Jansen-Jacobs, M.J. A Revision of the genus *Crucigera* (Polychaeta; Serpulidae); A proposed methodical approach to serpulids, with special reference to variation in *Serpula* and *Hydroides*. In *Proceedings of the First International Polychaete Conference*; Hutchings, P.A., Ed.; Linnean Society of New South West: Sydney, Australia, 1984; pp. 143–180.
516. Watanabe, M.E. The evolution of natural history collections: New research tools move specimens, data to center stage. *Bioscience* **2019**, *69*, 163–169. [CrossRef]
517. Annenkova, N.P. Polychaeta of the North Japan Sea and their horizontal and vertical distribution. USSR Hydrobiological Expedition in 1934 to the Japan Sea. *Trudy Daln. Filiala Akademii Nauk SSSR* **1938**, *1*, 81–230. (In Russian)
518. Oug, E.; Bakken, T.; Kongsrud, J.A. Original specimens and type localities of early described polychaete species (Annelida) from Norway, with particular attention to species described by O. F. Müller and M. Sars. *Mem. Mus. Vic.* **2014**, *71*, 217–236. [CrossRef]
519. Montagu, G. *Testacea Brittanica or Natural History of British Shells, Marine, Land and Fresh-Water, Including the Most Minute: Systematically Arranged and Embellished with Figures. Part II*; White: London, UK, 1803; pp. 293–606.
520. Von Siebold, C.T. Bericht Über die Leistungen in der Naturgeschichte der Würmer, Zoophyten und Protozoen Während der Jahre 1845, 1846 und 1847. *Arch. Naturgesch.* **1850**, *16*, 351–468.

diversity

MDPI

Review

On the Systematics and Biodiversity of the Opheliidae and Scalibregmatidae

Julio Parapar [1,*], Alejandro Martínez [2] and Juan Moreira [3]

1 Departamento de Bioloxía, Facultade de Ciencias, Universidade da Coruña, 15008 A Coruña, Spain
2 Molecular Ecology Group (MEG), Water Research Institute (IRSA), National Research Council of Italy (CNR), Largo Tonolli 5, 28922 Pallanza, Italy; alejandro.martinezgarcia@cnr.it
3 Departamento de Biología (Zoología) & Centro de Investigación en Biodiversidad y Cambio Global (CIBC-UAM), Facultad de Ciencias, Universidad Autónoma de Madrid, 28049 Madrid, Spain; juan.moreira@uam.es
* Correspondence: julio.parapar@udc.es

Abstract: In this paper we review the systematics, diversity, and ecology of two related annelid families: Opheliidae Malmgren, 1867 and Scalibregmatidae Malmgren, 1867. Opheliids are deposit-feeders and that are mainly found as burrowers in sandy sediments. Morphologically, opheliids are characterized by the smooth cuticle, as well as the presence of a conspicuous ventral groove, reduced parapodia, and a tubular-shaped structure often projecting from the posterior end. Scalibregmatids are also deposit-feeders, but compared to opheliids, they have a characteristic arenicoliform body, a T-shaped anterior end and a glandular, reticulated epidermis. For each family, we summarize the available information about the evolutionary relationships, taxonomic history, geographical distribution, ecological preferences and diversity of life strategies along with the techniques most commonly used for their study. By highlighting the main gaps in knowledge on each of these topics, this review ultimately aims at stimulating further research into members of these two families in the future.

Keywords: Opheliidae; Scalibregmatidae; diversity; taxonomy; anatomy; biology

Citation: Parapar, J.; Martínez, A.; Moreira, J. On the Systematics and Biodiversity of the Opheliidae and Scalibregmatidae. *Diversity* **2021**, *13*, 87. https://doi.org/10.3390/d13020087

Academic Editor: Luc Legal

Received: 30 December 2020
Accepted: 12 February 2021
Published: 18 February 2021

1. Introduction

Opheliidae Malmgren, 1867 is a well-known family of annelids distributed throughout the world mostly in sandy sediments [1–3]. Most of the described five to six genera and ca. 160 species of opheliids include elongate, deposit-feeding burrowing worms, which are easily recognized by the smooth cuticle and the presence of a conspicuous ventral groove along at least the posterior half of the body (Figure 1). Opheliids usually have a conical to pointed prostomium that lacks lateral antennae, whereas their pygidium often develops a tubular-shaped prolongation that may bear cirri and marginal papillae. Although some species may reach 100 mm in length, most opheliids range between 5–70 mm and their trunk comprise about 30–60 segments [4].

The knowledge on opheliid taxonomy and systematics has been substantially improved in the last two decades, including the delineation of subfamilies and phylogenetic affinities [1]. However, further work is still needed in order to assess the validity of the genus *Ammotrypanella* McIntosh, 1879 and some species of *Ophelia* Savigny, 1822 and *Ophelina* Örsted, 1843, as well as the status of the many synonymies attributed to the presumably cosmopolitan *Polyophthalmus pictus* (Dujardin, 1839). The opheliid fauna of some geographic areas is well known (e.g., North Atlantic, California) whereas other regions remain clearly understudied and may potentially hold many undescribed species (e.g., Tropical Atlantic, Indo-Pacific and Australasia). The biology, ecology, and burrowing behavior of some species were studied in detail due to their ecological importance in the intertidal and shallow subtidal of sandy beaches at temperate and tropical latitudes (e.g., [5–10]).

Some of these shallow water opheliids represent promising bioindicator species and have been even the target of experimental toxicological studies [11,12]. In contrast, we know virtually nothing on the biology of the opheliid species found at greater depths, despite their numerical importance in many macrofaunal assemblages in the deep-sea [13].

Traditionally, opheliid taxonomy has been based on conspicuous morphological characters, such as the number of branchiate chaetigers and different features associated to the anal tube. However, the branchiae and the anal tube are easily detached or damaged, leading to the wrong assessment of their absence or presence during species descriptions and identification and producing too much taxonomic confusion in the past (e.g., [1,2,14,15]). On the other hand, recent studies based on scanning electron microscopy (SEM) have revealed that the extended presence of lateral organs as well as a variety of nuchal organs features [1,15] may represent reliable taxonomic characters in those animals with simple bodies, reduced parapodia, and apparently similar simple chaetae. The internal anatomy of several opheliids has been studied in detail during the first half of the 20th century [16,17], when much attention was paid, for instance, to the structure of the sensory organs (e.g., [18,19]) and the arrangement of the body musculature (e.g., [20,21]). Methodological approaches such as the use of microcomputed X-ray tomography (Micro-CT) may update some of the results from these studies and provide further morphological support for the described genera (e.g., features of the digestive tract) by revealing new phylogenetically informative characters.

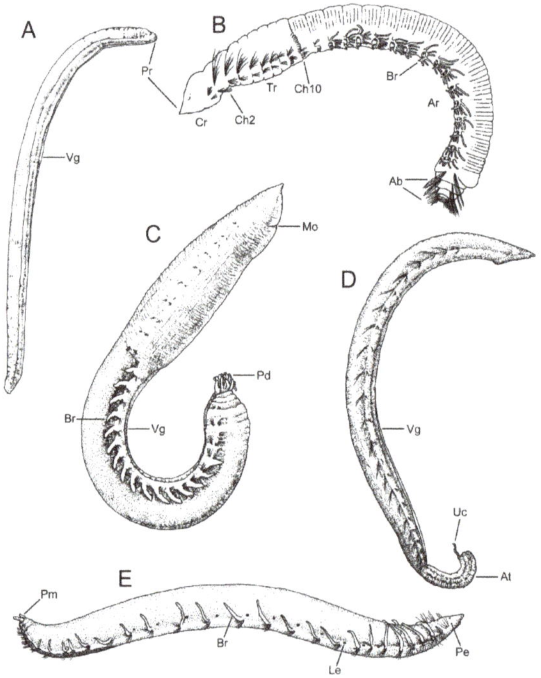

Figure 1. Stylized drawings of opheliids of the subfamily Opheliinae (**B**,**C**) and Ophelininae (**A**,**D**,**E**). (**A**) *Polyophthalmus pictus* in latero-ventral view; (**B**) *Thoracophelia japonica* in lateral view (chaetiger numbers mark limit between body regions); (**C**) *Ophelia bicornis* in lateral view; (**D**) *Ophelina abranchiata* in lateral view; (**E**) *Armandia cirrhosa* in lateral view. (**A**,**C**–**E**) redrawn after Parapar [4]; (**B**) modified after Misaka and Sato [22]. Abbreviations: Ab—abranchiate chaetigers; Ar—abdominal region; At—anal tube; Br—branchia; Ch—chaetiger; Cr—cephalic region; Le—lateral eye; Mo—mouth; Pd—pygidium dorsal papillae; Pe—prostomial subdermal eye; Pm—pygidium marginal papillae; Pr—prostomium; Tr—thoracic region; Uc—unpaired anal cirrus; Vg—ventral groove.

Scalibregmatidae is a worldwide distributed family of sedentary annelids currently including ca. 70 described species classified in 14 genera (see below) [3,23]. Most species are subsurface deposit-feeders and prefer muddy bottoms at considerable depths or in high latitudes. Typically, they range between 5–70 mm in body length, exhibiting a vividly red pigmentation and a relatively simple external morphology [24]. Traditionally, the body shape has been categorized either as arenicoliform, i.e., more or less elongated and tapering towards the posterior end, or as maggotlike, i.e., relatively short and stout [25]. The epidermis is thick and glandular, and each trunk segment is often divided in one to six annulated rows of elevated pads that give the body a characteristic tesselate appearance. The prostomium is usually small and forms a pair of lateral or frontal prostomial appendages, which give the anterior end a characteristic T-shaped appearance. The pygidium is typically simple and possesses a variable number of cirri. However, there are several exceptions to this body plan within morphologically divergent species classified in the genera *Axiokebuita*, *Speleobregma*, and *Scalibregmella* [26,27].

Scalibregmatids have been known for a relatively long period of time, and indeed, quite extensive monographs on the group were already published during the 19th and the early 20th centuries [28–30]. However, despite this early interest, the phylogenetic position of the family as well as the relationships amongst its genera remain poorly understood. This is despite the several taxonomic revisions that the family has undergone during the last few decades, notably involving the rearrangement of several genera [25,31,32] and the transference of the genus *Travisia* to the newly erected family Travisiidae [33]. While most Scalibregmatidae has been described from the Northern Atlantic [34,35], the family is unusually diverse in the Antarctic Ocean, from where 16 species have been described so far [31,36,37]. Most of those Northern Atlantic and Antarctic species have been recorded from muddy bottoms, where they might become very abundant and even locally dominate the benthic community. Records of scalibregmatids in lower latitudes are scarcer but often come from a wider range of environments, including sandy bottoms, *Posidonia* and *Zostera* seagrass meadows [38], corals and sponges [25,39], mussel beds [40], or even marine and anchialine cave systems [26].

Despite that the internal anatomy of Scalibregmatidae has long been known [28,41], no recent studies have revisited these early anatomical studies using modern imaging techniques. This has hampered our understanding of both the phylogenetic position of the family as well as its internal relationships insofar as the homology of many scalibregmatid characters in relation to other annelids [36,42,43], as well as the character evolution within the group remain obscure. Consequently, both the family Scalibregmatidae as well as many of its genera are diagnosed without any synapomorphies [24,44], but rather based on combinations of few external morphological characters [23,24,30,44] whose inter- and intraspecific variability remain, in general, poorly understood. The fact that many scalibregmatids have been described from limited or fragmented material has aggravate this situation [27,40], also because many traditional characters vary substantially across life stages of the same species [26,36]. This situation can be improved integrating different microscopical techniques in future taxonomic descriptions. This approach has already been followed by recent studies, which have successfully included previously overlooked characters, such as arrangement of ciliary bands, glands, or patterns of the epidermal ornamentation, in the diagnoses of several new species [26,36].

In this contribution, an updated revision of the current biodiversity knowledge of the families Opheliidae and Scalibregmatidae is provided, and an update in taxonomy, classification, and systematics of the members of both taxa, highlighting where major gaps in knowledge lie and where future efforts could be made.

2. Methods

Published literature on opheliids and scalibregmatids was reviewed thoroughly aiming for information on diversity, ecology, and distribution. The World Register of Marine Species [3] database was mostly used as the basis for systematic arrangement, synonymies and valid genera and species, as well as Blake and Maciolek [1] for Opheliidae and Blake [23] for Scalibregmatidae. Furthermore, brief accounts on systematics and general morphology of these families are also provided as well as tables with valid nominal species including type locality, depth (from original description) and marine realms (*sensu* [45]) (Tables A1 and A2 in Appendix A).

3. Results

3.1. Opheliidae Malmgren, 1867

3.1.1. Systematics

Until recently, the Opheliidae comprised three subfamilies: Opheliinae Hartman-Schröder, 1971, Ophelininae Hartman-Schröder, 1971 and Travisiinae Hartmann-Schröder, 1971. The latter only included the genus *Travisia* Johnston, 1840 that differed from other opheliids in having a grublike appearance and a papillated cuticle. Indeed, recent molecular phylogenetic analyses have demonstrated the monophyly of opheliids if the Travisiinae are excluded [46], subsequently motivating the establishment of Travisiinae as a family by Blake and Maciolek [33]. In fact, this possibility had been already proposed by Blake [47], Bleidorn et al. [48] and Hall et al. [49]. The morphological differences between Travisiinae and the other two subfamilies were further supported by Belova and Zhadan [50]. These authors suggested that the presence of several shared anatomical and ultrastructural features of the gills amongst several opheliid genera but absent in *Travisia*, might support the exclusion of the latter from Opheliidae and would constitute synapomorphies of the Opheliinae and Ophelininae. The Travisiidae is now considered the sister group to the Scalibregmatidae, while molecular analyses have highlighted the affinities of opheliids to capitellids and echiuroids [51,52] and to other "sedentary" families as well (e.g., Arenicolidae). Therefore, today Opheliidae includes only the subfamilies Opheliinae and Ophelininae [1,46]. The two subfamilies are represented only by species with elongated bodies and smooth cuticle, all sharing the presence of a conspicuous ventral groove [1].

According to Blake and Maciolek [1], Opheliidae comprises five genera distributed in the subfamilies Opheliinae (*Ophelia* and *Thoracophelia* Ehlers, 1897) and Ophelininae (*Armandia* Filippi, 1861, *Ophelina* and *Polyophthalmus* Quatrefages, 1850). However, there has been much confusion with the generic arrangement within the Opheliidae (e.g., synonymies and changes in diagnosis of genera). Sene-Silva [53] performed a cladistic morphological analysis of the family that have led to a redefinition of the previously established genera. In this context, *Lobochesis* Hutchings and Murray, 1984 was synonymized with *Thoracophelia*, which subsequently replaced *Euzonus* Grube, 1866 (a homonym of the diplopod myriapod *Euzonus* Menge, 1854 [54]). According to Blake and Maciolek [1], the genera *Tachytrypane* McIntosh, 1879 and *Ammotrypanella* would fall within the current diagnosis of *Ophelina* because the presence and distribution of branchiae seemed much variable within the latter. However, *Ammotrypanella* was retained by Wiklund et al. [2] who also amended the redefinition of the genus as given by Schüller [55]. Furthermore, the abranchiate *Antiobactrum* Chamberlin, 1919 is regarded as a valid genus in the World Register of Marine Species [3] but considered, in turn, as a synonym of *Ophelina* by Blake and Maciolek [1]. In this context, Paul et al. [46] have provided a phylogenetic analysis of the family but considering only a limited number of species; therefore, an analysis based on molecular and morphological characters including a greater taxa sampling would be desirable to assess the actual definition of genera.

3.1.2. Taxonomic History

The first described species was *Ophelia bicornis* Savigny, 1822. The number of newly described taxa increased gradually during the second half of the 19th century and along the first two decades of the 20th century. After WWII new species were described at a rate of about 10 per decade, whereas in the last decade (2010–2019) 30 new species were added to the family from all around the globe (Figure 2A). This overall tendency closely resembles those exhibited by each of the most speciose genera (Figure 3).

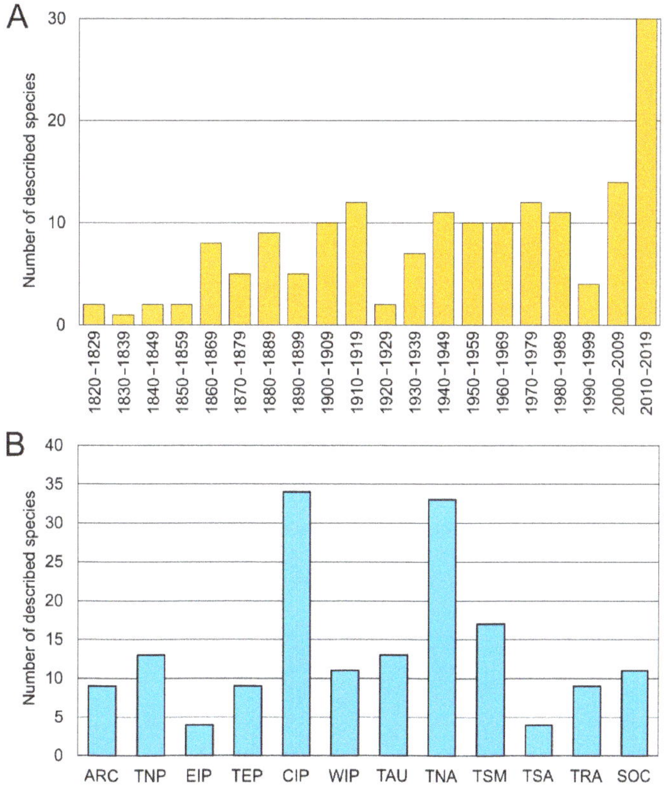

Figure 2. (**A**) number of opheliid species described per decade; (**B**) number of valid opheliid species listed under the bioregion (*sensu* Spalding et al. [45]) according to type locality. Abbreviations: ARC—Arctic; TNP—Temperate North Pacific; EIP—Eastern Indo-Pacific; TEP—Tropical Eastern Pacific; CIP—Central Indo-Pacific; WIP—Western Indo-Pacific; TAU—Temperate Australasia; TNA—Temperate Northern Atlantic; TSM—Temperate South America; TSA—Temperate South Africa; TRA—Tropical Atlantic; SOC—Southern Ocean.

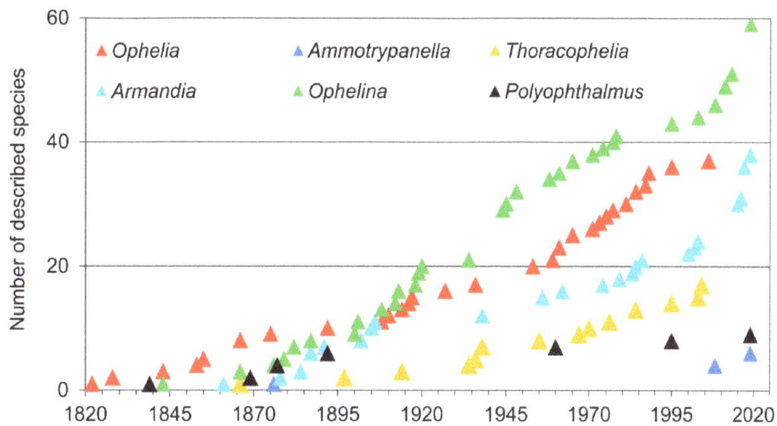

Figure 3. Number of described species (accumulated) of each opheliid genus (including *Ammotrypanella*) from 1820 to 2020.

There are a number of identification keys for Opheliidae from regions such as South Africa [56], California [47], the United Kingdom [54], and the Iberian Peninsula [4]. Some papers also provided tables that compile morphological features for species of the genus *Thoracophelia* (as *Euzonus* and *Lobochesis*; [57]), *Ophelina* from Australia [58], *Armandia* [59] and *Polyophthalmus* [60], and identification keys for *Ophelina* from NE Atlantic [61] and *Armandia* from Australasia and Central Indo-Pacific [62,63].

3.1.3. Taxonomic Characters and External Morphology

The opheliid body is usually elongated and divided into a defined number of segments, usually ranging between 30–70. The anterior end is inflated in the Opheliinae, but typically sleek and more elongate in the Ophelininae [1,4]. The trunk may be entire, as in Ophelininae (Figure 1A,D,E), or divided in two (e.g., *Ophelia*) (Figure 1C) or three regions (e.g., *Thoracophelia*) (Figure 1B). In *Thoracophelia*, the modified chaetiger 10 marks the limit between the thoracic and the abdominal region (Figure 1B). A conspicuous ventral groove is always present, but it may extend continuously throughout the trunk, as in Ophelininae (Figure 1A,D and Figure 4B,H,I), or be restricted to its posterior half, as in Opheliinae (Figure 1C). Some species present two additional longitudinal lateral grooves, one on each side of the body (Figure 4H). Paired lateral branchiae attached dorsally to the parapodia are present in many species, either along the entire trunk or limited to its posterior 1/2–2/3 portion. Branchiae are always absent in last few chaetigers (Figure 1B,E, Figure 4I,J and Figure 5A,C,H). Branchiae are bifurcate or pectinate in some *Thoracophelia*, but simple and cirriform in the remaining genera (Figure 1B).

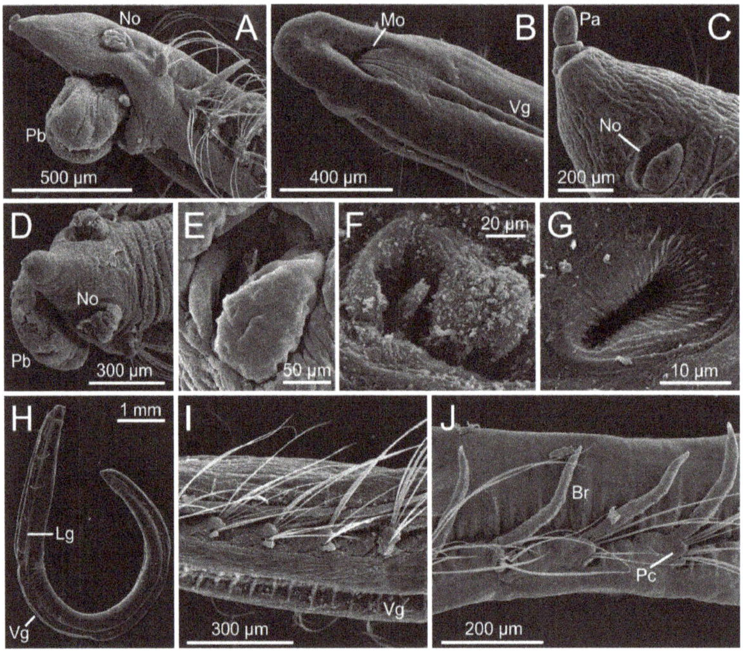

Figure 4. SEM micrographs of several Opheliidae showing main diagnostic characters. (**A**) *Ophelina breviata*, anterior end in lateral view, showing pointed prostomium; (**B**) *Polyophthalmus pictus*, anterior end in ventral view, showing the distally rounded prostomium; (**C**) *Ophelina helgolandiae*, anterior end in lateral view; (**D**) *O. breviata*, anterior end in dorsal view; (**E**) *O. helgolandiae*, nuchal organ; (**F**) *Armandia buccina*, nuchal organ; (**G**) *Ophelina abranchiata*, nuchal organ; (**H**) *O. abranchiata* in lateral view; (**I**) *A. buccina*, anterior chaetigers in lateral view; (**J**) *Armandia opisthoculata*, mid-body chaetigers in lateral view. Abbreviations: Br—branchia; Lg—lateral groove; Mo—mouth; No—nuchal organ; Pa—palpode; Pb—proboscis; Pc—prechaetal lobe; Vg—ventral groove.

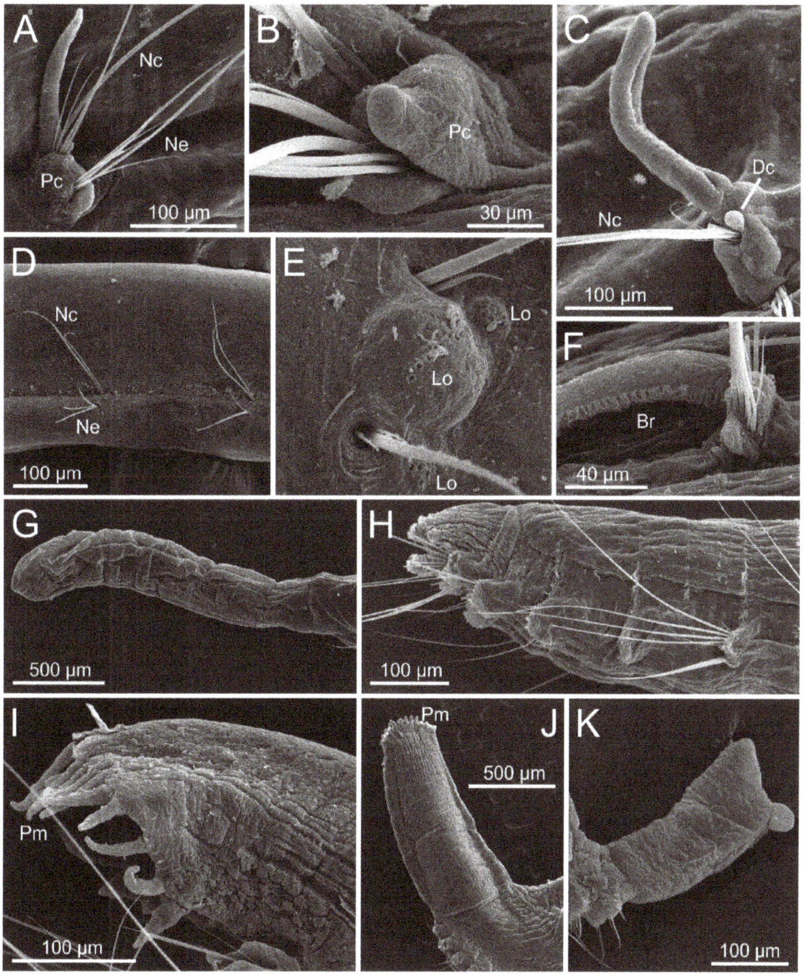

Figure 5. SEM micrographs of several Opheliidae showing main diagnostic characters. (**A**) *Ophelina basicirra*, parapodium and branchia; (**B**) *Armandia laminosa*, parapodium; (**C**) *Ophelina helgolandiae*, parapodium and branchia; (**D**) *Ophelina abranchiata*, mid-body parapodia; (**E**) *Polyophthalmus pictus*, lateral organs; (**F**) *Armandia paraintermedia*, parapodium and branchia; (**G**) *O. abranchiata*, anal tube in lateral view; (**H**) *Armandia parva*, posterior end in lateral view; (**I**) *Armandia tubulata*, anal tube in lateral view; (**J**) *Ophelina bowitzi*, anal tube in lateral view; (**K**) *Ophelina cylindricaudata*, anal tube in lateral view. Abbreviations: Br—branchia; Dc—"dorsal cirrus"; Lo—lateral organs; Nc—notochaetae; Ne—neurochaetae; Pc—prechaetal lobe; Pm—pygidium marginal papillae.

The prostomium is elongated, tapered, or conical in most species (Figure 4A,C), but rounded in *Polyophthalmus* (Figure 4B). It lacks lateral appendages, but a terminal palpode (sometimes biarticulated) is present in *Armandia* and several *Ophelina* species (Figure 4A,C). The proboscis is often an axial, nonmuscular eversible structure (Figure 4A,D and Figure 6A,B), but it might consist of several retractable ciliated tentacles in some species of *Armandia* [63,64]. Nuchal organs are eversible and represented by one pair of conspicuous ciliated pits/slits of various shapes depending on the species [65] (Figure 4A,C–G). Exceptionally, two pairs of nuchal organs are present in *Polyophthalmus* spp. and *Armandia polyophthalma* Kükenthal, 1887 (see [65]), often slightly pigmented [2]. Subdermal pigmented eyes (two to three) are present in several species [66] (Figure 1E); these simple eyes are present in larvae and may be retained in the adult [67]. Additional pairs of segmentally arranged pigmented eyes are present in *Armandia* and *Polyophthalmus*, at least on several midbody segments (Figure 1E).

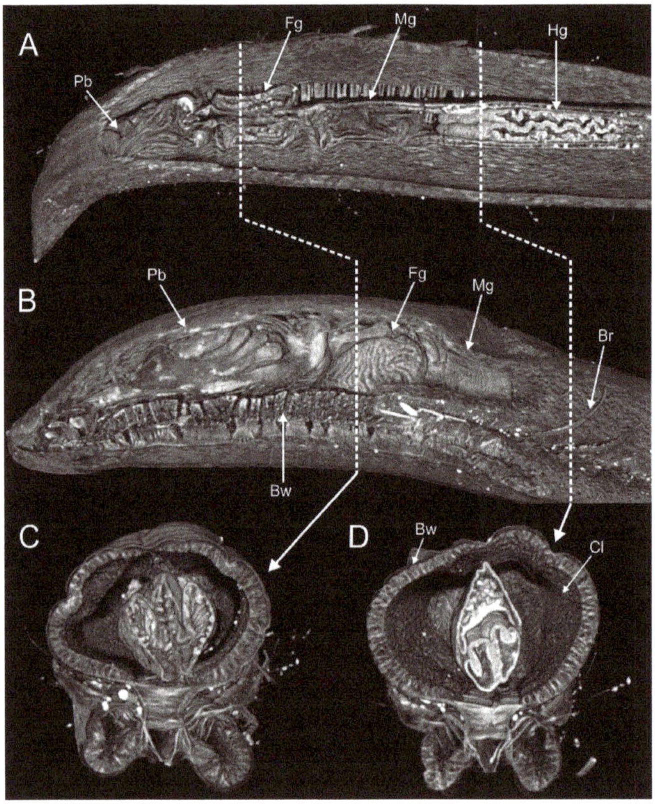

Figure 6. Microcomputed tomography (μCT) sections of *Ophelina acuminata* from Iceland. (**A**) frontal, (**B**) right sagittal and (**C**,**D**) transversal body sections showing internal anatomy. White discontinuous lines in (**A**,**B**) marking regions showed in (**C**,**D**). Abbreviations. Br—branchia; Bw—body wall musculature; Cl—coelomic space; Fg—foregut; Hg—hindgut; Mg—midgut; Pb—proboscis.

Parapodia are biramous and consist of small lobes or tori provided with simple capillary chaetae (Figure 4I,J and Figure 5A–F). A ventral cirrus is present and a small spherical projection may be also found dorsally on the prechaetal lobe in *Armandia* (termed as "dorsal cirrus": [15,62]; Figure 5C). Parapodial ciliated sensory organs were reported on the prechaetal lobe in several species of *Armandia* [62]. Lateral organs are usually present as ciliated pits in between noto- and neuropodia and may also occur in the anterior achaetous segments [15] (Figure 5E).

The last segments may be achaetous and are often retractile. The pygidium typically prolongs into a tubular funnellike structure (termed anal cone, funnel, or tube) that may be quite long in comparison to body length in some species of Ophelininae (Figures 1D and 5G–K). The shape of the funnel, as well as the presence of accessory structures, such as dorsal/marginal papillae and unpaired/paired cirri, diagnoses few genera and species (Figure 1C–E and Figure 5G–K).

Most opheliids have a relatively simple body, reduced parapodia, and simple chaetae. Therefore, the taxonomy of the family has traditionally relied on the limited number of available external characters. This is particularly evident among *Polyophthalmus*, a genus in which most described species are nearly identical morphologically (e.g., [19]). Opheliid genera are defined according to whether the body is divided in distinct regions or not, the extension of the ventral groove, as well as the presence of branchiae and lateral eyes. Species are instead diagnosed based on several parapodial features (e.g., shape of prechaetal lobe, ventral cirrus and presence of "dorsal cirrus", relative length of chaetae across body), the number of lateral eyes (if present), the length of the branchiae, as well as the number of branchiate segments and pygidial cirri. Features of the anal tube are mainly relevant to identify species of Ophelininae, and include its shape, length relative to last chaetigers, number and shape of marginal anal cirri/papillae, as well as presence, shape, and position of the unpaired ventral cirrus and the paired basal cirri [58]. Unfortunately, the fact that branchiae and anal tube are easily detached has generated much confusion regarding the taxonomic status and identifications of some opheliid taxa (e.g., [2]). The many species described based on single/damaged specimens have contributed to worsen the situation [58], together with the lack of information on the intraspecific variation exhibited by some characters, such as number and presence of lateral eyes and anal tube papillae, which may change through different ontogenetic stages in the same species [66]. Examination of a sufficient number of specimens of several sizes is therefore crucial to alleviate this situation in the future [2,58,66].

Parapar et al. [15] suggested that features of nuchal and lateral organs might represent useful characters to diagnose species in the future, in spite that the latter, for instance, can be easily overlooked or is difficult to examine due to state of preservation [2]. In this context, the use of SEM for examination of properly fixed specimens seems mandatory to fully assess features of parapodia, as well as nuchal and lateral organs (e.g., [2,58,60,62,63,68]).

3.1.4. Internal Morphology

The internal anatomy of opheliids has been studied mostly in several intertidal species [16,17,69], including later detailed accounts on the structure of the proboscis [64], body musculature [20,21], respiratory system [50], and sensory organs (see below).

Opheliids lack circular muscle fibers, but they possess bands of longitudinal muscles protruding along the body surface [20], as well as oblique muscles that insert into the midventral line thereby contributing to shape the typical opheliid ventral groove. The structure of the proboscis varies greatly among taxa, corresponding to several of the types described by Tzetlin and Zhadan [64]: type 1, symmetrical, bubblelike, and ciliated as found in the Opheliinae; type 3, asymmetrical, dorsal-lobed (e.g., *Ophelina*, *Polyophthalmus*); type 4, formed instead by several retractable ciliated tentacles (some species of *Armandia*). Exceptionally, the proboscis of *Armandia amukusaensis* Saito, Tamaki and Imajima, 2000 has been reported as flanked by several "filaments" [66]. The digestive tract, and particularly the intestine, might be regionalized in certain species [70] (Figure 6).

The circulatory system is closed [17]. Gills appear as body wall protrusions containing coelom or vessels connected to blood sinuses [50]. Metanephridia are present in several species [71], although protonephridia have been reported in *Thoracophelia mucronata* (Treadwell, 1914) by McConnaughey and Fox [17]. The ultrastructure of sensory organs has been described thoroughly in several opheliids, including the nuchal organs in *Ophelia bicornis* [72] and *Ophelia rathkei* McIntosh, 1908 [65]), the subdermal eyes in *Armandia brevis* (Moore, 1906) [18] and the juveniles of *O. rathkei* [67], as well as the lateral eyes in *A. brevis* [73], *P. pictus*, and *Polyophthalmus qingdaoensis* Purschke, Ding and Müller, 1995 [19].

In this sense, the consistent differences in the ultrastructure of lateral eyes in *Polyophthalmus* (e.g., size and number of cells, number, and dimensions of cellular elements) seem also useful to distinguish species [19]. Thus, future ultrastructural studies might provide phylogenetically informative morphological characters, perhaps further illuminating the delineation of genera. In the same line, the use of micro-CT seems a promising source for phylogenetically informative characters insofar as it offers a comparatively easy overview of the internal anatomy and produces a minimum damage to the examined specimen (e.g., [74]) (Figure 6). It therefore represents a useful tool to compare, for instance, the regionalization of the digestive tract as well as the organization of the circulatory system across genera and/or species.

3.1.5. Species Diversity and Distribution

The most speciose genera are *Ophelina* (about 59 species, excluding *Ammotrypanella*), *Armandia* (38) and *Ophelia* (37); *Thoracophelia* comprises 17 species. Depending on the sources, *Polyophthalmus* is composed of four [3] to nine [60] species, highlighting the need for further morphological and molecular work in order to assess its actual diversity as well as a fully review the synonyms and material attributed worldwide to *Polyophthalmus pictus* [4,60]. Finally, six species are classified into *Ammotrypanella* by those authors who consider the genus as valid [2,15,55].

Opheliids have been reported or described from the poles to the equator across all the 12 marine ecoregion realms defined by Spalding et al. [45] (Figure 2B). Similar distribution patterns are found in the genera *Armandia*, *Ophelia*, and *Ophelina;* whereas *Thoracophelia* is mostly restricted to the temperate realms (14 out of 17 species). Many opheliid species have been described from Temperate Northern Atlantic and Central Indo-Pacific (33 and 34, respectively) in comparison to other regions (ranging from 4 to 17). The type localities of half of the known species of *Ophelia* (16) are in the Temperate Northern Atlantic and about one third of each *Armandia* and *Ophelina* are found in the Central Indo-Pacific. These numbers, however, may be explained by the greater sampling effort historically performed in those areas and the subsequent more detailed knowledge that we have on their annelid faunas of the NW and NE Atlantic, California, and some areas of the Pacific Ocean [47,75]. Indeed, recent work done in unexplored Pacific areas has yielded many new taxa. For instance, Magalhães et al. [60] have described five new species from several western Pacific islands and Wiklund et al. [2] eight new species of *Ammotrypanella* and *Ophelina* plus other still formally undescribed taxa from the eastern Clarion-Clipperton Zone (central Pacific). Furthermore, Parapar and Moreira [62] and Moreira and Parapar [63] have described eleven new species of *Armandia* from Lizard Island (Great Barrier Reef) whereas only two valid species of this genus are present in the comparatively better-known Western Europe. These findings suggest that the actual diversity in other temperate and tropical regions may be greater, including other Pacific areas as well as Temperate Australia (only 13 species described so far) and Tropical Atlantic (nine species).

A wide geographic distribution has been reported for species such as *P. pictus*, *Armandia intermedia* Fauvel, 1902, *Ophelina acuminata* Örsted, 1843 and *O. abranchiata* Støp-Bowitz, 1948. However, these taxa might represent complex of cryptic species as suggested by recent molecular analyses of several populations previously attributed to *O. abranchiata* [2,76]. On the contrary, many taxa have not been reported after original description thus making it difficult to assess their distribution patterns. Finally, reports of species far away from their type locality should be considered with caution because of the lack of knowledge of local faunas (see [60]).

3.1.6. Biology and Ecology

Most *Ophelia* species inhabit clean sandy sediments from the intertidal fringe to the shallow subtidal down to depths of about 100 m [75]. The exception is *Ophelia profunda* Hartman, 1965 and *Ophelia pulchella* Tebble, 1953 that prefer, in turn, muddy bottoms; the former being reported down to 1700 m depth. Species of *Armandia*, *Polyophthalmus*, and *Thoracophelia* prefer coastal areas, the only remarkable exception being *Thoracophelia profunda* (Hartman, 1967) (4000 m). *Polyophthalmus translucens* Hartman, 1960 has been reported at depths of 900 m but Sene-Silva [53] suggested that this species may correspond to the genus *Ophelina*. Indeed, *Ophelina* shows a wider range of ecological preferences, with some species restricted to intertidal-shallow depths while others show wide bathymetric ranges (subtidal/shelf depths down to 2000–3000 m), or, alternatively, are limited to the deep-sea (at depths below 1000 m). *Ammotrypanella* species are distributed at depths below 400 m, more than reaching the abyssal realm.

The majority of opheliids burrow in coarse to fine sand or in muddy sediments. Ecology of several intertidal species of *Armandia*, *Ophelia*, and *Thoracophelia* have been extensively studied when compared to deep-sea species [13]. Some opheliids such as *Thoracophelia furcifera* Ehlers, 1897 and *T. mucronata* may reach high abundances in the intertidal of sandy beaches (2000–40,000 individuals per m^2 [8,17]). Spatial variations in abundance have been related to beach morphodynamics, granulometry, and organic content (e.g., [10]). Experimental work has suggested that the abundance of *A. brevis* is correlated negatively with proliferation of tube-building infaunal species [77]. In general, opheliids are found within well-oxygenated sediments but some *Ophelina* species thrive in muddy sediments with low oxygen content [50] or a high concentration of heavy metals [58]. On the other hand, *P. pictus* usually dwells among intertidal algae, reaching densities that surpass 5000 ind. per m^2 in *Cystoseira* mats where is also present all the year round [78]; *Polyophthalmus* is also found among fouling communities in artificial habitats [79].

Opheliids show two strategies to burrow into the sediment, i.e., peristalsis based on oblique muscular fibers acting in conjunction with cuticular fibers (e.g., *Thoracophelia*) resulting in a dual anchor burrowing mechanism [9,21] or, rather, by undulatory movements (e.g., *Armandia*). Regarding the latter, *A. brevis* lacks circular musculature and therefore relies on bands of oblique muscles that act antagonistically to longitudinal muscles. This muscular arrangement allows for lateral bending and undulating movements that rearrange the sediment grains around by creating a burrow [80]. *Armandia brevis* and other Ophelininae species display a similar pattern of movement when swimming in water; while there is no report of such behavior in Opheliinae.

These burrowing abilities facilitate the migration of intertidal species of *Thoracophelia* downwards or upwards into the sediment to cope with wave turbulence or avoid of low levels of oxygen in the interstitial water [81]. *Thoracophelia* is also capable to migrate horizontally seaward or landward into the sediment in response to changing beach morphodynamics in high-energy environments [7]. Vertical migration in *Ophelia* has also been related to the release of gametes/eggs near the sediment surface [82] or to the avoidance of interspecific competition [83]. Tamaki [84] reports that specimens of *Armandia* sp. migrate in offshore direction as they grow. Giangrande et al. [10] has suggested that the spatial migration in *Ophelia barquii* Fauvel, 1927, from the upper intertidal to upper infralittoral zones, may occur as a response to seasonal changes in hydrodynamics. Because of their burrowing activity, opheliids are important agents in sediment bioturbation [6].

Opheliids are nonselective deposit-feeders by swallowing sediment with the everted proboscis [85]. Feeding behavior has been studied in several species of *Ophelia*, *Ophelina*, and *Thoracophelia*; intertidal and shallow-water species show high ingestion rates [86]. On the contrary, *P. pictus* has been suggested to be a selective feeder [87].

In general, opheliids are mostly dioecious and synchronously release large amounts of gametes or eggs [88]. Life cycle and reproduction of several *Ophelia* species has been studied in North Atlantic and the Mediterranean. Life span extends from one to six years and reproduction occurs from spring to autumn. In general, species breed once a year. On the other hand, adults of *A. brevis* and *P. pictus* experience an epitokous planktonic phase in which they swimming into water to release their gametes [89,90]. Presence of similar epitokous phases has also been suggested for *A. polyophthalma* at least in aquarium conditions. Epitokous specimens show longer chaetae on the posterior five chaetigers that are in turn slightly compressed laterally [90].

Larvae might go through a short lecithotrophic planktonic stage of 4–12 days [5,10,82] or a longer planktonic life thus allowing for a greater dispersal ability (e.g., *A. brevis*; [89,91]). Larvae of *Ophelia*, *Thoracophelia*, *Armandia cirrhosa* Filippi, 1861, and *A. polyophthalma* consist only of two to five chaetigers right before settlement whereas those of *A. brevis* may have up to 20 segments [87,89,90,92,93]. Miner et al. [91] described the feeding mechanisms of the larvae of *A. brevis* that includes action by ciliary bands and direct ingestion with the mouth. After this pelagic phase, the larva settles on the substrate and the body enlarges to become a juvenile worm. Wilson [94] has demonstrated, after several experiments that the settlement of *O. bicornis* larvae is conditioned by the presence of bacteria in the sand grains rather than by the grain size itself.

Polyophthalmus pictus is among the few polychaetes unable to regenerate body segments although it may show wound healing of posterior segments [95].

Deep-sea opheliids are known to be the hosts of two parasitic cyclopoid copepod species of the genus *Ophelicola* [96]. Opheliids are also consumed by several fishes and crabs; for example, *Ophelia limacina* (Rathke, 1843) has been found in the digestive tract of demersal fishes [97] and *P. pictus* in *Trachurus mediterraneus* (Steindachner, 1868) [98]. Kicklighter and Hay [99] also suggested that *A. agilis* may have some chemical deterrents that make it unpalatable for some fishes.

Some opheliids have been the subject of a number of ecotoxicological studies by exposing them in experimental conditions to contaminants (e.g., heavy metals) or antifouling compounds (e.g., tributyltin) in sediments. For instance, *O. bicornis* has been demonstrated to be sensitive to cadmium [12] whereas the exposition of *A. brevis* to TBT resulted in changes in body growth rates [11]. *Armandia agilis* (Andrews, 1891) has been suggested as an appropriate target species to discriminate between clean and contaminated sediments [100] and *Armandia cyprophilia* Neave and Glasby, 2013 is abundant in sediments with high concentrations of copper in otherwise depauperated polychaete assemblages [58]. Therefore, the use of opheliids as indicators of marine pollution seems a promising field of study.

3.2. Scalibregmatidae Malmgren, 1867

3.2.1. Systematics

The first described scalibregmatid was *Scalibregma inflatum* Rathke, 1843 [30], originally classified as an allied to the genus *Arenicola* Lamarck, 1801 [24], until Malmgren [101] established the family Scalibregmatidae in 1867. Later classifications considered Scalibregmatidae as part of the suborder Opheliida [97,102]. This placement was congruent with the results of subsequent morphological analyses, which nested Scalibregmatidae within the clade Scolecida as sister group of Opheliidae, although without any synapomorphy [44]. In contrast, molecular data have more frequently favored a sister-group relationship between Scalibregmatidae and Arenicolidae, often including the genus *Travisia* Johnston, 1840 [46,103], nowadays classified as a separated family [23]. However, the placement of Scalibregmatidae must be considered unresolved, as those analyses were limited to few molecular markers and did not include morphological information. Despite phylogenomic information is available for at least one species in the family [104], Scalibregmatidae has never been included in broad phylogenomic analyses [105].

Scalibregmatidae comprises about 68 described species and 14 valid genera [23]. However, there has been much confusion regarding the species composition of several of them, hampered by the fact that many scalibregmatid species have been described based on incomplete specimens or limited material [27,40]. Scalibregmatids have been traditionally categorized as arenicoliform or maggotlike depending on their overall body shape, although without assigning to these groups any systematic value. Arenicoliform species are typically elongated, inflated in the anterior end, and tapering towards the pygidium; whereas maggotlike species are shorter and stouter [31]. This distinction has been progressively abandoned partly because we know that these differences often rely on preservation artefacts and post mortem contraction; but mostly because intermediate forms also exist and this character even changes during the development of certain species [26,36].

There have been no attempts to resolve the internal relationships of Scalibregmatidae apart from few studies aiming at placing a few specific taxa [2,26,46,103], so the character evolution within the group remains unknown [43].

3.2.2. Taxonomic History

The study of Scalibregmatidae received a notable attention during the 19th century. By the beginning of the 20th century, many common European species were already described [30,106–110] including also a few species from Australia [111], New Zealand [112], Cuba [113], and South Africa [111] (Table A2). This level of attention did not decline during the 20th century, when new species of Scalibregmatidae were described nearly every decade (Figures 7A and 8).

Figure 7. (**A**) number of scalibregmatid species described per decade; (**B**) number of valid scalibregmatid species listed under the bioregion (*sensu* Spalding et al. [45]) according to type locality. Abbreviations: ARC—Arctic; CIP—Central Indo-Pacific; SOC—Southern Ocean; TAU—Temperate Australasia; TEP—Tropical Eastern Pacific; TNA—Temperate Northern Atlantic; TNP—Temperate North Pacific; TRA—Tropical Atlantic; TSA—Temperate South Africa; TSM—Temperate South America; WIP—Western Indo-Pacific.

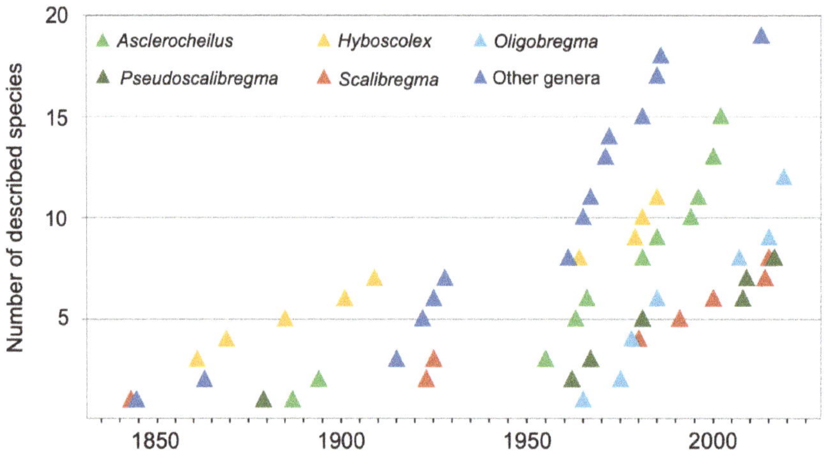

Figure 8. Number of described species (accumulated) of the main scalibregmatid genera from 1820 to 2020. Genera with less than five described species have been grouped under "Other genera".

The first major revision for the family was published in 1925 [114], followed by the work by Kudenov and Blake [38], Kudenov [25] and Blake [31,36,47]. There have also been important works focused on individual genera, such as *Axiokebuita* Pocklington and Fournier, 1987 [26,61], *Oligobregma* Kudenov and Blake, 1978 [37], and *Scalibregma* [35,36]. The status of the systematics of the family has been recently reviewed by Blake [23], who has summarized and amended the diagnoses of all the currently valid genera.

3.2.3. Taxonomic Characters and External Morphology

Members of Scalibregmatidae are relatively large annelids with few taxonomically informative characters. The prostomium is generally rounded or triangular, lacking antennae but often bearing a pair of lateral or frontal extensions, whose homology with palps remains unclear [43] (Figures 9 and 10B–E). Due to the presence of these structures, the prostomium has been often described as T-shaped [see 23]. Prostomial extensions are well developed in the species of *Axiokebuita* and *Speleobregma* Bertelsen, 1983, where they are separated from the prostomium by a basal furrow and bear longitudinal bands of motile ciliary bands capable of producing water currents (Figure 10D,E) [26]. Prostomial appendages are also long in the enigmatic *Scalibregmella antennata* Hartman and Fauchald, 1971, only known from its original collection off New England at 4800–5000 m depth [27], although the presence and arrangement of ciliary bands remain unknown (Figure 9G). In contrast, in the species of the genera *Asclerocheilus* Ashworth, 1901, *Oligobregma*, *Scalibregma*, and *Sclerobregma* Hartman, 1965 prostomial appendages consist of stiff hornlike prolongations, lacking ciliation and a basal furrow (Figure 9A,B) [31]. Despite these morphological differences, the fact that prostomial appendages follow a similar development in all investigated scalibregmatids suggests their homology across the family [26,36]. Epidermal eyes are sometimes present as simple ocelli (Figure 9F) or more complex structures composed of multiple ocelli (Figure 9B). Nuchal organs are usually small and often found retracted into grooves that extend transversally between the prostomium and the peristomium (Figure 10B,E). When they are everted, they resemble expanded bulbous vesicles [23]. Nuchal organs are associated with additional transverse bands of motile cilia in *Axiokebuita cavernicola* Martínez, Di Domenico and Worsaae, 2013 and *Speleobregma lanzaroteum* Bertelsen, 1983 [26] (Figure 9G, Figure 10D,E and Figure 11D).

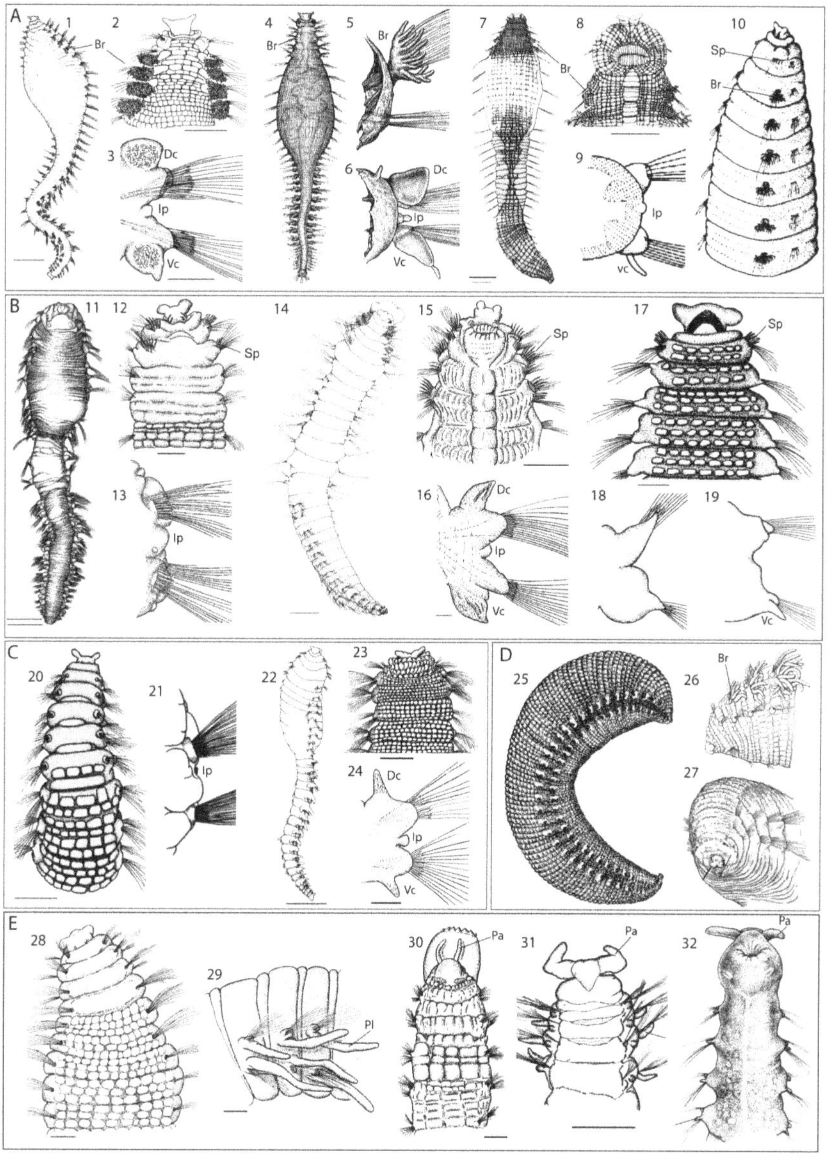

Figure 9. Stylized drawings showing the main taxonomic characters of different genera in Scalibregmatidae, artificially grouped according to the most conspicuous external traits. (**A**) Arenicoliform scalibregmatid genera with branchiae; genus *Scalibregma*, (1) *S. hanseni*, dorsal view, (2) *S. inflatum*, anterior end in dorsal view, (3) *S. hanseni*, left parapodium of chaetiger 21 in posterior view; genus *Sclerobregma*, (4) *S. branchiatum*, dorsal view, (5) chaetiger 3 in posterior view and (6) abdominal parapodium in anterior view; genus *Cryptosclerocheilus*, (7) *C. baffinensis*, dorsal view, (8) anterior end in ventral view; genus *Parasclerocheilus*, (9) *P. capensis*, chaetiger 40 and (10) anterior end in lateral view. (**B**) Arenicoliform scalibregmatids genera without branchiae and with spines; genus *Asclerocheilus*, (11) *A. tasmanius*, dorsal view, (12) *A. kudenovi*, anterior end dorsal view, (13) *A. beringianus*, chaetiger 15 in anterior view; genus *Oligobregma* (14) *O. quadrispinosa*, anterior view, (15) *O. mucronata*, anterior end in ventral view, (16) and posterior parapodium in anterior view; genus *Sclerocheilus*, (17) *S. unoculus*, anterior end in dorsal view, (18) chaetiger 16 in posterior view, and (19) chaetiger 29 in posterior view. (**C**) Arenicoliform scalibregmatid genera without branchiae and spines; genus *Hyboscolex*, (20) *H. quadricincta*, anterior end in dorsal view, (21)

H. pacificus, median parapodium in anterior view; genus *Pseudoscalibregma*, (22) *P. papilia*, dorsal view, (23) *P. usarpium*, anterior end in dorsal view, (24) *P. hartmanae*, posterior chaetigers in anterior view. (**D**) Maggotlike scalibregmatids genera; genus *Polyphysia*, (25) *P. crassa*, lateral view and (26) anterior end in anterior view; genus *Lipobranchius*, (27), *L. jeffreysi*, frontal view. (**E**) Morphologically divergent genera; genus *Scalibregmides*, (28) *S. peruanus*, anterior end in dorsal view, (29) *S. chilensis*; genus *Scalibregmella*, (30) *S. antennata*, anterior end in dorsal view; genus *Speleobregma*, (31) *S. lanzaroteum*, anterior end in dorsal view; genus *Axiokebuita*, (32) *A. minuta*, anterior end in ventral view. Abbreviations: Br—branchiae; Dc—dorsal cirri; Ip—interramal papillae or ciliation; Pa—prostomial appendages; Pl—parapodial lobe; Sp—spines; Vc—ventral cirri. Modified from (1,3) Bakken et al. [35]; (2) Mackie [34], (4–6) Hartman [115], (7–8) Blake [116], (9–10) Day, [117], (11) Kirkegaard [118], (12) Blake [119], (13) Imajima [120], (14) Schüller and Hilbig [37], (15–16) Blake [36], (17–20) Kudenov [25], (21) Imajima [121], (22) Schüller [55], (23–24, 28–29) Blake [31], (25) Støp-Bowitz, [122], (26) Hartmann-Schröder [97], (27) Wesenberg-Lund [123], (30) Blake [23], (31) Bertelsen [124], (32) Parapar et al. [61].

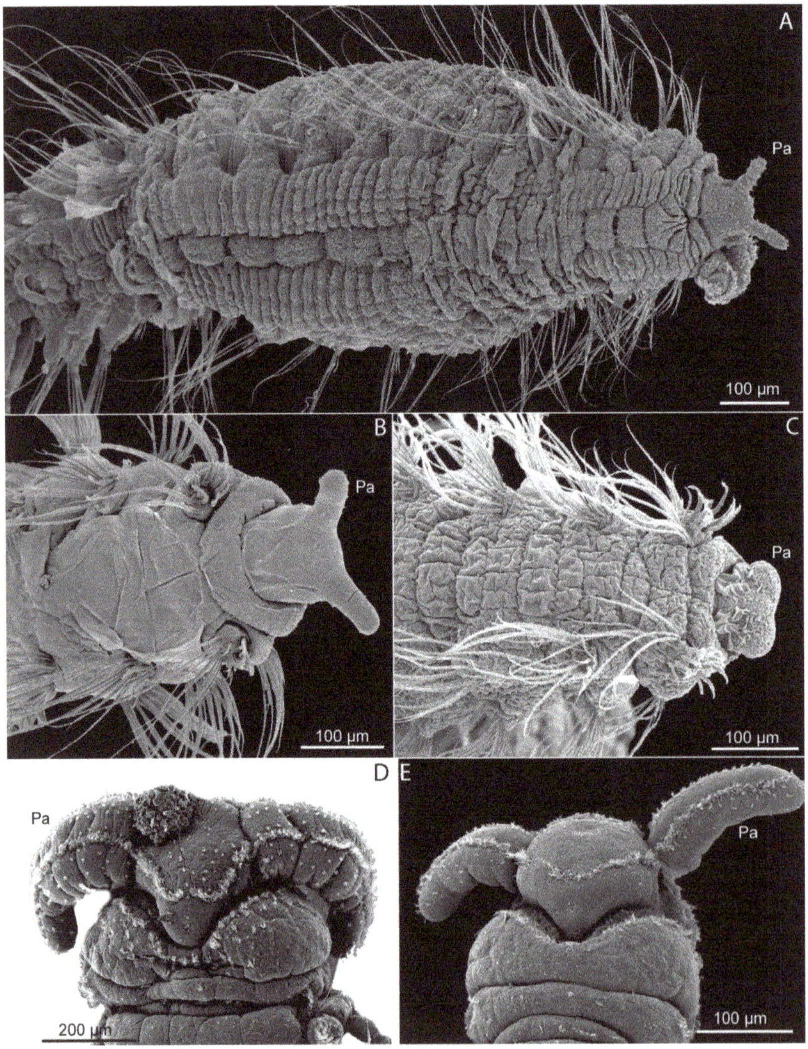

Figure 10. SEM micrographs of several Scalibregmatidae showing main prostomial diagnostic characters. (**A**) *Pseudoscalibregma* sp., Canary Islands, anterior end in ventral view, showing the pattern formed by the pads as well as the structure of the parapodia; (**B**) *Asclerocheilus* sp., Canary Islands, anterior end in dorsal view; (**C**) *Asclerocheilus* sp., northwestern Spain,

anterior end in dorsal view; (**D**) *S. lanzaroteum*, anterior end in dorsal view; (**E**) *A. cavernicola*, anterior end in dorsal view. Notice the difference in the prostomial shape and appendages amongst (**B–E**), as well the presence of different development of the peristomium, and the presence of different types of chaetae. Abbreviation: Pa—prostomial appendages.

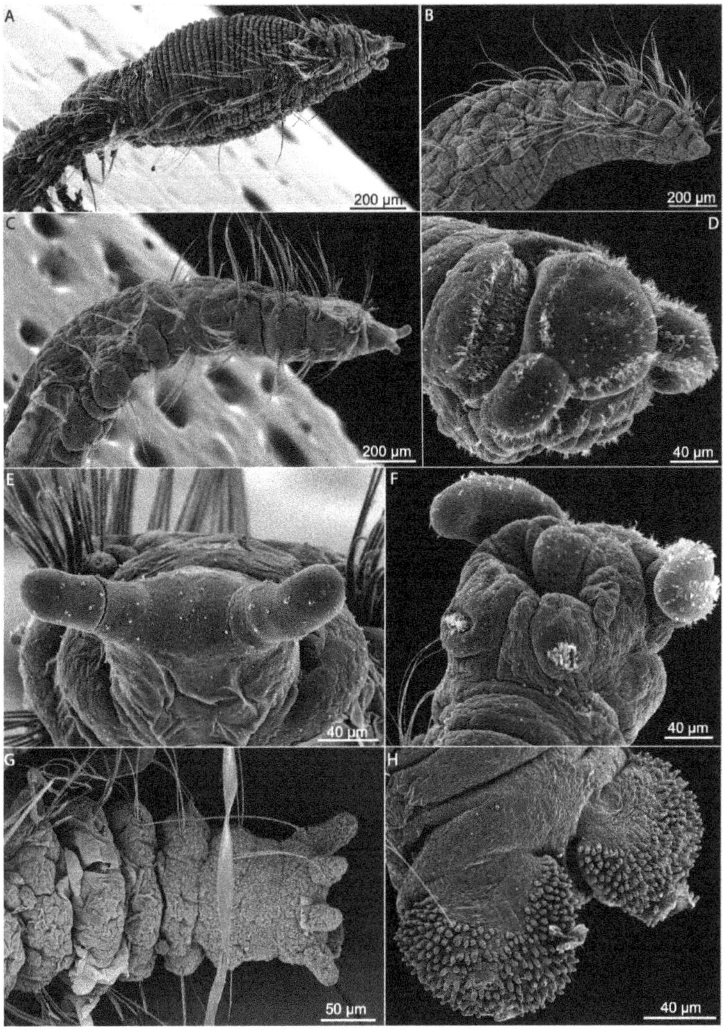

Figure 11. SEM micrographs of several Scalibregmatidae showing main diagnostic characters. (**A**) *Pseudoscalibregma* sp., Canary Islands, anterior end in lateral view; (**B**) *Asclerocheilus* sp., northwestern Spain, anterior end in lateral view; (**C**) *Asclerocheilus* sp., Canary Islands, anterior end in lateral view; notice the different morphology and epidermal pattern found on the anterior end on (**A–C**). (**D**) *A. cavernicola*, anterior end in lateral view; (**E**) *Asclerocheilus* sp., Canary Islands, anterior end in frontal view, compare the arrangement of the ciliary patterns between (**D,E**); (**F**) *A. cavernicola*, anterior end in ventral view, showing the ventral ciliary pads pn the peristomium; (**G**) *Asclerocheilus* sp., northwestern Spain, posterior end in dorsal view, showing a typical shape and arrangement of the pygidium in Scalibregmatidae; (**H**) *A. cavernicola*, posterior end in dorsal view, showing the adhesive pygidium typical of the genera *Axiokebuita* and *Speleobregma*.

The peristomium typically consist of one dorsal and one to three ventral rings, merging into the upper and lower lips of the mouth. Unfortunately, detailed morphological descriptions of the peristomium have only been provided for a few species [23,36] (Figure 10C–E). The ventral mouth is connected to an axial proboscis, which is multilobed when everted and divided into proximal unciliated and distal ciliated zones [64]. The posterior part of peristomium possesses a pair of rounded ciliated areas of unknown function in Axiokebuita and Speleobregma, with potential taxonomic value (Figure 11F) [26].

The trunk includes up to 60 segments, each of them typically bearing one to six rows of elevated pads giving the worms an areolate appearance (Figures 9C and 10A–C). The number of these rows, as well as the number and size of the pads that form each of them, varies across different species and body regions. The pattern formed by the pads has been used to diagnose certain species, suggesting that these patterns might be species-specific in some genera [36]. Epidermal papillae are absent in Scalibregmatidae. A midventral groove is present in most genera, extending from the mouth towards the pygidium along the longitudinal body axis (Figure 10A). It is not clear, though, whether this structure bears systematic information or if its appearance depends on the post mortem contraction of the trunk musculature [23]. Transverse bands of presumably motile cilia have been described on *S. lanzaroteum* and *A. cavernicola* [26] (Figure 10D,E).

Branchiae have been considered as an important taxonomic character. The presence of branchiae in the anterior segment characterizes the genera Scalibregma, Sclerobregma, Cryptosclerocheilus Blake, 1972, and Parasclerocheilus Fauvel, 1928 (Figure 9A), in which they are attached to the notopodium from segment 2 up to segment 5–7. Branchiae are arborescent in most species, branching dichotomously a variable number of times; but can also be pectinate, with individual branchial filaments arising from an elongate flattened lamella, as in Sclerobregma branchiatum Hartman, 1965 (Figure 9A) [23]. However, recent studies suggest that their number and arrangement might vary ontogenetically within the same species [23,36]. This has raised concerns about the validity of certain species identification, particularly when few small individuals have been studied, and growth series are not incorporated into species descriptions. More information on the ontogeny of other species of Scalibregmatidae can be found elsewhere [23,26,36].

Parapodia are biramous in all scalibregmatids. The development of each ramus largely varies across species and body regions, but they are typically smaller anteriorly and more elongated towards the posterior body end. Parapodial structures, such as interramal papillae and parapodial cirri have been described in some species, holding useful taxonomic information. Interramal papillae are retractile and ciliated in *S. inflatum* [28] and Asclerocheilus (Figure 12B); whereas species of Oligobregma present interramal ciliated areas (Figure 12C). Interramal papillae in *A. cavernicola* and *S. lanzaroteum* project from the body wall and bear terminal ciliation [26,61] (Figure 12A,D). Nonciliated glandular papillae have been observed in *S. minutus* Grube, 1863 [41], and *P. palmeri* Blake, 2015 [36]. Parapodial dorsal and ventral cirri may help discriminating amongst species. Cirri are filiform in Axiokebuita and Speleobregma (Figure 9G), and leaf-shaped in Oligobregma, Pseudoscalibregma Ashworth, 1901, Scalibregma, and Sclerobregma. Cirri often exhibit glands, which are tubular in some species of Scalibregma, Oligobregma, and Pseudoscalibregma; but circular in Axiokebuita and Speleobregma. Parapodial lobes or lamellae are described in Asclerocheilus californicum and in the two species of the genus Scalibregmides (Figure 9F) [31,40].

The arrangement of chaetae is a very important taxonomic characteristic in Scalibregmatidae. Chaetae are always simple and might include long capillaries (Figure 10A), geniculated (Figure 12I), lyrate (Figure 12E,G), short spinous (Figure 12H), and acicular (Figure 12F,H). Simple capillary chaetae are present in all described species, while the presence or absence of other types of chaetae is an important character to diagnose genera. The absence of lyrate chaetae characterizes the genera Speleobregma and Axiokebuita, whereas the morphology of these chaetae is useful to diagnose species in genera such as Hyboscolex and Asclerocheilus, amongst others. Spinous chaetae are small and typically arranged as a single row restricted to the anterior most body segments. Since they occupy similar position to the lyrate chaetae, they are presumed as homologous to the former and rarely used in taxonomy. The presence of acicular chaetae, in contrast, is very useful and characterizes the genera Sclerobregma, Parasclerocheilus, Asclerocheilus, Sclerocheilus, and Oligobregma. Acicular chaetae are large and conspicuous, typically sickle-shaped or curved, and covered with fibrils visible in the scanning electron microscope. They are restricted to the anterior most segments and their arrangement is useful for species diagnoses. They can extend through a variable number of segments either on the notopodia or in both rami. Finally, geniculate chaetae are only found in *S. lanzaroteum* [124].

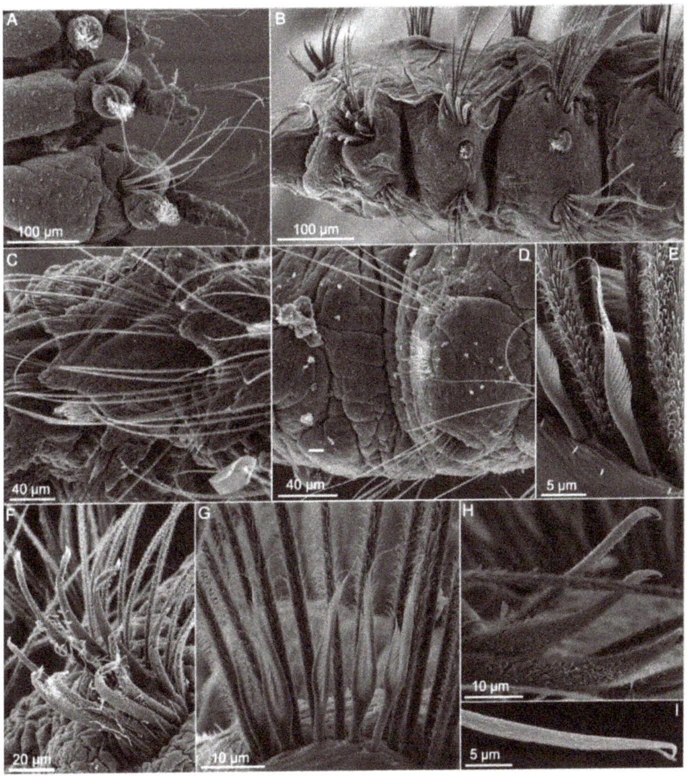

Figure 12. SEM micrographs of several Scalibregmatidae showing main diagnostic characters. (**A**) *Speleobregma lanzaroteum*, parapodia on the anterior segments in dorsal view, showing the presence of cirri; (**B**) *Asclerocheilus* sp., Canary Islands, parapodia on the anterior segments in lateral view; (**C**) *Asclerocheilus* sp., northwestern Spain, mid-body parapodium in frontal view; (**D**) *Axiokebuita cavernicola*, mid-body parapodium in frontal view; (**E**), *Asclerocheilus* sp., Canary Islands, lyrate chaetae on anterior segments in lateral view; (**F**) *Asclerocheilus* sp., northwestern Spain, spines on segment 1; (**G**) *Asclerocheilus* sp., Canary Islands, lyrate chaetae on mid-body segment; (**H**) *Asclerocheilus* sp., Canary Islands, spines; (**I**) *S. lanzaroteum*, geniculate chaetae.

The pygidium is quite variable across different scalibregmatids. However, since scalibregmatids are found lacking the posterior end in most samples, the usefulness of this character is limited. In most species, the pygidium is simple and bears a typically terminal anus and surrounded by a variable number of cirri (Figure 11G) whose arrangement, length, and number are potentially useful to identify species. Species of Axiokebuita and Speleobregma possess two enlarged rounded pygidial lobes covered with adhesive papillae (Figure 11H).

3.2.4. Internal Morphology

The internal morphology of Scalibregmatidae was thoroughly investigated during the early 20th century, particularly in the species *S. inflatum* [28] and *S. minutus* [41] mostly based on histological sections. Unfortunately, after these early works, very few studies have been undertaken using more modern microscopical techniques.

The body wall consists of the epidermis, which comprises elongated columnar cells and mucous secreting cells, as well as a muscular layer of circular muscles surrounding dorsal and ventral longitudinal muscular bundles [28]. Narrow oblique muscles are also present, arising ventrally from each side of the nerve cord and inserting into the body wall near the notopodial chaetal sacs. Parapodial musculature is limited to the chaetal sacs as well as the parapodial retractor muscles [28]. There is also a relatively strong mouth and pharyngeal musculature, with retractor muscles attached to the proboscis and two short muscles supplying the nuchal organs [28].

A thin epithelium delineates the coelomic cavity, which is well developed and spacious. As an adaptation for burrowing, septa are reduced along most of the body [28,41]. The gut is linear and attaches to the body cavity by few strands of muscular tissue in *S. inflatum*. The esophagus is straight and covered with secretory glands; whereas the midgut is wider and curled, and the hindgut is short, linear, and opens directly into the anus. Several blood sinuses are associated with the stomach in *S. inflatum* and *S. minutus* [28,41].

There is also a well-developed vascular system [28,41], consisting of dorsal and ventral vessels and their derivatives. The dorsal vessel extends along the alimentary canal supplying it with capillary vessels. It forms a blood reservoir near the anterior end of the stomach and a conical heart-like bulb before branching off to supply the pharynx, the peristomium, and the brain. The ventral blood vessel originates near the mouth and continues posteriorly, extending dorsally along the nerve cord. In *S. inflatum*, it supplies the branchiae, the stomach, and nephridia, as well as the chaetal sacs and their adjacent tissues.

A pair of metanephridia occurs in each chaetigerous segments, except for those most anterior. Gonads are associated with each metanephridium and are formed by the proliferation of cells covering the septum by which the nephrostome is attached to the body wall [28]. The gametes are released from the gonad at an early stage and complete their maturation in the coelom. Male gonads form sperm platelets bearing spermatids in *S. australis* and *O. mucronata*, and they mature into ect-aquasperm [36].

The brain has an anterior lobe associated with the prostomium and two posterior lobes associated with the nuchal organs. The prostomial appendages are innervated by a pair of nerves originating from the anterior lobe of the brain, whereas the esophageal connectives and the nerves innervating the nuchal organs arise from the middle and posterior lobes, respectively. The palps are innervated by one ventral and one dorsal nerve, corresponding to the fourth and ninth pairs respectively [42].

3.2.5. Species Diversity and Distribution

The most species rich scalibregmatid genus is Asclerocheilus, with 14 described species, followed by Oligobregma (12 species), Hyboscolex Schmarda, 1861 (10 species), Pseudoscalibregma (eight species), Scalibregma (eight species) and Sclerocheilus (four species). The remaining genera are less diverse, including Polyphysia Quatrefages, 1866 (three species), Axiokebuita (two or three species depending on the sources), Parasclerocheilus (two species), Scalibregmides Hartmann-Schröder, 1965 (two species), and the monotypic Cryptosclerocheilus, Lipobranchius Cunningham and Ramage, 1888, Sclerobregma, and Speleobregma.

From a geographical point of view, scalibregmatids have been reported throughout the world and are present in all marine ecoregions [45] (Figure 7B). Most of the species have been described from the Temperate Northern Atlantic (17 species) and the Southern Ocean ecoregions (14 species), which together host nearly the half of the scalibregmatid type localities (Figure 7B). However, while the abundance of described species in the Temperate Northern Atlantic might just reflect the higher attention that historically has been paid to the fauna of this region, the presence of so many scalibregmatids in Antarctica is somehow unusual and might respond to unidentified ecological or historical processes. This is particularly remarkable given that nearly all Antarctic species seem to be endemic from that area, although this endemism might be exacerbated by the lack of studies in surrounding deep-sea areas. The remaining type localities are distributed across the Temperate Northern Pacific (nine species), Temperate Australasia (seven species), Tropical Atlantic (six species), Tropical Eastern Pacific, Temperate South America, Arctic, and Temperate Southern Africa (all with four species), Western Indo-Pacific (three species), and Central Indo-Pacific (two species). However, given the fragmentary information available on the family, this pattern most likely reflects the different attention that the group has received across the world than any other biological meaningful factors.

Many scalibregmatids seem to have relatively broad distribution ranges. A remarkable example is A. minuta, which has been recorded both in Arctic and Antarctic latitudes, as well as hydrothermal vents in the Pacific Ocean and in the Galician Bank, off Northwest Spain (but see [23]); or *S. inflatum*, recorded from Northern Europe as well as from South Africa [117], Australia [38], Chile [40], and Japan [121]. However, many of these records are exclusively based on morphological data often evaluated from few specimens, generally preserved in suboptimal conditions. Therefore, one might expect that more detailed morphological examinations and the inclusion of molecular data will reveal that these records actually correspond to complex of species with narrower distributions and better-defined ecological preferences. For example, the re-examination of material originally attributed to *S. inflatum* has already revealed several different species with more restricted distribution. This includes the recent description of *S. australis* Blake, 2015 based on the detailed examination of growth series of Antarctic material [36], *S. californicum* Blake, 2000 from California [119], as well as *S. celticum* Mackie, 1991 and S. hanseni Bakken, Oug and Kongsrud, 2014 from Europe [34,35]. Remarkably, these last species show sympatric occurrence with *S. inflatum*. Such discoveries, even in the relatively well explored waters of Europe, highlight once again that our knowledge on the diversity of the Scalibregmatidae is still very limited. Therefore, most discussions on the distribution patterns of the scalibregmatid species remain speculative.

In contrast to those species with large distribution areas, other species are exclusively known from a few localities. This is the case of the species Scalibregmides chilensis Hartmann-Schröder, 1965, recorded only once from Puerto Aguirre (Chile) [40] and *S. peruanus* Blake, 1981 from Callao (Peru) [31]; a few species of the genus Oligobregma, such as *O. whaleyi* Wiklund, Neal, Glover, Drennan, Rabone and Dahlgren, 2019 from a single deep-sea locality in the Pacific [2], or *O. oculata* Kudenov and Blake, 1978 and *O. simplex* Kudenov and Blake, 1978 each known from a single locality around Victoria (Australia) [38]. However, once again, the actual endemic status of these species remains doubtful due to our limited knowledge.

The endemism of *S. lanzaroteum* and *A. cavernicola* deserves a separate comment since the species are restricted to two different volcanic lava tubes in the Canary Islands [26]. *Speleobregma lanzaroteum* is an elusive species exclusively known from La Corona lava tube, in Lanzarote, an anchialine cave penetrating the saline aquifer of the island and characterised by the presence of a highly distinct fauna [125]. The species was described based on a single specimen collected in 1981 and only observed again after 27 years, when two more individuals were recovered in two successive dives [26,125]. The fact that these are the only three records of the species is remarkable because the cave has been regularly sampled over the last 40 years by well-trained cave divers who were explicitly sampling the fauna [126–129]. Therefore, the scarcity of records for *Speleobregma lanzaroteum* is more likely attributed to the low population densities described for many other cave species, than to an actual lack of sampling efforts [130]. Axiokebuita cavernicola is, in contrast, limited to a specific gravelly patch found in Los Cerebros cave in Tenerife [26], while it is absent in the muddy or sandy sediments found elsewhere in the cave [131]. The fact that both species are found in specific areas inside caves supports the idea that they may be actually endemic from these cave localities [132].

3.2.6. Biology and Ecology

Most scalibregmatids prefer muddy sediments at depths greater than 100 m. This seems to be the case, at least, for species in the genus Scalibregma, Oligobregma, Polyphysia, Lipobranchius, and Pseudoscalibregma, which are considered subsurface deposit feeders capturing food particles with their eversible multilobulated proboscis [85]. In particular, *S. inflatum* and *S. californicum* burrow by pushing the sediment to the sides of the body by lateral movements of the prostomium and afterwards moving forward by producing peristaltic waves [133]. The presence of the prostomial appendages probably increases the efficiency of this so-called shoveling process, while the absence of septa makes the production of waves more efficient. Polyphysia crassa (Örsted, 1843) burrows in a similar way, lacking prostomial horn, but also possessing reduced septa and a glandular epidermis to increase the efficiency of the peristaltic movements [85,134,135]. Because of the burrowing behavior, scalibregmatids play an important ecological role in soft bottoms bringing burrowed particles near to the surface, as it has been showed in the Cape Hatteras area [136] where they can be present quite deep in the sediment column [137]. These burrowing species can become very abundant or even dominate the benthic communities, as it has been shown for *S. australis* at the east side of the Antarctic Peninsula [36], and *S. inflatum* in Cape Hatteras between 550–1500 m depth [138]. Indirect evidence from various sources suggests that these are not isolated cases, but rather that species of these scalibregmatid genera might dominate soft bottom assemblages in many areas in high latitudes [23,35].

However, there are other species of scalibregmatids that seem to exhibit different habitat preferences. For example, Oligobregma brasirae Wiklund, Neal, Glover, Drennan, Rabone and Dahlgren, 2019, *O. whaleyi*, and *O. tanyi* Wiklund, Neal, Glover, Drennan, Rabone and Dahlgren, 2019, are exclusively known from the polymetallic nodule exploration areas in the eastern Clarion-Clipperton Zone [2], and there are many records of species of Axiokebuita from gravel, deep Desmophyllum Ehrenberg, 1834 coral reefs [26,61,103], or even rock crevices near hydrothermal vents at the Pacific Antarctic Ridge [139]. In shallow waters scalibregmatids are not uncommon in hard substrates. For example, Scalibregmides chilensis has been recorded from a mytilid bank [40], Hyboscolex quadricincta Kudenov, 1985, Asclerocheilus tropicus Blake, 1981, and A. mexicanus Kudenov, 1985 have been collected from dead corals and sponges [25,39]; Asclerocheilus acirratus (Hartman, 1966) and Hyboscolex verrucosa Hartmann-Schröder, 1979 are known from algae in hard substrates [140,141], and Asclerocheilus kudenovi Blake, 2000 and H. oculatus (Ehlers, 1901) are recorded from unspecified hard, rocky substrates [119,142]. Axiokebuita cavernicola has only been reported from gravel sediments in the middle section of Los Cerebros lava tube in Tenerife, where there is an active water movement produced by waves driving a notable input of organic matter into the gravelly bed [26]. Adults of this species attach to the gravel particles using the adhesive papillae of the pygidium, while they collect suspended food particles using the water currents produced by the ciliation on their palps. Upon perturbation, they can also swim short distances using undulatory movements of the trunk. In contrast, juveniles of *A. cavernicola* lack palps and adhesive pygidium, and usually are found actively crawling and ciliary swimming in the petri dishes [26]. In contrast, *S. lanzaroteum* lives in La Corona lava tube, an anchialine cave system where food is limited to the organic matter carried by tidal currents [143]. Remarkably, *S. lanzaroteum* has only been found swimming in the water column using undulatory body movements and gentle movement of the parapodia. Similar life strategies have been discovered in other annelids exclusively reported in their isolated cave systems [126,144,145], suggesting that drifting in the water column might be the optimal life strategy in these type of cave environments. The capability of swimming is not unique in this cave-adapted scalibregmatid, since adults of several typically benthic species, such as S.

inflatum and *L. jeffreysii* (McIntosh, 1869) have been occasionally reported swarming in the plankton [146–148], although in all these cases, individuals possess specialized swimming chaetae.

The reproduction of scalibregmatids is largely unknown, and detailed studies are only available for a few selected species [23,149]. Fertilization is unknown but spawning might take place in the water column. This is presumed given the presence of ect-sperm and large oocytes in many species, as well as the observation of adult individuals of *S. inflatum*, *L. jeffreysii* and possibly *S. celticum*, swimming in the water column of the ocean, sometimes provided with long natatory chaetae [34,146–148]. Finally, despite nothing is known about the embryonic development, we known the postembryonic development of *S. australis*, *O. mucronata* and *A. cavernicola* [26,36] from the description of series of individuals of different size.

4. Conclusions and Future Perspectives

The Opheliidae are well known in some parts of the world such as the northern Atlantic; some areas (e.g., Pacific, Atlantic Africa), however, remain clearly understudied. The status of cosmopolitan species and several species not reported after original description should be reassessed. Furthermore, proper evaluation of some taxonomic characters needs to consider ontogenetic variability and preservation artefacts. Regarding Scalibregmatidae, the knowledge of the species richness and distribution is often fragmentary and strongly biased by the unbalanced sampling effort across the world. Indeed, except for the Antarctic and the northwestern Atlantic, the remaining marine areas have been poorly studied when it comes to Scalibregmatidae. Furthermore, both the position of this family within Annelida as well as its internal evolutionary relationships and systematics remain unresolved, warranting further assessment combining different sources of data; this also applies to Opheliidae because a full phylogenetic analysis of this family is still lacking as well.

Finally, as it still happens with other annelid families, current knowledge on the internal anatomy, life cycles, ecology, and behavior of opheliids and scalibregmatids has been obtained from a few studies on some common species. In this sense, traditional taxonomic approaches coupled with modern microscopy imaging techniques (e.g., micro-CT, SEM) and molecular methods (e.g., molecular phylogenies and species delimitation analyses) are needed; this will be paramount to assess properly intraspecific diversity issues that have hampered the taxonomy of these families in the past. This may be especially useful in finding appropriate, robust characters with systematic value in these morphologically homogenous taxa, aiding in an effective assessment of their current species diversity, and, ultimately, their distribution patterns and ecological preferences.

Author Contributions: Conceptualization and supervision, J.P., A.M. and J.M.; methodology, investigation, writing—original draft preparation, review and editing—and illustrations J.P., A.M. and J.M.; and funding acquisition, J.P. All authors have read and agreed to the published version of the manuscript.

Funding: This research has been partially funded by FAUNA IBÉRICA research project Polychaeta VII, Palpata, Canalipalpata II (PGC2018–095851–B–C64) sponsored by the Agencia Estatal de Investigación and coordinated by JP. Pat Hutchings (Australian Museum) which has also provided some funds to cover the publication costs.

Acknowledgments: Authors would like to thank Ada Castro and Catalina Sueiro (Servizos de Apoio á Investigación, Universidade da Coruña) for SEM assistance, to María Candás (Estación de Bioloxía Mariña da Graña–Ferrol, Universidade de Santiago de Compostela, Spain) for assistance with the micro-CT and to Antón Taboada for line drawings. We are also grateful to Jorge Núñez for sharing with us the material of *Pseudoscalibregma* sp. from the Canary Islands. Authors deeply thank reviewers' comments on the manuscript and Maria Bogomolova for language revision.

Conflicts of Interest: The authors declare no conflict of interest.

Appendix A

Table A1. List of valid species of Opheliidae (after Blake and Maciolek [1,3]) with their type locality, realms *sensu* Spalding et al. [45] and depth ranges of type locality. N.d. = no data.

Species	Type Locality	Realms	Depth (m)
Ammotrypanella arctica (McIntosh, 1879)	North Atlantic Ocean	Temperate Northern Atlantic	2014–5023
Ammotrypanella cirrosa Schüller, 2008	Antarctic Peninsula; Weddell Sea	Southern Ocean	2014–4817
Ammotrypanella keenani Wiklund, Neal, Glover, Drennan, Rabone and Dahlgren, 2019	Clarion-Clipperton Zone, Pacific Ocean	Tropical Eastern Pacific	404–4302
Ammotrypanella kersteni Wiklund, Neal, Glover, Drennan, Rabone and Dahlgren, 2019	Clarion-Clipperton Zone, Pacific Ocean	Tropical Eastern Pacific	4128
Ammotrypanella mcintoshi Schüller, 2008	off South Africa	Southern Ocean	1047–4720
Ammotrypanella princessa Schüller, 2008	Antarctic Peninsula, Weddell Sea	Southern Ocean	2014–4720
Armandia agilis (Andrews, 1891)	North Carolina, USA	Temperate Northern Atlantic	N.d.
Armandia amakusaensis Saito, Tamaki and Imajima, 2000	Western Kyushu, Japan	Temperate Northern Pacific	Intertidal
Armandia andamana Eibye-Jacobsen, 2002	Andaman Sea	Central Indo-Pacific	42–63
Armandia bifida Parapar and Moreira, 2015	Lizard Island, Australia	Central Indo-Pacific	Intertidal
Armandia bilobata Hartmann-Schröder, 1986	South Australia	Temperate Australasia	Intertidal
Armandia bipapillata Hartmann-Schröder, 1974	Inhambane, Mozambique	Western Indo-Pacific	Intertidal
Armandia brevis (Moore, 1906)	Icy Cape, Alaska, USA	Arctic	Intertidal to shallow subtidal
Armandia broomensis Hartmann-Schröder, 1979	Broome, Australia	Central Indo-Pacific	Intertidal
Armandia buccina Moreira and Parapar, 2017	Lizard Island, Australia	Central Indo-Pacific	0–12
Armandia casuarina Moreira and Parapar, 2017	Lizard Island, Australia	Central Indo-Pacific	Intertidal
Armandia circumpapillata Magalhães, Rizzo and Bailey-Brock, 2019	Oahu, Hawaii	Eastern Indo-Pacific	Shallow subtidal to 82 m
Armandia cirrhosa Filippi, 1861	Cagliari, Sardinia, Mediterranean	Temperate Northern Atlantic	Shallow subtidal
Armandia dolio Parapar and Moreira, 2015	Lizard Island, Australia	Central Indo-Pacific	5–14
Armandia exigua Kükenthal, 1887	China	Central Indo-Pacific	15
Armandia filibranchia Parapar and Moreira, 2015	Lizard Island, Australia	Central Indo-Pacific	9
Armandia garretti Magalhães, Rizzo and Bailey-Brock, 2019	Pearl Harbor, Hawaii	Eastern Indo-Pacific	0–20
Armandia hossfeldi Hartmann-Schröder, 1956	Brazil	Temperate South America	9.5–14
Armandia ilhabelae Hartmann-Schröder, 1956	Santos, Brazil	Temperate South America	Intertidal
Armandia intermedia Fauvel, 1902	River Casamance estuary, Senegal	Tropical Atlantic	Intertidal to subtidal
Armandia laminosa Parapar and Moreira, 2015	Lizard Island, Australia	Central Indo-Pacific	0–15
Armandia lanceolata Willey, 1905	South of Manaar Island, Sri Lanka	Western Indo-Pacific	15–16
Armandia leptocirris (Grube, 1878)	Philippines	Central Indo-Pacific	0–18
Armandia loboi Elías and Bremec, 2003	off Mar del Plata, Argentina	Temperate South America	5–13
Armandia maculata (Webster, 1884)	Bermuda	Tropical Atlantic	9–38
Armandia mariacapae Moreira and Parapar, 2017	Lizard Island, Australia	Central Indo-Pacific	0–12
Armandia melanura Gravier, 1905	Djibouti, Gulf of Aden	Western Indo-Pacific	Intertidal
Armandia nonpapillata Jones, 1962	Kingston Harbour, Jamaica	Tropical Atlantic	N.d.
Armandia opisthoculata Moreira and Parapar, 2017	Lizard Island, Australia	Central Indo-Pacific	5–15
Armandia paraintermedia Parapar and Moreira, 2015	Lizard Island, Australia	Central Indo-Pacific	0–24
Armandia parva Moreira and Parapar, 2017	Lizard Island, Australia	Central Indo-Pacific	0–15
Armandia polyophthalma Kükenthal, 1887	Gulf of Naples, Italy	Temperate Northern Atlantic	0–20
Armandia salvadoriana Hartmann-Schröder, 1956	El Salvador	Tropical Eastern Pacific	Intertidal
Armandia sampadae Gopal, Jaleel, Parameswaran and Vijayan, 2016	Andaman Islands	Western Indo-Pacific	52–57
Armandia secundariopapillata Hartmann-Schröder, 1984	SW Australia	Temperate Australasia	Intertidal
Armandia sinaitica Amoureux, 1983	Gulf of Akaba, Red Sea	Western Indo-Pacific	Intertidal
Armandia simodaensis Takahashi, 1938	Japan	Temperate Northern Pacific	N.d.
Armandia tubulata Parapar and Moreira, 2015	Lizard Island, Australia	Central Indo-Pacific	1–10
Armandia weissenbornii Kükenthal, 1887	Perim Island, Red Sea	Western Indo-Pacific	N.d.
Ophelia africana Tebble, 1953	Table Bay, South Africa	Temperate Southern Africa	Intertidal to shallow subtidal
Ophelia agulhana Day, 1961	False Bay, South Africa	Temperate Southern Africa	Intertidal to shallow subtidal

Table A1. *Cont.*

Species	Type Locality	Realms	Depth (m)
Ophelia algida Maciolek and Blake, 2006	Off Macquarie Island, Southern Ocean	Southern Ocean	112–124
Ophelia amoureuxi Bellan and Costa, 1988	Côte d'Azur	Temperate Northern Atlantic	90–95
Ophelia anomala Day, 1961	False Bay, South Africa	Temperate Southern Africa	15–80
Ophelia ashworthi Fauvel, 1917	Gulf of St Vincent and Spencer	Temperate Australasia	Subtidal
Ophelia assimilis Tebble, 1953	Central California	Temperate Northern Pacific	Intertidal to shallow subtidal
Ophelia barquii Fauvel, 1927	Agay, Var, France	Temperate Northern Atlantic	Intertidal
Ophelia bicornis Savigny, 1822	La Rochelle, France	Temperate Northern Atlantic	Intertidal to subtidal
Ophelia bipartita Monro, 1936	S Chile	Temperate South America	35
Ophelia borealis Quatrefages, 1866	Greenland	Arctic	N.d.
Ophelia bulbibranchiata Hartmann-Schröder and Parker, 1995	Pearson Island, Australia	Temperate Australasia	not recorded
Ophelia capensis Kirkegaard, 1959	Table Bay, South Africa	Temperate Southern Africa	50–80
Ophelia celtica Amoureux and Dauvin, 1981	Atlantic France	Temperate Northern Atlantic	<100
Ophelia dannevigi Benham, 1916	St. Francis Island, Australia	Temperate Australasia	~60
Ophelia denticulata Verrill, 1875	Maine, USA	Temperate Northern Atlantic	N.d.
Ophelia elongata Hutchings and Murray, 1984	Burwood Beach, Australia	Temperate Australasia	Subtidal
Ophelia formosa (Kinberg, 1866)	La Plata, Argentina (?)	Temperate South America	N.d.
Ophelia glabra Stimpson, 1853	East Canada	Temperate Northern Atlantic	N.d.
Ophelia kirkegaardi Intes and Le Loeuff, 1977	Off Abidjan, Ivory Coast	Tropical Atlantic	20–40
Ophelia koloana Gibbs, 1971	Solomon Islands	Central Indo-Pacific	2
Ophelia laubieri Bellan and Costa, 1988	Estuârio do Sado, Portugal	Temperate Northern Atlantic	Low intertidal to subtidal
Ophelia limacina (Rathke, 1843)	Norway	Temperate Northern Atlantic	0–500
Ophelia magna (Treadwell, 1914)	California, USA	Temperate Northern Pacific	N.d.
Ophelia multibranchia Hutchings and Murray, 1984	Botany Bay, Australia	Temperate Australasia	3–4
Ophelia neglecta Schneider, 1892	France	Temperate Northern Atlantic	Intertidal to subtidal
Ophelia peresi Bellan and Picard, 1965	Tuléar, Madagascar	Western Indo-Pacific	15–20
Ophelia praetiosa (Kinberg, 1866)	Cabo Vírgenes, Argentina	Temperate South America	13–95
Ophelia profunda Hartman, 1965	Off New England, USA	Temperate Northern Atlantic	1000–1700
Ophelia pulchella Tebble, 1953	Southern California, USA	Temperate Northern Pacific	Subtidal
Ophelia radiata (Delle Chiaje, 1828)	Gulf of Naples, Italy	Temperate Northern Atlantic	Intertidal to subtidal
Ophelia rathkei McIntosh, 1908	UK	Temperate Northern Atlantic	Intertidal to subtidal
Ophelia roscoffensis Augener, 1910	Roscoff, France	Temperate Northern Atlantic	Subtidal
Ophelia rullieri Bellan, 1975	Gaspésie, Quebec	Temperate Northern Atlantic	Intertidal to shallow subtidal
Ophelia simplex Leidy, 1855	Rhode Island, USA	Temperate Northern Atlantic	N.d.
Ophelia translucens (Katzmann, 1973)	Zlarin, Croatia	Temperate Northern Atlantic	20–40
Ophelia verrilli Riser, 1987	New England, USA	Temperate Northern Atlantic	Intertidal to shallow subtidal
Ophelina abranchiata Støp-Bowitz, 1948	Greenland	Arctic	90–4500
Ophelina acuminata Örsted, 1843	Øresund Strait, Hveen Island, Sweden	Temperate Northern Atlantic	Shelf and slope
Ophelina adamantea (Kinberg, 1866)	Rio de Janeiro, Brazil	Temperate South America	N.d.
Ophelina alata Elías, Bremec, Lana and Orensanz, 2003	SE Brazil	Temperate South America	Subtidal
Ophelina ammotrypanella Schüller, 2008	Antarctic peninsula, Weddell Sea	Southern Ocean	1970–3050
Ophelina aulogastrella (Hartman and Fauchald, 1971)	off New England, USA	Temperate Northern Atlantic	196–5023
Ophelina basicirra Parapar, Moreira and Helgason, 2011	NW Iceland	Temperate Northern Atlantic	23–2298
Ophelina bimensis (Caullery, 1944)	Indonesia	Central Indo-Pacific	N.d.
Ophelina bowitzi Parapar, Moreira and Helgason, 2011	Southern Iceland	Temperate Northern Atlantic	1897–2709
Ophelina brasiliensis Hansen, 1882	Brazil	Temperate South America	Subtidal
Ophelina brattegardi Kongsrud, Bakken and Oug, 2011	off East Greenland	Arctic	1600
Ophelina breviata (Ehlers, 1913)	Wilhelm II Coast, Antarctic Ocean	Southern Ocean	20–3000
Ophelina brevibranchiata (Caullery, 1944)	Indonesia	Central Indo-Pacific	N.d.
Ophelina buitendijki (Horst, 1919)	Java Sea	Central Indo-Pacific	N.d.
Ophelina chaetifera (Hartman, 1965)	off New England, USA	Temperate Northern Atlantic	1330–5007
Ophelina cordiformis (Caullery, 1944)	Indonesia	Central Indo-Pacific	N.d.

Table A1. *Cont.*

Species	Type Locality	Realms	Depth (m)
Ophelina curli Wiklund, Neal, Glover, Drennan, Rabone and Dahlgren, 2019	Clarion-Clipperton Zone, Pacific Ocean	Tropical Eastern Pacific	4026
Ophelina cylindricaudata (Hansen, 1879)	off West Norway	Temperate Northern Atlantic	7–4663
Ophelina cyprophilia Neave and Glasby, 2013	Darwin Harbour, Australia	Central Indo-Pacific	0–10
Ophelina delapidans (Kinberg, 1866)	Valparaiso, Chile	Temperate South America	Shallow subtidal
Ophelina dubia (Caullery, 1944)	Indonesia	Central Indo-Pacific	N.d.
Ophelina ehlersi (Horst, 1919)	Jedan, Aroe Isles, Indonesia	Central Indo-Pacific	N.d.
Ophelina fauveli (Caullery, 1944)	Indonesia	Central Indo-Pacific	10–21
Ophelina ganae Wiklund, Neal, Glover, Drennan, Rabone and Dahlgren, 2019	Clarion-Clipperton Zone, Pacific Ocean	Tropical Eastern Pacific	4076–4302
Ophelina gaucha Elías, Bremec, Lana and Orensanz, 2003	SE Brazil	Temperate South America	0–18
Ophelina hachaensis Augener, 1934	Riohacha, Colombia	Tropical Atlantic	6
Ophelina helgolandiae Augener, 1912	Spitsbergen, Norway	Arctic	562–2710
Ophelina gigantea (Rullier, 1965)	Moreton Bay, Australia	Temperate Australasia	N.d.
Ophelina grandis (Pillai, 1961)	Tambalagam Bay, Sri Lanka	Western Indo-Pacific	1.8–7.3
Ophelina groenlandica Støp-Bowitz, 1948	East Greenland	Arctic	Shelf and slope
Ophelina gymnopyge (Ehlers, 1908)	Kerguelen Islands	Southern Ocean	13–199
Ophelina jeffreysi (McIntosh in Jeffreys, 1876)	Labrador Sea, North Atlantic Ocean	Arctic	1066–3200
Ophelina juhazi Wiklund, Neal, Glover, Drennan, Rabone and Dahlgren, 2019	Clarion-Clipperton Zone, Pacific Ocean	Tropical Eastern Pacific	4100
Ophelina kampeni (Horst, 1919)	Jedan, Aroe Isles, Indonesia	Central Indo-Pacific	N.d.
Ophelina kinbergii Hansen, 1882	Río de Janeiro, Brazil	Temperate South America	Subtidal
Ophelina kohni Magalhães, Rizzo and Bailey-Brock, 2019	Guam	Central Indo-Pacific	Shallow subtidal
Ophelina kuekenthali (McIntosh, 1908)	North Atlantic Ocean	Temperate Northern Atlantic	~1400
Ophelina langii (Kükenthal, 1887)	Philippines	Central Indo-Pacific	20
Ophelina longicaudata (Caullery, 1944)	Indonesia	Central Indo-Pacific	N.d.
Ophelina longicephala Hartmann-Schröder, 1977	Off Portugal	Temperate Northern Atlantic	77
Ophelina longicirrata Hartmann-Schröder and Parker, 1995	South Australia	Temperate Australasia	N.d.
Ophelina manana Magalhães, Rizzo and Bailey-Brock, 2019	Oahu, Hawaii	Eastern Indo-Pacific	400–500
Ophelina martinezarbizui Wiklund, Neal, Glover, Drennan, Rabone and Dahlgren, 2019	Clarion-Clipperton Zone, Pacific Ocean	Tropical Eastern Pacific	4026–4425
Ophelina meyerae Wiklund, Neal, Glover, Drennan, Rabone and Dahlgren, 2019	Clarion-Clipperton Zone, Pacific Ocean	Tropical Eastern Pacific	4302
Ophelina minima Hartmann-Schröder, 1974	Skagerrak	Temperate Northern Atlantic	230–645
Ophelina modesta Støp-Bowitz, 1958	Oslo, Norway	Temperate Northern Atlantic	100–200
Ophelina nematoides (Ehlers, 1913)	Antarctic Ocean	Southern Ocean	246–2725
Ophelina norvegica Støp-Bowitz, 1945	East Norway	Temperate Northern Atlantic	Subtidal
Ophelina nunnallyi Wiklund, Neal, Glover, Drennan, Rabone and Dahlgren, 2019	Clarion-Clipperton Zone, Pacific Ocean	Tropical Eastern Pacific	4425–4302
Ophelina nybelini (Eliason, 1951)	Azores Islands	Temperate Northern Atlantic	4540–4600
Ophelina opisthobranchiata Wirén, 1901	Spitsbergen, Norway	Arctic	800–3900
Ophelina profunda (Caullery, 1944)	Indonesia	Central Indo-Pacific	N.d.
Ophelina pygocirrata (Ehlers, 1920)	Indonesia	Central Indo-Pacific	N.d.
Ophelina remigera (Ehlers, 1918)	Aru Islands, Sungi Manumbai, Indonesia	Central Indo-Pacific	N.d.
Ophelina robusta Schüller, 2008	Antarctic Peninsula, Weddell Sea	Southern Ocean	2668–3050
Ophelina scaphigera (Ehlers, 1900)	Magellan Strait	Temperate South America	18–3382
Ophelina setigera (Hartman, 1978)	Weddell Sea	Southern Ocean	3111
Ophelina sibogae (Caullery, 1944)	Java	Central Indo-Pacific	79–81
Ophelina syringopyge (Ehlers, 1901)	South Georgia	Southern Ocean	9–876
Ophelina tessellata Neave and Glasby, 2013	Melville Bay, Australia	Central Indo-Pacific	0–10
Polyophthalmus australis Grube, 1869	Cape York, Australia	Central Indo-Pacific	N.d.
Polyophthalmus ceylonensis Kükenthal, 1887	Sri Lanka	Western Indo-Pacific	N.d.
Polyophthalmus collaris Michaelsen, 1892	Sri Lanka	Western Indo-Pacific	N.d.
Polyophthalmus longisetosus Michaelsen, 1892	Sri Lanka	Western Indo-Pacific	N.d.
Polyophthalmus mauliola Magalhães, Rizzo and Bailey-Brock, 2019	Mamala Bay, Hawaii	Eastern Indo-Pacific	27–56
Polyophthalmus pictus (Dujardin, 1839)	France	Temperate Northern Atlantic	Intertidal
Polyophthalmus qingdaoensis Purschke, Ding and Müller, 1995	Qingdao, Yellow Sea	Temperate Northern Pacific	N.d.

Table A1. *Cont.*

Species	Type Locality	Realms	Depth (m)
Polyophthalmus striatus Kükenthal, 1887	Hong Kong	Central Indo-Pacific	N.d.
Polyophthalmus translucens Hartman, 1960	Southern California, USA	Temperate Northern Pacific	914
Thoracophelia arctica (Grube, 1866)	Arctic Ocean	Arctic	N.d.
Thoracophelia bibrancha (Hutchings and Murray, 1984)	Merimbula, Australia	Temperate Australasia	Intertidal
Thoracophelia dillonensis (Hartman, 1938)	Dillon Beach, California, USA	Temperate Northern Pacific	Intertidal
Thoracophelia ezoensis Okuda, 1936	Hokkaido, Japan	Temperate Northern Pacific	N.d.
Thoracophelia flabellifera Ziegelmeier, 1955	German Bight, North Sea	Temperate Northern Atlantic	13
Thoracophelia furcifera Ehlers, 1897	Punta Arenas, Magellan Strait	Temperate South America	Intertidal
Thoracophelia heterocirra (Rozbaczylo and Zamorano, 1970)	El Tabo, Chile	Temperate South America	Intertidal
Thoracophelia japonica (Misaka and Sato, 2003)	Oura Bay, Japan	Temperate Northern Pacific	0–16
Thoracophelia longiseta (Hutchings and Murray, 1984)	Ocean Beach, Australia	Temperate Australasia	Intertidal
Thoracophelia mammillata (Santos, Nonato and Petersen, 2004)	Rio de Janeiro, Brazil	Temperate South America	22–45
Thoracophelia mucronata (Treadwell, 1914)	La Jolla, Southern California	Temperate Northern Pacific	Intertidal
Thoracophelia otagoensis (Probert, 1976)	Otago Peninsula, New Zealand	Temperate Australasia	Intertidal
Thoracophelia papillata (Santos, Nonato and Petersen, 2004)	Abaís beach, Brazil	Tropical Atlantic	Intertidal
Thoracophelia profunda (Hartman, 1967)	Cape Horn, Chile	Temperate South America	4008
Thoracophelia yasudai Okuda, 1934	Kainawa, Japan	Temperate Northern Pacific	Intertidal
Thoracophelia williamsi (Hartman, 1938)	Dillon Beach, California, USA	Temperate Northern Pacific	Intertidal
Thoracophelia zeidleri (Hartmann-Schröder and Parker, 1995)	Haystack Beach, Australia	Temperate Australasia	Intertidal

Table A2. List of valid species of Scalibregmatidae (after Blake [3,23]) with their type locality, realms *sensu* Spalding et al. [45] and depth ranges of type locality. N.d. = no data.

Species	Type Locality	Realms	Depth (m)
Asclerocheilus acirratus (Hartman, 1966)	White Cove, Southern California, USA	Temperate Northern Pacific	0–3
Asclerocheilus ashworthi Blake, 1981	Elephant Island, Antarctica	Southern Ocean	223–397
Asclerocheilus beringianus Uschakov, 1955	Bering Sea	Arctic	986–2005
Asclerocheilus californicus Hartman, 1963	Santa Monica, Redondo and San Pedro valley, California, USA	Temperate Northern Pacific	542–890
Asclerocheilus capensis Day, 1963	South Africa	Temperate Southern Africa	9–26
Asclerocheilus elisabethae Eibye-Jacobsen, 2002	Thailand, Andaman Sea	Western Indo-Pacific	70–76
Asclerocheilus glabrus (Ehlers, 1887)	Cuba	Tropical Atlantic	320
Asclerocheilus intermedius (Saint-Joseph, 1894)	Dinard, France	Temperate Northern Atlantic	96–1830
Asclerocheilus kudenovi Blake, 2000	Point Arguello, California, USA	Temperate Northern Pacific	91.5–123
Asclerocheilus mexicanus Kudenov, 1985	Florida, Gulf of Mexico	Tropical Atlantic	2.4–76.2
Asclerocheilus shanei Hartmann-Schröder, 1994	Scamander, Tasmania	Temperate Australia	122
Asclerocheilus shanonae Eibye-Jacobsen, 2002	Thailand, Andaman Sea	Western Indo-Pacific	70–76
Asclerocheilus tasmanicus Kirkegaard, 1996	Tasman Sea, W of New Zealand	Temperate Australia	3710–3830
Asclerocheilus tropicus Blake, 1981	NW off Guayaquil, Ecuador	Tropical Eastern Pacific	3–10
Asclerocheilus victoriensis Blake, 2000	Victoria, Australia	Temperate Australia	6–22
Axiokebuita cavernicola Martínez, Di Domenico and Worsaae, 2013	Los Cerebros cave, Tenerife, Canary Islands	Temperate Northern Atlantic	8–15
Axiokebuita minuta (Hartman, 1967)	Antarctica	Southern Ocean	180–3685
Cryptosclerocheilus baffinensis Blake, 1972	Southern Baffin Bay	Arctic	1830
Hyboscolex dicranochaetus (Schmarda, 1861)	Table Bay, Cape New Hope, South Africa	Temperate Southern Africa	0.5–19.8
Hyboscolex equatorialis Blake, 1981	NW of Guayaquil, Ecuador	Temperate Southern America	8–9
Hyboscolex homochaetus (Schmarda, 1861)	New Zealand	Temperate Southern Africa	N.d.
Hyboscolex longisetus Schmarda, 1861	Table Bay, Cape New Hope, South Africa	Temperate Southern Africa	9–110

Table A2. *Cont.*

Species	Type Locality	Realms	Depth (m)
Hyboscolex oculatus (Ehlers, 1901)	Tumbes Peninsula, near Talcahuano, Chile	Temperate Southern America	ca. 0.5–10
Hyboscolex pacificus (Moore, 1909)	Santa Monica, California, USA	Temperate Northern Pacific	200
Hyboscolex quadricincta Kudenov, 1985	Florida, Gulf of Mexico	Tropical Atlantic	0.6–31
Hyboscolex reticulatus (McIntosh, 1885)	Queen Charlotte Sound, New Zealand	Temperate Australia	2011
Hyboscolex verrucosus Hartmann-Schröder, 1979	Port Hedland, Western Australia	Central Indo-Pacific	0–5
Lipobranchius jeffreysii McIntosh, 1869	Hebrides and Shetland Islands	Temperate Northern Atlantic	22–1194
Oligobregma aciculatum (Hartman, 1965)	New England, abyssal	Temperate Northern Atlantic	1925–4825
Oligobregma brasierae Wiklund, Neal, Glover, Drennan, Rabone and Dahlgren, 2019	Clarion-Clipperton Zone, Pacific Ocean	Tropical Eastern Pacific	4425
Oligobregma collare (Levenstein, 1975)	Drake Passage, Antarctica	Southern Ocean	3733–3806
Oligobregma lonchochaeta Detinova, 1985	Reykjanes Ridge, off Iceland	Temperate Northern Atlantic	2930–2951
Oligobregma mucronata Blake, 2015	Greenpeace Trough, East Antarctic Peninsula	Southern Ocean	323–912
Oligobregma notiale Blake, 1981	Palmer Archipelago, Antarctic Peninsula	Southern Ocean	18–923
Oligobregma oculata Kudenov and Blake, 1978	East of Saint Maurice island, New Caledonia	Central Indo-Pacific	57
Oligobregma pseudocollare Schüller and Hilbig, 2007	Scotia Sea, Antarctica	Southern Ocean	2889–2892
Oligobregma quadrispinosa Schüller and Hilbig, 2007	Scotia Sea, Antarctica	Southern Ocean	2258–2313
Oligobregma simplex Kudenov and Blake, 1978	Western Port, Victoria, Australia	Temperate Australia	11
Oligobregma tani Wiklund, Neal, Glover, Drennan, Rabone and Dahlgren, 2019	Clarion-Clipperton Zone, Pacific Ocean	Tropical Eastern Pacific	4137
Oligobregma whaleyi Wiklund, Neal, Glover, Drennan, Rabone and Dahlgren, 2019	Clarion-Clipperton Zone, Pacific Ocean	Tropical Eastern Pacific	4425
Parasclerocheilus branchiatus Fauvel, 1928	Shingle Island, Gulf of Manaar, India	Western Indo-Pacific	N.d.
Parasclerocheilus capensis Day, 1961	Langebaan Lagoon, South Africa	Temperate Southern Africa	26
Polyphysia caulleryi (McIntosh, 1922)	Inland Sea of Japan	Temperate Northern Pacific	47
Polyphysia crassa (Örsted, 1843)	Denmark	Temperate Northern Atlantic	0–1755
Polyphysia hystricis (McIntosh, 1922)	Channel slope, Antarctica	Temperate Northern Atlantic	4701261
Pseudoscalibregma bransfieldium (Hartman, 1967)	E. Bransfield Strait, Antarctica	Southern Ocean	323–916
Pseudoscalibregma hartmanae Blake, 1981	Weddell Sea, Antarctica	Southern Ocean	585
Pseudoscalibregma orientalis Imajima, 2009	Japan	Temperate Northern Pacific	373–1005
Pseudoscalibregma pallens Levenstein, 1962	Kermadec Trench	Temperate Australia	8928–9174
Pseudoscalibregma palmeri Blake, 2015	Weddell Sea, Off Lindenberg Island, Antarctica	Southern Ocean	385–768
Pseudoscalibregma papilia Schüller, 2008	South Sandwich Islands, Antarctica	Southern Ocean	2258–2313
Pseudoscalibregma parvum (Hansen, 1878)	North Sea	Temperate Northern Atlantic	53–1802
Pseudoscalibregma usarpium Blake, 1981	Ross Sea, Antarctica	Southern Ocean	2143
Scalibregma australis Blake, 2015	Greenpeace Trough, East Antarctic Peninsula	Southern Ocean	12–978
Scalibregma californicum Blake, 2000	Santa Maria Basin, off Point Sal, California	Temperate Northern Pacific	90–2710
Scalibregma celticum Mackie, 1991	Milford Haven, Dyfed, Wales	Temperate Northern Atlantic	6–21
Scalibregma hanseni Bakken, Oug and Kongsrud, 2014	Egga, west of Nordland County, Norway	Temperate Northern Atlantic	765
Scalibregma inflatum Rathke, 1843	Norway	Temperate Northern Atlantic	1–3690
Scalibregma robustum Zachs, 1925	White Sea, Russia	Artic	N.d.
Scalibregma stenocerum (Bertelsen and Weston, 1980)	Daytona Beach, Florida	Tropical Atlantic	17–65
Scalibregma wireni Furreg, 1925	Kaiser Joseph Fjord, East Greenland	Artic	3–9
Scalibregmella antennata Hartman and Fauchald, 1971	Bermuda, abyssal	Temperate Northern Atlantic	4833–5023
Scalibregmides chilensis Hartmann-Schröder, 1965	Puerto Aguirre, Chile	Temperate Southern America	10
Scalibregmides peruanus Blake, 1981	island near Pucusana, south of Callao, Peru	Temperate Southern America	0–5
Sclerobregma branchiatum Hartman, 1965	New England	Temperate Northern Atlantic	1330–2022
Sclerocheilus antarcticus Ashworth, 1915	Petermann Island, Antarctica	Southern Ocean	45–311
Sclerocheilus deriugini Zachs, 1925	Kola Fjord, Northern Russia	Temperate Northern Atlantic	N.d.
Sclerocheilus minutus Grube, 1863	Nerizine, Mali Losinj, Croatia	Temperate Northern Pacific	N.d.
Sclerocheilus unoculus Kudenov, 1985	Florida, Gulf of Mexico	Tropical Atlantic	18–37
Speleobregma lanzaroteum Bertelsen, 1986	La Corona lava tube, Lanzarote, Canary Islands	Temperate Northern Atlantic	5–25

References

1. Blake, J.A.; Maciolek, N.J. 7.6.1 Opheliidae Malmgren, 1867. In *Handbook of Zoology. Annelida: Pleistoannelida, Sedentaria II*; Schmidt-Rhaesa, A., Ed.; De Gruyter: Berlin, Germany, 2020.
2. Wiklund, H.; Neal, L.; Glover, A.G.; Drennan, R.; Rabone, M.; Dahlgren, T.G. Abyssal fauna of polymetallic nodule exploration areas, eastern Clarion-Clipperton Zone, central Pacific Ocean: Annelida: Capitellidae, Opheliidae, Scalibregmatidae, and Travisiidae. *Zookeys* **2019**, *883*, 1–82. [CrossRef]
3. WoRMS Editorial Board. Register of Marine Species. 2020. Available online: http://www.marinespecies.org/ (accessed on 15 February 2021).
4. Parapar, J. Familia Opheliidae. In *Fauna Ibérica—Annelida Polychaeta III*; Parapar, J., Alós, C., Núñez, J., Moreira, J., López, E., Aguirrezabalaga, F., Besteiro, C., Martínez, A., Ramos, M.A., Eds.; Museo Nacional de Ciencias Naturales, CSIC: Madrid, Spain, 2012; Volume 36, pp. 284–332.
5. Wilson, D.P. The Larval Development of *Ophelia bicornis* Savigny. *J. Mar. Biol. Assoc. UK* **1948**, *27*, 540–553. [CrossRef]
6. Dales, R.P. The larval development and ecology of *Thoracophelia mucronata* (Treadwell). *Biol. Bull.* **1952**, *102*, 232–242. [CrossRef]
7. Seike, K. Burrowing behaviour inferred from feeding traces of the opheliid polychaete *Euzonus* sp. as response to beach morphodynamics. *Mar. Biol.* **2008**, *53*, 1199–1206. [CrossRef]
8. Otegui, M.P.B.; Blankensteyn, A.; Pagliosa, P.R. Population structure, growth and production of *Thoracophelia furcifera* (Polychaeta: Opheliidae) on a sandy beach in Southern Brazil. *Helgol. Mar. Res.* **2012**, *66*, 479–488. [CrossRef]
9. Dorgan, K.M. Kinematics of burrowing by peristalsis in granular sands. *J. Exp. Biol.* **2018**, *221*, jeb167759. [CrossRef] [PubMed]
10. Giangrande, A.; Gambino, I.; Tundo, M.; Pasqua, M.; Licciano, M.; Fanini, L.; Pinna, M. Reproductive biology of *Ophelia barquii* (Annelida, Opheliidae) along the Salento Peninsula (Mediterranean Sea, South Italy). *Mar. Biodiv.* **2020**, *50*, 3. [CrossRef]
11. Meador, J.P.; Rice, C.A. Impaired growth in the polychaete *Armandia brevis* exposed to tributyltin in sediment. *Mar. Environ. Res.* **2001**, *51*, 113–129. [CrossRef]
12. Bat, L.; Şahin, F.; Öztekin, A. Acute toxicity of cadmium on *Ophelia bicornis* Savigny, 1822. *Acta Aquat. Turc.* **2019**, *15*, 289–297. [CrossRef]
13. Vanreusel, A.; Cosson-Sarradin, N.; Gooday, A.J.; Paterson, G.L.J.; Galéron, J.; Sibuet, M.; Vincx, M. Evidence for episodic recruitment in a small opheliid polychaete species from the abyssal NE Atlantic. *Prog. Oceanogr.* **2001**, *50*, 285–301. [CrossRef]
14. Kongsrud, J.A.; Bakken, T.; Oug, E. Deep-water species of the genus *Ophelina* (Annelida, Opheliidae) in the Nordic Seas, with the description of *Ophelina brattegardi* sp. nov. *Ital. J. Zool.* **2011**, *78*, 95–111. [CrossRef]
15. Parapar, J.; Moreira, J.; Helgason, G. Distribution and diversity of Opheliidae (Annelida, Polychaeta) on the continental shelf and slope of Iceland, with a review of the genus *Ophelina* in Northeast Atlantic waters and description of two new species. *Org. Divers. Evol.* **2011**, *11*, 83–105. [CrossRef]
16. Brown, R.S. The anatomy of the polychaete *Ophelia cluthensis* McGuire 1935. *Proc. R. Soc. Edinb.* **1938**, *58*, 135–160. [CrossRef]
17. McConnaughey, B.H.; Fox, D.L. The Anatomy and Biology of the Marine Polychaete *Thoracophelia mucronata* (Treadwell) Opheliidae. *Univ. Calif. Publ. Zool.* **1949**, *47*, 319–340.
18. Hermans, C.O.; Cloney, R.A. Fine structure of the prostomial eyes of *Armandia brevis* (Polychaeta: Opheliidae). *Z. Zellforsch. Mikrosk. Anat.* **1966**, *72*, 583–596. [CrossRef] [PubMed]
19. Purschke, G.; Ding, Z.; Müller, M.C. Ultrastructural differences as a taxonomic marker: The segmental ocelli of *Polyophthalmus pictus* and *Polyophthalmus qingdaoensis* sp. n. Polychaeta, Opheliidae). *Zoomorphology* **1995**, *115*, 229–241. [CrossRef]
20. Tzetlin, A.B.; Filippova, A.V. Muscular system in polychaetes (Annelida). *Hydrobiologia* **2005**, *535*, 113–126. [CrossRef]
21. Law, C.J.; Dorgan, K.M.; Rouse, G.W. Relating divergence in polychaete musculature to different burrowing behaviors: A study using Opheliidae. *J. Morphol.* **2014**, *275*, 548–571. [CrossRef] [PubMed]
22. Misaka, T.; Sato, M. A new species of *Euzonus* (Polychaeta: Opheliidae) from subtidal zones in Japan. *Zool. Sci.* **2003**, *20*, 1171–1177. [CrossRef]
23. Blake, J.A. 7.6.3 Scalibregmatidae Malmgren, 1867. In *Handbook of Zoology. Annelida: Pleistoannelida, Sedentaria II*; Schmidt-Rhaesa, A., Ed.; De Gruyter: Berlin, Germany, 2020.
24. Rouse, G.W.; Pleijel, F. *Polychaetes*; Oxford University Press: Oxford, UK, 2001.
25. Kudenov, J.D. Four new species of Scalibregmatidae (Polychaeta) from the Gulf of Mexico, with comments on the familial placement of *Mucibregma* Fauchald and Hancock, 1981. *Proc. Biol. Soc. Wash.* **1985**, *98*, 332–340.
26. Martínez, A.; Di Domenico, M.; Worsaae, K. Evolution of cave *Axiokebuita* and *Speleobregma* (Scalibregmatidae, Annelida). *Zool. Scr.* **2013**, *42*, 623–636. [CrossRef]
27. Hartman, O.; Fauchald, K. Deep-water benthic polychaetous annelids off New England to Bermuda and other North Atlantic areas. Part II. *Allan Hancock Monogr. Mar. Biol.* **1971**, *6*, 1–327.
28. Ashworth, J.H. Memoirs: The Anatomy of *Scalibregma inflatum*, Rathke. *J. Cell Sci.* **1901**, *45*, 237–309.
29. Ashworth, J.H. On a New Species of *Sclerocheilus*, with a revision of the genus. *Trans. R. Soc. Edinb.* **1915**, *50*, 405–422. [CrossRef]
30. Rathke, H. *Beiträge Zur Fauna Norwegens*; Für die Akademie in Eduard Weber's Buchhandlng in Bonn: Bonn, Germany, 1843; Volume 1, pp. 1–264.

31. Blake, J.A. The Scalibregmatidae (Annelida: Polychaeta) from South America and Antarctica collected chiefly during the cruises of the R/V Anton Bruun, R/V Hero and USNS Eltanin. *Proc. Biol. Soc. Wash.* **1981**, *94*, 1131–1162.
32. Pleijel, F.; Fauchald, K. *Scalispinigera oculata* Hartman, 1967 (Scalibregmatidae: Polychaeta): Senior synonym of *Lacydonia antarctica* (Lacydoniidae) Hartmann-Schröder & Rosenfeldt, 1988. *Proc. Biol. Soc. Wash.* **1993**, *106*, 673–677.
33. Blake, J.A.; Maciolek, N.J. 7.6.2 Travisiidae Hartmann-Schröder, 1971, new family status. In *Handbook of Zoology. Annelida: Pleistoannelida, Sedentaria II*; Schmidt-Rhaesa, A., Ed.; De Gruyter: Berlin, Germany, 2020; pp. 302–311.
34. Mackie, Y. *Scalibregma celticum* new species (Polychaeta: Scalibregmatidae) from Europe, with a redescription of *Scalibregma inflatum* Rathke, 1843 and comments on the genus *Sclerobregma* Hartman, 1965. *Bull. Mar. Sci.* **1991**, *48*, 268–276.
35. Bakken, T.; Oug, E.; Kongsrud, J.A. Occurrence and distribution of *Pseudoscalibregma* and *Scalibregma* (Annelida, Scalibregmatidae) in the deep Nordic Seas, with the description of *Scalibregma hanseni* n. sp. *Zootaxa* **2013**, *3753*, 101–117. [CrossRef] [PubMed]
36. Blake, J.A. New species of Scalibregmatidae (Annelida, Polychaeta) from the East Antarctic Peninsula including a description of the ecology and post-larval development of species of *Scalibregma* and *Oligobregma*. *Zootaxa* **2015**, *4033*, 57–93. [CrossRef]
37. Schüller, M.; Hilbig, B. Three new species of the genus *Oligobregma* (Polychaeta, Scalibregmatidae) from the Scotia and Weddell Seas (Antarctica). *Zootaxa* **2007**, *1391*, 35–45. [CrossRef]
38. Kudenov, D.; Blake, J.A. A review of the genera and species of the Scalibregmidae (Polychaeta) with descriptions of one new genus and three new species from Australia. *J. Nat. Hist.* **1978**, *12*, 427–444. [CrossRef]
39. Nogueira, J.M.M. *Asclerocheilus tropicus* Blake, 1981 (Polychaeta: Scalibregmatidae): Redescription Based on Brazilian specimens. *Proc. Biol. Soc. Wash.* **2002**, *115*, 323–332.
40. Hartmann-Schröder, G. Zur Kenntnis des Sublitorals der Chilenischen Küste unter besonderer Berücksichtigung der Polychaeten und Ostracoden. II Die Polychaeten des Sublitorals. *Mitt. Zool. Mus. Hamburg* **1965**, *62*, 59–305.
41. Dehorne, A.; Dehorne, L. Recherches sur *Sclerocheilus minutus* (polychète de la famille des Scalibregmides). Morphologie, yeux, néphridie et pavillon. *Arch. Zool. Exp. Gen.* **1913**, *53*, 61–137.
42. Orrhage, L.; Müller, M. Morphology of the Nervous System of Polychaeta (Annelida). *Hydrobiologia* **2005**, *535/536*, 79–111. [CrossRef]
43. Martínez, A.; Di Domenico, M.; Worsaae, K. Gain of palps within a lineage of ancestrally burrowing annelids (Scalibregmatidae). *Acta Zool.* **2014**, *95*, 421–429. [CrossRef]
44. Fauchald, K.; Rouse, G. Polychaete systematics: Past and present. *Zool. Scri.* **1997**, *26*, 71–138. [CrossRef]
45. Spalding, M.D.; Fox, H.E.; Allen, G.R.; Davidson, N.; Ferdaña, Z.A.; Finlayson, M.; Halpern, B.S.; Jorge, M.A.; Lombana, A.; Lourie, S.A.; et al. Marine Ecoregions of the World: A Bioregionalization of Coastal and Shelf Areas. *BioScience* **2007**, *57*, 573–583. [CrossRef]
46. Paul, C.; Halanych, K.M.; Tiedemann, R.; Bleidorn, C. Molecules reject an opheliid affinity for *Travisia* (Annelida). *Syst. Biodivers.* **2010**, *8*, 507–512. [CrossRef]
47. Blake, J.A. Family Opheliidae Malmgren, 1867. In *Taxonomic Atlas of the Benthic Fauna of the Santa Maria Basin and the Western Santa Barbara Channel. Vol. 7. The Annelida. Part 4: Polychaeta: Flabelligeridae to Sternaspidae*; Blake, J.A., Hilbig, B., Scott, P.V., Eds.; Santa Barbara Museum of Natural History: Santa Barbara, CA, USA, 2000; Volume 7, pp. 145–168.
48. Bleidorn, C.; Vogt, L.; Bartolomaeus, T. New Insights into Polychaete Phylogeny (Annelida) Inferred from 18S RDNA Sequences. *Mol. Phylogenet. Evol.* **2003**, *29*, 279–288. [CrossRef]
49. Hall, K.A.; Hutchings, P.A.; Colgan, D.J. Further phylogenetic studies of the Polychaeta using 18S RDNA sequence data. *J. Mar. Biol. Asspc. UK* **2004**, *84*, 949–960. [CrossRef]
50. Belova, P.A.; Zhadan, A.E. Comparative morphology and ultrastructure of the respiratory system in four species of the Opheliidae family. *Biol. Bull.* **2014**, *41*, 752–772. [CrossRef]
51. Weigert, A.; Helm, C.; Meyer, M.; Nickel, B.; Arendt, D.; Hausdorf, B.; Santos, S.R.; Halanych, K.M.; Purschke, G.; Bleidorn, C.; et al. Illuminating the Base of the Annelid Tree Using Transcriptomics. *Mol. Biol. Evol.* **2014**, *31*, 1391–1401. [CrossRef]
52. Andrade, S.C.S.; Novo, M.; Kawauchi, G.Y.; Worsaae, K.; Pleijel, F.; Giribet, G.; Rouse, G.W. Articulating "Archiannelids": Phylogenomics and Annelid Relationships, with Emphasis on Meiofaunal Taxa. *Mol. Biol. Evol.* **2015**, *32*, 2860–2875. [CrossRef]
53. Sene-Silva, G. Filogenia de Opheliidae (Annelida: Polychaeta). Ph.D. Thesis, Universidade Federal do Paraná, Curitiba, Brazil, 2007.
54. Blake, J.A. Revalidation of the genus *Thoracophelia* Ehlers, 1897, replacing *Euzonus* Grube, 1866 (Polychaeta: Opheliidae), junior homonym of *Euzonus* Menge, 1854 (Arthropoda: Diplopoda), together with a literature summary and updated listing of *Thoracophelia* species. *Zootaxa* **2011**, *2807*, 65–68. [CrossRef]
55. Schüller, M. New Polychaete Species Collected during the Expeditions ANDEEP I, II, and III to the Deep Atlantic Sector of the Southern Ocean in the Austral Summers 2002 and 2005—Ampharetidae, Opheliidae, and Scalibregmatidae. *Zootaxa* **2008**, *1705*, 51–68. [CrossRef]
56. Day, J.H. *A Monograph on the Polychaeta of Southern Africa. Part 2, Sedentaria*; British Museum (Natural History): London, UK, 1967; Volume 2.
57. Santos, C.S.G.; Nonato, E.F.; Petersen, M.E. Two New Species of Opheliidae (Annelida: Polychaeta): *Euzonus papillatus* sp. n. from a Northeastern Brazilian Sandy Beach and *Euzonus mammillatus* sp. n. from the Continental Shelf of Southeastern Brazil. *Zootaxa* **2004**, *478*, 1–12. [CrossRef]

58. Neave, M.J.; Glasby, C.J. New Species of *Ophelina* (Annelida: Opheliidae: Ophelininae) from Northern Australia. *Org. Divers. Evol.* **2013**, *13*, 331–347. [CrossRef]
59. Gopal, A.; Jaleel, A.K.U.; Parameswaran, U.V.; Vijayan, A.-K. *Armandia sampadae*, a new species of polychaete (Opheliidae) from Andaman Sea, Northern Indian Ocean. *J. Mar. Biol. Assoc. UK* **2016**, *96*, 1625–1632. [CrossRef]
60. Magalhães, W.F.; Rizzo, A.E.; Bailey-Brock, J.H. Opheliidae (Annelida: Polychaeta) from the Western Pacific Islands, including five new species. *Zootaxa* **2019**, *4555*, 209–235. [CrossRef]
61. Parapar, J.; Gambi, M.C.; Rouse, G.W. A revision of the deep-sea genus *Axiokebuita* Pocklington and Fournier, 1987 (Annelida: Scalibregmatidae). *Ital. J. Zool.* **2011**, *78*, 148–162. [CrossRef]
62. Parapar, J.; Moreira, J. Six new species of the genus *Armandia* Filippi, 1861 (Polychaeta, Opheliidae) from Lizard Island (Great Barrier Reef, Australia). *Zootaxa* **2015**, *4019*, 577–603. [CrossRef]
63. Moreira, J.; Parapar, J. New data on the Opheliidae (Annelida) from Lizard Island (Great Barrier Reef, Australia): Five new species of the genus *Armandia* Filippi, 1861. *Zootaxa* **2017**, *4290*, 483–502. [CrossRef]
64. Tzetlin, A.; Zhadan, A. Morphological variation of axial non-muscular proboscis types in the Polychaeta. *Zoosymposia* **2009**, *2*, 415–427. [CrossRef]
65. Purschke, G. Ultrastructure of Nuchal Organs in Polychaetes (Annelida)—New Results and Review. *Acta Zool.* **1997**, *78*, 123–143. [CrossRef]
66. Saito, H.; Tamaki, A.; Imajima, M. Description of a new species of *Armandia* (Polychaeta: Opheliidae) from Western Kyushu, Japan, with character variations. *J. Nat. Hist.* **2000**, *34*, 2029–2043. [CrossRef]
67. Bartolomaeus, T. Different photoreceptors in juvenile *Ophelia rathkei* (Annelida, Opheliida). *Microfauna Mar.* **1993**, *8*, 99–114.
68. Parapar, J.; Moreira, J. Sobre la presencia del género *Ophelina* Ørsted, 1843 (Polychaeta, Opheliidae) en el litoral de la península Ibérica. *Nova Acta Cient. Compostel. Biol.* **2008**, *17*, 117–134.
69. Hartmann-Schröder, G. Zur Morphologie der Opheliiden (Polychaeta Sedentaria). *Z. Wiss. Zool.* **1958**, *161*, 84–143.
70. Penry, D.L.; Jumars, P.A. Gut architecture, digestive constraints and feeding ecology of deposit-feeding and carnivorous polychaetes. *Oecologia* **1990**, *82*, 1–11. [CrossRef]
71. Bartolomaeus, T.; Quast, B. Structure and development of nephridia in Annelida and related taxa. *Hydrobiologia* **2005**, *535*, 139–165. [CrossRef]
72. West, D.L. Comparative Ultrastructure of juvenile and adult nuchal organs of an annelid (Polychaeta, Opheliidae). *Tissue Cell* **1978**, *10*, 243–257. [CrossRef]
73. Hermans, C.O. Fine structure of the segmental ocelli of *Armandia brevis* (Polychaeta: Opheliidae). *Z. Zellforsch. Mikrosk. Anat.* **1969**, *96*, 361–371. [CrossRef]
74. Parapar, J.; Candás, M.; Cunha-Veira, X.; Moreira, J. Exploring annelid anatomy using micro-computed tomography: A taxonomic approach. *Zool. Anz.* **2017**, *270*, 19–42. [CrossRef]
75. Bellan, G.; Dauvin, J.C. Phenetic and biogeographic relationships in *Ophelia* (Polychaeta, Opheliidae). *Bull. Mar. Sci.* **1991**, *48*, 544–558.
76. Neal, L.; Taboada, S.; Woodall, L.C. Slope-Shelf Faunal Link and Unreported Diversity off Nova Scotia: Evidence from Polychaete Data. *Deep Sea Res. Part I Oceanogr. Res. Pap.* **2018**, *138*, 72–84. [CrossRef]
77. Woodin, S.A. Polychaete Abundance Patterns in a Marine Soft-Sediment Environment: The Importance of Biological Interactions. *Ecol. Monog.* **1974**, *44*, 171–187. [CrossRef]
78. Kuş, S.; Kurt-Şahin, G. Temporal Changes in the Polychaeta (Annelida) Community Associated with *Cystoseira* Beds of Sinop Peninsula (Southern Black Sea). *Turk. J. Fish. Aquat. Sci.* **2016**, *16*, 61–68. [CrossRef]
79. Hadiyanto, H. Fouling Polychaetes in Tanjung Priok Port of Jakarta, Indonesia. *ASEAN J. Sci. Technol. Dev.* **2018**, *35*, 79–87. [CrossRef]
80. Dorgan, K.M.; Law, C.J.; Rouse, G.W. Meandering worms: Mechanics of undulatory burrowing in muds. *Proc. R. Soc. B* **2013**, *280*, 20122948. [CrossRef] [PubMed]
81. Dafoe, L.T.; Gingras, M.K.; Pemberton, S.G. Determining *Euzonus mucronata* Burrowing Rates with Application to Ancient *Macaronichnus segregatis* Trace-makers. *Ichnos* **2008**, *15*, 78–90. [CrossRef]
82. Riser, N.W. Observations on the genus *Ophelia* (Polychaeta: Opheliidae) with the description of a new species. *Ophelia* **1987**, *28*, 11–29. [CrossRef]
83. Castelli, A.; Lardicci, C.; Castellani, C.; Finocchiaro, C.; Genovesi, S.; Tataranni, M.; Maltagliati, F. Analysis of vertical distribution of two polychaetes (genus *Ophelia*) in sandy substrate along the Pisan Coast. *Biol. Mar. Mediterr.* **2006**, *13*, 160–161.
84. Tamaki, A. Zonation by size in the *Armandia* sp. (Polychaeta: Opheliidae) population on an intertidal sand flat. *Mar. Ecol. Prog. Ser.* **1985**, *27*, 123–133. [CrossRef]
85. Fauchald, K.; Jumars, P.A. The diet of worms: A study of polychaete feeding guilds. *Oceanogr. Mar. Biol. Ann. Rev.* **1979**, *17*, 193–284.
86. Kemp, P.F. Direct uptake of detrital carbon by the deposit feeding polychaete *Euzonus mucronata* (Treadwell). *J. Exp. Mar. Biol. Ecol.* **1986**, *99*, 49–61. [CrossRef]
87. Guérin, J.-P. Modalités d'élevage et description des stades larvaires de *Polyophthalmus pictus* Dujardin (Annélide Polychète). *Vie Milieu* **1971**, *22*, 143–152.

88. Giangrande, A. Polychaete reproductive patterns, life cycles and life histories: An overview. *Oceanogr. Mar. Biol. Ann. Rev.* **1997**, *35*, 323–386.
89. Hermans, C.O. Metamorphosis in the opheliid polychaete *Armandia brevis*. In *Settlement and Metamorphosis of Marine Invertebrate Larvae*; Chia, F., Rice, M.E., Eds.; Elsevier: New York, NY, USA, 1978; pp. 113–126.
90. Rivain, V. Contribution à L'étude Dynamique et Fonctionnelle des Peuplements de Sables fins du Golfe Normano-Breton. Ph.D Thesis, Université Pierre et Marie Curie, Paris, France, 1983.
91. Miner, B.G.; Sanford, E.; Strathmann, R.R.; Pernet, B.; Emlet, R.B. Functional and Evolutionary Implications of Opposed Bands, Big Mouths, and Extensive Oral Ciliation in Larval Opheliids and Echiurids (Annelida). *Biol. Bull.* **1999**, *197*, 14–25. [CrossRef]
92. Guérin, J.-P. Le développement larvaire d'*Armandia cirrhosa* Filippi (Annélide Polychète). *Tethys* **1972**, *4*, 969–974.
93. Parke, S.R. Biological Aspects of Speciation in Three Sympatric *Euzonus* Species at Dillon Beach, California (Polychaeta: Opheliidae). Master's Thesis, University of the Pacific, Stockton, CA, USA, 1973.
94. Wilson, D.P. The role of micro-organisms in the settlement of *Ophelia bicornis* Savigny. *J. Mar. Biol. Assoc. UK* **1955**, *34*, 531–543. [CrossRef]
95. Bely, A.E. Distribution of segment regeneration ability in the Annelida. *Integr. Comp. Biol.* **2006**, *46*, 508–518. [CrossRef]
96. Conradi, M.; Bandera, M.E.; Marin, I.; Martin, D. Polychaete-parasitizing copepods from the deep-sea Kuril–Kamchatka Trench (Pacific Ocean), with the description of a new *Ophelicola* species and comments on the currently known annelidicolous Copepods. *Deep Sea Res. II* **2015**, *111*, 147–165. [CrossRef]
97. Hartmann-Schröder, G. *Annelida, Borstenwürmer, Polychaeta: Tierwelt Deutschlands 58 (Die Tierwelt Deutschlands und der Angrenzenden Meeresteile)*; Gustav Fischer: Jena, Germany, 1996; ISBN 978-3-437-35038-2.
98. Georgieva, Y.G.; Daskalov, G.M.; Klayn, S.L.; Stefanova, K.B.; Stefanova, E.S. Seasonal Diet and Feeding Strategy of Horse Mackerel *Trachurus mediterraneus* (Steindachner, 1868) (Perciformes: Carangidae) in the South-Western Black Sea. *Acta Zool. Bulg.* **2019**, *71*, 201–210.
99. Kicklighter, C.E.; Hay, M.E. Integrating Prey Defensive Traits: Contrasts of Marine Worms from Temperate and Tropical Habitats. *Ecol. Monogr.* **2006**, *76*, 195–215. [CrossRef]
100. Saes, R.V.S.T.; Moreira, L.B.; Davanso, M.B.; Perina, F.C.; Abess, D.M.S. Developing a protocol whole sediment toxicity testing with the polychaete *Armandia agilis*. *Ecotoxicol. Environ. Contam.* **2018**, *13*, 85–97. [CrossRef]
101. Malmgren, A.J. *Annulata Polychaeta: Spetsbergiae, Groenlandiae, Islandiae et Scandinaviae. Hactenus Cognita*; Ex Officina Frenckelliana: Helsinki, Finland, 1867.
102. Fauchald, K. Polychaetes from intertidal areas in Panama, with a review of previous shallow-water records. *Smithson. Contr. Zool.* **1977**, *221*, 1–81. [CrossRef]
103. Persson, J.; Pleijel, F. On the phylogenetic relationships of *Axiokebuita*, *Travisia* and Scalibregmatidae (Polychaeta). *Zootaxa* **2005**, *998*, 1–14. [CrossRef]
104. Helm, C.; Beckers, P.; Bartolomaeus, T.; Drukewitz, S.H.; Kourtesis, I.; Weigert, A.; Purschke, G.; Worsaae, K.; Struck, T.H.; Bleidorn, C. Convergent evolution of the ladder-like ventral nerve cord in Annelida. *Front. Zool.* **2018**, *15*, 36. [CrossRef]
105. Weigert, A.; Bleidorn, C. Current status of annelid phylogeny. *Org. Divers. Evol.* **2016**, *16*, 345–362. [CrossRef]
106. Örsted, A.S. *Annulatorum Danicorum Conspectus. Fasc. I. Maricolæ*; Librarie Wahliannae: Copenhagen, Denmark, 1843.
107. Grube, A.E. Beschreibung Neuer Oder Wenig Bekannter Anneliden. Sechster Beitrag. *Arch. Nat.* **1863**, *29*, 37–69.
108. Saint-Joseph, A.D. Les Annélides Polychètes Des Côtes de Dinard. Troisième Partie. *Ann. Sci. Nat.* **1849**, *17*, 1–395.
109. Hansen, G.A. Annelider Fra Den Norske Nordhavsexpedition i 1876. *Nyt Mag. Naturvid.* **1879**, *24*, 1–17.
110. McIntosh, W.C. On the Structure of the British Nemerteans, and Some New British Annelids. *Trans. R. Soc. Edinb.* **1869**, *25*, 305–433. [CrossRef]
111. Schmarda, L.K. Neue Wirbellose Thiere: Beobachted und Gesammelt auf einer Reise um die Erdr 1853 bis 1857. In *Turbellarien, Rotatorien und Anneliden*; Erster Band, Zweite Hälfte; Verlag von Wilhelm Engelmann: Leipzig, Germany, 1861.
112. McIntosh, W.C. *Report on the Scientific Results of the Voyage of H.M.S. Challenger during the Years 1873–76, under the Command of Captain George S. Nares and the Late Captain Frank Tourle Thomson: A Summary of the Scientific Results*; Printed for H.M.S.O.: London, UK, 1885.
113. Ehlers, E. Reports on the Results of Dredging, Under the Direction of L.F. Pourtalès, During the Years 1868–1870, and of Alexander Agassiz in the Gulf of Mexico (1877–78) and in the Caribbean Sea (1878–79), in the US Coast. Survey Steamer "Blake" Lieut.-Com. CD Sigsbee, USN, and Commander JR Bartlett, USN, Commanding. XXXI. Report on the Annelids. *Mem. Mus. Comp. Zool. Harv. Coll.* **1887**, *15*, 1–335.
114. Furreg, E. Zur Systematik Der Polychätenfamilie Scalibregmidae. *Zool. Jahrb. Abt. Syst. Geogr. Biol. Tiere* **1925**, *50*, 123–190.
115. Hartman, O. Deep-Water Benthic Polychaetous Annelids off New England to Bermuda and Other North Atlantic Areas. *Allan Hancock Found. Publ. Occas. Pap.* **1965**, *28*, 1–384.
116. Blake, J.A. Two new species of polychaetous annelid worms from Baffin Bay and the Davis Strait. *Bull. South. Calif. Acad. Sci.* **1972**, *71*, 127–132.
117. Day, J.H. The Polychaet [Sic] Fauna of South Africa. Part 6. Sedentary Species Dredged off Cape Coasts with a Few New Records from the Shore. *J. Linn. Soc.* **1961**, *44*, 463–560. [CrossRef]
118. Kirkegaard, J.B. Bathyal and Abyssal Polychaetes (Sedentary Species I). *Galathea Rep.* **1996**, *17*, 57–77.

119. Blake, J. Family Scalibregmatidae Malmgren, 1867. In *Taxonomic Atlas of the Benthic Fauna of the Santa Maria Basin and the Western Santa Barbara Channel. Vol. 7. The Annelida. Part 4: Polychaeta: Flabelligeridae to Sternaspidae*; Blake, J., Hilbig, B., Scott, P., Eds.; Santa Barbara Museum of Natural History: Santa Barbara, CA, USA, 2000; Volume 4, pp. 129–144.
120. Imajima, M. Deep-Sea Benthic Polychaetes off Pacific Coast of the Northern Honshu, Japan. *Natl. Mus. Nat. Sci. Monogr.* **2009**, *39*, 39–192.
121. Imajima, M. Occurrence of *Oncoscolex pacificus* (Moore), a Species of Family Scalibregmidae (Polychaeta), in Japan. *J. Hokkaido Gakugei Univ.* **1961**, *12*, 11–13.
122. Støp-Bowitz, C. Les Scalibregmiens de Norvége. *Medd. Zool. Mus. Oslo* **1945**, *55*, 63–87.
123. Wesenberg-Lund, E. Lesser Antillean Polychaetes, Chiefly from Brackish Water with a Survey and a Bibliography of Fresh and Brackish–Water Polychaetes. Studies on the Fauna of Curaçao and Other Caribbean Islands 8.1. *Stud. Fauna Curacao Caribb. Isl.* **1958**, *8*, 1–41.
124. Bertelsen, R.D. *Speleobregma lanzaroteum*, a New Genus and Species of Scalibregmatidae (Polychaeta) from a Marine Cave in the Canary Islands. *Proc. Biol. Soc. Wash.* **1983**, *99*, 375–379.
125. Martínez, A.; Gonzalez, B.C.; Worsaae, K.; Wilkens, H.; Núñez, J.; Oromí, P.; Iliffe, T.M. *Guide to the Anchialine Ecosystems of Jameos Del Agua and Túnel de La Atlántida*; Medio Ambiente; Cabildo de Lanzarote: Arrecife, Spain, 2016. [CrossRef]
126. Martínez, A.; Palmero, A.M.; Brito, M.D.C.; Núñez, J.; Worsaae, K. Anchialine fauna of the Corona lava tube (Lanzarote, Canary Islands): Diversity, endemism and distribution. *Mar. Biodiver.* **2009**, *39*, 169–187. [CrossRef]
127. Martínez, A.; Gonzalez, B.C. Volcanic Anchialine Habitats of Lanzarote. In *Cave Ecology. Ecological Studies (Analysis and Synthesis)*; Moldovan, O.T., Kovac, L., Halse, S., Eds.; Springer: Cham, Switzerland, 2019.
128. García-Valdecasas, A. Estudio Faunístico de La Cueva Submarina "Túnel de La Atlántida", Jameos Del Agua, Lanzarote. *Nat. Hisp.* **1985**, *27*, 1–56.
129. Martínez, A.; Di Domenico, M.; Leasi, F.; Curini-Galletti, M.; Todaro, M.A.; Zotto, M.D.; Gobert, S.; Artois, T.; Norenburg, J.; Jörger, K.M.; et al. Patterns of Diversity and Endemism of Soft-Bodied Meiofauna in an Oceanic Island, Lanzarote, Canary Islands. *Mar. Biodiv.* **2019**, *49*, 2033–2055. [CrossRef]
130. Mammola, S.; Lunghi, E.; Bilandžija, H.; Cardoso, P.; Grimm, V.; Schmidt, S.I.; Hesselberg, T.; Martínez, A. Collecting Eco-Evolutionary Data in the Dark: Impediments to Subterranean Research and How to Overcome Them. *EcoEvorXiv* **2020**. [CrossRef]
131. Riera, R.; Monterroso, Ó.; Núñez, J.; Martínez, A. Distribution of meiofaunal abundances in a marine cave complex with secondary openings and freshwater filtrations. *Mar. Biodivers.* **2018**, *48*, 203–215. [CrossRef]
132. Gonzalez, B.C.; Martínez, A.; Borda, E.; Iliffe, T.M.; Fontaneto, D.; Worsaae, K. Genetic spatial structure of an anchialine cave annelid indicates connectivity within–but not between–islands of the Great Bahama Bank. *Mol. Phylogenet. Evol.* **2017**, *109*, 259–270. [CrossRef] [PubMed]
133. Dorgan, K.M.; Jumars, P.A.; Johnson, B.D.; Boudreau, B.P. Macrofaunal Burrowing: The Medium Is the Message. *Oceanogr. Mar. Biol. Ann. Rev.* **2006**, *44*, 5–121.
134. Elder, H.Y. Direct Peristaltic Progression and the Functional Significance of the Dermal Connective Tissues during Burrowing in the Polychaete *Polyphysia crassa* (Oersted). *J. Exp. Biol.* **1973**, *58*, 637–655.
135. Hunter, R.D.; Elder, H.L. Burrowing Dynamics and Energy Cost of Transport in the Soft-Bodied Marine Invertebrates *Polyphysia crassa* and *Priapulus caudatus*. *J. Zool.* **1983**, *218*, 209–222. [CrossRef]
136. Cahoon, L.B.; Laws, R.A.; Thomas, C.J. Viable diatoms and chlorophyll*a* in continental slope sediments off Cape Hatteras, North Carolina. *Deep Sea Res. Part II Top. Stud. Oceanogr.* **1994**, *41*, 767–782. [CrossRef]
137. Aller, J.Y.; Aller, R.C.; Green, M.A. Benthic faunal assemblages and carbon supply along the continental shelf/shelf break-slope off Cape Hatteras, North Carolina. *Deep Sea Res. Part II Top. Stud. Oceanogr.* **2002**, *49*, 4599–4625. [CrossRef]
138. Blake, J.A.; Hilbig, B. Dense infaunal assemblages on the continental slope off Cape Hatteras, North Carolina. *Deep Sea Res. Part II Top. Stud. Oceanogr.* **1994**, *41*, 875–899. [CrossRef]
139. Worsaae, K.; Rouse, G.W. *Mesonerilla neridae* sp. nov. (Nerillidae): First meiofaunal annelid from deep-sea hydrothermal vents. *Zoosymposia* **2009**, *2*, 297–303. [CrossRef]
140. Hartman, O. Quantitative survey of the benthos of San Pedro Basin, southern California. Part II. Final results and conclusions. *Allan Hancock Pac. Exped.* **1966**, *19*, 187–455.
141. Hartmann-Schröder, G.; Hartmann, G. Zur Kenntnis Des Eulitorals Der Australischen Küsten Unter Besonderer Berücksichtigung Der Polychaeten Und Ostracoden (Extract *Hyboscolex*). *Mitt. Hambg. Zool. Mus. Inst.* **1979**, *76*, 75–218.
142. Hartman, O. Submarine canyons of southern California. Polychaetous Annelids. *Allan Hancock Pac. Exped.* **1963**, *27*, 1–93.
143. Wilkens, H.; Iliffe, T.M.; Oromí, P.; Martínez, A.; Tysall, T.N.; Koenemann, S. The Corona Lava Tube, Lanzarote: Geology, Habitat Diversity and Biogeography. *Mar. Biodiv.* **2009**, *39*, 155–167. [CrossRef]
144. Martínez, A.; Kvindebjerg, K.; Iliffe, T.M.; Worsaae, K. Evolution of cave suspension feeding in Protodrilidae (Annelida). *Zool. Scr.* **2017**, *46*, 214–226. [CrossRef]
145. Worsaae, K.; Gonzalez, B.C.; Kerbl, A.; Nielsen, S.H.; Jørgensen, J.T.; Armenteros, M.; Iliffe, T.M.; Martínez, A. Diversity and evolution of the stygobitic *Speleonerilla* nom. nov. (Nerillidae, Annelida) with description of three new species from anchialine caves in the Caribbean and Lanzarote. *Mar. Biodiv.* **2019**, *49*, 2167–2192. [CrossRef]
146. Ditlevsen, H. Annelids from the Danmark Expedition. *Medd. Grønland Kobenhaven* **1911**, *45*, 411–432.
147. Clark, R. Pelagic Swarming of Scalibregmidae (Polychaeta). *Annu. Rep. Scott. Mar. Biol. Assoc.* **1952**, *53*, 20–22.

148. Fage, L.; Legendre, R. Pêches Planctoniques à La Lumière, Effectuées à Banyuls-Sur-Mer et à Concarneau. I. Annélides Polychètes. *Arch. Zool. Exp. Gen.* **1927**, *67*, 23–222.
149. Blake, J.A. Life history analysis of five dominant infaunal polychaete species from the continental slope off North Carolina. *J. Mar. Biol. Assoc. UK* **1993**, *73*, 123–141. [CrossRef]

diversity MDPI

Review

The Terebelliformia-Recent Developments and Future Directions

Pat Hutchings [1,2,*], Orlemir Carrerette [3], João M. M. Nogueira [4], Stephane Hourdez [5] and Nicolas Lavesque [6]

1 Australian Museum Research Institute, Australian Museum, Sydney, NSW 2010, Australia
2 Biological Sciences, Macquarie University, North Ryde, NSW 2109, Australia
3 Laboratório de Ecologia e Evolução de Mar Profundo, Instituto Oceanográfico, Universidade de São Paulo, Cidade Universitária, São Paulo 05508-090, Brazil; orlemir@gmail.com
4 Laboratório de Poliquetologia, Departamento de Zoologia, Instituto de Biociências, Universidade de São Paulo, Rua do Matão, Travessa 14, n. 101, São Paulo 05508-090, Brazil; nogueira@ib.usp.br
5 CNRS, Sorbonne Université, LECOB, UMR 8222, Observatoire Océanologique de Banyuls, 1 Avenue Pierre Fabre, 66650 Banyuls-sur-Mer, France; hourdez@obs-banyuls.fr
6 CNRS, Université Bordeaux, EPOC, UMR 5805, Station Marine d'Arcachon, 33120 Arcachon, France; nicolas.lavesque@u-bordeaux.fr
* Correspondence: pat.hutchings@australian.museum

Abstract: Terebelliformia comprises a large group of sedentary polychaetes which live from the intertidal to the deep sea. The majority live in tubes and are selective deposit feeders. This study synthesises the current knowledge of this group, including their distribution, in the different biogeographic regions. We highlight the new methodologies being used to describe them and the resolution of species complexes occurring in the group. The main aim of this review is to highlight the knowledge gaps and to stimulate research in those directions, which will allow for knowledge of their distribution and abundances to be used by ecologists and managers.

Keywords: Annelida; polychaetes; biodiversity assessment; geographical distribution; methods; knowledge gaps

check for updates

Citation: Hutchings, P.; Carrerette, O.; Nogueira, J.M.M.; Hourdez, S.; Lavesque, N. The Terebelliformia-Recent Developments and Future Directions. *Diversity* **2021**, *13*, 60. https://doi.org/10.3390/d13020060

Academic Editor: Bert W. Hoeksema
Received: 31 December 2020
Accepted: 28 January 2021
Published: 3 February 2021

1. Introduction

This review of the diversity of the Terebelliformia deals with the taxa previously considered as subfamilies of the Terebellidae Johnston, 1846, namely Polycirridae Malmgren, 1866, Terebellidae Johnston, 1846 (previously referred to as the Amphitritinae) and Thelepodidae Hessle, 1917, together with the closely related family Trichobranchidae Malmgren, 1866, and the recently described family Telothelepodidae Nogueira, Fitzhugh and Hutchings, 2013. For a detailed discussion of the elevation of the subfamilies of the Terebellidae sensu lato (s.l.) to family level, see Nogueira et al. [1] and Hutchings et al. [2]. As well, we include Alvinellidae Desbruyères and Laubier, 1986, Pectinariidae Johnston, 1865 and Ampharetidae Malmgren, 1866, which are all included within the Terebelliformia.

Terebelliformia are common worldwide, including the polar regions, and may be abundant in some areas [3–5]. While some genera are highly speciose, others are represented by few species or only by a single one (for details of genera and numbers of species, see [2] for terebellids, see [6] for pectinariids, see [7] for alvinellids and [8] for ampharetids).

Members of this diverse group are characterised by the presence of multiple grooved buccal tentacles used for selective deposit feeding. Although it is still debatable whether those structures are homologous among all the families of Terebelliformia, we assume they are [1,9] and, therefore, all are of prostomial origin. Due to the extensible characteristic of these structures, they can be easily recognized around their tubes or galleries, rendering these animals the name "spaghetti worms" (Figures 1–3). Typically, the tentacles are smooth, but some polycirrids have papillose tentacles and ampharetids may also have grooved, smooth or pinnate tentacles. In general, these tentacles are not retractable into the mouth, except in ampharetids and alvinellids, which are able to fully retract them (Figure 1e–g).

Figure 1. Diversity of Terebelliformia: Pectinariidae (PE), Ampharetidae (AM), Alvinellidae (AL) and Polycirridae (PO): (**a**) *Petta investigatoris* (PE), tube; (**b**) *Amphictene auricoma* (PE): entire worm, left lateral view; (**c**) *Petta pusilla* (PE): entire worms, ventral (left) and dorsal (right) views; (**d**) *Petta investigatoris* (PE): entire worm, dorsal view stained in methyl green; (**e**,**i**) *Amphicteis dalmatica* (paratype AM W.11667) (AM): anterior end, ventral and dorsal views, respectively; (**f**) *Amythas membranifera* (AM): entire worm, ventral view; (**g**,**h**) *Alvinella pompejana* (AM W.29585) (AL): anterior end, ventral and dorsal views, respectively; (**j**) *Polycirrus oculeus* (paratype AM W.44612) (PO): entire live worm, dorso-lateral view; (**k**) *Polycirrus rubrointestinalis* (PO): entire worm live, dorsal view; (**l**) *Hauchiella tentaculata* (holotype NTM W.023154) (PO): entire live worm, dorsal ventral view. Photos: (**d**)—E. Wong; (**f**)—Gabriel Monteiro; (**j**–**l**)—A. Semenov.

Figure 2. Diversity of Terebelliformia: Telothelepodidae (TE), Thelepodidae (TH) and Trichobranchidae (TR): (**a**,**b**) *Telothelepus capensis* (topotype NHMUK ANEA 1955.12.30.1) (TE): anterior end, dorsal and ventral views, respectively; (**c**,**d**) *Rhinothelepus mexicanus* (holotype LACM-AHF Poly 1449) (TE): anterior end, dorsal and ventral views, respectively; (**e**,**f**) *Thelepus paiderotos* (AM W.44600 and AM W.44283, respectively) (TH): entire live worms in right lateral and ventro-lateral views, respectively; (**g**) *Streblosoma curvus* (paratype AM W.44287) (TH): entire live worm (incomplete), dorsal view; (**h**,**i**) *Terebellides akares* (paratype AM W.45450) (TR): ventral and left dorso-lateral views, respectively, of live animals; (**j**) *Trichobranchus hirsutus* (AM W.45444) (TR): complete live worm, left lateral view. Photos: (**e**–**j**)—A. Semenov.

Figure 3. (**a**,**b**) Diversity of Terebelliformia: Terebellidae *s.l.* (TER) *Pistella franciscana*: complete live worm, right lateral views; (**c**,**d**) *Loimia tuberculata* (holotype AM W.44280): complete live worm, ventral and right lateral views, respectively; (**e**,**f**) *Pista chloroplokamia* (holotype AM W.44613): entire live worm, female, left and right lateral views; (**g**,**h**) *Loimia pseudotriloba* (holotype AM W.47810): entire live worm, right and left lateral views; (**i**) *Reteterebella lirrf* (paratype AM W.44545): entire live worm, dorso-lateral view. All animals removed from their tubes. Photos: (**a**–**i**)—A. Semenov.

In this paper, we discuss the current status of our knowledge of Terebelliformia, considering all the modern techniques available, which allows for much deeper analyses and observations, including at the molecular level, to document the diversity of the group. We also discuss the major gaps in our knowledge of Terebelliformia and their phylogeny, including some taxonomic issues, and point to directions to solve them, as well as highlighting other issues which need to be addressed.

The aims of this paper are (1) to present the taxonomic history of these worms, (2) their morphology, (3) the recent studies on their phylogenetic relationships, (4) their roles in the ecosystem and their distribution around the world, (5) the evolution of the methods used to describe them, (6) the knowledge gaps and challenges for the future, with focus on species complexes and taxonomic issues and, finally, (7) how such data can be used in marine park management as well as comments regarding the importance of using correct names.

2. Materials and Methods

This study provides a literature review of the Terebelliformia, including a list of valid species and their distribution according to biogeographical regions and their depth ranges (see Supplementary Material). This is based on the literature as well the World Register of Marine Species (WoRMS, http://www.marinespecies.org) to assess the number of currently valid taxa and analyses of species richness.

The citation of authors and date, and type localities policy: the original author(s) and date of a name of all taxa here included are cited the first time they appear in the text. However, due to the large number of taxa cited in this paper, we have not included all these citations in the references. Instead, they can be found in WoRMS as well as details of type localities and synonymies. We also discuss various genera for which diagnostic characters still need to be evaluated. Biodiversity information is referred to the realms proposed by Spalding et al. [10].

3. Terebelliforms

3.1. Taxonomic History of the Terebelliformia

The discovery of Terebelliformia began in 1766 (Figure 4), with the description of three species from the Dutch Sea, by Pallas: *Lanice conchilega* (Pallas, 1766) (Terebellidae), *Pectinaria belgica* (Pallas, 1766) and *P. capensis* (Pallas, 1766) (Pectinariidae). Since then, more than 1100 species of Terebelliformia have been described by 162 different first authors (Supplementary Material). During this period, four peaks were identified (Figure 4). The initial phase lasted for almost 100 years, from 1766 to 1859, and it was not the most productive, as only 46 species were described. The first peak occurred from 1860 to 1889 when 185 species were described by few taxonomists (Figure 4), as noted by Pamungkas et al. [11]. This productive period can be explained by the publication of important monographs by Europe-based polychaetologists: Grube (47) species) (e.g., [12,13]), Kinberg (12 species) [14], Malmgren (19 species) [15], McIntosh (36 species) [16] and Schmarda (13 species) [17]. Malmgren [15] launched the foundations for the modern taxonomy of Terebelliformia, describing most families of the group and a large number of genera.

By that time, most, if not all, of the researchers were European scientists, working on European material, but frequently no types were deposited in museums or zoological collections, and those species were later often reported from far-away locations. This has led to great taxonomic confusion, which in some cases threatens the stability of important genera (see below). Redescriptions and designation of neotypes from the type localities of some of these early described genera are urgently needed, such as *Amphitrite* O.F. Müller 1771, *Nicolea* Malmgren 1866, *Pista* Malmgren 1866 and *Terebella* Linnaeus 1767, for example.

The second phase of discovery occurred from 1900 to 1919, with 142 new species identified (Figure 4). This period corresponds, once again, to few active taxonomists, such as Augener (12 species) [18], Caullery (8 species) [19], Chamberlin (18 species) [20], Hessle (25 species) [21], Gravier (9 species) (e.g., [22]) and Moore (17 species) (e.g., [23]). It was not until 1970–1989 that the third phase took place, with the description of 165 species, by 34 different first authors. This peak corresponds mainly to the description of new species from Australia by Hutchings and collaborators (59 species) (e.g., [24–30]), but also to the description of the new family Alvinellidae by Desbruyères and Laubier (12 species, all from deep-sea environments and hydrothermal vents) [31]).

Finally, the years 2000–2019 were the most prolific, with 258 species described by 38 different first authors (Figure 4). Among them, Hutchings, Nogueira and Carrerette were the most productive taxonomists (Table 1), with descriptions of 85 species of Terebellidae s.l., mostly from Brazil and Australia (e.g., [32–40]); Ampharetidae were also well studied during this period, with 32 species described [41–43].

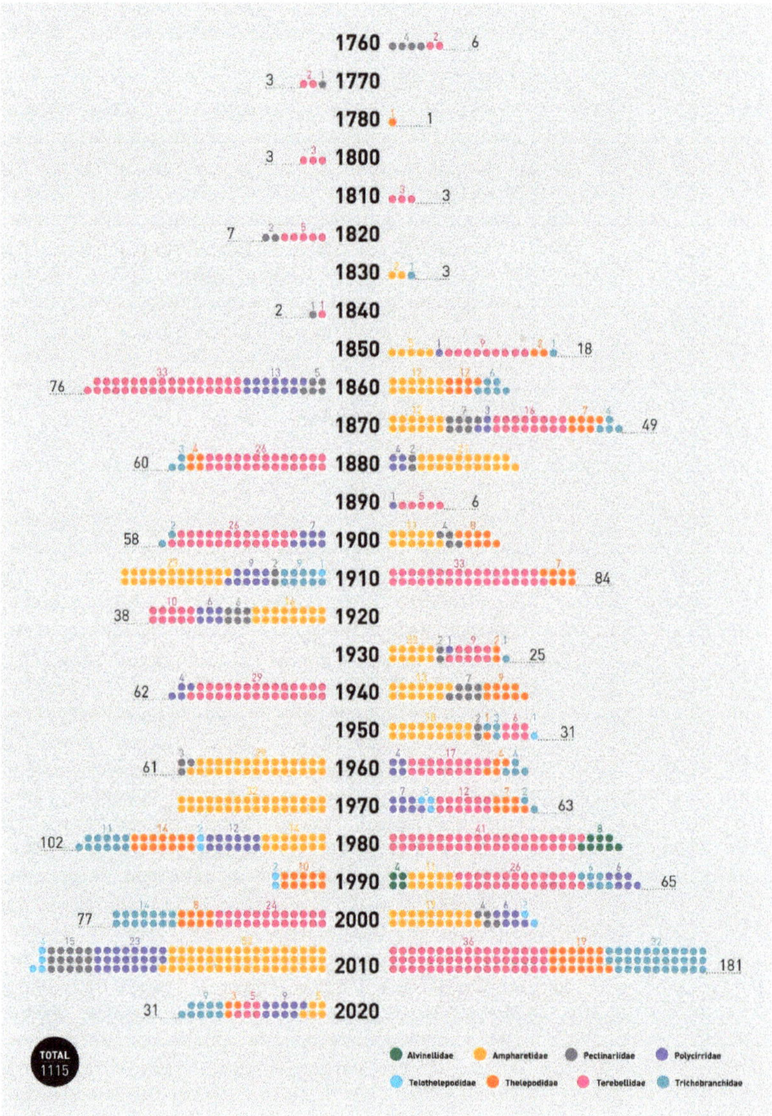

Figure 4. Number of Terebelliformia species described per period of ten years.

Table 1. The top 30 most prolific authors along with their numbers of Terebelliformia species described, first and last discoveries, and other polychaetes species described. Names in bold refer to active taxonomists.

Taxonomist	Terebelliformia Species Described	First Record	Last Record	Non Terebelliformia Species Described
P.A. Hutchings	217	1974	2020	152
J.M.N. Nogueira	74	2010	2020	54
A.E. Grube	58	1855	1878	409
M. Caullery	57	1915	1944	40
O. Carrerette	55	2013	2020	2
C.J. Glasby	49	1986	2014	43
O. Hartman	45	1941	1978	435
W.C. McIntosh	43	1869	1924	247
M.H. Londoño-Mesa	38	2003	2020	0
M. Reuscher	33	2009	2017	2
D. Fiege	31	2009	2016	39
G. Hartmann-Schröder	29	1962	1992	476
I.A. Jirkov	29	1985	2020	11
J.P. Moore	28	1904	1923	196
C. Hessle	27	1917	1917	5
M. Imajima	26	1964	2015	221
K. Fauchald	25	1971	1991	228
J. Parapar	24	1997	2020	45
N. Lavesque	23	2017	2020	3
R.V. Chamberlin	21	1919	1920	107
A.J. Malmgren	21	1865	1867	46
J. Moreira	20	2011	2020	36
J.H. Day	20	1934	1973	171
A.E. Verrill	18	1873	1901	102
P. Fauvel	17	1908	1959	125
H. Augener	15	1906	1926	197
T. Holthe	15	1985	2002	1
M. Schüller	15	2008	2013	8
D. Desbruyères	14	1977	1996	24
J.G.H. Kinberg	14	1866	1867	188

3.2. Morphology of Terebelliforms

Pectinariids are unique among terebelliforms, and among all polychaetes, by having rigid ice-cream cone-shaped tubes [6] (Figure 1a), which disintegrate once the animal dies. These animals are also unique among terebelliforms in having the prostomium and peristomium fused as a cephalic veil, of mixed prostomial and peristomial origin, together with a pair of rows of paleae at the anterior end, and the posterior end modified into a sucker-like scaphe (Figures 1b–d and 5b,i,j,m) [6].

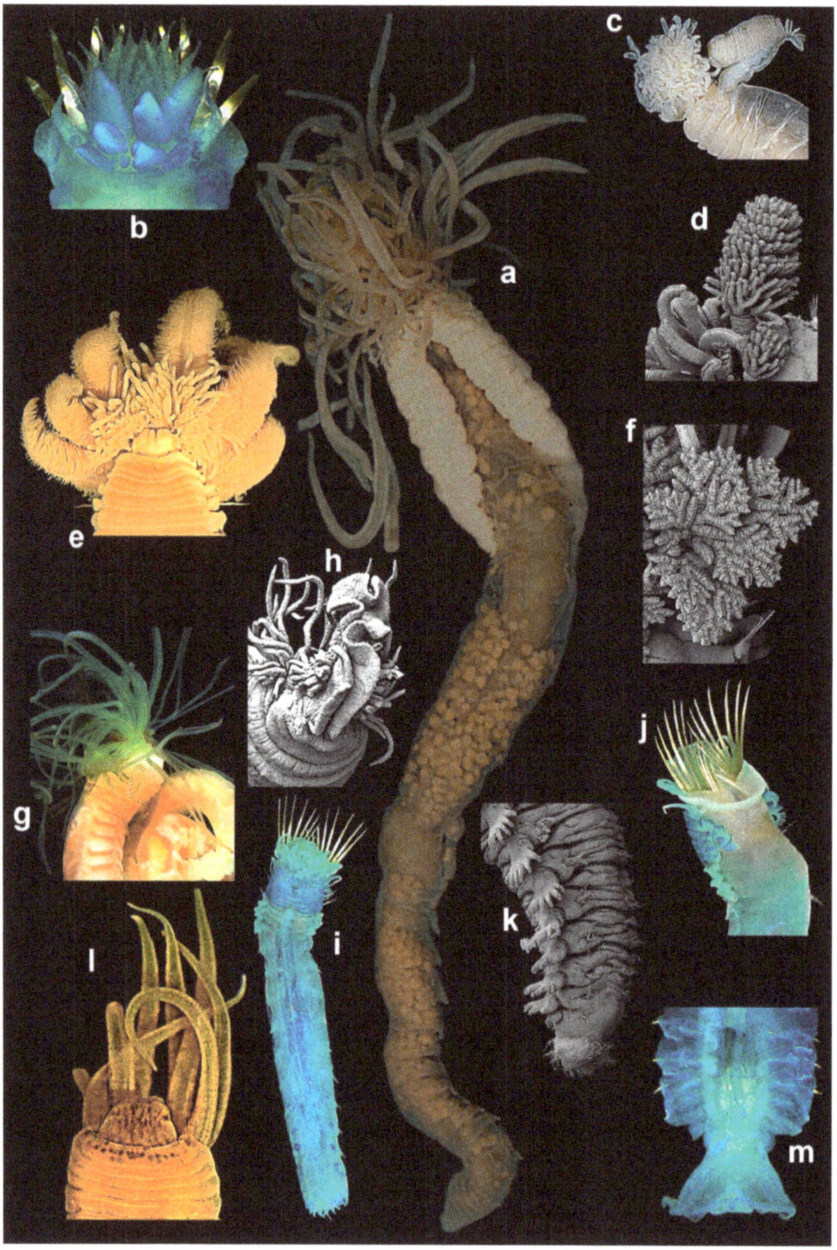

Figure 5. Diagnostic characters of terebelliforms: (**a**) *Nicolea lazowasemi* (holotype YPM 40593) (TER): entire worm, a gravid female, dorsal view; (**b,m**) *Pectinaria antipoda* (stained in methyl green) (PE): anterior and posterior ends, ventral views, respectively; (**c**) *Terebellides akares* (NTM W.023143) (TR): left lateral view; (**d**) *Pistella franciscana* (paratype AM W.44593) (TER): detail of branchiae (SEM); (**e**) *Alvinella pompejana* (AM W.29585) (AL): anterior end, ventral view; (**f,g**) *Pista chloroplokamia* (TER): detail of a branchia (SEM) and anterior end of live animal, right lateral view, respectively; (**h**) *Rhinothelepus occabus* (paratype AM W.201904 (TE): detail of oral area as shown by SEM; (**i,j**) *Petta investigatoris* (stained in methyl green) (PE): entire worm, ventral view, and anterior end, left dorso-lateral view, respectively; (**k**) *Trichobranchus hirsutus* (paratype AM W.47510) (TR): posterior end examined under SEM; (**l**) *Amphicteis dalmatica* (AM): anterior end, ventral view.

Alvinellids and ampharetids are more closely related because members of both families have buccal tentacles fully retractable into the mouth and branchiae originating from segments II–V, but arising as free filaments from segments II–III in ampharetids (following Reuscher et al. [44]) [1,8,9,43] and associated to segments III–IV in alvinellids (Figures 1e–i and 5e,l) [7].

In ampharetids, the body regions are well marked, with notopodia restricted to the anterior part of the body (together with neuropodia, frequently called the "thorax"; see below), and posterior abdominal region with neuropodia only (Figure 6l). The shape of the prostomium can vary with the degree of the extension of the tentacles [44,45] but is typically spatulate and swollen, tri-lobed, frequently with paired glandular ridges; these latter structures are also interpreted as nuchal organs [8,43]. Eyespots may be present in ampharetids, and the peristomium is represented by a ring without appendages or chaetae. The first chaetiger is segment II, often with differentiated notochaetae (also referred to as "paleae"), directed upwards (Figures 1e, 5l and 6c), which may be modified or even absent [8,41]. Other thoracic segments usually bear limbate capillary notochaetae (Figure 6d), but some groups present modifications to the anterior segments, including the presence of hook-like chaetae (Melinninae), different sizes and thicknesses of chaetae and notopodia. Notopodia are absent in the abdominal region, although notopodial rudiments may be present (Figure 6p) [9,42,44]. Neuropodia in ampharetids are sessile tori on thoracic segments, forming pinnules after the end of notopodia (Figure 6m); both regions typically bear short uncini, which usually vary in shape and number of teeth between anterior and posterior regions.

In alvinellids, the first chaetae (notopodial only) appear on segment III in *Paralvinella* Desbruyères and Laubier, 1982 and VI in *Alvinella* Desbruyères and Laubier, 1980 (Figure 1g). Neuropodia (uncini) are sessile and start as early as segment VI (chaetiger 4 in *Paralvinella*) but sometimes much later on the body for some species (ca. chaetiger 40). They occur until the end of the body and their morphology does not change markedly in anterior and posterior regions. As a result, body regions are not marked [1,9]. Chaetiger 7 (*Paralvinella*) or 4 and 5 (*Alvinella*) have strong hooks (Figure 1h). Both prostomium and peristomium are devoid of appendages and bear no eyes. All members have four pairs of branchiae, emerging as strong stems bearing lamellae (*Alvinella*) (Figure 1g,h and Figure 5e) or cylindrical extensions (*Paralvinella*). In addition to the typically grooved tentacles, males of alvinellids also possess a pair of short, thick modified tentacles, possibly involved in pseudocopulation.

Terebellidae s.l. is a group of five families previously considered as subfamilies of a single family, Terebellidae (= Terebellidae s.l.), which Nogueira et al. [1] showed to have originated independently in the evolution of Terebelliformia, raising each of those to family level, and describing a new one, the Telothelepodidae. Animals belonging to these families all have prostomium at the dorsal side of the upper lip, with buccal tentacles originating from the distal part of prostomium, therefore out of the mouth and not retractable into it (Figure 1j,l, Figures 2a–j and 3a–i). In addition, all these animals have up to three pairs of branchiae, usually from segment II, although several forms are abranchiate, including the entire family Polycirridae (Figures 1j–l and 6a); notopodia bearing distally winged (="smooth") (Figure 6d,f,g,i) or serrated capillaries (Figure 6e,h), frequently restricted to the anterior region of the body; neuropodia, extending until near pygidium, bearing uncini (Figure 6j–o,q) [2]. Members of these families, however, are distinguished from each other, mostly by the morphology of the upper lip, the branchiae, the ventral glandular areas of anterior segments, and neuropodia, and by the morphology and arrangement of the uncini of anterior neuropodia, if in single or double rows [2].

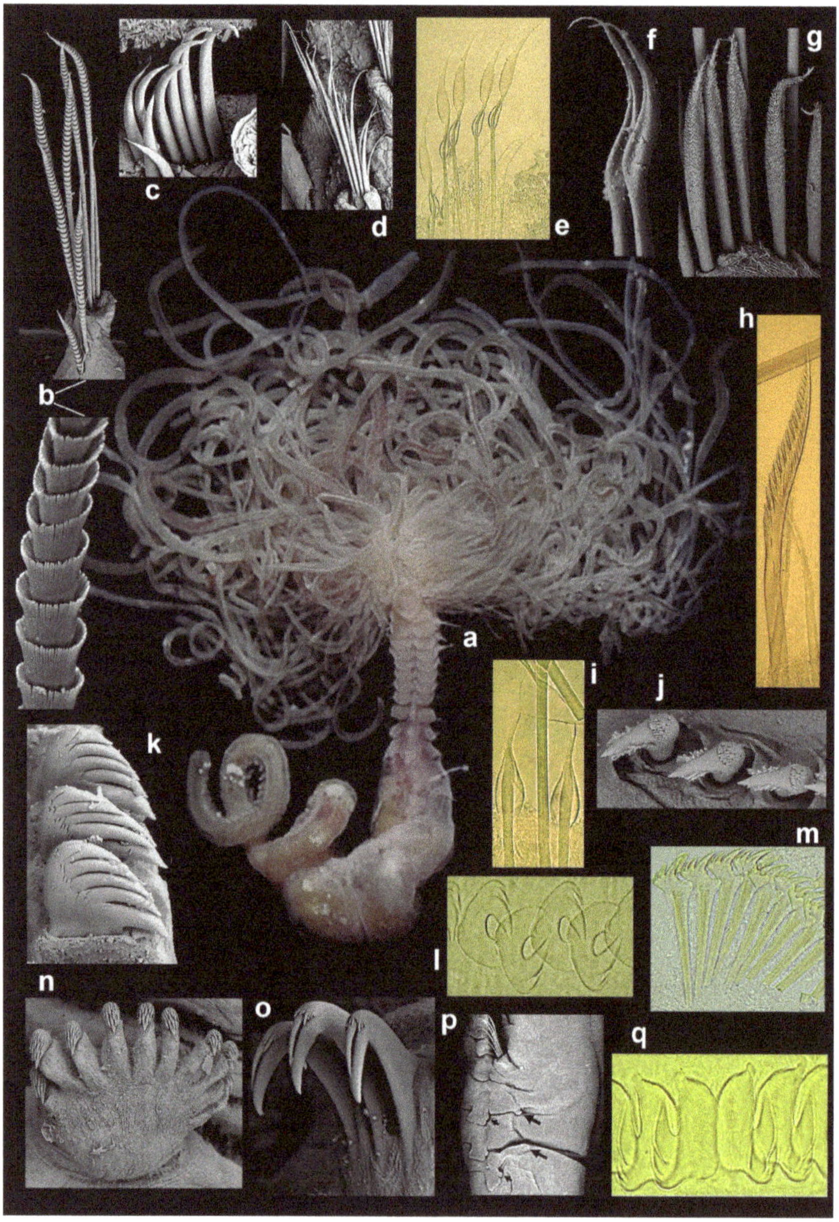

Figure 6. Diagnostic characters of terebelliforms: (**a**) *Polycirrus changbunker* (ZUEC 21354) (PO): entire worm, ventral view; (**b**,**o**) *Polycirrus papillatus* (PO): notochaetae, of two magnifications, and abdominal uncini, respectively (SEM); (**c**,**d**) *Amphicteis dalmatica* (AM): paleae and notochaetae of anterior segment, respectively; (**e**) *Spinosphaera barega* (TER): posterior thoracic notochaetae; (**f**) *Pista anneae* (TER): notochaetae, segment X; (**g**) *Pistella franciscana* (TER): notochaetae, segment VIII; (**h**) *Alvinella pompejana* (AL): notochaetae, anterior segment; (**i**) *Leaena ebranchiata* (TER): notochaetae of anterior row, anterior segment; (**j**) *Trichobranchus hirsutus* (TR): uncini, segment VI (SEM); (**k**) *Loimia pseudotriloba* (TER): abdominal uncini (SEM); (**l**) *Nicolea vaili* (TER): uncini, segment 10; (**m**) *Lanicides rubra* (TER): uncini, segment 5; (**n**) *Trichobranchus hirsutus* (TR): neuropodium, segment XXI; (**p**) *Amphicteis dalmatica* (AM): last thoracic and first two abdominal segments; large arrows point to abdominal rudimental notopodia, short arrows point to neuropodial dorsal papillae; (**q**) *Thelepus paiderotos* (TH): uncini, segment VII.

Polycirrids, in addition to being all abranchiate, have a circular upper lip, and the buccal tentacles are of two types, with the long ones often distally modified (Figures 1j–l and 6a). The body may be highly papillated and the anterior glandular areas of anterior segments are typically well developed, with paired mid-ventral shields, separated from each other within pairs by the mid-ventral longitudinal groove, extending from ~segment II or III to the pygidium (Figures 1l and 6a) [1,2,46]. In addition, there is a tendency for a reduction in parapodia in these animals, as members of some genera lack either notopodia (*Biremis* Polloni, Rowe and Teal, 1973), neuropodia (*Enoplobranchus* Verrill, 1879 and *Lysilla* Malmgren, 1866), or lacking all chaetae (*Hauchiella* Levinsen, 1893) (Figure 1l) [2,9].

Members of both Telothelepodidae and Thelepodidae have branchiae as multiple unbranched filaments, originating independently from the body wall on either side of the pairs, 2–3 pairs in thelepodids, on segments II–III or II–IV, always 3 pairs among telothelepodids, on segments II–IV (Figure 2a–c,e–g and Figure 5h). Members of these families are distinguished from each other because telothelepodids have a narrow and proportionally an elongate upper lip, frequently convoluted, very poorly developed ventral glandular areas on anterior segments and distinctly poorly developed neuropodia throughout the body, as low ridges on the anterior body (Figure 2a–d), where notochaetae are also present, and almost sessile pinnules after notopodia terminate. In contrast, members of Thelepodidae have a hood-like, almost circular upper lip and very well developed ventral glandular surfaces of anterior segments, although discrete ventral shields are not observed among these animals; fleshy, well developed neuropodia throughout, the posterior body neuropodial pinnules are frequently well raised from the body (Figure 2e–f) [1,2,9,32,39,40].

Trichobranchids are a group of three genera only, *Octobranchus* Marion and Bobretzki, 1875, *Terebellides* Sars, 1835 and *Trichobranchus* Malmgren, 1866, sharing the character of having neurochaetae on anterior segments as long-handled acicular uncini (Figure 6j), instead of avicular uncini, as in members of all other families, and also poorly developed ventral glandular areas on anterior segments and neuropodia almost sessile on the region with both noto- and neuropodia, and as developed neuropodial pinnules after notopodia terminate (Figures 2h–j and 6n). These animals have a circular, usually flaring upper lip, peristomial lobes are common and the anterior body segments present lobes as low collars of even length around the body, or only ventrally (Figures 2h–j and 5c). An eversible ventral process is present in *Trichobranchus*, in segment 1 [36]. Body regions are well marked in these animals, with notopodia extending only until ~segment XIX or XX, but beginning on segments III–VI, depending on the genus [1,2,9,32]. In *Terebellides*, branchiae are fused into a single structure with two paired lobes that bear lamellae and arise on segments II–IV (Figure 2h,i and Figure 5c) [46]. In contrast, in *Trichobranchus*, branchiae arise from segments II–IV but remain as three pairs of distinct organs (Figure 2j). In *Octobranchus*, there are four pairs present, on segments II–V, which may be digitiform or arranged as a foliaceous structure. Finally, terebellids sensu stricto (s.s.) are unique among Terebellidae s.l. in having neuropodial uncini arranged in double rows on at least some anterior segments (Figure 6l), while animals of all the other families of this group always have uncini in single rows. Terebellids s.s. also have well developed glandular ventral areas of anterior segments, with discrete, unpaired, rectangular to trapezoidal mid-ventral shields, and branchiae, whenever present, originate from a main stalk or at least a single point on the body wall on either side of pairs, and the branchial filaments may be unbranched or branching in a variety of ways (Figures 3a–i and 6a,c,d,f,g) [1,2,9,32,46].

3.3. Phylogenetic Relationships within the Group

A detailed discussion on the hypotheses for the position of Terebelliformia within Annelida through time was provided by [2,47,48]. The latest phylogenetic studies, mostly based on molecular data, suggest terebelliforms are a sister taxon to Arenicolidae Johnston, 1835, and the clade is sister to Clitellata, sometimes with Capitellida, Echiurida and Opheliida, also included in the group [49–51]. This contrasts with the traditional morphological hypotheses, which proposed a sister–taxon relationship between Terebelliformia and Cirratuliformia, grouped together in the taxon Terebellida, which is closely related to Sabellida and Spionida [52,53].

Many of these molecular phylogenies are based on a small number of taxa, and a small number of sequenced genes. Weigert et al. [51], for example, only included two species of alvinellids and one pectinariid, while Zrzavy et al. [50] used one alvinellid, three ampharetids, one pectinariid and two terebellids s.s. This limited number of taxa does not cover the range of morphologies present in the group and often differs from morphological phylogenetic studies. Future studies need a better representation of molecular data from all the families of Terebelliformia, especially of the type species of the genera to continue to resolve the phylogeny of this diverse group of polychaetes.

The most comprehensive phylogenetic study on the phylogenetic relationships within Terebelliformia was performed by Nogueira et al. [1], but was based exclusively on morphological data. The aim of that work was to study the relationships within the Terebellidae s.l., but representatives of the other families of Terebelliformia were also included, as well as three non-terebelliform species, one cirratulid, one spionid and one sabellariid. The authors examined 118 characters in members of 82 species of terebelliforms, including the type species of nearly all genera of Terebellidae s.l., plus the three outgroups, and noticed that all the groups previously considered as subfamilies of Terebellidae had originated independently along the Terebelliformia lineage. As a result, all these groups were raised to the familial level, together with a new family, Telothelepodidae, described therein [1]. According to Nogueira et al. [1], Trichobranchidae is monophyletic, but deeply nested within the Terebellidae s.l., sister to a clade in which Terebellidae s.s. is sister to Alvinellidae/Ampharetidae/Pectinariidae together. All those families originated along the Terebelliformia lineage as follows: Polycirridae (Telothelepodidae (Thelepodidae (Trichobranchidae (Terebellidae s.s. (Alvinellidae (monophyletic Pectinariidae and paraphyletic Ampharetidae)))))). However, the authors stressed that the study was totally focused on Terebellidae s.l., using characters and terminals especially selected for terebellids, but not representative of the diversity of alvinellids, ampharetids and pectinariids; therefore, the relationships between these latter three families had not been properly evaluated [1].

Prior to the study by [1], sister taxa relationships have been suggested between (1) Trichobranchidae and Alvinellidae, the group sister to Pectinariidae, and Ampharetidae and Terebellidae s.l. [54,55]; (2) Alvinellidae and Ampharetidae, and Pectinariidae and Terebellidae s.l., with a plesiomorphic Trichobranchidae, sister to all other terebelliforms [55]; (3) monophyletic Alvinellidae, all other families polyphyletic, except for Trichobranchidae, with a single species included in the study; Pectinariidae is also monophyletic, but out of Terebelliformia [56]. The relationships within Terebellidae s.l. had never been investigated before Nogueira et al. [1], except by Garraffoni and Lana [57,58], who found Trichobranchidae nested within Terebellidae s.l. In their analysis of Terebellidae s.l., Garraffoni and Lana [58] found polycirrids nested within telothelepodids + thelepodids, rendering paraphyletic the traditional Thelepodinae (including species of both thelepodids and telothelepodids, which were regarded as a single family until 2013), and Trichobranchidae sister to Terebellidae.

More recently, a phylogenetic study combining both morphological and molecular data by Stiller et al. [59] suggested a different arrangement for the internal groups of Terebelliformia. The authors first studied transcriptomes of one outgroup plus 20 terebelliform representatives, which included 1 Pectinariidae, 5 Ampharetidae (4 Ampharetinae and 1 Melinninae), 6 Alvinellidae, 2 Trichobranchidae and 6 Terebellidae s.l. (1 Polycirridae, 4 Terebellidae and 1 Thelepodidae), totalling 12,674 orthologous genes, to generate the "backbone" to a more general analysis, with 132 species of terebelliforms (13 Alvinellidae, 49 Ampharetidae (29 Ampharetinae, 5 Amaginae, 8 Amphicteinae and 7 Melinninae), 7 Pectinariidae, 47 Terebellidae s.l. (10 Polycirridae, 27 Terebellidae s.s., 1 Telothelepodidae and 9 Thelepodidae), and 16 Trichobranchidae), combining five genes (three nuclear and two mitochondrial, and not including any of those used for the first analysis) and 90 morphological characters. As a result of the combined analyses, the authors moved the newly erected families of Terebellidae by Nogueira et al. [1] back into the Terebellidae s.l., most of them as subfamilies and found a sister taxon relationship between Terebellidae and Melinninae, raising the latter to familial level, and also between the remaining Ampharetidae and Alvinellidae. In regard to the Terebellidae s.l., the authors found Polycirridae nested within Terebellidae s.s., and synonymised Telothelepodidae with Thelepodidae, keeping the subfamily status of Thelepodinae and Terebellinae, and suggesting the subdivision of the latter into four tribes, Lanicini, Polycirridi, Procleini and Terebellini. However, although the sampling for the combined analysis is very comprehensive, the one used for the first analysis, which was used as a "backbone" to direct the second study, only included 20 species, of which pectinariids, melinnins, polycirrids and thelepodids were all represented by a single species each, and no telothelepodids were included. In addition, Fitzhugh [60–62] thoroughly discussed the philosophical issues in comparing phylogenetic hypotheses generated by different datasets of characters, as made by Stiller et al. [59] to combine the "backbone" with the main analysis. Fitzhugh also argued against the combination of morphological and molecular data, as well as against molecular phylogenies per se, also due to philosophical issues [62]. We consider that these major changes still need to be re-evaluated, given that only one species of Melinninae was included and the limited sampling of species of Telothelepodidae and Thelepodidae and the validity of plotting morphological characters onto genetic trees. For those reasons, we prefer to follow herein the classification proposed by Nogueira et al. [1], which was subsequently confirmed in the phylogenetic analyses of Polycirridae [46], and Telothelepodidae [47]. However, this may change as additional species are added to the dataset after sequencing.

Another phylogenetic study on the relationships within Terebellidae s.s. was performed by Jirkov and Leontovich [63], which focused on the animals with large lateral lobes only, which they suggest form a monophyletic clade in the family, although the reasons for this were not given. The authors included 93 taxa with large lateral lobes and a single "outgroup" species without lobes, *Terebella lapidaria* Linnaeus, 1767, the type species of the family. They also considered the presence of short-handled or long-handled anterior uncini as a specific character, rather than generic, as had traditionally been considered. As a result, the authors considered only the following genera with large lateral lobes as valid: *Axionice* Malmgren, 1866, *Lanicides* Hessle, 1917, *Lanicola* Hartmann-Schröder, 1986, *Pista* Malmgren, 1866 and *Scionella* Moore, 1963, and synonymised under *Axionice* the genera *Betapista* Banse, 1980, *Eupistella* Chamberlin, 1919, *Euscione* Chamberlin, 1919, *Lanice* Malmgren, 1866, *Loimia* Malmgren, 1866, and *Paraxionice* Fauchald, 1972. The authors also changed the traditional diagnoses of *Axionice* and *Pista* (see below), but these changes have not been adopted by other workers.

In summary, the phylogenetic relationships within the group are still being debated as well as the boundaries of some genera. Hutchings et al. [2] provide a synthesis of the phylogeny of the group prior to the studies by Nogueira et al. [1].

3.4. Biological and Ecological Notes on Terebelliforms

3.4.1. Role of Terebelliforms in the Ecosystem

The majority of terebelliforms are tubiculous, living in robust tubes made of sand and sediment grains, which may be within the sediment or more commonly attached to rocks, algae or shells (Figure 7a–f). A few, such as some polycirrids and some alvinellids, lack tubes, instead covered in a mucous sheath. Pectinariids produce very characteristic cone-shaped tubes, using very well calibrated sediment grains (Figure 1a). The alvinellids build tubes on the walls of the vent chimneys, in basaltic cracks with venting (Figure 7f), or live in mucus sheaths at the base of vestimentiferan tubes. In all cases, once the animal dies, the tubes, which are constantly being maintained, tend to break apart, as the mucus binding the shell fragments and sediment particles degenerates.

Figure 7. Some typical habitats of terebelliforms: (**a**) wave-dominated coastline, Cape Leewin, WA, Australia. Photo: Pat Hutchings; (**b**) mangrove area in front of *Rhizophora* roots at Lizard Island. Photo: Gary Cranitch; (**c**) base of *Posidonia australis* beds. Photo: Clay Bryce; (**d**) soft mud, here collected by Van Veen grab. Photo: Pat Hutchings; (**e**) fine sand sampled by dredge. Photo: Jeurgen Freund; (**f**) deep-sea hydrothermal vents, tubes of alvinellids. Chimney wall surface at Tu'i Malila, Lau Basin. Copyright: Chubacarc cruise/Ifremer.

In general, these animals appear to have reduced mobility; however, members of the polycirrid *Biremis blandi* Polloni, Rowe and Teal, 1973 have been seen swimming in mid-water, at depths of 411–597 m, in the Florida Strait and Bahamas [64]. Other species have been observed to swim for a short time when removed from their tubes, presumably an avoidance reaction (Hutchings, pers. obs.; Nogueira, pers. obs.). On the other hand, species such as *Amphisamytha galapagensis* Zottoli, 1983 can apparently live free of the tubes when the material for their construction is scarce in the environment [65]. They then use fibrous structures such as byssal threads from mussels or setae on crabs to host them.

Some species occur in dense aggregations; for example, the ampharetid *Melinna palmata* Grube, 1870 occurs in aggregations up to 9000 ind./m^2 in Arcachon Bay, France [4]. The alvinellid *Alvinella* spp. can also form high-density aggregations on hydrothermal vent chimney walls where it affects the chemical conditions [66] or *Lanice conchilega* (Pallas, 1766), which is also considered as an ecosystem engineer for forming reef-like structures in intertidal sandy substrates, by the aggregation of their tubes [67]. Other terebelliforms, in contrast, form small aggregations or are found as single, solitary individuals.

All the terebelliforms are selective surface deposit feeders [68] gathering food particles with the buccal tentacles, and then conveying these to the mouth, through the ciliated longitudinal tentacular groove. This trophic mode largely modifies marine benthic environments by reworking large amounts of sediments [69] and directly affects their physical and chemical properties [70,71]. Particularly, terebelliforms have a great impact on the amount of organic matter at the water–sediment interface, modifying local hydrodynamics and sediment cohesion [72]. Finally, terebelliforms can influence the structures of benthic communities through tube-building [70].

The Alvinellidae, in addition, have been reported to supplement their deposit feeding diet by collecting particles suspended in the water, by filtering water through their branchiae, as well as feeding on the bacterial residents of the worm tubes [73]. Evidence of such supplementation, however, is lacking and gut contents only revealed mineral particles and bacterial cells gathered from the environment [74]. Although both species of *Alvinella* bear epibiotic bacteria, these do not appear to contribute to the nutrition of the worm hosts. Both species, however, produce structures that allow for the settlement of these bacteria and the association must be beneficial to the host [74].

Most Terebelliformia are dioecious with no morphological differences between males and females, except at the time of spawning when the mature gametes colour the body, where females may be pinkish or greenish, and males are typically cream coloured. In alvinellids, however, reproduction appears continuous; males bear a pair of modified buccal tentacles and females have genital pores [31]. In some taxa, the genital papillae may vary between sexes, as well as the distribution of glandular areas (Figure 6a) [36].

To date, no evidence of asexual reproduction has been observed, although all are capable of regenerating posterior ends, branchiae and buccal tentacles. Gametes are proliferated from the germinal epithelium, often associated with the nephridia, and released into the coelomic cavity, where vitellogenesis and spermatogenesis occur. Synchronised spawning occurs through the nephridia, and spawning varies from only one or two days to discrete periods over several months.

In alvinellids, the presence of sperm ducts, spermathecae and oviducts have been reported, lending support to continuous gamete production, episodic release, pseudocopulation, and internal fertilization [75]. Compared to other studied Terebelliformia, alvinellid sperm cells are highly modified entaquasperm, devoid of acrosome and sometimes of flagella, providing further support for internal fertilization in this family [76].

Among the other terebelliforms, mass spawning occurs in some taxa, while others produce a lecithotrophic larva, with varying planktonic larval phase durations (PLDs), and at least one species has a direct development within a cocoon, with larvae released at the 15 chaetiger stage [74]. Although few species have been studied, most of them produce large yolky eggs, and the embryo probably does not feed in the plankton [8,74], except in pectinariids, which may have a planktotrophic larva, capable of feeding through a capture system involving the generation of a current and the production of mucus [77]. In *Alvinella pompejana* Desbruyères and Laubier, 1980, the conditions the adults experience in their environment are actually harmful to the developing larvae and they need to encounter milder conditions to survive and develop properly [75]. Erpochaete larvae of *Paralvinella grasslei* Desbruyères and Laubier, 1982 as young as 13 segments (with a single pair of branchiae) have been captured near adults of this species, suggesting a very early recruitment following a planktonic phase [7].

Few species of ampharetids have been studied with regard to their reproduction (e.g., [65,78–85]. Some shallow water species reproduce annually, and all produce large yolky eggs, which spend only a few days in the plankton. *Melinna palmata*, for example, may spend 6 days living in the plankton before settling and building a tube at the 3-chaetiger stage [8]. *Hobsonia florida* (Hartman, 1951) has been reported as having larval development in the maternal tube and a 2-chaetiger stage leaves the tube, settles on the nearby sediment and builds its own tube [8]. Studies conducted with deep-sea species, from both hydrothermal vents and organic falls, suggest continuous reproduction and rapid maturation, possibly as a reflection of the ephemeral conditions of these chemosynthetic habitats [65,84,86].

So, in summary, among terebelliforms, we have a variety of reproductive strategies (see references in [2,7,8]).

3.4.2. Distribution and Biogeography

Historically, species were described from Europe with most of the type species of the 137 genera sampled from these waters, except for the Alvinellidae, which is restricted to deep-sea hydrothermal vents, and was erected in 1986 (although the first species was described in 1980, as an aberrant Ampharetidae *Alvinella pompejana*).

In the 1980s, the centre of gravity moved to the southern hemisphere with taxonomists based in Australia (Glasby, Hutchings), South America (Carrerette, Londoño-Mesa, Nogueira) and those involved in Pacific expeditions (Fiege, Imajima, Reuscher) (Table 1). In these regions, polychaete workers had to review earlier expeditions, which were mainly housed by European institutions, where the material was deposited, and over time some has been lost or damaged. In some cases, material from an expedition was deposited in several institutions and locating this material can be challenging. All the scientists working on the material collected during these expeditions were based in Europe and they tended to identify much of this material as European species even though they had been collected thousands of kilometres away in very different habitats and temperature regimes. This led to the idea that many polychaete species were cosmopolitan and certainly later European workers such as Fauvel [86] reinforced this view and recorded the widespread distributions of many species. Later this was reinforced by the catalogues of polychaetes produced by Hartman [87] and by Day [88]. An example of this is provided by Hutchings and Glasby [89] who analysed the species list of terebellids s.l. produced by Day and Hutchings [90] in their checklist of the polychaetes recorded from Australia and New Zealand, which was based entirely on the literature and listed 32 species in 17 genera. Hutchings and Glasby [89] showed that only 14 of these occurred in Australia, the rest having been misidentified as European species. They further analysed the diversity of Australian terebellids, as they were known in 1991, which was represented by 78 species in 27 genera, and of these 67 (85.89%) species were Australian endemics, and of the remaining 11, nine were restricted to the Indo-Pacific and only two were found in the northern hemisphere. Since that analysis in 1991, the remaining two species recorded from the northern hemisphere have been described as Australian endemics as well [91]. Since that study, many more species have been described from Australia. Similar patterns are also evident in other polychaete families, so this is certainly not solely a characteristic of terebellids but widespread across all polychaete families found in Australia.

More recently, it has been recognised that most species actually have discrete distributions, unless proven otherwise [92], and while many genera are widely distributed, it is at the species level that discrete distributions occur. However, in many parts of the world, taxonomists and ecologists still identify their material using the well-illustrated monographs of Faune de France ([86,93], and Southern Africa [88], despite their samples being collected many thousands of kilometres away from France or Southern Africa. This has tended to reinforce the concept that polychaete species are cosmopolitan in their distribution. While this has been shown not to be true—for example, *Terebellides stroemii* Sars, 1835 is now known to represent a highly speciose group—as the nominal species is restricted to a very small area in Western Norway [5,94]. In many cases, this is also because no regional keys exist in many parts of the world, and so a student has little option but to use keys from other regions. Once those names become enshrined in the local fauna, then subsequent workers just repeat them, explaining why species such as *T. stroemii* have been so widely reported.

Our current knowledge on the diversity of Terebelliformia shows great variation from some regions of the world to others. While places such as Europe and North America have been investigated for centuries, others are still virtually unknown, as is most of the African coast and the Eastern Indo-Pacific. This is largely due to the presence of more researchers based in Europe and North America than in other regions of the world, and also for the availability of financial resources available for biological research in these regions.

However, it should be stressed that even regions where the fauna has been studied, it is now being re-examined with molecular tools, as numerous complexes of species have been found, resulting in a much greater number of species than previously considered. For example, French coastal waters are well-known areas, studied for several centuries by early taxonomists and benthic ecologists. However, studying numerous terebelliforms (spaghetti worms), within the collaborative Spaghetti Project, using modern tools, such as the scanning electron microscope and molecular analysis, has revealed the existence of more than 20 species new for science [95–99]. We assume that this marked discrepancy in our knowledge of the diversity of polychaetes in many parts of the world is common for most if not all polychaete families.

To facilitate a review of the distribution of terebelliforms, we chose to look at various biogeographic schemes which have been suggested over the years ([100,101] and we are following Spalding et al. [10]. In an effort to strategically plan exploitation and marine conservation measures, Spalding et al. [10] suggested a classification for the marine biogeographic regions, the Marine Ecoregions of the World (MEOW), dividing coastal and shelf areas into 12 Realms, 62 Provinces and 232 Ecoregions (Figure 8). As said above, our analysis of the geographic distribution of terebelliforms follows that biogeographic classification. We have compiled a list of all terebelliforms described and, just using their type locality, allocated them to each of these regions and they are plotted in Figure 8. Obviously, these numbers are influenced by the number of taxonomists working in each area, which has varied over time, and the resources available.

As expected, the most diverse realm is the Central Indo-Pacific, with 233 species of terebelliforms described from the region (Table 2, Figure 8). This realm, corresponding to the area from the South China Sea, through the Pacific side of Indochina Peninsula, Philippines, Indonesia, Papua, Melanesia and Micronesia islands, Northwestern, Northern and Northeastern Australia, including the northern Great Barrier Reef (Figure 8), is considered as the world biodiversity hotspot for many groups of marine animals and is referred to as the Coral Triangle [102]. The following most diverse realms match the observations discussed above, as the regions with more polychaete taxonomists and economic resources are also the most diverse, Temperate Northern Atlantic (210 species), Temperate Northern Pacific (175 species), Tropical Atlantic (95 species), Southern Ocean (82 species), Temperate Australasia (76 species) and Temperate South America (67 species) (Table 2 and Figure 8).

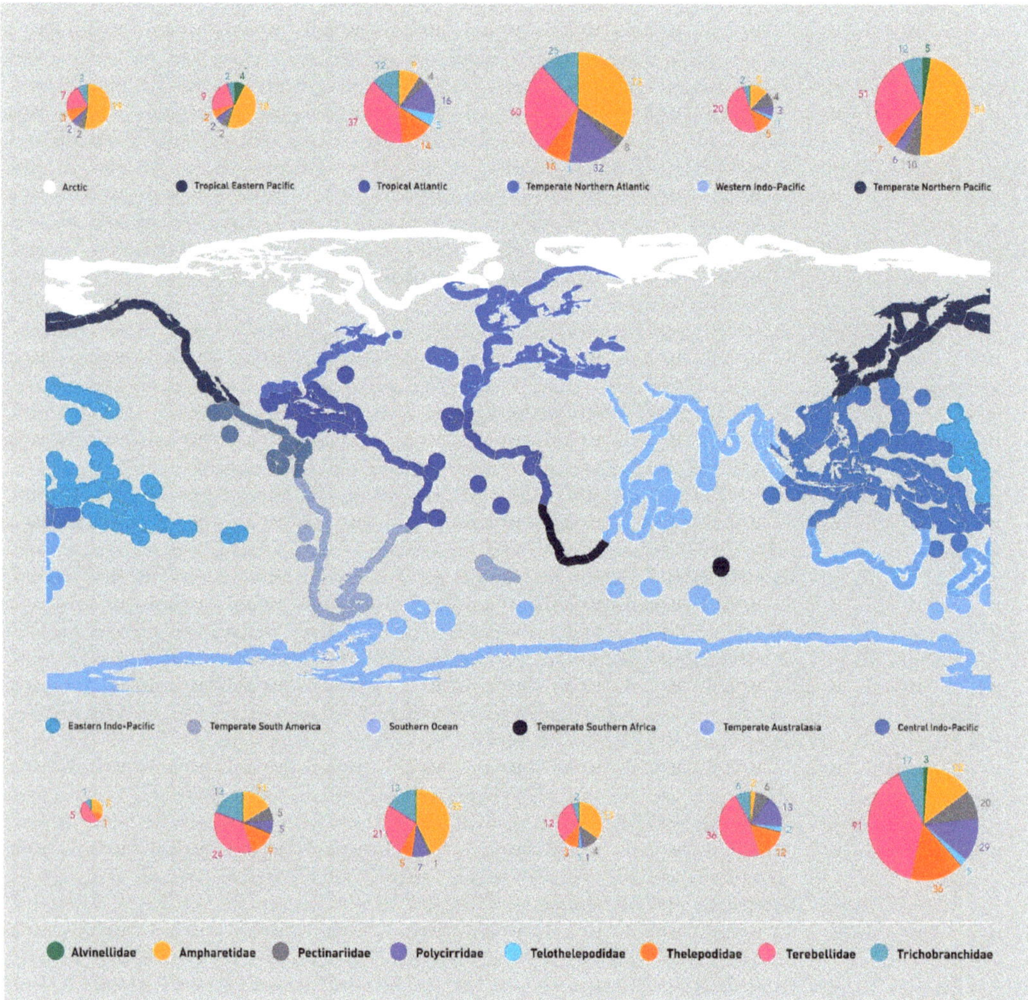

Figure 8. Marine Ecoregions of the World following [10] with number of species of terebelliforms described from each realm.

The fauna from Europe, corresponding to part of the Temperate Northern Atlantic realm, has been thoroughly investigated since Linnean times. Despite this long history, many new species are still being found [95–99], as discussed above for French terebelliforms. North America, corresponding to the remainder of the Temperate Northern Atlantic realm, and parts of the Temperate Northern Pacific and Tropical Atlantic realms, certainly had the most taxonomists and economic resources during the 20th century. Temperate Northern Pacific and Tropical Atlantic realms also include some countries which have dedicated resources to intensively study invertebrate taxonomy in the last few decades, such as Russia, Japan and China in the first case, and Mexico, Brazil and Colombia in the latter. The Southern Ocean was investigated by earlier expeditions, but as many countries established scientific bases in Antarctica, this has led to more taxonomic studies. Temperate Australasia and Temperate South America also include countries which have put an effort on the study of marine fauna in the last few decades, with large projects carried out, such as in Australia and New Zealand, in the case of Temperate Australasia, and Chile, Argentina and Brazil, in Temperate South America.

Table 2. Distribution of Terebelliformia around the world, following the marine regionalization created by Spalding et al. [10], and bathymetric variation of the terebelliform families, as well as the deepest records of each group.

	Alvinellidae	Ampharetidae	Pectinariidae	Polycirridae	Telothelepodidae	Thelepodidae	Terebellidae	Trichobranchidae
Realms by Spalding et al (2007)								
Arctic		19	2	2		3	7	3
Central Indo-Pacific	3	32	20	29	5	36	91	17
Eastern Indo-Pacific		3				1	5	1
Suthern Ocean		35	1	7		5	21	13
Temperate Australasia		2	6	13	2	12	36	6
Temperate Northern Atlantic		73	8	32	1	16	60	25
Temperate Northern Pacific	5	84	10	6		7	51	12
Temperate South America		11	5	5		9	24	13
Temperate SouternAfrica		13	4	1	1	3	12	2
Tropical Atlantic		9	4	16	5	14	37	12
Tropical Eastern Pacific	4	18	2	2		2	9	2
Western Indo-Pacific		5	4	3	1	5	20	2
Bathymetric distribution								
Inter tidal to 100 m		72	20	66	14	57	87	34
100–500 m		52	5	9		6	13	21
500–1000		42	2	3		4	1	3
1000–2000	4	38		1		4	1	5
2000–3000	7	23	2			2		5
3000–4000	1	21				1	1	
4000–5000		25					1	3
5000–6000		9						4
6000–7000		1						
8000–9000		1						
9000–10,000		1						
Deeper records	*Alvinella pompejana*, 2593 m	*Anobothrus auriculantus*,9584 m	*Petta assimilis*, −3000 m	*Polycurrus nonatoi* 1904 m	*Prathelepus anomalus* and *Rhinothelepus mexicanus*, 91 m (for both)	*Streblosoma chilensis*, 3950 m	*Pista torcuata* 4540 m	*Terebellides bulbosa* and *T. ginkgo*, −5200 m (for both)

On the other hand, the least diverse realms are also those with fewer taxonomists and frequently fewer economic resources. Western Indo-Pacific (including eastern Africa, Red Sea, Persian Gulf, Pakistan, India, Sri Lanka and the Indian side of the Indochina Peninsula) tally only 40 species. Temperate Southern Africa tallies only 37 species, in spite of the efforts by Day [90], but this last author reported many European species for South Africa, as did Fauvel [103], for the region of India and Sri Lanka. Tropical Eastern Pacific, which includes the Pacific side of Tropical America, comprises 38 species only and this is attributable to Mexican and Colombian researchers. Only 36 species are described from the Arctic, which is somewhat surprising, considering the Scandinavian and Russian scientists who have been working in the region since the 19th century, although many Northern European and North American species are reported for this region. Additionally, the Eastern Indo-Pacific realm, the poorest of all, including the region from Hawaii and Marshall Islands through Polynesia and the Mariana Islands to Easter Island, with only 10 species, but also with many records from the West Indo-Pacific (Table 2; Figure 8). We suggest that some of these patterns of diversity may just reflect the lack of sampling rather than a reflection of their true diversity.

As discussed below, many genera of terebelliforms are monotypic, several of which have never been sampled since they were first collected. These descriptions are very brief and do not mention several characters currently considered important for the taxonomy of the group. This is further complicated by the loss of the type of material or it is damaged in such a way that those characters cannot be assessed. The uncertainty about the identity of those genera obviously imposes several problems in regard to the knowledge on the distribution of those animals and several genera are considered as endemic to certain regions which may change as more studies are carried out.

Most of the non-monotypic genera of terebelliforms are widespread through [10] realms. One non-monotypic genus which apparently has a more restricted distribution is *Reteterebella* Hartman, 1963, with three species. The type species, *R. queenslandia* Hartman, 1963, described from Heron Island, Great Barrier Reef (Central Indo-Pacific), but apparently restricted to that region [37], *R. aloba* Hutchings and Glasby, 1988, from South Eastern Australia (Temperate Australasia) and *R. lirrf* Nogueira, Hutchings and Carrerette, 2015, described from Lizard Island, also on the Great Barrier Reef. The habitats in which *R. lirrf* and *R. queenslandia* occur are very different, the first being found in crevices deep down in boulders, whereas *R. queenslandia* occurs on reef flat with its flimsy tube attached to the underside of boulders. However, it should be stressed that reefs between Heron and Lizard have not been well sampled.

Another genus *Hadrachaeta* Hutchings, 1977 is only known by the type species and has only been found in the front of mangroves in Broken Bay, NSW and Moreton Bay, Queensland and, despite extensive sampling in these habitats along the east coast of Australia, no other material of this species has been found (Hutchings, pers., comm.).

3.4.3. Distribution of Terebelliforms with Depth

There is no generally accepted definition of the deep sea. One can consider depths below the euphotic layer (i.e., 300 m) as a natural upper limit to the deep sea. Overall, the deep-sea remains poorly explored outside of specific areas, such as cold seeps, hydrothermal vents, and organic falls. The typical lifestyle of terebelliforms also makes their capture unlikely by the gear typically used to sample the deep sea. In particular, species that live buried in the sediment or attached to rocks are often missed by dredges and beam trawls used by most recent general study programmes. This was clearly demonstrated by Gunton et al. [104] who studied the polychaete fauna from depths off the east coast of Australia (1000–4000 m), and while ampharetids were very well represented, with over 300 specimens belonging to more than six species, 10 specimens and 2 species of pectinariids were also present and described (*Petta investigatoris* Zhang, Hutchings and Kupriyanova, 2019 and *P. williamsonae* Zhang, Hutchings and Kupriyanova, 2019), and far fewer specimens of Terebellidae s.l. were collected, representing four genera but all in poor condition.

The deepest record among Terebelliformia comes from a species of ampharetid *Anobothrus auriculatus* Alalykina and Polyakova, 2020, found at 9584 m depth (Table 2). Ampharetids are well represented in the abyss and in different deep-sea habitats, with more than half of the known ampharetid species occurring below 500 m deep (Table 2). Several ampharetids are exclusively found in the deep-sea, in addition to some specialised representatives associated with chemosynthetic environments, such as some known species of the genera *Amage* (1 species., at cold seep), *Amphisamytha* (7 species., at cold seeps and hydrothermal vents), *Anobothrus* (3 species., at cold seeps and hydrothermal vent), *Decemunciger* (1 species., on decaying wood), *Endecamera* (1 species., on decaying wood), *Glyphanostomum* (2 species, at cold seep and sedimented hydrothermal vents), *Grassleia* (1 species at sedimented vents and cold seeps), *Paramytha* (2 species on decaying bones and sedimented hydrothermal vents), and *Pavelius* (3 species at cold seep and sedimented hydrothermal vents) [105].

The alvinellids are restricted to hydrothermal vents. All the species are exclusively found at hydrothermal vents in the Eastern and Western Pacific (Table 2), but a species was recently reported from vents in the Indian Ocean [7]. As a result, alvinellids are exclusively found at depths greater than 1500 m and can reach ~3600 m (Table 2).

Hydrothermal vents and cold seeps are also home to some terebellid species described recently (e.g., *Neoamphitrite hydrothermalis* Reuscher et al. 2012, and *Streblosoma kaia* Reuscher, Fiege and Wehe, 2012, for hydrothermal vents, and *Pista shizugawaensis* Nishi and Tanaka, 2006 for cold seeps; see [106] (Table 2). The Telothelepodidae, in contrast, so far have only been found at shallow depths, the deepest records (~91 m) for *Parathelepus anomalus* (Londoño-Mesa, 2009) and *Rhinothelepus mexicanus* (Glasby and Hutchings, 1986) (Table 2). In general, most polycirrids are found in intertidal to shallow water habitats, the deepest record being for *Polycirrus nonatoi* Carrerette and Nogueira, 2013, found from ~30–1900 m deep (Table 2). Trichobranchidae are also well represented in the deep sea, frequently by a large number of species (Table 2), despite sometimes being considered low in abundance, belonging to the genus *Terebellides*; the deepest records come from *Terebellides bulbosa* Schüller and Hutchings, 2012 and *T. gingko* Schüller and Hutchings, 2012, ~5200 m deep for both, from animals collected at the Brazil Basin [107] (Table 2). This genus is highly speciose, with many endemic species, while others can have wide distributions, indicating dispersion over long distances [107]. Both Pectinariidae and Thelepodidae are much more diverse intertidally to ~100 m, but in both families a few species adapted to the deep sea have been described, the deepest records being *Streblosomma chilensis* (McIntosh, 1885), for thelepodids, registered at ~4000 m deep off Chile, and the pectinariid *Petta assimilis* McIntosh, 1885, found ~3000 m deep, off Crozet Islands (Table 2).

3.5. Evolution of Methods Used to Describe Species

Earlier taxonomists in the 18th and 19th centuries worked with very rudimentary optical instruments, sometimes only a little more than a magnifying glass, capable of low magnifications. Nevertheless, albeit with limited resources, those authors did an amazing job. Except for the Alvinellidae, Telothelepodidae and Thelepodidae, all other families, 35% of the genera and 23% of the currently valid species of Terebelliformia, were described in the 18th and 19th centuries (see above). Those descriptions are frequently criticized for their simplicity, but they reflect the state of knowledge at those times, when the authors considered enough to define species characters which, nowadays, frequently do not allow even for the recognition of the genus. Additionally, it is noteworthy how, in spite of these instruments, some of those earlier descriptions included minutely detailed drawings of chaetae, showing structures which can only be clearly seen under the SEM, a technology that was obviously developed much later.

A great improvement on taxonomists' instruments came in the 20th century, first with more powerful compound optical microscopes, with techniques such as phase contrast, allowing for a much better visualization of chaetal ornamentation, together with better software to capture the images directly from the microscopes and process them, largely replacing traditional line drawings (Figure 9).

Then, from the end of the 20th century through to today, Scanning Electron Microscopy (SEM) provides a much better view of the surface of microscopic structures, such as chaetal ornamentation (Figure 9); Transmission Electron Microscopy (TEM) brought information on cellular ultrastructure; confocal microscopy allowed us to peer deep into the tissues and highlight specific organs; molecular tools became available to distinguish the taxa at the genetic level (DNA and RNA), providing much more detailed and complete descriptions.

All this has greatly increased our knowledge on the diversity of terebelliforms, as for all other polychaetes, with many more morphological and molecular characters available to characterise the taxa, allowing for the recognition of complexes of cryptic species and alien species, for example, opposing the traditional view of species with wide distributions or even cosmopolitan [5,92].

Molecular data are extremely useful to delimit new species or even identify valid species, however it must be accompanied by voucher specimens and preferably be obtained from animals from the type localities of the species, in the case of those already described. Molecular data of misidentified species can generate much confusion. Additionally, in most cases, type species of the genera were not sequenced yet and cannot be included in the resulting phylogenies, compromising all the results obtained. Molecular studies on Terebelliformia so far have resulted in 222,406 sequences available for Alvinellidae in Genbank (mostly transcriptomics and phylogenetic markers), 1011 for Ampharetidae, 1298 for Pectinariidae, 2588 for Terebellidae s.l., and 1476 for Trichobranchidae, considering mitochondrial and nuclear gene markers (Figure 10) (Table 3).

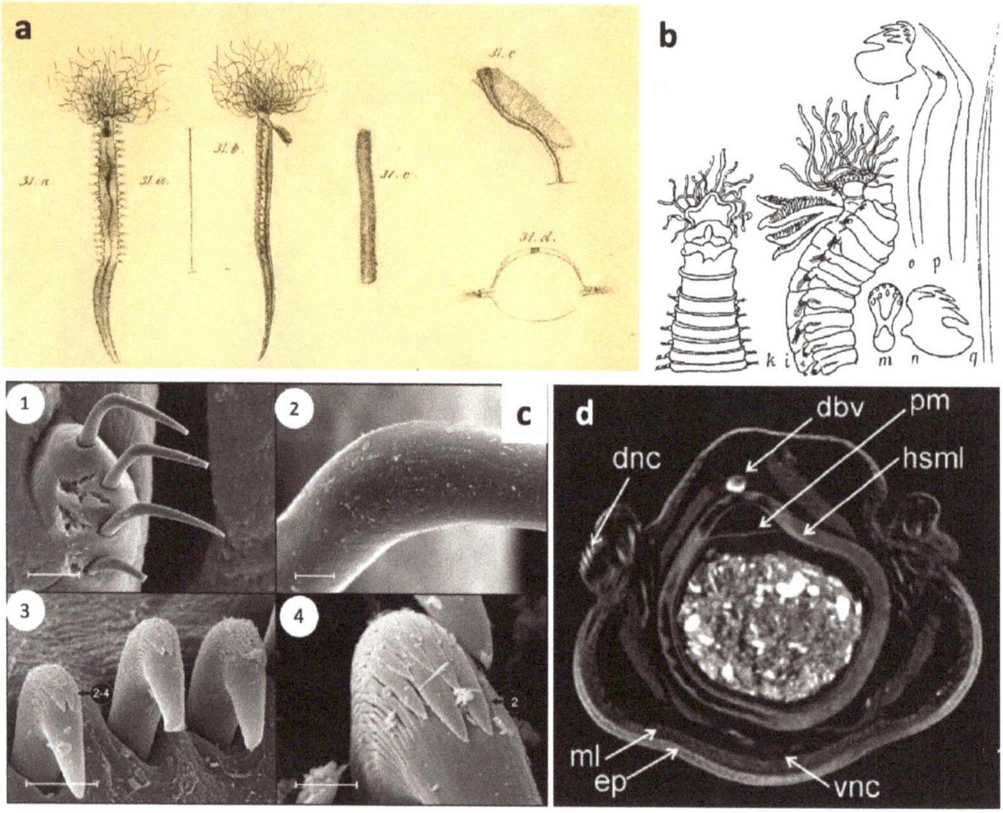

Figure 9. Evolution of appreciation of morphological traits important in taxonomy and techniques used, with *Terebellides stroemii* as an example. (**a**) Illustrations of the original description by Sars [108] (part of plate 13). (**b**) Later observations focused on optical microscopy of the chaetae (setae), here summarized by Fauvel [86]. Key: i side view of anterior part; k ventral view; l–n, uncini front and side view; o, thoracic ventral hook; p, geniculate chaeta; q, dorsal chaeta. (**c**) Scanning Electron Microscopy of geniculate chaetae (1) and detail of the bend (2), thoracic uncini (3) and detail of the teeth (4) (Parapar et al. [109]). (**d**) Micro-computed tomography (μCT) allows cross sections to look at the anatomical level. Here, section at the level of thoracic chaetiger 9. dnc, dorsal notochaetae; dbv, dorsal blood vessel; hsml, hind stomach muscle layer; ep, epidermis; ml, muscle layer; pm, peritrophic membrane; vnc, ventral nerve chord. (Modified after Parapar and Hutchings [110]). Scale bars a 50 μm, b 5 μm, c 10 μm, and d 3 μm.

Morphology-based polychaete taxonomy is largely based on external characters, particularly in Terebelliformia. Hessle [21] suggested a classification based on the structure of nephridia, however this requires dissection of the specimens, which is not feasible with museum material. However, a technique recently developed, computerized microtomography (μCT), scans the specimens and gives amazing 3D images of their internal anatomy (Figure 9), not causing any damage to the specimens, thus allowing type material to be examined. So far, few terebelliforms have been examined under the μCT, but as more are subjected to such analyses [111], additional morphological characters will certainly be found, increasing our knowledge on these animals.

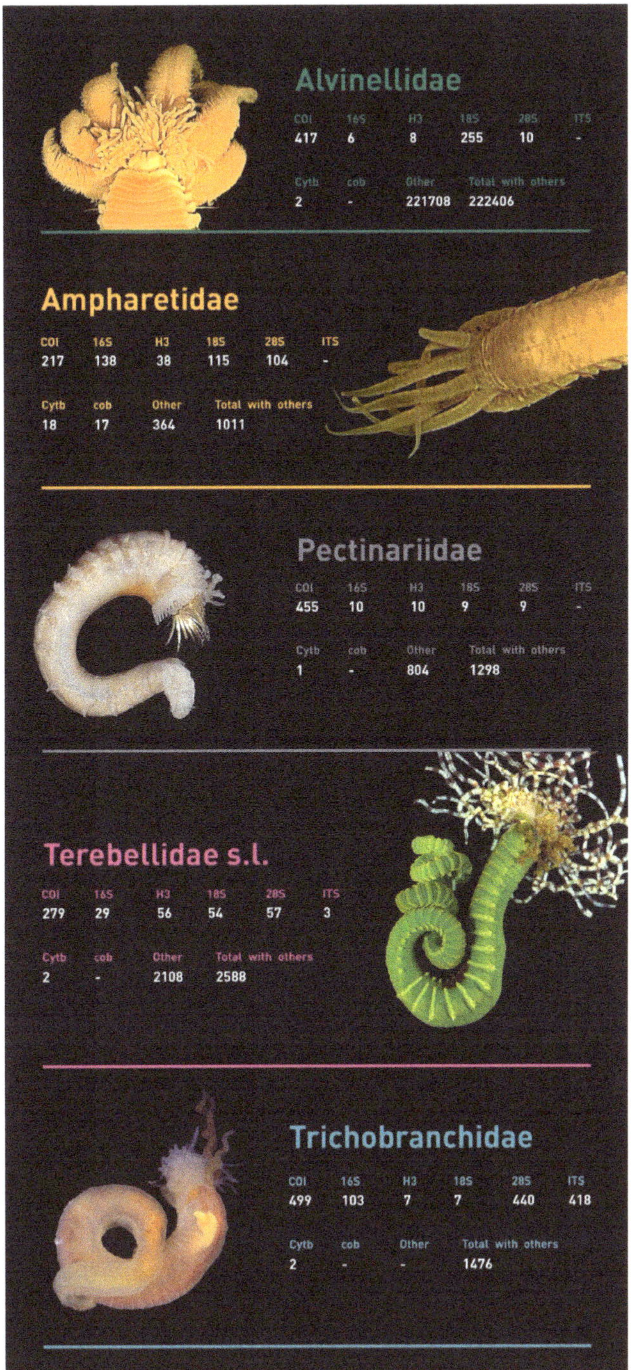

Figure 10. DNA sequences for Terebelliformia.

Table 3. Sequences available in Genbank for each main group of Terebelliformia.

	COI	16S	H3	18S	28S	ITS	Cytb	cob	Other	Total	% Sequences with Voucher
Alvinellidae	417	6	8	255	10	0	2	0	221,708	222,406	0.0
Ampharetidae	217	138	38	115	104	0	18	17	364	1011	53.9
Pectinariidae	455	10	10	9	9	0	1	0	804	1298	7.6
Terebellidae s.l.	279	29	56	54	57	3	2	0	2108	2588	19.7
Trichobranchidae	499	103	7	7	440	418	2	0	0	1476	793

3.6. Knowledge Gaps and Challenges for the Future

3.6.1. Poorly Known Regions of the World

As discussed above, some regions of the world have their local fauna of terebelliforms poorly known, as reflected by the low number of species originally described from those areas. In most cases, they correspond to poorly investigated areas of the world, such as the African coast (except for the Mediterranean part), Southern and Southeastern Asia, the western side of tropical America, corresponding to the Pacific Latin America shore, and all the Eastern Indo-Pacific realm, including the region from Hawaii and the Marshall Islands through Polynesia and the Mariana Islands to Easter Island. Those areas in most cases correspond to developing countries and/or with few institutions investigating invertebrate biodiversity. In some, however, the areas have been sampled and studied, but the identification of the specimens was made based on traditional monographs from other regions of the world, such as France [86,93], or South Africa [88], and resulting in many so called "cosmopolitan" species being recorded [92,112], whereas in fact they actually represent undescribed species. Even worse is that these names become incorporated into the ecological literature with no discussion as to the likelihood that a European species is present in China, for example (see [113]). With taxonomic studies of the fauna of Africa, India, China and other countries from SE Asia, and the Pacific side of America, the number of new species will certainly increase in the next decade or so and will mirror the tendency of the last decade (Figure 4).

Overall, the deep-sea also needs to be better explored, especially in areas that are not influenced by chemosynthetic local primary production. Better adapted gear for the sampling of sediment in which some species live may also be designed or adapted from other existing equipment.

Our knowledge of the deep-sea representatives of the terebelliforms, however, will soon expand as programmes are being conducted by many institutions around the world, such as the "Tropical Deep-Sea Benthos" series of cruises carried by the Museum National d'Histoire Naturelle, Paris, which have recently accessioned a large number of specimens to their collections. The use of tools, such as remotely operated vehicles (ROVs), has also allowed the targeted collection of deep-sea samples, and these are making their way to taxonomists around the world.

3.6.2. Species Complexes

Recently, with the rejection of cosmopolitanism [92] and the wide use of modern tools such as SEM imaging and molecular analysis, scientists have re-examined well-known species from well-known areas in Europe, resulting in the description of several cryptic species, as new to science. Consequently, the number of terebelliform species continues to increase, and many species previously considered widely distributed have become restricted to smaller areas. One of the best examples is *Terebellides stroemii*, reported from all around the world but almost certainly restricted to Norwegian waters [5]. These authors, using molecular data, showed the presence of more than 25 species in the Northeastern Atlantic alone, hidden behind this so-called "cosmopolitan" species. Parapar et al. [94] has just formally described five of these species identified by Nygren et al. [5]. By launching the Spaghetti project, Lavesque and collaborators are revising all French species of Terebellidae s.l. This project has allowed them to describe nine new species of Trichobranchidae [99], three species of Thelepodidae [98] and eight species of Polycirridae [97] from French waters, an area historically well studied by early polychaetes workers (Audouin, Caullery, Fauvel, Gravier, Quatrefages, Rullier, Saint-Joseph, Savigny, etc.). A subsequent paper will document the diversity of Terebellidae from French waters (Lavesque et al. in prep.).

3.6.3. Taxonomic Issues Which Need to Be Resolved

Several genera of terebelliforms are monotypic, known only from the original descriptions, which do not include many characters important for the taxonomy of these groups, and type material is lost, damaged or cannot be located. In many cases, the material was collected by earlier expeditions and corresponds to species described, for example, by Grube, Müller, Lamarck, McIntosh, Chamberlin and Caullery. In some cases, the descriptions and illustrations are such that it is impossible to define the genus and, in these cases, they must be declared as *nomen dubium*, or indeterminable, at least until more material from the type of locality is collected and a neotype designated. Currently the genera *Paralanice* Caullery, 1944, *Opisthopista* Caullery, 1944 and *Spiroverma* Uchida, 1968, in the Terebellidae s.s., cannot be defined. Ebbe and Purschke [8] also list the monotypic genera, *Aryandes* Kinberg, 1866, *Rytocephalus* Quatrefages, 1866 and *Uschakovius* Laubier, 1973, as of doubtful affiliation. In some of the other cases, genera are not well known and Nogueira et al. [1] list those which could not be included in their phylogenetic study as the type material was either poorly preserved or too incomplete for scoring.

Another example is *Hadrachaeta* Hutchings, 1977. Although the type of locality has been extensively sampled through the years, since the original description, no additional specimens of *H. aspeta* Hutchings, 1977 have been obtained (Hutchings, pers. obs. [1]), and the type of material has been dissected several times, removing important diagnostic characters.

Another issue is whether some characters should be regarded as generic or species characters. These include the number of pairs of branchiae; in some genera, such as *Nicolea* Malmgren, 1866, they have two pairs, whereas in other genera the number of pairs is used to distinguish between species, such as in *Pista* Malmgren (2–3) and *Terebella* Linnaeus, 1767 (2–3, although the segment on which they occur can vary).

In *Pista*, the type of branching of the branchiae is a specific character. However, the genus *Pistella* which has only one pair of branchiae resembling some *Pista* species has recently been synonymised with *Pista* by Jirkov and Leontovich [63] but lacks the long-handled uncini characteristic of *Pista*. This is complicated by the type species of *Pista* (*Amphitrite cristata* Müller 1776) which was described as having one pair of branchiae, while Malmgren who erected the genus *Pista* and designated *P. cristata* (Müller 1776) as the type species, stated it has two pairs of branchiae, and no type material exists. However, this synonymy has between *Pista* and *Pistella* has not been accepted by other workers, and Hutchings et al. [2] record 76 species currently assigned to *Pista*, whereas the genus *Pistella* has four species.

Another issue which needs to be resolved is the development of long-handled uncini on thoracic neuropodia, which occur in several terebellid genera and their actual structure. Jirkov and Leontovich [63] have also suggested that all genera with long-handled uncini be synonymised with *Axionice* and that such structures are specific and not generic characters. This hypothesis has not been accepted but highlights the need for more developmental studies to actually study the development of the branchiae and the chaetae as the larvae settle and become juveniles. Similarly, the development and homologies of the peristomium and prostomium needs to be carefully investigated by developmental studies. Finally, the development of the anterior lateral lobes needs to be examined in detail, as their shape, orientation and the segment on which they occur appear to be very useful specific characters in many genera, although Jirkov and Leontovich [63] have suggested that all genera with large lateral lobes be synonymised, although they do not explain why this should happen.

A final issue concerns the genus *Pseudothelepus* Augener, 1918. Augener described this genus for *P. nyanganus* Augener, 1918, from the Tropical Atlantic coast of Africa. Later, Hartman [74] incorrectly synonymised *P. nyanganus* with *Sabellides oligocirra* Schmarda, 1861, described from the Caribbean, keeping the validity of the genus *Pseudothelepus* and changing the type species to *P. oligocirrus*. Unaware that the name was preoccupied, Hutchings [26] described an unusual thelepodid from Houtman Abrolhos Islands, Western Australia, as a new genus and species, which she named *Pseudothelepus binara* Hutchings, 1997. One of us (J.M.M.N.) examined the type of material of the three species and verified that all three are separate, valid species, rejecting the synonymy between *P. nyanganus* and *S. oligocirra*. However, both *P. nyanganus* and *S. oligocirra* are species of *Streblosoma*, and therefore *Pseudothelepus* is not valid. *Pseudothelepus binara*, in contrast, is a very different species, which justifies the erection of a new genus, since the original name is preoccupied, although that still requires phylogenetic confirmation.

So, in summary, not only will new species continue to be described around the world, but a more robust discussion needs to be had on the way in which generic and specific characters are defined, as well as better descriptions of those type species, which are currently inadequate. Ideally these descriptions will be based on neotypes and ideally with associated molecular data.

4. Discussion

As our taxonomic knowledge of this large group of polychaetes (both in terms of diversity and abundance) continues to increase, we need to develop online resources to make these data widely available to the wider biological community. Currently, online keys to the families are being developed and will be uploaded when completed (Kupriyanova et al. in prep.), which include all annelid families and genera as well as Australian species. Similar guides need to be developed for other parts of the world and the views that old monographs such as [86,93] and [88] should not be used and instead retained as historical documents [112] should become widely accepted.

An initiative in Australia could be developed elsewhere. The Atlas of Living Australia (https://www.ala.org.au/) is regularly updated by all the State natural history museums who upload their registered collection onto ALA. These data are all specimen based and you can interrogate the data and download distribution maps, as shown in Figure 11, which shows all the terebellid s.l. data from around Australia and indicates the number of species recorded around the coast which have all been checked by Hutchings and her colleagues. Similar analyses could be carried out in other parts of the world, but one needs to check the validity of the original identifications.

For example, if such data from terebellids as a selective deposit feeding group are combined with other polychaete families, which are filter feeders, such as the sabellids/serpulids, and opportunistic feeders, such as nereidids, for which the taxonomic data are good, one would be able to characterise benthic communities. Such data would be invaluable when developing zoning plans for marine national parks, which currently are often based on physical parameters, such as depth, sediment type, surrogates, such as seagrass beds, sponge gardens, coral reefs, and with limited biological data, such as fisheries data. Yet, the benthic communities dominate these parks and play a crucial role in the marine ecosystem and are barely considered. The sort of data which can be extracted from ALA could provide valuable data to improve the representativeness of marine parks and help develop monitoring programs to ensure that such plans are effectively conserving their biodiversity. Critically important is the fact that climate change is impacting our marine communities.

If we, taxonomists, can provide this sort of data to ecologists, marine managers, this may enhance our ability to attract funds to continue our research and to facilitate the training and mentoring of the next generation of taxonomists.

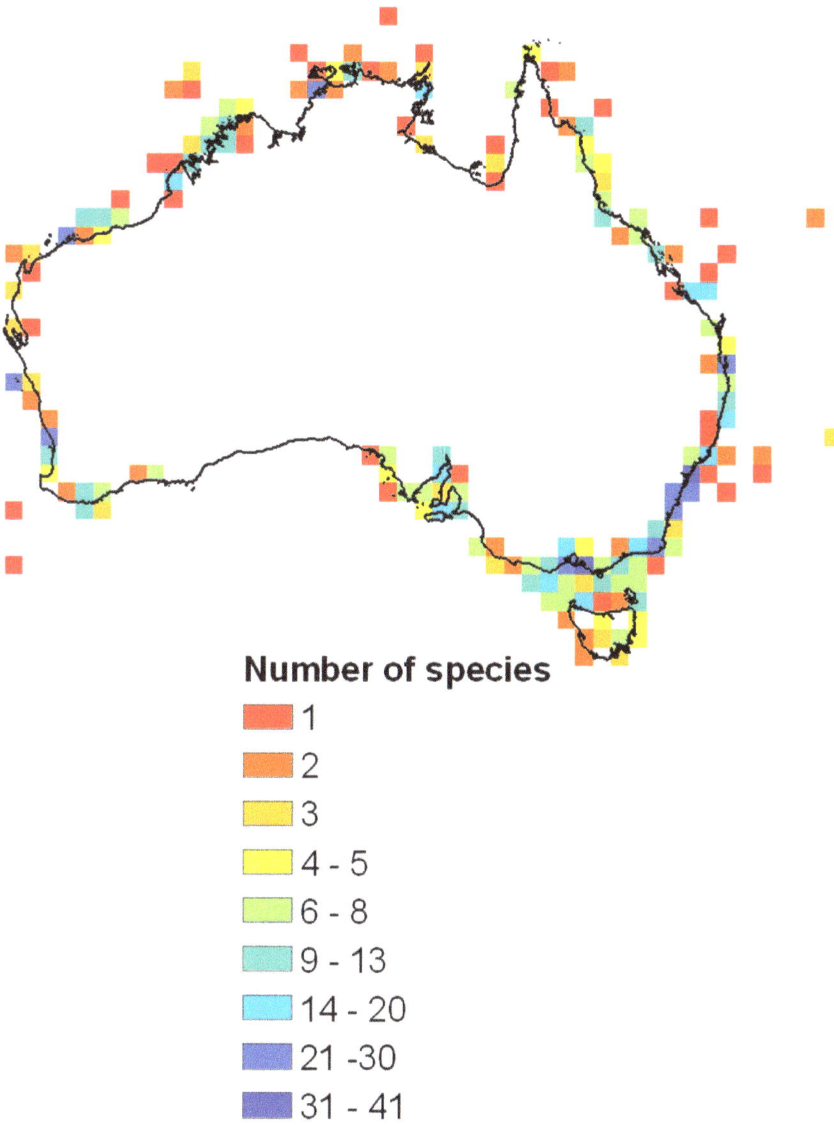

Figure 11. The distribution of terebellids ss. species around Australia based on data from ALA.

Supplementary Materials: The following are available online at https://www.mdpi.com/1424-2818/13/2/60/s1.

Author Contributions: Conceptualization, P.H., J.M.M.N. and N.L.; validation, P.H., J.M.M.N., O.C., N.L. and S.H.; writing—original draft preparation P.H., J.M.M.N., O.C., N.L. and S.H.; writing—review and editing, P.H., J.M.M.N., O.C., N.L. and S.H.; visualization, J.M.M.N., O.C., N.L. and S.H.; coordination; P.H., J.M.M.N., O.C., N.L. and S.H.; funding acquisition, P.H. All authors have read and agreed to the published version of the manuscript.

Funding: O.C. receives a post-doctoral fellowship from FUSP and Shell Brasil; J.M.M.N. receives a productivity grant from CNPq, "Conselho Nacional de Desenvolvimento Científico e Tecnológico,

Brazil, level 2; N.L. has received financial support from the French State in the frame of the "Investments for the future" Programme IdEx Bordeaux, reference ANR-10-IDEX-03-02.

Acknowledgments: Alisson Ricardo da Silva (Museu da Casa Brasileira, Brazil) helped with the figures.

Conflicts of Interest: The authors declare no conflict of interest. The funders had no role in the design of the study; in the collection, analyses, or interpretation of data; in the writing of the manuscript, or in the decision to publish the results.

References

1. Nogueira, J.M.M.; Fitzhugh, K.; Hutchings, P. The Continuing Challenge of Phylogenetic Relationships in (Annelida: Polychaeta). *Invert. Syst.* **2013**, *27*, 186–238. [CrossRef]
2. Hutchings, P.; Nogueira, J.M.N.; Carrerette, O. Terebellidae Johnston, 1846. In *Handbook of Zoology. A Natural History of the Phyla of the Animal Kingdom*; Schmidt-Rhaesa, A., Beutel, R.G., Glaubrecht, M., Kristensen, N.P., Prendini, L., Purschke, G., Richter, S., Westheide, W., Leschen, R.Z.E., Eds.; Walter de Gruyter & Co: Berlin, Germany, 2021; pp. 68–144. [CrossRef]
3. Holthe, T. Evolution, systematics and distribution of the polychaeta terebellomorpha, with a catalogue of the taxa and a bibliography. *Gunneria* **1986**, *55*, 1–236.
4. Bernard, G.; Delgard, M.L.; Maire, O.; Ciutat, A.; Lecroart, P.; Deflandre, B.; Duchêne, J.C.; Grémare, A. Comparative study of sediment particle mixing in a *Zostera noltei* meadow and a bare sediment mudflat. *Mar. Ecol. Prog. Ser.* **2014**, *514*, 71–86. [CrossRef]
5. Nygren, A.; Parapar, J.; Pons, J.; Meißner, K.; Bakken, T.; Kongsrud, J.A.; Oug, E.; Gaeva, D.; Sikorski, A.; Johansen, R.A.; et al. Megacryptic species complex hidden among one of the most common annelids in the North East Atlantic. *PLoS ONE* **2018**, *13*, e0198356. [CrossRef] [PubMed]
6. Hutchings, P.; Carrerette, O.; Nogueira, J.M.N. Pectinariidae de Quatrefages, 1866. In *Handbook of Zoology—A Natural History of the Phyla of the Animal Kingdom*; Sedentaria 111, Errantia 1; Purschke, G., Boggemann, M., Westheide, W., Eds.; De Grutyer: Berlin, Germany, 2021; Volume 3, pp. 34–49. [CrossRef]
7. Jollivet, D.; Hourdez, S. Alvinellidae desbruyeres & laubier, 1986. In *Handbook of Zoology—A Natural History of the Phyla of the Animal Kingdom*; Sedentaria 111, Errantia 1; Purschke, G., Boggemann, M., Westheide, W., Eds.; De Grutyer: Berlin, Germany, 2021; Volume 3, pp. 145–162. [CrossRef]
8. Ebbe, B.; Purschke, G. Ampharetidae Malmgren, 1866. In *Handbook of Zoology. A Natural History of the Phyla of the Animal Kingdom*; Sedentaria 111, Errantia 1; Purschke, G., Boggemann, M., Westheide, W., Eds.; De Grutyer: Berlin, Germany, 2021; Volume 3, pp. 50–67. [CrossRef]
9. Nogueira, J.M.; Hutchings, P.A.; Fukuda, M.F. Morphology of Terebelliform Polychaetes (Annelida: Polychaeta: Terebelliformia). *Zootaxa* **2010**, *2460*, 1–185. [CrossRef]
10. Spalding, M.D.; Fox, H.E.; Allen, G.R.; Davidson, N.; Ferdaña, Z.A.; Finlayson, M.; Halpern, B.S.; Jorge, M.A.; Lombana, A.L.; Lourie, S.A.; et al. Marine ecoregions of the world: A bioregionalization of coastal and shelf areas. *BioScience* **2007**, 573. [CrossRef]
11. Pamungkas, J.; Glasby, C.J.; Read, G.; Wilson, S.P.; Costello, M.J. Progress and perspectives in the discovery of polychaete. Worms (Annelida) of the world. *Helgol. Mar. Res.* **2019**, *73*, 3424. [CrossRef]
12. Grube, A.E. Beschreibungen neuer oder weniger Bekannter von Hrn. Ehrenberg gesammelter anneliden des rothen meeres. *Mon. K. Preuss. Akad. Wiss. Berl.* **1869**, *1870*, 484–521.
13. Grube, A.E. Annulata semperiana. Beiträge zur Kenntniss der annelidenfauna der Philippinen nach den von Herrn Prof. Semper. Mitgebrachten Sammlungen. *Mém. l'Acad. Imp. Sci. St.-Pétersbourg* **1878**, *25*, 1–300.
14. Kinberg, J.G.H. Annulata Nova. (Continuatio). *Öfvers. K. Vetensk. Akad. För.* **1866**, *23*, 337–357.
15. Malmgren, A.J. Nordiska hafs-annulater. *Öfvers. K. Vetensk. Akad För.* **1866**, *22*, 355–410.
16. McIntosh, W.C. Report on the annelida polychaeta collected by H.M.S. Challenger during the years 1873–1876. Report on the scientific results of the voyage of H.M.S. challenger during the years 1873–1876. *Zoology* **1885**, *12*, 1–554.
17. Schmarda, L.K. Neue Wirbellose Thiere: Beobachted Und Gesammelt Auf Einer Reise Um Die Erdr 1853 Bis 1857. In *Turbellarien, Rotatorien Und Anneliden*; Band, E., Hälfte, Z., Eds.; Verlag von Wilhelm Engelmann: Leipzig, Germany, 1861.
18. Augener, H. Beitrage zur Kenntnis der Meeresfauna Westafrikas. *Polychaeta* **1918**, *2*, 67–625.
19. Caullery, M. Sur les terebelliens du genre *Pista* mgn. et en particulier sur les uncini de ces annelides. *Bull. Soc. Zool. Fr.* **1915**, *40*, 68–78.
20. Chamberlin, R.V. New polychaetous annelids from Laguna Beach, California. *J. Entomol. Zool. Pomona Coll.* **1919**, *11*, 1–23.
21. Hessle, C. Zur kenntnis der terebellomorphen polychaeten. *Zool. Bidr. Upps.* **1917**, *5*, 39–258.
22. Gravier, C. Expédition antarctique Française Du "Pourquoi-Pas", dirigee par le dr. J.-B. Charcot (1908–1910). Espèces nouvelles d'annélides polychetes. *Bull. Mus. Natl. Hist. Nat. Paris* **1911**, *17*, 310–316.
23. Moore, J.P. Some polychaetous annelids of the Northern Pacific coast of North America. *Proc. Acad. Nat. Sci. USA* **1908**, *60*, 321–364.
24. Hutchings, P.A. Polychaeta of Wallis Lake, N.S.W. *Proc. Linn. Soc. NSW* **1974**, *98*, 175–196.

25. Hutchings, P.A. The Terebelliform polychaeta from Australia, chiefly from Moreton Bay, Queensland. *Rec. Aust Mus.* **1977**, *31*, 1–39. [CrossRef]
26. Hutchings, P.A. New species of the family terebellidae (Polychaeta) from Abrohlos island, Western Australia. In *Proceedings of the 7th International Marine Biological Workshop—The Marine Flora and Fauna of Abrohlos Island, Western Australia*; Wells, F.E., Ed.; Western Australian Museum: Perth, Australia, 1997; Volume 2, pp. 459–502.
27. Hutchings, P.A. New species of the family terebellidae (polychaeta) from Darwin harbour Australia. In *Proceedings of the 6th International Marine Biological Workshop—The Marine Flora and Fauna of Darwin Harbour, Northern Territory*; Museum & Art Gallery of the N.T.: Darwin, Australia, 1997; pp. 133–161.
28. Hutchings, P.A.; Murray, A. Taxonomy of polychaetes from the Hawkesbury river and the Southern Estuaries of New South Wales, Australia. *Rec. Aust Mus. Suppl.* **1984**, *3*, 1–118. [CrossRef]
29. Hutchings, P.A.; Glasby, C.J. The polycirrinae (*F. terebellidae*) of Australia. *Rec. Aust Mus.* **1986**, *38*, 319–350. [CrossRef]
30. Hutchings, P.A.; Glasby, C.J. The Amphitritinae (*F. terebellidae*) of Australia. *Rec. Aust Mus.* **1988**, *40*, 1–60. [CrossRef]
31. Desbruyères, D.; Laubier, L. Les Alvinellidae, une famille nouvelle d'annélides polychetes infeodees aux sources hydrothermales sous-marines: Systématique, Biologie et Écologie. *Can. J. Zool.* **1986**, *64*, 2227–2245. [CrossRef]
32. Hutchings, P.; Nogueira, J.M.M.; Carrerette, O. Telothelepodidae, Thelepodidae and Trichobranchidae (Annelida, Terebelliformia) from the Lizard Island group, Great Barrier Reef, Australia. *Zootaxa* **2015**, *4019*, 240–274. [CrossRef] [PubMed]
33. Carrerette, O.; Nogueira, J.M.M. Four new species of *Polycirrus* grube, 1850 (Polychaeta: Terebellidae) from Campos Basin, Southeastern Brazil. *Zootaxa* **2013**, *3626*, 146–172. [CrossRef] [PubMed]
34. Nogueira, J.M.M.; Amaral, A.C.Z. New terebellids (Polychaeta: Terebellidae) living in colonies of a stony coral in the state of São Paulo, Brazil. *Proc. Biol. Soc. Wash.* **2001**, *114*, 285–296.
35. Nogueira, J.M.M. A New species of *Paraeupolymnia* Young and Kritzler, 1986 (Polychaeta: Terebellidae: Terebellinae) from Brazil. *Sci. Mar.* **2003**, *67*, 403–411. [CrossRef]
36. Nogueira, J.M.M. Review of some terebelliform polychaetes (Polychaeta: Terebelliformia) at the Yale Peabody Museum. *Bull. Peabody Mus. Nat. Hist.* **2008**, *49*, 209–234. [CrossRef]
37. Nogueira, J.M.M.; Hutchings, P.A.; Carrerette, O. Terebellidae (Annelida, Terebelliformia) from Lizard Island, Great Barrier Reef, Australia. *Zootaxa* **2015**, *4019*, 484–576. [CrossRef]
38. Nogueira, J.M.M.; Hutchings, P.; Carrerette, O. Polycirridae (Annelida, Terebelliformia) from Lizard Island Group, Great Barrier Reef, Australia. *Zootaxa* **2015**, *4091*, 437–483. [CrossRef] [PubMed]
39. Noguiera, J.M.M.; Fitzhugh, K.; Hutchings, P.; Carrerette, O. Phylogenetic analysis of the family Telothelepodinae Nogueira, Fitzhugh & Hutchings 2013 (Annelida: Terebelliformia). *Mar. Biol. Res.* **2017**, *13*, 671–692. [CrossRef]
40. Carrerette, O.; Nogueira, J.M.M.; Hutchings, P. The genus *Thelepus* leuckart, 1849 (Annelida, Thelepodidae) in Brazil, with redescription of the holotype of *T. setosus* (Quatrefages, 1866). *Zootaxa* **2017**, *4250*, 587–599. [CrossRef]
41. Imajima, M.; Reuscher, M.; Fiege, D. Ampharetidae (Annelida: Polychaeta) from Japan. Part II: Genera with Elevated and Modified Notopodia. *Zootaxa* **2013**, *3647*, 137–166. [CrossRef] [PubMed]
42. Wang, W.; Sui, J.; Li, X.; Hutchings, P.; Nogueira, J.M.d.M. A new species of the genus *Amphicteis* grube, 1850 (Annelida, Ampharetidae) from the Yellow Sea, China, together with a redescription of *A. dalmatica* Hutchings & Rainer, 1979. *ZooKeys* **2020**, *988*, 1–15. [CrossRef]
43. Hilbig, B. Family Ampharetidae Malmgren, 1867. In *Taxonomic Atlas of the Benthic Fauna of the Santa Maria Basin and Western Santa Barbara Channel 7*; The Annelida Part 4. Polychaeta: Fabelligeridae to Sternaspidae; Santa Barbara Museum of Natural History: Santa Barbara, CA, USA, 2000; Volume 4, pp. 169–230.
44. Reuscher, M.G.; Fiege, D.; Wehe, T. Four new species of ampharetidae (Annelida: Polychaeta) from Pacific hot vents and cold seeps, with a key and synoptic table of characters for all genera. *Zootaxa* **2009**, *2191*, 1–40. [CrossRef]
45. Day, J.H. A Review of the family Ampharetidae (Polychaeta). *Ann. S. Afr. Mus.* **1964**, *48*, 97–120.
46. Jouin-Toulmond, C.; Hourdez, S. Morphology, ultrastructure and functional anatomy of the branchial organ of *Terebellides stroemii* (Polychaeta: Trichobranchidae) and remarks on the systematic position of the genus *Terebellides*. *Cah. Biol. Mar.* **2006**, *47*, 287–299. [CrossRef]
47. Fitzhugh, K.; Nogueira, J.M.M.; Carrerette, O.; Hutchings, P. An assessment of the status of polycirridae genera (Annelida: Polychaeta: Terebelliformia). *Zool. J. Linn. Soc.* **2015**. [CrossRef]
48. Nogueira, J.M.M.; Carrerette, O.; Hutchings, P.; Fitzhugh, K. Systematic review of the species of the family Telothelepodidae Nogueira, Fitzhugh & Hutchings, 2013 (Annelida, Terebelliformia), with descriptions of three new species. *Mar. Biol. Res.* **2018**, *14*, 217–257.
49. Struck, T.H.; Nesnidal, M.P.; Purschke, G.; Halanych, K.M. Detecting possibly saturated positions in 18S and 28S sequences and their influence on phylogenetic reconstruction of annelida (Lophotrochozoa). *Mol. Phylogenet. Evol.* **2008**, *48*, 628–645. [CrossRef] [PubMed]
50. Zrzavý, J.; Riha, P.; Pialek, L.; Janouskovec, J. Phylogeny of annelida (Lophotrochozoa): Total-evidence analysis of morphology and six genes. *BMC Evol. Biol.* **2009**, *9*. [CrossRef] [PubMed]
51. Weigert, A.; Helm, C.; Meyer, M.; Nickel, B.; Arendt, D.; Hausdorf, B.; Santos, S.R.; Halanych, K.M.; Purschke, G.; Bleidorn, C.; et al. Illuminating the base of the annelid tree using transcriptomics. *Mol. Biol. Evol.* **2014**, *31*, 1391–1401. [CrossRef]
52. Rouse, G.W.; Fauchald, K. Cladistics and polychaetes. *Zool. Scr.* **1997**, *26*, 139–204. [CrossRef]

53. Rouse, G.W.; Pleijel, F. *Polychaetes*; Oxford University Press: Oxford, UK, 2001.
54. Rousset, V.; Rouse, G.W.; Feral, J.P.; Desbruyeres, D.; Pleijel, F. Molecular and morphological evidence of Alvinellidae relationships(Terebelliformia, Polychaeta. *Zool. Scr.* **2003**, *32*, 185–197. [CrossRef]
55. Glasby, C.J.; Hutchings, P.A.; Hall, K. Phylogeny of the polychaete order Terebellomorpha (Polychaeta: Terebellidae) Based on morphology. *J. Mar. Biolog. Assoc. UK* **2004**, *84*, 961–971. [CrossRef]
56. Rousset, V.; Pleijel, F.; Rouse, G.W.; Erseus, C.; Siddall, M.E. A molecular phylogeny of annelids. *Cladistics* **2006**, *23*, 41–63. [CrossRef]
57. Garraffoni, A.R.S.; Lana, P.C. Cladistic analysis of the subfamily Trichobranchinae (Polychaeta: Terebellidae). *J. Mar. Biolog. Assoc. UK* **2004**, *84*, 973–982. [CrossRef]
58. Garraffoni, A.R.S.; Lana, P.C. Phylogenetic Relationships within the Terebellidae (Polychaeta: Terebellida) Based on Morphological Characters. *Invertebr. Syst.* **2008**, *22*, 605–626. [CrossRef]
59. Stiller, J.; Tilic, E.; Rousset, V.; Pleijel, F.; Rouse, G.W. Spaghetti to a tree: A robust phylogeny for Terebelliformia (Annelida) based on transcriptomes, molecular and morphological data. *Biology* **2020**, *9*, 73. [CrossRef]
60. Fitzhugh, K. The limits of understanding in biological systematics. *Zootaxa* **2012**, *3435*, 40–67. [CrossRef]
61. Fitzhugh, K. Defining 'species', 'biodiversity', and 'conservation' by their transitive relations. In *The Species Problem—Ongoing Problems*; Pavlinov, I.Y., Ed.; InTech: New York, NY, USA, 2013; pp. 93–130.
62. Fitzhugh, K. Sequence data, phylogenetic inference, and implications of downward causation. *Acta Biotheor.* **2016**, *64*, 133–160. [CrossRef] [PubMed]
63. Jirkov, I.A.; Leontovitch, M.K. Review of genera within the *Axionice/Pista* complex (Polychaeta, Terebellidae), with discussion of the taxonomic definition of other terebellidae with large lateral lobes. *J. Mar. Biolog. Assoc. UK* **2017**, *97*, 911–934. [CrossRef]
64. Londoño-Mesa, M.H. Terebellidae (Polychaeta: Terebellida) from the Grand Caribbean Region. *Zootaxa* **2009**, *2320*, 1–93. [CrossRef]
65. McHugh, D.; Tunnicliffe, V. Ecology and reproductive biology of the hydrothermal vent polychaete *Amphisamytha galapagensis* (Ampharetidae). *Mar. Ecol. Prog. Ser.* **1994**, *106*, 111–120. [CrossRef]
66. Le Bris, N.; Zbinden, M.; Gaill, F. Processes controlling the physico-chemical micro-environments associated with pompeii worms. *Deep Sea Res. Part 1* **2005**, *52*, 1071–1083. [CrossRef]
67. Rabaut, M.; Guilini, K.; Van Hoey, G.; Vincx, M.; Degraer, S. A bio-engineered softbottom environment: The impact of *Lanice conchilega* on the benthic species specific densities and community structure. *Estuar. Coast. Shelf Sci.* **2007**, *75*, 525–536. [CrossRef]
68. Jumars, P.A.; Dorgan, K.M.; Lindsay, S. Diet of worms emended: An update of polychaete feeding guilds. *Ann Rev Mar Sci.* **2015**, *7*, 497–520. [CrossRef]
69. Fauchald, K.; Jumars, P.A. The diet of worms: A study of polychaetes feeding guilds. *Oceanogr. Mar Biol.* **1979**, *17*, 193–284.
70. Maire, O.; Duchêne, J.C.; Amouroux, J.M.; Grémare, A. Activity patterns in the terebellid polychaete *Eupolymnia nebulosa* assessed using a new image analysis system. *Mar. Biol.* **2007**, *151*, 737–749. [CrossRef]
71. Ziegler, A.F.; Cape, M.; Lundesgaard, Ø.; Smith, C.R. Intense deposition and rapid processing of seafloor phytodetritus in a glaciomarine Fjord, Andvord Bay (Antarctica). *Prog. Oceanogr.* **2020**, *187*. [CrossRef]
72. Jumars, P.A.; Nowell, A.R.M. Effects of benthos on sediment transport: Difficulties with functional grouping. *Cont. Shelf Res.* **1984**, *3*, 115–130. [CrossRef]
73. Desbruyères, D.; Laubier, L. Systematics, phylogeny, ecology and distribution of the Alvinellidae (Polychaeta) from deep-sea hydrothermal vents. *Ophelia Suppl.* **1991**, *5*, 31–45.
74. Desbruyères, D.; Chevaldonné, P.; Alayse, A.-M.; Jollivet, D.; Lallier, F.H.; Jouin-Toulmond, C.; Zal, F.; Sarradin, P.-M.; Cosson, R.; Caprais, J.-C.; et al. Biology and ecology of the pompei worm (*Alvinella pompejana* Desbruyères and Laubier), a normal dweller on an extreme deep-sea environment: A synthesis of current knowledge and recent developments. *Deep Sea Res. Part II* **1998**, *45*, 383–422. [CrossRef]
75. Pradillon, F.; Gaill, F. Adaptation to deep-sea hydrothermal vents: Some molecular and developmental aspects. *J. Mar. Sci. Technol.* **2007**, *1*, 37–53.
76. Jouin-Toulmond, C.; Mozzo, M.; Hourdez, S. Ultrastructure of spermatozoa in four species of Alvinellidae (Annelida: Polychaeta). *Cah. Biol. Mar.* **2002**, *43*, 391–394. [CrossRef]
77. Rouse, G.W. Chapter 6. Annelid Larval Morphology. In *Reproductive Biology and Phylogeny of Annelida*; Rouse, F., Pleijel, G.W., Eds.; Science Publishers Inc.: Enfield, NH, USA, 2006.
78. Nyholm, K.G. Contributions to the life history of the ampharetid *Melinna cristata*. *Zool. Bidr. Uppsala* **1951**, *29*, 79–91.
79. Hutchings, P. Gametogenesis in a Northumberland population of the polychaete *Melinna cristata*. *Mar. Biol.* **1973**, *18*, 199–211. [CrossRef]
80. Hutchings, P. Age structure and spawning of a Northumberland population of *Melinna cristata* (Polychaeta: Ampharetidae). *Mar. Biol.* **1973**, *18*, 218–227. [CrossRef]
81. Zottoli, R.A. Reproduction and larval development of the ampharetid polychaete *Amphicteis floridus*. *Trans. Am. Microsc. Soc.* **1974**, *93*, 78–89. [CrossRef]
82. Zottoli, R.A. *Amphisamytha galapagensis*, a new species of ampharetid polychaete from the vicinity of abyssal hydrothermal vents in the Galapagos rift, and the role of this species in rift ecosystems. *Proc. Biol. Soc. Wash.* **1983**, *96*, 379–391.

83. Grehan, A.; Retière, C.; Keegan, B. Larval development in the ampharetid *Melinna palmata* grube (Polychaeta). *Ophelia Suppl.* **1991**, *5*, 321–332.
84. Blake, E.A.; Van Dover, C.L. The reproductive biology of *Amathys luti*, an ampharetid polychaete from hydrothermal vents on the mid-Atlantic ridge. *Invertebr. Biol.* **2005**, *124*, 254–264. [CrossRef]
85. Queirós, J.P.; Ravara, A.; Eilertsen, M.H.; Kongsrud, J.A.; Hilário, A. *Paramytha ossicola* sp. nov. (Polychaeta, Ampharetidae) from mammal bones: Reproductive biology and population structure. *Deep Sea Res. Part II Top Stud. Oceanogr.* **2017**, *137*, 349–358. [CrossRef]
86. Fauvel, P. *Polychètes Sédentaires. Addenda Aux Errantes, Archiannélides. Myzostomaires*; Faune de France: Paris, France, 1927; Volume 16, p. 494.
87. Hartman, O. Catalogue of the polychaetous annelids of the world. *Occ. Pap. Allan Hancock Fdn.* **1965**, *23*, 1–628.
88. Day, J.H. *A Monograph on the Polychaeta of Southern Africa*; British Museum (Natural History): London, UK, 1967.
89. Hutchings, P.A.; Glasby, C.J. Phylogenetic implications of the biogeography of Australian terebellidae. *Ophelia Suppl.* **1991**, *5*, 565–572.
90. Day, J.H.; Hutchings, P.A. An annotated checklist of Australian and New Zealand Polychaeta and Myzostomidae. *Rec. Aust. Mus.* **1979**, *32*, 80–161. [CrossRef]
91. Nogueira, J.M.M.; Carrerette, O.; Hutchings, P.A. Review of *Amaeana* Hartman, 1959 (Annelida, Terebelliformia, Polycirridae), with descriptions of seven new species. *Zootaxa* **2015**, *3994*, 1–52. [CrossRef]
92. Hutchings, P.; Kupriyanova, E. Cosmopolitan polychaetes—Fact or fiction? Personal and historical perspectives. *Invertebr. Syst.* **2018**, *32*, 1–9. [CrossRef]
93. Fauvel, P. *Polychètes Errantes*; Faune de France: Paris, France, 1923; Volume 5, p. 488.
94. Parapar, J.; Capa, M.; Nygren, A.; Moreira, J. To Name but a few: Descriptions of five new species of terebellides (Annelida, Trichobranchidae) from the North East Atlantic. *ZooKeys* **2020**, *992*, 1–58. [CrossRef]
95. Labrune, C.; Lavesque, N.; Bonifácio, P.; Hutchings, P.A. New species of *Pista* Malmgren, 1866 (Annelida, Terebellidae) from the Western Mediterranean sea. *ZooKeys* **2019**, *838*, 71–83. [CrossRef] [PubMed]
96. Lavesque, N.; Bonifácio, P.; Londoño-Mesa, M.H.; Le Garrec, V.; Grall, J. *Loimia ramzega* sp. nov. a New Giant Species of Terebellidae (Polychaeta) from French Waters (Brittany, English Channel). *J. Mar. Biolog. Assoc. UK* **2017**, *97*, 935–942. [CrossRef]
97. Lavesque, N.; Hutchings, P.; Daffe, G.; Londoño-Mesa, M.H. Revision of the French Polycirridae (Annelida, Terebelliformia), with descriptions of eight species. *Zootaxa* **2020**, *4869*, 151–186. [CrossRef] [PubMed]
98. Lavesque, N.; Londoño-Mesa, M.H.; Daffe, G.; Hutchings, P.A. Revision of the French Telothelepodidae and Thelepodidae (Annelida, Terebelliformia), with descriptions of three species and first European record of a non-indigenous species. *Zootaxa* **2020**, *4810*, 305–327. [CrossRef]
99. Lavesque, N.; Hutchings, P.; Daffe, G.; Nygren, A.; Londoño-Mesa, M.H. A revision of the French Trichobranchidae (Polychaeta), with descriptions of nine new species. *Zootaxa* **2019**, *4869*, 151–190. [CrossRef] [PubMed]
100. Kelleher, G.; Bleakley, C.; Wells, S. *A Global Representative System of Marine Protected Areas*; Great Barrier Reef Marine Park Authority, World Bank, IUCN: Washington, DC, USA, 1995.
101. Olson, D.M.; Dinerstein, E. The global 200: Priority ecoregions for conservation. *Ann. Missouri Bot. Gar.* **2002**, *89*, 199–224. [CrossRef]
102. Asaad, I.; Lundquist, C.J.; Erdmann, M.E.; Costello, M.J. Delineating priority areas for marine biodiversity conservation in the Coral Triangle. *Biol. Conserv.* **2018**, *222*, 198–211. [CrossRef]
103. Fauvel, P. *Annelida Polychaeta; The Fauna of India Including Pakistan, Ceylon, Burma and Malaya*; The Indian Press Ltd.: Allahabad, India, 1953; Volume 12.
104. Gunton, L.M.; Kupriyanova, E.; Alvestad, T.; Avery, L.; Blake, J.A.; Biriukova, O.; Böggemann, M.; Borisova, P.; Budaeva, N.; Burghardt, I.; et al. Annelids of the Eastern Australian abyss collected by the 2017 RV investigator voyage. *ZooKeys* **2021**, in press.
105. Eilertsen, M.H.; Kongsrud, J.A.; Alvestad, T.; Stiller, J.; Rouse, G.W.; Rapp, T.H. Do ampharetids take sedimented steps between vents and seeps? Phylogeny and habitat-use of Ampharetidae (Annelida, Terebelliformia) in chemosynthesis-based ecosystems. *BMC Evol. Biol.* **2017**, *17*, 222. [CrossRef]
106. Reuscher, M.G.; Fiege, D.; Wehe, T. Terebellomorph polychaetes from hydrothermal vents and cold seeps with the description of two new species of Terebellidae (Annelida: Polychaeta) representing the first records of the family from deep-sea vents. *J. Mar. Biolog. Assoc.* **2012**, *92*, 997–1012. [CrossRef]
107. Schuller, M.; Hutchings, P.A. New species of *Terebellides* (Polychaeta: Trichobranchidae) indicate long–distance dispersal between Western South Atlantic deep–sea basins. *Zootaxa* **2012**, *2395*, 1–16. [CrossRef]
108. Sars, M. *Beskrivelser Og Iagttagelser over Nogle Maerkelige Eller Nye i Havet Ved Den Bergenske Kyst Levende Dyr Af Polypernes, Acalephernes, Radiaternes, Annelidernes Og Molluskernes Classer, Med En Kort Oversigt over de Hidtil Af Forfatteren Sammesteds Fundne Arter Og Deres Forekommen*; Thorstein Hallegers Forlag: Bergen, Norway, 1835.
109. Parapar, J.; Mikac, B.; Fiege, D. Diversity of the genus *Terebellides* (Polychaeta: Trichobranchidae) in the Adriatic Sea with the description of a new species. *ZooKeys* **2013**, *3691*, 333–350. [CrossRef]
110. Parapar, J.; Hutchings, P. Redescription of *Terebellides stroemii* (Polychaeta, Trichobranchidae) and designation of a neotype. *J. Mar. Biol. Assoc. UK* **2014**, *95*, 323–337. [CrossRef]

111. Parapar, J.; Moreira, J.; Gil, J.; Martin, D. A new species of the genus *Terebellides* (Polychaeta, Trichobranchidae) from the Iranian coast. *Zootaxa* **2016**, *4117*, 321–340. [CrossRef] [PubMed]
112. Hutchings, P.; Lavesque, N. I know who you are, but do others know? Why correct scientific names are so important for the biological sciences. *Zoosymposia* **2020**, *19*, 151–163. [CrossRef]
113. Zhang, J.; Hutchings, P. Taxonomy and distribution of *Terebellides* (Polychaeta: Trichobranchidae) in the Northern South China Sea, with description of three new species. *Zootaxa* **2018**, *4377*, 387–411. [CrossRef]

diversity

Review

Cryptic Clitellata: Molecular Species Delimitation of Clitellate Worms (Annelida): An Overview

Svante Martinsson * **and Christer Erséus**

Systematics and Biodiversity, Department of Biological and Environmental Sciences, University of Gothenburg, Box 463, SE-405 30 Göteborg, Sweden; christer.erseus@bioenv.gu.se
* Correspondence: svante.martinsson@bioenv.gu.se

Abstract: Methods for species delimitation using molecular data have developed greatly and have become a staple in systematic studies of clitellate worms. Here we give a historical overview of the data and methods used to delimit clitellates from the mid-1970s to today. We also discuss the taxonomical treatment of the cryptic species, including the recommendation that cryptic species, as far as possible, should be described and named. Finally, we discuss the prospects and further development of the field.

Keywords: DNA-barcoding; Oligochaeta; cryptic species

Citation: Martinsson, S.; Erséus, C. Cryptic Clitellata: Molecular Species Delimitation of Clitellate Worms (Annelida): An Overview. *Diversity* **2021**, *13*, 36. https://doi.org/10.3390/d13020036

Received: 17 December 2020
Accepted: 18 January 2021
Published: 20 January 2021

1. Introduction

Species delimitation, i.e., the process of determining species boundaries and discovering species, is a field that has developed quickly since the introduction of genetic data [1,2]. The development has been both on the data side, from protein patterns to large genomic datasets, and on the analytical side, from clustering and measures of genetic distances to complex analyses based on coalescent theory. These advances have led to an increase in the discovery of cryptic species, i.e., species that are morphologically similar and, therefore, have been classified as the same nominal species [3]. Cryptic species are found all over the animal kingdom (e.g., [4,5]), including annelids (e.g., [6,7]) and, despite morphological similarities, they may differ in ecologically and physiologically important aspects (see, e.g., [8,9]). Species are basic biological units and entities of generalisation, and, therefore, the basis of most studies. A number of clitellate species are used as models in several fields, e.g., ecotoxicology, neurobiology and soil ecology [10,11] and, in several of the species used, taxonomical problems have been found [12–16]. In this kind of work, it is important to know the true identity of the organisms to be able to compare the results between studies, and to correctly generalise the findings to species level, and to understand the functional differences between the taxa in question.

Clitellata is a large "class" of segmented worms, comprising about one third of all known annelid species. It is placed within "subclass" Sedentaria (e.g., [17,18]), which is often thought of as a (major) polychaete group. Clitellates seem to have evolved in the transitional zones between marine and continental waters [19], and a majority of the species live within soil or aquatic sediment [20]. Unlike polychaetous annelids, they lack parapodia, and their prostomium lacks appendages. The monophyly of Clitellata is strongly supported by their unique mode of reproduction. Clitellates are hermaphrodites and characterized by the "clitellum", an epidermal structure, secreting a protective cocoon for the embryos, which develop without a larval stage (see, e.g., [21]). The external morphology of clitellates is rather stable and offers few characters trustworthy of the taxonomic separation of taxa. The shape, position and number of gonads have historically been of fundamental importance for the classification [22]. The burrowing and interstitial habitats of most clitellates are likely to be the reason for their conserved morphology, as the evolutionary pressures in these environments may favour morphological stasis [3,9,23]. Due to the lack of externally discernible characters, many clitellates are hard to delimit and identify without the aid of molecular markers, and their species diversity has, in many cases, been underestimated when based on morphology alone (many examples will be given below). This fact has led to the rise of molecular approaches to separate species, which we will explore in this review.

Species delimitation can be divided into two steps, species discovery and species validation [24]. In the first step, the researchers form hypotheses about the species boundaries, which are then tested in the second step. In the species discovery phase, typically a single data source, e.g., morphology or DNA-barcoding, is used. Testing these hypotheses in the species validation step are often based on additional data and more sophisticated analyses. In most studies, this division between species discovery and validation is not explicitly stated, but rather implied.

The definition of cryptic species varies between researchers. Some use a relaxed definition. They count all cases as cryptic where species fall within the morphological variation of the same nominal species, even when there are minor differences between them, e.g., [3]. Others use a stricter definition and distinguish between true cryptic and pseudo-cryptic species, where the first refer to species between which no morphological differences are observed, while the latter are species that do show some differences, but still are so similar that they would be classified as the same nominal species based on morphology (e.g., [25]). In this paper, we apply the broader definition of cryptic species. Moreover, we use a liberal definition of molecular species delimitation. We include papers that explore molecular data to support species also discriminated morphologically, even if the authors do not explicitly test species limits.

In this paper, we aim to give an overview of the research field of species delimitation, and cryptic species, in Clitellata. We will examine the development of methods and the new data used in delimitation of clitellate species and discuss some of the problems arising when describing cryptic species. Finally, we will consider possible directions for this field.

2. History of the Field

Here we present an historical overview over the field of molecular species delimitation of clitellate worms, from a first publication in the 1970s to papers published in 2020. In total, 104 studies where found (Figure 1, Table S1). We identified four categories of data studied and have structured the overview accordingly, dividing this section into methods categorized as: (1) gel electrophoresis of proteins; (2) non-sequenced DNA; (3) Sanger-sequencing of a limited number of DNA fragments; (4) High-Throughput Sequencing (HTS) of a large number of DNA fragments. This classification is somewhat arbitrary, and methods from more than one category have been used together in many instances and is schematically shown in Figure 1.

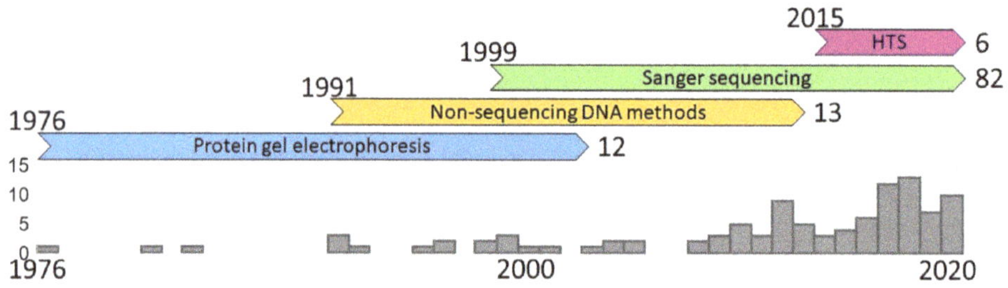

Figure 1. Historical timeline of the development of molecular species delimitation in Clitellata, showing the year of the first study, and the total number of studies, of the four major categories of methods referenced in this paper (see Table S1 for details). The histogram shows the total number of studies (all categories) per year.

2.1. Protein Gel Electrophoresis

The first publications on the species delimitation of clitellates, by means of molecular data, explored variation in proteins revealed by gel electrophoresis. In these molecular methods, proteins encoded by alleles at some locus (alloenzymes), or proteins with the same function but encoded by separate genes at different loci (isoenzymes) are separated on gels, and the pattern observed is used to infer the separation of populations. The first works by using protein gel electrophoresis to explicitly test species hypotheses of clitellates that occurred in the 1970s and 1980s, (e.g., [26–28]), although the implication of this method was discussed already by Milbrink and Nyman [29], who saw it mainly as a supplement to morphological identification of species in ecological studies. Isoenzymes and alloenzymes continued to be used, often in combination with other methods (e.g., [30–35]). Another gel-based method is the study of general protein patterns, where a mix of proteins extracted from a specimen is run on a gel, producing a banding pattern that is then compared between individuals. The pattern produced is assumed to be species specific and an index based on protein patterns was suggested [36], which was then mainly used for studies of the family Enchytraeidae [30,33,35,37,38]. Crossed immunoelectrophoresis (CIE) is another method that, to our knowledge, was only tested once in clitellate systematics—i.e., to separate populations of *Enchytraeus* (Enchytraeidae) [39]. In general, these methods seem to have worked well, as the re-examination of the same groups using more modern methods has given similar results.

2.2. Non-Sequencing DNA Methods

Restriction Fragment Patterns [40] was an early DNA-based method for the separation of species, where restriction enzymes are used to digest specific markers and the variation in restriction fragments is visualised on a gel. It was used to separate species in the genus *Enchytraeus* (Enchytraeidae) [41]. A number of other methods that generate data on the presence/absence of amplification or length variation in markers, to estimate genetic variation, both within and between species, have been used in clitellate studies. These include Arbitrary Primers PCR (AP-PCR) [42], which uses a set of primers to amplify arbitrary genetic markers, and the presence or absence of amplification is scored and used as a measure of genetic distance. This method was applied by Koperski et al. [43] in a study on the leech *Erpobdella octoculata* (Erpobdellidae). The Random Amplification of Polymorphic DNA (RAPD) method [44] also amplifies random segments of DNA, but with several shorter primers. The amplified patterns are visualised on a gel and scored. This method was used in some studies [45–49]. In the Amplified Fragment Length Polymorphism (AFLP) method [50], DNA is digested by restriction enzymes, followed by the amplification of the fragments, which are then separated and visualised on a gel, and scored as absent/present. AFLP has been used in some papers [51–53]. Lastly, microsatellites [54–57] are short repetitive regions of DNA with a high mutation rate, and the variation within them can be studied both with and without sequencing. Microsatellites have been used to study gene flow between possible cryptic species in a few studies on lumbricid earthworms [58,59].

2.3. Sanger Sequencing

When proper DNA sequencing, i.e., the Sanger-sequencing method [60,61], became more affordable, it started to be used for the species delimitation of clitellates. The first studies (e.g., [62–64]) used a single mitochondrial marker and tried to find clusters of sequences divided by large genetic distances. Studies using a single marker are continuously published [65–81]. These studies still have their merits, especially when the analysis of single gene data is integrated with the examination of morphology or other independent information. Most of the single marker studies have either (1) been distance-based, identifying clusters of sequences with short genetic distances within each cluster, but greater distances between clusters, the so-called "barcoding gap", i.e., a distinct gap in the distribution of genetic distances between low, i.e., intraspecific, distances and higher, i.e., interspecific, distances (see [82]), or (2) they have been tree-based, where a phylogeny is estimated, and used to identify well separated (monophyletic) clades, which are then being interpreted as potential species.

Today, however, studies based on more than one locus are becoming more and more common. In some analyses using multiple markers (e.g., [16,83–92]), the different sequence alignments are concatenated and a tree is estimated, and terminal clades are then identified and interpreted as species. Another approach is to estimate separate gene trees, or haplotype networks, and then identify congruent clades or network groups. Terminal clades (or specimen groups) found in all trees (or networks) are then interpreted as species, whereas conflicts between trees and groups are taken as support for gene-flow, and thus speak against speciation [13,93–115]. Several studies use a combination of the two approaches.

There is a plethora of software for dividing the individuals into species, as well as for testing species hypotheses. The most commonly used automated methods to divide single-marker datasets into species are Automated Barcode Gap Discovery, (ABGD) [116], and General Mixed Yule Coalescent (GMYC) [117]. ABGD delimits genetic clusters by detecting a significant gap in the pairwise distance distribution, and it uses genetic distances as the input. The method has been used in several studies [78,80,95,113,118–126]. GMYC, on the other hand, identifies a transition between the speciation and coalescence processes, by the identification of a shift in the branching patterns; the principle is that there are several short branches within species, but fewer and longer branches between species. It uses an ultrametric tree as input, i.e., a rooted tree where all terminal taxa are equidistant from the root; there is also a Bayesian implementation of the method (bGMYC), which applies Bayesian methodology, to account for uncertainty by sampling multiple trees [127]. This method has also been used for delimiting species of clitellates [78,113,118,126,128–131]. Another method is Bayesian Poisson Tree Processes (bPTP) [132]. It identifies significant changes in the pace of branching events on an input tree, using the number of substitutions between branching events, and it has been used in a few studies [118,131,133]. There are also a set of analyses in the Barcode of Life Database System (BOLD) [134], i.e., Barcode Gap Analysis (BGA) and the Refined Single Linkage (RESL) algorithm, the latter of which is the base of the Barcode Index Number (BIN) system [135]. These analyses have been used by Tiwari et al. [80] and Jeratthitikul et al. [118]. Haplowebs is a method that builds on the fields for recombination, i.e., sets of haplotypes connected by heterozygous individuals [136], where haplotype networks are constructed, and haplotypes that are found within the same heterozygous individual are connected to each other [137]. This method has been applied by Martinsson et al. [122] and Martin et al. [126].

To more formally test species hypotheses, both single and multi-locus approaches have been developed. Some of the single-locus methods are the statistical tests Rosenberg's P_{AB} [138] and $P_{(Randomly\ Distinct)}$ [139], which both test the distinctness of clades, and are implemented as a plugin in the software Geneious [140]. These tests have been used by some authors [119,121,123,124,131]. All of the methods mentioned in the previous two paragraphs are used on a single marker, and results from several loci have to be kept separate and each result interpreted as independent evidence. There are also explicit multi-locus species delimitation methods, and the most commonly used are based on the multispecies coalescent (MSC) model. In this model, genes evolve inside a species phylogeny where the branches are species and the properties of the branches restrict the gene trees. One of these restrictions is that the divergence times between species have to be more recent than the coalescent times for any genes shared between them, assuming no genetic transfer after speciation [141], and it can be used for the statistical testing of species assignments [2,142]. Different applications of MSC have been used in clitellate research, the most popular being the software BPP [143,144] used in several studies [12,113,122–125,145–147]. DISSECT (Division of Individuals into Species using Sequences and Epsilon-Collapsed Trees) [148], which is run within the software BEAST [149], is another species delimitation analysis based on the MSC and was used by Klinth et al. [119].

2.4. High-Throughput Sequencing (HTS)

An array of sequencing methods with a much higher throughput than Sanger sequencing have been developed today, and these methods are collectively known as Next-generation sequencing (NGS) or High-Throughput Sequencing (HTS). The techniques involved make the generation of genomic data possible, even for large samples of specimens, and HTS has made its way into species delimitation studies, also of clitellate worms. So far, four different methods have been used: (1) Restriction-Site-Associated DNA Sequencing (RAD-seq) [150,151] and (2) Genotyping by Sequencing (GBS) [152]; both work by using restriction enzymes for the digestion of the DNA, followed by the sequencing of short fragments from the restriction sites. This produces a dataset of DNA fragments from across the genome, which can either be used directly, or a set of Single Nucleotide Polymorphisms (SNP) and be extracted from the data and used for downstream analyses. The two methods differ mainly in RAD-seq implementing a fragment size selection step and more enzymatic and purification steps than GBS [152]. There are several variants of RAD-seq, and the double digest RAD-seq (ddRAD-seq) [153], which differs from the standard RAD-seq in that it lacks the random shearing and end repair of genomic DNA, but instead uses a double restriction enzyme digest, which reduces the cost of the library preparation, was used by Giska et al. [154]. On the other hand, Anderson et al. [155] use the standard RAD-seq protocol. Both of these studies are on the *Lumbricus rubellus* complex (Lumbricidae). GBS was used by Marchán et al. [156], to study the genus *Carpetania* (Hormogastridae). (3) In Transcriptome Sequencing, the transcribed mRNA is being sequenced, and this generates a dataset consisting of expressed protein coding genes, which are then used for further analyses. Transcriptomes were used by Shekhovtsov et al. [157] and Shekhovtsov et al. [158] to study the *Eisenia nordenskiold* complex (Lumbricidae), but also in some larger phylogenomic studies [19,159,160]. (4) Anchored Hybrid Enrichment (AHE) [161] enriches the target region by using a probe for conserved anchor regions. This captures both the highly conserved anchor regions and the more variable flanking regions and enriches them in the sample before sequencing. AHE was used by Taheri et al. [147] to study *Pontoscolex corethrurus* (Rhinodrilidae), and Phillips et al. [162] to test hypotheses of leech evolution. The Whole genome sequencing of clitellates is still rare, and sequenced genomes only exists for a couple of species [163–165], and no phylogenomic studies focusing on Clitellata have used whole genomes.

3. Taxonomical Treatment of Delimited Species

As many nominal species have been found to actually be species complexes, each consisting of more than one species, the question arises, how should these species be treated taxonomically? Our opinion is that the species should as far as it is possible be described as such, and given a binominal name in the context of the traditional Linnean nomenclature. In many cases, delimited species have been described, either in the paper delimited them (e.g., [12,87,93,107,108,121,166,167]), or in subsequent papers with or without additional analyses [168–173]. However, we understand that this is not always possible, due to limited material, and nomenclatorial issues, etc. that prevent a description at the moment. One obstacle to overcome when revising a cryptic species complex is to determine which of the species should keep the original name, i.e., which species is identical with the type material used in the original description. This also needs to be done for any synonyms, as these names may be applied to other species in the complex. This work may be hard but is important for taxonomic stability. In cases where type material is missing, a neotype can be designated, and this has been done for some species (e.g., [12,71,81,113,169,174]). The problem with how to treat cryptic species has been discussed for Enchytraeidae [175], and the recommendations in that paper are largely valid across Clitellata (as well as for many other organismal groups) and are briefly summarised here. The main point is that description of new species should include a good morphological description, following the standard within the specific taxonomic group, if possible, combined with at least two genetic markers that are informative at the species level—e.g., 16S, COI, H3, or ITS—

and at least one type specimen, preferably the holotype, should be sequenced. Further, specimens that are the basis for re-descriptions, including neotypes when appropriate, for nomenclatorial stability, should also be sequenced.

If species are delimited by genetic data in a study, and regardless of whether they are formally resolved, taxonomically or not, it is important that vouchers of the specimens used are deposited in natural history museums. This will enable the morphological re-examination of the specimens, to resolve possible conflicts between different datasets, as well as formal taxonomic description and revision.

4. Future Development of the Field

As we have shown in this overview, there is a great variation in the molecular methods used for species delimitation of clitellate worms, and we predict that the field will continue to grow and develop in the future. The recent introduction of High-Throughput Sequencing (HTS) methods in the systematics of clitellates has opened up a promising perspective, and we believe this will be commonplace in the near future. With continued methodological developments, we do not see a standardisation of methods used any time soon. However, there is a suggestion of using a standardised set of single-copy nuclear protein coding genes for species delimitation [176], which is an interesting suggestion, and perhaps, this will be developed and used in the future. It has the benefit of it being easier to re-use and combine data from more studies. We also see the great potential of Genotyping by Sequencing (GBS) as a relatively cheap method to generate genomic datasets for species delimitations—this method has already been used successfully for a group of hormogastrid earthworms [156]—and more studies using it will surely follow in the coming years. Finally, we hope that more of the delimited species will be formally described.

5. Summary and Conclusions

We hope this review has given a fair and inclusive description of how clitellate species have been delimited in recent years, thanks to a wide range of new data sources and methods, and also how we think delimited species should be handled and described from now on. Molecular species delimitation of clitellate worms is a research field in constant movement, evolving with molecular systematics at large that of course is universal to all groups of organisms, and we see no signs for this development to slow down. We hope this paper will give inspiration to further studies and the exploration of new methods.

With the continued testing of the many species hypotheses in Clitellata, characterized by a population genetics approach rather than traditional analyses of similarities and differences, we will get a better understanding of the species taxonomy of this species-rich and common annelid group. This will improve other fields of clitellate biology, especially with regard to phylogeny (evolutionary history) and classification, and it may stimulate studies on more applied aspects of their biology and function in various ecosystems (as suggested by [9]).

Supplementary Materials: The following are available online at https://www.mdpi.com/1424-2818/13/2/36/s1, Table S1: List of studies using molecular data to delimit species of Clitellata, listed chronologically.

Author Contributions: Conceptualization, S.M. and C.E.; writing—original draft preparation, S.M.; writing—review and editing, S.M. and C.E.; funding acquisition, C.E. All authors have read and agreed to the published version of the manuscript.

Funding: The study was funded by the Swedish Taxonomy Initiative, Swedish Species Information Centre (SLU ArtDatabanken); the Norwegian Taxonomy Initiative, Norwegian Biodiversity Information Centre (NTNU Artsdatabanken); the Swedish EPA's Environmental Research Fund in collaboration with the Swedish Agency for Marine and Water Management.

Institutional Review Board Statement: Not applicable.

Informed Consent Statement: Not applicable.

Data Availability Statement: Not applicable.

Conflicts of Interest: The authors declare no conflict of interest. The funders had no role in the design of the study; in the collection, analyses, or interpretation of data; in the writing of the manuscript, or in the decision to publish the results.

References

1. Sites, J.W.; Marshall, J.C. Delimiting species: A Renaissance issue in systematic biology. *Trends Ecol. Evol.* **2003**, *18*, 462–470. [CrossRef]
2. Fujita, M.K.; Leache, A.D.; Burbrink, F.T.; McGuire, J.A.; Moritz, C. Coalescent-based species delimitation in an integrative taxonomy. *Trends Ecol. Evol.* **2012**, *27*, 480–488. [CrossRef] [PubMed]
3. Bickford, D.; Lohman, D.J.; Sodhi, N.S.; Ng, P.K.; Meier, R.; Winker, K.; Ingram, K.K.; Das, I. Cryptic species as a window on diversity and conservation. *Trends Ecol. Evol.* **2007**, *22*, 148–155. [CrossRef] [PubMed]
4. Pfenninger, M.; Schwenk, K. Cryptic animal species are homogeneously distributed among taxa and biogeographical regions. *BMC Evol. Biol.* **2007**, *7*, 121. [CrossRef]
5. Perez-Ponce de Leon, G.; Poulin, R. Taxonomic distribution of cryptic diversity among metazoans: Not so homogeneous after all. *Biol. Lett.* **2016**, *12*. [CrossRef]
6. Erséus, C.; Gustafsson, D. Cryptic speciation in clitellate model organisms. In *Annelids in Modern Biology*; Shain, D.H., Ed.; John Wiley & Sons, Inc.: Hoboken, NJ, USA, 2009; pp. 31–46. [CrossRef]
7. Nygren, A. Cryptic polychaete diversity: A review. *Zool. Scr.* **2014**, *43*, 172–183. [CrossRef]
8. Feckler, A.; Schulz, R.; Bundschuh, M. Cryptic lineages-same but different? *Integr. Environ. Assess. Manag.* **2013**, *9*, 172–173. [CrossRef]
9. Marchán, D.F.; Díaz Cosín, D.J.; Novo, M. Why are we blind to cryptic species? Lessons from the eyeless. *Eur. J. Soil Biol.* **2018**, *86*, 49–51. [CrossRef]
10. Halanych, K.M.; Borda, E. Developing models for Lophotrochozoan and annelid Biology. In *Annelids in Modern Biology*; Shain, D.H., Ed.; John Wiley & Sons: Hoboken, NJ, USA, 2009; pp. 1–12.
11. Römbke, J.; Egeler, P. Oligochaete worms for ecotoxicological assessment of soils and sediments. In *Annelids in Modern Biology*; Shain, D.H., Ed.; Wiley-Blackwell: Hoboken, NJ, USA, 2009; pp. 228–241. [CrossRef]
12. Erséus, C.; Klinth, M.J.; Rota, E.; De Wit, P.; Gustafsson, D.R.; Martinsson, S. The popular model annelid *Enchytraeus albidus* is only one species in a complex of seashore white worms (Clitellata, Enchytraeidae). *Org. Divers. Evol.* **2019**, *19*, 105–133. [CrossRef]
13. Gustafsson, D.R.; Price, D.A.; Erséus, C. Genetic variation in the popular lab worm *Lumbriculus variegatus* (Annelida: Clitellata: Lumbriculidae) reveals cryptic speciation. *Mol. Phylogenet Evol.* **2009**, *51*, 182–189. [CrossRef]
14. Kille, P.; Andre, J.; Anderson, C.; Ang, H.N.; Bruford, M.W.; Bundy, J.G.; Donnelly, R.; Hodson, M.E.; Juma, G.; Lahive, E.; et al. DNA sequence variation and methylation in an arsenic tolerant earthworm population. *Soil Biol. Biochem.* **2013**, *57*, 524–532. [CrossRef]
15. Römbke, J.; Aira, M.; Backeljau, T.; Breugelmans, K.; Dominguez, J.; Funke, E.; Graf, N.; Hajibabaei, M.; Perez-Losada, M.; Porto, P.G.; et al. DNA barcoding of earthworms (*Eisenia fetida/andrei* complex) from 28 ecotoxicological test laboratories. *Appl. Soil Ecol.* **2016**, *104*, 3–11. [CrossRef]
16. Trontelj, P.; Utevsky, S.Y. Celebrity with a neglected taxonomy: Molecular systematics of the medicinal leech (genus *Hirudo*). *Mol. Phylogenet Evol.* **2005**, *34*, 616–624. [CrossRef] [PubMed]
17. Struck, T.H.; Golombek, A.; Weigert, A.; Franke, F.A.; Westheide, W.; Purschke, G.; Bleidorn, C.; Halanych, K.M. The evolution of annelids reveals two adaptive routes to the interstitial realm. *Curr. Biol.* **2015**, *25*, 1993–1999. [CrossRef] [PubMed]
18. Struck, T.H.; Paul, C.; Hill, N.; Hartmann, S.; Hosel, C.; Kube, M.; Lieb, B.; Meyer, A.; Tiedemann, R.; Purschke, G.; et al. Phylogenomic analyses unravel annelid evolution. *Nature* **2011**, *471*, 95–98. [CrossRef]
19. Erséus, C.; Williams, B.W.; Horn, K.M.; Halanych, K.M.; Santos, S.R.; James, S.W.; des Chatelliers, M.C.; Anderson, F.E. Phylogenomic analyses reveal a Palaeozoic radiation and support a freshwater origin for clitellate annelids. *Zool. Scr.* **2020**, *49*, 614–640. [CrossRef]
20. Timm, T. Life forms in Oligochaeta: A literature review. *Zool. Middle East* **2012**, *58*, 71–82. [CrossRef]
21. Kuo, D.H. The polychaete-to-clitellate transition: An EvoDevo perspective. *Dev. Biol.* **2017**. [CrossRef]
22. Brinkhurst, R.O.; Jamieson, B.G.M. *Aquatic Oligochaeta of the World*; Oliver and Boyd: Edinburgh, UK, 1971; p. 860.
23. Cerca, J.; Meyer, C.; Purschke, G.; Struck, T.H. Delimitation of cryptic species drastically reduces the geographical ranges of marine interstitial ghost-worms (*Stygocapitella*; Annelida, Sedentaria). *Mol. Phylogenet Evol.* **2019**. [CrossRef]
24. Carstens, B.C.; Pelletier, T.A.; Reid, N.M.; Satler, J.D. How to fail at species delimitation. *Mol. Ecol.* **2013**, *22*, 4369–4383. [CrossRef]
25. Mann, D.G.; Evans, K.M. The species concept and cryptic diversity. In Proceedings of the 12th International Conference on Harmful Algae, Copenhagen, Denmark, 4–8 September 2006; Moestrup, Ø., Doucette, G., Enevoldsen, H., Godhe, A., Hallegraeff, G., Luckas, B., Lundholm, N., Lewis, J., Rengefors, K., Sellner, K., et al., Eds.; International Society for the Study of Harmful Algae and Intergovernmental Oceanographic Commission of UNESCO: Copenhagen, Denmark, 2008.
26. Christensen, B.; Jelnes, J. Sibling species in the oligochaete worm *Lumbricillus rivalis* (Enchytraeidae) revealed by enzyme polymorphisms and breeding experiments. *Hereditas* **1976**, *83*, 237–243. [CrossRef]

27. Øien, N.; Stenersen, J. Esterases of earthworms—III. Electrophoresis reveals that *Eisenia fetida* (Savigny) is two species. *Comp. Biochem. Physiol. Part C Comp. Pharmacol.* **1984**, *78*, 277–282. [CrossRef]
28. Jaenike, J. "*Eisenia foetida*" is two biological species. *Megadrilogica* **1982**, *4*, 6–8.
29. Milbrink, G.; Nyman, L. Protein Taxonomy of Aquatic Oligochaetes and Its Ecological Applications. *Oikos* **1973**, *24*, 473. [CrossRef]
30. Brockmeyer, V. Isozymes and general protein patterns for use in discrimination and identification of *Enchytraeus* species (Annelida, Oligochaeta). *Z. Zool. Syst. Evol.* **1991**, *29*, 343–361. [CrossRef]
31. Christensen, B.; Hvilsom, M.; Pedersen, B.V. Genetic variation in coexisting sexual diploid and parthenogenetic triploid forms of *Fridericia galba* (Enchytraeidae, Oligochaeta) in a heterogeneous environment. *Hereditas* **1992**, *117*, 153–162. [CrossRef]
32. Holmstrup, M.; Simonsen, V. Genetic and physiological differences between two morphs of the lumbricid earthworm *Dendrodrilus rubidus* (Savigny, 1826). *Soil Biol. Biochem.* **1996**, *28*, 1105–1107. [CrossRef]
33. Schmelz, R. Separation of sympatric *Fridericia* species (Enchytraeidae, Oligochaeta) with isozyme and general protein patterns. *Newsl. Enchytraeidae* **1995**, *4*, 97–104.
34. Collado, R.; Hass-Cordes, E.; Schmelz, R.M. Microtaxonomy of fragmenting *Enchytraeus* species using molecular markers, with a comment on species complexes in enchytraeids. *Turk. J. Zool.* **2012**, *36*, 85–94. [CrossRef]
35. Schmelz, R.M.; Collado, R.; Myohara, M. A Taxonomic Study of *Enchytraeus japonensis* (Enchytraeidae, Oligochaeta): Morphological and Biochemical Comparisons with *E. bigeminus*. *Zool. Sci.* **2000**, *17*, 505–516. [CrossRef]
36. Westheide, W.; Brockmeyer, V. Suggestions for an index of enchytraeid species (Oligochaeta) based on general protein patterns. *Z. Zool. Syst. Evol.* **1992**, *30*, 89–99. [CrossRef]
37. Schmelz, R.M. Species separation and identification in the Enchytraeidae (Oligochaeta, Annelida): Combining morphology with general protein data. *Hydrobiologia* **1996**, *334*, 31–36. [CrossRef]
38. Schmelz, R.M. Taxonomy of *Fridericia* (Oligochaeta, Enchytraeidae). Revision of species with morphological and biochemical methods. *Abh. Nat. Ver. Hambg. (Neue Folge)* **2003**, *38*, 1-415, 73 figs.
39. Gabrich, A.; Jaros, P.P.; Brockmayer, V. Application of immunological methods for the taxonomic study of two selected animal taxa: *Tisbe* (Crustacea, Copepoda) and *Enchytraeus* (Annelida, Oligochaeta). *Z. Zool. Syst. Evol.* **1991**, *29*, 381–392. [CrossRef]
40. Avise, J.; Lansman, R.; Shade, R. The use of restriction endonucleases to measure mitochondrial DNA sequence relatedness in natural populations. I. Population structure and evolution in the genus *Peromyscus*. *Genetics* **1979**, *92*, 279–295.
41. Schlegel, M.; Steinbrück, G.; Kramer, M.; Brockmeyer, V. Restriction fragment patterns as molecular markers for species identification and phylogenetic analysis in the genus *Enchytraeus* (Oligochaeta). *Z. Zool. Syst. Evol.* **1991**, *29*, 362–372. [CrossRef]
42. Welsh, J.; McClelland, M. Fingerprinting Genomes Using PCR with Arbitrary Primers. *Nucleic Acids Res.* **1990**, *18*, 7213–7218. [CrossRef]
43. Koperski, P.; Milanowski, R.; Krzyk, A. Searching for cryptic species in *Erpobdella octoculata* (L.) (Hirudinea: Clitellata): Discordance between the results of genetic analysis and cross-breeding experiments. *Contrib. Zool.* **2011**, *80*, 85–94. [CrossRef]
44. Williams, J.G.K.; Kubelik, A.R.; Livak, K.J.; Rafalski, J.A.; Tingey, S.V. DNA Polymorphisms Amplified by Arbitrary Primers Are Useful as Genetic-Markers. *Nucleic Acids Res.* **1990**, *18*, 6531–6535. [CrossRef]
45. Bielecki, A.; Polok, K. Genetic variation and species identification among selected leeches (Hirudinea) revealed by RAPD markers. *Biologia* **2012**, *67*. [CrossRef]
46. Schirmacher, A.; Schmidt, H.; Westheide, W. RAPD-PCR investigations on sibling species of terrestrial *Enchytraeus* (Annelida: Oligochaeta). *Biochem. Syst. Ecol.* **1998**, *26*, 35–44. [CrossRef]
47. Trontelj, P.; Sotler, M.; Verovnik, R. Genetic differentiation between two species of the medicinal leech, *Hirudo medicinalis* and the neglected *H. verbana*, based on random-amplified polymorphic DNA. *Parasitol. Res.* **2004**, *94*, 118–124. [CrossRef] [PubMed]
48. Verovnik, R.; Trontelj, P.; Sket, B. Genetic differentiation and species status within the snail leech *Glossiphonia complanata* aggregate (Hirudinea: Glossiphoniidae) revealed by RAPD analysis. *Arch. Hydrobiol.* **1999**, *144*, 327–338. [CrossRef]
49. Dyer, A.R.; Fowler, J.C.S.; Baker, G.H. Detecting genetic variation in exotic earthworms, *Aporrectodea spp.* (Lumbricidae), in Australian soils using RAPD markers. *Soil Biol. Biochem.* **1998**, *30*, 159–165. [CrossRef]
50. Vos, P.; Hogers, R.; Bleeker, M.; Reijans, M.; van de Lee, T.; Hornes, M.; Frijters, A.; Pot, J.; Peleman, J.; Kuiper, M.; et al. AFLP: A new technique for DNA fingerprinting. *Nucleic Acids Res.* **1995**, *23*, 4407–4414. [CrossRef] [PubMed]
51. King, R.A.; Tibble, A.L.; Symondson, W.O. Opening a can of worms: Unprecedented sympatric cryptic diversity within British lumbricid earthworms. *Mol. Ecol.* **2008**, *17*, 4684–4698. [CrossRef]
52. Andre, J.; King, R.A.; Sturzenbaum, S.R.; Kille, P.; Hodson, M.E.; Morgan, A.J. Molecular genetic differentiation in earthworms inhabiting a heterogeneous Pb-polluted landscape. *Environ. Pollut.* **2010**, *158*, 883–890. [CrossRef]
53. Govedich, F.R.; Blinn, D.W.; Hevly, R.H.; Keim, P.S. Cryptic radiation in erpobdellid leeches in xeric landscapes: A molecular analysis of population differentiation. *Can. J. Zool.* **1999**, *77*, 52–57. [CrossRef]
54. Spritz, R.A. Duplication/deletion polymorphism 5′-to the human beta globin gene. *Nucleic Acids Res.* **1981**, *9*, 5037–5047. [CrossRef]
55. Tautz, D. Hypervariability of Simple Sequences as a General Source for Polymorphic DNA Markers. *Nucleic Acids Res.* **1989**, *17*, 6463–6471. [CrossRef]
56. Weber, J.L.; May, P.E. Abundant Class of Human DNA Polymorphisms Which Can Be Typed Using the Polymerase Chain-Reaction. *Am. J. Hum. Genet.* **1989**, *44*, 388–396. [PubMed]

57. Litt, M.; Luty, J.A. A hypervariable microsatellite revealed by in vitro amplification of a dinucleotide repeat within the cardiac muscle actin gene. *Am. J. Hum. Genet.* **1989**, *44*, 397–401. [PubMed]
58. Donnelly, R.K.; Harper, G.L.; Morgan, A.J.; Orozco-Terwengel, P.; Pinto-Juma, G.A.; Bruford, M.W. Nuclear DNA recapitulates the cryptic mitochondrial lineages of *Lumbricus rubellus* and suggests the existence of cryptic species in an ecotoxological soil sentinel. *Biol. J. Linn. Soc.* **2013**, *110*, 780–795. [CrossRef]
59. Dupont, L.; Lazrek, F.; Porco, D.; King, R.A.; Rougerie, R.; Symondson, W.O.C.; Livet, A.; Richard, B.; Decaëns, T.; Butt, K.R.; et al. New insight into the genetic structure of the *Allolobophora chlorotica* aggregate in Europe using microsatellite and mitochondrial data. *Pedobiologia* **2011**, *54*, 217–224. [CrossRef]
60. Sanger, F.; Nicklen, S.; Coulson, A.R. DNA sequencing with chain-terminating inhibitors. *Proc. Natl. Acad. Sci. USA* **1977**, *74*, 5463–5467. [CrossRef] [PubMed]
61. Sanger, F.; Coulson, A.R. A rapid method for determining sequences in DNA by primed synthesis with DNA polymerase. *J. Mol. Biol.* **1975**, *94*, 441–448. [CrossRef]
62. Beauchamp, K.A.; Kathman, R.D.; McDowell, T.S.; Hedrick, R.P. Molecular phylogeny of tubificid oligochaetes with special emphasis on *Tubifex tubifex* (Tubificidae). *Mol. Phylogenet Evol.* **2001**, *19*, 216–224. [CrossRef]
63. Heethoff, M.; Etzold, K.; Scheu, S. Mitochondrial COII sequences indicate that the parthenogenetic earthworm *Octolasion tyrtaeum* (Savigny 1826) constitutes of two lineages differing in body size and genotype. *Pedobiologia* **2004**, *48*, 9–13. [CrossRef]
64. Sturmbauer, C.; Opadiya, G.B.; Niederstätter, H.; Riedmann, A.; Dallinger, R. Mitochondrial DNA reveals cryptic oligochaete species differing in cadmium resistance. *Mol. Biol. Evol.* **1999**, *16*, 967–974. [CrossRef]
65. Cech, G.; Boros, G.; Dózsa-Farkas, K. Revision of *Bryodrilus glandulosus* (Dózsa-Farkas, 1990) and *Mesenchytraeus kuehnelti* Dózsa-Farkas, 1991 (Oligochaeta: Enchytraeidae) using morphological and molecular data. *Zool. Anz.* **2012**, *251*, 253–262. [CrossRef]
66. Dózsa-Farkas, K.; Porco, D.; Boros, G. Are *Bryodrilus parvus* Nurminen, 1970 and *Bryodrilus librus* (Nielsen and Christensen, 1959) (Annelida: Enchytraeidae) really different species? A revision based on DNA barcodes and morphological data. *Zootaxa* **2012**, *3276*, 38–50. [CrossRef]
67. Rota, E.; Martinsson, S.; Erséus, C.; Petushkov, V.N.; Rodionova, N.S.; Omodeo, P. Green light to an integrative view of *Microscolex phosphoreus* (Dugès, 1837) (Annelida: Clitellata: Acanthodrilidae). *Zootaxa* **2018**, *4496*, 175–189. [CrossRef] [PubMed]
68. Srinivasan, S.; Martinsson, S.; Naveed, M.I. On the identity and phylogenetic position of *Dero indica* (Clitellata: Naididae). *Biologia* **2020**, *75*, 1685–1689. [CrossRef]
69. Chang, C.H.; Lin, S.M.; Chen, J.H. Molecular systematics and phylogeography of the gigantic earthworms of the *Metaphire formosae* species group (Clitellata, Megascolecidae). *Mol. Phylogenet Evol.* **2008**, *49*, 958–968. [CrossRef] [PubMed]
70. Novo, M.; Almodóvar, A.; Díaz-Cosín, D.J. High genetic divergence of hormogastrid earthworms (Annelida, Oligochaeta) in the central Iberian Peninsula: Evolutionary and demographic implications. *Zool. Scr.* **2009**, *38*, 537–552. [CrossRef]
71. James, S.W.; Porco, D.; Decaens, T.; Richard, B.; Rougerie, R.; Erséus, C. DNA barcoding reveals cryptic diversity in *Lumbricus terrestris* L., 1758 (Clitellata): Resurrection of *L. herculeus* (Savigny, 1826). *PLoS ONE* **2010**, *5*, e15629. [CrossRef]
72. Paoletti, M.G.; Blakemore, R.J.; Csuzdi, C.; Dorigo, L.; Dreon, A.L.; Gavinelli, F.; Lazzarini, F.; Manno, N.; Moretto, E.; Porco, D.; et al. Barcoding *Eophila crodabepis* sp. nov. (Annelida, Oligochaeta, Lumbricidae), a Large Stripy Earthworm from Alpine Foothills of Northeastern Italy Similar to *Eophila tellinii* (Rosa, 1888). *PLoS ONE* **2016**, *11*, e0151799. [CrossRef]
73. Smythe, A.B.; Forgrave, K.; Patti, A.; Hochberg, R.; Litvaitis, M.K. Untangling the Ecology, Taxonomy, and Evolution of *Chaetogaster limnaei* (Oligochaeta: Naididae) Species Complex. *J. Parasitol.* **2015**, *101*, 320–326. [CrossRef]
74. Szederjesi, T.; Pop, V.V.; Márton, O.; Krízsik, V.; Csuzdi, C. The *Allolobophora sturanyi* species group revisited: Integrated taxonomy and new taxa (Clitellata: Megadrili). *Opusc. Zool. (Budap.)* **2016**, *47*, 87–92. [CrossRef]
75. Csuzdi, C.; Szederjesi, T.; Marchán, D.F.; Sosa, I.; Gavinelli, F.; Dorigo, L.; Pamio, A.; Dreon, A.L.; Fusaro, S.; Moretto, E.; et al. DNA barcoding of the Italian anecic *Octodrilus* species in rural (vineyard) and forested areas with description of *Octodrilus zicsiniello* sp. nov. (Clitellata, Megadrili). *Zootaxa* **2018**, *4496*, 43–64. [CrossRef]
76. Seesamut, T.; Sutcharit, C.; Jirapatrasilp, P.; Chanabun, R.; Panha, S. Morphological and molecular evidence reveal a new species of the earthworm genus *Pontodrilus* Perrier, 1874 (Clitellata, Megascolecidae) from Thailand and Peninsular Malaysia. *Zootaxa* **2018**, *4496*, 218–237. [CrossRef]
77. Nxele, T.C.; Plisko, J.D.; Mwabvu, T.; Zishiri, O.T. Molecular phylogeny of *Kazimierzus* Plisko, 2006 (Clitellata, Kazimierzidae) from the Western and Northern Cape Province inferred from mitochondrial DNA sequences. *Afr. Invertebr.* **2020**, *61*, 83–92. [CrossRef]
78. Prantoni, A.L.; Belmonte-Lopes, R.; Lana, P.C.; Erséus, C. Genetic diversity of marine oligochaetous clitellates in selected areas of the South Atlantic as revealed by DNA barcoding. *Invertebr. Syst.* **2018**, *32*, 524–532. [CrossRef]
79. Saglam, N.; Kutschera, U.; Saunders, R.; Saidel, W.M.; Balombini, K.L.W.; Shain, D.H. Phylogenetic and morphological resolution of the *Helobdella stagnalis* species-complex (Annelida: Clitellata: Hirudinea). *Zootaxa* **2018**, *4403*. [CrossRef] [PubMed]
80. Tiwari, N.; Lone, A.R.; Thakur, S.S.; Yadav, S. Interrogation of earthworm (Clitellata: Haplotaxida) taxonomy and the DNA sequence database. *J. Asia-Pac. Biodivers.* **2020**. [CrossRef]
81. Iwama, R.E.; Oceguera-Figueroa, A.; De Carle, D.; Manglicmot, C.; Erseus, C.; Miles, N.M.; Siddall, M.E.; Kvist, S. Broad geographic sampling and DNA barcoding do not support the presence of *Helobdella stagnalis* (Linnaeus, 1758) (Clitellata: Glossiphoniidae) in North America. *Zootaxa* **2019**, *4671*, 1–25. [CrossRef]

82. Meyer, C.P.; Paulay, G. DNA barcoding: Error rates based on comprehensive sampling. *PLoS Biol.* **2005**, *3*, e422. [CrossRef]
83. Perez-Losada, M.; Ricoy, M.; Marshall, J.C.; Dominguez, J. Phylogenetic assessment of the earthworm *Aporrectodea caliginosa* species complex (Oligochaeta: Lumbricidae) based on mitochondrial and nuclear DNA sequences. *Mol. Phylogenet Evol.* **2009**, *52*, 293–302. [CrossRef]
84. Buckley, T.R.; James, S.; Allwood, J.; Bartlam, S.; Howitt, R.; Prada, D. Phylogenetic analysis of New Zealand earthworms (Oligochaeta: Megascolecidae) reveals ancient clades and cryptic taxonomic diversity. *Mol. Phylogenet Evol.* **2011**, *58*, 85–96. [CrossRef]
85. Kvist, S.; Sarkar, I.N.; Erséus, C. Genetic variation and phylogeny of the cosmopolitan marine genus *Tubificoides* (Annelida: Clitellata: Naididae: Tubificinae). *Mol. Phylogenet Evol.* **2010**, *57*, 687–702. [CrossRef]
86. Matamoros, L.; Rota, E.; Erséus, C. Cryptic diversity among the achaetous *Marionina* (Annelida, Clitellata, Enchytraeidae). *Syst. Biodivers.* **2012**, *10*, 509–525. [CrossRef]
87. Marchán, D.F.; Fernández, R.; Novo, M.; Díaz Cosín, D. New light into the hormogastrid riddle: Morphological and molecular description of *Hormogaster joseantonioi* sp. n. (Annelida, Clitellata, Hormogastridae). *ZooKeys* **2014**, *414*, 1–17. [CrossRef] [PubMed]
88. Saglam, N.; Saunders, R.; Lang, S.A.; Shain, D.H. A new species of *Hirudo* (Annelida: Hirudinidae): Historical biogeography of Eurasian medicinal leeches. *BMC Zool.* **2016**, *1*, 5. [CrossRef]
89. De Sosa, I.; Marchán, D.F.; Novo, M.; Almodóvar, A.; Díaz Cosín, D.J. Bless this phylogeographic mess–Comparative study of *Eiseniella tetraedra* (Annelida, Oligochaeta) between an Atlantic area and a continental Mediterranean area in Spain. *Eur. J. Soil Biol.* **2017**, *78*, 50–56. [CrossRef]
90. Marchán, D.F.; Fernández, R.; de Sosa, I.; Díaz Cosín, D.J.; Novo, M. Pinpointing cryptic borders: Fine-scale phylogeography and genetic landscape analysis of the *Hormogaster elisae* complex (Oligochaeta, Hormogastridae). *Mol. Phylogenet Evol.* **2017**, *112*, 185–193. [CrossRef]
91. Anderson, K.; Braoudakis, G.; Kvist, S. Genetic variation, pseudocryptic diversity, and phylogeny of *Erpobdella* (Annelida: Hirudinida: Erpobdelliformes), with emphasis on Canadian species. *Mol. Phylogenet Evol.* **2019**. [CrossRef]
92. Timm, T.; Arslan, N.; Rüzgar, M.; Martinsson, S.; Erséus, C. Oligochaeta (Annelida) of the profundal of Lake Hazar (Turkey), with description of *Potamothrix alatus hazaricus* n. ssp. *Zootaxa* **2013**, *3716*, 144–156. [CrossRef]
93. De Wit, P.; Erséus, C. Genetic variation and phylogeny of Scandinavian species of *Grania* (Annelida: Clitellata: Enchytraeidae), with the discovery of a cryptic species. *J. Zool. Syst. Evol. Res.* **2010**, *48*, 285–293. [CrossRef]
94. Pérez-Losada, M.; Eiroa, J.; Mato, S.; Domínguez, J. Phylogenetic species delimitation of the earthworms *Eisenia fetida* (Savigny, 1826) and *Eisenia andrei* Bouché, 1972 (Oligochaeta, Lumbricidae) based on mitochondrial and nuclear DNA sequences. *Pedobiologia* **2005**, *49*, 317–324. [CrossRef]
95. Martinsson, S.; Erséus, C. Cryptic diversity in the well-studied terrestrial worm *Cognettia sphagnetorum* (Clitellata: Enchytraeidae). *Pedobiologia* **2014**, *57*, 27–35. [CrossRef]
96. Achurra, A.; Rodriguez, P.; Erséus, C. Pseudo-cryptic speciation in the subterranean medium: A new species of *Stylodrilus* Claparède, 1862, with a revision of the status of *Bichaeta* Bretscher, 1900 (Annelida, Clitellata, Lumbriculidae). *Zool. Anz.* **2015**, *257*, 71–86. [CrossRef]
97. Envall, I.; Gustavsson, L.M.; Erseus, C. Genetic and chaetal variation in *Nais* worms (Annelida, Clitellata, Naididae). *Zool. J. Linn. Soc.* **2012**, *165*, 495–520. [CrossRef]
98. Achurra, A.; Erséus, C. DNA barcoding and species delimitation: The *Stylodrilus heringianus* case (Annelida : Clitellata : Lumbriculidae). *Invertebr. Syst.* **2013**, *27*, 118–128. [CrossRef]
99. Dozsa-Farkas, K.; Felföldi, T. Unexpected occurrence of *Hemifridericia bivesiculata* Christensen & Dozsa-Farkas, 2006 in Hungary, a species presumed to be endemic to Devon Island, Canada, and its comparative analysis with *H. parva* Nielsen & Christensen, 1959 (Enchytraeidae, Oligochaeta). *Zootaxa* **2015**, *3914*, 185–194. [CrossRef] [PubMed]
100. Prantoni, A.L.; De Wit, P.; Erséus, C. First reports of *Grania* (Clitellata: Enchytraeidae) from Africa and South America: Molecular phylogeny and descriptions of nine new species. *Zool. J. Linn. Soc.* **2016**, *176*, 485–510. [CrossRef]
101. Shekhovtsov, S.V.; Berman, D.I.; Bazarova, N.E.; Bulakhova, N.A.; Porco, D.; Peltek, S.E. Cryptic genetic lineages in *Eisenia nordenskioldi pallida* (Oligochaeta, Lumbricidae). *Eur. J. Soil. Biol.* **2016**, *75*, 151–156. [CrossRef]
102. Dózsa-Farkas, K.; Csitári, B.; Felföldi, T. A new *Cernosvitoviella* species (Clitellata: Enchytraeidae) and its comparison with other *Cernosvitoviella* species from *Sphagnum* mires in Hungary. *Zootaxa* **2017**, *4254*, 322. [CrossRef]
103. Dozsa-Farkas, K.; Felfoldi, T. Comparative morphological and molecular taxonomic study of six *Achaeta* species (Clitellata: Enchytraeidae) with the description of a new *Achaeta* species from Koszeg Mountains, Hungary. *Zootaxa* **2017**, *4273*, 177–194. [CrossRef]
104. Martinsson, S.; Klinth, M.; Erséus, C. A new Scandinavian *Chamaedrilus* species (Clitellata: Enchytraeidae), with additional notes on others. *Zootaxa* **2018**, *4521*, 417–429. [CrossRef]
105. Shekhovtsov, S.V.; Golovanova, E.V.; Peltek, S.E. Genetic diversity of the earthworm *Octolasion tyrtaeum* (Lumbricidae, Annelida). *Pedobiologia* **2014**, *57*, 245–250. [CrossRef]
106. Shekhovtsov, S.V.; Rapoport, I.B.; Poluboyarova, T.V.; Geraskina, A.P.; Golovanova, E.V.; Peltek, S.E. Morphotypes and genetic diversity of *Dendrobaena schmidti* (Lumbricidae, Annelida). *Vavilov J. Genet. Breed.* **2020**, *24*, 48–54. [CrossRef]

107. Nagy, H.; Felföldi, T.; Dózsa-Farkas, K. Morphological and molecular distinction of two *Fridericia* species (Clitellata, Enchytraeidae) having same spermatheca type. *Zootaxa* **2018**, *4496*, 111–123. [CrossRef] [PubMed]
108. Rota, E.; Martinsson, S.; Erséus, C. Two new bioluminescent *Henlea* from Siberia and lack of molecular support for *Hepatogaster* (Annelida, Clitellata, Enchytraeidae). *Org. Divers. Evol.* **2018**, *18*, 291–312. [CrossRef]
109. Nagy, H.; Dózsa-Farkas, K.; Hong, Y.; Felföldi, T. Extending the geographic distribution of *Bryodrilus ehlersi* (Annelida, Enchytraeidae): Morphological and molecular comparison of korean and european specimens. *Acta Zool. Acad. Sci. Hung.* **2020**, *66*, 345–360. [CrossRef]
110. Novo, M.; Almodovar, A.; Fernandez, R.; Trigo, D.; Diaz Cosin, D.J. Cryptic speciation of hormogastrid earthworms revealed by mitochondrial and nuclear data. *Mol. Phylogenet Evol.* **2010**, *56*, 507–512. [CrossRef] [PubMed]
111. Shekhovtsov, S.V.; Golovanova, E.V.; Peltek, S.E. Cryptic diversity within the Nordenskiold's earthworm, *Eisenia nordenskioldi subsp. nordenskioldi* (Lumbricidae, Annelida). *Eur. J. Soil. Biol.* **2013**, *58*, 13–18. [CrossRef]
112. Latif, R.; Malek, M.; Aminjan, A.R.; Pasantes, J.J.; Briones, M.J.I.; Csuzdi, C. Integrative taxonomy of some Iranian peregrine earthworm species using morphology and barcoding (Annelida: Megadrili). *Zootaxa* **2020**, *4877*, 163–173. [CrossRef]
113. Liu, Y.; Fend, S.V.; Martinsson, S.; Erséus, C. Extensive cryptic diversity in the cosmopolitan sludge worm *Limnodrilus hoffmeisteri* (Clitellata, Naididae). *Org. Divers. Evol.* **2017**, *17*, 477–495. [CrossRef]
114. Torii, T.; Erséus, C.; Martinsson, S.; Ito, M. Morphological and genetic characterization of the first species of *Thalassodrilides* (Annelida: Clitellata: Naididae: Limnodriloidinae) from Japan. *Species Divers.* **2016**, *21*, 117–125. [CrossRef]
115. Felföldi, T.; Dózsa-Farkas, K.; Nagy, H.; Hong, Y. Three new enchytraeid species (Enchytraeidae, Annelida) from mountain soils of Korea and ten species new for the country. *Zootaxa* **2020**, *4896*, 1–45. [CrossRef]
116. Puillandre, N.; Lambert, A.; Brouillet, S.; Achaz, G. ABGD, Automatic Barcode Gap Discovery for primary species delimitation. *Mol. Ecol.* **2012**, *21*, 1864–1877. [CrossRef] [PubMed]
117. Pons, J.; Barraclough, T.G.; Gomez-Zurita, J.; Cardoso, A.; Duran, D.P.; Hazell, S.; Kamoun, S.; Sumlin, W.D.; Vogler, A.P. Sequence-based species delimitation for the DNA taxonomy of undescribed insects. *Syst. Biol.* **2006**, *55*, 595–609. [CrossRef] [PubMed]
118. Jeratthitikul, E.; Bantaowong, U.; Panha, S. DNA barcoding of the Thai species of terrestrial earthworms in the genera *Amynthas* and *Metaphire* (Haplotaxida: Megascolecidae). *Eur. J. Soil Biol.* **2017**, *81*, 39–47. [CrossRef]
119. Klinth, M.J.; Martinsson, S.; Erséus, C. Phylogeny and species delimitation of North European *Lumbricillus* (Clitellata, Enchytraeidae). *Zool. Scr.* **2017**, *46*, 96–110. [CrossRef]
120. Latif, R.; Malek, M.; Csuzdi, C. When morphology and DNA are discordant: Integrated taxonomic studies on the *Eisenia fetida/andrei* complex from different parts of Iran (Annelida, Clitellata: Megadrili). *Eur. J. Soil Biol.* **2017**, *81*, 55–63. [CrossRef]
121. Martinsson, S.; Achurra, A.; Svensson, M.; Erséus, C. Integrative taxonomy of the freshwater worm *Rhyacodrilus falciformis* s.l. (Clitellata: Naididae), with the description of a new species. *Zool. Scr.* **2013**, *42*, 612–622. [CrossRef]
122. Martinsson, S.; Klinth, M.; Erséus, C. Testing species hypotheses for *Fridericia magna*, an enchytraeid worm (Annelida: Clitellata) with great mitochondrial variation. *BMC Evol. Biol.* **2020**, *20*, 1–12. [CrossRef]
123. Martinsson, S.; Rhodén, C.; Erséus, C. Barcoding gap, but no support for cryptic speciation in the earthworm *Aporrectodea longa* (Clitellata: Lumbricidae). *Mitochondrial DNA* **2017**, *28*, 147–155. [CrossRef]
124. Martin, P.; Martinsson, S.; Wuillot, J.; Erséus, C. Integrative species delimitation and phylogeny of the branchiate worm *Branchiodrilus* (Clitellata, Naididae). *Zool. Scr.* **2018**, *47*, 727–742. [CrossRef]
125. Martinsson, S.; Erséus, C. Cryptic diversity in supposedly species-poor genera of Enchytraeidae (Annelida: Clitellata). *Zool. J. Linn. Soc.* **2018**, *183*, 749–762. [CrossRef]
126. Martin, P.; Sonet, G.; Smitz, N.; Backeljau, T. Phylogenetic analysis of the *Baikalodrilus* species flock (Annelida: Clitellata: Naididae), an endemic genus to Lake Baikal (Russia). *Zool. J. Linn. Soc.* **2019**, *187*, 987–1015. [CrossRef]
127. Reid, N.M.; Carstens, B.C. Phylogenetic estimation error can decrease the accuracy of species delimitation: A Bayesian implementation of the general mixed Yule-coalescent model. *BMC Evol. Biol.* **2012**, *12*, 196. [CrossRef] [PubMed]
128. Fernández, R.; Almodóvar, A.; Novo, M.; Simancas, B.; Díaz Cosín, D.J. Adding complexity to the complex: New insights into the phylogeny, diversification and origin of parthenogenesis in the *Aporrectodea caliginosa* species complex (Oligochaeta, Lumbricidae). *Mol. Phylogenet Evol.* **2012**, *64*, 368–379. [CrossRef] [PubMed]
129. Novo, M.; Almodovar, A.; Fernandez, R.; Trigo, D.; Diaz-Cosin, D.J.; Giribet, G. Appearances can be deceptive: Different diversification patterns within a group of Mediterranean earthworms (Oligochaeta, Hormogastridae). *Mol. Ecol.* **2012**, *21*, 3776–3793. [CrossRef] [PubMed]
130. Novo, M.; Fernandez, R.; Marchan, D.F.; Monica, G.; Cosin, D.J. Compilation of morphological and molecular data, a necessity for taxonomy: The case of *Hormogaster abbatissae* sp. n. (Annelida, Clitellata, Hormogastridae). *ZooKeys* **2012**, *242*, 1–16. [CrossRef] [PubMed]
131. Jirapatrasilp, P.; Backeljau, T.; Prasankok, P.; Chanabun, R.; Panha, S. Untangling a mess of worms: Species delimitations reveal morphological crypsis and variability in Southeast Asian semi-aquatic earthworms (Almidae, *Glyphidrilus*). *Mol. Phylogenet Evol.* **2019**, *139*, 106531. [CrossRef]
132. Zhang, J.; Kapli, P.; Pavlidis, P.; Stamatakis, A. A general species delimitation method with applications to phylogenetic placements. *Bioinformatics* **2013**, *29*, 2869–2876. [CrossRef]

133. De Carle, D.; Oceguera-Figueroa, A.; Tessler, M.; Siddall, M.E.; Kvist, S. Phylogenetic analysis of *Placobdella* (Hirudinea: Rhynchobdellida: Glossiphoniidae) with consideration of COI variation. *Mol. Phylogenet Evol.* **2017**, *114*, 234–248. [CrossRef]
134. Ratnasingham, S.; Hebert, P.D. BOLD: The Barcode of Life Data System (http://www.barcodinglife.org). *Mol. Ecol. Notes* **2007**, *7*, 355–364. [CrossRef]
135. Ratnasingham, S.; Hebert, P.D. A DNA-based registry for all animal species: The barcode index number (BIN) system. *PLoS ONE* **2013**, *8*, e66213. [CrossRef]
136. Doyle, J.J. The Irrelevance of Allele Tree Topologies for Species Delimitation, and a Non-Topological Alternative. *Syst. Bot.* **1995**, *20*, 574–588. [CrossRef]
137. Flot, J.F.; Couloux, A.; Tillier, S. Haplowebs as a graphical tool for delimiting species: A revival of Doyle's "field for recombination" approach and its application to the coral genus *Pocillopora* in Clipperton. *BMC Evol. Biol.* **2010**, *10*, 372. [CrossRef] [PubMed]
138. Rosenberg, N.A. Statistical tests for taxonomic distinctiveness from observations of monophyly. *Evolution* **2007**, *61*, 317–323. [CrossRef]
139. Rodrigo, A.; Bertels, F.; Heled, J.; Noder, R.; Shearman, H.; Tsai, P. The perils of plenty: What are we going to do with all these genes? *Philos. Trans. R. Soc. Lond. B Biol. Sci.* **2008**, *363*, 3893–3902. [CrossRef] [PubMed]
140. Masters, B.C.; Fan, V.; Ross, H.A. Species Delimitation—a Geneious plugin for the exploration of species boundaries. *Mol. Ecol. Resour.* **2011**, *11*, 154–157. [CrossRef]
141. Rannala, B.; Yang, Z. Bayes estimation of species divergence times and ancestral population sizes using DNA sequences from multiple loci. *Genetics* **2003**, *164*, 1645–1656.
142. Rannala, B. The art and science of species delimitation. *Curr. Zool.* **2015**, *61*, 846–853. [CrossRef]
143. Yang, Z. The BPP program for species tree estimation and species delimitation. *Curr. Zool.* **2015**, *61*, 854–865. [CrossRef]
144. Yang, Z.; Rannala, B. Bayesian species delimitation using multilocus sequence data. *Proc. Natl. Acad. Sci. USA* **2010**, *107*, 9264–9269. [CrossRef]
145. Martinsson, S.; Erseus, C. Cryptic speciation and limited hybridization within *Lumbricus* earthworms (Clitellata: Lumbricidae). *Mol. Phylogenet Evol.* **2017**, *106*, 18–27. [CrossRef]
146. Martinsson, S.; Erséus, C. Hybridisation and species delimitation of Scandinavian *Eisenia* spp. (Clitellata: Lumbricidae). *Eur. J. Soil Biol.* **2018**, *88*, 41–47. [CrossRef]
147. Taheri, S.; James, S.; Roy, V.; Decaens, T.; Williams, B.W.; Anderson, F.; Rougerie, R.; Chang, C.H.; Brown, G.; Cunha, L.; et al. Complex taxonomy of the 'brush tail' peregrine earthworm *Pontoscolex corethrurus*. *Mol. Phylogenet Evol.* **2018**, *124*, 60–70. [CrossRef] [PubMed]
148. Jones, G.; Aydin, Z.; Oxelman, B. DISSECT: An assignment-free Bayesian discovery method for species delimitation under the multispecies coalescent. *Bioinformatics* **2015**, *31*, 991–998. [CrossRef] [PubMed]
149. Drummond, A.J.; Rambaut, A. BEAST: Bayesian evolutionary analysis by sampling trees. *BMC Evol. Biol.* **2007**, *7*, 214. [CrossRef]
150. Miller, M.R.; Dunham, J.P.; Amores, A.; Cresko, W.A.; Johnson, E.A. Rapid and cost-effective polymorphism identification and genotyping using restriction site associated DNA (RAD) markers. *Genome Res.* **2007**, *17*, 240–248. [CrossRef]
151. Davey, J.W.; Blaxter, M.L. RADSeq: Next-generation population genetics. *Brief. Funct. Genom.* **2010**, *9*, 416–423. [CrossRef]
152. Elshire, R.J.; Glaubitz, J.C.; Sun, Q.; Poland, J.A.; Kawamoto, K.; Buckler, E.S.; Mitchell, S.E. A robust, simple genotyping-by-sequencing (GBS) approach for high diversity species. *PLoS ONE* **2011**, *6*, e19379. [CrossRef]
153. Peterson, B.K.; Weber, J.N.; Kay, E.H.; Fisher, H.S.; Hoekstra, H.E. Double digest RADseq: An inexpensive method for de novo SNP discovery and genotyping in model and non-model species. *PLoS ONE* **2012**, *7*, e37135. [CrossRef]
154. Giska, I.; Sechi, P.; Babik, W. Deeply divergent sympatric mitochondrial lineages of the earthworm *Lumbricus rubellus* are not reproductively isolated. *BMC Evol. Biol.* **2015**, *15*, 217. [CrossRef]
155. Anderson, C.; Cunha, L.; Sechi, P.; Kille, P.; Spurgeon, D. Genetic variation in populations of the earthworm, *Lumbricus rubellus*, across contaminated mine sites. *BMC Genet.* **2017**, *18*, 97. [CrossRef]
156. Marchán, D.F.; Novo, M.; Sanchez, N.; Dominguez, J.; Diaz Cosin, D.J.; Fernández, R. Local adaptation fuels cryptic speciation in terrestrial annelids. *Mol. Phylogenet Evol.* **2020**, *146*, 106767. [CrossRef] [PubMed]
157. Shekhovtsov, S.V.; Ershov, N.I.; Vasiliev, G.V.; Peltek, S.E. Transcriptomic analysis confirms differences among nuclear genomes of cryptic earthworm lineages living in sympatry. *BMC Evol. Biol.* **2019**, *19*. [CrossRef] [PubMed]
158. Shekhovtsov, S.V.; Shipova, A.A.; Poluboyarova, T.V.; Vasiliev, G.V.; Golovanova, E.V.; Geraskina, A.P.; Bulakhova, N.A.; Szederjesi, T.; Peltek, S.E. Species Delimitation of the *Eisenia nordenskioldi* Complex (Oligochaeta, Lumbricidae) Using Transcriptomic Data. *Front. Genet.* **2020**, *11*, 01508. [CrossRef] [PubMed]
159. Anderson, F.E.; Williams, B.W.; Horn, K.M.; Erseus, C.; Halanych, K.M.; Santos, S.R.; James, S.W. Phylogenomic analyses of Crassiclitellata support major Northern and Southern Hemisphere clades and a Pangaean origin for earthworms. *BMC Evol. Biol.* **2017**, *17*, 123. [CrossRef]
160. Novo, M.; Fernández, R.; Andrade, S.C.S.; Marchán, D.F.; Cunha, L.; Díaz Cosín, D.J. Phylogenomic analyses of a Mediterranean earthworm family (Annelida: Hormogastridae). *Mol. Phylogenet Evol.* **2016**, *94*, 473–478. [CrossRef] [PubMed]
161. Lemmon, A.R.; Emme, S.A.; Lemmon, E.M. Anchored hybrid enrichment for massively high-throughput phylogenomics. *Syst. Biol.* **2012**, *61*, 727–744. [CrossRef] [PubMed]

162. Phillips, A.J.; Dornburg, A.; Zapfe, K.L.; Anderson, F.E.; James, S.W.; Erseus, C.; Moriarty Lemmon, E.; Lemmon, A.R.; Williams, B.W. Phylogenomic Analysis of a Putative Missing Link Sparks Reinterpretation of Leech Evolution. *Genome Biol. Evol.* **2019**, *11*, 3082–3093. [CrossRef] [PubMed]
163. Kvist, S.; Manzano-Marin, A.; de Carle, D.; Trontelj, P.; Siddall, M.E. Draft genome of the European medicinal leech *Hirudo medicinalis* (Annelida, Clitellata, Hirudiniformes) with emphasis on anticoagulants. *Sci. Rep.* **2020**, *10*, 9885. [CrossRef] [PubMed]
164. Zwarycz, A.S.; Nossa, C.W.; Putnam, N.H.; Ryan, J.F. Timing and Scope of Genomic Expansion within Annelida: Evidence from Homeoboxes in the Genome of the Earthworm *Eisenia fetida*. *Genome Biol. Evol.* **2015**, *8*, 271–281. [CrossRef]
165. Simakov, O.; Marletaz, F.; Cho, S.J.; Edsinger-Gonzales, E.; Havlak, P.; Hellsten, U.; Kuo, D.H.; Larsson, T.; Lv, J.; Arendt, D.; et al. Insights into bilaterian evolution from three spiralian genomes. *Nature* **2013**, *493*, 526–531. [CrossRef]
166. Dozsa-Farkas, K.; Felföldi, T.; Nagy, H.; Hong, Y. New enchytraeid species from Mount Hallasan (Jeju Island, Korea) (Enchytraeidae, Oligochaeta). *Zootaxa* **2018**, *4496*, 337–381. [CrossRef] [PubMed]
167. Dózsa-Farkas, K.; Nagy, H.; Felföldi, T. Two new species of *Fridericia* (Annelida: Enchytraeidae) from Hungarian caves. *Eur. J. Taxon.* **2019**, *553*. [CrossRef]
168. Klinth, M.J.; Rota, E.; Erséus, C. Taxonomy of North European *Lumbricillus* (Clitellata, Enchytraeidae). *ZooKeys* **2017**, *703*, 15–96. [CrossRef] [PubMed]
169. Martinsson, S.; Rota, E.; Erséus, C. Revision of *Cognettia* (Clitellata, Enchytraeidae): Re-establishment of *Chamaedrilus* and description of cryptic species in the *sphagnetorum* complex. *Syst. Biodivers* **2015**, *13*, 257–277. [CrossRef]
170. Martinsson, S.; Rota, E.; Erséus, C. On the identity of *Chamaedrilus glandulosus* (Michaelsen, 1888) (Clitellata, Enchytraeidae), with the description of a new species. *ZooKeys* **2015**, *501*, 1–14. [CrossRef]
171. Marchán, D.F.; Fernández, R.; Domínguez, J.; Díaz Cosín, D.J.; Novo, M. Genome-informed integrative taxonomic description of three cryptic species in the earthworm genus *Carpetania* (Oligochaeta, Hormogastridae). *Syst. Biodivers.* **2020**, 1–13. [CrossRef]
172. Marchán, D.F.; Fernández, R.; Sánchez, N.; de Sosa, I.; Díaz Cosín, D.J.; Novo, M. Insights into the diversity of Hormogastridae (Annelida, Oligochaeta) with descriptions of six new species. *Zootaxa* **2018**, *4496*, 65–95. [CrossRef]
173. Kvist, S.; Erséus, C. Two new European species of the marine genus *Tubificoides* (Annelida: Clitellata: Naididae) with notes on the morphology of *T. pseudogaster* (Dahl, 1960). *Zootaxa* **2018**, *4433*, 561. [CrossRef]
174. Kvist, S.; Oceguera-Figueroa, A.; Siddall, M.E.; Erseus, C. Barcoding, types and the *Hirudo* files: Using information content to critically evaluate the identity of DNA barcodes. *Mitochondrial DNA* **2010**, *21*, 198–205. [CrossRef]
175. Schmelz, R.M.; Beylich, A.; Boros, G.; Dózsa-Farkas, K.; Graefe, U.; Hong, Y.; Römbke, J.; Schlaghamersky, J.; Martinsson, S. How to deal with cryptic species in Enchytraeidae, with recommendations on taxonomical descriptions. *Opusc. Zool.* **2017**, *48*, 45–51. [CrossRef]
176. Eberle, J.; Ahrens, D.; Mayer, C.; Niehuis, O.; Misof, B. A Plea for Standardized Nuclear Markers in Metazoan DNA Taxonomy. *Trends Ecol. Evol.* **2020**. [CrossRef] [PubMed]

diversity

MDPI

Review

Annelids in Extreme Aquatic Environments: Diversity, Adaptations and Evolution

Christopher J. Glasby [1,2,*], Christer Erséus [3] and Patrick Martin [4]

[1] Natural Sciences, Museum & Art Gallery Northern Territory, PO Box 4646, Darwin, NT 0801, Australia
[2] Research Institute for the Environment and Livelihoods, Charles Darwin University, Casuarina, NT 0810, Australia
[3] Systematics and Biodiversity, Department of Biological and Environmental Sciences, University of Gothenburg, Box 463, SE-405 30 Göteborg, Sweden; christer.erseus@bioenv.gu.se
[4] Royal Belgian Institute of Natural Sciences, Taxonomy and Phylogeny, 29 rue Vautier, B-1000 Brussels, Belgium; patrick.martin@naturalsciences.be
* Correspondence: chris.glasby@nt.gov.au

Abstract: We review the variety of morphological, physiological and behavioral modifications that annelids have acquired to cope with environments either unsuitable for, or on the limits of, survival for most animals. We focus on polychaetes (excluding sipunculans and echiurans) and clitellates (oligochaetes and leeches) and source information mostly from the primary literature. We identified many modifications common to both polychaetes and clitellates, and others that are specific to one or the other group. For example, certain land-adapted polychaetes show reduction in nuchal organs, epidermal ciliation and receptor cells, and other coastal polychaetes use adhesive glands and glue-reinforced tubes to maintain position in surf zones, while oligochaetes, with their simple body plans, appear to be 'pre-adapted' to life underground. Modifications common to both groups include the ability to construct protective cocoons, make cryoprotective substances such as antifreeze and heat shock proteins, develop gills, transform their bodies into a home for symbiotic chemoautotrophic bacteria, metabolize contaminants, and display avoidance behaviors. Convergent evolution in both directions has enabled annelids to transition from salt water to freshwater, sea to land via beaches, freshwater to soil, and surface water to subterranean water. A superficially simple worm-like body and a mostly benthic/burrowing lifestyle has facilitated radiation into every conceivable environment, making annelids among the most common and diverse animal groups on the planet.

Keywords: adaptation; habitat; environment; morphology; physiology; behavior; invertebrate; Annelida

check for updates

Citation: Glasby, C.J.; Erséus, C.; Martin, P. Annelids in Extreme Aquatic Environments: Diversity, Adaptations and Evolution. *Diversity* **2021**, *13*, 98. https://doi.org/10.3390/d13020098

Academic Editors: Michael Wink, Maria Capa and Pat Hutchings

Received: 14 January 2021
Accepted: 17 February 2021
Published: 23 February 2021

1. Introduction

Annelida has traditionally included Polychaeta, Oligochaeta and Hirudinea, the latter two together in the Clitellata. However, recent phylogenomic studies have shown that the monophyletic clade Clitellata is nested within the Polychaeta, rendering the latter paraphyletic and its taxonomic scope the equivalent of Annelida [1–3]. However, maintaining a distinction between the "polychaetes" and the "clitellates" remains evolutionary significant, as the polychaete–clitellate transition is associated with the loss of a larval stage and the emergence of direct development as one of several key adaptations to freshwater and terrestrial habitats [4]. In this chapter, and for the sake of convenience, we will use the vernacular term "polychaetes" to discriminate non-clitellate annelids from Clitellata. As pointed out by Weigert and Bleidorn [2], this is common practice for paraphyletic groups such as "fishes", "reptiles" "crustaceans".

In this overview, we consider annelids living in extreme aquatic environments (i.e., extremophiles) outside the typical range of conditions experienced by the group. Most examples we provide are free-living forms; commensals and ectoparasites, such as hirudineans and myzostomatids, are largely excluded as these groups combined represent only approximately 2% of the total Annelida species diversity estimated at 30,000 [5], and as such would have little impact on our general conclusions. The focus of this chapter is on aquatic polychaetes and oligochaetes, as these two groups have many more examples of extremophiles than any other annelid group.

2. What Is a Polychaete?

Polychaetes encompass the traditional 80 or so families of the former class Polychaeta, plus the former phyla Echiura (now sister to the polychaete family Capitellidae), Sipuncula (uncertain position), and Pogonophora and Vestimentifera (the latter two combined as the polychaete family Siboglinidae) ([6] and references therein). Polychaetes are characterized by several morphological characters including segmentation via a pre-pygidial growth zone, a dorsal brain and ventral nerve cord, nuchal organs, longitudinal muscle bands and the structural composition of the capillary chaetae [7,8], although some of these features may be absent. Nuchal organs for example have been lost in clitellates and a few traditional polychaete families [7,8]. In keeping with their paraphyletic status, there are no evolutionary novelties (autapomorphies) that separate uniquely 'polychaetes' from clitellates. However, polychaetes tend to have separate sexes (clitellates are hermaphrodites), appendages on the prostomium, peristomium or pygidium (all lacking in clitellates), and parapodia bearing complex and numerous chaetae (clitellates lack parapodia and most of them have few, simple chaetae; leeches lack chaetae altogether).

Annelids comprise a single large clade, Pleistoannelida, which includes the vast majority of taxa including Clitellata, Siboglinidae and Echiura; this large clade is comprised of two smaller clades, Sedentaria (including Clitellata) and Errantia [6]. A basal grade including the sipunculans and a group of small polychaete families represents an early annelid radiation [2,6,9]. Fossil Annelida date back to the Cambrian, although forms we recognize today mostly arose at the end of the Carboniferous, approximately 300 Ma [10].

3. What Is a Clitellate?

Clitellata is a large taxon comprising approximately one-third of all known annelid species. A recent estimate places its divergence from sedentarian polychaetes in the mid-Palaeozoic Era, at least approximately 400 Ma, with all major extant clitellate lineages arising over the next ~150 million years [11]. The vast majority of the limnetic and terrestrial annelids belong to Clitellata [8], but the estimated rise of the highly diverse earthworm lineage (largely the taxon Crassiclitellata) was not until approximately 200 Ma [11]. The origin of the family Enchytraeidae, with a majority of its species on land today, was probably approximately 250 Ma [11].

Hirudinea [comprising Hirudinida (true leeches), Acanthobdellida and Branchiobdellida] are predatory and parasitic clitellates previously considered as sister group to Oligochaeta. However, morphological and molecular studies support that they originated near the oligochaete family Lumbriculidae, making Oligochaeta paraphyletic [12,13]. In analogy to the polychaete–clitellate case mentioned above, we will use the vernacular term "oligochaetes" only to discriminate non-leech clitellates from Hirudinea.

All sexually mature clitellates possess a clitellum, i.e., a modification of the epidermis into a glandular "girdle" near the female pores. This structure secretes a mucous cocoon, in which eggs are laid and fertilized, and embryonal development then proceeds directly without any polychaete-type larval stage. Clitellates are distinguished from polychaetes not only morphologically (see above) but they show a unique combination of other distinctive features such as hermaphroditism, the organization of their reproductive system and sperm ultrastructure [14–18].

Considering their universal adaptations to freshwater and terrestrial life, clitellates can be said to be extreme annelids by default; Kuo [4] reviewed the evo-devo background to these adaptations. Although it is clear that Clitellata has its root in Sedentaria, its sister group is yet unknown [2], and it is unsettled whether the earliest ancestor with a clitellum and a typical clitellate reproductive mode arose before or after the colonialization of freshwater. Molecular evidence supports that the borders between sea, freshwater and land have been crossed several times in the evolutionary history of Clitellata [11].

4. What Is an Extreme Environment? What Is an "Extremophile"?

An organism that thrives in an extreme environment is an "extremophile". Given that all physical and geochemical factors of the environment in which annelids live are on a continuum, extremes in any one factor may make it difficult for an organism to function [19–21]. Whether a species is an extremophile will depend on the annelid group in question. For the predominantly marine polychaetes, freshwater and terrestrial conditions count as 'extreme' environments; for oligochaetes, which have had a long evolutionary history in continental environments, the sea may hold more possibilities for extreme conditions. Extreme living in environments that are on the limits for survival has evolved convergently and repeatedly in annelids. The list of examples in this chapter is not exhaustive, but chosen to illustrate, especially, the adaptations to a variety of extreme conditions that are on the limits for survival of most other organisms.

5. Terrestrial Freshwater

Annelids are commonly found living in terrestrial freshwater, from the sediments of rivers and lakes to moist soils. Freshwater can be considered to be extreme for the marine derived polychaetes and the oligochaetes that have transitioned from marine to terrestrial waters. Polychaetes struggle to survive in freshwater because most are osmoconformers; when exposed to low salinities, they swell uncontrollably. Certain intertidal and estuarine species can withstand short periods of stress caused by changing salinities, but fundamental osmotic modifications are required for longer periods. Those polychaetes that have been able to penetrate and live permanently in freshwater are most probably osmoregulators as adults, although how they accomplish this is unknown as the osmoregulation is thought to develop after the larval stages. Protection of embryos and larvae from osmotic stress can occur with or without modification of the most common reproductive mode among polychaetes, i.e., free spawning with planktotrophic larvae. Species retaining the common mode may simply migrate seaward to spawn in more saline coastal waters, while modified strategies include various forms of brooding either inside the parent body, inside their tube, or encapsulation of embryos in a gelatinous mass (below).

The most successful freshwater family among the polychaetes by far is the Nereididae, with approximately 60 named species able to tolerate freshwater conditions, more than the combined number of species in the next three most successful families, Spionidae (26), Fabriciidae (13), and Sabellidae (11; Table 1) [22]. Almost half of the terrestrial nereidids belong to the subfamily Namanereidinae, which are found in every continent except Antarctica. Some of the more unusual habitats include tree leaf axils and plant container habitats (*Namalycastis hawaiiensis, Namanereis catarractarum*), rice fields in SE Asia (*Namalycastis rhodochorde*), and cisterns (*Namalycastis indica*) (Figure 1). The subfamily includes the highest reported elevation for a polychaete, 1600 m a.s.l. in cave pools in Mexico [23]. Glasby and Timm [24] suggested that the ancestor of *Namanereis* colonized freshwater as a result of a single colonization event prior to the breakup of Gondwana, although two invasions, of Gondwanan and Tethyan ages, were postulated in a later study [25]. The nereidid, *Simplisetia limnetica*, a freshwater polychaete found in the upper freshwater reaches of the Hawkesbury River, Sydney, Australia, may protect its developing young from osmotic stress by brooding them in its sedimentary galleries, which are more saline than overlying waters [26]. The larvae of other estuarine nereidids, particularly *Hediste* species, use tides to the wash larvae seaward and provide better chances of survival.

Freshwater Sabellidae include seven species of the highly modified genus *Caobangia*, shell-boring symbionts of gastropods and bivalves restricted to south and southeast Asia [22]. The closely allied Fabriciidae are represented by three sympatric species of *Manayunkia* in Lake Baikal: *M. baicalensis*, *M. zenkewitschii* and *M. godlewskii*; each occupies a slightly different niche [27]. Sitnikova [28] claimed that a fourth species was present in the deep-water zone of the lake (at approximately 350 m depth) but it was never described and has not been mentioned again in later literature (see also below). Possibly, this 'fourth' species is a vagrant, as huge masses of water can suddenly sink into the deeper parts of the lake, probably bringing with them animals that normally do not live at such depths. A similar situation exists with the presence of *Nais* species (Naidinae, Naididae) under the dimictic layer of the lake, which is also unexpected given that species of this genus have a diet based on microalgae (and there is no light at such depths).

Among the oligochaetes, the Lake Baikal species *Baikalodrilus discolor* is characterized by unusual low osmotic concentration in its hemolymph, by comparison with *Spirosperma ferox*, a close Palaearctic relative, which is considered as an adaptation to low electrolyte concentrations of Baikal waters (see [29] and above). Another species of *Baikalodrilus*, *B. digitatus*, has peculiar epidermal projections on the body wall the function of which remains unknown (Figure 2A).

Table 1. Extremophile annelids and their adaptations. P = polychaete, O = oligochaete, H = hirudinid, and A = aphanoneuran.

Extreme Environment	Adaptation	Annelid Examples
Freshwater	osmoregulation	Namanereidinae (P), Fabriciidae (P), Sabellidae (P). *Baikalodrilus discolor* (O)
Land (for polychaetes)	reduced (or an absence) of nuchal organs; epidermis lacking ciliation; fewer types of receptor cells; embryos develop directly inside cocoons	Hrabeiellidae (P), Parergodrilidae (P), Nerillidae (P)
Deep-sea brine pools	avoidance	*Methanoaricia* sp. (P)
Hypersalinity, aridity	encysting; asexual reproduction	*Manayunkia athalassia* (P), *Lamprodrilus mrazeki* (O), some *Aeolosoma* species (A)
Frozen terrane	blood 'antifreeze' and other cryoprotective substances; increase internal energy levels; dehydration to lower mean supercooling point	Dinophilidae (P), *Dendrobaena octaedra* (O), *Enchytraeus albidus* (O), *Mesenchytraeus solifugus* (O)
High temperatures	protective tubes and cocoons, heat shock proteins	*Eisenia* spp. and *Amynthas* spp. (O), Siboglinidae (P) and Alvinellidae (P)
High-energy coasts and fast-flowing streams	small body, flat body, adhesive glands, strong tubes	many polychaete families; *Trichodrilus* sp. (O)
Subterranean	loss of eyes, pigmentation, elongation of sensory appendages (and legs), ability to cope with limited food resources	*Namanereis* spp. (P), *Marifugia cavatica* (P), *Croatobranchus mestrovi* (H), *Trichodrilus* sp. (O)
Low oxygen	gills, unsegmented posterior end, high O_2 affinity respiratory pigments, diel and seasonal vertical migration, oxygen debts	*Capitella* spp. (P), *Rhyacodrilus hiemalis* (O), *Croatobranchus mestrovi* (H), Phallodrilinae (O), *Aulodrilus* sp. (O)
Deep-sea vents and cold seeps	symbiotic autotrophic bacteria, host body-cavity 'homes' for symbiotic bacteria, long tubes	Siboglinidae, Alvinellidae (P), genera *Olavius* spp. (O) *Inanidrilus* spp. (O), Phallodrilinae (O)
Pollutants (organic and inorganic)	metabolism of PAHs; assimilation of metals into hard structures, e.g., jaws; biotransformation of aromatic hydrocarbons	*Capitella* spp. (P). Namanereidinae (P), *Thalassodrilides* cf. *briani* (O)

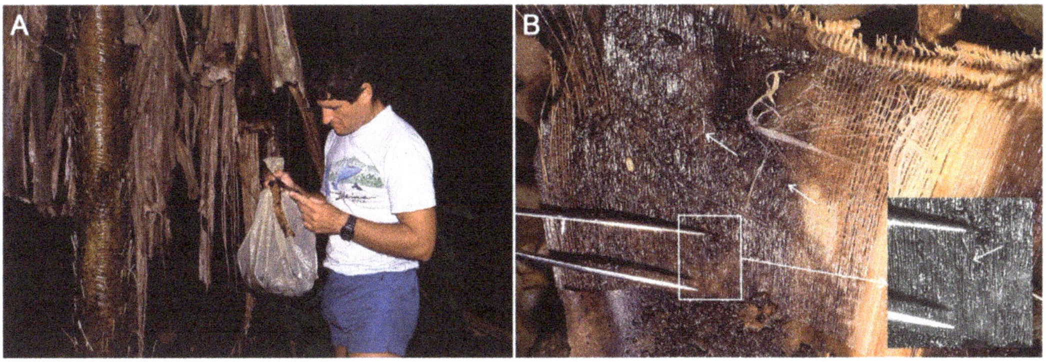

Figure 1. Habitat of *Namanereis catarractarum* (Nereididae) in the moist dead leaf axils of *Pandanus vitiensis* beside Savuro Creek near Suva, Fiji. (**A**) Collecting dead leaves for examination in the lab. (**B**) Worms (almost transparent, arrows indicating position) found at the base of dead leaves in the moist layer between the plant fibers. Photographs: C. Glasby.

6. Hypersalinity and Aridity

Hypersaline conditions occur on land and in the sea. On land, such conditions are often associated with aridity. The Australian fabriciid *Manayunkia athalassia* (Figure 2B) has been recorded from hypersaline ephemeral lakes adjacent to the Coorong Lagoon in South Australia and other undescribed species of the genus are known from records in similar habitats in Western Australia [30]. *Manayunkia athalassia* can tolerate salinities of up to three times that of normal seawater (27–95‰). The worms also appear to be resistant to desiccation over the hot, dry summer months in southern Australia, as they can be revived from dried mud samples with the addition of distilled water (Table 1) [31]. In the deep-sea of the Gulf of Mexico, the orbiniid *Methanoaricia* sp. can tolerate short stays in hypersaline brine pools associated with hydrocarbon seeps, but usually avoids the most extreme conditions by aggregating on top of the associated mussel beds (Table 1) [32].

Marine Naididae are common in shallow tropical seas (including the intertidal zones), sometimes even in habitats with hypersaline conditions, with documented records of the genera *Heterodrilus* and *Ainudrilus* from up to approximately 54‰ in Saudi Arabia and Kenya [33,34]. From the Kenyan site, a high-intertidal beach in a mangrove, Healy [35] also described a species of *Marionina* (family Enchytraeidae).

In North Africa, water evapotranspiration is important due to the semi-arid to arid climate. This often results in high water conductivity or even high chlorosities, which is probably a factor explaining why the oligochaete fauna is impoverished in these regions [36]. The species present are known to prefer highly mineralized or brackish waters; they include *Paranais birsteini* var. *maghrebensis* and *Nais elinguis* (Naidinae, Nainididae) [37], *Epirodrilus* spp. (Rhyacodrilinae, Naididae), forms of *Tubifex tubifex* (Tubificinae, Naididae), and *Cernosvitoviella immota* (Enchytraeidae) [38]. One species, *Aktedrilus yacoubii* (Naididae; Figure 3A), belongs to a mainly marine subfamily Phallodrilinae [39].

Among aquatic oligochaetes, encysting is an extreme adaptative response to the drying up of the animal's environment. *Lamprodrilus mrazeki* (Lumbriculidae) is a rare example of encysting in aquatic oligochaetes when pools dry up in summertime (Table 1). Architomy, i.e., asexual reproduction by fragmentation, then occurs in the cysts [40,41]. The ubiquitous freshwater *Lumbriculus variegatus* and *Tubifex tubifex* have been observed dormant in a mucous cyst in the sediment of temporary pools [17].

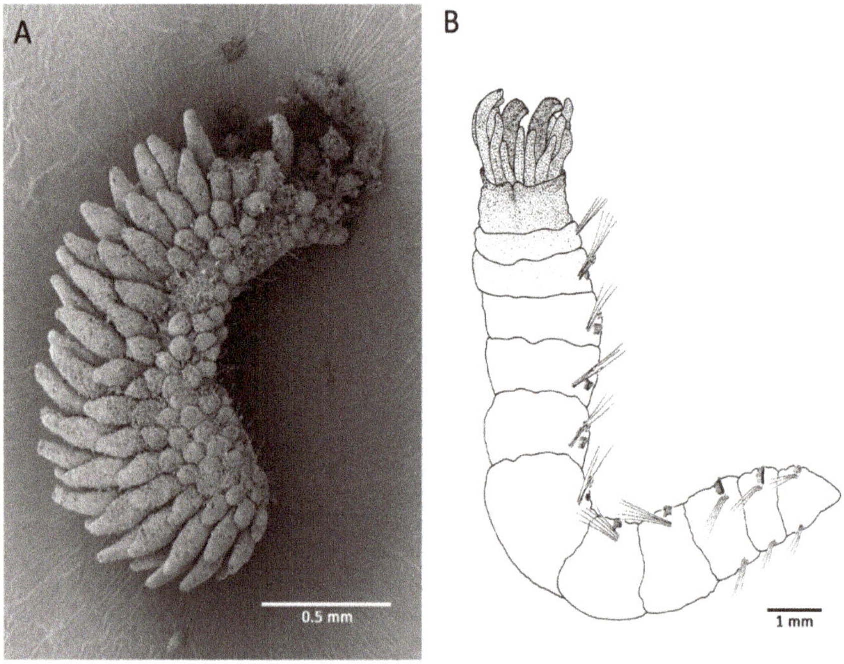

Figure 2. (**A**) The genus *Baikalodrilus* (Tubificinae, Naididae) is one of the most characteristic faunistic elements of the oligochaete community in Lake Baikal and constitutes a species flock in the lake with no less than 24 morphospecies known to date. *Baikalodrilus digitatus* is a species with epidermal projections of the body wall, giving the worm a resemblance to some nudibranchs. Photograph: P. Martin. (**B**), *Manayunkia athalassia* (Fabriciidae) from an ephemeral land-locked lake adjacent to Coorong, South Australia. After Hutchings et al. (1981: Figure 1a [31]).

A terrestrial habitat is considered extreme for polychaetes as, unlike oligochaetes, most typically lack protection for their eggs, and have vulnerable larval stages (but see below). There are only two entirely terrestrial families of polychaetes. Both are species poor and members live among other meiofauna in damp environments. The oligochaete-like *Hrabeiella periglandulata* (Hrabeiellidae) occurs in fresh, slightly acidic and well-drained soils in meadows and different types of forests (beech, fir, spruce) across Europe and northern Asia [42,43]. The Parergodrilidae are represented by two genera, the monotypic *Parergodrilus (P. heideri)*, and *Stygocapitella* which is represented by 11 cryptic species. *Parergodrilus heideri* is found terrestrially in the humus-rich upper layers of soil in the Palearctic Region and *Stygocapitella* species are found worldwide, excluding the tropics, in the supralittoral zone of sandy beaches (i.e., semi-terrestrial) [44–49]. Both families have reduced (or an absence of) nuchal organs, the epidermis shows less (or no) ciliation, there are fewer types of receptor cells, and in Parergodrilidae embryos develop directly inside cocoons (Table 1) [45,49,50].

Most members of the non-clitellate Aeolosomatidae and Potamodrilidae, together comprising the taxon Aphanoneura, have an aquatic oligochaete lifestyle but some *Aeolosoma* can form desiccation-resistant cysts (a hardened membrane of mucus secreted by the worm) that allow them to survive adverse conditions and possibly enabling transport between water bodies by animals (Table 1) [51]. Herlant-Meewis [52] studied the encystment of *Aeolosoma hemprichi* in Canada, and concluded that it was induced by low temperatures, enabling hibernation during the winter season.

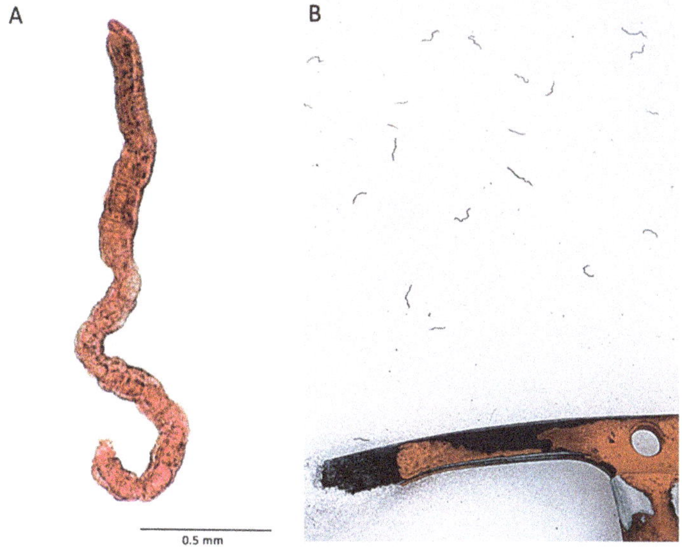

Figure 3. (**A**) *Aktedrilus yacoubii* (Phallodrilinae, Naididae), a stygobiotic oligochaete species from a resurgence in Morocco, illustrated as a freshwater representative of a mainly marine subfamily. Photograph: P. Martin. (**B**) The highly-pigmented aquatic "ice worms", *Mesenchytraeus solifugus*, live in the melted ice cracks of North-American glaciers, tolerating only a narrow ±7 °C range near freezing. Photograph: Dick Culbert, Stikine Icecap between Alaska and British Columbia; image cropped, otherwise unmodified, licensed under cc-by-2.0.

7. Extreme Temperatures, and Vents and Seeps

Aqueous environments are inherently less variable in temperature than subaerial environments. Therefore, it is not surprising to find more temperature extremophiles in terrestrial and supralittoral areas. Although many polychaete families show high diversity in sediments of polar seas, one appears to be particularly noteworthy. Dinophilidae, particularly *Trilobodrilus* species, inhabit interstitial coarse sands in the arctic, subarctic and boreal coasts, which can experience even colder temperatures than polar bottom waters, which largely remain above zero. In the laboratory, *T. axi* can withstand supercooled water without ice formation, with individuals of this species showing no damage after 5 h at −12 °C [53].

Holmstrup [54], referring to Zachariassen [55], noted that terrestrial oligochaete worms may have developed different strategies for survival in water-freezing temperatures. Sensitive species, likely to die if frozen, may promote supercooling by removal of ice-nucleating agents and proliferation of polyhydric alcohols and antifreeze proteins. Other species are more freeze tolerant and able to establish a controlled, protective freezing by dehydration of their extracellular fluids at high subzero temperatures by production of ice-nucleating agents. Holmstrup [54] studied the physiological mechanisms underlying the cold resistance of cocoons and juveniles of the earthworm *Dendrobaena octaedra* and found them not to produce cryoprotective, low molecular weight substances, but instead to use extensive dehydration to lower the mean supercooling point.

Sømme and Birkemoe [56] provided evidence that some arctic terrestrial Enchytraeidae also avoid freezing by dehydration at subzero temperatures and in this way survive temperatures as low as −15 °C. On the other hand, Slotsbo et al. [57] showed that the enchytraeid *Enchytraeus albidus*, which occurs in the supralittoral zone of arctic beaches, do use glycose as a cryoprotective substance (Table 1).

The pigmented "ice worms", *Mesenchytraeus solifugus*, live in ice cracks of some North-American glaciers, grazing on microalgae on ice surfaces on summer nights [58,59]; Figure 3B. They are aquatic rather than terrestrial animals, and they tolerate only a narrow ±7 °C range near freezing [60]. The species is the only psychrophilic annelid, i.e., it lives and reproduces in the ice/snow of glaciers. It lacks cryoprotective agents but is able to increase its internal energy level when temperature falls, and all its biological processes are maintained even at 0 °C [61].

At the other end of the temperature scale, some earthworms, especially *Eisenia*, are well known for the ability to thrive in the high temperatures of organic compost, but the upper limit for survival is approximately 33 °C [62]. Studies of tropical earthworms have shown that cocoons of species of *Amynthas* can survive up to 38 °C (Table 1) [63].

In the marine realm, high temperatures are muted, although compared to surrounding conditions they may be regarded as 'locally extreme'. Vestimentiferan siboglinids and alvinellids inhabiting deep-sea hydrothermal vents have the highest reported temperature tolerances for an aquatic annelid. The latter have developed strategies to cope with high (and variable) temperatures in these areas. Some species prefer temperatures between 40 and 50 °C, tolerating 55 °C for short periods [64]. Their tubes insulate the worm from high temperatures [65]. To cope with high temperatures, the alvinellids *Alvinella pompejana* and *Paralvinella sulfincola* have developed adaptations including a protective tube/cocoon made of positively and negatively twisted polymers layers comprising a stable glycoprotein matrix and elemental sulfur, a range of heat shock proteins and stress oxidative enzymes that remain stable and active at temperatures greater than 50 °C and, production of a thermally stable collagen (Table 1) [66–70]. To our knowledge, there are no records of oligochaetes from deep-sea hydrothermal vents with high temperatures (see McHugh and Tunnicliffe [71]), but naidid clitellates (in *Tubificoides* and *Limnodriloides*) have been described in close proximity to hydrocarbon seeps on the continental slopes (at 540 to >2000 m depth) of the Gulf of Mexico [72] and off the US coast of the North-West Atlantic Ocean [73].

Many polychaetes living in "extreme" habitats, in particular elevated temperatures, occur in high abundances, have high growth rates, and may be numerically dominant. Several polychaete species living in the abyss display this behavior including the ampharetid, *Amphisamytha galapagensis*, which reaches densities of approximately 2200–3000 individuals per m^2 near hydrothermal vents in the Juan de Fuca area [71,74]. Hydrothermal vents are relatively short lived, subject to major disturbances and waxing and waning of fluid flow. Thus, the vestimentiferan siboglinid *Riftia pachyptila* grows extremely fast (the fastest growing invertebrate), but only in areas with relatively vigorous diffuse vent flow, and is relatively short lived [75,76]. Mass colonization of vestimentiferan siboglinids of new sites has been found, with growth to massive adult size completed in two years [77]. Other species living in extreme habitats live in very low densities, and presumably have also had to adjust their reproductive strategy to be able to reproduce. Cold seeps are relatively long lived and stable, so invertebrates living there such as frenulate siboglinids grow extremely slowly and are extremely long-lived [78–80]. Large individuals of *Lamellibrachia luymesi* with tube lengths over 2 m of are estimated to be older than 200 yr [79].

Hydrothermal vents are known in Lake Baikal, in Frolikha Bay, at approximately 420 m deep. Sediment temperatures under the bacterial mats that develop there are greater than 16 °C, a 'local extreme' compared to an ambient temperature of 3.47 °C [81]. On the contrary, the Frolikha vent houses rich freshwater benthic communities, which are built on ancient carbon and depend on in situ bacterial biochemosynthesis [82,83]. Densities of oligochaete species living near the vent have been found higher than at similar water depths and approximately nearly half of these species are only found in the dimictic zone of the lake, which extends up to 250 m deep in Lake Baikal [84,85]. Their presence at approximately 420 m depth, near the Frolikha vent, suggests that their bathymetric distribution is restricted by availability of food. Lake Baikal is indeed strongly oligotrophic and nutrient elements in the surface layer are recycled on average approximately four times before being removed to deep waters [86]. Hydrothermal vents in Lake Baikal would instead appear as "food refuges" in an extreme environment from the point of view of food availability. In this respect, Lake Baikal shares similarities with the groundwater environment (see below).

8. Extreme Depths, and Other Depth-Related Factors

Great water depths are often associated with the marine environment, where depths can exceed 10 km. The pressure in deep water is extremely high, but this is largely unproblematic for soft-tissue animals (lacking free gases), as water volume is only slightly affected by it. Persistent low temperature and scarcity of food, however, are more challenging. The predominantly marine polychaetes are common in the deep-sea, where they occur in the water column and in the sediment of the deepest trenches, and as such are not dealt with in detail in this section. Many species of oligochaetes belonging to families Naididae (especially subfamily Phallodrilinae) and Enchytraeidae (genus *Grania*) occur in the deep-sea (species are listed in Erséus and Rota [87]), with the deepest record of a naidid from approximately 7700 m depth (*Bathydrilus hadalis*, in the Aleutian Trench, North Pacific Ocean [88]), and of an enchytraeid from approximately 2900 m (*Grania papillinasus*, off France in the North Atlantic [89]). For *B. hadalis*, even its clitellate-type cocoon has been described ([90] Figure 5).

Lake Baikal is one place on land where annelids can encounter extreme depths. The lake is unique because the water circulation carries oxygen to its deepest point (1642 m), which makes it the only freshwater lake in the world with an inhabitable abyssal area [91–93]. Oligochaetes are present at all depths although with an exponential decline of abundance and a decrease in sizes of animals with depth. No genuine deep-water oligochaete fauna seems to exist in Lake Baikal and the bathymetric distribution of oligochaetes is rather better explained by lower food availability at depth than by hydrostatic pressure ([85]; see also above).

Polychaetes (*Manayunkia* mentioned above) also occur in Lake Baikal, but they are primarily littoral species (3–20 m), although they are also present, but patchy, in the deep-water zone, below 250 m [27,28,94]. In the latter case, their presence is probably accidental because these sestonophagous, filter-feeding animals require an abundance of food particles which are probably limited at depth. Thus, the low salinity is more challenging for polychaetes (see above) than the extreme bathymetry of the lake. The strikingly low electrolyte content of Baikalian waters (not more than 150 μmol/L sodium and 12 μmol/L chloride) [95] can be considered as an extreme environment even for freshwater oligochaetes. *Baikalodrilus discolor*, which is one of the more characteristic faunal elements of the oligochaete community in Lake Baikal [96], presents a hemolymph with extremely low osmotic concentrations, which are considered as a physiological adaptation to the low electrolyte concentration of Baikal waters [29].

9. Subterranean and Cave Environments

The subterranean environment presents a difficult combination of extreme conditions including darkness, scarcity of food, limited variations in temperature, high physical fragmentation and, for the underground aquatic environment, low oxygen [97,98]. Only a few lineages are able to colonize and adapt to these inhospitable habitats [99]. Certain polychaete families have adapted to living in the water column of marine/anchialine caves, including the normally benthic Protodrilidae, Nerillidae, and Scalibregmatidae [100–102], while others live benthically and have secondary troglomorphic traits such as loss of eyes, pigmentation, and elongation of the sensory appendages [103]; Figure 4A–F. *Marifugia cavatica* (Serpulidae) is eyeless and lacks pigmentation, and is also considered to be a Tertiary relict of marine origin (Table 1) [104]; Figure 5B. The aptly named genus *Troglochaetus* (Nerillidae) is also found in the hyporheic zone of caves, wells and springs, with *T. beranecki* currently thought to occur in both Europe and North America. *Troglochaetus* has rudimentary parapodial cirri and, like all nerillids, development is direct, either via an external brood or in cocoons (Table 1) [105].

When polychaetes transition to underground environments, their typical sense organs (eyes, cirri) also tend to be reduced or absent, while other organs may be enlarged to compensate—for example, elongated parapodial cirri and antennae presumably both increase sensory awareness. Nereidid polychaetes adapted to living in the water column of marine/anchialine caves have been shown to have secondary troglomorphic traits such as loss of eyes or elongation of the sensory appendages (Table 1) [25,103].

Aquatic clitellates have a habitus that makes them pre-adapted to live in the subterranean environment [106]. In addition, they do not exhibit any peculiar morphological adaptations to subterranean life that can be seen in other subterranean organisms (loss of eyes, elongation of appendages and body, loss of pigmentation, increase in sensory structure). Therefore, the stygobiotic (i.e., obligatory hypogean) nature of the species can only be inferred from their exclusive presence in the subterranean environment. Despite their morphological pre-adaptation, only approximately 100 species of aquatic oligochaetes are exclusively present in groundwater out of approximately 1700 aquatic oligochaete species and 1100 freshwater species known to date [14,106]. However, biological and ecological traits that predispose oligochaetes and leeches to successfully colonize underground habitats are still mostly unknown. The ability to cope with limited food resources is probably an important adaptive factor.

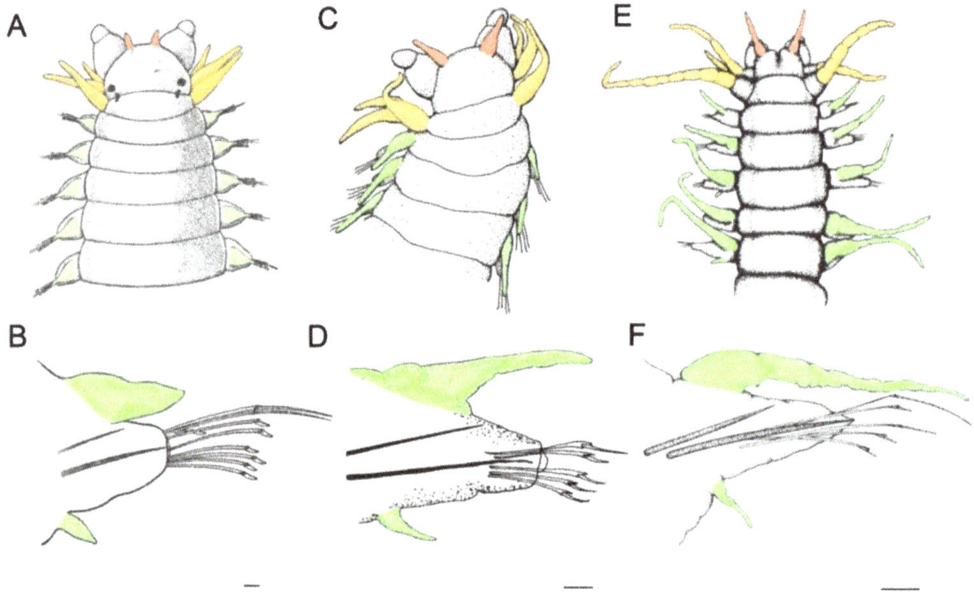

Figure 4. Typical morphology of *Namanereis* (Nereididae) exemplified by *Namanereis littoralis* (**A**,**B**), compared to the secondary troglomorphic traits in *Namanereis beroni* (**C**,**D**) and *Namanereis araps* (**E**,**F**), including loss of eyes, elongation of antennae, tentacular and parapodial cirri, and increase in length of the terminal blades of the articulated falcigerous chaetae (bar showing relative length). Adapted from Glasby et al. (2014: Figure 7 [25]).

On land, cave systems sometimes harbor physiologically or morphologically modified clitellates. An extreme example is the leech *Croatobranchus mestrovi*, originally described by Kerovec et al. [107], but also known as *Erpobdella mestrovi* [108], which lives in deep shaft-like caves in the Velebit karst area in Croatia. It has been recorded as deep as 1320 m below surface in waters of 4–6 °C, and it walks around on the cave walls with leg-like appendages, which probably also function as gills (Table 1) [108] (Figure 5A).

Figure 5. Two examples of subterranean annelids. (**A**) The leech *Croatobranchus mestrovim* lives in deep shaft-like caves in the Velebit karst area in Croatia. It has been recorded as deep as 1320 m below the surface, and it walks around on the cave walls with leg-like appendages. (**B**) The serpulid polychaete *Marifugia cavatica* attaches its calcareous tube to the walls of freshwater caves found on the eastern Adriatic coast and further inland; it is eyeless and lacks pigmentation. Photographs: Jana Bedek, archive of the Croatian Biospeleological Society.

Since photosynthesis is absent in the subsurface, life in this environment is highly dependent on trophic resources that can be imported from the surface environment. Yet food is not so much scarce as it is unevenly distributed. Moreover, among other ecosystems, the subsurface most resembles the hypolimnia of lakes (e.g., Lake Baikal) and the deep-sea in that, in this permanent and omnipresent darkness, the food that sinks from the surface into these habitats is patchily distributed. [99,106]. In Lake Baikal, food is clearly a limiting factor for some annelid species, as evidenced by the presence of some species near hydrothermal vents, where chemoautotrophic production is a substitute to primary photosynthesis production, located at depths where such species are never found [85]. As said earlier, the hypolimnion of Lake Baikal and groundwater are similar in environmental characteristics, which led Timm [109] to hypothesize that founders of some Baikalian oligochaete species flocks originated from the groundwater fauna.

Ironically, groundwater, as an extreme environment, turns out to be a refuge for annelids living in areas where conditions on the surface are even more extreme. In desert areas (Morocco, Arabian Peninsula), subterranean waters provide refuge for a diversified annelid fauna, including species of the oligochaete family Phreodrilidae whose presence in the northernmost part of an original geographical distribution assumed pan-Gondwanian is considered relictual, and the bifid-jawed polychaetes *Namanereis araps* and *N. socotrensis* whose present-day distributions on the Arabian Peninsula and Socotra Archipelago are suggestive of speciation following closure of the Tethys Sea [25,36,110,111]. These observations are interesting in that they demonstrate that an extreme environment loses its "extreme" nature once pre-adaptations, or exaptations, have allowed a population to pass the filter imposed by key ecological factors (in this case, the barrier of darkness) and actively colonize their new habitat [99].

10. High-Energy Habitats: Mobile Sediments and Surf Zone

Several families of polychaetes are specialized for living in unstable sedimentary environments of surf beaches. Species of the interstitial genera *Protodrilus* (Protodrilidae) and *Saccocirrus* (Saccocirridae) are swash zone specialists that are able to cling to sand grains using their caudal appendages and sticky mucous-exuding skin, and feed on plankton and carrion using elongated, highly motile ciliated feeding palps [112–115]. Larger-bodied forms such as Australian beachworms *Australonuphis* species (Onuphidae) also scavenge for dead morsels of food in the swash zone of surf beaches. Their great length, up to 2 m long, and muscular body help them to maintain station in the unstable substrate, and their prolonged anterior parapodia enable them to dig effectively and move rapidly toward stranded dead animals [116].

Other families also contain species adapted to high-energy conditions—the scalibregmatid *Axiokebuita cavernicola* occurs only in coarse sands and gravel sediments in areas where there is active water movement produced by waves and tides and where the water turbulence precludes particle deposition [100]. Scalibregmatid polychaetes anchor themselves to coarse sediment by the adhesive papillae of the pygidium. These same worms use undulatory movements to swim for short distances.

Many species of Sabellariidae are adapted for life in the surf zone or shallow depths where current is strong for the animals to feed and capture sand grains for tube building [117]. They actively build tubes made of sand, shell fragments or other suitable particles glued together by a dark layer of mucoprotein secretions [118–120]. The tubes are attached to a variety of substrata, including rocks, seaweeds or invertebrates or other sabellariid tubes [30,121,122].

Several pholoin sigalionids (i.e., *Pholoe, Laubierpholoe*) are also known to inhabit interstitial sands of the surf zone including also *Imajimapholoe parva*, an intertidal member of the interstitial fauna, which is found in the nearshore surf zone in northern Japan. *Imajimapholoe parva* produces a sticky substance from their adhesive glands [123], which presumably helps them to stay within the coarse sediment.

In the surf zone, adhesive glands and a small flat body are adaptations to interstitial life [124]. The interstitial can be seen as an extreme environment due to the tiny size of the voids in which animals are constrained to live. Reduction in body size is a common morphological adaptation to this habitat (Table 1). In addition, as the marine littoral zone may function as a transitional zone between the marine and continental subterranean environments [106], adaptation to life in interstitial habitats can be considered for many marine meiobenthic species as a pre-adaptation to life in ground water, another extreme environment. Miniaturization and progenesis have both played a significant role in the evolution of annelids in interstitial environments [125–127].

The size of the marine interstitial *Psammodriloides fauveli* Swedmark (Psammodrilidae), which is 25 times smaller than *Psammodrilus balanoglossoides* Swedmark, is not accompanied by a reduction in cell size but by anatomical changes and the absence of some organs as a result of progenesis [124].

The torrential regime of streams also constitutes in itself an "extreme" environment (water turbulence and sediment movement). In Maghreb countries (northwest Africa), it is responsible for strong variations in physico-chemical water parameters (temperature, salinity, dissolved oxygen, sedimentary load, etc.) [36]. As a result, streambeds are often an inhospitable environment for the oligochaete fauna, which is forced to find refuge beneath and alongside the riverbed (hyporheic zone). Due to greater environmental stability, groundwater (underflow, groundwater tables) probably plays an important role in maintaining a diverse aquatic fauna. In the high mountain Pyrenean torrents, the Enchytraeidae and the genus *Trichodrilus* (Lumbriculidae) are the only oligochaetes to resist the velocity of the current and the coldest temperatures [128]. Martínez-Ansemil and Collado [129] also showed that the Enchytraeidae dominate in streams with high current velocities associated with a mineral substrate of large granulometry. In the Maghreb countries, *Trichodrilus* is also commonly present and has been collected in streams where the temperature can reach

38 °C [130]. It thus appears that the ability of this genus to persist in streams with torrential flow depends less on its resistance to extreme temperatures than on its ability to withstand the power of fast-flowing water over a coarse, gravelly bottom (Table 1). The fact that *Trichodrilus* is frequently found in the hyporheic zone of the oueds is probably a strategy that allows the genus to cope with these extreme living conditions [36].

11. Low Oxygen: Organic Enrichment, Sulfidic, Methanic Sediments

Some groups of oligochaetes and polychaetes thrive in environments where levels of oxygen are low and hydrogen sulfide and methane are high. These conditions are common in coastal and terrestrial waters where there is organic enrichment, and in the deep-sea where there is mineral enrichment resulting from, for example, hydrocarbon seeps. Perhaps the best-known examples among polychaetes are the capitellids. *Capitella* spp. are often abundant in organically rich sediments (up to 47,000 ind.m^{-2} in *C. telata*, see [131]). In the latter category, *Capitella* spp. can thrive in organically rich sediments by their fast response to large influxes of organic material and the resultant anoxia due to the lack of competition by less-stress-tolerant species as well as their early maturation and high reproductive potential [132]. Forbes et al. [133] suggested that *Capitella* spp. populations were strongly influenced by a physiology well adapted to exploit low oxygen and respond rapidly when more favorable oxygen conditions returned. Other intertidal capitellid species are able to tolerate periods of low oxygen levels or reduced salinities during low tide by building up oxygen debts which are repaid on the incoming tide (Table 1) [134]. Such physiology may allow some species to live in highly anaerobic sediments.

Cirratulids are often the dominant group in soft, nearshore sediments having reduced oxygen levels often associated with organically enriched sediments. For example, *Raricirrus beryli* inhabits upper slope sediments, sometimes near oil rigs, at continental shelf depths down to 500 m in the North Sea [135,136], and is a colonizer of organically enriched sediments in experiments near mud volcanos in the Bay of Cádiz, Spain [137].

Among Errantia, members of Namanereidinae (Nereididae) and Eunicidae are perhaps the best known for living in reduced sedimentary environments. For example, the *Marphysa sanguinea* complex is one of the few polychaetes able to endure the high organic loads of marine fish farms in the tropics [138] and among the nereidids, the intertidal *Namalycastis* species are commonly one of the few annelids found in the polluted, foul-smelling mangrove muds.

In aquatic oligochaetes, the ability of many species to thrive in heavily organically enriched sediments has long been known [139]. This ability has been widely exploited in research on pollution biology and its applications, for which oligochaete worms have proved useful. Rodriguez and Reynoldson [140] give a recent and detailed overview for many species. Other interesting examples are some marine naidids, e.g., members of the genera *Tubificoides* (Tubificinae) (e.g., [141]) and *Thalassodrilides* (Limnodriloidinae) (e.g., [142], which are well known for their occurrence in eutrophic, often polluted, coastal sediments enriched with hydrogen sulfide. Of particular interest are also the marine gutless Phallodrilinae, which actively migrate between the lower sulfidic and the upper oxygenated sediments and have evolved a mutualistic obligatory symbiosis with chemoautotrophic bacteria (Table 1) (see [143] and below).

A physiological strategy common to both oligochaetes and polychaetes to cope with reduce oxygen levels is modification of their respiration pigments, including hemoglobins. The orbiniid *Methanoaricia* shows a high hemoglobin–oxygen affinity and is able to withstand extended periods of anoxia [144,145]. The oligochaetes *Potamothrix hammoniensis* and *Tubifex tubifex* living in an oxygen-poor sediment or even periodically anoxic are known to be rich in erythrocruorin [17]. In many annelids, erythrocruorins provide large oxygen-carrying capacity through their high number of oxygen binding sites per complex and their high concentration in the blood [146]. Erythrocruorin's high affinity for oxygen is usually seen as an adaptation to the hypoxic conditions prevailing in the habitat of animals using this protein.

Rhyacodrilus hiemalis is a freshwater naidid species endemic to Japan. It is known for its remarkable seasonal vertical migration in the sediment of Lake Biwa. During the cool season, it is a dominant littoral component of the oligochaete communities. During the summer, however, it can migrate up to 90 cm deep into the anoxic sediment, where temperatures are milder than in the littoral zone. In the absence of diel vertical migration, *R. hiemalis* is assumed to use anaerobic respiration when aestivating in this extreme environment, which could be based on glycolysis of glycogen [147].

The family Siboglinidae is specialized for life at the border of aerobic and anaerobic environments, including deep-sea vents and cold seeps. Siboglinids have a high concentration of hemoglobin in solution, which binds and transports both oxygen and sulfide, the latter in a non-toxic form that can be released to symbiotic chemoautotrophic bacteria [148]. Siboglinids utilize the organic material synthesized by the chemoautotrophic bacteria in their modified mid gut (trophosome) to produce their own organic compounds and derive energy. *Osedax* (Siboglinidae) utilize chemoheterotrophic bacteria to source their energy from whale bones and other bones on the seafloor.

While vestimentiferan siboglinids are arguably the most distinctive members of the faunal assemblage of hydrothermal vents, the polychaete family Alvinellidae is also well adapted to live in this harsh environment. Alvinellids differ in having functional feeding tentacles and a well-developed gut. They derive nutritional products from chemoautotrophic bacteria, which attach either to the inner part of the tube and/or mucous threads on the body secreted by the worm, and are ingested by the worm [149–151]. Further, the hemoglobins of Alvinellidae exhibit very high affinities for oxygen counterbalanced by a pronounced Bohr effect, which allows for an enhanced release of oxygen under low-pH conditions [152–154].

In both families, the relationship between the worm and its symbiotic bacteria has led to morphological adaptations in the worm. In siboglinids, the symbiotic bacteria live in a trophosome, a modification of the larval midgut; in alvinellids such as *A. pompejana* the epibiotic bacteria are housed in dorsal or posterior end expansions bearing modified geniculate setae [155].

Among the oligochaetes, one particular lineage (genera *Olavius* and *Inanidrilus*) of marine gutless Phallodrilinae has evolved a mutualistic obligatory symbiosis with subcuticular chemoautotrophic bacteria, resulting in the complete loss of the worms' mouth and alimentary canal (e.g., [143,156]). These gutless phallodrilines are particularly common in calcareous fine sands of shallow coral reef areas, but they may be found also in various other marine habitats, from the intertidal zone to several hundred meters depth. The symbiotic bacteria obviously provide nutrients to their hosts, which may migrate between the lower sulfidic and the upper oxygenated sediments in the sediment, implying "that sulfide is taken up in the anoxic deeper sediment layers, oxidized to sulfur by either nitrate from the environment or oxygen from the worm's hemoglobin, and the sulfur stored in the bacteria until the worms migrate to upper sediment layers where the sulfur could be fully oxidized to sulfate" ([143]: p. 256). In some cases, however, gutless worms may, in addition to the sulfide-oxidizing bacteria, have sulfate-reducing symbionts producing sulfide; these oligochaetes thus have established their own sulfur circuit by recycling sulfide internally [157].

Living in low oxygen sediments has produced one similar morphological response among gut-bearing oligochaetes and polychaetes: a proliferation of gills on the head or the posterior region, depending on which end of the body is closest to the oxygenated sediment/water-air interface. For example, the large, bright red gills of vestimentiferan siboglinid tube worms provide the huge surface areas required to extract oxygen from surrounding, often oxygen-poor, deep-sea waters. The animals extend their red branchial crown and white obturaculum into the water, leaving the rest of the body, from the collar backwards, inside the tube. The posterior end of both oligochaetes and polychaetes may be prolonged, or flattened, to increase the body surface area in order to maximize oxygen absorption.

Polychaetes, such as Sabellidae, Fabriciidae and Serpulidae, can also take up oxygen through their multi-functional, pinnulated tentacles surrounding the head, which also serve to collect food. In addition to modified posterior ends, gills and tentacles, parapodial structures can be modified (usually by way of enlarged and/or flattened lobes) to provide the additional surface area for oxygen uptake. For example, namanereidin polychaetes (*Namalycastis* species) inhabiting the decaying vegetation of mangrove zone use their highly vascularized, leaf-like posterior parapodial cirri for oxygen uptake [158]. Opheliids, capitellids and cirratulids are often abundant in low oxygen environments of the coastal zone; they use gills, and sometimes modified posterior ends to facilitate oxygen uptake.

Among oligochaetes, three morphological adaptations can be seen to cope with low oxygen content in the environment. Gills can sometimes be observed on some aquatic oligochaetes, but this is not a widespread adaptation. In the Naididae, they are known around the proctodeum in *Dero* and *Aulophorus*, or as external gills supported internally by hair chaetae in *Branchiodrilus* (Figure 6), or as finger-like gills dorsally and ventrally on the hind body of *Branchiura sowerbyi* [16]. In the Phreodrilidae, *Phreodrilus branchiatus* has finger-like gills present laterally [159]. Another type of morphological adaptation is the absence in species of the genus *Aulodrilus* (Naididae), of segmentation of their posterior end which is used as a respiratory organ (Table 1) [40]. In the Lumbriculidae, *Lumbriculus variegatus* and some *Trichodrilus* species, the circulatory blood vessel has developed networks of branched and blind lateral blood vessels that probably allow better irrigation of the segments and, hence, better oxygenation [17].

Figure 6. *Branchiodrilus* sp. (Naidinae, Naididae), a branchiate oligochaete showing external gills supported internally with hair chaetae (illustrated specimen from Cambodia). Photograph: P. Martin.

Another morphological strategy to maximize oxygen uptake exclusively employed by polychaetes is to live in very long tubes. Frenulate siboglinids use their long permeable tubes to reach oxygenated surface waters when the deeper layers are reduced. The animal's great length allows it to bridge the redox discontinuity, with oxygen uptake in the anterior body transported by its blood, bound to hemoglobin, to the symbiotic bacteria in the lower part of its trunk where sulfide or thiosulfate diffuse in from the sediment [148].

Sediment pore water and overlying sea water provide dissolved carbon dioxide, and organic nitrogen may be obtained as dissolved organic compounds or ammonia from the pore water. Similarly, *Spiochaetoperus* species (Chaetopteridae), which can be found in sediments from coastal to deep-sea hydrothermal vents, appear to use their long straw-like tubes to survive in the reduced muddy sediments adjacent a former bauxite refinery in Gove Harbour, northern Australia; they are the only macroinvertebrate able to tolerate the anoxic, alkaline muds resulting from waste water derived from the use of caustic soda to extract alumina from bauxite, although precisely how they do so is unknown (see also Extreme pH below; CJG pers. obs.).

12. Inorganic and Organic Contaminated Sediments

A few groups of polychaetes can tolerate heavy metal pollution. At Port Pirie in South Australia, sediments near an iron ore smelter operating since the beginning of the 1900s have extremely high levels of heavy metals—they are almost exclusively colonized by polychaetes. Ward and Hutchings [160] found that three or four species of polychaete were the most invertebrates at the most contaminated sites adjacent to the outflow pipes. One of these, a species of *Capitella*, occurred almost exclusively at the most contaminated site at a density of 322 worms per m^2, indicating that the species is self-sustaining and can tolerate very high concentrations of metals. Dense populations of *Capitella* can also build up rapidly in oil contaminated sediments [161], and members of this genus and *Ophryotrocha* (Dorvilleidae) are commonly encountered in the organically-enriched sediments of aquaculture facilities [132].

Heavy metals are a common component in sewage outfall discharges and a positive relationship between the concentrations of these components in the sediment and abundance of capitellids is shown in several studies. Densities of more than 300 ind. m^{-2} were found to occur in a self-sustaining population in proximity to a south Australian lead-zinc smelting facility, where high concentrations of Pb (up to 5270 $\mu g.g^{-1}$) and Zn (up to 16,700 $\mu g.g^{-1}$) occurred in the sediment [160].

Among the oligochaetes, the marine Limnodriloidinae *Thalassodrilides* cf. *briani* (Naididae) was found to survive in polluted sediments with polycyclic aromatic hydrocarbons, in particular 1-nitronaphtalene, a toxic and carcinogenic chemical, and to biotransform them into substances that are not toxic to fish (Table 1) [162,163]. *Thalassodrilides* cf. *briani* increases the CYP (cytochrome P450) gene expression when exposed to polluted sediments with polycyclic aromatic hydrocarbons (PAH). Cytochrome P450 enzymes play important roles in the metabolism of exogenous compounds such as PAHs [163].

13. Extreme pH

Lake Van in Eastern Turkey is the largest soda lake in the world [164], with a pH of 9.5–9.9, a salinity of 21–24‰, and an extreme alkalinity for lakes with 155 m eq/L [165]. Its impoverished fauna includes a single clitellate, *Enchytraeus polatdemiri*, which occurs at profundal (8–115 m) depths in the lake [166]. This taxon is a member of the *Enchytraeus albidus* species complex, where other members are characteristic of high-intertidal habitats in seashores [167]. Other *Enchytraeus* species considered closely related to *E. polatdemiri* are also known from brackish-water lakes (see, e.g., [166,168,169], giving further support to a marine rather than limnic origin of the Lake Van species.

14. Concluding Remarks

Polychaetes, clitellates, and the less diverse non-clitellate aphanoneurans present their own unique adaptations to cope with particular extreme environments, reflecting their existing body plans and morphological and physiological capabilities (Table 1). For example, some polychaetes construct protective tubes to cope with high-energy environments, oligochaetes are pre-adapted to life underground, and one leech is capable of growing legs to find its next meal. However, it is the shared similar strategies for coping with these adversities that is perhaps the most notable, particularly between polychaetes and clitellates—

both groups show the ability to construct protective cocoons, manufacture cryoprotective substances such as antifreeze and heat shock proteins, develop gills (greater capacity in polychaetes), transform their bodies into a home for symbiotic chemoautotrophic bacteria, metabolize contaminants, and avoid the most extreme of extreme conditions as much as possible by their local migrations and reproductive behaviors.

Annelids have largely retained a superficially simple worm-like body plan along with a mostly benthic/burrowing lifestyle, and yet they have been capable of immense radiation, both morphologically and physiologically. Over and over again, they have, by convergent evolution and in both directions, overcome and crossed the transitions from salt water to freshwater, sea to land via beaches, freshwater to soil, and surface water to subterranean water. Our overview has shown that these steps include a plethora of adaptations to extreme conditions and habitats not suitable for other animal life. Conceivably, it is the simplified "annelid" bauplan that has made it possible to cope with a multitude of living conditions. As one of the oldest group of invertebrates, annelids are indeed survivors and deserve their position among the most common and diverse animal groups on the planet.

Author Contributions: All authors contributed to the conceptualization, investigation, writing and revision in the following order: C.J.G., P.M., C.E. All authors have read and agreed to the published version of the manuscript.

Funding: This research received no external funding.

Acknowledgments: We thank the guest editors of this special edition, Maria Capa and Pat Hutchings, for inviting us to contribute this article. C.J.G. also thanks Olga Biriukova for preparation of figures, and the Australian Biological Resources Study for supporting Olga as part of the 'Annelid Key' project. C.E. is supported by the Swedish EPA's Environmental Research Fund in collaboration with the Swedish Agency for Marine and Water Management.

Conflicts of Interest: The authors declare no conflict of interest.

References

1. Struck, T.H.; Purschke, G.; Dordel, J.; Hösel, C.; Nesnidal, M.P.; Diersing, F.; Bleidorn, C.; Paul, C.; Hill, N.; Tiedemann, R.; et al. Phylogeny and evolution of Annelida based on molecular data. In *Deep Metazoan Phylogeny: The Backbone of the Tree of Life. New Insights from Analyses of Molecules, Morphology, and Theory of Data Analysis*; Wägele, J.W., Bartolomaeus, T., Eds.; De Gruyter: Berlin, Germany, 2014; pp. 143–160. [CrossRef]
2. Weigert, A.; Bleidorn, C. Current status of annelid phylogeny. *Org. Divers. Evol.* **2016**, *16*, 345–362. [CrossRef]
3. Struck, T.H.; Paul, C.; Hill, N.; Hartmann, S.; Hosel, C.; Kube, M.; Lieb, B.; Meyer, A.; Tiedemann, R.; Purschke, G.; et al. Phylogenomic analyses unravel annelid evolution. *Nature* **2011**, *471*, 95–98. Available online: http://www.nature.com/nature/journal/v471/n7336/abs/10.1038-nature09864-unlocked.html#supplementary-information (accessed on 20 December 2020). [CrossRef]
4. Kuo, D.H. The polychaete-to-clitellate transition: An EvoDevo perspective. *Dev. Biol.* **2017**, *427*, 230–240. [CrossRef]
5. Chapman, A.D. *Numbers of Living Species in Australia and the World*, 2nd ed.; Australian Biological Resources Study: Canberra, Austrilia, 2009; p. 84.
6. Struck, T.H. 2. Phylogeny. In *Handbook of Zoology. Annelida. Basal Groups and Pleistoannelida, Sedentaria I*; Purschke, G., Böggemann, M., Wilfried, W., Eds.; De Gruyter: Berlin, Germany, 2019; pp. 37–68. [CrossRef]
7. Purschke, G. On the ground pattern of Annelida. *Org. Divers. Evol.* **2002**, *2*, 181–196. [CrossRef]
8. Purschke, G.; Bleidorn, C.; Struck, T. Systematics, evolution and phylogeny of Annelida—a morphological perspective. *Mem. Mus. Vic.* **2014**, *71*, 247–269. [CrossRef]
9. Struck, T.H. Direction of evolution within Annelida and the definition of Pleistoannelida. *J. Zool. Syst. Evol. Res.* **2011**, *49*, 340–345. [CrossRef]
10. Rouse, G.W.; Pleijel, F. *Polychaetes*; Oxford University Press: Oxford, UK, 2001; p. xiii + 354.
11. Erséus, C.; Williams, B.W.; Horn, K.M.; Halanych, K.M.; Santos, S.R.; James, S.W.; Châtelliers, M.C.; Anderson, F.E. Phylogenomic analyses reveal a Paleozoic radiation and support a freshwater origin for clitellate annelids. *Zool. Scr.* **2020**, *49*, 614–640, in press. [CrossRef]
12. Erséus, C. Phylogeny of oligochaetous Clitellata. *Hydrobiologia* **2005**, *535–536*, 357–372.
13. Ferraguti, M.; Erséus, C. Sperm types and their use for a phylogenetic analysis of aquatic clitellates. *Hydrobiologia* **1999**, *402*, 225–237. [CrossRef]
14. Martin, P.; Martínez-Ansemil, E.; Pinder, A.; Timm, T.; Wetzel, M.J. Global diversity of oligochaetous clitellates ("Oligochaeta"; Clitellata) in freshwater. *Hydrobiologia* **2008**, *595*, 117–127. [CrossRef]

15. Purschke, G.; Westheide, W.; Rhode, D.; Brinkhurst, R.O. Morphological reinvestigation and phylogenetic relationship of *Acanthobdella peledina* (Annelida, Clitellata). *Zoomorphology* **1993**, *113*, 91–101. [CrossRef]
16. Rouse, G.W.; Fauchald, K. The articulation of annelids. *Zool. Scr.* **1995**, *24*, 269–301. [CrossRef]
17. Timm, T.; Martin, P.J. Chapter 21—Clitellata: Oligochaeta. In *Thorp and Covich's Freshwater Invertebrates*, 4th ed.; Thorp, J.H., Rogers, D.C., Eds.; Academic Press: Boston, MA, USA, 2015; pp. 529–549. [CrossRef]
18. Westheide, W. The direction of evolution within the Polychaeta. *J. Nat. Hist.* **1997**, *31*, 1–15. [CrossRef]
19. Rothschild, L.J.; Mancinelli, R.L. Life in extreme environments. *Nature* **2001**, *409*, 1092–1101. [CrossRef]
20. Gómez, F. Extreme Environment. In *Encyclopedia of Astrobiology*; Gargaud, M., Amils, R., Quintanilla, J.C., Cleaves, H.J., Irvine, W.M., Pinti, D.L., Viso, M., Eds.; Springer: Berlin/Heidelberg, Germany, 2011; pp. 570–572. [CrossRef]
21. Macelroy, R.D. Some comments on the evolution of extremophiles. *Biosystems* **1974**, *6*, 74–75. [CrossRef]
22. Glasby, C.J.; Timm, T.; Muir, A.I.; Gil, J. Catalogue of non-marine Polychaeta (Annelida) of the world. *Zootaxa* **2009**, *2070*, 1–52. [CrossRef]
23. Solís-Weiss, V.; Espinasa, L. *Lycastilla cavernicola*, a new freshwater nereidid from an inland Mexican cave (Polychaeta: Nereididae: Namanereidinae). *Proc. Biol. Soc. Wash.* **1991**, *104*, 631–639.
24. Glasby, C.J.; Timm, T. Global diversity of polychaetes (Polychaeta; Annelida) in freshwater. In *Freshwater Animal Diversity, Assessment*; Balian, E.V., Lévêque, C., Segers, H., Martens, K., Eds.; Springer: Dordrecht, The Netherlands, 2008; pp. 107–115. [CrossRef]
25. Glasby, C.J.; Fiege, D.; Van Damme, K. Stygobiont polychaetes: Notes on the morphology and the origins of groundwater *Namanereis* (Annelida: Nereididae: Namanereidinae), with a description of two new species. *Zool. J. Linn. Soc.* **2014**, *171*, 22–37. [CrossRef]
26. Glasby, C.J. Population structure and reproductive biology of *Ceratonereis limnetica* (Polychaeta: Nereididae) at Lower Portland, Hawkesbury River, Australia. *Mar. Biol.* **1986**, *90*, 589–595. [CrossRef]
27. Pudovkina, T.A.; Sitnikova, T.Y.; Matveyev, A.N.; Shcherbakov, D.Y. The history of *Manayunkia* [Polychaeta: Sedentaria: Sabellidae] propagation in North Eastern Asia. *Ecol. Genet.* **2014**, *12*, 32–43. [CrossRef]
28. Sitnikova, T.Y. Chapter 14: Polychaete worms (Annelida: Polychaeta) [In Russian]. In *Index of Animal Species Inhabiting Lake Baikal and Its Catchment Area. Volume 1: Lake Baikal*; Timoshkin, O.A., Proviz, V.I., Sitnikova, T.Y., Eds.; Russian Academy of Sciences: Novosibirsk, Russia, 2004; Volume 1, pp. 428–431.
29. Zerbst-Boroffka, I. Osmotic Adaptation of the Endemic Fauna to the Ancient Freshwater Lake Baikal. *Naturwissenschaften* **1999**, *86*, 330–333. [CrossRef]
30. Hutchings, P.A. Part C. Natural History. In *Polychaetes & Allies. The Southern Synthesis. Fauna of Australia Vol 4A Polychaeta, Myzostomida, Pogonophora, Echiura, Sipuncula*; Beesley, P.L., Ross, G.J.B., Glasby, C.J., Eds.; CSIRO Publishing: Melbourne, Australia, 2000; pp. 30–37.
31. Hutchings, P.A.; De Deckker, P.; Geddes, M.C. A new species of *Manayunkia* (Polychaeta) from ephemeral lakes near the Coorong, South Australia. *Trans. R. Soc. S. Aust.* **1981**, *105*, 25–28.
32. Van Gaest, A.L.; Young, C.M.; Young, J.J.; Helms, A.R.; Arellano, S.M. Physiological and behavioral responses of *Bathynerita naticoidea* (Gastropoda: Neritidae) and *Methanoaricia dendrobranchiata* (Polychaeta: Orbiniidae) to hypersaline conditions at a brine pool cold seep. *Mar. Ecol.* **2007**, *28*, 199–207. [CrossRef]
33. Erséus, C. Marine Tubificidae (Oligochaeta) of the Arabian Gulf Coast of Saudi Arabia (part 4). *Fauna Saudi Arab.* **1988**, *9*, 19–22.
34. Erséus, C. Marine Tubificidae (Oligochaeta) from a mangrove habitat in Kenya. *Trop. Zool.* **1999**, *12*, 137–143. [CrossRef]
35. Healy, B. *Marionina schrijversi* (Oligochaeta Enchytraeidae), a new species from a mangrove in Kenya. *Trop. Zool.* **1997**, *10*, 129–132. [CrossRef]
36. Martin, P.; Aït Boughrous, A. *Guide Taxonomique des Oligochètes Aquatiques du Maghreb*; Belgian National Focal Point to the Global Taxonomy Initiative; Royal Belgian Institute of Natural Sciences: Brussels, Belgium, 2012; Volume 12, p. 186.
37. Boumaiza, M.; Martínez-Ansemil, E.; Giani, N. Les Oligochètes et Aphanoneura des eaux courantes de Tunisie. I—Données faunistiques. *Ann. Limnol.* **1986**, *22*, 231–237. [CrossRef]
38. Schmelz, R.M.; Collado, R. A guide to European terrestrial and freshwater species of Enchytraeidae (Oligochaeta). *Soil Org.* **2010**, *82*, 1–176.
39. Erséus, C. A generic revision of the Phallodrilinae (Oligochaeta, Tubificidae). *Zool. Scr.* **1992**, *21*, 5–48. [CrossRef]
40. Brinkhurst, R.O.; Jamieson, B.G.M.G. *Aquatic Oligochaeta of the World*; Oliver and Boyd: Edinburgh, Scotland, 1971; p. 860.
41. Timm, T. A guide to the freshwater Oligochaeta and Polychaeta of Northern and Central Europe. *Lauterbornia* **2009**, *66*, 1–235.
42. Jans, W.; Römbke, J. Funde eines terrestrischen Polychaeten (Annelida) in Wäldern Baden-Württembergs. *Carolinea* **1989**, *47*, 158–162.
43. Dózsa-Farkas, K.; Hong, Y. Three new *Hemienchytraeus* species (Enchytraeidae, Oligochaeta, Annelida) from Korea, with first records of other enchytraeids and terrestrial polychaetes (Annelida). *Zootaxa* **2010**, *2406*, 29–56. [CrossRef]
44. Schmidt, P. Zonation of the interstitial polychaete *Stygocapitella subterranea* (Stygocapitellidae) in European sandy beaches. *Mar. Biol.* **1970**, *7*, 319–323. [CrossRef]
45. Purschke, G. Terrestrial polychaetes—models for the evolution of the Clitellata (Annelida) ? *Hydrobiologia* **1999**, *406*, 87–99. [CrossRef]
46. Westheide, W. *Interstitial Families*; The Linnean Society of London: London, UK, 2008; p. 169.

47. Rota, E.; de Jong, Y. Fauna Europaea: Annelida—Terrestrial Oligochaeta (Enchytraeidae and Megadrili), Aphanoneura and Polychaeta. *Biodivers. Data J.* **2015**, *3*, e5737. [CrossRef] [PubMed]
48. Struck, T.H.; Koczula, J.; Stateczny, D.; Meyer, C.; Purschke, G. Two new species in the annelid genus *Stygocapitella* (Orbiniida, Parergodrilidae) with comments on their biogeography. *Zootaxa* **2017**, *4286*, 301–332. [CrossRef]
49. Meca, M.A.; Zhadan, A.; Struck, T.H. The early branching group of Orbiniida *sensu* Struck et al., 2015: Parergodrilidae and Orbiniidae. *Diversity* **2021**, *13*, 29. [CrossRef]
50. Rota, E.; Lupetti, P. An ultrastructural investigation of *Hrabeiella* Pizl & Chalupsky, 1984 (Annelida). I. Chaetae and body wall organization. *Hydrobiologia* **1996**, *334*, 229–239.
51. Stout, J.D. Aquatic oligochaetes occurring in forest litter.-I. *Trans. R. Soc. N. Z.* **1956**, *84*, 97–102.
52. Herlant-Meewis, H. Cyst-formation in *Aeolosoma hemprichi* (Ehr). *Biol. Bull.* **1950**, *99*, 173–180. [CrossRef] [PubMed]
53. Westheide, W.; von Basse, M. Chilling and freezing resistance of two interstitial polychaetes from a sandy tidal beach. *Oecologia* **1978**, *33*, 45–54. [CrossRef]
54. Holmstrup, M. Cold hardiness strategy in cocoons of the lumbricid earthworm *Dendrobaena octaedra* (Savigny). *Comp. Biochem. Physiol. Part A Physiol.* **1992**, *102*, 49–54. [CrossRef]
55. Zachariassen, K.E. Physiology of cold tolerance in insects. *Physiol. Rev.* **1985**, *65*, 799–832. [CrossRef] [PubMed]
56. Sømme, L.; Birkemoe, T. Cold tolerance and dehydration in Enchytraeidae from Svalbard. *J. Comp. Physiol. B Biochem. Syst. Environ. Physiol.* **1997**, *167*, 264–269. [CrossRef]
57. Slotsbo, S.; Maraldo, K.; Malmendal, A.; Nielsen, N.C.; Holmstrup, M. Freeze tolerance and accumulation of cryoprotectants in the enchytraeid Enchytraeus albidus (Oligochaeta) from Greenland and Europe. *Cryobiology* **2008**, *57*, 286–291. [CrossRef]
58. Goodman, D. *Ecological Investigations of Ice Worms on Casement Glacier, Southeastern Alaska*; Ohio State University Research Foundation: Columbus, OH, USA, 1971.
59. Shain, D.H.; Mason, T.A.; Farrell, A.H.; Michalewicz, L.A. Distribution and behavior of ice worms (*Mesenchytraeus solifugus*) in South-central Alaska. *Can. J. Zool.* **2001**, *79*, 1813–1821. [CrossRef]
60. Edwards, J.S. How small ectotherms thrive in the cold without really trying. *Cryo Lett.* **1986**, *6*, 388–390.
61. Napolitano, M.J.; Nagele, R.G.; Shain, D.H. The ice worm, *Mesenchytraeus solifugus*, elevates adenylate levels at low physiological temperature. *Comp. Biochem. Physiol.* **2004**, *137*, 227–235. [CrossRef] [PubMed]
62. Miles, H.B. Heat-death temperature in *Allolobophora terrestris* (Sav.) forma *longa* (Ude) and *Eisenia foetida* (Sav.). *Nature* **1963**, *199*, 826. [CrossRef]
63. Johnston, M.R.; Herrick, B.M. Cocoon Heat Tolerance of Pheretimoid Earthworms Amynthas tokioensis and Amynthas agrestis. *Am. Midl. Nat.* **2019**, *181*, 299–309. [CrossRef]
64. Girguis, P.R.; Lee, R.W. Thermal preference and tolerance of alvinellids. *Science* **2006**, *312*, 231. [CrossRef] [PubMed]
65. Gaill, F.; Bouligand, Y. Supercoil of collagen fibrils in the integument of *Alvinella*, an abyssal annelid. *Tissue Cell* **1987**, *19*, 625–642. [CrossRef]
66. Gaill, F.; Hunt, S. Tubes of deep sea hydrothermal vent worms *Riftia pachyptila* (Vestimentifera) and *Alvinella pompejana* (Annelida). *Mar. Ecol. Prog. Ser.* **1986**, *34*, 267–274. [CrossRef]
67. Jollivet, D.; Desbruyères, D.; Ladrat, C.; Laubier, L. Evidence for differences in the allozyme thermostability of deep-sea hydrothermal vent polychaetes (Alvinellidae):a possible selection by habitat. *Mar. Ecol. Prog. Ser.* **1995**, *123*, 125–136. [CrossRef]
68. Sicot, F.X.; Mesnage, M.; Masselot, M.; Exposito, J.Y.; Garrone, R.; Deutsch, J.; Gaill, F. Molecular adaptation to an extreme environment: Origin of the thermal stability of the Pompeii worm collagen. *J. Mol. Biol.* **2000**, *302*, 811–820. [CrossRef]
69. Rinke, C.; Lee, R.W. Pathways, activities and thermal stability of anaerobic and aerobic enzymes in thermophilic vent paralvinellid worms. *Mar. Ecol. Prog. Series* **2009**, *382*, 99–112. [CrossRef]
70. Dilly, G.F.; Young, C.R.; Lane, W.S.; Pangilinan, J.; Girguis, P.R. Exploring the limit of metazoan thermal tolerance via comparative proteomics: Thermally induced changes in protein abundance by two hydrothermal vent polychaetes. *Proc. Biol. Sci.* **2012**, *279*, 3347–3356. [CrossRef]
71. McHugh, D.; Tunnicliffe, V. Ecology and reproductive biology of the hydrothermal vent polychaete *Amphisamytha galapagensis* (Ampharetidae). *Mar. Ecol. Prog. Ser.* **1994**, *106*, 111–120. [CrossRef]
72. Erséus, C.M.; Milligan, M.R. Three new species of Tubificidae (Oligochaeta) from an oil seepage area on the continental slope of the northern Gulf of Mexico. *Proc. Biol. Soc. Wash.* **1989**, *102*, 887–893.
73. Kvist, S.; Dreyer, J.; Erséus, C. Two new species of *Tubificoides* (Annelida: Clitellata: Naididae) from the Blake Ridge methane seep in the northwest Atlantic Ocean. *Proc. Biol. Soc. Wash.* **2008**, *121*, 531–540. [CrossRef]
74. Petrecca, R.F.; Grassle, J.F. *Notes on Fauna from Several Deep-Sea Hydrothermal Vent and Cold Seep Soft-Sediment Communities*; Springer: New York, NY, USA, 1990; pp. 279–283.
75. Fisher, C.R.; Childress, J.J.; Arp, A.J.; Brooks, J.M.; Distel, D.; Favuzzi, J.A.; Macko, S.A.; Newton, A.; Powell, M.A.; Somero, G.N.; et al. Physiology, morphology, and biochemical composition of *Riftia pachyptila* at Rose Garden in 1985. *Deep Sea Res. Part A Oceanogr. Res. Pap.* **1988**, *35*, 1745–1758. [CrossRef]
76. Hessler, R.R.; Smithey, W.M.; Boudrias, M.A.; Keller, C.H.; Lutz, R.A.; Childress, J.J. Temporal change in megafauna at the Rose Garden hydrothermal vent (Galapagos Rift; eastern tropical Pacific). *Deep Sea Res. Part A Oceanogr. Res. Pap.* **1988**, *35*, 1681–1709. [CrossRef]

77. Southward, E.C.; Tunnicliffe, V.; Black, M.B.; Dixon, D.R.; Dixon, L.R.J. Ocean-ridge segmentation and vent tubeworms (Vestimentifera) in the NE Pacific. *Geol. Soc. Lond. Spec. Publ.* **1996**, *118*, 211–224. [CrossRef]
78. Fisher, C.R.; Urcuyo, I.A.; Simpkins, M.A.; Nix, E. Life in the slow lane: Growth and longevity of cold-seep vestimentiferans. *Mar. Ecol.* **1997**, *18*, 83–94. [CrossRef]
79. Bergquist, D.C.; Williams, F.M.; Fisher, C.R. Longevity record for deep-sea invertebrate. *Nature* **2000**, *403*, 499–500. [CrossRef] [PubMed]
80. Cordes, E.E.; Bergquist, D.C.; Redding, M.L.; Fisher, C.R. Patterns of growth in cold-seep vestimenferans including Seepiophila jonesi: A second species of long-lived tubeworm. *Mar. Ecol.* **2007**, *28*, 160–168. [CrossRef]
81. Crane, K.; Hecker, B.; Golubev, V. Hydrothermal vents in Lake Baikal. *Nature* **1991**, *350*, 281. [CrossRef]
82. Grachev, M.; Flalkov, V.; Nakamura, T.; Ohta, T.; Kawai, T. Extant fauna of ancient carbon. *Nature* **1995**, *374*, 123–124. [CrossRef]
83. Sitnikova, T.Y.; Shirokaya, A.A. New data on deep water Baikal limpets found in hydrothermal vents and oil-seeps. *Arch. Molluskenkd. Int. J. Malacol.* **2013**, *142*, 257–278. [CrossRef]
84. Kaygorodova Irina, A. Clitellate communities in anomalous region of Lake Baikal, Eastern Siberia (Russia). *Biologia* **2016**, *71*, 31–37. [CrossRef]
85. Martin, P.; Martens, K.; Goddeeris, B. Oligochaeta from the abyssal zone of Lake Baikal (Siberia, Russia). *Hydrobiologia* **1999**, *406*, 165–174. [CrossRef]
86. Weiss, R.F.; Carmack, E.C.; Koropalov, V.M. Deep-water renewal and biological production in Lake Baikal. *Nature* **1991**, *349*, 665–669. [CrossRef]
87. Erséus, C.; Rota, E. New findings and overview of the oligochaetous Clitellata (Annelida) of the North Atlantic deep sea. *Proc. Biol. Soc. Wash.* **2003**, *116*, 892–900.
88. Erséus, C. Taxonomic revision of the marine genera *Bathydrilus* Cook and *Macroseta* Erséus (Oligochaeta, Tubificidae), with descriptions of six new species and subspecies. *Zool. Scr.* **1979**, *8*, 139–151. [CrossRef]
89. Rota, E.; Erséus, C. New records of *Grania* (Clitellata, Enchytraeidae) in the Northeast Atlantic (from Tromso to the Canary Islands), with descriptions of seven new species. *Sarsia* **2003**, *88*, 210–243. [CrossRef]
90. Erséus, C. Taxonomic revision of the Marine Genus *Phallodrilus* Pierantoni (Oligochaeta, Tubificidae), with description of thirteen new species. *Zool. Scr.* **1979**, *8*, 187–208. [CrossRef]
91. Kozhova, O.M.; Izmest'eva, L.R. *Lake Baikal. Evolution and Biodiversity*; Backhuys Publishers: Leiden, The Netherlands, 1998; p. 447.
92. Martin, P.; Granina, L.; Martens, K.; Goddeeris, B. Oxygen concentration profiles in sediments of two ancient lakes: Lake Baikal (Sideria, Russia) and Lake Malawi (East Africa). *Hydrobiologia* **1998**, *367*, 163–174. [CrossRef]
93. Sherstyankin, P.P.; Alekseev, S.P.; Abramov, A.M.; Stavrov, K.G.; De Batist, M.; Hus, R.; Canals, M.; Casamor, J.L. Computer-based bathymetric map of Lake Baikal. *Dokl. Earth Sci.* **2006**, *408*, 564–569. [CrossRef]
94. Pudovkina, T.A.; Sitnikova, T.Y.; Matveyev, A.N.; Shcherbakov, D.Y. Kindred Relations of Baikal polychaete of the *Manayunkia* Genus [Polychaeta: Sedentaria: Sabellidae] according to CO1 and settlement history analysis. *Russ. J. Genet. Appl. Res.* **2016**, *6*, 129–137. [CrossRef]
95. Falkner, K.K.; Measures, C.I.; Herbelin, S.E.; Edmond, J.M.; Weiss, R.F. The major and minor element geochemistry of Lake Baikal. *Limnol. Oceanogr.* **1991**, *36*, 413–423. [CrossRef]
96. Martin, P.; Sonet, G.; Smitz, N.; Backeljau, T. Phylogenetic analysis of the *Baikalodrilus* species flock (Annelida: Clitellata: Naididae), an endemic genus to Lake Baikal (Russia). *Zool. J. Linn. Soc.* **2019**, *187*, 987–1015. [CrossRef]
97. Culver, D.C.; White, W.B. *Encyclopedia of Caves*, 2nd ed.; Academic Press: Waltham, MA, USA, 2011; p. 945.
98. Ginet, R.; Decou, V. *Initiation à la Biologie et à L'écologie Souterraines*; Jean-Pierre Delarge: Paris, France, 1977; p. 345.
99. Gibert, J.; Deharveng, L. Subterranean Ecosystems: A Truncated Functional Biodiversity. *BioScience* **2002**, *52*, 473–481. [CrossRef]
100. Martínez, A.; Di Domenico, M.; Worsaae, K. Evolution of cave *Axiokebuita* and *Speleobregma* (Scalibregmatidae, Annelida). *Zool. Scr.* **2013**, *42*, 623–636. [CrossRef]
101. Martínez, A.; Gonzales, B.C.; Worsaae, K.; Wilkens, H.; Núñez Jiménez, A.; Oromi, P.; Iliffe, T.M. *Guide to the Anchialine Ecosystems of Jameos del Agua and Túnel de la Atlántida*; Cabildo de Lanzarote, Medio Ambiente: Arrecife, Lanzarote, 2016.
102. Worsaae, K.; Sterrer, W.; Iliffe, T.M. *Longipalpa saltatrix*, a new genus and species of the meiofaunal family Nerillidae (Annelida: Polychaeta) from an anchihaline cave in Bermuda. *Proc. Biol. Soc. Wash.* **2004**, *117*, 346–362.
103. Alves, P.R.; Glasby, C.J.; Santos, C.S.G. On the use of troglomorphic characters in Namanereidinae (Annelida; Nereididae) systematics. *Zootaxa* **2018**, *4531*, 195–210. [CrossRef]
104. Matjašič, J.; Sket, B. Développement larvaire du Serpulien cavernicole *Marifugia cavatica* Absolon et Hrabe (Polychaeta, Sedentaria). *Speleology* **1966**, *2*, 9–16. [CrossRef]
105. Worsaae, K. 2021. 7.9 Errantia incertae sedis: Nerillidae Levinsen. In *Handbook of Zoology, Pleistoannelida, Sedentaria III and Errantia I*; Purschke, G., Böggemann, M., and W. Westheide, W., Eds.; De Gruyter: Berlin, Germany, 1883; pp. 215–227. [CrossRef]
106. Creuzé des Châtelliers, M.; Juget, J.; Lafont, M.; Martin, P. Subterranean aquatic Oligochaeta. *Freshw. Biol.* **2009**, *54*, 678–690. [CrossRef]
107. Kerovec, M.; Kučinić, M.; Jalžic, B. *Croatobranchus mestrovi* sp. n. -predstavnik nove endemske podzemne vrste pijavica. *Spelolog* **1999**, *44–45*, 35–36.
108. Sket, B.; Dovc, P.; Jalzic, B.; Kerovec, M.; Kucinic, M.; Trontelj, P. A cave leech (Hirudinea, Erpobdellidae) from Croatia with unique morphological features. *Zool. Scr.* **2001**, *30*, 223–229. [CrossRef]

109. Timm, T. Profundal oligochaete assemblages in Palearctic lakes. *Turk. J. Zool.* **2012**, *36*, 121–131. [CrossRef]
110. Giani, N.; Martin, P.; Juget, J. A new species of Phreodrilidae (Oligochaeta), *Astacopsidrilus naceri* sp. nov., from Morocco (North Africa), with notes on the biogeography of the family. *Can. J. Zool.* **1995**, *73*, 2375–2381. [CrossRef]
111. Martínez-Ansemil, E.; Giani, N.; Sambugar, B. Oligochaetes from underground waters of Oman with descriptions of two new species of Phreodrilidae (Oligochaeta): *Antarctodrilus arabicus* n. sp. and *Phreodrilus stocki* n. sp. *Contrib. Zool.* **2002**, *71*, 147–158. [CrossRef]
112. Martínez, A.; Di Domenico, M.; Jörger, K.; Norenburg, J.; Worsaae, K. Description of three new species of *Protodrilus* (Annelida, Protodrilidae) from Central America. *Mar. Biol. Res.* **2013**, *9*, 676–691. [CrossRef]
113. Martínez, A.; Di Domenico, M.; Rouse, G.W.; Worsaae, K. Phylogeny and systematics of Protodrilidae (Annelida) inferred with total evidence analyses. *Cladistics* **2015**, *31*, 250–276. [CrossRef]
114. Di Domenico, M.; da Cunha Lana, P.; Garraffoni, A.R.S. Distribution patterns on interstitial polychaetes in sandy beaches of southern Brazil. *Mar. Ecol.* **2008**, *30*, 47–62. [CrossRef]
115. Di Domenico, M.; Martínez, A.; Lana, P.C.; Worsaae, K. *Protodrilus* (Protodrilidae, Annelida) from the southern and southeastern Brazilian coasts. *Helgol. Mar. Res.* **2013**, *67*, 733–748. [CrossRef]
116. Paxton, H. Taxonomy and aspects of the life history of Australian beachworms (Polychaeta: Onuphidae). *Aust. J. Mar. Freshw. Res.* **1979**, *30*, 265–294. [CrossRef]
117. Kirtley, D.W. Intertidal Reefs of Sabellariidae, Annelida Polychaeta, along the Coast of Florida. Ph.D. Thesis, The Florida State University, Tallahassee, FL, USA, 1966.
118. Vovelle, J. Le tube de *Sabellaria alveolata* (L.) annélide polychète Hermellidae et son ciment étude ecologique, expérimentale, histologique et histochimique. *Arch. Zool. Exp. Gén.* **1965**, *106*, 1–187.
119. Gaill, F.; Hunt, S. Tubes. In *The Ultrastructure of Polychaeta*; Westheide, W., Hermans, C.O., Eds.; John Wiley & Sons: Hoboken, NJ, USA, 1988; Volume 4, pp. 61–70.
120. Wang, C.S.; Svendsen, K.K.; Stewart, R.J. Morphology of the adhesive system in the sandcastle worm, *Phragmatopoma californica*. In *Biological Adhesive Systems: From Nature to Technical and Medical Application*; von Byern, J., Grunwald, I., Eds.; Springer: Vienna, Austria, 2010; pp. 169–179. [CrossRef]
121. Morgado, E.H.; Tanaka, M.O. The macrofauna associated with the bryozoan *Schizoporella unicornis* in southeastern Brazil. *Sci. Mar.* **2007**, *65*, 173–181. [CrossRef]
122. Pérez, C.D.; Vila-Nova, D.A.; Santos, A.M. Associated community with the zoanthid *Palythoa caribaeorum* (Duchassaing & Michelotti, 1860) (Cnidaria, Anthozoa) from littoral of Pernambuco, Brazil. *Hydrobiologia* **2005**, *548*, 207–215. [CrossRef]
123. Yamanishi, R. Notes on *Pholoe parva* Imajima & Hartman, 1964 (Annelida: Polychaeta: Sigalionidae). *Bull. Osaka Mus. Nat. Hist.* **1985**, *38*, 17–22.
124. Swedmark, B. The interstitial fauna of marine sand. *Biol. Rev.* **1964**, *39*, 1–42. [CrossRef]
125. Worsaae, K.; Kristensen, R.M. Evolution of interstitial Polychaeta (Annelida). *Hydrobiologia* **2005**, *535–536*, 319–340.
126. Westheide, W. Progenesis as a principle in meiofauna evolution. *J. Nat. Hist.* **1987**, *21*, 843–854. [CrossRef]
127. Struck, T.H.; Golombek, A.; Weigert, A.; Franke, F.A.; Westheide, W.; Purschke, W.; Bleidorn, C.; Halanych, K.M. The evolution of annelids reveals two adaptive routes to the interstitial realm. *Curr. Biol.* **2015**, *25*, 1993–1999. [CrossRef]
128. Giani, N.; Lavandier, P. Les oligochètes du torrent d'Estaragne (Pyrenées Centrales). *Bull. Soc. Hist. Nat. Toulouse* **1977**, *113*, 234–243.
129. Martínez-Ansemil, E.; Collado, R. Distribution patterns of aquatic oligochaetes inhabiting watercourses in the Northwestern Iberian Peninsula. *Hydrobiologia* **1996**, *334*, 73–83. [CrossRef]
130. Gagneur, J.; Giani, N.; Martínez-Ansemil, E. Les oligochètes aquatiques d'Algérie. *Bull. Soc. Hist. Nat. Toulouse* **1986**, *122*, 119–124.
131. Tsutsumi, H. Population dynamics of *Capitella capitata* (Polychaeta; Capitellidae) in an organically polluted cove. *Mar. Ecol. Prog. Ser.* **1987**, *36*, 139–149. [CrossRef]
132. Dean, H.K. The use of polychaetes (Annelida) as indicator species of marine pollution: A review. *Rev. Biol. Trop.* **2008**, *56*, 11–38.
133. Forbes, T.L.; Forbes, V.E.; Depledge, M.H. Individual physiological responses to environmental hypoxia and organic enrichment: Implications for early soft-bottom community succession. *J. Mar. Res.* **1994**, *52*, 1081–1100. [CrossRef]
134. Warren, L.M. Respiratory adaptations to temporary hypoxia by the polychaete *Cirriformia tentaculata*. *Comp. Biochem. Physiol. Part A Physiol.* **1981**, *69*, 321–324. [CrossRef]
135. Petersen, M.E.; George, J.D. A new species of *Raricirrus* from northern Europe, with notes on its biology and a discussion of the affinities of the genus (Polychaeta: Ctenodrilidae). In *Systematics, Biology and Morphology of World Polychaeta (Ophelia Supplement 5), Proceedings of the Second International Polychaete Conference, Copenhagen, Denmark, 18–23 August 1986*; Petersen, M.E., Kirkegaard, J.B., Eds.; Brill: Leiden, The Netherlands, 1991; pp. 185–208.
136. Ugland, K.I.; Bjørgesæter, A.; Bakke, T.; Fredheim, B.; Gray, J.S. Assessment of environmental stress with a biological index based on opportunistic species. *J. Exp. Mar. Biol. Ecol.* **2008**, *366*, 169–174. [CrossRef]
137. Cunha, M.R.; Rodrigues, C.F.; Génio, L.; Hilário, A.; Ravara, A.; Pfannkuche, O. Macrofaunal assemblages from mud volcanoes in the Gulf of Cadiz: Abundance, biodiversity and diversity partitioning across spatial scales. *Biogeosciences* **2013**, *10*, 2553–2568. [CrossRef]

138. Glasby, C.J.; Hutchings, P.A. A new species of *Marphysa* Quatrefages, 1865 (Polychaeta: Eunicida: Eunicidae) from northern Australia and a review of similar taxa from the Indo-west Pacific, including the genus *Nauphanta* Kinberg, 1865. *Zootaxa* **2010**, *2352*, 29–45. [CrossRef]
139. Brinkhurst, R.O. Distribution and Ecology. 1. Microdriles. In *Aquatic Oligochaeta of the World*; Brinkhurst, R.O., Jamieson, A.J., Eds.; Oliver & Boyd: Edinburgh, Scotland, 1971; pp. 104–146.
140. Rodriguez, P.; Reynoldson, T.B. *The Pollution Biology of Aquatic Oligochaetes*; Springer: Dordrecht, The Netherlands, 2011; p. 265.
141. Giere, O.; Preusse, J.-H.; Dubilier, N. *Tubificoides benedii* (Tubificidae, Oligochaeta)—a pioneer in hypoxic and sulfidic environments. An overview of adaptative pathways. *Hydrobiologia* **1999**, *406*, 235–241. [CrossRef]
142. Ito, K.; Ito, M.; Hano, T.; Onduka, T.; Mochida, K.; Shono, N.; Nakamura, R. Marine sediment conservation using benthic organisms. *Bull. Jpn. Fish. Res. Educ. Agency* **2019**, *49*, 45–51.
143. Dubilier, N.; Blazejak, A.; Rühland, C. Symbioses between bacteria and gutless marine oligochaetes. In *Molecular Basis of Symbiosis*; Overmann, J., Ed.; Springer: Berlin/Heidelberg, Germany, 2005; pp. 251–275. [CrossRef]
144. Hourdez, S.; Frederick, L.-A.; Schernecke, A.; Fisher, C.R. Functional respiratory anatomy of a deep-sea orbiniid polychaete from the Brine Pool NR-1 in the Gulf of Mexico. *Invertebr. Biol.* **2005**, *120*, 29–40. [CrossRef]
145. Hourdez, S.; Weber, R.E.; Green, B.N.; Kenney, J.M.; Fisher, C.R. Respiratory adaptations in a deep-sea orbiniid polychaete from Gulf of Mexico brine pool NR-1: Metabolic rates and hemoglobin structure/function relationships. *J. Exp. Biol.* **2002**, *205*, 1669–1681. [PubMed]
146. Weber, R.E.; Vinogradov, S.N. Nonvertebrate hemoglobins: Functions and molecular adaptations. *Physiol. Rev.* **2001**, *81*, 569–628. [CrossRef]
147. Narita, T. Seasonal vertical migration and eastivation of *Rhyacodrilus hiemalis* (Tubificidae, Clitellata) in the sediment of Lake Biwa, Japan. *Hydrobiologia* **2006**, *564*, 87–93. [CrossRef]
148. Southward, E.C. Class Pogonophora. In *Polychaetes & Allies. The Southern Synthesis. Fauna of Australia Vol 4A Polychaeta, Myzostomida, Pogonophora, Echiura, Sipuncula*; Beesley, P.L., Ross, G.J.B., Glasby, C.J., Eds.; CSIRO Publishing: Melbourne, Australia, 2000; pp. 331–351.
149. Desbruyères, D.; Laubier, L. Les Alvinellidae, une famille nouvelle d'annélides polychètes inféodées aux sources hydrothermales sous-marines: Systématique, biologie et écologie. *Can. J. Zool.* **1986**, *64*, 2227–2245. [CrossRef]
150. Tunnicliffe, V.; Desbruyères, D.; Jollivet, D.; Laubier, L. Systematic and ecological characteristics of *Paralvinella sulfincola* Desbruyères and Laubier, a new polychaete (family Alvinellidae) from northeast Pacific hydrothermal vents. *Can. J. Zool.* **1993**, *71*, 286–297. [CrossRef]
151. Bris, N.L.; Gaill, F. Microbial habitats associated with deep-sea hydrothermal vent invertebrates: Insights from microanalysis and geochemical modeling. In *The Vent and Seep Biota: Aspects from Microbes to Ecosystems*; Kiel, S., Ed.; Springer: Dordrecht, The Netherlands, 2010; pp. 51–71. [CrossRef]
152. Terwilliger, N.B.; Terwilliger, R.C. Hemoglobin from the "pompeii worm," *Alvinella pompejana*, an annelid from a deep sea hot hydrothermal vent environment. *Mar. Biol. Lett.* **1984**, *5*, 191–201.
153. Toulmond, A.; Slitine, F.E.I.; De Frescheville, J.; Jouin, C. Extracellular hemoglobins of hydrothermal vent annelids: Structural and functional characteristics in three alvinellid species. *Biol. Bull.* **1990**, *179*, 366–373. [CrossRef]
154. Hourdez, S.; Weber, R.E. Molecular and functional adaptations in deep-sea hemoglobins. *J. Inorg. Biochem.* **2005**, *99*, 130–141. [CrossRef]
155. Desbruyères, D.; Chevaldonné, P.; Alayse, A.M.; Jollivet, D.; Lallier, F.H.; Jouin-Toulmond, C.; Zal, F.; Sarradin, P.M.; Cosson, R.; Caprais, J.C.; et al. Biology and ecology of the "Pompeii worm" (*Alvinella pompejana* Desbruyères and Laubier), a normal dweller of an extreme deep-sea environment: A synthesis of current knowledge and recent developments. *Deep Sea Res. Part II Top. Stud. Oceanogr.* **1998**, *45*, 383–422. [CrossRef]
156. Zimmermann, J.; Wentrup, C.; Sadowski, M.; Blazejak, A.; Gruber-Vodicka, H.R.; Kleiner, M.; Ott, J.A.; Cronholm, B.; De Wit, P.; Erseus, C.; et al. Closely coupled evolutionary history of ecto- and endosymbionts from two distantly related animal phyla. *Mol. Ecol.* **2016**, *25*, 3203–3223. [CrossRef]
157. Dubilier, N.; Mulders, C.; Ferdelman, T.; de Beer, D.; Pernthaler, A.; Klein, M.; Wagner, M.; Erseus, C.; Thiermann, F.; Krieger, J.; et al. Endosymbiotic sulphate-reducing and sulphide-oxidizing bacteria in an oligochaete worm. *Nature* **2001**, *411*, 298–302. [CrossRef]
158. Glasby, C.J. The Namanereidinae (Polychaeta: Nereididae). Part 1. Taxonomy and phylogeny. *Rec. Aust. Mus. Suppl.* **1999**, *25*, 1–129. [CrossRef]
159. Pinder, A.M.; Brinkhurst, R.O. Review of the Phreodrilidae (Annelida: Oligochaeta: Tubificida) of Australia. *Invertebr. Taxon.* **1997**, *11*, 443–523. [CrossRef]
160. Ward, T.J.; Hutchings, P.A. Effects of trace metals on infaunal species composition in polluted intertidal and subtidal marine sediments near a lead smelter, Spencer Gulf, South Australia. *Mar. Ecol. Prog. Ser.* **1996**, *135*, 123–135. [CrossRef]
161. Grassle, J.F.; Grassle, J.P. Opportunistic life histories and genetic systems in marine benthic polychaetes. *J. Mar. Res.* **1974**, *32*, 253–284.
162. Ito, K.; Ito, M.; Onduka, T.; Ohta, K.; Torii, T.; Hano, T.; Mochida, K.; Ohkubo, N.; Miura, T.; Fujii, K. Differences in the ability of two marine annelid species, Thalassodrilides sp. and Perinereis nuntia, to detoxify 1-nitronaphthalene. *Chemosphere* **2016**, *151*, 339–344. [CrossRef]

163. Ito, M.; Ito, K.; Ohta, K.; Hano, T.; Onduka, T.; Mochida, K. Transcription of a novel P450 gene varies with some factors (pollutant exposure, temperature, time, and body region) in a marine oligochaete (Thalassodrilides sp.). *Mar. Pollut. Bull.* **2016**, *109*, 344–349. [CrossRef]
164. Kempe, S.; Khoo, F.; Guryelik, Y. Hydrography of Lake Van and its drainage area. In *The Geology of Lake Van*; Degens, E.T., Kurtman, F., Eds.; MTA: Ankara, Turkey, 1978; pp. 30–44.
165. Litt, T.; Krastel, S.; Sturm, M.; Kipfer, R.; Örcen, S.; Heumann, G.; Franz, S.O.; Ülgen, U.B.; Niessen, F. 'PALEOVAN', International Continental Scientific Drilling Program (ICDP): Site survey results and perspectives. *Quat. Sci. Rev.* **2009**, *28*, 1555–1567. [CrossRef]
166. Arslan, N.; Timm, T.; Rojo, V.; Vizcalno, A.; Schmelz, R.M. A new species of *Enchytraeus* (Enchytraeidae, Oligochaeta) from the profundal of Lake Van, the world's largest soda Lake (Turkey, East Anatolia). *Zootaxa* **2018**, *4382*, 367–380. [CrossRef] [PubMed]
167. Erséus, C.; Klinth, M.J.; Rota, E.; De Wit, P.; Gustafsson, D.R.; Martinsson, S. The popular model annelid Enchytraeus albidus is only one species in a complex of seashore white worms (Clitellata, Enchytraeidae). *Org. Divers. Evol.* **2019**, *19*, 105–133. [CrossRef]
168. Hrabě, S. Die Oligochaeten des Issykkulsees. (The oligochaetes of Lake Issyk-Kul. In Russian with German summary). *Tr. Kirg. Kompleks. Exped. Akad. Nauk SSSR* **1935**, *3*, 73–85.
169. Timm, T. Enchytraeids (Oligochaeta, Enchytraeidae) as lake-dwellers. In *Newsletter on Enchytraeidae No 5, Proceedings of the Symposium on Enchytraeidae, Budapest, Hungary, 24–25 June 1996*; Dózsa-Farkas, K., Ed.; Budapest, Hungary, 1996; pp. 47–55.

diversity

MDPI

Article

Interstitial Annelida

Katrine Worsaae [1,*], Alexandra Kerbl [1,2], Maikon Di Domenico [1,3], Brett C. Gonzalez [1,4], Nicolas Bekkouche [1,5] and Alejandro Martínez [1,6]

1 Marine Biological Section, Department of Biology, University of Copenhagen, Universitetsparken 4, 2100 Copenhagen, Denmark; alexandra.kerbl@uni-hamburg.de (A.K.); maik2dd@gmail.com (M.D.D.); gonzalezb@si.edu (B.C.G.); nicolas.bekkouche@gmail.com (N.B.); alejandro.martinezgarcia@cnr.it (A.M.)
2 Center of Natural History–Zoological Museum, Universität Hamburg, Martin-Luther-King-Platz 3, 20146 Hamburg, Germany
3 Center for Marine Studies, Federal University of Paraná, Av. Beira Mar s/n, 83255-976, Pontal do Sul, Pontal do Paraná, Paraná 83255-976, Brazil
4 Smithsonian Institution, National Museum of Natural History, Department of Invertebrate Zoology, P.O. Box 37021, Washington, DC 20013-7012, USA
5 Institut de Systématique, Evolution, Biodiversité, ISYEB–UMR 7205 MNHN CNRS UPMC EPHE, Sorbonne Universités, 45 rue Buffon, 75005 Paris, France
6 Molecular Ecology Group (MEG), Water Research Institute (IRSA), National Research Council of Italy (CNR), 28922 Verbania, Italy
* Correspondence: kworsaae@bio.ku.dk

Abstract: Members of the following marine annelid families are found almost exclusively in the interstitial environment and are highly adapted to move between sand grains, relying mostly on ciliary locomotion: Apharyngtidae n. fam., Dinophilidae, Diurodrilidae, Nerillidae, Lobatocerebridae, Parergodrilidae, Polygordiidae, Protodrilidae, Protodriloididae, Psammodrilidae and Saccocirridae. This article provides a review of the evolution, systematics, and diversity of these families, with the exception of Parergodrilidae, which was detailed in the review of Orbiniida by Meca, Zhadan, and Struck within this Special Issue. While several of the discussed families have previously only been known by a few described species, recent surveys inclusive of molecular approaches have increased the number of species, showing that all of the aforementioned families exhibit a high degree of cryptic diversity shadowed by a limited number of recognizable morphological traits. This is a challenge for studies of the evolution, taxonomy, and diversity of interstitial families as well as for their identification and incorporation into ecological surveys. By compiling a comprehensive and updated review on these interstitial families, we hope to promote new studies on their intriguing evolutionary histories, adapted life forms and high and hidden diversity.

Keywords: systematics; identification; meiobenthos; annelids

check for updates

Citation: Worsaae, K.; Kerbl, A.; Domenico, M.D.; Gonzalez, B.C.; Bekkouche, N.; Martínez, A. Interstitial Annelida. *Diversity* **2021**, *13*, 77. https://doi.org/10.3390/d13020077

Academic Editors: Michael Wink, Maria Capa and Pat Hutchings

Received: 22 December 2020
Accepted: 2 February 2021
Published: 12 February 2021

1. Introduction

"To see the world in a grain of sand ... " (William Blake) reaches another meaning when it comes to the amazing diversity of animals revealed upon examining a handful of sand. Interstices between sand grains constitute a generally protected and well-oxygenated environment, rich in trapped organic matter and benthic microalgae [1]. This environment houses a great diversity of microscopic metazoans [2], particularly among harpacticoid copepods and the worm-like taxa Nematoda, Acoela, Gnathostomulida, Gastrotricha, Platyhelminthes, and Annelida. Within Annelida, interstitial forms have evolved a significantly high number of times from larger ancestors, resulting in more than 400 species that are distributed across 14 macrofaunal families and 13 interstitial families [3,4]. Whereas the two meiofaunal freshwater families Aeolosomatidae Levinsen, 1884 and Potamodrilidae Bunke, 1967 are found in various environments, the following 11 marine families are exclusively interstitial: Apharyngtidae n. fam., Dinophilidae Macalister, 1876, Diurodrilidae Kristensen and Niilonen, 1982, Lobatocerebridae Rieger, 1980, Nerillidae Levinsen, 1883,

Parergodrilidae Reisinger, 1925, Polygordiidae Czerniavsky, 1881, Protodrilidae Hatschek, 1888, Protodriloididae Purschke and Jouin, 1988, Psammodrilidae Swedmark, 1952 and Saccocirridae Bobretzky, 1872.

Many interstitial species are considered meiofaunal, a term that today is generally applied to species that pass through a 500 μm mesh size but are retained on a 42 μm mesh [1]. This definition implies that not all meiofaunal species are microscopic, since a long "meiofaunal" species might be able to squeeze itself through a 500 μm mesh. The term "interstitial" allows for a broader size range, just referring to animals capable of moving through the interstices without displacing the sediment particles. As a result, there are evident differences in size between interstitial annelids, e.g., from the minute *Diurodrilus minimus* Remane, 1925, (ca. 250–450 μm long) to the comparatively enormous *Saccocirrus major* Pierantoni, 1907 (up to 70 mm long). Moreover, a few of the largest "interstitial" annelids may actually perform muscular burrowing, displacing the sand grains, rather than gliding in between them. Acknowledging these inconsistencies, we use the term interstitial, which best fits the majority of annelids addressed in this article. Indeed, most of these families share a common set of adaptations to the interstitial environment, such as ventral motile ciliation (for gliding), adhesive glands, and small and/or slender bodies. Historically, the marine families Dinophilidae, Histriobdellidae (not interstitial), Nerillidae, Polygordiidae, Protodrilidae, and Saccocirridae (but not Parergodrilidae and Psammodrilidae) were regarded as part of the now abandoned group "Archiannelida" [5]. The concept of Archiannelida originated from Hatschek's studies on *Polygordius* [6], possessing a superficially simple adult morphology but a highly advanced trochophore-like larva from which, he concluded, all other annelids with a trochophore larva might have derived. Nowadays, the archiannelid concept has been abandoned and all the interstitial marine families have been shown to be secondarily small, generally unrelated, highly derived lineages [7–13]. Nonetheless, the exact phylogenetic positions of many of these families remain debated, even after the analysis of large transcriptomic datasets, e.g., [8–13], and extensive morphological revisions based on state-of-the-art microscopy and imagining technology, e.g., [9,14–22]. Despite these challenges and incongruences, we summarize the phylogenetic positions of the eleven marine interstitial families based on the most recent phylogenomic analyses (Figure 1), while specific problems and alternative placements are further discussed in the subchapter of each family. Nonetheless, Figure 1 illustrates that the marine interstitial families represent at least five independent lineages: (1) Psammodrilidae is nested in a group of macrofaunal annelids, next to Apistobranchidae [9]. (2) Dinophilidae and Lobatocerebridae were recently proposed to constitute a clade called Dinophiliformia, sister to Pleistoannelida [12], but see also [10,11]. (3) Protodrilidae, Saccocirridae, and Protodriloidae form a well-supported clade within Errantia in all phylogenomic analyses, e.g., [11], sometimes recovered next to Polygordiidae. (4) Nerillidae has been recovered nested among either errantian, e.g., [13], or sedentarian taxa [11], sometimes closely related to other interstitial families [11]. (5) Diurodrilidae and Apharyngtidae n. fam. group together within the larger sedentarian clade Orbiniida, which also contains Parergodrilidae, e.g., [11,12].

Perhaps correlated with this phylogenetic and morphological disparity, these interstitial families show a wide range of ecological preferences. Individuals of some genera have never been found outside of the interstitial environment and show notable pharyngeal, glandular, and ciliary specializations (e.g., *Trilobodrilus* Remane, 1925; Psammodrilidae; *Lobatocerebrum* Rieger, 1980; Diurodrilidae; and Protodriloididae). Some of these species colonize the phreatic coastal waters through the upper beach zone (e.g., in Diurodrilidae), protected from waves and currents, and have secondarily lost some of their adhesive structural adaptations found in their close relatives [23–25]. Other species (e.g., in Dinophilidae and Nerillidae) are epibenthic, grazing on biofilms growing on sediments, gravel, and seaweeds e.g., [26,27].

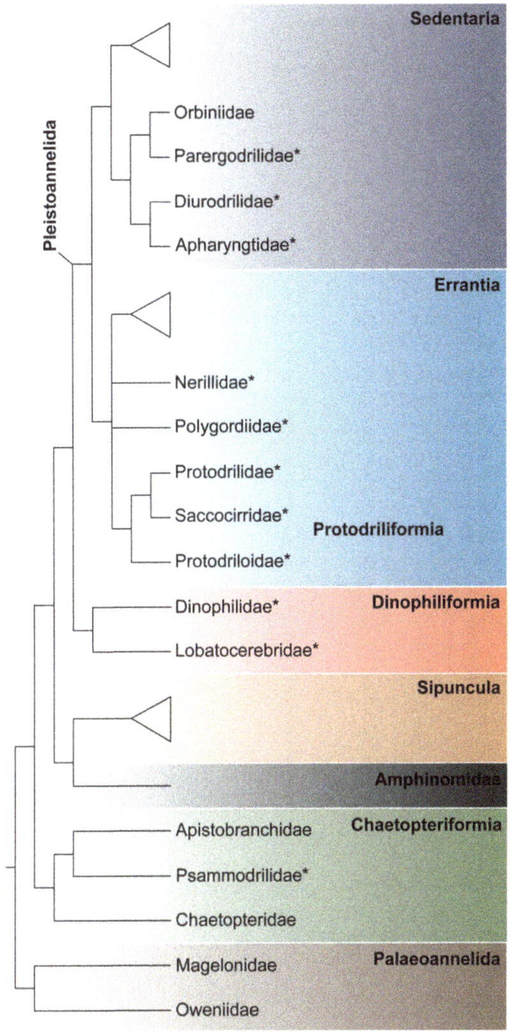

Figure 1. Selected hypotheses on the evolutionary relationship of interstitial annelid families (*), summarizing some of the most recent phylogenomic analyses [3,7–9]. Problems and alternatives to the depicted positions are discussed in the subchapter of each family (e.g., Nerillidae has been suggested to nest both with Errantian and Sedentarian taxa). Branches ending as a triangle indicate a cluster of multiple families.

Large species of Polygordiidae and Protodrilidae prefer the flocculent organic matter accumulated on top of the sediment, whereas the so-called surfing-species of *Saccocirrus*, *Protodrilus*, and *Megadrilus* drift along with the waves in the highly energetic zones of sandy beaches [28,29]. Some protodrilids are even semi-sessile suspension feeders [28,30], while the aberrant *Astomus taenioides* Jouin, 1979 lacks a functional mouth and gut, taking up nutrients through its body wall [31]. However, the highest ecological ubiquity is found amongst Nerillidae [32], the members of which are specialized to a wide range of habitats, such as mud e.g., [33], intertidal algae [27], groundwater [34], and anchialine caves e.g., [35–37], as well as deep-sea hydrothermal vents [38] and bacterial mats [39]. Two cave exclusive lineages of interstitial annelids, i.e., *Speleonerilla* Worsaae, Sterrer, and Iliffe, in Worsaae et al., 2018 and *Megadrilus pelagicus* Martínez, Kvindebjerg, Iliffe, and Worsaae, 2017, have even colonized the water columns of anchialine caves from interstitial ancestors, convergently gaining new traits to feed on suspended organic matter [40–42]. This vast range of life strategies, somewhat neglected in the literature, opens up new questions regarding early ecological radiations within these groups, emphasizing their potential as models for understanding general eco-evolutionary processes, including ecological radiations and adaptive morphological change as well as putative responses of marine organisms to future global and local climate changes [29,41,43,44].

It is difficult to make general statements on the biogeography and species diversity of interstitial annelids mostly due to two reasons; firstly, the available records for most species are fragmentary and therefore the distribution of most interstitial groups more likely reflect the unbalanced sampling effort across the world rather than any biological meaningful factors [45,46]; and secondly, many of these records are based on morphological identification, which largely underestimate species diversity in most of the investigated interstitial annelid lineage [20,37,47,48], leading to an inflated number of "cosmopolitan" species. Many of these "cosmopolitan" species have been shown to represent species complexes and a profound hidden diversity when examined using molecular approaches or when more detailed morphological studies are performed (e.g., [49]). In that regard, it is important to consider that not only may light microscopy observations fail to distinguish species, but sometimes even detailed morphological characterization combining measurements with confocal laser and scanning electron microscopy are insufficient to identify otherwise well-defined molecular lineages (e.g., [26,37,42]). Moreover, interstitial annelids represent a polyphyletic assemblage of animals with long evolutionary histories and different phylogenetic affinities, morphological traits, and ecological preferences [3]. Therefore, it is unlikely that their current distribution patterns have been affected by comparable processes and can thus be collectively discussed. However, it may be easier to extract meaningful biogeographical patterns when focusing on specific lineages such as single genera or species complex. For example, a single species of *Dinophilus* has recently been documented to be distributed across the Northern Atlantic using molecular data [26], whereas individuals belonging to the *Astomus taenioides* Jouin, 1979 species complex are restricted to the Pacific [50], and many cave species of *Mesonerilla* Remane, 1949 and *Speleonerilla,* as well as *Megadrilus pelagicus,* are endemic to individual cave systems (e.g., [41,42]). Of course, these patterns might still be considered with caution, since further sampling might render them spurious. This was illustrated by species of *Pharyngocirrus* Di Domenico, Martínez, Lana, and Worsaae, 2014, which were believed to be restricted to the Indopacific and Western Atlantic but were recently widely recorded throughout the Mediterranean and the Eastern Atlantic [51].

The goal of this paper was to provide an update and comprehensive review of our current knowledge on the eleven families of exclusively interstitial annelids. After a section devoted towards specialized methods applied to their study, we allocate a sub-chapter to each of these families. For each of them, we provide an overview of their current systematic placement along with the diversity and most relevant morphological features used in the identification of each genera. We complete each section with a review of their ecology and distribution patterns as well as a list of the most relevant literature. In order to increase the readability of the text, a reference to taxonomic authors will be limited in the taxonomic section under each family, along with the citation to each of the papers. We hope this review will stimulate further research on these somehow neglected Annelida, not only providing crucial elements to understand their character evolution within the phylum, but also as a potentially useful model for addressing broad eco-evolutionary questions across the marine realm.

2. Materials and Methods

2.1. Extraction Methods

One of the main challenges related to the study of interstitial annelids is that contrary to many other annelid groups, they are best investigated alive. Therefore, several extraction methods have been developed over the years in order to carefully extract the fragile animals from the substrate they live in without harming or breaking them. Methods widely used in other annelids or meiofaunal species, such as freshwater shock-treatment, harsh mixing, or formalin bulk-fixations are not recommended for interstitial annelids, as they will often destroy individuals or recover them in poor condition for subsequent morphological identification [2,52,53]. For the same reason, the use of density gradients, such as colloidal silica polymer (Ludox-TM) [54] or centrifugation methods, is disregarded.

Extraction of meiofauna from sandy sediments routinely involves the decantation of previously anesthetized samples through a mesh. Typically, large sediment samples must rest in the lab for a few hours or days after collection, so that the animals migrate to the uppermost two to five centimeters of the sediment. This layer is then scooped into a separate container with a 1:1 mix of sea water and isotonic $MgCl_2$-solution (or $MgS0_4$), gently stirred, and then left for 10–20 min in order to anesthetize the animals. After that time, the sediment is gently mixed again and the supernatant is decantated through a 30–100 µm mesh, often using a funnel-shaped sewn mesh, playfully referred to as a "mermaid bra" amongst meiobenthologists. This process might be repeated three to four times to ensure a total extraction of the fauna. The material retained within the "mermaid bra" is then transferred directly into petri dishes containing sea water, from which the animals are sorted out using a dissecting scope. Alternatively, the filtered material can be placed into small secondary 30–100 µm mesh sieves, which are then placed inside a petri dish with seawater. Over time, meiofaunal animals will squeeze through the mesh and accumulate in the underlying petri dish, making their sorting easier since most of the debris and larger individuals are retained in the mesh [2,52]. Once in the petri dish, animals are carefully picked up using plastic or glass Pasteur pipettes with a narrow opening. The fine tips create a rapid flux, making the animal collection more efficient.

Extraction from other substrates, such as silt, mud, and macroalgae, does not require anesthetization, since the animals in these habitats lack adhesive glands. Instead, for example, the mud sample (or the top layers of this) is resuspended in a large bucket of seawater, and left to settle for a minute or so, whereafter the surface layers are screened with a 100–200 µm aquarium net. The net is thereafter rinsed into a finer cone mesh ("mermaid bra"), transferred to a Petri dish and sorted [2]. Extraction from algae can be simply done by squeezing and rinsing a number of algae pieces, or parts of larger algae, onto a fine mesh.

2.2. Fixation and Preservation Methods

Preservation for DNA extraction is usually done using molecular grade ethanol (> 95%), although special buffers (i.e., RNAlater) or snap freezing in liquid nitrogen are necessary for RNA-extraction and/or give higher DNA yields, which is essential for transcriptome- or genome analyses, respectively. Preserved samples must be stored at −20–80 °C. Samples preserved for molecular analyses are not suitable for morphological investigation, and solutions proposed to be versatile for morphological and molecular studies such as DESS (20% DMSO, 0.25M disodium EDTA, saturated with NaCL, pH 8.0) [55] or HistoChoice Tissue Fixative (Amresco, patent #5,429,797, Solon, OH, USA) [56] do not work well with these soft-bodied, ciliated animals.

The selection of different fixatives and reagents for morphological analyses depends on the intended use of the samples. Fixation should be done on anesthetized animals and works better if fixatives, buffers, and samples are kept at the same temperature and osmolarity. Glutaraldehyde, trialdehyde [57] or any other mixture of paraformaldehyde and glutaraldehyde (i.e., Trumps (e.g., from product nr. 18030, Electron Microscopy Sciences, Hatfield, PA, USA) [58], SPAFG (3% glutaraldehyde, 1% paraformaldehyde, 7.5% picric acid saturated solution, 0.45M sucorse, 70mM cacodylate buffer) [59]) offer the best morphological fixation and are ideal for light microscopy, histology, and electron microscopy. For aldehyde-based fixatives, results can be enhanced by postfixation with osmium tetroxide at low concentrations (<1%). Direct fixation in 1% osmium tetroxide provides optimal fixation results for scanning electron microscopy in some groups. Since glutaraldehyde irreversibly binds proteins, fixation in <4% paraformaldehyde is preferred in immunohistochemical studies. Sufficient preservation by simultaneous epitope accessibility requires rather short fixation times and several rinses in appropriate buffer solutions (i.e., phosphate buffered saline, PBS), followed by storage in this buffer with fungicides (e.g., 0.05% NaN_3). Vouchers or museum specimens should be progressively transferred into 70–75% alcohol (in sealed vials), or whole mounted in glycerine on permanent slides sealed with resin or nail polish for long term storage

2.3. *Morphological and Molecular Methods for Species Identification*

Morphological identification requires combined light and electron microscopical observations, often at high magnification. Differential interference contrast (DIC) helps when examining epidermal structures such as cilia and glands, whereas phase contrast enhances hard structures such as jaws, chaetae, stylets, and scales. Description of coloration, glandular structures or epidermal patterns, as well as some internal structures (e.g., nephridia), demands observations on live individuals. External ciliary structures (i.e., ciliary bands, ciliary tufts) are better studied using scanning electron microscopy (SEM) [37,42,60], whereas internal ciliated structures such as nephridia and gonoducts are better revealed using confocal laser scanning microscopy (CLSM) and immunolabelling [42,61].

Descriptions and identification should be accompanied by molecular studies. Although next generation sequencing is getting more affordable, and protocols manage to produce good results even with a limited input of RNA or DNA, Sanger sequencing of conserved and fast evolving genes, predominantly the nuclear ribosomal markers (18S rRNA and 28S rRNA) as well as the mitochondrial markers 16S rRNA, COI, and CytB, are sufficient to resolve both inter- and intraspecific relationships (e.g., [26,48,62–68]).

3. Results

3.1. *Apharyngtidae n. fam*

3.1.1. Phylogenetic Affinities

Apharyngtidae n. fam. is a monotypic family of annelids that includes the single species *Apharyngtus punicus*. This species was originally described as a member of Dinophilidae based on its transverse ciliation, dorso-ventrally flattened body and the lack of appendages and chaetae [69]. Subsequently, *A. punicus* was transferred (along with Dinophilidae) to Dorvilleidae following the results of a morphologically based cladistic analysis [69–71] (see below). However, phylogenomic investigations [11,12] do not support a close relationship of *Apharyngtus* with Dorvilleidae, and only in some analyses, find them within Dinophilidae. However, a sister group relationship between *Apharyngtus* and Diurodrilidae within the clade Orbiniida was found in several analyses [11,12]. The nesting within Orbiniida suggests a progenetic origin for *A. punicus*, further supported by the presence of transverse ciliary bands on the prostomium and surrounding the mouth segment, resembling those found in the polytrochous larvae of Orbiniidae [11]. However, the morphological distinctiveness and newly available molecular evidence for *Apharyngtus*, representing a separate evolutionary lineage, highlights the need for a new family designation. Therefore, Apharyngtidae n. fam. is herein formally established and is in line with previous findings [71].

3.1.2. Morphology

Since this group is only represented by one species, the morphological characteristics are treated in the taxonomy section. Apharyngtidae resembles Diurodrilidae in the small size, lack of appendages and chaetae, and by the presence of paired posterior gonopores.

3.1.3. Taxonomy

Apharyngtidae n. fam.

ZooBank Number: urn:lsid:zoobank.org:pub:018F0060-208A-44DF-9D2F-3334D736AB3A.

Diagnosis: A microscopic annelid lacking appendages and chaetae, having multiple indistinct segments (Figure 2). The prostomium, peristomium, and pygidium are well delineated by ciliary bands and epidermal constrictions. A muscular pharyngeal bulb is absent. Ventrally, three dense longitudinal bands of locomotory cilia extend along the trunk. Internally, the densely ciliated mouth opens ventrally on the peristomium, continuing into a heavily ciliated esophagus, mid- and hindgut. The anus is located dorsally on the pygidium. A minimum of three pairs of segmental protonephridia are present. Both coelom and a blood vascular system were undetected, but coelenchyme cells are scattered throughout the body. Gonochoristic and sexually monomorphic. Females carry oocytes in the posterior segments and a pair of ventral gonopores near the pygidium. Males carry filiform spermatozoa, an unpaired copulatory organ and several ciliated glandular pores near the pygidium. A larval stage seems to be lacking and direct development is assumed.

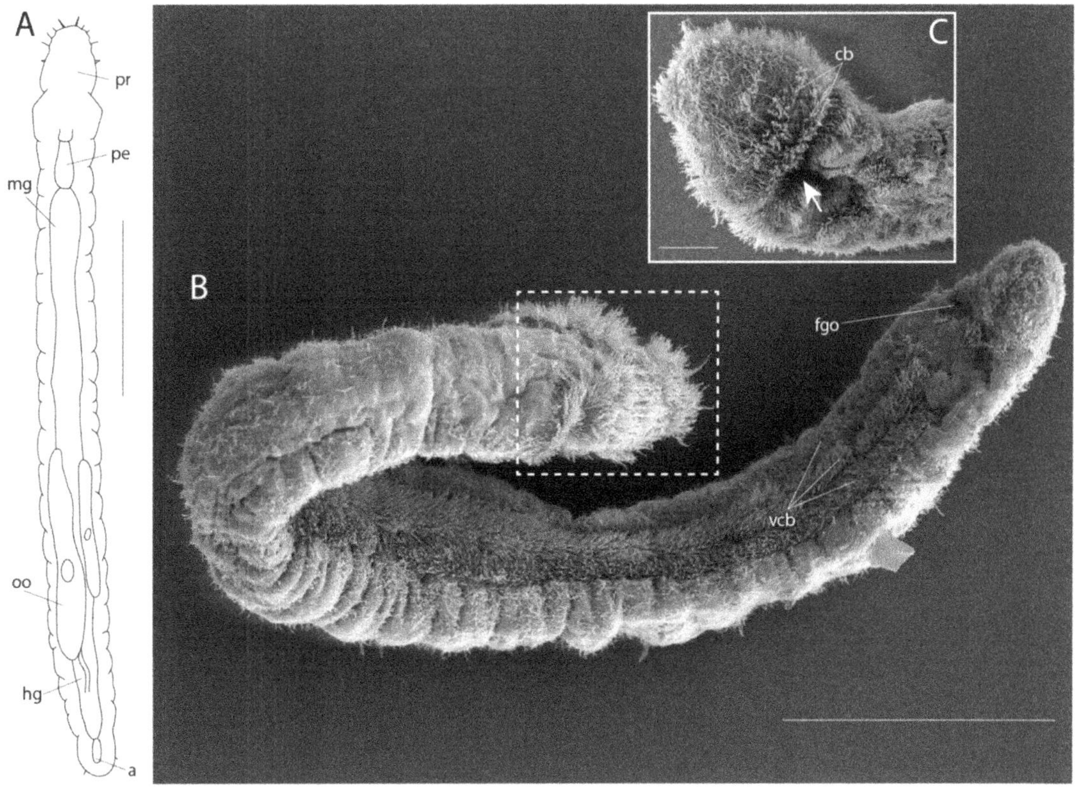

Figure 2. *Apharyngtus punicus* Westheide, 1971. (**A**) Dorsal overview of a mature female, redrawn from Westheide [72]. Scale = 200 μm; (**B**) scanning electron micrograph of a mature female in semi-curled state, note the three ventral longitudinal bands of locomotory cilia. Prostomium enclosed by the dashed box. Scale = 100 μm; (**C**) scanning electron micrograph of the ventral side of the prostomium. White arrow indicates mouth opening. Scale = 20 μm. Micrographs courtesy of Günter Purschke. Abbreviations: a, anus; cb, ciliary bands; fgo, female genital opening; hg, hindgut; mg, midgut; oo, oocyte; pe, posterior esophagus; pr, prostomium; vcb, ventral ciliary bands.

Apharyngtus Westheide, 1971 (Figure 2)

Diagnosis (modified from [72]): The only described member is 610–940 µm long and less than 100 µm in width. The elongated body supposedly has 18–20 externally indistinct segments. Three incomplete transverse ciliary bands are present on the prostomium, and scattered ciliation covers the trunk segments.

Monotypic. Type species: *Apharyngtus punicus* Westheide, 1971

3.1.4. Distribution

Apharyngtus punicus is likely a microphagous feeder, grazing on diatoms, bacteria, and detritus [71,73]. *Apharyngtus* was originally described in Tunisia and has been subsequently recorded in Corsica and the North Island of New Zealand [71,72,74], possibly representing undescribed species. *Apharyngtus* has been only found intertidally in the upper shoreline in fine sandy sediments between 5 and 15 cm depths [3,71].

3.1.5. Major Revisions and Most Important Literature

The main and most recent review on the family is by Westheide (2019) [71].

3.2. *Dinophilidae Macalister, 1878*

3.2.1. Phylogenetic Affinities

The family Dinophilidae consists of eighteen described microscopic species, all having six trunk segments (with the exception of the dwarf males) and inhabiting biofilms, coarse sediments, or living on macroalgae [53,75]. Dinophilids were first described as Platyhelminthes before being recognized as an annelid family. Previous morphological studies suggested Dinophilidae to be the last step in a miniaturization sequence within Dorvilleidae [69], but this hypothesis was rebutted by later molecular analyses [76]. Recent phylogenomic studies recovered conflicting relationships for Dinophilidae, either unresolved [10], or forming a clade within Orbiniida, along with Nerillidae and Diurodrilidae [11], or in the latest well supported analyses as a sister group to Lobatocerebridae [12] within Dinophiliformia; a sister group to Pleistoannelida in most analyses [12]. Dinophilidae and Lobatocerebridae share morphological characters such as widely separated ventral nerve cords, an unpaired medioventral nerve and a particularly broad range of epidermal glands partly condone the otherwise stark differences in brain organization, ciliation patterns, and segmentation.

Starting with Remane's description of the genus *Trilobodrilus* in 1925 [77] up to the morphological and molecular revision in 2019 [26], the family was long been thought to contain only two genera, *Dinophilus* and *Trilobodrilus*. *Dinophilus* then included both monomorphic species with long life cycles as well as species with strong sexual dimorphism and a short life cycle [53,75]. However, a recent phylogenetic analysis [26] shows the monomorphic *Dinophilus* and *Trilobodrilus* to form a clade, sister to a clade containing the sexually dimorphic species and then named *Dimorphilus* (Figure 3A). Developmental studies of *Trilobodrilus* are warranted for comparison with the two other genera, otherwise showing highly similar morphology [78–80].

3.2.2. Morphology

All Dinophilidae are microscopic in size, ranging from the diminutive 50 µm-long dwarf males of *D. gyrociliatus*, to the approximately 3 mm-long adult *D. vorticoides* [26,53,75,81]. They all have an elongated, cigar-shaped body, being plump in *Dinophilus*- and *Dimorphilus*-species, while slender in *Trilobodrilus*. A slight constriction demarcates the head from the six poorly delineated body segments, followed by a tapering pygidium (Figures 3B–D and 4A,D,G). Dinophilids also have a dense and broad ventral ciliary tract that is used for locomotion, and, in its anterior part, aids the transport of food particles into the Y-shaped mouth (Figures 3B–D and 4A,D,G).

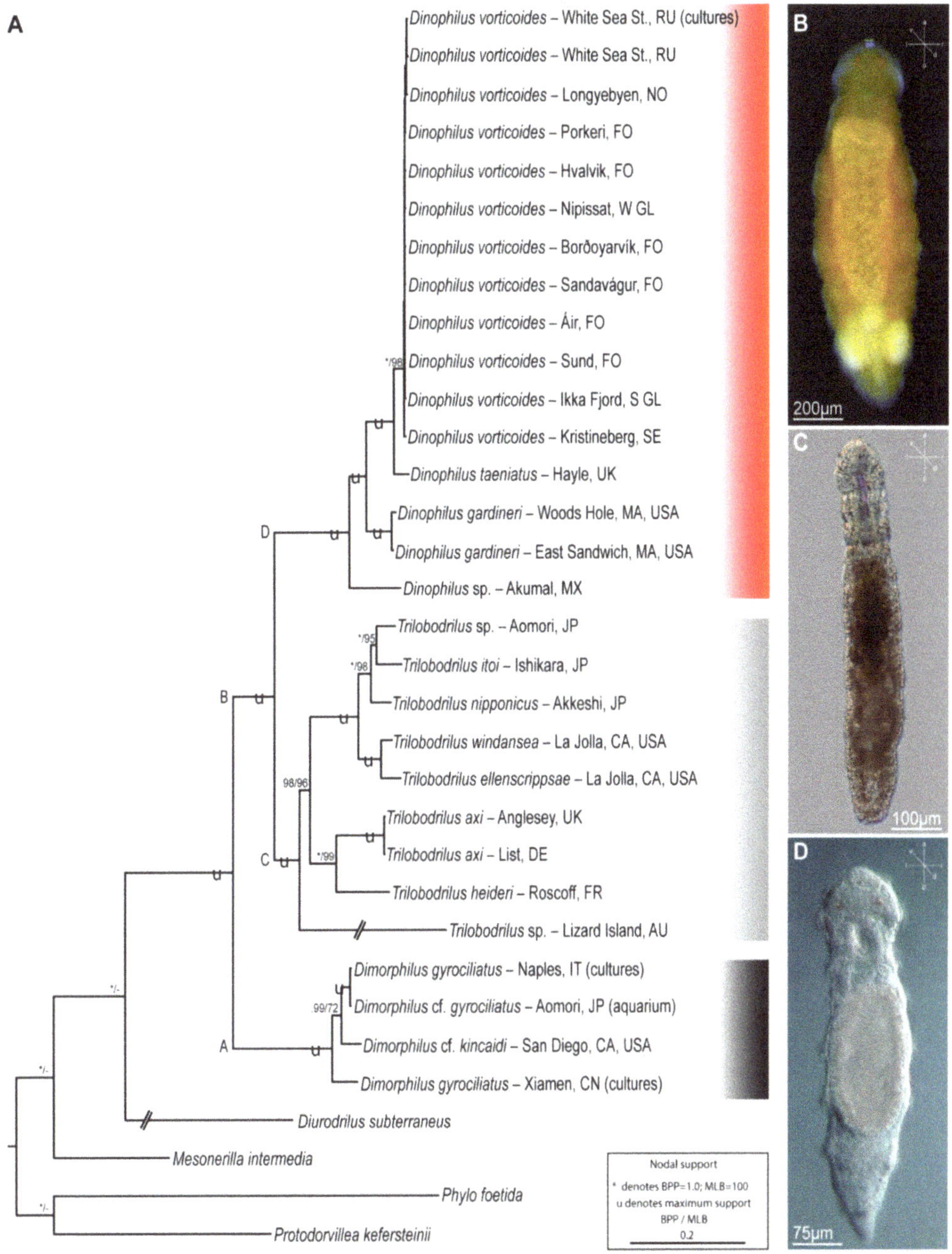

Figure 3. Phylogenetic relationships of Dinophilidae using combined gene analyses (18s rRNA, 28s rRNA, 16s rRNA, COI, CytB); (**A**) tree topology based on Bayesian analyses (BA) of combined gene datasets, nodal support indicated with Bayesian posterior probabilities (BPP) and maximum likelihood bootstrapping (MLB). Only nodal support above BPP = 0.5 or MLB = 50 shown, with those falling below threshold represented by a dash (–). "u" indicates maximum support (BPP = 1.0, MLB = 100). Color bars on right margin indicate three recovered clades: red, *Dinophilus*; (**B**) grey, *Trilobodrilus*; (**C**) black, *Dimorphilus*; (**D**). Small crosshairs indicate orientation. Images modified from [26].

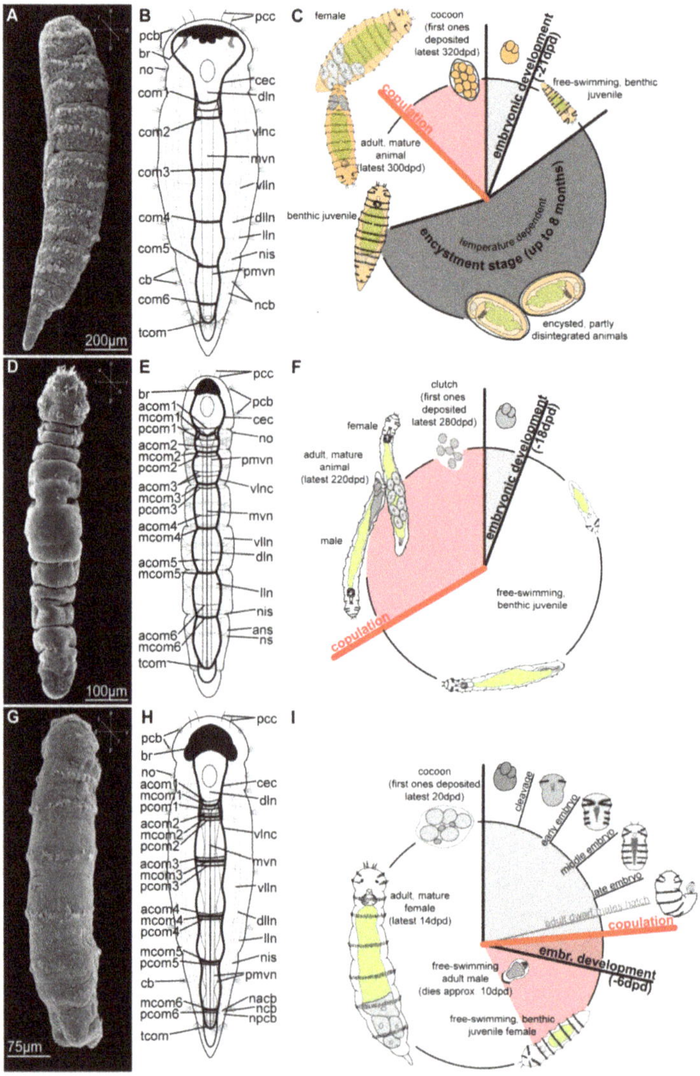

Figure 4. General external morphology, nervous system architecture and life cycle pattern of representatives of each of the three genera within Dinophilidae. (**A–C**). *Dinophilus vorticoides* (Faroe Islands); (**D–F**) *Trilobodrilus axi* (Sylt, Germany), (**G–I**) female *Dimorphilus gyrociliatus*. (**A**,**D**,**G**) Scanning electron micrographs illustrating external ciliation patterns (small crosshairs indicating orientation); (**B**,**E**,**H**) schematic reconstruction of the nervous system based on immunohistochemical labelling with anti-acetylated α-tubulin and confocal laser scanning microscopy (CLSM) in ventral view; (**C**,**F**,**I**) schematic summary of the life cycles of *D. vorticoides*, *T. axi* and *D. gyrociliatus* based on literature review and personal observations, developmental time given in days post deposition (dpd). Abbreviations: a, anterior; acom, anterior commissures; ans, angled segmental nerve; br, brain; cb, ciliary band; cec, circumesophageal connective; com, commissure; d, dorsal; dlln, dorsolateral longitudinal nerve; dln, dorsal longitudinal nerve; drcc, dorsal root of the circumesophageal connective; l, left; lln, lateral longitudinal nerve; mcom, median commissure; mvn, medioventral nerve; nacb, nerve anterior to the ciliary band; ncb, nerve of the ciliary band; nis, intersegmental nerve; no, nuchal organ; npcb, nerve posterior to the ciliary band; ns, segmental nerve; p, posterior; pcb, prostomial ciliary band; pcc, prostomial compound cilia; pcom, posterior commissure; pmvn, paramedian nerve(s); r, right; tcom, terminal commissure; v, ventral; vlln, ventrolateral longitudinal nerve; vlnc, ventrolateral nerve cord. Images modified from [26,79].

All dinophilids have two pairs of stiff compound cilia located anteroterminally on the prostomium, and one pair of densely ciliated nuchal organs positioned laterally on the "neck" region (Figure 4A,D,G). *Dinophilus* and female *Dimorphilus* have a pair of bean-shaped pigmented cup-type eyes (Figure 3B,D) [82,83]. *Trilobodrilus* lack such eyes but have anteriorly positioned ciliated organs underneath the prostomial epidermis [79,81], which most likely serve as light sensing organs. The epidermis has a range of glands, some of which show species characteristics in their vesicular or granular content, location, shape, and light refraction [24,60,84,85].

The distribution of transverse ciliary bands or dorsal ciliation is generally genus-specific (Figure 4A,D,G). For instance, while other dinophilids have two incomplete ciliary bands on their prostomium (broken around the eyes), only species of *Trilobodrilus* lack dorsolateral ciliary bands on their trunk segments, with the exception of some lateral ciliary tufts in *T. nipponicus* or *T. ellenscrippsae*, for example (Figure 4D, [54,61,76,85]). *Dimorphilus* females have a single continuous transverse ciliary band per trunk segment (Figure 4G, [27,86,87]). Different types of dorsal ciliation are found within the genus *Dinophilus*, ranging from two continuous transverse ciliary bands on each trunk segment, with the last two occasionally being incomplete (e.g., *D. vorticoides* (Figure 4A) and *D. taeniatus*), to almost complete dorsoanterior ciliation with additional ciliary tufts between the ciliary bands in the posterior body in *D. gardineri* [26,87].

Mature females can be identified by the presence of yolky eggs in the posterior body region, while unpaired male copulatory organs can be best observed in *Dinophilus* and *Dimorphilus* due to the refraction of their stylet glands [88]. An unpaired muscular copulatory organ is also present in *Trilobodrilus*, but less obvious.

Internally, all dinophilids have a thin layer of body wall musculature, which, depending on their size, varies between a more or less continuous layer of longitudinal muscles in the larger species of *Dinophilus* and ventrolaterally concentrated muscle bundles in *Trilobodrilus* and *Dimorphilus*. The longitudinal musculature is complemented by a thin outer layer of approximately equally spaced circular muscles [75,78,79,88]. The intestine is also surrounded by a thin muscle grid. All dinophilid species have a massive pharyngeal muscle bulb posterior to their mouth opening (Figure 3B–D), which is used to scrape or push off and transport biofilm from the substrate into the digestive tract [85,89]. All dinophilids have an anterodorsal brain with an internal neuropil and a surrounding somata-layer in the prostomium, as well as ventral nerve cords extending throughout the trunk (Figure 4B,E,H). The latter consists of a single pair of longitudinal ventrolateral nerve cords, one ventromedian nerve, one to two pairs of paramedian nerves, and different configurations of transverse commissures. *Dimorphilus* females have three transverse commissures in most segments (Figure 4H), *Dinophilus* has a single commissure per segment (Figure 4B), and *Trilobodrilus* has one prominent commissure complemented by a varying number of thin neurite bundles in each segment (Figure 3E, [75,78–81,88,90–96]).

The individual genera can be identified based on their size, coloration, and ciliation pattern using light microscopy (Figures 3B–D and 4A,D,G). Species identification requires the additional use of scanning electron microscopy to examine the detailed external morphology (e.g., to distinguish between *Trilobodrilus*-species, Figure 4A,D,G) and molecular analyses (e.g., for the distinction between *D. vorticoides* and *D. taeniatus*, [26]).

3.2.3. Taxonomy

Dinophilus O. Schmidt, 1848 (Figures 3B and 4A–C)

Diagnosis: All species are monomorphic, 1–3 mm long, brightly yellow to orange-brown and with cigar-shaped bodies that exhibit a broad ventral ciliary band and at least two transverse ciliary bands per segment (e.g., *D. vorticoides*, *D. gigas*, *D. taeniatus* and *D. jaegersteni*, [53,97,98]) or with complete dorsoanterior ciliation (*D. gardineri*). The life cycle of *Dinophilus* is the longest of the family, consisting of approximately three weeks to one month of embryonic development, with obligate, prolonged encystment stages lasting up to eight months in *D. vorticoides*, *D. taeniatus*, and *D. gardineri* [97,99].

Five species: Type species: *Dinophilus vorticoides* O. Schmidt, 1848; *D. gigas* Weldon, 1886, *D. taeniatus* Harmer, 1889; *D. gardineri* Moore, 1900; and *D. jaegersteni* Jones and Ferguson, 1957. *Dinophilus caudatus* Levinsen, 1880 and *D. metameroides* Hallez, 1879 were previously synonymized with *D. vorticoides* (Figures 3D and 4G–I, [100]).

Dimorphilus Worsaae, Kerbl, Vang and Gonzalez, 2019 (Figures 3B and 4A–C)

Diagnosis: Strong sexual dimorphism. *Dimorphilus* females are about 1 mm long with hyaline bodies, having a single transverse ciliary band per segment. Males are about 50 μm long, extremely reduced in size and complexity, e.g., lacking a digestive system and mainly containing testes, gametes and a muscular copulatory organ [78,88,101,102]. Dwarf males are well-studied in *D. gyrociliatus*; however, they have not been observed in *D. kincaidi* [103]. Fertilized eggs are deposited in gelatinous cocoons, and the embryonic development takes roughly one week (slightly less in males, upon hatching immediately starting mating). Given their fast life cycle, *Dimorphilus* species can rapidly colonize new (and artificial) habitats and are often found in aquaria systems.

Two species. Type species: *Dimorphilus gyrociliatus* (O. Schmidt, 1857) and *D. kincaidi* (Jones and Ferguson, 1957). *Dimorphilus apartis* (Korschelt, 1882) and *D. conklini* (Nelson, 1907), were previously synonymized with *D. gyrociliatus* (see e.g., [104]). *Dimorphilus pygmaeus* (Verrill, 1892), should probably be synonymized with *D. gyrociliatus*, too, yet more detailed analyses are needed. *Dimorphilus borealis* (Diesing, 1862), *D. simplex* (Verrill, 1892) and *D. rostratus* (Schultz, 1902) were also reported, yet the latter two were morphologically assigned to Turbellaria and Rhabdocoela, respectively, and a platyhelminth affiliation was also suggested for *D. borealis* [53,75]. It is furthermore not possible to validate the taxonomic status of *D. sphaerocephalus* Schmarda, 1861, due to the inadequate description.

Trilobodrilus Remane, 1925 (Figures 3C and 4D–F)

Diagnosis: All species are monomorphic, but have a more elongated, slender, hyaline body than *Dinophilus*. In contrast to the other two genera, *Trilobodrilus* has reduced lateral and dorsal ciliation, and lacks pigmented eyes. While little is known about the life cycle of the subtidal species (*T. heideri* Remane, 1925 and *T. ellenscrippsae*), intertidal species have a life cycle of approximately one year with reproductive periods between April and July, and embryonic development taking between two and four weeks within a gelatinous egg clutch. *Trilobodrilus* lacks an encystment stage [105].

Eight species: Type species: *Trilobodrilus heideri* Remane, 1925; *T. axi* Westheide, 1967; *T. indicus* Rao, 1973; *T. hermaphroditus* Riser, 1999; *T. nipponicus* Uchida and Okuda, 1943; *T. itoi* Kajihara, Ikoma, Yamasaki and Hiruta, 2015; *T. ellenscrippsae* Kerbl, Vereide, Gonzalez, Rouse and Worsaae, 2018; *T. windansea* Kerbl, Vereide, Gonzalez, Rouse and Worsaae, 2018).

3.2.4. Distribution and Diversity

Integration of molecular, developmental and morphological studies have helped to further unravel the relationships and distribution of species [26,60]. However, with limited sampling and taxonomic efforts a substantial cryptic and hidden diversity is expected, e.g., an undescribed species was recently discovered off the coast of the Yucatán Peninsula in México [26]. The highest species number is found in the genus *Trilobodrilus*, while *Dinophilus* species have the broadest distribution range within the family [26,53,75]. *Dinophilus* species are restricted to shallow waters and the intertidal areas of rocky or sandy shores. Most *Dinophilus* species inhabit the cold waters throughout the Atlantic, with *D. vorticoides* having the broadest distribution range, spanning from the west coast of Greenland to the White Sea, Russia [26]. In contrast, *D. taeniatus* has only been found along the west coast of the United Kingdom [106], and *D. gardineri* appears to be limited to the coast off Massachusetts [87]. Explanations for these varying distribution patterns are mainly speculation. However, it is likely that the lack of larval dispersal stages and a limited ability to migrate over long distances, as well as the temperature optimums during the different life cycle stages, has hampered a broad distribution in most species. On the other hand, stages of lengthy encapsulation, such as the encysted juveniles in *Dinophilus*-species or eggs deposited in gelatinous "cocoons", might increase dispersal abilities via rafting on algae, sediment, or debris, being at the whims of prominent currents.

Trilobodrilus prefers coarse, well-sorted sandy sediments from the eulittoral zone down to several meters depth [60,75,107]. Bathymetric ranges seem to be clearly demarcated, resulting in eu- and sublittoral species occurring at the same beach in close proximity to each other [60,84].

Very little is known about the distribution pattern of *Dimorphilus*, yet preliminary analyses found geographically widely separated populations to be genetically closely related [5]. However, most of these specimens came from old aquarium cultures, which could have been mixed over time, as the geographical origin cannot be verified. A "real" global distribution pattern of one species across Brazil, USA, Italy, Germany, Russia, and Japan as indicated by [5,26,108] warrants further studies on wild caught material [26].

3.2.5. Major Revisions and Most Important Literature

The most recent reviews of the family were given by Westheide [75] and Worsaae et al. [26]. The latter study [26] revised the genus *Dinophilus* and especially the relationship between the morphologically similar *D. vorticoides*, described in the Faroe Islands, and *D. taeniatus* described in the United Kingdom. *D. vorticoides* was here recognized as a valid taxon with a remarkably broad distribution in the boreal North Atlantic, while *D. taeniatus* was only found near its type location [26]. Populations previously reported along the French coast of the English Channel remain of particular interest, since their collection and identity will allow for interpretations of the ecological, developmental and physiological limits of the distribution ranges between *D. vorticoides* and *D. taeniatus*. Molecular analyses of *Trilobodrilus* species collected from several locations worldwide recovered taxa adapted to intertidal and subtidal sediments, respectively, for each geographical region [27,61,85]: *T. axi—T. heideri* in the Northwest Atlantic, *T. itoi—T. nipponicus* around Japan, *T. indicus* along the Indian coast and *T. windansea—T. ellenscrippsae* along the west coast of the United States (Figure 3A). Although specimens of *Dimorphilus* cf. *gyrociliatus* were collected from different locations in Europe (Naples, laboratory aquaria in Russia, Sweden and Denmark) as well as from Israel, USA, and Japan, their identity has only been analyzed superficially so far (Figure 3A, [26]).

3.3. *Diurodrilidae Kristensen and Niilonen, 1982*

3.3.1. Phylogenetic Affinities

The phylogenetic position of the microscopic members of Diurodrilidae has long been debated due to their lack of significant annelid characteristics, such as chaetae, head appendages, parapodia, nuchal organs and obvious segmentation (Figure 5). The first described species of *Diurodrilus* was assigned to Dinophilidae by Remane (1925) as part of the now-abandoned "Archiannelida" [7,109]. However, diurodrilids lack the characteristic continuous midventral ciliary band of most interstitial annelids. Their ventral side instead carries specialized multiciliated cells, called ciliophores. Diurodrilids also possess a ventral bowl-shaped muscular pharynx that differs from that of Dinophilidae, yet these animals have a reduced cuticle, showing some resemblance to other interstitial and juvenile annelids [14,24]. Their unique morphology was acknowledged by Kristensen and Niilonen [110] when they erected Diurodrilidae Kristensen and Niilonen, 1982, then by Westheide [111] in erecting Diurodrilida, and finally Worsaae and Rouse [14] questioned their annelid affinity based on a phylogenetic study of 18S rRNA and 28S rDNA data, which placed them outside Annelida. Moreover, diurodrilids have several traits in common with other meiofaunal metazoans, and particularly Gnathifera, such as the presence of trunk ciliophores with long ciliary rootlets, adhesive head and toe glands, spermatozoa with mushroom bodies, dorsal plates and a ventral muscular pharynx with large central glands [14]. However, a later mitochondrial genome study [112] and three comprehensive phylogenomic studies [8,11,12] found Diurodrilidae to nest within annelids. Although their exact position is not fully resolved, the latter two studies grouped *Diurodrilus* with another meiofaunal annelid taxon, *Apharyngtus* (within Orbinida), which at least shows some superficial morphological resemblance to Diurodrilidae [53,71]. The very small size, aberrant morphology and poorly segmented nervous system of *Diurodrilus* have therefore been discussed to possibly reflect an extreme case of pedomorphosis within Annelida [7,11,14,25].

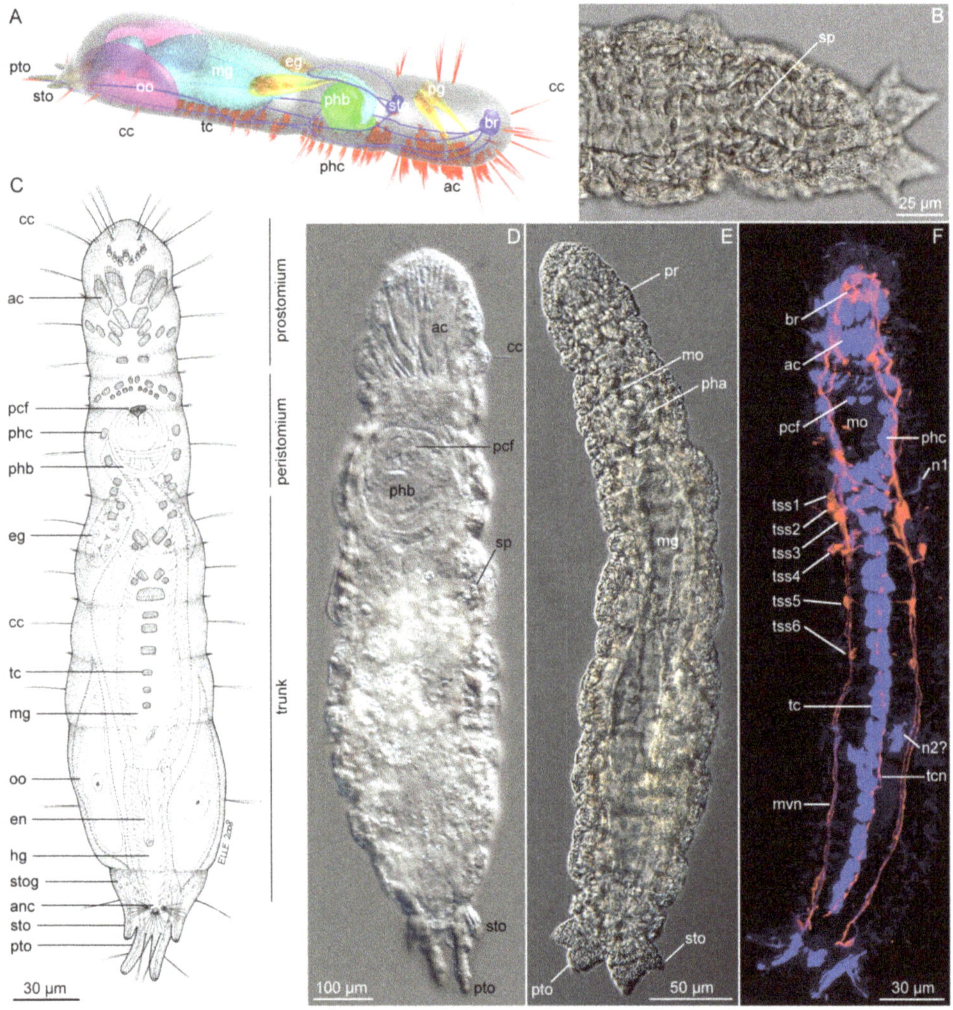

Figure 5. (**A**) Schematic illustration of *Diurodrilus* sp. from South Australia, Australia; (**B**) light micrograph of posterior trunk of *D. subterraneus* male with sperm. (**C**) Drawing of *Diurodrilus* sp. from Queensland, Australia, ventral side. (**D**) Light micrograph of male *Diurodrilus* sp. from Aomori, Japan. (**E**) Light micrograph of female *D. subterraneus*. (**F**). Confocal laser scanning microscopy (CLSM), maximum intensity projection of Z-stack images of *D. subterraneus* showing anti-acetylated α-tubulin like immunoreactivity (LIR) (blue) and anti-serotonin LIR (red). (**C**,**E**,**F**)—modified from [25]. Abbreviations: ac, anterior head ciliophore; anc, anal ciliary field; br, brain; cc, compound cilia; eg, esophageal gland; en, enteronephridium; hg, hindgut; mg, midgut; mo, mouth opening; mvn, main ventral nerve; n1–2, nephridium 1 and 2; pcf, prepharyngeal ciliary field; pg, prostomial gland; pha, pharynx; phb, pharyngeal muscle bulb (bowl-shaped); phc, peripharyngeal ciliophore; pr, prostomium; pto, primary toe; oo, oocyte; oon, oocyte nucleus; sp, spermatozoa; sto, secondary toe; stog, secondary toe gland; stg, stomatogastric ganglion; tc, trunk ciliophore; tcn, trunk ciliophore nerve; tss1–6, first to sixth pairs of trunk anti-serotonin LIR somata.

Only a single genus, *Diurodrilus*, has been described, reflecting a rather similar morphology across the seven described species. A phylogenetic study of the family that included three gene fragments from three species [113] did not group the two species from the East Pacific, but instead grouped the two upper littoral species *D. kunii* and *D. subterraneus*, which also show closer morphological resemblance. An ongoing phylogenetic study of the family across multiple species (Worsaae et al. unpublished) aims to test the degree of endemism among the European populations and whether possible adaptations to subtidal versus littoral habitats may be reflected in their phylogenetic relationships, despite their presumably more restricted dispersal potential of intertidal species.

3.3.2. Morphology

All members of Diurodrilidae are microscopic, dorso-ventrally flattened, hyalin and fast moving. Their 300–500 µm-long body comprises an elongated head region and a short, seemingly unsegmented coelomate trunk with two to four pygidial lobes (toes) and sometimes an anal cone (Figure 5). The prostomium carries long, presumably sensory, compound cilia that are also found along the lateral trunk. The peristomium has a ventral mouth opening and a bowl-shaped muscular pharynx with central paired glands. Along the entire ventral surface are the characteristic ciliophores (multiciliated cells with long ciliary rootlets), where the cilia of each cell beat in unison. The ciliophores are large and ovoid on the prostomium but rectangular on the peristomium and trunk. The ciliophores surrounding the mouth opening continue along the trunk as transverse rows of rectangular ciliophores, forming a discontinuous midventral band (Figure 5) [7,14,25,110].

Paired, long-necked adhesive glands open ventrally on the prostomium and at the tip of the pygidial toes, while two large salivary (esophageal) glands extend posteriorly to the muscular bulb. Diurodrilids glide quickly by way of ciliary beating of the ciliophores, intermittently adhering (and releasing) the head and/or the toes to the substrate, somewhat resembling the motility pattern of gastrotrichs. Their rapid release from substrate (post adhesion) indicates a duo-gland function of the diverging gland types found in the primary and secondary toes [7,14,25,110].

Diurodrilids possess an unsegmented, grid-like body wall musculature composed of two main and several thinner circular muscles. Inner circular musculature surrounds the intestine, some of which may act as sphincter muscles between the mid and hindgut as well as around the anus [14].

The only detailed study of the nervous system [14] showed an anterior bilobed brain and only a few anterior ganglia along the widely separated two main, and four minor ventral nerves, hereby defying the previous externally assessed interpretation of the trunk consisting of five segments. Likewise, only two pairs of protonephridia are found in the anterior and middle trunk [7,14,110,113]. Although their presence and lateral openings have been documented in four species using TEM, CLSM and SEM, their exact configuration and composition are still not fully understood and may vary slightly among species [26]. A third pair of densely ciliated ducts, presumably representing gonoducts, is found opening ventrally in the posteriormost trunk [25]. However, the paired ovaries or testes seem to disappear during development, with the gametes consequently lying freely in the coelomic cavity, being most prominent in the dorsal part of trunk and lacking an obvious peritoneal lining [7,14,110,114]. Diurodrilids are seemingly all direct developers and gonochoristic, with males producing specialized spermatozoa with large acrosomes and mushroom-shaped bodies [114]. An unpaired, ciliated, blind-ending enteronephridium extends along the hindgut from the dorso-posterior midgut in *Diurodrilus* sp. from Brisbane, Australia [25].

3.3.3. Taxonomy

The different species of *Diurodrilus* are distinguished by variation in ciliophore patterns, glandular patterns, absence/presence of cuticular plates, length of the toes and shape of the spermatozoa. Accompanying molecular barcoding and perhaps even population genetics might prove necessary in order to describe the vast hidden diversity of Diurodrilidae. The systematically important cilliophore patterns are best examined using anti-α-tubulin staining and CLSM, and alternatively, by meticulous high-resolution light microscopy on live animals. Key features of described species are listed in Table 1.

Table 1. Diagnostic characters of described species of *Diurodrilus*.

Species of *Diurodrilus*	Intertidal (I)/Subtidal (S)	Primary Toes Longer than Secondary (2°) Toes	Shape of Primary Toes	Shape of Secondary Toes	Anal Cone	Ciliophores on Anterior Head	Dorsal Cuticular Plates
D. ankeli	I	yes	cylindrical	cone shaped	absent	4 pairs + 1	present
D. benazzii	I	no 2° toes	bottle-shaped	no 2° toes	absent	3 pairs	absent
D. dohrni	I + S	yes	bottle-shaped	cone shaped	absent	unknown	unknown
D. kunii	I	yes	cylindrical	cone shaped	absent	3 pairs	absent
D. minimus	I + S	equal	cylindrical	cone shaped	small	unknown	absent
D. subterrraneus	I	yes, slightly	cone shaped	cone shaped	absent	5 pairs	present
D. westheidei	S	yes	cylindrical	cylindrical	large	3 pairs	absent

Diurodrilus Remane, 1925

Seven described species. Type species: *Diurodrilus minimus* Remane, 1925; *D. subterraneus* Remane, 1934; *D. benazzii* Gerlach, 1952; *D. dohrni* Gerlach, 1953; *D. ankeli* Ax, 1967; *D. westheidei* Kristensen and Niilonen, 1982; *D. kunii* Kajihara, Ikoma, Yamasaki and Hiruta, 2019. Ten unidentified species of *Diurodrilus* have additionally been found along the Atlantic coast of the USA [25], Galapagos Islands [115], New Zealand ([74], two species), northeast and southern Australia ([14,116]; two species), Tobago and Trinidad (K. Worsaae and R. M. Kristensen, unpublished), Brazil (M. Di Domenico, pers. comm.), Northern Cuba (K. Worsaae, unpublished), Aomori, Japan (K. Worsaae, unpublished), and Amsterdam Island, southern Indian Ocean (K. Worsaae, unpublished). Moreover, some of the multiple sampled populations of *Diurodrilus* cf. *minimus* and *D.* cf. *subterraneus* in the North Atlantic and *D.* cf. *dohrni* in Canary Island waters and the Mediterranean Sea may represent new cryptic species (Worsaae, unpublished).

3.3.4. Distribution and Diversity

Diurodrilids are only recorded from intertidal and shallow subtidal waters (less than 60 m depth). They prefer fine to coarse, well-sorted, oxygenated sediment. Most records are from European waters, but they are found in all major oceans worldwide, from polar to tropical regions ([25] and references herein).

The limited number of easily distinguishable external characteristics has most likely led to the arrest in the description of new species of Diurodrilidae in recent years. However, unpublished molecular data (K. Worsaae et al.) indicate a high hidden diversity similar to what is seen in other interstitial annelid families, with different species existing even within short geographical distances.

3.3.5. Major Revisions and Most Important Literature

The morphological diversity of Diurodrilidae has mainly been addressed by Kristensen and Niilonen [110], Villora-Moreno [117], Worsaae and Rouse [14], Westheide [53] and Worsaae [3], whereas the study by Kajihara et al. [113] was the first to compare molecular sequences among species of Diurodrilidae.

3.4. *Lobatocerebridae Rieger, 1980*

3.4.1. Phylogenetic Affinities

Lobatocerebridae is a family of rare and inconspicuous meiobenthic annelids [15,118,119]. Superficially, they do not share any morphological traits exclusively with annelids. However, their phylogenetic position among Annelida has been settled thanks to phylogenomics [8]. They are filiform, cylindrical, and completely ciliated worms lacking appendages, with a length ranging between 1 and 3 mm and a width of 40–100 μm (Figure 6A,F). The first described species, *Lobatocerebrum psammicola*, was assigned to Annelida. However, this phylogenetic position has been repeatedly questioned given their ambiguous morphological characters, leading to the subsequent erection of Lobatocerebromorpha as phylum [120,121]. It was only in 2015 that transcriptomic-based phylogenetic studies confirmed its affinities with Annelida and found it as a sister group to Sipuncula (albeit with low support) [8]. More recently, new studies placed *Lobatocerebrum* as the sister group to Dinophilidae, forming the clade Dinophiliformia [12], reciprocally monophyletic to the clade Pleistoannelida (Errantia—Sedentaria) (Figure 1).

3.4.2. Morphology

Lobatocerebridae are elongated animals, 1–3 mm long and 40–110 μm wide, with a body circular in cross-section and completely ciliated [15,118]. All lobatocerebrids lack segmentation and appendages. The densely ciliated pharynx (Figure 6D) is followed by an unciliated gut and a ciliated hindgut (Figure 6D,F) that terminates at a dorsal, subterminal anus. All species have a large, transparent, multilobed brain (hence the etymology of "*Lobatocerebrum*"; Figure 6C) positioned posteriorly in the rostrum, anterior to the mouth (Figure 6F). The highly glandular epidermis gives the animal a slightly greenish hue. The longitudinally elongated ventral mouth (Figure 6D,F) marks the border of the proportionally long rostrum (ca. 20–30% of the body length) and the trunk (Figure 6A,F).

Lobatocerebrum is hermaphroditic. The anterior-most reproductive structure is an unpaired testis, positioned approximately halfway along the body, containing elongated filamentous sperm cells. A pair of spermioducts (100–200 μm long) extend anteriorly from the testis and open in an unpaired, antero-dorsal gonopore surrounded by numerous elongated glands (Figure 6F). Posteriorly, approximately two thirds along the body, up to four oocytes can be found, which increase in volume and length posteriorly ($\leq$couple of hundred micrometers). No ovarium, oviduct, or female opening has been described [15,118]. Anterior to the hindgut, one to several 20–30 μm-wide seminal receptacles with ventrolateral openings were found, containing curled up sperm cells (Figure 6F). *Lobatocerebrum* supposedly has direct development, but observations and studies to confirm this are lacking.

The nervous system consists of a relatively large brain with a prominent neuropil, two pairs of segmentally arranged ganglia, anterior and posterior longitudinal nerves emerging from the neuropil (Figure 6C,F), five commissures and a peripheral nervous system [15,118]. The brain comprises three pairs of lobes (and sublobes hereof): anterior major rostral lobes and posterior pairs of minor and caudal lobes, respectively. Originating at the brain, four paired and one unpaired nerve extend through the rostrum, possibly innervating sensory cells and glands in the anterior-most tip of the animal. Two ventrolateral nerves extend posteriorly from the brain connected by the commissures of two pairs of post-pharyngeal ganglia localized posterior to the mouth opening. Each ganglion of the anterior-most pair supplies an additional nerve extending ventromedially. These nerves fuse medially with their contralateral partner at the level of the second commissure, forming an unpaired midventral nerve and extending posteriorly to the posterior-most (fifth) commissure alongside the two ventrolateral nerves. The third, fourth, and fifth posterior commissures are not associated with any ganglia. Additionally, a stomatogastric nerve ring is found encircling the mouth.

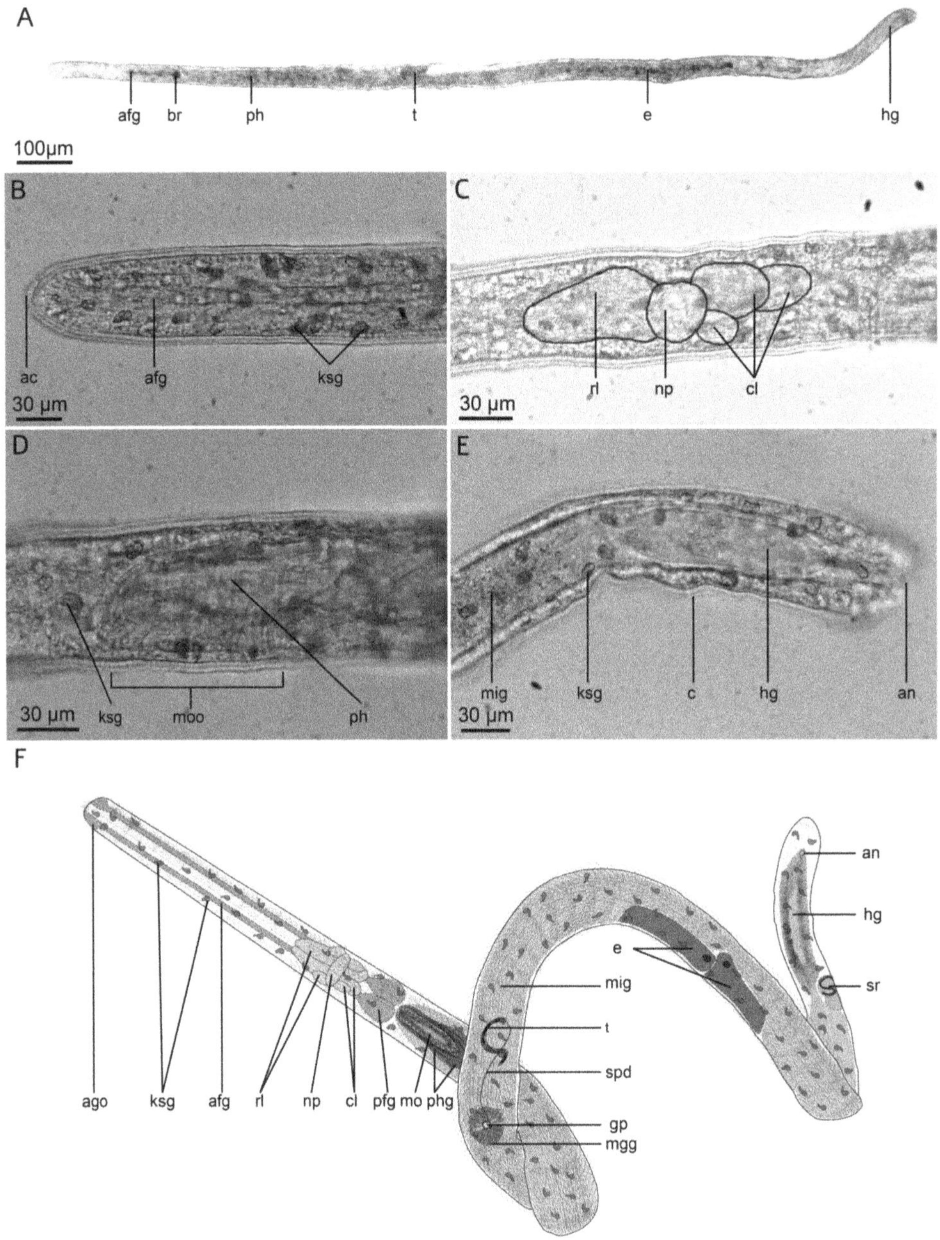

Figure 6. Morphology of Lobatocerebridae with illustrations of *Lobatocerebrum riegeri* (b–f, modified from [15]). Anterior is to the left and posterior is to the right. (**A**,**B**,**F**) Dorsal view. (**C**,**D**) Lateral view, ventral is down, and dorsal is up. (**A**–**E**) Light micrographs of *L. riegeri.* (**A**) Habitus of *L. riegeri.* (**B**) Close up of the tip of the rostrum. (**C**) Close up of the brain. (**D**) Close up of the mouth area. (**E**) Close up of the posterior end. (**F**) Schematic drawing of *L. riegeri.* Abbreviations: ac, anterior ciliation; afg, anterior frontal glands; ago; anterior frontal gland opening; an, anus; br, brain; c, body ciliation; cl, caudal lobe(s) of the brain; e, egg; gp, gonopore; hg, hindgut; ksg, kidney shaped glands; mgg, male gonopore glands; mig, midgut; mo, mouth; moo, mouth opening; np, neuropil; pfg, posterior frontal glands; ph, pharynx; rl, rostral lobe(s) of the brain; spd, spermioduct; sr, seminal receptacle; t, testes.

The muscular system consists of the body wall and gut musculature. The body wall musculature consists of six pairs of longitudinal muscles extending along the entire body length [15,122]. Regularly spaced muscle ring complexes are associated with these longitudinal muscles in the trunk, consisting of transverse muscle fibers extending perpendicularly from one longitudinal muscle to the next one, together giving the impression of unusual internal positioned circular musculature. This organization more closely resembles the transverse muscles of other annelids than their normally externally positioned circular musculature [123]. In the rostrum, these transverse muscles cross centrally to form a star-shaped grid of muscles in cross section. The musculature of the intestinal system comprises a muscular grid of longitudinal and perpendicularly arranged circular muscle fibers lining the entire digestive tract. Five circular sphincters are found supporting the pharynx, and a sixth sphincter is located anterior to the anus.

Lobatocerebrum possesses four types of unicellular gland and three types of multicellular gland [15,118,119,124]. Unicellular glands are characterized as: (1) regular scattered mucus glands, which are the largest unicellular glands with thick microvilli around the opening. The nucleus is found at the basal end and the cell body is densely packed with spherical vesicles. A long cell projection, several time longer than cell body, extends along the basal membrane. (2) Tubular glands, which are elongated and flask-shaped, with a basal projection along the basal lamina. The glandular content consists of small, rod-like granules. They are randomly distributed throughout the body, at least in *L. riegeri*. (3) Kidney-shaped gland cells, which are densely packed with spherical vesicles, but do not have a basal projection (Figure 6B,D,F). The dense packing of the granules affects the nucleus, which takes on a characteristic sickle shape. (4) Adhesive glands, which possess a ciliary ring around the opening, which is encircled by an anchor cell. The granules of these cells differ between the two described species in having rod-like electron-dense inclusions in *L. psammicola*, and granule-shaped (shorter) inclusions in *L. riegeri*.

Multicellular gland systems include: (1) two pairs of frontal glands (Figure 6A,B,F); an anterior pair of frontal glands lying anterior to the brain with elongated rod-shaped granules, and a pair of posterior frontal glands with spherical granules situated between the brain and the pharynx (Figure 6F). The ducts of these glands extend ventrolaterally throughout the rostrum and seem to release the glandular secrete mainly at the tip of the rostrum. (2) Pharyngeal glands, constituted by multiple epidermal glands whose duct openings encircle the mouth (Figure 6F). (3) Male gonopore glands, comprised of two different gland types in *L. psammicola*, and apparently only one in *L. riegeri* (Figure 6F). These cells resemble the pharyngeal glands in shape, size, and electron-density, but are arranged around the gonopore.

Lobatocerebrum psammicola was suggested to have three pairs of U-shaped protonephridia based on squeezed preparations and live observations of cyrtocyte-like structures [118,125]; however, only one nephridium, located posterior to the testis, was reported for *L. riegeri* (as *Lobatocerebrum* sp. II [118]).

3.4.3. Taxonomy

Lobatocerebrum Rieger, 1980

Two described species: Type species, *Lobatocerebrum psammicola* Rieger, 1980; *L. riegeri* Kerbl, Bekkouche, Worsaae, Sterrer, 2015. Species of *Lobatocerebrum* are diagnosed based on the proportional measurements of the body [15]: *Lobatocerebrum psammicola* is larger than *L. riegeri* in total length (2–3 mm vs. 1–1.5 mm, repectively) and diameter (40–60 μm vs. 70–110 μm, repectively), but has a shorter rostrum (15% vs. 20% of the body length, respectively), and a more anterior brain (at 10% vs. 18% of the body length, respectively) and mouth (at 14% vs. 20% of the body length, respectively). Other differences concern the inclusion in the granules of unicellular adhesive glands, which are rod-shaped in *L. psammicola* and more spherical (shorter) in *L. riegeri* [15,119]. So far, only specimens collected at Bocas del Toro, Panama have been sequenced for transcriptomic analyses, morphologically closely resembling *L. psammicola* or a cryptic species hereof [8]. Few additional sightings in the North Atlantic near Gran Canaria (Spain), the Mediterranean Sea near Elba (Italy), and possibly a location off Elsinore (Denmark) may represent additional new cryptic species (W. Sterrer, R. M. Kristensen, K. Worsaae, pers. comm.).

3.4.4. Distribution and Diversity

Lobatocerebrum is found in very low densities and abundances in different kinds of sediments: *Lobatocerebrum psammicola* is found in heterogenous medium-coarse sandy sediment off North Carolina but has also been reported in coral rubble (not well-sorted, mixed fine and coarse sediment) at Bocas del Toro, Panama [8]. *Lobatocerebrum riegeri* is found in coarse calcareous sand in Eilat, Israel. Dr. Wolfgang Sterrer has also found lobatocerebrids in fine sandy sediments underneath *Zostera* meadows around Gran Canaria (Canary Islands, Spain) and in southern Italy (Mediterranean Sea [119]).

3.4.5. Major Revisions and Most Important Literature

A recent morphological study [15] as well as [119] reviewed and summarized most of the extensive TEM and histology studies done previously [118,122,124,125].

3.5. Nerillidae Levinsen, 1883

3.5.1. Phylogenetic Affinities

Nerillidae contains 14 valid genera with 59 meiofaunal species, ranging in size from 300 µm to 2.1 mm and comprising seven to nine segments (Figure 7). Nerillids possess several morphological traits normally considered apomorphic for Errantia, including compound chaetae, one pair of pygidial cirri, a muscular ventral pharyngeal organ, prostomial antennae and short (except in *Speleonerilla*), non-grooved ventrolateral palps (Figure 8). The meiofaunal sizes and resemblance to early juvenile stages of Syllidae and Eunicidan taxa support a progenetic origin for the family, possibly from an ancestor within Errantia [7,126,127]. Early molecular studies likewise found support for their relatedness to errantian families, although the exact position remained debated [128]. However, one phylogenomic study recovered Nerillidae close to Orbiniidae within Sedentaria [11]. Recent morphological [13] and ongoing phylogenomic studies (K. Halanych, pers. comm.), however, continue to support a position within Errantia.

The genera-to-species ratio is quite high in Nerillidae, with many of the 14 valid genera being monotypic, yet, several genera have already been synonymized, including *Afronerilla* herein with *Nerillidium*, see below. Nerillids easily shed their appendages during the fixation processes, necessitating live examination and complicates gathering sufficient information for taxonomic descriptions. Furthermore, species may be diagnosed based on subtle differences in ciliary and glandular patterns, morphometrics of the body, appendages and chaetae (preferably on living individuals), as well as on the configuration of both nephridia and gonoducts. Taxonomic descriptions thus necessitate a broad range of fixation and examination techniques, including light- and scanning electron microscopy as well as immunolabelling and confocal laser scanning microscopy. In recent years, molecular studies have revealed a high diversity of cryptic species that can only be resolved by sequencing multiple genes, preferably from several specimens within each population, e.g., [37].

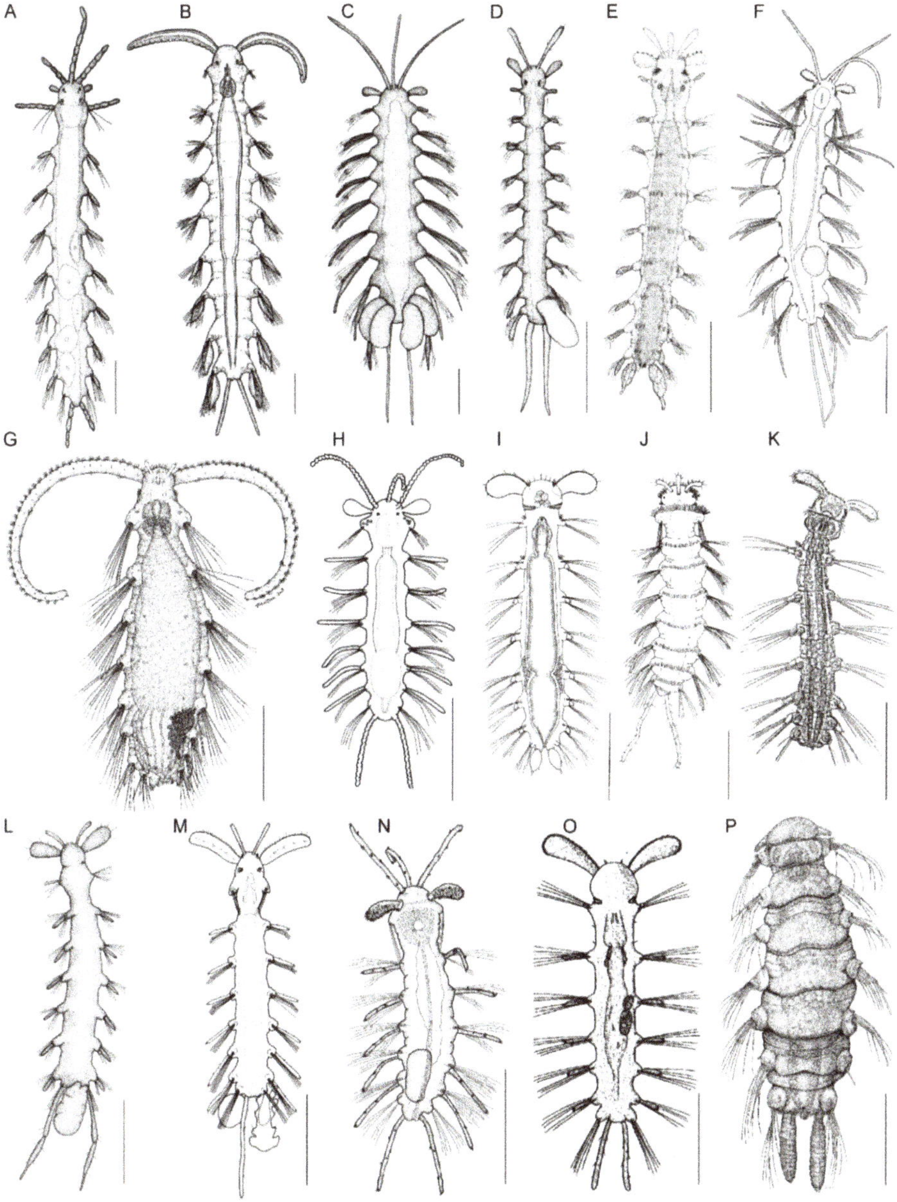

Figure 7. Schematic drawings of Nerillidae genera. (**A**) Nerilla antennata. (**B**) Meganerilla swedmarki. (**C**) Mesonerilla intermedia. (**D**) Mesonerilla biantennata. (**E**) Mesonerilla armoricana. (**F**) Leptonerilla diplocirrata. (**G**) Speleonerilla saltatrix. (**H**) Micronerilla minuta. (**I**) Thalassochaetus palpifoliaceus. (**J**) Trochonerilla mobilis. (**K**) Troglochaetus beranecki. (**L**) Nerillidium mediterraneum. (**M**) Nerillidopsis hyalina. (**N**) Aristonerilla brevis. (**O**) Psammoriedlia ruperti. (**P**) Paranerilla cilioscutata. Scale bars 200 μm; all dorsal view except for N (ventral view). Redrawn or modified from: (**A**,**C**,**D**,**L**), [129]; (**B**), [130], (**E**), [36]; (**F**), [127]; (**G**), [40]; (**H**,**M**), [53]; (**I**), [131]; (**J**), [132]; (**K**), [133]; (**N**), [134]; (**O**), [135]; (**P**), [33]—acknowledging copyright permissions granted by the publishers.

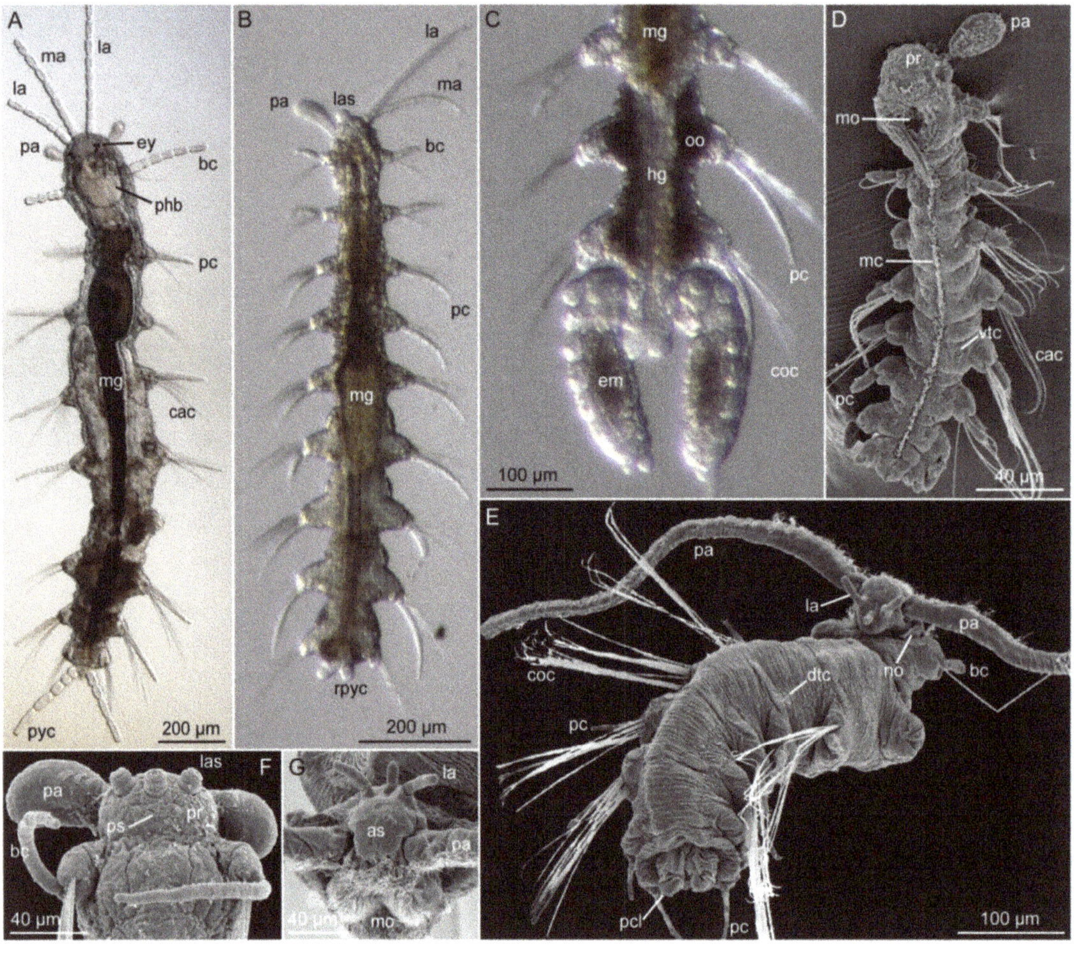

Figure 8. Morphology of Nerillidae (a–c, light micrographs; d–g, scanning electron micrographs). (**A**) *Nerilla* cf. *antennata* (from Roscoff, France), dorsal view (modified from [3]). (**B**,**C**). *Mesonerilla* cf. *luederitzi* (from Iriomote Island, Japan), dorsal views; entire male with sperm in posterior segments (b), posterior end of female with oocytes and embryos (c). (**D**). *Nerillidium* sp. (from Jeju Island, South Korea), ventral view. (**E**). *Speleonerilla calypso*, dorsal view (modified from [129]). (**F**). *Leptonerilla* cf. *diplocirrata* (from Jeju Island, South Korea), anterior end in dorsal view. (**G**) *Speleonerilla calypso*, anterior end, antero-ventral view. Abbreviations: as, anterior field of sensory cilia; bc, buccal (=segment 1) cirrus; cac, capillary chaetae; coc, compound chaetae; dtc, dorsal tuft of cilia; emb, embryo; ey, eye; hg, hindgut; la, lateral antenna; las, scar from lateral antennae; ma, median antenna; mc, midventral ciliary band; mg, midgut; mo, mouth opening; no, nuchal organ; oo, oocyte; pa, palp; pc, parapodial (interramal) cirrus; pcl, pygidial ciliated lobe; phb, pharyngeal muscular bulb; pr, prostomium; ps, posterior field of sensory cilia; pyc, pygidial cirrus; rpyc, regenerating pygidial cirrus; vtc, ventral tuft of cilia.

3.5.2. Morphology

Nerillids comprise a prostomium and seven to nine body segments, of which the first (buccal) segment may contain a peristomium limited to the central mouth region and a pygidium. The prostomium carries two palps (or two horns in *Paranerilla*), maximum three antennae and zero, two, or four eyes. The biramous parapodia possess soft outgrowths (parapodial or interramal cirri) uniquely positioned in between the dorsal and ventral bundles of capillary or compound chaetae. The pygidium likewise carries a pair of appendages, which are easily shed and lost or in a stage of regeneration upon collection. In

Speleonerilla, the pygidium furthermore carries two heavily ciliated lobes that aid with its unique swimming locomotion [129].

A midventral ciliary band extending from the densely ciliated mouth region to the pygidium is used for gliding locomotion. In some species, this band is accompanied by additional dorsal, lateral, and ventral ciliary fields that may facilitate swimming in the water column (i.e., *Trochonerilla*, *Speleonerilla*), or burrowing in soft sediment (i.e., *Paranerilla*) [34,62,136]. Configuration and number of dorsal, lateral and ventral ciliary tufts on the palps and body is relevant for species (and sometimes genus) characterization (i.e., cilia missing, cilia/tufts distributed in one or two transverse rows per segment or dense ciliated fields) (e.g., [33,37,38,43]). In addition, presumed sensory cilia are scattered across the body and the appendages; an anterior and posterior field of sensory cilia as well as two antero-lateral bands are always present on the prostomium. Two ciliated nuchal organs are found postero-laterally of the palps, on the border of segment one [129,137]. Detailed mapping of the external ciliation warrants scanning electron microscopy studies on carefully fixed individuals.

During locomotion, all nerillids are capable of twisting and turning their body using their two dorsal and two ventral bundles of longitudinal muscles, which are supported by both transverse and diagonal muscles [136]. All nerillids can also perform an escape reaction, rapidly undulating their longitudinal muscles to swim a short distance. The relatively small (muscular) parapodia seemingly do not aid the swimming or gliding locomotion (hanging passively along the body) but may assist with maneuvering and attaching the chaetae within the interstitial pore spaces [129]. A ventral pharyngeal bulb muscle aids in the processing of food particles within the mouth cavity and can in some species be extended to "lick" up particles or be accompanied by a protrusible muscular "tongue" (e.g., [138–142]).

Their behavior is coordinated by a relatively elaborate nervous system, overall resembling the ganglionated, subepidermal nervous system found in most macrofaunal errantian annelids, though having a clearly separated pentaneural, ganglionated nerve cord [61,129,143].

A range of glandular structures may be found, such as nuchal, pharyngeal, esophageal, parapodial, epidermal and integumental glands. Their patterns and coloration are most easily observed on live specimens in compound microscopes (with high magnification and DIC) and are occasionally used for species characterization [36,37,40].

Most nerillid species are gonochoristic and monomorphic except for organs or features related to reproduction [82]. However, some species, or genera, are found to be (simultaneously) hermaphroditic (see Table 2). Fertilization strategies may vary, but externalfertilization is generally presumed due to the absence of male genitalia, record of external sperm pouches in some species [144], and direct observation of male *Nerilla antennata* laying benthic spermatophores [145]. A select number of species (i.e., some *Mesonerilla*, *Nerillidium*, *Nerillidopsis*, and *Troglochaetus*) brood their offspring, attaching them to the posteriormost segments [146–151]. Nerillids are direct developers except for *Paranerilla* (and possibly also some species of *Meganerilla*), which possess a lecitotrophic benthic larvae [53,145,149,152]. Distribution and configuration of spermioducts holds systematic significance for genera or species groups (see Table 2) and can be observed on live specimens in high resolution light microscopy (e.g., [148,149]) or more easily with confocal microscopy of alpha-tubulin stained ciliated ducts ([42,61]). Similar techniques can be used to map the segmented nephridia and so called enteronephridia. Whereas most genera possess protonephridia, *Nerilla* has metanephridia and *Paranerilla* has mixonephridia (e.g., [61,153] and references herein, [42]). Blind-ending enteronephridia, extending from the posteriormost midgut and lining the hind gut, are found in species-specific numbers in all examined species [42,61,141]. Although so far not routinely examined, their number may add valuable information to Nerillidae systematics alongside the configuration of segmental nephridia and gonoducts.

Table 2. Diagnostic characters of the nerillid genera (and major clades within). Cell shading to distinguish differences. Ann, annulated; ant, antenna; br, brooding; capil, capillary; cf, cirriform; cil, ciliary; club, club-shaped; comp, compound; gono, gonochoristic; herma, hermaphroditic; IC, interramal cirri; mv, microvillar; nc, neurochaetal; pigm, pigmented; rudi, rudimental; s, shaped; segm, segment; wr, wrinkled; **, unpublished data, #, number.

Genus	# segm	Palp Shape	Lateral Antennae	Median Ant	Chaetae	Chaetae segm I	IC segm I	Other segm with IC	IC Shape	Pygidial cirri	Eyes	Reproduction	Spermioduct Openings (* Fused) in segm #	# of Species
Nerilla	9	short-club	long-ann	long-ann	capil	+	long-ann	2–8	short or long-cf	long ann	0 or 4 pigm	gono	6 and 7 and 8 *	11
Meganerilla	9	long-cf	0	0 or short	capil	+	+/−	2–9	short-leaf-s	short leaf-s or long-cf	0 or 2	gono	7	5
Mesonerilla "luederitzi-intermedia" group	9	club	long	long	comp	+	+	2–9	long-bottle-s	long-cf	0 or 2 cil	gono (+/- br-hood)	5 * or 5 * and 6 *	10
Mesonerilla "biantennata" group	9	club	short	0	comp	+	+	2–9	short-cf	short-bottle-s	0	gono	5 *	2
Mesonerilla "roscovita" group	9	club	long	short	comp	+/−	+	2–9	short-cf	long-cf, long-bottle-s	0	herma	6 and 7	3
Leptonerilla	9	club	long	long	comp	+	+	2–9,	long-cf, double	long-cf	2 pigm	gono	8	3
Speleonerilla isa	9	very long-cf	short	short	comp	+	+, rudi	2–9 + nc-lobes	short, bottle-s	long-cf + cil. lobes	0	herma	6 * and 7*	1
Speleonerilla (excl. *S. isa*)	8	very long-cf	short	short	comp	+	−/+, rudi	3–8 + nc-lobes	short, bottle-s	long-cf + cil. lobes	0	herma	7	3
Genus 1 Japan**	8	long-cf	long	long	comp	+	+	2–9	long-cf	long-cf	0	gono	7 and 8	-
Genus 2 Japan**	8	long-cf	long	long	comp	+	−	2–7?	very long-cf	long-cf?	0	herma	7	-
Micronerilla	8	club	long-wr	long-wr	comp	+	−	2–7	long-cf, +/− double	long-cf-wr	2 pigm	gono	7 and 8	1
Thalassochaetus	8	club	0	0	comp	+	−	2–7	rudi	short-bottle-s	0	?	?	1
Trochonerilla	8	club	short	short	capil	+	−	2–7	long-cf	long-cf-wr	2 pigm	gono	7 and 8	1
Troglochaetus	8	club	0	0	capil	+/−	+/−	2–7 or 2–8	rudi	short lobes or bulbs	0	herma	6 *	2
Nerillidium	8	club	short-medium	0	capil	+	+/−	2–7	rudi or short-cf	long-cf or bottle-s	0 or 2 mv	herma	6 or 6 *	10
Nerillidopsis	8	long-cf	long	long	capil and comp	+	−	2–7	long-cf	long-cf	0	herma	6 *	1
Aristonerilla	7	club	long-wr	long-wr	comp	+	−	2–7	long-cf	long-cf-wr	2 red-black	gono	7	1
Psammoriedlia	7	club	0	0	capil	+	+	2–6/7	short-cf	long-cf	0	?	?	2
Paranerilla	7	lateral horns	0	0	capil	+	+	2–7	rudi	long-cf	0	gono	5	2

3.5.3. Taxonomy

Besides molecular phylogenetics, combinations of the following main morphological characteristics define the different genera: number of segments, type of chaetae, number and shape of appendages (palps, antennae, parapodial and pygidial cirri), and type of reproduction (gonochoristic or hermaphroditic), as well as number and position of spermioducts. Characteristics are provided for each genus in Table 2 (illustrated in Figure 7) and the valid species of each genus listed below. Table 2 furthermore includes data on two undescribed genera, recently found in Japan (Worsaae, Hansen et al. unpublished) as well as on three groups of *Mesonerilla*, two of which (*M. biantenerilla*-group and *M. roscovita*-group) are indicated in previous and ongoing analyses to represent separate genera ([42], K. Worsaae et al., unpublished).

Nine-Segmented Genera

Nerilla E. O. Schmidt, 1848 (Figure 7A)

Eleven species: Type species: *Nerilla antennata* E. O. Schmidt, 1848 (includes *Dujardinia* Quatrefages, 1866); *N. rotifera* (Quatrefages, 1866); *N. mediterranea* Schlieper, 1925; *N. australis* Willis, 1951; *N. digitata* Wieser, 1957; *N. stygicola* Ax, 1957; *N. inopinata* Gray, 1968; *N. marginalis* Tilzer, 1970; *N. parva* Schmidt and Westheide, 1977; *N. jouini* Saphonov and Tzetlin, 1988; *N. taurica* Skulari, 1997.

Meganerilla Boaden, 1961 (Figure 7B)

Five species, including synonymized *Xenonerilla* Müller, Bernhard and Jouin-Toulmond, 2001. Type species: *Meganerilla swedmarki* Boaden, 1961; *M. clavata* Magagnini, 1966; *M. penicillicauda* Riser, 1988; *M. bactericola* (Müller, Bernhard, and Jouin-Toulmond, 2001) (as *Xenonerilla bactericola*); *M. cesari* Worsaae, Martínez, and Núñez, 2009.

Mesonerilla Remane, 1949 (Figure 7C–E)

Fifteen species. Type species: *Mesonerilla luederitzi* Remane, 1949; *M. intermedia* Wilke, 1953; *M. roscovita* Lévi, 1953; *M. armoricana* Swedmark, 1959; *M. fagei* Swedmark, 1959; *M. biantennata* Jouin, 1963; *M. pacifica* Jouin, 1970; *M. equadoriensis* Schmidt and Westheide, 1977; *M. neridae* Worsaae and Rouse, 2009; *M. arya*, *M. laerkae*, *M. katharinae*, *M. peteri*, *M. runae*, *M. xurxoi* Worsaae, Mikkelsen, and Martínez, 2019.

Leptonerilla Westheide and Purschke, 1996 (Figure 7F)

Three species. Type species: *Leptonerilla diplocirrata* Westheide and Purschke, 1996; *L. prospera* (Sterrer and Iliffe, 1982) (as *Mesonerilla prospera*); *L. diatomeophaga* (Núñez, 1997 in Núñez, Ocaña, and Brito 1997) (as *Mesonerilla diatomeophaga*).

Eight-Segmented Genera

Speleonerilla Worsaae, Sterrer and Iliffe, 2018 (Figure 7G)

Four species. Described as *Longipalpa* Worsaae, Sterrer and Iliffe, 2004. Type species: *Speleonerilla saltatrix* (Worsaae, Sterrer, and Iliffe, 2004); *S. calypso*, *S. isa*, *S. salsa* Worsaae, Gonzalez, Armenteros, Iliffe, Kerbl, Holdflod, Terp, and Martínez, 2018.

Micronerilla Jouin, 1970b (Figure 7H)

Monotypic. Type species: *Micronerilla minuta* (Swedmark, 1959) (as *Mesonerilla minuta*).

Thalassochaetus Ax, 1954 (Figure 7I)

Monotypic. Type species: *Thalassochaetus palpifoliaceus* Ax, 1954

Trochonerilla Tzetlin and Saphonov, 1992 (Figure 7J)

Monotypic. Type species: *Trochonerilla mobilis* Tzetlin and Saphonov, 1992.

Troglochaetus Delachaux, 1921 (Figure 7K)

Two species. Type species: *Troglochaetus beranecki* Delachaux, 1921; *T. simplex* (Levi, 1953) (as *Nerillidium simplex*).

Nerillidium Remane, 1925 (Figure 7L)

Ten species, including synonymized *Afronerilla* Faubel, 1978, *Akessoniella* Tzetlin and Larionov, 1988, *Bathynerilla* Faubel, 1978. Type species: *Nerillidium gracile* Remane, 1925; *N. troglochaetoides* Remane, 1925; *N. mediterraneum* Remane, 1928; *N. levetzovi* Remane, 1949; *N. macropharyngeum* Jouin, 1970; *N. renaudae* Jouin, 1970; *N. lothari* Schmidt and Westheide,

1977; *N. marinum* (Faubel, 1978) (as *Bathynerilla marinum*) and *N. hartwigi* (Faubel, 1978) (as *Afronerilla hartwigi*), *N. orientalis* (Tzetlin and Larionov, 1988) (as *Akessoniella orientalis*). We have here chosen to also refer *Afronerilla hartwigi* Faubel, 1978 to *Nerillidium*, since its description is obviously based on poorly preserved material, and the lack of antennae and cirri most likely reflects losses rather than diagnostic differences to *Nerillidium*. This synonymization does not affect the diagnosis of *Nerillidium*.

Nerillidopsis Jouin, 1966 (Figure 7M)

Monotypic. Type species: *Nerillidopsis hyalina* Jouin, 1966.

Seven-Segmented Genera

Aristonerilla Müller, 2002 (Figure 7N)

Monotypic. Type species: *Aristonerilla brevis* (Saphonov and Tzetlin, 1997) (as *Micronerilla brevis*).

Psammoriedlia Kirsteuer, 1966 (Figure 7O)

Two species. Includes synonymized *Bathychaetus* Faubel, 1978. Type species: *Psammoriedlia ruperti* Kirsteuer, 1966; *P. heptapous* (Faubel, 1978) (as *Bathychaetus heptapous*). Descriptions based on poor material possibly most likely having lost antennae and cirri and *B. heptapous* possibly even representing *Nerillidium* juveniles (with seven instead of eight segments).

Paranerilla Jouin and Swedmark, 1965 (Figure 7P)

Two species. Type species: *Paranerilla limicola* Jouin and Swedmark, 1965; *P. cilioscutata* Worsaae, and Kristensen, 2003.

3.5.4. Distribution and Diversity

Nerillidae is represented in all oceans across a large diversity of habitats, including brackish open waters, anchialine caves, and fresh groundwater habitats ([32] and references herein). *Troglochaetus beranecki* is found in (primarily subterranean) limnic and hyporheic environments of Europe and USA (e.g., [133,154–156]). Nerillid depth distribution ranges from the deep sea to shallow coastal waters, with the greatest diversity seen in fully marine, well-oxygenated, sandy to gravelly sediment (e.g., [27,32,36,129] and references herein).

The more species-rich genera *Mesonerilla*, *Meganerilla*, *Nerilla*, and *Nerillidium* are found worldwide [32]. In contrast, *Paranerilla* prefers colder Atlantic waters [33,152], and *Leptonerilla*, *Psammoriedlia*, *Speleonerilla*, and *Trochonerilla* have more often been recorded in warmer waters. Most geographical regions outside Europe and the US East Coast as well as deeper waters worldwide are generally undersampled. Further sampling is necessary to predict a possible geographical delimitation of some of these warm-water genera as well as of the more geographically restricted *Micronerilla*, *Aristonerilla*, *Thalassochaetus*, *Nerillidopsis* and *Troglochaetus*.

Nerillids lack a pelagic larval stage (except for species of *Paranerilla*) and any other dispersal stages. Molecular studies have revealed new (sometimes cryptic) species within relatively short geographical distances (e.g., among Caribbean islands; [37,42,45]). In fact, every new meiofaunal survey on sandy sediment in coastal tropical and sub-tropical regions has revealed undescribed nerillids. The species diversity of Nerillidae is therefore expected to multiply, raising with every taxonomic study in previously uninvestigated regions. The species number has already increased drastically over the last two decades despite the limited number of taxonomists working with this group. However, since the morphological disparity of new species is often limited, potential morphological apomorphies can only be documented through time consuming and detailed microscopical examinations. Moreover, any taxonomic studies should be accompanied by molecular sequences to ensure the future identification of cryptic species. In some cases, molecular taxonomy is the only or the fastest way forward [37,157], yet the cryptic diversity may only be fully unraveled by likewise challenging, extensive population genetic studies [38].

Surprisingly, new genera or species with highly diverging morphology have in recent decades only been found in highly diverging environments such as anchialine caves or the deep sea ([39,40], K. Worsaae, unpublished). The worldwide distribution of several genera and the lack of discoveries of new genera in "similar-type" shallow sandy sediments point to an old origin of the family as is also predicted from more recent and ongoing phylogenomic analyses ([13]; K. Halanych, G. Rouse, pers comm.).

3.5.5. Major Revisions and Most Important Literature

Multiple studies have addressed the diversity and systematics of nerillids, but some of the most recent and larger revisions include Müller et al. [39], Müller [134], Worsaae and Müller [61], Worsaae [3,32,42,129,158], Westheide [53], Worsaae et al. [37], as well as two ongoing phylogenetic studies (K. Worsaae et al., unpublished).

3.6. *Polygordiidae Czerniavsky, 1881*

3.6.1. Phylogenetic Affinities

Polygordiidae contains the genus *Polygordius*, with 18 described species and two subspecies [48]. *Polygordius* species are remarkable in their body simplicity, lacking parapodia and chaetae. Due to this lack of typical annelid characters, they were considered by early authors as a primitive form, closely resembling the common ancestor of Annelida [6]. Due to this, the genus *Polygordius* holds a tremendous historical importance towards the early theoretical studies on the evolution of Metazoa and Annelida, dating back to 1843. The early life stages of *Polygordius* were described prior to the adults. The endolarva was described by Lovèn [159], highlighting their dramatic metamorphosis. Agassiz [160] described the development and metamorphosis of the larva from the western Atlantic, being similar to Lovèn's. Schneider [161] found them so unique that he created an order for them, the Gymnotoma. Two years later, Schneider [162] described *Polygordius* adults from the west coast of Helgoland and the development of the larvae from an undescribed species from the Mediterranean. Finally, Hatschek [6] studied the larval development of *Polygordius*, describing it as the most primitive annelid genus, nearest to the generalized stem of the annelids. In his study, Hatschek also proposed the great phylogenetic significance of the trochophore larva, where groups possessing this larval type had evolved from a common stem form, the "Trochozoan." The "Trochophore Theory" gained wide acceptance and *Polygordius*, with its supposedly primitive anatomy and its highly developed trochophore larva, was regarded as the most primitive annelid (for a detailed historical description, see [5]).

Nowadays, the simplicity of *Polygordius* is interpreted as secondary, having evolved as an adaptation to life in interstitial habitats. The family is closely related to Protodrilida [11] or Phyllodocida [10] as part of Errantia. A molecular and morphological phylogeny of the family focusing on species from European Atlantic regions recovered six valid species: *Polygordius appendiculatus*; *P. lacteus*; *P. neapolitanus*; *P. triestinus*; *P. jouinae*; and *P. eschaturus*. Both *P. erythrophthalmus* and *P. villoti* are considered as invalid species, synonymous with *P. lacteus* [163]. After the first molecular phylogeny of the genus, Tustison et al. [47] described four new species by including data from the Pacific and Caribbean, increasing the number to the now 18 valid species [47,164].

3.6.2. Morphology

Polygordiidae have a thin, slender body, which is cylindrical in cross-section with shallow ventral and ventro-lateral grooves along the body in certain species (length 10–100 mm; width < 1 mm). Trunk segments are poorly delineated externally and the animals are characterized by an iridescent cuticle that resembles that of nematodes in their appearance due to the smooth body surface. The cuticle is comparatively thick and formed by several stacked layers of prominent collagen fibers arranged in parallel within a fine fibrillar matrix, giving way to the iridescent appearance [24]. Unlike many interstitial annelid species, external ciliation is usually absent, except for *P. jouinae*, which has a ciliated pygidium [165]. Parapodia and chaetae are absent.

The prostomium is rounded or conical with frontally orientated paired palps (Figure 9A,B) [166]. The homology of these appendages with palps has been long debated (see [167,168]). However, studies using transmission electron microscopy (TEM) have shown the same innervation patterns as palps in other annelids [166]. The palps are relatively rigid and lacks ciliation, vessels, coelomic cavities and musculature, and thus they have been interpreted as purely sensoric and not involved in feeding [164]. Pigmented eyes are absent, but unpigmented rhabdomeric and ciliary receptor cells might be present in front of the brain [166]. Red pigment spots are present in the prostomium of *P. lacteus* [164].

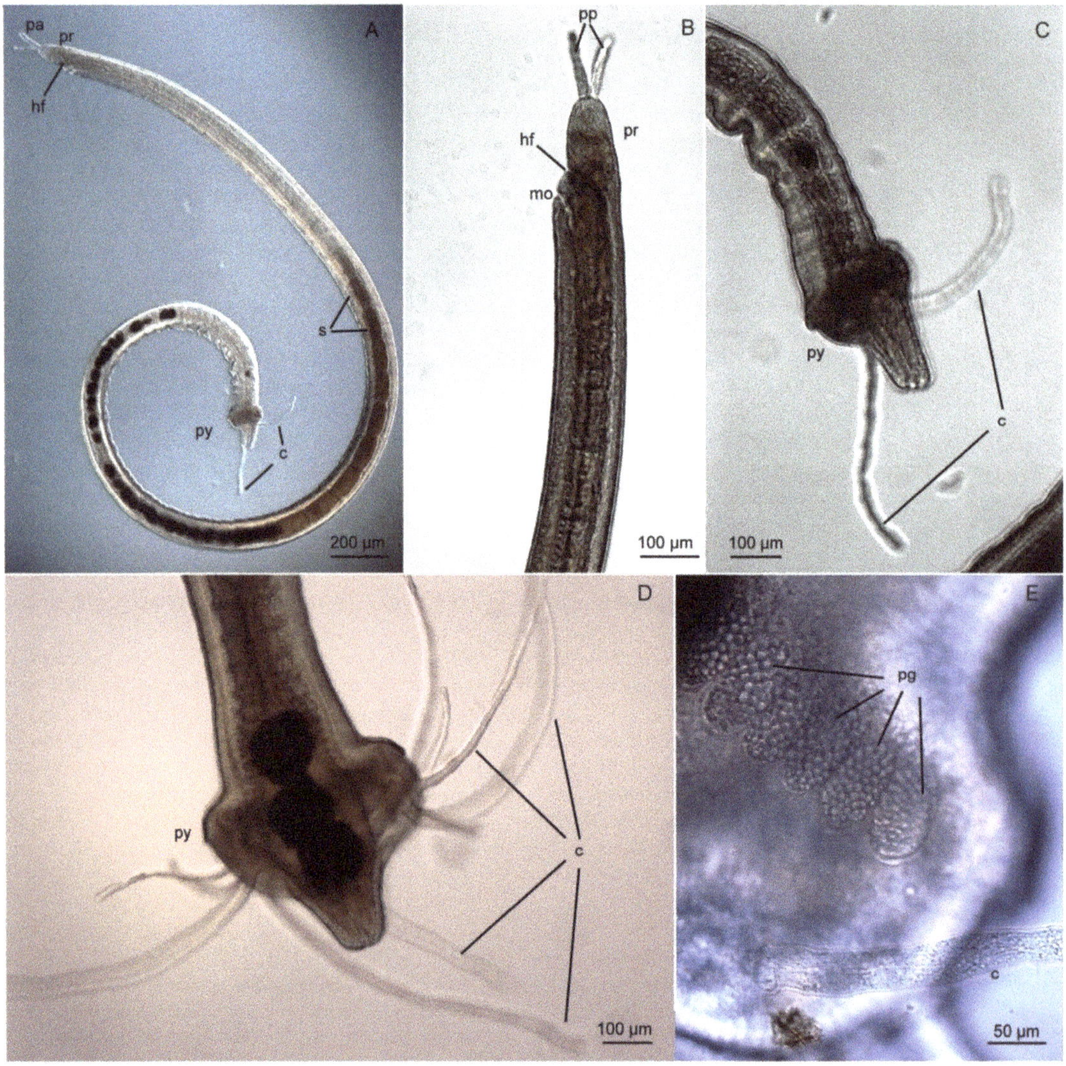

Figure 9. Morphology of *Polygordius*. (**A–C**) *Polygordius* sp. (São Sebastião Island, São Paulo, Brazil). (**A**) Entire body. (**B**) Anterior end. (**C**) Posterior end. (**D**,**E**) *Polygordius leo* (São Sebastião Island, São Paulo, Brazil). (**D**) Details of pygidium (**E**) Adhesive pygidial glands. Abbreviations: c, cirrus; hf, head fold; mo, mouth opening; pa, palps; pg, pygidial glands; pp, paired palps, pr, prostomium; py, pygidial lobe; s, segment. muscular bulb.

The peristomium is separated from the prostomium by the head fold, which raises in front of the ventral, slit-shaped mouth (Figure 9A,B [168]). The nuchal organs are oval and densely ciliated, extending dorso-laterally between the prostomium and the peristomium [137].

The trunk consists of 200 or more segments, followed by a pygidium that may be inflated or cylindrical depending on the species. The pygidium may be encircled by adhesive pygidial glands [169] that vary in size, shape, and number depending on the species (Figure 9E, [168]). Pygidial cirri may also be present, either terminally or subterminally, forming distinctive anal lobes at the tip of the pygidium.

The musculature of *Polygordius* resembles other interstitial annelids and is arranged in four groups of longitudinal fibers, numerous segmentally arranged oblique muscles and weakly developed circular fibers. The gut is a straight tube. The mouth cavity presents prominent densely ciliated protrudable dorsolateral folds, which continue into the pharynx, also containing a ventral pharyngeal sac directly posterior to the mouth. The pharynx opens dorsally into the esophagus, followed by the foregut with a characteristic ventral ridge carrying longer cilia. The lower epithelium of the hindgut comprises longitudinal folds and lacks glands. Coelomic cavities, mesenteries, and muscular septa are well developed throughout the trunk. The circulatory system is closed and well-developed. Excretory organs are segmentally arranged metanephridia; the first pair formed by fusion of the second pair of larval protonephridia with the first pair of metanephridia. Sexes are separated and sexual products occur in a variable number of fertile segments. The spermatozoa are typically of the ecto-aquasperm type (sense [170]), while oocytes are relatively small and occur in large numbers, completely filling the coelom of sexually mature females. Sexual products are probably released by rupture of the body wall, since no genital ducts are present.

3.6.3. Taxonomy

Identifying *Polygordius* species based on morphological characters alone can be challenging. Their long cylindrical bodies appear relatively similar to one another under visual inspection, and the distinguishing features useful for morphology-based discrimination of species are small, requiring examination with scanning electron microscopy (SEM) [48,164]. The list of the 18 valid species and the two subspecies is listed below.

Polygordius Schneider, 1868

Eighteen species, two subspecies: Type species, *Polygordius lacteus* Schneider, 1868; *P. appendiculatus* Fraipont, 1887; *P. antarcticus* Rota and Carchini, 1999; *P. arafura* Avery, Ramey and Wilson, 2009; *P. erikae* Tustison, Ramey-Balci and Rouse, 2020; *P. eschaturus* Du Bois-Reymond Marcus, 1948; *P. eschaturus brevipapillosus* Jouin and Rao, 1987; *P. ijimai* Izuka, 1903; *P. jouinae* Ramey, Fiege and Leander, 2006; *P. kiarama* Avery, Ramey and Wilson, 2009; *P. kurthcarolae* Tustison, Ramey-Balci and Rouse, 2020; *P. kurthsusanae* Tustison, Ramey-Balci and Rouse, 2020; *P. jenniferae* Tustison, Ramey-Balci and Rouse, 2020; *P. leo* Du Bois-Reymond Marcus, 1955 (Figure 9D,E); *P. madrasensis* Aiyar and Alikunhi, 1944; *P. neapolitanus* Fraipont, 1887; *P. pacificus* Uchida, 1935; *P. pacificus floreanensis* Schmidt and Westheide, 1977; *P. triestinus* Hempelmann, 1906; *P. triestinus sensu* Jouin, 1970; *P. uroviridis* Aiyar and Alikunhi, 1944.

3.6.4. Distribution and Diversity

Polygordiidae are often included as part of the meiofaunal or interstitial annelid literature because they spend their life living interstitially among the coarse sand grains. However, their size makes them part of the macrofaunal community as it is usually defined as organisms retained on a 500 µm mesh sieve [171]. Although they occur in several interstitial habitats, they have a strong affinity for highly energetic systems with coarse sediments [172,173]. They are found worldwide in intertidal, shallow subtidal, and also in continental slope sediments at depths between 1000 and 5000 m in Antarctic waters [164]. The limited knowledge on the behavior, feeding strategies, and general ecology of Polygordiidae comes from studies of *P. jouinae* from the inner continental shelf off of New Jersey, USA [14,172,173].

A recent molecular phylogeny illustrated the lack of a clear biogeographic pattern for the genus [47]. The Atlantic, European and Mediterranean terminals were placed in three regions of the tree and none were close to the Caribbean *P. jenniferae*, which showed a low COI divergence from *Polygordius* sp. from California. The Australian/French Polynesian species did form a discrete clade [47].

3.6.5. Major Revisions and Most Important Literature

Distribution patterns of the described species of Polygordiidae were first summarized in Rota and Carchini [170], but more recently, a large effort integrating morphology and molecular data is presented by [47,163,164].

3.7. *Protodrilidae Hatschek, 1888*

3.7.1. Phylogenetic Affinities

Protodrilidae contains six valid genera with 38 described species ranging in size from 1 to 13 mm and comprising up to 60 segments [174]. The presence of non-grooved ventral filiform palps with internal canals and the possession of a bilobed adhesive pygidium support their relationship to Saccocirridae and Protodriloididae. These three families are often recovered together in molecular analyses, even when few markers are employed [175]. However, due to their comparatively simple external morphology across all members of these three families, their phylogenetic position within Annelida shifted historically, from the basally branching clade "Archiannelida" (subsequently shown as polyphyletic) to the currently derived position as a sister clade of Errantia [10,11].

The names Protodrilidae and *Protodrilus*, which refer to this external simplicity, were first introduced by Czerniavsky [176] to accommodate *Protodrilus mirabilis*. However, since his short description is largely incomplete and lacks supporting type material, *P. mirabilis* is today considered an invalid species and Protodrilidae sensu Czerniavsky [176] is *nomen dubium*. Therefore, the genus *Protodrilus* was first erected by Hatschek (1881) to name the species *Protodrilus leuckartii* from a coastal lagoon in Sicily, Italy [177]. The same author afterwards erected Protodrilidae to accommodate this species [178], as well as *Lindrilus flavocapitatus*, originally described as a *Polygordius* species [179].

Protodrilidae sensu Hatschek, 1888 traditionally consisted of the genera *Protodrilus* (36 species), *Astomus* (one species), and *Protodriloides* (two species) [50]. However, *Protodrilus* was originally diagnosed based on plesiomorphic characters, as highlighted by cladistic analyses that first motivated the transfer of *Protodriloides* into a separate family (see section on Protodriloididae), and later, the systematic rearrangement of *Protodrilus* [50,180]. This last systematic rearrangement involved the re-description of *Protodrilus* and *Astomus*, and the description of four new genera consistent with each of the major monophyletic clades recovered in these analyses. Protodrilidae is currently divided into six genera, with *Lindrilus* and *Protodrilus* branching off as a sister to the clade including (1) *Astomus* and *Megadrilus*, and (2) *Meiodrilus* and *Claudrilus* [173,174] (Figure 10). Anecdotally, Lam [181] introduced a monotypic protodrilid genus for the species *Protannelis meyeri* Lam, 1922, currently considered as invalid, as it was described from a single, fragmented specimen lacking any of the typical protodrilid features.

3.7.2. Morphology

The external morphology of Protodrilidae is rather simple. The prostomium, generally rounded, is followed by a comparatively large peristomium (Figure 11A–E), a long cylindrical trunk lacking chaetae and parapodia (Figure 10G,L), and a bilobed adhesive pygidium (Figure 11N) [50]. The most conspicuous external characteristics are the paired filiform palps that arise anteroventrally from the prostomium and are provided with an internal coelomic channel connected behind the brain (Figure 10A,F) [180]. Exceptionally, *Astomus taenioides* has a characteristically festooned trunk (Figure 10F) [182,183], while the cave-exclusive *Megadrilus pelagicus* bears a dorsal keel which aides their stabilization while drifting in the water column of flooded lava tubes [41]. Although the family is almost exclusively interstitial, its members can barely be considered meiofaunal. In fact, the plesiomorphic state for the body size in Protodrilidae is rather large, decreasing evolutionarily in the lineages corresponding to the genera *Astomus, Claudrilus* and *Meiodrilus* [50]. This has been interpreted as an adaptation to the epidermal uptake of nutrients in *Astomus* (i.e., as it reduced the ratio between body surface and volume) and a consequence of a more specialized lifestyle in *Claudrilus—Meiodrilus*. In fact, species of *Claudrilus* and *Meiodrilus* often have shorter palps and are provided with segmentally arranged adhesive glands, which might facilitate squeezing through the tight interstitial spaces while providing additional attachment points in this genuinely turbulent environment [184].

In order to explore the interstitial spaces, all protodrilids glide by means of their midventral motile ciliary band, which consists of four to five rows of multiciliate cells arranged along a shallow groove [185]. If other ciliary bands are present, they are typically sensory or involved in the production of water currents [28,174]. The latter can be found on the head or arranged segmentally on the trunk, in both cases holding important systematic value. *Megadrilus pelagicus* uses ciliary bands to drift in the water column of anchialine caves [41]. The adhesive pygidium also plays a role in locomotion [184,186], both by providing attachment to the substrate and by facilitating changes in direction or fast retraction of the body in combination with trunk musculature.

The presence of a conspicuous ventral pharynx occupying the posterior half of the peristomium is another distinctive feature of Protodrilidae (Figure 11A). This structure is probably optimized for grazing on the biofilms and as well as for collecting and grating different types of deposited organic particles at low Reynolds numbers [174]. The pharynx consists of a prominent ventral muscular bulb formed by transverse muscular fibers and a few interstitial cells and surrounded by the sagittal muscle, which attaches dorsally to the so-called tongue-like organ [187], despite the structure having nothing to do with the tongue as described in vertebrates, but rather resembles the radula, as described in many gastropod molluscs. This organ is situated in the dorsoposterior part of the buccal cavity between the muscular bulb and the esophagus and bears a thick cuticular plate that is likely involved with grating and dislodging food particles [188]. The pharyngeal apparatus is vividly red in species of *Lindrilus* and *Megadrilus*, as well as in *Protodrilus ciliates*, while it retains dark pigmentation in *Claudrilus helgolandicus* and *C. hypoleucus* [49,174,189]. Segmentally arranged salivary glands extend from the pharynx to the posterior end of the body, often in a species-specific number of segments [190].

The presence of different types of sensory organs is another useful diagnostic feature in Protodrilidae. Ventral pigmented eyes characterize the genus *Lindrilus* (Figures 10A and 11A), whereas dorsal pigmented eyes are only found in adults of *Protodrilus oculifer* (Figure 10B). Large, rounded, unpigmented ciliary receptors are a synapomorphy of the clade *Lindrilus—Protodrilus* and are retained in all described species (Figures 10A–C and 11F), while smaller, oval phaosomal receptors are more common in *Meiodrilus* and *Claudrilus* (Figure 11G) [20]. No light-sensitive sensory structures are present in *Megadrilus*; however, they bear large nuchal organs that extend as transverse ciliated furrows on both sides of the prostomium (Figure 10D) [49].

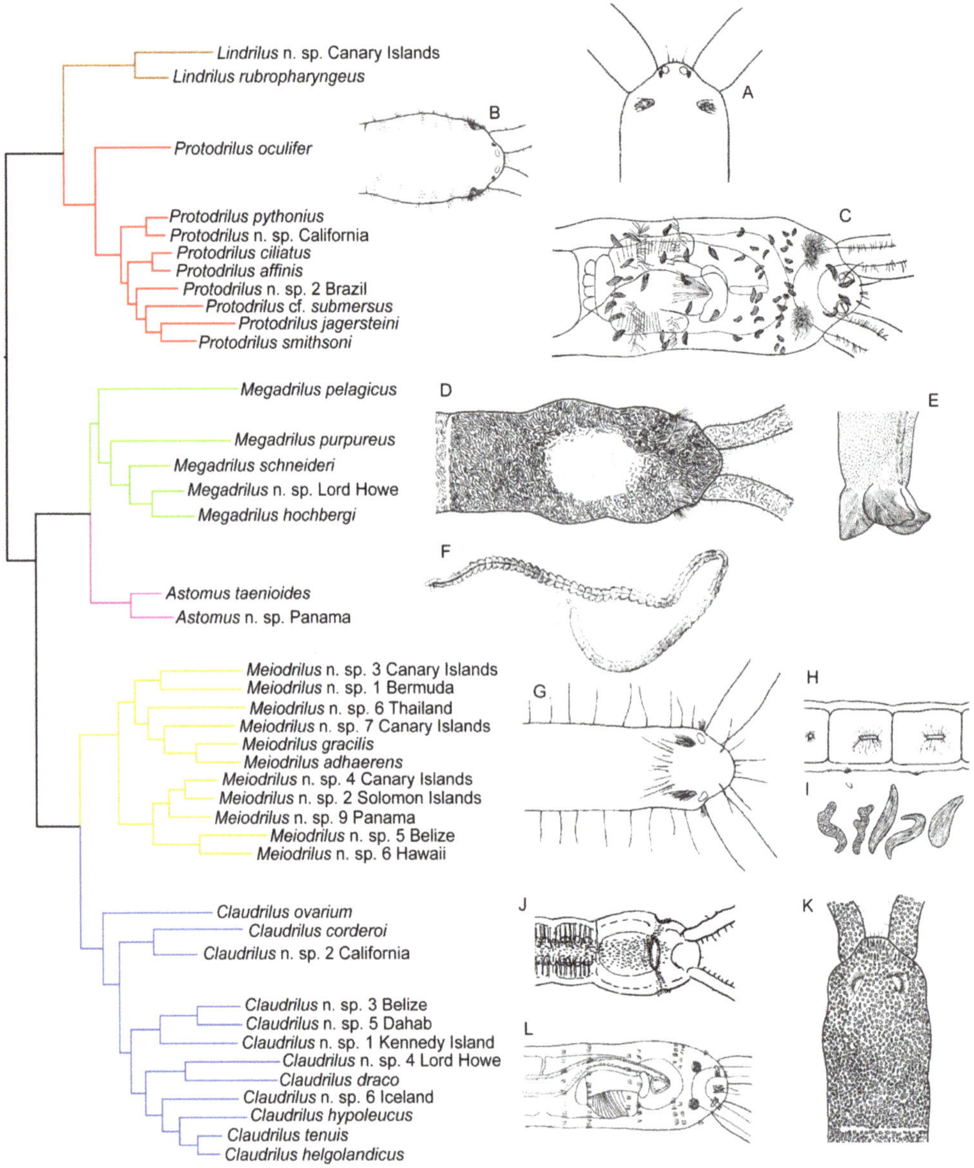

Figure 10. Diagram showing the phylogenetic relationships of Protodrilidae based on four molecular markers (18S rRNA, 28 rRNA, COI, H3) and morphology analyzed with maximum likelihood methods (Redrawn from [50], along with illustrations representing the diagnostic characters for each genus. (**A**) *Lindrilus rubropharyngeus,* anterior end showing the presence of eyes, ciliated organs and the shape of the nuchal organs (**B**) *Protodrilus oculifer,* anterior end showing the presence of eyes and ciliated unpigmented organs. (**C**) *Protodrilus albicans,* anterior end. (**D**) *Megadrilus purpureus,* anterior end. (**E**) *Megadrilus purpureus* and its characteristic trilobed pygidium. (**F**) *Astomus taenioides,* showing the festooned trunk segments. (**G**) *Meiodrilus adhaerens,* anterior end. (**H**) *Meiodrilus adhaerens,* lateral organs. (**I**) *M. adhaerens,* cocoon glands. (**J**) *Claudrilus corderoi,* anterior end. (**K**) *Claudrilus hypoleucus,* anterior end showing the epidermal glads. (**L**) *Claudrilus hypoleucus,* anterior end. A, redrawn from [191]; (**C,L**), redrawn from [28], (**D,K**), from [53]; (**F**), from [31], (**B,G–I**), from [192], (**J**) from [108].

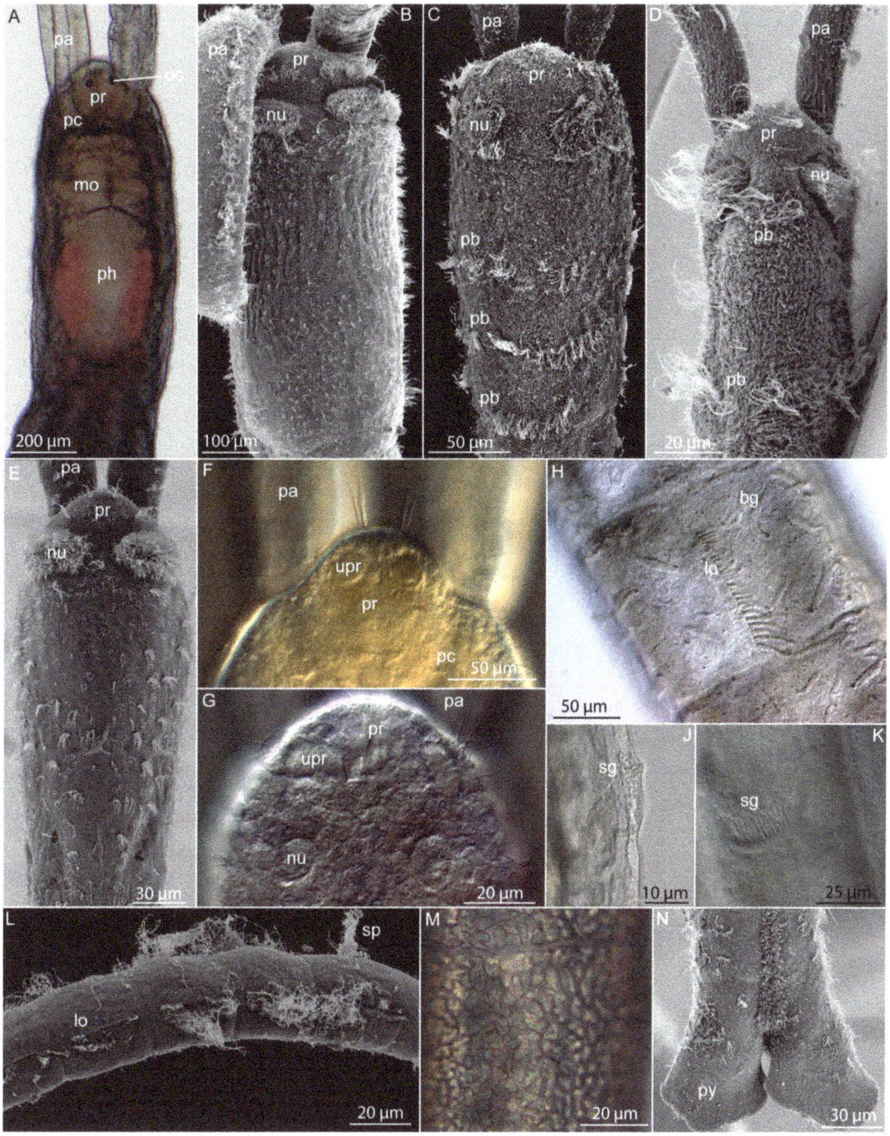

Figure 11. Morphology of Protodrilidae. (**A**,**F**,**N**) *Lindrilus* sp. (La Palma, Canary Islands). (**B**) *Megadrilus schneideri* (Capo Caccia, Sardegna, Italy). (**C**,**M**) *Claudrilus* cf. *hipoleucus* (Tenerife, Canary Islands). (**D**) *Meiodrilus* sp. (Phuket, Thailand). (**E**) *Protodrilus smithsoni* (Bocas del Toro, Panama). (**G**) *Claudrilus hypoleucus* (La Maddalena, Sardegna, Italy). (**H**,**J**,**K**) *Meiodrilus adhaerens* (Svinbaden, Helsingør, Denmark). (**L**) *Claudrilus ovarium* (Parana, Brasil). (**A**) Light micrographs of the anterior region. (**B–E**) Scanning electron micrographs of the anterior region. (**F–G**) Light micrographs of the prostomium showing different types of photoreceptors. (**H–K**) Light micrographs of the trunk segments showing adhesive and bacillary glands. (**L**) Scanning electron micrograph of the lateral organs and sperm cells. (**M**) Light micrograph of a trunk segment showing the epidermal glands. (**N**) Scanning electron micrograph of the pygidium. Abbreviations: bg, bacillary gland; lo, lateral organs; nu, nuchal organ; oc, ocelli; pa, palps; pb, prostomial ciliary band; pc, palp canal; ph, pharynx; pr, prostomium; mo, mouth; py, pygidium; sg, segmental adhesive glands; sp, sperm cells; upr, unpigmented prostomial receptor. All images reproduced with permission from [174].

All Protodrilidae are gonochoristic and have a fixed number of fertile trunk segments, using the coelomic cavity to store spermatids, spermatozoa, or oocytes [28]. There are two types of slender and filiform spermatozoa in males, the euspermatozoa, which are involved in the fertilization of the oocytes, and the paraspermatozoa, which play a role in the enzymatic opening of the spermatophore and the female epidermis during sperm transfer [193]. While the ultrastructure of sperm might hold phylogenetic information at the genus level, it seems too conserved to diagnose species. Similarly, the size and abundance of oocytes in females are useful characteristics to identify the different genera [174]. Externally, all protodrilid males present lateral organs in the anterior part of the body (Figure 11L) [194], consisting of paired invaginations of the epidermis that form a wide furrow into which numerous gland cells open and cilia project [193]. Lateral organs are involved in the production of the spermatophore and generally appear on a species-specific number of anterior segments. Therefore, the number, position and morphology of the lateral organs have been traditionally used as the main characteristics for species delineation [174,186]. The posteriormost pairs of lateral organs are associated with a specific number of gonoducts, which is generally conserved within each genus and therefore useful in their diagnoses and identification [194]. Females, in contrast, generally lack any conspicuous external reproductive structures. Exceptions include dorsal organs described in *Lindrilus rubropharyngeus* and *L. flavocapitatus*, consisting of segmentally arranged rosette-like structures extending backwards from segments 19–22, and the ciliary furrows described in *Protodrilus leuckartii* and *P. ciliatus*, consisting of a long structure similar to the lateral organs of males [195]. Both of these structures are involved in sperm transfer. Oviducts are described only in a few species (see [28,194]).

3.7.3. Taxonomy

The following main morphological features (in various combinations) define the different genera: length of the body, pharyngeal pigmentation, number and shape of prostomial sensory organs (ciliated receptors, phaosomal receptors, pigmented eyes, nuchal organs), extension of the salivary glands, external ciliary patterns (number of transverse bands on the head and trunk) as well as number and position of male lateral organs.

Protodrilus Hatschek, 1881

Diagnosis: Body opaque white. Prostomium with ciliated palps, with or without pigmented eyes and with rounded, unpigmented ciliary receptors. Nuchal organs oval, extending dorsolaterally between the prostomium and peristomium. Salivary glands of variable length, normally from segments 1 to 5–15. Two pygidial lobes. Males with continuous or segmented lateral organs and three to four pairs of spermioducts; females without oviducts and about ten small oocytes per segment.

Fifteen species: Type species: *Protodrilus leuckartii* Hatschek, 1880; *P. hatscheki* Pierantoni, 1908; *P. oculifer* Pierantoni, 1908; *P. ciliatus* Jägersten, 1952; *P. robustus* Jägersten, 1952; *P. albicans* Jouin, 1970; *P. brevis* Jouin, 1970; *P. infundibuliformis* Schmidt and Westheide, 1977; *P. huanghaiensis* Wu, Sun, and Chen, 1980; *P. jagersteni* von Nordheim, 1989; *P. litoralis* von Nordheim, 1989; *P. submersus* von Nordheim, 1989; *P. gelderi* Riser, 1997; *P. pythonius* Di Domenico, Martínez, Lana, and Worsaae, 2013; *P. smithsoni* Martínez, Di Domenico, Jörger, Norenburg, and Worsaae, 2013.

Astomus Jouin, 1979

Diagnosis: Body whitish, with festooned segments. Prostomium with ciliated palps, unpigmented receptors present, but no pigmented eyes. Nuchal organs rounded dorsal, positioned dorsally on the prostomium. Mouth and pharyngeal organ absent; ciliated gut residual all along the body, comprising few ciliated cells with large vacuoles containing sphaerocrystals. Salivary glands reduced. Two pygidial lobes. Males with lateral organs on segments 13–22 that extend over two consecutive segments and with two short ciliated grooves on segments 9–10; one pair of spermioducts in segment 17.

Monotypic: *Astomus taenioides* Jouin, 1979, but at least two undescribed species are known [50].

Lindrilus Martínez, Di Domenico, Rouse and Worsaae, 2015

Diagnosis. Body large and pigmented, pharynx red. Prostomium with long ciliated palps, pigmented eyes and rounded, large, and medial unpigmented receptors. Nuchal organs dorsolateral and rounded. Salivary glands extending further than segment 15. Two pygidial lobes. Lateral organs in males segmented and equal in size, connected to one pair of spermioducts; females with oviducts and abundant, small oocytes in each segment; dorsal organs in two species.

Three species: Type species: *Lindrilus rubropharyngeus* (Jägersten, 1940); *L. flavocapitatus* (Uljanin, 1877); *L. haurakiensis* (von Nordheim, 1989).

Megadrilus Martínez, Di Domenico, Rouse and Worsaae, 2014

Diagnosis: Body large and pigmented, pharynx reddish; prostomium with long ciliated palps, no eyes, small and rounded unpigmented receptors, sometimes absent. Nuchal organs long and transversely oriented between the prostomium and peristomium. Three pygidial lobes. Males with lateral organs from segment six, first lateral organ rounded and smaller; females with oviducts and abundant, small oocytes in each segment.

Four species: Type species: *Megadrilus purpureus* (Schneider, 1868); *M. schneideri* (Langerhans, 1880); *M. hochbergi* (Martínez, Di Domenico, Jörger, Norenburg and Worsaae, 2013); *M. pelagicus* (Martínez, Kvindebjerg, Iliffe and Worsaae, 2017).

Meiodrilus Martínez, Di Domenico, Rouse and Worsaae, 2015

Diagnosis: Body short and translucent, pharynx unpigmented. Prostomium with short and poorly ciliated palps. Eyes absent, but oval lateral unpigmented receptor present in most species. Nuchal organs rounded and dorsal. Salivary glands in segments 1–5, paired segmented adhesive glands ventrally on trunk segments. Two pygidial lobes. Males with continuous or segmented lateral organs; females with cocoon glands and a few comparatively large oocytes in each segment.

Four species: Type species: *Meiodrilus adhaerens* (Jägersten, 1952); *M. indicus* (Aiyar and Alikhuni, 1943); *M. gracilis* (von Nordheim, 1989); *M. jouinae* (von Nordheim, 1989).

Claudrilus Martínez, Di Domenico, Rouse and Worsaae, 2015

Diagnosis: Body whitish, pharynx unpigmented or grayish. Prostomium with palps, sometimes with motile cilia; no eyes; paired, oval, unpigmented ciliary receptors present laterally in most species. Nuchal organs dorsal and rounded, or indistinct. Salivary glands at least from segment 1 to 8 but may extend to segment 20. Epidermis sometimes with abundant glasslike vacuolar glands. Two pygidial lobes. Lateral organs variable across species; females with a few large oocytes per segment.

Ten species: Type species: *Claudrilus hypoleucus* (Armenante, 1903); *C. pierantonii* (Aiyar and Alikunhi, 1944); *C. corderoi* (du Bois-Reymond Marcus, 1948); *C. flabelliger* (Wieser, 1957); *C. minutus* (Kirsteuer, 1966); *C. similis* (Jouin, 1970c); *C. tenuis* (Jouin, 1970c); *C. helgolandicus* (von Nordheim, 1983); *C. draco* (Martínez, Di Domenico, Jörger, Norenburg and Worsaae, 2013); *C. ovarium* (Di Domenico, Martínez, Lana and Worsaae, 2013).

3.7.4. Distribution and Diversity

Protodrilids are known from all over the world, except for the Antarctic [174]. All genera are cosmopolitan, with the exception of *Astomus*, which seems to exclusively live in organic-matter rich sediments associated with Pacific coral reefs [196]. A few species of protodrilids are known from a single locality, whereas those described with wide distribution ranges may represent cryptic or pseudocryptic species [28,196,197]. Certain groups of species seem to be restricted to Europe (e.g., *Claudrilus hypoleucus* and *C. helgolandicus*, *Protodrilus ciliatus* and *P. affinis*), or represent clades with a Western Atlantic–Indopacific distribution (e.g., *Protodrilus jagersteni*, *P. smithsoni*, and *P. submersus*) [50]. However, these patterns are likely to be spurious and arise from poor species identification combined with a geographically biased sampling effort rather than accurately reflecting the real distribution of the group [174].

In regard to habitats, protodrilids are mostly recorded from shallow water marine sediments, where they prefer coarse, well-sorted sands between 0 and 10 m depth [14].

Records below 50 m depth are rare [30]. *Protodrilus spongioides* was once recorded from freshwater sediments at the Anton Dohrn Marine Station in Italy [198], although the rather scarce description raises doubts about the validity of this taxon. Other protodrilids exhibit a more or less loose preference for certain marine habitats, although whether those preferences are real or arise from an ecologically biased sampling effort warrants supplementary research. For example, *Protodrilus leuckartii* prefers brackish waters and has been commonly found in coastal lagoons or estuarine areas, mostly in Italy and the south of France [28,177]. Small species, with segmental adhesive glands, such as the European *Meiodrilus adhaerens* and *M. similis* or the South American *Claudrilus corderoi* [77,199,200], are often found deep in the sediment layers at the upper intertidal zone of sandy beaches. In contrast, the larger species are rather common at the surface of coarse or gravelly sediments. Species of *Lindrilus* seem to prefer the slope of exposed beaches [191,201], whereas a quite diverse assemblage of protodrilids can be found in coarse and gravelly subtidal sediments accumulated at the surf zone, including *Protodrilus albicans* and *M. schneideri* in the Atlanto-Mediterranean area [28,45,46,202] and *Protodrilus smithsoni, P. pythonius,* and *Megadrilus hochbergi* around the Caribbean and Brazil [49,200,203]. These species cope with the turbulence in this highly hydrodynamic zone by exhibiting swimming behaviors facilitated by the well-developed body wall musculature and large size [28,49].

Only *Megadrilus pelagicus* and *Protodrilus puniceus* are known from non-interstitial environments. *Megadrilus pelagicus* is endemic to the La Corona lava tube, a 1600 m long anchialine lava tube in Lanzarote, Canary Islands [204], where it hoovers in the water column using its specialized ciliation in combination with its dorsal keel, feeding on suspended organic matter with its specialized long palps [205]. This species can also swim by producing muscular waves along its entire trunk, favored by the lateral compression of the body and the presence of a dorsal keel [41]. *Protodrilus puniceus* was described from whale falls off Cape Noma Misaki, Kyushu Island (Japan) at 219–254 m depth. This species forms dense aggregations in the countless small pores of whale bones. Some individuals completely bury themselves in these bones, whereas others were found partially emerged [30]. In aquaria, *P. puniceus* were observed to use their palps and midventral ciliary bands to gather small organic particles around the mouth. A similar protodrilid was reported from whale falls in Monterrey Bay, California (USA) [206]. A suspension feeding behavior has also been described for the species *Protodrilus brevis,* which lives interstitially or inside empty mollusk shells, using its palps to feed [28].

3.7.5. Major Revisions and Most Important Literature

The family Protodrilidae attracted comparatively a lot of attention during the end of the nineteenth century with the publication of several studies [162,177,207,208], with a major review of the family by Umberto Pierantoni [189], who described most of the Mediterranean species as well as several details on the anatomy and larval development [189]. This research was followed up on by a revision by Gösta Jägersten from the Kristineberg Marine Research Station, who described the species diversity of the group in northern Europe and provided detailed studies on their larval development [191,192,209]. This work, along with a few papers targeting the description of new species from India [210,211], Washington (USA) [212], Brazil [108], and Egypt [135] was extensively reviewed by Claude Jouin [28,213], who established the systematics of the family by describing the genera *Astomus* and *Protodriloides,* although the latter was later trasferred to the family Protodriloididae [180]. The last major morphological review of the family corresponds to the series of articles published by Hennig von Nordheim [186,214], who described several new species from Europe and New Zealand, providing a very useful overview of the family.

The current systematics of the family was established after an integrative cladistic analyses that included all described species of Protodrilidae, including those few described after the works by von Nordheim (e.g., [41,49,196,200]). This study split the family into six genera, each corresponding to a monophyletic clade [50].

3.8. Protodriloididae Jouin, 1966

3.8.1. Phylogenetic Affinities

Protodriloididae Jouin, 1966 is a family of interstitial annelids with two valid species reaching up to 13 mm in length. Both species were originally described as protodrilid, despite the conspicuous morphological differences between these species and family Protodrilidae. Indeed, *Protodriloides symbioticus* [215] was originally described as an aberrant species of *Protodrilus* due to the presence of greenish epidermal inclusions, initially interpreted as symbiotic algae [215], while *Protodriloides chaetifer* was described a few years later from Helgoland (Germany), mainly distinguished from the former by the presence of hooked chaetae [216]. The genus *Protodriloides* was erected by Jouin [217] following a detailed morphological analysis that revealed fundamental differences from that of *Protodrilus* [218]. *Protodriloides* was finally placed in Protodriloididae (originally spelled as Protodriloidae, modified by [167]) as a result of a morphological phylogenetic analysis of *Astomus* Jouin, 1979, *Protodrilus* Hatschek, 1881, *Saccocirrus* Bobretzky, 1872 and *Protodriloides* Jouin, 1966 [217].

3.8.2. Morphology

Protodriloididae are very characteristic and their overall aspect can be fairly described as sticky noodle-like worms—somewhat comparable to overcooked spaghetti. They have a pair of anterior palps that are short and stiff (Figure 12B). These palps resemble anterior extensions of the prostomium [180] and differ from those of the related Protodrilidae and Saccocirridae [50,51] by lacking internal coelomic canals or basal ampullae (Figure 12C). The palps do, however, bear longitudinal ventral bands of motile cilia extending from near the base to the tip (Figure 12G,H).

Unlike members of Saccocirridae and Protodrilidae, species of *Protodriloides* have a small prostomium and lack pigmentation. They possess two pairs of unpigmented ciliary receptors, unique for the family [219], as well as a dorsal and a ventral ciliary band extending transversely near the border of the peristomium [220] (Figure 12A,C,G). The peristomium is longer than the prostomium and possesses a longitudinal slit-like mouth, a pair of lateral ciliary bands and discoid, densely ciliated nuchal organs (Figure 12G,H). Dorsal nuchal organs are well-developed in *P. chaetifer*, extending slightly obliquely to the longitudinal body axis, anterior to the peristomial lateral ciliary bands (Figure 12G), but they are reduced to inconspicuous dorsal papillae in *Protodriloides symbioticus* [20].

The trunk is elongated and compressed dorsoventrally. It consists of 45–50 segments in *P. chaetifer*, but only around 15–20 in *P. symbioticus*. Segments are indistinct in both species, delimited by incomplete septa (Figure 12B), and approximately equal in length, except for the shorter first segment [218,221]. The mid-ventral ciliary band extends from the mouth to the last four to five segments (Figure 12G,H), and, unlike in other families, covers approximately two thirds of the ventral side of each trunk segment [46]. Segmentally arranged sensory ciliated organs are present in *P. symbioticus* and *P. chaetifer*, consisting of groups of multiple cilia projecting from small papillae (Figure 12E). Segmentally arranged adhesive organs are exclusive for *P. symbioticus* [222], typically consisting of groups of 10–25 glandular cells arranged in dorsal, lateral and ventral pairs, which are aligned throughout the body, forming characteristic longitudinal bands [218]. A pair of bifid sigmoid chaetae is found exclusively on *P. chaetifer*, which additionally bears a curved and robust rostrum-like structure (Figure 12F) [216,218] that somewhat resembles those found across Clitellata or in the orbinid genus *Questa*.

The pygidium possesses a pair of lobes which is well-developed and rounded in *P. chaetifer* but somewhat smaller and less swollen in *P. symbioticus* (Figure 12D), but always smaller and thicker than that what is found in Protodrilidae and Saccocirridae [174,220].

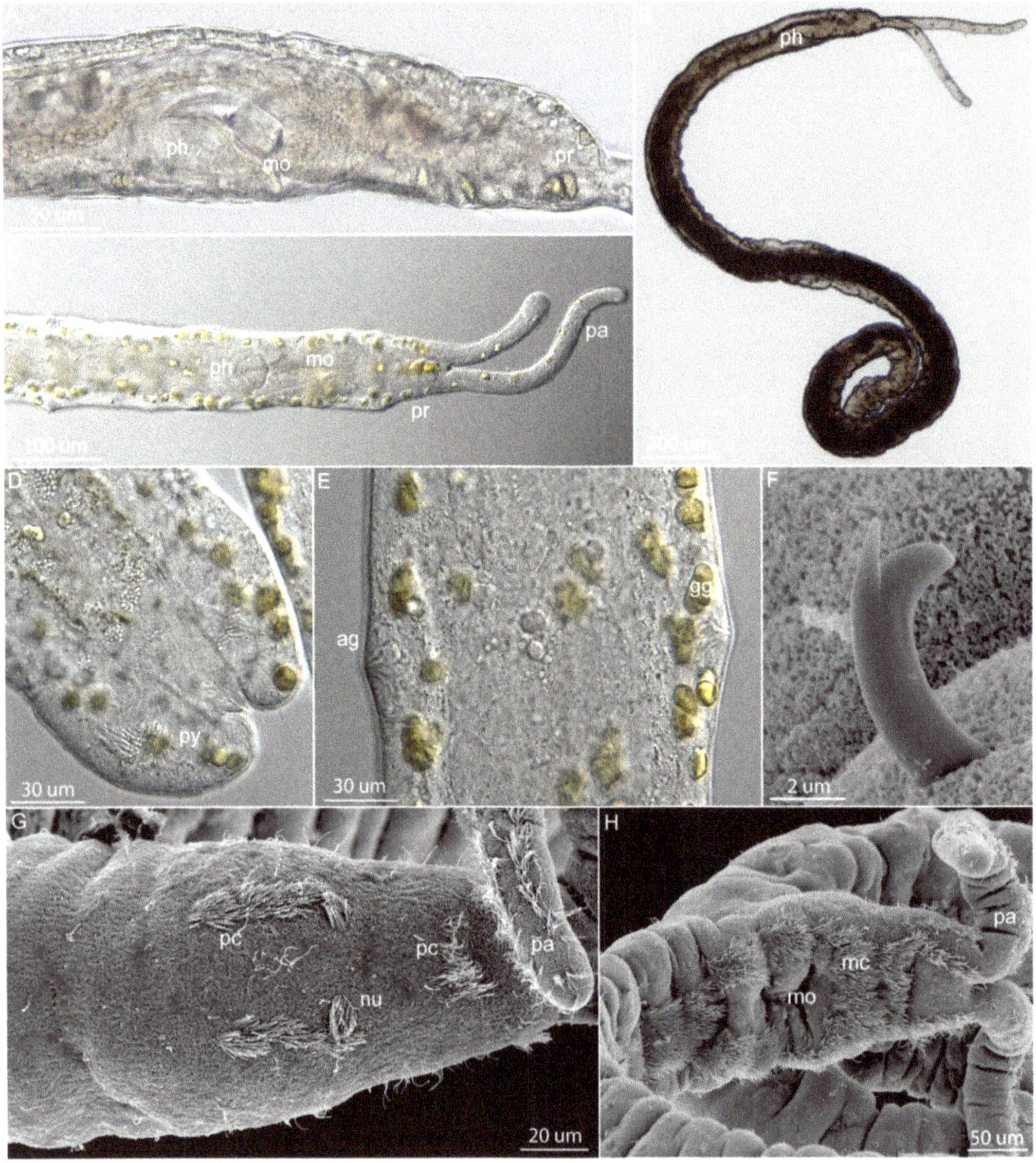

Figure 12. Morphology of Protodriloididae. (**A**–**E**) *Protodriloides symbioticus* (Roscoff, France), light micrographs. (**F**–**H**) *Protodriloides chaetifer* (Flakkerhuk, Greenland), scanning electron micrographs. (**A**) Lateral view of the anterior end. (**B**) Lateral overview of the body. (**C**) Dorsal view of the anterior end. (**D**) Dorsal view of the pygidium. (**E**) Dorsal view of a trunk segment. (**F**) Detail of the chaeta. (**G**) Dorsal view of the head. (**H**) Ventral view of the head. Abbreviations: ag, adhesive glands, gg green glands, mc mouth ciliation, mo mouth, nu nuchal organs, pa palp, pc prostomial ciliation, ph pharynx, pr prostomium, py pygidium. Reproduced with permission from [220].

3.8.3. Taxonomy

The only two described species of *Protodriloides* are mainly diagnosed by the presence or absence of chaeta, as well as by the arrangement of the epidermal glands along the trunk.

Protodriloides Jouin, 1966

Diagnosis: body elongated and flattened with a midventral ciliary band. Prostomium with two anterior palps lacking internal canals. Unpigmented prostomial receptors (so-called statocysts) and pigmented eyes absent. Two pairs of ciliary photoreceptor-like sensory organs are positioned behind the brain. Pharyngeal organ with bulbous and sagittal muscles as well as interstitial cells, but without a tongue-like organ. Salivary glands indistinct on the esophageal epithelium; not visible with light microscopy. Epidermis with green vacuolar glands. Pygidium with two rounded adhesive lobes. Fertile segments in the posterior half of the body. Males without lateral organs and with spermioducts in each fertile segment. Spermatozoa round and aflagellate. Large yolky eggs laid in cocoons produced by female cocoon glands. Fertilization and direct development inside the cocoon [220].

Two species. Type species: *Protodriloides symbioticus* Giard, 1904; *P. chaetifer* Remane, 1922.

3.8.4. Distribution and Diversity

The two described species of Protodriloididae have been recorded worldwide, except for Antarctica. Most records are from European coastal waters, reflecting a higher sampling effort here [220]. *Protodriloides symbioticus* was described in Ambleteuse, France [215] and has been mostly found in Northeastern Atlantic waters thereafter [199,216–218,222–228], although with a few records in Northern America and the Mediterranean Sea (Curini-Galletti et al. accepted; [189,229,230]). In contrast, *Protodriloides chaetifer* was originally described from Helgoland, Germany [216] and has subsequently been mainly recorded from Europe [199,218,227,228,231–235], but also occasionally in the Arctic [236,237] and Pacific oceans [74,212]. Most of these records are from intertidal to shallow subtidal medium-coarse sediments; often associated with the presence of groundwater submarine discharge [220]. Although RAFLP data [157] indicate that the *P. chaetifer* records comprise a complex of at least four cryptic species, the widely distributed records are still in need of careful review using integrative taxonomical studies, combining detailed morphological examinations with molecular methods. In contrast, the possibility of *P. chaetifer* dispersing due to stochastic catastrophic phenomena has recently been suggested in the area of Kerala (India), which was recently affected by a tsunami in 2006 [238].

3.8.5. Major Revisions and Most Important Literature

Main reviews of the family were provided by Jouin [218] and Purschke and Jouin [217], and recently updated by Martínez, Worsaae, and Purschke [220].

3.9. Psammodrilidae Swedmark, 1952

3.9.1. Phylogenetic Affinities

The eight described species of Psammodrilidae range in length from 1 to 8 mm and exhibit synapomorphic features such as thoracic motile cirri (with internal aciculae), a muscular collar region with unciliated "warty" epidermal cells and an almost entirely ciliated body, making them well-adapted to ciliary gliding between sand grains. Despite this aberrant morphology, their specialized hooked chaetae (uncini with barbules) resemble those of Arenicolidae and Maldanidae (e.g., [239,240]), causing them to ally with uncini-bearing families in Sedentaria in morphological phylogenies [241,242]. However, this hypothesis is rejected in phylogenomic analyses based on transcriptomic data [9], instead placing Psammodrilidae together with Apistobranchidae and Chaetopteridae (lacking uncini) in the basally branching Chaetopteriformia.

Morphologically, psammodrilids were previously divided into two genera: *Psammodrilus*, including the larger semi-sessile species that possesses a proposedly mucus-secreting collar region and *Psammodriloides*, which included the small-sized *P. fauveli* that lacks this collar and moves freely among the interstices [7]. However, the discovery of intermediate forms and life stages led to the synonymization of *Psammodriloides* with *Psammodrilus* [243]. This was further supported by a phylogenetic analysis of the family relying on increased taxon sampling [21], which showed that several small-sized species [239,243,244] have evolved secondarily and several times within the family from larger species [21] (Figure 13).

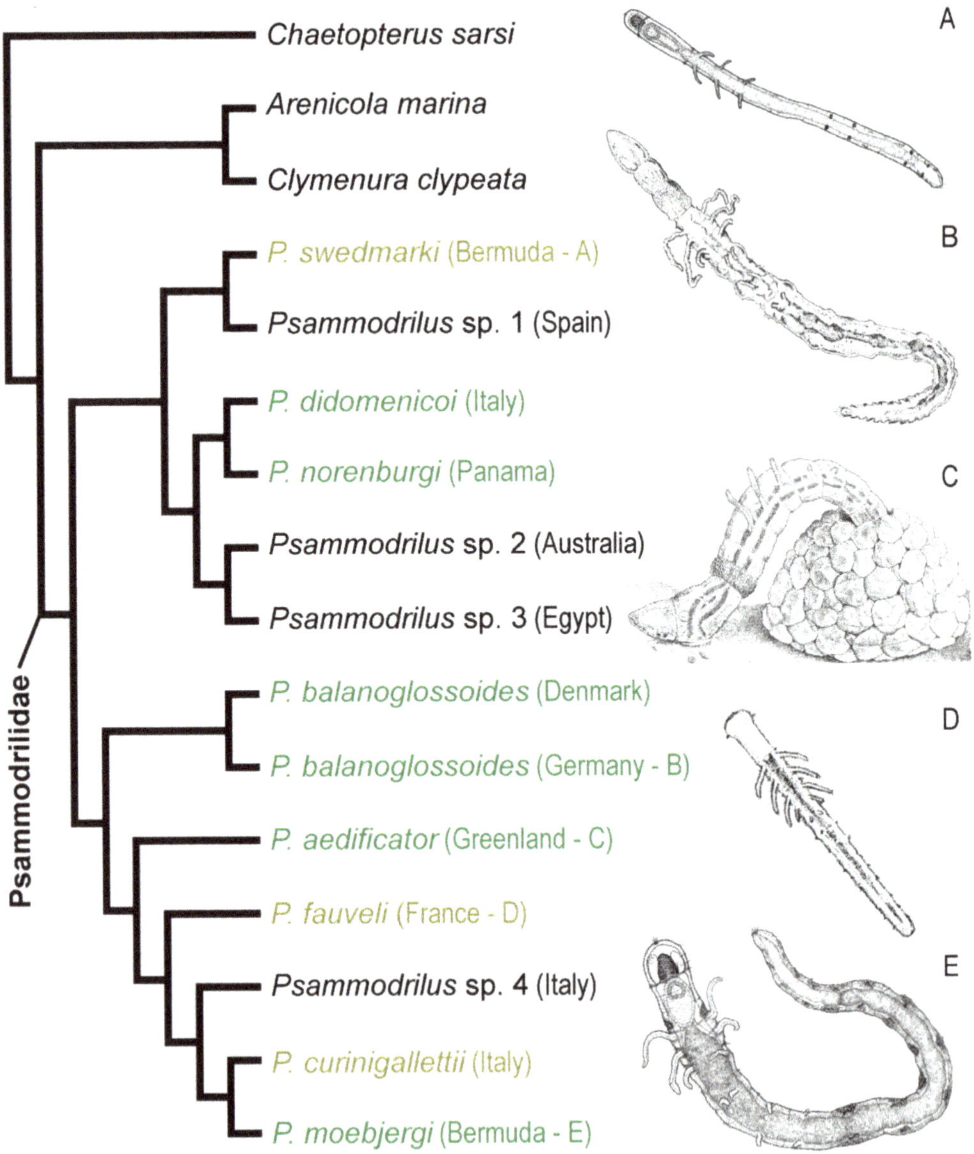

Figure 13. Phylogenetic tree of Psammodrilidae based on direct optimization parsimony analyses of combined molecular and morphological data (modified from [21]). Small-sized, collar-lacking species in olive green font two of which illustrated in (**A**,**D**), large-sized species in dark green font three of which illustrated in (**B**,**C**,**E**), psammodrilids of unknown size in black font.

These specializations are likely to have occurred through progenesis, since the small-sized species show great resemblance to juvenile stages of large-sized species, not yet having developed a collar, pharyngeal muscular diaphragms or as many chaetae and dorsal cirri [21,243]. The convergent and varying degree of regressive adaptations to a vagile, interstitial life form not only indicates that the development and morphology of Psammodrilidae is genetically susceptible to progenesis, causing the family to diversify worldwide throughout its long history [21,243], but also supports the suggestion that the family evolved from a larger, tube-building macrofaunal ancestor, as indicated by its position in Chaetopteriformia [9].

3.9.2. Morphology

The cylindrical coelomate body of psammodrilids comprises three different regions: a head (prostomium+peristomium), a thorax with six well defined segments, and an abdomen of varying length carrying uncini (Figure 14). The prostomium has an anterior apical sensory organ with compound cilia, lacking appendages and eyes and includes more or less well-developed prostomial coeloms with diaphragm sacs facilitating changes in prostomial length and width. The elongated peristomium bears an anterior ventral mouth opening and a posterior muscular pharyngeal apparatus depicting various levels of complexity. The larger species possess a complex suctorial pharynx containing muscular diaphragms and two coelomic muscle sacs within a collar of non-ciliated polygonal epidermal cells (in *P. moebjergi* limited to two lateral areas of collar cells). It is not yet clear whether the collar region constitutes the posterior part of the peristomium or an individual segment [245]. In the smaller species, the musculature is reduced, and the collar cells are missing.

The thorax carries one to six pairs of dorsal muscularized cirri of differing length and with internal pliable aciculae proximally nested within the ventral musculature. The abdomen lacks parapodia but carries paired groups of 1 to 16 lateroventral sigmoid uncini per segment (greatest number found in posterior segments) as well as a pygidium with a dorsal to terminal anus and compound sensory cilia. The epidermis is completely ciliated except in the collar region and the dorsal midline of thorax and abdomen. Rings of longer cilia delineate the prostomium, peristomium, and thoracic segments.

A range of unicellular glands are scattered throughout the epidermis, with numerous mucus glands on the head and ventral thorax seemingly aiding ciliary locomotion [239]. Numerous adhesive glands on the abdomen and cirri have such adhesive power that the animals easily get stuck in a petri dish and may rip them self apart (e.g., [236,243]). Segmental, multicellular, diamond-shaped glands are found dorso-laterally in the thorax and abdomen of *P. curinigallettii* [244]. The collar cells have well-developed Golgi complexes, and it has been proposed that they secrete mucus (possibly forming tubes) along the intercellular spaces where they interdigitate [236].

Body wall musculature is relatively simple with multiple longitudinal muscles sometimes forming two ventral and two dorsal bundles, as well as oblique muscles supporting the aciculae and thin outer circular muscles [244]. However, the unique suctional pharynx of larger psammodrilids is supported by several layers of both circular and longitudinal muscles, which together with an anterior and a posterior transverse muscular diaphragm aid in contracting or expanding the pharynx and the intake of detritus [245,246]. When moving through the sediment, a combination of muscles and coelomic cavities support the high flexibility of the prostomial shape (from anchor shaped to pointed).

Psammodrilids possess a variable number of paired metanephridia, being more developed with additional cilia in the larger species. The intraepidermal nervous system is poorly studied but comprises a dorsal brain and ganglionated paired ventro-lateral nerve cords [9,94,245].

Psammodrilids are gonochoristic, except for the hermaphroditic *P. curinigallettii* and *P. moebjergi*, producing both oocytes and sperm. Spermatozoa have ellipsoidal to elongate heads and long flagellate tails [243,244]. Fertilization is expectedly external but spermathecae were found in *P. swedmarki*, which suggests that some species might copulate [243]. Development is direct [245].

3.9.3. Taxonomy

Large-sized psammodrilids are identified based on the number of rows of cells in the collar, length and number of thoracic cirri, reproductive and glandular structures andshape and number of chaetae per segment. However, small-sized species have highly similar (and sometimes convergent) morphology, often only differing in combinations of absent features and variation in the number and length of structures rather than by the presence of species-specific autapomorphies. This makes taxonomy of Psammodrilidae extremely challenging and the presence of sexually mature stages crucial for morphological characterization. Molecular barcoding and comparison are therefore generally warranted to assure correct identification of psammodrilids.

Psammodrilus Swedmark, 1952

Eight described species, including synonymized *Psammodriloides* Swedmark, 1958. Type species: *Psammodrilus balanoglossoides* Swedmark, 1952; *P. fauveli* (Swedmark, 1958)

(as *Psammodriloides fauveli*); *P. aedificator* Kristensen and Nørrevang, 1982; *P. moebjergi*, *P. swedmarki* Worsaae and Sterrer, 2006; *P. curinigallettii* Worsaae, Kvindebjerg, and Martínez, 2015; *P. didomenicoi*, *P. norenburgi* Worsaae and Martínez, 2018. See Table 3 for morphological characteristics and distribution.

Table 3. List of main characters for the eight described species of Psammodrilidae (modified from [21]). Abbreviations: abdom., abdominal; max., maximum; rud., rudimental; segm., segment; #, number.

Species	Locality	Max. Body Length (mm)	Max. Body Width (mm)	Prostomial Sacs	Pharyngeal Musculature	Polygonal Collar Cells	Thoracic Cirri (Pairs)	Max. # Abdom. Segm.	Max. # Uncini/ Ramus
P. aedificator	Disko, W Greenland	8.1	0.19	Present	Well developed	ca. 10 rows	3 long, 3 short	20	3(5)
P. balanoglossoides	North Atlantic; White Sea; New Zealand	5–6	0.15	Present	Well developed	>25 rows	2 long, 1 medium, 3 short	31	16
P. didomenicoi	Napoli, Italy	1.9	0.13	Present	Well developed	ca. 7–8 rows	2 short, 3 rud.	11	5
P. fauveli	NE Atlantic	1	0.13	Absent	Absent	absent	3 medium, 1 short, 2 rud. or none	10	1
P. moebjergi	Bermuda	2.2	0.11	Present	Absent	2 lateral clusters	2 medium, 1 short, 3 rud.	11	4
P. norenburgi	Bocas del Toro, Panama	3.5	0.18	Present	Well developed	ca. 25 rows	3 long (or 2 long + 1 medium), 2–3 short	?	5
P. swedmarki	Bermuda	1.6	0.06	Absent	Absent	Absent	3 long, 1–2 rud. or none	7	2

Furthermore, at least nine potentially new species may exist for which details are lacking. Sequences are published from four undescribed species found off Lanzarote (Canary Islands), Lizard Island (NE Australia), Dahab (Egypt), and Campania (Italy) (Figure 1 in [21]). Recently, two seemingly undescribed species were found off Iriomote, Japan and off Amsterdam Island, French Southern and Antarctic Lands (Worsaae, unpublished). Moreover, *Psammodrilus balanoglossoides*, originally described off Roscoff, France, has been reported from widely separated geographic areas, including the White Sea [247], Barents Sea [237] and off Florida (USA) [243] and New Zealand [74]. Since even the two only sequenced populations of *P. balanoglossoides* from the North Sea and Baltic Sea, respectively, showed significant differences in 18S rRNA [21], the disparate records of *P. balanoglossoides* will most likely reveal several cryptic species.

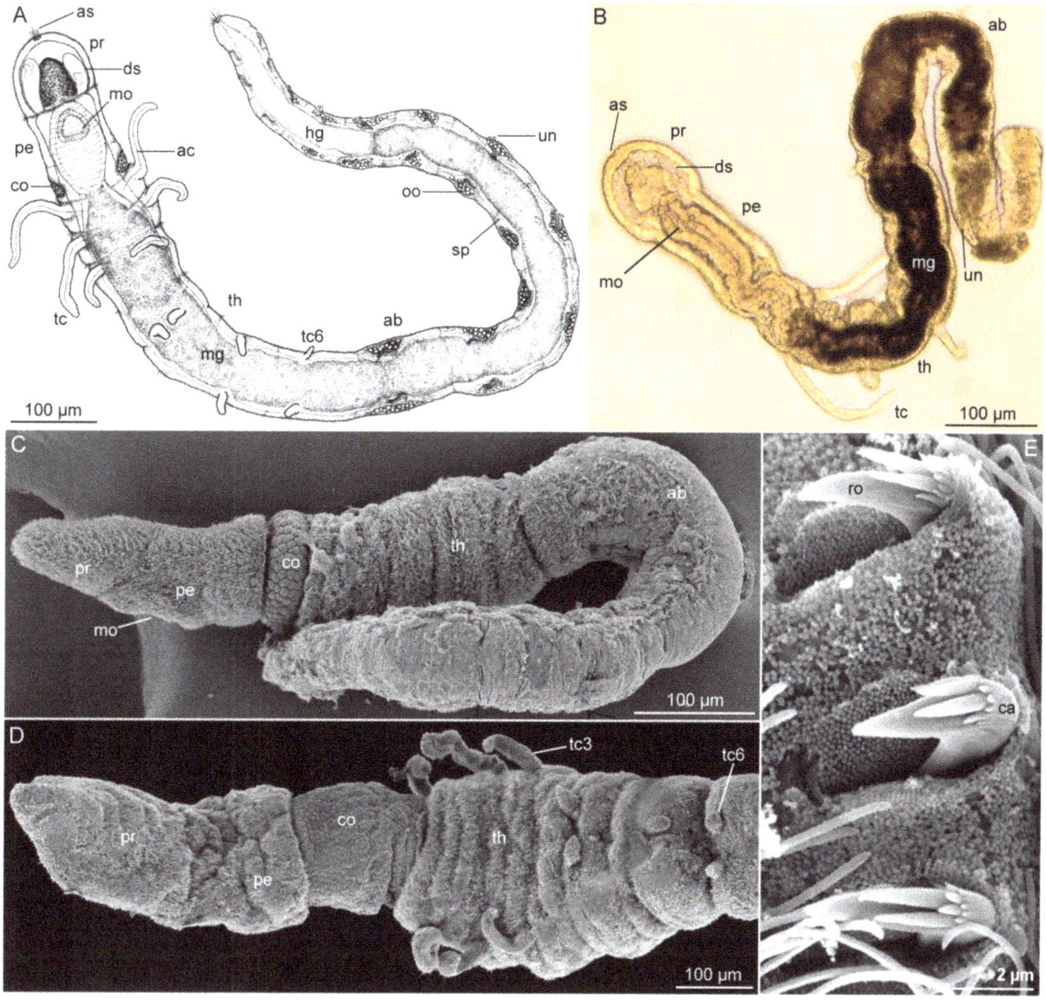

Figure 14. Morphology of Psammodrilidae. (**A**) Schematic illustration of *Psammodrilus moebjergi*. (**B**) Light micrograph of *P. moebjergi*, hind end lost. (**C–E**) Scanning electron micrographs. (**C**) *Psammodrilus didomenico*. (**D**) Anterior end of *P. balanoglossoides* from Denmark. (**E**) Uncini of *P. moebjergi*. (**A**,**B**,**E**) modified from [248]; (**C**), modified from [21]. Abbreviations: ab, abdomen; ac, acicula; as, anterior sensory cilia; ca, capitium; co, collar; ds, prostomial diaphragm sac; hg, hindgut; mg, midgut; mo, mouth opening; oo, oocyte; pe, peristomium; pr, prostomium; ro, rostrum; sp, sperm; tc, thoracic cirrus; tc3, tc6, third and sixth thoracic cirri; th, thorax; un, uncinus.

3.9.4. Distribution and Diversity

Psammodrilids are extremely fragile with their nearly entire ciliated epidermis and reduced cuticle, often breaking or disintegrating easily during collection and handling. This, in combination with an elusive and juvenile appearance of the small species, may be the reason why they are easily overlooked in meiofauna surveys and generally only found in low numbers, despite living in shallow subtidal, sandy-gravelly sediments. However, in the numerous new findings during recent meiofauna surveys ([21,245], Worsaae, unpublished) have expanded their distribution to most major oceans, suggesting a worldwide distribution and a high number of cryptic species [245].

3.9.5. Major Revisions

The evolution, morphology, and biology of psammodrilids are treated in the following major papers and references herein: Swedmark [246], Worsaae and Kristensen [7], Helm et al. [9], Worsaae et al. [21], and Worsaae [245].

3.10. *Saccocirridae Bobretzky, 1872*

3.10.1. Phylogenetic Affinities

Saccocirridae contains two valid genera with a total of 23 interstitials species. A combined phylogenetic analysis by Di Domenico et al. [51] using molecular and morphological data revealed that each of the 23 described species could be separated into two genera, namely *Saccocirrus* and *Pharyngocirrus*. These two clades were previously acknowledged as the "papillocercus" (= *Saccocirrus*) species complex and the "krusadensis" (= *Pharyngocirrus*) species complex in several earlier studies [53,217,248–251].

The systematic position of Saccocirridae within Annelida appears to be better resolved: Protodrilida (the clade Saccocirridae belongs to) plus Polygordiidae are united in a clade called Protodriliformia, which is a sister group to Phyllodocida and Eunicida within Errantia [10,11,252]. Saccocirridae are thought to have evolved by gradual miniaturization from macroscopic ancestors rather than by progenesis, as proposed for several other interstitial annelids, supported not only by their comparatively large size, but also by certain other characteristics, such as a life with trochophore larva and the presence of parapodia and chaetae [11].

3.10.2. Morphology

Species of Saccocirridae show body lengths ranging from 0.3 to 8 cm and comprise up to about 200 segments [51,53,248,250,253,254]. Compared to other interstitial and meiofaunal animals, saccocirrids have a large body. Since saccocirrids inhabit coarse sands, some species, such as *Saccocirrus major*, have macrofaunal body size dimensions. All saccocirrid species are characterized by the presence of two long and highly flexible muscular palps, small cylindrical uniramous parapodia with simple, retractable chaetae and a bilobed adhesive pygidium (Figures 15A,B and 16A,B,D). Depending on the species, the ventral palps are between 0.2 and 1.5 mm long and supplied with an internal coelomic canal that fuses with its corresponding canal from the opposite side behind the brain, and a pair of basal sac-like structures (= ampullae) that extend longitudinally until the first or third segment in all species.

Nuchal organs are present, and their size varies greatly among species. The eyes are situated anteriorly on the prostomium [51,83,249]. Whereas the small, pigmented eyes are easily seen with the light microscope, other, usually unpigmented photoreceptive sensory organs may be present as well, which, as a rule, can only be detected by TEM.

The digestive system of Saccocirridae comprises a ventral mouth, a buccal cavity with or without a muscular pharynx, a long esophagus, intestine, and a terminal anus [53,217,255,256]. The pharyngeal region of *S. papillocercus* and other *Saccocirrus* spp. bears dorsolateral ciliary folds but lacks a muscular ventral pharynx, whereas in *Pharyngocirrus* spp. possess a muscular pharynx (Figures 15A and 16A) [51,250,253,257–259].

Parapodia are uniramous, cylindrical and without lobes or cirri, bearing 5 to 10 chaetae (Figure 16C,F). Small ciliary tufts are present on the ventral side of the parapodia. Three types of chaetae are usually identified in Saccocirridae: long, medium and short [250] (Figure 16C,F). The longest chaetae in *Saccocirrus* species are robust and forked, whereas *Pharyngocirrus* species have delicate fan-shaped lyrate chaetae [51], which can be either asymmetrical or bilateraly symmetric. The medium length chaetae are spatulated to oar-shaped or have a smooth apex in *Saccocirrus*, or bifid with two equally long prongs in

Pharyngocirrus. The shortest (and thinnest) chaetae are spatulated with a smooth tip in *Saccocirrus* and bifid with a notched apex in *Pharyngocirrus*.

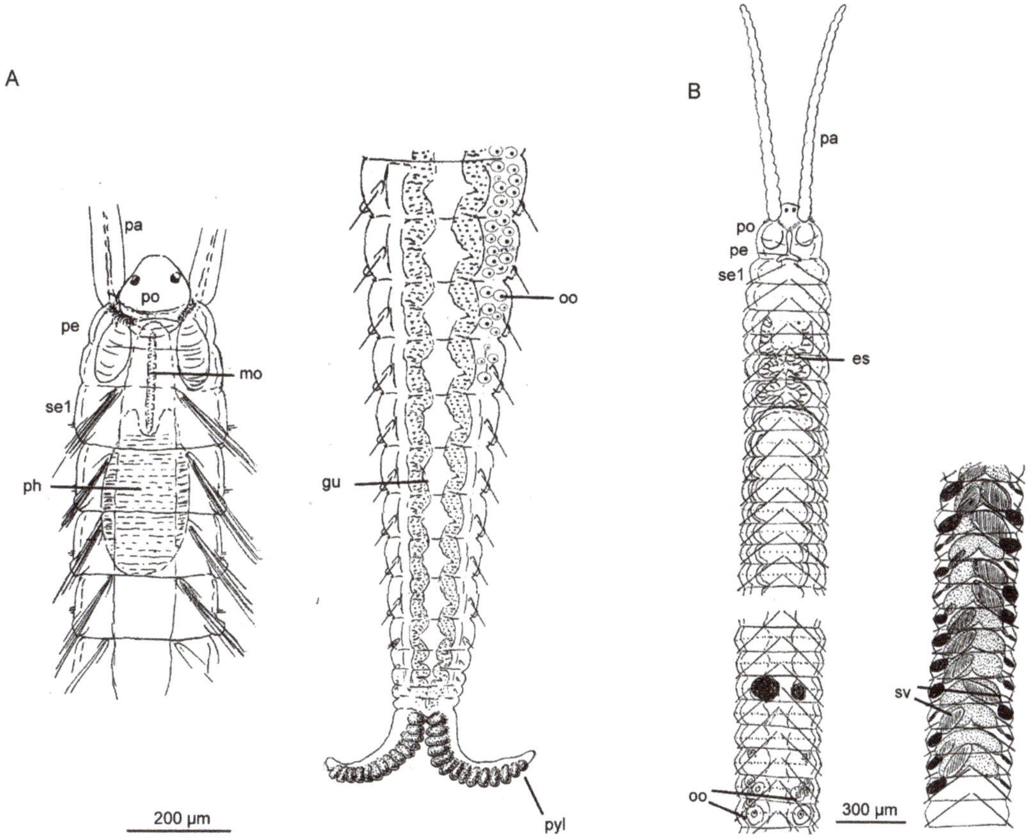

Figure 15. Saccocirridae species. (**A**) Drawing of *Pharyngocirrus gabriellae*, showing a female with an unpaired reproductive system and oocytes arranged on the right side of the gut (dorsal view). (**B**) Drawing of *Saccocirrus pussicus* showing the paired reproductive system of a female (oocytes) and a male (seminal vesicule). Abbreviations: es, esophagus; gu, gut; mo, mouth opening; oo, oocytes; pa, palps; pe, peristomium; ph, pharynx; po, prostomium; pyl, pygidial lobe; se1, segment 1; sv, seminal vesicle. References: Drawing (**A**) is modified from [260], and drawing (**B**) is modified from [108], acknowledging copyright permissions from publishers.

Saccocirridae usually possess adhesive glands that open on the pygidial lobes and on the trunk segments. These are generally well-developed and most likely necessitated by the turbulent environment most species live in. However, in a few species, the adhesive glands are less developed, as in *Saccocirrus minor* or *Pharyngocirrus jouinae*. The adhesive glands open in transverse rows with usually three gland cells opening in a common small papilla in the trunk region (Figure 16E).

The reproductive system is unpaired in *Pharyngocirrus* species and paired among *Saccocirrus* species [51,248]. Within each genus, the species differ in their distinct number of fertile segments, presence or absence of ovaries, size of mature oocytes, number of segments with oocytes, and position of the testes [51,248].

Figure 16. Morphology of Saccocirridae. (**A–C**) *Pharyngocirrus*. (**D–F**) *Saccocirrus*. (**A**) Light microscopy of *Pharyngocirrus* sp. (Bocas del Toro, Panamá). (**B**) Scanning electron micrograph of *Pharyngocirrus* sp. (Bird Rock, CA, USA). (**C**) Scanning electron micrograph of *Pharyngocirrus* sp. (Bocas del Toro, Panamá). (**D**) Light microscopy of *Saccocirrus* sp. (Bermuda), dorsal view. (**E**) *Saccocirrus slateri* (Abades, Tenerife, Canary Island). (**F**) *Saccocirrus* sp. (Mala, Lanzarote, Canary Island). Abbreviations: am, ampullae; ar, adhesive ridges; epg, epidermal adhesive glands; es, esophagus, ey, eye (ocellus); ls, longest setae; mo, mouth opening; ms, medium setae; pa, palps; pe, peristomium; ph, pharynx; po, prostomium; pyl, pygidial lobe; s, setae; se, setiger; ss, shortest setae; vcp, ventral (mouth) ciliary patch.

3.10.3. Taxonomy

The following characteristics are the main morphological features used to identify Saccocirridae species: number of segments, type of chaetae, number and shape of appendages (palps, antennae, parapodial and pygidial cirri), and type of reproductive organs (paired or unpaired), as well as size, number, and position of seminal vesicles and oocytes. A list of valid species for each genus is listed below (see Table 4).

Table 4. Meristic and morphometric characters of described species of *Saccocirrus* and *Pharyngocirrus*. Abbreviations: L, length; W, width; Max, maximum; #, number; pyg. adh., pygidial adhesive; Fem, Female; segm., segment; ?, unknown.

Species	Max L. (mm)	Max W. (μm)	Max # Segm.	Max # pyg. adh. Ridges	Pharyngeal Bulb	Gonads Fem.	# Fertile Segm.	Gonads Males	Ciliary Groove	Ciliary Patches Mouth	Longest Chaetae	Prongs Length
Saccocirrus												
S. slateri	25	730	155	22	absent	bilateral	72	bilateral	absent	absent	forked	equal
S. papi-locercus	30	400	150	8	absent	bilateral	120	bilateral	absent	absent	forked	equal
S. major	70	1000	200	14	absent	bilateral	175	bilateral	absent	absent	forked	equal
S. minor	15	200	100	absent	absent	bilateral	40	bilateral	absent	absent	forked	equal
S. orien-talis	12	?	170	4	absent	bilateral	60	bilateral	absent	absent	forked	equal
S. pussicus	30	400	120	12	absent	bilateral	36	bilateral	absent	absent	forked	unequal
S. hete-rochaetus	9	300	74	absent	absent	bilateral	20	bilateral	absent	absent	forked	unequal
S. parvus	3	280	70	absent	absent	bilateral	?	bilateral	absent	absent	forked	unequal
S. oahuen-sis	10.5	400	119	6	absent	bilateral	?	bilateral	absent	absent	forked	unequal
S. wa-ianaen-sis	10	450	210	absent	absent	bilateral	?	bilateral	absent	absent	forked	unequal
S. cirratus	45	?	200	absent	present	bilateral	115	bilateral	absent	present	lyrid	unequal
Pharyngocirrus												
P. archi-boldi	6	200	84	absent	present	?	?	?	absent	present	spatulated	?
P. kru-sadensis	25	400	150	9	present	left	80	left	absent	present	lyrid	unequal
P. gabriel-lae	30	400	160	15	present	right	100	left	present	present	lyrid	equal
P. eroticus	22	300	125	22	present	right	110	left	present	present	lyrid	equal
P. labilis	14	250	133	9	present	left	100	left	present	present	lyrid	unequal
P. sono-macus	25	330	140	12	present	right	55	left	absent	present	lyrid	equal
P. jouinae	20	550	120	?	present	unilateral	?	unilateral	present	present	lyrid	unequal
P. tri-dentiger	20	600	100	14	present	unilateral	?	unilateral	to segm. 8	present	lyrid	unequal
P. uchidai	20	350	146	23	present	left	100	left	present	present	lyrid	equal
P. goodrichi	15	300	130	7	present	left	35	unilateral	absent	absent	lyrid	unequal
P. alan-hongi	3.4	300	47	6	present	unilateral	?	unilateral	segm.1 to 3	present	lyrid	unequal
P. burchell	20	1000	200	10	present	unilateral	180	unilateral	absent	present	absent?	unequal

Pharyngocirrus Di Domenico, Martínez, Lana, and Worsaae 2014 (Figures 15A and 16A–C, Table 4)

Diagnosis: Brown body. Prostomium with two pigmented eyes and long filiform palps. Presence of prostomial transverse ciliary band. Mouth surrounded by ciliary patches consisting of paired longitudinal bands. Mid-ventral ciliary band may be present.

Ventral muscular pharynx present. Uniramous parapodia with three types of chaetae: (1) long capillary chaetae lyrate (equal or unequal sides) with a small median tooth; (2) medium bifid chaetae with equal lateral prongs; and (3) short chaetae with notched apex. Females have unilateral ovaries on the right or left side of the gut. Males have unilateral seminal vesicles on the right or left side of the gut.

Thirteen described species. Type species: *Pharyngocirrus gabriellae* (Du Boys-Reymond Marcus, 1946); *P. krusadensis* (Alikunhi, 1946); *P. eroticus* (Gray, 1969); *P. labilis* (Yamanishi, 1973); *P. archiboldi* (Kirsteuer, 1967); *P. sonomacus* (Martin, 1977); *P. jouinae* (Brown, 1981); *P. tridentiger* (Brown, 1981); *P. uchidai* (Sasaki, 1981); *P. goodrichi,* (Jouin-Toulmond and Gambi, 2007); *P. burchelli* (Silberbauer, 1969); *P. alanhongi* (Baley-Brock, Dreyer, and Brock, 2003).

Saccocirrus Bobretzky, 1872, (Figures 15B and 16D–F, Table 4)

Diagnosis: Brown body. Prostomium with two pigmented eyes and long filiform palps. Uniramous parapodia with three types of chaetae: (1) one to two long chaetae, robust and forked with equal or unequal prongs; (2) two to three medium spatuled chaetae; and (3) two to three short spatuled chaetae with notched apex. Females have bilateral ovaries. Males have bilateral seminal vesicles.

Eleven described species. Type species: *Saccocirrus papillocercus* Bobretzky, 1872; *S. major* Pierantoni, 1907; *S. orientalis* Alikunhi, 1946; *S. minor* Aiyar and Alikunhi, 1944; *S. pussicus* Du Bois-Reymond Marcus, 1948; *S. heterochaetus* Jouin, 1975; *S. parvus* Gerlach, 1953; *S. oahuensis* Baley-Brock, Dreyer and Brock, 2003; *S. waianaensis* Bailey-Brock, Dreyer and Brock, 2003; *S. cirratus* Aiyar and Alikunhi, 1944; *S. slateri* Di Domenico, Martínez and Worsaae, 2019.

3.10.4. Distribution and Diversity

Saccocirridae are found in medium to coarse sediments and gravel, from intertidal to subtidal regions. Records from the deeper sublittoral zone and the deep sea are lacking. These species are usually recorded from sandy beach environments [53,108,251,260–262]. A revision of the literature showed that *Pharyngocirrus* occurs intertidally on sheltered beaches, bays, or coves, as well as between rocks in tidal pools or subtidally. Generally, these animals occur in coarse sand with a well-defined redox layer. Species of *Saccocirrus* live intertidally in well-oxygenated coarse sand of exposed beaches and can often be found in the surf zone where they are hunting for prey or animals damaged by wave action. These annelids cling to sand grains or shells using their caudal appendages and sticky skin (caused by mucous produced by adhesive glands), particularly in high hydrodynamic environments [53,185,251,254,260].

Both genera are usually found worldwide in the tropical to temperate zones, but most species are described in warmer waters [51,248], except *Pharyngocirrus eroticus,* described near Orcas Island, on the west coast of North America [263]. The first reported species of Saccocirridae was *Saccocirrus papillocercus*, described in 1871 at Sebastopol Bay, Crimea peninsula, Black Sea [264]. Since then, *S. papillocercus* has been reported by several authors from the Mediterranean Sea [250,265,266] and North Atlantic [267,268]. The deepest record of the family is *Saccocirrus waianaensis*, found between 30 and 35 m off Oahu Island, Hawaii [269]. So far, records from boreal and polar latitudes are lacking. Geographic analyses yielded a well-supported diversity gradient for Saccocirridae, with a maximal diversity estimated at 20° N, and with 95% of the diversity being found between 0° and 30° N [254]. Distribution patterns of the described species of Saccocirridae have been summarized in Di Domenico et al. [51].

3.10.5. Major Revisions and Most Important Literature

Multiple studies have addressed the diversity and systematics of saccocirrids, but some of the most recent larger revisions include Jouin-Toulmond and Gambi 2007, Westheide 2008, Di Domenico et al. 2014a, and Di Domenico et al. 2019 [51,53,250,251]. Di Domenico et al. [270] provided the most comprehensive and recent revision of the family.

Author Contributions: K.W., A.K. and A.M. conceptualized and edited the manuscript. K.W. and A.M. drafted the abstract and introduction. A.K. prepared the first draft and plates for the methods and Dinophilidae; A.M. for Protodrilidae and Protodriloididae, B.C.G. for Apharyngtidae; K.W. for Diurodrilidae, Nerillidae and Psammodrilidae; M.D.D. for Polygordiidae and Saccocirridae; and N.B. for Lobatocerebridae. All authors have read and agreed to the published version of the manuscript.

Funding: This research received no external funding.

Institutional Review Board Statement: Not applicable.

Informed Consent Statement: Not applicable.

Data Availability Statement: Not applicable.

Acknowledgments: We greatly acknowledge our friends and family/colleagues and students that have joined our studies of these wonderful worms. We furthermore thank the following publishers for granting us permission to reuse, redraw, or copy selected images: CBM-Cahiers de Biologie Marine/Station Biologique de Roscoff; Javier González, Director of the Museo Nacional de Historia Natural from Montevideo; John Wiley & Sons; Linnean Society London; and Springer Nature. Lastly, we are very grateful for the careful eyes and helpful advices of the anonymous reviewers.

Conflicts of Interest: The authors declare no conflict of interest. The funders had no role in the design of the study; in the collection, analyses, or interpretation of data; in the writing of the manuscript, or in the decision to publish the results.

References

1. Giere, O. *Meiobenthology*; Springer Science & Business Media: Berlin/Heidelberg, Germany, 2013.
2. Schmidt-Rhaesa, A. (Ed.) *Guide to the Identification of Marine Meiofauna*; Dr. Friedrich Pfeil Verlag: Munich, Germany, 2020.
3. Worsaae, K. Annelida (excluding Clitellata and Sipuncula). In *Guide to the Identification of Marine Meiofauna*; Schmidt-Rhaesa, A., Ed.; Dr. Friedrich Pfeil Verlag: Munich, Germany, 2020; pp. 239–270.
4. Bunke, D. Aeolosomatidae and Potamodrilidae. In *Introduction to the Study of Meiofauna*; Higgins, R.P., Thiel, H., Eds.; Smithsonian Institution: Washington, DC, USA, 1988; pp. 345–348.
5. Hermans, C.O. The systematic position of the Archiannelida. *Syst. Biol.* **1969**, *18*, 85–102. [CrossRef]
6. Hatschek, B. Studien über die Entwicklungsgeschichte der Anneliden: Ein Beitrag zur Morphologie der Bilaterien. *Arb. Zool. Inst. Univ. Wien.* **1878**, *1*, 277–404.
7. Worsaae, K.; Kristensen, R.M. Evolution of interstitial Polychaeta (Annelida). *Hydrobiologia* **2005**, *535/536*, 319–340. [CrossRef]
8. Laumer, C.E.; Bekkouche, N.; Kerbl, A.; Goetz, F.; Neves, R.C.; Sørensen, M.V.; Kristensen, R.M.; Hejnol, A.; Dunn, C.W.; Giribet, G.; et al. Spiralian phylogeny informs the evolution of microscopic lineages. *Curr. Biol.* **2015**, *25*, 2000–2006. [CrossRef]
9. Helm, C.; Beckers, P.; Bartolomaeus, T.; Drukewitz, S.H.; Kourtesis, I.; Weigert, A.; Purschke, G.; Worsaae, K.; Struck, T.H.; Bleidorn, C. Convergent evolution of the ladder-like ventral nerve cord in Annelida. *Front. Zool.* **2018**, *15*, 36–17. [CrossRef] [PubMed]
10. Andrade, S.C.S.; Novo, M.; Kawauchi, G.Y.; Worsaae, K.; Pleijel, F.; Giribet, G.; Rouse, G.W. Articulating "Archiannelids": Phylogenomics and annelid relationships, with emphasis on meiofaunal taxa. *Mol. Biol. Evol.* **2015**, *32*, 2860–2875. [CrossRef] [PubMed]
11. Struck, T.H.; Golombek, A.; Weigert, A.; Franke, F.A.; Westheide, W.; Purschke, G.; Bleidorn, C.; Halanych, K.M. The evolution of annelids reveals two adaptive routes to the interstitial realm. *Curr. Biol.* **2015**, *25*, 1993–1999. [CrossRef] [PubMed]
12. Martín-Durán, J.M.; Vellutini, B.C.; Marletaz, F.; Cetrangolo, V.; Cvetesic, N.; Thiel, D.; Henriet, S.; Grau-Bové, X.; Carrillo-Baltodano, A.M.; Gu, W.; et al. Conservative route to extreme genome compaction in a miniature annelid. *Nat. Ecol. Evol.* **2020**.
13. Chen, H.; Parry, L.A.; Vinther, J.; Zhai, D.; Hou, X.; Ma, X. A Cambrian crown annelid reconciles phylogenomics and the fossil record. *Nature* **2020**, *583*, 249–252. [CrossRef] [PubMed]
14. Worsaae, K.; Rouse, G.W. Is *Diurodrilus* an annelid? *J. Morph.* **2008**, *269*, 1426–1455. [CrossRef]
15. Kerbl, A.; Bekkouche, N.; Sterrer, W.; Worsaae, K. Detailed reconstruction of the nervous and muscular system of Lobatocerebridae with an evaluation of its annelid affinity. *BMC Evol. Biol.* **2015**, *15*, 277. [CrossRef] [PubMed]
16. Helm, C.; Vöcking, O.; Kourtesis, I.; Hausen, H. *Owenia fusiformis*—A basally branching annelid suitable for studying ancestral features of annelid neural development. *BMC Evol. Biol.* **2016**, *16*, 129. [CrossRef]
17. Müller, M.C. Polychaete nervous systems: Ground pattern and variations—cLS microscopy and the importance of novel characteristics in phylogenetic analysis. *Integr. Comp. Biol.* **2006**, *46*, 125–133. [CrossRef]
18. Miyamoto, N.; Shinozaki, A.; Fujiwara, Y. Neuroanatomy of the vestimentiferan tubeworm Lamellibrachia satsuma provides insights into the evolution of the polychaete nervous system. *PLoS ONE* **2013**, *8*, 1–9. [CrossRef]
19. Purschke, G. Male genital organs, spermatogenesis and spermatozoa in the enigmatic terrestrial polychaete *Parergodrilus heideri* (Annelida, Parergodrilidae). *Zoomorphology* **2002**, *121*, 125–138. [CrossRef]
20. Purschke, G. Comparative electron microscopic investigation of the nuchal organs in *Protodriloides*, *Protodrilus*, and *Saccocirrus* (Annelida, Polychaeta). *Can. J. Zool.* **1990**, *68*, 325–338. [CrossRef]
21. Worsaae, K.; Giribet, G.; Martínez, A. The role of progenesis in the diversification of the interstitial annelid lineage Psammodrilidae. *Invertebr. Syst.* **2018**, *32*, 774–793. [CrossRef]
22. Bleidorn, C. The role of character loss in phylogenetic reconstruction as exemplified for the Annelida. *J. Zool. Syst. Evol. Res.* **2007**, *45*, 299–307. [CrossRef]

23. Mock, H. Zur Kenntnis von *Diurodrilus subterraneus* (Polychaeta, Dinophilidae) aus dem Sandhang der Nordseeinsel Sylt. *Helgol. Meeresunters.* **1981**, *34*, 329–335. [CrossRef]
24. Rieger, R.M.; Rieger, G.E. Fine structure of the archiannelid cuticle and remarks on the evolution of the cuticle within the Spiralia. *Acta Zool.* **1976**, *57*, 53–68. [CrossRef]
25. Worsaae, K.; Kristensen, R.M. Diurodrilidae Kristensen & Niilonen, 1982. In *Handbook of Zoology: Annelida*; Purschke, G., Böggemann, M., Westheide, W., Eds.; De Gruyter: Berlin, Germany, 2020; Volume 3.
26. Worsaae, K.; Kerbl, A.; Vang, Á.; Gonzalez, B.C. Broad North Atlantic distribution of a meiobenthic annelid—Against all odds. *Sci. Rep.* **2019**, *9*, 15497–13. [CrossRef]
27. Gelder, S.R. A review of the zoogeography and habitat data of the genus *Nerilla* Schmidt, 1848 (Annelida: Archiannelida). *J. Nat. Hist.* **1974**, *8*, 631–643. [CrossRef]
28. Jouin, C. Recherches sur les Protodrilidae (Archiannélides). I. Étude morphologique et systématique du genre *Protodrilus*. *Cah. Biol. Mar.* **1970**, *11*, 367–434.
29. Di Domenico, M.; Martínez, A.; Almeida, T.C.M.; Martins, M.O.; Worsaae, K.; Lana, P.C. Response of the meiofaunal annelid *Saccocirrus pussicus* (Saccocirridae) to sandy beach morphodynamics. *Hydrobiologia* **2014**, *734*, 1–16. [CrossRef]
30. Sato-Okoshi, W.; Okoshi, K.; Fujiwara, Y. A new species of *Protodrilus* (Annelida, Protodrilidae), covering bone surfaces bright red, in whale-fall ecosystems in the northwest Pacific. *Biol. Bull.* **2015**, *229*, 209–219. [CrossRef] [PubMed]
31. Jouin, C. The ultrastructure of a gutless annelid, *Parenterodrilus* gen. nov. *taenioides* (=*Astomus taenioides*) (Polychaeta, Protodrilidae). *Can. J. Zool.* **1992**, *70*, 1833–1848. [CrossRef]
32. Worsaae, K. Systematics of Nerillidae (Polychaeta, Annelida). *Meiofauna Marina* **2005**, *14*, 49–74.
33. Worsaae, K.; Kristensen, R.M. A new species of *Paranerilla* (Polychaeta: Nerillidae) from Northeast Greenland Waters, Arctic Ocean. *Cah. Biol. Mar.* **2003**, *44*, 23–39.
34. Morselli, I.; Sarto, M.; Mari, M. *Troglochaetus beranecki* Delachaux (Annelida, Polychaeta): Collecting methods and microscopy techniques for SEM and in vivo observations. *Hydrobiologia* **1998**, *379*, 213–216. [CrossRef]
35. Sterrer, W.; Iliffe, T.M. *Mesonerilla prospera*, a new archiannelid from marine caves in Bermuda. *Proc. Biol. Soc. Wash.* **1982**, *95*, 509–514.
36. Worsaae, K.; Martinez, A.; Núñez, J. Nerillidae (Annelida) from the Corona lava tube, Lanzarote, with description of *Meganerilla cesari* n. sp. *Mar. Biodivers.* **2009**, *39*, 195–207. [CrossRef]
37. Worsaae, K.; Mikkelsen, M.D.; Martinez, A. Description of six new species of *Mesonerilla* (Nerillidae, Annelida) and an emended description of *M. intermedia* Wilke, 1953, from marine and cave environments. *Mar. Biodivers.* **2019**, *49*, 2141–2165. [CrossRef]
38. Worsaae, K.; Rouse, G.W. *Mesonerilla neridae*, n. sp. (Nerillidae): First meiofaunal annelid from deep-sea hydrothermal vents. *Zoosymposia* **2009**, *2*, 297–303. [CrossRef]
39. Müller, M.C.M.; Bernhard, J.M.; Jouin-Toulmond, C. A new member of Nerillidae (Annelida: Polychaeta), *Xenonerilla bactericola* gen. et sp. nov., collected off California, USA. *Cah. Biol. Mar.* **2001**, *42*, 203–217.
40. Worsaae, K.; Sterrer, W.; Iliffe, T.M. *Longipalpa saltatrix*, a new genus and species of the meiofaunal family Nerillidae (Annelida: Polychaeta) from an anchialine cave in Bermuda. *Proc. Biol. Soc. Wash.* **2004**, *117*, 346–362.
41. Martinez, A.; Kvindebjerg, K.; Iliffe, T.M.; Worsaae, K. Evolution of cave suspension feeding in Protodrilidae (Annelida). *Zool. Scr.* **2017**, *46*, 214–226. [CrossRef]
42. Worsaae, K.; Gonzalez, B.C.; Kerbl, A.; Nielsen, S.H.; Jørgensen, J.T.; Armenteros, M.; Iliffe, T.M.; Martinez, A. Diversity and evolution of the stygobitic *Speleonerilla* nom. nov. (Nerillidae, Annelida) with description of three new species from anchialine caves in the Caribbean and Lanzarote. *Mar. Biodivers.* **2019**, *49*, 2167–2192. [CrossRef]
43. Cerca, J.; Meyer, C.; Stateczny, D.; Siemon, D.; Wegbrod, J.; Purschke, G.; Dimitrov, D.; Struck, T.H. Deceleration of morphological evolution in a cryptic species complex and its link to paleontological stasis. *Evolution* **2020**, *74*, 116–131. [CrossRef] [PubMed]
44. De Souza-Santos, L.P. An experimental approach to the ecophysiology of the interstitial polychaete *Polygordius eschaturus* (Annelida: Polygordiidae). *Scimar* **2017**, *81*, 509–515. [CrossRef]
45. Curini-Galletti, M.; Artois, T.; Delogu, V.; De Smet, W.H.; Fontaneto, D.; Jondelius, U.; Leasi, F.; Martinez, A.; Meyer-Wachsmuth, I.; Nilsson, K.S.; et al. Patterns of diversity in soft-bodied meiofauna: Dispersal ability and body size matter. *PLoS ONE* **2012**, *7*, e33801. [CrossRef]
46. Martinez, A.; Di Domenico, M.; Leasi, F.; Curini-Galletti, M.; Todaro, M.A.; Zotto, M.D.; Gobert, S.; Artois, T.; Norenburg, J.; Jörger, K.M.; et al. Patterns of diversity and endemism of soft-bodied meiofauna in an oceanic island, Lanzarote, Canary Islands. *Mar. Biodivers.* **2019**, *49*, 2033–2055. [CrossRef]
47. Tustison, C.A.; Ramey-Balci, P.A.; Rouse, G.W. More Knot Worms: Four new *Polygordius* (Annelida) species from the Pacific and Caribbean. *Diversity* **2020**, *12*, 146. [CrossRef]
48. Cerca, J.; Meyer, C.; Purschke, G.; Struck, T.H. Delimitation of cryptic species drastically reduces the geographical ranges of marine interstitial ghost-worms (*Stygocapitella*; Annelida, Sedentaria). *Mol. Phylogenet. Evol.* **2020**, *143*, 106663. [CrossRef] [PubMed]
49. Martinez, A.; Di Domenico, M.; Jörger, K.; Norenburg, J.; Worsaae, K. Description of three new species of *Protodrilus* (Annelida, Protodrilidae) from Central America. *Mar. Biol. Res.* **2013**, *9*, 676–691. [CrossRef]
50. Martinez, A.; Di Domenico, M.; Rouse, G.W.; Worsaae, K. Phylogeny and systematics of Protodrilidae (Annelida) inferred with total evidence analyses. *Cladistics* **2015**, *31*, 250–276. [CrossRef]

51. Di Domenico, M.; Martínez, A.; Lana, P.C.; Worsaae, K. Molecular and morphological phylogeny of Saccocirridae (Annelida) reveals two cosmopolitan clades with specific habitat preferences. *Mol. Phylogenet. Evol.* **2014**, *75*, 202–218. [CrossRef]
52. Higgins, R.P.; Thiel, H. *Introduction to the Study of Meiofauna*, 1st ed.; Higgins, R.P., Thiel, H., Eds.; Smithsonian Institution: Washington, DC, USA, 1988.
53. Westheide, W. *Polychaetes: Interstitial Families*, 2nd ed.; Crothers, J.H., Ed.; The Linnean Society of London (Field Studies Council Shrewsbury): London, UK, 2008.
54. De Jonge, V.N.; Bouwman, L.A. A simple density separation technique for quantitative isolation of meiobenthos using the colloidal silica Ludox-TM. *Mar. Biol.* **1977**, *42*, 143–148. [CrossRef]
55. Yoder, M.; De Ley, I.T.; King, I.W.; Mundo-Ocampo, M.; Mann, J.; Blaxter, M.; Poiras, L.; De Ley, P. DESS: A versatile solution for preserving morphology and extractable DNA of nematodes. *Nematology* **2006**, *8*, 367–376. [CrossRef]
56. Raikova, O.I.; Meyer-Wachsmuth, I.; Jondelius, U. The plastic nervous system of Nemertodermatida. *Org. Divers. Evol.* **2016**, *16*, 85–104. [CrossRef]
57. Kalt, M.R.; Tandler, B. A study of fixation of early amphibian embryos for electron microscopy. *J. Ultrastruct. Res.* **1971**, *36*, 633–645. [CrossRef]
58. Cheng, P.C.; Walden, D.B.; Greyson, R.I. Improved plant microtechnique for TEM, SEM and LM specimen preparation. *Natl. Sci. Counc. Month. Rep. China* **1979**, *7*, 1000–1007.
59. Westheide, W.; Purschke, G. Organism processing. In *Introduction to the Study of Meiofauna*; Higgins, R.P., Thiel, H., Eds.; Smithsonian Institution: Washington, DC, USA, 1988; pp. 146–160.
60. Kerbl, A.; Vereide, E.H.; Gonzalez, B.C.; Rouse, G.W.; Worsaae, K. Two new meiofaunal species of *Trilobodrilus* (Dinophilidae, Annelida) from California, USA. *Eur. J. Taxon.* **2018**, *421*, 1–18. [CrossRef]
61. Worsaae, K.; Müller, M.C. Nephridial and gonoduct distribution patterns in Nerillidae (Annelida: Polychaeta) examined by tubulin staining and cLSM. *J. Morphol.* **2004**, *261*, 259–269. [CrossRef]
62. Jörger, K.M.; Norenburg, J.L.; Wilson, N.G.; Schrödl, M. Barcoding against a paradox? Combined molecular species delineations reveal multiple cryptic lineages in elusive meiofaunal sea slugs. *BMC Evol. Biol.* **2012**, *12*, 1–18. [CrossRef]
63. Meyer-Wachsmuth, I.; Curini-Galletti, M.; Jondelius, U. Hyper-cryptic marine meiofauna: Species complexes in Nemertodermatida. *PLoS ONE* **2014**, *9*, e107688. [CrossRef] [PubMed]
64. Fontaneto, D.; Flot, J.-F.; Tang, C.Q. Guidelines for DNA taxonomy, with a focus on the meiofauna. *Mar. Biodivers.* **2015**, *45*, 433–451. [CrossRef]
65. Fujisawa, T.; Barraclough, T.G. Delimiting Species Using Single-Locus Data and the Generalized Mixed Yule Coalescent Approach: A Revised Method and Evaluation on Simulated Data Sets. *Syst. Biol.* **2013**, *62*, 707–724. [CrossRef]
66. Kapli, P.; Lutteropp, S.; Zhang, J.; Kobert, K.; Pavlidis, P.; Stamatakis, A.; Flouri, T. Multi-rate Poisson tree processes for single-locus species delimitation under maximum likelihood and Markov chain Monte Carlo. *Bioinformatics* **2017**, *33*, 1630–1638. [CrossRef] [PubMed]
67. Zhang, J.; Kapli, P.; Pavlidis, P.; Stamatakis, A. A general species delimitation method with applications to phylogenetic placements. *Bioinformatics* **2013**, *29*, 2869–2876. [CrossRef]
68. Puillandre, N.; Lambert, A.; Brouillet, S.; Achaz, G. ABGD, Automatic Barcode Gap Discovery for primary species delimitation. *Mol. Ecol.* **2012**, *21*, 1864–1877. [CrossRef] [PubMed]
69. Eibye-Jacobsen, D.; Kristensen, R.M. A new genus and species of Dorvilleidae (Annelida, Polychaeta) from Bermuda, with a phylogenetic analysis of Dorvilleidae, Iphitimidae and Dinophilidae. *Zool. Scr.* **1994**, *23*, 107–131. [CrossRef]
70. Struck, T.H. Phylogeny. In *Handbook of Zoology: Annelida*; Purschke, G., Böggemann, M., Westheide, W., Eds.; De Gruyter: Berlin, Germany, 2019; Volume 1, pp. 37–68.
71. Westheide, W. Apharyngtus Westheide, 1971. In *Handbook of Zoology: Annelida*; Purschke, G., Böggemann, M., Westheide, W., Eds.; De Gruyter: Berlin, Germany, 2019; Volume 1, pp. 234–237.
72. Westheide, W. *Apharyngtus punicus* nov. gen. nov. spec., an aberrant archiannelid from the mesopsammal of the Tunesian Coast of the Mediterranean. *Mikrofauna Meeresboden* **1971**, *6*, 233–248.
73. Jumars, P.A.; Dorgan, K.M.; Lindsay, S.M. Diet of worms emended: An update of polychaete feeding guilds. *Ann. Rev. Mar. Sci.* **2015**, *7*, 497–520. [CrossRef] [PubMed]
74. Riser, N.W. General observations on the intertidal interstitial fauna of New Zealand. *Tane* **1984**, *30*, 239–249.
75. Westheide, W. Dinophilidae Verrill, 1892. In *Handbook of Zoology: Annelida*; Purschke, G., Böggemann, M., Westheide, W., Eds.; De Gruyter: Berlin, Germany, 2019; Volume 1, pp. 217–233.
76. Struck, T.H.; Westheide, W.; Purschke, G. Progenesis in Eunicida ("Polychaeta," Annelida)—Separate evolutionary events? Evidence from molecular data. *Mol. Phylogenet. Evol.* **2002**, *25*, 190–199. [CrossRef]
77. Remane, A. Diagnosen neuer Archianneliden. *Zool. Anz.* **1925**, *65*, 15–17.
78. Kerbl, A.; Fofanova, E.G.; Mayorova, T.D.; Voronezhskaya, E.E.; Worsaae, K. Comparison of neuromuscular development in two dinophilid species (Annelida) suggests progenetic origin of *Dinophilus gyrociliatus*. *Front. Zool.* **2016**, *13*, 49. [CrossRef]
79. Kerbl, A.; Conzelmann, M.; Jekely, G.; Worsaae, K. High diversity in neuropeptide immunoreactivity patterns among three closely related species of Dinophilidae (Annelida). *J. Comp. Neurol.* **2017**, *525*, 3596–3635. [CrossRef]
80. Fofanova, E.G.; Mayorova, T.D.; Voronezhskaya, E.E. Paradoxical effect of serotonin on ciliary locomotion of the adult archiannelid worms. *Zool. Bespozvon.* **2017**, *14*, 114–120.

81. Müller, M.C.M.; Westheide, W. Comparative analysis of the nervous systems in presumptive progenetic dinophilid and dorvilleid polychaetes (Annelida) by immunohistochemistry and cLSM. *Acta Zool.* **2002**, *83*, 33–48. [CrossRef]
82. Purschke, G.; Arendt, D.; Hausen, H.; Müller, M.C.M. Photoreceptor cells and eyes in Annelida. *Arthropod Struct. Dev.* **2006**, *35*, 211–230. [CrossRef] [PubMed]
83. Eakin, R.M.; Martin, G.G.; Reed, C.T. Evolutionary significance of fine structure of archiannelid eyes. *Zoomorphology* **1977**, *88*, 1–18. [CrossRef]
84. Kajihara, H.; Ikoma, M.; Yamasaki, H.; Hiruta, S.F. *Trilobodrilus itoi* sp. nov., with a re-description of *T. nipponicus* (Annelida: Dinophilidae) and a molecular phylogeny of the genus. *Zool. Sci.* **2015**, *32*, 405–417. [CrossRef]
85. Purschke, G. On the ground pattern of Annelida. *Org. Divers. Evol.* **2002**, *2*, 181–196. [CrossRef]
86. Fofanova, E.G.; Nezlin, L.P.; Voronezhskaya, E.E. Ciliary and nervous structures in juvenile females of the annelid *Dinophilus gyrociliatus* (O. Schmidt, 1848) (Annelida: Polychaeta). *Russ. J. Mar. Biol.* **2014**, *40*, 43–52. [CrossRef]
87. Moore, A. *Dinophilus gardineri* (sp. nov.). *Biol. Bull.* **1899**, *1*, 15–18. [CrossRef]
88. Kerbl, A.; Tolstrup, E.W.; Worsaae, K. Nerves innervating copulatory organs show common FMRFamide, FVRIamide, MIP and serotonin immunoreactivity patterns across Dinophilidae (Annelida) indicating their conserved role in copulatory behaviour. *BMC Zool.* **2019**, *4*, 8. [CrossRef]
89. Jennings, J.B.; Gelder, S.R. Feeding and digestion in *Dinophilus gyrociliatus* (Annelida: Archiannelida). *J. Zool.* **1969**, *158*, 441–451. [CrossRef]
90. Beniash, E.A.; Yerlikova, N.N.; Yevdonin, L.A. Some characteristics of the *Dinophilus vorticoides* anatomy of the nervous system. In *Explorations of the Fauna of the Seas*; Bushinskaya, G.N., Ed.; Russian Academy of Sciences: St. Petersburg, Russia, 1992; Volume 43, pp. 5–9.
91. Repiakhov, V.M. K anatomii i istorii razvitiia *Dinophilus gyrociliatus*, O. Schmidt. *Zap. Novoross. Obshchestva Estestvoispyt.* **1886**, 77p.
92. Mamkaev, Y.V. Dinophilida as a primitive group of Trochozoa. *Investig. Mar. Fauna* **1985**, *34*, 99–103.
93. Müller, M.C.M.; Westheide, W. Das Nervensystem parapodienloser Polychaeten: Orthogonale Strukturen des Nervensystems juveniler Stadien und progenetischer Arten. *Verh. Dtsch. Zool. Ges.* **1997**, *90*, 209.
94. Purschke, G. Annelida: Basal groups and Pleistoannelida. In *Structure and Evolution of Invertebrate Nervous Systems*; Schmidt-Rhaesa, A., Harzsch, S., Purschke, G., Eds.; Oxford University Press: Oxford, UK, 2015; pp. 254–312.
95. Kerbl, A.; Martín-Durán, J.M.; Worsaae, K.; Hejnol, A. Molecular regionalization in the compact brain of the meiofaunal annelid *Dinophilus gyrociliatus* (Dinophilidae). *EvoDevo* **2016**, *7*, 20. [CrossRef] [PubMed]
96. Fofanova, E.G.; Voronezhskaya, E.E. The structure of archiannelid *Dinophilus gyrociliatus* ventral nerve cords. *Acta Biol. Hung.* **2012**, *63*, 88–90. [CrossRef]
97. Jägersten, G. Zur Kenntnis der Morphologie, Enzystierung und Taxonomie von *Dinophilus*. *Kungl. Svenska Vetenskapsakad. Handl.* **1944**, *21*, 1–50.
98. Jennings, J.B.; Donworth, P.J. Observations on the life cycle and nutrition of *Dinophilus taeniatus* Harmer 1889 (Annelida: Polychaeta). *Ophelia* **1986**, *25*, 119–137. [CrossRef]
99. Jägersten, G. Life cycle of *Dinophilus*, with special reference to the encystment and its dependence on temperature. *Oikos* **1951**, *3*, 143–165. [CrossRef]
100. Donworth, P.J. A reappraisal and validation of the species *Dinophilus taeniatus* Harmer 1889 and of taxonomically significant features in monomorphic dinophilids (Annelida: Polychaeta). *Zool. Anz.* **1985**, *216*, 32–38.
101. Windoffer, R.; Westheide, W. The nervous system of the male *Dinophilus gyrociliatus* (Annelida: Polychaeta). I. Number, types and distribution pattern of sensory cells. *Acta Zool.* **1988**, *69*, 55–64. [CrossRef]
102. Windoffer, R.; Westheide, W. The nervous system of the male *Dinophilus gyrociliatus* (Polychaeta, Dinophilidae): II. Electron microscopical reconstruction of nervous anatomy and effector cells. *J. Comp. Neurol.* **1988**, *272*, 475–488. [CrossRef]
103. Fauvel, P. Polychètes sédentaires. Addenda aux errantes, Arachniannélides, Myzostomaires. In *Faune de France*; Volume 16, Paul Lechevalier: Paris, France, 1927.
104. Jones, E.R.; Ferguson, F.F. The genus *Dinophilus* (Archiannelida) in the United States. *Am. Midl. Nat.* **1957**, *57*, 440–449. [CrossRef]
105. Ax, P. Das Fortpflanzungsverhalten von *Trilobodrilus* (Archiannelida, Dinophilidae). *Mar. Biol.* **1968**, *1*, 330–335. [CrossRef]
106. Harmer, S.F. Notes on the anatomy of *Dinophilus*. *Proc. Cambridge Philos. Soc.* **1889**, *6*, 119–143. [CrossRef]
107. Westheide, W. Die Gattung *Trilobodrilus* (Archiannelida, Polychaeta) von der deutschen Nordseeküste. *Helgol. Meeresunters.* **1967**, *16*, 207–215. [CrossRef]
108. Du Bois-Reymond Marcus, E. Further archiannelids from Brazil. *Com. Zool. Mus.* **1948**, *48*, 1–17.
109. Westheide, W.; Nordheim, H. Interstitial Dorvilleidae (Annelida, Polychaeta) from Europe, Australia and New Zealand. *Zool. Scr.* **1985**, *14*, 183–199. [CrossRef]
110. Kristensen, R.M.; Niilonen, T. Structural Studies on *Diurodrilus* Remane (Diurodrilidae fam.n.), with description of *Diurodrilus westheidei* sp.n. from the Arctic Interstitial Meiobenthos, W. Greenland. *Zool. Scr.* **1982**, *11*, 1–12. [CrossRef]
111. Westheide, W. The systematic position of the Dinophilidae and the archiannelid problem. In *The Origins and Relationships of Lower Invertebrates*; Conway Morris, S., George, J.D., Gibson, R., Platt, H.M., Eds.; Systematic Association: Oxford, UK, 1985; Volume 28, pp. 310–326.

112. Golombek, A.; Tobergte, S.; Nesnidal, M.P.; Purschke, G.; Struck, T.H. Mitochondrial genomes to the rescue—Diurodrilidae in the myzostomid trap. *Mol. Phylogenet. Evol.* **2013**, *68*, 312–326. [CrossRef] [PubMed]
113. Kajihara, H.; Ikoma, M.; Yamasaki, H.; Hiruta, S.F. *Diurodrilus kunii* sp. nov. (Annelida: Diurodrilidae) and a molecular phylogeny of the genus. *Zool. Sci.* **2019**, *36*, 250–258. [CrossRef] [PubMed]
114. Kristensen, R.M.; Eibye-Jacobsen, D. Ultrastructure of spermiogenesis and spermatozoa in *Diurodrilus subterraneus* (Polychaeta, Diurodrilidae). *Zoomorphology* **1995**, *115*, 117–132. [CrossRef]
115. Schmidt, P.; Westheide, W. Interstitielle Fauna von Galàpagos. XVII. Polygordiidae, Saccocirridae, Protodrilidae, Nerillidae, Dinophilidae (Polychaeta). *Mikrofauna Meeresbod.* **1977**, *62*, 1–38.
116. Paxton, H. Family Diurodrilidae. In *Polychaetes & Allies: The Southern Synthesis*; Beesley, P.L., Ross, G.J.B., Glasby, C.J., Eds.; Csiro Publishing: Clayton, Australia, 2000; pp. 104–105.
117. Villora-Moreno, S. *Diurodrilus benazzi* Gerlach, 1952 (Diurodrilida) and *Dinophilus gyrociliatus* O. Schmidt, 1857 (Dinophilida): First record of two orders of interstitial polychaetes on the Iberian Peninsula. *Oceanogr. Lit. Rev.* **1998**, *7*, 1186.
118. Rieger, R.M. A new group of interstitial worms, Lobatocerebridae nov. fam. (Annelida) and its significance for metazoan phylogeny. *Zoomorphology* **1980**, *95*, 41–84. [CrossRef]
119. Kerbl, A.; Worsaae, K. Lobatocerebridae Rieger, 1980. In *Handbook of Zoology: Annelida*; Purschke, G., Böggemann, M., Westheide, W., Eds.; De Gruyter: Berlin, Germany, 2019; Volume 1, pp. 201–216.
120. Haszprunar, G.; Rieger, R.M.; Schuchert, P. Extant "problematica" within or near the Metazoa. In *The Early Evolution of Metazoa and the Significance of Problematic Taxa*; Simonetta, A.M., Conway Morris, S., Eds.; Cambridge University Press: Cambridge, UK, 1991; pp. 99–105.
121. Zrzavý, J. Gastrotricha and metazoan phylogeny. *Zool. Scr.* **2003**, *32*, 61–81. [CrossRef]
122. Rieger, R.M. Comparative ultrastructure and the Lobatocerebridae: Keys to understand the phylogenetic relationship of Annelida and the acoelomates. In *The Ultrastructure of Polychaeta*; Westheide, W., Hermans, C.O., Eds.; John Wiley & Sons: Stuttgart, Germany, 1988; Volume 4, pp. 373–382.
123. Tzetlin, A.B.; Filippova, A.V. Muscular system in polychaetes (Annelida). *Hydrobiologia* **2005**, *535/536*, 113–126. [CrossRef]
124. Rieger, R.M. Fine structure of the body wall, nervous system, and digestive tract in the Lobatocerebridae Rieger and the organization of the gliointerstitial system in Annelida. *J. Morph.* **1981**, *167*, 139–165. [CrossRef]
125. Rieger, R.M. Neue Organisationstypen aus der Sandlückenraumfauna: Die Lobatocerebriden und *Jennaria pulchra*. *Verh. Dtsch. Zool. Ges.* **1991**, *84*, 247–259.
126. Westheide, W. Progenesis as a principle in meiofauna evolution. *J. Nat. Hist.* **1987**, *21*, 843–854. [CrossRef]
127. Westheide, W.; Purschke, G. *Leptonerilla diplocirrata*, a new genus and species of interstitial polychaetes from the island of Hainan, south China (Nerillidae). *Proc. Biol. Soc. Wash.* **1996**, *109*, 586–590.
128. Worsaae, K.; Nygren, A.; Rouse, G.W.; Giribet, G.; Persson, J.; Sundberg, P.; Pleijel, F. Phylogenetic position of Nerillidae and Aberrantia (Polychaeta, Annelida), analyzed by direct optimization of combined molecular and morphological data. *Zool. Scr.* **2005**, *34*, 313–328. [CrossRef]
129. Boaden, P.J.S. *Meganerilla swedmarki*, nov. gen., nov. spec., an archiannelid of the family Nerillidae. *Arkiv Zool.* **1961**, *13*, 553–559.
130. Ax, P. *Thalassochaetus palpifoliaceus* nov. gen., nov. spec. (Archiannelida, Nerillidae), ein marine Verwandter von *Troglochaetus beranecki* Delachaux. *Zool. Anz.* **1954**, *153*, 64–75.
131. Tzetlin, A.B.; Saphonov, M.V. *Trochonerilla mobilis* gen. et sp. n., a meiofaunal nerillid (Annelida, Polychaeta) from a marine aquarium in Moscow. *Zool. Scr.* **1992**, *21*, 251–254. [CrossRef]
132. Delachaux, T. Un polychète d'eau douce cavernicole: *Troglochaetus beranecki* nov. gen. nov. spec. *Bull. Soc. Neuchatel. Sci. Nat.* **1921**, *45*, 1–11.
133. Müller, M.C.M. *Aristonerilla*: A new genus (Annelida: Polychaeta) with description of *Aristonerilla* (*Micronerilla*) *brevis* comb. nov. from a seawater aquarium. *Cah. Biol. Mar.* **2002**, *43*, 131–139.
134. Kirsteuer, E. Zur Kenntnis der Archianneliden des Roten Meeres. *Zool. Anz.* **1966**, *177*, 288–296.
135. Worsaae, K. Nerillidae Levinsen, 1883. In *Handbook of Zoology: Annelida*; Purschke, G., Böggemann, M., Westheide, W., Eds.; De Gruyter: Berlin, Germany, 2020; Volume 3.
136. Müller, M.C.M.; Worsaae, K. CLSM analysis of the phalloidin-stained muscle system in *Nerilla antennata*, *Nerillidium* sp. and *Trochonerilla mobilis* (Polychaeta; Nerillidae). *J. Morph.* **2006**, *267*, 885–896. [CrossRef] [PubMed]
137. Purschke, G. Ultrastructure of nuchal organs in polychaetes (Annelida)—New results and review. *Acta Zool.* **1997**, *78*, 123–143. [CrossRef]
138. Gelder, S.R.; Uglow, R.F. Feeding and gut structure in *Nerilla antennata* (Annelida: Archiannelida). *J. Zool.* **1973**, *171*, 225–237. [CrossRef]
139. Purschke, G. Anatomy and ultrastructure of ventral pharyngeal organs and their phylogenetic importance in Polychaeta (Annelida). *Zoomorphology* **1985**, *105*, 223–239. [CrossRef]
140. Tzetlin, A.B.; Zarvarzina, E.G.; Saphonov, M.V. Functional morphological analysis of the pharynx in some annelids. *Dokl. Akad. Nauk SSSR* **1987**, *294*, 1008–1011.
141. Tzetlin, A.B.; Purschke, G.; Westheide, W.; Saphonov, M.V. Ultrastructure of enteronephridia and general description of the alimentary canal in *Trochonerilla mobilis* and *Nerillidium troglochaetoides* (Polychaeta, Nerillidae). *Acta Zool.* **1992**, *73*, 163–176. [CrossRef]

142. Worsaae, K. Functional morphology of two mud-dwelling meiofana polychaetes of the family Nerillidae (Annelida). In *Arctic Biology Field Course, Qeqertarsuaq*; University of Copenhagen: Copenhagen, Denmark, 2002; pp. 119–128.
143. Müller, M.C.M. Das Nervensystem der Polychaeten Untersuchungen an ausgewählten taxa. Ph.D. Thesis, University of Osnabrück, Osnabrück, Germany, 1999; pp. 1–207.
144. Jouin, C. Sexualité et biologie de la reproduction chez *Mesonerilla* Remane et *Meganerilla* Boaden. *Cah. Biol. Mar.* **1968**, *9*, 31–52.
145. Magagnini, G. Reproduction in *Nerilla antennata* O. Schmidt (Archiannelida Nerillidae): Induction of spawning. *Boll. Zool.* **1982**, *49*, 283–286. [CrossRef]
146. Fransen, M.E. Fine structure of the brooding apparatus of the archiannelid *Mesonerilla intermedia*: Maternal connections to brooded eggs. *Trans. Am. Microsc. Soc.* **1983**, *102*, 25. [CrossRef]
147. Jouin, C. *Mesonerilla biantennata* n. sp. nouvelle archiannélide Nerillidae de la region de Roscoff. *C. R. Acad. Sci.* **1963**, *257*, 4057–4060.
148. Jouin, C. Hermaphrodisme chez *Nerillidopsis hyalina* n. g., n. sp. et chez *Nerillidium* Remane, Archiannélides. *C. R. Acad. Sci.* **1966**, *263*, 412–415.
149. Jouin, C. Sexualité chez *Meganerilla* Boaden et *Mesonerilla* Remane (Archiannélides Nerillidae) et modalités de reproduction chez ce dernier genre. *C. R. Acad. Sci.* **1967**, *265*, 150–153.
150. Jouin, C. Étude morphologique et anatomique de *Nerillidopsis hyalina* Jouin et de quelques *Nerillidium* Remane (Archiannelides Nerillidae). *Arch. Zool. Exp. Gen.* **1967**, *108*, 97–110.
151. Jouin, C. Nouvelles données sur *Troglochaetus beranecki* Delachaux (Archiannelida Nerillidae). *Ann. Spéliol.* **1973**, *28*, 575–579.
152. Jouin, C.; Swedmark, B. *Paranerilla limicola* n. g., n. sp., Archiannélide Nerillidae du benthos vaseux marin. *Cah. Biol. Mar.* **1965**, *6*, 201–218.
153. Goodrich, E.S. *Nerilla* an archiannelid. *Quart. Journ. Micr. Sci.* **1912**, *57*, 397–425.
154. Pennak, R.W. A fresh-water archiannelid from the Colorado Rocky Mountains. *Trans. Am. Microsc. Soc.* **1971**, *90*, 372–375. [CrossRef]
155. Pennak, R.W.; Ward, J.V. Interstitial faunal communities of the hyporheic and adjacent groundwater biotopes of a Colorado mountain stream. *Arch. Hydrobiol.* **1986**, *74*, 356–396.
156. Särkkä, J.; Mäkelä, J. *Troglochaetus beranecki* Delachaux (Polychaeta, Archiannelida) in esker groundwaters of Finland: A new class of limnic animals for northern Europe. *Hydrobiologia* **1998**, *379*, 17–21. [CrossRef]
157. Schmidt, H.; Westheide, W. RAPD-PCR experiments confirm the distinction between three morphologically similar species of *Nerilla (Polychaeta: Nerillidae)*. *Zool. Anz.* **1998**, *236*, 277–285.
158. Worsaae, K. Phylogeny of Nerillidae (Polychaeta, Annelida) as inferred from combined 18S rDNA and morphological data. *Cladistics* **2005**, *21*, 143–162. [CrossRef]
159. Lovén, S. Observations on the Metamorphosis of an Annelide. *Ann. Mag. Nat.* **1843**, *111*, 43–45. [CrossRef]
160. Agassiz, A. On the young stages of a few annelids. *Ann. Mag. Nat. Hist.* **1867**, *19*, 203–218. [CrossRef]
161. Schneider, A. *Monographie der Nematoden*; Georg Reimer: Berlin, Germany, 1866.
162. Schneider, A. Über Bau und Entwicklung von *Polygordius*. *Arch. Anat. Physiol. Wiss. Med.* **1868**, 51–60.
163. Ramey-Balcı, P.; Fiege, D.; Struck, T.H. Molecular phylogeny, morphology, and distribution of *Polygordius* (Polychaeta: Polygordiidae) in the Atlantic and Mediterranean. *Mol. Phylogenet. Evol.* **2018**, *127*, 919–930. [CrossRef]
164. Ramey-Balci, P.A.; Fiege, D.; Purschke, G. Polygordiidae Czerniavsky, 1881. In *Handbook of Zoology: Annelida, V*; Purschke, G., Böggemann, M., Westheide, W., Eds.; De Gruyter: Berlin, Germany, 2020; Volume 3.
165. Ramey, P.A.; Fiege, D.; Leander, B.S. A new species of *Polygordius* (Polychaeta: Polygordiidae): From the inner continental shelf and in bays and harbours of the north-eastern United States. *J. Mar. Biol. Assoc. UK* **2006**, *86*, 1025–1034. [CrossRef]
166. Wilkens, V.; Purschke, G. Central nervous system and sense organs, with special reference to photoreceptor-like sensory elements, in *Polygordius appendiculatus* (Annelida), an interstitial polychaete with uncertain phylogenetic affinities. *Invertebr. Biol.* **2009**, *128*, 46–64. [CrossRef]
167. Rouse, G.W.; Fauchald, K. Cladistics and polychaetes. *Zool. Scr.* **1997**, *26*, 139–204. [CrossRef]
168. Rota, E.; Carchini, G. A new *Polygordius* (Annelida: Polychaeta) from the Terra Nova Bay, Ross Sea, Antarctica. *Polar Biol.* **1999**, *21*, 201–213. [CrossRef]
169. Avery, L.; Ramey, P.A.; Wilson, R.S. New Polygordiidae (Polychaeta) from the Australian region. *Zootaxa* **2009**, *2068*, 59–68. [CrossRef]
170. Jamieson, B.G.M.; Rouse, G.W. The spermatozoa of the polychaeta (annelida): An ultrastructural review. *Biol. Rev.* **1989**, *64*, 93–157. [CrossRef]
171. Grassle, J.F.; Ramey, P.A.; Petrecca, R.F. Temporal and spatial variation in infaunal community structure in physically active continental shelf sediments at a long-term ecosystem observatory (LEO-15) off New Jersey, USA. *J. Mar. Res.* **2009**, *67*, 869–897. [CrossRef]
172. Ramey, P.A. Life history and population dynamics of a dominant polychaete, *Polygordius jouinae*, in inner continental shelf sands of the Mid-Atlantic Bight, USA. *Mar. Biol.* **2008**, *154*, 443–452. [CrossRef]
173. Ramey, P.A.; Bodnar, E. Selection by a deposit-feeding polychaete, *Polygordius jouinae*, for sands with relatively high organic content. *Limnol. Oceanogr.* **2008**, *54*, 1512–1520. [CrossRef]

174. Martínez, A.; Purschke, G.; Worsaae, K. Protodrilidae Hatschek, 1888. In *Handbook of Zoology: Annelida*; Purschke, G., Böggemann, M., Westheide, W., Eds.; De Gruyter: Berlin, Germany, 2020; Volume 3.
175. Zrzavý, J.; Říha, P.; Piálek, L.; Janouškovec, J. Phylogeny of Annelida (Lophotrochozoa): Total-evidence analysis of morphology and six genes. *BMC Evol. Biol.* **2009**, *9*, 1–14. [CrossRef]
176. Czerniavsky, V. Materialia ad Zoographiam Ponticam comparatam Fasc. III. Vermes. *Bull. Soc. Nat. Moscou* **1881**, *55*, 213–334.
177. Hatschek, B. *Protodrilus leuckartii*. Eine neue Gattung Archianneliden. *Arb. Zool. Inst. Univ. Wien.* **1881**, *3*, 79–92.
178. Hatschek, B. *Lehrbuch der Zoologie: Eine Morphologische Übersicht des Thierreiches zur Einführung in das Studium Dieser Wissenschaft*; Gustav Fischer: Jena, Germany, 1888.
179. Uljanina, V.N. Observations on *Polygordius* living in Sevastopol Bay (In Russian). *Bull. Soc. Nat. Moscou* **1877**, *52*, 53–96.
180. Purschke, G. Structure of the prostomial appendages and the central nervous system in the Protodrilida (Polychaeta). *Zoomorphology* **1993**, *113*, 1–20. [CrossRef]
181. Lam, H.J. Über den Bau und die Verwandschaft der *Protoannelis meyeri*, nov. gen., nov. spec., eine neue Archiannelide. *Tijdschr. Nederl. Dierk. Ver.* **1922**, *18*, 44–84.
182. Jouin, C. Description of a free-living Polychaete without gut: *Astomus taenioides* n.gen., n.sp. (Protodrilidae, Archiannelida). *Can. J. Zool.* **1979**, *57*, 2448–2456. [CrossRef]
183. Jouin-Toulmond, C.; Purschke, G. Ultrastructure of the spermatozoa of *Parenterodrilus taenioides* (Protodrilida: "Polychaeta") and its phylogenetic significance. *Zoomorphology* **2004**, *123*, 139–146. [CrossRef]
184. Jägersten, G. On the locomotion and attachment of *Protodrilus*: With remarks on the function of locomotory cilia. *Zool. Bidr. Fran Upps.* **1954**, *31*, 315–320.
185. Martin, G.G. Ciliary gliding in lower invertebrates. *Zoomorphology* **1978**, *91*, 249–261. [CrossRef]
186. Von Nordheim, H. Six new species of *Protodrilus* (Annelida, Polychaeta) from Europe and New Zealand, with a concise presentation of the genus. *Zool. Scr.* **1989**, *18*, 245–268. [CrossRef]
187. Jouin, C. Anatomical and ultrastructural study of the pharyngeal bulb in *Protodrilus* (Polychaeta, Archiannelida). I. Muscles and myo-epithelial junctions. *Tissue Cell* **1978**, *10*, 269–287. [CrossRef]
188. Jouin, C. Anatomical and ultrastructural study of the pharyngeal bulb in *Protodrilus* (Polychaeta, Archiannelida) II. The stomodeal and its cuticle. *Tissue Cell* **1978**, *10*, 289–301. [CrossRef]
189. Pierantoni, U. Genus Protodrilus. *Fauna Flora Golf. Neapel.* **1908**, *31*, 1–226.
190. Pierantoni, U. Organi genitali e glandole salivari nei Protodrili. *Boll. Soc. Nat. Napoli* **1906**, *20*, 154–157.
191. Jägersten, G. Zur Kenntnis der äusseren Morphologie, Entwicklung und Ökologie von *Protodrilus rubropharyngeus* n. sp. *Ark. Zool.* **1940**, *32A*, 1–19.
192. Jägersten, G. Studies on the morphology, larval development and biology of *Protodrilus*. *Zool. Bidr. Fran Upps.* **1952**, *29*, 427–511.
193. Von Nordheim, H. Vergleichende Ultrastrukturuntersuchungen der Eu-und Paraspermien von 13 *Protodrilus*-Arten (Polychaeta, Annelida) und ihre taxonomische und phylogenetische Bedeutung. *Helgol. Meeresunters.* **1989**, *43*, 113–156. [CrossRef]
194. Von Nordheim, H. Ultrastructure and functional morphology of the female reproductive organs in *Protodrilus* (Polychaeta, Annelida). *Helgol. Meeresunters.* **1991**, *45*, 465–485. [CrossRef]
195. Von Nordheim, H. Ultrastructure and functional morphology of male genital organs and spermatophore formation in *Protodrilus* (Polychaeta, Annelida). *Zoomorphology* **1991**, *111*, 81–94. [CrossRef]
196. Bailey-Brock, J.H.; Jouin-Toulmond, C.; Brock, R.E. Protodrilidae (Annelida: Polychaeta) from the Hawaiian Islands and Comparison with Specimens from French Polynesia. *Pac. Sci.* **2010**, *64*, 463–472. [CrossRef]
197. Kirsteuer, E. Bredin-Archbold-Smithsonian Biological Survey of Dominica, 3: Marine Archiannelids from Dominica. *Proc. U. S. Natl. Mus.* **1967**, *123*, 1–6. [CrossRef]
198. Pierantoni, U. Sopra un nuovo *Protodrilus* a'acqua dolce. *Monit. Zool. Ital.* **1903**, *14*, 324–327.
199. Von Nordheim, H. Life histories of subtidal interstitial polychaetes of the families Polygordiidae, Protodrilidae, Nerillidae, Dinophilidae and Diurodrilidae from Helgoland (North Sea). *Helgol. Meeresunters.* **1984**, *38*, 1–20.
200. Di Domenico, M.; Martinez, A.; da Cunha Lana, P.; Worsaae, K. *Protodrilus* (Protodrilidae, Annelida) from the southern and southeastern Brazilian coasts. *Helgol. Meeresunters.* **2013**, *67*, 733–748. [CrossRef]
201. Kopiy, V. Some aspects of the biology and the present state of the population of *Protodrilus flavocapitatus* (Polychaeta: Protodrilidae) in the coastal zone of Crimea (the Black Sea). *J. Black Sea Medit. Envirn.* **2013**, *19*, 162–168.
202. Martinez, A.; Eckert, E.M.; Artois, T.; Careddu, G.; Casu, M.; Curini-Galletti, M.; Gazale, V.; Gobert, S.; Ivanenko, V.N.; Jondelius, U.; et al. Human access impacts biodiversity of microscopic animals in sandy beaches. *Commun. Biol.* **2020**, *3*, 1–9. [CrossRef]
203. Lagos, A.M.; Leon, M.V.; Quiroga, S.Y.; Martinez, A.; Lagos, A.M.; Leon, M.V.; Quiroga, S.Y. Interstitial annelids from the Caribbean Coast of Colombia. *Rev. Biol. Trop.* **2018**, *66*, 658–673. [CrossRef]
204. Martinez, A.; Gonzalez, B.C. Volcanic Anchialine Habitats of Lanzarote. In *Cave Ecology*; Analysis and Synthesis; Springer: Cham, Switzerland, 2018; Volume 235, pp. 399–414.
205. Martinez, A.; Gonzalez, B.C.; Worsaae, K.; Wilkens, H.; Núñez, J.; Oromi, P.; Iliffe, T.M. *Guide to the Anchialine Ecosystems of Túnel de la Atlántida*; Medio Ambiente, Cabilde de Lanzarote: Lanzarote, Spain, 2016.
206. Braby, C.E.; Rouse, G.W.; Johnson, S.B.; Jones, W.J.; Vrijenhoek, R.C. Bathymetric and temporal variation among *Osedax* boneworms and associated megafauna on whale-falls in Monterey Bay, California. *Deep Sea Res. Part I Oceanogr. Res. Pap.* **2007**, *54*, 1773–1791. [CrossRef]

207. Langerhans, P. Über einige kanarische Anneliden. *Nov. Act. Leop.* **1881**, *42*, 93–124.
208. Armenante, Z. Protodrilus hypoleucus n. *sp. Monit. Zool. Ital.* **1903**, *14*, 221–222.
209. Jägersten, G. Die Abhängigkeit der Metamorphose vom Substrat des Biotops bei *Protodrilus*. *Ark. Zool.* **1940**, *32*, 1–12.
210. Aiyar, G.; Alikunhi, K.H. On some archiannelids of the Madras coast. *Proc. Indian Acad. Sci. Sect. B* **1943**, *67*, 113–140.
211. Alikunhi, K.H. On some archiannelids of the Krusadai Island. *Proc. Nat. Inst. Sci. India* **1948**, *14*, 373–383.
212. Wieser, W. Archiannelids from the intertidal of Puget Sound. *Trans. Am. Microsc. Soc.* **1957**, *76*, 275–285. [CrossRef]
213. Jouin, C. Archiannélides Interstitielles de Nouvelle-Caledónie. *Expédition Francaise sur les Récifs Coralliens de la Nouvelle-Calédonie* **1970**, *4*, 147–167.
214. Von Nordheim, H. Systematics and ecology of *Protodrilus helgolandicus* sp.n., an Interstitial polychaete (Protodrilidae) from subtidal sands off Helgoland, German Bight. *Zool. Scr.* **1983**, *12*, 171–177. [CrossRef]
215. Giard, M.A. Sur une faunule caractéristique des sables a diatomées d'ambleteuse (Pas-de-Calais). *C. R. Seances Soc. Biol. Fil.* **1904**, *56*, 295–298.
216. Remane, A. Protodrilidae aus Ost- und Nordsee. *Zool. Anz.* **1926**, *67*, 119–125.
217. Purschke, G.; Jouin, C. Anatomy and ultrastructure of the ventral pharyngeal organs of *Saccocirrus* and *Protodriloides* with remarks on the phylogenetic relationships within the Protodrilida (Annelida: Polychaeta). *J. Zool.* **1988**, *215*, 405–432. [CrossRef]
218. Jouin, C. Morphologie et anatomie comparee de *Protodrilus chaetifer* Remane et *Protodrilus symbioticus* Giard; creation du nouveau genre *Protodriloides* (Archiannelides). *Cah. Biol. Mar.* **1966**, *7*, 139–155.
219. Purschke, G.; Müller, M.C.M. Structure of prostomial photoreceptor-like sense organs in *Protodriloides* species (Polychaeta, Protodrilida). *Cah. Biol. Mar.* **1996**, *37*, 205–219.
220. Martínez, A.; Worsaae, K.; Purschke, G. Protodriloididae Purschke & Jouin, 1988. In *Handbook of Zoology; Annelida*; Purschke, G., Böggemann, M., Westheide, W., Eds.; De Gruyter: Berlin, Germany, 2020; Volume 3.
221. Fransen, M.E. Ultrastructure of coelomic organization in annelids I. Archiannelids and other small polychaetes. *Zoomorphology* **1980**, *95*, 235–249. [CrossRef]
222. Swedmark, B. Étude du développement larvaire et remarques sur la morphologie de *Protodrilus symbioticus* Giard (Archiannelides). *Ark. Zool.* **1954**, *6*, 511–522.
223. Gray, J.S. The behaviour of *Protodrilus symbioticus* (Giard) in temperature gradients. *J. Anim. Ecol.* **1965**, *34*, 455–461. [CrossRef]
224. Gray, J.S. Factors controlling the localizations of populations of *Protodrilus symbioticus* Giard. *J. Anim. Ecol.* **1966**, *35*, 435. [CrossRef]
225. Gray, J.S. The response of *Protodrilus symbioticus* (Giard) (Archiannelida) to light. *J. Anim. Ecol.* **1966**, *35*, 55–64. [CrossRef]
226. Sterrer, W. *Paranerilla limicola* Jouin & Swedmark (Archiannelida) von der Norwegischen und Adriatischen Küste. *Sarsia* **2011**, *36*, 65–68.
227. Boaden, P.J.S. Soft meiofauna of sand from the Delta region of the Rhine, Meuse and Scheldt. *Neth. J. Sea Res.* **1976**, *10*, 461–471. [CrossRef]
228. Besteiro, C.; Eugénio, W.; Carvalho, L.H.; Veiga, P.; Rubal, M. Contributions to the autoecology and biogeography of some psammic species of Annelida Polychaeta from Galicia (North-West Iberian Peninsula). *Cah. Biol. Mar.* **2016**, *57*, 253–259.
229. Hamond, R. The non-polychaetous annelids of Norfolk, England, with additional notes on polychaetes. *Cah. Biol. Mar.* **1972**, *13*, 341–350.
230. Westheide, W. La faune des polychètes et des archiannelides dans les plages sableuses a ressac de la cote Méditerranéenne de la Tunisie. *Bull. Inst. Oceanogr. Peche* **1972**, *2*, 449–468.
231. Boaden, P.J.S. 1962 Colonization of graded sand by an interstitial fauna. *Cah. Biol. Mar.* **1962**, *3*, 245–248.
232. Jouin, C. Le développement larvaire de *Protodrilus chaetifer* Remane (Archiannelida). *C. R. Acad. Sci. Paris* **1962**, *255*, 3065–3067.
233. Bellan, G. Remarques au sujet de la faune annélidienne "épibiote mobile" de quelques biotopes marins des côtes de Provence. *Rapp. Proc. Verb. Réun.* **1965**, *18*, 93–98.
234. Renaud-Debyser, J. Note sur la faune interstitielle du bassin d"Arcachon et description d"un gastrotriche nouveau. *Cah. Biol. Mar.* **1964**, *5*, 111–123.
235. Westheide, W. Zur Polychaetenfauna des Eulitorals der Nordseeinsel Sylt. *Helgol. Meeresunters.* **1966**, *13*, 203–209. [CrossRef]
236. Kristensen, R.M.; Nørrevang, A. Description of *Psammodrilus aedificator* sp.n. (Polychaeta), with notes on the Arctic Interstitial Fauna of Disko Island, W. Greenland. *Zool. Scr.* **1982**, *11*, 265–279. [CrossRef]
237. Mastepanova, E.A. Interstitial polychaetes of the seas of Russia. *Zool. Bespozvon.* **2004**, *1*, 59–64.
238. Priyalakshmi, G.; Menon, N.R. Sediment transport and bioinvasion—Possible impact of Tsunami—*Protodriloides chaetifer* an example. *J. Mar. Biol. Ass. India* **2007**, *49*, 86–88.
239. Swedmark, B. *Psammodriloides fauveli* n. gen. n. sp. et la famille des Psammodrilidae (Polychaeta, Sedentaria). *Arkiv. Zool.* **1958**, *2*, 55–65.
240. Bartolomaeus, T.; Meyer, K. Development and phylogenetic significance of hooked setae in Arenicolidae (Polychaeta, Annelida). *Invertebr. Biol.* **1997**, *116*, 227. [CrossRef]
241. Rouse, G.; Pleijel, F. *Polychaetes*; Oxford Univdersity Press: Oxford, UK, 2001.
242. Parry, L.A.; Edgecombe, G.D.; Eibye-Jacobsen, D.; Vinther, J. The impact of fossil data on annelid phylogeny inferred from discrete morphological characters. *Proc. R. Soc. B* **2016**, *283*, 20161378–9. [CrossRef]

243. Worsaae, K.; Sterrer, W. Description of two new interstitial species of Psammodrilidae (Annelida) from Bermuda. *Mar. Biol. Res.* **2006**, *2*, 431–445. [CrossRef]
244. Worsaae, K.; Kvindebjerg, K.; Martinez, A. Morphology of a new interstitial *Psammodrilus* (Psammodrilidae, Annelida) from Sardinia, Italy. *Zool. Anz.* **2015**, *259*, 13–21. [CrossRef]
245. Worsaae, K. Psammodrilidae Swedmark, 1952. In *Handbook of Zoology*; Purschke, G., Böggemann, M., Westheide, W., Eds.; De Gruyter: Berlin, Germany, 2019; pp. 143–155.
246. Swedmark, B. Recherches sur la morphologie le developpement et la biologie de *Psammodrilus balanoglossoides*, polychète sedentaire de la microfaune des sables. *Arch. Zool. Exp. Gén.* **1955**, *92*, 141–220.
247. Tzetlin, A.B.; Saphonov, M.V. Interstitial polychaetes (Annelida) from the Kandalaksha Bay of the White Sea. *Zool. Zhurnal.* **2002**, *81*, 899–908.
248. Brown, R. Saccocirridae (Annelida: Archiannelida) from the central coast of New South Wales. *Austr. J. Mar. Freshw. Res.* **1981**, *32*, 439–456. [CrossRef]
249. Purschke, G. Ultrastructural investigation of presumed photoreceptive organs in two *Saccocirrus* species (Polychaeta, Saccocirridae). *J. Morphol.* **1992**, *211*, 7–21. [CrossRef] [PubMed]
250. Jouin-Toulmond, C.; Gambi, C. Description of *Saccocirrus goodrichi* sp. nov. (Annelida: Polychaeta: Saccocirridae), a new Mediterranean species and new data on the chaetae of *S. papillocercus* and *S. Major*. *Cah. Biol. Mar.* **2007**, *48*, 381–390.
251. Di Domenico, M.; Martínez, A.; Worsaae, K. Saccocirridae (Annelida) from the Canary Islands with a description of *Saccocirrus slateri* sp. nov. *Mar. Biodivers.* **2019**, *49*, 2125–2139. [CrossRef]
252. Weigert, A.; Bleidorn, C. Current status of annelid phylogeny. *Org. Divers. Evol.* **2016**, *16*, 345–362. [CrossRef]
253. Jouin, C. Status of the knowledge of the systematics and ecology of Archiannelida. *Smithson. Contr. Zool.* **1971**, *76*, 47–56.
254. Di Domenico, M.; Martínez, A.; Amaral, A.C.Z.; Lana, P.C.; Worsaae, K. Saccocirridae (Annelida) from the southern and southeastern Brazilian coasts. *Mar. Biodivers.* **2014**, *44*, 313–325. [CrossRef]
255. Goodrich, E.S. On the structure and affinities of *Saccocirrus*. *Q. J. Microsc. Sci.* **1901**, *44*, 413–428.
256. Salensky, W. Morphogenetische Studien an Würmern. *Mem. Acad. Sci. St. Petersbourg* **1907**, *19*, 1–349.
257. Brown, S.; Rouse, G.; Hutchings, P.; Colgan, D. Assessing the usefulness of histone H3, U2 snRNA and 28S rDNA in analyses of polychaete relationships. *Aust. J. Zool.* **1999**, *47*, 499–516. [CrossRef]
258. Purschke, G.; Tzetlin, A.B. Dorsolateral ciliary folds in the polychaete foregut: Structure, prevalence and phylogenetic significance. *Acta Zool.* **1996**, *77*, 33–49. [CrossRef]
259. Tzetlin, A.B.; Purschke, G. Pharynx and intestine. *Hydrobiologia* **2005**, *535/536*, 199–225. [CrossRef]
260. Du Bois-Reymond Marcus, E. On a new archiannelid, *Saccocirrus gabriellae*, from Brazil. *Com. Zool. Mus.* **1946**, *37*, 1–11.
261. Villora-Moreno, S.; Capaccioni-Azzati, R.; Garcia-Carrascosa, A.M. Meiobenthos of sandy beaches from the Gulf of Valencia (Western Mediterranean): Ecology of interstitial polychaetes. *Bull. Mar. Sci.* **1991**, *48*, 376–385.
262. Di Domenico, M.; Lana, P.C.; Garraffoni, A.R.S. Distribution patterns of interstitial polychaetes in sandy beaches of southern Brazil. *Mar. Ecol.* **2009**, *30*, 47–62. [CrossRef]
263. Gray, J.S.A. A new species of *Saccocirrus* (Archiannelida) from the West Coast of North America. *Pac. Sc.* **1969**, *23*, 238–251.
264. Bobretzky, N.V. *Saccocirrus papillocerus*, n. gen., n. sp. Tip' novago semeistra annelid. Sravmitel'no-Anatomichskii obcherk. *Kiev Odschestva Estest Zap.* **1871**, *2*, 211–259.
265. Pierantoni, U. Observazioni suilo sviluppo embrionale e larvale del *Saccocirrus papillocercus* Bobr. *Mitt. Zool. Stat. Neapel* **1906**, *18*, 46–72.
266. Pierantoni, U. Il genere *Saccocirrus* Bobretzky e le sue specie. *Ann. Ist. Mus. Zool. Univ. Napoli* **1907**, *2*, 1–11.
267. Boaden, P.J.S. The interstitial fauna of some North Wales beaches. *J. Mar. Biol. Assoc. UK* **1963**, *43*, 79–96. [CrossRef]
268. Núñez, J.; Brito, M.C.; Docoito, J.R. Anélidos poliquetos de Canarias: Catálogo de especies, distribución y hábitats. *Vieraea* **2005**, *33*, 297–321.
269. Bailey-Brock, J.H.; Dreyer, J.; Brock, R.E. Three new species of *Saccocirrus* (Polychaeta: Saccocirridae) from Hawai'i. *Pac. Sc.* **2003**, *57*, 463–478. [CrossRef]
270. Di Domenico, M.; Worsaae, K.; Purschke, G. Saccocirridae Czerniavsky, 1881. In *Handbook of Zoology: Annelida*; Purschke, G., Böggemann, M., Westheide, W., Eds.; De Gruyter: Berlin, Germany, 2020; Volume 3.

MDPI
St. Alban-Anlage 66
4052 Basel
Switzerland
Tel. +41 61 683 77 34
Fax +41 61 302 89 18
www.mdpi.com

Diversity Editorial Office
E-mail: diversity@mdpi.com
www.mdpi.com/journal/diversity

www.ingramcontent.com/pod-product-compliance
Ingram Content Group UK Ltd.
Pitfield, Milton Keynes, MK11 3LW, UK
UKHW060121010826
14089UKWH00006B/97

9 783036 513904